Second Edition

Handbook of

Water and Wastewater Treatment Plant Operations

Second Edition

Handbook of
Water and Wastewater Treatment Plant Operations

Frank R. Spellman

CRC Press
Taylor & Francis Group
Boca Raton London New York

CRC Press is an imprint of the
Taylor & Francis Group, an **informa** business

CRC Press
Taylor & Francis Group
6000 Broken Sound Parkway NW, Suite 300
Boca Raton, FL 33487-2742

Library of Congress Cataloging-in-Publication Data

Spellman, Frank R.
    Handbook of water and wastewater treatment plant operations / Frank R. Spellman. -- 2nd ed.
       p. cm.
    "A CRC title."
    Includes bibliographical references and index.
    ISBN 978-1-4200-7530-4 (alk. paper)
    1. Water treatment plants--Handbooks, manuals, etc. 2. Sewage disposal plants--Handbooks, manuals, etc. 3. Water--Purification--Handbooks, manuals, etc. 4. Sewage--Purification--Handbooks, manuals, etc. I. Title.

TD434.S64 2008
628.1'62--dc22
    2008018725

Visit the Taylor & Francis Web site at
http://www.taylorandfrancis.com

and the CRC Press Web site at
http://www.crcpress.com

Contents

Preface

Water is the new oil.

Water is a carrier of things it picks up as it passes through—it carries the good and the bad.

When I wrote the first edition to this text and it was published, I had no idea it would be so well received and become an instant bestseller. Hailed on its first publication as a real-world practical handbook for general readers, students, and water/wastewater practitioners, the *Handbook of Water and Wastewater Treatment Plant Operations* continues to make the same basic point in its second edition: Water and wastewater operators must be Jacks or Jills of many trades; that is, they must have basic skill sets best described as being all encompassing in nature. In light of the need for practitioners in the field who are well rounded in the sciences, cyber operations, math, mechanics, and technical aspects of water treatment, the second edition picks up where the original left off.

Based on constructive criticism of the first edition, the second edition has been upgraded and expanded from beginning to end. Many reviewers appreciated the candid portrayal of regulatory, privatization, management, and other ongoing current issues within the water/wastewater industry. With regard to privatization, this book recognizes that critics of privatization insist that water/wastewater operations are too important to be left to the mercies of private enterprise. It is my premise, however, that water/wastewater operations are too important *not* to be subject to market forces. In my opinion, the debate should not be about taking the supply, treatment, reuse, and distribution of water/wastewater away from perhaps dysfunctional public service agencies and implementing pricing, property rights, and private, competent enterprise instead, but over how best to do so. Based on actual personal experience in water/wastewater public sector operations, one thing seems certain to me: I have found that engineers, biologist, chemists, environmental scientists, hydrologists, and other public-service technical officials generally know little, and usually care even less, about markets and ratepayers. In a nutshell, even though the original edition took a hard-line, no-nonsense look at problems engendered by current management and various dysfunctional management styles unique to the industry, the second edition takes an even more penetrating look at these problems. This is especially the case with regard to management problems within the public sector entities involved in water and wastewater operations. Simply, it is the author's view,

based on personal experience, that, when it comes to water and wastewater public service managers, relevance and common sense are rare commodities.

In addition to a no-holds-barred look at current management issues, the second edition includes the latest security information pertinent to protecting public assets from the indiscriminate and destructive hand of terrorism. As a result of the events of September 11, 2001, things have changed for all freedom-loving people everywhere. How many of us thought security was a big deal prior to 9/11? Some of us did, but some of us didn't give it any thought at all. Today, we must adjust or fall behind. In the current climate, falling behind on keeping our potable water supplies secure is not an option. We must aggressively protect our precious water sources and those ancillaries that are critical to maintaining and protecting water quality. Because water and wastewater operations are listed by the Department of Homeland Security as one of this nation's 13 critical infrastructures, water and wastewater practitioners must realize and understand that the threat of terrorism is real. Accordingly, extensive coverage of security needs is provided in this edition.

In addition to more in-depth coverage of management aspects and security, the second edition includes a new chapter covering the basics of blueprint reading. This is a critical area of expertise that is important for maintenance operators and others. Also, the chapter on water and wastewater mathematics has been tripled in size and now contains an additional 200 example problems and 350 math system operational problems with solutions. These examples and operational math problems are typical of those seen on water and wastewater licensure examinations used throughout the United States. Every chapter has been upgraded to include emerging technologies pertinent to the content presented. Practical hands-on information necessary for proper plant operation is provided that will also help readers obtain passing scores on licensure examinations.

The text follows a pattern that is nontraditional; that is, the paradigm (i.e., model or prototype) used here is based on real-world experience and proven parameters—not on theoretical gobbledygook. Clearly written and user friendly, this timely revision of the handbook builds on the remarkable success of the first edition. Still intended to serve as an information source, this text is not limited in its potential for other uses. This work can be utilized by water/wastewater practitioners to gain valuable insight into the substance they work so hard to collect, treat, supply, reuse, or discharge for its intended purpose, but it can just as easily provide important information for policymakers who may be tasked with making decisions concerning water or wastewater resource utilization. Consequently, this book serves a varied audience: students, lay personnel, regulators, technical experts, attorneys, business leaders, and concerned citizens.

This text is not about the planning, designing, or construction of water and wastewater treatment facilities. Although these tasks are of paramount importance during the conception and construction of facilities and infrastructures, many excellent texts are already available that address these topics. This text is not about engineering at all. Instead, this handbook is about operations and is designed for the plant manager and plant operator. We often forget the old axiom: "Someone must build it, but once built, someone must operate it." It is the operation of *it* that concerns us here.

With regard to plant managers and operators, most texts ignore, avoid, or pay cursory attention to such important areas as the multiple-barrier concept, maintaining infrastructure, benchmarking, plant security, operator roles, water hydraulics, microbiology, water ecology, math operations, basic electrical principles, pumping, conveyance, flow measurement, basic water chemistry, water quality issues, biomonitoring, sampling and testing, water sources, and watershed protection. All of these important topics are thoroughly discussed in the second edition of the *Handbook of Water and Wastewater Treatment Plant Operations*.

To maximize the usefulness of the material contained in the text, it has been presented in plain English in a simplified and concise format. Many tables have been developed, using a variety of sources. Moreover, to ensure its relevance to modern practice and design, illustrative problems are presented in terms of commonly used operational parameters.

Each chapter ends with chapter review questions to help evaluate the reader's mastery of the concepts presented. Before going on to the next chapter, work through the questions, compare your answers to the key provided in Appendix A, and review the pertinent information for any problems you missed. If you miss many items, review the entire chapter.

This text is accessible to those who have no experience with water or wastewater operations. If you work through the text systematically, an understanding of and skill in water/wastewater operations can be acquired—adding a critical component to your professional knowledge.

Frank R. Spellman
Old Dominion University
Norfolk, VA

To the Reader

While reading this text, you are going to spend some time following water on its travels.

Even after being held in bondage, sometimes for eons, eventually water moves.

Do you have any idea where this water has been?

Where this water is going?

What changes it has undergone, during all the long ages that the water has lain on and under the face of the Earth?

Sometimes we can look at this water ... analyze this water ... test this water to find out where it has been.

Water, because it is the universal solvent, has a tendency to pick up materials through which it flows.

When this happens, we must sometimes treat the water before we consume it.

Whether this is the case or not, water continues its endless cycle.

And for us this is the best of news.

So, again, do you have any idea where water has been?

More importantly, where is the water going?

If we could first know where we are and wither we are tending, we could better judge what we do and how to do it.

—Abraham Lincoln

Author

Frank R. Spellman, Ph.D., is assistant professor of Environmental Health at Old Dominion University, Norfolk, VA, and the author of 55 books. Dr. Spellman's book topics range from concentrated animal feeding operations (CAFOs) to all areas of environmental science and occupational health. Many of Dr. Spellman's texts are listed on Amazon.com and Barnes and Noble. Several of his texts have been adopted for classroom use at major universities throughout the United States, Canada, Europe, and Russia; two of them are currently being translated into Spanish for South American markets.

Dr. Spellman has been cited in more than 400 publications. He serves as a professional expert witness for three law groups and as an accident investigator for a northern Virginia law firm. He also consults on homeland security vulnerability assessments (VAs) for critical infrastructure, including water/wastewater facilities nationwide. Dr. Spellman has joined many well-recognized experts in contributing to texts in several scientific fields; for example, he is a contributor to the second edition of the prestigious text *The Engineering Handbook* (CRC Press).

Dr. Spellman lectures on sewage treatment, water treatment, homeland security, and health and safety topics throughout the country and teaches water/wastewater operator short courses at Virginia Tech (Blacksburg, VA). He has earned a bachelor of arts in Public Administration, bachelor of science in Business Management, master of business administration, master of science in Environmental Engineering, and doctorate in Environmental Engineering.

Part I

Water and Wastewater Operations:
An Overview

1 Current Issues in Water and Wastewater Treatment Operations

The failure to provide safe drinking water and adequate sanitation services to all people is perhaps the greatest development failure of the twentieth century.

Gleick (1998, 2000)

1.1 INTRODUCTION

Although not often thought of as a commodity (or, for that matter, not thought about at all), water is a commodity—a very valuable, vital commodity. We consume water, waste it, discard it, pollute it, poison it, and relentlessly modify the hydrological cycles (natural and urban cycles), with total disregard to the consequences: "Too many people, too little water, water in the wrong places and in the wrong amounts. The human population is burgeoning, but water demand is increasing twice as fast" (De Villiers, 2000). It is our position that, with the passage of time, potable water will become even more valuable. Moreover, with the passage of even more time, potable water will be even more valuable than we might ever imagine—possibly (likely) comparable in pricing, gallon for gallon, to what we pay for gasoline or even more. From urban growth to infectious disease and newly identified contaminants in water, greater demands are being placed on our planet's water supply (and other natural resources). As the global population continues to grow, people will place greater and greater demands on our water supply (Anon., 2000). The fact is—simply, profoundly, without a doubt in the author's mind—water is the new oil.

Earth was originally allotted a finite amount of water; we have no more or no less than that original allotment today—they are not making any more of it. Thus, it logically follows that, in order to sustain life as we know it, we must do everything we can to preserve and protect our water supply. Moreover, we also must purify and reuse the water we currently waste (i.e., wastewater).

Did You Know?

More than 50% of Americans drink bottled water occasionally or as their major source of drinking water—an astounding fact given the high quality and low cost of U.S. tapwater.

1.2 THE PARADIGM SHIFT

Historically, the purpose of water supply systems has been to provide pleasant drinking water that is free of disease organisms and toxic substances. In addition, the purpose of wastewater treatment has been to protect the health and well-being of our communities. Water/wastewater treatment operations have accomplished this goal by: (1) preventing disease and nuisance conditions; (2) avoiding contamination of water supplies and navigable waters; (3) maintaining clean water for survival of fish, bathing, and recreation; and (4) generally conserving water quality for future use.

The purpose of water supply systems and wastewater treatment processes has not changed; however, the paradigm has shifted, primarily because of new regulations that include: (1) protecting against protozoan and virus contamination; (2) implementing the multiple-barrier approach to microbial control; (3) new requirements of the Ground Water Disinfection Rule (GWDR), the Total Coliform Rule (TCR) and Distribution System (DS), and the Lead and Copper (Pd/Cu) Rule; (4) regulations for trihalomethanes (THMs) and disinfection byproducts (DBPs); and (5) new requirements to remove even more nutrients (nitrogen and phosphorus) from wastewater effluent. We will discuss this important shift momentarily, but first it is important to abide by Voltaire's advice; that is, "If you wish to converse with me, please define your terms."

For those not familiar with the term *paradigm*, it can be defined in the following ways. A paradigm is the consensus of the scientific community: "concrete problem solutions that the profession has come to accept" (Holyningen-Huene, 1993). Thomas Kuhn coined the term *paradigm*; he outlined it in terms of the scientific process and felt that "one sense of paradigm is global, embracing all the shared commitments of a scientific group; the other isolates a particularly important sort of commitment and is thus a subset of the first" (Holyningen-Huene, 1993). The concept of paradigm has two general levels. The first is the encompassing whole, the summation of parts. It consists of the theories, laws, rules, models, concepts, and definitions that go into a generally accepted fundamental theory of science. Such a paradigm is "global" in character. The other level of paradigm is that it can also be just one of these laws, theories, models, etc. that combine to formulate a global paradigm. These have the property of being "local." For example, Galileo's theory that the Earth rotated around the sun became a paradigm in itself—namely, a generally accepted law in astronomy. Yet, on the other hand, his theory combined with other local paradigms in areas such as religion and politics to transform culture. Paradigm can also be defined as a pattern or point of view that determines what is seen as reality. We use this definition in this text.

A paradigm shift is defined as a major change in the way things are thought about, especially scientifically. Once a problem can no longer be solved in the existing paradigm, new laws and theories emerge to form a new paradigm, overthrowing the old if the new one is accepted. Paradigm shifts are the "occasional, discontinuous, revolutionary changes in tacitly shared points of view and preconceptions" (Daly, 1980). Simply, a paradigm shift represents "a profound change in the thoughts, perceptions, and values that form a particular vision of reality" (Capra, 1982). For our purposes, we use the term *paradigm shift* to refer to a change in the way things are understood and done.

1.2.1 A Change in the Way Things Are Understood and Done

In water supply systems, the historical focus, or traditional approach, has been to control turbidity, iron and manganese, taste and odor, color, and coliforms. New regulations provided new focus and thus a paradigm shift. Today, the traditional approach is no longer sufficient. Providing acceptable water has become more sophisticated and costly. To meet the requirements of the new paradigm, a systems approach must be employed. In the systems approach, all components are interrelated. What affects one impacts others. The focus has shifted to multiple requirements (e.g., new regulations require the process to be modified or the plant upgraded).

To illustrate the paradigm shift in the operation of water supply systems, let us look back at the traditional approach of disinfection. Disinfection was used in water to destroy harmful organisms. Currently, disinfection is still used in water to destroy harmful organisms but is now only one part of the multiple-barrier approach. Moreover, disinfection has traditionally been used to treat for coliforms only, but, because of the paradigm shift, disinfection now (and in the future) is used against coliforms, *Legionella*, *Giardia*, *Cryptosporidium*, and others. (Note: To effectively remove the protozoa *Giardia* and *Cryptosporidium*, filtration is required; disinfection is not effective against the oocysts of *Cryptosporidium*.) Another example of traditional vs. current practices is seen in the traditional approach to particulate removal in water to lessen turbidity and improve aesthetics. Current practice is still to decrease turbidity to improve aesthetics but now microbial removal plus disinfection is practical.

Another significant factor that contributed to the paradigm shift in water supply systems was the introduction of the Surface Water Treatment Rule (SWTR) in 1989. SWTR requires water treatment plants to achieve 99.9% (3 log) removal activation/inactivation of *Giardia* and 99.99% (4 log) removal or inactivation of viruses. SWTR applies to all surface water and groundwater under the direct influence of surface water (GWUDI).

As mentioned earlier, removal of excess nutrients such as nitrogen and phosphorus in wastewater effluent is now receiving more attention from regulators (e.g., U.S. Environmental Protection Agency) and others. One of the major concerns is the appearance of dead zones in various water bodies, where excess nutrients cause oxygen-consuming algae to grow and thus create oxygen-deficient dead zones. In recent years, for example, it has not been uncommon to find several dead zone locations in the Chesapeake Bay region; consider the case study below.

■ CASE STUDY 1.1. Chesapeake Bay Cleanup

The following newspaper article, written by the author, appeared in the January 2, 2005, issue of *The Virginian-Pilot*. It is an Op-Ed rebuttal to the article referenced in the text below. It should be pointed out that this piece was well received by many, but a few stated that it was nothing more than a rhetorical straw man. Of course, in contrast, I felt that the organizational critics were using the rhetorical Tin Man approach; that is, when you need to justify your cause and your organization's existence and you need more grease, you squawk. The grease that many of these organizations require, however, is grease that is the consistency of paper-cloth and is colored green; thus, they squawk quite often. You be the judge.

Chesapeake Bay Cleanup: Good Science vs. "Feel Good" Science

In your article, "Fee to help Bay faces anti-tax mood" (*The Virginian-Pilot*, 1/2/05), you pointed out that environmentalists call it the "Virginia Clean Streams Law." Others call it a "flush tax." I call the environmentalists' (and others') view on this topic a rush to judgment, based on "feel good" science vs. good science. The environmentalists should know better.

Consider the following:

Environmental policymakers in the Commonwealth of Virginia came up with what is called the Lower James River Tributary Strategy on the subject of nitrogen (a nutrient) from the Lower James River and other tributaries contaminating the Lower Chesapeake Bay Region. When in excess, nitrogen is a pollutant. Some "theorists" jumped on nitrogen as being the cause of a decrease in the oyster population in the Lower Chesapeake Bay Region. Oysters are important to the local region. They are important for economical and other reasons. From an environmental point of view, oysters are important to the Lower Chesapeake Bay Region because they have worked to maintain relatively clean Bay water in the past. Oysters are filter-feeders. They suck in water and its accompanying nutrients and other substances. The oyster sorts out the ingredients in the water and uses those nutrients it needs to sustain its life. Impurities (pollutants) are aggregated into a sort of ball that is excreted by the oyster back into the James River.

You must understand that there was a time, not all that long ago (maybe 50 years ago), when oysters thrived in the Lower Chesapeake Bay. Because they were so abundant, these filter-feeders were able to take in turbid Bay water and turn it almost clear in a matter of three days. (How could anyone dredge up, clean, and then eat such a wonderful natural vacuum cleaner?)

Of course, this is not the case today. The oysters are almost all gone. Where did they go? Who knows?

The point is that they are no longer thriving, no longer colonizing the Lower Chesapeake Bay Region in numbers they did in the past. Thus, they are no longer providing economic stability to watermen; moreover, they are no longer cleaning the Bay.

Ah! But don't panic! The culprit is at hand; it has been identified. The "environmentalists" know the answer. They say it has to be nutrient contamination—namely, nitrogen is the culprit. Right?

Not so fast.

A local sanitation district and a university in the Lower Chesapeake Bay region formed a study group to formally, professionally, and scientifically study this problem. Over a five-year period, using Biological Nutrient Removal (BNR) techniques at a local wastewater treatment facility, it was determined that the effluent leaving the treatment plant and entering the Lower James River consistently contained below 8 mg/L nitrogen (a relatively small amount) for five consecutive years.

The first question is: Has the water in the Chesapeake Bay become cleaner, clearer because of the reduced nitrogen levels leaving the treatment plant?

The second question is: Have the oysters returned?

Answers to both questions, respectively: no; not really.

Wait a minute. The environmentalists, the regulators, and other well-meaning interlopers stated that the problem was nitrogen. If nitrogen levels have been reduced in the Lower James River, shouldn't the oysters start thriving, colonizing, and cleaning the Lower Chesapeake Bay again?

You might think so, but they are not. It is true that the nitrogen level in the wastewater effluent was significantly lowered through treatment. It is also true that a major point source contributor of nitrogen was reduced with a corresponding decrease in the nitrogen level in the Lower Chesapeake Bay.

If the nitrogen level has decreased, then where are the oysters?

A more important question is: What is the real problem?

The truth is that no one at this point in time can give a definitive answer to this question.

Back to the original question: Why has the oyster population decreased?

One theory states that because the tributaries feeding the Lower Chesapeake Bay (including the James River) carry megatons of sediments into the bay (stormwater runoff, etc.), they are adding to the Bay's turbidity problem. When waters are highly turbid, oysters do the best they can to filter out the sediments but eventually they decrease in numbers and then fade into the abyss.

Is this the answer? That is, is the problem with the Lower Chesapeake Bay and its oyster population related to turbidity?

Only solid, legitimate, careful scientific analysis may provide the answer.

One thing is certain; before we leap into decisions that are ill-advised, that are based on anything but sound science, and that "feel" good, we need to step back and size up the situation. This sizing-up procedure can be correctly accomplished only through the use of scientific methods.

Don't we already have too many dysfunctional managers making too many dysfunctional decisions that result in harebrained, dysfunctional analysis—and results?

Obviously, there is no question that we need to stop the pollution of Chesapeake Bay.

However, shouldn't we replace the timeworn and frustrating position that "we must start somewhere" with good common sense and legitimate science?

The bottom line: We shouldn't do anything to our environment until science supports the investment. Shouldn't we do it right?

Frank R. Spellman

1.3 MULTIPLE-BARRIER CONCEPT

On August 6, 1996, during the Safe Drinking Water Act Reauthorization signing ceremony, President Bill Clinton stated:

> A fundamental promise we must make to our people is that the food they eat and the water they drink are safe.

No rational person could doubt the importance of the promise made in this statement.

The Safe Drinking Water Act (SDWA), passed in 1974, amended in 1986, and (as stated above) reauthorized in 1996, gives the U.S. Environmental Protection Agency (USEPA) the authority to set drinking water standards.

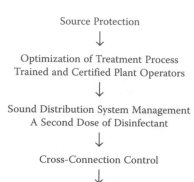

Source Protection

↓

Optimization of Treatment Process
Trained and Certified Plant Operators

↓

Sound Distribution System Management
A Second Dose of Disinfectant

↓

Cross-Connection Control

↓

Continuous Monitoring and Testing

FIGURE 1.1 Multiple-barrier approach.

This document is important for many reasons but is even more important because it describes how USEPA establishes these standards.

Drinking water standards are regulations that USEPA sets to control the level of contaminants in the nation's drinking water. These standards are part of the Safe Drinking Water Act's *multiple-barrier approach* to drinking water protection. As shown in Figure 1.1, the multiple-barrier approach includes the following elements:

- *Assessing and protecting drinking water sources*, which means doing everything possible to prevent microbes and other contaminants from entering water supplies. Minimizing human and animal activity around our watersheds is one part of this barrier.
- *Optimizing treatment processes*, which provides a second barrier. This usually means filtering and disinfecting the water. It also means making sure that the people who are responsible for our water are properly trained and certified and knowledgeable of the public health issues involved.
- *Ensuring the integrity of distribution systems*, which consists of maintaining the quality of water as it moves through the system on its way to the customer's tap.
- *Effecting correct cross-connection control procedures*, which is a critical fourth element in the barrier approach. It is critical because the greatest potential hazard in water distribution systems is associated with cross-connections to nonpotable waters. Many connections exist between potable and nonpotable systems (every drain in a hospital constitutes such a connection), but cross-connections are those through which backflow can occur (Angele, 1974).
- *Continuous monitoring and testing of the water before it reaches the tap* are critical elements in the barrier approach and should include specific procedures to follow should potable water ever fail to meet quality standards.

With the involvement of USEPA, local governments, drinking water utilities, and citizens, these multiple-barriers ensure that the tap water in the United States and territories is safe to drink. Simply, in the multiple-barrier concept, we employ a holistic approach to water management that begins at the source and continues with treatment through disinfection and distribution.

1.3.1 MULTIPLE-BARRIER APPROACH: WASTEWATER OPERATIONS

Not shown in Figure 1.1 is the fate of the used water. What happens to the wastewater produced? Wastewater is treated via the *multiple-barrier treatment train*, which is a combination of unit processes used in the system. The primary mission of the wastewater treatment plant (and the operator/practitioner) is to treat the wastestream to a level of purity acceptable to return it to the environment or for immediate reuse (e.g., at the present time for reuse in such applications as irrigation of golf courses).

Water and wastewater professionals maintain a continuous urban water cycle on a daily basis. B.D. Jones (1980) summed this up as follows:

> Delivering services is the primary function of municipal government. It occupies the vast bulk of the time and effort of most city employees, is the source of most contacts that citizens have with local governments, occasionally becomes the subject of heated controversy, and is often surrounded by myth and misinformation. Yet, service delivery remains the "hidden function" of local government.

Did You Know?

Artificially generated water cycles or urban water cycles consist of: (1) source (surface or groundwater), (2) water treatment and distribution, (3) use and reuse, and (4) wastewater treatment and disposition, as well as the connection of the cycle to the surrounding hydrological basins (see Figure 1.2 and Figure 1.3).

In the *Handbook of Water and Wastewater Treatment Plant Operations*, 2nd ed., we focus on sanitary (or environmental) services (excluding solid-waste disposal)—water and wastewater treatment—because they have been and remain indispensable for the functioning and growth of cities. Water is the most important (next to air) life-sustaining product on Earth. Yet, it is its service delivery (and all that it entails) that remains a hidden function of local government (Jones, 1980). This hidden function is what this text is all about. We present our discussion in a completely new and unique dual manner—in what we consider a new paradigm shift in water management and in the concept of the multiple-barrier

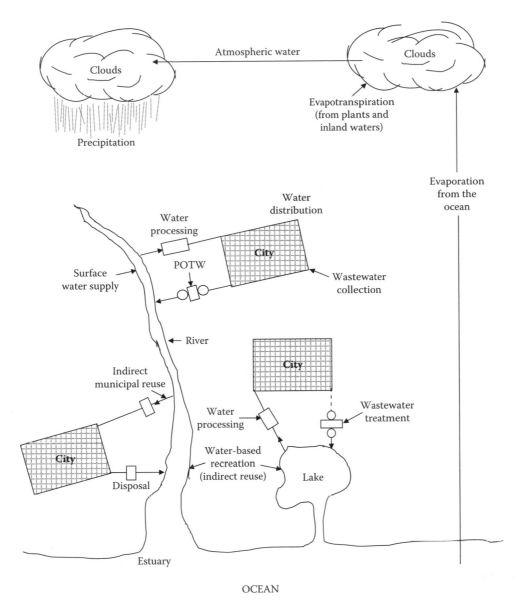

FIGURE 1.2 Urban water cycle.

approach. Essentially, in blunt, plain English, the *Handbook* exposes the "hidden" part of services delivered by water and wastewater professionals.

Water service professionals provide water for typical urban domestic and commercial uses, eliminate wastes, protect the public health and safety, and help control many forms of pollution. Wastewater service professionals treat the urban wastestream to remove pollutants before discharging the effluent into the environment. Water and wastewater treatment services are the urban circulatory system, the hidden circulatory system. In addition, like the human circulatory system, the urban circulatory system is less than effective if flow is not maintained. In a practical sense, we must keep both systems plaque free and freeflowing.

Maintaining flow is what water and wastewater operations is all about. This seems easy enough: Water has been flowing for literally eons, emerging from mud, rocks, silt. It is the very soul of moving water to carve a path, to pick up its load, to forge its way to the open arms of a waiting sea.

This is not to say that water and wastewater operations are not without problems or challenges. After we survived the Y2K fiasco (were you surrounded by dysfunctional managers running about helter-skelter waiting until midnight as I was?), the dawn of the 21st century brought with it, for many of us, aspirations of good things ahead in the constant struggle to provide quality food and water for humanity. However, the only way in which we can hope to accomplish this is to stay on the cutting edge of technology

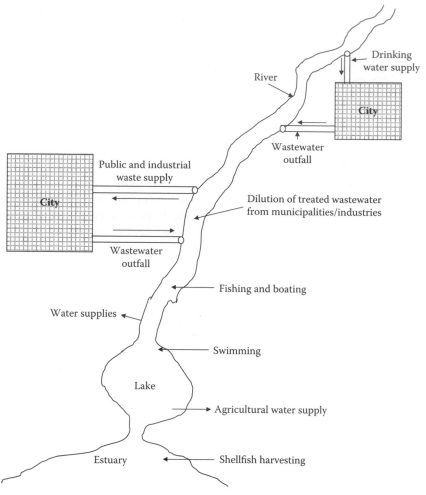

FIGURE 1.3 Indirect water reuse process.

and to face all challenges head on. Some of these other challenges are addressed in the following sections.

1.4 MANAGEMENT PROBLEMS FACING WATER AND WASTEWATER OPERATIONS

Problems come and go, shifting from century to century, decade to decade, year to year, and from site to site. They range from problems caused by natural forces (storms, earthquakes, fires, floods, and droughts) to those caused by social forces, currently including terrorism. In general, five areas are of concern to many water and wastewater management personnel:

1. Complying with regulations and coping with new and changing regulations
2. Maintaining infrastructure
3. Privatization and/or reengineering
4. Benchmarking
5. Upgrading security (see Chapter 3)

1.4.1 COMPLIANCE WITH NEW, CHANGING, AND EXISTING REGULATIONS

Note: This section is from the notes of F.R. Spellman provided to J.E. Drinan (2001).

Adapting the workforce to the challenges of meeting changing regulations and standards for both water and wastewater treatment is a major concern. As mentioned, drinking water standards are regulations that USEPA sets to control the level of contaminants in the nation's drinking water. Again, these standards are part of the SDWA's multiple-barrier approach to drinking water protection. There are two categories of drinking water standards:

1. *National primary drinking water regulation (primary standard)*. This is a legally enforceable standard that applies to public water systems. Primary standards protect drinking water quality by limiting the levels of specific contaminants that can adversely affect public health and are known or anticipated to occur in

water. They take the form of Maximum Contaminant Levels or Treatment Techniques.

2. *National secondary drinking water regulation (secondary standard).* This is a non-enforceable guideline regarding contaminants that may cause cosmetic effects (such as skin or tooth discoloration) or aesthetic effects (such as taste, odor, or color) in drinking water. USEPA recommends secondary standards to water systems but does not require systems to comply; however, states may choose to adopt them as enforceable standards. This information focuses on national primary standards.

Drinking water standards apply to public water systems, which provide water for human consumption through at least 15 service connections or which regularly serve at least 25 individuals. Public water systems include municipal water companies, homeowner associations, schools, businesses, campgrounds, and shopping malls.

More recent requirements (e.g., the Clean Water Act Amendments that went into effect in February of 2001) require water treatment plants to meet tougher standards; they present new problems for treatment facilities to deal with but offer some possible solutions to the problems of meeting the new standards. These regulations provide for communities to upgrade existing treatment systems by replacing an aging and outdated infrastructure with new process systems. Their purpose is to ensure that facilities are able to filter out higher levels of impurities from drinking water, thus reducing the health risk from bacteria, protozoa, and viruses, and that they are able to decrease levels of turbidity and reduce concentrations of chlorine byproducts in drinking water.

With regard to wastewater collection and treatment, the National Pollutant Discharge Elimination System (NPDES) program established by the Clean Water Act issues permits that control wastewater treatment plant discharges. Meeting permit is always a concern for wastewater treatment managers because the effluent discharged Into water bodies affects those downstream of the release point. Individual point-source dischargers must use the best available technology (BAT) to control the levels of pollution in the effluent they discharge into streams. As systems age and the best available technology changes, meeting permit with existing equipment and unit processes becomes increasingly difficult.

1.4.2 MAINTAINING INFRASTRUCTURE

During the 1950s and 1960s, the U.S. government encouraged the prevention of pollution by providing funds for the construction of municipal wastewater treatment plants, water pollution research, and technical training and assistance. New processes were developed to treat sewage,

analyze wastewater, and evaluate the effects of pollution on the environment. In spite of these efforts, however, expanding population and industrial and economic growth caused pollution and health problems to increase.

In response to the need to make a coordinated effort to protect the environment, the National Environmental Policy Act (NEPA) was signed into law on January 1, 1970. In December of that year, a new independent body, the U.S. Environmental Protection Agency, was created to bring under one roof all of the pollution-control programs related to air, water, and solid wastes. In 1972, the Water Pollution Control Act Amendments expanded the role of the federal government in water pollution control and significantly increased federal funding for construction of wastewater treatment plants.

Many of the wastewater treatment plants in operation today are the result of federal grants made over the years; for example, because of the 1977 Clean Water Act Amendment to the Federal Water Pollution Control Act of 1972 and the 1987 Clean Water Act reauthorization bill, funding for wastewater treatment plants was provided. Many large sanitation districts, with multiple plant operations, and even a larger number of single plant operations in smaller communities in operation today are a result of these early environmental laws. As a result of these laws, the federal government provided grants of several hundred million dollars to finance the construction of wastewater treatment facilities throughout the country.

Many of these locally or federally funded treatment plants are aging; based on my experience, I rate some as dinosaurs. The point is that many facilities are facing problems caused by aging equipment, facilities, and infrastructure. Complicating the problems associated with natural aging is the increasing pressure on inadequate older systems to meet the demands of increased population and urban growth. Facilities built in the 1960s and 1970s are now 30 to 40 years old; not only are they showing signs of wear and tear, but they simply were not designed to handle the level of growth that has occurred in many municipalities.

Regulations often lead to a need to upgrade. By matching funds or providing federal money to cover some of the costs, municipalities can take advantage of a window of opportunity to improve their facility at a lower direct cost to the community. Those federal dollars, of course, do come with strings attached; they are to be spent on specific projects in specific areas. On the other hand, many times new regulatory requirements are put in place without the financial assistance required to implement them. When this occurs, either local communities ignore the new requirements (until caught and forced to comply) or they face the situation and implement them through local tax hikes to pay for the cost of compliance.

An example of how a change in regulations can force the issue is demonstrated by the demands made by the Occupational Safety and Health Administration (OSHA) and USEPA in their Process Safety Management (PSM)/ Risk Management Planning (RMP) regulations (29 CFR 1910.119). These regulations put the use of elemental chlorine (and other listed hazardous materials) under scrutiny. Moreover, because of these regulations, plant managers throughout the country are forced to choose which side of a double-edged sword cuts their way the most. One edge calls for full compliance with the regulations (analogous to stuffing the regulation through the eye of a needle). The other edge calls for substitution— that is, replacing elemental chlorine with a nonlisted chemical (e.g., hypochlorite) or with a physical (ultraviolet irradiation, UV) disinfectant. Either way, it is a very costly undertaking.

Note: Many of us who have worked in water and wastewater treatment for years characterize PSM and RMP as the elemental chlorine killer. You have probably heard the old saying: "If you can't do away with something in one way, then regulate it to death."

Note: Changes resulting because of regulatory pressure sometimes mean replacing or changing existing equipment or increased chemical costs (e.g., substituting hypochlorite for chlorine typically increases costs threefold) and could easily involve increased energy and personnel costs. Equipment condition, new technology, and financial concerns are all considerations when upgrades or new processes are chosen. In addition, the safety of the process must be considered, of course, because of the demands made by USEPA and OSHA. The potential of harm to workers, the community, and the environment is under study, as are the possible long-term effects of chlorination on the human population.

■ CASE STUDY 1.2. Chesapeake Bay and Nutrients: A Modest Proposal

In Case Study 1.1, the technical problems and political controversy regarding nutrient contamination of the Chesapeake Bay were discussed. Nutrient pollution in Chesapeake Bay and other water bodies is real and ongoing, and the controversy over what the proper mitigation procedures might be is intense and neverending (and very political). Nutrients are substances that all living organisms need for growth and reproduction. Two major nutrients, nitrogen and phosphorus, occur naturally in water, soil, and air. Nutrients are present in animal and human waste and in chemical fertilizers. All organic material such as leaves and grass clippings contains nutrients. These nutrients cause algal growth and depletion of oxygen in the Bay, which leads to the formation of dead zones lacking in oxygen and aquatic life.

The U.S. Fish and Wildlife Service (USFWS, 2007) pointed out that nutrients can find their way to the Bay from anywhere within the 64,000-square-mile Chesapeake Bay watershed, and that is the problem. All streams, rivers, and storm drains in this huge area eventually lead to the Chesapeake. The activities of over 13.6 million people in the watershed have overwhelmed the Bay with excess nutrients. Nutrients come from a wide range of sources, including sewage treatment plants (20 to 22%), industry, agricultural fields, and even the atmosphere. Nutrient inputs are divided into two general categories: point sources and nonpoint sources.

Sewage treatment plants, industries, and factories are the major point sources. These facilities discharge wastewater containing nutrients directly into a waterway. Although each facility is regulated for the amount of nutrients that can be legally discharged, at times violations still occur.

In this text, it is the wastewater treatment plant, a point-source discharger, that is of concern to us. It should be pointed out that wastewater treatment plants (approximately 350 units outfalling effluent to nine major rivers and other locations all flowing into the Chesapeake Bay region) discharge somewhere between 20 and 22% of total nutrients into the Bay. Many target these point-source dischargers as being the principal causes of oxygen depletion and creators of dead zones within the Bay. If this is true, one needs to ask the question: Why is wastewater outfalled into the Chesapeake Bay in the first place? Water is the new oil, and if we accept this as fact we should therefore preserve and use our treated wastewater with great care and even greater utility. Thus, it makes good sense (to me) to take the wastewater that is currently being discharged into Chesapeake Bay and reuse it. This recycling of water saves raw water supplies in reservoirs and aquifers and limits the amount of wastewater that is discharged from wastewater treatment plants into public waterways, such as Chesapeake Bay. Properly treated wastewater could be used for many other purposes that raw water now serves, such as irrigating lawns, parks, gardens, golf courses, and farms; fighting fires; washing cars; controlling dust; cooling industrial machinery and towers and nuclear reactors; making concrete; and cleaning streets. It is my contention, of course, that when we get thirsty enough we will find another use for properly treated and filtered wastewater, and when this occurs we certainly will not use this treated wastewater for any purpose other than quenching our thirst. In reality, we are doing this already, but this is the subject for another forthcoming book: *Water Is the New Oil: Sustaining Freshwater Supplies.*

Let's get back to the Chesapeake Bay problem. The largest source of nutrients dumped into the Bay are from nonpoint sources. These nonpoint sources pose a greater threat to the Chesapeake ecosystem, as they are much more difficult to control and regulate. It is my view that, because of the difficulty of controlling runoff from agricultural fields, the lack of political will, and the technical difficulty of preventing such flows, wastewater treatment plants and other end-of-pipe dischargers have become the targets of convenience for the regulators. The problem is that the regulators are requiring the expenditure of hundreds of millions to billions of dollars to upgrade wastewater treatment to biological nutrient removal (BNR), tertiary treatment, or the combining of microfiltration membranes with a biological process to produce superior quality effluent—these requirements are commendable, interesting, and achievable but not at all necessary.

What is the alternative, the answer to the dead zone problem—the lack of oxygen in various locations in Chesapeake Bay? Putting it simply, take a portion of the hundreds of millions of dollars earmarked for upgrading wastewater treatment plants (which is a total waste, in my opinion) and build mobile, floating platforms containing electromechanical aerators or mixers. These platforms should be outfitted with diesel generators and accessories to provide power to the mixers. The mixer propellers will be adjustable and will be able to mix at a water depth of as little as 10 feet, or they can be extended to mix at a depth of 35 feet. Again, these platforms are mobile. When a dead zone appears in the Bay, the mobile platforms with their mixers are moved to a center portion of the dead zone area and energized at the appropriate depth. These mobile platforms are anchored to the Bay bottom and so arranged to accommodate shipping to ensure that maritime traffic is not disrupted. The idea is to churn the dead zone water and sediment near the benthic zone and force a geyser-like effect above the surface to aerate the Bay water in the dead zone regions. Nothing adds more oxygen to water than natural or artificial aeration. Of course, while aerating and forcing oxygen back into the water, bottom sediments containing contaminants will also be stirred up and sent to the surface, and temporary air pollution problems will occur around the mobile platforms. Some will view this turning up of contaminated sediments as a bad thing, not a good thing. I, in contrast, suggest that removing contaminants from the Bay by volatizing them is a very good thing.

How many of these mobile mixer platforms will be required? It depends on the number of dead zones. Enough platforms should be constructed to handle the average number of dead zones that appear in the Bay in the warm season.

Will this modest proposal—using aerators to eliminate dead zones—actually work? I do not have a clue. This proposal makes more sense to me, though, than spending billions of dollars on upgrading wastewater treatment plants and effluent quality when they account for only 20 to 22% of the actual problem. The regulators and others do not have the political will or insight to go after the runoff that is the real culprit in contaminating Chesapeake Bay with nutrient pollution.

Recall that it was that great mythical hero Hercules, the world's first environmental engineer, who said that "dilution is the solution to pollution." I agree with this; however, in this text, the solution to preventing dead zones in Chesapeake Bay is to prohibit the discharge of wastewater from point sources (i.e., reuse to prevent abuse) and to aerate the dead zones.

1.4.3 Privatization and/or Reengineering

As mentioned, water and wastewater treatment operations are undergoing a new paradigm shift. I explained that this paradigm shift focused on the holistic approach to treating water; however, the shift is more inclusive. It also includes thinking outside the box. To remain efficient and therefore competitive in the real world of operations, water and wastewater facilities have either bought into the new paradigm shift or have been forcibly shifted to doing other things (often these other things have little to do with water or wastewater operations) (Johnson and Moore, 2002).

Experience has shown that few words conger up more fear among plant managers than *privatization* or *reengineering*. *Privatization* means allowing private enterprise to compete with the government in providing public services such as water and wastewater operations. Privatization is often proposed as one solution to the numerous woes facing water and wastewater utilities, including corruption, inefficiencies (dysfunctional management), and the lack of capital for needed service improvements and infrastructure upgrades and maintenance. Existing management, on the other hand, can accomplish reengineering, internally, or it can be used (and usually is) during the privatization process. *Reengineering* is the systematic transformation of an existing system into a new form to realize quality improvements in operation, system capability, functionality, performance, or evolvability at a lower cost, schedule, or risk to the customer.

Many on-site managers consider privatization and reengineering schemes threatening. In the worst-case scenario, a private contractor could bid the entire staff out of their jobs. In the best case, privatization and reengineering are often a very real threat that forces on-site managers into workforce cuts (the Waldrop Syndrome, explained in Case Study 1.3), improving efficiency, and reducing costs. Using fewer workers, on-site managers must work to ensure that the community receives safe drinking water and that the facility meets standards and permits, without injury to workers, the facility, or the environment.

Local officials take a hard look at privatization and reengineering for a number of reasons:

- *Decaying infrastructures*—Many water and wastewater operations include water and wastewater infrastructures that date back to the early 1900s. The most recent systems were built with federal funds during the 1970s, and even these now require upgrading or replacing. USEPA recently estimated that the nation's 75,000+ drinking water systems alone will require more than $100 billion in investments over the next 20 years. Wastewater systems will require a similar level of investment.

- *Mandates*—The federal government has reduced its contributions to local water and wastewater systems over the past 30 years while at the same time imposing stricter water quality and effluent standards under the Clean Water Act and Safe Drinking Water Act. Moreover, as previously mentioned, new unfunded mandated safety regulations, such as OSHA's Process Safety Management and USEPA's Risk Management Planning, are expensive to implement using local sources of revenues or state revolving loan funds.
- *Hidden function*—Earlier it was stated that much of the work of water and wastewater treatment is a "hidden function." Because of this lack of visibility, it is often difficult for local officials to commit to making the necessary investments in community water and wastewater systems. Simply, the local politicians lack the political will, as water pipes and interceptors are not visible and not perceived as immediately critical for adequate funding. It is easier for elected officials to ignore them in favor of expenditures on more visible services, such as police and fire. Additionally, raising water and sewage rates to cover operations and maintenance costs is not always possible because it is an unpopular move for elected officials to make. This means that water and sewer rates do not adequately cover the actual cost of providing services in many municipalities.

In many locations throughout the United States, expenditures on water and wastewater services are the largest problem facing local governments today. (This is certainly the case for those municipalities struggling to implement the latest stormwater and nutrient reduction requirements.) Thus, this area presents a great opportunity for cost savings. Through privatization, water/wastewater companies can take advantage of advanced technology, more flexible management practices, and streamlined procurement and construction practices to lower costs and make the critical improvements more quickly.

With regard to privatization, the view taken in this text is that ownership of water resources, treatment plants, and wastewater operations should be maintained by the public (local government entities) to prevent a "Tragedy of the Commons"; that is, free access and unrestricted demand for water (or other natural resource) ultimately dooms the resource through overexploitation by private interests. However, because management is also a hidden function of many public service operations (e.g., water and wastewater operations), privatization may be a better alternative to prevent creating a home for dysfunctional managers and for ROAD gangs (see Case Study 1.3).

The bottom line: Water and wastewater are commodities, the quantity and quality of which are much too important to leave to the whims of public authorities.

■ CASE STUDY 1.3. Waldrop Syndrome

In 1995, G. ("don't call me George") Daniel Waldrop, Director of Operations for one of the nation's largest sanitation districts, had a dilemma. The commission that governed his operation was bandying about the idea of privatizing the organization. The commissioners reasoned that public ratepayers would be better served by a private, efficient operation. In their view, private operation would not only save money but also bring in talented top managers. The commissioners felt that public service entities do not have a stellar history of attracting quality managers; for example, they viewed their own operation as a repository for a bunch of good ol' boys belonging to the ROAD (Retired on Active Duty) gang. Unfortunately for the commissioners (and anyone else for that matter), once they become entrenched, it is almost impossible, excepting for criminal behavior bordering on murder, to get rid of these no-loads.

Waldrop understood that one of the first moves any privatizer would likely make would be to replace him and many of his cronies with proven private, non-public service professionals. This was, of course, unacceptable to him. Waldrop, though, had an ace in the hole: He knew that the commission would delay making the decision to privatize or not until after they conducted a study (a pilot study; engineers just love the words "pilot study") to determine if privatization actually made sense in their operation.

Waldrop understood that the results of the pilot study would help to determine whether privatization would actually occur. Knowing that the metrics involved with the commissioners' pilot study would focus on staffing, value-added operation, and total costs, Waldrop decided to nip the situation in the bud, so to speak. He knew that each of the 11 wastewater treatment plants for which he was responsible was overstaffed. Each plant had a workforce of approximately 44 personnel. Of this number, each plant had 4 to 6 plant operator assistants. Many of these plant operator assistants were licensed operators at the lowest professional certification levels (i.e., Class 3 or 4 Wastewater Operator, with Class 4 being the lowest certification level). Annual salaries for each assistant plant operator averaged around $32,000, depending on longevity. In reviewing assistant operator pay and experience levels,

Waldrop was surprised to find that their average continuous time served on the job was 12 years. And, although Waldrop understood that these operator assistants performed important functions at each of his treatment plants, he figured they simply had to go. He reasoned that reducing plant staff levels to less than 40 personnel would save a large portion of the annual outlay of funds for salaries, benefits, and personnel equipment (uniforms, safety shoes, training expenses, etc.). To address benefits (company-paid life insurance, retirement, and total medical care), Waldrop multiplied the cost of these benefits (1.4 times annual salary) and determined that on average each of the 56 plant operator assistants cost his operation $44,800 per annum. He then did the math: 56 × $44,800 = $2,508,800. Waldrop was surprised at this total cost figure; he could not fathom how a bunch of assistant operators could cost so much!

Before the upcoming budget submission deadline of July 1, 1996, Waldrop subtracted the cost of the operator assistants ($2,508,800), generated and distributed a memo giving all assistant treatment plant operators notice that their services were no longer required after June 30, 1996 (last day of the physical year), and sat back in his chair chain-smoking his cigarettes, actually quite pleased with himself.

Waldrop's feeling good about his actions and himself was short term, however. As it turned out, after the new physical year began July 1, 1996, the governor replaced the commission chairperson with a professional public administrator who had no intention (no personal motivation) to privatize the district. So, the privatization effort evaporated before it ever got started.

The results of Waldrop's downsizing of the assistant operators did not sit well with the 11 plant managers who had lost their assistants. Waldrop, being the dysfunctional public service manager that he was, had not bothered to discuss his downsizing plans with anyone, let alone with the plant managers—the ones most affected by the downsizing. If Waldrop had taken the time and the common decency (in the view of the plant managers) to explain his downsizing plan with them, they would have adamantly and forcefully argued against such a short-sighted move; for example, the plant managers would have explained to Waldrop that the assistant operators formed a pool of fully trained personnel to draw upon instantly whenever it became necessary to replace full-time treatment plant operators (the average turnover rate of fulltime operators was approximately three per year per plant).

In addition to losing their pool of trained assistant operators to draw upon to fill vacant positions, the lack of fully qualified assistant operators caused another glaring problem. The district had a very generous annual leave policy that allowed employees with more than 20 years of service to take 6 to 8 weeks of annual leave, depending on pay grade and length of service. When the assistants were removed from the payroll, fill-in personnel for those

employees wanting to use their annual leave each year were no longer available. Thus, it became increasingly more difficult for plant management personnel to allow key operators to take their annual leave each year.

Another unforeseen problem with doing away with the assistant operators was that, when an operator became ill or injured or encountered personal family problems, it placed a huge burden on those plant personnel who had to fill the shoes of those who were unable to work. When the plants had a cadre of assistant plant operators to draw upon to fill in for these employees, their absences had not a problem.

In summarizing the Waldrop Syndrome, it can be said that it adds meaning to Albert Einstein's thought in the following: "You cannot solve a problem within the mindset that created it."

1.4.4 Benchmarking

As shown in Case Study 1.3, it is primarily out of self-preservation (to retain their lucrative positions) that many utility directors work against the trend to privatize water, wastewater, and other public operations. Usually the real work to prevent privatization is delegated to the individual managers in charge of each specific operation because they also have a stake in making sure that their relatively secure careers are not affected by privatization. It can easily be seen that working against privatization by these local managers is in their own self-interest and in the interest of their workers whose jobs may be at stake.

The question is, of course, how do these managers go about preventing their water and wastewater operations from being privatized? The answer is rather straightforward and clear: Efficiency must be improved at reduced cost of operations. In the real world, this is easier said than done but is not impossible; for example, for those facilities properly implementing Total Quality Management (TQM) the process can be much easier. The advantage that TQM can offer plant managers is the variety of tools provided to help plan, develop, and implement water and wastewater efficiency measures. These tools include self-assessments, statistical process control, International Organization for Standards (ISO) 9000 and 14000 certification, process analysis, quality circle, and benchmarking (see Figure 1.4).

In this text, the focus is on use of the *benchmarking* tool to improve the efficiency of water and wastewater operations. Benchmarking is a process for rigorously measuring your performance vs. best-in-class operations and using the analysis to meet and exceed the best in class.

What benchmarking is:

1. Benchmarking vs. best practices gives water and wastewater operations a way to evaluate their operations overall:

a. How effective?

b. How cost effective?

2. Benchmarking shows plants both how well their operations stack up and how well those operations are implemented.
3. Benchmarking is an objective-setting process.
4. Benchmarking is a new way of doing business.
5. Benchmarking forces an external view to ensure correctness of objective setting.
6. Benchmarking forces internal alignment to achieve plant goals.
7. Benchmarking promotes teamwork by directing attention to those practices necessary to remain competitive.

Potential results of benchmarking:

1. Benchmarking may indicate the direction of required change rather than specific metrics:
   a. Costs must be reduced.
   b. Customer satisfaction must be increased.
   c. Return on assets must be increased.
   d. Maintenance must be improved.
   e. Operational practices must be improved.
2. Best practices are translated into operational units of measure.

Targets:

1. Consideration of available resources converts benchmark findings to targets.
2. A target represents what can realistically be accomplished within a given time frame.
3. Progress toward benchmark practices and metrics can be demonstrated.
4. Quantification of precise targets is based on achieving the benchmark.

Note: Benchmarking can be performance based, process based, or strategic based and can compare financial or operational performance measures, methods or practices, and strategic choices.

1.4.4.1　Benchmarking: The Process

When forming a benchmarking team, the goal should be to provide a benchmark that evaluates and compares privatized and reengineered water and wastewater treatment operations to your operation to be more efficient, remain competitive, and make continual improvements. It is important to point out that benchmarking is more than simply setting a performance reference or comparison; it is a way to facilitate learning for continual improvements. The key to the learning process is looking outside one's own plant to other plants that have discovered better ways of achieving improved performance.

Start → Plan → Research → Observe → Analyze → Adapt

FIGURE 1.4 Benchmarking process.

1.4.4.1.1　*Benchmarking Steps*

As shown in Figure 1.4, the benchmarking process consists of five steps:

1. *Planning*—Managers must select a process (or processes) to be benchmarked. A benchmarking team should be formed. The process of benchmarking must be thoroughly understood and documented. The performance measure for the process should be established (i.e., cost, time, and quality).
2. *Research*—Information on the best-in-class performer must be determined through research. The information can be derived from the industry's network, industry experts, industry and trade associations, publications, public information, and other award-winning operations.
3. *Observation*—The observation step is a study of the performance level, processes, and practices of the benchmarking subject that have achieved those levels, as well as other enabling factors.
4. *Analysis*—In this phase, performance levels among facilities are compared. The root causes for any performance gaps are studied. To make accurate and appropriate comparisons, the comparison data must be sorted, controlled for quality, and normalized.
5. *Adaptation*—This phase is putting what is learned throughout the benchmarking process into action. The findings of the benchmarking study must be communicated to gain acceptance, functional goals must be established, and a plan must be developed. Progress should be monitored and, as required, corrections in the process made.

Note: Benchmarking should be interactive. It should also recalibrate performance measures and improve the process itself.

■ CASE STUDY 1.4. Benchmarking: An Example

To gain a better understanding of the benchmarking process, the following limited example is provided. (It is in outline and summary form only, as discussion of a full-blown study is beyond the scope of this text.)

Rachel's Creek Sanitation District

Introduction

In January 1997, Rachel's Creek Sanitation District formed a benchmarking team with the goal of providing a benchmark that evaluates and compares privatized and reengineered wastewater treatment operations to those of Rachel's Creek in order to be more efficient and remain competitive. After 3 months of evaluating wastewater facilities using the benchmarking tool, our benchmarking is complete. This report summarizes our findings and should serve as a benchmark by which to compare and evaluate Rachel's Creek Sanitation District operations.

Facilities

41 wastewater treatment plants throughout the United States

Target Areas

The benchmarking team focused on the following target areas for comparison:

1. Reengineering
2. Organization
3. Operations and maintenance
   a. Contractual services
   b. Materials and supplies
   c. Sampling and data collection
   d. Maintenance
4. Operational directives
5. Utilities
6. Chemicals
7. Technology
8. Permits
   a. Water quality
   b. Solids quality
   c. Air quality
   d. Odor quality
9. Safety
10. Training and development
11. Process
12. Communication
13. Public relations
14. Reuse
15. Support services
    a. Pretreatment
    b. Collection systems
    c. Procurement
    d. Finance and administration
    e. Laboratory
    f. Human resources

Summary of Findings

Our overall evaluation of Rachel's Creek Sanitation District as compared to our benchmarking targets is a good one; that is, we are in good standing as compared to the 41 target facilities we benchmarked against. In the area of safety, we compare quite favorably. Only plant 34, with its own full-time safety manager, appeared to be better than we are. We were very competitive with the privatized plants in our usage of chemicals and far ahead of many public plants. We were also competitive in the use of power. Our survey of what other plants are doing to cut power costs showed that we had clearly identified those areas of improvement and that our current effort to further reduce power costs is on track. We were far ahead in the optimization of our unit processes, and we were leaders in the area of odor control.

We also found areas where we need to improve. To the Rachel's Creek employee, reengineering applies only to the treatment department and has been limited to cutting staff while plant practices and organizational practices remain outdated and inefficient. Under the reengineering section of this report, we have provided a summary of reengineering efforts at the reengineered plants visited. The experiences of these plants can be used to improve our own reengineering effort. The next area we examined was our organization and staffing levels. A private company could reduce the entire treatment department staff by about 18 to 24%, based on the number of employees and not costs. In the organization section of this report, organizational models and their staffing levels are provided as guidelines to improving our organization and determining optimum staffing levels. The last big area where we need to improve is in the way we accomplish the work we perform. Our people are not used efficiently because of outdated and inefficient policies and work practices. Methods to improve the way we do work are found throughout this report. We noted that efficient work practices used by private companies allow plants to operate with small staffs.

Overall, Rachel's Creek Sanitation District's treatment plants are performing much better than other public service plants. Although some public plants may have better equipment, better technology, and cleaner effluents, their labor and materials costs are much higher than ours. Several of the public plants were in bad condition. Contrary to popular belief, the privately operated plants had good to excellent operations. These plants met permit, complied with safety regulations, maintained plant equipment, and kept the plant clean. Due to their efficiency and low staff, we feel that most of the privately operated plants are performing better than we are. We agree that this needs to be changed. Using what we learned during our benchmarking effort, we can be just as efficient as a privately operated plant and still maintain our standards of quality.

The bottom line on privatization: Privatization is becoming of greater and greater concern. Governance boards, such as state commissions, see privatization as a potential way to shift liability and responsibility from the municipality's shoulders, with the attractive bonus of cutting costs. Water and wastewater facilities face constant pressure to work more efficiently and more cost effectively with fewer workers and to produce a higher quality product; that is, all functions must be value added. Privatization is increasing, and many municipalities are seriously considering outsourcing parts or all of their operations to contractors (Drinan, 2001).

1.5 WATER: THE NEW OIL?

Even though over 70% of the Earth is covered with water, only 3% is fit for human consumption, of which two thirds is comprised of frozen and largely uninhabited ice caps and glaciers, leaving 1% available for consumption. The remaining 97% is saltwater, which cannot be used for agriculture or drinking. If all of the Earth's water fit in a quart jug, available freshwater would not equal a teaspoon; thus, some (including the author) would consider water the new oil. Let's take a moment to make a few important points about water, the new oil.

Unless you are thirsty, in real need of refreshment, when you look at a glass of water you might wonder what could be more boring. The curious might want to know what physical and chemical properties of water make the water in the glass so unique and necessary for living things. Again, when you look at a glass of water, when you taste it and smell it, well, what could be more boring? Pure water is virtually colorless and has no taste or smell, but the hidden qualities of water make it a most interesting subject.

When the uninitiated become initiated to the wonders of water, one of the first surprises is that the total quantity of water on Earth is much the same now as it was more than 3 or 4 billion years ago, when the 320+ million cubic miles of it were first formed—there is no more freshwater on Earth today than there was millions of years ago. The water reservoir has gone round and round, building up, breaking down, cooling, and then warming. Water is very durable but remains difficult to explain because it has never been isolated in a completely undefiled state.

Have you ever wondered about the nutritive value of water? Well, the fact is that water has no nutritive value. It has none; yet, it is the major ingredient of all living things. Consider yourself, for example. Think about what you need to survive—just to survive. Food? Air? PS-3? MTV? Water? Naturally, the focus of this text is on water. Water is of major importance to all living things; in some organisms, up to 90% of their body weight comes from water. In humans, up to 60% of their body weight is water, the brain is composed of 70% water, the lungs are nearly 90% water, and about 83% of our blood is water. Water

also helps us digest our food, transport waste, and control body temperature. Each day humans must replace 2.4 liters of water, some through drinking and the remainder by the foods we eat.

There wouldn't be any you, me, or Lucy the dog without the existence of an ample liquid water supply on Earth. The unique qualities and properties of water are what make it so important and basic to life. The cells in our bodies are full of water. The excellent ability of water to dissolve so many substances allows our cells to use valuable nutrients, minerals, and chemicals in biological processes. Water's "stickiness" (from surface tension) plays a part in the body's ability to transport these materials throughout our bodies. The carbohydrates and proteins that out bodies use as food are metabolized and transported by water in the bloodstream. No less important is the ability of water to transport waste material out of our bodies.

Water facts:

- Water is used to fight forest fires; yet, we spray water on coal in a furnace to make it burn better.
- Chemically, water is basically hydrogen oxide. Upon more advanced analysis, however, it can be a mixture of more than 30 possible compounds. In addition, all of its physical constants are abnormal (strange).
- At a temperature of 2900°C, some substances that contain water cannot be forced to part with it; yet, others that do not contain water will liberate it when even slightly heated.
- When liquid, water is virtually incompressible; as it freezes, it expands by an 11th of its volume.

For all of these reasons, and for many others, we can truly say that water is special, strange, and different.

1.5.1 CHARACTERISTICS OF WATER

To this point in this text many things have been said about water; however, it has not been said (nor will it be) that water is plain. Nowhere in nature is plain water to be found. Here on Earth, with a geologic origin dating back over 3 to 5 billion years, water found in even its purest form is composed of many constituents. You probably know that the chemical description of water is H_2O—that is, one atom of oxygen bound to two atoms of hydrogen. The hydrogen atoms are attached to one side of the oxygen atom, resulting in a water molecule having a positive charge on the side where the hydrogen atoms are and a negative charge on the other side, where the oxygen atom is. Because opposite electrical charges attract, water molecules tend to attract each other, making water kind of "sticky," as the hydrogen atoms (positive charge) attract the oxygen side (negative charge) of other water molecules.

Important Point: All of these water molecules attracting each other mean they tend to clump together. This is why water drops are, in fact, drops! If not for some of Earth's forces, such as gravity, a drop of water would be ball shaped—a perfect sphere. Even if it does not form a perfect sphere on Earth, we should be happy that water is sticky.

Along with H_2O molecules, hydrogen (H^+), hydroxyl (OH^-), sodium, potassium, and magnesium, other ions and elements are present. Additionally, water contains dissolved compounds, including various carbonates, sulfates, silicates, and chlorides. Rainwater, often assumed to be the equivalent of distilled water, is not immune to contamination that is collected as it descends through the atmosphere. The movement of water across the face of land contributes to its contamination when the water takes up dissolved gases, such as carbon dioxide and oxygen, and a multitude of organic substances and minerals leached from the soil. Do not let that crystal-clear lake or pond fool you. These bodies of water are not filled with water alone but are composed of a complex medium of chemical ingredients far exceeding the brief list presented here; it is a special medium in which highly specialized life can occur.

How important is water to life? To answer this question all we need do is to take a look at the common biological cell, as it easily demonstrates the importance of water to life. Living cells are comprised of a number of chemicals and organelles within a liquid substance (the cytoplasm), and survival of the cell may be threatened by changes in the proportion of water in the cytoplasm. This change in proportion of water in the cytoplasm can occur through desiccation (evaporation), oversupply, or the loss of either nutrients or water to the external environment. A cell that is unable to control and maintain homeostasis (i.e., the correct equilibrium/proportion of water) in its cytoplasm may be doomed; it may not survive.

Important Point: As mentioned, water is called the *universal solvent* because it dissolves more substances than any other liquid. This means that wherever water goes, either through the ground or through our bodies, it takes along valuable chemicals, minerals, and nutrients.

1.5.2 WATER USE

In the United States, rainfall averages approximately 4250×10^9 gallons a day. About two thirds of this rainfall returns to the atmosphere through evaporation directly from the surface of rivers, streams, and lakes and transpiration from plant foliage. This leaves approximately 1250×10^9 gallons a day to flow across or through the Earth to the sea.

The U.S. Geological Survey (USGS, 2004) reported that estimates in the United States indicate that about 408 billion gallons per day (1000 million gallons per day, abbreviated Bgal/day) were withdrawn from all uses during the year 2000. This total has varied less than 3% since 1985 as withdrawals have stabilized for the two largest uses—thermoelectric power and irrigation. Fresh groundwater withdrawals (83.3 Bgal/day) during 2000 were 14% more than during 1985. Fresh surface-water withdrawals for 2000 were 262 Bgal/day, varying less than 2% since 1985.

About 195 Bgal/day, or 8% of all freshwater and saline-water withdrawals for 2000, were used for thermoelectric power. Most of this water was derived from surface water and used for once-through cooling at power plants. About 52% of fresh surface-water withdrawals and about 96% of saline-water withdrawals were for thermoelectric-power use. Withdrawals for thermoelectric power have been relatively stable since 1985.

Irrigation totaled 137 Bgal/day and represented the largest use of freshwater in the United States in 2000. Since 1950, irrigation has accounted for about 65% of total water withdrawals, excluding those for thermoelectric power. Historically, more surface water than groundwater has been used for irrigation; however, the percentage of total irrigation withdrawals from groundwater has continued to increase, from 23% in 1950 to 42% in 2000. Total irrigation withdrawals were 2% more for 2000 than for 1995, because of a 16% increase in groundwater withdrawals and a small decrease in surface-water withdrawals. Irrigated acreage more than doubled between 1950 and 1980, then remained constant before increasing nearly 7% between 1995 and 2000. The number of acres irrigated with sprinkler and microirrigation systems has continued to increase, and they now account for more than one half of the total irrigated acreage.

Public-supply withdrawals were more than 43 Bgal/day for 2000 compared to public-supply withdrawals in 1950 of 14 Bgal/day. During 2000, about 85% of the population in the United States obtained drinking water from public suppliers, compared to 62% during 1950. Surface water provided 63% of the total during 2000, whereas surface water provided 74% during 1950.

Self-supplied industrial withdrawals totaled nearly 20 Bgal/day in 2000, or 12% less than in 1995; compared to 1985, industrial self-supported withdrawals declined by 24%. Estimates of industrial water use in the United State were largest during the years from 1965 to 1980, but usage estimates for 2000 were at the lowest level since reporting began in 1950. Combined withdrawals for self-supplied domestic, livestock, aquaculture, and mining were less than 13 Bgal/day for 2000 and represented about 3% of total withdrawals. California, Texas, and Florida accounted for one fourth of all water withdrawals for 2000. States with the largest surface-water withdrawals were California,

Texas, and Nebraska, all of which had large withdrawals for irrigation.

All of this factual information is interesting. Well, it is interesting to those of us who are admirers, purveyors, or students of water. Obviously, these are the folks who read and use a book like this one. However, the question is what does all of this information have to do with water being the new oil?

Water is the new oil because there is no more freshwater on Earth today than there was millions of years ago. Yet, at the present time, more than 6 billion people share it. Since the 1950s, the world population has doubled, but water use has tripled. A simple extrapolation of today's water usage compared to projected usage in the future shows that water will become a much more important commodity than it is today. Earlier it was suggested that the day is coming when a gallon of water will be comparable in value to (or even more expensive than) a gallon of gasoline. There are those who will read this and shake their heads in doubt and think: "Water is everywhere. Water belongs to no one; water belongs to everyone; no one owns the water. Water pours freely from the sky. Water has no real value. Certainly water is nowhere near as valuable as gasoline, nowhere near as valuable as gold or diamonds."

Water has no real value? Really?

With regard to water and diamonds and which of the two is more valuable, consider the following. In his epic book, *The Wealth of Nations*, Adam Smith, the 18th-century philosopher credited with laying the foundation of modern economics, described the paradox of diamonds and water. Smith asked how could it be that water, so vital to life, was so cheap, while diamonds, used only for adornment, were very costly? Smith pointed out that, when it came to value, a container full of diamonds was exponentially more valuable than an equal amount of water. In today's value system, this is still true. It is true unless you are dying of thirst. While on the verge of dying of thirst, what value would you place on that same container of diamonds? On the container of water? If you were offered one or the other, which would you choose? Which would you give up everything you own for?

That is my point. Although Adam Smith used the paradox for his own pedagogical purposes, explaining the basic concepts of supply and demand and showing that prices reflect relative scarcity, today the paradox provides a troubling description of the way water is treated in our economy. Water may be critical to life itself, but we do not have a clue as to its true value (USEPA, 2003). We have not reached that point yet. However, with the majority of the world's population being relatively thirsty and many dying of thirst or dying from drinking filthy, pathogen-contaminated water, the dawn of new understanding is just around the corner.

Moreover, as our population continues to grow and degradation of the world's supply worsens and global climate changes accelerate, it is my view that the diamond and water paradox will flip-flop. Diamonds will lose some value when compared to safe, potable drinking water. This will occur because when it comes to sustaining life and quenching our thirst, all of the diamond-encrusted drinking glasses filled to the brim with diamond-clear water will be just what the doctor ordered, thank you very much!

Years ago, when I first stated that water will be more valuable than an equal amount of gasoline, many folks (reviewers) asked me what part of the planet Mars was I from? Well, I have not been to Mars and have not changed my opinion on the ever-increasing value of water—and this same realization will soon confront us all. By the way, with regard to that water we flush down our toilets and drains, the day is coming, in my opinion, when we will have direct pipe-to-pipe connection from wastewater treatment plants to our municipal potable water supply systems. Why? Because water is the new oil. Furthermore, have you heard about the recent discovery of the presence of water on ancient Mars? My guess is that if we do not protect our water supplies, the Mars of today may be the Earth of tomorrow. This is a thought to keep close at hand, close at heart, very close to the brain cells as a reminder of what really matters.

If you do not accept the premise that water is the new oil, maybe you are willing to accept the possibility that we can use water to make oil. I am not talking about converting hydrogen from water into fuel; instead, consider that we can turn algae into fuel. Scientists at Old Dominion University (ODU), Norfolk, Virginia, for example, are conducting successful research on growing algae in treated sewage and extracting fatty oils from the weedy slime, then converting the oils into cleaner-burning fuel. As part of the research project, the algae is grown in tanks at a wastewater treatment plant in Norfolk and then converted to biofuel at an ODU facility. It should be pointed out that this wastewater-grown-algae-to-oil-to-fuel-process has already proven itself in New Zealand (Harper, 2007).

The bottom line: The day is drawing near when water becomes new oil. This day is closer than we may be willing to readily acknowledge.

Did You Know?

Growing algae in wastewater will soak up nutrients in wastewater at the wastewater plant, thus helping receiving water bodies that are suffering from excessive nutrients discharged by such treatment plants.

1.6 TECHNICAL MANAGEMENT VS. PROFESSIONAL MANAGEMENT

Water treatment operations management is management that is directed toward providing water of the right quality, in the right quantity, at the right place, at the right time, and at the right price to meet various demands. Wastewater treatment management is directed toward providing treatment of incoming raw influent (no matter what the quantity), at the right time, at the right level to meet regulatory requirements, and at the right price. The techniques of management are manifold both in water resource management and wastewater treatment operations. In water treatment operations, for example, management techniques may include (Mather, 1984):

> Storage to detain surplus water available at one time of the year for use later, transportation facilities to move water from one place to another, manipulation of the pricing structure for water to reduce demand, use of changes in legal systems to make better use of the supplies available, introduction to techniques to make more water available through watershed management, cloud seeding desalination of saline or brackish water, or area-wide educational programs to teach conservation or reuse of water.

Many of the management techniques employed in water treatment operations are also employed in wastewater treatment. In addition, wastewater treatment operations employ management techniques that may include upgrading existing systems for nutrient removal, reuse of process residuals in an Earth-friendly manner, and area-wide educational programs to teach proper domestic and industrial waste disposal practices.

Whether managing a waterworks or wastewater treatment plant, the expertise of the manager must include being a well-rounded, highly skilled individual. No one questions the need for incorporating these highly trained practitioners—well-versed in the disciplines and practice of sanitary engineering, biology, chemistry, hydrology, environmental science, safety principles, accounting, auditing, technical aspects, and operations—in both professions. Based on years of experience in the water and wastewater profession and personal experience dealing with high-level public service managers, however, engineers, biologists, chemists, and others with no formal management training and no proven leadership expertise can often be limited in their ability to solve the complex management problems currently facing both industries. I admit that my view is biased in this regard because my experience in public service has, unfortunately, exposed me to more dysfunctional than functional managers.

So what is dysfunctional management? How is it defined? Consider Case Study 1.5. Maybe it will provide an answer.

■ CASE STUDY 1.5. Dysfunctional Management

Earlier, in Case Study 1.2, G. Daniel Waldrop, Chief Operations Officer for a large, well-known sanitation district took various steps to downgrade his workforce in an effort to economize operational expenses. The need to economize, in his mind anyway, was driven by a threat to privatize his operation. Such a move would have been a direct threat to Waldrop because he assumed that a private operation would immediately replace him and his cronies with proven management personnel who were competent and less expensive. Simply put, privatization of his organization was viewed by Waldrop as the ultimate kiss of death—*the* career killer.

It is interesting to note that when Waldrop was downsizing his operation by doing away with all the operator assistant positions he was also looking at other, higher up positions that could be downgraded to further reduce costs; in his view, he simply had to make deeper cuts in personnel, the operation's largest expense. When it came time to deliver the bad news, however, especially to those engineers and other managers who might be the target of such cost-cutting measures, Waldrop was not the type to address the problem face-to-face. Waldrop suffered from chronic jellyfish syndrome in that, like the stately jellyfish, Waldrop simply did not possess a backbone.

Waldrop hired an outside professional management firm with extensive experience in auditing organizations and determining where to make cuts in the organization to economize operations. The cost-cutter (a.k.a., hatchet man) assigned to assess Waldrop's operation was highly skilled in his work; he had more than 20 years of experience in the field. It is interesting to note that the cost-cutter needed no more than 10 minutes at each of Waldrop's 11 facilities to come up with the solution to Waldrop's problem.

Two weeks after being hired, the cost-cutter met with Waldrop to deliver his findings and recommendations. The cost-cutter began his presentation by telling Waldrop that he could immediately save his organization approximately $1,355,200 per year simply by doing away with plant manager positions. Approximately 12 years earlier, Waldrop had created plant manager positions to handle management of each plant. The main qualification for each of these plant managers was that they had to be college graduates and engineers. Prior to hiring the engineers, the plants had been managed for years by plant superintendents. The superintendents were blue-collar employees who had worked their way to the top and had managed the plants successfully for several years; they had proven themselves to be competent managers. Waldrop found that he had difficulty talking to and relating to these superintendents, the blue-collar types. Simply, he had nothing in common with any of them, and they did not communicate at his level, his super-educated, superior-intellect level. Thus, the plant manager position was created so Waldrop could deal with people who understood him. Further, Waldrop, a dysfunctional manager, but a smart one, knew that the only thing protecting him from blame and termination for anything that went wrong at a plant was the plant manager, to whom he could readily point the finger of blame. Waldrop was one of those dysfunctional managers who are expert in personal survival techniques. He kept a worn, faded, tattered, dog-eared copy of *The Prince* within close reach at all times. Niccolo Machiavelli would have been proud.

As Waldrop listened intently, with jaw dropped, the hired cost-cutter explained that each plant manager earned an annual salary of $88,000 per annum plus 1.4 times that for benefits, equaling $123,200 per annum. Multiplying that figure by 11 (the total number of managers) produced a grand total of $1,355,200, a considerable cost that could be turned into considerable savings.

The cost-cutter ignored the shocked look on Waldrop's face and continued. "You currently have both a plant manager and a plant superintendent at each facility. This is overkill. Your upper management is top heavy with plant managers, plant superintendents, chief operators, and lead operators—too much costly management. When I asked what function the plant managers performed at each site, I pretty much got the same answer from each." The cost-cutter stopped to check his notes and Waldrop, feeling severe chest pains and an irritable bowel problem at the same time, sat there like a potted plant, wilting with each of the cost-cutter's words. "The plant managers all basically say the same thing," the cost-cutter continued. "Basically, they stated that their job is to oversee operations but never to interfere with the plant superintendents, unless the superintendents are incoherent or out of control for one reason or another. When I asked them if that was their total function, they all said, no, they were assigned to various organizational teams studying various organizational problems in line with the organization's TQM program."

After swallowing hard and finally finding his voice, Waldrop said, "Okay, thanks for your information. Please forward your invoice to me for payment. We no longer need your help. Oh, and thank you, thank you very much." And, with those words of absolute dismissal, the cost-cutter departed, never to return. Waldrop just sat there in his chair shaking his head in disgust. "It will be a cold day in hell before I ever get rid of my brother and sister engineers, no matter how much money it would save," he told himself.

In my Industrial Environmental Management, Risk Management, and Occupational Safety and Health Management undergraduate and graduate classes at Old Dominion University, my students are required to study and complete research projects on many aspects of professional management. Much of their research is based on their experiences while interning at public service entities where they spend their summers learning on-the-job skills and earning income to help pay their way through college. When they have completed their internships and return to the classroom, I am always amazed at how pumped up the students are. They cannot wait to relate their internship experience to their classmates, which is fortunate because all intern students are required to formally present their experiences to their peers in my classes.

Over the last 8 years, one of the interesting trends I have noticed about these returning students is that almost all of them have portrayed the managers they were exposed to in a negative light. I find this trend intriguing. I usually give the students a chance to say what they have to say (in my view, there is no such thing as an incorrect opinion), and then eventually I ask: "Can you say anything positive about the folks you worked for?" They are usually surprised that I would ask such a question, but eventually they decide that they did have a few good experiences here and there. Moreover, it is interesting that most of them have given the same answer to the following question: "What did you find was the biggest problem with the managers you worked with and for?" Almost every respondent has answered: "The managers I worked with seemed to have difficulty in making decisions. They simply did not want to make a decision."

After listening to the students' presentations, I finally get around to my main question: "Do you think the organization you worked for during the summer is functional or dysfunctional?" Most reply that the organization is functional but that the managers seemed dysfunctional. Year after year, I ask students: "How do you define the dysfunctional manager?" The answers I receive are usually vague, ambiguous, and definitely incomplete (like most OSHA regulations). One student told me that she would rather define beauty (which is impossible, according to her) than define the dysfunctional manager.

Eventually, the students, as all students eventually do, turn to their teacher (me) for a definition of the dysfunctional manager. The first time I was asked to come up with a definition, I was tongued tied. I stumbled over explaining that the dysfunctional manager cannot lead … that he cannot foresee the future … that he is not a problem solver … that he simply lacks that which is not so common—common sense. A few of the really good students accused me of copping out; they were correct, of course. I simply could not answer this question without having some time to think about it.

Keeping in mind that identifying the dysfunctional is relatively easy (especially if you have the misfortune to work for one) but that describing one is much more difficult, I thought about it and Case Study 1.6 is the result.

■ CASE STUDY 1.6. Defining the Dysfunctional Manager

Have you ever felt the need (the absolute, overwhelming, crushing, pulsating need) to get close and personal with your manager? That is, have you ever wanted to get in your manager's face and laugh, scream, howl, shout, spit, throttle, or just stare that stare that sums it all up? Maybe you are one of those straightforward folks who would take a different approach. Maybe you would just walk up to your manager and coolly, calmly, and with intense purpose reach out and place your resignation on the manager's junk heap of a desk, then march off smartly to the office door, make an about face, and issue your former oppressor the ultimate *coup de grace* whose meaning cannot be misunderstood.

Some daring folks (maybe you are one) might take a more dramatic approach. Those with ice water flowing through their veins might terminate a disturbing and distressing relationship with the boss by walking right into his office, picking up his trash can, moving it over to the nearest wall, inserting one foot into the can, donning a gas mask, and then placing an index finger against the wall. At this point in time, the manager is either ignoring you or is breathless with anticipation of your next move. This is what you want, of course; you now have the manager's full, undivided attention. When you have an attentive audience, you simply yell out as loud as you can: "*Elevator going up—you want to ride along, nerd-breath?*"

Of course, people have devised many other ways to end relationships with their bosses, their employers, their managers, their harassers—some classic, some more imaginative. The underlying question, though, is what was it that drove these people to end it, to quit, to self-terminate? Was an employee's terminal visit to face the manager brought about by feelings of anxiety, panic, frustration, depression, grief, guilt, shame, worry, anger, jealousy, or belligerence? Or maybe what it boiled down to was that the employee felt like he was "going to pieces." Certainly, some particular event or series of events triggers such a reaction, and those who reach this state may decide that it is time to put the pieces back together again. Thus, such actions are brought about for one simple reason (that same old standard reason): You have had all you are going to take, and you are not going to take it any more.

Well, whatever it is that initiates this "fed up with it all" attitude, I understand. The fact is that most of the rest of us out here in the real-world workplace also understand, although it is probably true that a small minority of workers out there (a very small group, too small to count) does not understand what it is that would drive any person to act in the manner described above. These unenlightened individuals probably view their managers (and management in general) as their guiding lights, their knights in shining white armor, their Horatio at the bridge, the very personification of Mother Teresa, their father or mother figure, or something even more stellar. This type of thinking is, of course, quite sad. You might even characterize it as dysfunctional thinking, but that is another story, another topic. What I am addressing in this particular treatise (diatribe) is dysfunctional management in the public service sector, or any sector of employment, although it should be quite obvious to all readers that dysfunctional thinking does have something to do with inept or dysfunctional management.

Don't you agree?

Let's get back to the reasons why workers (aren't we all classified in this terrible, ignoble category) suddenly decide that they have had enough; that is, why has the worker gone over the edge or thrown off the jellyfish label. (Note that no disrespect or disparagement is intended or directed toward the amorphous, silky, stately jellyfish; instead, the intent here is to point out that jellyfish are not equipped with backbones. They are gelatinous masses that tend toward quivering and quavering with each rustle of the wind, with each lapping of the water surface, with each undulation of the watery mass. Simply stated, jellyfish go with the flow; they have little choice. They are not equipped with backbones.)

In the process of throwing off the jellyfish label, some particular situation, some single event occurs—an initiator that causes or literally drives a suddenly enlightened worker to metamorphose from jellyfish to 600-pound gorilla mode. Why the worker finally finds the backbone and energy to sort things out, to right millions of wrongs, to change to something better is worth our consideration.

Don't you agree?

Exactly what generates the hurricane force wind at the back of the worker who makes this metamorphosis is what this treatise is all about. More specifically, this treatise is an account of the types of dysfunctional management practices that have driven environmental compliance workers in public service to react, to respond, to leave their places of employment.

Actually, when you get right down to it, is it not the practice of dysfunctional management that motivates us to react at all times—in one way or another? Pain stimulus is a highly effective motivational technique, whether it is called behavior modification or torture. (B.F. Skinner, go back to sleep; you have nothing positive to add to this discussion.)

At this point in the presentation a few individuals out there (less than 3% of the readers, probably) are asking themselves: What exactly is a dysfunctional manager? To ensure that all readers have a clear understanding (an absolutely clear understanding) of what a dysfunctional manager is, a definition or explanation is called for. Remember, it was that genius Voltaire who stated, "If you wish to converse with me, please define your terms." Simply stated, if management is considered to be the glue that holds an organization together, then functional management is the super glue. On the other hand, dysfunctional management is that agent (however nebulous it might be) that works to break the cohesive bond.

You don't care for analogy? You still want a more precise definition? Okay.

A more accurate, correct, and precise explanation of what a dysfunctional manager really is can be made more clearly by explaining what a dysfunctional manager *is not*. With this in mind, consider what I call the *dysfunctional dozen:*

1. A dysfunctional manager *is not* qualified to prevent turnover costs and hassles by using specific hiring and interviewing techniques.
2. A dysfunctional manager *is not* qualified to blend differing personality types, backgrounds, and age groups into a smooth-running, productive team.

Are you starting to get the idea? Yeah, I thought so. Let's move on.

3. A dysfunctional manager *is not* qualified to supervise former peers and friends without losing their respect.
4. A dysfunctional manager *is not* qualified to establish boundaries for supervisor/subordinate relationships that will not be misunderstood.
5. A dysfunctional manager *is not* qualified to quickly identify difficult employees and redirect them with swiftness and ease.
6. A dysfunctional manager *is not* qualified to relay constructive criticism without it being taken personally—even by the least sensitive employee.
7. A dysfunctional manager *is not* qualified to originate project plans and set goals that the staff will buy into.
8. A dysfunctional manager *is not* qualified to control absenteeism and tardiness (they have enough trouble controlling their own). They don't even try … why should they?
9. A dysfunctional manager *is not* qualified to fire or take corrective action—or to learn the legal implications for each.
10. A dysfunctional manager *is not* qualified to work under pressure.
11. A dysfunctional manager *is not* qualified to organize people, projects, and schedules on an ongoing basis (actually, not on any basis).
12. A dysfunctional manager *is not* qualified to keep top performers at their maximum level without burning out.

After having reviewed the dysfunctional dozen, even those who are normally confused should now have an understanding of what I am talking about in this treatise: *dysfunctional management* and the *dysfunctional public service manager*. In case there is still some lingering doubt, I will sum it up for you quite succinctly: The dysfunctional manager is without conscience, without heart, without feeling, without direction, without discretion, without motivation, without scruples, and without leadership ability.

Before moving on from the tenets of the "dozen," I would like to point out that, while some managers may fit cleanly into any one division of the dysfunctional dozen, most will display additional dysfunctional behaviors at one time or another, sooner or later. It's a given. But, keep in mind that it takes competence in only one aspect of the "dozen" to fully qualify someone as a dysfunctional manager.

Have you ever worked for a manager who meets the criteria listed in *every* member of the "dozen"? If you have, you are not alone. Perhaps those of us who have should start a support group of Dysfunctionary Survivors Anonymous (DSA).

Right about now, readers may be asking themselves: "If a manager consistently displays any of the shortcomings listed in the dysfunctional dozen, why not replace the manager? Why isn't such a loser fired?" These are, of course, logical questions, and they identify the proper solution, but the solution is not easy to implement. Three major problems stand in the way.

The first problem is the manager. Keep in mind that dysfunctional managers are dysfunctional but not necessarily stupid. In fact, because these managers are dysfunctional, they have learned through experience (as a tenet to their own self-preservation) to hide their incompetence and disability from those who have the power to take proper corrective action. They cover their dysfuntionalism with ambition. Ambition? Yes, but remember what Oscar Wilde said about ambition: "Ambition is the last resort of failure."

The second problem has to do with connections. Some dysfunctional managers owe their management positions to connections with those who own the company. The owners may or may not be aware of the dysfunctional manager's inherent problems and may not care. Their friendship or association priorities may rank above (believe it or not) their concern for the welfare, well-being, or mental health of "lower caste" employees. Even if the manager's shortcomings are known, the owner may still choose not to replace even the most dysfunctional manager. If the department is making a profit, upper management falsely believes that nothing is wrong. This, of course, is the drawback to the "if it ain't broke, don't fix it" rule. The dysfunctional manager/employee relationship *is* "broke," just not financially. Besides, keep in mind that a particular dysfunctional manager may be working for another higher up dysfunctional manager, and you know what that means: Birds of a feather tend to roost together.

The third problem has to do with the organization's culture. For example, some organizations have incorporated Total Quality Management (TQM)—sometimes referred to as Managing Total Quality (MTQ)—into their organizations. TQM is often implemented because an organization experiences a crisis; however, TQM is more often implemented at the instigation of some bored-to-tears top manager who has read or heard about it and thinks it seems like a good idea, an idea that will lead to the manager winning the Baldrige National Quality Award or, at the very least, receiving high approval ratings from the higher-ups (i.e., the brown-noser syndrome).

As mentioned earlier, I hold the view that TQM is important; however, in the hands of a dysfunctional manager, TQM can be used for purposes not intended. The author (and this text) supports the view that when employees are empowered (when employees closest to the work are empowered to correct problems or defects on their own) the middle manager is often taken out of the management scheme; the manager becomes redundant and not necessary. To dysfunctional managers, TQM provides a shield, a facade that they can hide behind. What I am saying here is that the dysfunctional manager can be a very cunning, adroit, sly, crafty, and well-connected person who has the uncanny ability to survive; TQM aids this process. TQM is just another management fad, a so-called panacea for all the ills that plague management. It was dreamed up by management theorists and gurus who have never run anything in their lives but have only thought about it from the half-baked point of view of human resource management. Let's face it. When you empower employees to make their own decisions, why do you need managers? TQM is not the silver bullet; effective management (i.e., leadership) is.

To aid the majority of dysfunctional managers who might read this text (with wonder, surprise, and guilt, if they have the intelligence and nerve to perform self-examination) and who might have difficulty understanding simple ideas and concepts, the term *dysfunctional manager* can be replaced with the term *dysfunctionary*.

I know what you might be thinking. Dysfunctionary? Dysfunctionary is not a word. It does not matter whether or not this word can be found in your standard college dictionary or even the *Oxford English Dictionary*. The words we all use every day to describe our own personal relationships with dysfunctionaries are most commonly found in dictionaries of slang—or are simply not suitable to be printed here (or anywhere else). A rose by any other name would smell as sweet, and a skunk still stinks even when called a pole-cat. Whether I call these burdens to the working world dysfunctional managers or dysfunctionaries, you will know what—and who—I mean ... that is, unless you are totally dysfunctional.

CHAPTER REVIEW QUESTIONS

Answers to chapter review questions can be found in Appendix A.

1.1 Define *paradigm* as used in this text.
1.2 Define *paradigm shift* as used in this text.
1.3 List five elements of the multiple-barrier approach.
1.4 "Water service delivery remains one of the 'hidden functions' of local government." Explain.
1.5 _____ Drinking Water Standards are not enforceable.
1.6 Explain the difference between privatization and reengineering.
1.7 Define *benchmarking*.
1.8 List the five benchmarking steps.

THOUGHT-PROVOKING QUESTION (ANSWERS WILL VARY)

Given the assignment and the proper resources, how would you clean up Chesapeake Bay?

REFERENCES AND SUGGESTED READING

Angele, Sr., F.J. 1974. *Cross Connections and Backflow Protection*, 2nd ed. Denver: American Water Association.

Anon. 2000. USGS says water supply will be one of challenges in coming century. *U.S. Water News Online*, March (http://www.uswaternews.com/archives/arcsupply/tusgsay3.html).

Capra, F. 1982. *The Turning Point: Science, Society and the Rising Culture*. New York: Simon & Schuster, p. 30.

Daly, H.E. 1980. Introduction to steady-state economics. In *Ecology, Ethics: Essays Toward a Steady-State Economy*. New York: W.H. Freeman & Company.

De Villiers, M. 2000. *Water: The Fate of our Most Precious Resource*. Boston: Mariner Books.

Drinan, J.E. 2001. *Water & Wastewater Treatment: A Guide for the Non-Engineering Professional*. Boca Raton, FL: CRC Press.

Garcia, M.L. 2001. *The Design and Evaluation of Physical Protection Systems*. London: Butterworth-Heinemann.

Gleick, P.H. 1998. *The World's Water 1998–1999: The Biennial Report on Freshwater Resources*. Washington, D.C.: Island Press.

Gleick, P.H. 2000. *The World's Water 2000–2001: The Biennial Report on Freshwater Resources*. Washington, D.C.: Island Press.

Gleick, P.H. 2004. *The World's Water 2004–2005: The Biennial Report on Freshwater Resources*. Washington, D.C.: Island Press.

Harper, S. 2007. Virginia grants to fuel green research. *The Virginian-Pilot*, June 30.

Holyningen-Huene, P. 1993. *Reconstructing Scientific Revolutions*. Chicago: University of Chicago, p. 134.

IBWA. 2004. *Bottled Water Safety and Security*. Alexandria, VA: International Bottled Water Association.

Johnson, R. and Moore, A. 2002. *Opening the Floodgates: Why Water Privatization Will Continue*, Policy Brief 17. Los Angeles, CA: Reason Public Policy Institute (www rppi.org.pbrief17).

Jones, B.D. 1980. *Service Delivery in the City: Citizen Demand and Bureaucratic Rules*. New York: Longman, p.2.

Jones, F.E. 1992. *Evaporation of Water*. Chelsea, MI: Lewis Publishers.

Lewis, S.A. 1996. *The Sierra Club Guide to Safe Drinking Water*. San Francisco, CA: Sierra Club Books.

Mather, J.R. 1984. *Water Resources: Distribution, Use, and Management*. New York: John Wiley & Sons.

McGhee, T.J. 1991. *Water Supply and Sewerage*, 6th ed. New York: McGraw-Hill.

Meyer, W.B. 1996. *Human Impact on Earth*. New York: Cambridge University Press.

Peavy, H.S. et al. 1985. *Environmental Engineering*. New York: McGraw-Hill.

Pielou, E.C. 1998. *Fresh Water*. Chicago: University of Chicago Press.

Powell, J.W. 1904. *Twenty-Second Annual Report of the Bureau of American Ethnology to the Secretary of the Smithsonian Institution, 1900–1901*. Washington, D.C.: U.S. Government Printing Office.

Spellman, F.R. 2003. *Handbook of Water and Wastewater Treatment Plant Operations*. Boca Raton, FL: Lewis Publishers.

Turk, J. and Turk, A. 1988. *Environmental Science*, 4th ed. Philadelphia, PA: Saunders College Publishing.

USEPA. 2003. Providing advice on how to pay for environmental protection: diamonds and water. *EFAB Newslett.*, 3(2) (www.epa.gov/efinpage/efab/newslaters/newsletters6.htm).

USEPA. 2005. *Water and Wastewater Security Product Guide*. Washington, D.C.: U.S. Environmental Protection Agency (http://cfpub.epa.gov.safewater/watersecurity/guide).

USEPA. 2006. *Watersheds*. Washington, D.C.: U.S. Environmental Protection Agency (http://www.epa.gov/owow/watershed/ whatis.html).

USFWS. 2007. *Nutrient Pollution*. Washington, D.C.: U.S. Fish and Wildlife Service (http://www.fws.gov/chesapeakebay/nutrient.htm).

USGS. 2004. *Estimated Use of Water in the United States in 2000*. Washington, D.C.: U.S. Geological Survey.

USGS. 2006. *Water Science in Schools*. Washington, D.C.: U.S. Geological Survey. Geological Survey.

2 Water/Wastewater Operators

Our Planet is shrouded in water, and yet 8 million children under the age of five will die this year from lack of safe water.

United Nations Environmental Program

If you are short of water, the choices are stark: conservation, treatment and reuse of wastewater, technological invention, or the politics of violence. It is not technology (engineering and science) that will mitigate the looming water crisis. Instead, providing safe drinking water to the masses can only be accomplished through politics and management. The irony is that both politics and management are at the heart of the water crisis.

2.1 INTRODUCTION

To begin our discussion of water and wastewater operators, it is important that we point out a few significant factors:

- Employment as a water and wastewater operator is concentrated in local government and private water supply and sanitary services companies.
- Postsecondary training is increasingly becoming an asset as the number of regulated contaminants grows and treatment unit processes become more complex.
- Operators must pass examinations certifying that they are capable of overseeing various treatment processes.
- Plants operate 24/7; therefore, plant operators must be willing to work shifts.
- Operators have a relatively high incidence of on-the-job injuries.

To properly operate a water treatment and distribution or a wastewater treatment and collection system usually requires a team of highly skilled personnel filling a variety of job classifications. Typical positions include plant manager/plant superintendent, chief operator, lead operator, operator, maintenance operator, distribution or interceptor system technicians, assistant operators, laboratory professionals, and clerical personnel, to list just a few.

Beyond the distinct job classification titles, over the years those operating water and wastewater plants have been given a variety of various titles. These include water jockey, practitioner of water, purveyor of water, sewer rat, or just plain water or wastewater operator. Based on our experience we have come up with a title that perhaps more closely characterizes what the water/wastewater operator really is: a Jack or Jill of all trades. This characterization seems only fitting when we take into account the knowledge and skills required of operators to properly perform their assigned duties. Moreover, operating the plant or distribution/collection system is one thing; taking samples, operating equipment, monitoring conditions, and determining settings for chemical feed systems and high-pressure pumps, along with performing laboratory tests and recording the results in the plant daily operating log, are quite another.

It is, however, the nontypical functions, the diverse functions, and the off-the-wall functions that lead to our describing operators as Jacks or Jills of all trades. For example, in addition to their normal, routine daily operating duties, operators may be called upon to make emergency repairs to systems (e.g., making a welding repair to a vital piece of machinery to keep the plant or unit process online); perform material handling operations; make chemical additions to process flow; respond to hazardous materials emergencies; make confined space entries; perform site landscaping duties; and carry out several other assorted functions. Remember, the plant operator's job is to keep the plant running and to make permit. Keeping the plant running, the flow flowing, and making permit—no matter what—requires not only talent but also the performance of a wide range of functions, many of which are not mentioned in written job descriptions.

Did You Know?

Water and wastewater treatment plant and system operators held about 94,000 jobs in 2004. Almost 4 in 5 operators worked for local governments. Others worked primarily for private water, sewage, and other systems utilities and for private waste treatment and disposal and waste management services companies. Private firms are increasingly providing operation and management services to local governments on a contract basis (BLS, 2006).

2.2 SETTING THE RECORD STRAIGHT

Based on experience, I have found that either most people have preconceived notions as to what water and wastewater operations are all about or they have nary a clue. On the one hand, most of us understand that clean water is essential for everyday life. Moreover, we have at least a vague concept that water treatment plants and water operators treat water to make it safe for consumption. On the other hand, when it comes to wastewater treatment and system operations, many of us have an ingrained image of toilets flushing and a sewer system managed and run by a bunch of sewer rats. Others give wastewater and its treatment and the folks who treat it no thought at all (that is, unless they are irate rate payers upset at a back-flushed toilet or the cost of wastewater service).

Typically, the average person has other misconceptions about water and wastewater operations; for example, very few people can identify the exact source of their drinking water. Is it pumped from wells, rivers, or streams to water treatment plants? Similarly, where is it treated and distributed to customers? The average person is clueless as to the ultimate fate of wastewater. Once the toilet is flushed, it is out of sight and out of mind.

Beyond the few functions we have pointed out to this point, what exactly is it that those water and wastewater operators—the 90,000+ Jacks or Jills of all trades in the United States—do? Operators in both water and wastewater treatment systems control unit processes and equipment to remove or destroy harmful materials, chemical compounds, and microorganisms from the water. They also control pumps, valves, and other processing equipment (including a wide array of computerized systems) to convey the water or wastewater through the various treatment processes (unit processes) and dispose (or reuse) of the removed solids (waste materials, such as sludge or biosolids). Operators also read, interpret, and adjust meters and gauges to make sure plant equipment and processes are working properly. They operate chemical-feeding devices, take samples of the water or wastewater, perform chemical and biological laboratory analyses, and adjust the amount of chemicals, such as chlorine, in the water/wastestream. They use a variety of instruments to sample and measure water quality and common hand and power tools to make repairs and adjustments. Operators also make minor repairs to valves, pumps, basic electrical equipment (note that electrical work should only be accomplished by qualified personnel), and other equipment.

As mentioned, water and wastewater system operators increasingly rely on computers to help monitor equipment, store sampling results, make process-control decisions, schedule and record maintenance activities, and produce reports. Computer-operated automatic sampling devices are beginning to gain widespread acceptance and use in both industries, especially at the larger facilities. When a system malfunction occurs, operators may use system computers to determine the cause and the solution to the problem.

2.3 THE COMPUTER-LITERATE JACK OR JILL

At many modern water/wastewater treatment plants, operators are required to perform skilled treatment plant operations work and to monitor, operate, adjust, and regulate a computer-based treatment process. In addition, the operator is also required to operate and monitor electrical, mechanical, and electronic processing and security equipment through central and remote terminal locations in a solids processing, water purification, or wastewater treatment plant. In those treatment facilities that are only partially automated or computer controlled, facility computers are used in other applications such as in clerical applications or computer maintenance management systems (CMMSs). The operator must be qualified to operate and navigate such computer systems; for example, a computer-literate operator typically:

- Monitors, adjusts, starts, and stops automated water treatment processes and emergency response systems to maintain a safe and efficient water treatment operation; monitors treatment plant processing equipment and systems to identify malfunctions and their probable causes following prescribed procedures; places equipment in or out of service or redirects processes around failed equipment; following prescribed procedures, monitors and starts process-related equipment, such as boilers, to maintain process and permit objectives; refers difficult equipment maintenance problems and malfunctions to a supervisor; monitors the system through a process-integrated control terminal or remote station terminal to ensure that control devices are making proper treatment adjustments; operates the central control terminal keyboard to perform backup adjustments to such treatment processes as influent and effluent pumping, chemical feed, sedimentation, and disinfection; monitors specific treatment processes and security systems at assigned remote plant stations; observes and reviews the terminal screen display of graphs, grids, charts, and digital readouts to determine process efficiency; responds to visual and audible alarms and indicators that indicate deviations from normal treatment processes and chemical hazards; identifies false alarms and other indicators that do not require immediate response; alerts remote-control locations to respond to alarms indicating trouble in that area; performs alarm investigations.

- Switches over to semiautomatic or manual control when the computer control system is not properly controlling the treatment process; off-scans a malfunctioning field sensor point and inserts data obtained from the field to maintain computer control; controls automated mechanical and electrical treatment processes through the computer keyboard when computer programs have failed; performs field tours to take readings when problems cannot be corrected through the computer keyboard; makes regular field tours of the plant to observe physical conditions; manually controls processes when necessary.

- Determines and changes the amount of chemicals to be added for the amount of water, wastewater, or biosolids to be treated; takes periodic samples of treated residuals, biosolids processing products and byproducts, and clean water or wastewater for laboratory analysis; receives, stores, handles, and applies chemicals and other supplies necessary for operation of the assigned station; maintains inventory records of suppliers on hand and quantities used; prepares and submits daily shift operational reports; records daily activities in the plant operation log or computer database or from a computer terminal; changes chemical feed tanks, chlorine cylinders, and feed systems; flushes clogged feed and sampling lines.

- Notes any malfunctioning equipment; makes minor adjustments when required; reports major malfunctions to higher level operators; enters maintenance and related task information into a computerized maintenance management system; processes work requests for skilled maintenance personnel.

- Performs routine mechanical maintenance such as packing valves, adjusting belts, and replacing shear pins and air filters; lubricates equipment by applying grease and adding oil; changes and cleans strainers; drains condensate from pressure vessels, gearboxes, and drip traps; performs minor electrical maintenance such as replacing bulbs and resetting low-voltage circuit switches; prepares equipment for maintenance crews by unblocking pipelines and pumps and isolating and draining tanks; checks equipment as part of a preventive and predictive maintenance program; reports more complex mechanical–electrical problems to supervisors.

- Responds, in a safe manner, to chlorine leaks and chemical spills in compliance with the requirements of OSHA's Hazardous Waste Operations and Emergency Response Standard (HAZWOPER; 29 CFR 1910.120) and with plant-specific emergency response procedures; participates in chlorine and other chemical emergency-response drills.

- Prepares operational and maintenance reports as required, including flow and treatment information; changes charts and maintains recording equipment; utilizes system and other software packages to generate reports, charts, and graphs of flow and treatment status and trends; maintains workplace housekeeping.

2.4 PLANT OPERATORS AS EMERGENCY RESPONDERS

As mentioned, occasionally operators must work under emergency conditions. Sometimes these emergency conditions are operational and not necessarily life threatening. A good example occurs during a rain event when there may be a temporary loss of electrical power and large amounts of liquid waste flow into sewers, exceeding the treatment capacity of a plant. Emergencies can also be caused by conditions inside a plant, such as oxygen deficiency within a confined space or exposure to toxic or explosive off-gases such as hydrogen sulfide and methane. To handle these conditions, operators are trained to make an emergency management response and use special safety equipment and procedures to protect coworkers, the public health, the facility, and the environment. During emergencies, operators may work under extreme pressure to correct problems as quickly as possible. These periods may create dangerous working conditions; operators must be extremely careful and cautious.

Operators who must aggressively respond to hazardous chemical leaks or spills (e.g., enter a chlorine gas-filled room and install chlorine repair kit B on a damaged 1-ton cylinder to stop the leak) must possess a HAZMAT Emergency Response Technician 24-hour certification. Additionally, many facilities where elemental chlorine is used for disinfection, odor control, or other process applications require operators to possess an appropriate certified pesticide applicator training completion certificate. Because of OSHA's specific confined-space requirement (i.e., a standby rescue team for entrants must be available), many plants require operators to hold and maintain CPR/first aid certification.

Note: It is important to point out that many wastewater facilities have substituted elemental chlorine with sodium or calcium hypochlorite, ozone, or ultraviolet irradiation because of the stringent requirements of OSHA's Process Safety Management

Standard (29 CFR 1910.119) and the Risk Management Program of the U.S. Environmental Protection Agency (USEPA). This is not the case in most water treatment operations, however. In water treatment systems, elemental chlorine is still employed because it provides chlorine residual that is important in maintaining safe drinking water supplies, especially throughout lengthy distribution systems.

2.5 OPERATOR DUTIES, NUMBERS, AND WORKING CONDITIONS

The specific duties of plant operators depend on the type and size of plant. In smaller plants, one operator may control all machinery, perform sampling and lab analyses, keep records, handle customer complaints, troubleshoot and make repairs, or perform routine maintenance. In some locations, operators may handle both water treatment and wastewater treatment operations. On the other hand, in larger plants with many employees, operators may be more specialized and only monitor one unit process (e.g., a solids handling operator who operates and monitors an incinerator). Along with treatment operators, plant staffing may include environmentalists, biologists, chemists, engineers, laboratory technicians, maintenance operators, supervisors, clerical help, and various assistants.

In the United States, notwithstanding a certain amount of downsizing brought on by privatization activities, employment opportunities for water/wastewater operators have increased in number. The number of operators has increased because of the ongoing construction of new water/wastewater and solids handling facilities. In addition, operator jobs have increased because of water pollution standards that have become increasingly more stringent since adoption of two major federal environmental regulations: the Clean Water Act of 1972 (and subsequent amendments), which implemented a national system of regulation on the discharge of pollutants, and the Safe Drinking Water Act of 1974, which established standards for drinking water.

Operators are often hired in industrial facilities to monitor or pretreat wastes before discharge to municipal treatment plants. These wastes must meet certain minimum standards to ensure that they have been adequately pretreated and will not damage municipal treatment facilities. Municipal water treatment plants also must meet stringent drinking water standards. This often means that additional qualified staff members must be hired to monitor and treat or remove specific contaminants. Complicating the problem is the fact that the list of contaminants regulated by these regulations has grown over time; for example, the 1996 Safe Drinking Water Act Amendments include standards for monitoring *Giardia* and *Cryptospo-*

ridium, two biological organisms (protozoa) that cause health problems. Operators must be familiar with the guidelines established by federal regulations and how they affect their plant. In addition to federal regulations, operators must be aware of any guidelines imposed by the state or locality in which the treatment process operates.

Another unique factor related to water/wastewater operators is their working conditions. Water and wastewater treatment plant operators work indoors and outdoors in all kinds of weather. Operators' work is physically demanding and often is performed in unclean locations (hence, the descriptive but inappropriate title of "sewer rat"). They are exposed to slippery walkways, vapors, odors, heat, dust, and noise from motors, pumps, engines, and generators. They work with hazardous chemicals. In water and wastewater plants, operators may be exposed to many bacterial and viral contaminants. As mentioned, dangerous gases such as methane and hydrogen sulfide could be present so they need to use proper safety gear.

Operators generally work a 5-day, 40-hour week; however, many treatment plants are in operation 24/7, and operators may have to work nights, weekends, holidays, or rotating shifts. Some overtime is occasionally required in emergencies.

Over the years, statistical reports have related historical evidence showing that the water/wastewater industry is an extremely unsafe occupational field. This less than stellar safety performance has continued to deteriorate even in the age of the Occupational Safety and Health Act, which was passed in 1970.

Why are the water/wastewater treatment industry on-the-job injury rates so high? Several reasons help to explain the high injury rates. First, all of the major classifications of hazards exist at water/wastewater treatment plants (with the typical exception of radioactivity):

- Oxygen deficiency
- Physical injuries
- Toxic gases and vapors
- Infections
- Fire
- Explosion
- Electrocution

Along with all of the major classifications of hazards, other factors that contribute to the high incidence of injury in the water/wastewater industry include:

- Complex treatment systems
- Shift work
- New employees
- Liberal workers' compensation laws
- Absence of safety laws
- Absence of safe work practices and safety programs

Experience has shown that a lack of well-managed safety programs and safe work practices is a major cause of the water/wastewater industry's high incidence of on-the-job injuries (Spellman, 2001).

2.6 OPERATOR CERTIFICATION AND LICENSURE

A high-school diploma or its equivalency usually is required as the entry-level credential for becoming a water or wastewater treatment plant operator-in-training. Operators need mechanical aptitude and should be competent in basic mathematics, chemistry, and biology. They must have the ability to apply data to formulas of treatment requirements, flow levels, and concentration levels. Some basic familiarity with computers also is necessary because of the current trend toward the use of computer-controlled equipment and more sophisticated instrumentation. Certain operator positions—particularly in larger cities—are covered by civil service regulations. Applicants for these positions may be required to pass a written examination testing mathematics skills, mechanical aptitude, and general intelligence.

Because treatment operations are becoming more complex, completion of an associates degree or 1-year certificate program in water quality and wastewater treatment technology is highly recommended. These credentials improve an applicant's chances for both employment and promotion. Advanced training programs are offered throughout the country. They provide good general through advanced training on water and wastewater treatment processes, as well as basic preparation for becoming a licensed operator. They also offer a wide range of computer training courses.

New water and wastewater operators-in-training typically start out as attendants or assistants and learn the practical aspects of their job under the direction of an experienced operator. They learn by observing, show-and-tell, and performing routine tasks such as recording meter readings, taking samples of liquid waste and sludge, and performing simple maintenance and repair work on pumps, electrical motors, valves, and other plant or system equipment. Larger treatment plants generally combine this on-the-job training with formal classroom or self-paced study programs. Some large sanitation districts operate their own 3- to 4-year apprenticeship schools. In some of these programs, each year of apprenticeship school completed not only prepares the operator for the next level of certification or licensure but also satisfies a requirement for advancement to the next higher pay grade.

The Safe Drinking Water Act Amendments of 1996, enforced by USEPA, specify national minimum standards for certification (licensure) and recertification of operators of community and nontransient, noncommunity water systems. As a result, operators must pass an examination to certify that they are capable of overseeing water/wastewater treatment operations. The various levels of certification depend on the operator's experience and training. Higher certification levels qualify the operator for a wider variety of treatment processes. Certification requirements vary by state and by size of treatment plants. Although relocation may mean having to become certified in a new location, many states accept the certifications of other states.

In an attempt to ensure that operators' training and qualifications are current and to improve their skills and knowledge, most state drinking water and water pollution control agencies offer ongoing training courses. These courses cover principles of treatment processes and process control methods, laboratory practices, maintenance procedures, management skills, collection system operation, general safe work practices, chlorination procedures, sedimentation, biological treatment, sludge/biosolids treatment, biosolids land application and disposal, and flow measurements. Correspondence courses covering both water/wastewater operations and preparation for state licensure examinations are provided by various state and local agencies. Many employers provide tuition assistance for formal college training.

Whether received from formal or informal sources, training provided for or obtained by water and wastewater operators must include coverage of very specific subject areas. Table 2.1 and Table 2.2 list many of the specialized topics that waterworks and wastewater operators are expected to have a fundamental knowledge of.

Note: It is important to note that both water and wastewater operators must have fundamental knowledge of basic science and math operations.

Note: For many water/wastewater operators, crossover training or overlapping training is common practice.

TABLE 2.1
Specialized Topics for Waterworks Operators

Chemical Addition	Hydraulics—Math
Chemical Feeders	Laboratory Practices
Chemical Feeders—Math	Measuring and Control
Clarification	Piping and Valves
Coagulation–Flocculation	Public Health
Corrosion Control	Pumps
Disinfection	Recordkeeping
Disinfection—Math	General Science
Basic Electricity and Controls	Electric Motors
Filtration	Finances
Filtration—Math	Storage
Fluoridation	Leak Detection
Fluoridation—Math	Hydrants
General Safe Work Practices	Cross-Connection Control
Bacteriology	and Backflow
	Stream Ecology

TABLE 2.2
Specialized Topics for Wastewater Operators

Wastewater Math	Fecal Coliform Testing
Troubleshooting Techniques	Recordkeeping
Preliminary Treatment	Flow Measurement
Sedimentation	Sludge Dewatering
Ponds	Drying Beds
Trickling Filters	Centrifuges
Rotating Biological Contactors	Vacuum Filtration
Activated Sludge	Pressure Filtration
Chemical Treatment	Sludge Incineration
Disinfection	Land Application of Biosolids
Solids Thickening	Laboratory Procedures
Solids Stabilization	General Safety

CHAPTER REVIEW QUESTIONS

2.1   Briefly, explain the causal factors behind the high incidence of on-the-job injuries for water/wastewater operators.

2.2   Why is computer literacy so important in operating a modern water/wastewater treatment system?

2.3   Define CMMS.

2.4   What is the necessary training requirement for HAZMAT responders?

2.5   Specifies national minimum standards for certification (licensure) and recertification for water/wastewater operators.

REFERENCES AND SUGGESTED READING

BLS. 2006. *Water and Liquid Waste Treatment Plant and System Operators*. Washington, D.C.: Bureau of Labor Statistics (http://www.bls.gov.oco/229.htm).

Spellman, F.R. 2001. *Safe Work Practices for Wastewater Treatment Plants*. Boca Raton, FL: CRC Press.

Spellman, F.R. 2007. *The Science of Water*, 2nd ed. Boca Raton, FL: CRC Press.

3 Upgrading Security

You may say Homeland Security is a Y2K problem that doesn't end January 1 of any given year.

Governor Tom Ridge

Worldwide conflicts are ongoing and seem neverending. One of the most important conflicts of our time, the ongoing Israeli–Palestinian conflict, is in fact conflict over scarce but vital water resources. This conflict over water, unfortunately, may be a harbinger of things to come.

3.1 INTRODUCTION

According to the U.S. Environmental Protection Agency (USEPA, 2004), approximately 160,000 public water systems (PWSs) are located in the United States, each of which regularly supplies drinking water to at least 25 persons or 15 service connections. Of the total U.S. population, 84% is served by PWSs, while the remainder is served primarily by private wells. PWSs are divided into community water systems (CWSs) and non-community water systems (NCWSs). Examples of CWSs include municipal water systems that serve mobile home parks or residential developments, and examples of NCWSs include schools, factories, churches, commercial campgrounds, hotels, and restaurants.

Did You Know?

As of 2003, community water systems served by far the largest proportion of the U.S. population—273 million out of a total population of 290 million (USEPA, 2004).

Because drinking water is consumed directly, health effects associated with contamination have long been major concerns. In addition, interruption or cessation of the drinking-water supply can disrupt society, impacting human health and critical activities such as fire protection. Although they have no clue as to its true economic value and its future worth, the general public correctly perceives drinking water as being central to the life of an individual and of society; however, many know very little about wastewater treatment and the fate of its end product.

Wastewater treatment is important for preventing disease and protecting the environment. Wastewater is treated by publicly owned treatment works (POTW) and by private facilities such as industrial plants. There are approximately 2.3 million miles of distribution system pipes and approximately 16,255 POTW in the United States. About 75% of the total U.S. population is served by POTW with existing flows of less than 1 million gallons per day (MGD) which are considered small; they number approximately 13,057 systems. For purposes of determining population served, 1 MGD equals approximately 10,000 persons served.

Disruption of a wastewater treatment system or service can cause loss of life, economic impacts, and severe public health incidents. If structural damage occurs, wastewater systems can become vulnerable to inadequate treatment. The public is much less sensitive to wastewater as an area of vulnerability than it is to drinking water; however, wastewater systems do provide opportunities for terrorist threats.

Federal and state agencies have long been active in addressing these risks and threats to water and wastewater utilities by implementing regulations, technical assistance, and research and outreach programs. As a result, an extensive system of regulations governing maximum contaminant levels of 90 conventional contaminants (most established by USEPA), construction and operating standards (implemented primarily by the states), monitoring, emergency response planning, training, research, and education has been developed to better protect the nation's drinking-water supply and receiving waters. Since the events of 9/11, USEPA has been designated as the sector-specific agency responsible for infrastructure protection activities for the nation's drinking-water and wastewater system. USEPA is utilizing its position within the water sector and working with its stakeholders to provide information to help protect the nation's drinking-water supply from terrorism or other intentional acts.

3.2 CONSEQUENCES OF 9/11

One consequence of the events of September 11, 2001, was USEPA's directive to establish a Water Protection Task Force to ensure that activities to protect and secure the country's water supply and wastewater treatment infrastructure are comprehensive and carried out expeditiously. Another consequence is a heightened concern among citizens in the United States over the security of their critical water and wastewater infrastructure. The nation's water and wastewater infrastructure consists of several thousand publicly owned water/wastewater treatment plants, more than 100,000 pumping stations, hundreds of thousands of miles of water distribution and sanitary sewers, and another 200,000 miles of storm sewers. It is one of America's most valuable resources, and its treatment and distribution/collection systems are valued at more than $2.5 trillion. Wastewater treatment operations alone include the sanitary and storm sewers that form an extensive network running near or beneath key buildings and roads and contiguous to many communication and transportation networks. Significant damage to the nation's wastewater facilities or collection systems would result in loss of life; catastrophic environmental damage to rivers, lakes, and wetlands; contamination of drinking-water supplies; long-term public health impacts; reduced fish and shellfish production; and disruption to commerce, the economy, and our normal way of life.

Governor Tom Ridge (Henry, 2002) pointed out the security role for the public professional (I interpret this to include water and wastewater professionals):

Americans should find comfort in knowing that millions of their fellow citizens are working every day to ensure our security at every level—federal, state, county, municipal. These are dedicated professionals who are good at what they do. I've seen it up close, as Governor of Pennsylvania ... but there may be gaps in the system. The job of the Office of Homeland Security will be to identify those gaps and work to close them.

Eliminating these gaps in the system has driven many water and wastewater facilities to increase their security. Moreover, in its *Water Protection Task Force Alert #IV: What Wastewater Utilities Can Do Now to Guard Against Terrorist and Security Threats* (USEPA, 2001), USEPA made several recommendations to increase security and reduce threats from terrorism. The recommendations include:

1. Guard against unplanned physical intrusion (water/wastewater).
   a. Lock all doors and set alarms at your office, pumping stations, treatment plants, and vaults, and make it a rule that doors are locked and alarms are set.
   b. Limit access to facilities and control access to pumping stations and chemical and fuel storage areas, giving close scrutiny to visitors and contractors.
   c. Post guards at treatment plants, and post "Employee Only" signs in restricted areas.
   d. Control access to storm sewers.
   e. Secure hatches, metering vaults, manholes, and other access points to the sanitary collection system.
   f. Increase lighting in parking lots, treatment bays, and other areas with limited staffing.
   g. Control access to computer networks and control systems, and change the passwords frequently.
   h. Do not leave keys in equipment or vehicles at any time.
2. Make security a priority for employees.
   a. Conduct background security checks on employees at hiring and periodically thereafter.
   b. Develop a security program with written plans and train employees frequently.
   c. Be sure that all employees are aware of communications protocols with relevant law enforcement, public health, environmental protection, and emergency response organizations.
   d. Be sure that employees are fully aware of the importance of vigilance and the seriousness of breaches in security; make note of unaccompanied strangers on the site, and immediately notify designated security officers or local law enforcement agencies.
   e. Consider varying the timing of operational procedures, if possible, so someone watching will realize that the pattern changes.
   f. Upon the dismissal of an employee, change passcodes and make sure keys and access cards are returned.
   g. Provide customer service staff with training and checklists of how to handle a threat if one is called in.
3. Coordinate actions for an effective emergency response.
   a. Review existing emergency response plans, and be sure they are current and relevant.
   b. Make sure employees have necessary training in emergency operating procedures.
   c. Develop clear protocols and chains of command for reporting and responding to threats with relevant emergency, law enforcement, environmental, public health officials, consumers, and the media. Practice the emergency protocols regularly.
   d. Be sure that key utility personnel (both on and off duty) have access to crucial telephone numbers and contact information at all times. Keep the call list up to date.
   e. Develop close relationships with local law enforcement agencies, and make sure they know where critical assets are located.

Request that they add your facilities to their routine rounds.

f. Work with local industries to ensure that their pretreatment facilities are secure.

g. Report to county or state health officials any illness among the employees that might be associated with wastewater contamination.

h. Report criminal threats, suspicious behavior, or attacks on wastewater utilities immediately to law enforcement officials and the relevant field office of the FBI.

4. Invest in security and infrastructure improvements.

a. Assess the vulnerability of the collection/distribution system, major pumping stations, water and wastewater treatment plants, chemical and fuel storage areas, outfall pipes, and other key infrastructure elements.

b. Assess the vulnerability of the stormwater collection system. Determine where large pipes run near or beneath government buildings, banks, commercial districts, or industrial facilities, or are contiguous with major communication and transportation networks.

c. Move as quickly as possible to make the most obvious and cost-effective physical improvements, such as perimeter fences, security lighting, tamper-proof manhole covers and valve boxes, etc.

d. Improve computer system and remote operational security.

e. Use local citizen watches.

f. Seek financing for more expensive and comprehensive system improvements.

Ideally, in a perfect world, water and wastewater infrastructure would be secured in a layered fashion (i.e., the *multibarrier approach*). Layered security systems are vital. Utilizing the protection-in-depth approach, which requires that adversaries must defeat several protective barriers or security layers to accomplish their goal, can make water and wastewater infrastructures more secure. *Protection in depth* is a term commonly used by the military to describe security measures that reinforce one another and mask the defense mechanisms from view of intruders, thus allowing the defender time to respond to intrusion or attack.

A prime example of the use of the multibarrier approach to ensure security and safety is demonstrated by the practices of the bottled water industry. In the aftermath of 9/11, the increased emphasis on homeland security has led to a new paradigm of national security and vulnerability awareness. Recall that, in the immediate aftermath of the 9/11 tragedies, emergency responders and others responded quickly and worked to exhaustion. In addition to the emergency responders, bottled water companies

responded immediately by donating several million bottles of water to the crews at the crash sites in New York, at the Pentagon, and in Pennsylvania. The International Bottled Water Association reported that "within hours of the first attack, bottled water was delivered where it mattered most: to emergency personnel on the scene who required ample water to stay hydrated as they worked to rescue victims and clean up debris" (IBWA, 2004, p. 2).

Bottled water companies continued to provide bottled water to responders and rescuers at the 9/11 sites for the duration. These patriotic actions by the bottled water companies, however, beg the question: How do we ensure the safety and security of the bottled water provided to anyone? IBWA's answer is to use a multibarrier approach, along with other defense principles, to enhance the safety and security of bottled water. IBWA (2004, p. 3) described its multibarrier approach as follows:

A multibarrier approach—Bottled water products are produced utilizing a multibarrier approach, from source to finished product, that helps prevent possible harmful contaminants (physical, chemical, or microbiological) from adulterating the finished product as well as storage, production, and transportation equipment. Measures in a multibarrier approach may include source protection, source monitoring, reverse osmosis, distillation, filtration, ozonation, or ultraviolet (UV) light. Many of the steps in a multibarrier system may be effective in safeguarding bottled water from microbiological and other contamination. Piping in and out of plants as well as storage silos and water tankers are also protected and maintained through sanitation procedures. In addition, bottled water products are bottled in a controlled, sanitary environment to prevent contamination during the filling operation.

In water and wastewater infrastructure security, *protection in depth* is used to describe a layered security approach. A protection-in-depth strategy uses several forms of security techniques and devices against an intruder and does not rely on a single defensive mechanism to protect the infrastructure. By implementing multiple layers of security, a hole or flaw in one layer is covered by the other layers, and an intruder will have to break through each layer without being detected. This layered approach means that, no matter how they attempt to accomplish their goal, intruders will encounter a rigorous physical protection system.

In the following sections, various security hardware and devices are described. These devices serve the main purpose of providing security against physical or digital intrusion; that is, they are designed to delay and deny intrusion and are normally coupled with detection and assessment technology. Keep in mind, however, that when it comes to trying to make something absolutely secure from intrusion or attack there is no absolute silver bullet.

3.3 SECURITY HARDWARE/DEVICES

USEPA (2005) groups water/wastewater infrastructure security devices and products into four general categories:

- Physical asset monitoring and control devices
- Water monitoring devices
- Communication and integration devices
- Cyber protection devices

3.3.1 PHYSICAL ASSET MONITORING AND CONTROL DEVICES

3.3.1.1 Aboveground Outdoor Equipment Enclosures

Water and wastewater systems consist of multiple components spread over a wide area and typically include a centralized treatment plant, as well as distribution or collection system components that are usually distributed at multiple locations throughout the community. In recent years, however, distribution and collection system designers have favored placing critical equipment—especially assets that require regular use and maintenance—above ground. A primary reason for doing so is that locating this equipment above ground eliminates the safety risks associated with confined space entry, which is often required for the maintenance of equipment located below ground. In addition, space restrictions often limit the amount of equipment that can be located inside, and there are concerns that some types of equipment (such as backflow-prevention devices) can, under certain circumstances, discharge water that could flood pits, vaults, or equipment rooms; therefore, many pieces of critical equipment are located outdoors and above ground. Examples of the many different system components that can be installed outdoors and above ground include:

- Backflow-prevention devices
- Air release and control valves
- Pressure vacuum breakers
- Pumps and motors
- Chemical storage and feed equipment
- Meters
- Sampling equipment
- Instrumentation

Much of this equipment is installed in remote locations or in areas where the public can access it.

One of the most effective security measures for protecting aboveground equipment is to place it inside a building. Where this is not possible, enclosing the equipment or parts of the equipment using some sort of commercial or homemade add-on structure may help to prevent tampering with the equipment. Equipment enclosures can generally be categorized into one of four main configurations:

- One-piece, drop-over enclosures
- Hinged or removable top enclosures
- Sectional enclosures
- Shelters with access locks

Other security features implemented on aboveground, outdoor equipment enclosures include locks, mounting brackets, tamper-resistant doors, and exterior lighting.

3.3.1.2 Alarms

An *alarm system* is a type of electronic monitoring system that is used to detect and respond to specific types of events, such as unauthorized access to an asset, or a possible fire. In water and wastewater systems, alarms are also used to alert operators when process operating or monitoring conditions go out of preset parameters (i.e., process alarms). These types of alarms are primarily integrated with process monitoring and reporting systems (e.g., SCADA systems). Note that this discussion does not focus on alarm systems that are not related to the processes of a utility.

Alarm systems can be integrated with fire detection systems, intrusion detection systems, access control systems, or closed-circuit television (CCTV) systems such that these systems automatically respond when the alarm is triggered. A smoke detector alarm, for example, can be set up to automatically notify the fire department when smoke is detected, or an intrusion alarm can automatically trigger cameras to turn on in a remote location so personnel can monitor that location.

An alarm system consists of sensors that detect different types of events; an arming station that is used to turn the system on and off; a control panel that receives information, processes it, and transmits the alarm; and an annunciator, which generates a visual or audible response to the alarm. When a sensor is tripped, it sends a signal to a control panel, which triggers a visual or audible alarm or notifies a central monitoring station. A more complete description of each of the components of an alarm system is provided below.

Detection devices (also called *sensors*) are designed to detect a specific type of event (such as smoke or intrusion). Depending on the type of event they are designed to detect, sensors can be located inside or outside of the facility or other asset. When an event is detected, the sensors use some type of communication method (such as wireless radio transmitters, conductors, or cables) to send signals to the control panel to generate the alarm; for example, a smoke detector sends a signal to a control panel when it detects smoke.

An *arming station*, which is the main user interface with the security system, allows the user to arm (turn on), disarm (turn off), and communicate with the system. How a specific system is armed will depend on how it is used. Although intrusion detection systems can be armed for continuous operation (24 hours a day), they are usually armed and disarmed according to the work schedule at a specific location so personnel going about their daily activities do not set off the alarms. In contrast, fire protection systems are typically armed 24 hours a day.

The *control panel* receives information from the sensors and sends it to an appropriate location, such as to a central operations station or to a 24-hour monitoring facility. When the alarm signal is received at the central monitoring location, personnel monitoring for alarms can respond (such as by sending security teams to investigate or by dispatching the fire department).

The *annunciator* responds to the detection of an event by emitting a signal. This signal may be visual, audible, or electronic, or a combination of these three; for example, fire alarm signals will always be connected to audible annunciators, whereas intrusion alarms may not be.

Alarms can be reported locally, remotely, or both locally and remotely. A *local alarm* emits a signal at the location of the event (typically using a bell or siren). A local-only alarm emits a signal at the location of the event but does not transmit the alarm signal to any other location (i.e., it does not transmit the alarm to a central monitoring location). Typically, the purpose of a local-only alarm is to frighten away intruders and possibly to attract the attention of someone who might notify the proper authorities. Because no signal is sent to a central monitoring location, personnel can only respond to a local alarm if they are in the area and can hear or see the alarm signal.

Fire alarm systems must have local alarms, including both audible and visual signals. Most fire alarm signal and response requirements are codified in the National Fire Alarm Code, National Fire Protection Association (NFPA) 72. NFPA 72 discusses the application, installation, performance, and maintenance of protective signaling systems and their components. In contrast to fire alarms, which require a local signal when fire is detected, many intrusion detection systems do not have a local alert device, because monitoring personnel do not wish to inform potential intruders that they have been detected. Instead, these types of systems silently alert monitoring personnel that an intrusion has been detected, thus allowing monitoring personnel to respond.

In contrast to systems that are set up to transmit local-only alarms when the sensors are triggered, systems can also be set up to transmit signals to a central location, such as to a control room or guard post at the utility or to a police or fire station. Most fire and smoke alarms are set up to signal both at the location of the event and at a fire station or central monitoring station. Many insurance companies require that facilities install certified systems that include alarm communication to a central station; for example, systems certified by the Underwriters Laboratory (UL) require that the alarm be reported to a central monitoring station.

The main differences between alarm systems lie in the types of event detection devices used in different systems. Intrusion sensors, for example, consist of two main categories: perimeter sensors and interior (space) sensors. *Perimeter intrusion sensors* are typically applied on fences, doors, walls, windows, etc. and are designed to detect intruders before they gain access to a protected asset (e.g., perimeter intrusion sensors are used to detect intruders attempting to enter through a door or window). In contrast, *interior intrusion sensors* are designed to detect an intruder who has already accessed the protected asset (i.e., interior intrusion sensors are used to detect intruders when they are already within a protected room or building). These two types of detection devices can be complementary, and they are often used together to enhance security for an asset. A typical intrusion alarm system, for example, might employ a perimeter glass-break detector that protects against intruders accessing a room through a window, as well as an ultrasonic interior sensor that detects intruders that have gotten into the room without using the window.

Fire detection and fire alarm systems consist of various combinations of fire detection devices and fire alarm systems. These systems may detect fire, heat, or smoke, or a combination of any of these. A typical fire alarm system might consist only of heat sensors located throughout a facility that detect high temperatures or a certain change in temperature over a fixed time period, whereas a different system might be outfitted with both smoke and heat detection devices.

When a sensor in an alarm system detects an event, it must communicate an alarm signal. The two basic types of alarm communication systems are *hardwired* and *wireless*. Hardwired systems rely on wire that is run from the control panel to each of the detection devices and annunciators. Wireless systems transmit signals from a transmitter to a receiver through the air—primarily using radio or other waves. Hardwired systems are usually lower cost, more reliable (they are not affected by terrain or environmental factors), and significantly easier to troubleshoot than wireless systems; however, a major disadvantage of hardwired systems is that it may not be possible to hardwire all locations (e.g., it may be difficult to hardwire remote locations). In addition, running wires to their required locations can be both time consuming and costly. The major advantage to using wireless systems is that they can often be installed in areas where hardwired systems are not feasible; however, wireless components can be much more expensive when compared to hardwired systems. In addition, in the past it has been difficult to perform self-diagnostics on wireless systems to confirm that they are communicating properly with the controller. Currently, the majority of wireless systems incorporate supervising circuitry, which allows the subscriber to recognize immediately any problem with the system (such as a broken detection device or a low battery) or if a protected door or window has been left open.

3.3.1.3 Backflow-Prevention Devices

As their name suggests, backflow-prevention devices are designed to prevent backflow, which is the reversal of the normal and intended direction of water flow in a water system. Backflow is a potential problem in a water system because it can spread contaminated water back through a distribution system. For example, pollution or backflow at uncontrolled cross-connections (any actual or potential connection between the public water supply and a source of contamination) can allow pollutants or contaminants to enter the potable water system. More specifically, backflow from private plumbing systems, industrial areas, hospitals, and other hazardous contaminant-containing systems into public water mains and wells poses serious public health risks and security problems. Cross-contamination from private plumbing systems can contain biological hazards (such as bacteria or viruses) or toxic substances that can contaminate and sicken an entire population in the event of backflow. The majority of historical incidences of backflow have been accidental, but growing concern that contaminants could be intentionally backfed into a system is prompting increased awareness among private homeowners, businesses, industries, and areas most vulnerable to intentional strikes; therefore, backflow prevention is a major component of water system protection.

Backflow may occur under two types of conditions: backpressure and backsiphonage. *Backpressure* is a reversal of normal flow direction within a piping system resulting from the downstream pressure being higher than the supply pressure. These reductions in the supply pressure occur whenever the amount of water being used exceeds the amount of water supplied, such as during water-main flushing, fire fighting, or breaks in water mains. *Backsiphonage* is the reversal of normal flow direction within a piping system that is caused by negative pressure in the supply piping (i.e., the reversal of normal flow in a system caused by a vacuum or partial vacuum within the water supply piping). Backsiphonage can occur due to high velocity in a pipeline, a line repair or break that is lower than a service point, or a lowered main pressure caused by a high water withdrawal rate, such as during fire fighting or water-main flushing.

To prevent backflow, various types of backflow preventers are appropriate for use. The primary types of backflow preventers are:

- Air gap drains
- Double check valves
- Reduced pressure principle assemblies
- Pressure vacuum breakers

3.3.1.4 Barriers

3.3.1.4.1 Active Security Barriers (Crash Barriers)

Active security barriers (or crash barriers) are large structures that are placed in roadways at entrance and exit points of protected facilities to control vehicle access to these areas. These barriers are placed perpendicular to traffic to block the roadway, and traffic can only pass the barrier if it is moved out of the roadway. These types of barriers are typically constructed from sturdy materials, such as concrete or steel, so vehicles cannot penetrate them. They are also situated at such a height off the roadway that vehicles cannot go over or under them.

The key difference between active security barriers, which include wedges, crash beams, gates, retractable bollards, and portable barricades, and passive security barriers, which include immovable bollards, jersey barriers, and planters, is that active security barriers are designed so they can be easily raised and lowered or moved out of the roadway to allow authorized vehicles to pass them. Many of these types of barriers are designed so they can be opened and closed automatically (i.e., mechanized gates, hydraulic wedge barriers), while others are easy to open and close manually (swing crash beams, manual gates). In contrast to active barriers, passive barriers are permanent, immovable barriers that are typically used to protect the perimeter of a protected facility, such as sidewalks and other areas that do not require vehicular traffic to pass them. Several of the major types of active security barriers such as wedge barriers, crash beams, gates, bollards, and portable/removable barricades are described below.

Wedge barriers are plated, rectangular steel buttresses approximately 2 to 3 feet high that can be raised and lowered from the roadway. When they are in the open position, they are flush with the roadway, and vehicles can pass over them; however, when they are in the closed (armed) position, they project up from the road at a 45° angle, with the upper end pointing toward the oncoming vehicle and the base of the barrier away from the vehicle. Generally, wedge barriers are constructed from heavy-gauge steel or concrete that contains an impact-dampening iron rebar core that is resistant to breaking or cracking, thereby allowing the barrier to withstand the impact of a vehicle attempting to crash through it. In addition, both of these materials help to transfer the energy of the impact over the entire volume of the barrier, thus helping to prevent the barrier from being sheared off its base. Also, because of the angle of the barrier, the force of any vehicle impacting the barrier is distributed over the entire surface of the barrier and is not concentrated at the base, which helps prevent the barrier from breaking off at the base. Finally, any vehicles attempting to drive over the barrier will most likely be hung up on it.

Wedge barriers can be fixed or portable. Fixed wedge barriers can be mounted on the surface of the roadway (surface-mounted wedges) or in a shallow mount in the surface of the road, or they can be installed completely below the road surface. Surface-mounted wedge barricades operate by rising from a flat position on the surface of the roadway, whereas shallow-mount wedge barriers rise from their resting position just below the road surface. In contrast, below-surface wedge barriers operate by rising from beneath the road surface. Both the shallow-mounted and surface-mounted barriers require little or no excavation and thus do not interfere with buried utilities. All three types of barriers project above the road surface and block traffic when they are raised into the armed position. When they are disarmed and lowered, they are flush with the road, thereby allowing traffic to pass. Portable wedge barriers are moved into place on wheels that are removed after the barrier has been set into place.

Installing rising wedge barriers requires preparation of the road surface. Installing surface-mounted wedges does not require that the road be excavated; however, the road surface must be intact and strong enough to allow the bolts anchoring the wedge to the road surface to attach properly. Shallow-mount and below-surface wedge barricades require excavation of a pit that is large enough to accommodate the wedge structure, as well as any arming/disarming mechanisms. Generally, the bottom of the excavation pit is lined with gravel to allow for drainage. A gravity drain or self-priming pump can be installed in areas not sheltered from rain or surface runoff.

Crash beam barriers consist of aluminum beams that can be opened or closed across the roadway. Although crash beam designs vary, every crash beam system consists of an aluminum beam that is supported on each side by a solid footing or buttress, which is usually constructed from concrete, steel, or some other strong material. Beams typically contain an interior steel cable (at least 1 inch in diameter) to give the beam added strength and rigidity. The beam is connected by a heavy-duty hinge or other mechanism to one of the footings so it can swing or rotate out of the roadway when it is open and can swing back across the road when it is in the closed (armed) position, blocking the road and inhibiting access by unauthorized vehicles. The nonhinged end of the beam can be locked into its footing, thus providing anchoring for the beam on both sides of the road and increasing the resistance of the beam to vehicles attempting to penetrate through it. In addition, if the crash beam is hit by a vehicle, the aluminum beam transfers the impact energy to the interior cable, which in turn transfers the impact energy through the footings and into their foundation, thereby minimizing the chance that the impact will snap the beam and allow the intruding vehicle to pass through.

Crash beam barriers can employ drop-arm, cantilever, or swing beam designs. Drop-arm crash beams operate by raising and lowering the beam vertically across the road. Cantilever crash beams are projecting structures that are opened and closed by extending the beam from the hinge buttress to the receiving buttress located on the opposite side of the road. In the swing beam design, the beam is hinged to the buttress such that it swings horizontally across the road. Generally, swing beam and cantilever designs are used at locations where a vertical lift beam is impractical; for example, the swing beam or cantilever designs are utilized at entrances and exits with overhangs, trees, or buildings that would physically block the operation of the drop-arm beam design. Installing any of these crash beam barriers involves the excavation of a pit approximately 48 inches deep for both the hinge and the receiver footings. Due to the depth of excavation, the site should be inspected for underground utilities before digging begins.

In contrast to wedge barriers and crash beams, which are typically installed separately from a fence line, *gates* are often integrated units of a perimeter fence or wall around a facility. Gates are basically movable pieces of fencing that can be opened and closed across a road. When the gate is in the closed (armed) position, the leaves of the gate lock into steel buttresses that are embedded in a concrete foundation located on both sides of the roadway, thereby blocking access to the roadway. Generally, gate barricades are constructed from a combination of heavy-gauge steel and aluminum that can absorb the impact from vehicles attempting to ram through them. Any remaining impact energy not absorbed by the gate material is transferred to the steel buttresses and their concrete foundation.

Gates can utilize a cantilever, linear, or swing design. Cantilever gates are projecting structures that operate by extending the gate from the hinge footing across the roadway to the receiver footing. A linear gate is designed to slide across the road on tracks via a rack-and-pinion drive mechanism. Swing gates are hinged so they can swing horizontally across the road. Installation of the cantilever, linear, or swing gate designs involves the excavation of a pit approximately 48 inches deep for both the hinge and receiver footings to which the gates are attached. Again, due to the depth of excavation, the site should be inspected for underground utilities before digging begins.

Bollards are vertical barriers at least 3 feet tall and 1 to 2 feet in diameter that are typically set 4 to 5 feet apart from each other so they block vehicles from passing between them. Bollards can be fixed in place, removable, or retractable. Fixed and removable bollards are passive barriers that are typically used along building perimeters or on sidewalks to prevent vehicle access while still allowing pedestrians to pass through them. In contrast to passive bollards, retractable bollards are active security barriers that can easily be raised and lowered to allow vehicles to pass between them; thus, they can be used in driveways or on roads to control vehicular access. When the bollards are raised, they project above the road surface and block the roadway; when they are lowered, they sit flush with the road surface and allow traffic to pass over them. Retractable bollards are typically constructed from steel or other materials that have a low weight-to-volume ratio so they require low power to raise and lower. Steel is also more resistant to breaking than is a more brittle material such as concrete, and it is better able to withstand direct vehicular impact without breaking apart.

Retractable bollards are installed in a trench dug across a roadway, typically at an entrance or a gate. Installing retractable bollards requires preparing the road surface. Depending on the vendor, bollards can be installed either in a continuous slab of concrete or in individual excavations with concrete poured in place. The required excavation for a bollard is typically slightly wider and slightly deeper than the bollard height when it is extended above ground. The bottom of the excavation is typically lined with gravel to allow drainage. The bollards are then connected to a control panel that controls the raising and lowering of the bollards. Installation typically requires mechanical, electrical, and concrete work; if utility personnel with these skills are available, then the utility can install the bollards themselves.

Portable or removable barriers, which can include removable crash beams and wedge barriers, are mobile obstacles that can be moved in and out of position on a roadway; for example, a crash beam may be completely removed and stored off-site when it is not needed. An additional example would be wedge barriers that are equipped with wheels that can be removed after the barricade is towed into place.

When portable barricades are needed, they can be moved into position rapidly. To provide them with added strength and stability, they are typically anchored to buttress boxes that are located on either side of the road. These buttress boxes, which may or may not be permanent, are usually filled with sand, water, cement, gravel, or concrete to make them heavy and aid in stabilizing the portable barrier. In addition, these buttresses can help dissipate any impact energy from vehicles crashing into the barrier itself.

Because these barriers are not anchored into the roadway, they do not require excavation or other related construction for installation, and they can be assembled and made operational in a short period of time. The primary shortcoming to this type of design is that these barriers may move if they are hit by vehicles; therefore, it is important to carefully assess the placement and anchoring of these types of barriers to ensure that they can withstand the types of impacts that may be anticipated at that location.

Because the primary threat to active security barriers is that vehicles will attempt to crash through them, their most important attributes are their size, strength, and crash resistance. Other important features for an active security barrier are the mechanisms by which the barrier is raised and lowered to allow authorized vehicle entry, as well as such other factors as weather resistance and safety features.

3.3.1.4.2 Passive Security Barriers

One of the most basic threats facing any facility is from intruders accessing the facility with the intention of causing damage to its assets. These threats may include intruders actually entering the facility, as well as intruders attacking the facility from outside without actually entering it (e.g., detonating an explosive near enough to the facility to cause damage within its boundaries). Security barriers are one of the most effective ways to counter the threat of intruders accessing a facility or the facility perimeter. Security barriers are large, heavy structures used to control access through a perimeter by either vehicles or personnel. They can be used in many different ways depending on how they are installed or where they are located at the facility; for example, security barriers can be used on or along driveways or roads to direct traffic to a checkpoint (e.g., placement of jersey barriers to direct traffic in particular direction). Other types of security barriers (e.g., crash beams, gates) can be installed at the checkpoint so guards can regulate which vehicles can access the facility. Finally, security barriers (e.g., bollards or security planters) can be used along the facility perimeter to establish a protective buffer area between the facility and approaching vehicles. Establishing such a protective buffer can help in mitigating the effects of an explosive by potentially absorbing some of the blast and by increasing the stand-off distance between the blast and the facility. The force of an explosion is reduced as the shock wave travels away from the source; thus, the greater the distance between the target and an explosion, the less damage will be incurred.

Security barriers can be either active or passive. *Active security barriers*, which include gates, retractable bollards, wedge barriers, and crash barriers, are readily movable and are typically used in areas where they must be moved often to allow vehicles to pass—such as in roadways at entrances and exits to a facility. In contrast to active security barriers, *passive security barriers*, which include jersey barriers, bollards, and security planters, are not designed to be moved on a regular basis and are typically used in areas where access is not required or allowed—such as along building perimeters or in traffic-control areas. Passive security barriers are typically large, heavy structures that are usually several feet high, and they are designed so even heavy-duty vehicles cannot go over or though them. They can be placed in a roadway parallel to the flow of traffic so they direct traffic in a particular direction (such as to a guardhouse, a gate, or some other sort of checkpoint) or perpendicular to traffic such that they prevent a vehicle from using a road or approaching a building or area.

3.3.1.5 Biometric Security Systems

Biometrics involves measuring the unique physical characteristics or traits of the human body. Any aspect of the body that is measurably different from person to person—for example, fingerprints or eye characteristics—can serve as unique biometric identifiers for individuals. Biometric systems recognizing fingerprints, palm shape, eyes, face, voice, and signature comprise the bulk of current biometric systems. Biometric security systems use biometric technology combined with some type of locking mechanisms to control access to specific assets. To access an asset controlled by a biometric security system, an individual's biometric trait must be matched with an existing profile stored in a database. If a match between the two is identified, the locking mechanism (e.g., a physical lock at a doorway or an electronic lock at a computer terminal) is disengaged, and the individual is given access to the asset. A biometric security system is typically comprised of the following components:

- A sensor measures and records a biometric characteristic or trait.
- A control panel serves as the connection point between various system components. The control panel communicates information back and forth between the sensor and the host computer and controls access to the asset by engaging or disengaging the system lock based on internal logic and information from the host computer.
- A host computer processes and stores the biometric trait in a database.
- Specialized software compares an individual image taken by the sensor with stored profiles.
- A locking mechanism is controlled by the biometric system.
- A power source supplies power to the system.

3.3.1.5.1 Biometric Hand and Finger Geometry Recognition

Hand and finger geometry recognition is the process of identifying an individual through the unique geometry (e.g., shape, thickness, length, width) of that individual's hand or fingers. Hand geometry recognition has been employed since the early 1980s and is among the most widely used biometric technologies for controlling access to important assets. It is easy to install and use and is appropriate for any location requiring highly accurate biometric security; for example, it is currently used in numerous workplaces, daycare facilities, hospitals, universities, airports, and power plants.

A newer option within hand geometry recognition technology is finger geometry recognition (not to be confused with fingerprint recognition). Finger geometry recognition relies on the same scanning methods and technologies as does hand geometry recognition, but the scanner scans only two of the user's fingers, as opposed to the entire hand. Finger geometry recognition has been in commercial use since the mid-1990s and is mainly used in time and attendance applications (i.e., to track when individuals have entered and exited a location). To date, the only large-scale commercial application of two-finger geometry for controlling access is at Disney World, where season-pass holders use the geometry of their index and middle finger to gain access to the facilities.

Hand and finger geometry recognition systems can be used in several different types of applications, including access control and time and attendance tracking. Although time and attendance tracking can be used for security purposes, it is primarily used in operations and payroll areas (e.g., clocking in and clocking out). In contrast, access control applications are more likely to be security related. Biometric systems are widely used for access control and can be used to protect various types of assets, including entryways, computers, and vehicles. Because of their size, however, hand and finger recognition systems are primarily limited to use in entryway access control applications.

3.3.1.5.2 Iris Recognition

The iris is the colored or pigmented area of the eye surrounded by the sclera (the white portion of the eye); it is a muscular membrane that controls the amount of light entering the eye by contracting or expanding the pupil (the dark center of the eye). The dense, unique patterns of connective tissue in the human iris were first noted in 1936, but it was not until 1994, when algorithms for iris recognition were created and patented, that commercial applications using biometric iris recognition began to be used extensively. Currently, only two vendors are producing iris recognition technology—the original developer of these algorithms as well as another company that has developed and patented a different set of algorithms for iris recognition.

The iris is an ideal characteristic for identifying individuals because it is formed *in utero*, and its unique patterns stabilize around eight months after birth. No two irises are alike—neither an individual's right and left irises nor the irises of identical twins. The iris is protected by the cornea (the clear covering over the eye); therefore, it is not subject to the aging or physical changes (and potential variation) that are common to some other biometric measures, such as the hand, fingerprints, and the face. Although some limited changes can occur naturally over time, these changes generally occur in the melanin of the iris and therefore affect only the eye's color, not its unique patterns. Because iris scanning uses only black and white images, color changes do not affect the effectiveness of the scan. Barring specific injuries or surgeries directly affecting the iris, the unique patterns of the iris remain relatively unchanged over an individual's lifetime.

Iris recognition systems employ a monochromatic, or black and white, video camera that uses both visible and near-infrared light to take a video of an individual's iris. Video is used rather than still photography as an extra security procedure. The video is used to confirm the normal continuous fluctuations of the pupil as the eye focuses, which ensures that the scan is of a living human being and not a photograph or some other attempted hoax. A high-resolution image of the iris is then captured or extracted from the video using a device often referred to as a *frame grabber*. The unique characteristics identified in this image are then converted into a numeric code, which is stored as a template for that user.

3.3.1.6 Card Identification and Access and Tracking Systems

A card reader system is a type of electronic identification system that is used to read a card and then perform an action associated with that card. Depending on the system, the card may identify where a person is or where the person was at a certain time, or it may authorize another action such as disengaging a lock. Security guards, for example, may use their cards at card readers located throughout a facility to indicate that they have checked certain locations at certain times. The reader will store the information or send it to a central location where the data can be checked later to verify that various areas in the facility have been patrolled. Other card reader systems can be associated with a lock, such that cardholders must have their cards read and accepted by the reader before the lock disengages. A complete card reader system typically consists of the following components:

- Access cards carried by the user
- Card readers, which read the card signals and send the information to control units
- Control units, which control the response of the card reader to the card
- A power source

Numerous card reader systems are available. All card systems are similar with regard to how the card reader and control unit interact; however, they differ in how data are encoded on the cards and are transferred between the cards and the card readers, which determines the types of applications for which they are best suited. Several types of technologies are available for card reader systems:

- Proximity
- Wiegand
- Smart cards
- Magnetic stripe
- Bar code
- Infrared

- Barium ferrite
- Hollerith
- Mixed technologies

The level of security required (low, moderate, or high) affects the choice of card technology (e.g., how simple is it to duplicate a particular technology and thus bypass the security system). Vulnerability ratings are based on how easily the card reader can be damaged due to frequent use or challenging working conditions (e.g., weather conditions if the reader is located outside). Often, the vulnerability of a system is influenced by the number of moving parts in the system—the more moving parts, the greater the potential susceptibility to damage. Life-cycle ratings are based on the durability of a given card reader system over its entire operational period. Systems requiring frequent physical contact between the reader and the card often have a shorter life cycle due to the wear and tear to which the equipment is exposed. For many card reader systems, the vulnerability rating and life-cycle ratings have a reciprocal relationship; for example, if a given system has a high vulnerability rating it will almost always have a shorter life cycle.

Card reader technology can be implemented for facilities of any size and with any number of users; however, because individual systems vary in the complexity of their technology and the level of security they can provide to a facility, individual users must determine the appropriate system for their needs. Some important features to consider when selecting a card reader system include:

- What level of technological sophistication and security does the card system have?
- How large is the facility, and what are its security needs?
- How frequently will the card system be used? For systems that will experience a high frequency of use, it is important to consider a system that has a longer life cycle and lower vulnerability rating, thus making it more cost effective to implement.
- Under what conditions will the system be used? (Will it be installed on the interior or exterior of buildings? Does it require light or humidity controls?) Most card reader systems can operate under normal environmental conditions; therefore, this would be a mitigating factor only in extreme conditions.
- What are the system costs?

3.3.1.7 Fences

A fence is a physical barrier that can be set up around the perimeter of an asset. Fences often consist of individual pieces (such as individual pickets in a wooden fence or individual sections of a wrought iron fence) that are fastened together. Individual sections of the fence are fastened together using posts, which are sunk into the ground to provide stability and strength for the sections of the fence hung between them. Gates are installed between individual sections of the fence to allow access inside the fenced area.

Fences are often used as decorative architectural features to separate physical spaces from each other and to mark the location of a boundary (such as a fence installed along a properly line); however, a fence can also serve as an effective means for preventing intruders from gaining access to a water or wastewater asset. Many utilities install fences around their primary facilities, around remote pump stations, or around hazardous materials storage areas or sensitive areas within a facility. Access to the area can be controlled through security located at gates or doors in the fence (e.g., posting a guard at the gate or locking it). To gain access to the asset, unauthorized persons would have to find a way either around or through the fence.

Fences are often compared with walls when determining the appropriate system for perimeter security. Both fences and walls can provide adequate perimeter security, and fences are often easier and less expensive to install than walls; however, they do not usually provide the same physical strength that walls do. In addition, many types of fences have gaps between the individual pieces that make up the fence (e.g., the spaces between chain links in a chain-link fence or the space between pickets in a picket fence); thus, many types of fences allow the interior of the fenced area to be seen. This may allow intruders to gather important information about the locations or defenses of vulnerable areas within the facility.

Numerous types of materials are used to construct fences, including chain link, aluminum, wood, or wire. Some types of fences, such as split rails or pickets, may not be appropriate for security purposes because they are traditionally low fences and are not physically strong; for example, the rails in a split-rail fence may easily be broken. Potential intruders may be able to easily defeat fences by jumping or climbing over them or by breaking through them.

Important security features of a fence include the height to which it can be constructed, the strength of the material comprising the fence, the method used to attach the individual sections of the fence together at the posts, and the ability of the fence to restrict the view of the assets inside the fence. Additional considerations include the ease of installing the fence and the ease of removing and reusing sections of the fence.

Some fences can include additional measures to delay, or even detect, potential intruders. Such measures may include the addition of barbed wire, razor wire, or other deterrents at the top of the fence. Barbed wire employed at the base of a fence can also impede a would-be intruder's access to the fence. Fences can be fitted with security cameras to provide visual surveillance of the perimeter, and some facilities have installed motion sensors along their fences to detect movement on the fence. Several manufacturers have combined these multiple perimeter security features into one product and offer alarms and other security features.

The correct implementation of a fence can make it a much more effective security measure. Security experts recommend the following when a facility constructs a fence:

- The fence should be at least 7 to 9 feet high.
- Any outriggers, such as barbed wire, that are affixed on top of the fence should be angled out and away from the facility, not in toward the facility. This will make climbing the fence more difficult and will prevent the placement of ladders against the fence.
- Other types of hardware that can increase the security of the fence include installing concertina wire along the fence (this can be done in front of the fence or at the top of the fence) or adding intrusion sensors, camera, or other hardware to the fence.
- All undergrowth should be cleared for several feet (typically 6 feet) on both sides of the fence. This will allow for a clearer view of the fence by any patrols in the area.
- Any trees with limbs or branches hanging over the fence should be trimmed so intruders cannot use them to go over the fence. Also, it should be noted that fallen trees can damage fences, so management of trees around the fence can be important. This can be especially important in areas where the fence runs through a remote area.
- Fences that do not block the view from outside the fence to inside the fence allow patrols to see inside the fence without having to enter the facility.
- "No Trespassing" signs posted along a fence can be a valuable tool in prosecuting any intruders who claim that the fence was broken and that they did not enter through the fence illegally. Adding signs that highlight the local ordinances against trespassing can further dissuade simple troublemakers from illegally climbing the fence.

3.3.1.8 Films for Glass Shatter Protection

Most water and wastewater utilities have numerous windows on the outside of buildings, in doors, and in interior offices. In addition, many facilities have glass doors or other glass structures, such as glass walls or display cases.

These glass objects are potentially vulnerable to shattering when heavy objects are thrown or launched at them, when explosions occur near them, or when there are high winds. If the glass is shattered, intruders may potentially enter an area. In addition, shattered glass projected into a room from an explosion or from an object being thrown through a door or window can injure and potentially incapacitate personnel in the room. Materials that prevent glass from shattering can help to maintain the integrity of the door, window, or other glass object and can delay an intruder from gaining access. These materials can also prevent flying glass and thus reduce potential injuries.

Materials designed to prevent glass from shattering include specialized films and coatings. These materials can be applied to existing glass objects to improve their strength and their ability to resist shattering. The films have been tested against many scenarios that could result in glass breakage, including penetration by blunt objects, bullets, high winds, and simulated explosions. They are tested against simulated weather scenarios (including high winds and the force of objects blown into the glass), as well as criminal or terrorist scenarios where the glass could be subject to explosives or bullets. Many vendors provide information on the results of these types of tests so potential users can compare different product lines to determine which products best suit their needs. The primary considerations with regard to films for shatter protection are:

- The materials from which the film is made
- The adhesive that bonds the film to the glass surface
- The thickness of the film

3.3.1.9 Fire Hydrant Locks

Fire hydrants are installed at strategic locations throughout a community's water distribution system to supply water for fire fighting. Because the many hydrants in a system are often located in residential neighborhoods, industrial districts, and other areas where they cannot be easily observed or guarded, they are potentially vulnerable to unauthorized access. Many municipalities, states, and USEPA regions have recognized this potential vulnerability and have instituted programs to lock hydrants; for example, USEPA Region 1 includes locking hydrants as the seventh item in its Drinking Water Security and Emergency Preparedness "top ten" list for small groundwater suppliers.

A hydrant lock is a physical security device designed to prevent unauthorized access to the water supply through a hydrant. Such locks can ensure water and water pressure availability to fire fighters and prevent water theft and associated lost water revenue. These locks have been used successfully in numerous municipalities and in various climates and weather conditions.

Fire hydrant locks are basically steel covers or caps that are locked in place over the operating nut of a fire hydrant. The lock prevents unauthorized persons from accessing the operating nut and opening the fire hydrant valve. The lock also makes it more difficult to remove the bolts from the hydrant and access the system that way. Finally, hydrant locks shield the valve from being broken off. Should a vandal attempt to breach the hydrant lock by force and succeed in breaking the hydrant lock, the vandal will only succeed in bending the operating valve. If the operating valve of the hydrant is bent, the hydrant will not be operational, but the water asset remains protected and inaccessible to vandals; however, the entire hydrant will have to be replaced.

The locking mechanisms for fire hydrant locking systems ensure that hydrants can only be accessed by authorized personnel who have the special key wrench required to operate a hydrant without removing the lock. These specialized wrenches are generally distributed to the fire department, public works department, and other authorized persons so they can access the hydrants as needed. An inventory of wrenches and their serial numbers is generally kept by a municipality so the location of all wrenches is known. These operating key wrenches may only be purchased by registered lock owners.

The most important features of hydrant are their strength and the security of their locking systems. The locks must be strong so they cannot be broken off. Hydrant locks are constructed from stainless or alloyed steel. Stainless-steel locks are stronger and are ideal for all climates; however, they are more expensive than alloy locks.

3.3.1.10 Hatch Security

A hatch is basically a door installed on a horizontal plane (such as in a floor, a paved lot, or a ceiling), instead of on a vertical plane (such as in a building wall). Hatches are usually used to provide access to assets that are located underground (e.g., in basements or in underground storage areas) or above ceilings (such as emergency roof exits). At water and wastewater facilities, hatches are typically used to provide access to underground vaults containing pumps, valves, or piping or to the interior of water tanks or covered reservoirs. Securing a hatch by locking it or upgrading materials to give the hatch added strength can help to delay unauthorized access to any asset behind the hatch. Like all doors, a hatch consists of a frame anchored to the horizontal structure, a door or doors, hinges connecting the door to the frame, and a latching or locking mechanism that keeps the hatch door closed.

It should be noted that improving hatch security is straightforward and that hatches with upgraded security features can be installed new or they can be retrofit for existing applications. Many municipalities already have specifications for hatch security at their water and waste-water utility assets.

Depending on the application, the primary security-related attributes of a hatch are the strength of the door and frame, its resistance to the elements and corrosion, its ability to be sealed against water or gas, and its locking features. Hatches must be both strong and lightweight so they can withstand typical static loads (such as people or vehicles walking or driving over them) while still being easy to open. In addition, because hatches are typically installed at outdoor locations, they are usually fabricated out of corrosion-resistant metal that can withstand the elements such as high-gauge steel or lightweight aluminum.

The hatch locking mechanism is perhaps the most important component of hatch security. A number of locks can be implemented for hatches, including:

- Slam locks (internal locks located within the hatch frame)
- Recessed cylinder locks
- Bolt locks
- Padlocks

3.3.1.11 Intrusion Sensors

An exterior intrusion sensor is a detection device used in an outdoor environment to detect intrusions into a protected area. These devices are designed to detect an intruder and then communicate an alarm signal to an alarm system. The alarm system can respond to the intrusion in many different ways, such as by triggering an audible or visual alarm signal or by sending an electronic signal to a central monitoring location that notifies security personnel of the intrusion. Intrusion sensor can be used to protect many kinds of assets. Intrusion sensors that protect physical space are classified according to whether they protect indoor, or *interior*, spaces (e.g., an entire building or room within a building), or outdoor, or *exterior*, spaces (e.g., a fence line or perimeter). Interior intrusion sensors are designed to protect the interior space of a facility by detecting an intruder who is attempting to enter or who has already entered a room or building. In contrast, exterior intrusion sensors are designed to detect an intrusion into a protected outdoor/exterior area. Exterior protected areas are typically arranged as zones or exclusion areas placed so the intruder is detected early in the intrusion attempt before gaining access to more valuable assets (e.g., into a building located within the protected area). Early detection creates additional time for security forces to respond to the alarm.

3.3.1.7.1 Buried Exterior Intrusion Sensors

Buried sensors are electronic devices designed to detect potential intruders. The sensors are buried along the perimeters of sensitive assets and are able to detect intruder activity both above and below ground. Some of these systems are composed of individual, stand-alone sensor units, while other sensors consist of buried cables.

3.3.1.12 Ladder Access Control

Water and wastewater utilities have a number of assets that are raised above ground level, including raised water tanks, raised chemical tanks, raised piping systems, and roof access points into buildings. In addition, communications equipment, antennae, or other electronic devices may be located on the top of these raised assets. Typically, these assets are reached by ladders that are permanently anchored to the asset; for example, raised water tanks typically are accessed by ladders that are bolted to one of the legs of the tank. Controlling access to these raised assets by controlling access to the ladder can increase security at a water or wastewater utility.

A typical ladder access control system consists of some type of cover that is locked or secured over the ladder. The cover can be a casing that surrounds most of the ladder or a door or shield that covers only part of the ladder. In either case, several rungs of the ladder (the number of rungs depends on the size of the cover) are made inaccessible by the cover, and these rungs can only be accessed by opening or removing the cover. The cover is locked so only authorized personnel can open or remove it and use the ladder. Ladder access controls are usually installed at several feet above ground level, and they usually extend several feet up the ladder so they cannot be circumvented by someone accessing the ladder above the control system.

The covers are constructed from aluminum or some type of steel. This should provide adequate protection from being pierced or cut through. The metals are corrosion resistant so they will not corrode or become weakened due to extreme weather conditions in outdoor applications. The bolts used to install each of these systems are galvanized steel. In addition, the bolts for each cover are installed on the inside of the unit so they cannot be removed from the outside. The important features of ladder access control are the size and strength of the cover and the ability to lock or otherwise secure the cover from unauthorized access.

3.3.1.13 Locks

A lock is a type of physical security device that can be used to delay or prevent the opening, moving, or operation of a door, window, manhole, filing-cabinet drawer, or some other physical feature. Locks typically operate by connecting two pieces together, such as connecting a door to a door jamb or a manhole to its casement. Every lock has two modes: engaged (or locked) and disengaged (or opened). When a lock is disengaged, the asset on which the lock is installed can be accessed by anyone, but when the lock is engaged, only access to the locked asset.

Locks are excellent security features because they have been designed to function in many ways and to work on many different types of assets. Locks can also provide different levels of security depending on how they are designed and implemented. The security provided by a lock is dependent on several factors, including its ability to withstand physical damage (e.g., being cut off, broken, or otherwise physically disabled) as well as its requirements for supervision or operation (e.g., combinations may have to be changed frequently so they are not compromised and the locks remain secure). Although no agreed-upon rating of lock security exists, locks are often designated as being minimum, medium, or maximum security. Minimum-security locks are those that can be easily disengaged (or "picked") without the correct key or code or those that can be disabled easily (such as small padlocks that can be cut with bolt cutters). Higher security locks are more complex and thus are more difficult to pick, or they are sturdier and more resistant to physical damage.

Many locks only have to be unlocked from one side; for example, most door locks (single-cylinder locks) can be opened on the outside by inserting a key in the lock. On the inside, a person can unlock the same lock by pushing a button or turning a knob or handle. Double-cylinder locks require a key to be locked or unlocked from both sides.

3.3.1.14 Manhole Intrusion Sensors

Manholes are located at strategic locations throughout most municipal water, wastewater, and other underground utility systems. Manholes are designed to provide access to underground utilities, and they represent potential entry points to a system; for example, manholes in water or wastewater systems may provide access to sewer lines or vaults containing on/off or pressure-reducing water valves. Because many utilities run under other infrastructure (roads, building), manholes also provide potential access points to critical infrastructure as well as water and wastewater assets. In addition, because the portion of the system to which manholes provide entry is primarily located underground, access to a system through a manhole increases the chances that an intruder will not be seen; therefore, protecting manholes can be a critical component of guarding an entire community.

The various methods for protecting manholes are designed to prevent unauthorized personnel from physically accessing the manhole or to detect attempts at unauthorized access to the manhole. A manhole intrusion sensor is a physical security device designed to detect unauthorized access to the utility through a manhole. Monitoring a manhole that provides access to a water or wastewater system can mitigate two distinct types of threats. First, monitoring a manhole may detect access of unauthorized personnel to water or wastewater systems or assets through the manhole. Second, monitoring manholes may also allow the detection of the introduction of hazardous substances into the water system.

Several different technologies have been used for manhole intrusion sensors, including mechanical systems, magnetic systems, and fiberoptic and infrared sensors. Some of these intrusion sensors have been specifically designed for manholes, while others consist of standard, off-the-shelf intrusion sensors that have been implemented in a system specifically designed for application in a manhole.

3.3.1.15 Manhole Locks

A manhole lock is a physical security device designed to delay unauthorized access to the utility through a manhole. Locking a manhole that provides access to a water or wastewater system can mitigate two distinct types of threats. First, locking a manhole may delay access of unauthorized personnel to water or wastewater systems through the manhole. Second, locking manholes may also prevent the introduction of hazardous substances into the wastewater or stormwater system.

3.3.1.16 Radiation Detection Equipment for Monitoring Personnel and Packages

A major potential threat faced by water and wastewater facilities is contamination by radioactive substances. Radioactive substances brought on-site at a facility could be used to contaminate the facility, thereby preventing workers from safely entering the facility to perform necessary water treatment tasks. In addition, radioactive substances brought on-site at a water treatment plant could be discharged into the water source or into the distribution system, contaminating the downstream water supply; therefore, the detection of radioactive substances being brought on-site can be an important security enhancement.

Various radionuclides have unique properties, and different equipment is required to detect the different types of radiation; however, it is impractical and potentially unnecessary to monitor for specific radionuclides. Instead, for security purposes, it may be more useful to monitor for gross radiation as an indicator of unsafe substances. To protect against radioactive materials being brought on-site, a facility may set up monitoring sites outfitted with radiation detection instrumentation at entrances to the facility. Depending on the specific types of equipment chosen, this equipment would detect radiation emitted from people, packages, or other objects being brought through an entrance.

One of the primary differences among the various types of detection equipment is the means by which the equipment reads the radiation. Radiation may be detected by direct measurement or through sampling. Direct radiation measurement involves measuring radiation through an external probe on the detection instrumentation. Some direct measurement equipment detects radiation emitted into the air around the monitored object. Because this equipment detects radiation in the air, it does not require that the monitoring equipment make physical contact with the monitored object. Direct means for detecting radiation include using either a walk-through, portal-type monitor that detects elevated radiation levels on a person or in a package or a hand-held detector, which would be moved or swept over individual objects to locate an radioactive source.

Some types of radiation, such as alpha or low-energy beta radiation, have a short range and are easily shielded by various materials. These types of radiation cannot be measured through direct measurement. Instead, they must be measured through sampling. Sampling involves wiping the surface to be tested with a special filter cloth and then exposing the cloth to a special counter; for example, specialized smear counters measure alpha and low-energy beta radiation.

3.3.1.17 Reservoir Covers

Reservoirs are used to store raw or untreated water. They can be located underground (buried), at ground level, or on an elevated surface. Reservoirs can vary significantly in size; small reservoirs can hold as little as 1000 gallons, and larger reservoirs may hold many millions of gallons. Reservoirs can be either natural or manmade. Natural reservoirs can include lakes or other contained water bodies, whereas manmade reservoirs usually consist of some sort of engineered structure, such as a tank or other impoundment structure. In addition to the water containment structure itself, reservoir systems may also include associated water treatment and distribution equipment, including intakes, pumps, pump houses, piping systems, chemical treatment, and chemical storage areas.

Drinking-water reservoirs are of particular concern because they are potentially vulnerable to contamination of the stored water, through direct contamination of the storage area or via infiltration of the equipment, piping, or chemicals associated with the reservoir. Because many drinking-water reservoirs are designed as aboveground, open-air structures, they are potentially vulnerable to airborne deposition, bird and animal wastes, human activities, and dissipation of chlorine or other treatment chemicals; however, one of the most serious potential threats to the system is direct contamination of the stored water through the dumping of contaminants into the reservoir. Utilities have taken various measures to mitigate this type of threat, including fencing off the reservoir, installing cameras to monitor for intruders, and monitoring for changes in water quality. Another option for enhancing security is covering the reservoir using some type of manufactured cover to prevent intruders from gaining physical access to the stored water. Implementing a reservoir cover may or may not be practical depending on the size of the reservoir; covers are not typically used on natural reservoirs because they are too large for the cover to be technically feasible and cost effective. This section focuses on drinking-water reservoir covers, where and how they are typically implemented, and how they can be used to reduce the threat of contamination of the stored water. Although covers can enhance the security of a reservoir, it should be noted that covering a reservoir typically changes the operational requirements of the reservoir; for example, vents must be installed in the cover to provide gas exchange between the stored water and the atmosphere.

A reservoir cover is a structure installed on or over the surface of the reservoir to minimize water quality degradation. The three basic design types of reservoir covers are:

- Floating
- Fixed
- Air-supported

A variety of materials are used to manufacture a cover, including reinforced concrete, steel, aluminum, polypropylene, chlorosulfonated polyethylene, or ethylene interpolymer alloys. Several factors affect the effectiveness of a reservoir cover and thus its ability to protect the stored water, including:

- The location, size, and shape of the reservoir
- The ability to lay or support a foundation (e.g., footing, soil, and geotechnical support conditions)
- The length of time reservoir can be removed from service for cover installation or maintenance
- Aesthetic considerations
- Economic factors, such as capital and maintenance costs

It may not be practical, for example, to install a fixed cover over a reservoir if the reservoir is too large or if the local soil conditions cannot support a foundation. A floating or air-supported cover may be more appropriate for these types of applications.

In addition to the practical considerations for installation of these types of covers, a number of operations and maintenance (O&M) concerns can affect the utility of a cover for specific applications, including how the various cover materials available will withstand local climatic conditions, what types of cleaning and maintenance will be required for each particular type of cover, and how these factors will affect the covers lifespan and its ability to be repaired when it is damage.

The primary feature affecting the security of a reservoir cover is its ability to maintain its integrity. Any type of cover, no matter what its construction material, will provide good protection from contamination by rainwater or atmospheric deposition, as well as from intruders attempting to access the stored water with the intent of causing intentional contamination. The covers are large and heavy, and it is difficult to circumvent them to get into the reservoir. At the very least, it would take a determined intruder, as opposed to a vandal, to defeat the cover.

3.3.1.18 Side-Hinged Door Security

Doorways are the main access points to a facility or to rooms within a building. They are used on the exterior or in the interior of buildings to provide privacy and security for the areas behind them. Different types of doorway security systems may be installed in various doorways depending on the requirements of the building or room. For example, exterior doorways tend to have heavier doors to withstand the elements and to provide some security to the entrance of the building. Interior doorways in office areas may have lighter doors that may be primarily designed to provide privacy rather than security; these doors may be made of glass or lightweight wood. Doorways in industrial areas may have sturdier doors than do other interior doorways and may be designed to provide protection or security for areas behind the doorway; for example, fireproof doors may be installed in chemical storage areas or in other areas where there is a danger of fire. Because they are the main entries into a facility or a room, doorways are often prime targets for unauthorized entry into a facility or an asset; therefore, securing doorways may be a major step in providing security at a facility. A doorway includes four main components:

- The *door*, which blocks the entrance. The primary threat to the actual door is breaking or piercing through it; therefore, the primary security features of doors are their strength and resistance to various physical threats, such as fire or explosions.
- The *door frame*, which connects the door to the wall. The primary threat to a door frame is that the door can be pried away from the frame; therefore, the primary security feature of a door frame is its resistance to prying.
- The *hinges*, which connect the door to the door frame. The primary threat to door hinges is that

they can be removed or broken, which will allow intruders to remove the entire door; therefore, security hinges are designed to be resistant to breaking. They may also be designed to minimize the threat of removal from the door.
- The *lock*, which connects the door to the door frame. Use of the lock is controlled through various security features, such as keys, combinations, etc., such that only authorized personnel can open the lock and go through the door. Locks may also incorporate other security features, such as software to track overall use of the door or to track individuals using the door.

Each of these components is integral to providing security for a doorway. Upgrading the security of only one of these components while leaving the other components unprotected may not improve the overall security of the doorway. Many facilities upgrade door locks as a basic step toward increasing their security, but if a facility does not also modify the door hinges or the door frame then the door may remain vulnerable to being removed from its frame, thus defeating the purpose of installing the new door lock.

The primary attribute for the security of a door is its strength. Many security doors are 4- to 20-gauge hollow metal doors consisting of steel plates over a hollow cavity reinforced with steel stiffeners to give the door extra stiffness and rigidity. This increases resistance to any blunt force applied in an attempt to penetrate through the door. The space between the stiffeners may be filled with specialized materials to provide fire, blast, or bullet resistance to the door. The Window and Door Manufacturers Association has developed a list of performance attributes for doors:

- Structural resistance
- Forced-entry resistance
- Hinge-style screw resistance
- Split resistance
- Hinge resistance
- Security rating
- Fire resistance
- Bullet resistance
- Blast resistance

The first five bullets relate to the resistance of a door to standard physical breaking and prying attacks. Tests are used to evaluate the strength of the door and the resistance of the hinges and the frame in a standardized way. The *rack load test* simulates a prying attack on a corner of the door. A test panel is restrained at one end, and a third corner is supported; loads are applied and measured at the fourth corner. The *door impact test* uses a steel pendulum to simulate a battering attack on a door and frame with impacts of 200 ft-lb; the door must remain fully operable after the test. It should be noted that door glazing is also rated for resistance to shattering, etc. Manufacturers will be able to provide security ratings for these features of a door, as well.

Door frames are an integral part of doorway security because they anchor the door to the wall. Door frames are typically constructed from wood or steel, and they are installed such that they extend for several inches over the doorway that has been cut into the wall. For added security, frames can be designed to have varying degrees of overlap or wrapping. This overlap can make prying the frame from the wall more difficult. A frame formed from a continuous piece of metal (as opposed to a frame constructed from individual metal pieces) will prevent prying between pieces of the frame.

Many security doors can be retrofit into existing frames; however, many security door installations require replacement of the door frame as well as the door itself. Bullet resistance per the Underwriters Laboratory (UL) 752 standard encompasses the resistance of the door and frame assembly both; thus, replacing the door only would not meet UL 752 requirements.

3.3.1.19 Valve Lockout Devices

Valves are utilized as control elements in water and wastewater process piping networks. They regulate the flow of both liquids and gases by opening, closing, or obstructing a flow passageway. Valves are typically located where flow control is necessary; they can be located inline or at pipeline and tank entrance and exit points. Valves serve multiple purposes in a process pipe network, including:

- Redirecting and throttling flow
- Preventing backflow
- Shutting off flow to a pipeline or tank (for isolation purposes)
- Releasing pressure
- Draining extraneous liquid from pipelines or tanks
- Introducing chemicals into the process network
- Serving as access points for sampling process water

Valves are located at critical junctures throughout water and wastewater systems, both on-site at treatment facilities and off-site within water distribution and wastewater collection systems. They may be located either above or below ground. Because many valves are located within the community, it is critical to provide protection against valve tampering. Tampering with a pressure-relief valve, for example, could result in a pressure buildup and potential explosion in the piping network. On a larger scale, addition of a pathogen or chemical to the water distribution system through an unprotected valve could result in the release of that contaminant to the general population.

Various security products are available to protect aboveground vs. belowground valves; for example, valve lockout devices can protect valves and valve controls located above ground. Vaults containing underground valves can be locked to prevent access to these valves. Valve-specific lockout devices are available in a variety of colors, which can be useful in distinguishing different valves. Different colored lockouts can be used to distinguish the type of liquid passing through the valve (e.g., treated, untreated, potable, chemical) or to identify the party responsible for maintaining the lockout. Implementing a system of colored locks on operating valves can increase system security by reducing the likelihood of an operator inadvertently opening the wrong valve and causing a problem in the system.

3.3.1.20 Vent Security

Vents are installed in aboveground, covered water reservoirs and in underground reservoirs to allow ventilation of the stored water. Specifically, vents permit the passage of air that is being displaced from, or drawn into, the reservoir as the water level in the reservoir rises and falls due to system demands. Small reservoirs may require only one vent, whereas larger reservoirs may have multiple vents throughout the system.

The specific vent design for any given application will vary depending on the design of the reservoir, but every vent consists of an open-air connection between the reservoir and the outside environment. These air-exchange vents are an integral part of covered or underground reservoirs, but they also represent a potential security threat. Improving vent security by making the vents tamper resistant or by adding other security features, such as security screens or security covers, can enhance the security of the entire water system. Many municipalities already have specifications for vent security at their water assets. These specifications typically include the following requirements:

- Vent openings are to be angled down or shielded to minimize the entrance of surface water or rainwater into the vent through the opening.
- Vent designs are to include features that exclude insects, birds, animals, and dust.
- Corrosion-resistant materials are to be used to construct the vents.

Some states have adopted more specific requirements for vent security at their water utility assets. The State of Utah's Department of Environmental Quality, Division of Drinking Water, Division of Administrative Rules (DAR), provide specific requirements for public drinking-water storage tanks. The rules for drinking-water storage tanks as they apply to venting are set forth in R309-545-15 (Venting) and include the following requirements:

- Drinking-water storage tank vents must have an open discharge on buried structures.
- The vents must be located 24 to 36 inches above the earthen covering.
- The vents must be located and sized to avoid blockage during winter conditions.

In a second example, the Washington State Department of Health (DOH, 2007) requires that vents must be protected to prevent the water supply from being contaminated. Noncorrodible No. 4 mesh may be used to screen vents on elevated tanks, but vent openings for storage facilities located underground or at ground level should be 24 to 36 inches above the roof or ground and must be protected with a No. 24 mesh noncorrodible screen. New Mexico's administrative code also specifies that vents must be covered with No. 24 mesh (NMAC Title 20, Chapter 7, Subpart I, 208.E). Washington and New Mexico, as well as many other municipalities, require vents to be screened using a noncorrodible mesh to minimize the entry of insects, other animals, and rain-borne contamination into the vents. When selecting the appropriate mesh size, it is important to identify the smallest mesh size that meets both the strength and durability requirements for that application.

3.3.1.21 Visual Surveillance Monitoring

Visual surveillance is used to detect threats through continuous observation of important or vulnerable areas of an asset. The observations can also be recorded for later review or use (e.g., in court proceedings). Visual surveillance system can be used to monitor various parts of collection, distribution, or treatment systems, including the perimeter of a facility, outlying pumping stations, or entry or access points into specific buildings. These systems are also useful in recording individuals who enter or leave a facility, thereby helping to identify unauthorized access. Images can be transmitted live to a monitoring station, where they can be monitored in real time, or they can be recorded and reviewed later. Many facilities have found that a combination of electronic surveillance and security guards provides an effective means of facility security. Visual surveillance is provided through a closed-circuit television (CCTV) system, in which the capture, transmission, and reception of an image are localized within a closed circuit. This is different than other broadcast images, such as over-the-air television, which is broadcast over the air to any receiver within range. At a minimum, a CCTV system consists of:

- One or more cameras
- A monitor for viewing the images
- A system for transmitting the images from the camera to the monitor

3.3.2 Water Monitoring Devices

(*Note:* The following information is adapted from Spellman, F.R., *Water Infrastructure Protection and Homeland Security*, Government Institutes Press, Lanham, MD, 2007.)

Earlier it was pointed out that proper security preparation really comes down to a three-legged approach: *detect, delay, respond.* The third leg of security—detect—is discussed in this section; specifically, this section deals with the monitoring of water samples to detect toxicity or contamination. Many of the major monitoring tools that can be used to identify anomalies in process streams or finished water that may represent potential threats are discussed, including:

- Sensors for monitoring chemical, biological, and radiological contamination
- Chemical sensors—arsenic measurement system
- Chemical sensors—adapted biochemical oxygen demand (BOD) analyzer
- Chemical sensors—total organic carbon analyzer
- Chemical sensors—chlorine measurement system
- Chemical sensors—portable cyanide analyzer
- Portable field monitors to measure volatile organic compounds (VOCs)
- Radiation detection equipment
- Radiation detection equipment for monitoring water assets
- Toxicity monitoring and toxicity meters

Water quality sensors may be used to monitor key elements of water or wastewater treatment processes (such as influent water quality, treatment processes, or effluent water quality) to identify anomalies that may indicate threats to the system. Some sensors, such as sensors for biological organisms or radiological contaminants, measure potential contamination directly, while others, particularly some chemical monitoring systems, measure surrogate parameters that may indicate problems in the system but do not identify sources of the contamination directly. In addition, sensors can provide more accurate control of critical components in water and wastewater systems and may provide a means of early warning so the potential effects of certain types of attacks can be mitigated. One advantage of using chemical and biological sensors to monitor for potential threats to water and wastewater systems is that many utilities already employ sensors to monitor potable water (raw or finished) or influent/effluent for Safe Drinking Water Act (SDWA) or Clean Water Act (CWA) water quality compliance or process control.

Chemical sensors that can be used to identify potential threats to water and wastewater systems include inorganic monitors (e.g., chlorine analyzer), organic monitors (e.g., total organic carbon analyzer), and toxicity meters. Radiological meters can be used to measure concentrations of several different radioactive species. Monitors that use biological species can be used as sentinels for the presence of contaminants of concern, such as toxics. At the present time, biological monitors are not in widespread use and very few biomonitors are used by drinking-water utilities in the United States.

Monitoring can be conducted using either portable or fixed-location sensors. Fixed-location sensors are usually used as part of a continuous, online monitoring system. Continuous monitoring has the advantage of allowing immediate notification when there is an upset; however, the sampling points are fixed, and only certain points in the system can be monitored. In addition, the number of monitoring locations required to capture the physical, chemical, and biological complexity of a system can be prohibitive. The use of portable sensors can overcome this problem of monitoring many points in the system. Portable sensors can be used to analyze grab samples at any point in the system but have the disadvantage that they provide measurements only at one point in time.

3.3.2.1 Sensors for Monitoring Chemical, Biological, and Radiological Contamination

Toxicity tests measure water toxicity by monitoring adverse biological effects on test organisms. Toxicity tests have traditionally been used to monitor wastewater effluent streams for National Pollutant Discharge Elimination System (NPDES) permit compliance or to test water samples for toxicity; however, this technology can also be used to monitor drinking-water distribution systems or other water/wastewater streams for toxicity. Currently, several types of biosensors and toxicity tests are being adapted for use in the water/wastewater security field. The keys to using biomonitoring or biosensors for drinking-water or other water/wastewater asset security are rapid response and the ability to use the monitor at critical locations in the system, such as in water distribution systems downstream of pump stations or prior to the biological process in a wastewater treatment plant. Several different organisms that can be used to monitor for toxicity (including bacteria, invertebrates, and fish), but bacteria-based biosensors are ideal for use as early-warning screening tools for drinking-water security because bacteria usually respond to toxins in a matter of minutes. In contrast to methods using bacteria, toxicity

screening methods that use higher level organisms such as fish may require several days to produce a measurable result. Bacteria-based biosensors have recently been incorporated into portable instruments, making rapid response and field testing more practical. These portable meters detect decreases in biological activity (e.g., decreases in bacterial luminescence), which are highly correlated with increased levels of toxicity.

At the present time, few utilities are using biologically based toxicity monitors to monitor water/wastewater assets for toxicity, and very few products are now commercially available. Several new approaches to the rapid monitoring of microorganisms for security purposes (e.g., microbial source tracking) have been identified; however, most of these methods are still in the research and development phase.

3.3.2.2 Chemical Sensors: Arsenic Measurement System

Arsenic is an inorganic toxin that occurs naturally in soils. It can enter water supplies from many sources, including erosion of natural deposits, runoff from orchards, runoff from glass and electronics production wastes, or leaching from products treated with arsenic, such as wood. Synthetic organic arsenic is also used in fertilizer. Arsenic toxicity is primarily associated with inorganic arsenic ingestion and has been linked to cancerous health effects, including cancer of the bladder, lungs, skin, kidney, nasal passages, liver, and prostate. Arsenic ingestion has also been linked to noncancerous cardiovascular, pulmonary, immunological, and neurological, endocrine problems. According to USEPA's Safe Drinking Water Act Arsenic Rule, inorganic arsenic can exert toxic effects after acute (short-term) or chronic (long-term) exposure. Toxicological data for acute exposure, which is typically give as an LD_{50} value (the dose that would be lethal to 50% of the test subjects in a given test), suggests that the LD_{50} of arsenic ranges from 1 to 4 milligrams arsenic per kilogram (mg/kg) of body weight. This dose would correspond to a lethal dose range of 70 to 280 mg for 50% of adults weighing 70 kg. At nonlethal, but high, acute doses, inorganic arsenic can cause gastroenterological effects, shock, neuritis (continuous pain), and vascular effects in humans. USEPA has set a maximum contaminant level (MCL) goal of 0 for arsenic in drinking water. In 2006, the enforceable maximum contaminant level for arsenic was lowered from 0.050 mg/L to 0.010 mg/L.

The SDWA requires arsenic monitoring for public water systems. The Arsenic Rule indicates that surface-water systems must collect one sample annually, and groundwater systems must collect one sample in each compliance period (once every 3 years). Samples are collected at entry points to the distribution system, and analysis is usually done in the lab using one of several methods approved by USEPA, including inductively coupled plasma–mass spectroscopy (ICP–MS) and a number of atomic absorption (AA) methods. Several different technologies, including colorimetric test kits and portable chemical sensors, are currently available for monitoring inorganic arsenic concentrations in the field. These technologies can provide a quick estimate of arsenic concentrations in a water sample, and they may prove useful for spot-checking various parts of a drinking-water system (e.g., reservoirs, isolated areas of distribution systems) to ensure that the water is not contaminated with arsenic.

3.3.2.3 Chemical Sensors: Adapted BOD Analyzer

One manufacturer has adapted a BOD analyzer to measure oxygen consumption as a surrogate for general toxicity. The critical element in the analyzer is the bioreactor, which is used to continuously measure the respiration of the biomass under stable conditions. As the toxicity of the sample increases, the oxygen consumption in the sample decreases. An alarm can be programmed to sound if oxygen reaches a minimum concentration (i.e., if the sample is strongly toxic). The operator must then interpret the results into a measure of toxicity. Note that, at the current time, it is difficult to directly define the sensitivity or the detection limit of toxicity measurement devices because limited data are available regarding the specific correlation of decreased oxygen consumption and increased toxicity of the sample.

3.3.2.4 Chemical Sensors: Total Organic Carbon Analyzer

Total organic carbon (TOC) analysis is a well-defined and commonly used methodology that measures the carbon content of dissolved and particulate organic matter present in water. Many water utilities monitor TOC to determine raw water quality or to evaluate the effectiveness of processes designed to remove organic carbon. Some wastewater utilities also employ TOC analysis to monitor the efficiency of the treatment process. In addition to these uses for TOC monitoring, measuring changes in TOC concentrations can be an effective surrogate for detecting contamination from organic compounds (e.g., petrochemicals, solvents, pesticides). Thus, although TOC analysis does not give specific information about the nature of the threat, identifying changes in TOC can be a good indicator of potential threats to a system. TOC analysis includes inorganic carbon removal, oxidation of organic carbon into CO_2, and quantification of the CO_2. The primary differences among various online TOC analyzers lie in the methods used for oxidation and CO_2 quantification.

The oxidation step can be high or low temperature. The determination of the appropriate analytical method (and thus the appropriate analyzer) is based on the expected characteristics of the wastewater sample (TOC concentrations and the individual components making up the TOC fraction). In general, high-temperature (combustion) analyzers achieve more complete oxidation of the carbon fraction than do low-temperature (wet chemistry/ultraviolet) analyzers. This can be important both in distinguishing different fractions of the organics in a sample and in achieving a precise measurement of the organic content of the sample. Three different methods are available for the detection and quantification of CO_2 produced in the oxidation step of a TOC analyzer:

- Nondispersive infrared (NDIR) detector
- Colorimetric methods
- Aqueous conductivity methods

The most common detector that online TOC analyzers use for source-water and drinking-water analysis is the nondispersive infrared detector.

Although the differences among analytical methods employed by various TOC analyzers may be important in compliance or process monitoring, high levels of precision and the ability to distinguish specific organic fractions from a sample may not be required for detection of a potential chemical threat. Instead, gross deviations from normal TOC concentrations may be the best indication of a chemical threat to the system.

The detection limit for organic carbon depends on the measurement technique used (high- or low-temperature) and the type of analyzer. Because TOC concentrations are simply surrogates that can indicate potential problems in a system, gross changes in these concentrations are the best indicators of potential threats; therefore, high-sensitivity probes may not be required for security purposes. The following detection limits can be expected:

- High-temperature method (between 680 and 950°C or higher in a few special cases, best possible oxidation), 1 mg/L carbon
- Low-temperature method (below 100°C, limited oxidation potential), 0.2 mg/L carbon

The response time of a TOC analyzer may vary depending on the manufacturer's specifications, but it usually takes from 5 to 15 minutes to get a stable, accurate reading.

3.3.2.5 Chemical Sensors: Chlorine Measurement System

Residual chlorine is one of the most sensitive and useful indicator parameters in water distribution system monitoring. All water distribution systems monitor for residual chlorine concentrations as part of their Safe Drinking Water Act requirements, and procedures for monitoring chlorine concentrations are well established and accurate. Chlorine monitoring ensures proper residual chlorine levels at all points in the system, helps pace rechlorination when required, and quickly and reliably signals any unexpected increase in disinfectant demand. A significant decline or loss of residual chlorine could be an indication of potential threats to the system. Several key points regarding residual chlorine monitoring for security purposes are provided below:

- Residual chlorine can be measured using continuous online monitors at fixed points in the system or by taking grab samples at any point in the system and using chlorine test kits or portable sensors to determine chlorine concentrations.
- Correct placement of residual chlorine monitoring points within a system is crucial to early detection of potential threats. Although dead ends and low-pressure zones are common trouble spots that can show low residual chlorine concentrations, these zones are generally not of great concern for water security purposes because system hydraulics will limit the circulation of any contaminants present in these areas of the system.
- Monitoring points and monitoring procedures for SDWA compliance vs. system security purposes may be different, and utilities must determine the best use of online, fixed monitoring systems vs. portable sensors or test kits to balance their SDA compliance and security needs.

Various portable and online chlorine monitors are commercially available. These range from sophisticated online chlorine monitoring systems to portable electrode sensors to colorimetric test kits. Online systems can be equipped with control, signal, and alarm systems that notify the operator of low chlorine concentrations, and some may be tied into feedback loops that automatically adjust chlorine concentrations in the system. In contrast, the use of portable sensors or colorimetric test kits requires technicians to take a sample and read the results. The technician then initiates required actions based on the results of the test.

Several methods are currently available to measure chlorine in water samples, including:

- *N,N*-diethyl-*p*-phenylenediamine (DPD) colorimetric method
- Iodometric method
- Amperometric electrodes
- Polarographic membrane sensors

It should be noted that these methods differ with regard to the specific type of analyte, their range, and their accuracy. In addition, these methods have different operations and maintenance requirements; for example, DPD systems require periodic replenishment of buffers, whereas polarographic systems do not. Users may want to consider these requirements when choosing appropriate sensors for their systems.

3.3.2.6 Chemical Sensors: Portable Cyanide Analyzer

Portable cyanide detection systems are designed to be used in the field to evaluate for potential cyanide contamination of a water asset. These detection systems use one of two distinct analytical methods—either a colorimetric method or an ion-selective method—to provide a quick, accurate cyanide measurement that does not require laboratory evaluation. Aqueous cyanide chemistry can be complex. Various factors, including the pH and redox potential of the water, can affect the toxicity of cyanide in that asset. Although personnel using these cyanide detection devices do not need to have advanced knowledge of cyanide chemistry to successfully screen a water asset for cyanide, understanding aqueous cyanide chemistry can help users to interpret whether the cyanide concentration represents a potential threat. For this reason, a short summary of aqueous cyanide chemistry, including a discussion of cyanide toxicity, is provided below. For more information, the reader is referred to Greenburg et al. (1999).

Cyanide (CN^-) is a toxic carbon–nitrogen organic compound that is the functional portion of the lethal gas hydrogen cyanide (HCN). The toxicity of aqueous cyanide varies depending on its form. At near-neutral pH, *free cyanide* (commonly designated as CN^-, although it is actually defined as the total of HCN and CN^-) is the predominant cyanide form in water. Free cyanide is potentially toxic in its aqueous form, although the primary concern regarding aqueous cyanide is that it could volatilize. Free cyanide is not highly volatile (it is less volatile than most VOCs, but its volatility increases as the pH decreases below 8). When free cyanide does volatilize, it volatilizes in its highly toxic gaseous form (gaseous HCN). As a general rule, metal–cyanide complexes are much less toxic than free cyanide because they do not volatilize unless the pH is low.

Analyses for cyanide in public water systems are often conducted in certified labs using various USEPA-approved methods, such as the preliminary distillation procedure with subsequent analysis by colorimetric, ion-selective electrode, or flow injection methods. Lab analyses using these methods require careful sample preservation and pretreatment procedures and are generally expensive and time consuming. Using these methods, several cyanide fractions are typically defined:

- *Total cyanide* includes free cyanide (CN⁻ + HCN) and all metal-complexed cyanide.
- *Weak acid dissociable (WAD) cyanide* includes free cyanide (CN⁻ + HCN) and weak cyanide complexes that could potentially be toxic due to hydrolysis to free cyanide in the pH range 4.5 to 6.0.
- *Amendable cyanide* includes free cyanide (CN⁻ + HCN) and weak cyanide complexes that can release free cyanide at high pH (11 to 12); this fraction gets its name because it includes the measurement of cyanide from complexes that are amendable to oxidation by chlorine at high pH. To measure amendable cyanide, the sample is split into two fractions. One of the fractions is analyzed for total cyanide; the other fraction is treated with high levels of chlorine for approximately an hour, dechlorinated, and distilled per the method for total cyanide. Amendable cyanide is determined by the difference in the cyanide concentrations in these two fractions.
- *Soluble cyanide* is measured by using the preliminary filtration step, followed by a total cyanide analysis.

As discussed above, these different methods yield various cyanide measurements that may or may not give a complete picture of that potential toxicity of the sample. The total cyanide method, for example, includes cyanide complexed with metals, some of which will not contribute to cyanide toxicity unless the pH is out of the normal range. In contrast, the WAD cyanide measurement includes metal-complexed cyanide that could become free cyanide at low pH, and amendable cyanide measurements include metal-complexed cyanide that could become free cyanide at high pH. Personnel using these kits should therefore be aware of the potential differences in actual cyanide toxicity vs. the cyanide potential differences in actual cyanide toxicity vs. the cyanide measured in the sample under different environmental conditions.

Ingestion of aqueous cyanide can result in numerous adverse health effects and may be lethal. USEPA's maximum contaminant level (MCL) for cyanide in drinking water is 0.2 µg/L (0.2 parts per million, or ppm). This MCL is based on free cyanide analysis per the amendable cyanide method. Note that USEPA has recognized that very stable metal–cyanide complexes, such as the iron–cyanide complex, are nontoxic (unless exposed to significant UV radiation); therefore, these fractions are not considered when defining cyanide toxicity. Ingestion of free cyanide at concentrations in excess of the MCL causes both acute effects (e.g., rapid breathing, tremors, and neurological symptoms) and chronic effects (e.g., weight loss, thyroid effects, and nerve damage). Under the current primary drinking-water standards, public water systems are required to monitor their systems to minimize public exposure to cyanide levels in excess of the MCL.

Hydrogen cyanide gas is also toxic, and the Occupational Safety and Health Administration (OSHA) has set a permissible exposure limit (PEL) of 10 ppmv for HCN inhalation. HCN also has a strong, bitter, almond-like smell and an odor threshold of approximately 1 ppmv. Considering the fact that HCN is relatively nonvolatile, a slight cyanide odor emanating from a water sample suggests very high aqueous cyanide concentrations—greater than 10 to 50 mg/L, which is in the range of a lethal or near-lethal dose with the ingestion of one pint of water.

3.3.2.7 Portable Field Monitors to Measure VOCs

Volatile organic compounds (VOCs) are a group of highly utilized chemicals that have widespread applications, including use as fuel components, solvents, and cleaning and liquefying agents in degreasers, polishes, and dry-cleaning solutions. VOCs are also used in herbicides and insecticides for agriculture applications. Laboratory-based methods for analyzing VOCs are well established; however, analyzing VOCs in the lab is time consuming. Obtaining a result may require several hours to several weeks, depending on the specific method. Faster commercially available methods for analyzing VOCs quickly in the field include the use of a portable gas chromatograph (GC), mass spectrometer (MS), or gas chromatograph/mass spectrometer (GC/MS), all of which can be used to obtain VOC concentration results within minutes. These instruments can be useful in rapid confirmation of the presence of VOCs in an asset or for monitoring an asset on a regular basis. In addition, portable VOC analyzers can analyze for a wide range of VOCs, such as toxic industrial chemicals (TICs), chemical warfare agents (CWAs), drugs, explosives, and aromatic compounds. Several easy-to-use, portable VOC analyzers currently on the market are effective in evaluating VOC concentrations in the field. These instruments utilize gas chromatography, mass spectroscopy, or a combination of both methods, to provide near-laboratory-quality analysis for VOCs.

3.3.2.8 Radiation Detection Equipment

Radioactive substances (radionuclides) are known health hazards that emit energetic waves or particles that can cause both carcinogenic and noncarcinogenic health effects. Radionuclides pose unique threats to source water supplies and water treatment, storage, or distribution systems because radiation emitted from radionuclides in water systems can affect individuals through several pathways—by direct contact with the contaminated water or by ingestion of, inhalation of, or external exposure to the contaminated water. Radiation can occur naturally in some cases due to the decay of some minerals, but the intentional or unintentional release of manmade radionuclides into water systems is also a realistic threat.

Threats to water and wastewater facilities from radioactive contamination could involve two major scenarios. First, the facility or its assets could be contaminated, preventing workers from accessing and operating the facility. Second, at drinking-water facilities, the water supply could be contaminated, and tainted water could be distributed to users downstream. These two scenarios require different threat reduction strategies. The first scenario requires that facilities monitor for radioactive substances being brought on-site; the second requires that water assets be monitored for radioactive contamination. The effects of radioactive contamination are basically the same under both types of threats, but each of these threats requires different types of radiation monitoring and equipment.

3.3.2.9 Radiation Detection Equipment for Monitoring Water Assets

Most water systems are required to monitor for radioactivity and certain radionuclides and to meet maximum contaminant levels for these contaminants, to comply with the Safe Drinking Water Act. Currently, USEPA requires drinking water to meet MCLs for beta/photon emitters (includes gamma radiation), alpha particles, combined radium 226/228, and uranium; however, this monitoring is required only at entry points into the system. In addition, after the initial sampling requirements, only one sample is required every 3 to 9 years, depending on the contaminant type and the initial concentrations. This is adequate to monitor for long-term protection from overall radioactivity and specific radionuclides in drinking water, but it may not be adequate to identify short-term spikes in radioactivity, such as from spills, accidents, or intentional releases. In addition, compliance with the SDWA requires analyzing water samples in a laboratory, which results in a delay in receiving results. In contrast, security monitoring is more effective when results can be obtained quickly in the field. In addition, monitoring for security purposes does not necessarily require that the specific radionuclides causing the contamination be identified. Thus, for security purposes, it may be more appropriate to monitor for non-radionuclide-specific radiation using either portable field meters, which can be used as necessary to evaluate grab samples, or online systems, which can provide continuous monitoring of a system.

Ideally, measuring radioactivity in water assets in the field would involve minimal sampling and sample preparation; however, the physical properties of specific types of radiation combined with the physical properties of water make evaluating radioactivity in water assets in the field somewhat difficult. Alpha particles, for example, can only travel short distances, and they cannot penetrate through most physical objects; therefore, instruments designed to evaluate alpha emissions must be specially designed to capture emissions at a short distance from the source, and they must not block alpha emissions from entering the detector. Gamma radiation does not have the same types of physical properties, so it can be measured using different detectors.

Measuring different types of radiation is further complicated by the relationship between the intrinsic properties of the radiation and the medium in which the radiation is being measured. Gas-flow proportional counters are typically used to evaluate gross alpha and beta radiation from smooth, solid surfaces, but because water is not a smooth surface and because alpha and beta emissions are relatively short range and can be attenuated within the water, these types of counters are not appropriate for measuring alpha and beta activity in water. An appropriate method for measuring alpha and beta radiation in water is using a liquid scintillation counter; however, this requires mixing an aliquot of water with a liquid scintillation cocktail. The liquid scintillation counter is a large, sensitive piece of equipment and is not appropriate for field use; therefore, measurements for alpha and beta radiation from water assets are not typically made in the field.

Unlike the problems associated with measuring alpha and beta activity in water in the field, the properties of gamma radiation allow it to be measured relatively well in water samples in the field. The standard instrumentation used to measure gamma radiation from water samples in the field is a sodium iodide (NaI) scintillator.

Although the devices outlined above are the most commonly used for evaluating total alpha, beta, and gamma radiation, other methods and other devices can be used. In addition, local conditions (e.g., temperature, humidity) or the properties of the specific radionuclides emitting the radiation may make other types of devices or other methods more optimal to achieve the goals of the survey than the devices noted above. Experts or individual vendors should be consulted to determine the appropriate measurement device for any specific application.

An additional factor to consider when developing a program to monitor for radioactive contamination in water assets is whether to take regular grab samples or to sample continuously. Portable sensors can be used to analyze grab samples at any point in the system, but they have the disadvantage that they provide measurements only at one point in time. On the other hand, fixed-location sensors are usually used as part of a continuous, online monitoring system. These systems continuously monitor a water asset and could be outfitted with some type of alarm system that would alert operators if radiation increased above a certain threshold; however, the sampling points are fixed, and only certain points in the system can be monitored. In addition, the number of monitoring locations required to capture the physical and radioactive complexity of a system can be prohibitive.

3.3.2.10 Toxicity Monitoring/Toxicity Meters

Toxicity measurement devices measure general toxicity to biological organisms, and detection of toxicity in any water/wastewater asset can indicate a potential threat to the treatment process (in the case of influent toxicity), to human health (in the case of drinking-water toxicity), or to the environment (in the case of effluent toxicity). Currently, whole effluent toxicity (WET) tests, in which effluent samples are tested against test organisms, are required of many National Pollutant Discharge Elimination System (NPDES) discharge permits. The WET tests are used as a complement to the effluent limits on physical and chemical parameters to assess the overall effects of the discharge on living organisms or aquatic biota. Toxicity tests may also be used to monitor wastewater influent streams for potential hazardous contamination, such as organic heavy metals (arsenic, mercury, lead, chromium, and copper) that might upset the treatment process.

The ability to get feedback on sample toxicity from short-term toxicity tests or toxicity meters can be valuable in estimating the overall toxicity of a sample. Online real-time toxicity monitoring is still being actively researched and developed, but several portable toxicity measurement devices are currently available. They can be divided into categories based on the way they measure toxicity:

- Meters measuring direct biological activity (e.g., luminescent bacteria) and correlating decreases in this direct biological activity with increased toxicity
- Meters measuring oxygen consumption and correlating decrease in oxygen consumption with increased toxicity

3.3.3 Communication and Integration

This section discusses those devices necessary for communication and the integration of water and wastewater system operations, such as electronic controllers, two-way radios, and wireless data communications. Electronic controllers are used to automatically activate security equipment (such as lights, surveillance cameras, audible alarms, or locks) when they are triggered. Triggering could be the result of the tripping of an alarm or a motion sensor, a window or glass door breaking, variation in vibration sensor readings, or simply input from a timer. Two-way wireless radios allow two or more users who have their radios tuned to the same frequency to communicate instantaneously with each other without the radios being physically connected with wires or cables. Wireless data communications devices are used to enable transmission of data between computer systems or between a SCADA server (see Section 3.4) and its sensing devices, without individual components being physically linked together via wires or cables. In water and wastewater utilities, these devices are often used to link remote monitoring stations (e.g., SCADA components) or portable computers to computer networks without using physical wiring connections.

3.3.3.1 Electronic Controllers

An electronic controller is a piece of electronic equipment that receives incoming electric signals and uses preprogrammed logic to generate electronic output signals based on the incoming signals. Electronic controllers can be implemented for any application that involves inputs and outputs (such as to control equipment in a factory), but in a security application they essentially act as the system's brain and can respond to specific security-related inputs with preprogrammed output responses. These systems combine the control of electronic circuitry with a logic function such that circuits are opened and closed (and equipment is turned on and off) through some preprogrammed logic. The basic principle behind the operation of an electrical controller is that it receives electronic inputs from sensors or any device generating an electrical signal (e.g., electrical signals from motion sensors) and then uses its preprogrammed logic to produce electrical outputs (e.g., these outputs could turn on power to a surveillance camera or an audible alarm). Thus, these systems automatically generate a preprogrammed, logical response to a preprogrammed input scenario.

The three major types of electronic controllers are timers, electromechanical relays, and programmable logic controllers (PLCs), which are often called *digital relays*. Timers use internal signals or inputs (in contrast to externally generated inputs) to generate electronic output signals at certain times. More specifically, timers control electric current flow to any application to which they are connected and can turn the current on or off on a schedule prespecified by the user. The typical timer range (amount of time that can be programmed to elapse before the timer activates linked equipment) is from 0.2 seconds to 10 hours, although some of the more advanced timers have ranges of up to 60 hours. Timers are useful in fixed applications that do not require frequent schedule changes; for example, a timer can be used to turn on the lights in a room or a building at a certain time every day. Timers are usually connected to their own power supply (usually 120 to 240 V).

In contrast to timers, which have internal triggers operating on a regular schedule, electromechanical relays and PLCs have both external inputs and external outputs; however, PLCs are more flexible and more powerful than electromechanical relays and are the predominant technology for security-related electronic control applications. Electromechanical relays are simple devices that use a magnetic field to control a switch. Voltage applied to the input coil of the relay creates a magnetic field that attracts an internal metal switch. This causes the contacts of the relay to touch, thus closing the switch and completing the electrical circuit. This activates any linked equipment. These types of systems are often used for high-voltage applications, such as in some automotive and other manufacturing processes.

3.3.3.2 Two-Way Radios

Two-way radios, as discussed here, are limited to direct unit-to-unit radio communication, either via single unit-to-unit transmission and reception or via multiple hand-held units to a base station radio contact and distribution system. Radiofrequency spectrum limitations apply to all hand-held units and are directed by the Federal Communications Commission (FCC). This discussion also distinguishes between a hand-held unit and a base station or base station unit, such as those used by amateur (ham) radio operators, which operate under different wavelength parameters.

Two-way radios allow a user to contact another user or group of users instantly on the same frequency and to transmit voice or data without the need for wires. They use *half-duplex* communication, which means that they cannot transmit and receive data simultaneously. In other words, only one person may talk while other personnel with radios can only listen. To talk, the user depresses the talk button and speaks into the radio. The audio then transmits the voice wirelessly to the receiving radios. When the speaker has finished speaking and the channel has cleared, users on any of the receiving radios can transmit, either to answer the first transmission or to begin a new conversation. In addition to carrying voice data, many types of wireless radios also allow the transmission of digital data, and these radios may be interfaced with computer networks that can use or track these data. Some two-way radios can send information such as global positioning system (GPS) data or the ID of the radio, and others can send data through a SCADA system.

Wireless radios broadcast these voice or data communications over the airwaves from the transmitter to the receiver. This can be an advantage in that the signal emanates in all directions and does not require a direct physical connection to be received at the receiver, but it can also make the communications vulnerable to being blocked, intercepted, or otherwise altered. Additional security features are available, however, to ensure that the communications are not tampered with.

3.3.3.3 Wireless Data Communications

A wireless data communication system consists of two components—a *wireless access point* (WAP) and a *wireless network interface card* (sometimes also referred to as a *client*)—which work together to complete the communications link. These wireless systems can link electronic devices, computers, and computer systems together using radio-

waves, thus eliminating the need for the individual components to be connected physically via wires. Wireless data communications have found widespread application in water and wastewater systems, but they also have limitations. First, wireless data connections are limited by the distance between components (radiowaves scatter over a long distance and cannot be received efficiently, unless directional antennas are used). Second, these devices only function if the individual components are in direct line of sight with each other, because radiowaves are affected by interference from physical obstructions. In some cases, however, repeater units can be used to amplify and retransmit wireless signals to circumvent these problems. The two components of wireless devices are discussed in more detail below.

The wireless access point provides the wireless data communication service. It usually consists of a housing (constructed from plastic or metal, depending on the environment in which it will be used) that contains a circuit board, as well as flash memory that holds the necessary software, one of two external ports to connect to existing wired networks, a wireless radio transmitter/receiver, and one or more antenna connections. Typically, the WAP requires a one-time user configuration to allow the device to interact with the local area network (LAN). This configuration is usually accomplished using web-driven software accessed via a computer.

The wireless network interface card or client is a piece of hardware that plugs into a computer and enables that computer to make a wireless network connection. The card consists of a transmitter, functional circuitry, and a receiver for the wireless signal, all of which work together to enable communication between the computer, its wireless transmitter/receiver, and its antenna connection. Wireless cards are installed in a computer through a variety of connections, including USB adapters or laptop PCMCIA Cardbus or desktop PCI peripheral cards. As for the WAP, software is loaded onto the user's computer, allowing configuration of the card so it may operate over the wireless network.

Two of the primary applications for wireless data communications systems are to enable mobile or remote connections to a LAN and to establish wireless communications links between SCADA remote terminal units (RTUs) and sensors in the field. Wireless card connections are usually used for LAN access from mobile computers. Wireless cards can also be incorporated into RTUs to allow them to communicate with sensing devices that are located remotely.

3.3.4 CYBER PROTECTION DEVICES

Various cyber protection devices are currently available for use in protecting utility computer systems. These protection devices include antivirus and pest eradication software, firewalls, and network intrusion hardware/software. These products are discussed in this section.

3.3.4.1 Antivirus and Pest-Eradication Software

Antivirus programs are designed to detect, delay, and respond to programs or pieces of code that are specifically designed to harm computers. These programs are known as *malware*. Malware can include computer viruses, worms, and Trojan horse programs (programs that appear to be benign but which have hidden harmful effects). Pest-eradication tools are designed to detect, delay, and respond to *spyware* (strategies that websites use to track user behavior, such as by sending cookies to the user's computer) and hacker tools that track keystrokes (keystroke loggers) or reveal passwords (password crackers).

Viruses and pests can enter a computer system through the Internet or through infected floppy discs or CDs. They can also be placed onto a system by insiders. Some of these programs, such as viruses and worms, then move throughout the drives and files of a computer or among networked computers. This malware can deliberately damage files, utilize memory and network capacity, crash application programs, and initiate transmissions of sensitive information from a PC. The specific mechanisms of these programs differ, but they all can infect files and affect even the basic operating program of the computer.

The most important features of an antivirus program are its abilities to identify potential malware and to alert a user before infection occurs, as well as its ability to respond to a virus already resident on a system. Most of these programs provide a log so the user can see what viruses have been detected and where they were detected. After detecting a virus, the antivirus software may delete the virus automatically, or it may prompt the user to delete the virus. Some programs will also fix files or programs damaged by the virus.

Various sources of information are available to inform the general public and computer system operators about new viruses being detected. Because antivirus programs use signatures (or snippets of code or data) to detect the presence of a virus, periodic updates are required to identify new threats. Many antivirus software providers offer free updates that are able to detect and respond to the latest viruses.

3.3.4.2 Firewalls

A firewall is an electronic barrier designed to keep computer hackers, intruders, or insiders from accessing specific data files and information on a utility's computer network or other electronic/computer systems. Firewalls operate by evaluating and then filtering information coming through a public network (such as the Internet) into the utility's computer or other electronic system. This evaluation can include identifying the source or destina-tion addresses and ports and allowing or denying access based on this identification. Two methods are used by firewalls to limit access to a utility's computers or other electronic systems from the public network:

- The firewall may deny all traffic unless it meets certain criteria.
- The firewall may allow all traffic through unless it meets certain criteria.

A simple example of the first method is screening requests to ensure that they come from an acceptable (i.e., previously identified) domain name and Internet protocol address. Firewalls may also use more complex rules that analyze the application data to determine if the traffic should be allowed through; for example, the firewall may require user authentication (i.e., use of a password) to access the system. How a firewall determines what traffic to let through depends on which network layer it operates within and how it is configured. Firewalls may be a piece of hardware, a software program, or an application card that contains both.

Advanced features that can be incorporated into firewalls allow for the tracking of attempts to log onto the local area network system; for example, a report of successful and unsuccessful log-in attempts may be generated for the computer specialist to analyze. For systems with mobile users, firewalls allow remote access to the private network via secure log-on procedures and authentication certificates. Most firewalls have a graphical user interface for managing the firewall. In addition, new Ethernet firewall cards that fit in the slot of an individual computer bundle additional layers of defense (such as encryption and permit/deny) for individual computer transmissions to the network interface function. The cost of these new cards is only slightly higher than for traditional network interface cards.

3.3.4.3 Network Intrusion
Hardware and Software

Network intrusion detection and prevention system are software- and hardware-based programs designed to detect unauthorized attacks on a computer network system. Whereas other applications such as firewalls and antivirus software share similar objectives with network intrusion systems, network intrusion systems provide a deeper layer of protection beyond the capabilities of these other systems because they evaluate patterns of computer activity rather than specific files. It is worth noting that attacks may come from either outside or within the system (i.e., from an insider) and that network intrusion detection systems may be more applicable for detecting patterns of suspicious activity from inside a facility (e.g., accessing sensitive data) than other information technology solutions. Network intrusion detection systems employ a variety of mechanisms to evaluate potential threats. The types of search and detection mechanisms are dependent upon the level of sophistication of the system. Some of the available detection methods include:

- *Protocol analysis*—Protocol analysis is the process of capturing, decoding, and interpreting electronic traffic. The protocol analysis method of network intrusion detection involves the analysis of data captured during transactions between two or more systems or devices and the evaluation of these data to identify unusual activity and potential problems. When a problem has been isolated and recorded, potential threats can be linked to pieces of hardware or software. Sophisticated protocol analysis will also provide statistics and trend information on the captured traffic.

- *Traffic anomaly detection*—Traffic anomaly detection identifies potential threatening activity by comparing incoming traffic to normal traffic patterns and identifying deviations. It does this by comparing user characteristics against thresholds and triggers defined by the network administrator. This method is designed to detect attacks that span a number of connections, rather than a single session.

- *Network honeypot*—This method establishes nonexistent services to identify potential hackers. A network honeypot impersonates services that do not exist by sending fake information to people scanning the network. It identifies attackers when they attempt to connect to the service. There is no reason for legitimate traffic to access these resources because they do not exist; therefore, any attempt to access them constitutes an attack.

- *Anti-intrusion detection system evasion techniques*—These methods are designed to detect attackers who may be trying to evade intrusion detection system scanning. They include such methods as IP defragmentation, TCP streams reassembly, and deobfuscation.

These detection systems are automated, but they can only indicate patterns of activity, and a computer administer or other experienced individual must interpret the activities to determine whether or not they are potentially harmful. Monitoring the logs generated by these systems can be time consuming, and there may be a learning curve with regard to determining a baseline of normal traffic patterns from which to distinguish potential suspicious activity.

3.4 SCADA

In Queensland, Australia, on April 23, 2000, police stopped a car on the road to Deception Bay and found a stolen computer and radio transmitter inside. Using commercially available technology, a disgruntled former employee had turned his vehicle into a pirate command center for sewage treatment along Australia's Sunshine Coast.

The former employee's arrest solved a mystery that had plagued the Maroochy Shire wastewater system for two months. Somehow the system was leaking hundreds of thousands of gallons of putrid sewage into parks, rivers, and the manicured grounds of a Hyatt Regency hotel … marine life died, the creek water turned black, and the stench was unbearable for residents. Until the former employee's capture—during his 46th successful intrusion—the utility's managers did not know why.

Specialists in cyber terrorism have studied this case because it is the only one known in which someone used a digital control system deliberately to wreak harm. Details of the former employee's intrusion show how easily he broke in—and how restrained he was with his power.

To sabotage the system, he set the software on his laptop to identify itself as a pumping station and then suppressed all alarms. The former employee was the central control station during his intrusions, with unlimited command of 300 SCADA nodes governing sewage and drinking water alike.

Gellman (2002)

The bottom line: As serious as the former employee's intrusions were, they pale in comparison with what he could have done to the freshwater system—he could have done anything he liked.

In 2000, the Federal Bureau of Investigation (FBI) identified and listed threats to critical infrastructure. These threats are listed in Table 3.1. In the past few years, especially since 9/11, it has been somewhat routine for us to pick up a newspaper or magazine or to view a television news program where a major topic of discussion is cyber security or the lack thereof. Many of the cyber intrusion incidents we read or hear about have added new terms or new uses for old terms to our vocabulary; for example, old terms such as Trojan horse, worms, and viruses have taken on new connotations with regard to cyber security issues. Relatively new terms such as scanners, Windows hacking tools, ICQ hacking tools, mail bombs, sniffer, logic bomb, nukers, dots, backdoor Trojan, key loggers, hackers' Swiss knife, password crackers, and BIOS crackers are now commonly encountered.

TABLE 3.1
Threats to Critical Infrastructure Observed by the FBI

Threat	Description
Criminal groups	The use of cyber intrusions by criminal groups who attack systems for purposes of monetary gain has increased.
Foreign intelligence services	Foreign intelligence services use cyber tools as part of their information-gathering and espionage activities.
Hackers	Hackers sometimes crack into networks for the thrill of the challenge or for bragging rights in the hacker community. Whereas remote cracking once required a fair amount of skill or computer knowledge, hackers can now download attack scripts and protocols from the Internet and launch them against victim sites. Thus, although attack tools have become more sophisticated, they have also become easier to use.
Hacktivists	*Hacktivism* refers to politically motivated attacks on publicly accessible Web pages or e-mail servers. Such groups and individuals overload e-mail servers and hack into Web sites to send a political message.
Information warfare	Several nations are aggressively working to develop information warfare doctrine, programs, and capabilities. Such capabilities enable a single entity to have a significant and serious impact by disrupting the supply, communications, and economic infrastructures that support military power—impacts that, according to the Director of Central Intelligence, can affect the daily lives of Americans across the country.
Inside threat	The disgruntled organization insider is a principal source of computer crimes. Insiders may not need a great deal of knowledge about computer intrusions because their knowledge of a victim system often allows them to gain unrestricted access to cause damage to the system or to steal system data. The insider threat also includes outsourcing vendors.
Virus writers	Virus writers are posing an increasingly serious threat. Several destructive computer viruses and worms have harmed files and hard drives; examples include the Melissa macrovirus, the Explore.Zip worm, the CIH (Chernobyl) virus, Nimda, and Code Red.

Source: FBI, *Threat to Critical Infrastructure*, Federal Bureau of Investigation, Washington, D.C., 2000.

Not all relatively new and universally recognizable cyber terms have sinister connotations or meanings, of course. Consider, for example, the following digital terms: backup, binary, bit, byte, CD-ROM, CPU, database, e-mail, HTML, icon, memory, cyberspace, modem, monitor, network, RAM, Wi-Fi, record, software, World Wide Web—none of these terms normally generates thoughts of terrorism in most of us. There is, however, one digital term, SCADA, that most people have not heard of. This is not the case, however, for those who work with the nation's critical infrastructure, including water/wastewater. SCADA is an acronym for *supervisory control and data acquisition* and sometimes referred to as *digital control systems* or *process control systems*, and it plays an important role in computer-based control systems. Many water/wastewater systems use computer-based systems to remotely control sensitive processes and system equipment previously controlled manually. These SCADA systems allow a water/wastewater utility to collect data from sensors and control equipment located at remote sites. Common water/wastewater system sensors measure elements such as fluid level, temperature, pressure, water purity, water clarity, and pipeline flow rates. Common water/wastewater system equipment includes valves, pumps, and mixers for mixing chemicals in the water supply.

3.4.1 WHAT IS SCADA?

Simply, SCADA is a computer-based system that remotely controls processes previously controlled manually. SCADA allows an operator using a central computer to supervise (control and monitor) multiple networked computers at remote locations. Each remote computer can control mechanical processes (pumps, valves, etc.) and collect data from sensors at its remote location, thus the phrase *supervisory control and data acquisition*. The central computer is the *master terminal unit* (MTU). The operator interfaces with the MTU using software referred to as the *human–machine interface* (HMI). The remote computer is the *programmable logic controller* (PLC) or *remote terminal unit* (RTU). The RTU activates a relay (or switch) that turns mechanical equipment on and off. The RTU also collects data from sensors.

Initially, utilities ran wires (hardwire or land lines) from the central computer (MTU) to the remote computers (RTUs). Because remote locations can be located hundreds of miles from the central location, utilities began to use public phone lines and modems, leased telephone company lines, and radio and microwave communication. More recently, they have also begun to use satellite links, the Internet, and newly developed wireless technologies.

Because the sensors of SCADA systems provide valuable information, many utilities established connections between their SCADA systems and their business system. This allowed utility management and other staff access to valuable statistics, such as water usage. When utilities later connected their systems to the Internet, they were able to provide stakeholders with water/wastewater statistics on the utility's web pages.

3.4.2 SCADA Applications in Water/Wastewater Systems

As stated above, SCADA systems can be designed to monitor a variety of equipment operating conditions and parameters, such as volumes, flow rates, or water quality, as well as to respond to changes in those parameters either by alerting operators or by modifying system operation through a feedback loop system without having personnel physically visit each process or piece of equipment on a daily basis to check it to ensure that it is functioning properly. SCADA systems can also be used to automate certain functions, so they can be performed without having to be initiated by an operator (e.g., injecting chlorine in response to periodic low chlorine levels in a distribution system or turning on a pump in response to low water levels in a storage tank). In addition to process equipment, SCADA systems can also integrate specific security alarms and equipment, such as cameras, motion sensors, lights, data from card-reading systems, etc., thereby providing a clear picture of what is happening at areas throughout a facility. Finally, SCADA systems also provide constant, real-time data on processes, equipment, location access, etc., so the necessary response can be made quickly. This can be extremely useful during emergency conditions, such as when distribution mains break or when potentially disruptive BOD spikes appear in wastewater influent.

Because these systems can monitor multiple processes, equipment, and infrastructure and then provide quick notification of or response to problems or disruptions, SCADA systems typically provide the first line of detection for atypical or abnormal conditions. For example, a real-time, customized operator interface screen can display critical system monitoring parameters, while a SCADA system connected to sensors that measure the water quality parameters can report findings measured outside of a specific range. The system can transmit warning signals back to the operators, such as by initiating a call to a personal pager. This might allow the operators to initiate actions to prevent contamination and disruption of the water supply. Further automation of the system could ensure that the system initiates measures to rectify the problem. Preprogrammed control functions (e.g., shutting a valve, controlling flow, increasing chlorination, or adding other chemicals) can be triggered and operated based on SCADA utility.

3.4.3 SCADA Vulnerabilities

According to USEPA (2005), SCADA networks were developed with little attention paid to security, thus the security of these systems can often be weak. Studies have found that, although technological advancements introduced vulnerabilities, many water/wastewater utilities have spent little time securing their SCADA networks, many of which may be susceptible to attacks and misuse. Remote monitoring and supervisory control of processes was initially developed in the early 1960s and has since adopted many technological advancements. The advent of minicomputers made it possible to automate a vast number of once manually operated switches. Advancements in radio technology reduced the communication costs associated with installing and maintaining buried cable in remote areas. SCADA systems continued to adopt new communication methods, including satellite and cellular. As the price of computers and communications dropped, it became economically feasible to distribute operations and to expand SCADA networks to include even smaller facilities.

Advances in information technology and the need for improved efficiency have resulted in an increasingly automated and interlinked infrastructure and created new vulnerabilities due to potential equipment failure, human error, weather and other natural causes, and physical and cyber attacks. Some examples of SCADA vulnerabilities include:

- *Humans*—People can be tricked or corrupted and may commit errors.
- *Communications*—Messages can be fabricated, intercepted, changed, deleted, or blocked.
- *Hardware*—Security features are not easily adapted to small self-contained units with limited power supplies.
- *Physical threats*—Intruders can break into a facility to steal or damage SCADA equipment.
- *Natural threats*—Tornadoes, floods, earthquakes, and other natural disasters can damage equipment and connections.
- *Software*—Programs can be poorly written.

A survey among water utilities found that they were doing little to secure their SCADA network vulnerabilities (Ezell, 1998); for example, many respondents reported that they had remote access, which can allow an unauthorized person to access the system without being physically present. More than 60% of the respondents believed that their systems were not safe from unauthorized access and use, and 20% of the respondents reported known attempts and successful unauthorized access to their systems. Yet, 22 of 43 respondents reported that they did not spend any time ensuring the safety of their networks, and 18 of 43 respondents reported that they spent less than 10% of their time ensuring network safety.

SCADA system computers and their connections are susceptible to a variety of information system attacks and misuse, such as system penetration and unauthorized access to information. The Computer Security Institute and the FBI conduct annual Computer Crime and Security surveys (FBI, 2004). A recent survey addressed ten types of attacks or misuse and reported that viruses and denial of service had the greatest negative economic impact. The same study also found that 15% of the respondents reported abuse of wireless networks, which can be components of a SCADA system. On average, respondents from all sectors did not believe that their organization invested enough in security awareness. Utilities as a group reported a lower average computer security expenditure/investment per employee than many other sectors such as transportation, telecommunications, and financial.

Sandia National Laboratories' *Common Vulnerabilities in Critical Infrastructure Control Systems* described some of the common problems it has identified in the following five categories (Stamp et al., 2003):

1. *System data*—Important data attributes for security include availability, authenticity, integrity, and confidentiality. Data should be categorized according to their sensitivity, and their ownership and responsibility must be assigned; however, SCADA data are often not classified at all, making it difficult to identify where security precautions are appropriate.
2. *Security administration*—Vulnerabilities can emerge when systems lack properly structured security policies, equipment and system implementation guides, configuration management, training, and enforcement and compliance auditing.
3. *Architecture*—Many common practices negatively affect SCADA security; for example, although it is convenient to use SCADA capabilities for other purposes such as fire and security systems, these practices create single points of failure. Also, the connection of SCADA networks to other automation systems and business networks introduces multiple entry points for potential adversaries.
4. *Network* (including communication links)— Legacy system hardware and software have very limited security capabilities, and the vulnerabilities of contemporary systems (based on modern information technology) are publicized. Wireless and shared links are susceptible to eavesdropping and data manipulation.
5. *Platforms*—Many platform vulnerabilities exist, including default configurations still being in place, poor password practices, shared accounts, inadequate protection for hardware,

and nonexistent security monitoring controls. In most cases, important security patches are not installed, often due to concern about negatively impacting system operation; in some cases, technicians are contractually forbidden from updating systems by their vendor agreements.

The following incident helps to illustrate some of the risks associated with SCADA vulnerabilities:

During the course of conducting a vulnerability assessment, a contractor stated that personnel from his company penetrated the information system of a utility within minutes. Contractor personnel drove to a remote substation and noticed a wireless network antenna. Without leaving their vehicle, they plugged in their wireless radios and connected to the network within 5 minutes. Within 20 minutes they had mapped the network, including SCADA equipment, and accessed the business network and data.

This illustrates what a cyber security advisor from Sandia National Laboratories specializing in SCADA stated— that utilities are moving to wireless communication without understanding the added risks.

3.4.4 The Increasing Risk

According to the GAO (2003), historically, security concerns about control systems (SCADA included) were related primarily to protecting against physical attack and misuse of refining and processing sites or distribution and holding facilities. More recently, however, there has been a growing recognition that control systems are now vulnerable to cyber attacks from numerous sources, including hotel governments, terrorist groups, disgruntled employees, and other malicious intruders. In addition to control system vulnerabilities mentioned earlier, several factors have contributed to the escalation of risk to control systems, including: (1) the adoption of standardized technologies with known vulnerabilities, (2) the connectivity of control systems to other networks, (3) constraints on the implementation of existing security technologies and practices, (4) insecure remote connections, and (5) the widespread availability of technical information about control systems.

3.4.5 Adoption of Technologies with Known Vulnerabilities

When a technology is not well known, not widely used, not understood, or not publicized, it is difficult to penetrate it and thus disable it. Historically, proprietary hardware, software, and network protocols made it difficult to understand how control systems operated—and therefore how to hack into them. Today, however, to reduce costs and improve performance, organizations have been transitioning from proprietary systems to less expensive, standardized technologies such as Microsoft's Windows and Unix-like operating systems and the common networking protocols used by the Internet. These widely used standardized technologies have commonly known vulnerabilities, and sophisticated and effective exploitation tools are widely available and relatively easy to use. As a consequence, both the number of people with the knowledge to wage attacks and the number of systems subject to attack have increased. Also, common communication protocols and the emerging use of Extensible Markup Language (XML) can make it easier for a hacker to interpret the content of communications among the components of a control system.

Control systems are often connected to other networks, as enterprises often integrate their control systems with their enterprise networks. This increased connectivity has significant advantages, including providing decision makers with access to real-time information and allowing engineers to monitor and control the process control system from different points on the enterprise network. In addition, the enterprise networks are often connected to the networks of strategic partners and to the Internet. Further, control systems are increasingly using wide area networks and the Internet to transmit data to their remote or local stations and individual devices. This convergence of control networks with public and enterprise networks potentially exposes the control systems to additional security vulnerabilities. Unless appropriate security controls are deployed in the enterprise network and the control system network, breaches in enterprise security can affect the operation of controls system. According to industry experts, the use of existing security technologies, as well as strong user authentication and patch management practices, are generally not implemented in control systems because control systems operate in real time, typically are not designed with cybersecurity in mind, and usually have limited processing capabilities.

Existing security technologies such as authorization, authentication, encryption, intrusion detection, and filtering of network traffic and communications require more bandwidth, processing power, and memory than control system components typically have. Because controller stations are generally designed to do specific tasks, they use low-cost, resource-constrained microprocessors. In fact, some devices in the electrical industry still use the Intel 8088 processor first introduced in 1978; consequently, it is difficult to install existing security technologies without seriously degrading the performance of the control system.

Further, complex passwords and other strong password practices are not always used to prevent unauthorized access to control systems, in part because this could hinder a rapid response to safety procedures during an emergency. As a result, according to experts, weak passwords that are easy to guess, shared, and infrequently changed are reportedly common in control systems, as well as the use of default passwords or even no password at all.

In addition, although modern control systems are based on standard operating systems, they are typically customized to support control system applications; consequently, vendor-provided software patches are generally either incompatible or cannot be implemented without compromising service by shutting down always-on systems or affecting interdependent operations.

Potential vulnerabilities in control systems are exacerbated by insecure connections. Organizations often leave access links, such as dial-up modems, open for remote diagnostics, maintenance, and examination of system status. Such links may not be protected with authentication or encryption which increases the risk that hackers could use these insecure connections to break into remotely controlled systems. Also, control systems often use wireless communications systems, which are especially vulnerable to attack, or leased lines that pass through commercial telecommunications facilities. Without encryption to protect data as it flows through these insecure connections or authentication mechanisms to limit access, there is limited protection for the integrity of the information being transmitted.

Public information about infrastructures and control systems is available to potential hackers and intruders. The relatively easy availability of such data was demonstrated by a university graduate student, whose dissertation reportedly mapped every business and industrial sector in the American economy to the fiberoptic network that connects them—using material that was available publicly on the Internet, none of which was classified. Many of the electric utility officials who were interviewed for the National Security Telecommunications Advisory Committee's Information Assurance Task Force's Electric Power Risk Assessment expressed concern over the amount of information about their infrastructure that is readily available to the public.

In the electric power industry, open sources of information, such as product data and educational videotapes from engineering associations, can be used to understand the basics of the electrical grid. Other publicly available information, including filings of the Federal Energy Regulatory Commission (FERC), industry publications, maps, and material available on the Internet, is sufficient to allow someone to identify the most heavily loaded transmission lines and the most critical substations in the power grid.

In addition, significant information on control systems is publicly available, including design and maintenance documents, technical standards for the interconnection of control systems and remote terminal units, and standards for communication among control devise—all of which could assist hackers in understanding the systems and how to attack them. Moreover, numerous former employees, vendors, support contractors, and other end users of the same equipment worldwide have inside knowledge of the operation of control systems.

3.4.6 Cyber Threats to Control Systems

There is a general consensus—and increasing concern—among government officials and experts on control systems regarding potential cyber threats to the control systems that govern our critical infrastructures. As components of control systems increasingly make critical decisions that were once made by humans, the potential effect of a cyber threat becomes more devastating. Such cyber threats could come from numerous sources, ranging from hostile governments and terrorist groups to disgruntled employees and other malicious intruders. Based on interviews and discussions with representatives throughout the electric power industry, the Information Assurance Task Force of the National Security Telecommunications Advisory Committee concluded that an organization with sufficient resources, such as a foreign intelligence service or a well-supported terrorist group, could conduct a structured attack on the electric power grid electronically, with a high degree of anonymity and without having to set foot in the target nation.

In July 2002, the National Infrastructure Protection Center (NIPC) reported that the potential for compound cyber and physical attacks, referred to as *swarming attacks*, is an emerging threat to the U.S. critical infrastructure. As NIPC reports, the effects of a swarming attack include slowing or complicating any response to the physical attack. An example would be a cyber attack that disabled the water supply or the electrical system in conjunction with a physical attack that denied emergency services the necessary resources to manage the consequences, such as controlling fires, coordinating actions, and generating light.

Control systems, such as SCADA, can be vulnerable to cyber attacks. Entities or individuals with malicious intent might take one or more of the following actions to successfully attack control systems:

- Disrupt the operation of control systems by delaying or blocking the flow of information through control networks, thereby denying availability of the networks to control system operations.
- Make unauthorized changes to programmed instructions in PLCs, RTUs, or distributed control system (DCS) controllers; change alarm thresholds; or issue unauthorized commands to control equipment, which could potentially result in damage to equipment (if tolerances are exceeded), premature shutdown of processes (such as prematurely shutting

down transmission lines), or even disabling control equipment.
- Send false information to control system operators to disguise unauthorized changes or to initiate inappropriate actions by system operators.
- Modify the control system software, producing unpredictable results.
- Interfere with the operation of safety systems.

In addition, in control systems that cover a wide geographic area, the remote sites are often unstaffed and may not be physically monitored. If such remote systems are physically breached, the attackers could establish a cyber connection to the control network.

3.4.7 Securing Control Systems

Several challenges must be addressed to effectively secure control systems against cyber threats. These challenges include: (1) limitations of current security technologies with regard to securing control systems, (2) the perception that securing control systems may not be economically justifiable, and (3) conflicting priorities within organizations regarding the security of control systems. A significant challenge in effectively securing control systems is the lack of specialized security technologies for these systems. The computing resources in control systems that are required to perform security functions tend to be quite limited, making it very difficult to use security technologies within control system networks without severely hindering performance. Securing control systems may not be perceived as economically justifiable. Experts and industry representatives have indicated that organizations may be reluctant to spend more money to secure control systems. Hardening the security of control systems would require industries to expend more resources, including acquiring more personnel, providing training for personnel, and potentially prematurely replacing current systems that typically have a lifespan of about 20 years. Finally, several experts and industry representatives have indicated that the responsibility for securing control systems typically includes two separate groups: IT security personnel and control system engineers and operators. IT security personnel tend to focus on securing enterprise systems, while control system engineers and operators tend to be more concerned with the reliable performance of their control systems. Further, they indicate that, as a result, those two groups do not always fully understand each other's requirements and do not collaborate to implement secure control systems.

3.4.8 Steps to Improve SCADA Security

The President's Critical Infrastructure Protection Board and the Department of Energy (DOE) have developed the steps outlined below to help organizations improve the security of their SCADA networks. These steps are not meant to be prescriptive or all inclusive; however, they do address essential actions to be taken to improve the protection of SCADA networks. The steps are divided into two categories: specific actions to improve implementation and actions to establish essential underlying management processes and policies (DOE, 2001).

■ 21 Steps to Increase SCADA Security

The following steps focus on specific actions to be taken to increase the security of SCADA networks:

1. *Identify all connections to SCADA networks.* Conduct a thorough risk analysis to assess the risk and necessity of each connection to the SCADA network. Develop a comprehensive understanding of all connections to the SCADA network and how well those connections are protected. Identify and evaluate the following types of connections:
   - Internal local area and wide area networks, including business networks
   - The Internet
   - Wireless network devices, including satellite uplinks
   - Modem or dial-up connections
   - Connections to business partners, vendors, or regulatory agencies

2. *Disconnect unnecessary connections to the SCADA network.* To ensure the highest degree of security of SCADA systems, isolate the SCADA network from other network connections to as great a degree as possible. Any connection to another network introduces security risks, particularly if the connection creates a pathway from or to the Internet. Although direct connections with other networks may allow important information to be passed efficiently and conveniently, insecure connections are simply not worth the risk; isolation of the SCADA network must be a primary goal to provide needed protection. Strategies such as the utilization of demilitarized zones (DMZs) and data warehousing can facilitate the secure transfer of data from the SCADA network to business networks; however, they must be designed and implemented properly to avoid the introduction of additional risk through improper configuration.

3. *Evaluate and strengthen the security of any remaining connections to the SCADA networks.* Conduct penetration testing or vulnerability analysis of any remaining connections to the SCADA network to evaluate the protection pos-

ture associated with these pathways. Use this information in conjunction with risk-management processes to develop a robust protection strategy for any pathways to the SCADA network. Because the SCADA network is only as secure as its weakest connecting point, it is essential to implement firewalls, intrusion detection systems (IDSs), and other appropriate security measures at each point of entry. Configure firewall rules to prohibit access from and to the SCADA network, and be as specific as possible when permitting approved connections. For example, an Independent System Operator (ISO) should not be granted blanket network access simply because of a need for a connection to certain components of the SCADA system. Strategically place IDSs at each entry point to alert security personnel of potential breaches of network security. Organization management must understand and accept the responsibilities and risks associated with any connection to the SCADA network.

4. *Harden SCADA networks by removing or disabling unnecessary services.* SCADA control servers built on commercial or open-source operating systems can be exposed to attack default network services. To the greatest degree possible, remove or disable unused services and network demons to reduce the risk of direct attack. This is particularly important when SCADA networks are interconnected with other networks. Do not permit a service or feature on a SCADA network unless a thorough risk assessment of the consequences of allowing the service or feature shows that the benefits of the service or feature far outweigh the potential for vulnerability exploitation. Examples of services to remove from SCADA networks include automated meter reading/remote billing systems, e-mail services, and Internet access. An example of a feature to disable is remote maintenance. Refer to the National Security Agency's series of security guides. Additionally, work closely with SCADA vendors to identify secure configurations and coordinate any and all changes to operational systems to ensure that removing or disabling services does not cause downtime, interruption of service, or loss of support.

5. *Do not rely on proprietary protocols to protect your system.* Some SCADA systems are unique, proprietary protocols for communications between field devices and servers. Often the security of SCADA systems is based solely on the secrecy of these protocols. Unfortunately, obscure protocols provide very little real

security. Do not rely on proprietary protocols or factory default configuration settings to protect your system. Additionally, demand that vendors disclose any backdoors or vendor interfaces to your SCADA systems, and expect them to provide systems that are capable of being secured.

6. *Implement the security features provided by device and system vendors.* Older SCADA systems (most systems in use) have no security features whatsoever. SCADA system owners must insist that their system vendors implement security features in the form of product patches or upgrades. Some newer SCADA devices are shipped with basic security features, but these are usually disabled to ensure ease of installation. Analyze each SCADA device to determine whether security features are present. Additionally, factory default security settings (such as in computer network firewalls) are often set to provide maximum usability but minimal security. Set all security features to provide the maximum security only after a thorough risk assessment of the consequences of reducing the security level.

7. *Establish strong controls over any medium that is used as a backdoor into the SCADA network.* Where backdoors or vendor connections do exist in SCADA systems, strong authentication must be implemented to ensure secure communications. Modems, wireless, and wired networks used for communications and maintenance represent a significant vulnerability to the SCADA network and remote sites. Successful war dialing or war driving attacks could allow an attacker to bypass all of the controls and to have direct access to the SCADA network or resources. To minimize the risk of such attacks, disable inbound access and replace it with some type of callback system.

8. *Implement internal and external intrusion detection systems and establish 24-hour-a-day incident monitoring.* To be able to effectively respond to cyber attacks, establish an intrusion detection strategy that includes alerting network administrators of malicious network activity originating from internal or external sources. Intrusion detection system monitoring is essential 24 hours a day; this capability can be easily set up through a pager. Additionally, incident response procedures must be in place to allow an effective response to any attack. To complement network monitoring, enable logging on all systems and audit system logs daily to detect suspicious activity as soon as possible.

9. *Perform technical audits of SCADA devices and networks, and any other connected networks, to identify security concerns.* Technical audits of SCADA devices and networks are critical to ongoing security effectiveness. Many commercial and open-sourced security tools are available that allow system administrators to conduct audits of their systems and networks to identify active services, patch level, and common vulnerabilities. The use of these tools will not solve systemic problems but will eliminate the paths of least resistance that an attacker could exploit. Analyze identified vulnerabilities to determine their significance, and take corrective actions as appropriate. Track corrective actions and analyze this information to identify trends. Additionally, retest systems after corrective actions have been taken to ensure that vulnerabilities were actually eliminated. Scan nonproduction environments actively to identify and address potential problems.

10. *Conduct physical security surveys and assess all remote sites connected to the SCADA network to evaluate their security.* Any location that has a connection to the SCADA network is a target, especially unmanned or unguarded remote sites. Conduct a physical security survey and inventory access points at each facility that has a connection to the SCADA system. Identify and assess any source of information, including remote telephone/computer network/fiberoptic cables, that could be tapped; radio and microwave links that are exploitable computer terminals that could be accessed; and wireless local area network access points. Identify and eliminate single points of failure. The security of the site must be adequate to detect or prevent unauthorized access. Do not allow live network access points at remote, unguarded sites simply for convenience.

11. *Establish SCADA "Red Teams" to identify and evaluate possible attack scenarios.* Establish a "Red Team" to identify potential attack scenarios and evaluate potential system vulnerabilities. Use a variety of people who can provide insight into weaknesses of the overall network, SCADA system, physical systems, and security controls. People who work on the system every day have great insight into the vulnerabilities of your SCADA network and should be consulted when identifying potential attack scenarios and possible consequences. Also, ensure that the risk from a malicious insider is fully evaluated, given that this represents one of the greatest threats to an organization. Feed information

resulting from the "Red Team" evaluation into risk-management processes to assess the information and establish appropriate protection strategies.

The following steps focus on management actions to establish an effective cyber security program:

12. *Clearly define cyber security roles, responsibilities, and authorities for managers, system administrators, and users.* Organization personnel need to understand the specific expectations associated with protecting information technology resources through the definition of clear and logical roles and responsibilities. In addition, key personnel need to be given sufficient authority to carry out their assigned responsibilities. Too often, good cyber security is left up to the initiative of the individual, which usually leads to inconsistent implementations and ineffective security. Establish a cyber security organizational structure that defines roles and responsibilities and clearly identifies how cyber security issues are escalated and who is notified in an emergency.

13. *Document network architecture and identify systems that serve critical functions or contain sensitive information that require additional levels of protection.* Develop and document a robust information security architecture as part of a process to establish an effective protection strategy. It is essential that organizations design their network with security in mind and continue to have a strong understanding of their network architecture throughout its lifecycle. Of particular importance, an in-depth understanding of the functions that the systems perform and the sensitivity of the stored information is required. Without this understanding, risk cannot be properly assessed and protection strategies may not be sufficient. Documenting the information security architecture and its components is critical to understanding the overall protection strategy and identifying single points of failure.

14. *Establish a rigorous, ongoing risk-management process.* A thorough understanding of the risks to network computing resources from denial-of-service attacks and the vulnerability of sensitive information to compromise is essential to an effective cyber security program. Risk assessments from the technical basis of this understanding and are critical to formulating effective strategies to mitigate vulnerabilities and preserve the integrity of computing resources. Ini-

tially, perform a baseline risk analysis based on current threat assessment to use for developing a network protection strategy. Due to rapidly changing technology and the emergence of new threats on a daily basis, an ongoing risk-assessment process is also needed so routine changes can be made to the protection strategy to ensure it remains effective. Fundamental to risk management is identification of residual risk with a network protection strategy in place and acceptance of that risk by management.

15. *Establish a network protection strategy based on the principle of defense-in-depth.* A fundamental principle that must be part of any network protection strategy is defense-in-depth. Defense-in-depth must be considered early in the design phase of the development process and must be an integral consideration in all technical decision-making associated with the network. Utilize technical and administrative controls to mitigate threats from identified risks to as great a degree as possible at all levels of the network. Single points of failure must be avoided, and cyber security defense must be layered to limit and contain the impact of any security incidents. Additionally, each layer must be protected against other systems at the same layer. For example, to protect against the inside threat, restrict users to accessing only those resources necessary to perform their job functions.

16. *Clearly identity cyber security requirements.* Organizations and companies need structured security programs with mandated requirements to establish expectations and allow personnel to be held accountable. Formalized policies and procedures are typically used to establish and institutionalize a cyber security program. A formal program is essential to establishing a consistent, standards-based approach to cyber security through an organization and eliminates sole dependence on individual initiative. Policies and procedures also inform employees of their specific cyber security responsibilities and the consequences of failing to meet those responsibilities. They also provide guidance regarding actions to be taken during a cyber security incident and promote efficient and effective actions during a time of crisis. As part of identifying cyber security requirements, include user agreements and notification and warning banners. Establish requirements to minimize the threat from malicious insiders, including the need for conducting background checks and limiting network privileges to those absolutely necessary.

17. *Establish effective configuration management processes.* A fundamental management process needed to maintain a secure network is configuration management. Configuration management must cover both hardware configurations and software configurations. Changes to hardware or software can easily introduce vulnerabilities that undermine network security. Processes are required to evaluate and control any change to ensure that the network remains secure. Configuration management begins with well-tested and documented security baselines for your various systems.

18. *Conduct routine self-assessments.* Robust performance evaluation processes are needed to provide organizations with feedback on the effectiveness of cyber security policy and technical implementation. A sign of a mature organization is one that is able to identify issues, conduct root-cause analyses, and implement effective corrective actions that address individual and systemic problems. Self-assessment processes that are normally part of an effective cyber security program include routine scanning for vulnerabilities, automated auditing of the network, and self-assessments of organizational and individual performance.

19. *Establish system backups and disaster recovery plans.* Establish a disaster recovery plan that allows for rapid recover from any emergency (including a cyber attack). System backups are an essential part of any plan and allow rapid reconstruction of the network. Routinely exercise disaster recovery plans to ensure that they work and that personnel are familiar with them. Make appropriate changes to disaster recovery plans based on lessons learned from exercises.

20. *Encourage senior organizational leadership to establish expectations for cyber security performance and hold individuals accountable for their performance.* Effective cyber security performance requires commitment and leadership from senior managers in the organization. It is essential that senior management establish an expectation for strong cyber security and communicate this to their subordinate managers throughout the organization. It is also essential that senior organizational leadership establish a structure for implementation of a cyber security program. This structure will promote consistent implementation and the ability to sustain a strong cyber security program. It is then important for individuals to be held accountable for their performance as it relates to cyber security. This includes managers, system administrators, technicians, and users/operators.

21. *Establish policies and conduct training to minimize the likelihood that organizational personnel will inadvertently disclose sensitive information regarding SCADA system design, operations, or security controls.* Release data related to the SCADA network only on a strict, need-to-know basis and only to persons explicitly authorized to receive such information. "Social engineering," the gathering of information about a computer or computer network via questions to naïve users, is often the first step in a malicious attack on computer networks. The more information revealed about a computer or network, the more vulnerable the computer or network is. Never divulge data revealed to a SCADA network, including names and contact information about the system operators and administrators, computer operating systems, or physical and logical locations of computers and network systems over telephones or to personnel unless they are explicitly authorized to receive such information. Any requests for information by unknown persons should be sent to a central network security location for verification and fulfillment. People can be a weak link in an otherwise secure network. Conduct training and information awareness campaigns to ensure that personnel remain diligent in guarding sensitive network information, particularly their passwords.

THE BOTTOM LINE ON SECURITY

Again, when it comes to the security of our nation and even of water/wastewater treatment facilities, few have summed it up better than Governor Ridge (Henry, 2002):

> Now, obviously, the further removed we get from September 11, the natural tendency is to let down our guard. Unfortunately, we cannot do that. …The government will continue to do everything we can to find and stop those who seek to harm us. And I believe we owe it to the American people to remind them that they must be vigilant, as well.

CHAPTER REVIEW QUESTION

Thought-provoking question:

3.1   Do you feel that water and/or wastewater facilities are realistic targets for terrorism? Why? (Answers will vary.)

REFERENCES AND SUGGESTED READING

DOE. 2001. *21 Steps to Improve Cyber Security of SCADA Networks*. Washington, D.C.: Department of Energy.

DOH. 2007. *Drinking Water Tech Tips: Sanitary Protection of Reservoirs—Vents*, DOH Publ. No. 331-250. Olympia: Washington State Department of Health, Division of Environmental Health, Office of Drinking Water.

Ezell, B.C. 1998. *Risks of Cyber Attack to Supervisory Control and Data Acquisition*. Charlottesville: University of Virginia.

FBI. 2000. *Threat to Critical Infrastructure*. Washington, D.C.: Federal Bureau of Investigation.

FBI. 2004. *Ninth Annual Computer Crime and Security Survey*. Washington, D.C.: Computer Crime Institute and Federal Bureau of Investigations.

GAO. 2003. *Critical Infrastructure Protection: Challenges in Securing Control System*. Washington, D.C.: U.S. General Accounting Office.

Gellman, B. 2002. Cyber attacks by Al Qaeda feared: terrorists at threshold of using Internet as tool of bloodshed, experts say. *Washington Post*, June 26, p. A01.

Greenburg, A. et al. 1999. *Standard Methods for the Examination of Water and Wastewater*, 20th ed. Denver, CO: American Water Works Association.

Henry, K. 2002. New face of security. *Government Security*, April, pp. 30–31.

IBWA. 2004. *Bottled Water Safety and Security*. Alexandria, VA: International Bottled Water Association.

NIPC. 2002. *National Infrastructure Protection Center Report*. Washington, D.C.: National Infrastructure Protection Center.

Stamp, J. et al. 2003. *Common Vulnerabilities in Critical Infrastructure Control Systems*, 2nd ed. Albuquerque, NM: Sandia National Laboratories.

USEPA. 2001. *Water Protection Task Force Alert #IV: What Wastewater Utilities Can Do Now to Guard Against Terrorist and Security Threats*. Washington, D.C.: U.S. Environmental Protection Agency.

USEPA. 2004. *Water Security: Basic Information*. Washington, D.C.: U.S. Environmental Protection Agency (http://cfpub. epa.gov/safewater/watersecurity/basicinformation.cfm).

USEPA. 2005. EPA needs to determine what barriers prevent water systems from securing known SCADA vulnerabilities. In *Final Briefing Report*, Harris, J., Ed., Washington, D.C.: U.S. Environmental Protection Agency.

4 Water/Wastewater References, Models, and Terminology

Living things depend on water but water does not depend on living things. It has a life of its own.

Pielou (1998)

Did You Know?

It takes between 250 and 600 gallons of water to grow a pound of rice. That is more water than many households use in a week. For just a bag of rice (Pearce, 2006).

4.1 SETTING THE STAGE

This handbook is a compilation or summary of information available in many expert sources. Although every attempt has been made to cover all aspects of water and wastewater treatment system operation, it should be pointed out that no one single handbook has *all* the information or all the answers. Moreover, because of the physical limits of any written text, some topics are given only cursory exposure and limited coverage. For those individuals seeking a more in-depth treatment of specific topics germane to water and wastewater treatment system operations, we recommend consulting one or more of the references listed in Table 4.1 and any of the many other outstanding references appearing throughout this text.

4.2 TREATMENT PROCESS MODELS

Figure 4.1 shows a basic schematic of the water treatment process. Other unit processes used in the treatment of water (fluoridation, for example) are not represented in Figure 4.1; however, we discuss many of the other unit processes in detail within the handbook. Figure 4.2 shows a basic schematic or model of a wastewater treatment process that provides primary and secondary treatment using the *activated sludge* process. In secondary treatment, which provides biochemical oxygen demand (BOD) removal beyond what is achievable by simple sedimentation, three approaches are commonly used: trickling filter, activated sludge, and oxidation ponds. These systems are discussed in detail later in the text, as are biological nutrient removal (BNR) and standard tertiary or advanced wastewater treatment. The purpose of the models shown in Figure 4.1 and

Figure 4.2 is to allow readers to visually follow the water and wastewater treatment process step-by-step as they are presented in this text. The figures will help the reader understand how all the various unit processes sequentially follow and tie into each other. This format simply provides a pictorial presentation along with pertinent written information to enhance the learning process.

4.3 KEY TERMS USED IN WATER AND WASTEWATER OPERATIONS

To learn water/wastewater treatment operations (or any other technology for that matter), you must master the language associated with the technology. Each technology has its own terms with their own definitions. Many of the terms used in water/wastewater treatment are unique; others combine words from many different technologies and professions. One thing is certain: Water/wastewater operators without a clear understanding of the terms related to their profession are ill equipped to perform their duties in the manner required. Usually, a handbook or text like this one includes a glossary of terms at the end of the work. In this *Handbook*, however, we list and define many of the terms used right up front. Experience has shown that an early introduction to keywords is a benefit to readers. An up-front introduction to key terms facilitates a more orderly, logical, systematic learning activity. Those terms not defined in this section are defined as they appear in the text. A short quiz on many of the following terms follows the end of this chapter.

Absorb—To take in; many things absorb water.
Acid rain—The acidic rainfall that results when rain combines with sulfur-oxide emissions from the combustion of fossil fuels (coal, for example).
Acre-feet (acre-foot)—An expression of water quantity. One acre-foot will cover 1 acre of ground 1 foot deep. An acre-foot contains 43,560 cubic feet, 1233 cubic meters, or 325,829 gallons (U.S). Also abbreviated as ac-ft.
Activated carbon—Derived from vegetable or animal materials by roasting in a vacuum furnace. Its porous nature gives it a very high surface area per unit mass—as much as 1000 square meters per gram, which is 10 million times the surface area of 1 gram of water in an open container. Used in

TABLE 4.1
Recommended Reference Material

1. Kerri, K., *Small Water System Operation and Maintenance*, California State University, Sacramento, 1990.
2. Kerri, K., *Water Distribution System Operation and Maintenance*, 2nd ed., California State University, Sacramento, 1989.
3. Kerri, K., *Water Treatment Plant Operation*, Vols. 1 and 2, 6th ed. California State University, Sacramento, CA, 2004.
4. CDC, *Basic Mathematics*, Publ. No. 3011-G, Centers for Disease Control, Atlanta, GA, 1987.
5. CDC, *Waterborne Disease Control*, Publ. No. 3014-G, Centers for Disease Control, Atlanta, GA, 1987.
6. CDC, *Water Fluoridation*, Publ. No. 3017-G, Centers for Disease Control, Atlanta, GA, 1987.
7. *Introduction to Water Sources and Transmission*, Vol. 1, American Water Works Association, Denver, CO, 1983.
8. *Introduction to Water Treatment*, Vol. 2, American Water Works Association, Denver, CO, 1984.
9. *Introduction to Water Distribution: Principles and Practices of Water Supply Operations*, Vol. 3, American Water Works Association, Denver, CO, 1986.
10. *Introduction to Water Quality Analysis: Principles and Practices of Water Supply Operations*, Vol. 4, American Water Works Association, Denver, CO, 1982.
11. *Basic Science Concepts and Applications: Principles and Practices of Water Supply Operations*, American Water Works Association, Denver, CO, 2003.
12. *Handbook of Water Analysis*, 2nd ed., HACH Chemical Company, Loveland, CO, 1992.
13. *Methods for Chemical Analysis of Water and Wastes*, EPA-6000/4-79-020, Environmental Monitoring Systems Laboratory, U.S. Environmental Protection Agency, Cincinnati, OH, 1979/1983.
14. *Standard Methods for the Examination of Water and Wastewater*, 21st ed., American Public Health Association, Washington, D.C., 2005.
15. Price, J.K., *Basic Math Concepts for Water and Wastewater Plant Operators*, CRC Press, Boca Raton, FL, 1991.
16. Spellman, F.R., *Spellman's Standard Handbook for Wastewater Operators*, Vols. 1–3, Technomic, Lancaster, PA, 1999–2000.
17. Spellman, F.R. and Drinan, J., *The Handbook for Waterworks Operator Certification*, Vols. 1–3, Technomic, Lancaster, PA, 2001.
18. Spellman, FR. and Drinan, J., *Fundamentals for the Water and Wastewater Maintenance Operator Series: Electricity; Electronics; Pumping; Water Hydraulics; Piping and Valves; Blueprint Reading*, Technomic, Lancaster, PA, 2000–2002.
19. Qasim, S.R., *Wastewater Treatment Plants: Planning, Design, and Operation*, 2nd ed., Technomic Lancaster, PA, 1999.
20. Haller, E., *Simplified Wastewater Treatment Plant Operations*, Technomic, Lancaster, PA, 1999.
21. Kerri, K., *Operation of Wastewater Treatment Plants: A Field Study Program*, Vols. I and II, 4th ed., California State University, Sacramento, CA, 1992.

adsorption (see definition), activated carbon adsorbs substances that are not or are only slightly adsorbed by other methods.

Activated sludge—The solids formed when microorganisms are used to treat wastewater using the activated sludge treatment process. It includes organisms, accumulated food materials, and waste products from the aerobic decomposition process.

Adsorption—The adhesion of a substance to the surface of a solid or liquid. Adsorption is often used to extract pollutants by causing them to attach to such adsorbents as activated carbon or silica gel. *Hydrophobic* (water-repulsing) adsorbents are used to extract oil from waterways in oil spills.

Advanced wastewater treatment—Treatment technology used to produce an extremely high-quality discharge.

Aeration—The process of bubbling air through a solution, sometimes cleaning water of impurities by exposure to air.

Aerobic—Condition in which free, elemental oxygen is present; also used to describe organisms, biological activity, or treatment processes that require free oxygen.

Agglomeration—Floc particles colliding and gathering into a larger settleable mass.

Air gap—The air space between the free-flowing discharge end of a supply pipe and an unpressurized receiving vessel.

Algae bloom—A phenomenon whereby excessive nutrients within a river, stream, or lake causes an explosion of plant life that results in the depletion of oxygen in the water needed by fish and other aquatic life. Algae bloom is usually the result of urban runoff (lawn fertilizers, etc.). A potential consequence is a fish kill, where the stream life dies *en masse*.

Alum—Aluminum sulfate, a standard coagulant used in water treatment.

Ambient—The expected natural conditions that occur in water unaffected or uninfluenced by human activities.

Anaerobic—Conditions in which no oxygen (free or combined) is available; also used to describe organisms, biological activity, or treatment processes that function in the absence of oxygen.

Anoxic—Condition in which no free, elemental oxygen is present; the only source of oxygen is combined

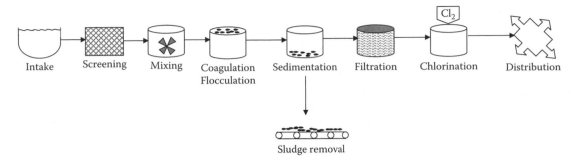

FIGURE 4.1 Unit processes: water treatment.

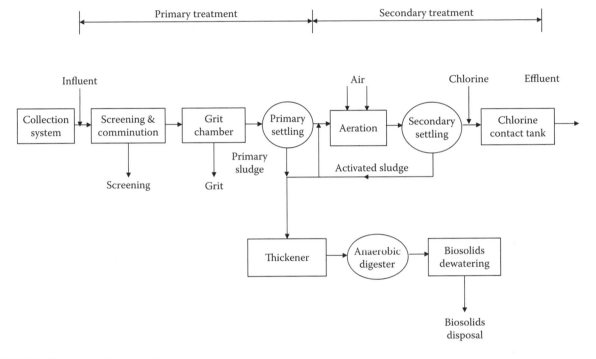

FIGURE 4.2 Conventional wastewater treatment.

oxygen, such as that found in nitrate compounds. Also used to describe biological activity of treatment processes that function only in the presence of combined oxygen.

Aquifer—A water-bearing stratum of permeable rock, sand, or gravel.

Aquifer system—A heterogeneous body of introduced permeable and less permeable material that acts as a water-yielding hydraulic unit of regional extent.

Artesian water—A well tapping a confined or artesian aquifer in which the static water level stands above the top of the aquifer. The term is sometimes used to include all wells tapping confined water. Wells with water level above the water table are said to have positive artesian head (pressure), and those with water level below the water table have negative artesian head.

Average monthly discharge limitation—The highest allowable discharge over a calendar month.

Average weekly discharge limitation—The highest allowable discharge over a calendar week.

Backflow—Reversal of flow when pressure in a service connection exceeds the pressure in the distribution main.

Backwash—Fluidizing filter media with water, air, or a combination of the two so individual grains can be cleaned of the material that has accumulated during the filter run.

Bacteria—Any of a number of one-celled organisms, some of which cause disease.

Bar screen—A series of bars formed into a grid used to screen out large debris from influent flow.

Base—A substance that has a pH value between 7 and 14.

Basin—A groundwater reservoir defined by the overlying land surface and underlying aquifers that contain water stored in the reservoir.

Beneficial use of water—The use of water for any beneficial purpose, such as for domestic consumption, irrigation, recreation, fish and wildlife, fire protection, navigation, power generation, and industrial use. The benefit varies from one location to another and by custom. What constitutes beneficial use is often defined by statute or court decisions.

Biochemical oxygen demand (BOD)—The oxygen used to meet the metabolic needs of aerobic microorganisms in water that is rich in organic matter.

Biosolids*—Solid organic matter recovered from a sewage treatment process and used especially as fertilizer or soil amendment; usually referred to in the plural (*Merriam-Webster's Collegiate Dictionary*, 10th ed., 1998).

Biota—All the species of plants and animals indigenous to a certain area.

Boiling point—The temperature at which a liquid boils; the temperature at which the vapor pressure of a liquid equals the pressure on its surface. If the pressure of the liquid varies, the actual boiling point varies. The boiling point of water is 212° Fahrenheit or 100° Celsius.

Breakpoint—Point at which chlorine dosage satisfies chlorine demand.

Breakthrough—In filtering, when unwanted materials start to pass through the filter.

Buffer—A substance or solution that resists changes in pH.

Calcium carbonate—Compound principally responsible for hardness.

Calcium hardness—Portion of total hardness caused by calcium compounds.

Carbonaceous biochemical oxygen demand (CBOD)—The amount of biochemical oxygen demand that can be attributed to carbonaceous material.

Carbonate hardness—Caused primarily by compounds containing carbonate.

Chemical oxygen demand (COD)—The amount of chemically oxidizable materials present in the wastewater.

Chlorination—Disinfection of water using chlorine as the oxidizing agent.

Clarifier—A device designed to permit solids to settle or rise and be separated from the flow; also known as a settling tank or sedimentation basin.

Coagulation—Neutralization of the charges of colloidal matter.

Coliform—A type of bacteria used to detect possible human or animal contamination of water.

Combined sewer—A collection system that carries both wastewater and stormwater flows.

Comminution—A process to shred solids into smaller, less harmful particles.

Composite sample—A combination of individual samples taken in proportion to flow.

Connate water—Pressurized water trapped in the pore spaces of sedimentary rock at the time it was deposited; it is usually highly mineralized.

Consumptive use—(1) The quantity of water absorbed by crops and transpired or used directly in the building of plant tissue, together with the water evaporated from the cropped area. (2) The quantity of water transpired and evaporated from a cropped area or the normal loss of water from the soil by evaporation and plant transpiration. (3) The quantity of water discharged to the atmosphere or incorporated into the products of the process in connection with vegetative growth, food processing, or an industrial process.

Contamination (water)—Damage to the quality of water sources by sewage, industrial waste, or other material.

Cross-connection—A connection between a storm drain system and a sanitary collection system, a connection between two sections of a collection system to handle anticipated overloads of one system, or a connection between drinking (potable) water and an unsafe water supply or sanitary collection system.

Daily discharge—The discharge of a pollutant measured during a calendar day or any 24-hour period that reasonably represents a calendar day for the purposes of sampling. Limitations expressed as weight are total mass (weight) discharged over the day; limitations expressed in other units are average measurement of the day.

Daily maximum discharge—The highest allowable values for a daily discharge.

Darcy's law—An equation for the computation of the quantity of water flowing through porous media. Darcy's law assumes that the flow is laminar and that inertia can be neglected. The law states that the rate of viscous flow of homogenous fluids through isotropic porous media is proportional to, and in the direction of, the hydraulic gradient.

Detention time—The theoretical time water remains in a tank at a give flow rate.

Dewatering—The removal or separation of a portion of water present in a sludge or slurry.

Diffusion—The process by which both ionic and molecular species dissolved in water move from areas of higher concentration to areas of lower concentration.

Discharge monitoring report (DMR)—The monthly report required by the treatment plant's National Pollutant Discharge Elimination System (NPDES) discharge permit.

---

* In this text, the term *biosolids* is used in many places to replace the standard term *sludge* (activated sludge being the exception). It is the opinion of the author that the term *sludge* is an ugly four-letter word inappropriate for describing biosolids. Biosolids are a product that can be reused; they have some value. Because biosolids have value, they certainly should not be classified as a waste product, and when the topic of biosolids for beneficial reuse is addressed, it is made clear that they are not a waste product.

Disinfection—Water treatment process that kills pathogenic organisms.

Disinfection byproducts (DBPs)—Chemical compounds formed by the reaction of disinfectant with organic compounds in water.

Dissolved oxygen (DO)—The amount of oxygen dissolved in water or sewage. Concentrations of less than 5 parts per million (ppm) can limit aquatic life or cause offensive odors. Excessive organic matter present in water because of inadequate waste treatment and runoff from agricultural or urban land generally causes low DO levels.

Dissolved solids—The total amount of dissolved inorganic material contained in water or wastes; excessive dissolved solids make water unsuitable for drinking or industrial uses.

Domestic consumption (use)—Water used for household purposes such as washing, food preparation, and showers. The quantity (or quantity per capita) of water consumed in a municipality or district for domestic uses or purposes during a given period, it sometimes encompasses all uses, including the quantity wasted, lost, or otherwise unaccounted for.

Drawdown: Lowering the water level by pumping; it is measured in feet for a given quantity of water pumped during a specified period or after the pumping level has become constant.

Drinking water standards—Established by state agencies, the U.S. Public Health Service, and the U.S. Environmental Protection Agency (USEPA) for drinking water in the United States.

Effluent—Something that flows out, usually a polluting gas or liquid discharge.

Effluent limitation—Any restriction imposed by the regulatory agency on quantities, discharge rates, or concentrations of pollutants discharged from point sources into state waters.

Energy—In scientific terms, the ability or capacity of doing work. Various forms of energy include kinetic, potential, thermal, nuclear, rotational, and electromagnetic. One form of energy may be changed to another, as when coal is burned to produce steam to drive a turbine, which produces electric energy.

Erosion—The wearing away of the land surface by wind, water, ice, or other geologic agents. Erosion occurs naturally from weather or runoff but is often intensified by human land use practices.

Eutrophication—The process of enrichment of water bodies by nutrients. Eutrophication of a lake normally contributes to its slow evolution into a bog or marsh and ultimately to dry land. Eutrophication may be accelerated by human activities, thereby speeding up the aging process.

Evaporation—The process by which water becomes a vapor at a temperature below the boiling point.

Facultative—Organisms that can survive and function in the presence or absence of free, elemental oxygen.

Fecal coliform—The portion of the coliform bacteria group that is present in the intestinal tracts and feces of warm-blooded animals.

Field capacity—The capacity of soil to hold water; it is measured as the ratio of the weight of water retained by the soil to the weight of the dry soil.

Filtration—The mechanical process that removes particulate matter by separating water from solid material, usually by passing it through sand.

Floc—Solids that join to form larger particles that will settle better.

Flocculation—Slow mixing process in which particles are brought into contact, with the intent of promoting their agglomeration.

Flume—A flow-rate measurement device.

Fluoridation—Chemical addition to water to reduce incidence of dental caries in children.

Food-to-microorganisms ratio (F/M)—An activated sludge process control calculation based on the amount of food (BOD or COD) available per pound of mixed liquor volatile suspended solids.

Force main—A pipe that carries wastewater under pressure from the discharge side of a pump to a point of gravity flow downstream.

Grab sample—An individual sample collected at a randomly selected time.

Graywater—Water that has been used for showering, clothes washing, and faucet uses; kitchen sink and toilet water is excluded. This water has excellent potential for reuse as irrigation for yards.

Grit—Heavy inorganic solids, such as sand, gravel, eggshells, or metal filings.

Groundwater—The supply of fresh water found beneath the surface of the Earth (usually in aquifers) often used for supplying wells and springs. Because groundwater is a major source of drinking water, concern is growing over areas where leaching agricultural or industrial pollutants or substances from leaking underground storage tanks (USTs) are contaminating groundwater.

Groundwater hydrology—The branch of hydrology that deals with groundwater, its occurrence and movements, its replenishment and depletion, the properties of rocks that control groundwater movement and storage, and the methods of investigation and use of groundwater.

Groundwater recharge—Inflow to a groundwater reservoir.

Groundwater runoff—A portion of runoff that has passed into the ground, has become groundwater, and has been discharged into a stream channel as spring or seepage water.

Hardness—The concentration of calcium and magnesium salts in water.

Head loss—Amount of energy used by water in moving from one point to another.

Heavy metals—Metallic elements with high atomic weights (e.g., mercury, chromium, cadmium, arsenic, and lead). They can damage living things at low concentrations and tend to accumulate in the food chain.

Holding pond—A small basin or pond designed to hold sediment-laden or contaminated water until it can be treated to meet water quality standards or used in some other way.

Hydraulic cleaning—Cleaning pipe with water under enough pressure to produce high water velocities.

Hydraulic gradient—A measure of the change in groundwater head over a given distance.

Hydraulic head—The height above a specific datum (generally sea level) that water will rise in a well.

Hydrologic cycle (water cycle)—The cycle of water movement from the atmosphere to the Earth and back to the atmosphere through various processes. These processes include precipitation, infiltration, percolation, storage, evaporation, transpiration, and condensation.

Hydrology—The science dealing with the properties, distribution, and circulation of water.

Impoundment—A body of water such as a pond, confined by a dam, dike, floodgate, or other barrier and used to collect and store water for future use.

Industrial wastewater—Wastes associated with industrial manufacturing processes.

Infiltration—The gradual downward flow of water from the surface into soil material.

Infiltration/inflow—Extraneous flows in sewers; simply, inflow is water discharged into sewer pipes or service connections from such sources as foundation drains, roof leaders, cellar and yard area drains, cooling water from air conditioners, and other clean-water discharges from commercial and industrial establishments. Defined by Metcalf & Eddy, Inc. (2003) as follows:

　Infiltration—Water entering the collection system through cracks, joints, or breaks.

　Steady inflow—Water discharged from cellar and foundation drains, cooling water discharges, and drains from springs and swampy areas. This type of inflow is steady and is identified and measured along with infiltration.

　Direct flow—Those types of inflow that have a direct stormwater runoff connection to the sanitary sewer and cause an almost immediate increase in wastewater flows. Possible sources are roof leaders, yard and areaway drains, manhole covers, cross connections from storm drains and catch basins, and combined sewers.

　Total inflow—The sum of the direct inflow at any point in the system plus any flow discharged from the system upstream through overflows, pumping station bypasses, and the like.

　Delayed inflow—Stormwater that may require several days or more to drain through the sewer system. This category can include the discharge of sump pumps from cellar drainage as well as the slowed entry of surface water through manholes in ponded areas.

Influent—Wastewater entering a tank, channel, or treatment process.

Inorganic chemicals/compounds—Chemical substances of mineral origin, not of a carbon structure. These include metals such as lead, iron (ferric chloride), and cadmium.

Ion exchange process—Used to remove hardness from water.

Jar test—Laboratory procedure used to estimate proper coagulant dosage.

Langelier saturation index (LSI)—A numerical index that indicates whether calcium carbonate will be deposited or dissolved in a distribution system.

Leaching—The process by which soluble materials in the soil such as nutrients, pesticide chemicals, or contaminants are washed into a lower layer of soil or are dissolved and carried away by water.

License—A certificate issued by the State Board of Waterworks/Wastewater Works Operators authorizing the holder to perform the duties of a wastewater treatment plant operator.

Lift station—A wastewater pumping station designed to lift the wastewater to a higher elevation. A lift station normally employs pumps or other mechanical devices to pump the wastewater and discharges into a pressure pipe called a *force main*.

Maximum contaminant level (MCL)—An enforceable standard for protection of human health.

Mean cell residence time (MCRT)—The average length of time a mixed liquor suspended solids particle remains in the activated sludge process; may also be referred to as *sludge retention time*.

Mechanical cleaning—Clearing a pipe by using equipment (bucket machines, power rodders, or hand rods) that scrapes, cuts, pulls, or pushes material out of the pipe.

Membrane process—A process that draws a measured volume of water through a filter membrane with small enough openings to take out contaminants.

Metering pump—A chemical solution feed pump that adds a measured amount of solution with each stroke or rotation of the pump.

Milligrams/liter (mg/L)—A measure of concentration equivalent to parts per million (ppm).

Mixed liquor volatile suspended solids (MLVSS)—The concentration of organic matter in the mixed liquor suspended solids.

Nephelometric turbidity unit (NTU)—Indicates the amount of turbidity in a water sample.

Nitrogenous oxygen demand (NOD)—A measure of the amount of oxygen required to biologically oxidize nitrogen compounds under specified conditions of time and temperature.

Nonpoint source (NPS) pollution—Pollution caused by surface runoff of sediment, nutrients, and organic and toxic substances originating from land use activities that are carried to lakes and streams. Nonpoint source pollution occurs when the rate of materials entering these waterbodies exceeds natural levels.

NPDES permit—National Pollutant Discharge Elimination System permit, which authorizes the discharge of treated wastes and specifies the conditions that must be met for discharge.

Nutrients—Substances required to support living organisms. Usually refers to nitrogen, phosphorus, iron, and other trace metals.

Organic chemicals/compounds—Animal- or plant-produced substances containing mainly carbon, hydrogen, and oxygen, such as benzene and toluene.

Parts per million (ppm)—The number of parts by weight of a substance per million parts of water; this unit is commonly used to represent pollutant concentrations. Large concentrations are expressed in percentages.

Pathogenic—Disease causing; a pathogenic organism is capable of causing illness.

Percolation—The movement of water through the subsurface soil layers, usually continuing downward to the groundwater or water table reservoirs.

pH—A way of expressing both acidity and alkalinity on a scale of 0 to 14, with 7 representing neutrality; numbers less than 7 indicate increasing acidity, and numbers greater than 7 indicate increasing alkalinity.

Photosynthesis—A process in green plants in which water, carbon dioxide, and sunlight combine to form sugar.

Piezometric surface—An imaginary surface that coincides with the hydrostatic pressure level of water in an aquifer.

Point source pollution—A type of water pollution resulting from discharges into receiving waters from easily identifiable points. Common point sources of pollution are discharges from factories and municipal sewage treatment plants.

Pollution—Alteration of the physical, thermal, chemical, or biological quality of, or the contamination of, any water to a state that renders the water harmful, detrimental, or injurious to humans, animal life, vegetation, property, or public health, safety, or welfare, or impairs the usefulness or the public enjoyment of the water for any lawful or reasonable purpose.

Porosity—Measure of the part of a rock containing pore spaces without regard to size, shape, interconnection, or arrangement of openings; it is expressed as percentage of total volume occupied by spaces.

Potable water—Water satisfactorily safe for drinking purposes from the standpoint of its chemical, physical, and biological characteristics.

Precipitate—A deposit of hail, rain, mist, sleet, or snow. The process by which atmospheric water becomes surface or subsurface water is *precipitation*, a term that is commonly used to designate the quantity of water precipitated.

Preventive maintenance (PM)—Regularly scheduled servicing of machinery or other equipment using appropriate tools, tests, and lubricants. This type of maintenance can prolong the useful life of equipment and machinery and increase its efficiency by detecting and correcting problems before they cause a breakdown of the equipment.

Purveyor—An agency or a person that supplies potable water.

Radon—A radioactive, colorless, odorless gas that occurs naturally in the earth; when radon is trapped in buildings, concentrations build up that can cause health hazards such as lung cancer.

Recharge—The addition of water into a groundwater system.

Reservoir—A pond, lake, tank, or basin (natural or manmade) where water is collected and used for storage. Large bodies of groundwater are called *groundwater reservoirs*; water behind a dam is also referred to as a *reservoir of water*.

Return activated sludge solids (RASS)—The concentration of suspended solids in the sludge flow being returned from the settling tank to the head of the aeration tank.

Reverse osmosis—Process in which almost pure water is passed through a semipermeable membrane.

River basin—A term used to designate the area drained by a river and its tributaries.

Sanitary wastewater—Wastes discharged from residences and from commercial, institutional, and similar facilities that include both sewage and industrial wastes.

Schmutzdecke—Layer of solids and biological growth that forms on top of a slow sand filter, allowing the filter to remove turbidity effectively without chemical coagulation.

Scum—The mixture of floatable solids and water removed from the surface of the settling tank.

Sediment—Transported and deposited particles derived from rocks, soil, or biological material.

Sedimentation—A process that reduces the velocity of water in basins so suspended material can settle out by gravity.

Seepage—The appearance and disappearance of water at the ground surface. Seepage designates movement of water in saturated material; it differs from *percolation*, which is predominantly the movement of water in unsaturated material.

Septic tanks—Tanks used to hold domestic wastes when a sewer line is not available to carry them to a treatment plant. The wastes are piped to underground tanks directly from a house. Bacteria in the wastes decompose some of the organic matter, the sludge settles on the bottom of the tank, and the effluent flows out of the tank into the ground through drains.

Settleability—A process control test used to evaluate the settling characteristics of the activated sludge. Readings taken at 30 to 60 minutes are used to calculate the settled sludge volume (SSV) and the sludge volume index (SVI).

Settled sludge volume (SSV)—The volume (in percent) occupied by an activated sludge sample after 30 to 60 minutes of settling. Normally written as SSV with a subscript to indicate the time of the reading used for calculation (e.g., SSV_{60} or SSV_{30}).

Sludge—The mixture of settleable solids and water removed from the bottom of the settling tank.

Sludge retention time (SRT)—See mean cell residence time.

Sludge volume index (SVI)—A process control calculation used to evaluate the settling quality of the activated sludge; requires the SSV_{30} and mixed liquor suspended solids test results to calculate.

Soil moisture (soil water)—Water diffused in the soil. It is found in the upper part of the zone of aeration from which water is discharged by transpiration from plants or by soil evaporation.

Specific heat—The heat capacity of a material per unit mass. The amount of heat (in calories) required to raise the temperature of 1 gram of a substance 1°C; the specific heat of water is 1 calorie.

Storm sewer—A collection system designed to carry only stormwater runoff.

Stormwater—Runoff resulting from rainfall and snowmelt.

Stream—A general term for a body of flowing water. In hydrology, the term is generally applied to the water flowing in a natural channel as distinct from a canal. More generally, it is applied to the water flowing in any channel, natural or artificial. Some types of streams include: (1) *ephemeral*, a stream that flows only in direct response to precipitation and whose channel is at all times above the water table; (2) *intermittent* or *seasonal*, a stream that flows only at certain times of the year when it receives water from springs, rainfall, or from surface sources such as melting snow; (3) *perennial*, a stream that flows continuously; (4) *gaining*, a stream or reach of a stream that receives water from the zone of saturation (an effluent stream); (5) *insulated*, a stream or reach of a stream that neither contributes water to the zone of saturation nor receives water from it and is separated from the zones of saturation by an impermeable bed; (6) *losing*, a stream or reach of a stream that contributes water to the zone of saturation—an influent stream; and (7) *perched*, a losing stream or an insulated stream that is separated from the underlying groundwater by a zone of aeration.

Supernatant—The liquid standing above a sediment or precipitate.

Surface tension—The free energy produced in a liquid surface by the unbalanced inward pull exerted by molecules underlying the layer of surface molecules.

Surface water—Lakes, bays, ponds, impounding reservoirs, springs, rivers, streams, creeks, estuaries, wetlands, marshes, inlets, canals, gulfs inside the territorial limits of the state, and all other bodies of surface water, natural or artificial, inland or coastal, fresh or salt, navigable or nonnavigable, and including the beds and banks of all watercourses and bodies of surface water that are wholly or partially inside or bordering the state or subject to the jurisdiction of the state. Waters in treatment systems that are authorized by state or federal law, regulation, or permit and which are created for the purpose of water treatment are not considered to be waters in the state.

Thermal pollution—The degradation of water quality by the introduction of a heated effluent. Primarily the result of the discharge of cooling waters from industrial processes (particularly from electrical power generation); waste heat eventually results from virtually every energy conversion.

Titrant—A solution of known strength of concentration; used in titration.

Titration—A process whereby a solution of known strength (titrant) is added to a certain volume of treated sample containing an indicator; a color change shows when the reaction is complete.

Titrator—An instrument, usually a calibrated cylinder (tube-form), used in titration to measure the amount of titrant being added to the sample.

Total dissolved solids—The amount of material (inorganic salts and small amounts of organic material) dissolved in water and commonly expressed as a concentration in terms of milligrams per liter.

Total suspended solids (TSS)—Total suspended solids in water, commonly expressed as a concentration in terms of milligrams per liter.

Toxicity—The occurrence of lethal or sublethal adverse affects on representative sensitive organisms due to exposure to toxic materials. Adverse effects caused by conditions of temperature, dissolved oxygen, or nontoxic dissolved substances are excluded from the definition of toxicity.

Transpiration—The process by which water vapor escapes from the living plant (principally the leaves) and enters the atmosphere.

Vaporization—The change of a substance from a liquid or solid state to a gaseous state.

Volatile organic compound (VOC)—Any organic compound that participates in atmospheric photochemical reactions except for those designated by the USEPA Administrator as having negligible photochemical reactivity.

Waste activated sludge solids (WASS)—The concentration of suspended solids in the sludge being removed from the activated sludge process.

Wastewater—The water supply of a community after it has been soiled by use.

Water cycle—The process by which water travels in a sequence from the air (condensation) to the Earth (precipitation) and returns to the atmosphere (evaporation); it is also referred to as the *hydrologic cycle*.

Water quality—A term used to describe the chemical, physical, and biological characteristics of water with respect to its suitability for a particular use.

Water quality standard—A plan for water quality management containing four major elements: water use, criteria to protect uses, implementation plans, and enforcement plans. An antidegradation statement is sometimes prepared to protect existing high-quality waters.

Water supply—Any quantity of available water.

Waterborne disease—A disease caused by a microorganism that is carried from one person or animal to another by water.

Watershed—The area of land that contributes surface run-off to a given point in a drainage system.

Weir—A device used to measure wastewater flow.

Zone of aeration—A region in the earth above the water table; water in the zone of aeration is under atmospheric pressure and would not flow into a well.

Zoogleal slime—The biological slime that forms on fixed-film treatment devices. It contains a wide variety of organisms essential to the treatment process.

CHAPTER REVIEW QUESTION

4.1   *Matching Exercise:* Match the definitions listed in Part A with the terms listed in Part B by placing the correct letter in the blank. (*Note:* After completing this exercise, check your answers with those provided in Appendix A.)

Part A

1. ____ A nonchemical turbidity removal layer in a slow sand filter
2. ____ Region in earth (soil) above the water table
3. ____ Compound associated with photochemical reaction
4. ____ Oxygen used in water rich in inorganic matter
5. ____ A stream that receives water from the zone of saturation
6. ____ The addition of water to a groundwater system
7. ____ The natural water cycle
8. ____ Present in intestinal tracts and feces of animals and humans
9. ____ Discharge from a factory or municipal sewage treatment plant
10. ____ Common to fixed-film treatment devices
11. ____ Identified water that is safe to drink
12. ____ The capacity of soil to hold water
13. ____ Used to measure acidity and alkalinity
14. ____ Rain mixed with sulfur oxides
15. ____ Enrichment of water bodies by nutrients
16. ____ Solution of known strength of concentration
17. ____ Water lost by foliage
18. ____ Another name for wastewater pumping station
19. ____ Plants and animals indigenous to an area
20. ____ The amount of oxygen dissolved in water
21. ____ A stream that flows continuously
22. ____ A result of excessive nutrients within a water body
23. ____ Change in groundwater head over a given distance
24. ____ Water trapped in sedimentary rocks
25. ____ Heat capacity of a material per unit mass
26. ____ A compound derived from material that once lived

Part B

a. pH	j. BOD	s. Gaining
b. Algae bloom	k. Field capacity	t. VOC
c. Zone of aeration	l. Transpiration	u. Potable
d. Hydrological cycle	m. Biota	v. Acid rain
e. Point source pollution	n. Specific heat	w. Titrant
f. Perennial	o. Schmutzdecke	x. Lift station
g. Organic	p. Recharge	y. DO
h. Connate water	q. Zoogleal slime	z. Hydraulic gradient
i. Fecal coliform	r. Eutrophication	

REFERENCES AND SUGGESTED READING

Metcalf & Eddy, Inc. 2003. *Wastewater Engineering: Treatment, Disposal, Reuse*, 4th ed. New York: McGraw-Hill.

Pearce, F. 2006. *When the Rivers Run Dry*. Boston, MA: Beacon Press.

Pielou, E.C. 1998. *Fresh Water*. Chicago, IL: University of Chicago.

Part II

Water/Wastewater Operations:
Math and Technical Aspects

5 Water/Wastewater Math Operations

To operate a waterworks or a wastewater treatment plant and to pass the examination for an operator's license, you must know how to perform certain mathematical operations. Do not panic. As Price (1991) pointed out, "Those who have difficulty in math often do not lack the ability for mathematical calculation; they merely have not learned, or have been taught, the 'language of math.'"

5.1 INTRODUCTION

Without the ability to perform mathematical calculations, it would be difficult for operators to properly operate water/wastewater systems. In reality, most of the calculations that operators have to perform are not difficult. Generally, math ability through basic algebra is all that is needed. Experience has shown that skill with math operations used in water/wastewater system operations is an acquired skill that is enhanced and strengthened with practice. In this chapter, we assume the reader is well grounded in basic math principles; thus, we do not cover basic operations such as addition, subtraction, multiplying, and dividing.

Note: Keep in mind that mathematics is a *language*, a universal language. Mathematical symbols have the same meaning to people speaking many different languages all over the globe. The key to learning mathematics is to learn the language—the symbols, definitions, and terms of mathematics that allow you to understand the concepts necessary to perform the operations.

5.2 CALCULATION STEPS

As with all math operations, many methods can be used successfully to solve water/wastewater system problems. We provide one of the standard methods of problem solving in the following:

1. If appropriate, make a drawing of the information in the problem.
2. Place the given data on the drawing.
3. Answer "What is the question?" This is the first thing you should do as you begin to solve the problem, as well as asking yourself "What are they really looking for?" Writing down exactly what is being looked for is always smart. Sometimes the answer has more than one unknown; for example, you may need to find x and then find y.
4. If the calculation calls for an equation, write it down.
5. Fill in the data in the equation; look to see what is missing.
6. Rearrange or transpose the equation, if necessary.
7. Use a calculator when necessary.
8. Always write down the answer.
9. Check any solution obtained.

5.3 EQUIVALENTS, FORMULAE, AND SYMBOLS

To work mathematical operations to solve problems (for practical application or for taking licensure examinations), it is essential to understand the language, equivalents, symbols, and terminology used. Because this handbook is designed for use in practical on-the-job applications, equivalents, formulae, and symbols are included as a ready reference in Table 5.1.

5.4 BASIC WATER/WASTEWATER MATH OPERATIONS

5.4.1 ARITHMETIC AVERAGE (OR ARITHMETIC MEAN) AND MEDIAN

During the day-to-day operation of a wastewater treatment plant, considerable mathematical data are collected. The data, if properly evaluated, can provide useful information for trend analysis and indicate how well the plant or unit process is operating; however, because much variation may be found in the data, it is often difficult to determine trends in performance. *Arithmetic average* refers to a statistical calculation used to describe a series of numbers such as test results. By calculating an average, a group of data is represented by a single number. This number may be considered typical of the group. The *arithmetic mean* is the most commonly used measurement of average value.

TABLE 5.1
Equivalents, Formulae, and Symbols

Equivalents

12 inches (in.)	=	1 foot (ft)
36 inches (in.)	=	1 yard (yd)
144 inches^2 (in.^{2})	=	1 foot^2 (ft^2)
9 feet^2 (ft^2)	=	1 yard^2 (yd^2)
43,560 ft^2	=	1 acre
1 foot^3 (ft^3)	=	1728 inch^3 (in.^{3})
1 foot^3 (ft^3) of water	=	7.48 gallons (gal)
1 foot^3 (ft^3) of water	=	62.4 pounds (lb)
1 gallon (gal) of water	=	8.34 pounds (lb)
1 liter (L)	=	1000 milliliters (mL)
1 gram (g)	=	1000 milligrams (mg)
1 million gallons per day (gpd)	=	694 gallons per minute (gpm)
	=	1.545 ft^3/sec
Average biochemical oxygen demand (BOD)/capita/day	=	.17 pound (lb)
Average suspended solids (SS)/capita/day	=	.20
Average daily flow	=	Assume 100 gal/capita/day

Symbols

A	=	Area
V	=	Velocity
t	=	Time
SVI	=	Sludge volume index
Vol	=	Volume
eff	=	Effluent
W	=	Width
D	=	Depth
L	=	Length
Q	=	Flow
r	=	Radius
π	=	pi (= 3.14)
WAS	=	Waste activated sludge
RAS	=	Return activated sludge
MLSS	=	Mixed liquor suspended solids
MLVSS	=	Mixed liquor volatile suspended solids

Formulae

$$\text{SVI} = \frac{\text{Volume}}{\text{Concentration}} \times 100$$

$$Q = A \times V$$

$$\text{Detention time} = \text{Volume}/Q$$

$$\text{Volume} = L \times W \times D$$

$$\text{Area} = W \times L$$

$$\text{Circular area} = \pi \times (\text{radius})^2 \ (= 0.785 \times \text{diameter}^2)$$

$$\text{Circumference} = \pi \times \text{diameter}$$

$$\text{Hydraulic loading rate} = Q/A$$

$$\text{Sludge age} = \frac{\text{MLSS (lb) in aeration tank}}{\text{SS in primary effluent (lb/day)}}$$

$$\text{Mean cell residence time} = \frac{\text{SS (lb) in secondary system (aeration tank + secondary clarifier)}}{\text{WAS (lb/day) + SS in effluent (lb/day)}}$$

$$\text{Organic loading rate} = \frac{\text{BOD (lb/day)}}{\text{Volume}}$$

Note: When evaluating information based on averages, remember that the "average" reflects the general nature of the group and does not necessarily reflect any one element of that group.

Arithmetic average is calculated by dividing the sum of all of the available data points (test results) by the number of test results:

$$\frac{\text{Test } 1 + \text{Test } 2 + \text{Test } 3 + \ldots + \text{Test } N}{\text{Number of tests performed } (N)} \quad (5.1)$$

■ EXAMPLE 5.1

Problem: Effluent biochemical oxygen demand (BOD) test results for the treatment plant during the month of September are shown below:

Test 1   20 mg/L
Test 2   31 mg/L
Test 3   22 mg/L
Test 4   15 mg/L

What is the average effluent BOD for the month of September?

Solution:

$$\text{BOD} = \frac{20 \text{ mg/L} + 31 \text{ mg/L} + 22 \text{ mg/L} + 15 \text{ mg/L}}{4}$$
$$= 22 \text{ mg/L}$$

■ EXAMPLE 5.2

Problem: For the primary influent flow, the following composite-sampled solids concentrations were recorded for the week:

Monday	300 mg/L SS
Tuesday	312 mg/L SS
Wednesday	315 mg/L SS
Thursday	320 mg/L SS
Friday	311 mg/L SS
Saturday	320 mg/L SS
Sunday	310 mg/L SS
Total	2188 mg/L SS

Solution:

$$\text{Average suspended solids} = \frac{\text{Sum of all measurements}}{\text{Number of measurements}}$$
$$= \frac{2188 \text{ mg/L suspended solids}}{7}$$
$$= 312.6 \text{ mg/L suspended solids}$$

TABLE 5.2
Daily Chlorine Residual Results

Day	Chlorine Residual (mg/L)
Monday	0.9
Tuesday	1.0
Wednesday	1.2
Thursday	1.3
Friday	1.4
Saturday	1.1
Sunday	0.9

■ EXAMPLE 5.3

Problem: A waterworks operator takes a chlorine residual measurement every day. We show part of the operating log in Table 5.2. Find the mean.

Solution: Add up the seven chlorine residual readings:

$$0.9 + 1.0 + 1.2 + 1.3 + 1.4 + 1.1 + 0.9 = 7.8$$

Next, divide by the number of measurements:

$$7.8 \div 7 = 1.11$$

The mean chlorine residual for the week was 1.11 mg/L.

Definition: The *median* is defined as the value of the central item when the data are arrayed by size. First, arrange all of the readings in either ascending or descending order, then find the middle value.

■ EXAMPLE 5.4

Problem: In our chlorine residual example, what is the median?

Solution: Arrange the values in ascending order:

0.9   0.9   1.0   1.1   1.2   1.3   1.4

The middle value is the fourth one (1.1); therefore, the median chlorine residual is 1.1 mg/L. (Usually, the median will be a different value than the mean). If the data contain an even number of values, you must add one more step because no middle value is present. In this case, find the two values in the middle and then find the mean of those two values.

■ EXAMPLE 5.5

Problem: A water system has 4 wells with the following capacities: 115 gallons per minute (gpm), 100 gpm, 125 gpm, and 90 gpm. What are the mean and the median pumping capacities?

Solution: The mean is:

$$\frac{115 \text{ gpm} + 100 \text{ gpm} + 125 \text{ gpm} + 90 \text{ gpm}}{4} = 107.5 \text{ gpm}$$

To find the median, arrange the values in order:

90 gpm   100 gpm   115 gpm   125 gpm

With four values, there is no single middle value, so we must take the mean of the two middle values:

$$\frac{100 \text{ gpm} + 115 \text{ gpm}}{2} = 107.5 \text{ gpm}$$

Note: At times, determining what the original numbers were like is difficult (if not impossible) when dealing only with averages.

■ EXAMPLE 5.6

Problem: A water system has four storage tanks. Three of them have a capacity of 100,000 gal each, while the fourth has a capacity of 1 million gal. What is the mean capacity of the storage tanks?

Solution: The mean capacity of the storage tanks is:

$$\frac{100{,}000 + 100{,}000 + 100{,}000 + 1{,}000{,}000}{4} = 325{,}000 \text{ gal}$$

Note: Notice that no tank in Example 5.6 has a capacity anywhere close to the mean. The median capacity requires us to take the mean of the two middle values; because they are both 100,000 gal, the median is 100,000 gal. Although three of the tanks have the same capacity as the median, these data offer no indication that one of these tanks holds 1 million gal, information that could be important for the operator to know.

5.4.2 UNITS AND CONVERSIONS

Most of the calculations made in the water/wastewater operations involve the use of *units*. Whereas the number tells us how many, the units tell us what we have. Examples of units include inches, feet, square feet, cubic feet, gallons, pounds, milliliters, milligrams per liter, pounds per square inch, miles per hour, and so on. *Conversions* are a process of changing the units of a number to make the number usable in a specific instance. Multiplying or dividing into another number to change the units of the number accomplishes conversions. Common conversions in water/wastewater operations are:

TABLE 5.3
Common Conversions

Linear Measurements	Weight
1 in. = 2.54 cm	1 ft³ of water = 62.4 lb
1 ft = 30.5 cm	1 gal = 8.34 lb
1 m = 100 cm = 3.281 ft = 39.4 in.	1 lb = 453.6 g
1 acre = 43,560 ft²	1 kg = 1000 g = 2.2 lb
1 yd = 3 ft	1 % = 10,000 mg/L
Volume	**Pressure**
1 gal = 3.78 L	1 ft of head = 0.433 psi
1 ft³ = 7.48 gal	1 psi = 2.31 ft of head
1 L = 1000 mL	
1 acre-foot (ac-ft) = 43,560 ft³	**Flow**
1 gal = 32 cups	1 cfs = 448 gpm
1 lb = 16 oz. dry wt.	1 gpm = 1440 gpd

- Gallons per minute (gpm) to cubic feet per second (cfs)
- Million gallons to acre-feet
- Cubic feet to acre-feet
- Cubic feet of water to weight
- Cubic feet of water to gallons
- Gallons of water to weight
- Gallons per minute (gpm) to million gallons per day (MGD)
- Pounds per square inch (psi) to feet of head (the measure of the pressure of water expressed as height of water in feet); 1 psi = 2.31 feet of head

In many instances, the conversion factor cannot be derived; it must be known. We use tables such as Table 5.3 to determine the common conversions.

Note: Conversion factors are used to change measurements or calculated values from one unit of measure to another. To convert from one unit to another, we must know two things:

1. The exact number that relates the two units
2. Whether to multiply or divide by that number

Most operators memorize some standard conversions. This happens due to their frequent use of the conversions, not because of attempts to memorize them.

5.4.2.1 Temperature Conversions

An example of a type of conversion typical in water and wastewater operations is provided in this section on temperature conversions.

Note: Operators should keep in mind that temperature conversions are only a small part of the many conversions that must be made in real-world systems operations.

Most water and wastewater operators are familiar with the formulas used for Fahrenheit and Celsius temperature conversions:

$$°C = 5/9(°F − 32°) \qquad (5.2)$$

$$°F = 5/9(°C + 32°) \qquad (5.3)$$

These conversions are not difficult to perform. The difficulty arises when we must recall these formulas from memory. Probably the easiest way to recall these important formulas is to remember three basic steps for both Fahrenheit and Celsius conversions:

1. Add 40°
2. Multiply by the appropriate fraction (5/9 or 9/5)
3. Subtract 40°

Obviously, the only variable in this method is the choice of 5/9 or 9/5 in the multiplication step. To make the proper choice, you must be familiar with two scales. On the Fahrenheit scale, the freezing point of water is 32°; it is 0° on the Celsius scale. The boiling point of water is 212° on the Fahrenheit scale and 100° on the Celsius scale.

Why is this important? Note, for example, that for the same temperatures, higher numbers are associated with the Fahrenheit scale and lower numbers with the Celsius scale. This important relationship helps us decide whether to multiply by 5/9 or 9/5. Let us look at a few conversion problems to see how the three-step process works.

■ **EXAMPLE 5.7**

Problem: Convert 220°F to Celsius.

Solution:

1. Add 40°:

$$220°F + 40°F = 260°F$$

2. 260°F must be multiplied by either 5/9 or 9/5. Because the conversion is to the Celsius scale, you will be moving to number *smaller* than 260. Through reason and observation, obviously we see that multiplying 260 by 9/5 would almost be the same as multiplying by 2, which would double 260, rather than make it smaller. On the other hand, multiplying by 5/9 is about the same as multiplying by 1/2, which would cut 260 in half. In this problem, we need to obtain a smaller number, so we should multiply by 5/9:

$$(5/9) × 260° = 144.4°C$$

3. Now subtract 40°:

$$144.4°C − 40.0°C = 104.4°C$$

Thus, 220°F = 104.4°C.

■ **EXAMPLE 5.8**

Problem: Convert 22°C to Fahrenheit.

1. Add 40°:

$$22°C + 40°C = 62°C$$

2. Because we are converting from Celsius to Fahrenheit, we are moving from a smaller to larger number, so we should use 9/5 in the multiplication:

$$(9/5) × 62° = 112°$$

3. Subtract 40°:

$$112°F − 40°F = 72°F$$

Thus, 22°C = 72°F.

Obviously, knowing how to make these temperature conversion calculations is useful; however, in practical (real-world) operations, we may prefer to use a temperature conversion table.

5.4.2.2 Milligrams Per Liter (Parts Per Million)

One of the most common terms for concentration is *milligrams per liter* (mg/L). If a mass of 15 mg of oxygen is dissolved in a volume of 1 L of water, for example, the concentration of that solution is expressed simply as 15 mg/L. Very dilute solutions are more conveniently expressed in terms of *micrograms per liter* (μg/L). A concentration of 0.005 mg/L, for example, is preferably written as its equivalent: 5 μg/L. Because 1000 μg = 1 mg, simply move the decimal point three places to the right when converting from mg/L to μg/L. Move the decimal three places to the left when converting from μg/L to mg/L; for example, a concentration of 1250 μg/L is equivalent to 1.25 mg/L.

A liter of water has a mass of 1 kg, and 1 kg is equivalent to 1000 g or 1,000,000 mg; therefore, if we dissolve 1 mg of a substance in 1 liter of water, we can say that there is 1 mg of solute per 1 million mg of water—or, in other words, *one part per million* (ppm).

Note: For comparative purposes, we like to say that 1 ppm is analogous to a full shotglass of water sitting in the bottom of a full standard-size swimming pool.

Neglecting the small change in the density of water as substances are dissolved in it, we can say that, in general, a concentration of 1 milligram per liter is equivalent to 1 part per million: 1 mg/L = 1 ppm. Conversions are very simple; for example, a concentration of 18.5 mg/L is identical to 18.5 ppm. The expression *mg/L* is preferred over *ppm*, just as the expression *μg/L* is preferred over its equivalent of *ppb*. Both types of units are still used, however, and the waterworks operator should be familiar with both of them.

5.4.3 Area and Volume

Water and wastewater operators are often required to calculate surface areas and volumes. *Area* is a calculation of the surface of an object. The length and the width of a water tank can be measured, but the surface area of the water in the tank must be calculated. An area is found by multiplying two length measurements, so the result is a square measurement; for example, when multiplying feet by feet, we get square feet (ft²). *Volume* is the calculation of the space inside a three-dimensional object and is calculated by multiplying three length measurements, or an area by a length measurement. The result is a cubic measurement, such as cubic feet (ft³).

5.4.4 Force, Pressure, and Head

Force, pressure, and head are important parameters in water and wastewater operations. Before we study calculations involving the relationship between force, pressure, and head, we must first define these terms:

- *Force*—The push exerted by water on any confining surface. Force can be expressed in pounds, tons, grams, or kilograms.
- *Pressure*—The force per unit area. The most common way of expressing pressure is in pounds per square inch (psi).
- *Head*—The vertical distance or height of water above a reference point. Head is usually expressed in feet. In the case of water, head and pressure are related.

Figure 5.1 illustrates these terms. A cubical container measuring 1 foot on each side can hold 1 cubic foot of water. A basic fact of science states that 1 cubic foot of water weighs 62.4 pounds. The force acting on the bottom of the container would be 62.4 pounds. The pressure acting on the bottom of the container would be 62.4 pounds per square foot (lb/ft²). The area of the bottom in square inches is:

$$1 \text{ ft}^2 = 12 \text{ in.} \times 12 \text{ in.} = 144 \text{ in.}^2 \quad (5.4)$$

Therefore, the pressure in pounds per square inch (psi) is:

$$\frac{62.4 \text{ lb/ft}^2}{1 \text{ ft}^2} = \frac{62.4 \text{ lb/ft}^2}{144 \text{ in.}^2/\text{ft}^2} = 0.433 \text{ lb/in.}^2 \text{ (psi)} \quad (5.5)$$

If we use the bottom of the container as our reference point, the head would be 1 foot. From this we can see that 1 foot of head is equal to 0.433 psi. Figure 5.2 illustrates some other important relationships between pressure and head.

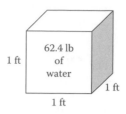

FIGURE 5.1 One cubic foot of water weighs 62.4 lb.

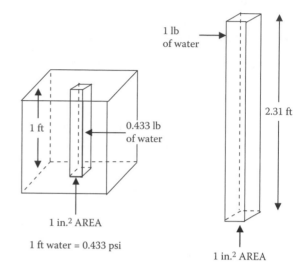

FIGURE 5.2 The relationship between pressure and head.

Note: In water/wastewater operations, 0.433 psi is an important parameter.

Note: Force acts in a particular direction. Water in a tank exerts force down on the bottom and out of the sides. Pressure, however, acts in all directions. A marble at a water depth of one foot would have 0.433 psi of pressure acting inward on all sides.

Key Point: 1 ft of head = 0.433 psi.

This is a valuable parameter that should be committed to memory. We should understand the relationship between pressure and feet of head; in other words, we should know how many feet of head 1 psi represents. This is determined by dividing 1 by 0.433:

$$\text{Feet of head} = 1 \text{ ft}/0.433 \text{ psi} = 2.31 \text{ ft/psi}$$

What we are saying here is that if a pressure gauge were reading 12 psi then the height of the water necessary to represent this pressure would be 12 psi × 2.31 ft/psi = 27.7 feet.

Again, the key points here are that 1 ft = 0.433 psi and 1 psi = 2.31 feet.

Having two conversions methods for the same thing is often confusing; thus, memorizing one and staying with it is best. The most accurate conversion is 1 ft = 0.433 psi, the standard conversion used throughout this handbook.

■ EXAMPLE 5.9

Problem: Convert 50 psi to feet of head.

Solution:

$$50 \times \frac{psi}{1} \times \frac{ft}{0.433 \ psi} = 115.5 \ ft$$

Problem: Convert 50 feet of head to psi.

Solution:

$$50 \times \frac{psi}{1} \times \frac{0.433 \ psi}{ft} = 21.7 \ psi$$

As the above examples demonstrate, when attempting to convert psi to feet, we divide by 0.433; when attempting to convert feet to psi, we multiply by 0.433. The above process can be most helpful in clearing up the confusion as to whether to multiply or divide. Another approach, however, may be more beneficial and easier for many operators to use. Notice that the relationship between psi and feet is almost 2 to 1. It takes slightly more than 2 feet to make 1 psi. When looking at a problem where the data are in pressure and the result should be in feet, the answer will be at least twice as large as the starting number. For example, if the pressure is 25 psi, we intuitively know that the head is over 50 ft; therefore, we must divide by 0.433 to obtain the correct answer.

■ EXAMPLE 5.10

Problem: Convert a pressure of 55 psi to feet of head.

Solution:

$$55 \times \frac{psi}{1} \times \frac{1 \ ft}{0.433 \ psi} = 127 \ ft$$

■ EXAMPLE 5.11

Problem: Convert 14 psi to feet.

Solution:

$$14 \times \frac{psi}{1} \times \frac{1 \ ft}{0.433 \ psi} = 32.3 \ ft$$

■ EXAMPLE 5.12

Problem: Between the top of a reservoir and the watering point, the elevation is 115 feet. What will the static pressure be at the watering point?

Solution:

$$115 \times \frac{ft}{1} \times \frac{0.433 \ psi}{1 \ ft} = 49.8 \ psi$$

Using the preceding information, we can develop Equation 5.6 and Equation 5.7 for calculating pressure and head:

$$\text{Pressure (psi)} = 0.433 \times \text{head (ft)} \qquad (5.6)$$

$$\text{Head (ft)} = 2.31 \times \text{pressure (psi)} \qquad (5.7)$$

■ EXAMPLE 5.13

Problem: Find the pressure (psi) in a tank 12 ft deep at a point 15 ft below the water surface.

Solution:

$$\text{Pressure (psi)} = 0.433 \times 5 \ ft = 2.17 \ psi \ (\text{rounded})$$

■ EXAMPLE 5.14

Problem: A pressure gauge at the bottom of a tank reads 12.2 psi. How deep is the water in the tank?

Solution:

$$\text{Head (ft)} = 2.31 \times 12.2 \ psi = 28.2 \ ft \ (\text{rounded})$$

■ EXAMPLE 5.15

Problem: What is the pressure (static pressure) 4 miles beneath the ocean surface?

Solution: Change miles to ft then to psi.

$$5380 \ ft/mile \times 4 = 21,120 \ ft$$

$$\frac{21,120 \ ft}{2.31 \ ft/psi} = 9143 \ psi \ (\text{rounded})$$

■ EXAMPLE 5.16

Problem: A 150-ft diameter cylindrical tank contains 2.0 MG of water. What is the water depth? At what pressure would a gauge at the bottom read in psi?

Solution:

1. Change MG to cubic feet:

    2,000,000 gal ÷ 7.48 gal/ft^3 = 267,380 ft^3

2. Using volume, solve for depth:

    Volume = 0.785 × (diameter)^2 × depth

    267,380 ft^3 = 0.785 × (150)^2 × depth

    Depth = 15.1 ft

■ EXAMPLE 5.17

Problem: The pressure in a pipe is 70 psi. What is the pressure in feet of water? What is the pressure in psf?

Solution:

1. Convert pressure to feet of water:

    70 psi × 2.31 ft/psi = 161.7 ft of water

2. Convert psi to psf:

    70 psi × 144 in.^{2}/ft^2 = 10,080 psf

■ EXAMPLE 5.18

Problem: The pressure in a pipeline is 6476 psf. What is the head on the pipe?

Solution:

Head on pipe = Feet of pressure

Pressure = Weight × height

6476 psf = 62.4 lb/ft^3 × height

Height = 104 ft (rounded)

5.4.5 FLOW

Flow is expressed in many different terms (English system of measurements). The most common flow terms are:

- Gallons per minute (gpm)
- Cubic feet per second (cfs)
- Gallons per day (gpd)
- Million gallons per day (MGD)

In converting flow rates, the most common flow conversions are 1 cfs = 448 gpm and 1 gpm = 1440 gpd. To convert gallons per day to MGD, divide the gpd by 1,000,000. For example, convert 150,000 gal to MGD:

$$\frac{150,000 \text{ gpd}}{1,000,000} = 0.150 \text{ MGD}$$

In some instances, flow is given in MGD but is needed in gpm. To make the conversion (MGD to gpm) requires two steps:

1. Convert the gpd by multiplying by 1,000,000.
2. Convert to gpm by dividing by the number of minutes in a day (1440 min/day).

■ EXAMPLE 5.19

Problem: Convert 0.135 MGD to gpm.

Solution:

1. Convert the flow in MGD to gpd:

    0.135 MGD × 1,000,000 = 135,000 gpd

2. Convert to gpm by dividing by the number of minutes in a day (24 hr per day × 60 min per hr = 1440 min/day):

    (135,000 gpd) ÷ (1440 min/day) = 93.8 or 94 gpm

To determine flow through a pipeline, channel or stream, we use the following equation:

$$Q = VA \qquad (5.8)$$

where:

Q = cubic feet per second (cfs).
V = velocity in feet per second (ft/sec).
A = area in square feet (ft^2).

■ EXAMPLE 5.20

Problem: Find the flow in cfs in an 8-in. line if the velocity is 3 ft/sec.

Solution:

1. Determine the cross-sectional area of the line in square feet. Start by converting the diameter of the pipe to inches.
2. The diameter is 8 in.; therefore, the radius is 4 in. = 4/12 of a foot, or 0.33 ft.
3. Find the area in square feet:

$$A = \pi r^2 \qquad (5.9)$$

    $A = \pi(0.33 \text{ ft})^2 = \pi × 0.109 \text{ ft}^2 = 0.342 \text{ ft}^2$

4. $Q = VA$:

    $Q = 3 \text{ ft/sec} × 0.342 \text{ ft}^2 = 1.03 \text{ cfs}$

■ **EXAMPLE 5.21**

Problem: Find the flow in gpm when the total flow for the day is 75,000 gpd.

Solution:

$$\frac{75,000 \text{ gpd}}{1440 \text{ min/day}} = 52 \text{ gpm}$$

■ **EXAMPLE 5.22**

Problem: Find the flow in gpm when the flow is 0.45 cfs.

Solution:

$$0.45 \times \frac{\text{cfs}}{1} \times \frac{448 \text{ gpd}}{1 \text{ cfs}} = 202 \text{ gpm}$$

5.4.6 FLOW CALCULATIONS

In water and wastewater treatment, one of the major concerns of the operator is not only to maintain flow but also to measure it. Normally, flow measurements are determined by metering devices. These devices measure water flow at a particular moment (instantaneous flow) or over a specified time (total flow). Instantaneous flow can also be determined mathematically. In this section, we discuss how to mathematically determine instantaneous and average flow rates and how to make flow conversions.

5.4.6.1 Instantaneous Flow Rates

In determining instantaneous flows rates through channels, tanks, and pipelines, we can use $Q = AV$ (Equation 5.8).

Note: It is important to remember that when using an equation such as $Q = AV$ that the units on the left side of the equation must match the units on the right side of the equation (A and V) with respect to volume (cubic feet or gallons) and time (seconds, minutes, hours, or days).

■ **EXAMPLE 5.23**

Problem: A channel 4 ft wide has water flowing to a depth of 2 ft. If the velocity through the channel is 2 feet per second (fps), what is the cubic feet per second (cfs) flow rate through the channel?

Solution:

$$Q = AV = (4 \text{ ft} \times 2 \text{ ft}) \times 2 \text{ fps} = 16 \text{ cfs}$$

5.4.6.2 Instantaneous Flow into and out of a Rectangular Tank

One of the primary flow measurements that water/wastewater operators are commonly required to calculate is flow through a tank. This measurement can be determined using the $Q = AV$ equation. For example, if the discharge valve to a tank were closed, the water level would begin to rise. If we time how fast the water rises, this would give us an indication of the velocity of flow into the tank. This information can be plugged into $Q = VA$ (Equation 5.8) to determine the flow rate through the tank. Let us look at an example.

■ **EXAMPLE 5.24**

Problem: A tank is 8 ft wide and 12 ft long. With the discharge valve closed, the influent to the tank causes the water level to rise 1.5 ft in 1 min. What is the gpm flow into the tank?

Solution:

1. Calculate the cfm flow rate:

$$Q = AV = (8 \text{ ft} \times 12 \text{ ft}) \times 1.5 \text{ fpm} = 144 \text{ cfm}$$

2. Convert cfm flow rate to gpm flow rate:

$$144 \text{ cfm} \times 7.48 \text{ gal/ft}^3 = 1077 \text{ gpm}$$

How do we compute flow rate from a tank when the influent valve is closed and the discharge pump remains on, lowering the wastewater level in the tank? First, we time the rate of this drop in wastewater level so the velocity of flow from the tank can be calculated. Then we use the $Q = AV$ equation to determine the flow rate out of the tank, as illustrated in Example 5.25.

■ **EXAMPLE 5.25**

Problem: A tank is 9 ft wide and 11 ft long. The influent valve to the tank is closed and the water level drops 2.5 ft in 2 min. What is the gpm flow from the tank?

Solution:

1. Drop rate = 2.5 ft/2 min = 1.25 ft/min.
2. Calculate the cfm flow rate:

$$Q = AV = (9 \text{ ft} \times 11 \text{ ft}) \times 1.25 \text{ fpm} = 124 \text{ cfm}$$

3. Convert cfm flow rate to gpm flow rate:

$$124 \text{ cfm} \times 7.48 \text{ gal/ft}^3 = 928 \text{ gpm}$$

5.4.6.3 Flow Rate into a Cylindrical Tank

We can use the same basic method to determine the flow rate when the tank is cylindrical in shape, as shown in Example 5.26.

■ **EXAMPLE 5.26**

Problem: The discharge valve to a cylindrical tank 25 ft in diameter is closed. If the water rises at a rate of 12 in. in 4 min, what is the gpm flow into the tank?

Solution:

1. Rise = 12 in. = 1 ft = 1 ft/4 min = 0.25 fpm.
2. Calculate the cfm flow rate:

$$Q = AV$$
$$= (0.785 \times 25 \text{ ft} \times 25 \text{ ft}) \times 0.25 \text{ fpm}$$
$$= 123 \text{ cfm}$$

3. Convert cfm flow rate to gpm flow rate:

$$123 \text{ cfm} \times 7.48 \text{ gal/ft}^3 = 920 \text{ gpm}$$

5.4.6.4 Flow through a Full Pipeline

Flow through pipelines is of considerable interest to water distribution operators and wastewater collection workers. The flow rate can be calculated using $Q = AV$ (Equation 5.8). The cross-sectional area of a round pipe is a circle, so the area (A) is represented by $0.785 \times (\text{diameter})^2$.

Note: To avoid errors in terms, it is prudent to express pipe diameters as feet.

■ **EXAMPLE 5.27**

Problem: The flow through a pipeline with an 8-in. diameter is moving at a velocity of 4 ft/sec. What is the cfs flow rate through the full pipeline?

Solution:

1. Convert 8 in. to feet:

$$8 \text{ in.}/12 \text{ in.} = 0.67 \text{ ft.}$$

2. Calculate the cfs flow rate:

$$Q = AV$$
$$= (0.785 \times 0.67 \text{ ft} \times 0.67 \text{ ft}) \times 4 \text{ fps}$$
$$= 1.4 \text{ cfs}$$

5.4.6.5 Velocity Calculations

To determine the velocity of flow in a channel or pipeline we use the $Q = AV$ equation; however, to use the equation correctly we must transpose it. We simply write into the equation the information given and then transpose for the unknown (V in this case), as illustrated in Example 5.28 for channels and Example 5.29 for pipelines.

■ **EXAMPLE 5.28**

Problem: A channel has a rectangular cross-section. The channel is 5 ft wide with wastewater flowing to a depth of 2 ft. If the flow rate through the channel is 8500 gpm, what is the velocity of the wastewater in the channel (ft/sec)?

Solution:

1. Convert gpm to cfs:

$$\frac{8500 \text{ gpm}}{7.48 \text{ gal} \times 60 \text{ sec}} = 18.9 \text{ cfs}$$

2. Calculate the velocity:

$$Q = AV$$
$$18.9 \text{ cfs} = (5 \text{ ft} \times 2 \text{ ft}) \times x \text{ (fps)}$$
$$\text{Velocity } (x) \text{ (fps)} = \frac{18.9}{5 \times 2} = 1.89 \text{ fps}$$

■ **EXAMPLE 5.29**

Problem: A full 8-in.-diameter pipe delivers 250 gpm. What is the velocity of flow in the pipeline (fps)?

Solution:

1. Convert 8 in. to feet:

$$8 \text{ in.}/12 \text{ in.} = 0.67 \text{ ft.}$$

2. Convert gpm to cfs flow:

$$\frac{250 \text{ gpm}}{(7.48 \text{ gal/ft}^3) \times (60 \text{ sec/min})} = 0.56 \text{ cfs}$$

3. Calculate the velocity:

$$Q = AV$$
$$0.56 \text{ cfs} = (0.785 \times 0.67 \text{ ft} \times 0.67 \text{ ft}) \times x \text{ (fps)}$$
$$\text{Velocity } (x) \text{ (fps)} = \frac{0.56 \text{ cfs}}{0.785 \times 0.67 \times 0.67} = 1.6 \text{ fps}$$

5.4.6.6 Average Flow Rate Calculations

Flow rates in water/wastewater systems vary considerably during the course of a day, week, month, or year; therefore, when computing flow rates for trend analysis or for other purposes, an *average flow rate* is used to determine the typical flow rate.

■ EXAMPLE 5.30

Problem: The following flows (MGD) were recorded for a week:

Monday	8.2 MGD
Tuesday	8.0 MGD
Wednesday	7.3 MGD
Thursday	7.6 MGD
Friday	8.2 MGD
Saturday	8.9 MGD
Sunday	7.7 MGD

What was the average daily flow rate for the week?

Solution:

$$\text{Average daily flow} = \frac{\text{Total of all sample flows}}{\text{Number of days}} \quad (5.10)$$

$$\text{Average daily flow} = \frac{55.9 \text{ MGD}}{7 \text{ days}} = 8.0 \text{ MGD}$$

5.4.6.7 Flow Conversion Calculations

One task involving calculations that the wastewater operator is typically called on to perform involves converting one expression of flow to another. The ability to do this is also a necessity for those preparing for licensure examinations. Probably the easiest way in which to accomplish flow conversions is to employ the box method illustrated in Figure 5.3. When using the box method, it is important to remember that moving from a smaller box to a larger box requires multiplication by the factor indicated. Moving from a larger box to a smaller box requires division by the factor indicated. From Figure 5.3 it should be obvious that memorizing the 9 boxes and the units in each box is not that difficult. The values of 60, 1440, 7.48, and 8.34 are not that difficult to remember either; it is a matter of remembering the exact placement of the units and the values. Once this is accomplished, we have a powerful tool that allows us to make flow conversions in a relatively easy manner.

5.4.7 DETENTION TIME

Detention time is the length of time that water is retained in a vessel or basin or the period from the time the water enters a settling basin until it flows out the other end. To

Flow Conversions Using the Box Method*

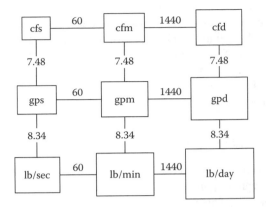

cfs = cubic feet per second
cfm = cubic feet per minute
cfd = cubic feet per day

gps = gallons per second
gpm = gallons per minute
gpd = gallons per day

*The factors shown in the diagram have the following units associated with them:

60 sec/min, 1440 min/day, 7.48 gal/ft³, and 8.34 lb/gal

FIGURE 5.3 Flow conversion diagram. (From Price, J.K., *Applied Math for Wastewater Plant Operators*, CRC Press, Boca Raton, FL, 1991, p. 32. With permission.)

calculate the detention period of a basin, the volume of the basin must first be obtained. Using a basin 70 ft long by 25 ft wide by 12 ft deep, the volume would be:

$$V = L \times W \times D = 70 \text{ ft} \times 25 \text{ ft} \times 12 \text{ ft} = 21,000 \text{ ft}^3$$

$$\text{Gallons} = V \times 7.48 \text{ gal/ft}^3 = 21,000 \times 7.48 = 157,080$$

If we assume that the plant filters 300 gpm, then we have 157,080/300 = 524 min (rounded) or roughly 9 hr of detention time. Stated another way, the detention time is the length of time theoretically required for the coagulated water to flow through the basin. If chlorine were added to the water as it entered the basin, the chlorine contact time would be 9 hr. To determine the contact time (CT) used to determine the effectiveness of chlorine, we must calculate detention time.

Key Point: True detention time is the "T" portion of the CT value.

Note: Detention time is also important when evaluating the sedimentation and flocculation basins of a water treatment plant.

Detention time is expressed in units of time. The most common are seconds, minutes, hours, and days. The simplest way to calculate detention time is to divide the volume of the container by the flow rate into the container. The theoretical detention time of a container is the same as the amount of time it would take to fill the container if it were empty. For volume, the most common units used are gallons; however, on occasion cubic feet may also be used. Time units will be whatever units are used to express the flow; for example, if the flow is in gpm, then the detention time will be in days. If, in the result, the detention time is in the wrong time units, simply convert to the appropriate units.

■ **EXAMPLE 5.31**

Problem: The reservoir for a community is 110,000 gal. The well will produce 60 gpm. What is the detention time in the reservoir in hours?

Solution:

$$\text{Detention time} = \frac{110,000 \text{ gal}}{60 \text{ gpm}} = 1834 \text{ min}$$

$$= \frac{1834 \text{ min}}{60 \text{ min/hr}} = 30.6 \text{ hr}$$

■ **EXAMPLE 5.32**

Problem: Find the detention time in a 55,000-gal reservoir if the flow rate is 75 gpm.

Solution:

$$\text{Detention time} = \frac{55,000 \text{ gal}}{75 \text{ gpm}} = 734 \text{ min}$$

$$= \frac{734 \text{ min}}{60 \text{ min/hr}} = 12 \text{ hr}$$

■ **EXAMPLE 5.33**

Problem: If the fuel consumption to the boiler is 30 gpd, how many days will the 1000-gal tank last?

Solution:

$$\text{Days} = 1000 \text{ gal} \div 30 \text{ gpd} = 33.3$$

5.4.8 HYDRAULIC DETENTION TIME

The term *hydraulic detention time* (HDT) refers to the average length of time (theoretical time) a drop of water, wastewater, or suspended particles remains in a tank or channel. It is calculated by dividing the water/wastewater in the tank by the flow rate through the tank. The units of flow rate used in the calculation are dependent on whether the detention time is to be calculated in seconds, minutes, hours, or days. Hydraulic detention time is used in con-

junction with various treatment processes, including sedimentation and coagulation–flocculation. Generally, in practice, the hydraulic detention time is associated with the amount of time required for a tank to empty. The range of detention time varies with the process; for example, in a tank used for sedimentation, the detention time is commonly measured in minutes. The calculation methods used to determine hydraulic detention time are illustrated in the following sections.

5.4.8.1 Hydraulic Detention Time in Days

Note: The general hydraulic detention time calculation is:

$$\text{Hydraulic detention time (HDT)} = \frac{\text{Tank volume}}{\text{Flow rate}} \quad (5.11)$$

This general formula is then modified based on the information provided or available and the "normal" range of detention times for the unit being evaluated:

$$\text{HDT (days)} = \frac{\text{Tank volume (ft}^3) \times 7.48 \text{ gal/ft}^3}{\text{Flow (gal/day)}} \quad (5.12)$$

■ **EXAMPLE 5.34**

Problem: An anaerobic digester has a volume of 2,200,000 gal. What is the detention time in days when the influent flow rate is 0.06 MGD?

Solution:

$$\text{HDT (days)} = \frac{2,200,000 \text{ gal}}{0.06 \text{ MGD} \times 1,000,000 \text{ gal/MG}} = 37 \text{ days}$$

5.4.8.2 Hydraulic Detention Time in Hours

$$\text{HDT (hr)} = \left[\frac{\begin{array}{c} \text{Tank volume (ft}^3) \\ \times 7.48 \text{ gal/ft}^3 \times 24 \text{ hr/day} \end{array}}{\text{Flow (gal/day)}} \right] \quad (5.13)$$

■ **EXAMPLE 5.35**

Problem: A settling tank has a volume of 40,000 ft³. What is the detention time in hours when the flow is 4.35 MGD?

Solution:

$$\text{HDT (hr)} = \frac{40,000 \text{ ft}^3 \times 7.48 \text{ gal/ft}^3 \times 24 \text{ hr/day}}{4.35 \text{ MGD} \times 1,000,000 \text{ gal/MG}}$$

5.4.8.3 Hydraulic Detention Time in Minutes

$$\text{HDT (min)} = \left[\frac{\text{Tank volume (ft}^3) \times 7.48 \text{ gal/ft}^3 \times 1440 \text{ min/day}}{\text{Flow (gal/day)}} \right] \quad (5.14)$$

■ **EXAMPLE 5.36**

Problem: A grit channel has a volume of 1240 ft³. What is the detention time in minutes when the flow rate is 4.1 MGD?

Solution:

$$\text{HDT (min)} = \left[\frac{1240 \text{ ft}^3 \times 7.48 \text{ gal/ft}^3 \times 1440 \text{ min/day}}{4,100,000 \text{ gal/day}} \right] = 3.26 \text{ min}$$

Note: The tank volume and the flow rate must be in the same dimensions before calculating the hydraulic detention time.

5.4.9 CHEMICAL DOSAGE CALCULATIONS

Chemicals are used extensively in wastewater treatment plant operations. Wastewater treatment plant operators add chemicals to various unit processes for slime-growth control, corrosion control, odor control, grease removal, BOD reduction, pH control, sludge-bulking control, ammonia oxidation, and bacterial reduction, among other reasons. To apply any chemical dose correctly it is important to be able to make certain dosage calculations. One of the most frequently used calculations in wastewater mathematics is the conversion of milligrams per liter (mg/L) concentration to pounds per day (lb/day) or pounds (lb) dosage or loading. The general types of mg/L to lb/day or lb calculations are for chemical dosage, BOD, chemical oxygen demand (COD), suspended solids (SS) loading/removal, pounds of solids under aeration, and waste activated sludge (WAS) pumping rate. These calculations are usually made using either of the following equations:

$$\text{mg/L} \times \text{flow (MGD)} \times 8.34 \text{ lb/gal} = \text{lb/day} \quad (5.15)$$

$$\text{mg/L} \times \text{volume (MG)} \times 8.34 \text{ lb/gal} = \text{lb} \quad (5.16)$$

Note: If mg/L concentration represents a concentration in a flow, then million gallons per day (MGD) flow is used as the second factor; however, if the concentration pertains to a tank or pipeline volume, then million gallons (MG) volume is used as the second factor.

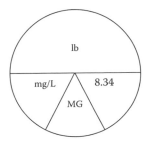

FIGURE 5.4 Dosage formula pie chart.

5.4.9.1 Dosage Formula Pie Chart

In converting pounds (lb) or mg/L, million gallons (MG) and 8.34 are key parameters. The pie chart shown in Figure 5.4 and the steps listed below can be helpful in finding lb or mg/L.

1. Determine what unit the question is asking you to find (lb or mg/L).
2. Physically cover or hide the area of the chart containing the desired unit. Write the desired unit down alone on one side of the equals sign to begin the necessary equation (e.g., lb =).
3. Look at the remaining uncovered areas of the circle. These *exactly* represent the other side of the equals sign in the necessary equation. If the unit above the center line is not covered, your equation will have a top (or numerator) and a bottom (or denominator), just like the pie chart. Everything above the center line goes in the numerator (the top of the equation) and everything below the center lines goes in the denominator (the bottom of the equation). Remember that all units below the line are *always multiplied together*; for example, if you are asked to find the dosage in mg/L, you would cover mg/L in the pie chart and write it down on one side of the equals sign to start your equation, like this:

$$\text{mg/L} =$$

The remaining portions of the pie chart are lb on top divided by MGD × 8.34 on the bottom and would be written like this:

$$\text{mg/L} = \frac{\text{lb}}{\text{MGD} \times 8.34 \text{ lb/gal}}$$

If the area above the center line is covered, the right side of your equation will be made up of only the units below the center line. Remember that all units below the line are always multiplied together.

If you are asked, for example, to find the number of pounds needed, you would cover lb in the pie chart and write it down on one side of the equals sign to start your equation, like this:

$$lb =$$

All of the remaining areas of the pie chart are together on one line (below the center line of the circle), multiplied together on the other side of the equals sign and written like this:

$$lb/day = mg/L \times MGD \times 8.34 \ lb/gal$$

5.4.9.2 Chlorine Dosage

Chlorine is a powerful oxidizer commonly used in water treatment for purification and in wastewater treatment for disinfection, odor control, bulking control, and other applications. When chlorine is added to a unit process, we want to ensure that a measured amount is added. The amount of chemical added or required can be specified in two ways:

- Milligrams per liter (mg/L)
- Pounds per day (lb/day)

To convert from mg/L (or ppm) concentration to lb/day, we use Equation 5.17:

$$mg/L \times MGD \times 8.34 \ lb/gal = lb/day \qquad (5.17)$$

Note: In previous years, it was normal practice to use the expression *parts per million* (ppm) as an expression of concentration, as 1 mg/L = 1 ppm; however, current practice is to use mg/L as the preferred expression of concentration.

■ **EXAMPLE 5.37**

Problem: Determine the chlorinator setting (lb/day) required to treat a flow of 8 MGD with a chlorine dose of 6 mg/L.

Solution:

$$mg/L \times MGD \times 8.34 \ lb/gal = lb/day$$

$$6 \ mg/L \times 8 \ MGD \times 8.34 \ lb/gal = 400 \ lb/day$$

■ **EXAMPLE 5.38**

Problem: What should the chlorinator setting be (lb/day) to treat a flow of 3 MGD if the chlorine demand is 12 mg/L and a chlorine residual of 2 mg/L is desired?

Note: The chlorine demand is the amount of chlorine used in reacting with various components of the wastewater such as harmful organisms and other organic

and inorganic substances. When the chlorine demand has been satisfied, these reactions stop.

$$mg/L \times MGD \times 8.34 \ lb/gal = lb/day$$

To find the unknown value of lb/day, we must first determine chlorine dose. To do this we must use Equation 5.18:

$$Chlorine \ dose \ (mg/L) = \begin{bmatrix} Chlorine \ demand \ (mg/L) \\ + \ Chlorine \ residual \ (mg/L) \end{bmatrix} (5.18)$$

$$Chlorine \ dose \ (mg/L) = 12 \ mg/L + 2 \ mg/L = 14 \ mg/L$$

Then we can make the mg/L to lb/day calculation:

$$12 \ mg/L \times 3 \ MGD \times 8.34 \ lb/gal = 300 \ lb/day$$

5.4.9.3 Hypochlorite Dosage

At many wastewater facilities, sodium hypochlorite or calcium hypochlorite is used instead of chlorine. The reasons for substituting hypochlorite for chlorine vary; however, due to the passage of stricter hazardous chemicals regulations by the Occupational Safety and Health Administration (OSHA) and the U.S. Environmental Protection Agency (USEPA), many facilities are deciding to substitute the hazardous chemical chlorine with nonhazardous hypochlorite. Obviously, the potential liability involved with using deadly chlorine is also a factor involved in the decision to substitute it with a less toxic chemical substance.

For whatever reason, when a wastewater treatment plant decides to substitute hypochlorite for chlorine, the wastewater operator needs to be aware of the differences between the two chemicals. Chlorine is a hazardous material. Chlorine gas is used in wastewater treatment applications as 100% available chlorine. This is an important consideration to keep in mind when making or setting chlorine feed rates. For example, if the chlorine demand and residual requires 100 lb/day chlorine, the chlorinator setting would be just that—100 lb/24 hr. Hypochlorite is less hazardous than chlorine; it is similar to strong bleach and comes in two forms: dry calcium hypochlorite (often referred to as HTH) and liquid sodium hypochlorite. Calcium hypochlorite contains about 65% available chlorine; sodium hypochlorite contains about 12 to 15% available chlorine (in industrial strengths).

Note: Because either type of hypochlorite is not 100% pure chlorine, more lb/day must be fed into the system to obtain the same amount of chlorine for disinfection. This is an important economical consideration for facilities thinking about substituting hypochlorite for chlorine. Some studies indicate that such a switch can increase overall operating costs by up to 3 times the cost of using chlorine.

To calculate the lb/day hypochlorite required, a two-step calculation is necessary:

$$mg/L \times MGD \times 8.34 \text{ lb/gal} = \text{lb/day} \quad (5.19)$$

$$\frac{\text{Chlorine (lb/day)}}{\% \text{ Available}} \times 100 = \text{Hypochlorite (lb/day)} \quad (5.20)$$

■ EXAMPLE 5.39

Problem: A total chlorine dosage of 10 mg/L is required to treat a particular wastewater. If the flow is 1.4 MGD and the hypochlorite has 65% available chlorine, how many lb/day of hypochlorite will be required?

Solution:

1. Calculate the lb/day chlorine required using the mg/L to lb/day equation:

$$mg/L \times MGD \times 8.34 \text{ lb/gal} = \text{lb/day}$$

$$10 \text{ mg/L} \times 1.4 \text{ MGD} \times 8.34 \text{ lb/gal} = 117 \text{ lb/day}$$

2. Calculate the lb/day hypochlorite required. Because only 65% of the hypochlorite is chlorine, more than 117 lb/day will be required:

$$\frac{117 \text{ lb/day chlorine}}{65 \text{ available chlorine}} \times 100 = 180 \text{ lb/day hypochlorite}$$

■ EXAMPLE 5.40

Problem: A wastewater flow of 840,000 gpd requires a chlorine dose of 20 mg/L. If sodium hypochlorite (15% available chlorine) is to be used, how many lb/day of sodium hypochlorite are required? How many gallons per day of sodium hypochlorite is this?

Solution:

1. Calculate the lb/day chlorine required:

$$mg/L \times MGD \times 8.34 \text{ lb/gal} = \text{lb/day}$$

$$20 \text{ mg/L} \times 0.84 \text{ MGD} \times 8.34 \text{ lb/gal} = 140 \text{ lb/day chlorine}$$

2. Calculate the lb/day sodium hypochlorite:

$$\frac{140 \text{ lb/day chlorine}}{15 \text{ available chlorine}} \times 100 = 933 \text{ lb/day hypochlorite}$$

3. Calculate gallons per day sodium hypochlorite:

$$\frac{933 \text{ lb/day}}{8.34 \text{ lb/gal}} = 112 \text{ gal/day hypochlorite}$$

■ EXAMPLE 5.41

Problem: How many pounds of chlorine gas are necessary to treat 5,000,000 gallons of wastewater at a dosage of 2 mg/L?

Solution:

1. Calculate the pounds of chlorine required:

$$(V \times 10^6 \text{ gal}) \times \text{chlorine conc. (mg/L)} \times 8.34 \text{ lb/gal}$$
$$= \text{lb chlorine}$$

2. Substitute:

$$(5 \times 10^6 \text{ gal}) \times 2 \text{ mg/L} \times 8.34 \text{ lb/gal} = 83 \text{ lb chlorine}$$

5.4.10 PERCENT REMOVAL

Percent removal is used throughout the wastewater treatment process to express or evaluate the performance of the plant and individual treatment unit processes. The results can be used to determine if the plant is performing as expected or in troubleshooting unit operations by comparing the results with those listed in the plant's operations and maintenance (O&M) manual. It can be used with either concentration or quantities:

For concentrations use:

$$\% \text{ Removal} = \frac{(\text{Influent conc.} - \text{effluent conc.}) \times 100}{\text{Influent conc.}} \quad (5.21)$$

For quantities use:

$$\% \text{ Removal} = \frac{(\text{Influent quant.} - \text{effluent quant.}) \times 100}{\text{Influent quant.}} \quad (5.22)$$

Note: The calculation used for determining the performance (percent removal) for a digester is different from that used for performance (percent removal) for other processes such as some process residuals or biosolids treatment processes. Be sure the correct formula is selected.

■ EXAMPLE 5.42

Problem: The plant influent contains 259 mg/L BOD$_5$ and the plant effluent contains 17 mg/L BOD$_5$. What is the % BOD$_5$ removal?

Solution:

$$\% \text{ Removal} = \frac{(259 \text{ mg/L} - 17 \text{ mg/L}) \times 100}{259 \text{ mg/L}} = 93.4\%$$

5.4.11 POPULATION EQUIVALENT OR UNIT LOADING FACTOR

When it is impossible to conduct a wastewater characterization study and other data are unavailable, population equivalent or unit per capita loading factors are used to estimate the total waste loadings to be treated. If the BOD contribution of a discharger is known, the loading placed on the wastewater treatment system in terms of equivalent number of people can be determined. The BOD contribution of a person is normally assumed to be 0.17 lb BOD/day:

$$\text{P.E. (people)} = \frac{\text{BOD}_5 \text{ contribution (lb/day)}}{0.17 \text{ lb BOD}_5 \text{ per person per day}} \quad (5.23)$$

■ EXAMPLE 5.43

Problem: A new industry wishes to connect to the city's collection system. The industrial discharge will contain an average BOD concentration of 349 mg/L and the average daily flow will be 50,000 gpd. What is the population equivalent of the industrial discharge?

Solution:

1. Convert flow rate to million gallons per day:

$$\text{Flow} = \frac{50,000 \text{ gpd}}{1,000,000 \text{ gal/MG}} = 0.050 \text{ MGD}$$

2. Calculate the population equivalent:

$$\text{P.E.} = \frac{349 \text{ mg/L} \times 0.050 \text{ MGD} \times 8.34 \text{ lb/mg/L/MG}}{0.17 \text{ lb BOD}_5 \text{ per person per day}}$$

$$= 856 \text{ people per day}$$

5.4.12 SPECIFIC GRAVITY

Specific gravity is the ratio of the density of a substance to that of a standard material under standard conditions of temperature and pressure. The standard material for gases is air, and for liquids and solids it is water. Specific gravity can be used to calculate the weight of a gallon of liquid chemical:

$$\text{Weight (lb/gal)} = \text{Water (lb/gal)} \times \text{specific gravity} \quad (5.24)$$

■ EXAMPLE 5.44

Problem: The label of the chemical states that the chemical has a specific gravity of 1.4515. What is the weight of 1 gal of solution?

Solution:

$$\text{Weight (lb/gal)} = 8.34 \text{ lb/gal} \times 1.4515 = 12.1 \text{ lb/gal}$$

5.4.13 PERCENT VOLATILE MATTER REDUCTION IN SLUDGE

The calculation used to determine *percent volatile matter reduction* is complicated because of the changes occurring during sludge digestion:

$$\% \text{ VM reduction} = \frac{(\% \text{ VM}_{in} - \% \text{ VM}_{out}) \times 100}{\% \text{ VM}_{in} - (\% \text{ VM}_{in} \times \% \text{ VM}_{out})} \quad (5.25)$$

where VM = volatile matter.

■ EXAMPLE 5.45

Problem: Determine the percent volatile matter reduction for the digester using the digester data provided below:

Raw sludge volatile matter	72%
Digested sludge volatile matter	51%

Solution:

$$\% \text{ VM reduction} = \frac{(0.72 - 0.51) \times 100}{0.72 - (0.72 \times 0.51)} = 59\%$$

5.4.14 CHEMICAL COAGULATION AND SEDIMENTATION

Chemical *coagulation* consists of treating the water with certain chemicals to bring nonsettleable particles together into larger heavier masses of solid material (called *floc*), which are then relatively easy to remove.

5.4.14.1 Calculating Feed Rate

The following equation is used to calculate the feed rate of chemicals used in coagulation:

$$\text{Feed rate (lb/day)} = \left[\begin{array}{l} \text{Dose (mg/L)} \times \text{flow (MGD)} \\ \times 8.34 \text{ lb/gal} \end{array} \right] \quad (5.26)$$

■ EXAMPLE 5.46

Problem: A water treatment plant operates at a rate of 5 MGD. The dosage of alum is 40 ppm (or mg/L). How many pounds of alum are used a day?

Solution:

$$\text{Feed rate (lb/day)} = \left[\begin{array}{l} \text{Dose (mg/L)} \times \text{flow (MGD)} \\ \times 8.34 \text{ lb/gal} \end{array} \right]$$

$$= 40 \text{ mg/L} \times 5 \text{ MGD} \times 8.34 \text{ lb/gal}$$

$$= 1668 \text{ lb/day of alum}$$

5.4.14.2 Calculating Solution Strength

Use the following procedure to calculate solution strength.

■ EXAMPLE 5.47

Problem: Eight pounds of alum are added to 115 lb of water. What is the solution strength?

Solution:

$$70\% = \frac{8}{8+115} \times 100 = 6.5\% \text{ solution}$$

We use this same concept in determining other solution strengths.

■ EXAMPLE 5.48

Problem: Twenty-five pounds of alum are added to 90 lb of water. What is the solution strength?

Solution:

$$\frac{25}{25+90} \times 100 = 22\% \text{ solution}$$

In the previous examples, we added pounds of chemicals to pounds of water. Recall that 1 gal of water = 8.34 lb. By multiplying the number of gallons by the 8.34 factor, we can find pounds.

■ EXAMPLE 5.49

Problem: 40 pounds of soda ash are added to 65 gal of water. What is the solution strength?

Solution: Units must be consistent, so convert gallons of water to pounds of water:

$$65 \text{ gal} \times 8.34 \text{ lb/gal} = 542.7 \text{ lb water}$$

$$\frac{40 \text{ lb}}{542.7 \text{ lb} + 40 \text{ lb}} \times 100 = 6.9\% \text{ solution}$$

5.4.15 FILTRATION

In waterworks operation (and to an increasing degree in wastewater treatment), the rate of flow through filters is an important operational parameter. Although flow rate can be controlled by various means or may proceed at a variable declining rate, the important point is that with flow suspended matter continuously builds up within the filter bed, affecting the rate of filtration.

5.4.15.1 Calculating the Rate of Filtration

■ EXAMPLE 5.50

Problem: A filter box (including sand area) is 20 ft × 30 ft. If the influent valve is shut, the water drops 3.0 in./min. What is the rate of filtration in MGD?

Solution:

Given:

Filter box = 20 ft × 30 ft
Water drops = 3 in./min

Find the volume of water passing through the filter:

$$\text{Volume} = \text{Area} \times \text{height}$$
$$\text{Area} = \text{Width} \times \text{length}$$

Note: The best way to perform calculations of this type is to systematically break down the problem into what is given and what is to be found.

1. Determine area:

$$\text{Area} = 20 \text{ ft} \times 30 \text{ ft} = 600 \text{ ft}^2$$

Convert 3 in. to feet:

$$(3 \text{ in.}) \div (12 \text{ in./ft}) = 0.25 \text{ ft}$$

Determine volume of water passing through the filter in 1 min:

$$\text{Volume} = 600 \text{ ft}^2 \times 0.25 \text{ ft} = 150 \text{ ft}^3$$

2. Convert cubic feet to gallons:

$$150 \text{ ft}^3 \times 7.48 \text{ gal/ft}^3 = 1122 \text{ gpm}$$

3. The problem asks for the rate of filtration in MGD. To find MGD, multiply the number of gallons per minute by the number of minutes per day:

$$1122 \text{ gpm} \times 1440 \text{ min/day} = 1.62 \text{ MGD}$$

5.4.15.2 Filter Backwash

In filter backwashing, one of the most important operational parameters to be determined is the amount of water in gallons required for each backwash. This amount depends on the design of the filter and the quality of the water being filtered. The actual washing typically lasts 5 to 10 min and uses amounts to 1 to 5% of the flow produced.

■ **EXAMPLE 5.51**

Problem: A filter has the following dimensions:

Length = 30 ft
Width = 20 ft
Depth of filter media = 24 in.

Assuming that a backwash rate of 15 gal/ft²/min is recommended and 10 min of backwash is required, calculate the amount of water in gallons required for each backwash.

Solution:

Given:

Length = 30 ft
Width = 20 ft
Depth of filter media = 24 in.
Rate = 15 gal/ft²/min
Backwash time = 10 min

Find the amount of water in gallons required:

1. Area of filter = 30 ft × 20 ft = 600 ft²
2. Gallons of water used per square foot of filter = 15 gal/ft²/min × 10 min = 15 gal/ft²
3. Gallons required for backwash = 150 gal/ft² × 600 ft² = 90,000 gal

5.4.16 WATER DISTRIBUTION SYSTEM CALCULATIONS

After water is adequately treated, it must be conveyed or distributed to the customer for domestic, commercial, industrial, and fire-fighting applications. Water distribution systems should be capable of meeting the demands placed on them at all times and at satisfactory pressures. Waterworks operators responsible for water distribution must be able to perform basic calculations for both practical and licensure purposes; such calculations deal with water velocity, rate of water flow, water storage tanks, and water disinfection.

5.4.16.1 Water Flow Velocity

The velocity of a particle (any particle) is the speed at which it is moving. Velocity is expressed by indicating the length of travel and how long it takes to cover the distance. Velocity can be expressed in almost any distance and time units.

$$\text{Velocity} = \frac{\text{Distance traveled}}{\text{Time}} \qquad (5.27)$$

Note that water flow that enters the pipe (any pipe) is the same flow that exits the pipe (under steady flow conditions). Water flow is continuous. Water is incompressible; it cannot accumulate inside. The flow at any given point

is the same flow at any other given point in the pipeline; therefore, a given flow volume may not change (it shouldn't), but the velocity of the water may change. At any given flow, velocity is dependent upon the cross-sectional area of the pipe or conduit. Velocity (the speed at which the flow is traveling) is an important parameter. Recall that when dealing with velocity of flow, again the most basic hydraulic equation is:

$$Q = AV$$

where:
  Q = flow.
  A = area; cross-sectional area of conduit = 0.785 × (diameter)².
  V = velocity.

■ **EXAMPLE 5.52**

Problem: A flow of 2 MGD occurs in a 10-in.-diameter conduit. What is the water velocity?

Solution: Change MGD to cfs and inches to feet, then solve for velocity using Equation 5.8:

$$Q = AV$$
$$2 \text{ MGD} \times 1.55 = (0.785 \times 0.832) \times V$$
$$3.1 = (0.785 \times 0.69) \times V$$
$$5.7 \text{ ft/sec} = V$$

■ **EXAMPLE 5.53**

Problem: A 24-in.-diameter pipe carries water at a velocity of 140 ft/min. What is the flow rate (gpm)?

Solution: Change ft/min to ft/sec and inches to feet, then solve for flow:

$$Q = AV$$
$$Q \text{ (cfs)} = (0.785 \times 2^2) \times 2.3$$
$$Q \text{ (cfs)} = 7.2 \text{ cfs}$$
$$Q \text{ (gpm)} = 7.2 \text{ cfs} \times 7.48 \text{ ft}^3 \times 60 \text{ min} = 3231$$

■ **EXAMPLE 5.54**

Problem: If water travels 700 ft in 5 min, what is the velocity?

Solution:

$$\text{Velocity} = \frac{\text{Distance traveled}}{\text{Time}}$$
$$= \frac{700 \text{ ft}}{5 \text{ min}} = 140 \text{ ft/min}$$

■ EXAMPLE 5.55

Problem: Flow in a 6-in. pipe is 400 gpm. What is the average velocity?

Solution:

1. Determine area (A) by first converting 6 in. to ft:

$$6 \text{ in.}/12 \text{ in./ft} = 0.5 \text{ ft}$$

Then,

$$\text{Area} = 0.785 \times (\text{diameter})^2$$
$$\text{Area} = 0.785 \times (0.5)^2$$
$$\text{Area} = 0.785 \times 0.25 = 0.196 \text{ ft}^2 \text{ (rounded)}$$

2. Determine flow (Q):

$$\text{Flow (cfs)} = \frac{\text{Flow (gpm)}}{7.48 \text{ gal} \times 60 \text{ sec/min}}$$

$$\text{Flow (cfs)} = \frac{400 \text{ gpm}}{7.48 \text{ gal} \times 60 \text{ sec/min})}$$

$$\text{Flow (cfs)} = \frac{400 \text{ ft}^3}{448.3 \text{ sec}} = 0.89 \text{ cfs}$$

3. Determine velocity (V):

$$\text{Velocity (ft/sec)} = \frac{\text{Flow (ft}^3\text{/sec)}}{\text{Area (ft}^2\text{)}}$$

$$\text{Velocity (ft/sec)} = \frac{0.89 \text{ ft}^3\text{/sec}}{0.196 \text{ ft}^2}$$

$$= 4.5 \text{ ft/sec}$$

■ EXAMPLE 5.56

Problem: Flow in a 2.0-ft-wide rectangular channel is 1.2 ft deep and measures 11.0 cfs. What is the average velocity?

Solution:

1. Transpose $Q = VA$ to $V = Q/A$.

   Given:

   Rate of flow (Q) = 11.0 cfs
   Width = 2.0 ft
   Depth = 1.2 ft

   Find average velocity.

2. Determine area:

$$\text{Area} = \text{Width} \times \text{depth}$$
$$\text{Area } (A) = 2.0 \text{ ft} \times 1.2 \text{ ft} = 2.4 \text{ ft}^2$$

3. Determine velocity:

$$\text{Velocity (ft/sec)} = \frac{\text{Flow (ft}^3\text{/sec)}}{\text{area (ft}^2\text{)}}$$

$$\text{Velocity } (V) = \frac{11.0 \text{ ft}^3\text{/sec}}{2.4 \text{ ft}^2}$$

$$= 4.6 \text{ ft/sec (rounded)}$$

5.4.16.2 Storage Tank Calculations

Water is stored at a waterworks to provide allowance for differences in water production rates and high-lift pump discharge to the distribution system. Water within the distribution system may be stored in elevated tanks, standpipes, covered reservoirs, or underground basins. The waterworks operator should be familiar with the basic storage tank calculation illustrated in the following example.

■ EXAMPLE 5.57

Problem: A cylindrical tank is 120 ft high by 25 ft in diameter. How many gallons of water will it contain?

Solution:

   Given:

   Height = 120 ft
   Diameter = 25 ft
   Cylindrical shape

   Find total gallons of water contained in the tank:

1. Find the volume in cubic feet:

$$\text{Volume} = 0.785 \times (\text{diameter})^2 \times \text{height}$$
$$= 0.785 \times (25 \text{ ft})^2 \times 120 \text{ ft}$$
$$= 0.785 \times 625 \text{ ft}^2 \times 120 \text{ ft} = 58,875 \text{ ft}^3$$

2. Find the number of gallons of water the cylindrical tank will contain:

$$\text{No. of gallons} = 58,875 \text{ ft}^3 \times 7.48 \text{ gal/ft}^3 = 440,385$$

5.4.16.3 Distribution System Disinfection Calculations

Delivering a clean, pathogen-free product to the customer is what water treatment operation is all about. Before being placed in service, all facilities and appurtenances associated with the treatment and distribution of water must be disinfected, because water may become tainted anywhere in the system. In the examples that follow, we demonstrate how to perform the necessary calculations for this procedure.

■ **EXAMPLE 5.58**

Problem: A waterworks has a tank containing water that needs to be disinfected using HTH 70% available chlorine. The tank is 100 ft high by 25 ft in diameter. The dose to use is 50 ppm. How many pounds of HTH are needed?

Solution:

Given:

> Height = 100 ft
> Diameter = 25 ft
> Chlorine dose = 50 ppm
> Available chlorine = 70%

Find pounds of HTH:

1. Find the volume of the tank:

> Radius (r) = Diameter/2 = 25/2 = 12.5 ft

> Volume = (3.14) × (r^2) × (height)

> Volume = 3.14 × (12.5)^2 × 100

> Volume = 3.14 × 156.25 × 100 = 49,062.5 ft^3

2. Convert cubic feet to million gallons (MG):

$$49{,}062.5 \text{ ft} \times \frac{7.48 \text{ gal}}{\text{ft}^3} = \frac{\text{MG}}{1{,}000{,}000 \text{ gal}} = 0.367 \text{ MG}$$

3. Determine available chlorine:

> Chemical weight (lb) = Chemical dose (mg/L)
> × water volume (MG)
> × 8.34 lb/gal

> Chlorine = 50 mg/L × 0.367 MG
> × 8.34 lb/gal
> = 153 lb

Note: The fundamental concept to keep in mind when computing hypochlorite calculations is that once we determine how many pounds of chlorine will be required for disinfection we will always need more pounds hypochlorite as compared to elemental chlorine.

4. Determine hypochlorite (HTH) required:

$$\text{HTH required} = \frac{\text{Available chlorine}}{\text{Chlorine fraction}}$$

$$\text{HTH required} = \frac{153 \text{ lb}}{0.7} = 218.6 \text{ lb (rounded)}$$

■ **EXAMPLE 5.59**

Problem: When treating 4000 ft of 8-in. water line by applying enough chlorine for an 80-ppm dosage, how many pounds of hypochlorite of 70% available chlorine are required?

Solution:

Given:

> Length = 4000 ft
> Available chlorine = 70%
> Diameter = 8 in.
> Chlorine dose = 80 ppm

Find pounds of hypochlorite required.

1. Find the volume of the pipe by first changing 8 in. to feet:

$$\text{Diameter (ft)} = \frac{8 \text{ in.}}{12 \text{ in./ft}} = 0.66 \text{ ft} = 0.70 \text{ ft (rounded)}$$

Then, determine radius (r):

> r = Diameter/2 = .70 ft/2 = .35 ft

Then, determine volume:

Volume = 3.14 × r^2 × height

Volume = 3.14 × (0.35 ft)^2 × 4000 ft

Volume = 3.14 × 0.1225 ft × 4000 ft = 1538.6 ft^3

2. Convert cubic feet (ft^3) to million gallons (MG):

$$1538.6 \text{ ft} \times \frac{7.48 \text{ gal/ft}^3}{\text{ft}^3} \times \frac{\text{MG}}{1{,}000{,}000 \text{ gal}} = 0.0115 \text{ MG}$$

3. Determine available chlorine:

> Chlorine = 80 mg/L × 0.0115 MG × 8.34 lb/gal
> = 7.67 lb

4. Determine hypochlorite (HTH) required:

> HTH required = 7.67 lb ÷ 0.70 = 11 lb (rounded)

5.4.17 Complex Conversions

Water and wastewater operators use complex conversions to convert, for example, laboratory test results to other units of measure which can be used to adjust or control the treatment process. Conversions such as these require the use of several measurements (e.g., concentration, flow rate, tank volume) and an appropriate conversion factor. The most widely used of these conversions are discussed in the following sections.

5.4.17.1 Concentration to Quantity

Concentration (milligrams/liter) to pounds

$$\text{Pounds} = \text{Concentration (mg/L)} \times \text{tank volume (MG)} \times 8.34 \text{ lb/MG/mg/L} \quad (5.28)$$

■ EXAMPLE 5.60

Problem: Given mixed liquor suspended solids (MLSS) = 2580 mg/L and aeration tank volume = 0.90 MG, what is the concentration in pounds?

Solution:

$$\text{Pounds} = 2580 \text{ mg/L} \times 0.90 \text{ MG} \times 8.34 \text{ lb/MG/mg/L}$$

$$= 19,366 \text{ lb}$$

Concentration (milligrams/liter) to pounds/day

$$\text{Pounds/day} = \text{Concentration (mg/L)} \times \text{flow (MGD)} \times 8.34 \text{ lb/MG/mg/L} \quad (5.29)$$

■ EXAMPLE 5.61

Problem: Given effluent BOD_5 = 23 mg/L and effluent flow = 4.85 MGD, what is the concentration in pounds per day?

Solution:

$$\text{Pounds/day} = 23 \text{ mg/L} \times 4.85 \text{ MGD} \times 8.34 \text{ lb/MG/mg/L}$$

$$= 930 \text{ lb/day}$$

Concentration (milligrams/liter) to kilograms/day

$$\text{Kilograms/day} = \begin{bmatrix} \text{Concentration (mg/L)} \\ \times \text{flow (MGD)} \\ \times 3.785 \text{ lb/MG/mg/L} \end{bmatrix} \quad (5.30)$$

■ EXAMPLE 5.62

Problem: Given effluent total suspended solids (TSS) = 29 mg/L and effluent flow = 11.5 MGD, what is the concentration in kilograms per day?

Solution:

$$\text{kg/day} = \begin{vmatrix} 29 \text{ mg/L} \times 11.5 \text{ MGD} \\ \times 3.785 \text{ lb/MG/mg/L} \end{vmatrix} = 1263 \text{ kg/day}$$

Concentration (milligrams/kilogram) to pounds/ton

$$\text{Pounds/ton} = \begin{bmatrix} \text{Concentration (mg/kg)} \\ \times 0.002 \text{ lb/ton/mg/kg} \end{bmatrix} \quad (5.31)$$

■ EXAMPLE 5.63

Problem: Given that the biosolids contain 0.97 mg/kg of lead, how many pounds of lead are being applied per acre if the current application rate is 5 dry tons of solids per acre?

Solution:

$$\text{Pounds/acre} = \begin{bmatrix} 0.97 \text{ mg/kg} \times 5 \text{ tons/ac} \\ \times 0.002 \text{ lb/ton/mg/kg} \end{bmatrix}$$

$$= 0.0097 \text{ lb/ac}$$

5.4.17.2 Quantity to Concentration

Pounds to concentration (milligrams/liter)

$$\text{Conc. (mg/L)} = \frac{\text{Quantity (lb)}}{\text{Volume (MG)} \times 8.34 \text{ lb/mg/L/MG}} \quad (5.32)$$

■ EXAMPLE 5.64

Problem: The aeration tank contains 73,529 lb of solids. The volume of the tank is 3.20 MG. What is the concentration of solids in the aeration tank in milligrams/liter?

Solution:

$$\text{Conc. (mg/L)} = \frac{73,529 \text{ lb}}{3.20 \text{ MG} \times 8.34 \text{ lb/mg/L/MG}}$$

$$= 2755 \text{ mg/L (rounded)}$$

Pounds/day to concentration (milligrams/liter)

$$\text{Conc. (mg/L)} = \frac{\text{Quantity (lb/day)}}{\text{Flow (MGD)} \times 8.34 \text{ lb/mg/L/MG}} \quad (5.33)$$

■ EXAMPLE 5.65

Problem: What is the chlorine dose in milligrams/liter when 490 lb/day of chlorine is added to an effluent flow of 11.0 MGD?

Solution:

$$\text{Dose (mg/L)} = \frac{490 \text{ lb/day}}{11.0 \text{ MGD} \times 8.34 \text{ lb/mg/L/MG}}$$

$$= 5.34 \text{ mg/L}$$

Kilograms/day to concentration (milligrams/liter)

$$\text{Conc. (mg/L)} = \frac{\text{Quantity (kg/day)}}{\text{Flow (MGD)} \times 3.785 \text{ kg/mg/L/MG}} \quad (5.34)$$

5.4.17.3 Quantity to Volume or Flow Rate

Pounds to tank volume (million gallons)

$$\text{Volume (mg/L)} = \frac{\text{Quantity (lb)}}{\text{Conc. (mg/L)} \times 8.34 \text{ lb/mg/L/MG}} \quad (5.35)$$

Pounds/day to flow (million gallons per day)

$$\text{Flow (MGD)} = \frac{\text{Quantity (lb/day)}}{\text{Conc. (mg/L)} \times 8.34 \text{ lb/mg/L/MG}} \quad (5.36)$$

■ EXAMPLE 5.66

Problem: You must remove 8485 lb of solids from the activated sludge process. The waste activated sludge solids concentration is 5636 mg/L. How many million gallons must be removed?

Solution:

$$\text{Flow (MGD)} = \frac{8485 \text{ lb/day}}{5636 \text{ mg/L} \times 8.34 \text{ lb/MG/mg/L}}$$

$$= 0.181 \text{ MGD}$$

Kilograms/day to flow (million gallons per day)

$$\text{Flow (MGD)} = \frac{\text{Quantity (kg/day)}}{\text{Conc. (mg/L)} \times 3.785 \text{ kg/MG/mg/L}} \quad (5.37)$$

5.5 APPLIED MATH OPERATIONS

5.5.1 MASS BALANCE AND MEASURING PLANT PERFORMANCE

The simplest way to express the fundamental engineering principle of *mass balance* is to say, "Everything has to go somewhere." More precisely, the *law of conservation of mass* says that when chemical reactions take place, matter is neither created nor destroyed. What this important concept allows us to do is track materials (e.g., pollutants, microorganisms, chemicals) from one place to another. The concept of mass balance plays an important role in treatment plant operations (especially wastewater treatment) where we assume that a balance exists between the material entering and leaving the treatment plant or a treatment process: "What comes in must equal what goes out." The concept is very helpful in evaluating biological

systems, sampling and testing procedures, and many other unit processes within the treatment system. In the following sections, we illustrate how the mass balance concept is used to determine the quantity of solids entering and leaving settling tanks and mass balance using BOD removal.

5.5.2 MASS BALANCE FOR SETTLING TANKS

The mass balance for the settling tank calculates the quantity of solids entering and leaving the unit.

Key Point: The two numbers—in (influent) and out (effluent)—must be within 10 to 15% of each other to be considered acceptable. Larger discrepancies may indicate sampling errors, increasing solids levels in the unit, or undetected solids discharge in the tank effluent.

To get a better feel for how the mass balance for settling tanks procedure is formatted for actual use, consider the steps below that are used in Example 5.67:

Step 1. Solids in = Pounds of influent suspended solids

Step 2. Pounds of effluent suspended solids

Step 3. Biosolids solids out = Pounds of biosolids solids pumped per day

Step 4. Balance = Solids in – (solids out + biosolids solids pumped)

■ EXAMPLE 5.67

Problem: The settling tank receives a daily flow of 4.20 MGD. The influent contains 252 mg/L suspended solids, and the unit effluent contains 140 mg/L suspended solids. The biosolids pump operates 10 min/hr and removes biosolids at the rate of 40 gpm. The biosolids are comprised of 4.2% solids. Determine if the mass balance for solids removal is within the acceptable 10 to 15% range.

Solution:

Step 1. Solids in = 252 mg/L × 4.20 MGD × 8.34
= 8827 lb/day

Step 2. Solids out = 140 mg/L × 4.20 MGD × 8.34
= 4904 lb/day

Step 3. Biosolids solids = 10 min/hr × 24 hr/day × 40 gpm × 8.34 × 0.042 = 3363 lb/day

Step 4. Balance = 8827 lb/day – (4904 lb/day + 3363 lb/day) = 560 lb, or 6.3%

5.5.3 MASS BALANCE USING BOD REMOVAL

The amount of BOD removed by a treatment process is directly related to the quantity of solids the process will generate. Because the actual amount of solids generated will vary with operational conditions and design, exact

TABLE 5.4
General Conversion Factors

Process Type	Conversion Factor (lb Solids/lb BOD Removal)
Primary treatment	1.7
Trickling filters	1.0
Rotating biological contactors	1.0
Activated biosolids with primary treatment	0.7
Activated biosolids without primary treatment	
Conventional	0.85
Extended air	0.65
Contact stabilization	1.0
Step feed	0.85
Oxidation ditch	0.65

figures must be determined on a case-by-case basis; however, research has produced general conversion factors for many of the common treatment processes. These values are given in Table 5.4 and can be used if plant-specific information is unavailable. Using these factors, the mass balance procedure determines the amount of solids the process is anticipated to produce. This is compared with the actual biosolids production to determine the accuracy of the sampling or the potential for solids buildup in the system or unrecorded solids discharges.

Step 1. BOD_{in} = Influent BOD × flow × 8.34

Step 2. BOD_{out} = Effluent BOD × flow × 8.34.

Step 3. BOD pounds removed = BOD_{in} − BOD_{out}

Step 4. Solids generated (lb) = BOD removed (lb) × factor

Step 5. Solids removed = Sludge pumped (gpd) × % solids × 8.34

Step 6. Effluent solids (mg/L) × flow (MGD) × 8.34

■ EXAMPLE 5.68

Problem: A conventional activated biosolids system with primary treatment is operating at the levels listed below. Does the mass balance for the activated biosolids system indicate that a problem exists?

Given:

Plant influent BOD = 250 mg/L

Primary effluent BOD = 166 mg/L

Activated biosolids system effluent BOD = 25 mg/L

Activated biosolids system effluent TSS = 19 mg/L

Plant flow = 11.40 MGD

Waste concentration = 6795 mg/L

Waste flow = 0.15 MGD

Solution:

BOD_{in} = 166 mg/L × 11.40 MGD × 8.34 = 15,783 lb/day

BOD_{out} = 25 mg/L × 11.40 MGD × 8.34 = 2377 lb/day

BOD removed = 15,783 lb/day − 2377 lb/day = 13,406 lb/day

Solids produced = 13,406 lb/day × 0.7 lb solids per lb BOD = 9384 lb solids per day

Solids removed = 6795 mg/L × 0.15 MGD × 8.34 = 8501 lb/day

Difference = 9384 lb/day − 8501 lb/day = 883 lb/day, or 9.4%

These results are within the acceptable range.

Key Point: We have demonstrated two ways in which mass balance can be used; however, it is important to note that the mass balance concept can be used for all aspects of wastewater and solids treatment. In each case, the calculations must take into account all of the sources of material entering the process and all of the methods available for removal of solids.

5.5.4 MEASURING PLANT PERFORMANCE

To evaluate how well a plant or unit process is performing, performance efficiency or percent removal is used. The results obtained can be compared with those listed in the plant's O&M manual to determine if the facility is performing as expected. In this section, sample calculations often used to measure plant performance or efficiency are presented. The *efficiency* of a unit process is its effectiveness in removing various constituents from the wastewater or water. Suspended solids and BOD removal are therefore the most common calculations of unit process efficiency. In wastewater treatment, the efficiency of a sedimentation basin may be affected by such factors as the types of solids in the wastewater, the temperature of the wastewater, and the age of the solids. Typical removal efficiencies for a primary sedimentation basin are as follows:

Settleable solids, 90–99%
Suspended solids, 40–60%
Total solids, 10–15%
BOD, 20–50%

5.5.4.1 Plant Performance/Efficiency

Key Point: The calculation used for determining the performance (percent removal) for a digester is different from that used for performance (percent removal) for other processes. Care must be taken to select the correct formula:

$$\% \text{ Removal} = \frac{(\text{Influent conc.} - \text{effluent conc.}) \times 100}{\text{Influent conc.}} \quad (5.38)$$

■ EXAMPLE 5.69

Problem: The influent BOD_5 is 247 mg/L, and the plant effluent BOD is 17 mg/L. What is the percent removal?

Solution:

$$\% \text{ Removal} = \frac{(247 \text{ mg/L} - 17 \text{ mg/L}) \times 100}{247 \text{ mg/L}} = 93\%$$

5.5.4.2 Unit Process Performance/Efficiency

Equation 5.42 is used to determine unit process efficiency. The concentration entering the unit and the concentration leaving the unit (e.g., primary, secondary) are used to determine the unit performance:

$$\% \text{ Removal} = \frac{(\text{Influent conc.} - \text{effluent conc.}) \times 100}{\text{Influent conc.}} \quad (5.39)$$

■ EXAMPLE 5.70

Problem: The primary influent BOD is 235 mg/L, and the primary effluent BOD is 169 mg/L. What is the percent removal?

Solution:

$$\% \text{ Removal} = \frac{(253 \text{ mg/L} - 169 \text{ mg/L}) \times 100}{235 \text{ mg/L}} = 28\%$$

5.5.4.3 Percent Volatile Matter Reduction in Sludge

The calculation used to determine *percent volatile matter reduction* is more complicated because of the changes occurring during biosolids digestion:

$$\% \text{ VM reduction} = \frac{(\% \text{ VM}_{in} - \% \text{ VM}_{out}) \times 100}{\% \text{ VM}_{in} - (\% \text{ VM}_{in} \times \% \text{ VM}_{out})} \quad (5.40)$$

■ EXAMPLE 5.71

Problem: Using the digester data provided below, determine the percent volatile matter reduction for the digester.

Given:

Raw biosolids volatile matter = 74%
Digested biosolids volatile matter = 54%

Solution:

$$\% \text{ VM reduction} = \frac{(0.74 - 0.54) \times 100}{0.74 - (0.74 \times 0.54)} = 59\%$$

5.6 WATER MATH CONCEPTS

5.6.1 WATER SOURCES AND STORAGE CALCULATIONS

Approximately 40 million cubic miles of water cover or reside within the Earth. The oceans contain about 97% of all water on Earth. The other 3% is freshwater: (1) snow and ice on the surface of the Earth contain about 2.25% of the water, (2) usable groundwater is approximately 0.3%, and (3) surface freshwater is less than 0.5%. In the United States, for example, average rainfall is approximately 2.6 ft (a volume of 5900 km³). Of this amount, approximately 71% evaporates (about 4200 km³), and 29% goes to stream flow (about 1700 km³).

Beneficial freshwater uses include manufacturing, food production, domestic and public needs, recreation, hydroelectric power production, and flood control. Stream flow withdrawn annually is about 7.5% (440 km³). Irrigation and industry use almost half of this amount (3.4%, or 200 km³/yr). Municipalities use only about 0.6% (35 km³/yr) of this amount. Historically, in the United States, water usage has been increasing (as might be expected); for example, in 1900, 40 billion gallons of freshwater were used. In 1975, usage increased to 455 billion gallons. Projected use in 2000 was about 720 billion gallons.

The primary sources of freshwater include the following:

- Captured and stored rainfall in cisterns and water jars
- Groundwater from springs, artesian wells, and drilled or dug wells
- Surface water from lakes, rivers, and streams
- Desalinized seawater or brackish groundwater
- Reclaimed wastewater

5.6.2 WATER SOURCE CALCULATIONS

Water source calculations covered in this section apply to wells and pond or lake storage capacity. Specific well calculations discussed include well drawdown, well yield, specific yield, well casing disinfection, and deep-well turbine pump capacity.

5.6.2.1 Well Drawdown

Drawdown is the drop in the level of water in a well when water is being pumped. Drawdown is usually measured in feet or meters. One of the most important reasons for measuring drawdown is to make sure that the source water is adequate and not being depleted. The data collected to calculate drawdown can indicate if the water supply is slowly declining. Early detection can give the system time to explore alternative sources, establish conservation measures, or obtain any special funding that may be needed to get a new water source. Well drawdown is the difference between the pumping water level and the static water level:

$$\text{Drawdown (ft)} = \begin{bmatrix} \text{Pumping water level (ft)} \\ - \text{ static water level (ft)} \end{bmatrix} \quad (5.41)$$

■ EXAMPLE 5.72

Problem: The static water level for a well is 70 ft. If the pumping water level is 90 ft, what is the drawdown?

Solution:

$$\text{Drawdown (ft)} = \begin{bmatrix} \text{Pumping water level (ft)} \\ - \text{ static water level (ft)} \end{bmatrix}$$

$$= 90 \text{ ft} - 70 \text{ ft} = 20 \text{ ft}$$

■ EXAMPLE 5.73

Problem: The static water level of a well is 122 ft. The pumping water level is determined using the sounding line. The air pressure applied to the sounding line is 4.0 psi, and the length of the sounding line is 180 ft. What is the drawdown?

Solution: First calculate the water depth in the sounding line and the pumping water level:

1. Water depth in sounding line = 4.0 psi × 2.31 ft/psi = 9.2 ft
2. Pumping water level = 180 ft – 9.2 ft = 170.8 ft

Then calculate drawdown as usual:

$$\text{Drawdown (ft)} = \begin{bmatrix} \text{Pumping water level (ft)} \\ - \text{ static water level (ft)} \end{bmatrix}$$

$$= 170.8 \text{ ft} - 122 \text{ ft} = 48.8 \text{ ft}$$

5.6.2.2 Well Yield

Well yield is the volume of water per unit of time that is produced from the well pumping. Usually, well yield is measured in terms of gallons per minute (gpm) or gallons per hour (gph). Sometimes, large flows are measured in cubic feet per second (cfs). Well yield is determined by using the following equation:

$$\text{Well yield (gpm)} = \frac{\text{Gallons produced}}{\text{Duration of test (min)}} \quad (5.42)$$

■ EXAMPLE 5.74

Problem: When the drawdown level of a well was stabilized, it was determined that the well produced 400 gal during a 5-min test. What was the well yield?

Solution:

$$\text{Well yield (gpm)} = \frac{\text{Gallons produced}}{\text{Duration of test (min)}}$$

$$= \frac{400 \text{ gal}}{5 \text{ min}} = 80 \text{ gpm}$$

■ EXAMPLE 5.75

Problem: During a 5-min test for well yield, a total of 780 gal was removed from the well. What was the well yield in gpm? In gph?

Solution:

$$\text{Well yield (gpm)} = \frac{\text{Gallons produced}}{\text{Duration of test (min)}}$$

$$= \frac{780 \text{ gal}}{5 \text{ min}}$$

$$= 156 \text{ gpm}$$

Then convert gpm flow to gph flow:

$$156 \text{ gpm} \times 60 \text{ min/hr} = 9360 \text{ gph}$$

5.6.2.3 Specific Yield

Specific yield is the discharge capacity of the well per foot of drawdown. The specific yield may range from 1 gpm/ft drawdown to more than 100 gpm/ft drawdown for a properly developed well. Specific yield is calculated using Equation 5.43:

$$\text{Specific yield (gpm/ft)} = \frac{\text{Well yield (gpm)}}{\text{Drawdown (ft)}} \quad (5.43)$$

■ **EXAMPLE 5.76**

Problem: A well produces 260 gpm. If the drawdown for the well is 22 ft, what is the specific yield in gpm/ft?

Solution:

$$\text{Specific yield (gpm/ft)} = \frac{\text{Well yield (gpm)}}{\text{Drawdown (ft)}}$$

$$= \frac{260 \text{ gpm}}{22 \text{ ft}} = 11.8 \text{ gpm/ft}$$

■ **EXAMPLE 5.77**

Problem: The yield for a particular well is 310 gpm. If the drawdown for this well is 30 ft, what is the specific yield in gpm/ft?

Solution:

$$\text{Specific yield (gpm/ft)} = \frac{\text{Well yield (gpm)}}{\text{Drawdown (ft)}}$$

$$= 310 \text{ gpm/30 ft}$$

$$= 10.3 \text{ gpm/ft}$$

5.6.2.4 Well Casing Disinfection

A new, cleaned, or a repaired well normally contains contamination that may remain for weeks unless the well is thoroughly disinfected. This may be accomplished by using ordinary bleach at a concentration of 100 parts per million (ppm) of chlorine. The amount of disinfectant required is determined by the amount of water in the well. The following equation is used to calculate the pounds of chlorine required for disinfection:

$$\text{Chlorine (lb)} = \begin{bmatrix} \text{Chlorine (mg/L)} \\ \times \text{casing volume (MG)} \\ \times 8.34 \text{ lb/gal} \end{bmatrix} \quad (5.44)$$

■ **EXAMPLE 5.78**

Problem: A new well is to be disinfected with chlorine at a dosage of 50 mg/L. If the well casing diameter is 8 in. and the length of the water-filled casing is 110 ft, how many pounds of chlorine will be required?

Solution: First calculate the volume of the water-filled casing:

$$0.785 \times 0.67 \times 67 \times 110 \text{ ft} \times 7.48 \text{ gal/ft}^3 = 290 \text{ gal}$$

Then determine the pounds of chlorine required using the mg/L to lb equation:

$$\text{Chlorine (lb)} = \begin{bmatrix} \text{Chlorine (mg/L)} \\ \times \text{casing volume (MG)} \times 8.34 \text{ lb/gal} \end{bmatrix}$$

$$= 50 \text{ mg/L} \times 0.000290 \text{ MG} \times 8.34 \text{ lb/gal}$$

$$= 0.12 \text{ lb}$$

5.6.2.5 Deep-Well Turbine Pumps

The deep-well turbine pump is used for high-capacity deep wells. The pump, usually consisting of more than one stage of centrifugal pump, is fastened to a pipe called the *pump column*; the pump is located in the water. The pump is driven from the surface through a shaft running inside the pump column. The water is discharged from the pump up through the pump column to surface. The pump may be driven by a vertical shaft, electric motor at the top of the well, or some other power source, usually through a right-angle gear drive located at the top of the well. A modern version of the deep-well turbine pump is the submersible type of pump, where the pump (as well as a close-coupled electric motor built as a single unit) is located below water level in the well. The motor is built to operate submerged in water.

5.6.2.6 Vertical Turbine Pump Calculations

The calculations pertaining to well pumps include head, horsepower, and efficiency calculations. *Discharge head* is measured to the pressure gauge located close to the pump discharge flange. The pressure (psi) can be converted to feet of head using the equation:

$$\text{Discharge head (ft)} = \text{Pressure (psi)} \times 2.31 \text{ ft/psi} \quad (5.45)$$

Total pumping head (*field head*) is a measure of the lift below the discharge head pumping water level (*discharge head*). Total pumping head is calculated as follows:

$$\text{Pumping head (ft)} = \begin{bmatrix} \text{Pumping water level (ft)} \\ + \text{ discharge head (ft)} \end{bmatrix} \quad (5.46)$$

■ **EXAMPLE 5.79**

Problem: The pressure gauge reading at a pump discharge head is 4.1 psi. What is this discharge head expressed in feet?

Solution:

$$4.1 \text{ psi} \times 2.31 \text{ ft/psi} = 9.5 \text{ ft}$$

■ **EXAMPLE 5.80**

Problem: The static water level of a pump is 100 ft. The well drawdown is 26 ft. If the gauge reading at the pump discharge head is 3.7 psi, what is the total pumping head?

Solution:

$$\text{Pumping head (ft)} = \begin{bmatrix} \text{Pumping water level (ft)} \\ + \text{discharge head (ft)} \end{bmatrix}$$

$$= (100 \text{ ft} + 26 \text{ ft}) + (3.7 \text{ psi} \times 2.31 \text{ ft/psi})$$

$$= 126 \text{ ft} + 8.5 \text{ ft} = 134.5 \text{ ft}$$

5.6.3 WATER STORAGE CALCULATIONS

Water storage facilities for water distribution systems are required primarily to provide for fluctuating demands of water usage (to provide a sufficient amount of water to average or equalize daily demands on the water supply system). In addition, other functions of water storage facilities include increasing operating convenience, leveling pumping requirements (to keep pumps from running 24 hours a day), decreasing power costs, providing water during power source or pump failure, providing large quantities of water to meet fire demands, providing surge relief (to reduce the surge associated with stopping and starting pumps), increasing detention time (to provide chlorine contact time and satisfy the desired contact time value requirements), and blending water sources. The storage capacity, in gallons, of a reservoir, pond, or small lake can be estimated using Equation 5.47:

$$\text{Capacity (gal)} = \begin{bmatrix} \text{Average length (ft)} \\ \times \text{average width (ft)} \\ \times \text{average depth (ft)} \times 7.48 \text{ gal/ft} \end{bmatrix} \quad (5.47)$$

■ **EXAMPLE 5.81**

Problem: A pond has an average length of 250 ft, an average width of 110 ft, and an estimated average depth of 15 ft. What is the estimated volume of the pond in gallons?

Solution:

$$\text{Volume (gal)} = \begin{bmatrix} \text{Average length (ft)} \times \text{average width (ft)} \\ \times \text{average depth (ft)} \times 7.48 \text{ gal/ft} \end{bmatrix}$$

$$= 250 \text{ ft} \times 110 \text{ ft} \times 15 \text{ ft} \times 7.48 \text{ gal/ft}^3$$

$$= 3,085,500 \text{ gal}$$

■ **EXAMPLE 5.82**

Problem: A small lake has an average length of 300 ft and an average width of 95 ft. If the maximum depth of the lake is 22 ft, what is the estimated gallons volume of the lake?

Note: For small ponds and lakes, the average depth is generally about 0.4 times the greatest depth; therefore, to estimate the average depth, measure the greatest depth and multiply that number by 0.4.

Solution: First, the average depth of the lake must be estimated:

$$\text{Est. average depth (ft)} = \text{Maximum depth (ft)} \times 0.4 \text{ ft}$$

$$= 22 \text{ ft} \times 0.4 \text{ ft} = 8.8 \text{ ft}$$

Then, the lake volume can be determined:

$$\text{Volume (gal)} = \begin{bmatrix} \text{Average length (ft)} \times \text{average width (ft)} \\ \times \text{average depth (ft)} \times 7.48 \text{ gal/ft} \end{bmatrix}$$

$$= 300 \text{ ft} \times 95 \text{ ft} \times 8.8 \text{ ft} \times 7.48 \text{ gal/ft}^3$$

$$= 1,875,984 \text{ gal}$$

5.6.4 COPPER SULFATE DOSING

Algal control by applying copper sulfate is perhaps the most common *in situ* treatment of lakes, ponds, and reservoirs; the copper ions in the water kill the algae. Copper sulfate application methods and dosages will vary depending on the specific surface water body being treated. The desired copper sulfate dosage may be expressed in mg/L copper, lb copper sulfate per ac-ft, or lb copper sulfate per acre. For a dose expressed as mg/L copper, the following equation is used to calculate lb copper sulfate required:

$$\text{Copper sulfate (lb)} = \frac{\text{Copper (mg/L)} \times \text{volume (MG)} \times 8.34 \text{ lb/gal}}{\% \text{ Available copper}/100} \quad (5.48)$$

■ **EXAMPLE 5.83**

Problem: For algae control in a small pond, a dosage of 0.5 mg/L copper is desired. The pond has a volume of 15 MG. How many pounds of copper sulfate will be required? (Note that copper sulfate contains 25% available copper.)

Solution:

$$\text{Copper sulfate (lb)} = \frac{\text{Copper (mg/L)} \times \text{volume (MG)} \times 8.34 \text{ lb/gal}}{\% \text{ Available copper}/100}$$

$$= \frac{0.5 \text{ mg/L} \times 15 \text{ MG} \times 8.34 \text{ lb/gal}}{25/100}$$

$$= 250 \text{ lb copper sulfate}$$

For calculating lb copper sulfate per ac-ft, use the following equation (assuming a desired copper sulfate dose of 0.9 lb/ac-ft):

$$\text{Copper sulfate (lb)} = \frac{0.9 \text{ lb copper sulfate} \times \text{ac-ft}}{1 \text{ ac-ft}} \quad (5.49)$$

■ EXAMPLE 5.84

Problem: A pond has a volume of 35 ac-ft. If the desired copper sulfate dose is 0.9 lb/ac-ft, how many lb of copper sulfate will be required?

Solution:

$$\text{Copper sulfate (lb)} = \frac{0.9 \text{ lb Copper sulfate} \times \text{ac-ft}}{1 \text{ ac-ft}}$$

$$\frac{0.9 \text{ lb Copper sulfate}}{1 \text{ ac-ft}} = \frac{x \text{ lb Copper sulfate}}{35 \text{ ac-ft}}$$

$$0.9 \times 35 = x$$

$$x = 31.5 \text{ lb copper sulfate}$$

The desired copper sulfate dosage may also be expressed in terms of lb copper sulfate per acre. The following equation is used to determine lb copper sulfate (assuming a desired copper sulfate dose of 5.2 lb/ac):

$$\text{Copper sulfate (lb)} = \frac{5.2 \text{ lb copper sulfate} \times \text{acres}}{1 \text{ ac}} \quad (5.50)$$

■ EXAMPLE 5.85

Problem: A small lake has a surface area of 6.0 ac. If the desired copper sulfate dose is 5.2 lb/ac, how many pounds of copper sulfate are required?

Solution:

$$\text{Copper sulfate (lb)} = \frac{5.2 \text{ lb copper sulfate} \times 6 \text{ ac}}{1 \text{ ac}}$$

$$= 31.2 \text{ lb copper sulfate}$$

5.6.5 COAGULATION AND FLOCCULATION

5.6.5.1 Coagulation

Following screening and the other pretreatment processes, the next unit process in a conventional water treatment system is a mixer where the first chemicals are added in what is known as *coagulation*. The exception to this situation occurs in small systems using groundwater, when chlorine or other taste and odor control measures are introduced at the intake and are the extent of treatment. The term *coagulation* refers to the series of chemical and mechanical operations by which coagulants are applied and made effective. These operations are comprised of two distinct phases: (1) rapid mixing to disperse coagulant chemicals by violent agitation into the water being treated, and (2) flocculation to agglomerate small particles into well-defined floc by gentle agitation for a much longer time. The coagulant must be added to the raw water and perfectly distributed into the liquid; such uniformity of chemical treatment is reached through rapid agitation or mixing.

Coagulation is a reaction caused by adding salts or iron or aluminum to the water. Common coagulants (salts) include:

- Alum (aluminum sulfate)
- Sodium aluminate
- Ferric sulfate
- Ferrous sulfate
- Ferric chloride
- Polymers

5.6.5.2 Flocculation

Flocculation follows coagulation in the conventional water treatment process. *Flocculation* is the physical process of slowly mixing the coagulated water to increase the probability of particle collision. Through experience, we see that effective mixing reduces the required amount of chemicals and greatly improves the sedimentation process, which results in longer filter runs and higher quality finished water. The goal of flocculation is to form a uniform, feather-like material similar to snowflakes—a dense, tenacious floc that traps the fine, suspended, and colloidal particles and carries them down rapidly in the settling basin. To increase the speed of floc formation and the strength and weight of the floc, polymers are often added.

5.6.5.3 Coagulation and Flocculation Calculations

Proper operation of the coagulation and flocculation unit processes requires calculations to determine chamber or basin volume, chemical feed calibration, chemical feeder settings, and detention time.

5.6.5.4.1 Chamber and Basin Volume Calculations

To determine the volume of a square or rectangular chamber or basin, we use Equation 5.51 or Equation 5.52:

$$\text{Volume (ft}^3) = \text{Length (ft)} \times \text{width (ft)} \times \text{depth (ft)} \quad (5.51)$$

$$\text{Volume (gal)} = \begin{bmatrix} \text{Length (ft)} \times \text{width (ft)} \\ \times \text{depth (ft)} \times 7.48 \text{ gal/ft}^3 \end{bmatrix} \quad (5.52)$$

■ EXAMPLE 5.86

Problem: A flash mix chamber is 4 ft square with water to a depth of 3 ft. What is the volume of water (in gallons) in the chamber?

Solution:

$$\text{Volume (gal)} = \begin{bmatrix} \text{Length (ft)} \times \text{width (ft)} \\ \times \text{depth (ft)} \times 7.48 \text{ gal/ft}^3 \end{bmatrix}$$

$$= 4 \text{ ft} \times 4 \text{ ft} \times 3 \text{ ft} \times 7.48 \text{ gal/ft}^3$$

$$= 359 \text{ gal}$$

■ EXAMPLE 5.87

Problem: A flocculation basin is 40 ft long by 12 ft wide with water to a depth of 9 ft. What is the volume of water (in gallons) in the basin?

Solution:

$$\text{Volume (gal)} = \begin{bmatrix} \text{Length (ft)} \times \text{width (ft)} \\ \times \text{depth (ft)} \times 7.48 \text{ gal/ft}^3 \end{bmatrix}$$

$$= 40 \text{ ft} \times 12 \text{ ft} \times 9 \text{ ft} \times 7.48 \text{ gal/ft}^3$$

$$= 32,314 \text{ gal}$$

■ EXAMPLE 5.88

Problem: A flocculation basin is 50 ft long by 22 ft wide and contains water to a depth of 11 ft, 6 in. How many gallons of water are in the tank?

Solution: First convert the 6-in. portion of the depth measurement to feet:

$$(6 \text{ in.}) \div (12 \text{ in./ft}) = 0.5 \text{ ft}$$

Then calculate basin volume:

$$\text{Volume (gal)} = \begin{bmatrix} \text{Length (ft)} \times \text{width (ft)} \\ \times \text{depth (ft)} \times 7.48 \text{ gal/ft}^3 \end{bmatrix}$$

$$= 50 \text{ ft} \times 22 \text{ ft} \times 11.5 \text{ ft} \times 7.48 \text{ gal/ft}^3$$

$$= 94,622 \text{ gal}$$

5.6.5.4.2 Detention Time

Because coagulation reactions are rapid, detention time for flash mixers is measured in seconds, whereas the detention time for flocculation basins is generally between 5 and 30 min. The equation used to calculate detention time is shown below:

$$\text{Detention time (min)} = \frac{\text{Volume of tank (gal)}}{\text{Flow rate (gpm)}} \quad (5.53)$$

■ EXAMPLE 5.89

Problem: The flow to a flocculation basin that is 50 ft long by 12 ft wide by 10 ft deep is 2100 gpm. What is the detention time in the tank (in minutes)?

Solution:

$$\text{Tank volume (gal)} = 50 \text{ ft} \times 12 \text{ ft} \times 10 \text{ ft} \times 7.48 \text{ gal/ft}^3$$

$$= 44,880 \text{ gal}$$

$$\text{Detention time (min)} = \frac{\text{Volume of tank (gal)}}{\text{Flow rate (gpm)}}$$

$$= \frac{44,880 \text{ gal}}{2100 \text{ gpm}}$$

$$= 21.4 \text{ min}$$

■ EXAMPLE 5.90

Problem: A flash mix chamber is 6 ft long by 4 ft with water to a depth of 3 ft. If the flow to the flash mix chamber is 6 MGD, what is the chamber detention time in seconds (assuming that the flow is steady and continuous)?

Solution: First, convert the flow rate from gpd to gps so the time units will match:

$$\frac{6,000,000}{1440 \text{ min/day} \times 60 \text{ sec/min}} = 69 \text{ gps}$$

Then calculate detention time:

$$\text{Detention time (sec)} = \frac{\text{Volume of tank (gal)}}{\text{Flow rate (gps)}}$$

$$= \frac{6 \text{ ft} \times 4 \text{ ft} \times 3 \text{ ft} \times 7.48 \text{ gal /ft}^3}{69 \text{ gps}}$$

$$= 7.8 \text{ sec}$$

5.6.5.4.3 Determining Dry Chemical Feeder Setting (lb/day)

When adding (dosing) chemicals to the water flow, a measured amount of chemical is called for. The amount of chemical required depends on such factors as the type of chemical used, the reason for dosing, and the flow rate being treated. To convert from mg/L to lb/day, the following equation is used:

$$\text{Chemical added (lb/day)} = \begin{bmatrix} \text{Chemical (mg/L)} \\ \times \text{flow (MGD)} \times 8.34 \text{ lb/gal} \end{bmatrix} \quad (5.54)$$

■ EXAMPLE 5.91

Problem: Jar tests indicate that the best alum dose for water is 8 mg/L. If the flow to be treated is 2,100,000 gpd, what should the lb/day settling be on the dry alum feeder?

Solution:

$$\text{Setting (lb/day)} = \begin{bmatrix} \text{Chemical (mg/L)} \\ \times \text{flow (MGD)} \times 8.34 \text{ lb/gal} \end{bmatrix}$$

$$= 8 \text{ mg/L} \times 2.10 \text{ MGD} \times 8.34 \text{ lb/gal}$$

$$= 140 \text{ lb/day}$$

■ EXAMPLE 5.92

Problem: Determine the desired lb/day setting on a dry chemical feeder if jar tests indicate an optimum polymer dose of 12 mg/L and the flow to be treated is 4.15 MGD.

Solution:

$$\text{Setting (lb/day)} = 12 \text{ mg/L} \times 4.15 \text{ MGD} \times 8.34 \text{ lb/gal}$$

$$= 415 \text{ lb/day}$$

5.6.5.4.4 Determining Chemical Solution Feeder Setting (gpd)

When solution concentration is expressed as pound chemical per gallon solution, the required feed rate can be determined using the following equations:

$$\text{Chemical (lb/day)} = \begin{bmatrix} \text{Chemical (mg/L)} \\ \times \text{flow (MGD)} \times 8.34 \text{ lb/gal} \end{bmatrix} \quad (5.55)$$

Then convert the lb/day dry chemical to gpd solution:

$$\text{Solution (gpd)} = \frac{\text{Chemical (lb/day)}}{\text{lb Chemical per gal solution}} \quad (5.56)$$

■ EXAMPLE 5.93

Problem: Jar tests indicate that the best alum dose for water is 7 mg/L. The flow to be treated is 1.52 MGD. Determine the gpd setting for the alum solution feeder if the liquid alum contains 5.36 lb of alum per gallon of solution.

Solution: First calculate the lb/day of dry alum required, using the mg/L to lb/day equation:

$$\text{Dry alum (lb/day)} = \begin{bmatrix} \text{Chemical (mg/L)} \\ \times \text{flow (MGD)} \times 8.34 \text{ lb/gal} \end{bmatrix}$$

$$= 7 \text{ mg/L} \times 1.52 \text{ MGD} \times 8.34 \text{ lb/gal}$$

$$= 89 \text{ lb/day}$$

Then calculate gpd solution required:

$$\text{Alum solution (gpd)} = \frac{89 \text{ lb/day}}{5.36 \text{ lb alum per gal solution}}$$

$$= 16.6 \text{ gpd}$$

5.6.5.4.5 Determining Chemical Solution Feeder Setting (mL/min)

Some chemical solution feeders dispense chemical as milliliters per minute (mL/min). To calculate the mL/min solution required, use the following procedure:

$$\text{Solution (mL/min)} = \frac{\text{gpd} \times 3785 \text{ mL/gal}}{1440 \text{ min/day}} \quad (5.57)$$

■ EXAMPLE 5.94

Problem: The desired solution feed rate was calculated to be 9 gpd. What is this feed rate expressed as mL/min?

Solution:

$$\text{Solution (mL/min)} = \frac{\text{gpd} \times 3785 \text{ mL/gal}}{1440 \text{ min/day}}$$

$$= \frac{9 \text{ gpd} \times 3785 \text{ mL/gal}}{1440 \text{ min/day}}$$

$$= 24 \text{ mL/min feed rate}$$

■ EXAMPLE 5.95

Problem: The desired solution feed rate has been calculated to be 25 gpd. What is this feed rate expressed as mL/min?

Solution:

$$\text{Solution (mL/min)} = \frac{\text{gpd} \times 3785 \text{ mL/gal}}{1440 \text{ min/day}}$$

$$= \frac{25 \text{ gpd} \times 3785 \text{ mL/gal}}{1440 \text{ min/day}}$$

$$= 65.7 \text{ mL/min feed rate}$$

Sometimes we will need to know the mL/min solution feed rate but we do not know the gpd solution feed rate. In such cases, calculate the gpd solution feed rate first, using the following the equation:

$$\text{gpd} = \frac{\text{Chemical (mg/L)} \times \text{flow (MGD)} \times 8.34 \text{ lb/gal}}{\text{Chemical (lb)/solution (gal)}} \quad (5.58)$$

5.6.5.4.6 Determining Percent of Solutions

The strength of a solution is a measure of the amount of chemical solute dissolved in the solution. We use the following equation to determine the percent strength of a solution:

$$\% \text{ Strength} = \frac{\text{Chemical (lb)}}{\text{Water (lb)} + \text{chemical (lb)}} \times 100 \quad (5.59)$$

■ EXAMPLE 5.96

Problem: If a total of 10 oz. of dry polymer is added to 15 gal of water, what is the percent strength (by weight) of the polymer solution?

Solution: Before calculating percent strength, the ounces of chemical must be converted to pounds of chemical:

$$(10 \text{ oz.}) \div (16 \text{ oz./lb}) = 0.625 \text{ lb chemical}$$

Now calculate percent strength:

$$\% \text{ Strength} = \frac{\text{Chemical (lb)}}{\text{Water (lb)} + \text{chemical (lb)}} \times 100$$

$$= \frac{0.625 \text{ lb chemical}}{(15 \text{ gal} \times 8.34 \text{ lb/gal}) + 0.625 \text{ lb}} \times 100$$

$$= \frac{0.625 \text{ lb chemical}}{125.7 \text{ lb solution}} \times 100$$

$$= 0.5\%$$

■ EXAMPLE 5.97

Problem: If 90 g (1 g = 0.0022 lb) of dry polymer is dissolved in 6 gal of water, what percent strength is the solution?

Solution: First, convert grams of chemical to pounds:

$$90 \text{ g polymer} \times 0.0022 \text{ lb/g} = 0.198 \text{ lb polymer}$$

Now calculate percent strength of the solution:

$$\% \text{ Strength} = \frac{\text{Polymer (lb)}}{\text{Water (lb)} + \text{polymer (lb)}} \times 100$$

$$= \frac{0.198 \text{ lb polymer}}{(6 \text{ gal} \times 8.34 \text{ lb/gal}) + 0.198 \text{ lb}} \times 100$$

$$= 4\% \text{ (rounded)}$$

5.6.5.4.7 Determining Percent Strength of Liquid Solutions

When using liquid chemicals to make up solutions (e.g., liquid polymer), a different calculation is required, as shown below:

$$\frac{\substack{\text{Liquid polymer (lb)} \\ \times \text{liquid polymer} \\ (\% \text{ strength})}}{100} = \frac{\substack{\text{Polymer solution (lb)} \\ \times \text{polymer solution} \\ (\% \text{ strength})}}{100} \quad (5.60)$$

■ EXAMPLE 5.98

Problem: A 12% liquid polymer is to be used in making up a polymer solution. How many pounds of liquid polymer should be mixed with water to produce 120 lb of a 0.5% polymer solution?

Solution:

$$\frac{\substack{\text{Liquid polymer (lb)} \\ \times \text{liquid polymer} \\ (\% \text{ strength})}}{100} = \frac{\substack{\text{Polymer solution (lb)} \\ \times \text{polymer solution} \\ (\% \text{ strength})}}{100}$$

$$\frac{x \text{ lb} \times 12}{100} = \frac{120 \text{ lb} \times 0.5}{100}$$

$$x = \frac{120 \times 0.005}{0.12} = 5 \text{ lb}$$

5.6.5.4.8 Determining Percent Strength of Mixed Solutions

The percent strength of solution mixture is determined using the following equation:

$$\% \text{ Strength of mix} = \left[\left(\frac{\frac{\text{Sol. 1 (lb)} \times \% \text{ strength Sol. 1}}{100}}{\text{Sol. 1 (lb)}} \right) + \left(\frac{\frac{\text{Sol. 2 (lb)} \times \% \text{ strength Sol. 2}}{100}}{\text{Sol. 2 (lb)}} \right) \right] \times 100 \quad (5.61)$$

■ **EXAMPLE 5.99**

Problem: If 12 lb of a 10% strength solution is mixed with 40 lb of a 1% strength solution, what is the percent strength of the solution mixture?

Solution:

$$\% \text{ Strength of mix} = \left[\frac{\left(\dfrac{\text{Sol. 1 (lb)} \times \% \text{ strength Sol. 1}}{100} \right)}{\text{Sol. 1 (lb)}} + \frac{\left(\dfrac{\text{Sol. 2 (lb)} \times \% \text{ strength Sol. 2}}{100} \right)}{\text{Sol. 2 (lb)}} \right] \times 100$$

$$= \frac{12 \text{ lb} \times 0.1}{12 \text{ lb}} + \frac{40 \text{ lb} \times 0.01}{40 \text{ lb}} \times 100$$

$$= \frac{1.2 \text{ lb} + 0.40 \text{ lb}}{52 \text{ lb}} \times 100 = 3.1\%$$

5.6.5.4.9 Dry Chemical Feeder Calibration

Occasionally, we need to perform a calibration calculation to compare the actual chemical feed rate with the feed rate indicated by the instrumentation. To calculate the actual feed rate for a dry chemical feeder, place a container under the feeder, weigh the container when empty, then weigh the container again after a specified length of time (e.g., 30 min). The actual chemical feed rate can be calculated using the following equation:

$$\text{Feed rate (lb/min)} = \frac{\text{Chemical applied (lb)}}{\text{Length of application (min)}} \quad (5.62)$$

If desired, the chemical feed rate can be converted to lb/day:

$$\text{Feed rate (lb/day)} = \text{Feed rate (lb/min)} \times 1400 \text{ min/day}$$

■ **EXAMPLE 5.100**

Problem: Calculate the actual chemical feed rate (lb/day) if a container is placed under a chemical feeder and a total of 2 lb is collected during a 30-min period.

Solution: First calculate the lb/min feed rate:

$$\text{Feed rate (lb/min)} = \frac{\text{Chemical applied (lb)}}{\text{Length of application (min)}}$$

$$= \frac{2 \text{ lb}}{30 \text{ min}}$$

$$= 0.06 \text{ lb/min feed rate}$$

Then calculate the lb/day feed rate:

$$\text{Feed rate} = 0.06 \text{ lb/min} \times 1440 \text{ min/day} = 86.4 \text{ lb/day}$$

■ **EXAMPLE 5.101**

Problem: Calculate the actual chemical feed rate (lb/day) if a container is placed under a chemical feeder and a total of 1.6 lb is collected during a 20-min period.

Solution: First calculate the lb/min feed rate:

$$\text{Feed rate (lb/min)} = \frac{\text{Chemical applied (lb)}}{\text{Length of application (min)}}$$

$$= \frac{1.6 \text{ lb}}{20 \text{ min}}$$

$$= 0.08 \text{ lb/min feed rate}$$

Then calculate the lb/day feed rate:

$$\text{Feed rate} = 0.08 \text{ lb/min} \times 1440 \text{ min/day} = 115 \text{ lb/day}$$

5.6.5.4.10 Chemical Solution Feeder Calibration

As with other calibration calculations, the actual chemical solution feed rate is determined and then compared with the feed rate indicated by the instrumentation. To calculate the actual chemical solution feed rate, first express the solution feed rate in MGD. When the MGD solution flow rate has been calculated, use the mg/L equation to determine chemical dosage in lb/day. If solution feed is expressed as mL/min, first convert mL/min flow rate to gpd flow rate:

$$\text{gpd} = \frac{(\text{mL/min}) \times (1440 \text{ min/day})}{3785 \text{ mL/gal}} \quad (5.63)$$

Then calculate chemical dosage:

$$\text{Chemical (lb/day)} = \left[\begin{array}{l} \text{Chemical (mg/L)} \\ \times \text{ flow (MGD)} \times 8.34 \text{ lb/day} \end{array} \right] \quad (5.64)$$

■ **EXAMPLE 5.102**

Problem: A calibration test is conducted for a chemical solution feeder. During a 5-min test, the pump delivered 940 mg/L of the 1.20% polymer solution. (Assume that the polymer solution weighs 8.34 lb/gal.) What is the polymer dosage rate in lb/day?

Solution: The flow rate must be expressed as MGD; therefore, the mL/min solution flow rate must first be converted to gpd and then MGD. The mL/min flow rate is calculated as:

$$\frac{940 \text{ mL}}{5 \text{ min}} = 188 \text{ mL/min}$$

Next convert the mL/min flow rate to gpd flow rate:

$$\frac{188 \text{ mL/min} \times 1440 \text{ min/day}}{3785 \text{ mL/gal}} = 72 \text{ gpd flow rate}$$

Then calculate the polymer feed rate:

$$12,000 \text{ mg/L} \times 0.000072 \text{ MGD} \times 8.34 \text{ lb/day} = 7.2 \text{ lb/day}$$

■ **EXAMPLE 5.103**

Problem: A calibration test is conducted for a chemical solution feeder. During a 24-hr period, the solution feeder delivers a total of 100 gal of solution. The polymer solution is a 1.2% solution. What is the lb/day feed rate? (Assume that the polymer solution weighs 8.34 lb/gal.)

Solution: The solution feed rate is 100 gal per day, or 100 gpd. Expressed as MGD, this is 0.000100 MGD. Use the mg/L to lb/day equation to calculate the actual feed rate:

$$\text{Chemical (lb/day)} = \begin{bmatrix} \text{Chemical (mg/L)} \\ \times \text{flow (MGD)} \times 8.34 \text{ lb/day} \end{bmatrix}$$

$$= \begin{bmatrix} 12,000 \text{ mg/L} \times 0.000100 \text{ MGD} \\ \times 8.34 \text{ lb/day} \end{bmatrix}$$

$$= 10 \text{ lb/day polymer}$$

The actual pumping rates can be determined by calculating the volume pumped during a specified time frame; for example, if 60 gal are pumped during a 10-min test, the average pumping rate during the test is 6 gpm. Actual volume pumped is indicated by the drop in tank level. By using the following equation, we can determine the flow rate in gpm:

$$\text{Flow rate (gpm)} = \frac{\begin{matrix} 0.785 \times (\text{diameter})^2 \\ \times \text{drop in level (ft)} \\ \times 7.48 \text{ gal/ft}^3 \end{matrix}}{\text{Duration of test (min)}} \quad (5.65)$$

■ **EXAMPLE 5.104**

Problem: A pumping rate calibration test is conducted for a 15-min period. The liquid level in the 4-ft-diameter solution tank is measured before and after the test. If the level drops 0.5 ft during the 15-min test, what is the pumping rate in gpm?

Solution:

$$\text{Pumping rate (gpm)} = \frac{\begin{matrix} 0.785 \times (\text{diameter})^2 \\ \times \text{drop in level (ft)} \times 7.48 \text{ gal/ft}^3 \end{matrix}}{\text{Duration of test (min)}}$$

$$= \frac{0.785 \times (4 \text{ ft})^2 \times 0.5 \text{ ft} \times 7.48 \text{ gal/ft}^3}{15 \text{ min}}$$

$$= 3.1 \text{ gpm}$$

5.6.6 Determining Chemical Usage

One of the primary functions performed by water operators is the recording of data. The lb/day or gpd chemical use is part of the data from which the average daily use of chemicals and solutions can be determined. This information is important in forecasting expected chemical use, comparing it with chemicals in inventory, and determining when additional chemicals will be required. To determine average chemical use, we use Equation 5.66 (lb/day) or Equation 5.67 (gpd):

$$\text{Average use (lb/day)} = \frac{\text{Total chemical used (lb)}}{\text{Number of days}} \quad (5.66)$$

$$\text{Average use (gpd/day)} - \frac{\text{Total chemical used (gal)}}{\text{Number of days}} \quad (5.67)$$

Then we can calculate the days supply in inventory:

$$\text{Days supply} = \frac{\text{Total chemical in inventory (lb)}}{\text{Average use (lb/day)}} \quad (5.68)$$

$$\text{Days supply} = \frac{\text{Total chemical in inventory (gal)}}{\text{Average use (gpd)}} \quad (5.69)$$

■ **EXAMPLE 5.105**

Problem: The chemical used for each day during a week is given below. Based on these data, what was the average lb/day chemical use during the week?

Monday	88 lb/day
Tuesday	93 lb/day
Wednesday	91 lb/day
Thursday	88 lb/day
Friday	96 lb/day
Saturday	92 lb/day
Sunday	86 lb/day

Solution:

$$\text{Average use (lb/day)} = \frac{\text{Total chemical used (lb)}}{\text{Number of days}}$$

$$= \frac{634 \text{ lb}}{7 \text{ days}} = 90.6 \text{ lb/day}$$

■ EXAMPLE 5.106

Problem: The average chemical use at a plant is 77 lb/day. If the chemical inventory is 2800 lb, how many days supply is this?

Solution:

$$\text{Days supply (lb/day)} = \frac{\text{Total chemical in inventory (lb)}}{\text{Average use (lb/day)}}$$

$$= \frac{2800 \text{ lb}}{77 \text{ lb / day}} = 36.4 \text{ days supply}$$

5.6.7 Sedimentation Calculations

Sedimentation, the solid–liquid separation by gravity, is one of the most basic processes of water and wastewater treatment. In water treatment, plain sedimentation, such as the use of a presedimentation basin for grit removal and a sedimentation basin following coagulation–flocculation, is the most commonly used approach. The two common tank shapes of sedimentation tanks are rectangular and cylindrical. The equations for calculating the volume for each type tank are shown below.

5.6.7.1 Calculating Tank Volume

For rectangular sedimentation basins, we use Equation 5.70:

$$\text{Volume (gal)} = \begin{bmatrix} \text{Length (ft)} \times \text{width (ft)} \\ \times \text{depth (ft)} \times 7.48 \text{ gal/ft}^3 \end{bmatrix} \quad (5.70)$$

For circular clarifiers, we use Equation 5.71:

$$\text{Volume (gal)} = \begin{bmatrix} 0.785 \times (\text{diameter})^2 \\ \times \text{depth (ft)} \times 7.48 \text{ gal/ft}^3 \end{bmatrix} \quad (5.71)$$

■ EXAMPLE 5.107

Problem: A sedimentation basin is 25 ft wide by 80 ft long and contains water to a depth of 14 ft. What is the volume of water in the basin, in gallons?

Solution:

$$\text{Volume (gal)} = \begin{bmatrix} \text{Length (ft)} \times \text{width (ft)} \\ \times \text{depth (ft)} \times 7.48 \text{ gal/ft}^3 \end{bmatrix}$$

$$= 80 \text{ ft} \times 25 \text{ ft} \times 14 \text{ ft} \times 7.48 \text{ gal/ft}^3$$

$$= 209,440 \text{ gal}$$

■ EXAMPLE 5.108

Problem: A sedimentation basin is 24 ft wide by 75 ft long. When the basin contains 140,000 gal, what would the water depth be?

Solution:

$$\text{Volume (gal)} = \begin{bmatrix} \text{Length (ft)} \times \text{width (ft)} \\ \times \text{depth (ft)} \times 7.48 \text{ gal/ft}^3 \end{bmatrix}$$

$$140,000 \text{ gal} = 75 \text{ ft} \times 24 \text{ ft} \times x \text{ ft} \times 7.48 \text{ gal/ft}^3$$

$$x \text{ ft} = \frac{140,000 \text{ gal}}{75 \text{ ft} \times 24 \text{ ft} \times 7.48 \text{ gal/ft}^3}$$

$$x \text{ ft} = 10.4 \text{ ft}$$

5.6.7.2 Detention Time

Detention time for clarifiers varies from 1 to 3 hr. The equations used to calculate detention time are shown below.

Basic detention time equation:

$$\text{Detention time (hr)} = \frac{\text{Volume of tank (gal)}}{\text{Flow rate (gph)}} \quad (5.72)$$

Rectangular sedimentation basin equation:

$$\text{Detention time (hr)} = \frac{\begin{array}{c} \text{Length (ft)} \times \text{width (ft)} \\ \times \text{depth (ft)} \times 7.48 \text{ gal/ft}^3 \end{array}}{\text{Flow rate (gph)}} \quad (5.73)$$

Circular basin equation:

$$\text{Detention time (hr)} = \frac{\begin{array}{c} 0.785 \times (\text{diameter})^2 \\ \text{depth (ft)} \times 7.48 \text{ gal/ft}^3 \end{array}}{\text{Flow rate (gph)}} \quad (5.74)$$

■ EXAMPLE 5.109

Problem: A sedimentation tank has a volume of 137,000 gal. If the flow to the tank is 121,000 gph, what is the detention time in the tank (in hours)?

Solution:

$$\text{Detention time (hr)} = \frac{\text{Volume of tank (gal)}}{\text{Flow rate (gph)}}$$

$$= \frac{137,000 \text{ gal}}{121,000 \text{ gph}} = 1.1 \text{ hr}$$

■ EXAMPLE 5.110

Problem: A sedimentation basin is 60 ft long by 22 ft wide and has water to a depth of 10 ft. If the flow to the basin is 1,500,000 gpd, what is the sedimentation basin detention time?

Solution: First, convert the flow rate from gpd to gph so the times units will match:

$$(1,500,000 \text{ gpd}) \div (24 \text{ hr/day}) = 62,500 \text{ gph}$$

Then calculate detention time:

$$\text{Detention time (hr)} = \frac{\text{Volume of tank (gal)}}{\text{Flow rate (gph)}}$$

$$= \frac{60 \text{ ft} \times 22 \text{ ft} \times 10 \text{ ft} \times 7.48 \text{ gal/ft}^3}{62,500 \text{ gph}}$$

$$= 1.6 \text{ hr}$$

5.6.7.3 Surface Overflow Rate

The surface overflow rate—similar to the hydraulic loading rate (flow per unit area)—is used to determine loading on sedimentation basins and circular clarifiers. Hydraulic loading rate, however, measures the total water entering the process, whereas surface overflow rate measures only the water overflowing the process (plant flow only).

Note: Surface overflow rate calculations do not include recirculated flows. Other terms used synonymously with surface overflow rate are *surface loading rate* and *surface settling rate*.

Surface overflow rate is determined using the following equation:

$$\text{Surface overflow rate} = \frac{\text{Flow (gpm)}}{\text{Area (ft}^2)} \quad (5.75)$$

■ EXAMPLE 5.111

Problem: A circular clarifier has a diameter of 80 ft. If the flow to the clarifier is 1800 gpm, what is the surface overflow rate in gpm/ft²?

Solution:

$$\text{Surface overflow rate} = \frac{\text{Flow (gpm)}}{\text{Area (ft}^2)}$$

$$= \frac{1800 \text{ gpm}}{0.785 \times 80 \text{ ft} \times 80 \text{ ft}}$$

$$= 0.36 \text{ gpm/ft}^2$$

■ EXAMPLE 5.112

Problem: A sedimentation basin 70 ft by 25 ft receives a flow of 1000 gpm. What is the surface overflow rate in gpm/ft²?

Solution:

$$\text{Surface overflow rate} = \frac{\text{Flow (gpm)}}{\text{Area (ft}^2)}$$

$$= \frac{1000 \text{ gpm}}{70 \text{ ft} \times 25 \text{ ft}}$$

$$= 0.6 \text{ gpm/ft}^2$$

5.6.7.4 Mean Flow Velocity

The measure of average velocity of the water as it travels through a rectangular sedimentation basin is known as *mean flow velocity*. Mean flow velocity is calculated using Equation 5.76:

$$Q = AV$$

$$\text{Flow } (Q) \text{ (ft}^3\text{/min)} = \begin{bmatrix} \text{Cross-sectional area } (A) \text{ (ft}^2) \\ \times \text{ volume } (V) \text{ (ft/min)} \end{bmatrix} (5.76)$$

■ EXAMPLE 5.113

Problem: A sedimentation basin is 60 ft long by 18 ft wide and has water to a depth of 12 ft. When the flow through the basin is 900,000 gpd, what is the mean flow velocity in the basin in ft/min?

Solution: Because velocity is desired in ft/min, the flow rate in the $Q = AV$ equation must be expressed in ft³/min (cfm):

$$\frac{900,000 \text{ gpd}}{1440 \text{ min/day} \times 7.48 \text{ gal/ft}^3} = 84 \text{ cfm}$$

Then, use $Q = AV$ to calculate velocity:

$$Q = AV$$

$$84 \text{ cfm} = (18 \text{ ft} \times 12 \text{ ft}) \times x \text{ fpm}$$

$$x \text{ fpm} = \frac{84 \text{ cfm}}{18 \text{ ft} \times 12 \text{ ft}}$$

$$= 0.4 \text{ fpm}$$

■ EXAMPLE 5.114

Problem: A rectangular sedimentation basin 50 ft long by 20 ft wide has a water depth of 9 ft. If the flow to the basin is 1,880,000 gpd, what is the mean flow velocity in ft/min?

Solution: Because velocity is desired in ft/min, the flow rate in the $Q = AV$ equation must be expressed in ft³/min (cfm):

$$\frac{1,880,000 \text{ gpd}}{1440 \text{ min/day} \times 7.48 \text{ gal/ft}^3} = 175 \text{ cfm}$$

Then, use the $Q = AV$ equation to calculate velocity:

$$Q = AV$$

$$175 \text{ cfm} = (20 \text{ ft} \times 9 \text{ ft}) \times x \text{ fpm}$$

$$x = \frac{175 \text{ cfm}}{20 \text{ ft} \times 9 \text{ ft}}$$

$$= 0.97 \text{ fpm}$$

5.6.7.5 Weir Loading Rate (Weir Overflow Rate)

Weir loading rate (weir overflow rate) is the amount of water leaving the settling tank per linear foot of weir. The result of this calculation can be compared with design. Normally, weir overflow rates of 10,000 to 20,000 gal/day/ft are used in the design of a settling tank. Typically, the weir loading rate is a measure of the flow in gallons per minute (gpm) over each foot of weir. The weir loading rate is determined using the following equation:

$$\text{Weir loading rate (gpm/ft)} = \frac{\text{Flow (gpm)}}{\text{Weir length (ft)}}$$

■ **EXAMPLE 5.115**

Problem: A rectangular sedimentation basin has a total of 115 ft of weir. What is the weir loading rate in gpm/ft when the flow of 1,110,000 gpd?

Solution:

$$\text{Flow} = \frac{1,110,000 \text{ gpd}}{1440 \text{ min/day}} = 771 \text{ gpm}$$

$$\text{Weir loading rate (gpm/ft)} = \frac{\text{Flow (gpm)}}{\text{Weir length (ft)}}$$

$$= \frac{771 \text{ gpm}}{115 \text{ ft}}$$

$$= 6.7 \text{ gpm/ft}$$

■ **EXAMPLE 5.116**

Problem: A circular clarifier receives a flow of 3.55 MGD. If the diameter of the weir is 90 ft, what is the weir loading rate in gpm/ft?

Solution:

$$\text{Flow} = \frac{3,550,000 \text{ gpd}}{1440 \text{ min/day}} = 2465 \text{ gpm}$$

$$\text{Feet of weir} = 3.14 \times 90 \text{ ft} = 283 \text{ ft}$$

$$\text{Weir loading rate (gpm/ft)} = \frac{\text{Flow (gpm)}}{\text{Weir length (ft)}}$$

$$= \frac{2465 \text{ gpm}}{283 \text{ ft}} = 8.7 \text{ gpm/ft}$$

5.6.7.6 Percent Settled Biosolids

The percent settled biosolids test (*volume over volume test*, or V/V test) is conducted by collecting a 100-mL slurry sample from the solids contact unit and allowing it to settle for 10 min. After 10 min, the volume of settled biosolids at the bottom of the 100-mL graduated cylinder is measured and recorded. The equation used to calculate percent settled biosolids is shown below:

$$\% \text{ Settled biosolids} = \frac{\text{Settled biosolids (mL)}}{\text{Total sample (mL)}} \times 100 \quad (5.77)$$

■ **EXAMPLE 5.117**

Problem: A 100-mL sample of slurry from a solids contact unit is placed in a graduated cylinder and allowed to set for 10 min. The settled biosolids at the bottom of the graduated cylinder after 10 min is 22 mL. What is the percent of settled biosolids of the sample?

Solution:

$$\% \text{ Settled biosolids} = \frac{\text{Settled biosolids (mL)}}{\text{Total sample (mL)}} \times 100$$

$$= \frac{22 \text{ mL}}{100 \text{ mL}} \times 100 = 19\%$$

■ **EXAMPLE 5.118**

Problem: A 100-mL sample of slurry from a solids contact unit is placed in a graduated cylinder. After 10 min, a total of 21 mL of biosolids settled to the bottom of the cylinder. What is the percent settled biosolids of the sample?

Solution:

$$\% \text{ Settled biosolids} = \frac{\text{Settled biosolids (mL)}}{\text{Total sample (mL)}} \times 100$$

$$= \frac{21 \text{ mL}}{100 \text{ mL}} \times 100$$

$$= 21\% \text{ settled biosolids}$$

5.6.7.7 Determining Lime Dosage (mg/L)

During the alum dosage process, lime is sometimes added to provide adequate alkalinity (HCO_3^-) in the solids contact clarification process for the coagulation and precipitation of the solids. To determine the lime dose required, in mg/L, three steps are required.

In Step 1, the total alkalinity required to react with the alum to be added and provide proper precipitation is determined using the following equation:

$$\text{Total alkalinity required (mg/L)} = \text{Alkalinity reacting with alum (mg/L)} + \text{alkalinity in the water (mg/L)} \quad (5.78)$$

(1 mg/L alum reacts with 0.45 mg/L alkalinity)

■ EXAMPLE 5.119

Problem: Raw water requires an alum dose of 45 mg/L, as determined by jar testing. If a residual 30-mg/L alkalinity must be present in the water to ensure complete precipitation of alum added, what is the total alkalinity required (in mg/L)?

Solution: First calculate the alkalinity that will react with 45 mg/L alum:

$$\frac{0.45 \text{ mg/L alkalinity}}{1 \text{ mg/L alum}} = \frac{x \text{ mg/L alkalinity}}{45 \text{ mg/L alum}}$$

$$x = 0.45 \times 45 = 20.25 \text{ mg/L}$$

Then calculate the total alkalinity required:

$$\text{Total alkalinity req. (mg/L)} = \text{Alkalinity reacting with alum (mg/L)} + \text{residual alkalinity (mg/L)}$$

$$= 20.25 \text{ mg/L} + 30 \text{ mg/L}$$

$$= 50.25 \text{ mg/L}$$

■ EXAMPLE 5.120

Problem: Jar tests indicate that 36 mg/L alum is optimum for a particular raw water. If a residual 30 mg/L alkalinity must be present to promote complete precipitation of the alum added, what is the total alkalinity required (in mg/L)?

Solution: First calculate the alkalinity that will react with 36 mg/L alum:

$$\frac{0.45 \text{ mg/L alkalinity}}{1 \text{ mg/L alum}} = \frac{x \text{ mg/L alkalinity}}{36 \text{ mg/L alum}}$$

$$x = 0.45 \times 36 = 16.2 \text{ mg/L}$$

Then calculate the total alkalinity required:

$$\text{Total alkalinity req. (mg/L)} = 16.2 \text{ mg/L} + 30 \text{ mg/L}$$

$$= 46.2 \text{ mg/L}$$

In Step 2, we make a comparison between required alkalinity and alkalinity already in the raw water to determine how many mg/L alkalinity should be added to the water. The equation used to make this calculation is shown below:

$$\text{Added alkalinity} = \left[\begin{array}{l} \text{Total alkalinity req. (mg/L)} \\ - \text{alkalinity in water (mg/L)} \end{array} \right] \quad (5.79)$$

■ EXAMPLE 5.121

Problem: A total of 44-mg/L alkalinity is required to react with alum and ensure proper precipitation. If the raw water has an alkalinity of 30 mg/L as bicarbonate, how much mg/L alkalinity should be added to the water?

Solution:

$$\text{Added alkalinity} = \left[\begin{array}{l} \text{Total alkalinity req. (mg/L)} \\ - \text{alkalinity in water (mg/L)} \end{array} \right]$$

$$= 44 \text{ mg/L} - 30 \text{ mg/L}$$

$$= 14 \text{ mg/L alkalinity to be added}$$

In Step 3, after determining the amount of alkalinity to be added to the water, we determine how much lime (the source of alkalinity) must be added. We accomplish this by using the ratio shown in Example 5.119.

■ EXAMPLE 5.122

Problem: It has been calculated that 16 mg/L alkalinity must be added to a raw water. How much mg/L lime will be required to provide this amount of alkalinity? (One mg/L alum reacts with 0.45 mg/L alkalinity and 1 mg/L alum reacts with 0.35 mg/L lime.)

Solution: First determine the mg/L lime required by using a proportion that relates bicarbonate alkalinity to lime:

$$\frac{0.45 \text{ mg/L alkalinity}}{0.35 \text{ mg/L lime}} = \frac{16 \text{ mg/L alkalinity}}{x \text{ mg/L lime}}$$

Then cross-multiply:

$$0.45x = 16 \times 0.35$$

$$x = \frac{16 \times 0.35}{0.45} = 12.4 \text{ mg/L lime}$$

In Example 5.123, we use all three steps to determine the lime dosage (mg/L) required.

■ EXAMPLE 5.123

Problem: Given the following data, calculate the lime dose required, in mg/L:

Alum dose required (determined by jar tests) = 52 mg/L
Residual alkalinity required for precipitation = 30 mg/L
1 mg/L alum reacts with 0.35 mg/L lime
1 mg/L alum reacts with 0.45 mg/L alkalinity
Raw water alkalinity = 36 mg/L

Solution: To calculate the total alkalinity required, we must first calculate the alkalinity that will react with 52 mg/L alum:

$$\frac{0.45 \text{ mg/L alkalinity}}{1 \text{ mg/L alum}} = \frac{x \text{ mg/L alkalinity}}{52 \text{ mg/L alum}}$$

$$x = 0.45 \times 52$$

$$x = 23.4 \text{ mg/L alkalinity}$$

The total alkalinity requirement can now be determined:

$$\text{Total alkalinity req. (mg/L)} = \begin{bmatrix} \text{Alkalinity reacting with} \\ \text{alum (mg/L)} \\ + \text{residual alkalinity (mg/L)} \end{bmatrix}$$

$$= 23.4 \text{ mg/L} + 30 \text{ mg/L}$$

$$= 53.4 \text{ mg/L}$$

Next calculate how much alkalinity must be added to the water:

$$\text{Added alkalinity} = \begin{bmatrix} \text{Total alkalinity req. (mg/L)} \\ - \text{alkalinity in water (mg/L)} \end{bmatrix}$$

$$= 53.4 \text{ mg/L} - 36 \text{ mg/L}$$

$$= 17.4 \text{ mg/L alkalinity to be added}$$

Finally, calculate the lime required to provide this additional alkalinity:

$$\frac{0.45 \text{ mg/L alkalinity}}{0.35 \text{ mg/L lime}} = \frac{17.4 \text{ mg/L alkalinity}}{x \text{ mg/L lime}}$$

$$0.45x = 17.4 \times 0.35$$

$$x = \frac{17.4 \times 0.35}{0.45} = 13.5 \text{ mg/L lime}$$

5.6.7.8 Determining Lime Dosage (lb/day)

After the lime dose has been determined in terms of mg/L, it is a fairly simple matter to calculate the lime dose in lb/day, which is one of the most common calculations in water and wastewater treatment. To convert from mg/L to lb/day lime dose, we use the following equation:

$$\text{Lime (lb/day)} = \begin{bmatrix} \text{Lime (mg/L)} \times \text{flow (MGD)} \\ \times 8.34 \text{ lb/gal} \end{bmatrix} \quad (5.80)$$

■ EXAMPLE 5.124

Problem: The lime dose for a raw water has been calculated to be 15.2 mg/L. If the flow to be treated is 2.4 MGD, how many lb/day lime will be required?

Solution:

$$\text{Lime (lb/day)} = \text{Lime (mg/L)} \times \text{flow (MGD)} \times 8.34 \text{ lb/gal}$$

$$= 15.2 \text{ mg/L} \times 2.4 \text{ MGD} \times 8.34 \text{ lb/gal}$$

$$= 304 \text{ lb/day lime}$$

■ EXAMPLE 5.125

Problem: The flow to a solids contact clarifier is 2,650,000 gpd. If the lime dose required is determined to be 12.6 mg/L, how many lb/day lime will be required?

Solution:

$$\text{Lime (lb/day)} = \text{Lime (mg/L)} \times \text{flow (MGD)} \times 8.34 \text{ lb/gal}$$

$$= 12.6 \text{ mg/L} \times 2.65 \text{ MGD} \times 8.34 \text{ lb/gal}$$

$$= 278 \text{ lb/day lime}$$

5.6.7.9 Determining Lime Dosage (g/min)

To convert from mg/L lime to g/min lime, use Equation 5.81:

Key Point: 1 lb = 453.6 g.

$$\text{Lime (g/min)} = \frac{\text{Lime (lb/day)} \times 453.6 \text{ g/lb}}{1440 \text{ min/day}} \quad (5.81)$$

■ EXAMPLE 5.126

Problem: A total of 275 lb/day lime will be required to raise the alkalinity of the water passing through a solids-contact clarification process. How many g/min lime does this represent?

Solution:

$$\text{Lime (g/min)} = \frac{\text{Lime (lb/day)} \times 453.6 \text{ g/lb}}{1440 \text{ min/day}}$$

$$= \frac{275 \text{ lb/day} \times 453.6 \text{ g/lb}}{1440 \text{ min/day}}$$

$$= 86.6 \text{ g/min lime}$$

■ EXAMPLE 5.127

Problem: A lime dose of 150 lb/day is required for a solids-contact clarification process. How many g/min lime does this represent?

Solution:

$$\text{Lime (g/min)} = \frac{\text{Lime (lb/day)} \times 453.6 \text{ g/lb}}{1440 \text{ min/day}}$$

$$= \frac{150 \text{ lb/day} \times 453.6 \text{ g/lb}}{1440 \text{ min/day}}$$

$$= 47.3 \text{ g/min lime}$$

5.6.8 FILTRATION CALCULATIONS

Water filtration is a physical process of separating suspended and colloidal particles from waste by passing the water through a granular material. The process of filtration involves straining, settling, and adsorption. As floc passes into the filter, the spaces between the filter grains become clogged, reducing this opening and increasing removal. Some material is removed merely because it settles on a grain of media. One of the most important processes is adsorption of the floc onto the surface of individual filter grains. In addition to removing silt and sediment, floc, algae, insect larvae, and any other large elements, filtration also contributes to the removal of bacteria and protozoa such as *Giardia lamblia* and *Cryptosporidium*. Some filtration processes are also used for iron and manganese removal.

The *Surface Water Treatment Rule* (SWTR) specifies four filtration technologies, although SWTR also allows the use of alternative filtration technologies (e.g., cartridge filters). These include slow sand filtration/rapid sand filtration, pressure filtration, diatomaceous earth filtration, and direct filtration. Of these, all but rapid sand filtration are commonly employed in small water systems that use filtration. Each type of filtration system has advantages and disadvantages. Regardless of the type of filter, however, filtration involves the processes of *straining* (where particles are captured in the small spaces between filter media grains), *sedimentation* (where the particles land on top of the grains and stay there), and *adsorption* (where a chemical attraction occurs between the particles and the surface of the media grains).

5.6.8.1 Flow Rate through a Filter (gpm)

Flow rate in gpm through a filter can be determined by simply converting the gpd flow rate, as indicated on the flow meter. The gpm flow rate can be calculated by taking the meter flow rate (gpd) and dividing by 1440 min/day as shown in Equation 5.82:

$$\text{Flow rate (gpm)} = \frac{\text{Flow rate (gpd)}}{1440 \text{ min/day}} \qquad (5.82)$$

■ EXAMPLE 5.128

Problem: The flow rate through a filter is 4.25 MGD. What is this flow rate expressed as gpm?

Solution:

$$\text{Flow rate (gpm)} = \frac{4.25 \text{ MGD}}{1440 \text{ min/day}}$$

$$= \frac{4,250,000 \text{ gpd}}{1440 \text{ min/day}} = 2951 \text{ gpm}$$

■ EXAMPLE 5.129

Problem: During a 70-hr filter run, a total of 22.4 million gal of water are filtered. What is the average flow rate through the filter in gpm during this filter run?

Solution:

$$\text{Flow rate (gpm)} = \frac{\text{Total gallons produced}}{\text{Filter run (min)}}$$

$$= \frac{22,400,000 \text{ gal}}{70 \text{ hr} \times 60 \text{ min/hr}} = 5333 \text{ gpm}$$

■ EXAMPLE 5.130

Problem: At an average flow rate of 4000 gpm, how long of a filter run (in hours) would be required to produce 25 MG of filtered water?

Solution: Write the equation as usual, filling in known data:

$$\text{Flow rate (gpm)} = \frac{\text{Total gallons produced}}{\text{Filter run (min)}}$$

$$4000 \text{ gpm} = \frac{25,000,000 \text{ gal}}{x \text{ hr} \times 60 \text{ min/hr}}$$

Then solve for *x:*

$$x = \frac{25,000,000 \text{ gal}}{4000 \text{ gpm} \times 60 \text{ min/hr}} = 104 \text{ hr}$$

■ **EXAMPLE 5.131**

Problem: A filter box is 20 ft by 30 ft (including the sand area). If the influent valve is shut, the water drops 3 in./min. What is the rate of filtration in MGD?

Solution:

Given:

> Filter box = 20 ft × 30 ft
> Water drops = 3 in./min

Find the volume of water passing through the filter:

$$\text{Volume} = \text{Area} \times \text{height}$$

$$\text{Area} = \text{Width} \times \text{length}$$

Note: The best approach to performing calculations of this type is a step-by-step one, breaking down the problem into what is given and what is to be found.

Step 1

> Area = 20 ft × 30 ft = 600 ft²
> Convert 3.0 in. into feet: 3/12 = 0.25 ft
> Volume = 600 ft² × 0.25 ft = 150 ft³ of water passing through the filter in one minute

Step 2
Convert cubic feet to gallons:

$$150 \text{ ft}^3 \times 7.48 \text{ gal/ft}^3 = 1122 \text{ gpm}$$

Step 3
The problem asks for the rate of filtration in MGD. To find MGD, multiply the number of gallons per minute by the number of minutes per day:

$$1122 \text{ gpm} \times 1440 \text{ min/day} = 1.62 \text{ MGD}$$

■ **EXAMPLE 5.132**

Problem: The influent valve to a filter is closed for 5 min. During this time, the water level in the filter drops 0.8 ft (10 in.). If the filter is 45 ft long and 15 ft wide, what is the gpm flow rate through the filter? Water drop equals 0.16 ft/min.

Solution: First calculate cfm flow rate using the $Q = AV$ equation:

$$Q \text{ (cfm)} = \text{Length (ft)} \times \text{width (ft)} \times \text{drop velocity (ft/min)}$$

$$= 45 \text{ ft} \times 15 \text{ ft} \times 0.16 \text{ ft/min} = 108 \text{ cfm}$$

Then convert cfm flow rate to gpm flow rate:

$$108 \text{ cfm} \times 7.48 \text{ gal/ft}^3 = 808 \text{ gpm}$$

5.6.8.2 Filtration Rate

One measure of filter production is filtration rate (generally ranging from 2 to 10 gpm/ft²). Along with filter run time, it provides valuable information for operation of filters. It is the gallons of water filtered per minute through each square foot of filter area. Filtration rate is determined using Equation 5.83:

$$\text{Filtration rate (gpm/ft}^2) = \frac{\text{Flow rate (gpm)}}{\text{Filter surface area (ft}^2)} \quad (5.83)$$

■ **EXAMPLE 5.133**

Problem: A filter 18 ft by 22 ft receives a flow of 1750 gpm. What is the filtration rate in gpm/ft²?

Solution:

$$\text{Filtration rate (gpm/ft}^2) = \frac{\text{Flow rate (gpm)}}{\text{Filter surface area (ft}^2)}$$

$$= \frac{1750 \text{ gpm}}{18 \text{ ft} \times 22 \text{ ft}} = 4.4 \text{ gpm/ft}^2$$

■ **EXAMPLE 5.134**

Problem: A filter 28 ft long by 18 ft wide treats a flow of 3.5 MGD. What is the filtration rate in gpm/ft²?

Solution:

$$\text{Flow rate} = \frac{3,500,000 \text{ gpd}}{1440 \text{ min/day}}$$

$$= 2431 \text{ gpm}$$

$$\text{Filtration rate (gpm/ft}^2) = \frac{\text{Flow rate (gpm)}}{\text{Filter surface area (ft}^2)}$$

$$= \frac{2431 \text{ gpm}}{28 \text{ ft} \times 18 \text{ ft}} = 4.8 \text{ gpm/ft}^2$$

■ **EXAMPLE 5.135**

Problem: A filter 45 ft long by 20 ft wide produces a total of 18 MG during a 76-hr filter run. What is the average filtration rate in gpm/ft² for this filter run?

Solution: First calculate the gpm flow rate through the filter:

$$\text{Flow rate (gpm)} = \frac{\text{Total gallons produced}}{\text{Filter run (min)}}$$

$$= \frac{18,000,000 \text{ gal}}{76 \text{ hr} \times 60 \text{ min/hr}}$$

$$= 3947 \text{ gpm}$$

Then calculate the filtration rate:

$$\text{Filtration rate} = \text{Flow rate (gpm)/filter area (ft}^2)$$

$$= 3947 \text{ gpm}/(45 \text{ ft} \times 20 \text{ ft}) = 4.4 \text{ gpm/ft}^2$$

■ EXAMPLE 5.136

Problem: A filter is 40 ft long by 20 ft wide. During a test of flow rate, the influent valve to the filter is closed for 6 min. The water level drop during this period is 16 in. What is the filtration rate for the filter in gpm/ft²?

Solution: First calculate the gpm flow rate using the $Q = AV$ equation:

$$Q \text{ (gpm)} = \left[\begin{array}{l} \text{Length (ft)} \times \text{width (ft)} \\ \times \text{drop velocity (ft/min)} \times 7.48 \text{ gal/ft}^3 \end{array} \right]$$

$$= \frac{40 \text{ ft} \times 20 \text{ ft} \times 1.33 \text{ ft} \times 7.48 \text{ gal/ft}^3}{6 \text{ min}} = 1326 \text{ gpm}$$

Then calculate the filtration rate:

$$\text{Filtration rate} = \frac{\text{Flow rate (gpm)}}{\text{Filter area (ft}^2)}$$

$$= 1316 \text{ gpm}/(40 \text{ ft} \times 20 \text{ ft})$$

$$= 1.7 \text{ g/ft}^2 \text{ (rounded)}$$

5.6.8.3 Unit Filter Run Volume

The unit filter run volume (UFRV) calculation indicates the total gallons passing through each square foot of filter surface area during an entire filter run. This calculation is used to compare and evaluate filter runs. UFRVs are usually at least 5000 gal/ft² and generally in the range of 10,000 gpd/ft². The UFRV value will begin to decline as the performance of the filter begins to deteriorate. The equation to be used in these calculations is shown below:

$$\text{Unit filter run volume} = \frac{\text{Total gallons filtered}}{\text{Filter surface area (ft}^2)} \quad (5.84)$$

■ EXAMPLE 5.137

Problem: The total water filtered during a filter run (between backwashes) is 2,220,000 gal. If the filter is 18 ft by 18 ft, what is the unit filter run volume in gal/ft²?

Solution:

$$\text{Unit filter run volume} = \frac{\text{Total gallons filtered}}{\text{Filter surface area (ft}^2)}$$

$$= \frac{2,220,000 \text{ gal}}{18 \text{ ft} \times 18 \text{ ft}} = 6852 \text{ gal/ft}^2$$

■ EXAMPLE 5.138

Problem: The total water filtered during a filter run is 4,850,000 gal. If the filter is 28 ft by 18 ft, what is the unit filter run volume in gal/ft²?

Solution:

$$\text{Unit filter run volume} = \frac{\text{Total gallons filtered}}{\text{Filter surface area (ft}^2)}$$

$$= \frac{4,850,000 \text{ gal}}{28 \text{ ft} \times 18 \text{ ft}} = 9623 \text{ gal/ft}^2$$

Equation 5.84 can be modified as shown in Equation 5.85 to calculate the unit filter run volume given filtration rate and filter run data:

$$\text{Unit filter run volume} = \left[\begin{array}{l} \text{Filtration rate (gpm/ft}^2) \\ \times \text{filter run time (min)} \end{array} \right] \quad (5.85)$$

■ EXAMPLE 5.139

Problem: The average filtration rate for a filter was determined to be 2 gpm/ft². If the filter run time was 4250 minutes, what was the unit filter run volume in gal/ft²?

Solution:

$$\text{Unit filter run volume} = \left[\begin{array}{l} \text{Filtration rate (gpm/ft}^2) \\ \times \text{filter run time (min)} \end{array} \right]$$

$$= 2 \text{ gpm/ft}^2 \times 4250 \text{ min}$$

$$= 8500 \text{ gal/ft}^2$$

The problem indicates that, at an average filtration rate of 2 gal entering each square foot of filter each minute, the total gallons entering during the total filter run is 4250 times that amount.

■ EXAMPLE 5.140

Problem: The average filtration rate during a particular filter run was determined to be 3.2 gpm/ft². If the filter run time was 61.0 hr, what was the UFRV in gal/ft² for the filter run?

Solution:

$$\text{Unit filter run volume} = \left[\begin{array}{l} \text{Filtration rate (gpm/ft}^2) \\ \times \text{filter run time (hr)} \times 60 \text{ min/hr} \end{array} \right]$$

$$= 3.2 \text{ gpm/ft}^2 \times 61.0 \text{ hr} \times 60 \text{ min/hr}$$

$$= 11,712 \text{ gal/ft}^2$$

5.6.8.4 Backwash Rate

In filter backwashing, one of the most important operational parameters to be determined is the amount of water in gallons required for each backwash. This amount depends on the design of the filter and the quality of the water being filtered. The actual washing typically lasts 5 to 10 min and usually amounts to 1 to 5% of the flow produced.

■ **EXAMPLE 5.141**

Problem: A filter has the following dimensions:

Length = 30 ft
Width = 20 ft
Depth of filter media = 24 in.

Assuming that a backwash rate of 15 gal/ft²/min is recommended and 10 min of backwash is required, calculate the amount of water in gallons required for each backwash.

Solution: Given the above data, find the amount of water in gallons required:

1. Area of filter = 30 ft × 20 ft = 600 ft²
2. Gallons of water used per square foot of filter
   = 15 gal/ft²/min × 10 min = 150 gal/ft²
3. Gallons required for backwash = 150 gal/ft² × 600 ft² = 90,000 gal

Typically, backwash rates will range from 10 to 25 gpm/ft². The backwash rate is determined by using Equation 5.86:

$$\text{Backwash rate} = \frac{\text{Flow rate (gpm)}}{\text{Filter area (ft}^2)} \quad (5.86)$$

■ **EXAMPLE 5.142**

Problem: A filter 30 ft by 10 ft has a backwash rate of 3120 gpm. What is the backwash rate in gpm/ft²?

Solution:

$$\text{Backwash rate} = \frac{\text{Flow rate (gpm)}}{\text{Filter area (ft}^2)}$$

$$= \frac{3120 \text{ gpm}}{30 \text{ ft} \times 10 \text{ ft}} = 10.4 \text{ gpm/ft}^2$$

■ **EXAMPLE 5.143**

Problem: A filter 20 ft by 20 ft has a backwash rate of 4.85 MGD. What is the filter backwash rate in gpm/ft²?

Solution:

$$\text{Flow rate} = \frac{4,850,000 \text{ gpd}}{1440 \text{ min/day}} = 3368 \text{ gpm}$$

$$\text{Backwash rate} = \frac{\text{Flow rate (gpm)}}{\text{Filter area (ft}^2)}$$

$$= \frac{3368 \text{ gpm}}{20 \text{ ft} \times 20 \text{ ft}} = 8.42 \text{ gpm/ft}^2$$

5.6.8.5 Backwash Rise Rate

Backwash rate is occasionally measured as the upward velocity of the water during backwashing, expressed as in./min rise. To convert from gpm/ft² backwash rate to an in./min rise rate, use either Equation 5.87 or Equation 5.88:

$$\begin{array}{c}\text{Backwash rate} \\ \text{(in./min)}\end{array} = \frac{\begin{array}{c}\text{Backwash rate (gpm/ft}^2) \\ \times 12 \text{ in./ft}\end{array}}{7.48 \text{ gal/ft}^3} \quad (5.87)$$

$$\begin{array}{c}\text{Backwash rate} \\ \text{(in./min)}\end{array} = \text{Backwash rate (gpm/ft}^2) \times 1.6 \quad (5.88)$$

■ **EXAMPLE 5.144**

Problem: A filter has a backwash rate of 16 gpm/ft². What is this backwash rate expressed as an in./min rise rate?

$$\text{Backwash rate (in./min)} = \frac{\begin{array}{c}\text{Backwash rate (gpm/ft}^2) \\ \times 12 \text{ in./ft}\end{array}}{7.48 \text{ gal/ft}^3}$$

$$= \frac{16 \text{ gpm/ft}^2 \times}{7.48 \text{ gal/ft}^3} = 25.7 \text{ in./min}$$

■ **EXAMPLE 5.145**

Problem: A filter 22 ft long by 12 ft wide has a backwash rate of 3260 gpm. What is this backwash rate expressed as an in./min rise?

Solution: First calculate the backwash rate as gpm/ft²:

$$\text{Backwash rate} = \frac{\text{Flow rate (gpm)}}{\text{Filter area (ft}^2)}$$

$$= \frac{3260 \text{ gpm}}{22 \text{ ft} \times 12 \text{ ft}} = 12.3 \text{ gpm/ft}^2$$

Then convert gpm/ft² to the in./min rise rate:

$$\text{Rise rate} = \frac{12.3 \text{ gpm/ft}^2 \times 12 \text{ in./ft}}{7.48 \text{ gal/ft}^3} = 19.7 \text{ in./min}$$

5.6.8.6 Volume of Backwash Water Required (gal)

To determine the volume of water required for backwashing, we must know both the desired backwash flow rate (gpm) and the duration of backwash (min):

$$\text{Water volume (gal)} = \begin{bmatrix} \text{Backwash flow rate (gpm)} \\ \times \text{ backwash duration (min)} \end{bmatrix} \quad (5.89)$$

■ EXAMPLE 5.146

Problem: For a backwash flow rate of 9000 gpm and a total backwash time of 8 min, how many gallons of water will be required for backwashing?

Solution:

$$\text{Water volume (gal)} = \begin{bmatrix} \text{Backwash flow rate (gpm)} \\ \times \text{ backwash duration (min)} \end{bmatrix}$$

$$= 9000 \text{ gpm} \times 8 \text{ min}$$

$$= 72,000 \text{ gal}$$

■ EXAMPLE 5.147

Problem: How many gallons of water would be required to provide a backwash flow rate of 4850 gpm for a total of 5 min?

Solution:

$$\text{Water volume (gal)} = \begin{bmatrix} \text{Backwash flow rate (gpm)} \\ \times \text{ backwash duration (min)} \end{bmatrix}$$

$$= 4850 \text{ gpm} \times 7 \text{ min}$$

$$= 33,950 \text{ gal}$$

5.6.8.7 Required Depth of Backwash Water Tank (ft)

The required depth of water in the backwash water tank is determined from the volume of water required for backwashing. To make this calculation, simply use Equation 5.90:

$$\text{Volume (gal)} = \begin{bmatrix} 0.785 \times (\text{diameter})^2 \\ \times \text{ depth (ft)} \times 7.48 \text{ gal/ft}^3 \end{bmatrix} \quad (5.90)$$

■ EXAMPLE 5.148

Problem: The volume of water required for backwashing has been calculated to be 85,000 gal. What is the required depth of water in the backwash water tank to provide this amount of water if the diameter of the tank is 60 ft?

Solution: Use the volume equation for a cylindrical tank, filling in known data, then solve for *x*:

$$\text{Volume (gal)} = \begin{bmatrix} 0.785 \times (\text{diameter})^2 \\ \times \text{ depth (ft)} \times 7.48 \text{ gal/ft}^3 \end{bmatrix}$$

$$85,000 \text{ gal} = 0.785 \times (60 \text{ ft})^2 \times x \text{ ft} \times 7.48 \text{ gal/ft}^3$$

$$x = \frac{85,000}{0.785 \times 60 \times 60 \times 7.48} = 4 \text{ ft}$$

■ EXAMPLE 5.149

Problem: A total of 66,000 gal of water will be required for backwashing a filter at a rate of 8000 gpm for a 9-min period. What depth of water is required if the backwash tank has a diameter of 50 ft?

Solution: Use the volume equation for cylindrical tanks:

$$\text{Volume (gal)} = \begin{bmatrix} 0.785 \times (\text{diameter})^2 \\ \times \text{ depth (ft)} \times 7.48 \text{ gal/ft}^3 \end{bmatrix}$$

$$85,000 \text{ gal} = 0.785 \times (60 \text{ ft})^2 \times x \text{ ft} \times 7.48 \text{ gal/ft}^3$$

$$x = \frac{85,000}{0.785 \times 60 \times 60 \times 7.48} = 4 \text{ ft}$$

5.6.8.8 Backwash Pumping Rate (gpm)

The desired backwash pumping rate (gpm) for a filter depends on the desired backwash rate in gpm/ft² and the ft² area of the filter. The backwash pumping rate (gpm) can be determined by using Equation 5.91:

$$\text{Pumping rate (gpm)} = \begin{bmatrix} \text{Backwash rate (gpm/ft}^2) \\ \times \text{ filter area (ft}^2) \end{bmatrix} \quad (5.91)$$

■ EXAMPLE 5.150

Problem: A filter is 25 ft long by 20 ft wide. If the desired backwash rate is 22 gpm/ft², what backwash pumping rate (gpm) will be required?

Solution: The desired backwash flow through each square foot of filter area is 20 gpm. The total gpm flow through the filter is therefore 20 gpm times the entire square foot area of the filter:

$$\text{Pumping rate (gpm)} = \begin{bmatrix} \text{Backwash rate (gpm/ft}^2) \\ \times \text{ filter area (ft}^2) \end{bmatrix}$$

$$= 20 \text{ gpm/ft}^2 \times (25 \text{ ft} \times 20 \text{ ft})$$

$$= 10,000 \text{ gpm}$$

■ **EXAMPLE 5.151**

Problem: The desired backwash pumping rate for a filter is 12 gpm/ft². If the filter is 20 ft long by 20 ft wide, what backwash pumping rate (gpm) will be required?

Solution:

$$\text{Pumping rate (gpm)} = \begin{bmatrix} \text{Backwash rate (gpm/ft}^2) \\ \times \text{filter area (ft}^2) \end{bmatrix}$$

$$= 12 \text{ gpm/ft}^2 \times (20 \text{ ft} \times 20 \text{ ft})$$

$$= 4800 \text{ gpm}$$

5.6.8.9 Percent Product Water Used for Backwashing

Along with measuring filtration rate and filter run time, another aspect of filter operation that is monitored for filter performance is the percent of product water used for backwashing. The equation for percent of product water used for backwashing calculations used is shown below:

$$\text{Backwash water (\%)} = \frac{\text{Backwash water (gal)}}{\text{Water filtered (gal)}} \times 100 \quad (5.92)$$

■ **EXAMPLE 5.152**

Problem: A total of 18,100,000 gal of water was filtered during a filter run. If backwashing used 74,000 gal of this product water, what percent of the product water was used for backwashing?

Solution:

$$\text{Backwash water (\%)} = \frac{\text{Backwash water (gal)}}{\text{Water filtered (gal)}} \times 100$$

$$= \frac{74,000 \text{ gal}}{18,100,000 \text{ gal}} \times 100 = 0.4\%$$

■ **EXAMPLE 5.153**

Problem: A total of 11,400,000 gal of water was filtered during a filter run. If backwashing used 48,500 gal of this product water, what percent of the product water was used for backwashing?

Solution:

$$\text{Backwash water (\%)} = \frac{\text{Backwash water (gal)}}{\text{Water filtered (gal)}} \times 100$$

$$= \frac{48,500 \text{ gal}}{11,400,000 \text{ gal}} \times 100 = 0.43\%$$

5.6.8.10 Percent Mud Ball Volume

Mud balls are heavier deposits of solids near the top surface of the medium that break into pieces during backwash, resulting in spherical accretions of floc and sand (usually less than 12 in. in diameter). The presence of mud balls in the filter media is checked periodically. The principal objection to mudballs is that they diminish the effective filter area.

■ **EXAMPLE 5.154**

Problem: A 3350-mL sample of filter media was taken for mud ball evaluation. The volume of water in the graduated cylinder rose from 500 mL to 525 mL when mud balls were placed in the cylinder. What is the percent mud ball volume of the sample?

Solution: First determine the volume of mud balls in the sample:

$$525 \text{ mL} - 500 \text{ mL} = 25 \text{ mL}$$

Then calculate the percent mud ball volume:

$$\text{Mud ball volume (\%)} = \frac{\text{Mud ball volume (mL)}}{\text{Total sample volume (mL)}} \times 100$$

$$= \frac{25 \text{ mL}}{3350 \text{ mL}} \times 100 = 0.75\%$$

■ **EXAMPLE 5.155**

Problem: A filter is tested for the presence of mud balls. The mud ball sample has a total sample volume of 680 mL. Five samples were taken from the filter. When the mud balls were placed in 500 mL of water, the water level rose to 565 mL. What is the percent mud ball volume of the sample?

Solution:

$$\text{Mud ball volume (\%)} = \frac{\text{Mud ball volume (mL)}}{\text{Total sample volume (mL)}} \times 100$$

The mud ball volume is the volume that the water rose:

$$565 \text{ mL} - 500 \text{ mL} = 65 \text{ mL}$$

Because 5 samples of media were taken, the total sample volume is 5 times the sample volume:

$$5 \times 680 \text{ mL} = 3400 \text{ mL}$$

% Mud ball volume = (65 mL/3400 mL) × 100 = 1.9%

5.6.8.11 Filter Bed Expansion

In addition to backwash rate, it is also important to expand the filter media during the wash to maximize the removal of particles held in the filter or by the media; that is, the efficiency of the filter wash operation depends on the expansion of the sand bed. Bed expansion is determined by measuring the distance from the top of the unexpanded media to a reference point (e.g., top of the filter wall) and from the top of the expanded media to the same reference. A proper backwash rate should expand the filter 20 to 25%. Percent bed expansion is given by dividing the bed expansion by the total depth of expandable media (i.e., media depth less support gravels) and multiplied by 100, as follows:

Expanded measurement = Depth to top of media during backwash (in.)

Unexpanded measurement = Depth to top of media before backwash (in.)

Bed expansion = Unexpanded measurement (in.) − expanded measurement (in.)

$$\text{Bed expansion (\%)} = \frac{\text{Bed expansion (in.)}}{\text{Total expandable media depth (in.)}} \times 100 \quad (5.93)$$

■ EXAMPLE 5.156

Problem: The backwashing practices for a filter with 30 in. of anthracite and sand are being evaluated. While at rest, the distance from the top of the media to the concrete floor surrounding the top of filter is measured to be 41 in. After the backwash has begun and the maximum backwash rate is achieved, a probe containing a white disk is slowly lowered into the filter bed until anthracite is observed on the disk. The distance from the expanded media to the concrete floor is measured to be 34.5 in. What is the percent bed expansion?

Solution:

Given:

Unexpanded measurement = 41 in.
Expanded measurement = 34.5 in.

Bed expansion = 41 in. − 34.5 in. = 6.5

% Bed expansion = (6.5 in./30 in.) × 100 = 22%

5.6.9 WATER CHLORINATION CALCULATIONS

Chlorine is the most commonly used substance for disinfection of water in the United States. The addition of chlorine or chlorine compounds to water is called *chlori-*

nation. Chlorination is considered to be the single most important process for preventing the spread of waterborne disease.

5.6.9.1 Chlorine Disinfection

Chlorine can destroy most biological contaminants by various mechanisms, including:

*   Damaging the cell wall
*   Altering the permeability of the cell (the ability to pass water in and out through the cell wall)
*   Altering the cell protoplasm
*   Inhibiting the enzyme activity of the cell so it is unable to use its food to produce energy
*   Inhibiting cell reproduction

Chlorine is available in a number of different forms: (1) as pure elemental gaseous chlorine (a greenish-yellow gas possessing a pungent and irritating odor that is heavier than air, nonflammable, and nonexplosive), which, when released to the atmosphere, is toxic and corrosive; (2) as solid calcium hypochlorite (in tablets or granules); or (3) as a liquid sodium hypochlorite solution (in various strengths). The choice of one form of chlorine over another for a given water system depends on the amount of water to be treated, configuration of the water system, the local availability of the chemicals, and the skill of the operator. One of the major advantages of using chlorine is the effective residual that it produces. A residual indicates that disinfection is completed and the system has an acceptable bacteriological quality. Maintaining a residual in the distribution system helps to prevent regrowth of those microorganisms that were injured but not killed during the initial disinfection stage.

5.6.9.2 Determining Chlorine Dosage (Feed Rate)

The expressions milligrams per liter (mg/L) and pounds per day (lb/day) are most often used to describe the amount of chlorine added or required. Equation 5.95 can be used to calculate either mg/L or lb/day chlorine dosage:

$$\text{Chlorine feed rate (lb/day)} = \begin{bmatrix} \text{Chlorine (mg/L)} \\ \times \text{flow (MGD)} \\ \times 8.34 \text{ lb/gal} \end{bmatrix} \quad (5.94)$$

■ EXAMPLE 5.157

Problem: Determine the chlorinator setting (lb/day) required to treat a flow of 4 MGD with a chlorine dose of 5 mg/L.

Solution:

$$\text{Chlorine feed rate (lb/day)} = \left[\begin{array}{l} \text{Chlorine (mg/L)} \\ \times \text{flow (MGD)} \times 8.34 \text{ lb/gal} \end{array} \right]$$

$$= \left[\begin{array}{l} 5 \text{ mg/L} \times 4 \text{ MGD} \\ \times 8.34 \text{ lb/gal} \end{array} \right]$$

$$= 167 \text{ lb/day}$$

■ EXAMPLE 5.158

Problem: A pipeline that is 12 in. in diameter and 1400 ft long is to be treated with a chlorine dose of 48 mg/L. How many pounds of chlorine will this require?

Solution: First determine the gallon volume of the pipeline:

$$\text{Volume (gal)} = \left[\begin{array}{l} 0.785 \times (\text{diameter})^2 \\ \times \text{length (ft)} \times 7.48 \text{ gal/ft}^3 \end{array} \right]$$

$$= 0.785 \times (1 \text{ ft})^2 \times 1400 \text{ ft} \times 7.48 \text{ gal/ft}^3$$

$$= 8221 \text{ gal}$$

Now calculate the pounds chlorine required:

$$\text{Chlorine (lb)} = \left[\begin{array}{l} \text{Chlorine (mg/L)} \times \text{volume (MG)} \\ \times 8.34 \text{ lb/gal} \end{array} \right]$$

$$= 48 \text{ mg/L} \times 0.008221 \text{ MG} \times 8.34 \text{ lb/gal}$$

$$= 3.3 \text{ lb}$$

■ EXAMPLE 5.159

Problem: A chlorinator setting is 30 lb per 24 hr. If the flow being chlorinated is 1.25 MGD, what is the chlorine dosage expressed as mg/L?

Solution:

$$\text{Chlorine (lb/day)} = \left[\begin{array}{l} \text{Chlorine (mg/L)} \times \text{flow (MGD)} \\ \times 8.34 \text{ lb/gal} \end{array} \right]$$

$$30 \text{ lb/day} = x \text{ mg/L} \times 1.25 \text{ MGD} \times 8.34 \text{ lb/gal}$$

$$x = \frac{30}{1.25 \text{ MGD} \times 8.34 \text{ lb/gal}}$$

$$= 2.9 \text{ mg/L}$$

■ EXAMPLE 5.160

Problem: A flow of 1600 gpm is to be chlorinated. At a chlorinator setting of 48 lb per 24 hr, what would be the chlorine dosage in mg/L?

Solution: Convert the gpm flow rate to MGD flow rate:

$$1600 \text{ gpm} \times 1440 \text{ min/day} = 2,304,000 \text{ gpd}$$

$$= 2.304 \text{ MGD}$$

Now calculate the chlorine dosage in mg/L:

$$\text{Chlorine (lb/day)} = \left[\begin{array}{l} \text{Chlorine (mg/L)} \times \text{flow (MGD)} \\ \times 8.34 \text{ lb/gal} \end{array} \right]$$

$$48 \text{ lb/day} = x \text{ mg/L} \times 2.304 \text{ MGD} \times 8.34 \text{ lb/gal}$$

$$x = \frac{48}{2.304 \text{ MGD} \times 8.34 \text{ lb/gal}}$$

$$= 2.5 \text{ mg/L}$$

5.6.9.3 Calculating Chlorine Dose, Demand, and Residual

Common terms used in chlorination include the following:

- *Chlorine dose*—The amount of chlorine added to the system. It can be determined by adding the desired residual for the finished water to the chlorine demand of the untreated water. Dosage can be either milligrams per liter (mg/L) or pounds per day (lb/day). The most common is mg/L:

$$\text{Chlorine dose (mg/L)} = \left[\begin{array}{l} \text{Chlorine demand (mg/L)} \\ + \text{chlorine (mg/L)} \end{array} \right]$$

- *Chlorine demand*—The amount of chlorine used by iron, manganese, turbidity, algae, and microorganisms in the water. Because the reaction between chlorine and microorganisms is not instantaneous, demand is relative to time; for example, the demand 5 min after applying chlorine will be less than the demand after 20 min. Demand, like dosage, is expressed in mg/L:

$$\text{Chlorine demand (mg/L)} = \left[\begin{array}{l} \text{Chlorine dose (mg/L)} \\ - \text{chlorine residual (mg/L)} \end{array} \right]$$

- *Chlorine residual*—The amount of chlorine (determined by testing) remaining after the demand is satisfied. Residual, like demand, is based on time. The longer the time after dosage, the lower the residual will be, until all of the demand has been satisfied. Residual, like dosage and demand, is expressed in mg/L. The presence of a *free residual* of at least 0.2 to 0.4 ppm usually provides a high degree of assurance that the disinfection of the water is complete. *Combined residual* is the result of combining free chlorine with nitrogen compounds. Combined residuals are also called *chloramines*. *Total chlorine residual* is the mathematical combination of free and combined residuals. Total residual can be determined directly with standard chlorine residual test kits.

The following examples illustrate the calculation of chlorine dose, demand, and residual using Equation 5.95:

$$\text{Chlorine dose (mg/L)} = \begin{bmatrix} \text{Chlorine demand (mg/L)} \\ + \text{ chlorine residual (mg/L)} \end{bmatrix} \quad (5.95)$$

■ EXAMPLE 5.161

Problem: A water sample is tested and found to have a chlorine demand of 1.7 mg/L. If the desired chlorine residual is 0.9 mg/L, what is the desired chlorine dose in mg/L?

Solution:

$$\text{Chlorine dose (mg/L)} = \begin{bmatrix} \text{Chlorine demand (mg/L)} \\ + \text{ chlorine residual (mg/L)} \end{bmatrix}$$

$$= 1.7 \text{ mg/L} + 0.9 \text{ mg/L}$$

$$= 2.6 \text{ mg/L}$$

■ EXAMPLE 5.162

Problem: The chlorine dosage for water is 2.7 mg/L. If the chlorine residual after 30 min of contact time is found to be 0.7 mg/L, what is the chlorine demand expressed in mg/L?

Solution:

$$\text{Chlorine dose (mg/L)} = \begin{bmatrix} \text{Chlorine demand (mg/L)} \\ + \text{ chlorine residual (mg/L)} \end{bmatrix}$$

$$2.7 \text{ mg/L} = x \text{ mg/L} + 0.6 \text{ mg/L}$$

$$x \text{ mg/L} = 2.7 \text{ mg/L} - 0.7 \text{ mg/L}$$

$$= 2.0 \text{ mg/L}$$

■ EXAMPLE 5.163

Problem: What should the chlorinator setting (lb/day) be to treat a flow of 2.35 MGD if the chlorine demand is 3.2 mg/L and a chlorine residual of 0.9 mg/L is desired?

Solution: Determine the chlorine dosage in mg/L:

$$\text{Chlorine dose (mg/L)} = \begin{bmatrix} \text{Chlorine demand (mg/L)} \\ + \text{ chlorine residual (mg/L)} \end{bmatrix}$$

$$= 3.2 \text{ mg/L} + 0.9 \text{ mg/L}$$

$$= 4.1 \text{ mg/L}$$

Calculate the chlorine dosage (feed rate) in lb/day:

$$\text{Chlorine (lb/day)} = \begin{bmatrix} \text{Chlorine (mg/L)} \times \text{flow (MGD)} \\ \times 8.34 \text{ lb/gal} \end{bmatrix}$$

$$= 4.1 \text{ mg/L} \times 2.35 \text{ MGD} \times 8.34 \text{ lb/gal}$$

$$= 80.4 \text{ lb/day}$$

To calculate the actual increase in chlorine residual that would result from an increase in chlorine dose, we use the mg/L to lb/day equation as shown below:

$$\begin{matrix} \text{Chlorine increase} \\ \text{(lb/day)} \end{matrix} = \begin{bmatrix} \text{Expected increase (mg/L)} \\ \times \text{ flow (MGD)} \\ \times 8.34 \text{ lb/gal} \end{bmatrix} \quad (5.96)$$

Key Point: The actual increase in residual is simply a comparison of new and old residual data.

■ EXAMPLE 5.164

Problem: A chlorinator setting is increased by 2 lb/day. The chlorine residual before the increased dosage was 0.2 mg/L. After the increased chlorine dose, the chlorine residual was 0.5 mg/L. The average flow rate being chlorinated is 1.25 MGD. Is the water being chlorinated beyond the breakpoint?

Solution: Calculate the expected increase in chlorine residual using the mg/L to lb/day equation:

$$\begin{matrix} \text{Chlorine increase} \\ \text{(lb/day)} \end{matrix} = \begin{bmatrix} \text{Expected increase (mg/L)} \\ \times \text{ flow (MGD)} \times 8.34 \text{ lb/gal} \end{bmatrix}$$

$$2 \text{ lb/day} = \begin{bmatrix} x \text{ mg/L} \times 1.25 \text{ MGD} \\ \times 8.34 \text{ lb/gal} \end{bmatrix}$$

$$x = \frac{2 \text{ lb/day}}{1.25 \text{ MGD} \times 8.34 \text{ lb/gal}}$$

$$= 0.19 \text{ mg/L}$$

The actual increase in residual chlorine is:

$$0.5 \text{ mg/L} - 0.19 \text{ mg/L} = 0.31 \text{ mg/L}$$

■ EXAMPLE 5.165

Problem: A chlorinator setting of 18 lb chlorine per 24 hr results in a chlorine residual of 0.3 mg/L. The chlorinator setting is increased to 22 lb per 24 hr. The chlorine residual increased to 0.4 mg/L at this new dosage rate. The average flow being treated is 1.4 MGD. On the basis of these data, is the water being chlorinated past the breakpoint?

Solution: Calculate the expected increase in chlorine residual:

$$\text{Chlorine increase} \atop \text{(lb/day)} = \left[\begin{matrix} \text{Expected increase (mg/L)} \\ \times \text{flow (MGD)} \times 8.34 \text{ lb/gal} \end{matrix} \right]$$

$$4 \text{ lb/day} = \left[\begin{matrix} x \text{ mg/L} \times 1.4 \text{ MGD} \\ \times 8.34 \text{ lb/gal} \end{matrix} \right]$$

$$x = \frac{4 \text{ lb/day}}{1.4 \text{ MGD} \times 8.34 \text{ lb/gal}}$$

$$= 0.34 \text{ mg/L}$$

The actual increase in residual chlorine is:

$$0.4 \text{ mg/L} - 0.3 \text{ mg/L} = 0.1 \text{ mg/L}$$

5.6.9.4 Calculating Dry Hypochlorite Rate

The most commonly used dry hypochlorite, calcium hypochlorite contains about 65 to 70% available chlorine, depending on the brand. Because hypochlorites are not 100% pure chorine, more lb/day must be fed into the system to obtain the same amount of chlorine for disinfection. The equation used to calculate the lb/day hypochlorite required for disinfection can be found using Equation 5.97:

$$\text{Hypochlorite (lb/day)} = \frac{\text{Chlorine (lb/day)}}{\left(\dfrac{\% \text{ Available chlorine}}{100} \right)} \quad (5.97)$$

■ EXAMPLE 5.166

Problem: A chlorine dosage of 110 lb/day is required to disinfect a flow of 1,550,000 gpd. If the calcium hypochlorite to be used contains 65% available chlorine, how many lb/day hypochlorite will be required for disinfection?

Solution: Because only 65% of the hypochlorite is chlorine, more than 110 lb of hypochlorite will be required:

$$\text{Hypochlorite (lb/day)} = \frac{\text{Chlorine (lb/day)}}{\left(\dfrac{\% \text{ Available chlorine}}{100} \right)}$$

$$= \frac{110 \text{ lb/day}}{\dfrac{65}{100}} = \frac{110 \text{ lb/day}}{0.65}$$

$$= 169 \text{ lb/day}$$

■ EXAMPLE 5.167

Problem: A water flow of 900,000 gpd requires a chlorine dose of 3.1 mg/L. If calcium hypochlorite (65% available chlorine) is to be used, how many lb/day of hypochlorite are required?

Solution: Calculate the lb/day chlorine required:

$$\text{Chlorine (lb/day)} = \left[\begin{matrix} \text{Chlorine (mg/L)} \times \text{flow (MGD)} \\ \times 8.34 \text{ lb/gal} \end{matrix} \right]$$

$$= 3.1 \text{ mg/L} \times 0.90 \text{ MGD} \times 8.34 \text{ lb/gal}$$

$$= 23 \text{ lb/day}$$

Calculate the lb/day hypochlorite:

$$\text{Hypochlorite (lb/day)} = \frac{\text{Chlorine (lb/day)}}{\left(\dfrac{\% \text{ Available chlorine}}{100} \right)}$$

$$= \frac{23 \text{ lb/day}}{0.65} = 169 \text{ lb/day}$$

■ EXAMPLE 5.168

Problem: A tank contains 550,000 gal of water and is to receive a chlorine dose of 2.0 mg/L. How many pounds of calcium hypochlorite (65% available chlorine) will be required?

Solution:

$$\text{Hypochlorite (lb)} = \frac{\text{Chlorine (mg/L)} \times \text{volume (MG)} \times 8.34 \text{ lb/gal}}{\left(\dfrac{\% \text{ Available chlorine}}{100} \right)}$$

$$= \frac{2.0 \text{ mg/L} \times 0.550 \text{ MG} \times 8.34 \text{ lb/gal}}{\dfrac{65}{100}}$$

$$= \frac{9.2 \text{ lb}}{0.65} = 14.2 \text{ lb}$$

■ **EXAMPLE 5.169**

Problem: A total of 40 lb of calcium hypochlorite (65% available chlorine) is used in a day. If the flow rate treated is 1,100,000 gpd, what is the chlorine dosage in mg/L?

Solution: Calculate the lb/day chlorine dosage:

$$\text{Hypochlorite (lb/day)} = \frac{\text{Chlorine (lb/day)}}{\dfrac{\% \text{ Available chlorine}}{100}}$$

$$40 \text{ lb/day} = \frac{x \text{ lb/day}}{0.65}$$

$$0.65 \times 40 = x \text{ lb/day}$$

$$x = 26 \text{ lb/day chlorine}$$

Then calculate mg/L chlorine using the mg/L to lb/day equation and filling in the known information:

$$26 \text{ lb/day chlorine} = \begin{bmatrix} x \text{ mg/L chlorine} \times 1.10 \text{ MGD} \\ \times 8.34 \text{ lb/gal} \end{bmatrix}$$

$$x = \frac{26 \text{ lb/day}}{1.10 \text{ MGD} \times 8.34 \text{ lb/gal}}$$

$$= 2.8 \text{ mg/L}$$

■ **EXAMPLE 5.170**

Problem: A flow of 2,550,000 gpd is disinfected with calcium hypochlorite (65% available chlorine). If 50 lb of hypochlorite are used in a 24-hr period, what is the mg/L chlorine dosage?

Solution: Calculate the lb/day chlorine dosage:

$$50 \text{ lb/day hypochlorite} = \frac{x \text{ lb/day chlorine}}{0.65}$$

$$x = 32.5 \text{ lb/day chlorine}$$

Calculate mg/L chlorine:

$$\begin{bmatrix} x \text{ mg/L chlorine} \times 2.55 \text{ MGD} \\ \times 8.34 \text{ lb/gal} \end{bmatrix} = 32.5 \text{ lb/day}$$

$$x = 1.5 \text{ mg/L chlorine}$$

5.6.9.5 Calculating Hypochlorite Solution Feed Rate

Liquid hypochlorite (i.e., sodium hypochlorite) is supplied as a clear, greenish-yellow liquid in strengths varying from 5.25 to 16% available chlorine. Often referred to as *bleach*,

it is, in fact, used for bleaching. Common household bleach is a solution of sodium hypochlorite containing 5.25% available chlorine. When calculating gallons per day (gpd) liquid hypochlorite, the lb/day hypochlorite required must be converted to gpd hypochlorite. This conversion is accomplished using Equation 5.98:

$$\text{Hypochlorite (gpd)} = \frac{\text{Hypochlorite (lb/day)}}{8.34 \text{ lb/gal}} \quad (5.98)$$

■ **EXAMPLE 5.171**

Problem: A total of 50 lb/day sodium hypochlorite is required for disinfection of a 1.5-MGD flow. How many gallons per day hypochlorite is this?

Solution: Because lb/day hypochlorite has already has been calculated, we simply convert lb/day to gpd hypochlorite required:

$$\text{Hypochlorite (gpd)} = \frac{\text{Hypochlorite (lb/day)}}{8.34 \text{ lb/gal}}$$

$$= \frac{50 \text{ lb/day}}{8.34 \text{ lb/gal}}$$

$$= 6.0 \text{ gpd}$$

■ **EXAMPLE 5.172**

Problem: A hypochlorinator is used to disinfect the water pumped from a well. The hypochlorite solution contains 3% available chlorine. A chlorine dose of 1.3 mg/L is required for adequate disinfection throughout the system. If the flow being treated is 0.5 MGD, how many gpd of the hypochlorite solution will be required?

Solution: Calculate the lb/day chlorine required:

$$\text{Chlorine (lb/day)} = \begin{bmatrix} 1.3 \text{ mg/L} \times 0.5 \text{ MGD} \\ \times 8.34 \text{ lb/gal} \end{bmatrix}$$

$$= 5.4 \text{ lb/day}$$

Calculate the lb/day hypochlorite solution required:

$$\text{Hypochlorite (lb/day)} = \frac{5.4 \text{ lb/day chlorine}}{0.03}$$

$$= 180 \text{ lb/day}$$

Calculate the gpd hypochlorite solution required:

$$\text{Hypochlorite (gpd)} = \frac{180 \text{ lb/day}}{8.34 \text{ lb/gal}} = 21.6 \text{ gpd}$$

5.6.9.6 Calculating Percent Strength of Solutions

If a teaspoon of salt is dropped into a glass of water it gradually disappears. The salt dissolves in the water, but a microscopic examination of the water would not show the salt. Only examination at the molecular level, which is not easily done, would show the salt and water molecules intimately mixed. If we taste the liquid, we would know that the salt is there. We can recover the salt by evaporating the water. In a solution, the molecules of the salt, the *solute,* are homogeneously dispersed among the molecules of water, the *solvent.* This mixture of salt and water is homogeneous on a molecular level. Such a homogeneous mixture is called a *solution.* The composition of a solution can be varied within certain limits. The three common states of matter are gas, liquid, and solid. In this discussion, of course, we are only concerned, at the moment, with the solid (calcium hypochlorite) and liquid (sodium hypochlorite) states.

5.6.9.7 Calculating Percent Strength Using Dry Hypochlorite

To calculate the percent strength of a chlorine solution, we use Equation 5.99:

$$\% \text{ Chlorine strength} = \frac{\text{Hypochlorite (lb)} \times \dfrac{\% \text{ available chlorine}}{100}}{\text{Water (lb)} + \left[\text{hypo- chlorite (lb)} \times \dfrac{\% \text{ available chlorine}}{100}\right]} \times 100 \quad (5.99)$$

■ EXAMPLE 5.173

Problem: If a total of 72 oz. of calcium hypochlorite (65% available chlorine) is added to 15 gal of water, what is the percent chlorine strength (by weight) of the solution?

Solution: Convert the ounces of hypochlorite to pounds of hypochlorite:

$$72 \text{ oz.} \div 16 \text{ oz./lb} = 4.5 \text{ lb chemical}$$

$$\% \text{ Chlorine strength} = \frac{\text{Hypochlorite (lb)} \times \dfrac{\% \text{ available chlorine}}{100}}{\text{Water (lb)} + \left[\text{hypo- chlorite (lb)} \times \dfrac{\% \text{ available chlorine}}{100}\right]} \times 100$$

$$= \frac{4.5 \text{ lb} \times 0.65}{(15 \text{ gal} \times 8.34 \text{ lb/gal}) + (4 \text{ lb} \times 0.65)} \times 100$$

$$= \frac{2.9 \text{ lb}}{125.1 \text{ lb} + 2.6 \text{ lb}} = \frac{2.9}{128} \times 100 = 2.3\%$$

5.6.10 CHEMICAL USE CALCULATIONS

In a typical plant operation, chemical use is recorded each day. Such data provide a record of daily use from which the average daily use of the chemical or solution can be calculated. To calculate average use in pounds per day (lb/day), we use Equation 5.100. To calculate average use in gallons per day (gpd), we use Equation 5.101:

$$\text{Average use (lb/day)} = \frac{\text{Total chemical used (lb)}}{\text{Number of days}} \quad (5.100)$$

$$\text{Average use (gpd)} = \frac{\text{Total chemical used (gal)}}{\text{Number of days}} \quad (5.101)$$

To calculate the days supply in inventory, we use Equation 5.102 or Equation 5.103:

$$\text{Days supply} = \frac{\text{Total chemical in inventory (lb)}}{\text{Average use (lb/day)}} \quad (5.102)$$

$$\text{Days supply} = \frac{\text{Total chemical in inventory (gal)}}{\text{Average use (gpd)}} \quad (5.103)$$

■ EXAMPLE 5.174

Problem: Calcium hypochlorite usage for each day during a week is given below. Based on these data, what was the average lb/day hypochlorite chemical use during the week?

Monday	50 lb/day
Tuesday	55 lb/day
Wednesday	51 lb/day
Thursday	46 lb/day
Friday	56 lb/day
Saturday	51 lb/day
Sunday	48 lb/day

Solution:

$$\text{Average use (lb/day)} = \frac{\text{Total chemical used (lb)}}{\text{Number of days}}$$

$$= \frac{357 \text{ lb}}{7 \text{ days}} = 51 \text{ lb/day}$$

■ EXAMPLE 5.175

Problem: The average calcium hypochlorite use at a plant is 40 lb/day. If the chemical inventory in stock is 1100 lb, how many days supply is this?

$$\text{Days supply} = \frac{\text{Total chemical in inventory (lb)}}{\text{Average use (lb/day)}}$$

$$= \frac{1100 \text{ lb in inventory}}{40 \text{ lb/day average use}} = 27.5 \text{ days}$$

5.6.11 FLUORIDATION

The key terms used in this section are defined as follows:

- *Fluoride*—Element found in many waters; also added to water systems to reduce tooth decay
- *Dental caries*—Tooth decay
- *Dental fluorosis*—Result of excessive fluoride content in drinking water causing mottled and discolored teeth

5.6.11.1 Water Fluoridation

As of 1989, fluoridation in the United States was being practiced in over 8000 communities serving more than 126 million people. More than 9 million residents of over 1800 additional communities were consuming water containing at least 0.7 mg/L fluoride from natural sources. Some key facts about fluoride include:

- Fluoride is seldom found in appreciable quantities in surface waters and appears in groundwater in only a few geographical regions.
- Fluoride is sometimes found in a few types of igneous or sedimentary rocks.
- Fluoride is toxic to humans in large quantities; it is also toxic to some animals.
- Based on human experience, fluoride, used in small concentrations (about 1.0 mg/L in drinking water), can be beneficial.

5.6.11.2 Fluoride Compounds

Theoretically, any compound that forms fluoride ions in water solution can be used for adjusting the fluoride content of a water supply; however, several practical considerations are involved in selecting compounds:

- The compound must have sufficient solubility to permit its use in routine water plant practice.
- The cation to which the fluoride ion is attached must not have any undesirable characteristics.
- The material should be relatively inexpensive and readily available in grades of size and purity suitable for their intended use.

Caution: Fluoride chemicals, like chlorine, caustic soda, and many other chemicals used in water treatment, can constitute a safety hazard for the water plant operator unless proper handling precautions are observed. It is essential that the operator be aware of the hazards associated with each individual chemical prior to its use.

The three commonly used fluoride chemicals should meet the American Water Works Associations (AWWA) standards for use in water fluoridation: sodium fluoride, B701-90; sodium fluorosilicate, B702-90; and fluorosilicic acid, B703-90.

TABLE 5.5
Solubility of Fluoride Chemicals

Chemical	Temperature	Solubility (g per 100 mL of H$_2$O)
Sodium fluoride	0.0	4.00
	15.0	4.03
	20.0	4.05
	25.0	4.10
	100.0	5.00
Sodium fluorosilicate	0.0	0.44
	25.0	0.76
	37.8	0.98
	65.6	1.52
	100.0	2.45
Fluorosilicic acid	Infinite at all temperatures	

Source: DHHS, *Water Fluoridation: A Manual for Water Plant Operators*, U.S. Department of Health and Human Services, Washington, D.C., 1994, p. 17.

5.6.11.2.1 Sodium Fluoride

The first fluoride compound used in water fluoridation was *sodium fluoride* (NaF). It was selected based on the above criteria and also because its toxicity and physiological effects had been so thoroughly studied. It has become the reference standard used in measuring fluoride concentration. Other compounds came into use, but sodium fluoride is still widely used because of its unique physical characteristics. Sodium fluoride is a white, odorless material available either as a powder or in the form of crystals of various sizes. It is a salt that in the past was manufactured by adding sulfuric acid to fluorspar and then neutralizing the mixture with sodium carbonate. It is now produced by neutralizing fluorosilicic acid with caustic soda (NaOH). Approximately 19 lb of sodium fluoride will add 1 ppm of fluoride to 1 million gal of water. The solubility of sodium fluoride is practically constant at 4 g per 100 mL in water at temperatures generally encountered in water treatment practice (see Table 5.5).

5.6.11.2.2 Sodium Fluorosilicate

Fluorosilicic acid can readily be converted into various salts. One of these, *sodium fluorosilicate* (Na$_2$SiF$_6$), also known as sodium silicofluoride, is widely used as a chemical for water fluoridation. As with most fluorosilicates, it is generally obtained as a byproduct from the manufacture of phosphate fertilizers. Phosphate rock is ground up and treated with sulfuric acid, thus forming a gas byproduct. This gas reacts with water and forms fluorosilicic acid. When neutralized with sodium carbonate, sodium fluorosilicate will precipitate out. The conversion of fluorosilicic acid to a dry material containing a high percentage of available fluoride results in a compound with most of the advantages of the acid and few of its disadvantages. When

TABLE 5.6
Properties of Fluorosilicic Acid

Acid (%)[a]	Specific Gravity	Density (lb/gal)
0 (water)	1.000	8.345
10	1.0831	9.041
20	1.167	9.739
23	1.191	9.938
25	1.208	10.080
30	1.250	10.431
35	1.291	10.773

[a] Based on the other percentage being distilled water.

Note: Actual densities and specific gravities will be slightly higher when distilled water is not used. Add approximately 0.2 lb/gal to density depending on impurities.

Source: DHHS, *Water Fluoridation: A Manual for Water Plant Operators*, U.S. Department of Health and Human Services, Washington, D.C., 1994, p. 19.

it was shown that fluorosilicates form fluoride ions in water solution as readily as do simple fluoride compounds and that there is no difference in the physiological effect, fluorosilicates were rapidly accepted for water fluoridation. In many cases, they have displaced the use of sodium fluoride, except in saturators. Sodium fluorosilicate is a white, odorless, crystalline powder. Its solubility varies (see Table 5.5). Approximately 14 lb of sodium fluorosilicate will add 1 ppm of fluoride to 1 million gal of water.

5.6.11.2.3 Fluorosilicic Acid

Fluorosilicic acid (H_2SiF_6), also known as *hydrofluorosilicic* or *silicofluoric acid*, is a 20 to 35% aqueous solution with a formula weight of 144.08. It is a straw-colored, transparent, fuming, corrosive liquid having a pungent odor and an irritating action on the skin. Solutions of 20 to 35% fluorosilicic acid exhibit a low pH (1.2) and at a concentration of 1 ppm can slightly depress the pH of poorly buffered potable waters. It must be handled with great care because it can cause a delayed burn on skin tissue. The specific gravity and density of fluorosilicic acid are given in Table 5.6. Approximately 46 lb (4.4 gal) of 23% acid are required to add 1 ppm of fluoride to 1 million gal of water. Two different processes, resulting in products with differing characteristics, are used to manufacture fluorosilicic acid. The largest source of the acid is phosphate fertilizer manufacture, for which it is a byproduct. Phosphate rock is ground up and treated with sulfuric acid to form a gas byproduct. Hydrofluoric acid (HF) is an extremely corrosive material. Its presence in fluorosilicic acid, whether from intentional addition (i.e., fortified acid) or from normal production processes, demands careful handling.

5.6.11.3 Optimal Fluoride Levels

The recommended optimal fluoride concentrations for fluoridated water supply systems are given in Table 5.7. These levels are based on the annual average of the maximum daily air temperature in the area of the involved school or community. In areas where the mean temperature is not shown on Table 5.7, the optimal fluoride level can be determined by the following formula:

$$\text{Fluoride (ppm)} = \frac{0.34}{E} \qquad (5.104)$$

TABLE 5.7
Recommended Optimal Fluoride Level

Annual Average of Maximum Daily Air Temperatures (°F)[a]	Recommended Fluoride Concentrations		Recommended Control Range			
	Community (ppm)	School[b] (ppm)	Community Systems		School Systems	
			0.1 Below	0.5 Above	20% Low	20% High
40.0–53.7	1.2	5.4	1.1	1.7	4.3	6.5
53.8–58.3	1.1	5.0	1.0	1.6	4.0	6.0
58.4–63.8	1.0	4.5	0.9	1.5	3.6	5.4
63.9–70.6	0.9	4.1	0.8	1.4	3.3	4.9
70.7–79.2	0.8	3.6	0.7	1.3	2.9	4.3
79.3–90.5	0.7	3.2	0.6	1.2	2.6	3.8

[a] Based on temperature data obtained for a minimum of 5 years.
[b] Based on 4.5 times the optimal fluoride level for communities.

Source: DHHS, *Water Fluoridation: A Manual for Water Plant Operators*, U.S. Department of Health and Human Services, Washington, D.C., 1994, p. 21.

where E is the estimated average daily water consumption for children through 10 years of age in ounces of water per pound of body weight. E is obtained from the formula:

$$E = 0.038 + 0.0062 \text{ avg. max. daily air temp. (°F)} \quad (5.105)$$

In Table 5.7, the recommended control range is shifted to the high side of the optimal fluoride level for two reasons:

1. It has become obvious that many water plant operators try to maintain the fluoride level in their community at the lowest level possible. The result is that the actual fluoride level in the water will vary around the lowest value in the range instead of around the optimal level.
2. Some studies have shown that suboptimal fluoride levels are relatively ineffective in preventing dental caries. Even a drop of 0.2 ppm below optimal levels can reduce dental benefits significantly.

Important Point: In water fluoridation, underfeeding is a much more serious problem than overfeeding.

5.6.11.4 Fluoridation Process Calculations

In this section, important process calculations are discussed.

5.6.11.4.1 Percent Fluoride Ion in a Compound

When calculating the percent fluoride ion present in a compound, we need to know the chemical formula for the compound (e.g., NaF) and the atomic weight of each element in the compound. The first step is to calculate the molecular weight of each element in the compound (number of atoms × atomic weight = molecular weight). Then we calculate the percent fluoride in the compound using Equation 5.106:

$$\% \text{ Fluoride} = \frac{\text{Molecular wt. of fluoride}}{\text{Molecular wt. of compound}} \times 100 \quad (5.106)$$

Important Point: Available fluoride ion concentration is abbreviated as AFI in the calculations that follow.

■ EXAMPLE 5.176

Problem: Given the following data, calculate the percent fluoride in sodium fluoride (NaF):

Element	No. of Atoms	Atomic Wt.	Molecular Wt.
Na	1	22.997	22.997
F	1	19.00	19.00

Molecular weight of NaF = 41.997

Solution: Calculate the percent fluoride in NaF:

$$\% \text{ Fluoride} = \frac{\text{Molecular wt. of fluoride}}{\text{Molecular wt. of compound}} \times 100$$

$$= \frac{19.00}{41.9997} \times 100$$

$$= 45.2\%$$

Key Point: The molecular weight of hydrofluorosilicic acid (H_2SiF_6) is 144.08, and the molecular weight of sodium silicofluoride (Na_2SiF_6) is 188.06.

5.6.11.4.2 Fluoride Feed Rate

Adjusting the fluoride level in a water supply to an optimal level is accomplished by adding the proper concentration of a fluoride chemical at a consistent rate. To calculate the fluoride feed rate for any fluoridation feeder in terms of pounds of fluoride to be fed per day, it is necessary to determine:

- Dosage
- Maximum pumping rate (capacity)
- Chemical purity
- Available fluoride ion concentration

The fluoride feed rate formula is a general equation used to calculate the concentration of a chemical added to water. It will be used for all fluoride chemicals except sodium fluoride when used in a saturator.

Important Point: mg/L is equal to ppm.

The fluoride feed rate (the amount of chemical required to raise the fluoride content to the optimal level) can be calculated as follows:

$$\text{Fluoride feed rate (lb/day)} = \frac{\text{Dosage (mg/L)} \times \text{capacity (MGD)} \times 8.34 \text{ lb/gal}}{\text{AFI} \times \text{chemical purity}} \quad (5.107)$$

If the capacity is in MGD, the fluoride feed rate will be in pounds per day. If the capacity is in gpm, the feed rate will be pounds per minute if a factor of 1 million is included in the denominator:

$$\text{Fluoride feed rate (lb/min)} = \frac{\text{Dosage (mg/L)} \times \text{capacity (gpm)} \times 8.34 \text{ lb/gal}}{1,000,000 \times \text{AFI} \times \text{chemical purity}} \quad (5.108)$$

■ EXAMPLE 5.177

Problem: A water plant produces 2000 gpm and the city wants to add 1.1 mg/L of fluoride. What would the fluoride feed rate be?

Solution:

$$2000 \text{ gpm} \times 1440 \text{ min/day} = 2{,}880{,}000 \text{ gpd}$$

$$2{,}880{,}000 \text{ gpd}/1{,}000{,}000 = 2.88 \text{ MGD}$$

$$\text{Feed rate (lb/day)} = \frac{1.1 \text{ mg/L} \times 2.88 \text{ MGD} \times 8.34 \text{ lb/gal}}{0.607 \times 0.985}$$

$$= 44.2 \text{ lb/day}$$

The fluoride feed rate is 44.2 lb/day. Some feed rates from equipment design data sheets are given in grams per minute. To convert to g/min, divide by 1440 min/day and multiply by 454 g/lb:

$$\text{Feed rate (g/min)} = \frac{44.2 \text{ lb/day}}{1440 \text{ min/day}} \times 454 \text{ g/lb}$$

$$= 13.9 \text{ g/min}$$

■ EXAMPLE 5.178

Problem: If it is known that the plant rate is 4000 gpm and the dosage needed is 0.8 mg/L, what is the fluoride feed rate in mL/min for 23% fluorosilicic acid?

Solution:

$$1{,}000{,}000 = 10^6$$

$$\text{Fluoride feed rate (lb/min)} = \frac{\text{Dosage (mg/L)} \times \text{capacity (gpm)} \times 8.34 \text{ lb/gal}}{10^6 \times \text{AFI} \times \text{chemical purity}}$$

$$= \frac{0.8 \text{ mg/L} \times 4000 \text{ gpm} \times 8.34 \text{ lb/gal}}{10^6 \times 0.79 \times 0.23}$$

$$= 0.147 \text{ lb/min}$$

Note: A gallon of 23% fluorosilicic acid weighs 10 lb, and there are 3785 mL/gal. The following formula can be used to convert the feed rate to mL/min:

$$\text{Feed rate (mL/min)} = \frac{0.147 \text{ lb/min}}{10 \text{ lb/gal}} \times 3785 \text{ mL/gal}$$

$$= 55.6 \text{ mL/min}$$

■ EXAMPLE 5.179

Problem: If a small water plant wishes to use sodium fluoride in a dry feeder and the water plant has a capacity (flow) of 180 gpm, what would be the fluoride feed rate? Assume that 0.1 mg/L natural fluoride and 1.0 mg/L are desired in the drinking water.

Important Point: Centers for Disease Control (CDC) recommends against using sodium fluoride in a dry feeder.

Solution:

$$\text{Fluoride feed rate (lb/min)} = \frac{\text{Dosage (mg/L)} \times \text{capacity (gpm)} \times 8.34 \text{ lb/gal}}{10^6 \times \text{AFI} \times \text{chemical purity}}$$

$$= \frac{(1.0 - 0.1) \text{ mg/L} \times 180 \text{ gpm} \times 8.34 \text{ lb/gal}}{10^6 \times 0.45 \times 0.98}$$

$$= 0.003 \text{ lb/min, or } 0.18 \text{ lb/hr}$$

Thus, sodium fluoride can be fed at a rate of 0.18 lb/hr to obtain 1.0 mg/L of fluoride in the water.

5.6.11.4.3 Fluoride Feed Rates for Saturator

A sodium fluoride saturator is unique in that the strength of the saturated solution formed is always 18,000 ppm. This is because sodium fluoride has a solubility that is practically constant at 4.0 g per 100 mL of water at temperatures generally encountered in water treatment. This means that each liter of solution contains 18,000 mg of fluoride ion: 40,000 mg/L × percent available fluoride (45%) = 18,000 mg/L. This simplifies calculations because it eliminates the need for weighing the chemicals. A meter on the water inlet of the saturator provides this volume, so all we need is the volume of solution added to the water:

$$\text{Fluoride feed rate (gpm)} = \frac{\text{Capacity (gpm)} \times \text{dosage (mg/L)}}{18{,}000 \text{ mg/L}} \quad (5.109)$$

The fluoride feed rate will have the same units as the capacity. If the capacity is in gallons per minute (gpm), the feed rate will also be in gpm. If the capacity is in gallons per day (gpd), the feed rate will also be in gpd.

Note: For the mathematician, the following derivation is given:

$$\text{Fluoride feed rate (lb/min)} = \frac{\text{Dosage (mg/L)} \times \text{capacity (gpm)} \times 8.34 \text{ lb/gal}}{10^6 \times \text{AFI} \times \text{chemical purity}} \quad (5.110)$$

To change the fluoride feed rate from pounds of dry feed to gallons of solution, divide by the concentration of sodium fluoride and the density of the solution (water).

Note: The chemical purity of the sodium fluoride in solution will be 4% × 8.34 lb/gal.

$$\text{Fluoride feed rate (gal/min)} = \frac{\text{Capacity (gpm)} \times \text{dosage (mg/L)} \times 8.34 \text{ lb/gal}}{10^6 \times \text{AFI} \times \text{chemical purity}}$$

$$= \frac{\text{Capacity (gpm)} \times \text{dosage (mg/L)} \times 8.34 \text{ lb/gal}}{10^6 \times 0.45 \times 4\% \times 8.34 \text{ lb/gal}} \quad (5.111)$$

$$= \frac{\text{Capacity (gpm)} \times \text{dosage (mg/L)}}{10^6 \times 0.45 \times .04}$$

$$= \frac{\text{Capacity (gpm)} \times \text{dosage (mg/L)}}{18,000 \text{ mg/L}}$$

■ EXAMPLE 5.180. Feed Rate for Saturator

Problem: A water plant produces 1.0 MGD and has less than 0.1 mg/L of natural fluoride. What would the fluoride feed rate be to obtain 1.0 mg/L in the water?

Solution:

$$\text{Fluoride feed rate (gpd)} = \frac{\text{Capacity (gpd)} \times \text{dosage (mg/L)}}{18,000 \text{ mg/L}}$$

$$= \frac{1,000,000 \text{ gpd} \times 1.0 \text{ mg/L}}{18,000 \text{ mg/L}}$$

$$= 55.6 \text{ gpd}$$

Thus, it takes approximately 56 gal of saturated solution to treat 1 MG of water at a dose of 1.0 mg/L.

5.6.11.4.4 Calculated Dosages

Some states require that records be kept regarding the amount of chemical used and that the theoretical concentration of chemical in the water be determined mathematically. To compute the theoretical concentration of fluoride, the calculated dosage must be determined. Adding the calculated dosage to the natural fluoride level in the water supply will yield the theoretical concentration of fluoride in the water. This number, the theoretical concentration, is calculated as a safety precaution to help ensure that an overfeed or accident does not occur. It is also an aid in solving troubleshooting problems. If the theoretical concentration is significantly higher or lower than the measured concentration, steps should be taken to determine the discrepancy. The fluoride feed rate formula can be changed to find the calculated dosage as follows:

$$\text{Dosage (mg/L)} = \frac{\text{Fluoride feed rate (lb/day)} \times \text{AFI} \times \text{chemical purity}}{\text{Capacity (MGD)} \times 8.34 \text{ lb/gal}} \quad (5.112)$$

When the fluoride feed rate is changed to fluoride fed and the capacity is changed to actual daily production of water in the water system, then the dosage becomes the calculated dosage. The units remain the same, except that fluoride feed goes from lb/day to lb and actual production goes from MGD to MG (the day units cancel).

Note: The amount of fluoride fed (lb) will be determined over a time period (e.g., day, week, month), and the actual production will be determined over the same time period.

$$\text{Calculated dosage (mg/L)} = \frac{\text{Fluoride fed (lb/day)} \times \text{AFI} \times \text{chemical purity}}{\text{Actual prod. (MGD)} \times 8.34 \text{ lb/gal}} \quad (5.113)$$

The numerator of the equation gives the pounds of fluoride ion added to the water, and the denominator gives million pounds of water treated. Pounds of fluoride divided by million pounds of water equals ppm or mg/L. The formula for calculated dosage for the saturator is as follows:

$$\text{Calculated dosage (mg/L)} = \frac{\text{Solution fed (lb/day)} \times 18,000 \text{ mg/L}}{\text{Actual prod. (MGD)}} \quad (5.114)$$

Determining the calculated dosage for an unsaturated sodium fluoride solution is based on the particular strength of the solution; for example, a 2% strength solution is equal to 8550 mg/L. The percent strength is based on the pounds of sodium fluoride dissolved into a certain amount of water. For example, find the percent solution if 6.5 lb of sodium fluoride are dissolved in 45 gal of water:

$$45 \text{ gal} \times 8.34 \text{ lb/gal} = 375 \text{ lb water}$$

$$(6.5 \text{ lb NaF}) \div (375 \text{ lb water}) = 1.7\% \text{ NaF solution}$$

This means that 6.5 lb of fluoride chemical dissolved in 45 gal of water will yield a 1.7% solution. To find the solution concentration of an unknown sodium fluoride solution, use the following formula:

$$\text{Solution concentration} = \frac{18,000 \text{ mg/L} \times \text{solution strength (\%)}}{4\%} \quad (5.115)$$

■ **EXAMPLE 5.181**

Problem: Assume that 6.5 lb of NaF is dissolved in 45 gal of water, as previously given. What would be the solution concentration? Solution strength is 1.7% (see above).

Solution:

$$\text{Solution conc.} = \frac{18{,}000 \text{ mg/L}}{4\%} \times \text{solution strength (\%)}$$

$$= \frac{18{,}000 \text{ mg/L} \times 1.7\%}{4\%} = 7650 \text{ mg/L}$$

Note: The calculated dosage formula for an unsaturated sodium fluoride solution is:

$$\text{Calculated dosage (mg/L)} = \frac{\text{Solution fed (lb/day)}}{\text{Actual production (gal)}} \times \text{solution concentration (mg/L)}$$

Caution: CDC recommends against the use of unsaturated sodium fluoride solution in water fluoridation.

5.6.11.4.5   Calculated Dosage Problems

■ **EXAMPLE 5.182. Sodium Fluorosilicate Dosage**

Problem: A plant uses 65 lb. of sodium fluorosilicate to treat 5,540,000 gal of water in 1 day. What is the calculated dosage?

Solution:

$$\text{Calculated dosage (mg/L)} = \frac{\text{Fluoride fed (lb/day)}}{\text{Actual prod. (MGD)} \times 8.34 \text{ lb/gal}} \times \text{AFI} \times \text{chemical purity}$$

$$= \frac{65 \text{ lb} \times 0.607 \times 0.985}{5540 \text{ MG} \times 8.34 \text{ lb/gal}} = 0.84 \text{ mg/L}$$

■ **EXAMPLE 5.183. Fluorosilicic Acid Dosage**

Problem: A plant uses 43 lb of fluorosilicic acid to treat 1,226,000 gal of water. Assume that the acid is 23% pure. What is the calculated dosage?

Solution:

$$\text{Calculated dosage (mg/L)} = \frac{\text{Fluoride fed (lb/day)}}{\text{Actual prod. (MGD)} \times 8.34 \text{ lb/gal}} \times \text{AFI} \times \text{chemical purity}$$

$$= \frac{43 \text{ lb} \times 0.792 \times 0.23}{1.226 \text{ MG} \times 8.34 \text{ lb/gal}} = 0.77 \text{ mg/L}$$

Note: The calculated dosage is 0.77 mg/L. If the natural fluoride level is added to this dosage, then it should equal what the actual fluoride level is in the drinking water.

■ **EXAMPLE 5.184. Sodium Fluoride (Dry) Dosage**

Problem: A water plant feeds sodium fluoride in a dry feeder. They use 5.5 lb of the chemical to fluoridate 240,000 gal of water. What is the calculated dosage?

Solution:

$$\text{Calculated dosage (mg/L)} = \frac{\text{Fluoride fed (lb/day)}}{\text{Actual prod. (MGD)} \times 8.34 \text{ lb/gal}} \times \text{AFI} \times \text{chemical purity}$$

$$= \frac{5.5 \text{ lb} \times 0.45 \times 0.98}{0.24 \text{ MG} \times 8.34 \text{ lb/gal}} = 1.2 \text{ mg/L}$$

■ **EXAMPLE 5.185. Sodium Fluoride—
Saturator Dosage**

Problem: A plant uses 10 gal of sodium fluoride from its saturator to treat 200,000 gal of water. What is the calculated dosage?

Solution:

$$\text{Calculated dosage (mg/L)} = \frac{\text{Solution fed (lb/day)}}{\text{Actual production (MGD)}} \times 18{,}000 \text{ mg/L}$$

$$= \frac{10 \text{ gal} \times 18{,}000 \text{ mg/L}}{200{,}000 \text{ gal}} = 0.9 \text{ mg/L}$$

■ **EXAMPLE 5.186. Sodium Fluoride—
Unsaturated Solution Dosage**

Problem: A water plant adds 93 gpd of a 2% solution of sodium fluoride to fluoridate 800,000 gpd. What is the calculated dosage?

Solution:

$$\text{Solution conc. (mg/L)} = \frac{18{,}000 \text{ mg/L}}{4\%} \times \text{solution strength (\%)}$$

$$= \frac{18{,}000 \text{ mg/L} \times 0.02}{.04} = 9000 \text{ mg/L}$$

$$\text{Calculated dosage (mg/L)} = \frac{\text{Solution fed (lb/day)}}{\text{Actual production (gal)}} \times \text{solution conc. (mg/L)}$$

$$= \frac{93 \text{ gal} \times 9000 \text{ mg/L}}{800{,}000 \text{ gal}} = 1.05 \text{ mg/L}$$

5.6.12 WATER SOFTENING

Hardness in water is caused by the presence of certain positively charged metallic ions in solution in the water. The most common of these hardness-causing ions are calcium and magnesium; others include iron, strontium, and barium. The two primary constituents of water that determine the hardness of water are calcium and magnesium. If the concentration of these elements in the water is known, the total hardness of the water can be calculated. To make this calculation, the equivalent weights of calcium, magnesium, and calcium carbonate must be known; the equivalent weights are given below:

Calcium (Ca)   20.04
Magnesium (Mg)   12.15
Calcium carbonate ($CaCO_3$)   50.045

5.6.12.1 Calculating Calcium Hardness as $CaCO_3$

The hardness (in mg/L as $CaCO_3$) for any given metallic ion is calculated using Equation 5.116:

$$\frac{\text{Calcium hardness (mg/L) as } CaCO_3}{\text{Equivalent wt. of } CaCO_3} = \frac{\text{Calcium (mg/L)}}{\text{Equivalent wt. of calcium}} \quad (5.116)$$

■ EXAMPLE 5.187

Problem: A water sample has calcium content of 51 mg/L. What is this calcium hardness expressed as $CaCO_3$?

Solution:

$$\frac{\text{Calcium hardness (mg/L) as } CaCO_3}{\text{Equivalent wt. of } CaCO_3} = \frac{\text{Calcium (mg/L)}}{\text{Equivalent wt. of calcium}}$$

$$\frac{x \text{ mg/L}}{50.045} = \frac{51 \text{ mg/L}}{20.45}$$

$$x = \frac{51 \times 50.045}{20.45} = 124.8 \text{ mg/L Ca as } CaCO_3$$

■ EXAMPLE 5.188

Problem: The calcium content of a water sample is 26 mg/L. What is this calcium hardness expressed as $CaCO_3$?

Solution:

$$\frac{\text{Calcium hardness (mg/L) as } CaCO_3}{\text{Equivalent wt. of } CaCO_3} = \frac{\text{Calcium (mg/L)}}{\text{Equivalent wt. of calcium}}$$

$$= \frac{x \text{ mg/L}}{50.045} = \frac{26 \text{ mg/L}}{20.04}$$

$$x = \frac{26 \times 50.045}{20.04} = 64.9 \text{ mg/L Ca as } CaCO_3$$

5.6.12.2 Calculating Magnesium Hardness as $CaCO_3$

To calculate magnesium harness, we use Equation 5.117:

$$\frac{\text{Magnesium hardness (mg/L) as } CaCO_3}{\text{Equivalent wt. of } CaCO_3} = \frac{\text{Magnesium (mg/L)}}{\text{Equivalent wt. of magnesium}} \quad (5.117)$$

■ EXAMPLE 5.189

Problem: A sample of water contains 24 mg/L magnesium. Express this magnesium hardness as $CaCO_3$.

Solution:

$$\frac{\text{Magnesium hardness (mg/L) as } CaCO_3}{\text{Equivalent wt. of } CaCO_3} = \frac{\text{Magnesium (mg/L)}}{\text{Equivalent wt. of magnesium}}$$

$$= \frac{x \text{ mg/L}}{50.045} = \frac{24 \text{ mg/L}}{12.15}$$

$$x = \frac{24 \times 50.045}{12.15} = 98.9 \text{ mg/L Mg as } CaCO_3$$

■ EXAMPLE 5.190

Problem: The magnesium content of a water sample is 16 mg/L. Express this magnesium hardness as $CaCO_3$.

Solution:

$$\frac{\text{Magnesium hardness (mg/L) as } CaCO_3}{\text{Equivalent wt. of } CaCO_3} = \frac{\text{Magnesium (mg/L)}}{\text{Equivalent wt. of magnesium}}$$

$$= \frac{x \text{ mg/L}}{50.045} = \frac{16 \text{ mg/L}}{12.15}$$

$$x = \frac{16 \times 50.045}{12.15} = 65.9 \text{ mg/L Mg as } CaCO_3$$

5.6.12.3 Calculating Total Hardness

Calcium and magnesium ions are the two constituents that are the primary cause of hardness in water. To find total hardness, we simply add the concentrations of calcium and magnesium ions, expressed in terms of calcium carbonate ($CaCO_3$), using Equation 5.118:

$$\begin{array}{c} \text{Total hardness} \\ \text{(mg/L)} \\ \text{as } CaCO_3 \end{array} = \left[\begin{array}{l} \text{Ca hardness (mg/L) as } CaCO_3 \\ + \text{Mg hardness (mg/L) as } CaCO_3 \end{array} \right] \quad (5.118)$$

■ **EXAMPLE 5.191**

Problem: A sample of water has calcium content of 70 mg/L as $CaCO_3$ and magnesium content of 90 mg/L as $CaCO_3$.

Solution:

$$\begin{array}{c} \text{Total hardness} \\ \text{(mg/L)} \\ \text{as } CaCO_3 \end{array} = \left[\begin{array}{l} \text{Ca hardness (mg/L) as } CaCO_3 \\ + \text{Mg hardness (mg/L) as } CaCO_3 \end{array} \right]$$

$$= 70 \text{ mg/L} + 90 \text{ mg/L}$$

$$= 160 \text{ mg/L as } CaCO_3$$

■ **EXAMPLE 5.192**

Problem: Determine the total hardness as $CaCO_3$ of a sample of water that has calcium content of 28 mg/L and magnesium content of 9 mg/L.

Solution: Express calcium and magnesium in terms of $CaCO_3$:

$$\frac{\text{Calcium hardness (mg/L) as } CaCO_3}{\text{Equivalent wt. of } CaCO_3}$$

$$= \frac{\text{Calcium (mg/L)}}{\text{Equivalent wt. of calcium}}$$

$$= \frac{x \text{ mg/L}}{50.045} = \frac{28 \text{ mg/L}}{20.04}$$

$$x = 69.9 \text{ mg/L Ca as } CaCO_3$$

$$\frac{\text{Magnesium hardness (mg/L) as } CaCO_3}{\text{Equivalent wt. of } CaCO_3}$$

$$= \frac{\text{Magnesium (mg/L)}}{\text{Equivalent wt. of magnesium}}$$

$$= \frac{x \text{ mg/L}}{50.045} = \frac{9 \text{ mg/L}}{12.15}$$

$$x = 37.1 \text{ mg/L Mg as } CaCO_3$$

Now, total hardness can be calculated:

$$\begin{array}{c} \text{Total hardness} \\ \text{(mg/L)} \\ \text{as } CaCO_3 \end{array} = \left[\begin{array}{l} \text{Ca hardness (mg/L) as } CaCO_3 \\ + \text{Mg hardness (mg/L) as } CaCO_3 \end{array} \right]$$

$$= 69.9 \text{ mg/L} + 37.1 \text{ mg/L}$$

$$= 107 \text{ mg/L as } CaCO_3$$

5.6.12.4 Calculating Carbonate and Noncarbonate Hardness

As mentioned, total hardness is comprised of calcium and magnesium hardness. When total hardness has been calculated, it is sometimes used to determine another expression of hardness—carbonate and noncarbonate. When hardness is numerically greater than the sum of bicarbonate and carbonate alkalinity, that amount of hardness equivalent to the total alkalinity (both in units of mg $CaCO_3$/L) is called the *carbonate hardness*; the amount of hardness in excess of this is the *noncarbonate hardness*. When the hardness is numerically equal to or less than the sum of carbonate and noncarbonate alkalinity, all hardness is carbonate hardness, and noncarbonate hardness is absent. Again, the total hardness is comprised of carbonate hardness and noncarbonate hardness:

$$\text{Total hardness} = \left[\begin{array}{l} \text{Carbonate hardness} \\ + \text{noncarbonate hardness} \end{array} \right] \quad (5.119)$$

When the alkalinity (as $CaCO_3$) is greater than the total hardness, all of the hardness is carbonate hardness:

$$\begin{array}{l} \text{Total hardness (mg/L) as } CaCO_3 \\ = \text{Carbonate hardness (mg/L) as } CaCO_3 \end{array} \quad (5.120)$$

When the alkalinity (as $CaCO_3$) is less than the total hardness, then the alkalinity represents carbonate hardness, and the balance of the hardness is noncarbonate hardness:

$$\begin{array}{c} \text{Total hardness} \\ \text{(mg/L)} \\ \text{as } CaCO_3 \end{array} = \left[\begin{array}{l} \text{Carbonate hardness (mg/L)} \\ \text{as } CaCO_3 \\ + \text{noncarbonate hardness (mg/L)} \\ \text{as } CaCO_3 \end{array} \right] \quad (5.121)$$

When carbonate hardness is represented by the alkalinity, we use Equation 5.122:

$$\begin{array}{c} \text{Total hardness} \\ \text{(mg/L)} \\ \text{as } CaCO_3 \end{array} = \left[\begin{array}{l} \text{Alkalinity (mg/L) as } CaCO_3 \\ + \text{noncarbonate hardness (mg/L)} \\ \text{as } CaCO_3 \end{array} \right] \quad (5.122)$$

■ **EXAMPLE 5.193**

Problem: A water sample contains 110 mg/L alkalinity as $CaCO_3$ and 105 mg/L total hardness as $CaCO_3$. What is the carbonate and noncarbonate hardness of the sample?

Solution: Because the alkalinity is greater than the total hardness, all of the hardness is carbonate hardness:

$$\begin{array}{l} \text{Total hardness} \\ \text{(mg/L)} \\ \text{as } CaCO_3 \end{array} = \left[\begin{array}{l} \text{Carbonate hardness (mg/L) as } CaCO_3 \\ + \text{noncarbonate hardness (mg/L)} \\ \text{as } CaCO_3 \end{array} \right]$$

105 mg/L as $CaCO_3$ = Carbonate hardness

No noncarbonate hardness is present in this water.

■ **EXAMPLE 5.194**

Problem: The alkalinity of a water sample is 80 mg/L as $CaCO_3$. If the total hardness of the water sample is 112 mg/L as $CaCO_3$, what is the carbonate and noncarbonate hardness in mg/L as $CaCO_3$?

Solution: Alkalinity is less than total hardness; therefore, both carbonate and noncarbonate hardness will be present in the hardness of the sample:

$$\begin{array}{l} \text{Total hardness} \\ \text{(mg/L)} \\ \text{as } CaCO_3 \end{array} = \left[\begin{array}{l} \text{Carbonate hardness (mg/L) as } CaCO_3 \\ + \text{noncarbonate hardness (mg/L)} \\ \text{as } CaCO_3 \end{array} \right]$$

$$112 \text{ mg/L} = 80 \text{ mg/L} - x \text{ mg/L}$$

$$x = 112 \text{ mg/L} - 80 \text{ mg/L}$$

$$= 32 \text{ mg/L noncarbonate hardness}$$

5.6.12.5 Alkalinity Determination

Alkalinity measures the acid-neutralizing capacity of a water sample. It is an aggregate property of the water sample and can be interpreted in terms of specific substances only when a complete chemical composition of the sample is also performed. The alkalinity of surface waters is primarily due to the carbonate, bicarbonate, and hydroxide content and is often interpreted in terms of the concentrations of these constituents. The higher the alkalinity, the greater the capacity of the water to neutralize acids; conversely, the lower the alkalinity, the less the neutralizing capacity. To detect the different types of alkalinity, the water is tested for phenolphthalein and total alkalinity, using Equation 5.123 and Equation 5.124:

$$\begin{array}{l} \text{Phenolphthalein alkalinity} \\ \text{(mg/L) as } CaCO_3 \end{array} = \frac{A \times N \times 50,000}{\text{mL of sample}} \quad (5.123)$$

$$\begin{array}{l} \text{Total alkalinity} \\ \text{(mg/L) as } CaCO_3 \end{array} = \frac{B \times N \times 50,000}{\text{mL of sample}} \quad (5.124)$$

where:

A	=	mL titrant used to titrate to pH 8.3.
N	=	normality of the acid (0.02N H_2SO_4 for this alkalinity test).
50,000	=	a conversion factor to change the normality into units of $CaCO_3$.
B	=	total mL of titrant used to titrate to pH 4.5.

■ **EXAMPLE 5.195**

Problem: A 100-mL water sample is tested for phenolphthalein alkalinity. If 1.3-mL titrant is used to pH 8.3 and the sulfuric acid solution has a normality of 0.02N, what is the phenolphthalein alkalinity of the water?

Solution:

$$\begin{array}{l} \text{Phenolphthalein alkalinity} \\ \text{(mg/L) as } CaCO_3 \end{array} = \frac{A \times N \times 50,000}{\text{mL of sample}}$$

$$= \frac{1.3 \text{ mL} \times 0.02N \times 50,000}{100 \text{ mL}}$$

$$= 13 \text{ mg/L}$$

■ **EXAMPLE 5.196**

Problem: A 100-mL sample of water is tested for alkalinity. The normality of the sulfuric acid used for titrating is 0.02N. If 0 mL is used to pH 8.3, and 7.6-mL titrant is used to pH 4.5, what are the phenolphthalein and total alkalinity of the sample?

Solution:

$$\begin{array}{l} \text{Phenolphthalein alkalinity} \\ \text{(mg/L) as } CaCO_3 \end{array} = \frac{A \times N \times 50,000}{\text{mL of sample}}$$

$$= \frac{0 \text{ mL} \times 0.02N \times 50,000}{100 \text{ mL}}$$

$$= 0 \text{ mg/L}$$

$$\begin{array}{l} \text{Total alkalinity} \\ \text{(mg/L) as } CaCO_3 \end{array} = \frac{B \times N \times 50,000}{\text{mL of sample}}$$

$$= \frac{7.6 \text{ mL} \times 0.02N \times 50,000}{100 \text{ mL}}$$

$$= 76 \text{ mg/L}$$

5.6.12.6 Calculation for Removal of Noncarbonate Hardness

Soda ash (Na_2CO_3) is used for precipitation and removal of noncarbonate hardness. To calculate the soda ash dosage required, we can use Equation 5.125 or Equation 5.126:

$$\begin{array}{c} \text{Total hardness} \\ \text{(mg/L)} \\ \text{as } CaCO_3 \end{array} = \begin{bmatrix} \text{Carbonate hardness (mg/L)} \\ \text{as } CaCO_3 \\ + \text{noncarbonate hardness (mg/L)} \\ \text{as } CaCO_3 \end{bmatrix} \quad (5.125)$$

$$\text{Soda ash (mg/L)} = \begin{bmatrix} \text{Noncarbonate hardness} \\ \text{(mg/L) as } CaCO_3 \end{bmatrix} \times \frac{106}{100} \quad (5.126)$$

■ EXAMPLE 5.197

Problem: A water sample has a total hardness of 250 mg/L as $CaCO_3$ and a total alkalinity of 180 mg/L. What soda ash dosage (mg/L) will be required to remove the noncarbonate hardness?

Solution: Determine the noncarbonate hardness:

$$\begin{array}{c} \text{Total hardness} \\ \text{(mg/L)} \\ \text{as } CaCO_3 \end{array} = \begin{bmatrix} \text{Carbonate hardness (mg/L)} \\ \text{as } CaCO_3 \\ + \text{noncarbonate hardness (mg/L)} \\ \text{as } CaCO_3 \end{bmatrix}$$

$$250 \text{ mg/L} = 180 \text{ mg/L} + x \text{ mg/L}$$

$$250 \text{ mg/L} - 180 \text{ mg/L} = x \text{ mg/L}$$

$$70 \text{ mg/L} = x$$

Calculate the soda ash required:

$$\text{Soda ash (mg/L)} = \begin{bmatrix} \text{Noncarbonate hardness} \\ \text{(mg/L) as } CaCO_3 \end{bmatrix} \times \frac{106}{100}$$

$$= 70 \text{ mg/L} \times \frac{106}{100}$$

$$= 74.2 \text{ mg/L}$$

■ EXAMPLE 5.198

Problem: Calculate the soda ash required (in mg/L) to soften water if the water has a total hardness of 192 mg/L and a total alkalinity of 103 mg/L.

Solution: Determine the noncarbonate hardness:

$$\begin{array}{c} \text{Total hardness} \\ \text{(mg/L)} \\ \text{as } CaCO_3 \end{array} = \begin{bmatrix} \text{Carbonate hardness (mg/L)} \\ \text{as } CaCO_3 \\ + \text{noncarbonate hardness (mg/L)} \\ \text{as } CaCO_3 \end{bmatrix}$$

$$192 \text{ mg/L} = 103 \text{ mg/L} + x \text{ mg/L}$$

$$192 \text{ mg/L} - 103 \text{ mg/L} = x \text{ mg/L}$$

$$89 \text{ mg/L} = x$$

Calculate the soda ash required:

$$\text{Soda ash (mg/L)} = \begin{bmatrix} \text{Noncarbonate hardness} \\ \text{(mg/L) as } CaCO_3 \end{bmatrix} \times \frac{106}{100}$$

$$= 89 \text{ mg/L} \times \frac{106}{100}$$

$$= 94 \text{ mg/L}$$

5.6.12.7 Recarbonation Calculation

Recarbonation involves the reintroduction of carbon dioxide into the water, either during or after lime softening, lowering the pH of the water to about 10.4. After the addition of soda ash, recarbonation lowers the pH of the water to about 9.8, promoting better precipitation of calcium carbonate and magnesium hydroxide. Equation 5.127 and Equation 5.128 are used to estimate carbon dioxide dosage:

$$\text{Excess lime (mg/L)} = (A + B + C + D) \times 0.15 \quad (5.127)$$

$$\begin{array}{c} \text{Total } CO_2 \\ \text{dosage (mg/L)} \end{array} = \begin{bmatrix} Ca(OH)_2 \text{ excess (mg/L)} \\ \times (44/74) \end{bmatrix} + \begin{bmatrix} Mg^{+2} \text{ residual (mg/L)} \\ \times (44/24.3) \end{bmatrix} \quad (5.128)$$

■ EXAMPLE 5.199

Problem: The A, B, C, and D factors of the excess lime equation have been calculated as follows: $A = 14$ mg/L, $B = 126$ mg/L, $C = 0$ mL, and $D = 66$ mg/L. If the residual magnesium is 5 mg/L, what is the carbon dioxide (in mg/L) required for recarbonation?

Solution: Calculate the excess lime concentration:

$$\text{Excess lime (mg/L)} = (A + B + C + D) \times 0.15$$

$$= \left(\begin{array}{c} 14 \text{ mg/L} + 126 \text{ mg/L} \\ + 0 \text{ mg/L} + 66 \text{ mg/L} \end{array} \right) \times 0.15$$

$$= 206 \text{ mg/L} \times 0.15$$

$$= 31 \text{ mg/L}$$

Determine the required carbon dioxide dosage:

$$\text{Total CO}_2 \atop \text{dosage (mg/L)} = \left(\begin{bmatrix} \text{Ca(OH)}_2 \text{ excess (mg/L)} \times \left(\frac{44}{74} \right) \end{bmatrix} \\ + \begin{bmatrix} \text{Mg}^{+2} \text{ residual (mg/L)} \times \left(\frac{44}{24.3} \right) \end{bmatrix} \right)$$

$$= \begin{bmatrix} 31 \text{ mg/L} \times \left(\frac{44}{74} \right) \end{bmatrix} + \begin{bmatrix} 5 \text{ mg/L} \times \left(\frac{44}{24.3} \right) \end{bmatrix}$$

$$= 18 \text{ mg/L} + 9 \text{ mg/L}$$

$$= 27 \text{ mg/L CO}_2$$

■ EXAMPLE 5.200

Problem: The A, B, C, and D factors of the excess lime equation have been calculated as $A = 10$ mg/L, $B = 87$ mg/L, $C = 0$ mg/L, and $D = 111$ mg/L. If the residual magnesium is 5 mg/L, what carbon dioxide dosage would be required for recarbonation?

Solution: The excess lime is:

$$\text{Excess lime (mg/L)} = (A + B + C + D) \times 0.15$$

$$= \begin{pmatrix} 10 \text{ mg/L} + 87 \text{ mg/L} \\ + 0 \text{ mg/L} + 111 \text{ mg/L} \end{pmatrix} \times 0.1$$

$$= 208 \text{ mg/L} \times 0.15$$

$$= 31 \text{ mg/L}$$

The required carbon dioxide dosage for recarbonation is:

$$\text{Total CO}_2 \atop \text{dosage (mg/L)} = \begin{bmatrix} 31 \text{ mg/L} \times \left(\frac{44}{74} \right) \end{bmatrix} + \begin{bmatrix} 5 \text{ mg/L} \times \left(\frac{44}{24.3} \right) \end{bmatrix}$$

$$= 18 \text{ mg/L} + 9 \text{ mg/L}$$

$$= 27 \text{ mg/L CO}_2$$

5.6.12.8 Calculating Feed Rates

The appropriate chemical dosage for various unit processes is typically determined by lab or pilot scale testing (e.g., jar testing, pilot plant), only monitoring, and historical experience. When the chemical dosage has been determined, the feed rate can be calculated using Equation 5.129. When the chemical feed rate is known, this value must be translated into a chemical feeder setting:

$$\text{Feed rate (lb/day)} = \begin{bmatrix} \text{Flow rate (MGD)} \\ \times \text{chemical dose (mg/L)} \\ \times 8.34 \text{ lb/gal} \end{bmatrix} \quad (5.129)$$

To calculate the lb/min chemical required, we use Equation 5.130:

$$\text{Chemical (lb/min)} = \frac{\text{Chemical (lb/day)}}{1440 \text{ min/day}} \quad (5.130)$$

■ EXAMPLE 5.201

Problem: Jar tests indicate that the optimum lime dosage is 200 mg/L. If the flow to be treated is 4.0 MGD, what should be the chemical feeder setting in lb/day and lb/min?

Solution: Calculate the lb/day feed rate:

$$\text{Feed rate (lb/day)} = \begin{bmatrix} \text{Flow rate (MGD)} \\ \times \text{chemical dose (mg/L)} \times 8.34 \text{ lb/gal} \end{bmatrix}$$

$$= 200 \text{ mg/L} \times 4.0 \text{ MGD} \times 8.34 \text{ lb/gal}$$

$$= 6672 \text{ lb/day}$$

Convert this feed rate to lb/min:

$$(6672 \text{ lb/day}) \div (1440 \text{ min/day}) = 4.6 \text{ lb/min}$$

■ EXAMPLE 5.202

Problem: What should the lime dosage setting be, in lb/day and lb/hr, if the optimum lime dosage has been determined to be 125 mg/L and the flow to be treated is 1.1 MGD?

Solution: The lb/day feed rate for lime is:

$$\text{Lime (lb/day)} = \text{Lime (mg/L)} \times \text{flow (MGD)} \times 8.34 \text{ lb/gal}$$

$$= 125 \text{ mg/L} \times 1.1 \text{ MGD} \times 8.34 \text{ lb/day}$$

$$= 1147 \text{ lb/day}$$

Convert this to a lb/hr feed rate, as follows:

$$(1147 \text{ lb/day}) \div (24 \text{ hr/day}) = 48 \text{ lb/hr}$$

5.6.12.9 Ion Exchange Capacity

An ion exchange softener is a common alternative to the use of lime and soda ash for softening water. Natural water sources contain dissolved minerals that dissociate in water to form charged particles called *ions*. Of main concern are the positively charged ions of calcium, magnesium, and sodium; bicarbonate, sulfate, and chloride are the normal negatively charged ions of concern. An ion exchange medium, called *resin*, is a material that will exchange a hardness-causing ion for another one that does not cause hardness, hold the new ion temporarily, and then release it when a regenerating solution is poured over the resin.

The removal capacity of an exchange resin is generally reported as grains of hardness removal per cubic foot of resin. To calculate the removal capacity of the softener, we use Equation 5.131:

$$\begin{array}{l} \text{Exchange} \\ \text{capacity} \\ \text{(grains)} \end{array} = \left[\begin{array}{l} \text{Removal capacity (grains/ft}^3) \\ \times \text{media volume (ft}^3) \end{array} \right] \quad (5.131)$$

■ EXAMPLE 5.203

Problem: The hardness removal capacity of an exchange resin is 24,000 grains/ft³. If the softener contains a total of 70 ft³ of resin, what is the total exchange capacity (grains) of the softener?

Solution:

$$\begin{array}{l} \text{Exchange capacity} \\ \text{(grains)} \end{array} = \left[\begin{array}{l} \text{Removal capacity (grains/ft}^3) \\ \times \text{media volume (ft}^3) \end{array} \right]$$

$$= 22,000 \text{ grains/ft}^3 \times 70 \text{ ft}^3$$

$$= 1,540,000 \text{ grains}$$

■ EXAMPLE 5.204

Problem: An ion exchange water softener has a diameter of 7 ft. The depth of resin is 5 ft. If the resin has a removal capacity of 22 kg/ft³, what is the total exchange capacity of the softener (in grains)?

Solution: Before the exchange capacity of a softener can be calculated, the ft³ resin volume must be known:

$$\text{Volume (ft}^3) = 0.785 \times (\text{diameter})^2 \times \text{depth (ft)}$$
$$= 0.785 \times (7 \text{ ft})^2 \times 5 \text{ ft} = 192 \text{ ft}^3$$

Calculate the exchange capacity of the softener:

$$\text{Exchange capacity (grains)} = \text{Removal capacity (grains/ft}^2 \\ \times \text{media volume (ft}^3)$$

$$= 22,000 \text{ grains/ft}^3 \times 192 \text{ ft}^3$$

$$= 4,224,000 \text{ grains}$$

5.6.12.10 Water Treatment Capacity

To calculate when the resin must be regenerated (based on volume of water treated), we know the exchange capacity of the softener and the hardness of the water. Equation 5.132 is used for this calculation:

$$\text{Treatment capacity (gal)} = \dfrac{\begin{array}{c}\text{Exchange capacity} \\ \text{(grains)}\end{array}}{\begin{array}{c}\text{Hardness} \\ \text{(grains/gal)}\end{array}} \quad (5.132)$$

■ EXAMPLE 5.205

Problem: An ion-exchange softener has an exchange capacity of 2,445,000 grains. If the hardness of the water to be treated is 18.6 grains/gal, how many gallons of water can be treated before regeneration of the resin is required?

Solution:

$$\text{Treatment capacity (gal)} = \dfrac{\text{Exchange capacity (grains)}}{\text{Hardness (grains/gal)}}$$

$$= \dfrac{2,455,000 \text{ grains}}{18.6 \text{ grains/gal}}$$

$$= 131,989 \text{ gal}$$

■ EXAMPLE 5.206

Problem: An ion exchange softener has an exchange capacity of 5,500,000 grains. If the hardness of the water to be treated is 14.8 grains/gal, how many gallons of water can be treated before regeneration of the resin is required?

Solution:

$$\text{Treatment capacity (gal)} = \dfrac{\text{Exchange capacity (grains)}}{\text{Hardness (grains/gal)}}$$

$$= \dfrac{5,500,000 \text{ grains}}{14.8 \text{ grains/gal}}$$

$$= 371,622 \text{ gal}$$

■ EXAMPLE 5.207

Problem: The hardness removal capacity of an ion exchange resin is 25 kilograins (kgrains) per ft³. The softener contains a total of 160 ft³ of resin. If the water to be treated contains 14.0 grains/gal hardness, how many gallons of water can be treated before regeneration of the resin is required?

Solution: Both the water hardness and the exchange capacity of the softener must be determined before the gallons water can be calculated:

$$\begin{array}{l} \text{Exchange capacity} \\ \text{(grains)} \end{array} = \left[\begin{array}{l} \text{Removal capacity (grains/ft}^3) \\ \times \text{media volume (ft}^3) \end{array} \right]$$

$$= 25,000 \text{ grains/ft}^3 \times 160 \text{ ft}^3$$

$$= 4,000,000 \text{ grains}$$

Calculate the gallons of water treated:

$$\text{Treatment capacity (gal)} = \frac{\text{Exchange capacity (grains)}}{\text{Hardness (grains/gal)}}$$

$$= \frac{4,000,000 \text{ grains}}{14.0 \text{ grains/gal}}$$

$$= 285,714 \text{ gal}$$

5.6.12.11 Treatment Time Calculation (Until Regeneration Required)

After calculating the total number of gallons water to be treated (before regeneration), we can also calculate the operating time required to treat that amount of water. Equation 5.133 is used to make this calculation.

$$\text{Operating time (hr)} = \frac{\text{Water treated (gal)}}{\text{Flow rate (gal/hr)}} \quad (5.133)$$

■ EXAMPLE 5.208

Problem: An ion exchange softener can treat a total of 642,000 gal before regeneration is required. If the flow rate treated is 25,000 gal/hr, how many hours of operation remain before regeneration is required?

Solution:

$$\text{Operating time (hr)} = \frac{\text{Water treated (gal)}}{\text{Flow rate (gal/hr)}}$$

$$= \frac{642,000 \text{ gal}}{25,000 \text{ gal/hr}}$$

$$= 25.7 \text{ hr before regeneration}$$

■ EXAMPLE 5.209

Problem: An ion exchange softener can treat a total of 820,000 gallons of water before regeneration of the resin is required. If the water is to be treated at a rate of 32,000 gph, how many hours of operation remain until regeneration is required?

Solution:

$$\text{Operating time (hr)} = \frac{\text{Water treated (gal)}}{\text{Flow rate (gal/hr)}}$$

$$= \frac{820,000 \text{ gal}}{32,000 \text{ gal/hr}}$$

$$= 25.6 \text{ hr before regeneration}$$

5.6.12.12 Salt and Brine Required for Regeneration

When calcium and magnesium ions replace the sodium ions in the ion exchange resin, the resin can no longer remove the hardness ions from the water. When this occurs, pumping a concentrated solution (10 to 14% sodium chloride solution) on the resin regenerates the resin. When the resin is completely recharged with sodium ions, it is ready for softening again. Typically, the salt dosage required to prepare the brine solution ranges from 5 to 15 lb salt/ft^3 resin. Equation 5.134 is used to calculate salt required (lb) and Equation 5.135 is used to calculate brine (gal):

$$\begin{aligned}\text{Salt req. (lb)} = &\text{ Salt required (lb/kgrains removed)}\\&\times \text{hardness removed (kgrains)}\end{aligned} \quad (5.134)$$

$$\text{Brine (gal)} = \frac{\text{Salt required (lb)}}{\text{Brine solution (lb salt per gal brine)}} \quad (5.135)$$

To determine the brine solution (lb salt per gal brine) to use in Equation 5.135, we must refer to the salt solutions table shown below:

Salt Solutions		
NaCl (%)	NaCl (lb/gal)	NaCl (lb/ft^3)
10	0.874	6.69
11	0.990	7.41
12	1.09	8.14
13	1.19	8.83
14	1.29	9.63
15	1.39	10.4

■ EXAMPLE 5.210

Problem: An ion exchange softener removes 1,310,000 grains hardness from the water before the resin must be regenerated. If 0.3 lb salt are required for each kilograin removed, how many pounds of salt will be required for preparing the brine to be used in resin regeneration?

$$\begin{aligned}\text{Salt req. (lb)} = &\text{ Salt required (lb/kgrains removed)}\\&\times \text{hardness removed (kgrains)}\\= &\text{ 0.3 lb salt/kgrain removed} \times 1310 \text{ kgrains}\\= &\text{ 393 lb}\end{aligned}$$

■ EXAMPLE 5.211

Problem: A total of 430 lb salt will be required to regenerate an ion exchange softener. If the brine solution is to be a 12% brine solution (see the Salt Solutions table to determine the lb salt per gal brine for a 12% brine solution), how many gallons brine will be required?

Solution:

$$\text{Brine (gal)} = \frac{\text{Salt required (lb)}}{\text{Brine solution (lb salt per gal brine)}}$$

$$= \frac{430 \text{ lb salt}}{1.09 \text{ lb salt per gal brine}}$$

$$= 394 \text{ gal of } 12\% \text{ brine}$$

Thus, it takes 430 lb salt to make up the 394 gal brine that will result in the desired 12% brine solution.

5.7 WASTEWATER MATH CONCEPTS

5.7.1 Preliminary Treatment Calculations

The initial stage of treatment in the wastewater treatment process (following collection and influent pumping) is *preliminary treatment*. Process selection normally is based on the expected characteristics of the influent flow. Raw influent entering the treatment plant may contain many kinds of materials (trash), and preliminary treatment protects downstream plant equipment by removing these materials, which could cause clogs, jams, or excessive wear in plant machinery. In addition, the removal of various materials at the beginning of the treatment train saves valuable space within the treatment plant.

Two of the processes used in preliminary treatment include screening and grit removal; however, preliminary treatment may also include other processes, each designed to remove a specific type of material that presents a potential problem for downstream unit treatment processes. These processes include shredding, flow measurement, preaeration, chemical addition, and flow equalization. Except in extreme cases, plant design will not include all of these items. In this chapter, we focus on and describe typical calculations used in two of these processes: screening and grit removal.

5.7.1.1 Screening

Screening removes large solids, such as rags, cans, rocks, branches, leaves, roots, etc. from the flow before the flow moves on to downstream processes.

5.7.1.1.1 Screenings Removal Calculations

Wastewater operators who are responsible for screenings disposal are typically required to keep a record of the amount of screenings removed from the flow. To keep and maintain accurate screening records, the volume of screenings withdrawn must be determined. Two methods are commonly used to calculate the volume of screenings withdrawn:

$$\text{Screenings removed (ft}^3/\text{day)} = \frac{\text{Screenings (ft}^3)}{\text{day}} \quad (5.136)$$

$$\text{Screenings removed (ft}^3/\text{MG)} = \frac{\text{Screenings (ft}^3)}{\text{Flow (MG)}} \quad (5.137)$$

■ EXAMPLE 5.212

Problem: A total of 65 gal of screenings is removed from the wastewater flow during a 24-hr period. What is the screenings removal reported as ft³/day?

Solution: First convert gallon screenings to cubic feet:

$$(65 \text{ gal}) \div (7.48 \text{ gal/ft}^3) = 8.7 \text{ ft}^3 \text{ screenings}$$

Then calculate screenings removed as ft³/day:

$$\text{Screenings removed (ft}^3/\text{day)} = \frac{8.7 \text{ ft}^3}{1 \text{ day}} = 8.7 \text{ ft}^3/\text{day}$$

■ EXAMPLE 5.213

Problem: During 1 week, a total of 310 gal of screenings was removed from the wastewater screens. What is the average screenings removal in ft³/day?

Solution: First convert gallon screenings to cubic feet:

$$(310 \text{ ft}^3) \div (7.48 \text{ gal/ft}^3) = 41.4 \text{ ft}^3 \text{ screenings}$$

Then calculate screenings removed as ft³/day:

$$\text{Screenings removed (ft}^3/\text{day)} = \frac{41.4 \text{ ft}^3}{7.48 \text{ gal/ft}^3} = 5.5 \text{ ft}^3/\text{day}$$

5.7.1.1.2 Screenings Pit Capacity Calculations

Recall that detention time may be considered the time required for flow to pass through a basin or tank or the time required to fill a basin or tank at a given flow rate. In screenings pit capacity problems, the time required to fill a screenings pit is being calculated. The equation used in screenings pit capacity problems is given below:

$$\text{Fill time (days)} = \frac{\text{Volume of pit (ft}^3)}{\text{Screenings removed (ft}^3/\text{day)}} \quad (5.138)$$

■ EXAMPLE 5.214

Problem: A screenings pit has a capacity of 500 ft³. (The pit is actually larger than 500 ft³ to accommodate soil for covering.) If an average of 3.4 ft³ of screenings is removed daily from the wastewater flow, in how many days will the pit be full?

Solution:

$$\text{Fill time (days)} = \frac{\text{Volume of pit (ft}^3)}{\text{Screenings removed (ft}^3/\text{day})}$$

$$= \frac{500 \text{ ft}^3}{3.4 \text{ ft}^3/\text{day}} = 147.1 \text{ days}$$

■ EXAMPLE 5.215

Problem: A plant has been averaging a screenings removal of 2 ft³/MG. If the average daily flow is 1.8 MGD how many days will it take to fill the pit with an available capacity of 125 ft³?

Solution: The filling rate must first be expressed as ft³/day:

$$\frac{2 \text{ ft}^3 \times 1.8 \text{ MGD}}{\text{MG}} = 3.6 \text{ ft}^3/\text{day}$$

Then,

$$\text{Fill time (days)} = \frac{\text{Volume of pit (ft}^3)}{\text{Screenings removed (ft}^3/\text{day})}$$

$$= \frac{125 \text{ ft}^3}{3.6 \text{ ft}^3/\text{day}} = 34.7 \text{ days}$$

■ EXAMPLE 5.216

Problem: A screenings pit has a capacity of 12 yd³ available for screenings. If the plant removes an average of 2.4 ft³ of screenings per day, in how many days will the pit be filled?

Solution: Because the filling rate is expressed as ft³/day, the volume must be expressed as ft³:

$$12 \text{ yd}^3 \times 27 \text{ ft}^3/\text{yd}^3 = 324 \text{ ft}^3$$

Now calculate fill time:

$$\text{Fill time (days)} = \frac{\text{Volume of pit (ft}^3)}{\text{Screenings removed (ft}^3/\text{day})}$$

$$= \frac{324 \text{ ft}^3}{2.4 \text{ ft}^3/\text{day}} = 135 \text{ days}$$

5.7.1.2 Grit Removal

The purpose of *grit removal* is to remove inorganic solids (sand, gravel, clay, egg shells, coffee grounds, metal filings, seeds, and other similar materials) that could cause excessive mechanical wear. Several processes or devices are used for grit removal, all based on the fact that grit is heavier than the organic solids, which should be kept in suspension for treatment in following unit processes. Grit removal may be accomplished in grit chambers or by the centrifugal separation of biosolids. Processes use gravity/velocity, aeration, or centrifugal force to separate the solids from the wastewater.

5.7.1.2.1 Grit Removal Calculations

Wastewater systems typically average 1 to 15 ft³ of grit per million gallons of flow (sanitary systems, 1 to 4 ft³/MG; combined wastewater systems, 4 to 15 ft³/MG of flow), with higher ranges during storm events. Generally, grit is disposed of in sanitary landfills. Because of this process and for planning purposes, operators must keep accurate records of grit removal. Most often, the data are reported as cubic feet of grit removed per million gallons for flow:

$$\text{Grit removed (ft}^3/\text{MG}) = \frac{\text{Grit volume (ft}^3)}{\text{Flow (MG)}} \quad (5.139)$$

Over a given period, the average grit removal rate at a plant (at least a seasonal average) can be determined and used for planning purposes. Typically, grit removal is calculated as cubic yards, because excavation is normally expressed in terms of cubic yards:

$$\text{Grit (yd}^3) = \frac{\text{Total grit (ft}^3)}{27 \text{ ft}^3/\text{yd}} \quad (5.140)$$

■ EXAMPLE 5.217

Problem: A treatment plant removes 10 ft³ of grit in one day. How many ft³ of grit are removed per million gallons if the plant flow was 9 MGD?

Solution:

$$\text{Grit removed (ft}^3/\text{MG}) = \frac{\text{Grit volume (ft}^3)}{\text{Flow (MG)}}$$

$$= \frac{10 \text{ ft}^3}{9 \text{ MG}} = 1.1 \text{ ft}^3/\text{MG}$$

■ EXAMPLE 5.218

Problem: The total daily grit removed for a plant is 250 gal. If the plant flow is 12.2 MGD, how many cubic feet of grit are removed per MG flow?

Solution: First, convert gallons grit removed to ft³:

$$(250 \text{ gal}) \div (7.48 \text{ gal/ft}^3) = 33 \text{ ft}^3$$

Next, complete the calculation of ft³/MG:

$$\text{Grit removed (ft}^3/\text{MG}) = \frac{\text{Grit volume (ft}^3)}{\text{Flow (MG)}}$$

$$= \frac{33 \text{ ft}^3}{12.2 \text{ MG}} = 2.7 \text{ ft}^3/\text{MG}$$

■ EXAMPLE 5.219

Problem: The monthly average grit removal is 2.5 ft³/MG. If the monthly average flow is 2,500,000 gpd, how many cubic yards must be available for grit disposal if the disposal pit is to have a 90-day capacity?

Solution: First, calculate the grit generated each day:

$$(2.5 \text{ ft}^3/\text{MG}) \times (2.5 \text{ MGD}) = 6.25 \text{ ft}^3 \text{ each day}$$

The ft³ grit generated for 90 days would be:

$$(6.25 \text{ ft}^3/\text{day}) \times 90 \text{ days} = 562.5 \text{ ft}^3$$

Convert ft³ to yd³ grit:

$$(562.5 \text{ ft}^3) \div (27 \text{ ft}^3/\text{yd}^3) = 21 \text{ yd}^3$$

5.7.1.2.2 Grit Channel Velocity Calculation

The optimum velocity in sewers is approximately 2 ft/sec (fps) at peak flow, because this velocity normally prevents solids from settling from the lines; however, when the flow reaches the grit channel, the velocity should decrease to about 1 fps to permit the heavy inorganic solids to settle. In the example calculations that follow, we describe how the velocity of the flow in a channel can be determined by the float and stopwatch method and by channel dimensions.

■ EXAMPLE 5.220. Velocity by Float and Stopwatch

$$\text{Velocity (fps)} = \frac{\text{Distance traveled (ft)}}{\text{Time required (sec)}} \quad (5.141)$$

Problem: A float requires 30 sec to travel 37 ft in a grit channel. What is the velocity of the flow in the channel?

Solution:

$$\text{Velocity (fps)} = 37 \text{ ft}/30 \text{ sec} = 1.2 \text{ fps}$$

■ EXAMPLE 5.221. Velocity by Flow and Channel Dimensions

This calculation can be used for a single channel or tank or for multiple channels or tanks with the same dimensions and equal flow. If the flow through each unit of the unit dimensions is unequal, the velocity for each channel or tank must be computed individually:

$$\text{Velocity (fps)} = \frac{\text{Flow (MGD)} \times 1.55 \text{ cfs/MGD}}{\substack{\text{Channels in service} \\ \times \text{ channel width (ft)} \\ \times \text{ water depth (ft)}}} \quad (5.142)$$

Problem: The plant is currently using two girt channels. Each channel is 3 ft wide and has a water depth of 1.3 ft. What is the velocity when the influent flow rate is 4.0 MGD?

Solution:

$$\text{Velocity (fps)} = \frac{40 \text{ MGD} \times 1.55 \text{ cfs/MGD}}{2 \text{ channels} \times 3 \text{ ft} \times 1.3 \text{ ft}}$$

$$= \frac{6.2 \text{ cfs}}{7.8 \text{ ft}^2}$$

$$= 0.79 \text{ fps}$$

Key Point: Because 0.79 is within the 0.7 to 1.4 level, the operator of this unit would not make any adjustments.

Key Point: The channel dimensions must always be in feet. Convert inches to feet by dividing by 12 inches per foot.

■ EXAMPLE 5.222. Required Settling Time

This calculation can be used to determine the time required for a particle to travel from the surface of the liquid to the bottom at a given settling velocity. To compute the settling time, the settling velocity in fps must be provided or determined by experiment in a laboratory.

$$\text{Settling time (sec)} = \frac{\text{Liquid depth (ft)}}{\text{Settling velocity (fps)}} \quad (5.143)$$

Problem: The plant's grit channel is designed to remove sand, which has a settling velocity of 0.080 fps. The channel is currently operating at a depth of 2.3 ft. How many seconds will it take for a sand particle to reach the channel bottom?

Solution:

$$\text{Settling time (sec)} = \frac{2.3 \text{ ft}}{0.080 \text{ fps}}$$

$$= 28.7 \text{ sec}$$

■ EXAMPLE 5.223. Required Channel Length

This calculation can be used to determine the length of channel required to remove an object with a specified settling velocity:

$$\text{Channel length} = \frac{\begin{array}{c}\text{Channel depth (ft)}\\ \times \text{flow velocity (fps)}\end{array}}{0.080 \text{ fps}} \quad (5.144)$$

Problem: The plant's grit channel is designed to remove sand, which has a settling velocity of 0.080 fps. The channel is currently operating at a depth of 3 ft. The calculated velocity of flow through the channel is 0.85 fps. The channel is 36 ft long. Is the channel long enough to remove the desired sand particle size?

Solution:

$$\text{Channel length} = \frac{3 \text{ ft} \times 0.85 \text{ fps}}{0.080 \text{ fps}} = 31.6 \text{ ft}$$

Yes, the channel is long enough to ensure that all of the sand will be removed.

5.7.2 PRIMARY TREATMENT CALCULATIONS

Primary treatment (primary sedimentation or clarification) should remove both settleable organic and floatable solids. Poor solids removal during this step of treatment may cause organic overloading of the biological treatment processes following primary treatment. Normally, each primary clarification unit can be expected to remove 90 to 95% of settleable solids, 40 to 60% of the total suspended solids, and 25 to 35% of BOD.

5.7.2.1 Process Control Calculations

As with many other wastewater treatment plant unit processes, several process control calculations may be helpful in evaluating the performance of the primary treatment process. Process control calculations are used in the sedimentation process to determine:

- Surface loading rate (surface settling rate)
- Weir overflow rate (weir loading rate)
- BOD and SS removed (lb/day)
- Percent removal
- Hydraulic detention time
- Biosolids pumping
- Percent total solids (%TS)

In the following sections, we take a closer look at a few of these process control calculations and provide example problems.

Key Point: The calculations presented in the following sections allow us to determine values for each function performed. Again, keep in mind that an optimally operated primary clarifier should have values in an expected range. Recall that the expected ranges of percent removal for a primary clarifier are:

- Settleable solids, 90–99%
- Suspended solids, 40–60%
- BOD, 20–50%

The expected range of hydraulic detention time for a primary clarifier is 1 to 3 hr. The expected range of surface loading/settling rate for a primary clarifier is 600 to 1200 gpd/ft² (ballpark estimate). The expected range of weir overflow rate for a primary clarifier is 10,000 to 20,000 gpd/ft.

5.7.2.2 Surface Loading Rate (Surface Settling Rate/Surface Overflow Rate)

Surface loading rate is the number of gallons of wastewater passing over 1 ft² of tank per day. This can be used to compare actual conditions with design. Plant designs generally use a surface loading rate of 300 to 1200 gpd/ft²:

$$\begin{array}{c}\text{Surface loading rate}\\ (\text{gpd/ft}^2)\end{array} = \frac{\text{Flow (gpd)}}{\text{Surface tank area (ft}^2)} \quad (5.145)$$

■ EXAMPLE 5.224

Problem: The circular settling tank has a diameter of 120 ft. If the flow to the unit is 4.5 MGD, what is the surface loading rate in gpd/ft²?

Solution:

$$\text{Surface loading rate} = \frac{4.5 \text{ MGD} \times 1,000,000 \text{ gal/MGD}}{0.785 \times 120 \text{ ft} \times 120 \text{ ft}}$$

$$= 398 \text{ gpd/ft}^2$$

■ EXAMPLE 5.225

Problem: A circular clarifier has a diameter of 50 ft. If the primary effluent flow is 2,150,000 gpd, what is the surface overflow rate in gpd/ft²?

Solution:

Key Point: Remember that area = 0.785 × 50 ft × 50 ft

$$\text{Surface overflow rate (gpd/ft}^2) = \text{Flow (gpd)/area (ft}^2)$$

$$= \frac{2,150,000 \text{ gpd}}{0.785 \times 50 \text{ ft} \times 50 \text{ ft}}$$

$$= 1096 \text{ gpd/ft}$$

■ EXAMPLE 5.226

Problem: A sedimentation basin 90 ft by 20 ft receives a flow of 1.5 MGD. What is the surface overflow rate in gpd/ft²?

Solution:

$$\text{Surface loading rate} \atop (\text{gpd/ft}^2) = \frac{\text{Flow (gpd)}}{\text{Surface tank area (ft}^2)}$$

$$= \frac{1,500,000 \text{ gpd}}{90 \text{ ft} \times 20 \text{ ft}} = 833 \text{ gpd/ft}^2$$

5.7.2.3 Weir Overflow Rate (Weir Loading Rate)

A weir is a device used to measure wastewater flow. *Weir overflow rate* (weir loading rate) is the amount of water leaving the settling tank per linear foot of water. The result of this calculation can be compared with design. Normally, weir overflow rates of 10,000 to 20,000 gpd/ft are used in the design of a settling tank:

$$\text{Weir overflow rate (gpd/ft)} = \frac{\text{Flow (gpd)}}{\text{Weir length (ft)}} \quad (5.146)$$

Key Point: To calculate weir circumference, use total feet of weir = 3.14 × weir diameter (ft).

■ EXAMPLE 5.227

Problem: The circular settling tank is 80 ft in diameter and has a weir along its circumference. The effluent flow rate is 2.75 MGD. What is the weir overflow rate in gallons per day per foot?

Solution:

$$\text{Weir overflow rate (gpd/ft)} = \frac{2.75 \text{ MGD} \times 1,000,000 \text{ gal}}{3.14 \times 80 \text{ ft}}$$

$$= 10,947 \text{ gpd/ft}$$

Key Point: Notice that 10,947 gpd/ft is above the recommended minimum of 10,000.

■ EXAMPLE 5.228

Problem: A rectangular clarifier has a total of 70 ft of weir. What is the weir overflow rate in gpd/ft when the flow is 1,055,000 gpd?

Solution:

$$\text{Weir overflow rate (gpd/ft)} = \frac{\text{Flow (gpd)}}{\text{Weir length (ft)}}$$

$$= \frac{1,055,000 \text{ gpd}}{70 \text{ ft}} = 15,071 \text{ gpd}$$

5.7.2.4 BOD and Suspended Solids Removed (lb/day)

To calculate the pounds of BOD or suspended solids removed each day, we need to know the mg/L BOD or SS removed and the plant flow, then we can use the mg/L to lb/day equation:

$$\text{SS removed} = \text{mg/L} \times \text{MGD} \times 8.34 \text{ lb/gal} \quad (5.147)$$

■ EXAMPLE 5.229

Problem: If 120 mg/L suspended solids are removed by a primary clarifier, how many lb/day suspended solids are removed when the flow is 6,250,000 gpd?

Solution:

$$\text{SS removed} = 120 \text{ mg/L} \times 6.25 \text{ MGD} \times 8.34 \text{ lb/gal}$$

$$= 6255 \text{ lb/day}$$

■ EXAMPLE 5.230

Problem: The flow to a secondary clarifier is 1.6 MGD. If the influent BOD concentration is 200 mg/L and the effluent BOD concentration is 70 mg/L, how many pounds of BOD are removed daily?

Solution:

$$\text{BOD removed (lb/day)} = 200 \text{ mg/L} - 70 \text{ mg/L} = 130 \text{ mg/L}$$

After calculating mg/L BOD removed, calculate lb/day BOD removed:

$$\text{BOD removed} = 130 \text{ mg/L} \times 1.6 \text{ MGD} \times 8.34 \text{ lb/gal}$$

$$= 1735 \text{ lb/day}$$

5.7.3 TRICKLING FILTER PROCESS

The *trickling filter process* is one of the oldest forms of dependable biological treatment for wastewater. By its very nature, the trickling filter has its advantages over other unit processes; for example, it is a very economical and dependable process for treatment of wastewater prior to discharge. Capable of withstanding periodic shock loading, process energy demands are low because aeration is a natural process.

Trickling filter operation involves spraying wastewater over solid media such as rock, plastic, or redwood slats (or laths). As the wastewater trickles over the surface of the media, a growth of microorganisms (bacteria, protozoa, fungi, algae, helminths or worms, and larvae) develops. This growth is visible as a shiny slime very similar to the slime found on rocks in a stream. As wastewater passes over this slime, the slime adsorbs the organic (food) matter. This organic matter is used for food by the microorganisms. At the same time, air moving through the open spaces in the filter transfers oxygen to the wastewater. This oxygen is then transferred to the slime to keep the outer layer aerobic. As the microorganisms use the food and oxygen, they produce more organisms, carbon dioxide, sulfates, nitrates, and other stable byproducts; these materials are then discarded from the slime back into the wastewater flow and are carried out of the filter.

5.7.3.1 Trickling Filter Process Calculations

Several calculations are useful in the operation of trickling filters; these include hydraulic loading, organic loading, and BOD and suspended solids removal. Each type of trickling filter is designed to operate with specific loading levels. These levels vary greatly depending on the filter classification. To operate the filter properly, filter loading must be within the specified levels. The main three loading parameters for the trickling filter are hydraulic loading, organic loading, and recirculation ratio.

5.7.3.2 Hydraulic Loading

Calculating the *hydraulic loading rate* is important in accounting for both the primary effluent as well as the recirculated trickling filter effluent. These are combined before being applied to the filter surface. The hydraulic loading rate is calculated based on filter surface area. The normal hydraulic loading rate ranges for standard rate and high rate trickling filters are:

Standard rate—25 to 100 gpd/ft² or 1 to 40 MGD/ac
High rate—100 to 1000 gpd/ft² or 4 to 40 MGD/ac

Key Point: If the hydraulic loading rate for a particular trickling filter is too low, septic conditions will begin to develop.

■ EXAMPLE 5.231

Problem: A trickling filter 80 ft in diameter is operated with a primary effluent of 0.588 MGD and a recirculated effluent flow rate of 0.660 MGD. Calculate the hydraulic loading rate on the filter in units of gpd/ft².

Solution: The primary effluent and recirculated trickling filter effluent are applied together across the surface of the filter; therefore, 0.588 MGD + 0.660 MGD = 1.248 MGD = 1,248,000 gpd:

$$\text{Circular surface area} = 0.785 \times (\text{diameter})^2$$
$$= 0.785 \times (80 \text{ ft})^2 = 5024 \text{ ft}^2$$

$$\frac{\text{Hydraulic loading}}{\text{rate (gpd/ft}^2)} = \frac{1,248,000 \text{ gpd}}{5024 \text{ ft}^2} = 248.4 \text{ gpd/ft}^2$$

■ EXAMPLE 5.232

Problem: A trickling filter 80 ft in diameter treats a primary effluent flow of 750,000 gpd. If the recirculated flow to the clarifier is 0.2 MGD, what is the hydraulic loading on the trickling filter?

Solution:

$$\frac{\text{Hydraulic loading}}{\text{rate (gpd/ft}^2)} = \frac{\text{Total flow gpd}}{\text{Area (ft}^2)}$$

$$= \frac{750,000 \text{ gpd}}{0.785 \times 80 \text{ ft} \times 80 \text{ ft}} = 149 \text{ gpd/ft}^2$$

■ EXAMPLE 5.233

Problem: A high-rate trickling filter receives a daily flow of 1.8 MGD. What is the dynamic loading rate in MGD/ac if the filter is 90 ft in diameter and 5 ft deep?

Solution:

$$0.785 \times 90 \text{ ft} \times 90 \text{ ft} = 6359 \text{ ft}^2$$

$$\frac{6359 \text{ ft}^2}{43,560 \text{ ft}^2/\text{ac}} = 0.146 \text{ ac}$$

$$\frac{\text{Hydraulic}}{\text{loading rate}} = \frac{1.8 \text{ MGD}}{0.146 \text{ ac}} = 12.3 \text{ MGD/ac}$$

Key Point: When hydraulic loading rate is expressed as MGD per acre, this is still an expression of gallon flow over surface area of trickling filter.

5.7.3.3 Organic Loading Rate

Trickling filters are sometimes classified by the *organic loading rate* applied. The organic loading rate is expressed as a certain amount of BOD applied to a certain volume of media. In other words, the organic loading is defined as the pounds of BOD_5 or chemical oxygen demand (COD) applied per day per 1000 ft³ of media—a measure of the amount of food being applied to the filter slime. To calculate the organic loading on the trickling filter, two things must be known: the pounds of BOD or COD being applied to the filter media per day and the volume of the filter media in 1000-ft³ units. The BOD and COD contribution of the recirculated flow is not included in the organic loading.

■ EXAMPLE 5.234

Problem: A trickling filter, 60 ft in diameter, receives a primary effluent flow rate of 0.440 MGD. Calculate the organic loading rate in units of pounds of BOD applied per day per 1000 ft³ of media volume. The primary effluent BOD concentration is 80 mg/L. The media depth is 9 ft.

Solution:

$$0.440 \text{ MGD} \times 80 \text{ mg/L} \times 8.34 \text{ lb/gal} = 293.6 \text{ lb BOD}$$

$$\text{Surface area} = 0.785 \times (60)^2 = 2826 \text{ ft}^2$$

$$\text{Area} \times \text{depth} = \text{Volume} = 2826 \text{ ft}^2 \times 9 \text{ ft} = 25,434 \text{ ft}^3$$

Key Point: To determine the pounds of BOD per 1000 ft³ in a volume of thousands of cubic feet, we must set up the equation as shown below:

$$\frac{293.6 \text{ lb BOD/day}}{25,434 \text{ ft}^3} \times \frac{1000}{1000}$$

Regrouping the numbers and the units together:

$$\frac{293.6 \text{ lb BOD/day} \times 1000}{25,434 \text{ ft}^3} \times \frac{\text{lb BOD/day}}{1000 \text{ ft}^3}$$

$$= 11.5 \times \frac{\text{lb BOD/day}}{1000 \text{ ft}^3}$$

5.7.3.4 BOD and Suspended Solids Removed

To calculate the pounds of BOD or suspended solids removed each day, we need to know the BOD (mg/L) and suspended solids (mg/L) removed and the plant flow.

■ **EXAMPLE 5.235**

Problem: If 120 mg/L suspended solids are removed by a trickling filter, how many lb/day suspended solids are removed when the flow is 4.0 MGD?

Solution:

$$\left[\begin{array}{l} \text{Suspended solids removed (mg/L)} \\ \times \text{flow (MGD)} \times 8.34 \text{ lb/gal} \end{array} \right] = \text{lb/day}$$

$$120 \text{ mg/L} \times 4.0 \text{ MGD} \times 8.34 \text{ lb/gal} = 4003 \text{ lb/day}$$

■ **EXAMPLE 5.236**

Problem: The 3,500,000-gpd influent flow to a trickling filter has a BOD content of 185 mg/L. If the trickling filter effluent has a BOD content of 66 mg/L, how many pounds of BOD are removed daily?

Solution:

$$\text{BOD removed} = 185 \text{ mg/L} - 66 \text{ mg/L} = 119 \text{ mg/L}$$

$$\text{BOD removed (mg/L)} \times \text{flow (MGD)} \times 8.34 \text{ lb/gal} = \text{lb/day}$$

$$119 \text{ mg/L} \times 3.5 \text{ MGD} \times 8.34 \text{ lb/gal} = 3474 \text{ lb/day}$$

5.7.3.5 Recirculation Flow

Recirculation in trickling filters involves the return of filter effluent back to the head of the trickling filter. It can level flow variations and assist in solving operational problems, such as ponding, filter flies, and odors. The operator must check the rate of recirculation to ensure that it is within design specifications. Rates above design specifications indicate hydraulic overloading; rates under design specifications indicate hydraulic underloading. The trickling filter recirculation ratio is the ratio of the recirculated trickling filter flow to the primary effluent flow. The trickling filter recirculation ratio may range from 0.5:1 (.5) to 5:1 (5); however, the ratio is often found to be 1:1 or 2:1:

$$\text{Recirculation ratio} = \frac{\text{Recirculated flow (MGD)}}{\text{Primary effluent flow (MGD)}} \quad (5.148)$$

■ **EXAMPLE 5.237**

Problem: A treatment plant receives a flow of 3.2 MGD. If the trickling filter effluent is recirculated at the rate of 4.50 MGD, what is the recirculation ratio?

Solution:

$$\text{Recirculation ratio} = \frac{\text{Recirculated flow (MGD)}}{\text{Primary effluent flow (MGD)}}$$

$$= \frac{4.5 \text{ MGD}}{3.2 \text{ MGD}} = 1.4$$

■ **EXAMPLE 5.238**

Problem: A trickling filter receives a primary effluent flow of 5 MGD. If the recirculated flow is 4.6 MGD, what is the recirculation ratio?

Solution:

$$\text{Recirculation ratio} = \frac{\text{Recirculated flow (MGD)}}{\text{Primary effluent flow (MGD)}}$$

$$= \frac{4.6 \text{ MGD}}{5 \text{ MGD}} = 0.92$$

5.7.4 Rotating Biological Contactors

The *rotating biological contactor* (RBC) is a variation of the attached growth idea provided by the trickling filter. Still relying on microorganisms that grow on the surface of a medium, the RBC is instead a *fixed-film* biological treatment device. The basic biological process, however, is similar to that occurring in trickling filters. An RBC consists of a series of circular plastic disks mounted side by side and closely spaced; they are typically about 11.5 ft in diameter. Attached to a rotating horizontal shaft, approximately 40% of each disk is submersed in a tank that contains the wastewater to be treated. As the RBC rotates, the attached biomass film (zoogleal slime) that grows on the surface of the disks moves into and out of the wastewater. While submerged in the wastewater, the microorganisms absorb organics; when they are rotated out of the wastewater, they are supplied with needed oxygen for aerobic decomposition. As the zoogleal slime reenters the wastewater, excess solids and waste products are stripped off the media as *sloughings*. These sloughings are transported with the wastewater flow to a settling tank for removal.

5.7.4.1 RBC Process Control Calculations

Several process control calculations may be useful in the operation of an RBC. These include soluble BOD, total media area, organic loading rate, and hydraulic loading. Settling tank calculations and biosolids pumping calculations may be helpful for evaluation and control of the settling tank following the RBC.

5.7.4.2 Hydraulic Loading Rate

The manufacturer normally specifies the RBC media surface area, and the hydraulic loading rate is based on the media surface area, usually in square feet (ft^2). Hydraulic loading is expressed in terms of gallons of flow per day per square foot of media. This calculation can be helpful in evaluating the current operating status of the RBC. Comparison with design specifications can determine if the unit is hydraulically over- or under-loaded. Hydraulic loading on an RBC can range from 1 to 3 gpd/ft^2.

■ EXAMPLE 5.239

Problem: An RBC treats a primary effluent flow rate of 0.244 MGD. What is the hydraulic loading rate in gpd/ft^2 if the media surface area is 92,600 ft^2?

Solution:

$$\text{Hydraulic loading rate} = \frac{\text{Flow (gpd)}}{\text{Media area (ft}^2)}$$

$$= \frac{244,000 \text{ gpd}}{92,000 \text{ ft}^2} = 2.63 \text{ ft}^2$$

■ EXAMPLE 5.240

Problem: An RBC treats a flow of 3.5 MGD. The manufacturer's data indicate a media surface area of 750,000 ft^2. What is the hydraulic loading rate on the RBC?

Solution:

$$\text{Hydraulic loading rate} = \frac{\text{Flow (gpd)}}{\text{Media area (ft}^2)}$$

$$= \frac{3,500,000 \text{ gpd}}{750,000 \text{ ft}^2} = 4.7 \text{ ft}^2$$

■ EXAMPLE 5.241

Problem: A rotating biological contactor treats a primary effluent flow of 1,350,000 gpd. The manufacturer's data indicate that the media surface area is 600,000 ft^2. What is the hydraulic loading rate on the filter?

Solution:

$$\text{Hydraulic loading rate} = \frac{\text{Flow (gpd)}}{\text{Media area (ft}^2)}$$

$$= \frac{1,350,000 \text{ gpd}}{600,000 \text{ ft}^2} = 2.3 \text{ ft}^2$$

5.7.4.3 Soluble BOD

The *soluble BOD* concentration of the RBC influent can be determined experimentally in the laboratory, or it can be estimated using the suspended solids concentration and the K factor. The K factor is used to approximate the BOD (particulate BOD) contributed by the suspended matter. The K factor must be provided or determined experimentally in the laboratory. The K factor for domestic wastes is normally in the range of 0.5 to 0.7:

$$\text{Soluble BOD}_5 = \text{Total BOD}_5 - (K \text{ factor} \times \text{total SS}) \quad (5.149)$$

■ EXAMPLE 5.242

Problem: The suspended solids concentration of a wastewater is 250 mg/L. If the K value at the plant is 0.6, what is the estimated particulate biochemical oxygen demand concentration of the wastewater?

Solution:

$$250 \text{ mg/L} \times 0.6 = 150 \text{ mg/L particulate BOD}$$

Key Point: The K value of 0.6 indicates that about 60% of the suspended solids are organic suspended solids (particulate BOD).

■ EXAMPLE 5.243

Problem: A rotating biological contactor receives a flow of 2.2 MGD with a BOD content of 170 mg/L and suspended solids concentration of 140 mg/L. If the K value is 0.7, how many pounds of soluble BOD enter the RBC daily?

Solution:

$$\text{Total BOD} = \text{Particulate BOD} + \text{soluble BOD}$$

$$170 \text{ mg/L} = (140 \text{ mg/L} \times 0.7) + x \text{ mg/L}$$

$$170 \text{ mg/L} = 98 \text{ mg/L} + x \text{ mg/L}$$

$$170 \text{ mg/L} - 98 \text{ mg/L} = x$$

$$x = 72 \text{ mg/L soluble BOD}$$

Now the lb/day soluble BOD may be determined:

$$\text{Soluble BOD (mg/L)} \times \text{flow (MGD)} \times 8.34 \text{ lb/gal} = \text{lb/day}$$

$$72 \text{ mg/L} \times 2.2 \text{ MGD} \times 8.34 \text{ lb/gal} = 1321 \text{ lb/day}$$

5.7.4.4 Organic Loading Rate

The *organic loading rate* can be expressed as total BOD loading in pounds per day per 1000 ft^2 of media. The actual values can then be compared with plant design specifications to determine the current operating condition of the system.

$$\text{Organic loading rate} = \frac{\begin{array}{c}\text{Soluble BOD} \times \text{flow (MGD)} \\ \times 8.34 \text{ lb/gal}\end{array}}{\text{Media area (1000 ft}^2)} \quad (5.150)$$

■ **EXAMPLE 5.244**

Problem: A rotating biological contactor has a media surface area of 500,000 ft^2 and receives a flow of 1,000,000 gpd. If the soluble BOD concentration of the primary effluent is 160 mg/L, what is the organic loading on the RBC in lb/day/1000 ft^2?

Solution:

$$\text{Organic loading rate} = \frac{\begin{array}{c}\text{Soluble BOD} \times \text{flow (MGD)} \\ \times 8.34 \text{ lb/gal}\end{array}}{\text{Media area (1000 ft}^2)}$$

$$= \frac{160 \text{ mg/L} \times 1.0 \text{ MGD} \times 8.34 \text{ lb/gal}}{500 \times 1000 \text{ ft}^2}$$

$$= 2.7 \text{ lb/day/1000 ft}^2$$

■ **EXAMPLE 5.245**

Problem: The wastewater flow to an RBC is 3,000,000 gpd. The wastewater has a soluble BOD concentration of 120 mg/L. The RBC consists of six shafts (each 110,000 ft^2), with two shafts comprising the first stage of the system. What is the organic loading rate in lb/day/1,000 ft^2 on the first stage of the system?

Solution:

$$\text{Organic loading rate} = \frac{\text{Soluble BOD (lb/day)}}{\text{Media area (1000 ft}^2)}$$

$$= \frac{120 \text{ mg/L} \times 3.0 \text{ MGD} \times 8.34 \text{ lb/gal}}{220 \times 1000 \text{ ft}^2}$$

$$= 13.6 \text{ lb/day/1000 ft}^2$$

5.7.4.5 Total Media Area

Several process control calculations for the RBC use the total surface area of all the stages within the train. As was the case with the soluble BOD calculation, plant design information or information supplied by the unit manufacturer must provide the individual stage areas (or the total train area), because physical determination of this would be extremely difficult:

$$\begin{array}{r}\text{Total area} = \text{1st stage area} + \text{2nd stage area} + \\ \ldots + n\text{th stage area}\end{array} \quad (5.151)$$

5.7.5 ACTIVATED BIOSOLIDS

The *activated biosolids process* is a manmade process that mimics the natural self-purification process that takes place in streams. In essence, we can state that the activated biosolids treatment process is a "stream in a container." In wastewater treatment, activated biosolids processes are used for both secondary treatment and complete aerobic treatment without primary sedimentation. *Activated biosolids* refers to biological treatment systems that use a suspended growth of organisms to remove BOD and suspended solids.

The basic components of an activated biosolids sewage treatment system include an aeration tank and a secondary basin, settling basin, or clarifier. Primary effluent is mixed with settled solids recycled from the secondary clarifier and is then introduced into the aeration tank. Compressed air is injected continuously into the mixture through porous diffusers located at the bottom of the tank, usually along one side.

Wastewater is fed continuously into an aerated tank, where the microorganisms metabolize and biologically flocculate the organics. Microorganisms (activated biosolids) are settled from the aerated mixed liquor under quiescent conditions in the final clarifier and are returned to the aeration tank. Left uncontrolled, the number of organisms would eventually become too great; therefore, some must periodically be removed (wasted). A portion of the concentrated solids from the bottom of the settling tank must be removed from the process (waste activated sludge, or WAS). Clear supernatant from the final settling tank is the plant effluent.

5.7.5.1 Activated Biosolids Process Control Calculations

As with other wastewater treatment unit processes, process control calculations are important tools used by the operator to control and optimize process operations. In the following sections, we review many of the most frequently used activated biosolids process calculations.

5.7.5.2 Moving Averages

When performing process control calculations, the use of a 7-day *moving average* is recommended. The moving average is a mathematical method for leveling the impact of any one test result. The moving average is determined by adding the test results collected during the past 7 days and dividing by the number of tests:

$$\frac{\text{Moving}}{\text{average}} = \frac{\text{Test } 1 + \text{Test } 2 + \text{Test } 3 + \ldots + \text{Test } 7}{\text{No. of tests performed in 7 days}} \quad (5.152)$$

■ EXAMPLE 5.246

Problem: Calculate the 7-day moving average for days 7, 8, and 9.

Day	MLSS	Day	MLSS
1	3340	6	2780
2	2480	7	2476
3	2398	8	2756
4	2480	9	2655
5	2558	10	2396

Solution:

1. Moving average, day 7 = (3340 + 2480 + 2398 + 2480 + 2558 + 2780 + 2476)/7 = 2645
2. Moving average, day 7 = (2480 + 2398 + 2480 + 2558 + 2780 + 2476 + 2756)/7 = 2561
3. Moving average, day 7 = (2398 + 2480 + 2558 + 2780 + 2476 + 2756 + 2655)/7 = 2586

5.7.5.3 BOD or COD Loading

When calculating BOD, COD, or SS loading on an aeration process (or any other treatment process), loading on the process is usually calculated as lb/day. The following equation is used:

$$\frac{\text{BOD, COD, or SS}}{\text{loading (lb/day)}} = \begin{bmatrix} \text{BOD, COD, or SS (mg/L)} \\ \times \text{ flow (MGD)} \times 8.34 \text{ lb/gal} \end{bmatrix} \quad (5.153)$$

■ EXAMPLE 5.247

Problem: The BOD concentration of the wastewater entering an aerator is 210 mg/L. If the flow to the aerator is 1,550,000 gpd, what is the lb/day BOD loading?

Solution:

$$\begin{aligned} \text{BOD (lb/day)} &= \text{BOD (mg/L)} \times \text{flow (MGD)} \times 8.34 \text{ lb/gal} \\ &= 210 \text{ mg/L} \times 1.55 \text{ MGD} \times 8.34 \text{ lb/gal} \\ &= 2715 \text{ lb/day} \end{aligned}$$

■ EXAMPLE 5.248

Problem: The flow to an aeration tank is 2750 gpm. If the BOD concentration of the wastewater is 140 mg/L, how many pounds of BOD are applied to the aeration tank daily?

Solution: First convert the gpm flow to gpd flow:

$$2750 \text{ gpm} \times 1440 \text{ min/day} = 3,960,000 \text{ gpd}$$

Then calculate lb/day BOD:

$$\begin{aligned} \text{BOD (lb/day)} &= \text{BOD (mg/L)} \times \text{flow (MGD)} \times 8.34 \text{ lb/gal} \\ &= 140 \text{ mg/L} \times 3.96 \text{ MGD} \times 8.34 \text{ lb/gal} \\ &= 4624 \text{ lb/day} \end{aligned}$$

5.7.5.4 Solids Inventory

In the activated biosolids process, it is important to control the amount of solids under aeration. The suspended solids in an aeration tank are called *mixed liquor suspended solids* (MLSS). To calculate the pounds of solids in the aeration tank, we need to know the mg/L MLSS concentration and the aeration tank volume, then lb MLSS can be calculated as follows:

$$\text{MLSS (lb)} = \begin{bmatrix} \text{MLSS (mg/L)} \times \text{volume (MG)} \\ \times 8.34 \text{ lb/gal} \end{bmatrix} \quad (5.154)$$

■ EXAMPLE 5.249

Problem: If the mixed liquor suspended solids concentration is 1200 mg/L, and the aeration tanks has a volume of 550,000 gal, how many pounds of suspended solids are in the aeration tank?

Solution:

$$\begin{aligned} \text{MLSS (lb)} &= \begin{bmatrix} \text{MLSS (mg/L)} \times \text{volume (MG)} \\ \times 8.34 \text{ lb/gal} \end{bmatrix} \\ &= 1200 \text{ mg/L} \times 0.550 \text{ MG} \times 8.34 \text{ lb/gal} \\ &= 5504 \text{ lb} \end{aligned}$$

5.7.5.5 Food-to-Microorganism Ratio

The food-to-microorganism ratio (F/M ratio) is a process control approach based on maintaining a specified balance between available food materials (BOD or COD) in the aeration tank influent and the aeration tank mixed liquor volatile suspended solids (MLVSS) concentration. The chemical oxygen demand (COD) test is sometimes used, because the results are available in a relatively short period of time. To calculate the F/M ratio, the following information is required:

- Aeration tank influent flow rate (MGD)
- Aeration tank influent BOD or COD (mg/L)
- Aeration tank MLVSS (mg/L)
- Aeration tank volume (MG)

$$\text{F/M ratio} = \frac{\begin{array}{c}\text{Primary effluent COD/BOD (mg/L)}\\ \times \text{flow (MGD)}\\ \times 8.34 \text{ lb/mg/L/MG}\end{array}}{\begin{array}{c}\text{MLVSS (mg/L)}\\ \times \text{aerator volume (MG)}\\ \times 8.34 \text{ lb/mg/L/MG}\end{array}} \quad (5.155)$$

Typical F/M ratios for activated biosolids process are shown in the following table:

	BOD (lb)/ MLVSS (lb)	COD (lb)/ MLVSS (lb)
Conventional	0.2–0.4	0.5–1.0
Contact stabilization	0.2–0.6	0.5–1.0
Extended aeration	0.05–0.15	0.2–0.5
Pure oxygen	0.25–1.0	0.5–2.0

■ EXAMPLE 5.250

Problem: The aeration tank influent BOD is 145 mg/L, and the aeration tank influent flow rate is 1.6 MGD. What is the F/M ratio if the MLVSS is 2300 mg/L and the aeration tank volume is 1.8 MG?

Solution:

$$\text{F/M ratio} = \frac{145 \text{ mg/L} \times 1.6 \text{ MGD} \times 8.34 \text{ lb/mg/L/MG}}{2300 \text{ mg/L} \times 1.8 \text{ MG} \times 8.34 \text{ lb/mg/L/MG}}$$

$$= 0.0.6 \text{ lb BOD per lb MLVSS}$$

Key Point: If the MLVSS concentration is not available, it can be calculated if the percent volatile matter (%VM) of the mixed liquor suspended solids is known:

$$\text{MLVSS} = \text{MLSS} \times \% \text{ volatile matter (\%VM)} \quad (5.156)$$

Key Point: The F value in the F/M ratio for computing loading to an activated biosolids process can be either BOD or COD. Remember, the reason for biosolids production in the activated biosolids process is to convert BOD to bacteria. One advantage of using COD over BOD for analysis of organic load is that COD is more accurate.

■ EXAMPLE 5.251

Problem: The aeration tank contains 2885 mg/L of MLSS. Lab tests indicate that the MLSS is 66% volatile matter. What is the MLVSS concentration in the aeration tank?

Solution:

$$\text{MLVSS (mg/L)} = 2885 \text{ mg/L} \times 0.66 = 1904 \text{ mg/L}$$

5.7.5.5.1 Required MLVSS Quantity (lb)

The pounds of MLVSS required in the aeration tank to achieve the optimum F/M ratio can be determined from the average influent food (BOD or COD) and the desired F/M ratio:

$$\text{MLVSS (lb)} = \frac{\begin{array}{c}\text{Primary effluent BOD/COD}\\ \times \text{flow (MGD)} \times 8.34 \text{ lb/gal}\end{array}}{\text{Desired F/M ratio}} \quad (5.157)$$

The required pounds of MLVSS determined by this calculation can then be converted to a concentration value by:

$$\text{MLVSS (mg/L)} = \frac{\text{Desired MLVSS (lb)}}{\begin{array}{c}\text{Aeration volume (MG)}\\ \times 8.34 \text{ lb/gal}\end{array}} \quad (5.158)$$

■ EXAMPLE 5.252

Problem: The aeration tank influent flow rate is 4.0 MGD and the influent COD is 145 mg/L. The aeration tank volume is 0.65 MG. The desired F/M ratio is 0.3 lb COD per lb MLVSS. How many pounds of MLVSS must be maintained in the aeration tank to achieve the desired F/M ratio?

Solution: Determine the required concentration of MLVSS in the aeration tank:

$$\text{MLVSS (lb)} = \frac{145 \text{ mg/L} \times 4.0 \text{ MGD} \times 8.34 \text{ lb/gal}}{0.3 \text{ lb COD per lb MLVSS}}$$

$$= 16,124 \text{ lb}$$

$$\text{MLVSS (mg/L)} = \frac{16,124 \text{ lb}}{0.65 \text{ MG} \times 8.34 \text{ lb/gal}}$$

$$= 2974 \text{ mg/L}$$

5.7.5.5.2 Calculating Waste Rates Using F/M Ratio

Maintaining the desired F/M ratio is accomplished by controlling the MLVSS level in the aeration tank. This may be achieved by adjustment of return rates; however, the most practical method is by proper control of the waste rate:

$$\text{Waste volume solids} = \text{Actual MLVSS (lb/day)} \\ - \text{desired MLVSS (lb/day)} \quad (5.159)$$

If the desired MLVSS is greater than the actual MLVSS, wasting is stopped until the desired level is achieved. Practical considerations require that the waste quantity be converted to a required volume of waste per day. This is accomplished by converting the waste pounds to flow rate in million gallons per day or gallons per minute:

$$\text{Waste (MGD)} = \frac{\text{Waste volatile (lb/day)}}{\text{Waste volatile conc. (mg/L)} \times 8.34 \text{ lb/gal}} \quad (5.160)$$

$$\text{Waste (gpm)} = \frac{\text{Waste (MGD)} \times 1,000,000 \text{ gpd/MGD}}{1440 \text{ min/day}} \quad (5.161)$$

Key Point: When the F/M ratio is used for process control, the volatile content of the waste activated sludge should be determined.

■ EXAMPLE 5.253

Problem: Given the following information, determine the required waste rate in gallons per minute to maintain an F/M ratio of 0.17 lb COD per lb MLVSS:

Primary effluent COD	140 mg/L
Primary effluent flow	2.2 MGD
MLVSS (mg/L)	3549 mg/L
Aeration tank volume	0.75 MG
Waste volatile concentrations	4440 mg/L

Solution:

$$\text{Actual MLVSS (lb/day)} = 3549 \text{ mg/L} \times 0.75 \text{ MG} \\ \times 8.34 \text{ lb/gal} \\ = 22{,}199 \text{ lb/day}$$

$$\text{Required MLVSS (lb/day)} = \frac{140 \text{ mg/L} \times 2.2 \text{ MGD} \times 8.34 \text{ lb/gal}}{0.17 \text{ lb COD per lb MLVSS}} \\ = 15{,}110 \text{ lb/day}$$

$$\text{Waste (lb)} = 22{,}199 \text{ lb/day} - 15{,}110 \text{ lb/day} = 7089 \text{ lb/day}$$

$$\text{Waste (MGD)} = \frac{7089 \text{ lb/day}}{4440 \text{ mg/L} \times 8.34 \text{ lb/gal}} = 0.19 \text{ MGD}$$

$$\text{Waste (gpm)} = \frac{0.19 \text{ MGD} \times 1{,}000{,}000 \text{ gpd/MGD}}{1440 \text{ min/day}} \\ = 132 \text{ gpm}$$

5.7.5.6 Gould Biosolids Age

Biosolids age refers to the average number of days a particle of suspend solids remains under aeration. It is a calculation used to maintain the proper amount of activated biosolids in the aeration tank. This calculation is sometimes referred to as *Gould biosolids age* so it is not confused with similar calculations such as the solids retention time (or mean cell residence time). When considering sludge age, in effect we are asking, "How many days of suspended solids are in the aeration tank?" For example, if 3000 lb of suspended solids enter the aeration tank daily, the aeration tank contains 12,000 lb of suspended solids when 4 days of solids are in the aeration tank—a sludge age of 4 days:

$$\text{Sludge age (days)} = \frac{\text{SS in tank (lb)}}{\text{SS added (lb/day)}} \quad (5.162)$$

■ EXAMPLE 5.254

Problem: A total of 2740 lb/day suspended solids enter an aeration tank in the primary effluent flow. If the aeration tank has a total of 13,800 lb of mixed liquor suspended solids, what is the biosolids age in the aeration tank?

Solution:

$$\text{Sludge age (days)} = \frac{\text{MLSS (lb)}}{\text{SS added (lb/day)}} \\ = \frac{13{,}800 \text{ lb}}{2740 \text{ lb/day}} = 5 \text{ days}$$

5.7.5.7 Mean Cell Residence Time

Mean cell residence time (MCRT), sometimes called *sludge retention time*, is another process control calculation used for activated biosolids systems. MCRT represents the average length of time an activated biosolids particle remains in the activated biosolids system. It can also be defined as the length of time required at the current removal rate to remove all the solids in the system:

$$\text{MCRT (days)} = \frac{\begin{array}{c}\text{MLSS (mg/L)} \\ \times (\text{aeration vol.} + \text{clarifier vol.}) \\ \times 8.34 \text{ lb/mg/L/MG}\end{array}}{\begin{array}{c}[\text{WAS (mg/L)} \times \text{WAS flow} \times 8.34] \\ + (\text{TSS out} \times \text{flow out} \times 8.34)\end{array}} \quad (5.163)$$

Key Point: MCRT can be calculated using only the aeration tank solids inventory. When comparing plant operational levels to reference materials, we must determine which calculation the reference manual uses to obtain its example values. Other methods are available to determine the clarifier solids concentrations; however, the simplest method assumes that the average suspended solids concentration is equal to the solids concentration of the aeration tank.

■ **EXAMPLE 5.255**

Problem: Given the following data, what is the MCRT?

Aerator volume	1,000,000 gal
Final clarifier	600,000 gal
Flow	5.0 MGD
Waste rate	0.085 MGD
MLSS	2500 mg/L
Waste	6400 mg/L
Effluent TSS	14 mg/L

Solution:

$$MCRT = \frac{2500 \text{ mg/L} \times (1.0 \text{ MG} + 0.60 \text{ MG}) \times 8.34}{\begin{array}{c} [6400 \text{ mg/L}] \times 0.085 \text{ MGD} \times 8.34] \\ + (14 \text{ mg/L} \times 5.0 \text{ mgd} \times 8.34) \end{array}}$$

$$= 6.5 \text{ days}$$

5.7.5.7.1 Waste Quantities/Requirements

Mean cell residence time for process control requires the determination of the optimum range for MCRT values. This is accomplished by comparison of the effluent quality with MCRT values. When the optimum MCRT is established, the quantity of solids to be removed (wasted) is determined by:

$$Water \text{ (lb/day)} = \left(\frac{\begin{array}{c} MLSS \times (\text{aeration vol.} \\ + \text{clarifier vol.}) \times 8.34 \end{array}}{Desired \text{ MCRT}} \right) \quad (5.164)$$
$$- (TSS_{out} \times flow \times 8.34)$$

■ **EXAMPLE 5.256**

$$\frac{\begin{array}{c} 3400 \text{ mg/L} \times (1.4 \text{ MG} \\ + 0.50 \text{ MG}) \times 8.34 \end{array}}{8.6 \text{ days}} - (10 \text{ mg/L} \times 5.0 \text{ MGD} \times 8.34)$$

Water quality (lb/day) = 5848 lb

5.7.5.7.2 Waste Rate in Million Gallons/Day

When the quantity of solids to be removed from the system is known, the desired waste rate in million gallons per day can be determined. The unit used to express the rate

(MGD, gpd, and gpm) is a function of the volume of waste to be removed and the design of the equipment:

$$Waste \text{ (MGD)} = \frac{Waste \text{ (lb/day)}}{WAS \text{ concentration (mg/L)}} \quad (5.165)$$
$$\times 8.34$$

$$Waste \text{ (gpm)} = \frac{\begin{array}{c} Waste \text{ (MGD)} \\ \times 1,000,000 \text{ gpd/MGD} \end{array}}{1440 \text{ min/day}} \quad (5.166)$$

■ **EXAMPLE 5.257**

Problem: Given the following data, determine the required waste rate to maintain a MCRT of 8.8 days:

MLSS	2500 mg/L
Aeration volume	1.20 MG
Clarifier volume	0.20 MG
Effluent TSS	11 mg/L
Effluent flow	5.0 MGD
Waste concentration	6000 mg/L

Solution:

$$Waste \text{ (lb/day)} = \frac{2500 \text{ mg/L} \times (1.20 + 0.20) \times 8.34}{8.8 \text{ days}}$$
$$- (11 \text{ mg/L} \times 5.0 \text{ MGD} \times 8.34)$$

$$= 3317 \text{ lb/day} - 459 \text{ lb/day}$$

$$= 2858 \text{ lb/day}$$

$$Waste \text{ (MGD)} = \frac{2858 \text{ lb/day}}{6000 \text{ mg/L} \times 8.34 \text{ gal/day}}$$

$$= 0.057 \text{ MGD}$$

$$Waste \text{ (gpm)} = \frac{0.057 \text{ MGD} \times 1,000,000 \text{ gpd/MGD}}{1440 \text{ min/day}}$$

$$= 40 \text{ gpm}$$

5.7.5.8 Estimating Return Rates from SSV$_{60}$

Many methods are available for estimation of the proper return biosolids rate. A simple method described in the *Operation of Wastewater Treatment Plants: Field Study Programs*, published by California State University, Sacramento, uses the 60-min percent settled sludge volume. The %SSV$_{60}$ test results can provide an approximation of the appropriate return activated biosolids rate. This calculation assumes that the SSV$_{60}$ results are representative of the actual settling occurring in the clarifier. If this is true, the return rate in percent should be approximately equal to the SSV$_{60}$. To determine the approximate return rate in million gallons per day (MGD), the influent flow rate, the current return rate, and the SSV_{60} must be known. The results of this calculation can then be adjusted based on sampling and visual observations to develop the optimum return biosolids rate.

Key Point: The $\%SSV_{60}$ must be converted to a decimal percent and total flow rate (wastewater flow and current return rate in million gallons per day must be used).

$$\text{Est. return} \atop \text{rate (MGD)} = \left[\begin{array}{c} \text{Influent flow (MGD)} \\ + \text{return flow (MGD)} \end{array} \right] \times \% \, SSV_{60}$$
(5.167)

$$\text{RAS rate} \atop \text{(gpm)} = \frac{\text{Biosolids return rate (gpd)}}{1440 \text{ min/day}}$$

where it is assumed that:

- $\%SSV_{60}$ is representative.
- Return rate in percent equals $\%SSV_{60}$.
- Actual return rate is normally set slightly higher to ensure organisms are returned to the aeration tank as quickly as possible.

The rate of return must be adequately controlled to prevent the following:

- Aeration and settling hydraulic overloads
- Low MLSS levels in the aerator
- Organic overloading of aeration
- Septic return activated biosolids
- Solids loss due to excessive biosolids blanket depth

■ **EXAMPLE 5.258**

Problem: The influent flow rate is 5.0 MGD, and the current return activated sludge flow rate is 1.8 MGD. The $\%SSV_{60}$ is 37%. Based on this information, what should be the return biosolids return rate in million gallons per day (MGD)?

Solution:

$$\text{Return (MGD)} = (5.0 \text{ MGD} + 1.8 \text{ MGD}) \times 0.37$$

$$= 2.5 \text{ MGD}$$

5.7.5.9 Sludge Volume Index

Sludge volume index (SVI) is a measure (an indicator) of the settling quality (a quality indicator) of the activated biosolids. As the SVI increases, the biosolids settle more slowly, do not compact as well, and are likely to result in an increase in effluent suspended solids. As the SVI decreases, the biosolids become denser, settling is more

rapid, and the biosolids age. SVI is the volume in milliliters occupied by 1 g of activated biosolids. For the settled biosolids volume (mL/L) and the mixed liquor suspended solids (MLSS) calculation, mg/L are required. The proper SVI range for any plant must be determined by comparing SVI values with plant effluent quality:

$$\text{Sludge volume index (SVI)} = \frac{\text{SSV (mL/L} \times 1000)}{\text{MLSS (mg/L)}} \quad (5.168)$$

■ **EXAMPLE 5.259**

Problem: The SSV_{30} is 365 mL/L, and the MLSS is 2365 mg/L. What is the SVI?

Solution:

$$\text{Sludge volume index (SVI)} = \frac{365 \text{ mL/L} \times 1000}{2365 \text{ mg/L}}$$

$$= 154.3$$

So, the SVI equals 154.3. What does this mean? It means that the system is operating normally with good settling and low effluent turbidity. How do we know this? We know this because we compare the 154.3 result with the parameters listed in the table below to obtain the expected condition (the result):

SVI	Expected Conditions
Less than 100	Old biosolids; possible pin floc Effluent turbidity increasing
100–250	Normal operation; good settling Low effluent turbidity
Greater than 250	Bulking biosolids; poor settling High effluent turbidity

5.7.5.10 Mass Balance: Settling Tank Suspended Solids

Solids are produced whenever biological processes are used to remove organic matter from wastewater. Mass balance for an anaerobic biological process must take into account both the solids removed by physical settling processes and the solids produced by biological conversion of soluble organic matter to insoluble suspended matter organisms. Research has shown that the amount of solids produced per pound of BOD removed can be predicted based on the type of process being used. Although the exact amount of solids produced can vary from plant to plant, research has developed a series of K factors that can be used to estimate the solids production for plants using a particular treatment process. These average factors provide a simple method to evaluate the effectiveness of a facility's process control program. The mass balance also provides an excellent mechanism to evaluate the validity of process control and effluent monitoring data generated. Recall that average K factors are listed in pounds of solids produced per pound of BOD removed for selected processes:

$$BOD_{in} \ (lb) = \left[\begin{array}{l} BOD \ (mg/L) \times flow \ (MGD) \\ \times \ 8.34 \ lb/gal \end{array} \right]$$

$$BOD_{out} \ (lb) = \left[\begin{array}{l} BOD \ (mg/L) \times flow \ (MGD) \\ \times \ 8.34 \ lb/gal \end{array} \right]$$

$$\text{Solids produced (lb/day)} = [BOD_{in} \ (lb) - BOD_{out} \ (lb)] \times K$$

$$TSS_{out} \ (lb/day) = \left[\begin{array}{l} TSS_{out} \ (mg/L) \times flow \ (MGD) \\ \times \ 8.34 \ lb/gal \end{array} \right]$$

$$\text{Waste (lb/day)} = \left[\begin{array}{l} \text{Waste (mg/L)} \times flow \ (MGD) \\ \times \ 8.34 \ lb/gal \end{array} \right]$$

$$\text{Solids removed (lb/day)} = TSS_{out} \ (lb/day) + \text{waste (lb/day)}$$

$$\underset{\text{balance}}{\% \ Mass} = \frac{(\text{Solids produced} - \text{solids removed})}{\text{Solids produced}} \times 100 \qquad (5.169)$$

5.7.5.11 Biosolids Waste Based on Mass Balance

$$\text{Waste rate (MGD)} = \frac{\text{Solids produced (lb/day)}}{\substack{\text{Waste concentration} \\ \times \ 8.34 \ lb/gal}} \qquad (5.170)$$

■ EXAMPLE 5.260

Problem: Given the data in the following table, determine the mass balance of the biological process and the appropriate waste rate to maintain current operating conditions.

Extended Aeration (No Primary)	
Influent flow	1.1 MGD
BOD	220 mg/L
TSS	240 mg/L
Effluent flow	1.5 MGD
BOD	18 mg/L
TSS	22 mg/L
Waste flow	24,000 gpd
TSS	8710 mg/L

Solution:

$$BOD_{in} = 220 \ mg/L \times 1.1 \ MGD \times 8.34 \ lb/gal = 2018 \ lb/day$$

$$BOD_{out} = 18 \ mg/L \times 1.1 \ MGD \times 8.34 \ lb/gal = 165 \ lb/day$$

$$\text{BOD removed} = 2018 \ lb/day - 165 \ lb/day = 1853 \ lb/day$$

$$\begin{array}{l} \text{Solids produced} = 1853 \ lb/day \times 0.65 \ lb/lb \ BOD \\ = 1204 \ lb/day \end{array}$$

$$\text{Solids out} = 22 \ mg/L \times 1.1 \ MGD \times 8.34 \ lb/gal = 202 \ lb/day$$

$$\begin{array}{l} \text{Sludge out} = 8710 \ mg/L \times 0.024 \ MGD \times 8.34 \ lb/gal \\ = 1743 \ lb/day \end{array}$$

$$\text{Solids removed} = 202 \ lb/day + 1743 \ lb/day = 1945 \ lb/day$$

$$\% \ \text{Mass balance} = \frac{1204 \ lb/day - 1945 \ lb/day) \times 100}{1204 \ lb/day} = 62\%$$

The mass balance indicates that:

1. The sampling points, collection methods, or laboratory testing procedures are producing nonrepresentative results.
2. The process is removing significantly more solids than is required. Additional testing should be performed to isolate the specific cause of the imbalance.

To assist in the evaluation, the waste rate based on the mass balance information can be calculated:

$$\text{Waste (gpd)} = \frac{\text{Solids produced (lb/day)}}{\text{Waste TSS (mg/L)} \times 8.34} \qquad (5.171)$$

$$\text{Waste (gpd)} = \frac{1204 \ lb/day \times 1,000,000}{8710 \ mg/L \times 8.34} = 16,575 \ gpd$$

5.7.5.12 Oxidation Ditch Detention Time

Oxidation ditch systems may be used where the treatment of wastewater is amendable to aerobic biological treatment and the plant design capacities generally do not exceed 1.0 mgd. The oxidation ditch is a form of aeration basin where the wastewater is mixed with return biosolids. The oxidation ditch is essentially a modification of a completely mixed activated biosolids system used to treat wastewater from small communities. This system can be classified as an extended aeration process and is considered to be a low loading rate system. This type of treatment facility can remove 90% or more of influent BOD. Oxygen requirements will generally depend on the maximum diurnal organic loading, degree of treatment, and suspended solids concentration to be maintained in the aerated channel mixed liquor suspended solids. Detention time is the length of time required for a given flow rate to pass through a tank. Detention time is not normally calculated for aeration basins, but it is calculated for oxidation ditches.

Key Point: When calculating detention time it is essential that the time and volume units used in the equation are consistent with each other.

$$\text{Detention time (hr)} = \frac{\text{Oxidation ditch volume (gal)}}{\text{Flow rate (gph)}} \quad (5.172)$$

■ EXAMPLE 5.261

Problem: An oxidation ditch has a volume of 160,000 gal. If the flow to the oxidation ditch is 185,000 gpd, what is the detention time in hours?

Solution: Because detention time is desired in hours, the flow must be expressed as gph:

$$(185{,}000 \text{ gpd}) \div (24 \text{ hr/day}) = 7708 \text{ gph}$$

Now calculate detention time:

$$\text{Detention time (hr)} = \frac{\text{Oxidation ditch volume (gal)}}{\text{Flow rate (gph)}}$$

$$= \frac{160{,}000 \text{ gal}}{7708 \text{ gph}} = 20.8 \text{ hr}$$

5.7.6 TREATMENT PONDS

The primary goals of wastewater treatment ponds focus on simplicity and flexibility of operation, protection of the water environment, and protection of public health. Moreover, ponds are relatively easy to build and manage, they accommodate large fluctuations in flow, and they can also provide treatment that approaches conventional systems (producing a highly purified effluent) at much lower cost. It is the cost (the economics) that drives many managers to decide on the pond option of treatment. The actual degree of treatment provided in a pond depends on the type and number of ponds used. Ponds can be used as the sole type of treatment, or they can be used in conjunction with other forms of wastewater treatment—that is, other treatment processes followed by a pond or a pond followed by other treatment processes. Ponds can be classified based on their location in the system, by the type of wastes they receive, and by the main biological process occurring in the pond.

5.7.6.1 Treatment Pond Parameters

Before we discuss process control calculations, it is important first to describe the calculations for determining the area, volume, and flow rate parameters that are crucial in making treatment pond calculations.

5.7.6.1.1 Pond Area in Inches

$$\text{Area (ac)} = \frac{\text{Area (ft}^2)}{43{,}560 \text{ ft}^2\text{/ac}} \quad (5.173)$$

5.7.6.1.2 Pond Volume in Acre-Feet

$$\text{Volume (ac-ft)} = \frac{\text{Volume (ft}^3)}{43{,}560 \text{ ft}^2\text{/ac-ft}} \quad (5.174)$$

5.7.6.1.3 Flow Rate in Acre-Feet/Day

$$\text{Flow (ac-ft/day)} = \text{Flow (MGD)} \times 3069 \text{ ac-ft/MG} \quad (5.175)$$

Key Point: Acre-feet (ac-ft) is a unit that can cause confusion, especially for those not familiar with pond or lagoon operations. One ac-ft is the volume of a box with a 1-ac top and 1 ft of depth—but the top does not have to be an even number of acres in size to use ac-ft.

5.7.6.1.4 Flow Rate in Acre-Inches/Day

$$\text{Flow (ac-in./day)} = \text{Flow (MGD)} \times 36.8 \text{ ac-in./MG} \quad (5.176)$$

5.7.6.2 Treatment Pond Process Control Calculations

Although there are no recommended process control calculations for the treatment pond, several calculations may be helpful in evaluating process performance or identifying causes of poor performance. These include hydraulic detention time, BOD loading, organic loading rate, BOD removal efficiency, population loading, and hydraulic loading rate. In the following, we provide a few calculations that might be helpful in pond performance evaluation and identification of causes of poor performance process along with other calculations and equations that may be helpful.

5.7.6.3 Hydraulic Detention Time (Days)

$$\begin{array}{l} \text{Hydraulic detention} \\ \text{time (days)} \end{array} = \frac{\text{Pond volume (ac-ft)}}{\text{Influent flow (ac-ft/day)}} \quad (5.177)$$

Key Point: Normally, hydraulic detention time ranges from 30 to 120 days for stabilization ponds.

■ EXAMPLE 5.262

Problem: A stabilization pond has a volume of 54.5 ac-ft. What is the detention time in days when the flow is 0.35 MGD?

Solution:

$$\text{Flow (ac-ft/day)} = 0.35 \text{ MGD} \times 3069 \text{ ac-ft/MG}$$

$$= 1.07 \text{ ac-ft/day}$$

$$\text{Detention time (days)} = \frac{54.5 \text{ ac-ft}}{1.07 \text{ ac-ft/day}} = 51 \text{ days}$$

5.7.6.4 BOD Loading

When calculating BOD loading on a wastewater treatment pond, the following equation is used:

$$\text{BOD (lb/day)} = \begin{bmatrix} \text{BOD (mg/L)} \times \text{flow (MGD)} \\ \times 8.34 \text{ lb/gal} \end{bmatrix} \quad (5.178)$$

■ **EXAMPLE 5.263**

Problem: Calculate the BOD loading (lb/day) on a pond if the influent flow is 0.3 MGD with a BOD of 200 mg/L.

Solution:

$$\text{BOD (lb/day)} = \text{BOD (mg/L)} \times \text{flow (MGD)} \times 8.34 \text{ lb/gal}$$

$$= 200 \text{ mg/L} \times 0.3 \text{ MGD} \times 8.34 \text{ lb/gal}$$

$$= 500 \text{ lb/day}$$

5.7.6.5 Organic Loading Rate

Organic loading can be expressed as pounds of BOD per acre per day (most common), pounds of BOD per acre-foot per day, or people per acre per day.

$$\text{Organic loading rate (lb BOD/ac/day)} = \frac{\begin{array}{c} \text{BOD (mg/L)} \\ \times \text{influent flow (MGD)} \\ \times 8.34 \end{array}}{\text{Pond area (ac)}} \quad (5.179)$$

Key Point: Normal range is 10 to 50 lb BOD per day per acre.

■ **EXAMPLE 5.264**

Problem: A wastewater treatment pond has an average width of 370 ft and an average length of 730 ft. The influent flow rate to the pond is 0.10 MGD with a BOD concentration of 165 mg/L. What is the organic loading rate to the pond in pounds per day per acre (lb/day/ac)?

Solution:

$$730 \text{ ft} \times 370 \text{ ft} \times \frac{1 \text{ ac}}{43,560 \text{ ft}^2} = 6.2 \text{ ac}$$

$$0.10 \text{ MGD} \times 165 \text{ mg/L} \times 8.34 \text{ lb/gal} = 138 \text{ lb/day}$$

$$\frac{138 \text{ lb/day}}{6.2 \text{ ac}} = 22.2 \text{ lb/day/ac}$$

5.7.6.6 BOD Removal Efficiency

As mentioned, the efficiency of any treatment process is its effectiveness in removing various constituents from the water or wastewater. BOD removal efficiency is therefore a measure of the effectiveness of the wastewater treatment pond in removing BOD from the wastewater.

$$\% \text{ BOD removed} = \frac{\text{BOD removed (mg/L)}}{\text{BOD total (mg/L)}} \times 100 \quad (5.180)$$

■ **EXAMPLE 5.265**

Problem: The BOD entering a waste treatment pond is 194 mg/L. If the BOD in the pond effluent is 45 mg/L, what is the BOD removal efficiency of the pond?

$$\% \text{ BOD removed} = \frac{\text{BOD removed (mg/L)}}{\text{BOD total (mg/L)}} \times 100$$

$$= \frac{149 \text{ mg/L}}{194 \text{ mg/L}} \times 100 = 77\%$$

5.7.6.7 Population Loading

$$\text{Population loading (people/ac/day)} = \frac{\begin{array}{c} \text{BOD (mg/L)} \\ \times \text{influent flow (MGD)} \\ \times 8.34 \end{array}}{\text{Pond area (ac)}} \quad (5.181)$$

5.7.6.8 Hydraulic Loading Rate (In./Day) (Hydraulic Overflow Rate)

$$\text{Hydraulic loading (in./day)} = \frac{\text{Influent flow (ac-in./day)}}{\text{Pond area (ac)}} \quad (5.182)$$

5.7.7 CHEMICAL DOSING

Chemicals are used extensively in wastewater treatment (and water treatment) operations. Plant operators add chemicals to various unit processes for slime growth control, corrosion control, odor control, grease removal, BOD reduction, pH control, biosolids bulking control, ammonia oxidation, and bacterial reduction, among other reasons. To apply any chemical dose correctly, it is important to make certain dosage calculations. One of the most frequently used calculations in wastewater/water mathematics is the dosage or loading. The general types of mg/L to lb/day or lb calculations are for chemical dosage, BOD, COD, SS loading/removal, pounds of solids under aeration, and WAS pumping rate. These calculations are usually made using either Equation 5.183 or Equation 5.184:

$$\left[\begin{array}{c} \text{Chemical (mg/L)} \times \text{flow (MGD)} \\ \times\, 8.34\ \text{lb/gal} \end{array}\right] = \text{lb/day} \quad (5.183)$$

$$\left[\begin{array}{c} \text{Chemical (mg/L)} \times \text{volume (MG)} \\ \times\, 8.34\ \text{lb/gal} \end{array}\right] = \text{lb} \quad (5.184)$$

Key Point: If mg/L concentration represents a concentration in a flow, then million gallons per day (MGD) flow is used as the second factor; however, if the concentration pertains to a tank or pipeline volume, then million gallons (MG) volume is used as the second factor.

Key Point: Typically, especially in the past, the expression parts per million (ppm) was used as an expression of concentration, because 1 mg/L = 1 ppm; however, current practice is to use mg/L as the preferred expression of concentration.

5.7.7.1 Chemical Feed Rate

In chemical dosing, a measured amount of chemical is added to the wastewater (or water). The amount of chemical required depends on the type of chemical used, the reason for dosing, and the flow rate being treated. The two expressions most often used to describe the amount of chemical added or required are:

- Milligrams per liter (mg/L)
- Pounds per day (lb/day)

A milligram per liter is a measure of concentration. As shown below, if a concentration of 5 mg/L is desired, then a total of 15 mg chemical would be required to treat 3 L:

$$\frac{5\ \text{mg} \times 3}{\text{L} \times 3} = \frac{15\ \text{mg}}{3\ \text{L}}$$

The amount of chemical required therefore depends on two factors:

- The desired concentration (mg/L)
- The amount of wastewater to be treated (normally expressed as MGD).

To convert from mg/L to lb/day, Equation 5.26 is used.

■ EXAMPLE 5.266

Problem: Determine the chlorinator setting (lb/day) required to treat a flow of 5 MGD with a chemical dose of 3 mg/L.

Solution:

$$\text{Chemical (lb/day)} = \left[\begin{array}{c} \text{Chemical (mg/L)} \times \text{flow (MGD)} \\ \times\, 8.34\ \text{lb/gal} \end{array}\right]$$

$$= 3\ \text{mg/L} \times 5\ \text{MGD} \times .34\ \text{lb/gal}$$

$$= 125\ \text{lb/day}$$

■ EXAMPLE 5.267

Problem: The desired dosage for a dry polymer is 10 mg/L. If the flow to be treated is 2,100,000 gpd, how many lb/day polymer will be required?

Solution:

$$\text{Polymer (lb/day)} = \left[\begin{array}{c} \text{Polymer (mg/L)} \times \text{flow (MGD)} \\ \times\, 8.34\ \text{lb/day} \end{array}\right]$$

$$= 10\ \text{mg/L} \times 2.10\ \text{MGD} \times 8.34\ \text{lb/day}$$

$$= 175\ \text{lb/day}$$

Key Point: To calculate chemical dose for tanks or pipelines, a modified equation must be used. Instead of MGD flow, MG volume is used:

$$\text{Chemical (lb)} = \left[\begin{array}{c} \text{Chemical (mg/L)} \\ \times \text{tank volume (MG)} \times 8.34\ \text{lb/gal} \end{array}\right] (5.185)$$

■ EXAMPLE 5.268

Problem: To neutralize a sour digester, 1 lb of lime is added for every pound of volatile acids in the digester biosolids. If the digester contains 300,000 gal of biosolids with a volatile acid level of 2200 mg/L, how many pounds of lime should be added?

Solution: Because the volatile acid concentration is 2200 mg/L, the lime concentration should also be 2200 mg/L, so:

$$\text{Lime required (lb)} = \left[\begin{array}{c} \text{Lime (mg/L)} \times \text{digester volume (MG)} \\ \times\, 8.34\ \text{lb/gal} \end{array}\right]$$

$$= 2200\ \text{mg/L} \times 0.30\ \text{MG} \times 8.34\ \text{lb/gal}$$

$$= 5504\ \text{lb}$$

5.7.7.2 Chlorine Dose, Demand, and Residual

Chlorine is a powerful oxidizer that is commonly used in wastewater and water treatment for disinfection and in wastewater treatment for odor and bulking control, among other applications. When chlorine is added to a unit process, we want to ensure that a measured amount is added, obviously. Chlorine dose depends on two considerations: the chlorine demand and the desired chlorine residual:

$$\text{Chlorine dose} = \begin{bmatrix} \text{Chlorine demand} \\ + \text{chlorine residual} \end{bmatrix} \quad (5.186)$$

5.7.7.2.1 Chlorine Dose

To describe the amount of chemical added or required, we use Equation 5.187:

$$\text{Chemical (lb/day)} = \begin{bmatrix} \text{Chemical (mg/L)} \\ \times \text{flow (MGD)} \times 8.34 \text{ lb/day} \end{bmatrix} (5.187)$$

■ EXAMPLE 5.269

Problem: Determine the chlorinator setting (lb/day) required to treat an 8-MGD flow with a chlorine dose of 6 mg/L.

Solution:

$$\text{Chemical (lb/day)} = \begin{bmatrix} \text{Chemical (mg/L)} \times \text{flow (MGD)} \\ \times 8.34 \text{ lb/day} \end{bmatrix}$$

$$= 6 \text{ mg/L} \times 8 \text{ MGD} \times 8.34 \text{ lb/gal}$$

$$= 400 \text{ lb/day}$$

5.7.7.2.2 Chlorine Demand

The chlorine demand is the amount of chlorine used in reacting with various components of the water such as harmful organisms and other organic and inorganic substances. When the chlorine demand has been satisfied, these reactions cease.

■ EXAMPLE 5.270

Problem: The chlorine dosage for a secondary effluent is 6 mg/L. If the chlorine residual after a 30-min contact time is found to be 0.5 mg/L, what is the chlorine demand expressed in mg/L?

Solution:

$$\text{Chlorine dose} = \text{Chlorine demand} + \text{chlorine residual}$$

$$6 \text{ mg/L} = x \text{ mg/L} + 0.5 \text{ mg/L}$$

$$6 \text{ mg/L} - 0.5 \text{ mg/L} = x \text{ mg/L}$$

$$x = 5.5 \text{ mg/L chlorine demand}$$

5.7.7.2.3 Chlorine Residual

Chlorine residual is the amount of chlorine remaining after the demand has been satisfied.

■ EXAMPLE 5.271

Problem: What should the chlorinator setting (lb/day) be to treat a flow of 3.9 MGD if the chlorine demand is 8 mg/L and a chlorine residual of 2 mg/L is desired?

Solution: First calculate the chlorine dosage in mg/L:

$$\text{Chlorine dose} = \text{Chlorine demand} + \text{chlorine residual}$$

$$= 8 \text{ mg/L} + 2 \text{ mg/L} = 10 \text{ mg/L}$$

Then calculate the chlorine dosage (feed rate) in lb/day:

$$\begin{bmatrix} \text{Chlorine (mg/L)} \times \text{flow (MGD)} \\ \times 8.34 \text{ lb/gal} \end{bmatrix} = \text{lb/day chlorine}$$

$$10 \text{ mg/L} \times 3.9 \text{ MGD} \times 8.34 \text{ lb/gal} = 325 \text{ lb/day chlorine}$$

5.7.7.3 Hypochlorite Dosage

Hypochlorite is less hazardous than chlorine; therefore, it is often used as a substitute chemical for elemental chlorine. Hypochlorite is similar to strong bleach and comes in two forms: dry calcium hypochlorite (often referred to as HTH) and liquid sodium hypochlorite. Calcium hypochlorite contains about 65% available chlorine; sodium hypochlorite contains about 12 to 15% available chlorine (in industrial strengths).

Key Point: Because either type of hypochlorite is not 100% pure chlorine, more lb/day must be fed into the system to obtain the same amount of chlorine for disinfection. This is an important economical consideration for those facilities thinking about substituting hypochlorite for chlorine. Some studies indicate that such a substitution can increase operating costs, overall, by up to three times the cost of using chlorine.

To calculate the lb/day hypochlorite required, a two-step calculation is required:

$$\begin{bmatrix} \text{Chlorine (mg/L)} \times \text{flow (MGD)} \\ \times 8.34 \text{ lb/gal} \end{bmatrix} = \text{lb/day chlorine}$$

$$\frac{\text{Chlorine (lb/day)}}{\dfrac{\text{\% Available}}{100}} = \text{Hypochlorite (lb/day)} \quad (5.188)$$

■ EXAMPLE 5.272

Problem: A total chlorine dosage of 10 mg/L is required to treat a particular wastewater. If the flow is 1.4 MGD and the hypochlorite has 65% available chlorine, how many lb/day of hypochlorite will be required?

Solution: First calculate the lb/day chlorine required using the mg/L to lb/day equation:

$$\left[\begin{array}{c} \text{Chlorine (mg/L)} \times \text{flow (MGD)} \\ \times \ 8.34 \text{ lb/gal} \end{array}\right] = \text{lb/day chlorine}$$

$$10 \text{ mg/L} \times 1.4 \text{ MGD} \times 8.34 \text{ lb/gal} = 117 \text{ lb/day chlorine}$$

Then calculate the lb/day hypochlorite required. Because only 65% of the hypochlorite is chlorine, more than 117 lb/day will be required:

$$\frac{117 \text{ lb/day chlorine}}{\dfrac{65 \text{ available chlorine}}{100}} = 180 \text{ lb/day hypochlorite}$$

■ EXAMPLE 5.273

Problem: A wastewater flow of 840,000 gpd requires a chlorine dose of 20 mg/L. If sodium hypochlorite (15% available chlorine) is to be used, how many lb/day of sodium hypochlorite are required? How many gallons per day of sodium hypochlorite is this?

Solution: First calculate the lb/day chlorine required:

$$\left[\begin{array}{c} \text{Chlorine (mg/L)} \times \text{flow (MGD)} \\ \times \ 8.34 \text{ lb/gal} \end{array}\right] = \text{lb/day chlorine}$$

$$20 \text{ mg/L} \times 0.84 \text{ MGD} \times 8.34 \text{ lb/gal} = 140 \text{ lb/day chlorine}$$

Then calculate the lb/day sodium hypochlorite:

$$\frac{140 \text{ lb/day chlorine}}{\dfrac{15 \text{ available chlorine}}{100}} = 933 \text{ lb/day hypochlorite}$$

Now calculate the gpd sodium hypochlorite:

$$\frac{933 \text{ lb/day}}{8.34 \text{ lb/gal}} = 112 \text{ gpd sodium hypochlorite}$$

■ EXAMPLE 5.274

Problem: How many pounds of chlorine gas are necessary to treat 5,000,000 gal of wastewater at a dosage of 2 mg/L?

Solution: First calculate the pounds of chlorine required:

$$\text{Volume (}10^6 \text{ gal)} = \text{Chlorine conc. (mg/L)} \times 8.34 \text{ lb/gal}$$

$$= \text{lb chlorine}$$

Then substitute:

$$(5 \times 10^6 \text{ gal}) \times 2 \text{ mg/L} \times 8.34 \text{ lb/gal} = 83 \text{ lb chlorine}$$

5.7.8 CHEMICAL SOLUTIONS

A *water solution* is a homogeneous liquid consisting of the *solvent* (the substance that dissolves another substance) and the *solute* (the substance that dissolves in the solvent). Water is the solvent. The solute (whatever it may be) may dissolve up to a certain point. This is called its *solubility*—that is, the solubility of the solute in the particular solvent (water) at a particular temperature and pressure. Remember, in chemical solutions, the substance being dissolved is called the *solute*, and the liquid present in the greatest amount in a solution (and that does the dissolving) is called the *solvent*. We should also be familiar with another term, *concentration*—the amount of solute dissolved in a given amount of solvent. Concentration is measured as:

$$\% \text{ Strength} = \frac{\text{Weight of solute}}{\text{Weight of solution}} \times 100$$

$$= \frac{\text{Weight of solute}}{\text{Weight of solute + solvent}} \times 100 \qquad (5.189)$$

■ EXAMPLE 5.275

Problem: If 30 lb of chemical is added to 400 lb of water, what is the percent strength (by weight) of the solution?

Solution:

$$\% \text{ Strength} = \frac{30 \text{ lb solute}}{400 \text{ lb water}} \times 100$$

$$= \frac{30 \text{ lb solute}}{30 \text{ lb solute + 400 lb water}} \times 100$$

$$= \frac{30 \text{ lb solute}}{430 \text{ lb solute + water}} = 7.0\%$$

Important to making accurate computations of chemical strength is a complete understanding of the dimensional units involved; for example, it is important to understand exactly what *milligrams per liter* (mg/L) signifies:

$$\text{Milligrams per liter (mg/L)} = \frac{\text{Milligrams of solute}}{\text{Liters of solution}} \qquad (5.190)$$

Another important dimensional unit commonly used when dealing with chemical solutions is *parts per million* (ppm):

$$\text{Parts per million (ppm)} = \frac{\text{Parts of solute}}{\text{Million parts of solution}} \qquad (5.191)$$

Key Point: A part is usually a weight measurement; for example:

$$8 \text{ ppm} = \frac{8 \text{ lb solids}}{1,000,000 \text{ lb solution}}$$

$$8 \text{ ppm} = \frac{8 \text{ mg solids}}{1,000,000 \text{ mg solution}}$$

5.7.8.1 Chemical Solution Feeder Setting (gpd)

Calculating the gallon-per-day feeder setting depends on how the solution concentration is expressed: lb/gal or percent. If the solution strength is expressed as lb/gal, use the following equation:

$$\text{Solution (gpd)} = \frac{\begin{array}{c}\text{Chemical (mg/L)} \times \text{flow (MGD)} \\ \times 8.34 \text{ lb/gal}\end{array}}{\text{Chemical solution (lb)}} \quad (5.192)$$

In water and wastewater operations, a standard, trial-and-error method known as *jar testing* is conducted to determine optimum chemical dosage. Jar testing has been the accepted bench testing procedure for many years. After jar testing results are analyzed to determine the best chemical dosage, the actual calculations are made, as demonstrated by the following example problems.

■ EXAMPLE 5.276

Problem: Jar tests indicate that the best liquid alum dose for a water is 8 mg/L. The flow to be treated is 1.85 MGD. Determine the gpd setting for the liquid alum chemical feeder if the liquid alum contains 5.30 lb of alum per gallon of solution.

Solution: First, calculate the lb/day of dry alum required, using the mg/L to lb/day equation:

$$\text{Alum (lb/day)} = \frac{\begin{array}{c}\text{Dose (mg/L)} \times \text{flow (MGD)} \\ \times 8.34 \text{ lb/gal}\end{array}}{\text{Chemical solution (lb)}}$$

$$= 8 \text{ mg/L} \times 1.85 \text{ MGD} \times 8.34 \text{ lb/gal}$$

$$= 123 \text{ lb/day}$$

Then, calculate the gpd solution required:

$$\text{Alum solution (gpd)} = \frac{123 \text{ lb/day alum}}{5.30 \text{ lb alum per gal solution}}$$

$$= 23 \text{ gpd}$$

The feeder setting, then, is 23 gpd alum solution. If the solution strength is expressed as a percent, we use the following equation:

$$\begin{bmatrix} \text{Chemical (mg/L)} \times \text{flow treated (MGD)} \\ \times 8.34 \text{ lb/gal} \end{bmatrix}$$
$$= \begin{bmatrix} \text{Solution (mg/L)} \times \text{solution flow (MGD)} \\ \times 8.34 \text{ lb/gal} \end{bmatrix} \quad (5.193)$$

■ EXAMPLE 5.277

Problem: The flow to a plant is 3.40 MGD. Jar testing indicates that the optimum alum dose is 10 mg/L. What should the gpd setting be for the solution feeder if the alum solution is a 52% solution?

Solution: A solution concentration of 52% is equivalent to 520,000 mg/L:

$$\text{Desired dose (lb/day)} = \text{Actual dose (lb/day)}$$

$$[\text{Chemical (mg/L)} \times \text{flow treated (MGD)} \times 8.34 \text{ lb/gal}]$$
$$= [\text{Solution (mg/L)} \times \text{solution flow (MGD)} \times 8.34 \text{ lb/gal}$$

$$10 \text{ mg/L} \times 3.40 \text{ MGD} \times 8.34 \text{ lb/gal}$$
$$= 520,00 \text{ mg/L} \times x \text{ MGD} \times 8.34 \text{ lb/gal}$$

$$x = \frac{10 \times 3.40 \times 8.34}{520,000 \times 8.34} = 0.0000653 \text{ MGD}$$

This can be expressed as gpd flow:

$$0.0000653 \text{ MGD} = 65.3 \text{ gpd flow}$$

5.7.8.2 Chemical Feed Pump: Percent Stroke Setting

Chemical feed pumps are generally positive-displacement pumps (also called *piston pumps*). This type of pump displaces, or pushes out, a volume of chemical equal to the volume of the piston. The length of the piston, called the *stroke*, can be lengthened or shortened to increase or decrease the amount of chemical delivered by the pump. As mentioned, each stroke of a piston pump displaces or pushes out chemical. When calculating the percent stroke setting, use the following equation:

$$\text{Stroke setting (\%)} = \frac{\text{Required feed (gpd)}}{\text{Maximum feed (gpd)}} \quad (5.194)$$

■ EXAMPLE 5.278

Problem: The required chemical pumping rate has been calculated as 8 gpm. If the maximum pumping rate is 90 gpm, what should the percent stroke setting be?

Solution: The percent stroke setting is based on the ratio of the gpm required to the total possible gpm:

$$\text{Stroke setting (\%)} = \frac{\text{Required feed (gpd)}}{\text{Maximum feed (gpd)}}$$

$$= \frac{8 \text{ gpm}}{90 \text{ gpm}} \times 100 = 8.9\%$$

5.7.8.3 Chemical Solution Feeder Setting (mL/min)

Some chemical solution feeders dispense chemical as milliliters per minute (mL/min). To calculate the mL/min solution required, use the following equation:

$$\text{Solution (mL/min)} = \frac{\text{gpd} \times 3785 \text{ mL/gal}}{1440 \text{ min/day}} \quad (5.195)$$

■ EXAMPLE 5.279

Problem: The desired solution feed rate was calculated to be 7 gpd. What is this feed rate expressed as mL/min?

Solution: Because the gpd flow has already been determined, the mL/min flow rate can be calculated directly:

$$\text{Solution (mL/min)} = \frac{\text{gpd} \times 3785 \text{ mL/gal}}{1440 \text{ min/day}}$$

$$= \frac{7 \text{ gpd} \times 3785 \text{ mL/gal}}{1440 \text{ min/day}}$$

$$= 18 \text{ mL/min}$$

5.7.8.4 Chemical Feed Calibration

Routinely, to ensure accuracy, we need to compare the actual chemical feed rate with the feed rate indicated by the instrumentation. To accomplish this, we use calibration calculations. To calculate the actual chemical feed rate for a dry chemical feed, place a container under the feeder, weigh the container when empty, and then weigh the container again after a specified length of time, such as 30 min. Then actual chemical feed rate can then be determined as:

$$\text{Feed rate (lb/min)} = \frac{\text{Chemical applied (lb)}}{\text{Length of application (min)}} \quad (5.196)$$

■ EXAMPLE 5.280

Problem: Calculate the actual chemical feed rate (lb/day) if a container is placed under a chemical feeder and a total of 2.2 lb is collected during a 30-min period.

Solution: First calculate the lb/min feed rate:

$$\text{Feed rate (lb/min)} = \frac{\text{Chemical applied (lb)}}{\text{Length of application (min)}}$$

$$= \frac{2.2 \text{ lb}}{30 \text{ min}} = 0.07 \text{ lb/min}$$

Then calculate the lb/day feed rate:

$$\text{Chemical feed rate (lb/day)} = 0.07 \text{ lb/min} \times 1440 \text{ min/day}$$

$$= 101 \text{ lb/day}$$

■ EXAMPLE 5.281

Problem: A chemical feeder is to be calibrated. The container to be used to collect chemical is placed under the chemical feeder and weighed (0.35 lb). After 30 min, the weight of the container and chemical is found to be 2.2 lb. Based on this test, what is the actual chemical feed rate in lb/day?

Key Point: The chemical applied is the weight of the container and chemical minus the weight of the empty container.

Solution: First calculate the lb/min feed rate:

$$\text{Chemical feed rate (lb/min)} = \frac{\text{Chemical applied (lb)}}{\text{Length of application (min)}}$$

$$= \frac{2.2 \text{ lb} - 0.35 \text{ lb}}{30 \text{ min}}$$

$$= \frac{1.85 \text{ lb}}{30 \text{ min}} = 0.062 \text{ lb/min}$$

Then calculate the lb/day feed rate:

$$0.062 \text{ lb/min} \times 1440 \text{ min/day} = 89 \text{ lb/day}$$

When the chemical feeder is for a solution, the calibration calculation is slightly more difficult than that for a dry chemical feeder. As with other calibration calculations, the actual chemical feed rate is determined and then compared with the feed rate indicated by the instrumentation. The calculations used for solution feeder calibration are as follows:

$$\text{Flow rate (gpd)} = \frac{(\text{mL/min}) \times 1440 \text{ min/day}}{3785 \text{ mL/gal}} = \text{gpd} \quad (5.197)$$

Then calculate chemical dosage (lb/day):

$$\text{Chemical (lb/day)} = \left[\begin{array}{l} \text{Chemical (mg/L)} \\ \times \text{flow (MGD)} \times 8.34 \text{ lb/day} \end{array} \right] \quad (5.198)$$

■ **EXAMPLE 5.282**

Problem: A calibration test is conducted for a chemical solution feeder. During 5 min, the solution feeder delivers a total of 700 mL. The polymer solution is a 1.3% solution. What is the lb/day feed rate? (Assume that the polymer solution weighs 8.34 lb/gal.)

Solution: The mL/min flow rate is calculated as:

$$700 \text{ mL} \div 5 \text{ min} = 140 \text{ mL/min}$$

Then convert the mL/min flow rate to a gpd flow rate:

$$\frac{140 \text{ mL/min} \times 1440 \text{ min/day}}{3785 \text{ mL/gal}} = 53 \text{ gpd}$$

Now calculate the lb/day feed rate:

$$\left[\begin{array}{c} \text{Chemical (mg/L)} \times \text{flow (MGD)} \\ \times\, 8.34 \text{ lb/day} \end{array} \right] = \text{Chemical (lb/day)}$$

$$\frac{13,000 \text{ mg/L}}{\times\, 0.000053 \text{ MGD} \times 8.34 \text{ lb/day}} = 5.7 \text{ lb/day polymer}$$

Actual pumping rates can be determined by calculating the volume pumped during a specified time frame; for example, if 120 gal are pumped during a 15-min test, the average pumping rate during the test is 8 gpm. The gallons pumped can be determined by measuring the drop in tank level during the timed test:

$$\text{Flow (gpm)} = \frac{\text{Volume pumped (gal)}}{\text{Duration of test (min)}} \quad (5.199)$$

The actual flow rate (gpm) is then calculated using:

$$\text{Flow rate (gpm)} = \frac{\begin{array}{c} 0.785 \times (\text{diameter})^2 \\ \times \text{drop in level (ft)} \times 7.48 \text{ gal/ft}^3 \end{array}}{\text{Duration of test (min)}} \quad (5.200)$$

■ **EXAMPLE 5.283**

Problem: A pumping rate calibration test is conducted for 5 min. The liquid level in the 4-ft-diameter solution tank is measured before and after the test. If the level drops 0.4 ft during the 5-min test, what is the pumping rate in gpm?

Solution:

$$\text{Pumping rate (gpm)} = \frac{\begin{array}{c} 0.785 \times (\text{diameter})^2 \\ \times \text{drop in level (ft)} \times 7.48 \text{ gal/ft}^3 \end{array}}{\text{Duration of test (min)}}$$

$$= \frac{\begin{array}{c} 0.785 \times (4 \text{ ft} \times 4 \text{ ft}) \\ \times\, 0.4 \text{ ft} \times 7.48 \text{ gal/ft}^3 \end{array}}{5 \text{ min}} = 38 \text{ gpm}$$

5.7.8.5 Average Use Calculations

During a typical shift, operators log in or record several parameter readings. The data collected are important in monitoring plant operation as they provide information on how to best optimize plant or unit process operation. One of the important parameters monitored each shift or each day is actual use of chemicals. From the recorded chemical use data, expected chemical use can be forecasted. These data are also important for inventory control, because when additional chemical supplies will be required can be estimated. To determine average chemical use, we first must determine the average chemical use:

$$\text{Average use (lb/day)} = \frac{\text{Total chemical used (lb)}}{\text{Number of days}} \quad (5.201)$$

or

$$\text{Average use (gpd)} = \frac{\text{Total chemical used (gal)}}{\text{Number of days}} \quad (5.202)$$

Then, we calculate the days supply in inventory:

$$\text{Days supply in inventory} = \frac{\begin{array}{c}\text{Total chemical}\\ \text{in inventory (lb)}\end{array}}{\text{Average use (lb/day)}} \quad (5.203)$$

or

$$\text{Days supply in inventory} = \frac{\begin{array}{c}\text{Total chemical}\\ \text{in inventory (gal)}\end{array}}{\text{Average use (gpd)}} \quad (5.204)$$

■ **EXAMPLE 5.284**

Problem: The chemical used for each day during a week is given below. Based on these data, what was the average lb/day chemical use during the week?

Monday	92 lb/day
Tuesday	94 lb/day
Wednesday	92 lb/day
Thursday	88 lb/day
Friday	96 lb/day
Saturday	92 lb/day
Sunday	88 lb/day

Solution:

$$\text{Average use (lb/day)} = \frac{\text{Total chemical used (lb)}}{\text{Number of days}}$$

$$= \frac{642 \text{ lb}}{7 \text{ days}} = 91.7 \text{ lb/day}$$

■ **EXAMPLE 5.285**

Problem: The average chemical use at a plant is 83 lb/day. If the chemical inventory in stock is 2600 lb, how many days supply is this?

Solution:

$$\text{Days supply in inventory} = \frac{\text{Total chemical in inventory (lb)}}{\text{Average use (lb/day)}}$$

$$= \frac{2600 \text{ lb}}{83 \text{ lb/day}}$$

$$= 31.3 \text{ days}$$

5.7.8.6 Process Residuals: Biosolids Production and Pumping Calculations

The wastewater unit treatment processes remove solids and biochemical oxygen demand from the wastestream before the liquid effluent is discharged to its receiving waters. What remains to be disposed of is a mixture of solids and wastes, called *process residuals*, more commonly referred to as *biosolids* (or *sludge*).

Key Point: *Sludge* is the commonly accepted term for wastewater residual solids; however, if wastewater sludge is used for beneficial reuse (i.e., as a soil amendment or fertilizer), it is commonly called *biosolids*. I choose to refer to process residuals as biosolids in this text.

The most costly and complex aspect of wastewater treatment can be the collection, processing, and disposal of biosolids. This is the case because the quantity of biosolids produced may be as high as 2% of the original volume of wastewater, depending somewhat on the treatment process being used. Because the biosolids can be as much as 97% water content and because cost of disposal will be related to the volume of biosolids being processed, one of the primary purposes or goals (along with stabilizing it so it is no longer objectionable or environmentally damaging) of biosolids treatment is to separate as much of the water from the solids as possible.

5.7.8.7 Primary and Secondary Solids Production Calculations

It is important to point out that when making calculations pertaining to solids and biosolids, the term *solids* refers to dry solids and the term *biosolids* refers to the solids and water. The solids produced during primary treatment depend on the solids that settle in, or are removed by, the primary clarifier. When making primary clarifier solids production calculations, we use the mg/L to lb/day equation for suspended solids as shown below:

$$\begin{array}{c} \text{SS removed} \\ \text{(lb/day)} \end{array} = \begin{bmatrix} \text{SS removed (mg/L)} \\ \times \text{flow (MGD)} \times 8.34 \text{ lb/gal} \end{bmatrix} \quad (5.205)$$

5.7.8.8 Primary Clarifier Solids Production Calculations

■ **EXAMPLE 5.286**

Problem: A primary clarifier receives a flow of 1.80 MGD with suspended solids concentrations of 340 mg/L. If the clarifier effluent has a suspended solids concentration of 180 mg/L, how many pounds of solids are generated daily?

Solution:

$$\begin{array}{c} \text{SS removed} \\ \text{(lb/day)} \end{array} = \begin{bmatrix} \text{SS removed (mg/L)} \\ \times \text{flow (MGD)} \times 8.34 \text{ lb/gal} \end{bmatrix}$$

$$= 160 \text{ mg/L} \times 1.80 \text{ MGD} \times 8.34 \text{ lb/gal}$$

$$= 2402 \text{ lb/day}$$

■ **EXAMPLE 5.287**

Problem: The suspended solids content of the primary influent is 350 mg/L and the primary influent is 202 mg/L. How many pounds of solids are produced during a day when the flow is 4,150,000 gpd?

Solution:

$$\begin{array}{c} \text{SS removed} \\ \text{(lb/day)} \end{array} = \begin{bmatrix} \text{SS removed (mg/L)} \\ \times \text{flow (MGD)} \times 8.34 \text{ lb/gal} \end{bmatrix}$$

$$= 148 \text{ mg/L} \times 4.15 \text{ MGD} \times 8.34 \text{ lb/gal}$$

$$= 5122 \text{ lb/day}$$

5.7.8.9 Secondary Clarifier Solids Production Calculations

Solids produced during secondary treatment depend on many factors, including the amount of organic matter removed by the system and the growth rate of the bacteria. Because precise calculations of biosolids production is complex, we use a rough estimate method of solids production that uses an estimated growth rate (unknown) value. We use the BOD removed (lb/day) equation shown below:

$$\begin{array}{c} \text{BOD removed} \\ \text{(lb/day)} \end{array} = \begin{bmatrix} \text{BOD removed (mg/L)} \\ \times \text{flow (MGD)} \times 8.34 \text{ lb/day} \end{bmatrix} \quad (5.206)$$

■ **EXAMPLE 5.288**

Problem: The 1.5-MGD influent to the secondary system has a BOD concentration of 174 mg/L. The secondary effluent contains 22 mg/L BOD. If the bacteria growth rate for this plant is 0.40 lb SS per lb BOD removed, how many pounds of dry biosolids solids are produced each day by the secondary system?

Solution:

$$\begin{aligned}\text{BOD removed} \atop \text{(lb/day)} &= \begin{bmatrix}\text{BOD removed (mg/L)}\\ \times \text{flow (MGD)} \times 8.34 \text{ lb/day}\end{bmatrix}\\ &= \begin{bmatrix}152 \text{ mg/L} \times 1.5 \text{ MGD}\\ \times 8.34 \text{ lb/gal}\end{bmatrix}\\ &= 1902 \text{ lb/day}\end{aligned}$$

Then use the unknown x value to determine lb/day solids produced:

$$\frac{0.44 \text{ lb SS produced}}{1 \text{ lb BOD removed}} = \frac{x \text{ lb SS produced}}{1902 \text{ lb/day BOD removed}}$$

$$\frac{0.44 \times 1902}{1} = x$$

837 lb/day solids produced $= x$

Key Point: Typically, for every pound of food consumed (BOD removed) by the bacteria, between 0.3 and 0.7 lb of new bacteria cells are produced; these are solids that have to be removed from the system.

5.7.8.10 Percent Solids

Biosolids are composed of water and solids. The vast majority of biosolids is water, usually in the range of 93 to 97%. To determine the solids content of a biosolids, a sample of biosolids is dried overnight in an oven at 103 to 105°F. The solids that remain after drying represent the total solids content of the biosolids. Solids content may be expressed as a percent or as a mg/L. Either of two equations is used to calculate percent solids:

$$\% \text{ Solids} = \frac{\text{Total solids (g)}}{\text{Biosolids sample (g)}} \times 100 \quad (5.207)$$

$$\% \text{ Solids} = \frac{\text{Solids (lb/day)}}{\text{Biosolids (lb/day)}} \times 100 \quad (5.208)$$

■ **EXAMPLE 5.289**

Problem: The total weight of a biosolids sample (sample only, not the dish) is 22 g. If the weight of the solids after drying is 0.77 g, what is the percent total solids of the biosolids?

Solution:

$$\% \text{ Solids} = \frac{\text{Total solids (g)}}{\text{Biosolids sample (g)}} \times 100$$

$$= \frac{0.77 \text{ g}}{22 \text{ g}} \times 100 = 3.5\%$$

5.7.8.11 Biosolids Pumping

While on shift, wastewater operators are often required to make various process control calculations. An important calculation involves biosolids pumping. The biosolids pumping calculations the operator may be required to make are covered in the following sections.

5.7.8.12 Estimating Daily Biosolids Production

The calculation for estimation of the required biosolids pumping rate provides a method to establish an initial pumping rate or to evaluate the adequacy of the current withdrawal rate:

$$\text{Est. pump rate} = \frac{\begin{array}{c}(\text{Influent TSS conc.}\\ - \text{effluent TSS conc.})\\ \times \text{flow} \times 8.34\end{array}}{\begin{array}{c}\% \text{ Solids in sludge} \times 8.34\\ \times 1440 \text{ min/day}\end{array}} \quad (5.209)$$

■ **EXAMPLE 5.290**

Problem: The biosolids withdrawn from the primary settling tank contain 1.4% solids. The unit influent contains 285 mg/L TSS, and the effluent contains 140 mg/L TSS. If the influent flow rate is 5.55 MGD, what is the estimated biosolids withdrawal rate in gallons per minute (assuming the pump operates continuously)?

Solution:

$$\text{Biosolids rate} = \frac{\begin{array}{c}(285 \text{ mg/L} - 140 \text{ mg/L})\\ \times 5.55 \times 8.34\end{array}}{0.014 \times 8.34 \times 1440 \text{ min/day}}$$

$$= 40 \text{ gpm}$$

5.7.8.13 Biosolids Production in Pounds per Million Gallons

A common method of expressing biosolids production is in pounds of biosolids per million gallons of wastewater treated.

$$\text{Biosolids (lb/MG)} = \frac{\text{Total biosolids production (lb)}}{\text{Total wastewater flow (MG)}} \quad (5.210)$$

■ **EXAMPLE 5.291**

Problem: Records show that a plant has produced 85,000 gal of biosolids during the past 30 days. The average daily flow for this period was 1.2 MGD. What was the plant's biosolids production in pounds per million gallons?

Solution:

$$\text{Biosolids (lb/MG)} = \frac{\text{Total biosolids production (lb)}}{\text{Total wastewater flow (MG)}}$$

$$= \frac{85{,}000 \text{ gal} \times 8.34 \text{ lb/gal}}{1.2 \text{ MGD} \times 30 \text{ days}}$$

$$= 19{,}692 \text{ lb/MG}$$

5.7.8.14 Biosolids Production in Wet Tons/Year

Biosolids production can also be expressed in terms of the amount of biosolids (water and solids) produced per year. This is normally expressed in wet tons per year:

$$\text{Biosolids (wet tons/yr)} = \frac{\begin{array}{c}\text{Biosolids prod. (lb/MG)}\\ \times \text{ average daily flow (MGD)}\\ \times 365 \text{ days/yr}\end{array}}{2000 \text{ lb/ton}} \quad (5.211)$$

■ EXAMPLE 5.292

Problem: The plant is currently producing biosolids at the rate of 16,500 lb/MG. The current average daily wastewater flow rate is 1.5 MGD. What will be the total amount of biosolids produced per year in wet tons per year?

Solution:

$$\text{Biosolids (wet tons/yr)} = \frac{\begin{array}{c}\text{Biosolids prod. (lb/MG)}\\ \times \text{ average daily flow (MGD)}\\ \times 365 \text{ days/yr}\end{array}}{2000 \text{ lb/ton}}$$

$$= \frac{16{,}500 \text{ lb/MG} \times 1.5 \text{ MGD} \times 365 \text{ days/yr}}{2000 \text{ lb/ton}}$$

$$= 4517 \text{ wet tons/yr}$$

5.7.8.15 Biosolids Pumping Time

The biosolids pumping time is the total time the pump operates during a 24-hr period in minutes:

$$\text{Pump operating time} = \begin{bmatrix}\text{Time/cycle (min)}\\ \times \text{ frequency (cycles/day)}\end{bmatrix} \quad (5.212)$$

The following information is used for Examples 5.293 to 5.297:

   Frequency = 24 times/day
   Pump rate = 120 gpm
   Solids = 3.70%
   Volatile matter = 66%

■ EXAMPLE 5.293

Problem: What is the pump operating time?

Solution:

$$\text{Pump operating time} = 15 \text{ min/hr} \times 24 \text{ cycles/day}$$

$$= 360 \text{ min/day}$$

5.7.8.16 Biosolids Pumped per Day in Gallons

$$\text{Biosolids (gpd)} = \begin{bmatrix}\text{Operating time (min/day)}\\ \times \text{ pump rate (gpm)}\end{bmatrix} \quad (5.213)$$

■ EXAMPLE 5.294

Problem: What is the biosolids pumped per day in gallons?

Solution:

$$\text{Biosolids (gpd)} = 360 \text{ min/day} \times 120 \text{ gpm} = 43{,}200 \text{ gpd}$$

5.7.8.17 Biosolids Pumped per Day in Pounds

$$\text{Sludge (lb/day)} = \begin{bmatrix}\text{Gallons of biosolids pumped}\\ \times 8.34 \text{ lb/gal}\end{bmatrix} \quad (5.214)$$

■ EXAMPLE 5.295

Problem: What is the biosolids pumped per day in pounds?

Solution:

$$\text{Biosolids (lb/day)} = 43{,}200 \text{ gpd} \times 8.34 \text{ lb/gal}$$

$$= 360{,}000 \text{ lb/day}$$

5.7.8.18 Solids Pumped per Day in Pounds

$$\text{Solids pumped (lb/day)} = \begin{bmatrix}\text{Biosolids pumped (lb/day)}\\ \times \% \text{ solids}\end{bmatrix} \quad (5.215)$$

■ EXAMPLE 5.296

Problem: What are the solids pumped per day?

Solution:

$$\text{Solids pumped (lb/day)} = 360{,}300 \text{ lb/day} \times 0.0370$$

$$= 13{,}331 \text{ lb/day}$$

5.7.8.19 Volatile Matter Pumped per Day in Pounds

$$\text{Volatile matter (lb/day)} = \begin{bmatrix}\text{Solids pumped (lb/day)}\\ \times \% \text{ volatile matter}\end{bmatrix} \quad (5.216)$$

■ **EXAMPLE 5.297**

Problem: What is the volatile matter in pounds per day?

Solution:

$$\text{Volatile matter (lb/day)} = 13{,}331 \text{ lb/day} \times 0.66$$

$$= 8798 \text{ lb/day}$$

5.7.8.20 Biosolids Thickening

Biosolids thickening (or concentration) is a unit process used to increase the solids content of the biosolids by removing a portion of the liquid fraction. In other words, biosolids thickening is all about volume reduction. Increasing the solids content allows more economical treatment of the biosolids. Biosolids thickening processes include the following:

- Gravity thickeners
- Flotation thickeners
- Solids concentrators

Biosolids thickening calculations are based on the concept that the solids in the primary or secondary biosolids are equal to the solids in the thickened biosolids. The solids are the same. It is primarily water that has been removed to thicken the biosolids and result in higher percent solids. In this unthickened biosolids, the solids might represent 1 to 4% of the total pounds of biosolids, but when some of the water is removed those same amount solids might represent 5 to 7% of the total pounds of biosolids.

Key Point: The key to biosolids thickening calculations is that solids remain constant.

5.7.8.21 Gravity/Dissolved Air Flotation Thickener Calculations

As mentioned, biosolids thickening calculations are based on the concept that the solids in the primary or secondary biosolids are equal to the solids in the thickened biosolids; that is, assuming a negligible amount of solids are lost in the thickener overflow, the solids are the same. Note that the water is removed to thicken the biosolids and results in higher percent solids.

5.7.8.21.1 Estimating Daily Biosolids Production

The calculation for estimation of the required biosolids pumping rate provides a method to establish an initial pumping rate or to evaluate the adequacy of the current pump rate:

$$\text{Est. pump rate} = \frac{\begin{array}{c}(\text{Influent TSS conc.}\\ - \text{ effluent TSS conc.})\\ \times \text{ flow} \times 8.34\end{array}}{\begin{array}{c}\% \text{ Solids in biosolids} \times 8.34\\ \times 1440 \text{ min/day}\end{array}} \quad (5.217)$$

■ **EXAMPLE 5.298**

Problem: The biosolids withdrawn from the primary settling tank contain 1.5% solids. The unit influent contains 280 mg/L TSS, and the effluent contains 141 mg/L TSS. If the influent flow rate is 5.55 MGD, what is the estimated biosolids withdrawal rate in gallons per minute (assuming that the pump operates continuously)?

Solution:

$$\text{Biosolids withdrawal rate} = \frac{\begin{array}{c}(280 \text{ mg/L} - 141 \text{ mg/L})\\ \times 5.55 \text{ MGD} \times 8.34\end{array}}{0.015 \times 8.34 \times 1440 \text{ min/day}}$$

$$= 36 \text{ gpm}$$

5.7.8.21.2 Surface Loading Rate (gal/day/ft²)

The surface loading rate (surface overflow rate or surface settling rate) is hydraulic loading—the amount of biosolids applied per square foot of a gravity thickener:

$$\text{Surface loading (gal/day/ft}^2) = \frac{\begin{array}{c}\text{Biosolids applied}\\ \text{to thickener (gpd)}\end{array}}{\text{Thickener area (ft}^2)} \quad (5.218)$$

■ **EXAMPLE 5.299**

Problem: A 70-ft-diameter gravity thickener receives 32,000 gpd of biosolids. What is the surface loading in gallons per square foot per day?

Solution:

$$\text{Surface loading} = \frac{32{,}000 \text{ gpd}}{0.785 \times 70 \text{ ft} \times 70 \text{ ft}} = 8.32 \text{ gpd/ft}^2$$

5.7.8.21.3 Solids Loading Rate (lb/day/ft²)

The solids loading rate is the pounds of solids per day being applied to 1 ft² of tank surface area. The calculation uses the surface area of the bottom of the tank. It assumes that the floor of the tank is flat and has the same dimensions as the surface:

$$\begin{array}{c}\text{Surface loading rate}\\ \text{(lb/day/ft}^2)\end{array} = \frac{\begin{array}{c}\% \text{ Biosolids solids}\\ \times \text{ biosolids flow (gpd)}\\ \times 8.34 \text{ lb/gal}\end{array}}{\text{Thickener area (ft}^2)} \quad (5.219)$$

■ EXAMPLE 5.300

Problem: The thickener influent contains 1.6% solids. The influent flow rate is 39,000 gpd. The thickener is 50 ft in diameter and 10 ft deep. What is the solids loading in pounds per day?

Solution:

$$\frac{\text{Surface loading rate}}{(\text{lb/day/ft}^2)} = \frac{0.016 \times 39{,}000 \text{ gpd} \times 8.34 \text{ lb/gal}}{0.785 \times 50 \text{ ft} \times 50 \text{ ft}}$$

$$= 2.7 \text{ lb/ft}^2$$

5.7.8.21.4 Concentration Factor (CF)

The concentration factor (CF) represents the increase in concentration resulting from the thickener; it is a means of determining the effectiveness of the gravity thickening process:

$$\text{CF} = \frac{\text{Thickened biosolids concentration (\%)}}{\text{Influent biosolids concentration (\%)}} \quad (5.220)$$

■ EXAMPLE 5.301

Problem: The influent biosolids contain 3.5% solids. The thickened biosolids solids concentration is 7.7%. What is the concentration factor?

Solution:

$$\text{CF} = \frac{7.7\%}{3.5\%} = 2.2$$

5.7.8.21.5 Air-to-Solids Ratio

The air-to-solids ratio is the ratio of the pounds of solids entering the thickener to the pounds of air being applied:

$$\text{Air/solids ratio} = \frac{\begin{array}{c}\text{Air flow (ft}^3\text{/min)}\\ \times 0.0785 \text{ lb/ft}^3\end{array}}{\begin{array}{c}\text{Biosolids flow (gpm)}\\ \times \% \text{ solids} \times 8.34 \text{ lb/gal}\end{array}} \quad (5.221)$$

■ EXAMPLE 5.302

Problem: The biosolids pumped to the thickener are comprised of 0.85% solids. The airflow is 13 cfm. What is the air-to-solids ratio if the current biosolids flow rate entering the unit is 50 gpm?

Solution:

$$\text{Air/solids ratio} = \frac{13 \text{ ft}^3\text{/min} \times 0.0785 \text{ lb/ft}^3}{50 \text{ gpm} \times 0.0085 \times 8.34 \text{ lb/gal}}$$

$$= 0.28$$

5.7.8.21.1 Recycle Flow in Percent

The amount of recycle flow can be expressed as a percent:

$$\text{Recycle \%} = \frac{\text{Recycle flow rate (gpm} \times 100)}{\text{Sludge flow (gpm)}} \quad (5.222)$$

■ EXAMPLE 5.303

Problem: The sludge flow to the thickener is 80 gpm. The recycle flow rate is 140 gpm. What is the percent recycle?

Solution:

$$\% \text{ Recycle} = \frac{140 \text{ gpm} \times 100}{80 \text{ gpm}} = 175\%$$

5.7.8.22 Centrifuge Thickening Calculations

A centrifuge exerts a force on the biosolids thousands of times greater than gravity. Sometimes polymer is added to the influent of the centrifuge to help thicken the solids. The two most important factors that affect the centrifuge are the volume of the biosolids put into the unit (gpm) and the pounds of solids put in. The water that is removed is the *centrate*. Normally, hydraulic loading is measured as flow rate per unit of area; however, because of the variety of sizes and designs, hydraulic loading to centrifuges does not include area considerations. It is expressed only as gallons per hour. The equations to be used if the flow rate to the centrifuge is given as gallons per day or gallons per minute are:

$$\text{Hydraulic loading (gph)} = \frac{\text{Flow (gpd)}}{24 \text{ hr/day}} \quad (5.223)$$

$$\text{Hydraulic loading (gph)} = \frac{\text{Flow (gpm)} \times 60 \text{ min}}{\text{hr}} \quad (5.224)$$

■ EXAMPLE 5.304

Problem: A centrifuge receives a waste activated biosolids flow of 40 gpm. What is the hydraulic loading on the unit in gal/hr?

Solution:

$$\text{Hydraulic loading (gph)} = \frac{\text{Flow (gpm)} \times 60 \text{ min}}{\text{hr}}$$

$$= \frac{40 \text{ gpm} \times 60 \text{ min}}{\text{hr}} = 2400 \text{ gph}$$

■ EXAMPLE 5.305

Problem: A centrifuge receives 48,600 gal of biosolids daily. The biosolids concentration before thickening is 0.9%. How many pounds of solids are received each day?

Solution:

$$\frac{48,600 \text{ gal}}{\text{day}} \times \frac{8.34 \text{ lb/gal}}{\text{gal}} \times \frac{0.9}{100} = 3648 \text{ lb/days}$$

5.7.8.23 Biosolids Digestion or Stabilization

A major problem in designing wastewater treatment plants is disposing of biosolids into the environment without causing damage or nuisance. Untreated biosolids are even more difficult to dispose of. Untreated raw biosolids must be stabilized to minimize disposal problems. In many cases, the term *stabilization* is considered synonymous with digestion.

Key Point: The *stabilization* of organic matter is accomplished biologically using variety of organisms. The microorganisms convert the colloidal and dissolved organic matter into various gases and into protoplasm. Because protoplasm has a specific gravity slightly higher than that of water, it can be removed from the treated liquid by gravity.

Biosolids digestion is a process in which biochemical decomposition of the organic solids occurs; in the decomposition process, the organics are converted into simpler and more stable substances. Digestion also reduces the total mass or weight of biosolids solids, destroys pathogens, and makes it easier to dry or dewater the biosolids. Well-digested biosolids have the appearance and characteristics of a rich potting soil.

Biosolids may be digested under aerobic or anaerobic conditions. Most large municipal wastewater treatment plants use anaerobic digestion. Aerobic digestion finds application primarily in small, package-activated biosolids treatment systems.

5.7.8.24 Aerobic Digestion Process Control Calculations

The purpose of *aerobic digestion* is to stabilize organic matter, to reduce volume, and to eliminate pathogenic organisms. Aerobic digestion is similar to the activated biosolids process. Biosolids are aerated for 20 days or more. Volatile solids are reduced by biological activity.

5.7.8.25 Volatile Solids Loading (lb/ft³/day)

Volatile solids (organic matter) loading for the aerobic digester is expressed in pounds of volatile solids entering the digester per day per cubic foot of digester capacity:

$$\frac{\text{Volatile solids loading}}{(\text{lb/day/ft}^3)} = \frac{\text{Volatile solids added (lb/day)}}{\text{Digester volume (ft}^3)} \quad (5.225)$$

EXAMPLE 5.306

Problem: An aerobic digester is 20 ft in diameter and has an operating depth of 20 ft. The biosolids that are added to the digester daily contain 1500 lb of volatile solids. What is the volatile solids loading in pounds per day per cubic foot?

Solution:

$$\frac{\text{Volatile solids loading}}{(\text{lb/day/ft}^3)} = \frac{1500 \text{ lb/day}}{0.785 \times 20 \text{ ft} \times 20 \text{ ft} \times 20 \text{ ft}}$$

$$= 0.24 \text{ lb/day/ft}^3$$

5.7.8.26 Digestion Time, Days

The theoretical time the biosolids remain in the aerobic digester is calculated by:

$$\text{Digestion time (days)} = \frac{\text{Digester volume (gal)}}{\text{Biosolids added (gpd)}} \quad (5.226)$$

EXAMPLE 5.307

Problem: The digester volume is 240,000 gal. Biosolids are added to the digester at the rate of 15,000 gpd. What is the digestion time in days?

Solution:

$$\text{Digestion time (days)} = \frac{240,000 \text{ gal}}{15,000 \text{ gpd}}$$

$$= 16 \text{ days}$$

5.7.8.27 pH Adjustment

In many instances, the pH of the aerobic digester will fall below the levels required for good biological activity. When this occurs, the operator must perform a laboratory test to determine the amount of alkalinity required to raise the pH to the desired level. The results of the lab test must then be converted to the actual quantity required by the digester:

$$\text{Chemical required (lb)} = \frac{\text{Lab test chemical (mg)} \times \text{digester volume} \times 3.785}{\text{Sample volume (L)} \times 454 \text{ g/lb} \times 1000 \text{ mg/g}} \quad (5.227)$$

EXAMPLE 5.308

Problem: 240 mg of lime will increase the pH of a 1-L sample of the aerobic digester contents to pH 7.1. The digester volume is 240,000 gal. How many pounds of lime will be required to increase the digester pH to 7.3?

Solution:

$$\text{Lime required (lb)} = \frac{240 \text{ mg} \times 240{,}000 \text{ gal} \times 3.785 \text{ L/gal}}{(1 \text{ L} \times 454 \text{ g/lb} \times 1000 \text{ mg/g})}$$

$$= 480 \text{ lb}$$

5.7.8.28 Anaerobic Digestion Process Control Calculations

The purpose of *anaerobic digestion* is the same as aerobic digestion: to stabilize organic matter, to reduce volume, and to eliminate pathogenic organisms. Equipment used in anaerobic digestion includes an anaerobic digester of either the floating or fixed cover type. These include biosolids pumps for biosolids addition and withdrawal, as well as heating equipment such as heat exchangers, heaters and pumps, and mixing equipment for recirculation. Typical ancillaries include gas storage, cleaning equipment, and safety equipment such as vacuum relief and pressure relief devices, flame traps, and explosion-proof electrical equipment. In the anaerobic process, biosolids enter the sealed digester where organic matter decomposes anaerobically. Anaerobic digestion is a two-stage process:

1. Sugars, starches, and carbohydrates are converted to volatile acids, carbon dioxide, and hydrogen sulfide.
2. Volatile acids are converted to methane gas.

Key anaerobic digestion process control calculations are covered in the sections that follow.

5.7.8.29 Required Seed Volume in Gallons

$$\text{Seed volume (gal)} = \begin{bmatrix} \text{Digester volume (gal)} \\ \times \% \text{ seed} \end{bmatrix} \quad (5.228)$$

■ EXAMPLE 5.309

Problem: The new digester requires 25% seed to achieve normal operation within the allotted time. If the digester volume is 280,000 gal, how many gallons of seed material will be required?

Solution:

$$\text{Seed volume (gal)} = 280{,}000 \times 0.25 = 70{,}000 \text{ gal}$$

5.7.8.30 Volatile Acids/Alkalinity Ratio

The ratio of volatile acids to alkalinity can be used to control an anaerobic digester:

$$\text{Ratio} = \frac{\text{Volatile acids concentration}}{\text{Alkalinity concentration}} \quad (5.229)$$

■ EXAMPLE 5.310

Problem: The digester contains 240 mg/L volatile acids and 1840 mg/alkalinity. What is the volatile acids/alkalinity ratio?

Solution:

$$\text{Volatile acids/alkalinity ratio} = \frac{240 \text{ mg/L}}{1840 \text{ mg/L}} = 0.13$$

Key Point: Increases in the ratio normally indicate a potential change in the operating condition of the digester.

5.7.8.31 Biosolids Retention Time

The length of time the biosolids remain in the digester is calculated as:

$$\text{Retention time} = \frac{\text{Digester volume (gal)}}{\text{Biosolids volume added (gpd)}} \quad (5.230)$$

■ EXAMPLE 5.311

Problem: Biosolids are added to a 520,000-gal digester at the rate of 12,600 gal/day. What is the biosolids retention time?

Solution:

$$\text{Biosolids retention time} = \frac{520{,}000 \text{ gal}}{12{,}600 \text{ gpd}} = 41.3 \text{ days}$$

5.7.8.32 Estimated Gas Production in Cubic Feet/Day

The rate of gas production is normally expressed as the volume of gas (ft^3) produced per pound of volatile matter destroyed. The total cubic feet of gas a digester will produce per day can be calculated by:

$$\begin{array}{l} \text{Gas production} \\ (ft^3/\text{day}) \end{array} = \begin{bmatrix} \text{Volatile matter in (lb/day)} \\ \times \% \text{ volatile matter reduction} \\ \times \text{ production rate } (ft^3/\text{lb}) \end{bmatrix} \quad (5.231)$$

Key Point: Multiplying the volatile matter added to the digester per day by the percent volatile matter reduction (in decimal percent) gives the amount of volatile matter being destroyed by the digestion process per day.

■ EXAMPLE 5.312

Problem: The digester reduces 11,500 lb of volatile matter per day. Currently, the volatile matter reduction achieved by the digester is 55%. The rate of gas production is 11.2 ft^3 of gas per pound of volatile matter destroyed.

Solution:

$$\text{Gas production (ft}^3/\text{day)} = \begin{bmatrix} 11{,}500 \text{ lb/day} \\ \times 0.55 \times 11.2 \text{ ft}^3/\text{lb} \end{bmatrix}$$

$$= 70{,}840 \text{ ft}^3/\text{day}$$

5.7.8.33 Percent Volatile Matter Reduction

Because of the changes occurring during biosolids digestion, the calculation used to determine percent volatile matter reduction is more complicated:

$$\% \text{ Reduction} = \frac{(\% \text{ VM}_{in} - \% \text{ VM}_{out}) \times 100}{\left[\% \text{ VM}_{in} - (\% \text{ VM}_{in} \times \% \text{ VM}_{out})\right]} \quad (5.232)$$

■ EXAMPLE 5.313

Problem: Using the digester data provided here, determine the percent volatile matter reduction for the digester with raw biosolids volatile matter of 71% and digested biosolids volatile matter of 54%.

Solution:

$$\% \text{ Volatile matter reduction} = \frac{0.71 - 0.54}{\left[0.71 - (0.71 \times 0.54)\right]}$$

$$= 52\%$$

5.7.8.34 Percent Moisture Reduction in Digested Biosolids

$$\% \text{ Moisture reduction} = \frac{\begin{bmatrix} (\% \text{ Moisture}_{in} - \% \text{ moisture}_{out}) \\ \times 100 \end{bmatrix}}{\begin{bmatrix} \% \text{ Moisture}_{in} - \\ (\% \text{ moisture}_{in} \times \% \text{ moisture}_{out}) \end{bmatrix}} \quad (5.233)$$

Key Point: % Moisture = 100% − percent solids.

■ EXAMPLE 5.314

Problem: Using the digester data provided below, determine the percent moisture reduction and percent volatile matter reduction for the digester.

    Raw biosolids
        % Solids   9%
        % Moisture 91% (100% − 9%)
    Digested biosolids
        % Solids   15%
        % Moisture 85% (100% − 15%)

Solution:

$$\% \text{ Moisture reduction} = \frac{(0.91 - 0.85) \times 100}{\left[0.91 - (0.91 \times 0.85)\right]} = 44\%$$

5.7.9 Biosolids Dewatering

The process of removing enough water from a liquid biosolids to change its consistency to that of damp solid is called *biosolids dewatering*. Although the process is also referred to as *biosolids drying*, the dry or dewatered biosolids may still contain a significant amount of water, often as much as 70%. But, at moisture contents of 70% or less, the biosolids no longer behave as a liquid and can be handled manually or mechanically. Several methods are available to dewater biosolids. The particular types of dewatering techniques or devices used best describe the actual processes used to remove water from biosolids and change their form from a liquid to damp solid. The commonly used techniques and devices include the following:

- Filter presses
- Vacuum filtration
- Sand drying beds

Key Point: Centrifugation is also used in the dewatering process; however, in this text we concentrate on those unit processes listed above that are traditionally used for biosolids dewatering.

Note that an ideal dewatering operation would capture all of the biosolids at minimum cost and the resultant dry biosolids solids or cake would be capable of being handled without causing unnecessary problems. Process reliability, ease of operation, and compatibility with the plant environment would also be optimized.

5.7.9.1 Pressure Filtration

In *pressure filtration*, the liquid is forced through the filter media by a positive pressure. Several types of presses are available, but the most commonly used types are plate and frame presses and belt presses.

5.7.9.2 Plate and Frame Press Calculations

The *plate and frame press* consists of vertical plates that are held in a frame and that are pressed together between a fixed and moving end. A cloth filter medium is mounted on the face of each individual plate. The press is closed, and the biosolids are pumped into the press at pressures up to 225 psi. They then pass through feed holes in the trays along the length of the press. Filter presses usually require a precoat material, such as incinerator ash or diatomaceous earth, to aid in solids retention on the cloth and to allow easier release of the cake. Performance factors for plate and frame presses include feed biosolids characteristics, type and amount of chemical conditioning, operating pressures, and the type and amount of precoat. Filter press calculations (and other dewatering calculations) typically used in wastewater solids handling operations include solids loading rate, net filter yield, hydraulic loading rate, biosolids feed rate, solids loading rate, flocculant feed rate, flocculant dosage, total suspended solids, and percent solids recovery.

5.7.9.2.1 Solids Loading Rate

The solids loading rate is a measure of the lb/hr solids applied per square foot of plate area, as shown in Equation 5.234:

$$\text{Solids loading rate}\ (\text{lb/hr/ft}^2) = \frac{\substack{\text{Biosolids (gph)} \times 8.34\ \text{lb/gal}\\ \times\ (\%\ \text{Solids}/100)}}{\text{Plate area (ft}^2)} \quad (5.234)$$

Key Point: The solids loading rate measures the lb/hr of solids applied to each ft² of plate surface area; however, this does not reflect the time when biosolids feed to the press is stopped.

5.7.9.2.2 Net Filter Yield

Operated in the batch mode, biosolids are fed to the plate and frame filter press until the space between the plates is completely filled with solids. The flow of biosolids to the press is then stopped and the plates are separated, allowing the biosolids cake to fall into a hopper or conveyor below. The net filter yield, measured in lb/hr/ft², reflects the runtime as well as the downtime of the plate and frame filter press. To calculate the net filter yield, simply multiply the solids loading rate (in lb/hr/ft²) by the ratio of filter runtime to total cycle time as follows:

$$\text{Net filter yield} = \frac{\text{Filter run time}}{\text{Total cycle time}} \quad (5.235)$$

■ EXAMPLE 5.315

Problem: A plate and frame filter press receives a flow of 660 gal of biosolids during a 2-hr period. The solids concentration of the biosolids is 3.3%. The surface area of the plate is 110 ft². If the downtime for biosolids cake discharge is 20 min, what is the net filter yield in lb/hr/ft²?

Solution: First calculate solids loading rate then multiply that number by the corrected time factor:

$$\text{Solids loading rate}\ (\text{lb/hr/ft}^2) = \frac{\substack{\text{Biosolids (gph)} \times 8.34\ \text{lb/gal}\\ \times\ (\%\ \text{solids}/100)}}{\text{Plate area (ft}^2)}$$

$$= \frac{330\ \text{gph} \times 8.34\ \text{lb/gal} \times (3.3/100)}{100\ \text{ft}^2}$$

$$= 0.83\ \text{lb/hr/ft}^2$$

Next, calculate net filter yield, using the corrected time factor:

$$\text{Net filter yield}\ (\text{lb/hr/ft}^2) = \frac{0.83\ \text{lb/hr/ft}^2 \times 2\ \text{hr}}{2.33\ \text{hr}}$$

$$= 0.71\ \text{lb/hr/ft}^2$$

5.7.9.3 Belt Filter Press Calculations

The *belt filter press* consists of two porous belts. The biosolids are sandwiched between the two porous belts. The belts are pulled tight together as they are passed around a series of rollers to squeeze water out of the biosolids. Polymer is added to the biosolids just before it gets to the unit. The biosolids are then distributed across one of the belts to allow for some of the water to drain by gravity. The belts are then put together with the biosolids in between.

5.7.9.3.1 Hydraulic Loading Rate

Hydraulic loading for belt filters is a measure of gpm flow per foot or belt width:

$$\text{Hydraulic loading rate}\ (\text{gpm/ft}) = \frac{\text{Flow (gpm)}}{\text{Belt width (ft)}} \quad (5.236)$$

■ EXAMPLE 5.316

Problem: A 6-ft-wide belt press receives a flow of 110 gpm of primary biosolids. What is the hydraulic loading rate in gpm/ft?

Solution:

$$\text{Hydraulic loading rate}\ (\text{gpm/ft}) = \frac{\text{Flow (gpm)}}{\text{Belt width (ft)}}$$

$$= \frac{110\ \text{gpm}}{6\ \text{ft}}$$

$$= 18.3\ \text{gpm/ft}$$

■ EXAMPLE 5.317

Problem: A belt filter press 5 ft wide receives a primary biosolids flow of 150 gpm. What is the hydraulic loading rate in gpm/ft²?

Solution:

$$\text{Hydraulic loading rate}\ (\text{gpm/ft}) = \frac{\text{Flow (gpm)}}{\text{Belt width (ft)}}$$

$$= \frac{150\ \text{gpm}}{5\ \text{ft}}$$

$$= 30\ \text{gpm/ft}$$

5.7.9.3.2 Biosolids Feed Rate

The biosolids feed rate to the belt filter press depends on several factors, including the biosolids (lb/day) that must be dewatered, the maximum solids feed rate (lb/hr) that will produce an acceptable cake dryness, and the number of hours per day the belt press is in operation. The equation used in calculating biosolids feed rate is:

$$\text{Biosolids feed rate (lb/hr)} = \frac{\text{Dewatered biosolids (lb/day)}}{\text{Operating time (hr/day)}} \quad (5.237)$$

■ EXAMPLE 5.318

Problem: The amount of biosolids to be dewatered by the belt filter press is 20,600 lb/day. If the belt filter press is to be operated 10 hr each day, what should the biosolids feed rate in lb/hr be to the press?

Solution:

$$\text{Biosolids feed rate (lb/hr)} = \frac{\text{Dewatered biosolids (lb/day)}}{\text{Operating time (hr/day)}}$$

$$= \frac{20,600 \text{ lb/day}}{10 \text{ hr/day}} = 2060 \text{ lb/hr}$$

5.7.9.3.3 Solids Loading Rate

The solids loading rate may be expressed as lb/hr or as tons/hr. In either case, the calculation is based on biosolids flow (or feed) to the belt press and percent of mg/L concentration of total suspended solids (TSS) in the biosolids. The equation used in calculating solids loading rate is:

$$\text{Solids loading rate (lb/hr)} = \begin{bmatrix} \text{Feed (gpm)} \times 60 \text{ (min/hr)} \\ \times 8.34 \text{ lb/gal} \times (\% \text{ TSS}/100) \end{bmatrix} \quad (5.238)$$

■ EXAMPLE 5.319

Problem: The biosolids feed to a belt filter press is 120 gpm. If the total suspended solids concentration of the feed is 4%, what is the solids loading rate in lb/hr?

Solution:

$$\text{Solids loading rate (lb/hr)} = \begin{bmatrix} \text{Feed (gpm)} \times 60 \text{ (min/hr)} \\ \times 8.34 \text{ lb/gal} \times (\% \text{ TSS}/100) \end{bmatrix}$$

$$= \begin{bmatrix} 120 \text{ gpm} \times 60 \text{ min/hr} \\ \times 8.34 \text{ lb/gal} \times (4/100) \end{bmatrix}$$

$$= 2402 \text{ lb/hr}$$

5.7.9.3.4 Flocculant Feed Rate

The flocculant feed rate may be calculated like all other mg/L to lb/day calculations and then converted to a lb/hr feed rate:

$$\text{Flocculant feed (lb/day)} = \frac{\text{Flocculant (mg/L)} \times \text{feed rate (MGD)} \times 8.34 \text{ lb/gal}}{24 \text{ hr/day}} \quad (5.239)$$

■ EXAMPLE 5.320

Problem: The flocculant concentration for a belt filter press is 1% (10,000 mg/L). If the flocculant feed rate is 3 gpm, what is the flocculant feed rate in lb/hr?

Solution: First calculate lb/day flocculant using the mg/L to lb/day calculation. Note that the gpm feed flow must be expressed as MGD feed flow:

$$\frac{3 \text{ gpm} \times 1440 \text{ min/day}}{1,000,000} = 0.00432 \text{ MGD}$$

$$\text{Flocculant feed (lb/day)} = \text{Flocculant (mg/L)}$$
$$\times \text{feed rate (MGD)} \times 8.34 \text{ lb/gal}$$

$$= 10,000 \text{ mg/L}$$
$$\times 0.00432 \text{ MGD} \times 8.34 \text{ lb/gal}$$

$$= 360 \text{ lb/day}$$

Then convert lb/day flocculant to lb/hr:

$$\frac{360 \text{ lb/day}}{24 \text{ hr/day}} = 15 \text{ lb/hr}$$

5.7.9.3.5 Flocculant Dosage

When the solids loading rate (tons/hr) and flocculant feed rate (lb/hr) have been calculated, the flocculant dose in lb/ton can be determined. The equation used to determine flocculant dosage is:

$$\text{Flocculant dosage (lb/ton)} = \frac{\text{Flocculant (lb/hr)}}{\text{Solids treated (ton/hr)}} \quad (5.240)$$

■ EXAMPLE 5.321

Problem: A belt filter has solids loading rate of 3100 lb/hr and a flocculant feed rate of 12 lb/hr. Calculate the flocculant dose in lb per ton of solids treated?

Solution: First convert lb/hr solids loading to tons/hr solids loading:

$$\frac{3100 \text{ lb/hr}}{2000 \text{ lb/ton}} = 1.55 \text{ tons/hr}$$

Now calculate pounds flocculant per ton of solids treated:

$$\text{Flocculant dosage (lb/ton)} = \frac{\text{Flocculant (lb/hr)}}{\text{Solids treated (ton/hr)}}$$

$$= \frac{12 \text{ lb/hr}}{1.55 \text{ tons/hr}} = 7.8 \text{ lb/ton}$$

5.7.9.3.6 Total Suspended Solids

The feed biosolids solids are comprised of two types of solids: suspended solids and dissolved solids. *Suspended solids* will not pass through a glass-fiber filter pad. Suspended solids can be further classified as total suspended solids (TSS), volatile suspended solids, or fixed suspended solids and can also be separated into three components based on settling characteristics: settleable solids, floatable solids, and colloidal solids. Total suspended solids in wastewater is normally in the range of 100 to 350 mg/L. *Dissolved solids* will pass through a glass-fiber filter pad. Dissolved solids can also be classified as total dissolved solids (TDS), volatile dissolved solids, and fixed dissolved solids. Total dissolved solids are normally in the range of 250 to 850 mg/L.

Two lab tests can be used to estimate the total suspended solids concentration of the feed biosolids to the filter press: *total residue test* (which measures both suspended and dissolved solids concentrations) and *total filterable residue test* (which measures only the dissolved solids concentration). By subtracting the total filterable residue from the total residue, we obtain the total nonfilterable residue (total suspended solids), as shown in Equation 5.241:

$$\begin{array}{l} \text{Total residue (mg/L)} \\ - \text{ total filterable residue (mg/L)} \end{array} = \begin{array}{l} \text{Total nonfilterable} \\ \text{residue (mg/L)} \end{array} \quad (5.241)$$

■ EXAMPLE 5.322

Lab tests indicate that the total residue portion of a feed biosolids sample is 22,000 mg/L. The total filterable residue is 720 mg/L. On this basis, what is the estimated total suspended solids concentration of the biosolids sample?

$$\begin{array}{l} \text{Total residue (mg/L)} \\ - \text{ total filterable residue (mg/L)} \end{array} = \begin{array}{l} \text{Total nonfilterable} \\ \text{residue (mg/L)} \end{array}$$

$$22{,}000 \text{ mg/L} - 720 \text{ mg/L} = 21{,}280 \text{ mg/L}$$

5.7.9.4 Rotary Vacuum Filter Dewatering Calculations

The *rotary vacuum filter* is a device used to separate solid material from liquid. The vacuum filter consists of a large drum with large holes in it covered with a filter cloth. The drum is partially submerged and rotated through a vat of conditioned biosolids. This filter is capable of excellent solids capture and high-quality supernatant or filtrate; solids concentrations of 15 to 40% can be achieved.

5.7.9.4.1 Filter Loading

The filter loading for vacuum filters is a measure of lb/hr of solids applied per square foot of drum surface area. The equation to be used in this calculation is shown below:

$$\text{Filter loading (lb/hr/ft}^2) = \frac{\text{Solids to filter (lb/hr)}}{\text{Surface area (ft}^2)} \quad (5.242)$$

■ EXAMPLE 5.323

Problem: Digested biosolids are applied to a vacuum filter at a rate of 70 gpm, with a solids concentration of 3%. If the vacuum filter has a surface area of 300 ft², what is the filter loading in lb/hr/ft²?

Solution:

$$\begin{array}{l} \text{Filter loading rate} \\ \text{(lb/hr/ft}^2) \end{array} = \frac{\begin{array}{l} \text{Biosolids (gpm)} \times 60 \text{ min/hr} \\ \times 8.34 \text{ lb/gal} \times (\% \text{ solids}/100) \end{array}}{\text{Surface area (ft}^2)}$$

$$= \frac{\begin{array}{l} 70 \text{ gpm} \times 60 \text{ min/hr} \\ \times 8.34 \text{ lb/gal} \times (3/100) \end{array}}{300 \text{ ft}^2}$$

$$= 3.5 \text{ lb/hr/ft}^2$$

5.7.9.4.2 Filter Yield

One of the most common measures of vacuum filter performance is filter yield. It is the lb/hr of dry solids in the dewatered biosolids (cake) discharged per square foot of filter area. It can be calculated using Equation 5.243:

$$\text{Filter yield (lb/hr/ft}^2) = \frac{\begin{array}{l} \text{Wet cake flow (lb/hr)} \\ \times (\% \text{ solids in cake}/100) \end{array}}{\text{Filter area (ft}^2)} \quad (5.243)$$

■ EXAMPLE 5.324

Problem: The wet cake flow from a vacuum filter is 9000 lb/hr. If the filter areas is 300 ft² and the percent solids in the cake is 25%, what is the filter yield in lb/hr/ft²?

Solution:

$$\text{Filter yield (lb/hr/ft}^2) = \frac{\begin{array}{l} \text{Wet cake flow (lb/hr)} \\ \times (\% \text{ solids in cake}/100) \end{array}}{\text{Filter area (ft}^2)}$$

$$= \frac{9000 \text{ lb/hr} \times (25/100)}{300 \text{ ft}^2}$$

$$= 7.5 \text{ lb/hr/ft}^2$$

5.7.9.4.3 Vacuum Filter Operating Time

■ **EXAMPLE 5.325**

Problem: A total of 4000 lb/day primary biosolids solids is to be processed by a vacuum filter. The vacuum filter yield is 2.2 lb/hr/ft². The solids recovery is 95%. If the area of the filter is 210 ft², how many hours per day must the vacuum filter remain in operation to process these solids?

Solution:

$$\text{Filter yield} \atop (\text{lb/hr/ft}^2) = \frac{\left(\dfrac{\text{Solids to filter (lb/day)}}{\text{Filter operation (lb/day)}}\right)}{\text{Filter area (ft}^2)} \times \frac{\% \text{ recovery}}{100}$$

$$2.2 \text{ lb/hr/ft}^2 = \frac{4000 \text{ lb/day}}{x \text{ hr/day}} \times \frac{1}{210 \text{ ft}^2} \times \frac{95}{100}$$

$$x = \frac{4000 \times 1 \times 95}{2.2 \times 210 \times 100} = 8.2 \text{ hr/day}$$

5.7.9.4.4 Percent Solids Recovery

As mentioned, the function of the vacuum filtration process is to separate the solids from the liquids in the biosolids being processed; therefore, the percent of feed solids recovered (sometimes referred to as the *percent solids capture*) is a measure of the efficiency of the process. Equation 5.244 is used to determine percent solids recovery:

$$\% \text{ Solids recovery} = \left[\frac{\begin{array}{c}\text{Wet cake flow (lb/hr)}\\ \times (\% \text{ solids in cake/100})\end{array}}{\begin{array}{c}\text{Biosolids feed (lb/hr)}\\ \times (\% \text{ solids in feed/100})\end{array}}\right] \times 100 \quad (5.244)$$

■ **EXAMPLE 5.326**

Problem: The biosolids feed to a vacuum is 3400 lb/day, with a solids content of 5.1%. If the wet cake flow is 600 lb/hr with a 25% solids content, what is the percent solids recovery?

Solution:

$$\% \text{ Solids recovery} = \left[\frac{\begin{array}{c}\text{Wet cake flow (lb/hr)}\\ \times (\% \text{ solids in cake/100})\end{array}}{\begin{array}{c}\text{Biosolids feed (lb/hr)}\\ \times (\% \text{ solids in feed/100})\end{array}}\right] \times 100$$

$$= \frac{600 \text{ lb/hr} \times (25/100)}{3400 \text{ lb/hr} \times (5.1/100)} \times 100$$

$$= \frac{150 \text{ lb/hr}}{173 \text{ lb/hr}} \times 100 = 87\%$$

5.7.9.5 Sand Drying Beds

Drying beds are generally used for dewatering well-digested biosolids. Biosolids drying beds consist of a perforated or open joint drainage system in a support media, usually gravel or wire mesh. Drying beds are usually separated into workable sections by wood, concrete, or other materials. Drying beds may be enclosed or opened to the weather. They may rely entirely on natural drainage and evaporation processes or may use a vacuum to assist the operation. *Sand drying beds* are the oldest biosolids dewatering technique. They consist of 6 to 12 in. of coarse sand underlain by layers of graded gravel ranging from 1/8 to 1/4 in. at the top and 3/4 to 1-1/2 in. at the bottom. The total gravel thickness is typically about 1 ft. Graded natural earth (4 to 6 in.) usually makes up the bottom with a web of drain tile placed on 20- to 30-ft centers. Sidewalls and partitions between bed sections are usually of wooden planks or concrete and extend about 14 in. above the sand surface. Typically, three calculations are used to monitor sand drying bed performance: total biosolids applied, solids loading rate, and biosolids withdrawal to drying beds.

5.7.9.5.1 Total Biosolids Applied

The total gallons of biosolids applied to sand drying beds may be calculated using the dimensions of the bed and depth of biosolids applied, as shown by Equation 5.245:

$$\text{Volume (gal)} = \left[\begin{array}{c}\text{Length (ft)} \times \text{width (ft)}\\ \times \text{depth (ft)} \times 7.48 \text{ gal/ft}^3\end{array}\right] \quad (5.245)$$

■ **EXAMPLE 5.327**

Problem: A drying bed is 220 ft long and 20 ft wide. If biosolids are applied to a depth of 4 in., how many gallons of biosolids are applied to the drying bed?

Solution:

$$\text{Volume (gal)} = \left[\begin{array}{c}\text{Length (ft)} \times \text{width (ft)}\\ \times \text{depth (ft)} \times 7.48 \text{ gal/ft}^3\end{array}\right]$$

$$= 220 \text{ ft} \times 20 \text{ ft} \times 0.33 \text{ ft} \times 7.48 \text{ gal/ft}^3$$

$$= 10,861 \text{ gal}$$

5.7.9.5.2 Solids Loading Rate

The biosolids loading rate may be expressed as lb/yr/ft². The loading rate is dependent on biosolids applied per application (lb), percent solids concentration, cycle length, and square feet of sand bed area. The equation for biosolids loading rate is given below:

$$\text{Solids loading} \atop \text{rate (lb/yr/ft}^2) = \frac{\left[\begin{array}{c}\left(\dfrac{\text{Biosolids applied (lb)}}{\text{Days of application}}\right)\\ \times (365 \text{ days/yr}) \times (\% \text{ solids/100})\end{array}\right]}{\text{Length (ft)} \times \text{width (ft)}} \quad (5.246)$$

■ **EXAMPLE 5.328**

Problem: A biosolids bed is 210 ft long and 25 ft wide. A total of 172,500 lb of biosolids is applied during each application of the sand drying bed. The biosolids have a solids content of 5%. If the drying and removal cycle requires 21 days, what is the solids loading rate in lb/yr/ft²?

Solution:

$$\text{Solids loading rate (lb/yr/ft}^2) = \frac{\left[\left(\dfrac{\text{Biosolids applied (lb)}}{\text{Days of application}}\right) \times \dfrac{365 \text{ days}}{\text{yr}} \times \dfrac{\% \text{ solids}}{100}\right]}{\text{Length (ft)} \times \text{width (ft)}}$$

$$= \frac{\left(\dfrac{172{,}500 \text{ lb}}{21 \text{ days}} \times \dfrac{365 \text{ days}}{\text{yr}} \times \dfrac{5}{100}\right)}{210 \text{ ft} \times 25 \text{ ft}}$$

$$= 37.5 \text{ lb/yr/ft}^2$$

5.7.9.5.3 Biosolids Withdrawal to Drying Beds

Pumping digested biosolids to drying beds is one method among many for dewatering biosolids, thus making the dried biosolids useful as a soil conditioner. Depending on the climate of a region, the drying bed depth may range from 8 to 18 in. Because the area covered by these beds may be substantial, the use of drying beds is more common for smaller plants than for larger plants. To calculate biosolids withdrawal to drying beds, use Equation 5.247:

$$\text{Biosolids withdrawn (ft}^3) = \left[\begin{array}{l} 0.785 \times (\text{diameter})^2 \\ \times \text{drawdown (ft)} \end{array}\right] \quad (5.247)$$

■ **EXAMPLE 5.329**

Problem: Biosolids are withdrawn from a digester that has a diameter of 40 ft. If the biosolids are drawn down 2 ft, how many ft³ will be sent to the drying beds?

Solution:

$$\text{Biosolids withdrawn (ft}^3) = \left[\begin{array}{l} 0.785 \times (\text{diameter})^2 \\ \times \text{drawdown (ft)} \end{array}\right]$$

$$= 0.785 \times (40 \text{ ft} \times 40 \text{ ft}) \times 2 \text{ ft}$$

$$= 2512 \text{ ft}^3$$

5.7.10 Biosolids Disposal

In the disposal of biosolids, land application, in one form or another, has become not only a necessity (because of the banning of ocean dumping in the United States in 1992 and the shortage of landfill space since then) but also quite popular as a beneficial reuse practice. Benefi-cial reuse means that the biosolids are disposed of in an environmentally sound manner by recycling nutrients and soil conditions. The application of biosolids is occurring throughout the United States on agricultural and forest lands. For use in land applications, the biosolids must meet certain conditions. Biosolids must comply with state and federal biosolids management and disposal regulations and must also be free of materials dangerous to human health (e.g., toxins, pathogenic organisms) or dangerous to the environment (e.g., pesticides, heavy metals). Biosolids can be land applied by direct injection, by application, by incorporation (plowing in), or by composting.

5.7.10.1 Land Application Calculations

Land application of biosolids requires precise control to avoid problems. Use of process control calculations is part of the overall process control process. Calculations include determining disposal cost, plant available nitrogen (PAN), application rates (dry tons and wet tons/acre), metals loading rates, maximum allowable applications based on metals loading, and site life based on metals loading.

5.7.10.1.1 Disposal Cost

The cost of disposal of biosolids can be determined by:

$$\text{Cost} = \left[\begin{array}{l} \text{Wet tons biosolids produced per year} \\ \times \% \text{ solids} \times \text{cost per dry ton} \end{array}\right] \quad (5.248)$$

■ **EXAMPLE 5.330**

Problem: The treatment system produces 1925 wet tons of biosolids for disposal each year. The biosolids are comprised of 18% solids. A contractor disposes of the biosolids for $28 per dry ton. What is the annual cost for biosolids disposal?

Solution:

$$\text{Cost} = 1925 \text{ wet tons per year} \times 0.18 \times \$28/\text{dry ton}$$

$$= \$9702$$

5.7.10.1.2 Plant Available Nitrogen

One factor considered when land applying biosolids is the amount of nitrogen in the biosolids available to the plants grown on the site. This includes ammonia nitrogen and organic nitrogen. The organic nitrogen must be mineralized for plant consumption. Only a portion of the organic nitrogen is mineralized per year. The mineralization factor (f^1) is assumed to be 0.20. The amount of ammonia nitrogen available is directly related to the time elapsed between applying the biosolids and incorporating (plowing) the biosolids into the soil. Volatilization rates are presented in the example below:

$$\text{PAN} \atop \text{(lb/dry ton)} = \left(\begin{bmatrix} \left(\text{Organic nitrogen (mg/kg)} \times f^1 \right) \\ + \left(\text{ammonia nitrogen (mg/kg)} \times V_1 \right) \end{bmatrix} \atop \times 0.002 \text{ lb/dry ton} \right) \quad (5.249)$$

where:

f^1 = Mineral rate for organic nitrogen (assume 0.20).

V_1 = Volatilization rate for ammonia nitrogen.
= 1.00 if biosolids are injected.
= 0.85 if biosolids are plowed in within 24 hr.
= 0.70 if biosolids are plowed in within 7 days.

■ **EXAMPLE 5.331**

Problem: Biosolids contain 21,000 mg/kg of organic nitrogen and 10,500 mg/kg of ammonia nitrogen. The biosolids is incorporated into the soil within 24 hr after application. What is the plant available nitrogen (PAN) per dry ton of solids?

Solution:

$$\text{PAN (lb/dry ton)} = \begin{bmatrix} (21{,}000 \text{ mg/kg} \times 0.20) \\ + (10{,}500 \times 0.85) \end{bmatrix} \times 0.002$$

$$= 26.3 \text{ lb/dry ton}$$

5.7.10.1.3 Application Rate Based on Crop Nitrogen Requirement

In most cases, the application rate of domestic biosolids to crop lands will be controlled by the amount of nitrogen that the crop requires. The biosolids application rate based on the nitrogen requirement is determined by the following:

1. Use an agriculture handbook to determine the nitrogen requirement of the crop to be grown.
2. Determine the amount of biosolids in dry tons required to provide this much nitrogen:

$$\text{Dry tons/ac} = \frac{\text{Plant nitrogen requirement (lb/ac)}}{\text{Plant available nitrogen (lb/dry ton)}} \quad (5.250)$$

■ **EXAMPLE 5.332**

Problem: The crop to be planted on the land application site requires 150 lb of nitrogen per acre. What is the required biosolids application rate if the PAN of the biosolids is 30 lb/dry ton?

Solution:

$$\text{Dry tons/ac} = \frac{150 \text{ lb/ac}}{30 \text{ lb/dry ton}} = 5 \text{ dry tons/ac}$$

5.7.10.1.4 Metals Loading

When biosolids are land applied, metals concentrations are closely monitored and the loading on land application sites is calculated:

$$\text{Loading (lb/ac)} = \begin{bmatrix} \text{Metal concentration (mg/kg)} \\ \times 0.002 \text{ lb/dry ton} \\ \times \text{application rate (dry tons/ac)} \end{bmatrix} \quad (5.251)$$

■ **EXAMPLE 5.333**

Problem: Biosolids contain 14 mg/kg of lead and are currently being applied to the site at a rate of 11 dry tons/ac. What is the metals loading rate for lead in pounds per acre?

Solution:

$$\text{Loading rate (lb/ac)} = \begin{bmatrix} 14 \text{ mg/kg} \times 0.002 \text{ lb/dry ton} \\ \times 11 \text{ dry tons} \end{bmatrix}$$

$$= 0.31 \text{ lb/ac}$$

5.7.10.1.5 Maximum Allowable Applications Based on Metals Loading

If metals are present, they may limit the total number of applications a site can receive. Metals loadings are normally expressed in terms of the maximum total amount of metal that can be applied to a site during its use:

$$\text{Application} = \frac{\text{Maximum allowable cumulative load (lb/ac)}}{\text{Metal loading (lb/ac)}} \quad (5.252)$$

■ **EXAMPLE 5.334**

Problem: The maximum allowable cumulative lead loading is 48 lb/ac. Based on the current loading of 0.35 lb/ac, how many applications of biosolids can be made to this site?

Solution:

$$\text{Application} = \frac{48.0 \text{ lb/ac}}{0.35 \text{ lb/ac}} = 137$$

5.7.10.1.6 Site Life Based on Metals Loading

The maximum number of applications based on metals loading and the number of applications per year can be used to determine the maximum site life:

$$\text{Site life (yr)} = \frac{\text{Maximum allowable applications}}{\text{Number of applications per year}} \quad (5.253)$$

■ **EXAMPLE 5.335**

Problem: Biosolids are currently being applied to a site twice annually. Based on the lead content of the biosolids, the maximum number of applications is determined to be 135 applications. Based on the lead loading and the applications rate, how many years can this site be used?

Solution:

$$\text{Site life (yr)} = \frac{135 \text{ applications}}{2 \text{ applications per year}} = 68 \text{ yr}$$

Key Point: When more than one metal is present, the calculations must be performed for each metal. The site life would then be the lowest value generated by these calculations.

5.7.10.2 Biosolids to Compost

The purpose of composting biosolids is to stabilize the organic matter, reduce volume, eliminate pathogenic organisms, and produce a product that can be used as a soil amendment or conditioner. Composting is a biological process. In a composting operation, dewatered solids are usually mixed with a bulking agent (e.g., hardwood chips) and stored until biological stabilization occurs. The composting mixture is ventilated during storage to provide sufficient oxygen for oxidation and to prevent odors. After stabilization of the solids, they are separated from the bulking agent. The composted solids are then stored for curing and are applied to farm lands or other beneficial uses. Expected performance of the composting operation for both percent volatile matter reduction and percent moisture reduction ranges from 40 to 60%.

Performance factors related to biosolids composting include moisture content, temperature, pH, nutrient availability, and aeration. The biosolids must contain sufficient moisture to support the biological activity. If the moisture level is too low (40% less), biological activity will be reduced or stopped. At the same time, if the moisture level exceeds approximately 60%, it will prevent sufficient airflow through the mixture. The composting process operates best when the temperature is maintained within an operating range of 130 to 140°F, but biological activities provide enough heat to increase the temperature well above this range. Forced air ventilation or mixing is used to remove heat and maintain the desired operating temperature range. The temperature of the composting solids, when maintained at the required levels, will be sufficient to remove pathogenic organisms. The influent pH can affect the performance of the process if extreme (less than 6.0 or greater than 11.0). The pH during composting may have some impact on the bio-logical activity but does not appear to be a major factor. Composted biosolids generally have a pH in the range of 6.8 to 7.5. The critical nutrient in the composting process is nitrogen. The process works best when the ratio of nitrogen to carbon is in the range of 26 to 30 carbon to one nitrogen. Above this ratio, composting is slowed. Below this ratio, the nitrogen content of the final product may be less attractive as compost. Aeration is essential to provide oxygen to the process and to control the temperature. In forced air processes, some means of odor control should be included in the design of the aeration system.

5.7.10.3 Composting Calculations

Pertinent composting process control calculations include determination of percent of moisture of compost mixture and compost site capacity. An important consideration in compost operation is the solids processing capability (fill time in lb/day or lb/wk). Equation 5.254 is used to calculate site capacity:

$$\frac{\text{Fill time}}{\text{(days)}} = \frac{\text{Total available capacity (yd}^3)}{\left(\dfrac{\text{Wet compost (lb/day)}}{\text{Compost bulk density (lb/yd}^3)}\right)} \quad (5.254)$$

■ **EXAMPLE 5.336**

Problem: A composting facility has an available capacity of 7600 yd³. If the composting cycle is 21 days, how many lb/day wet compost can be processed by this facility? Assume a compost bulk density of 900 lb/yd³.

Solution:

$$\frac{\text{Fill time}}{\text{(days)}} = \frac{\text{Total available capacity (yd}^3)}{\left(\dfrac{\text{Wet compost (lb/day)}}{\text{Compost bulk density (lb/yd}^3)}\right)}$$

$$= \frac{7600 \text{ yd}^3}{\left(\dfrac{x \text{ lb/day}}{900 \text{ lb/yd}^3}\right)}$$

$$21 \text{ days} = \frac{7600 \text{ yd}^3 \times 900 \text{ lb/yd}^3}{x \text{ lb/day}}$$

$$x \text{ lb/day} = \frac{7600 \text{ yd}^3 \times 900 \text{ lb/yd}^3}{21 \text{ days}}$$

$$x \text{ lb/day} = \frac{6,840,000 \text{ lb}}{21 \text{ days}}$$

$$= 325,714 \text{ lb/day}$$

5.8 WATER/WASTEWATER LABORATORY CALCULATIONS

Waterworks and wastewater treatment plants are sized to meet current needs, as well as those of the future. No matter the size of the treatment plant, some space or area within the plant is designated as the lab area, which can range from being the size of a closet to being fully equipped and staffed environmental laboratories. Water and wastewater laboratories usually perform a number of different tests. Lab test results provide the operator with the information necessary to operate the treatment facility at optimal levels. Laboratory testing usually includes determining service line flushing time, solution concentration, pH, COD, total phosphorus, fecal coliform count, chlorine residual, and BOD (seeded), to name a few. The standard reference for performing wastewater testing is contained in the *Standard Methods for Examination of Water & Wastewater* (Clesceri et al., 1999).

In this section, the focus is on standard water/wastewater lab tests that involve various calculations. Specifically, the focus is on calculations used to determine the proportioning factor for composite sampling, flow from a faucet estimation, service line flushing time, solution concentration, biochemical oxygen demand, molarity and moles, normality, settleability, settleable solids, biosolids total, fixed and volatile solids, suspended and volatile suspended solids, and biosolids volume index and biosolids density index.

5.8.1 Faucet Flow Estimation

On occasion, the waterworks sampler must take water samples from a customer's residence. In small water systems, the sample is usually taken from the customer's front yard faucet. A convenient flow rate for taking water samples is about 0.5 gpm. To estimate the flow from a faucet, use a 1-gallon container and record the time it takes to fill the container. To calculate the flow in gpm, insert the recorded information into Equation 5.255:

$$\text{Flow (gpm)} = \frac{\text{Volume (gal)}}{\text{Time (min)}} \qquad (5.255)$$

■ EXAMPLE 5.337

Problem: The flow from a faucet filled up the gallon container in 48 sec. What was the gpm flow rate from the faucet? Because the flow rate is desired in minutes the time should also be expressed as minutes:

$$(48 \text{ sec}) \div (60 \text{ sec/min}) = 0.80 \text{ min}$$

Solution: Calculate flow rate from the faucet:

$$\text{Flow (gpm)} = \frac{\text{Volume (gal)}}{\text{Time (min)}} = \frac{1 \text{ gal}}{0.80 \text{ min}} = 1.25 \text{ gpm}$$

■ EXAMPLE 5.338

Problem: The flow from a faucet filled up the gallon container in 55 sec. What was the gpm flow rate from the faucet?

Solution:

$$(55 \text{ sec}) \div (60 \text{ sec/min}) = 0.92 \text{ min}$$

Calculate the flow rate:

$$\text{Flow (gpm)} = \frac{\text{Volume (gal)}}{\text{Time (min)}} = \frac{1 \text{ gal}}{0.92 \text{ min}} = 1.1 \text{ gpm}$$

5.8.2 Service Line Flushing Time

To determine the quality of potable water delivered to the consumer, a sample is taken from the customer's outside faucet—water that is typical of the water delivered. To obtain an accurate indication of the system water quality, this sample must be representative. Further, to ensure that the sample taken is typical of water delivered, the service line must be flushed twice. Equation 5.256 is used to calculate flushing time:

$$\frac{\text{Flushing time}}{\text{(min)}} = \frac{\begin{array}{c}0.785 \times (\text{diameter})^2 \times \text{length (ft)} \\ \times 7.48 \text{ gal/ft}^3 \times 2\end{array}}{\text{Flow rate (gpm)}} \qquad (5.256)$$

■ EXAMPLE 5.339

Problem: How long (in min) will it take to flush a 40-ft length of 1/2-in.-diameter service line if the flow through the line is 0.5 gpm?

Solution: Calculate the diameter of the pump in feet:

$$(0.50 \text{ in.}) \div (12 \text{ in./ft}) = 0.04 \text{ ft}$$

Calculate the flushing time:

$$\frac{\text{Flushing time}}{\text{(min)}} = \frac{\begin{array}{c}0.785 \times (\text{diameter})^2 \times \text{length (ft)} \\ \times 7.48 \text{ gal/ft}^3 \times 2\end{array}}{\text{Flow rate (gpm)}}$$

$$= \frac{\begin{array}{c}0.785 \times (0.04 \text{ ft} \times 0.04 \text{ ft}) \times 40 \text{ ft} \\ \times 7.48 \text{ gal/ft}^3 \times 2\end{array}}{0.5 \text{ gpm}}$$

$$= 1.5 \text{ min}$$

■ EXAMPLE 5.340

Problem: At a flow rate of 0.5 gpm, how long (min and sec) will it take to flush a 60-ft length of 3/4-in. service line?

Solution:

$$3/4\text{-in. diameter} = 0.06 \text{ ft.}$$

$$\begin{aligned}\text{Flushing time} \atop \text{(min)} &= \frac{\begin{array}{c}0.785 \times (\text{diameter})^2 \times \text{length (ft)} \\ \times 7.48 \text{ gal/ft}^3 \times 2\end{array}}{\text{Flow rate (gpm)}} \\[2em] &= \frac{\begin{array}{c}0.785 \times (0.06 \text{ ft} \times 0.06 \text{ ft}) \times 60 \text{ ft} \\ \times 7.48 \text{ gal/ft}^3 \times 2\end{array}}{0.5 \text{ gpm}} \\[2em] &= 5.1 \text{ min}\end{aligned}$$

Convert the fractional part of a minute (0.1) to seconds:

$$0.1 \text{ min} \times 60 \text{ sec/min} = 6 \text{ sec}$$

$$5.1 \text{ min} = 5 \text{ min, 6 sec}$$

5.8.3 COMPOSITE SAMPLING

When preparing oven-baked food, a cook pays close attention in setting the correct oven temperature, usually setting the oven to the correct temperature and then moving on to some other chore. The oven thermostat maintains the correct temperature, and that is that. Unlike the cook, in water and wastewater treatment plant operations the operator does not have the luxury of setting a plant parameter and then walking off and forgetting about it. To optimize plant operations, various adjustments to unit processes must be made on an ongoing basis.

The operator makes unit process adjustments based on local knowledge (experience) and on lab test results; however, before lab tests can be performed, samples must be taken. The two basic types of samples are grab samples and composite samples. The type of sample taken depends on the specific test, the reason the sample is being collected, and requirements in the plant discharge permit.

A *grab sample* is a discrete sample collected at one time and one location. It is primarily used for any parameter whose concentration can change quickly (e.g., dissolved oxygen, pH, temperature, total chlorine residual) and is representative only of the conditions at the time of collection.

A *composite sample* consists of a series of individual grab samples taken at specified time intervals and in proportion to flow. The individual grab samples are mixed together in proportion to the flow rate at the time the sample was collected to form the composite sample. The composite sample represents the character of the water/wastewater over a period of time. Because knowledge of the procedure used in processing composite samples is important (a basic requirement) to the water/wastewater operator, the actual procedure used is covered in this section:

1. Determine the total amount of sample required for all tests to be performed on the composite sample.
2. Determine the average daily flow of the treatment system.

Key Point: Average daily flow can be determined by using several months of data, which will provide a more representative value.

3. Calculate a proportioning factor:

$$\begin{aligned}\text{Proportioning} \atop \text{factor (PF)} = \frac{\begin{array}{c}\text{Total sample volume} \\ \text{required (mm)}\end{array}}{\begin{array}{c}\text{No. of samples} \\ \times \text{average daily flow (MGD)}\end{array}}\end{aligned} \quad (5.257)$$

Key Point: Round the proportioning factor to the nearest 50 units (e.g., 50,100, 150) to simplify calculation of the sample volume.

4. Collect the individual samples in accordance with the schedule (e.g., once/hr, once/15 min).
5. Determine flow rate at the time the sample was collected.
6. Calculate the specific amount to add to the composite container:

$$\text{Required volume (mL)} = \left[\begin{array}{c}\text{Flow}_T \\ \times \text{proportioning factor}\end{array}\right] \quad (5.258)$$

where T = time sample was collected.

7. Mix the individual sample thoroughly, measure the required volume, and add to composite storage container.
8. Keep the composite sample refrigerated throughout the collection period.

■ EXAMPLE 5.341

Problem: The effluent testing will require 3825 mL of sample. The average daily flow is 4.25 MGD. Using the flows given below, calculate the amount of sample to be added at each of the times shown:

8 a.m.	3.88 MGD
9 a.m.	4.10 MGD
10 a.m.	5.05 MGD
11 a.m.	5.25 MGD
12 noon	3.80 MGD
1 p.m.	3.65 MGD
2 p.m.	3.20 MGD
3 p.m.	3.45 MGD
4 p.m.	4.10 MGD

Solution:

$$\text{Proportioning factor (PF)} = \frac{3825 \text{ mL}}{9 \text{ samples} \times 4.25 \text{ MGD}}$$

$$= 100$$

$\text{Volume}_{8\text{a.m.}} = 3.88 \times 100 = 388 \ (400 \text{ rounded}) \text{ mL}$
$\text{Volume}_{9\text{a.m.}} = 4.10 \times 100 = 410 \ (410 \text{ rounded}) \text{ mL}$
$\text{Volume}_{10\text{a.m.}} = 5.05 \times 100 = 505 \ (500 \text{ rounded}) \text{ mL}$
$\text{Volume}_{11\text{a.m.}} = 5.25 \times 100 = 525 \ (530 \text{ rounded}) \text{ mL}$
$\text{Volume}_{12\text{noon}} = 3.80 \times 100 = 380 \ (380 \text{ rounded}) \text{ mL}$
$\text{Volume}_{1\text{p.m.}} = 3.65 \times 100 = 365 \ (370 \text{ rounded}) \text{ mL}$
$\text{Volume}_{2\text{p.m.}} = 3.20 \times 100 = 320 \ (320 \text{ rounded}) \text{ mL}$
$\text{Volume}_{3\text{p.m.}} = 3.45 \times 100 = 345 \ (350 \text{ rounded}) \text{ mL}$
$\text{Volume}_{4\text{p.m.}} = 4.10 \times 100 = 410 \ (410 \text{ rounded}) \text{ mL}$

5.8.4 BIOCHEMICAL OXYGEN DEMAND CALCULATIONS

Biochemical oxygen demand (BOD_5) measures the amount of organic matter that can be biologically oxidized under controlled conditions (5 days at 20°C in the dark). Several criteria are used when selecting which BOD_5 dilutions should be used for calculating test results. Consult a laboratory testing reference manual (such as *Standard Methods*) for this information. Two basic calculations are used for BOD_5. The first is used for samples that have not been seeded, and the second must be used whenever BOD_5 samples are seeded. Both methods are introduced and examples are provided below.

5.8.4.1 BOD₅ Unseeded

$$\text{BOD}_5 \atop \text{(unseeded)} = \frac{\left(\begin{array}{c} [\text{DO}_{\text{start}} \ (\text{mg/L}) - \text{DO}_{\text{final}} \ (\text{mg/L})] \\ \times 300 \text{ mL} \end{array} \right)}{\text{Sample volume (mL)}} \quad (5.259)$$

■ EXAMPLE 5.342

Problem: The BOD_5 test is completed. Bottle 1 contained 120 mg/L of sample and had dissolved oxygen (DO) of 7.1 mg/L at the start of the test. After 5 days, bottle 1 had a DO of 2.9 mg/L. Determine the unseeded BOD_5.

Solution:

$$\text{BOD}_5 \ (\text{unseeded}) = \frac{\left(7.1 \text{ mg/L} - 2.9 \text{ mg/L}\right) \times 300 \text{ mL}}{120 \text{ mL}}$$

$$= 10.5 \text{ mg/L}$$

5.8.4.2 BOD₅ Seeded

If the BOD_5 sample has been exposed to conditions that could reduce the number of healthy, active organisms, the sample must be seeded with organisms. Seeding requires the use of a correction factor to remove the BOD_5 contribution of the seed material:

$$\text{Seed correction} = \frac{\begin{array}{c} \text{Seed material BOD}_5 \\ \times \text{ seed in dilution (mL)} \end{array}}{300 \text{ mL}} \quad (5.260)$$

$$\text{BOD}_5 \atop \text{(seeded)} = \frac{\left(\begin{array}{c} \left[\begin{array}{c} \text{DO}_{\text{start}} \ (\text{mg/L}) - \text{DO}_{\text{final}} \ (\text{mg/L}) \\ - \text{ seed correction} \end{array} \right] \\ \times 300 \text{ mL} \end{array} \right)}{\text{Sample volume (mL)}} \quad (5.261)$$

5.8.4.3 BOD 7-Day Moving Average

Because the BOD characteristic of wastewater varies from day to day, even hour to hour, operational control of the treatment system is most often accomplished based on trends in data rather than individual data points. The BOD 7-day moving average is a calculation of the BOD trend.

Key Point: The 7-day moving average is called a moving average because a new average is calculated each day, adding the new day's value to the six previous days' values.

$$\begin{array}{c} \text{7-day} \\ \text{average} \\ \text{BOD} \end{array} = \frac{\left(\begin{array}{c} \text{BOD}_{\text{day1}} + \text{BOD}_{\text{day2}} + \text{BOD}_{\text{day3}} \\ + \text{BOD}_{\text{day4}} + \text{BOD}_{\text{day5}} + \text{BOD}_{\text{day6}} \\ + \text{BOD}_{\text{day7}} \end{array} \right)}{7} \quad (5.262)$$

■ EXAMPLE 5.343

Problem: Given the following primary effluent BOD test results, calculate the 7-day average:

June 1	200 mg/L
June 2	210 mg/L
June 3	204 mg/L
June 4	205 mg/L
June 5	222 mg/L
June 6	214 mg/L
June 7	218 mg/L

$$\text{7-day average BOD} = \frac{\left(\begin{array}{c} 200 + 210 + 204 + 205 \\ + 222 + 214 + 218 \end{array} \right)}{7}$$

$$= 210 \text{ mg/L}$$

5.8.5 MOLES AND MOLARITY

Chemists have defined a very useful unit called the *mole*. Moles and molarity, a concentration term based on the mole, have many important applications in water/wastewater operations. A mole is defined as a gram molecular weight; that is, the molecular weight expressed as grams. For example, a mole of water is 18 g of water, and a mole of glucose is 180 g of glucose. A mole of any compound always contains the same number of molecules. The number of molecules in a mole is called *Avogadro's number* and has a value of 6.022×10^{23}.

Interesting Point: How big is Avogadro's number? An Avogadro's number of soft drink cans would cover the surface of the Earth to a depth of over 200 miles.

Key Point: Molecular weight is the weight of one molecule. It is calculated by adding the weights of all the atoms that are present in one molecule. The units are atomic mass units (amu). A mole is a gram molecular weight—that is, the molecular weight expressed in grams. The molecular weight is the weight of 1 molecule in daltons. All moles contain the same number of molecules (Avogadro's number), equal to 6.022×10^{23}. The reason all moles have the same number of molecules is because the value of the mole is proportional to the molecular weight.

5.8.5.1 Moles

As mentioned, a mole is a quantity of a compound equal in weight to its formula weight; for example, the formula weight for water can be determined using the Periodic Table of Elements:

$$\text{Hydrogen } (1.008) \times 2 = \ \ 2.016$$
$$+ \text{ Oxygen} = \underline{16.000}$$
$$\text{Formula weight of } H_2O = 18.016$$

Because the formula weight of water is 18.016, a mole is 18.016 units of weight. A *gram-mole* is 18.016 grams of water. A *pound-mole* is 18.016 pounds of water. For our purposes in this text, the term *mole* will be understood to mean *gram-mole*. The equation used to determine moles is shown below:

$$\text{Moles} = \frac{\text{Grams of chemical}}{\text{Formula weight of chemical}} \quad (5.263)$$

■ EXAMPLE 5.344

Problem: The atomic weight of a certain chemical is 66. If 35 g of the chemical are used in making up a 1-L solution, how many moles are used?

$$\text{Moles} = \frac{\text{Grams of chemical}}{\text{Formula weight of chemical}}$$

$$= \frac{66 \text{ g}}{35 \text{ g/mole}} = 1.9 \text{ moles}$$

The molarity of a solution is calculated by taking the moles of solute and dividing by the liters of solution:

$$\text{Molarity} = \frac{\text{Moles of solute}}{\text{Liters of solution}} \quad (5.264)$$

■ EXAMPLE 5.345

Problem: What is the molarity of 2 moles of solute dissolved in 1 L of solvent?

Solution:

$$\text{Molarity} = \frac{2 \text{ moles}}{1 \text{ L}} = 2 \ M$$

Key Point: Measurement in moles is a measurement of the amount of a substance. Measurement in molarity is a measurement of the concentration of a substance—the amount (moles) per unit volume (liters).

5.8.5.2 Normality

As mentioned, the *molarity* of a solution refers to its concentration (the solute dissolved in the solution). The *normality* of a solution refers to the number of *equivalents* of solute per liter of solution. The definition of chemical equivalent depends on the substance or type of chemical reaction under consideration. Because the concept of equivalents is based on the reacting power of an element or compound, it follows that a specific number of equivalents of one substance will react with the same number of equivalents of another substance. When the concept of equivalents is taken into consideration, it is less likely that chemicals will be wasted as excess amounts. Keeping in mind that normality is a measure of the reacting power of a solution (i.e., 1 equivalent of a substance reacts with 1 equivalent of another substance), we use the following equation to determine normality:

$$\text{Normality } (N) = \frac{\text{No. of equivalents of solute}}{\text{Liters of solution}} \quad (5.265)$$

■ EXAMPLE 5.346

Problem: If 2.0 equivalents of a chemical are dissolved in 1.5 L of solution, what is the normality of the solution?

Solution:

$$\text{Normality } (N) = \frac{\text{No. of equivalents of solute}}{\text{Liters of solution}}$$

$$= \frac{2.0 \text{ equivalents}}{1.5 \text{ L}} = 1.33 \ N$$

■ EXAMPLE 5.347

Problem: A 800-mL solution contains 1.6 equivalents of a chemical. What is the normality of the solution?

Solution: First convert 800 mL to liters:

$$800 \text{ mL} \div 1000 \text{ mL} = 0.8 \text{ L}$$

Then calculate the normality of the solution:

$$\text{Normality } (N) = \frac{\text{No. of equivalents of solute}}{\text{Liters of solution}}$$

$$= \frac{1.6 \text{ equivalents}}{0.8 \text{ L}} = 2 \text{ } N$$

5.8.6 SETTLEABILITY (ACTIVATED BIOSOLIDS)

The settleability test is a test of the quality of the activated biosolids solids—or activated sludge solids (mixed liquor suspended solids). Settled biosolids volume (SBV)—or settled sludge volume (SSV)—is determined at specified times during sample testing. For control, 30- and 60-minute observations are made. Subscripts (SBV$_{30}$ or SSV$_{30}$ and SBV$_{60}$ or SSV$_{60}$) indicate settling time. A sample of activated biosolids is taken from the aeration tank, poured into a 2000-mL graduated cylinder, and allowed to settle for 30 or 60 min. The settling characteristics of the biosolids in the graduated cylinder give a general indication of the settling of the MLSS in the final clarifier. From the settleability test, the percent settleable solids can be calculated using the following equation:

$$\% \text{ Settleable solids} = \frac{\text{Settled solids (mL)}}{2000\text{-mL sample}} \times 100 \quad (5.266)$$

■ EXAMPLE 5.348

Problem: The settleability test is conducted on a sample of MLSS. What is percent settleable solids if 420 mL settle in the 2000-mL graduate?

Solution:

$$\% \text{ Settleable solids} = \frac{\text{Settled solids (mL)}}{2000\text{-mL sample}} \times 100$$

$$= \frac{420 \text{ mL}}{2000 \text{ mL}} \times 100 = 21\%$$

■ EXAMPLE 5.349

Problem: A 2000-mL sample of activated biosolids is tested for settleability. If the settled solids are measured as 410 mL, what is the percent settled solids?

Solution:

$$\% \text{ Settleable solids} = \frac{\text{Settled solids (mL)}}{2000\text{-mL sample}} \times 100$$

$$= \frac{410 \text{ mL}}{2000 \text{ mL}} \times 100 = 20.5\%$$

5.8.7 SETTLEABLE SOLIDS

The settleable solids test is an easy, quantitative method to measure sediment found in wastewater. An Imhoff cone is filled with 1 L of sample wastewater, stirred, and allowed to settle for 60 min. The settleable solids test, unlike the settleability test, is conducted on samples from the sedimentation tank or clarifier influent and effluent to determine percent removal of settleable solids. The percent settleable solids is determined by:

$$\begin{array}{l}\% \text{ Settleable} \\ \text{solids removed}\end{array} = \frac{\begin{array}{c}\text{Settled solids} \\ \text{removed (mL/L)}\end{array}}{\begin{array}{c}\text{Settled solids} \\ \text{in influent (mL/L)}\end{array}} \times 100 \quad (5.267)$$

■ EXAMPLE 5.350

Problem: Calculate the percent removal of settleable solids if the settleable solids of the sedimentation tank influent is 15 mL/L and the settleable solids of the effluent is 0.4 mL/L.

Solution: First determine removed settleable solids:

$$15.0 \text{ mL/L} - 0.4 \text{ mL/L} = 14.6 \text{ mL/L}$$

Next, insert the parameters into Equation 5.267:

$$\% \text{ Settleable solids removed} = \frac{14.6 \text{ mL/L}}{15.0 \text{ mL/L}} \times 100 = 97\%$$

■ EXAMPLE 5.351

Problem: Calculate the percent removal of settleable solids if the settleable solids of the sedimentation tank influent are 13 mL/L and the settleable solids of the effluent are 0.5 mL/L.

Solution: First determine removed settleable solids:

$$13 \text{ mL/L} - 0.5 \text{ mL/L} = 12.5 \text{ mL/L}$$

Next, insert the parameters into Equation 5.267:

$$\% \text{ Settleable solids removed} = \frac{12.5 \text{ mL/L}}{13.0 \text{ mL/L}} \times 100 = 96\%$$

5.8.8 BIOSOLIDS TOTAL SOLIDS, FIXED SOLIDS, AND VOLATILE SOLIDS

Wastewater consists of both water and solids. The *total solids* may be further classified as either *volatile solids* (organics) or *fixed solids* (inorganics). Normally, total solids and volatile solids are expressed as percents, whereas suspended solids are generally expressed as mg/L. To calculate either percents or mg/L concentrations, certain concepts must be understood:

- *Total solids*—The residue left in the vessel after evaporation of liquid from a sample and subsequent drying in an oven at 103 to 105°C
- *Fixed solids*—The residue left in the vessel after a sample is ignited (heated to dryness at 550°C)
- *Volatile solids*—The weight loss after a sample is ignited (heated to dryness at 550°C)

Determinations of fixed and volatile solids do not distinguish precisely between inorganic and organic matter because the loss on ignition is not confined to organic matter. It includes losses due to decomposition or volatilization of some mineral salts.

Key Point: When the term *biosolids* is used, it may be understood to mean a semiliquid mass composed of solids and water. The term *solids*, however, is used to mean dry solids after the evaporation of water.

The percent total solids and percent volatile solids are calculated as follows:

$$\% \text{ Total solids} = \frac{\text{Total solids weight}}{\text{Biosolids sample weight}} \times 100 \quad (5.268)$$

$$\% \text{ Volatile solids} = \frac{\text{Volatile solids weight}}{\text{Total solids weight}} \times 100 \quad (5.269)$$

■ EXAMPLE 5.352

Problem: Given the information below, determine: (1) the percent solids in the sample, and (2) the percent volatile solids in the biosolids sample:

	Biosolids Sample	After Drying	After Burning (Ash)
Weight of sample and dish	73.43 g	24.88	22.98
Weight of dish (tare weight)	22.28 g	22.28	22.28

To calculate percent total solids, the grams total solids (solids after drying) and grams biosolids sample must be determined:

Total solids

24.88 g (weight of total solids and dish)
−22.28 g (weight of dish)
 2.60 g (weight of total solids)

Biosolids sample

73.43 g (weight of biosolids and dish)
−22.28 g (weight of dish)
 51.15 g (weight of biosolids sample)

$$\% \text{ Total solids} = \frac{\text{Total solids weight}}{\text{Biosolids sample weight}} \times 100$$

$$= \frac{2.60 \text{ g}}{51.15 \text{ g}} \times 100 = 5\%$$

To calculate the percent volatile solids, the grams total solids and grams volatile solids must be determined. Because total solids has already been calculated (above), only volatile solids must be calculated:

Volatile solids

24.88 g (weight of sample and dish before burning)
−22.98 g (weight of sample and dish after burning)
 1.90 g (weight of solids lost in burning)

$$\% \text{ Volatile solids} = \frac{\text{Volatile solids weight}}{\text{total solids weight}} \times 100$$

$$= \frac{1.90 \text{ g}}{2.60 \text{ g}} \times 100 = 73\%$$

5.8.9 WASTEWATER SUSPENDED SOLIDS AND VOLATILE SUSPENDED SOLIDS

Total suspended solids (TSS) are the amount of filterable solids in a wastewater sample. Samples are filtered through a glass-fiber filter. The filters are dried and weighed to determine the amount of total suspended solids in mg/L of sample. *Volatile suspended solids* (VSS) are those solids lost on ignition (heating to 500°C.). They are useful to the treatment plant operator because they give a rough approximation of the amount of organic matter present in the solid fraction of wastewater, activated biosolids, and industrial wastes. With the exception of the required drying time, the suspended solids and volatile suspended solids tests of wastewater are similar to those of the total and volatile solids performed for biosolids. Calculation of suspended solids and volatile suspended solids are demonstrated in the example below.

Key Point: The total and volatile solids of biosolids are generally expressed as percents, by weight. The biosolids samples are 100 mL and are unfiltered.

■ EXAMPLE 5.353

Problem: Given the following data regarding a primary effluent sample, calculate: (1) the mg/L suspended solids, and (2) the percent volatile suspended solids of the sample.

	After Drying (Before Burning)	After Burning (Ash)
Weight of sample and dish	24.6268 g	24.6232 g
Weight of dish (tare weight)	24.6222 g	24.6222 g

Sample volume = 50 mL

Solution: To calculate the milligrams suspended solids per liter of sample (mg/L), we must first determine grams suspended solids:

24.6268 g (weight of dish and suspended solids)
−24.6222 g (weight of dish)
00.0046 g (weight of suspended solids)

Next, we calculate mg/L suspended solids (using a multiplication factor of 20, a number that will vary with sample volume) to make the denominator equal to 1 L (1000 mL):

$$\frac{0.0046 \text{ g SS}}{50 \text{ mL}} \times \frac{1000 \text{ mg}}{1 \text{ g}} \times \frac{20}{20} = \frac{92 \text{ mg}}{1000 \text{ mL}} = 92 \text{ mg/L SS}$$

To calculate percent volatile suspended solids, we must know the weight of both total suspended solids (calculated above) and volatile suspended solids.

24.6268 g (weight of dish and SS *before* burning)
−24.6234 g (weight of dish and SS *after* burning)
0.0034 g (weight of solids lost in burning)

$$\% \text{ Volatile solids} = \frac{\text{Volatile solids weight}}{\text{Total solids weight}} \times 100$$

$$= \frac{0.0034 \text{ g}}{0.0046 \text{ g}} \times 100 = 70\%$$

5.8.10 Biosolids Volume Index and Biosolids Density Index

Two variables are used to measure the settling characteristics of activated biosolids and to determine what the return biosolids pumping rate should be. These are the *volume of the biosolids* (BVI) and the *density of the biosolids* (BDI) indices:

$$BVI = \frac{\% \text{ MLSS volume after 30 min}}{\% \text{ MLSS (mg/L MLSS)}}$$ (5.270)

= Settled biosolids (mL) × 1000

$$BDI = \frac{\% \text{ MLSS}}{\% \text{ MLSS volume after 30 min}} \times 100$$ (5.271)

These indices relate the weight of biosolids to the volume the biosolids occupies. They show how well the liquid–solids separation part of the activated biosolids system is performing its function on the biological floc that has been produced and is to be settled out and returned to the aeration tanks or wasted. The better the liquid–solids separation is, the smaller will be the volume occupied by the settled biosolids and the lower the pumping rate required to keep the solids in circulation.

■ EXAMPLE 5.354

Problem: A settleability test indicates that after 30 min 220 mL of biosolids settle in the 1-L graduated cylinder. If the mixed liquor suspended solids (MLSS) concentration in the aeration tank is 2400 mg/L, what is the biosolids volume?

Solution:

$$BVI = \frac{\text{Volume (determined by settleability test)}}{\text{Density (determined by MLSS conc.)}}$$

$$= \frac{220 \text{ mL/L}}{2400 \text{ mg/L}} = \frac{220 \text{ mL}}{2400 \text{ mg}} = \frac{220 \text{ mL}}{2.4 \text{ g}} = 92$$

The biosolids density index is also a method of measuring the settling quality of activated biosolids, yet it, like the BVI parameter, may or may not provide a true picture of the quality of the biosolids in question unless compared with other relevant process parameters. It differs from BVI in that the higher the BDI value, the better the settling quality of the aerated mixed liquor. Similarly, the lower the BDI, the poorer the settling quality of the mixed liquor. BDI is the concentration in percent solids that the activated biosolids will assume after settling for 30 min. BDI will range from 2.00 to 1.33, and biosolids with values of one or more are generally considered to have good settling characteristics. To calculate the BDI, we simply invert the numerator and denominator and multiply by 100.

■ EXAMPLE 5.355

Problem: The MLSS concentration in the aeration tank is 2500 mg/L. If the activated biosolids settleability test indicates 225 mL settled in the 1-L graduated cylinder, what is the biosolids density index?

Solution:

$$BDI = \frac{\text{Density (determined by MLSS conc.)}}{\text{Volume (determined by settleability test)}} \times 100$$

$$= \frac{2500 \text{ mg}}{225 \text{ mL}} \times 100 = \frac{2.5 \text{ g}}{225 \text{ mL}} \times 100 = 1.11 \text{ mL}$$

CHAPTER REVIEW QUESTIONS

GENERAL WASTEWATER TREATMENT PROBLEMS

5.1 Determine the chlorinator setting (lb/day) required to treat a flow of 5.5 MGD with a chlorine dose of 2.5 mg/L.

5.2 To dechlorinate a wastewater, sulfur dioxide is to be applied at a level 4 mg/L more than the chlorine residual. What should the sulfonator feed rate (lb/day) be for a flow of 4.2 MGD with a chlorine residual of 3.1 mg/L?

5.3 What should the chlorinator setting (lb/day) be to treat a flow of 4.8 MGD if the chlorine demand is 8.8 mg/L and a chlorine residual of 3 mg/L is desired?

5.4 A total chlorine dosage of 10 mg/L is required to treat the water in a unit process. If the flow is 1.8 MGD and the hypochlorite has 65% available chlorine, how many lb/day of hypochlorite will be required?

5.5 A storage tank is to be disinfected with 60 mg/L of chlorine. If the tank holds 86,000 gal, how many pounds of chlorine (gas) will be needed?

5.6 To neutralize a sour digester, 1 lb of lime is to be added for every pound of volatile acids in the digester liquor. If the digester contains 225,000 gal of sludge with a volatile acid level of 2220 mg/L, how many pounds of lime should be added?

5.7 A flow of 0.83 MGD requires a chlorine dosage of 8 mg/L. If the hypochlorite has 65% available chlorine, how many lb/day of hypochlorite will be required?

5.8 The suspended solids concentration of the wastewater entering the primary system is 450 mg/L. If the plant flow is 1,840,000 gpd, how many lb/day suspended solids enter the primary system?

5.9 Calculate the BOD loading (lb/day) on a stream if the secondary effluent flow is 2.90 MGD and the BOD of the secondary effluent is 25 mg/L.

5.10 The daily flow to a trickling filter is 5,450,000 gpd. If the BOD content of the trickling filter influent is 260 mg/L, how many lb/day BOD enter the trickling filter?

5.11 The flow to an aeration tank is 2540 gpm. If the COD concentration of the water is 144 mg/L, how many pounds of COD are applied to the aeration tank daily?

5.12 The daily flow to a trickling filter is 2300 gpm with a BOD concentration of 290 mg/L. How many pounds of BOD are applied to the trickling filter daily?

5.13 If a primary clarifier removes 152 mg/L suspended solids, how many lb/day suspended solids are removed when the flow is 5.7 MGD?

5.14 The flow to a primary clarifier is 1.92 MGD. If the influent to the clarifier has a suspended solids concentration of 310 mg/L and the primary effluent has 122 mg/L SS, how many lb/day suspended solids are removed by the clarifier?

5.15 The flow to a primary clarifier is 1.88 MGD. If the influent to the clarifier has a suspended solids concentration of 305 mg/L and the primary effluent has 121 mg/L SS, how many lb/day suspended solids are removed by the clarifier?

5.16 The flow to a trickling filter is 4,880,000 gpd. If the primary effluent has a BOD concentration of 150 mg/L and the trickling filter effluent has a BOD concentration of 25 mg/L, how many pounds of BOD are removed daily?

5.17 A primary clarifier receives a flow of 2.13 MGD with a suspended solids concentration of 367 mg/L. If the clarifier effluent has a suspended solids concentration of 162 mg/L, how many pounds of suspended solids are removed daily?

5.18 The flow to the trickling filter is 4,200,000 gpd with a BOD concentration of 210 mg/L. If the trickling filter effluent has a BOD concentration of 95 mg/L, how many lb/day BOD does the trickling filter remove?

5.19 The aeration tank has a volume of 400,000 gal. If the mixed liquor suspended solids concentration is 2230 mg/L, how many pounds of suspended solids are in the aerator?

5.20 The aeration tank of a conventional activated sludge plant has a mixed liquor volatile suspended solids concentration of 1890 mg/L. If the aeration tank is 115 ft long by 40 ft wide and has wastewater to a depth of 12 ft, how many pounds of MLVSS are under aeration?

5.21 The volume of an oxidation ditch is 23,800 ft^3. If the mixed liquor volatile suspended solids concentration is 3125 mg/L, how many pounds of volatile solids are under aeration?

5.22 An aeration tank is 110 ft long by 40 ft wide. The operating depth is 16 ft. If the mixed liquor suspended solids concentration is 2250 mg/L, how many pounds of mixed liquor suspended solids are under aeration?

5.23 An aeration tank is 105 ft long by 50 wide. The depth of wastewater in the tank is 16 ft. If the tank contains a mixed liquor suspended solids concentration of 2910 mg/L, how many pounds of MLSS are under aeration?

5.24   The WAS suspended solids concentration is 6150 mg/L. If 5200 lb/day solids are to be wasted, what must the WAS pumping rate be (in MGD)?

5.25   The WAS suspended solids concentration is 6200 mg/L. If 4500 lb/day solids are to be wasted, (a) what must the WAS pumping rate be in MGD, and (b) what is this rate expressed in gpm?

5.26   It has been determined that 6070 lb/day of solids must be removed from the secondary system. If the RAS suspended solids concentration is 6600 mg/L, what must the WAS pumping rate be in gpm?

5.27   The RAS suspended solids concentration is 6350 mg/L. If a total of 7350 lb/day solids is to be wasted, what should the WAS pumping rate be in gpm?

5.28   A total of 5750 lb/day of solids must be removed from the secondary system. If the RAS suspended solids concentration is 7240 mg/L, what must the WAS pumping rate be in gpm?

5.29   Determine the chlorinator setting (lb/day) required to treat a flow of 3,650,000 gpd with a chlorine dose of 2.5 mg/L.

5.30   Calculate the BOD loading (lb/day) on a stream if the secondary effluent flow is 2.10 MGD and the BOD of the secondary effluent is 17 mg/L.

5.31   The flow to a primary clarifier is 4.8 MGD. If the influent to the clarifier has a suspended solids concentration of 310 mg/L and the primary effluent suspended solids concentration is 120 mg/L, how many lb/day suspended solids are removed by the clarifier?

5.32   What should the chlorinator setting (lb/day) be to treat a flow of 5.5 MGD if the chlorine demand is 7.7 mg/L and a chlorine residual of 2 mg/L is desired?

5.33   The suspended solids concentration of the wastewater entering primary system is 305 mg/L. If the plant flow is 3.5 MGD, how many lb/day suspended solids enter the primary system?

5.34   A total chlorine dosage of 10 mg/L is required to treat water in a unit process. If the flow is 3.1 MGD and the hypochlorite has 65% available chlorine, how many lb/day of hypochlorite will be required?

5.35   A primary clarifier receives a flow of 3.44 MGD with a suspended solids concentration of 350 mg/L. If the clarifier effluent has a suspended solids concentration of 140 mg/L, how many pounds of suspended solids are removed daily?

5.36   A storage tank is to be disinfected with 60 mg/L of chlorine. If the tank holds 90,000 gal, how many pounds of chlorine gas will be needed?

5.37   An aeration tank is 110 ft long and 45 ft wide. This operating depth is 14 ft. If the mixed liquor suspended solids concentration is 2720 mg/L, how many pounds of MLSS are under aeration?

5.38   The WAS suspended solids concentration is 5870 mg/L. If 5480 lb/day solids are to be wasted, what must the WAS pumping rate be in MGD?

5.39   The flow to an aeration tank is 2300 gpm. If the COD concentration of the water is 120 mg/L, how many pounds COD enter the aeration tank daily?

5.40   The daily flow to a trickling filter is 2210 gpm. If the BOD concentration of the trickling filter influent is 240 mg/L, how many lb/day BOD are applied to the trickling filter?

5.41   The 1.7-MGD influent to the secondary system has a BOD concentration of 220 mg/L. The secondary effluent contains 24 mg/L BOD. How many pounds of BOD are removed each day by the secondary system?

5.42   The chlorine feed rate at a plant is 330 lb/day. If the flow is 5,300,000 gpd, what is this dosage expressed in mg/L?

5.43   It has been determined that 6150 lb/day solids must be removed from the secondary system. If the RAS suspended solids concentration is 5810 mg/L, what must the WAS pumping rate be in gpm?

5.44   A trickling filter 100 ft in diameter treats a primary effluent flow of 2.5 MGD. If the recirculated flow to the clarifier is 0.9 MGD, what is the hydraulic loading on the trickling filter in gpd/ft^2?

5.45   The flow to a 90-ft-diameter trickling filter is 2,850,000 gpd. The recirculated flow is 1,675,000 gpd. At this flow rate, what is the hydraulic loading rate in gpd/ft^2?

5.46   A rotating biological contactor treats a flow of 3.8 MGD. The manufacturer data indicates a media surface area of 870,000 ft^2. What is the hydraulic loading rate on the RBC in gpd/ft^2?

5.47   A pond receives a flow of 2,100,000 gpd. If the surface area of the pond is 16 ac, what is the hydraulic loading in in./day?

5.48   What is the hydraulic loading rate in gpd/ft^2 to a 90-ft-diameter trickling filter if the primary effluent flow to the tickling filter is 3,880,000 gpd and the recirculated flow is 1,400,000 gpd?

5.49   A 20-ac pond receives a flow of 4.4 ac-ft/day. What is the hydraulic loading on the pond in in./day?

5.50   A sedimentation tank 70 ft by 25 ft receives a flow of 2.05 MGD. What is the surface overflow rate in gpd/ft^2?

5.51   A circular clarifier has a diameter of 60 ft. If the primary effluent flow is 2.44 MGD, what is the surface overflow rate in gpd/ft²?

5.52   A sedimentation tank is 110 ft long by 50 ft wide. If the flow to the tank is 3.45 MGD, what is the surface overflow rate in gpd/ft²?

5.53   The primary effluent flow to a clarifier is 1.66 MGD. If the sedimentation tank is 25 ft by 70 ft long, what is the surface overflow rate of the clarifier in gpd/ft²?

5.54   The flow to a circular clarifier is 2.66 MGD. If the diameter of the clarifier is 70 ft, what is the surface overflow rate in gpd/ft²?

5.55   A filter 40 ft by 20 ft receives a flow of 2230 gpm. What is the filtration rate in gpm/ft²?

5.56   A filter 40 ft by 25 ft receives a flow rate of 3100 gpm. What is the filtration rate in gpm/ft²?

5.57   A filter 26 ft by 60 ft receives a flow of 2500 gpm. What is the filtration rate in gpm/ft²?

5.58   A filter 40 ft by 20 ft treats a flow of 2.2 MGD. What is the filtration rate in gpm/ft²?

5.59   A filter has a surface area of 880 ft². If the flow treated is 2850 gpm, what is the filtration rate?

5.60   A filter 14 ft by 14 ft has a backwash flow rate of 4750 gpm. What is the filter backwash rate in gpm/ft²?

5.61   A filter 20 ft by 20 ft has a backwash flow rate of 4900 gpm. What is the filter backwash rate in gpm/ft²?

5.62   A filter is 25 ft by 15 ft. If the backwash flow rate is 3400 gpm, what is the filter backwash rate in gpm/ft²?

5.63   A filter 25 ft by 30 ft backwashes at a rate of 3300 gpm. What is this backwash rate expressed as gpm/ft²?

5.64   The backwash flow rate for a filter is 3800 gpm. If the filter is 15 ft by 20 ft, what is the backwash rate expressed as gpm/ft²?

5.65   The total water filtered during a filter is 3,770,000 gal. If the filter is 15 ft by 30 ft, what is the unit filter run volume (UFRV) in gal/ft²?

5.66   The total water filtered during a filter run (between backwashes) is 1,860,000 gal. If the filter is 20 ft by 15 ft, what is the unit filter run volume (UFRV) in gal/ft²?

5.67   A filter 25 ft by 20 ft filters a total of 3.88 MG during the filter run. What is the unit filter run volume in gal/ft²?

5.68   The total water filtered between backwashes is 1,410,200 gal. If the length of the filter is 20 ft and the width is 14 ft, what is the unit filter run volume in gal/ft²?

5.69   A filter is 30 ft by 20 ft. If the total water filtered between backwashes is 5,425,000 gal, what is the unit filter run volume (UFRV) in gal/ft²?

5.70   A rectangular clarifier has a total of 163 ft of weir. What is the weir overflow rate in gpd/ft when the flow is 1,410,000 gpd?

5.71   A circular clarifier receives a flow of 2.12 MGD. If the diameter of the weir is 60 ft, what is the weir overflow rate in gpd/ft?

5.72   A rectangular clarifier has a total of 240 ft of weir. What is the weir overflow rate in gpd/ft when the flow is 2.7 MGD?

5.73   The flow rate to a clarifier is 1400 gpm. If the diameter of the weir is 80 ft, what is the weir overflow rate in gpd/ft?

5.74   A rectangular sedimentation basin has a total weir length of 189 ft. If the flow to the basin is 4.01 MGD, what is the weir loading rate in gpm/ft?

5.75   A trickling filter 80 ft in diameter with a media depth of 5 ft receives a flow of 2,450,000 gpd. If the BOD concentration of the primary effluent is 210 mg/L, what is the organic loading on the trickling filter in pounds BOD per day per 1000 ft³?

5.76   The flow to a 3.5-ac wastewater pond is 120,000 gpd. The influent BOD concentration is 170 mg/L. What is the organic loading to the pond in pounds BOD per day per ac?

5.77   An 85-ft-diameter trickling filter with a media depth of 6 ft receives a primary effluent flow of 2,850,000 gpd with a BOD of 120 mg/L. What is the organic loading on the trickling filter in pounds BOD per day per 1000 ft³?

5.78   A rotating biological contactor (RBC) receives a flow of 2.20 MGD. If the soluble BOD of the influent wastewater to the RBC is 140 mg/L and the surface area of the media is 900,000 ft², what is the organic loading rate in pounds BOD per day per 1000 ft²?

5.79   A 90-ft-diameter trickling filter with a media depth of 4 ft receives a primary effluent flow of 3.5 MGD. If the BOD concentration of the wastewater flow to the trickling filter is 150 mg/L, what is the organic loading rate in pounds BOD per day per 1000 ft³?

5.80   An activated sludge aeration tank receives a primary effluent flow of 3,420,000 gpd with a BOD of 200 mg/L. The mixed liquor volatile suspended solids concentration is 1875 mg/L, and the aeration tank volume is 420,000 gal. What is the current F/M ratio?

5.81   The volume of an aeration tank is 280,000 gal. The mixed liquor suspended solids concentration is 1710 mg/L. If the aeration tank receives a primary effluent flow of 3,240,000 gpd with a BOD of 190 mg/L, what is the F/M ratio?

5.82   The desired F/M ratio at a particular activated sludge plant is 0.9 lb COD per 1 lb mixed liquor volatile suspended solids. If the 2.25-MGD primary effluent flow has a COD of 151 mg/L, how many pounds of MLVSS should be maintained?

5.83   An activated sludge plant receives a flow of 2,100,000 gpd with a COD concentration of 160 mg/L. The aeration tank volume is 255,000 gal and the MLVSS concentration is 1900 mg/L. What is the current F/M ratio?

5.84   The flow to an aeration tank is 3,110,000 gpd, with a BOD content of 180 mg/L. If the aeration tank is 110 ft long by 50 ft wide and has wastewater to a depth of 16 ft, and the desired F/M ratio is 0.5, what is the desired MLVSS concentration (mg/L) in the aeration tank?

5.85   A secondary clarifier is 70 ft in diameter and receives a combined primary effluent and return activated sludge (RAS) flow of 3.60 MGD. If the MLSS concentration in the aerator is 2650 mg/L, what is the solids loading rate on the secondary clarifier in lb/day/ft²?

5.86   A secondary clarifier, 80 ft in diameter, receives a primary effluent flow of 3.10 MGD and a return sludge flow of 1.15 MGD. If the MLSS concentration is 2825 mg/L, what is the solids loading rate on the clarifier in lb/day/ft²?

5.87   The desired solids loading rate for a 60-ft-diameter clarifier is 26 lb/day/ft². If the total flow to the clarifier is 3,610,000 gpd, what is the desired MLSS concentration?

5.88   A secondary clarifier 60 ft in diameter receives a primary effluent flow of 2,550,000 gpd and a return sludge flow of 800,000 gpd. If the MLSS concentration is 2210 mg/L, what is the solids loading rate on the clarifier in lb/day/ft²?

5.89   The desired solids loading rate for a 60-ft-diameter clarifier is 20 lb/day/ft². If the total flow to the clarifier is 3,110,000 gpd, what is the desired MLSS concentration?

5.90   A digester receives a total of 12,110 lb/day volatile solids. If the digester volume is 33,100 ft³, what is the digester loading in volatile solids added per day per ft³?

5.91   A digester 60 ft in diameter with a water depth of 25 ft receives 124,000 lb/day raw sludge. If the sludge contains 6.5% solids with 70% volatile matter, what is the digester loading in pounds volatile solids added per day per ft³?

5.92   A digester 50 ft in diameter with a liquid level of 20 ft receives 141,000 lb/day sludge with 6% total solids and 71% volatile solids. What is the digester loading in pounds volatile solids added per day per ft³?

5.93   A digester 40 ft in diameter with a liquid level of 16 ft receives 21,200 gpd sludge with 5.5% solids and 69% volatile solids. What is the digester loading in pounds volatile solids per day per ft³? Assume the sludge weighs 8.34 lb/gal.

5.94   A digester 50 ft in diameter with a liquid level of 20 ft receives 22,000 gpd sludge with 5.3% total solids and 70% volatile solids. What is the digester loading in pounds volatile solids per day per ft³? Assume the sludge weighs 8.6 lb/gal.

5.95   A total of 2050 lb/day volatile solids is pumped to a digester. The digester sludge contains a total of 32,400 lb of volatile solids. What is the volatile solids loading on the digester in pounds volatile solids in the digester?

5.96   A digester contains a total of 174,600 lb of sludge that has a total solids content of 6.1% and volatile solids of 65%. If 620 lb/day volatile solids are added to the digester, what is the volatile solids loading on the digester in pounds volatile solids added per day per pound volatile solids in the digester?

5.97   A total of 63,200 lb/day sludge is pumped to an 115,000-gal digester. The sludge being pumped to the digester has a total solids content of 5.5% and a volatile solids content of 73%. The sludge in the digester has a solids content of 6.6% with a 59% volatile solids content. What is the volatile solids loading on the digester in pounds volatile solids added per day per pound volatile solids in the digester? Assume the sludge in the digester weighs 8.34 lb/gal.

5.98   A total of 110,000 gal of digested sludge is in a digester. The digested sludge contains 5.9% total solids and 58% volatile solids. If the desired volatile solids loading ratio is 0.08 lb volatile solids added per day per pound volatile solids under digestion, what is the desired rate of pound volatile solids per day to enter the digester? Assume that the sludge in the digester weighs 8.34 lb/gal.

5.99   A total of 7900 gpd sludge is pumped to the digester. The sludge has 4.8% solids with a volatile solids content of 73%. If the desired volatile solids loading ratio is 0.06 lb volatile solids added per day per pound volatile solids under digestion, how many pounds volatile solids should be in the digester for this volatile solids load? Assume that the sludge pumped to the digester weighs 8.34 lb/gal.

5.100   A 5.3-ac wastewater pond serves a population of 1733. What is the population loading on the pond in persons per acre?

5.101 A wastewater pound serves a population of 4112. If the pond is 10 ac, what is the population loading on the pond?

5.102 A 381,000-gpd wastewater flow has a BOD concentration of 1765 mg/L. Using an average of 0.2 lb/day BOD per person, what is the population equivalent of this wastewater flow?

5.103 A wastewater pound is designed to serve a population of 6000. If the desired population loading is 420 persons per acre, how many acres of pond will be required?

5.104 A 100,000-gpd wastewater flow has a BOD content of 2210 mg/L. Using an average of 0.2 lb/day BOD per person, what is the population equivalent of this flow?

5.105 A circular clarifier has a diameter of 80 ft. If the primary effluent flow is 2.25 MGD, what is the surface overflow rate in gpd/ft^2?

5.106 A filter has an area of 190 ft^2. If the flow rate to the filter is 2960 gpm, what is this filter backwash rate expressed as gpm/ft^2?

5.107 The flow rate to a circular clarifier is 2,100,000 gpd. If the diameter of the weir is 80 ft, what is the weir overflow rate in gpd/ft?

5.108 A trickling filter 90 ft in diameter treats a primary effluent flow of 2.8 MGD. If the recirculated flow to the clarifier is 0.5 MGD, what is the hydraulic loading on the trickling filter in gpd/ft^2?

5.109 The desired F/M ratio at an activated sludge plant is 0.7 lb BOD per day per pound mixed liquor volatile suspended solids. If the 2.1-MGD primary effluent flow has a BOD of 161 mg/L, how many pounds of MLVSS should be maintained in the aeration tank?

5.110 A digester contains a total of 182,000 lb sludge that has a total solids content of 6.4% and volatile solids of 67%. If 500 lb/day volatile solids are added to the digester, what is the volatile solids loading on the digester in lb/day volatile solids added per pound volatile solids in the digester?

5.111 A secondary clarifier is 80 ft in diameter and receives a combine primary effluent and return activated sludge (RAS) flow of 3.58 MGD. If the MLSS concentration in the aerator is 2760 mg/L, what is the solids loading rate on the secondary clarifier in lb/day/ft^2?

5.112 A digester 70 ft in diameter with a water depth of 21 ft receives 115,000 lb/day raw sludge. If the sludge contains 7.1% solids with 70% volatile solids, what is the digester loading in pounds volatile solids added per day per ft^3 volume?

5.113 A 25-ac pond receives a flow of 4.15 ac-ft/day. What is the hydraulic loading on the pond in in./day?

5.114 The flow to an aeration tank is 3,335,000 gpd with a BOD content of 174 mg/L. If the aeration tank is 80 ft long by 40 ft wide and has wastewater to a depth of 12 ft, and the desired F/M ratio is 0.5, what is the desired MLVSS concentration (mg/L) in the aeration tank?

5.115 A sedimentation tank 80 ft by 25 ft receives a flow of 2.0 MGD. What is the surface overflow rate in gpd/ft?

5.116 The total water filtered during a filter run (between backwashes) is 1,785,000 gal. If the filter is 25 ft by 20 ft, what is the unit filter run volume (UFRV) in gal/ft^2?

5.117 The volume of an aeration tank is 310,000 gal. The mixed liquor volatile suspended solids concentration is 1920 mg/L. If the aeration tank receives a primary effluent flow of 2,690,000 gpd with a COD of 150 mg/L, what is the F/M ratio?

5.118 A total of 24,500 gal of digested sludge is in a digester. The digested sludge contains 5.5% solids and 56% volatile solids. To maintain a desired volatile loading ratio of 0.09 lb volatile solids added per day per pound volatile solids under digestion, what is the desired pound volatile solids per day loading to the digester?

5.119 The flow to a filter is 4.44 MGD. If the filter is 40 ft by 30 ft, what is the filter loading rate in gpm/ft^2?

5.120 An 80-ft-diameter trickling filter with a media depth of 4 ft receives a primary effluent flow of 3.3 MGD with a BOD concentration of 115 mg/L. What is the organic loading on the filter in pounds BOD per day per 1000 ft^3?

5.121 A circular clarifier receives a flow of 2.56 MGD. If the diameter of the weir is 80 ft, what is the weir overflow rate in gpd/ft?

5.122 A 5.5-acre wastewater pond serves a population of 1900. What is the population loading on the pond (people per acre)?

5.123 A rotating biological contactor (RBC) receives a flow of 2.44 MGD. If the soluble BOD of the influent wastewater to the RBC is 140 mg/L and the surface area of the media is 750,000 ft^2, what is the organic loading rate in pounds soluble BOD per day per 1000 ft^2?

5.124 A filter 40 ft by 30 ft treats a flow of 4.15 MGD. What is the filter loading rate in gpm/ft^2?

5.125 A flocculation basin is 8 ft deep, 16 ft wide, and 30 ft long. If the flow through the basin is 1.45 MGD, what is the detention time in minutes?

5.126 The flow to a sedimentation tank 80 ft long by 20 ft wide by 12 ft deep is 1.8 MGD. What is the detention time in the tank in hours?

5.127 A basin 3 ft by 4 ft is to be filled to the 3-ft level. If the flow to the tank is 6 gpm, how long will it take to fill the tank (in hours)?

5.128 The flow rate to a circular clarifier is 5.20 MGD. If the clarifier is 80 ft in diameter with water to a depth of 10 ft, what is the detention time in hours?

5.129 A waste treatment pond is operated at a depth of 6 ft. The average width of the pond is 500 ft and the average length is 600 ft. If the flow to the pond is 222,500 gpd, what is the detention time in days?

5.130 An aeration tank has a total of 12,300 lb of mixed liquor suspended solids. If a total of 2750 lb/day suspended solids enters the aerator in the primary effluent flow, what is the sludge age in the aeration tank?

5.131 An aeration tank is 110 ft long by 30 ft wide with a wastewater to a depth of 20 ft. The mixed liquor suspended solids concentration is 2820 mg/L. If the primary effluent flow is 988,000 gpd with a suspended solids concentration of 132 mg/L, what is the sludge age in the aeration tank?

5.132 An aeration tank contains 200,000 gal of wastewater. The MLSS is 2850 mg/L. If the primary effluent flow is 1.52 MGD with a suspended solids concentration of 84 mg/L, what is the sludge age?

5.133 The 2.10-MGD primary effluent flow to an aeration tank has a suspended solids concentration of 80 mg/L. The aeration tank volume is 205,000 gal. If a sludge age of 6 days is desired, what is the desired MLSS concentration?

5.134 A sludge age of 5.5 days is desired. Assume that 1610 lb/day suspended solids enter the aeration tank in the primary effluent. To maintain the desired sludge age, how many pounds of MLSS must be maintained in the aeration tank?

5.135 An aeration tank has a volume of 320,000 gal, and the final clarifier has a volume of 180,000 gal. The MLSS concentration in the aeration tank is 3300 mg/L. If a total of 1610 lb/day suspended solids is wasted, and 340 lb/day suspended solids are in the secondary effluent, what is the solids retention time for the activated sludge system? Use the solids retention equation that utilizes combined aeration tank and final clarifier volumes to estimate system solids.

5.136 Determine the solids retention time (SRT) given the data below. Use the solids retention time equation that uses combined aeration tank and final clarifier volumes to estimate system solids.

Aerator volume	250,000 gal
MLSS	2750 mg/L
Final clarifier volume	110,000 gal
WAS SS	5410 mg/L
Population equivalent flow	2.35 MGD
Secondary effluent SS	16 mg/L
WAS pumping rate	19,200 gpd

5.137 Calculate the solids retention time given the data below. Use the SRT equation that uses combined aeration tank and final clarifier volumes to estimate system solids.

Aeration tank volume	1.4 MG
MLSS	2550 mg/L
Final clarifier volume	0.4 MG
WAS SS	6240 mg/L
Population equivalent flow	2.8 MGD
Secondary effluent SS	20 mg/L
WAS pumping rate	85,000 gpd

5.138 The volume of an aeration tank is 800,000 gal and the final clarifier is 170,000 gal. The desired solids retention time (SRT) for the plant is 8 days. The primary effluent flow is 2.6 MGD and the WAS pumping rate is 32,000 gpd. If the WAS SS concentration is 6340 mg/L and the secondary effluent SS concentration is 20 mg/L, what is the MLSS concentration in mg/L?

5.139 The flow to a sedimentation tank 75 ft long by 30 ft wide by 14 ft deep is 1,640,000 gpd. What is the detention time in the tank in hours?

5.140 An aeration tank has a total of 12,600 lb of mixed liquor suspended solids. If a total of 2820 lb/day suspended solids enter the aeration tank in the primary effluent flow, what is the sludge age in the aeration tank?

5.141 An aeration tank has a volume of 310,000 gal. The final clarifier has a volume of 170,000 gal. The MLSS concentration in the aeration tank is 3120 mg/L. If a total of 1640 lb/day suspended solids is wasted and 320 lb/day suspended solids are in the secondary effluent, what is the solids retention time for the activated sludge system?

5.142 The flow through a flocculation basin is 1.82 MGD. If the basin is 40 ft long, 20 ft wide, and 10 ft deep, what is the detention time in minutes?

5.143 Determine the solids retention time given the data below:

Aeration tank volume	220,000 gal
MLSS	2810 mg/L
Final clarifier volume	115,000 gal
WAS SS	6100 mg/L
Population equivalent flow	2,400,000 gpd
Secondary effluent SS	18 mg/L
WAS pumping rate	18,900 gpd

5.144 The mixed liquor suspended solids concentration in an aeration tank is 3250 mg/L. The aeration tank contains 330,000 gal. If the primary effluent flow is 2,350,000 gpd with suspended solids concentrations of 100 mg/L, what is the sludge age?

5.145 Calculate the solids retention time given the following data:

Aeration tank volume	1.5 MG
MLSS	2408 mg/L
Final clarifier volume	0.4 MG
WAS SS	6320 mg/L
Population equivalent flow	2.85 MGD
Secondary effluent SS	25 mg/L
WAS pumping rate	71,200 gpd

5.146 An aeration tank is 80 ft long by 25 ft wide with wastewater to a depth of 10 ft. The mixed liquor suspended solids concentration is 2610 mg/L. If the influent flow rate to the aeration tank is 920,000 gpd with a suspended solids concentration of 140 mg/L, what is the sludge age in the aeration tank?

5.147 A tank 6 ft in diameter is to be filled to the 4-ft level. If the flow to the tank is 12 gpm, how long will it take to fill the tank (in min)?

5.148 A sludge age of 6 days is desired. The suspended solids concentration of the 2.14-MGD influent flow to the aeration tank is 140 mg/L. To maintain the desired sludge age, how many pounds of MLSS must be maintained in the aeration tank?

5.149 The average width of a pond is 400 ft and the average length is 440 ft. The depth is 6 ft. If the flow to the pond is 200,000 gpd, what is the detention time in days?

5.150 The volume of an aeration tank is 480,000 gal and the volume of the final clarifier is 160,000 gal. The desired solids retention time for the plant is 7 days. The primary effluent flow is 2,920,000 gpd and the WAS pumping rate is 34,000 gpd. If the WAS suspended solids concentration is 6310 mg/L, and the secondary effluent suspended solids concentration is 12 mg/L, what is the desired MLSS concentration in mg/L?

5.151 The suspended solids concentration entering a trickling filter is 110 mg/L. If the suspended solids concentration in the trickling filter effluent is 21 mg/L, what is the suspended solids removal efficiency of the trickling filter?

5.152 The BOD concentration of the raw wastewater at an activated sludge plant is 230 mg/L. If the BOD concentration of the final effluent is 14 mg/L, what is the overall efficiency of the plant in BOD removal?

5.153 The influent flow to a waste treatment pond has a BOD content of 260 mg/L. If the pond effluent has a BOD content of 60 mg/L, what is the BOD removal efficiency of the pond?

5.154 The suspended solids concentration of the primary clarifier influent is 310 mg/L. If the suspended solids concentration of the primary effluent is 135 mg/L, what is the suspended solids removal efficiency?

5.155 A total of 3700 gal of sludge is pumped to a digester. If the sludge has a 4.9% solids content, how many lb/day solids are pumped to the digester? Assume the sludge weighs 8.34 lb/gal.

5.156 The total weight of a sludge sample is 12.87 g (sample only, not the dish). If the weight of the solids after drying is 0.87 g, what is the percent total solids of the sludge?

5.157 A total of 1450 lb/day suspended solids is removed from a primary clarifier and pumped to a sludge thickener. If the sludge has a solids content of 3.3%, how many lb/day sludge is this?

5.158 It is anticipated that 258 lb/day suspended solids will be pumped from the primary clarifier of a new plant. If the primary clarifier sludge has a solids content of 4.4%, how many gpd sludge will be pumped from the clarifier? Assume a sludge weight of 8.34 lb/gal.

5.159 A total of 291,000 lb/day sludge is pumped from a primary clarifier to a sludge thickener. If the total solids content of the sludge is 3.6%, how many lb/day total solids are sent to the thickener?

5.160 A primary sludge flow of 3100 gpd with a solids content of 4.4% is mixed with a thickened secondary sludge flow of 4100 gpd that has a solids content of 3.6%. What is the percent solids content of the mixed sludge flow? Assume that the density of both sludges is 8.34 lb/gal.

5.161 Primary and thickened secondary sludges are to be mixed and sent to the digester. The 8100-gpd primary sludge has a solids content of 5.1%, and the 7000-gpd thickened secondary sludge has a solids content of 4.1%. What would be the percent solids content of the mixed sludge? Assume that the density of both sludges is 8.34 lb/gal.

5.162 A 4750-gpd primary sludge has a solids content of 4.7%. The 5250-gpd thickened secondary sludge has a solids content of 3.5%. If the sludges were blended, what would be the percent solids content of the mixed sludge? Assume that the density of both sludges is 8.34 lb/gal.

5.163   A primary sludge flow of 8925 gpd with a solids content of 4.0% is mixed with a thickened secondary sludge flow of 11,340 gpd with 6.6% solids content. What is the percent solids of the combined sludge flow? Assume that the density of both sludges is 8.34 lb/gal.

5.164   If 3250 lb/day solids with a volatile solids content of 65% are sent to the digester, how many lb/day volatile solids are sent to the digester?

5.165   A total of 4120 gpd of sludge is to be pumped to the digester. If the sludge has 7% solids content with 70% volatile solids, how many lb/day volatile solids are pumped to the digester? Assume that the sludge weighs 8.34 lb/gal.

GENERAL WATER TREATMENT PROBLEMS

5.166   The static water level for a well is 91 ft. If the pumping water level is 98 ft, what is the well drawdown?

5.167   The static water level for a well is 110 ft. The pumping water level is 125 ft. What is the well drawdown?

5.168   Before the pump is started, the water level is measured at 144 ft. The pump is then started. If the pumping water level is determined to be 161 ft, what is the well drawdown?

5.169   The static water level of a well is 86 ft. The pumping water level is determined using the sounding line. The air pressure applied to the sounding line is 3.7 psi and the length of the sound line is 112 ft. What is the drawdown?

5.170   A sounding line is used to determine the static water level for a well. The air pressure applied to the sounding line is 4.6 psi and the length of the sounding line is 150 ft. If the pumping water level is 171 ft, what is the drawdown?

5.171   If the well yield is 300 gpm, and the drawdown is 20 ft, what is the specific capacity?

5.172   During a 5-min well yield test, a total of 420 gal was pumped from the well. What is the well yield in gpm?

5.173   When the drawdown of a well was stabilized, it was determined that the well produced 810 gal during a 5-min pumping test. What is the well yield in gpm?

5.174   During a test for well yield, a total of 856 gal was pumped from the well. If the well yield test lasted 5 min, what was the well yield in gpm? In gph?

5.175   A bailer is used to determine the approximate yield of a well. The bailer is 12 ft long and has a diameter of 12 in. If the bailer is placed in the well and removed a total of 12 times during a 5-min test, what is the well yield in gpm?

5.176   During a 5-min well yield test, a total of 750 gal of water was pumped from the well. At this yield, if the pump is operated a total of 10 hr each day, how many gallons of water are pumped daily?

5.177   The discharge capacity of a well is 200 gpm. If the drawdown is 28 ft, what is the specific yield in gpm/ft of drawdown?

5.178   A well produces 620 gpm. If the drawdown for the well is 21 ft, what is the specific yield in gpm/ft of drawdown?

5.179   A well yields 1100 gpm. If the drawdown is 41.3 ft, what is the specific yield in gpm/ft of drawdown?

5.180   The specific yield of a well is listed as 33.4 gpm/ft. If the drawdown for the well is 42.8 ft, what is the well yield in gpm?

5.181   A new well is to be disinfected with chlorine at a dosage of 40 mg/L. If the well casing diameter is 6 in. and the length of the water-filled casing is 140 ft, how may pounds of chlorine will be required?

5.182   A new well with a casing diameter of 12 in. is to be disinfected. The desired chlorine dosage is 40 mg/L. The casing is 190 ft long and the water level in the well is 81 ft from the top of the well. How many pounds of chlorine will be required?

5.183   An existing well has a total casing length of 210 ft. The top 180-ft section of the casing has a 12-in. diameter, and the bottom 40-ft section of the casing has an 8-in. diameter. The water level is 71 ft from the top of the well. How many pounds of chlorine will be required if a chlorine dosage of 110 mg/L is desired?

5.184   The water-filled casing of a well has a volume of 540 gal. If 0.48 lb of chlorine was used in disinfection, what was the chlorine dosage in mg/L?

5.185   A total of 0.09 lb of chlorine is required to disinfect a well. If sodium hypochlorite (5.25% available chlorine) is to be used, how many fluid ounces of sodium hypochlorite are required?

5.186   A new well is to be disinfected with calcium hypochlorite (65% available chlorine). The well casing diameter is 6 in. and the length of the water-filled casing is 120 ft. If the desired chlorine dosage is 50 g/L, how many ounces (dry measure) of calcium hypochlorite will be required?

5.187   How many pounds of chloride of lime (25% available chlorine) will be required to disinfect a well if the casing is 18 in. in diameter and 200 ft long with a water level 95 ft from the top of the well? The desired chlorine dosage is 100 mg/L.

5.188 The water-filled casing of a well has a volume of 240 gal. How many fluid ounces of sodium hypochlorite (5.25% available chlorine) are required to disinfect the well if a chlorine concentration of 60 mg/L is desired?

5.189 The pressure gauge reading at a pump discharge head is 4.0 psi. What is this discharge head expressed in feet?

5.190 The static water level of a well is 94 ft. The well drawdown is 24 ft. If the gauge reading at the pump discharge head is 3.6 psi, what is the field head?

5.191 A pond has an average length of 400 ft, an average width of 110 ft, and an estimated average depth of 14 ft. What is the estimated volume of the pond in gallons?

5.192 A pond has an average length of 400 ft and an average width of 110 ft. If the maximum depth of the pond is 30 ft, what is the estimated gallon volume of the pond?

5.193 A pond has an average length of 200 ft, an average width of 80 ft, and an average depth of 12 ft. What is the ac-ft volume of the pond?

5.194 A small pond has an average length of 320 ft, an average width of 170 ft, and a maximum depth of 16 ft. What is the ac-ft volume of the pond?

5.195 For algal control in a reservoir, a dosage of 0.5 mg/L copper is desired. The reservoir has a volume of 20 MG. How many pounds of copper sulfate pentahydrate (25% available copper) will be required?

5.196 The static water level for a well is 93.5 ft. If the pumping water level is 131.5 ft, what is the drawdown?

5.197 During a 5-min well yield test, a total of 707 gal was pumped from the well. What is the well yield in gpm? In gph?

5.198 A bailer is used to determine the approximate yield of a well. The bailer is 12 ft long and has a diameter of 12 in. If the bailer is placed in the well and removed a total of 8 times during a 5-min test, what is the well yield in gpm?

5.199 The static water level in a well is 141 ft. The pumping water level is determined using the sounding line. The air pressure applied to the sounding line is 3.5 psi and the length of the sounding line is 167 ft. What is the drawdown?

5.200 A well produces 610 gpm. If the drawdown for the well is 28 ft, what is the specific yield in gpm/ft of drawdown?

5.201 A new well is to be disinfected with a chlorine dose of 55 mg/L. If the well casing diameter is 6 in. and the length of the water-filled casing is 150 ft, how many pounds of chlorine will be required?

5.202 During a 5-min well yield test, a total of 780 gal of water was pumped from the well. At this yield, if the pump is operated a total of 8 hr each day, how many gallons of water are pumped daily?

5.203 The water-filled casing of a well has a volume of 610 gal. If 0.47 lb of chlorine was used for disinfection, what was the chlorine dosage in mg/L?

5.204 An existing well has a total casing length of 230 ft. The top 170-ft section of casing has a 12-in. diameter and the bottom 45-ft section of casing has an 8-in. diameter. The water level is 81 ft from the top of the well. How many pounds of chlorine will be required if a chlorine dosage of 100 mg/L is desired?

5.205 A total of 0.3 lb of chlorine is required for the disinfection of a well. If sodium hypochlorite is to be used (5.25% available chlorine), how many fluid ounces of sodium hypochlorite are required?

5.206 A flash mix chamber is 4 ft wide by 5 ft long with water to a depth of 3 ft. What is the volume of water (in gal) in the flash mix chamber?

5.207 A flocculation basin is 50 ft long by 20 ft wide with water to a depth of 8 ft. What is the volume of water (in gal) in the basin?

5.208 A flocculation basin is 40 ft long by 16 ft wide with water to a depth of 8 ft. How many gallons of water are in the basin?

5.209 A flash mix chamber is 5 ft square with water to a depth of 42 in. What is the volume of water (in gal) in the flash mixing chamber?

5.210 A flocculation basin is 25 ft wide by 40 ft long and contains water to a depth of 9 ft, 2 in. What is the volume of water (in gal) in the flocculation basin?

REFERENCES AND SUGGESTED READING

AWWA and ASCE. 1990. *Water Treatment Plant Design*, 2nd ed. New York: McGraw-Hill.

Clesceri, L.S., Greenberg, A.E., and Eaton, A.D., Eds. 1999. *Standard Methods for Examination of Waste & Wastewater*, 20th ed. Washington, D.C.: American Public Health Association.

DHHS. 1994. *Water Fluoridation: A Manual for Water Plant Operators*. Washington, D.C.: U.S. Department of Health and Human Services.

Price, J.M. 1991. *Applied Math for Wastewater Plant Operators*. Boca Raton, FL: CRC Press.

6 Blueprint Reading

You have heard the old saying, "A picture is worth a thousand words." This is certainly true when referring to water or wastewater unit processes, plant process machinery, or a plant electrical motor controller. It would be next to impossible for a maintenance supervisor (or any other knowledgeable person) to describe in words the shape, size, configuration, relations of the various components of a machine, or its operation in sufficient detail for a water or wastewater maintenance operator to troubleshoot the process or machine properly. Blueprints are the universal language used to communicate quickly and accurately the information necessary to understand process operations or to disassemble, service, and reassemble process equipment.

The original drawing is seldom used in the plant or field, but copies, commonly called *blueprints*, are made and distributed to maintenance operators who need them. These blueprints are used extensively in water and wastewater operations to convey the ideas relating to the design, manufacture, and operation of equipment and installations. Simply, blueprints are reproductions or copies of original drawings. Blueprints are made by a special process that produces a white image on a blue background from drawings having dark lines on a light background. In addition to understanding applicable blueprints, *schematic diagrams* are also important pictorial representations with which the maintenance operator should be familiar. A schematic is a line drawing made for a technical purpose that uses symbols and connecting lines to show how a particular system operates.

Blueprints and schematics are particularly important to a maintenance operator because they provide detailed information (or views) for troubleshooting; that is, they help familiarize the troubleshooter with the overall characteristics of systems and equipment. In this chapter, the focus is on blueprints and schematics representative of major water and wastewater plant support equipment, systems, and processes. Major support equipment and systems discussed include machine parts, machines, hydraulic and pneumatic systems, piping and plumbing systems, electrical systems, welding, and air conditioning and refrigeration (AC&R) systems.

6.1 BLUEPRINTS: THE UNIVERSAL LANGUAGE

Technical information about the shape and construction of a simple part, mechanism, or system may be conveyed from one person to another by the spoken or written word. As the addition of details makes the part, mechanism, or system more complex, the water and wastewater maintenance operator must have a precise method available that describes the object adequately. The methodology? Blueprints. Blueprints provide a universal language by which all information about a part, mechanism, or system is furnished to the operator and others.

■ CASE STUDY 6.1. Groping in the Dark

One evening in December, an automatic bar screen at the Rachel's Creek Wastewater Treatment Plant jammed, and its motor overloaded. The overload tripped the electrical switchgear main circuit breaker, cutting off power to the entire building complex. The lack of electricity left the on-duty plant operator unable to perform her normal duties and left her literally groping in the dark.

The sudden, unexpected interruption of electrical power and simultaneous shutdown of the bar screen, the lighting and ventilation in the building, and other systems brought a halt to all things electrical but did not stop the flow of influent into the plant. The plant operator, well trained for such contingencies, immediately contacted the on-call maintenance operator and donned a self-contained breathing apparatus (SCBA) to protect herself against high sulfide levels; the lack of ventilation immediately allowed off-gases (sulfides and methane) from raw influent to accumulate to dangerous levels in the bar screen room. Guided by the beam of a flashlight, the operator located the tripped circuit breaker. She reset the breaker, but electrical power was not restored. She correctly discerned that power to the entire switchgear was tripped.

Realizing that she would have to wait for the on-call maintenance operator to restore electrical power, she directed the flashlight beam along the floor leading to the manual bar screen. She raked debris from the screen until power was restored about 10 minutes later.

Important Point: The first treatment unit process for raw wastewater (influent) is coarse screening. The purpose of screening is to remove large solids (rags, cans, rocks, branches, leaves, roots, etc.) from the flow before the flow moves on to downstream processes. A *bar screen* traps debris as wastewater influent passes through. Typically, a bar screen consists of a series of parallel, evenly spaced long metal bars about 1 in. apart, or a perforated screen, and is placed in a channel. The screen may be coarse (2- to 4-in. openings) or fine (0.75 to 2.0-in. openings). The wastestream passes through the screen and the large solids (screenings) are retained (trapped) on the bars for later removal. Bar screens may be manually cleaned (bars or screens are placed at an angle of 30° for easier solids removal with a hand rake) or mechanically cleaned (bars are placed at a 45° to 60° angle to improve automatic mechanical cleaner operation).

In the meantime, the on-call maintenance operator arrived at the plant. The maintenance operator had electrical power restored within 5 minutes. The troublesome bar screen was back in operation the next morning. This was a good response and service from the maintenance operator. Obviously, maintenance operators who can provide such good service are valuable to plant operations. They advance up the promotional ladder quickly, and they usually make more money than the average plant operator. How did the maintenance operator at Rachel's Creek do such a professional job?

When the call came, the maintenance operator asked two questions: "How extensive is the power failure?" and "Do you know what caused it?" The operator said the whole building complex was down and that there was a loud grinding noise, then a louder popping sound from the bar screen when the trouble started.

Upon arrival at the plant site, the on-call maintenance operator, who had been cross-trained as an electrician, went to the electrical substation first to check for damage to the circuit breaker. After determining that the main circuit breaker appeared to be undamaged, she went to the local switchgear for the building. The maintenance operator cut the individual power switches to all equipment before resetting the main circuit breaker at the substation. Then the maintenance operator put the lighting, ventilation, and other equipment on line, one by one, avoiding resetting the bar screen breaker.

After restoring electrical power to lighting, ventilation, and other circuits, the maintenance operator checked out the bar screen and found that it was indeed jammed. The maintenance operator then examined the bar screen circuit breaker, which was built into the motor controller. This breaker should have tripped (opened), so the jammed bar screen would have been isolated and would be the only electrical device losing power. Something apparently had happened to this circuit breaker to make it malfunction.

The initial response to get the bar screen building complex (e.g., lighting, ventilation) back on line again was good. Even though the maintenance operator was highly skilled and had considerable local knowledge (experience), she needed a lot of information to get this job done, and she needed it quickly. The information she needed came from blueprints.

- The maintenance operator located the substation on a plant layout.
- At the substation, the maintenance operator used an electrical utilities plan showing what equipment was installed there and how it should look.
- Later the following day, the maintenance team that made the actual bar screen repairs relied on drawings showing how to disassemble the bar screen drive mechanism. The drawing also showed the identification number of the part that failed (i.e., the bearings on the drive mechanism) so a replacement could be obtained from stock.

Blueprints are used almost everywhere in water and wastewater treatment systems. The bar screen malfunction just described points out that blueprints are among the most important forms of communication among people involved in plant maintenance operations.

6.1.1 BLUEPRINT STANDARDS

To provide a universal language, blueprints must communicate ideas to many different people. It logically follows that, to facilitate this communication, all industrialized nations need to develop technical drawings according to universally adopted standards. Moreover, such drawing standards must also include symbols, technical data, and principles of graphic representation. Universal standardization practices allow blueprints to be uniformly interpreted throughout the globe. The standardization implication should be obvious: Parts, structures, machines, and all other products (designed according to the same system of measurement) may be actually manufactured and interchangeable.

Important Point: Universal standardization of blueprints and drawings is sometimes referred to as *drawing conventions*—that is, standard ways of drawing things so everyone understands the information being conveyed.

In this modern era, with its global economy, the interchangeability of manufactured parts is of increasing importance. Consider, for example, a machine manufactured in Europe that subsequently ends up being used in a factory in North America. Although such a machine is manufactured on one continent and is used on another, getting replacement repair parts normally is not a significant

TABLE 6.1
Blueprint Sheet Size

Standard USA Size (in.)	Nearest International Size (mm)
A, 8.5 × 11	A4, 210 × 297
B, 11 × 17	A3, 297 × 420
C, 17 × 22	A2, 420 × 594
D, 22 × 34	A1, 594 × 841
E, 34 × 44	AO, 841 × 1189

Source: ANSI, *ANSI Y14.1: Blueprint Sheet Size*, American National Standards Institute, New York, 1980.

problem; however, if the user company has its own machine shop or access to one, it may decide, for one reason or another, that it wants to manufacture its own replacement parts. Without standardized blueprints, such an operation would be very difficult to accomplish.

6.1.1.1 Standards-Setting Organizations

Two standards-setting organizations have developed drafting standards that are accepted and widely used throughout the globe: the *American National Standards Institute* (ANSI) and the *International Organization Standardization* (ISO; metric standards). Incorporated into these systems are other engineering standards generated and accepted by professional organizations dealing with specific branches of engineering, science, and technology. These organizations include the *American Society of Mechanical Engineers* (ASME), the *American Welders Society* (AWS), the *American Institute of Architects* (AIA), the U.S. military (MIL), and others. Moreover, to suit their own needs, some large corporations have adopted their own standards.

Important Point: All references to blueprints in this text closely follow the ANSI and ISO standards and current industrial practices.

6.1.1.2 ANSI Standards for Blueprint Sheets

The American National Standards Institute has established standards for the sheets onto which blueprints are printed (see Table 6.1).

6.1.2 Finding Information

In the previous section, the importance of using universal standards or drawing conventions with regard to the correct interpretation of blueprints was pointed out. In addition to knowing these conventions, we must also know where to look for information. In this section, we explain how to find the information provided in blueprints. Typically, designers use technical shorthand in their drawings; however, there generally is too much information to be

included in a single drawing sheet. For this reason, several blueprints are often assembled to make a set of *working drawings*. Working drawings (drawings that furnish all the information required to construct an object) consist of two basic types: (1) *detail drawings* for the parts produced, and (2) an *assembly drawing* for each unit or subunit to be put together.

6.1.2.1 Detail Drawings

A *detail drawing* is a working drawing that includes a great deal of data, such as the size and shape of the project, what kinds of materials should be used, how the finishing should be done, and what degree of accuracy is needed for a single part. Each detail must be given.

Important Point: Usually, a detail drawing contains only dimensions and information needed by the department for which it is made. The only part that may not have to be drawn is a *standard part*, one that can probably be purchased from an outside supplier more economically than it can be manufactured.

6.1.2.2 Assembly Drawings

Most machines and systems contain more than one part. Simply, *assembly drawings* show how these parts fit together. In addition to showing how the parts fit together, they provide an overall look of the construction and the dimensions required for installation. Assembly drawings also include a *parts list*, which identifies all the pieces comprising the item. A parts list is also called a *bill of materials*.

Important Point: Because assembly drawings show the working relationship of the various parts of a machine or structure as they fit together, each part in the assembly is usually numbered and listed in a table on the drawing.

6.1.2.3 Title Block

The first place to look for information on a blueprint is the *title block*, an outlined rectangular space located in the lower right corner of the sheet. The title block is placed in the lower right corner because, when the print is correctly folded, it may be seen for easy reference and for filing. The purpose of the title block is to provide supplementary information on the part or assembly to be made and to include in one section of the print information that aids in identification and filing of the print.

Although title blocks used by different organizations can vary, certain information is basic. Figure 6.1 shows a blank blueprint sheet, with its title block and other features. The following list describes the information usually found in a title block; the letters after each description refer to the example in Figure 6.1:

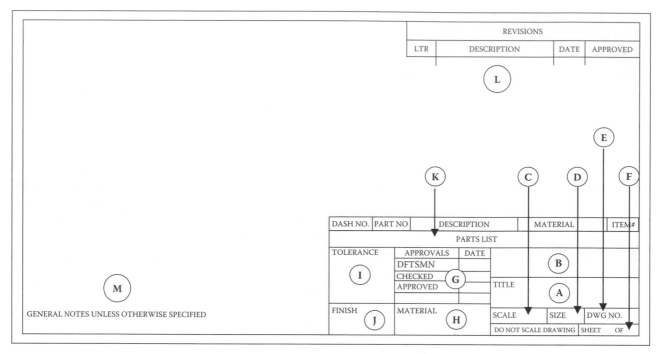

FIGURE 6.1 Blank blueprint sheet.

1. *Title of drawing*. This box identifies the part or assembly illustrated (A).
2. *Name of company and its location*. The space above the title is reserved for the name and location (complete address) of the designing or manufacturing firm (B).
3. *Scale*. The drawing scale indicates the relationship between the size of the image and the size of the actual object. Some parts are shown at actual size, whereas others might be either too big or too small to show conveniently at full size; for example, we cannot show a large machine full size on an ordinary sheet of paper. The designer has the choice of drawing a machine, mechanical part, or other object larger or smaller than actual size. Typical scale notations are: 1/2" = 1" (one half actual size); FULL (actual size); 1:1 (actual size); 2:1 (twice size); 3, 4, 5, etc. (3, 4, 5, etc. times true size). When the scale is indicated to be "as noted," it means that several scales have been used in making the drawing and each is indicated below the particular view to which it pertains (C).

Important Point: Measurements on a blueprint should never be used because the print may have been reduced or enlarged. Work only from the dimensions given on the print.

4. *Drawing size*. Drawings are prepared on standard-size sheets in multiples of 8.5 × 11 or 9 × 12 in. and are designated by a letter to indicate size (D).

5. *Drawing number*. The drawing number is used to identify and control the blueprint. It is also used to designate the part or assembly shown on the blueprint (i.e., it becomes the number of the part itself). The number is usually coded (to the particular industry, not universally applicable to all industries) to indicate department, model, group, serial number, and dash numbers. This number is also used to file the drawings, making it easier to locate them later on (E).

Important Point: A *dash number* is a number preceded by a dash after the drawing number; it indicates right- or left-hand parts as well as neutral parts or detail and assembly drawings.

6. *Sheet number*. Sheet numbering is used on multiple-sheet blueprints to indicate the consecutive order and total number of prints, and which one of the series this particular drawing happens to be (F).
7. *Approvals block*. This block is for the signatures and date of release by those who have responsibility for making or approving all or certain facets of the drawing or the manufacture of the part. The block may include signature and date blocks for the following:
   - Draftsperson
   - Checked (for the engineer who checked the drawing for completeness, accuracy, and clarity)

- Design (for the person responsible for the design of the part)
- Stress (for the engineer who ran the stress calculations for the part)
- Materials (for the person whose responsibility it is to see that the materials required to make the part are available)
- Production (for the engineer who approved the producibility of the part)
- Supervisor (for the person in charge of drafting, who indicates approval by signing and dating this block)
- Approved (any other required approvals)

Each person signs the document and fills in the date on the appropriate line when his or her portion of the work is finished or approved (G).

8. *Materials block.* This block specifies exactly what the part is made of (e.g., the type of steel to be used) and often includes the size of raw stock to be used (H).

9. *Tolerance block.* The tolerance block indicates the general tolerance limits for one-, two-, and three-place decimal and angular dimensions. The tolerance limits are often necessary because nothing can be made to the exact size specified on a drawing. Normal machining and manufacturing processes allow for slight deviations. These limits are applicable unless the tolerance is given along with the dimension callout (I).

Important Point: *Tolerance* is defined as the total amount of variation permitted from the design size of a part. Parts may have a tolerance given in fractions or decimal inches or decimal millimeters.

10. *Finish block.* The finish block gives information on how the part is to be finished (buffed, painted, plated, anodized, or other). Specific finish requirements would appear as a callout on the drawing with the word "NOTED" in the finish block (J).

11. *Parts list (bill of materials).* A parts list is a tabular form usually appearing right above the title block on the blueprint; it is used only on assembly and installation drawings. The purpose of the parts list is to provide specific information on the quantity and types of materials used in the manufacturing and assembling of parts of a machine or structure. The parts list allows a purchasing department to requisition the quantity of materials required to produce a given number of the assemblies. Individual component parts, their part numbers, and the quantity required for each unit are listed (K).

Important Point: The list is built from the bottom up. The columns are labeled at the bottom (just above the words "Parts List"), and parts are listed in reverse numerical order above these categories. This allows the list to grow into the blank space as additional parts are added. If a drawing is complicated, containing many parts, the parts list may be on a separate piece of paper that is attached to the drawing.

12. *Revision/change block.* On occasion, after a blueprint has been released, it is necessary to make design revisions or changes. The revision or change block is a separate block positioned in the upper right-hand corner of the drawing. It is used to note any changes that have been made to the drawing after its final approval. It is placed in a prominent position, because we need to know which revision we are using and what features have been revised or changed. We also need to know whether the changes have been approved, so the initials of the draftsperson making the change and those approving it are required. When a drawing revision/change notice has been prepared, the drawing is revised and the pertinent information recorded in the revision/change block (L). The following items are usually included in the revision/change block; the letters after the descriptions refer to Figure 6.2:

NOTICE OF CHANGE						
LTR	ZONE	DESCRIPTION	SERIAL	DATE	DR	APP
(A)	(B)	(C)	(D)	(E)	(F)	(G)

FIGURE 6.2 Revision/change block.

REVISIONS					
ZONE	LTR	DESCRIPTION	DATE	APPROVED	
E-3	B	INCORP. FEI I REVISED MARKING B18 WAS A4 N/A WAS 12303			

FIGURE 6.3 Change block with drawing change notice (DCN) recorded.

- *Sequence letter (LTR)*. A sequence letter is assigned to the change or revision and recorded in the change/revision block (A). This index letter is also referenced to the field of the drawing next to the changed effected—for example, "A1. BREAK ALL SHARP EDGES."
- *ZONE*. Used on larger sized prints, this column aids in locating changes (B).
- *DESCRIPTION*. This column provides a concise description of change; for example, when a note is removed from the drawing, the type of note is referred to in the description block (e.g., "PLATING NOTE REMOVED"), or a note is added when a dimension is changed (e.g., "WAS .975–1.002") (C).
- *Serial number (SERIAL)*. This column lists the serial number of the assembly or machine on which the change becomes effective (D).
- *DATE*. Entries in this column indicate the dates when changes were written (E).
- *Drafter (DR)*. This column is initialed by the drafter making the change (F).
- *Approved (APP)*. This column carries the initials or name of the engineer approving the change (G).

A few industries will include other items in their revision/change block to further document the changes made in the original blueprint. Some of these include:

- *Checked*. This column is for the initials or signature of the person who has checked and approved a revision or change.
- *Authority*. This column records the approved engineering change request number.
- *Change number*. This column lists drawing change notice (DCN) numbers.
- *Disposition*. This column carries coded numbers indicating the *disposition* of a change request.
- *Microfilm*. This column is used to indicate the date a revised drawing was placed on microfilm.

- *Effective on*. This column (sometimes a separate block) gives the serial number or ship number of the machine, assembly, or part on which the change becomes effective. The change may also be noted as becoming effective on a certain date.

Important Point: A drawing that has been extensively revised or changed may be redrawn and carry an entry to that effect in the revised/change block or the drawing number may carry a dash letter (-A) indicating a revised/changed drawing.

Important Point: Considerable variation exists among industries in the form of processing and recording changes in prints (see Figure 6.3). The information presented in this section will allow maintenance operators to develop an understanding of the change system in general.

6.1.2.4 Drawing Notes

Notes on drawings provide information and instructions that supplement the graphic presentation as well as the information in the title block and list of materials. Notes on drawings convey many kinds of information (e.g., the size of holes to be drilled, type fasteners to be used, removal of machining burrs). Specific notes like these are tied by *leaders* directly to specific features.

6.1.2.4.1 General Notes

General notes refer to the entire drawing. They are located at the bottom of the drawing, to the left of the title block. General notes are not referenced in the list of materials nor from specific areas of the drawing. Some examples of general notes are given in Figure 6.4.

1. Break sharp edges .030 r unless otherwise specified.
2. This part shall be purchased only from sources approved by the treatment department.
3. Finish all over.
4. Remove burrs.
5. Metallurgical inspection required before machining.

FIGURE 6.4 Examples of general notes on drawings.

FIGURE 6.5 Local note directed to feature.

FIGURE 6.6 Local note reference.

Important Point: When there are exceptions to general notes on the field of the drawing, the general note will usually be followed by the phrases "EXCEPT AS SHOWN" or "UNLESS OTHERWISE SPECIFIED." These exceptions will be shown by local notes or data on the field of the drawing.

6.1.2.4.2 Local Notes

Specific notes or *local notes* apply only to certain features or areas and are located near, and directed to, the feature or area by a leader (see Figure 6.5). Local notes may also be referenced from the field of the drawing or the list of materials by the note number enclosed in a *flag* (equilateral triangle; see Figure 6.6). Some examples of local notes are given in Figure 6.7.

■ CASE STUDY 6.2. The Maintenance Operator's Toolbox

When we place a service call for a heating, air conditioning, television, washing machine or dryer, or other household appliance or system malfunction, usually the repair person responds in short order. The usual practice is for the repair person to check out or look over the appliance or system to get a feel for what the problem is. After determining that the problem is more than just turning the "on" switch to the correct position and that the system is properly aligned for operation (e.g., valves opened or closed as per design), the repair person will usually open his or her tool box and begin troubleshooting to identify the problem. After zeroing in on the cause of the malfunction, the repair person makes the necessary repair or adjustment and then tests the unit to ensure proper operation.

There is nothing unusual about the scenario just described; it is nothing more than a routine practice that most of us are familiar with—unfortunately, some of us much more than others. In fact, this practice is so common and familiar that we do not give it much thought. We are trying to make an important point here, though. Let's take another look at the routine service call procedure described above:

- We have a problem with a home appliance or system.
- We place a service call.
- A repair person responds.
- The repair person checks out the appliance or system.

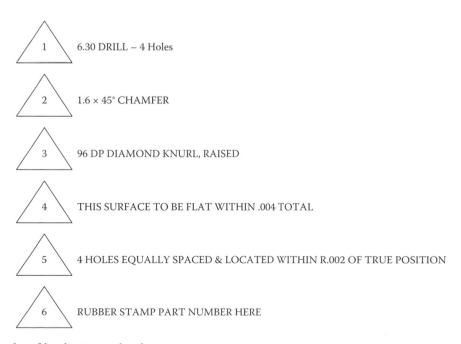

FIGURE 6.7 Examples of local notes on drawings.

- The repair person corrects the problem immediately, or …
- The repair person determines that the malfunction is a bit more complex.
- The repair person opens his or her toolbox and the troubleshooting process begins.
- Eventually, the fault is found and corrected (we hope).

The repair person opens his or her toolbox. This is the part of the routine repair procedure we want to focus on. Why? Good question.

The best way to answer this question is to provide an illustration. Whenever I have hired a repair person to repair a home appliance, a carpenter to repair a wooden deck, a plumber to unclog our pipes, or an electrician to install a new lighting fixture, these skilled technicians always responded with tool box in hand. Have you noticed this, too?

We once hired a carpenter to replace several windows with new ones. During this replacement process, we noticed that the carpenter lugged a huge, heavy, clumsy tool box from window to window. We also noticed that he was able to remove the old windows and install the new ones using only a few basic handtools—screwdrivers, chisel, and hammer. Why, then, the big, clumsy toolbox? Eventually, we got around to asking him that very question: "Why the big toolbox?" After looking at us like we were somewhat out to lunch, he replied as follows: "I lug this big toolbox around with me everywhere I go because I never know for sure what tool I will need to do the job at hand. It is easier to have my complete set of tools in easy reach so I can get the job done without running all over the place to get this or that tool. The way I see it, it just makes good common sense."

"I lug this big toolbox around with me everywhere I go because I never know for sure what tool I will need to do the job at hand" not only makes good common sense but also is highly practicable. For those of us who have worked in water or wastewater treatment plant maintenance operations, we have seen this same routine practiced over and over again by the plant maintenance operator who responds to a plant trouble call: "It just makes good common sense."

At this point the reader might be thinking, "So, beyond the obvious, what is the author's point?" Another good question. Simply put, maintenance operators must have an extensive assortment of tools in their toolboxes to perform many of the plant maintenance actions they are required to. "It just makes good common sense."

But, there is more to consider. We can choose from many types of toolboxes and a large variety of tools, and one thing is certain: If we fill a standard-sized portable toolbox with tools (the best tools that money can buy), they do us little good unless we know how to use them properly. Tools are designed to assist us in performing certain tasks. Like electricity, the internal combustion engine, and the computer, tools, when properly used, they are extremely helpful to us. They not only make tasks easier but also save much time. On the other hand, few would argue against the adage that any tool is only as good as the person's ability or skill in using it properly.

Ability and *skill* are also important tools. They are important tools in the sense that any water or wastewater maintenance operator without a certain amount of ability or skill is, no matter the sophistication of the tool in hand, just another unskilled user of the tool. If we agree that ability and skill are tools, we must also agree that they are tools kept in a different kind of toolbox. The toolbox we are referring to is obvious, of course. Moreover, this kind of toolbox is one we do not have to worry about forgetting to bring to any job we are assigned to perform; we automatically carry it with us everywhere we go. Ability and skill are not normally innate qualities; instead, they are characteristics that have to be learned. They are also general terms, their connotations wide and various.

Simply, ability and skill entail more than just knowing how to properly use a handsaw or other portable tool; for example, with a little practice, just about anyone can use a handsaw to cut a piece of lumber. But, what if we need to cut the lumber to a particular size or dimension? Obviously, to make such a cut to proper size or specification, we not only need to know how to use the cutting tool but also how to measure the stock properly. Moreover, to properly measure and determine the proper size of the cut to be made, we also need to know how to use basic math operations to determine how much has to be actually cut. "It just makes good common sense."

This chapter provides a brief review of the basic math we need and find most useful in reading blueprints. We are likely to find that we already know most of it. Some of it may be new, but only because we have not worked with blueprints before.

Although blueprints ordinarily give us sizes, we occasionally have to do some calculating to get the exact dimension of what we are particularly concerned with. Usually, we find it by adding and subtracting. At other times, we may have to calculate the number of pieces of a given length we can get from a particular piece of wood, pipe, or bar of steel or aluminum or other material. That is usually simply a matter of multiplying and dividing. We may also want to know how many square feet there are in a particular room, doorway, or roof area. In addition to basic math operations of adding, subtracting, multiplying, and dividing, the maintenance operator must also know how to work with fractions and decimals. A basic understanding of angles, areas of rectangles, and the radius of a circle is also important.

TABLE 6.2
Commonly Used Units and Conversions

Quantity	SI Units	SI Symbol	×Conversion Factor	USCS Units
Length	Meter	m	3.2808	ft
Area	Square meter	m²	10.7639	ft²
Volume	Cubic meter	m³	35.3147	ft³

6.2 UNITS OF MEASUREMENT

A basic knowledge of units of measurement and how to use them is essential. Water/wastewater maintenance operators should be familiar with both the U.S. Customary System (USCS) or English system and the International System of Units (SI). Table 6.2 gives conversion factors between the SI and USCS systems for three of the basic units that are encountered in blueprint reading.

The basic units used in blueprint reading are for straight line (linear) measurements; that is, most of the calculating we do is with numbers of yards, feet, and inches, or parts of them. What we actually do is find the distance between two or more points and then use numbers to express the answer in terms of yards, feet, and inches (or parts of them). 12 inches make 1 foot (ft), 3 ft make 1 yard (yd). The symbol indicating an inch is "; the symbol for a foot is '.

As technology has improved, so has the need for closer measurement. As we develop new improved measuring tools, it becomes possible to make parts to greater accuracy. Moreover, now that we are in the age of interchangeable parts, we have developed standards of various kinds. This, in turn, allows us to know what is needed and meant by a given specification. Various thread specifications are a good example, as are the conventions and symbols used on blueprints themselves.

Water and wastewater operations familiar to us today would not be possible without our ability to make close measurements and to do so accurately. This is why basic math operations are used and why it is important to be familiar with them. The basic unit of linear measurement in the United States is the yard, which we break down into feet and inches and parts of them. In our maintenance work, we are concerned with all of them. We may work more with yards and feet when using plant building drawings. At other times, when we work with plant or pumping station machinery, we may find them represented in inches or parts of inches.

6.2.1 FRACTIONS AND DECIMAL FRACTIONS

The number 8 divided by 4 gives an exact quotient of 2, which may be written 8/4 = 2; however, if we attempt to divide 5 by 6 we are unable to derive an exact quotient.

This division may be written 5/6 (read "five sixths"). This is called a *fraction*. The fraction 5/6 represents a number, but it is not a whole number. For this reason, our idea of numbers must be enlarged to include fractions.

In blueprint reading, we are specifically concerned with fractions of some unit involved with measurements; for example, a half inch is one of two parts that make up an inch. That could be written down as 1/2 inch, where the right-hand number (2, the denominator) indicates that it takes two parts to make up the whole unit, and the left-hand number (1, the numerator) indicates that we have one of those two parts. One quarter of an inch, which would be shown on a blueprint as 1/4", means that we need four parts to make up an inch. Because one is all we need, one is all that is shown. We could just as easily have 3/4, 5/8, or 11/16. The basic meaning of the numbers would still be that we need three quarters of an inch, or five eighths of an inch, or eleven sixteenths of an inch. In maintenance practice, the inch is further broken down into thirty-seconds (32nds) and sixty-fourths (64ths).

Important Point: A 64th is the smallest fraction we will use; it is the smallest fraction shown on a machinist's ruler, sometimes incorrectly referred to as a *scale*.

A *decimal fraction* is a fraction that may be written with 10, 100, 1000, 10,000, or some other multiple of 10 as its denominator; thus, 47/100, 4256/10,000, 77/1000, and 3437/1000, for example, are decimal fractions. When writing a decimal fraction, it is standard procedure to omit the denominator and instead merely indicate what that denominator is by placing a *decimal point* in the numerator so there are as many figures to the right of this point as there are zeros in the denominator. Thus, 47/100 is written 0.47; 4256/10,000 = 0.4256; 77/1000 = 0.077; and 3437/1000 = 3.437.

Important Point: The word "decimal" comes from the Latin word for tenth or tenth part.

Important Point: Called *decimal fractions* because they are small parts of a whole unit, these fractions are useful in shortening calculations. Most technologies now use the decimal fractions as a matter of course.

Table 6.3 lists all of the fractions we are likely to see on our plant machine prints. The figures at the right-hand side of each column mean exactly the same thing, except that the numbers are expressed as decimal parts of an inch.

With the passage of time and corresponding improvements in technology, greater accuracy became possible (measurements in fractions of an inch were no longer exact enough). Smaller parts of an inch were needed and were provided by dividing the inch into 1000 parts, the parts being referred to as "thousandths of an inch." One thousandth of an inch is written as ".001."

TABLE 6.3
Common Fractions and Their Decimal Equivalents

Fraction	Decimal	Fraction	Decimal
1/64	.015625	33/64	.515625
1/32	.03125	17/32	.53215
3/64	.046875	35/64	.546875
1/16	.0625	9/16	.5625
5/64	.078125	37/64	.578125
3/32	.09375	19/32	.59375
7/64	.109375	39/64	.609375
1/8	.125	5/8	.625
9/64	.140625	41/64	.640625
5/32	.15625	21/32	.65625
11/64	.171875	43/64	.671875
3/16	.1875	11/16	.6875
13/64	.203125	45/64	.703125
7/32	.21875	23/32	.71875
15/64	.234375	47/64	.734375
1/4	.25	3/4	.75
17/64	.265625	49/64	.765625
9/32	.28125	25/32	.78125
19/64	.296875	51/64	.796875
5/16	.3125	13/16	.8125
21/64	.328125	53/64	.828125
11/32	.34375	27/32	.84375
23/64	.359375	55/64	.859375
3/8	.375	7/8	.875
25/64	.390625	57/64	.890625
13/32	.40625	29/32	.90625
27/64	.421875	59/64	.921875
7/16	.4375	15/16	.9375
29/64	.453125	61/64	.953125
15/32	.46875	31/32	.96875
31/64	.484375	63/64	.984375
1/2	.5	1/1	1.0

Common measurements of an inch, for example, can be expressed as follows:

One inch	1.000"
One thousandth of an inch	.001"
One ten-thousandth	.0001"
One millionth	.000001"

6.3 ALPHABET OF LINES

■ **CASE STUDY 6.3. Just Lines**

Over the years we have heard seasoned maintenance operators claim that no blueprint or technical diagram exists that they could not use or understand. At one time, we might have considered this to be more braggadocio than truth, and we might have continued to think this way if we hadn't finally cornered one of these seasoned maintenance types and asked: "What makes you so confident that you can read and understand any technical blueprint or drawing?" At first, somewhat peeved that we would ask such an obviously dumb question, this maintenance operator answered in a condescending tone of voice: "I know that I can read and understand any print or drawing because prints and drawings are nothing more than a bunch of drawn lines. Even the components that the lines are hooked to are nothing more than lines shown in a different fashion. It all comes down to a bunch of lines, nothing more."

6.3.1 JUST A BUNCH OF DRAWN LINES?

Notwithstanding the summation provided by this seasoned maintenance operator, for the engineer, the designer, and the drafter, engineering-type drawings are more than "a bunch of drawn lines." No doubt lines are important; few would argue with this point. Moreover, to correctly interpret the blueprint when servicing a part or assembly, the maintenance operator and technician must recognize and understand the meaning of ten kinds of lines that are commonly used in engineering and technical drawings. These lines, known as the *alphabet of lines* (a list of line symbols), are universally used throughout industry. Each line has a definite form, shape, and width (thick, medium, or thin), and when they are combined in a drawing they convey information essential to understanding the blueprint (i.e., shape and size of an object).

Important Point: Each line on a technical drawing has a definite meaning and is drawn in a certain way. We use the line conventions, together with illustrations showing various applications, recommended by the American National Standards Institute (ANSI, 1973a) throughout this text.

With regard to the seasoned maintenance operator stating that technical drawings are "nothing more than a bunch of drawn lines," we agree, to a point, but feel it would be more accurate to say that the *line* is the basis of all technical drawings. The point is that, by combining lines of different thicknesses, types, and lengths, just about anything can be described graphically and in sufficient detail so persons with a basic understanding of blueprint reading can accurately visualize the shape of the component. To understand a blueprint, then, we must know and understand the alphabet of lines. The following sections explain and describe each of these lines.

6.3.2 VISIBLE LINES

The *visible line* (also called *object line*) is a thick (dark), continuous line that represents all edges and surfaces of an object that are visible in view. A visible line (see Figure 6.8) is always drawn thick and solid so the outline or shape of the object is clearly emphasized on the drawing.

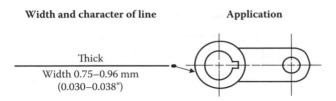

Width and character of line   **Application**

Thick
Width 0.75–0.96 mm
(0.030–0.038")

FIGURE 6.8 Visible line. (Adapted from Brown, W.C., *Blueprint Reading for Industry*, The Goodheart–Wilcox Company, South Holland, IL, 1989, pp. 14–18.).

Important Point: The visible line represents the outline of an object. The thickness of the line may vary according to the size and complexity of the part being described (Olivo and Olivo, 1999).

6.3.3 HIDDEN LINES

Hidden lines are thin, dark, medium-weight, short dashes used to show edges, surfaces, and corners that are not visible in a particular view (see Figure 6.9). Many of these lines are invisible to the observer because they are covered by other portions of the object. They are used when their presence helps to clarify a drawing and are sometimes omitted when the drawing seems to be clearer without them.

6.3.4 SECTION LINES

Usually drawn at an angle of 45°, *section lines* are thin lines used to indicate the cut surface of an object in a sectional view. In Figure 6.10A, the section lining is

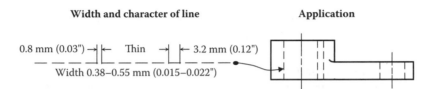

Width and character of line   **Application**

0.8 mm (0.03") →| |← Thin →| |← 3.2 mm (0.12")
Width 0.38–0.55 mm (0.015–0.022")

FIGURE 6.9 Hidden line. (Adapted from Brown, W.C., *Blueprint Reading for Industry*, The Goodheart–Wilcox Company, South Holland, IL, 1989, pp. 14–18.)

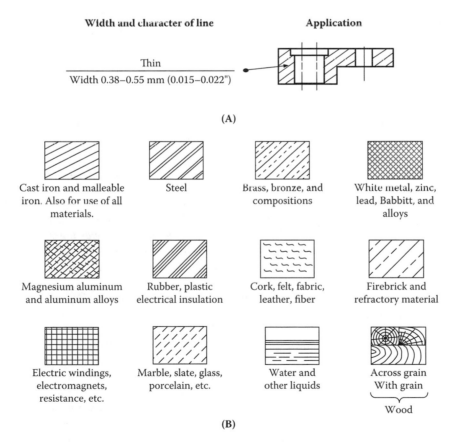

Width and character of line   **Application**

Thin
Width 0.38–0.55 mm (0.015–0.022")

(A)

| Cast iron and malleable iron. Also for use of all materials. | Steel | Brass, bronze, and compositions | White metal, zinc, lead, Babbitt, and alloys |

| Magnesium aluminum and aluminum alloys | Rubber, plastic electrical insulation | Cork, felt, fabric, leather, fiber | Firebrick and refractory material |

| Electric windings, electromagnets, resistance, etc. | Marble, slate, glass, porcelain, etc. | Water and other liquids | Across grain With grain — Wood |

(B)

FIGURE 6.10 (A) Section line. (B) Symbols for materials in section. (Adapted from Brown, W.C., *Blueprint Reading for Industry*, The Goodheart–Wilcox Company, South Holland, IL, 1989, pp. 14–18, 82.)

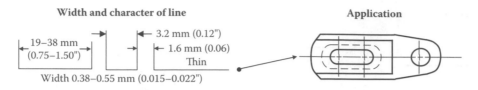

FIGURE 6.11 Center line. (Adapted from Brown, W.C., *Blueprint Reading for Industry*, The Goodheart–Wilcox Company, South Holland, IL, 1989, p. 83.)

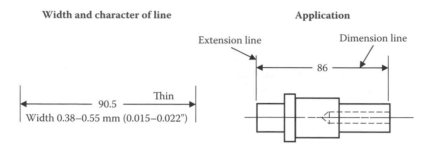

FIGURE 6.12 Dimension line, extension line and leaders. (Adapted from Brown, W.C., *Blueprint Reading for Industry*, The Goodheart–Wilcox Company, South Holland, IL, 1989, p. 84.)

composed of cast iron. This particular section lining is commonly used for other materials in the section unless the draftsperson wants to indicate the specific material in section. Figure 6.10B, for example, shows symbols for other specific materials.

6.3.5 CENTER LINES

Center lines are thin (light), broken lines consisting of alternating long and short dashes and are used to designate the centers of a whole or part of a circle, holes, arcs, and symmetrical objects (see Figure 6.11). The symbol L is often used with a center line. On some drawings, only one side of a part is drawn and the letters *SYM* are added to indicate that the other side is identical in dimension and shape. Center lines are also used to indicate paths of motion.

6.3.6 DIMENSION AND EXTENSION LINES

Dimension lines are thin, dark, solid lines broken at the dimension and terminated by arrowheads, which indicate the direction and extent of a dimension (see Figure 6.12). Fractional, decimal, and metric dimensions are used on drawings to give size dimensions. On machine drawings, the dimension line is broken, usually near the middle, to provide an open space for the dimension figure.

Important Point: The tips of arrowheads used on dimension lines indicate the exact distance referred to by a dimension placed at a break in the line. The tip of the arrowhead touches the extension line. The size of the arrow is determined by the thickness of the dimension line and the size of the drawing.

Extension lines are thin, dark, solid lines that extend from a point on the drawing to which a dimension refers. Simply, extension lines are used in dimensioning to show the size of an object (see Figure 6.12).

Important Point: A space of 1/16 in. is usually allowed between the object and the beginning of the extension line.

6.3.7 LEADERS

Leaders are thin inclined solid lines leading from a note or a dimension (see Figure 6.12) that terminate in an arrowhead or a dot touching the part to which attention is directed.

6.3.8 CUTTING PLANE OR VIEWING PLANE LINES

To obtain a sectional view, an imaginary cutting plane is passed through the object as shown in Figure 6.13. This *cutting plane line* or *viewing plane line* is either a thick (heavy) line with one long and two short dashes or a series of thick (heavy), equally spaced long dashes.

6.3.9 BREAK LINES

To break out sections for clarity (e.g., from behind a hidden surface) or to shorten parts of objects that are constant in detail and would be too long to place on a blueprint, *break lines* are used. Typically, three types of break lines are used. When the part to be broken requires a short line, the thick, wavy *short-break line* is used (see Figure 6.14). If the part to be broken is longer, the thin *long-break line* is used (see Figure 6.15). In round stock such as shafts or pipe, the thick *S break* is used (see Figure 6.16).

Width and character of line

Application

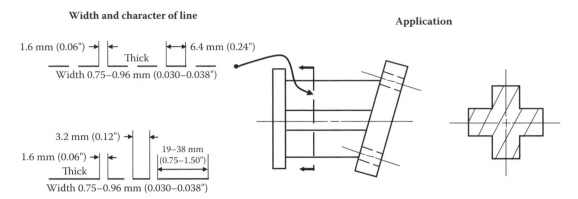

FIGURE 6.13 Cutting-plane or viewing-plane lines. (Adapted from Brown, W.C., *Blueprint Reading for Industry*, The Good-heart–Wilcox Company, South Holland, IL, 1989, p. 85.)

Width and character of line

Application

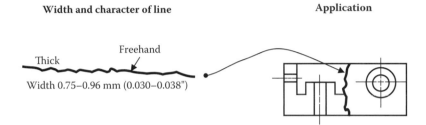

FIGURE 6.14 Short-break line. (Adapted from Brown, W.C., *Blueprint Reading for Industry*, The Goodheart–Wilcox Company, South Holland, IL, 1989, pp. 14–18.)

Width and character of line

Application

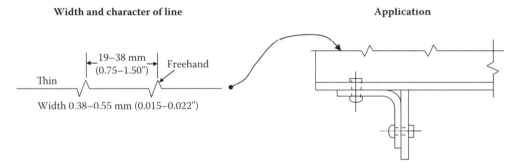

FIGURE 6.15 Long-break line. (Adapted from Brown, W.C., *Blueprint Reading for Industry*, The Goodheart–Wilcox Company, South Holland, IL, 1989, p. 80.)

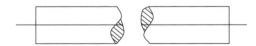

FIGURE 6.16 Cylindrical S break.

6.3.10 PHANTOM LINES

Limited almost entirely to detail drawings, *phantom lines* are thin lines composed of alternating long dashes and pairs of short dashes. They are used primarily to indicate: (1) alternative positions of moving parts, such as is shown in Figure 6.17 (right end); (2) adjacent positions of related parts such as an existing column (see Figure

6.18); and (3) objects having a series of identical features (repeated detail), as in screwed shafts and long springs (see Figure 6.19).

6.3.11 LINE GAUGE

The *line gauge*, used by draftspersons and shown in Figure 6.20, is convenient when referring to lines of various widths.

6.3.12 VIEWS

In the context of reading blueprints, when we speak of various views, they speak for themselves.

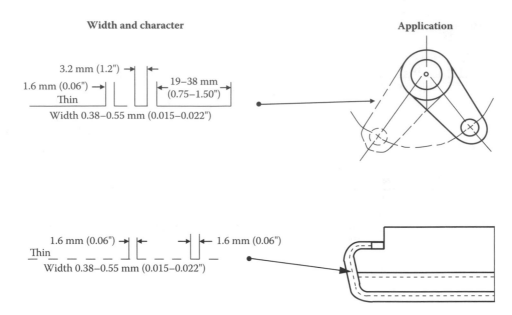

FIGURE 6.17 Phantom line. (Adapted from Brown, W.C., *Blueprint Reading for Industry*, The Goodheart–Wilcox Company, South Holland, IL, 1989, p. 79.)

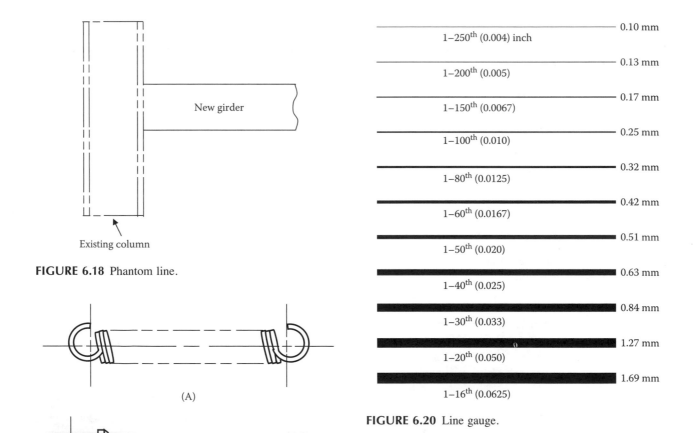

FIGURE 6.18 Phantom line.

FIGURE 6.19 Phantom lines: (A) spring; (B) screw shaft.

FIGURE 6.20 Line gauge.

6.3.12.1 Orthographic Projections

When a draftsperson sets pencil to paper to draw a particular object such as a machine part, a basic problem is faced because such objects are three dimensional; that is,

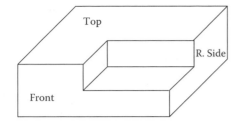

FIGURE 6.21 Pictorial view of a 3-D object.

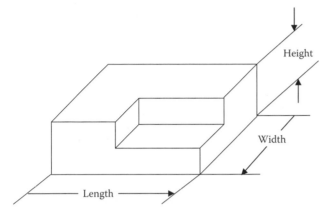

FIGURE 6.22 Dimension of an object.

they have height, width, and depth. No matter the skill of the draftsperson, an object can only be drawn in two dimensions on a flat two-dimensional (2-D) sheet of paper (height and width). A draftsperson typically works with three-dimensional (3-D) objects, but how can these 3-D objects be represented on a 2-D sheet of paper? One way to do this is with a *three-dimensional pictorial*. A 3-D pictorial is a drawing that displays three sides of an object. A pictorial view of a 3-D object is shown in Figure 6.21. In a pictorial view, the object is seen in such a way that three of the six sides of the object are visible. In this case, the *top*, *front*, and *right* sides of the object are visible. The other sides, the *bottom*, *rear*, and *left*, are not visible in this view. In addition to addressing of the sides of a 3-D

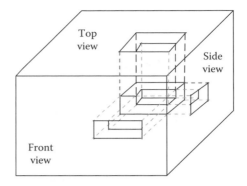

FIGURE 6.23 3-D object inside a glass box.

object, we must also refer to the three dimensions of an object: *length*, *width*, and *height*. The dimensions of an object are shown in Figure 6.22.

To get around the basic problem of drawing 3-D objects in such a manner as to make them usable in industry, *orthographic views* are used. It is often useful to choose the position from which an object is seen, or the *viewpoint*, so only one side (two dimensions) of the object is visible. This is an orthographic view of an object. A *top* (or *plan*) view shows the top side of the object, with the length and width displayed. A *front* view (or *front elevation*) shows the front side of the object, with the length and height displayed. A *right side* view (or *right elevation*) shows the right side of the object, with the width and height displayed.

Important Point: The object is usually drawn so its most important feature appears in the front view.

To create an orthographic view, imagine that the object shown in Figure 6.22 is inside a glass box (see Figure 6.23). The edges of the object are projected onto the glass sides. Next, imagine that the sides of the glass box are hinged so when it is opened the views are as shown in Figure 6.24. These are the orthographic projections of the 3-D object.

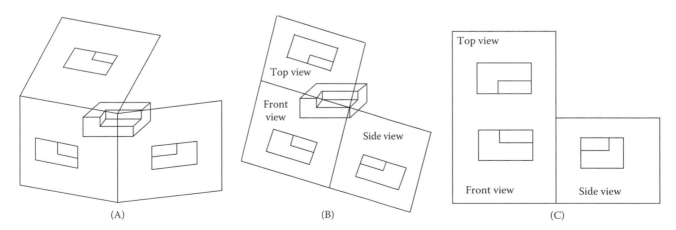

(A)   (B)   (C)

FIGURE 6.24 Various orthographic projection of the 3-D object.

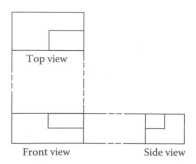

FIGURE 6.25 Proper alignment of orthographic projections.

Important Point: Notice that the views shown in Figure 6.24 are arranged so the top and front projections are in vertical alignment while the front and side views are in horizontal alignment (see Figure 6.25).

6.3.12.2   One-View Drawings

Frequently, a single view supplemented by a note or lettered symbols is sufficient to describe clearly the shape of a relatively simple object (i.e., *simple* here meaning parts that are uniform in shape); for example, cylindrical objects (shafts, bolts, screws, and similar parts) require only one view to describe them adequately. According to ANSI standards, when a one-view drawing of a cylindrical part is used (see Figure 6.26), the dimension for the diameter must be preceded by the symbol ø. In many cases, the older, but widely used, practice for dimensioning diameters is to place the letters *DIA* after the dimension. The main advantage of one-view drawings is the saving in drafting time; moreover, they simplify blueprint reading.

Important Point: In both applications, use of the symbol ø or the letters DIA and a center line indicate that the part is cylindrical.

The one-view drawing is also used extensively for flat parts. With the addition of notes to supplement the dimensions on the view, the one view furnishes all the necessary information for accurately describing the part (see Figure

6.27). (Note that in Figure 6.27 a note indicates the thickness as 3/8".)

6.3.12.3   Two-View Drawings

Often only two views are required to describe clearly the shape of simple, symmetrical flat objects and cylindrical parts; for example, sleeves, shafts, rods, and studs only require two views to give the full details of construction (see Figure 6.28). The two views usually include the front view and a right-side or left-side view, or a top or bottom view.

Important Point: If an object requires only two views, and the left-side and right-side views are equally descriptive, the right-side view is customarily chosen. Similarly, if an object requires only two views, and the top and bottom views are equally descriptive, the top view is customarily chosen. Finally, if only two views are necessary, and the top view and right-side view are equally descriptive, the combination chosen is that which spaces best on the paper.

In the front view shown in Figure 6.28, the center lines run through the axis of the part as a horizontal center line. If the rotor shaft is in a vertical position, the center line runs through the axis as a vertical center line. The second view of the two-view drawing shown in Figure 6.28 contains a horizontal and a vertical center line intersecting at the center of the circles which make up the part in the view. Some of the two-view combinations commonly used in industrial blueprints are shown in Figure 6.29. In many two-view drawings, hidden edge or invisible edge lines, such as shown in Figure 6.30, are common. A hidden detail may be straight, curved, or cylindrical.

6.3.12.4   Three-View Drawings

Regularly shaped flat objects that require only simple machining operations are often adequately described with notes on a one-view drawing. Moreover, any two related views will show all three dimensions, but two views may

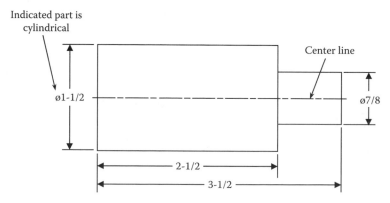

FIGURE 6.26 One-view drawing of a cylindrical shaft; ø or DIA indicate that the part is cylindrical. Center line shows that the part is symmetrical.

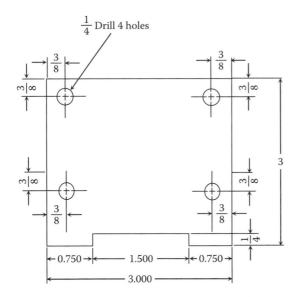

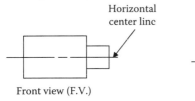

FIGURE 6.27 One-view drawing of a flat machine cover.

(A) Horizontal position

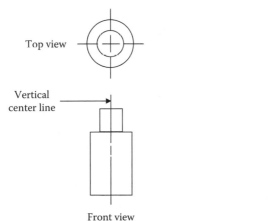

FIGURE 6.28 Examples of two-view drawings of a rotor shaft.

not show enough detail to make the intentions of the designer completely clear. In addition, when the shape of the object changes, when portions are cut away or relieved,

(A) Horizontal position

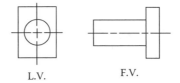

(B) Alternative horizontal position

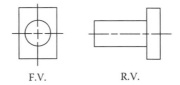

(C) Alternative vertical position views

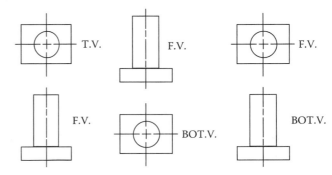

FIGURE 6.29 Several views for a two-view drawing of the same object.

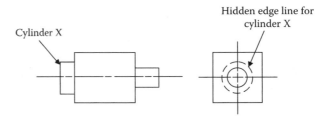

FIGURE 6.30 Two-view object with invisible edge lines.

or when complex machine or fabrication processes must be represented on a drawing, the one view may not be sufficient to describe the part accurately. For this reason, a set of three related views has been established as the usual standard for technical drawings. The combination of front, top, and right-side views represents the method most commonly used by draftspersons to describe simple objects (see Figure 6.31). The object is usually drawn so its most important feature appears in the front view.

Important Point: The number and selection of views is governed by the shape or complexity of the object. A view should not be drawn unless it makes a drawing easier to read or furnishes other information needed to describe the part clearly.

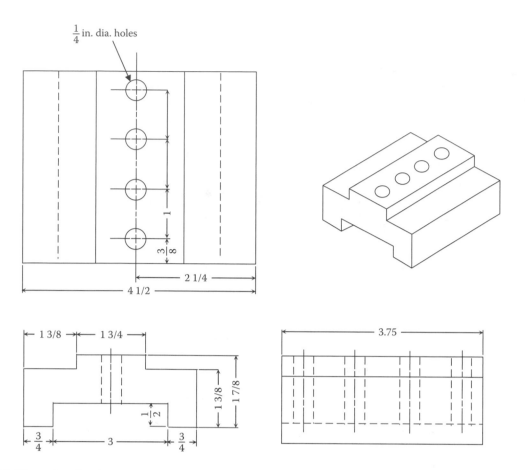

FIGURE 6.31 Three-view drawing.

6.3.12.5 Auxiliary Views

The purpose of the technical drawing is to show the size and shape of each surface. As long as all the surfaces of an object are parallel or at right angles to one another, they may be represented in one or more views. On occasion, however, even three views are not enough. To overcome this problem, draftspersons sometimes find it necessary to use auxiliary views of an object to show the shape and size of surfaces that cannot be shown in the regular view; that is, many objects are of such a shape that their principal faces cannot always be assumed to be parallel to the regular planes of projection. If an object has a surface that is not at a 90° angle from the other surfaces, the drawing will not show its true size and shape; thus, an auxiliary view is drawn to overcome this problem. Figure 6.32A shows an object with an inclined surface (surface cut at an angle). Figure 6.32B shows the three standard views of the same object; however, because of the inclined surface of the object, it is impossible to determine its true size and shape. To show its true size and shape, the draftsperson draws an auxiliary view of the inclined surface (see Figure 6.33). In Figure 6.33, the auxiliary view shows the inclined surface from a position perpendicular to the surface.

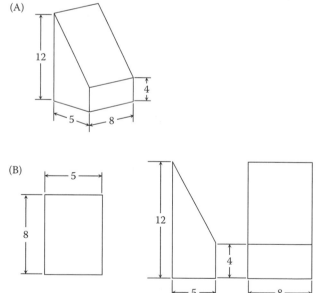

FIGURE 6.32 (A) Object with inclined surface; (B) three views of object with inclined surface.

Important Point: Auxiliary views may be projected from any view in which the inclined surface appears as a line.

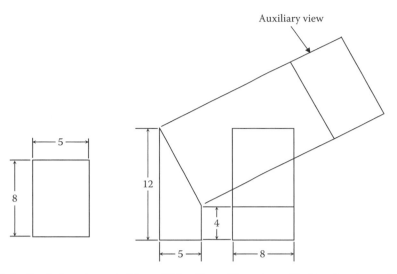

FIGURE 6.33 Object with inclined view shown in Figure 6.32 with auxiliary view added to give size and shape of inclined surface.

6.4 DIMENSIONS AND SHOP NOTES

Early on, *dimensioning* (or measuring using basic units of measurement) was rather simple and straightforward. For example, in the time of Noah and the Ark, a *cubit* was the length of a man's forearm, or about 18 in. In pre-industrialized England, an inch used to be "three barley corns, round and dry." More recently, we have all heard of "rule of thumb." At one time, an inch was defined as the width of a thumb, and a foot was simply the length of a man's foot. Although it is still somewhat common to hear some of these terms used today, dimensions are stated somewhat differently. One major difference is in the adoption of the standardized dimensioning units we currently use. This use came about because of the relatively recent rapid growth of worldwide science, technology, and commerce—all of which has fostered an international system of units (SI) we use today.

6.4.1 DIMENSIONING

As mentioned, technical drawings consist of several types of lines that are used singly or in combination with each other to describe the shape and internal construction of an object or mechanism; however, to rebuild a machine or remachine or reproduce a part, the blueprint or drawing must include dimensions that indicate exact sizes and locations of surfaces, indentations, holes, and other details. Stated differently, in addition to a complete shape description of an object, a technical drawing of the object must also give a complete size description; that is, it must be *dimensioned*.

In the early days of industrial manufacturing, products were typically produced under one roof, often by one individual, using parts and subassemblies manufactured on the premises. Today, most major industries do not manufacture all of the parts and subassemblies in their products. Frequently, the parts are manufactured by specialty industries to standard specifications or to specifications provided by the major industry.

Important Point: The key to successful operation of the various parts and subassemblies in the major product is the ability of two or more nearly identical duplicate parts to be used individually in an assembly and function satisfactorily.

The modern practice of utilizing interchangeable parts is the basis for the development of widely accepted methods for size description. Drawings today are dimensioned so machinists in widely separated places can make mating parts that will fit properly when brought together for final assembly in the factory or when replacement parts are used to make repairs to plant equipment. In today's modern wastewater treatment plant, the responsibility for size control has shifted from the maintenance operator to the draftsperson. The operator no longer exercises judgment in engineering matters, but only in the proper execution of instructions given on the drawings.

Technical drawings show the object in its completed condition and contain all necessary information to bring it to its final state. A properly dimensioned drawing takes into account the shop processes required to finish a piece and the function of the part in assembly. Moreover, shop drawings are dimensioned for convenience for the shop worker or maintenance operator. These dimensions are given so it is not necessary to calculate, scale, or assume any dimensions.

Designers and draftspersons provide dimensions that are neither irrelevant nor superfluous. Only those dimensions are given that are needed to produce and inspect the part exactly as intended by the designer. More importantly,

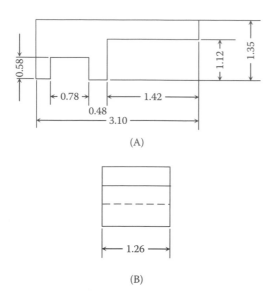

FIGURE 6.34 Size dimensions: (A) view with two size dimensions; (B) third side dimension on this view.

only those dimensions are given that are needed by the maintenance operator who may have to rely on the blueprint (usually as a last resort) to determine the exact dimensions of the replacement part.

The meaning of various terms and symbols and conventions used in shop notes as well as procedures and techniques relating to dimensioning are presented in this chapter to assist the operator in accurately interpreting plant blueprints; however, before defining these important terms we first discuss decimal and size dimensions.

6.4.2 Decimal and Size Dimensions

Dimensions may appear on blueprints as decimals, usually two-place decimals (normally given in even hundredths of an inch). In fact, it is common practice to use two-place decimals when the range of dimensional accuracy of a part is between 0.01 in. larger or smaller than nominal size (specified dimension). For more precise dimensions (e.g., dimensions requiring machining accuracies in thousandths or ten thousandths of an inch), three- and four-place decimal dimensions are used. Every solid object has three size dimensions: depth (or thickness), length (or width), and height. In the case of the object shown in Figure 6.34, two of the dimensions are placed on the principal view and the third dimension is placed on one of the other views.

6.4.3 Definition of Dimensioning Terms

An old Chinese proverb states: "The beginning of wisdom is to call things by their right names." This statement is quite fitting because to satisfactorily read and interpret blueprints it is necessary to understand the terms relating to conditions and applications of dimensioning.

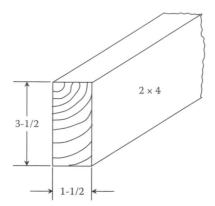

FIGURE 6.35 Nominal size of a construction 2 × 4.

6.4.3.1 Nominal Size

Nominal size is the designation that is used for the purpose of general identification. It may or may not express the true numerical size of the part or material. For example, the standard 2 × 4 stud used in building construction has an actual size of 1-1/2 × 3-1/2" (see Figure 6.35). In the case of the hole and shaft shown in Figure 6.36, however, the nominal size of both the hole and the shaft is 1-1/4", which would be 1.25" in a decimal system of dimensioning. So, again, it may be seen that the nominal size may or may not be the true numerical size of a material.

Important Point: When the term *nominal size* is used synonymously with *basic size*, we are to assume the exact or theoretical size from which all limiting variations are made.

6.4.3.2 Basic Size

Basic size (or dimension) is the size of a part determined by engineering and design requirements. More specifically, it is the theoretical size from which limits of size are derived by the application of allowances and tolerances; that is, it is the size from which limits are determined for the size, shape, or location of a feature. For example, strength and stiffness may require a 1-in.-diameter shaft. The basic 1-in. size (with tolerance) will most likely be applied to the hole size because allowance is usually applied to the shaft (see Figure 6.37).

6.4.3.3 Allowance

Allowance is the designed difference in the dimensions of mating parts to provide for different classes of fit. Simply, it is the minimum clearance space (or maximum interference) of mating parts; consequently, it represents the tightest permissible fit and is simply the smallest hole minus the largest shaft. Recall that in Figure 6.37 we allowed .003 on the shaft for clearance (1.000 − .003 = .997) (see Figure 6.38).

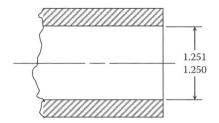

FIGURE 6.36 Nominal size 1.25".

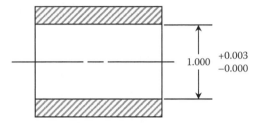

FIGURE 6.37 Basic size.

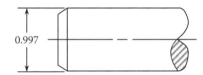

FIGURE 6.38 Design size (after application of allowance).

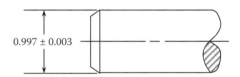

FIGURE 6.39 Design size (after allowance and tolerance are applied).

6.4.3.4 Design Size

Design size is the size of a part after an allowance for clearance has been applied and tolerances have been assigned. The design size of the shaft in Figure 6.38 is .997 after the allowance of .003 has been made. A tolerance of ±.003 is assigned after the allowance is applied (see Figure 6.39).

Important Point: After defining basic and design size, the reader may be curious as to what the definition of *actual size* is. Actual size is simply the measured size.

6.4.3.5 Limits

Limits are the maximum and minimum sizes indicated by a toleranced dimension; for example, the design size of a part may be 1.435. If a tolerance of plus or minus two thousandths (±.002) is applied, then the two limit dimensions are maximum limit 1.437 and minimum limit 1.433 (see Figure 6.40).

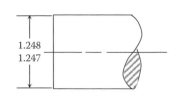

FIGURE 6.40 Tolerance expressed by limits.

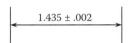

FIGURE 6.41 Design size with tolerance.

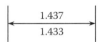

FIGURE 6.42 One tolerance value given.

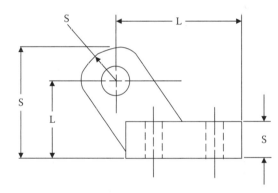

FIGURE 6.43 Size and location dimensions.

6.4.3.6 Tolerance

Tolerance is the total amount by which a given dimension may vary or the difference (variation) between limits (as shown in Figure 6.40). Tolerance should always be as large as possible, other factors considered, to reduce manufacturing costs. It can also be expressed as the design size followed by the tolerance (see Figure 6.41). Moreover, tolerance can be expressed when only one tolerance value is given, as the other value is assumed to be zero (see Figure 6.42). Tolerance is also applied to *location dimensions* for other features (e.g., holes, slots, surfaces) of a part (see Figure 6.43). (Note that location dimensions are usually made from either a center line or a finished surface; this practice is followed to overcome inaccuracies due to variations caused by surface irregularities.) Because

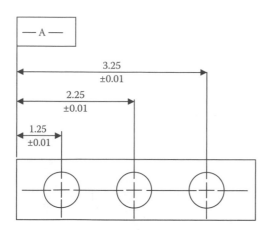

FIGURE 6.44 Dimensioning datum surface.

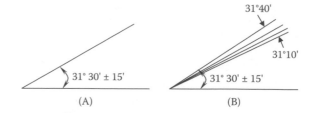

FIGURE 6.45 Angular dimensions and tolerance.

Important Point: In the construction and structural industries, linear dimensions are given in feet, inches, and common fractional parts of an inch.

6.4.4.2 Angular Dimensions

Angular dimensions are used on blueprints to indicate the size of angles in degrees (°) and fractional parts of a degree, minutes (') and seconds ("). Each degree is 1/360 a circle. There are 60 minutes (') in each degree. As mentioned, each minute may be divided into smaller units called seconds. There are 60 seconds (") in each minute. For example, 15 degrees, 12 minutes, and 45 seconds can be written 15°12'45". Current practice is to use decimalized angles. To convert angles given in whole degrees, minutes, and seconds, the following example should be followed.

■ **EXAMPLE 6.1**

Problem: Convert 15°12'45" into decimal degrees.

Solution:

1. Convert minutes into degrees by dividing by 60 (60' = 1°):

$$12 \div 60 = .20°$$

2. Convert seconds into degrees by dividing seconds by 3600:

$$(3600° = 1)$$

$$45" \div 3600 = .01°$$

3. Add whole degrees plus decimal degrees:

$$15° + .20° + .01° = 15.21°$$

Therefore, 15°12'45" = 15.21° decimal degrees

The size of an angle with the tolerance may be shown on the angular dimension itself (see Figure 6.45). The tolerance may also be given in a note on the drawing, as shown in Figure 6.46.

the size of the shaft shown in Figure 6.38 is 0.997 after the allowance has been applied, the tolerance applied must be below this size to ensure the minimum clearance (allowance) of .003. If a tolerance of ±.003 is permitted on the shaft, the total variation of .006 (+.003 and −.003) must occur between .997 and .994. Then, the design of the shaft is given a *bilateral tolerance*, where variation is permitted in both directions from the design size, as shown in Figure 6.41, vs. *unilateral tolerance*, where variation is permitted only in one direction from the design size, as shown in Figure 6.42.

Important Point: Tolerances may be specific and given with the dimension value or general and given by means of a printed note in or just above the title block.

6.4.3.7 Datum

A *datum* (a point, axis, or plane) identifies the origin of a dimensional relationship between a particular (designated) point or surface and a measurement; that is, it is assumed to be exact for purpose of reference and is the origin from which the location or geometric characteristic of features or a part are established. The datum is indicated by the assigned letter preceded and followed by a dash, enclosed in a small rectangle or box (see Figure 6.44).

6.4.4 Types of Dimensions

The types of dimensions include linear, angular, reference, tabular, and arrowless dimensions. Each of these is discussed in the following sections.

6.4.4.1 Linear Dimensions

Linear dimensions are typically used in aerospace, automotive, machine tool, sheet metal, electrical, electronics, and similar industries. Linear dimensions are usually given in inches for measurements of 72" and under and in feet and inches if greater than 72".

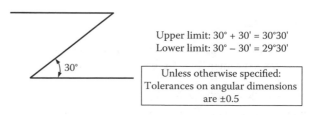

Upper limit: 30° + 30' = 30°30'
Lower limit: 30° − 30' = 29°30'

Unless otherwise specified:
Tolerances on angular dimensions are ±0.5

FIGURE 6.46 Tolerance specified as a note.

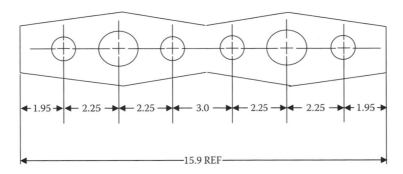

FIGURE 6.47 Dimensions for reference only.

Part No.	A	B	C	D	E
69-3705	0.650	1.00	1.250	0.158	0.890
69-3706	0.760	1.200	1.820	0.384	1.00
69-3707	0.800	1.300	2.160	0.496	1.115

FIGURE 6.48 Example table used for dimensioning a series of sizes.

6.4.4.3 Reference Dimensions

Reference dimensions are occasionally given on drawings for reference and checking purposes; they are given for information only. They are not intended to be measured and do not govern the shop operations. They represent the calculated dimensions and are often useful in showing the intended design size. Reference dimensions are marked by parentheses or followed by the notation "REF" (see Figure 6.47).

6.4.4.4 Tabular Dimensions

Tabular dimensions are used when a series of objects having like features but varying in dimensions may be represented by one drawing. Letters are substituted for dimension figures on the drawing, and the varying dimensions are given in tabular form (see Figure 6.48).

6.4.4.5 Arrowless Dimensions

Arrowless dimensions are frequently used on drawings that contain datum lines or planes (see Figure 6.49). This practice improves the clarity of the drawing by eliminating numerous dimension and extension lines.

6.4.5 Shop Notes

To convey the information the machinist needs to make a part, the draftsperson typically uses notes. Notes such as those used for reaming, counterboring, drilling, or countersinking holes are added to ordinary dimensions. The order of items in a note corresponds to the order of procedures in the shop for producing the hole. Two or more

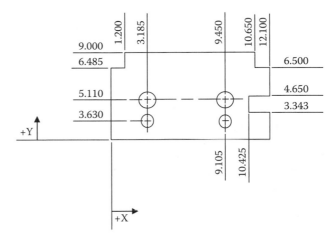

FIGURE 6.49 Application of arrowless dimensioning.

holes are dimensioned by a single note, the leader pointing to one of the holes. A note may consist of a very brief statement at the end of a leader, or it may be a complete sentence that gives an adequate picture of machining processes and all necessary dimensions. On drawings of parts to be produced in large quantities for interchangeable assembly, dimensions and notes may be given without specification of the shop process to be used. A note is placed on a drawing near the part to which it refers.

Important Point: Notes should always be lettered horizontally on the drawing paper, and guide lines should always be used.

6.5 MACHINE DRAWINGS

Your ability to work with machines depends on your ability to understand them. If water and wastewater maintenance operators know how the parts of a machine fit together and how they are intended to work together, they are better able to operate the machine, perform proper preventive maintenance on it, and repair it when it breaks down. At larger plants, no one can possibly know all the details of every machine or machine tool, as too many variations exist among machines of the same type. All centrifugal pumps, for example, do the same basic work, but the many different manufacturers of centrifugal pumps each has several different models and sizes. It would be difficult to understand all of the different centrifugal pumps without being able to read blueprints or basic machine drawings (such as those depicted in this chapter).

We have chosen, along with a standard centrifugal pump, a submersible pump, a turbine pump, and a simple solid packed stuffing box assembly as the machines to refer to for our discussion on how to read simple machine drawings. We chose these machines and the stuffing box assembly because they are among the most familiar of all machines and assemblies in water and wastewater treatment operations. The reader will see how the parts that make up various pumps and pump mechanisms are represented on simplified assembly drawings. Understanding these simplified drawings will add an important tool to the reader's toolbox.

6.5.1 The Centrifugal Pump Drawing (Simplified)

Figure 6.49 shows a simplified assembly drawing of a standard centrifugal pump used in water and wastewater treatment, collections, and distribution.

6.5.1.1 The Centrifugal Pump

The *centrifugal pump* is the most widely used type of pumping equipment in the water and wastewater industries. Pumps of this type are capable of moving high volumes of water in a relatively efficient manner. The centrifugal pump is very dependable, has relatively low maintenance requirements, and can be constructed out of a wide variety of materials. The centrifugal pump is available in a wide range of sizes with capacities ranging from

a few gallons per minute up to several thousand pounds per square inch (Cheremisinoff and Cheremisinoff, 1989). The centrifugal pump is considered to be one of the most dependable systems available for water and wastewater liquid transfer.

6.5.1.2 Centrifugal Pump: Description

The centrifugal pump consists of a rotating element (impeller) sealed in a casing (volute). The rotating element is connected to a drive unit or prime mover (motor/engine) that supplies the energy to spin the rotating element. As the impeller spins inside the volute casing, an area of low pressure is created in the center of the impeller. This pressure allows the atmospheric pressure on the water in the supply tank to force the water up to the impeller. (Note that use of the term *water* in our discussion includes both fresh, or potable, water and wastewater, unless otherwise specified.) Because the pump will not operate if no low-pressure zone is created at the center of the impeller, it is important that the casing be sealed to prevent air from entering the casing. To ensure that the casing is airtight, the pump includes some type of seal (mechanical or conventional packing) assembly at the point where the shaft enters the casing. This seal also includes some type of lubrication (water, grease, or oil) to prevent excessive wear.

When water enters the casing, the spinning action of the impeller transfers energy to the water. This energy is transferred to the water in the form of increased speed or velocity. The water is thrown outward by the impeller into the volute casing, where the design of the casing allows the velocity of the water to be reduced, which, in turn, converts the velocity energy (velocity head) to pressure energy (pressure head). The water then travels out of the pump through the pump discharge. The major components of the centrifugal pump are shown in Figure 6.50.

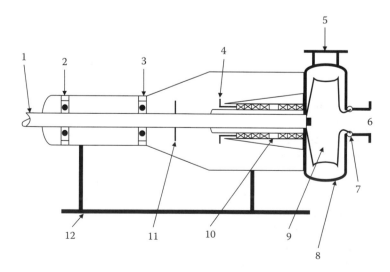

FIGURE 6.50 Major components of a centrifugal pump.

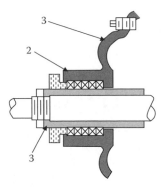

FIGURE 6.51 Solid packed stuffing box.

6.5.1.3 Centrifugal Pump: Components

Figure 6.50 shows a simplified representation of the major components that make up a standard centrifugal pump. This is the type of drawing typically used to train maintenance operators in the classroom on the centrifugal pump. It also serves as a basic shop assembly drawing showing how each of the components is related relative to each other. Figure 6.50 shows all the parts that are typically shown in contact with one another in their assembled positions. Notice that all of the key components shown in Figure 6.50 are numbered. For this type of drawing, this is standard practice. Usually, along with the view of the numbered components, a key is provided on the drawing that identifies each numbered part. For the purpose of simplification, each component of the centrifugal pump in Figure 6.50 is numbered as follows:

1. Shaft
2. Thrust bearing
3. Radial bearing
4. Packing gland
5. Discharge
6. Suction
7. Impeller wear ring
8. Volute
9. Impeller
10. Stuffing box
11. Slinger ring
12. Pump frame

6.5.2 PACKING GLAND DRAWING

The pump packing gland is part of the stuffing box and seal assembly. Such sealing devices are used on pumps to prevent water leakage along the pump driving shaft (component 1 in Figure 6.50 and component 3 in Figure 6.51). Shaft sealing devices must control water leakage without causing wear to the pump shaft. Two systems are available to accomplish this seal: the conventional stuffing box–packing assembly and the mechanical seal assembly.

Important Point: We have included an elementary drawing of a packing gland in this text (see Figure 6.51) because it is a critical pump component with which all maintenance operators become familiar sooner rather than later in their tenure.

The *stuffing box* of a centrifugal pump is a cylindrical housing, the bottom of which may be the pump casing, a separate throat bushing attached to the stuffing box, or a bottoming ring. Stuffing boxes for pumps used in water/wastewater treatment processes are available in a number of designs. Generally, at the top of the stuffing box is a *packing gland* (see Figure 6.51). The gland encircles the pump shaft sleeve and is cast with a flange that slips securely into the stuffing box. Stuffing box glands are manufactured as a single piece split in half and held together with bolts. The advantage of the split gland is being able to remove it from the pump shaft without dismantling the pump. In Figure 6.51, each numbered component is identified as follows:

1. Pump housing
2. Packing gland
3. Shaft sleeve

6.5.3 SUBMERSIBLE PUMP DRAWING (SIMPLIFIED)

The *submersible pump* is another machine familiar to most water or wastewater maintenance operators, as it is used extensively in both industries. The submersible pump is, as the name suggests, placed directly in deep wells and pumping station wet wells. In some cases, only the pump is submerged; in other cases, the entire pump–motor assembly is placed in the well or wet well. A simplified drawing of a typical submersible pump is shown in Figure 6.52, where each numbered component is identified as follows:

1. Electrical connection
2. Drop pipe
3. Inlet screen
4. Electric motor
5. Check valve
6. Inlet screen
7. Bowls and impellers

6.5.4 TURBINE PUMP DRAWING (SIMPLIFIED)

The *turbine pump* is another familiar type of pump. It consists of a motor, drive shaft, a discharge pipe of varying lengths, and one or more impeller–bowl assemblies. It is normally a vertical assembly in which the water enters at the bottom, passes axially through the impeller–bowl assembly where the energy transfer occurs, then moves upward through additional impeller–bowl assemblies to the discharge pipe. The length of this discharge pipe will vary with the distance from the wet well to the desired

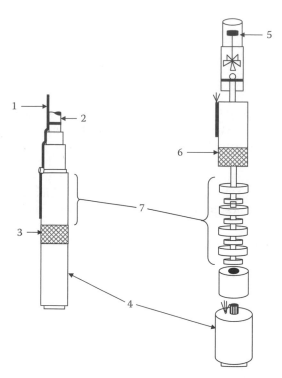

FIGURE 6.52 Submersible pump.

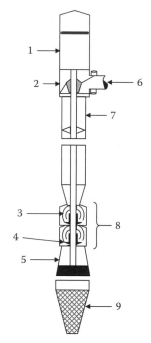

FIGURE 6.53 Vertical turbine pump.

point of discharge; see Figure 6.53, where each numbered component is identified as follows:

1. Hollow motor shaft
2. Stuffing box
3. Bowl
4. Impeller
5. Suction bell

6. Discharge head
7. Driving shaft
8. Pump unit
9. Screen

6.6 SHEET METAL DRAWINGS

Some water and wastewater maintenance work may involve making simple pieces of equipment for the plant. Much of this equipment is made from sheet metal. To use the sheet metal correctly, maintenance operators need to know how to read basic drawings.

6.6.1 SHEET METAL

Thin-gauged metals, such as sheet metal, which is made from sheet stock (metal that has been rolled into sheet stock), are used in water and wastewater treatment operations to fabricate many objects and devices. Examples include safety guards, shelves, machinery cover plates, brackets for tools and parts, and ducts for heating and air conditioning units.

Important Point: Wide rolls of sheet stock (steel, aluminum, copper, and brass) are called *coils*. When sheet stock is cut into rectangular sections, it is referred to as *sheet metal*.

Sheet metal is either plain metal with very little surface protection or metal covered with a thin protective coat of zinc (galvanized) to prevent rusting. Typically, in water/wastewater treatment operations, either galvanized or aluminum sheet metal is used. Sheet metal drawings tell maintenance operators what they need to know to fabricate various kinds of objects and devices. Typically, the calculations and layout on the drawings are exact. On many occasions, the metal is machined in a flat position and then folded or assembled, so the relationship and location of the resulting planes and features must be within their specified tolerances (see Figure 6.54).

6.6.2 DIMENSION CALCULATIONS

As mentioned, usually blueprints are dimensioned in the flat layout form by the draftsperson to achieve the desired dimension of the device and its features after it is folded or assembled. However, as most maintenance operators know, there are exceptions to this practice. In the event that dimensions must be calculated in the shop or field, it is helpful to know how to make these calculations (Brown, 1989).

6.6.2.1 Calculations for Allowances in Bend

Calculations for allowances in bend radii may be obtained in two ways: (1) directly from a set-back table or (2) by use of a mathematical formula. *Set-back* tables are available from sheet metal suppliers.

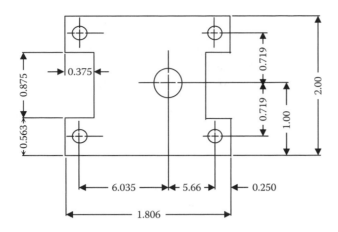

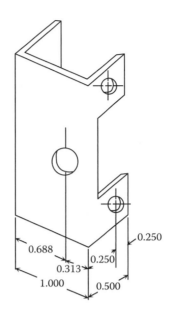

FIGURE 6.54 Sheet metal part.

Notes
1. Material 0.097 sheet metal
2. Tolerance ±0.010
3. Bend radii = 1.32

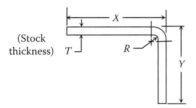

FIGURE 6.55 Diagram for calculating developed length from set-back table.

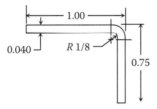

FIGURE 6.56 Illustration for Example 6.2.

6.6.2.2 Set-Back Table

The diagram (from which the formula $X + Y - Z$ is derived) shown in Figure 6.55 shows the application of the set-back value to calculate the length in the flat length (developed length; see note below) to produce desired folded size of a sheet metal part or device.

Important Point: Laying out a three-dimensional shape on a flat surface is a technique called *development*. To make something out of sheet metal, we must first lay out

the development on a flat sheet known as a *template*. We then place the template on a piece of sheet metal, cut the metal to the proper shape, and bend it to form the three-dimensional piece we want.

■ EXAMPLE 6.2

Problem: Using Figure 6.56, find the developed length.

Solution:

$$\text{Developed length} = X + Y - Z$$
$$= 1.00 + .75 - .106$$
$$= 1.75 - .106$$
$$\text{Developed length} = 1.64 \text{ (rounded)}$$

6.6.2.3 Formulae Used to Determine Developed Length

One of two formulae can be used to calculate the developed length. The formula used first calculates the *lineal length* of sheet metal parts depending on the size of the bend radius and the thickness of the metal, then the developed length is calculated. Figure 6.57 shows the diagram used for calculating the developed length. As mentioned, two formulae can be used to make this calculation.

When R is less than twice the stock thickness:

$$A = 1/2\pi(R + .4T) \tag{6.1}$$

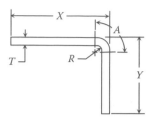

FIGURE 6.57 Diagram used for calculating developed length using formulae.

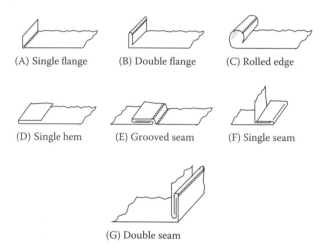

(A) Single flange   (B) Double flange   (C) Rolled edge

(D) Single hem   (E) Grooved seam   (F) Single seam

(G) Double seam

FIGURE 6.58 Examples of sheet metal hems and joints.

When the bend radius is more than twice the stock thickness:

$$A = 1/2\pi(R + .5T) \tag{6.2}$$

where:

A = lineal length of a 90° bend.
R = inside radius.
T = material thickness.

After determining A (lineal bend length), the developed length of the part or device is determined using the following formula (refer to the diagram in Figure 6.57):

$$\text{Developed length} = X + Y + A - (2R + 2T) \tag{6.3}$$

Important Point: The two formulae shown above are provided for informational purposes only. For our purposes, the same answer can be obtained in shorter time using a sheet metal set-back table. In the shop or field (i.e., in actual practice), the set-back table is most commonly used.

6.6.3 HEMS AND JOINTS

Keeping in mind that the development of a sheet metal surface is that surface laid out on a plane, a wide variety of hems and joints are used in the fabrication of these sheet metal developments (see Figure 6.58). *Hems* are used to eliminate the raw edge and also to stiffen the material. The need to eliminate raw edges and stiffen sheet metal can, for example, best be seen in the use of sheet metal to fabricate ventilation ducts. To ensure efficient ventilation operation, the ductwork through which the air is conveyed must be smooth to reduce turbulence and mechanically strong enough to stand up to internal and external forces. In the development of sheet metal surfaces, *joints* (seams) may be made by bending, welding, riveting, or soldering.

Important Point: In the fabrication of sheet metal developments, sufficient material as required for hems and joints must be added to the layout or development. The amount of allowance depends on the thickness of the material and the production equipment.

6.7 HYDRAULIC AND PNEUMATIC DRAWINGS

Hydraulic and pneumatic power systems are widely used in water and wastewater treatment operations. They operate small tools and large machines. Maintenance operators must know how these fluid power systems work to be able to repair and maintain them; however, before attempting to understand and service fluid power systems, it is necessary to also know how to read hydraulic and pneumatic drawings. Although it is not our purpose in this section to make fluid mechanics experts out of anyone (e.g., able to discuss such fluid principles as Pascal's law and multiplying forces), we do intend to explain the basics of hydraulic and pneumatic systems.

Many water and wastewater machines are operated by *hydraulic* and *pneumatic* power systems. These systems transmit forces through a *fluid* (defined as either a gas or a liquid). Power plungers, power bar screen assemblies, fluid-drive transmissions, hydraulic lifts and racks, air brakes, and various heavy-duty power tools are examples of hydraulic and pneumatic power systems. Air-powered drills and grinders are tools that operate on *compressed air*. Larger pneumatic devices are used on larger machines. Because all of these mechanisms must be maintained and repaired, we need to know how they operate. Moreover, we must also be able to read and interpret the drawings that show the construction of these systems.

6.7.1 STANDARD HYDRAULIC SYSTEM

A standard hydraulic system operates by means of a *liquid* (hydraulic fluid) under pressure. A basic hydraulic system has five components:

- *Reservoir*—Provides storage space for the liquid
- *Pump*—Provides pressure to the system
- *Piping*—Directs fluid through the system

- *Control valve*—Controls the flow of fluid
- *Actuating unit*—Reacts to the pressure and does some kind of useful work

Because the hydraulic fluid never leaves the system, it is a *closed system*. The hydraulic system on a plant forklift is an example of a hydraulic system.

6.7.2 STANDARD PNEUMATIC SYSTEM

A standard pneumatic system operates by means of a *gas* under pressure. The gas is usually dry air. A basic pneumatic system is very much like the hydraulic system described above and includes the following main components:

- *Atmosphere*—Serves the same function as the reservoir of the hydraulic system
- *Intake pipe and filter*—Provides a passage for air to enter the system
- *Compressor*—Compresses the air, putting it under pressure; its counterpart in the hydraulic system is the pump
- *Receiver*—Stores pressurized air until it is needed; it helps provide a constant flow of pressurized air in situations where air demand is high or varies
- *Relief valve*—Can be set to open and bleed off some of the air if the pressure becomes too high
- *Pressure-regulating valve*—Ensures that the air delivered to the actuating unit is at the proper pressure
- *Control valve*—Provides a path for the air to the actuating unit

Pneumatic systems are usually *open* systems (i.e., the air leaves the system after it is used).

6.7.3 HYDRAULIC AND PNEUMATIC SYSTEMS: SIMILARITIES AND DIFFERENCES

With regard to their similarities, both hydraulic and pneumatic systems use a pressure-building source, which can be either a pump or a compressor. They both also require either a reservoir or a receiver to store the fluid. In addition, they require valves and actuators and lines to connect these components in a system. The motion that results from the actuator may be either straight line (linear) or circular (rotary). With regard to their differences, we must note an important difference between a liquid (for hydraulic systems) and a gas (for pneumatic systems). A liquid is difficult to compress; water, for example, cannot be compressed into a space that is noticeably smaller in size. On the other hand, a gas is easy to compress; for example, a large volume of air from the atmosphere can be compressed into a much smaller volume.

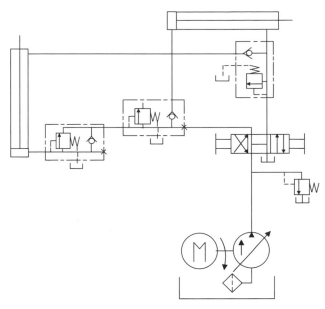

FIGURE 6.59 Graphic drawing for a fluid power system.

6.7.4 TYPES OF HYDRAULIC AND PNEUMATIC DRAWINGS

Several types of drawings are used in showing instrumentation, control circuits, and hydraulic or pneumatic systems. These include graphic, pictorial, cutaway, and combination drawings:

- *Graphic drawings* consist of graphic symbols joined by lines that provide an easy method of emphasizing functions of the system and its components (see Figure 6.59).
- *Pictorial drawings* are used when piping is to be shown between components (see Figure 6.60).
- *Cutaway drawings* consist of cutaway symbols of components and emphasize component function and piping between components (see Figure 6.61).
- *Combination drawings* utilize in one drawing the type of component illustration that best suits the purpose of the drawing (see Figure 6.62).

The emphasis in the following discussion is on graphic diagrams, as these are the most widely used in water and wastewater operations and the graphic symbols have been standardized.

6.7.5 GRAPHIC SYMBOLS FOR FLUID POWER SYSTEMS

You may or may not have had difficulty reading any of these four types of drawings shown in Figure 6.59 through Figure 6.62. In contrast, unless you were familiar with the basic symbols used in the drawings, you probably did have some difficulty understanding them.

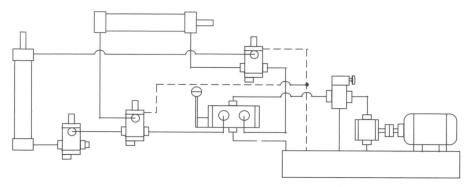

FIGURE 6.60 Pictorial drawing for a fluid power system.

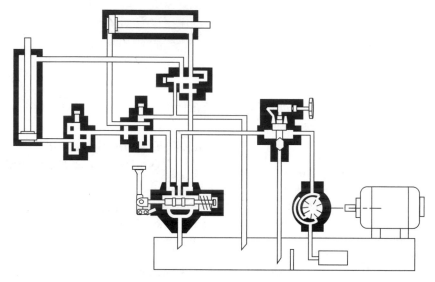

FIGURE 6.61 Cutaway drawing of a fluid power system.

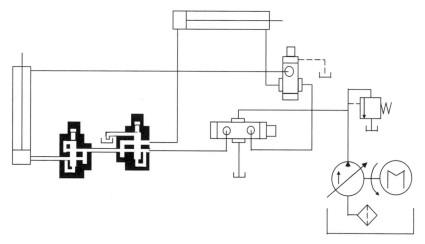

FIGURE 6.62 Combination drawing of a fluid power system.

6.7.5.1 Symbols for Methods of Operation (Controls)

Figure 6.63 shows the standard graphic symbols used in hydraulic and pneumatic system diagrams for methods of operation (controls).

6.7.5.2 Symbols for Rotary Devices

Figure 6.64 shows the standard graphic symbols used in hydraulic and pneumatic system diagrams for rotary devices, such as pumps, motors, oscillators, and internal combustion engines.

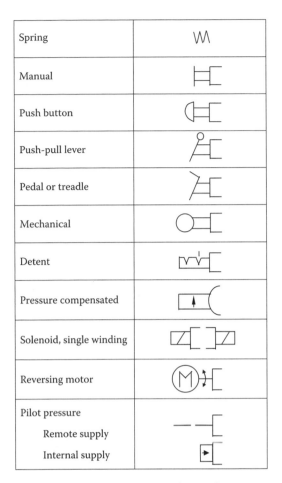

FIGURE 6.63 Symbols for methods of operation.

6.7.5.3 Symbols for Lines

Figure 6.65 shows the standard graphic symbols used in hydraulic and pneumatic system diagrams for system lines.

6.7.5.4 Symbols for Valves

Figure 6.66 shows the standard graphic symbols used in hydraulic and pneumatic system diagrams for valves.

6.7.5.5 Symbols for Miscellaneous Units

Figure 6.67 shows the standard graphic symbols used in hydraulic and pneumatic system diagrams for miscellaneous units, such as energy storage and fluid storage devices.

6.7.6 SUPPLEMENTARY INFORMATION ACCOMPANYING GRAPHIC DRAWINGS

Once we have become familiar with the symbols used in graphic drawings, the drawings are relatively easy to read and understand. In addition to the graphic drawing, prints of hydraulic and pneumatic systems usually include a

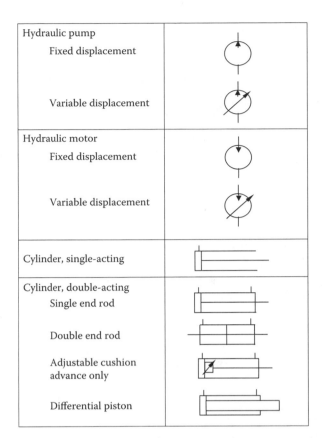

FIGURE 6.64 Symbols for rotary devices.

listing of the sequence of operations, solenoid chart, and parts used to facilitate understanding of the function and purpose of the system and its components.

6.7.6.1 Sequence of Operations

A listing of the sequence of operations is an explanation of the various functions of the system explained in order of occurrence. Each phase of the operation is numbered or lettered and a brief description is given of the initiating and resulting action.

Important Point: A listing of the sequence of operations is usually given in the upper part of the print or on an attached sheet.

6.7.6.2 Solenoid Chart

If solenoids are used in the instrumentation or control circuits of a hydraulic or pneumatic system, a chart is normally located in the lower left corner of the print to help explain the operation of the electrically controlled circuit.

Important Point: Solenoids are usually given a letter on the drawings and the chart shows where the solenoids are energized (+) or de-energized (−) at each phase of system operation.

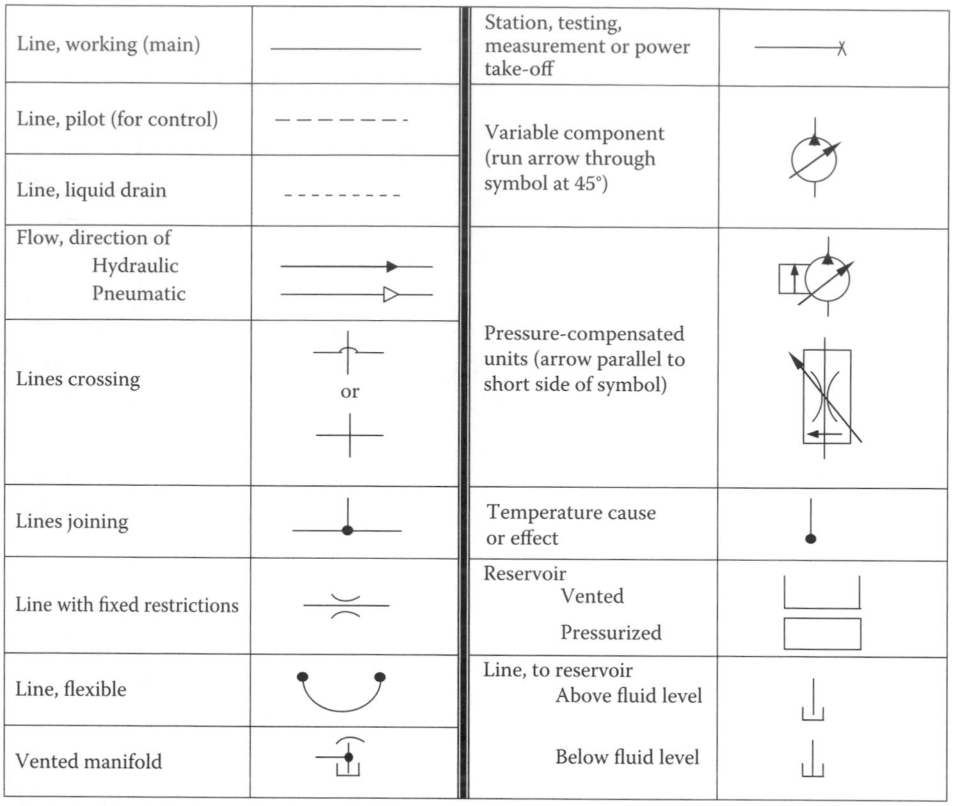

Line, working (main)		Station, testing, measurement or power take-off	
Line, pilot (for control)		Variable component (run arrow through symbol at 45°)	
Line, liquid drain			
Flow, direction of Hydraulic Pneumatic			
Lines crossing	or	Pressure-compensated units (arrow parallel to short side of symbol)	
Lines joining		Temperature cause or effect	
Line with fixed restrictions		Reservoir Vented Pressurized	
Line, flexible		Line, to reservoir Above fluid level	
Vented manifold		Below fluid level	

FIGURE 6.65 Symbols for lines.

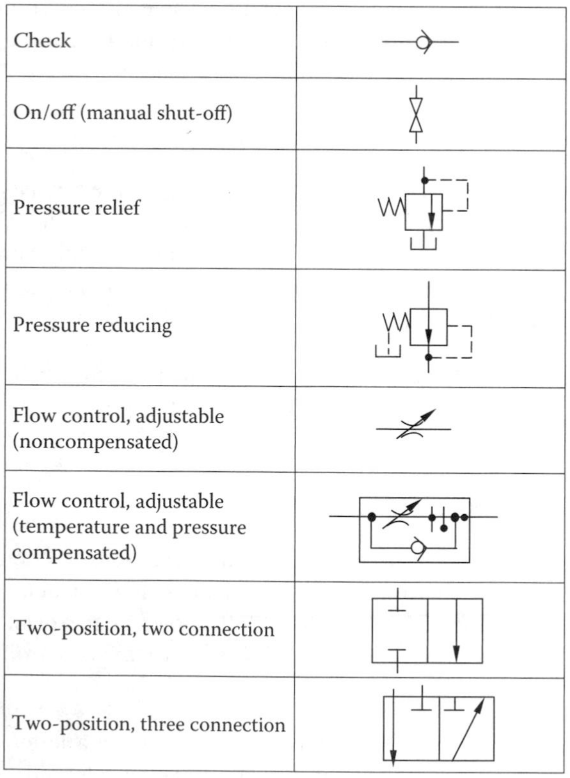

Check	
On/off (manual shut-off)	
Pressure relief	
Pressure reducing	
Flow control, adjustable (noncompensated)	
Flow control, adjustable (temperature and pressure compensated)	
Two-position, two connection	
Two-position, three connection	

FIGURE 6.66 Symbols for valves.

6.7.6.3 Bill of Materials

The bill of materials, sometimes referred to as a *component* or *parts list*, includes an itemized list of the several parts of a structure or device shown on a graphic detail drawing or a graphic assembly drawing. This parts list usually appears right above the title block; however, this list is also often given on a separate sheet.

Important Point: The title strip alone is sufficient on graphic detail drawings of only one part, but a parts list is necessary on graphic detail drawings of several parts.

Parts lists on machine drawings contain the part numbers or symbols, a descriptive title of each part, the quantity required, the material specified, and frequently other information such as pattern numbers, stock sizes of materials, and weights of parts.

Important Point: Parts are listed in general order of size or importance; for example, the main castings or forgings are listed first, parts cut form cold-rolled stock second, and standard parts such as bushings and roller bearings third.

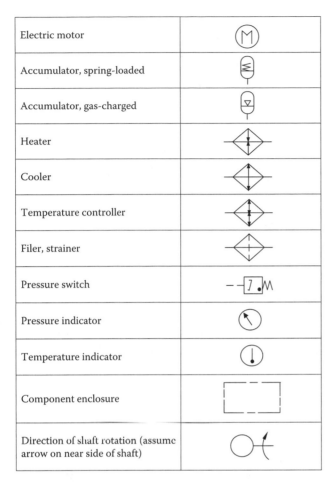

Electric motor	
Accumulator, spring-loaded	
Accumulator, gas-charged	
Heater	
Cooler	
Temperature controller	
Filer, strainer	
Pressure switch	
Pressure indicator	
Temperature indicator	
Component enclosure	
Direction of shaft rotation (assume arrow on near side of shaft)	

FIGURE 6.67 Miscellaneous symbols.

6.8 WELDING BLUEPRINTS AND SYMBOLS

Water and wastewater maintenance operators may be called upon to perform welding operations. In many cases, the operator/welder is required to do nothing more than tack pieces of metal together—a very basic operation; however, occasionally a welding job must be performed precisely according to the specifications of the designer. Obviously, the strength and durability of the piece to be welded depend on the ability of the operator/welder to make the welds properly.

The designer of a part communicates the welding specifications by means of blueprints or drawings. Special symbols on the drawing provide all the information required concerning the preparation of the parts to be welded, the kind of welding to be performed, which side to weld from, how to shape the weld, and how to finish the welded surface. Operator/welders who cannot understand all of the information contained in the symbol will not be able to perform the welds properly. Because welding is used so extensively in water/wastewater operations for so large a variety of purposes, it is essential to have an accurate method of showing the exact types, sizes, and locations of

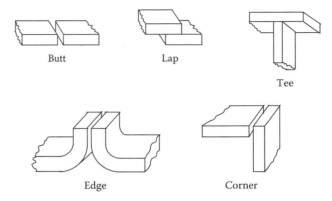

FIGURE 6.68 Basic weld joints.

welds on the working drawings of machines and plant equipment. In the past, many parts were cast in foundries. These parts are now being constructed by welding. To provide a means for placing complete welding information on the drawing in a simple manner, a system of welding symbols was developed by the ASW in conjunction with ANSI (ANSI/AWS, 1991). The welding symbols included in this chapter will assist in reading and interpreting blueprints and drawings involving welding processes.

6.8.1 WELDING PROCESSES

The American Welding Society defines *welding* as "a joining process that produces coalescence of materials by heating them to the welding temperature, with or without the application of pressure or by the application of pressure alone, and with or without the use of filler metal" (AWS, 1996). Three of the principal methods of welding are the oxyacetylene method, generally known as *gas welding*; the electric-arc method, generally known as *arc welding*; and *electric-resistance welding*. The first two are the most widely used welding processes for maintenance welding; the high temperatures necessary for fusion are obtained with a gas flame in oxyfuel welding and with an electric arc in arc welding. Electric-resistance welding, generally referred to as *resistance welding*, requires electricity and the application of pressure to make welded joints. Resistance welding is primarily a production welding operation; it is rarely used in water/wastewater maintenance operations.

6.8.2 TYPES OF WELDED JOINTS

A welded joint is the union of two or more pieces of metal by means of a welding process. Five basic types of welded joints are specified on drawings. Each type of joint is identified by the position of the parts to be joined together. Parts that are welded by using butt, corner, tee, lap, or edge joints are illustrated in Figure 6.68. Each of these joints has several variations.

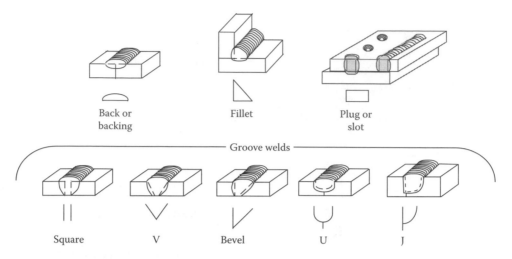

FIGURE 6.69 Arc and gas welds and symbols.

Important Point: Various types of welds are applicable to each type of joint, depending on the thickness of metal, the strength of joint required and other considerations.

6.8.2.1 Butt Joints

Butt joints join the edges of two metals that are placed against each other end to end in the same plane. The joint is reasonably strong in static tension but is not recommended when it is subjected to fatigue or impact loads, especially at low temperatures. The preparation of the joint is relatively simple as it requires only matching the edges of the plates; consequently, the cost of making the joint is low. These joints are frequently used for plate, sheet metal, and pipe work (see Figure 6.68).

6.8.2.2 Lap Joints

A *lap joint*, as the name implies, is made by connecting overlapping pieces of metal, which are often part of a structure or assembly. The lap joint is popular because it is strong and easy to weld. Moreover, special beveling or edge preparations are seldom necessary. For joint efficiency, an overlap greater than three times the thickness of the thinnest member is recommended. Lap joints are common in torch brazing processes, for which filler metal is drawn into the joint area by capillary action, and in sheet metal structures fabricated with the spot welder (see Figure 6.68).

6.8.2.3 Tee Joints

A *tee joint* (which, as the name implies, is T-shaped) is made by placing the edge of one piece of metal on the surface of the other piece at approximately a 90° angle. This joint is used for all ordinary plate thicknesses (see Figure 6.68).

6.8.2.4 Edge Joints

The *edge joint* is suitable for plate 1/4 in. or less in thickness and can sustain only light loads. These joints are made when one or more of the pieces to be connected is flared or flanged (see Figure 6.68).

6.8.2.5 Corner Joints

L-shaped *corner joints* have wide applications in joining sheet and plate metal sections where generally severe loads are not encountered. Boxes, trays, low-pressure tanks, and other objects are made with corner joints (see Figure 6.68).

6.8.3 Basic Weld Symbols

It is important to point out that the term *welding symbol* and *weld symbol* are different. The welding symbol consists of several elements that provide instructions to the welder. The weld symbol, on the other hand, indicates the type of weld only. In the following sections, we describe weld symbols.

6.8.3.1 Symbols for Arc and Gas Welds

The most commonly used arc and gas welds for fusing parts are shown in Figure 6.69. The four types of arc and gas welds are the *back* or *backing*, the *fillet*, the *plug* or *slot*, and *groove*. The groove welds are further classified as square, V, bevel, U, and J (see Figure 6.69) (ANSI/AWS, 1991).

6.8.3.2 Symbols for Resistance Welds

In resistance welding, the fusing temperature is produced in the particular area that is to be welded by applying force and passing electric current between two electrodes

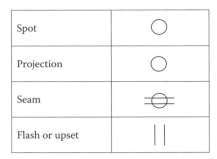

FIGURE 6.70 Symbols for resistance welds.

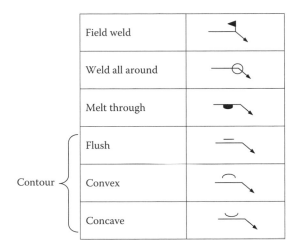

FIGURE 6.71 Supplementary weld symbols.

and the parts. The four basic resistance welds are the *spot*, *projection*, *seam*, and *flash* or *upset*. The symbols for the general types of resistance welds are given in Figure 6.70.

6.8.3.3 Symbols for Supplementary Welds

General supplementary weld symbols are shown in Figure 6.71. These symbols convey additional information about the extent of welding, the location, and the contour of the weld bead. The contour symbols are placed above or below the weld symbol.

6.8.4 THE WELDING SYMBOL

The complete welding symbol (see Figure 6.72) consists of six elements: reference line, arrowhead, weld symbol, dimensions, special symbols, and tail. Each element is described and shown in the following sections.

Important Point: Although welding symbols are often complex and carry a large amount of data, they may also be quite simple. Maintenance operators should study the various examples that follow and learn to read the symbols.

6.8.4.1 Reference Line

The basis of the welding symbol and all elements shown in Figure 6.73 is the *reference line*. As shown, it is the horizontal line (may appear vertically on some prints) portion of a welding symbol (see Figure 6.72). The reference line contains weld data regarding size, type, position, length, pitch, and strength. Data are written or drawn above, below, or on this line. Two or more reference lines may be used to specify steps to be performed in sequence. The first operation to be performed is the one specified on the reference line nearest the arrowhead. Additional operations are specified on subsequent lines as shown in Figure 6.74.

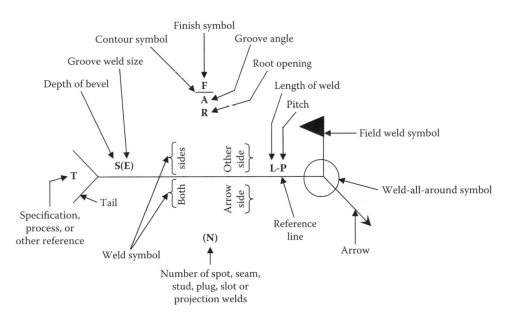

FIGURE 6.72 Standard welding symbol.

FIGURE 6.73 Reference line.

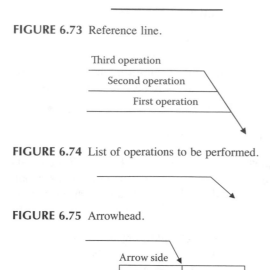

FIGURE 6.74 List of operations to be performed.

FIGURE 6.75 Arrowhead.

Arrow side

Other side

FIGURE 6.76 Arrowhead showing arrow side and other side.

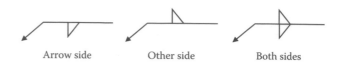

Arrow side   Other side   Both sides

FIGURE 6.77 Location of welds.

Square butt weld	⏸
Fillet weld	◺
Bevel weld	⋁
J-groove weld	⋃
V weld	⋁
U-groove weld	⋃

FIGURE 6.78 Weld symbols.

6.8.4.2 Arrowhead

An *arrowhead* is used to connect the welding symbol reference line to one side of the joint (see Figure 6.75). This is considered the arrow side of the joint (the surface that is in direct line of vision). The side opposite the arrow is considered the other side of the joint; that is, this side is the opposite surface of the joint (see Figure 6.76).

Important Point: A straight arrow pointing to a joint with a *chamfer* (shaped edge-groove) indicates that the chamfer is to be cut on both pieces. A broken arrow indicates that the chamfer is to be cut only on the piece that the arrow points toward. Two or more arrows from a single reference line indicate multiple locations for identical welds.

6.8.4.3 Weld Symbol

The *weld symbol* is attached to the reference line to show the kind of weld and the sides to be welded. The location of a weld is indicated by its placement on the reference line (see Figure 6.77). Weld symbols placed on the side of the reference line nearest the reader indicate welds on the arrow side of the joint. Weld symbols on the reference line side away from the reader indicate welds on the other side of the joint. Weld symbols on both sides of the reference line indicate welds on both sides of the joint. The symbols for some of the most common welds are shown in Figure 6.78.

6.8.4.4 Dimensions

Dimensions of welds are shown on the same side of the reference line as the weld symbol (see Figure 6.79). Dimensions of the chamfer and the weld cross-section

are written to the left of the weld symbol. The welding symbol at the right indicates that the fillet weld should be 3/8 in. high by 3/8 in. wide. Length dimensions are written to the right of the weld symbol. The welding symbol at the right indicates that the fillet weld should be 1.5 in. long. If an intermittent weld is required, the center-to-center spacing of the welds follows a hyphen after the length dimension. The welding symbol at the right indicates a series of 2-in. fillet welds 4 in. apart, measured center-to-center.

6.8.4.5 Special Symbols

Special symbols are used with the welding symbols to further specify the type of weld. These special symbols include contour, groove angle, spot welds, weld-all-around, field weld, melt-thru, and finish symbols.

6.8.4.6 Contour Symbol

As shown in Figure 6.80, a *contour symbol* is shown next to the weld symbol to indicate fillet welds that are to be flush (A), concave (B), or convex (C).

6.8.4.7 Groove Angle

A *groove angle* is shown on the same side of the reference line as the weld symbol. The size (depth) of groove welds is shown to the left of the weld symbol. The root opening of a groove weld is shown inside the weld symbol (see Figure 6.81).

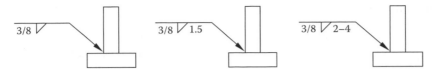

FIGURE 6.79 Dimensions.

A   B   C

FIGURE 6.80 Contour symbols.

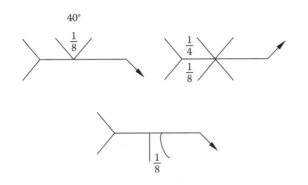

FIGURE 6.81 Groove angles.

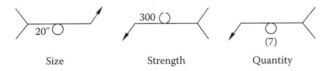

Size   Strength   Quantity

FIGURE 6.82 Spot weld symbols.

FIGURE 6.83 Weld-all-around symbol.

6.8.4.8 Spot Welds

Spot welds are dimensioned either by size or strength (see Figure 6.82). Size is designated as the diameter of the weld expressed in fractions, decimals, or millimeters and placed to the left of the symbol. The strength is also placed to the left of the symbol and expresses the required minimum shear strength in pounds per spot. If a joint requires a certain number of spot welds, the number is given in parentheses above or below the symbol.

6.8.4.9 Weld-All-Around

When a weld will extend completely around a joint, a small circle, the *weld-all-around* symbol, is placed where the arrow connects with the reference line (see Figure 6.83).

FIGURE 6.84 Field weld symbol.

FIGURE 6.85 Melt-thru symbol.

FIGURE 6.86 Finish symbol (M, machining).

FIGURE 6.87 Tail symbol.

6.8.4.10 Field Weld

The *field weld* symbol is used when welds are not to be made in the shop or at the place of initial construction. They are shown by a darkened triangular flag at the juncture of the reference line and arrow. The flag always points toward the tail of the arrow (see Figure 6.84).

6.8.4.11 Melt-Thru Welds

The *melt-thru* symbol indicates the welds where 100% joint or member penetration plus reinforcement is required in welds made from one side (see Figure 6.85). No dimension of melt-thru, except height of reinforcement, is shown on the weld symbol.

6.8.4.12 Finish Symbols

Welds that will be mechanically finished carry a *finish* symbol (C, chipping; G, grinding; M, machining; R, rolling; H, hammering) with the contour symbols (see Figure 6.86).

6.8.4.13 Tail

A tail may be added to the reference line to provide additional welding process information or specifications that are not otherwise shown by symbols (see Figure 6.87). These data are often in the form of symbols or abbreviations.

Important Point: Standard symbols and abbreviations and suffixes for optional use in applying welding and allied processes are given in tables provided by the American Welding Society.

6.9 ELECTRICAL DRAWINGS

Working drawings for the fabrication and troubleshooting of electrical machinery, switching devices, chassis for electronic equipment, cabinets, housings, and other mechanical elements associated with electrical equipment are based on the same principles as given earlier. To operate, maintain, and repair electrical equipment in the plant, the water and wastewater operator (qualified in electrical work) must understand electrical systems. The electrical operator must be able to read electrical drawings and determine what is wrong when electrical equipment fails to run properly. This section introduces electrical drawings, and the functions of important electrical components and how they are shown on drawings are explained.

6.9.1 TROUBLESHOOTING AND ELECTRICAL DRAWINGS

A water/wastewater maintenance operator who is qualified to perform electrical work is often assigned to repair or replace components of an electrical system. To repair anything, the first thing that must be done is to find the problem. The operator may solve the problem by simply restoring electrical power (i.e., resetting a circuit breaker or replacing a fuse). At other times, however, a good deal of troubleshooting may be required. Troubleshooting is like detective work: Find the culprit (the problem or what happened) and remedy (fix) the situation. Troubleshooting is a skill, but even the best troubleshooter would have some difficulty in troubleshooting many complex electrical machines without the proper electrical blueprint or wiring diagram.

6.9.2 ELECTRICAL SYMBOLS

Figure 6.88 shows some of the most common symbols used on electrical drawings. It is not necessary to memorize these symbols, but the maintenance operator should be familiar with them and be able to recognize them as an aid to reading electrical drawings.

6.9.2.1 Electrical Voltage and Power

Because of the force of its electrostatic field, an electric charge has the ability to do the work of moving another charge by attraction or repulsion. The force that causes free electrons (electricity) to move in a conductor as electric current may be referred to as follows:

- Electromotive force (emf)
- Difference in potential
- Voltage

6.9.2.2 What Is Voltage?

When a difference in potential exists between two charged bodies that are connected by a wire (conductor), electrons (current) will flow along the conductor. This flow is from the negatively charged body to the positively charged body until the two charges are equalized and the potential difference no longer exists.

Important Point: The basic unit of potential difference is the volt (V). The symbol for voltage is V, indicating the ability to do the work of forcing electrons (current flow) to move. Because the volt unit is used, potential difference is called *voltage*.

6.9.2.3 How Is Voltage Produced?

Voltage may be produced in many ways, but some methods are much more widely used than others. The following is a list of the six most common methods of producing voltage:

- *Friction*—Voltage produced by rubbing two materials together
- *Pressure*—Voltage produced by squeezing crystals of certain substances
- *Heat*—Voltage produced by heating the joint (junction) where two unlike metals are joined
- *Light*—Voltage produced by light striking photosensitive substances
- *Chemical action*—Voltage produced by chemical reaction in a battery cell
- *Magnetism*—Voltage produced in a conductor when the conductor moves through a magnetic field or a magnetic field moves through the conductor in such a manner as to cut the magnetic lines of force of the field

6.9.2.4 How Is Electricity Delivered to the Plant?

Electricity is delivered to the plant from a generating station. The electricity travels to the plant over high-voltage wires. It is important to point out that it is not the voltage *per se* that travels through the wires; instead, current travels through the wires to the plant. Voltage is the pressure or driving force that pushes the current through the wires. This high-voltage electricity must be reduced to a much lower voltage before it can be used to operate most plant electrical equipment. *Transformers* reduce the voltage. Again, it is *current* that actually flows through the wire. Current is measured in units called *amps or amperes*.

6.9.2.5 Electric Power

Power, whether electrical or mechanical, pertains to the rate at which work is being done. The electrical power consumption in a plant is related to current flow. A large

Ceiling fixture	○	Power transformer	T
Wall fixture	─○	Branch circuit concealed in ceiling or wall	————
Duplex outlet (grounded)	─⊖	Branch circuit concealed in floor	– – – – –
Single receptacle floor outlet (ungrounded)	⊟ UNG	Branch circuit exposed	- - - - - -
Street light	⊗	Feeders (note heavy line)	▬▬▬▬
Motor	Ⓜ	Number of wires in conduit (3)	─///─
Single-pole switch	─o⟍o─ S	Fuse	─[▭]─
Three-way switch	─o⟍ o─ S₃	Transformer	⫴
Circuit breaker	⟨ S_CB	Ground	⏚
Panel board and cabinet	▬▬▬	Normally open contacts	─| |─
Power panel	▨▨▨	Normally closed contacts	─|⧸|─
Motor controller	MC		

FIGURE 6.88 Common electrical symbols.

electric pump motor or air dryer consumes more power (and draws more current, particularly for motors at start) in a given length of time than, for example, an indicating light on a motor controller. *Work* is done whenever a force causes motion. If a mechanical force is used to lift or move a weight, work is done; however, force exerted *without* causing motion, such as the force of a compressed spring acting between two fixed objects, does not constitute work.

Key Point: Power is the rate at which work is done.

Electric power is measured in *watts*. One watt is a current of one ampere flowing at a voltage of one volt. In determining watts or wattage, we multiply the current (in amps) by the voltage (in volts).

Important Point: With regard to electric power, in a transformer, the number of watts coming in equals the number of watts going out.

6.9.2.6 Electrical Drawings

The two kinds of electrical drawings used for troubleshooting in water/wastewater treatment operations are *architectural drawings* and *circuit drawings*. An architectural drawing shows the physical locations of the electric lines in a plant building or between buildings. A circuit drawing shows the electrical loads served by each circuit.

Important Point: A circuit drawing does not indicate the physical location of any load or circuit.

6.9.2.7 Types of Architectural Drawings

Figure 6.89 shows three types of architectural drawings. The *plot plan* shows the electric distribution to all of the plant buildings. The *floor plan* shows where branch circuits are located in one building or pumping station, where equipment is located, and where outside and inside tie-ins to water, heat, and electric power are located. The *riser diagram* shows how the wiring runs to each floor of the building.

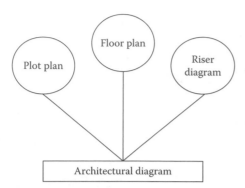

FIGURE 6.89 Types of architectural diagrams.

6.9.2.8 Circuit Drawings

A *circuit drawing* shows how a single circuit distributes electricity to various loads (e.g., pump motors, grinders, bar screens, mixers). Unlike an architectural drawing, a circuit drawing does not show the location of these loads. Figure 6.90 depicts a typical single-line circuit drawing and shows power distribution to 11 loads. The number in each circle indicates the power rating of the loads in horsepower.

Note: Electrical loads in all plants can be divided into two categories: critical and noncritical. Critical loads

are those that are essential to the operation of the plant and cannot be turned off (e.g., critical unit processes). Noncritical loads include those pieces of equipment that would not disrupt the operation of the plant or pumping station or compromise safety if they were turned off for a short period of time (e.g., air conditioners, fan systems, electric water heaters, and certain lighting systems).

The numbers in the rectangles show the current ratings of circuit breakers. The upper number is the current in amps that the circuit breaker will allow as a momentary surge. The lower number is the maximum current the circuit breaker will allow to flow continuously.

Note: Circuit breakers are typically equipped with surge protection for three-phase motors and other devices. When a three-phase motor is started, current demand is six to ten times normal value. After start, current flow decreases to its normal rated value. Surge protection is also provided to allow slight increases in current flow when the load varies or increases slightly. Most circuit breakers are also equipped with an instantaneous trip value for protection against short circuits.

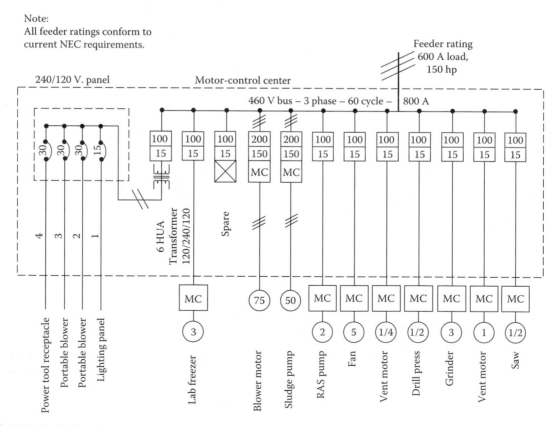

FIGURE 6.90 Single-line circuit diagram.

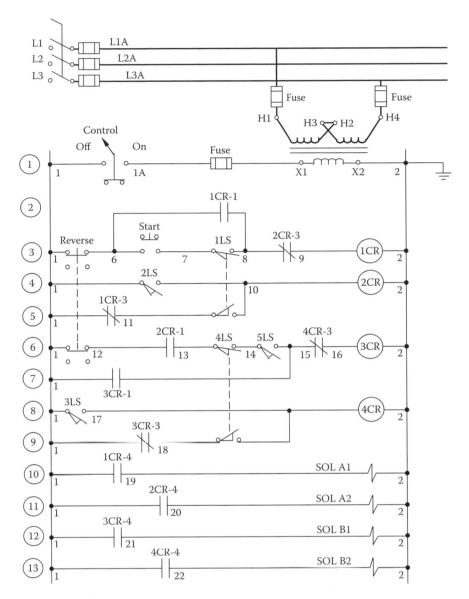

FIGURE 6.91 Typical ladder drawing.

6.9.2.9 Ladder Drawing

A *ladder drawing* is a type of schematic diagram that shows a control circuit. The parts of the control circuit lie on horizontal lines, like the rungs of a ladder. Figure 6.91 is an example of a ladder diagram.

Important Point: The size of electrical drawings is important. This can be understood first in the problem of storing drawings. If every size and shape were allowed, the task of systematic and protective filing of drawings could be tremendous. Pages that are 8-1/2 × 11" or 9 × 12" and multiples thereof are generally accepted. The drawing size can also be a problem for the troubleshooter or maintenance operator. If the drawing is too large, it is unwieldy to handle at the machine. If it is too small, it is difficult to read the schematic.

The purpose of a ladder drawing, such as the one shown in Figure 6.91, is to cut maintenance and troubleshooting time. This is accomplished when the designer uses certain guidelines in making electrical drawings and layouts. Let's take a closer look at ladder drawing for the control circuit shown in Figure 6.91. Note the numbering of elementary circuit lines. Normally closed contacts are indicated by a bar under the line number. Moreover, note that the line numbers are enclosed in a geometric figure to prevent mistaking the line numbers for circuit numbers.

All contacts and the conductors connected to them are properly numbered. Typically, numbering is carried throughout the entire electrical system. This may involve going through one or more terminal blocks. The incoming and outgoing conductors as well as the terminal blocks carry the proper electrical circuit numbers. When possible, connections to all electrical components are taken back to one common checkpoint.

All electrical elements on a machine should be correctly identified with the same markings as shown in the ladder drawing in Figure 6.91; for example, if a given solenoid is marked "solenoid A2" on the drawing, the actual solenoid on the machine should carry the marking, "solenoid A2."

The bottom line: An electrical drawing is made to show the relative location of each electrical component on the machine. The drawing is not normally drawn to scale, and it need not be; however, usually it is reasonably accurate in showing the location of parts relative to each other and in relative size.

6.10 AC&R DRAWINGS

In the not too distant past, central indoor climate control was reserved for a limited, privileged few. Today, central cooling in industrial plants is common in some locations. Although repairs often require the expertise of a trained specialist, water and wastewater maintenance operators may occasionally be required to perform certain maintenance operations on plant air conditioning and refrigeration (AC&R) systems. To perform certain kinds of service, maintenance operators need to understand the basic parts of air conditioning and refrigeration systems. To properly troubleshoot these systems, maintenance operators must also be able to read basic air conditioning and refrigeration drawings. Because almost all air conditioning systems cool the air inside a plant, the system must include refrigeration equipment. It logically follows, therefore, that to understand air conditioning we must first understand the principles of refrigeration.

6.10.1 Refrigeration

Refrigeration is the process of removing heat from an enclosed space or from a substance to lower the temperature. Before mechanical refrigeration systems were introduced, people cooled their food with ice transported from the mountains. Stored ice was the principal means of refrigeration until the beginning of the 20th century, and it is still used in some areas. Cooling caused by the rapid expansion of gases is the primary means of refrigeration today.

6.10.1.1 Basic Principles of Refrigeration

The basic principles of refrigeration are based on a few natural scientific facts; for example, it is a scientific (and often observed) fact that heat flows naturally from warm substances to cooler substances. For the heat to flow, it is

only necessary that the substances have different temperatures and that they make contact with one another. As the heat flows from one substance to the other, the temperature of the warmer substance falls and the temperature of the cooler substance rise. The flow of heat continues until the two substances become equal in temperature. The purpose of a refrigeration system is obvious: to reverse the natural flow of heat. The refrigeration system moves heat from a cool substance and delivers it to a warmer substance. In doing so, the temperature of the cooler substance falls, and the temperature of the warmer substance rises. The refrigeration system reverses heat flow by means of a circulating fluid. The fluid may be either a liquid or a gas, or it may change back and forth from one form to the other. If it is always a liquid or always a gas, the fluid is called a *coolant*. If it changes back and forth, it is called a *refrigerant*.

Refrigerants are the working fluid of air conditioning and refrigeration systems. Their purpose is to absorb heat by evaporation in one location, transport it to another location through the application of external work, then release it through condensation. A wide range of commercially available refrigerants is suitable for use in an equally wide range of applications. These refrigerants vary in chemical composition, boiling points, freezing points, critical temperatures, flammability, toxicity, chemical stability, chemical reactiveness, and cost.

Important Point: Selection of a particular refrigerant depends on the application and characteristics required. In most cases the choice will be a compromise between different properties.

A refrigeration system, where the fluid changes back and forth from liquid to gas and *vice versa*, works in such a way that the circulating fluid has a lower temperature than the cool substance at one point of the circulation. At another point, it has a higher temperature than the warm substance. The changing temperature of the refrigerant is important because it makes it possible for heat to flow out of the cool substance and into the warm substance. Although we are talking about an artificial cooling process, it is based on a natural heat flow process. The problem is that energy is required to make the system run. The amount of energy (electrical or mechanical) required is greater than the amount of heat energy moved from the cool place to the warm place.

6.10.1.2 Refrigeration System Components

A mechanical refrigeration system is an arrangement of components in a system that puts the theory of gases into practice to provide artificial cooling. To do this, we must provide the following:

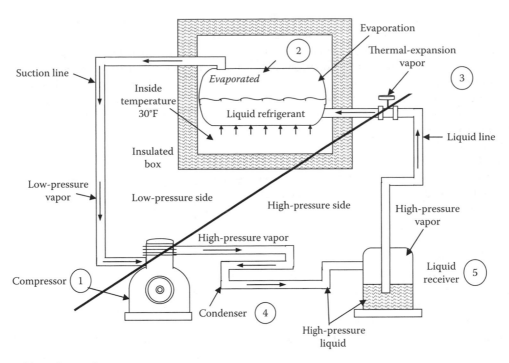

FIGURE 6.92 Refrigeration cycle.

- A metered supply of relatively cool liquid under pressure
- A device in the space to be cooled that operates at reduced pressure so when the cool, pressurized liquid enters it will expand, evaporate, and take heat from the space
- A means of repressurizing (compressing) the vapor
- A means of condensing it back into a liquid to remove its superheat, latent heat of vaporization, and some of its sensible heat

6.10.1.3 Refrigeration System Operation

Every mechanical refrigeration system operates at two different pressure levels. The dividing line is shown in Figure 6.92, which is also the type of simple system drawing with which maintenance operators need to be familiar. The line passes through the discharge valves of the compressor on one end and through the orifice of the metering device or expansion valve on the other. The high-pressure side of the refrigeration system is comprised of all of the components that operate at or above condensing pressure. These components are the discharge side of the compressor, the condenser, the receiver, and all interconnected tubing up to the metering device or expansion valve. The low-pressure side of a refrigeration system consists of all the components that operate at or below evaporating pressure. These components include the low-pressure side of the expansion valve, the evaporator, and all the interconnecting

tubing up to and including the low side of the compressor. Refrigeration maintenance operators refer to the pressure on the refrigerant low-pressure vapor drawn from the high-side discharge pressure as *head pressure*. On the low side, the pressure is compressed by the compressor to what is called a *suction pressure* or *low-side pressure*.

6.10.1.4 Using Refrigeration Drawings in Troubleshooting

The maintenance operator would use the basic drawing shown in Figure 6.92 to troubleshoot a typical refrigeration system. Moreover, the troubleshooter who is interested in making repairs or adjustments to the system to restore correct operation obviously needs to understand system operation. In this case, Figure 6.92 is helpful.

Note: The following explanation is rather detailed to provide insight on how valuable system drawings can be in assisting the troubleshooter first to understand system operation and then to determine the cause of the fault. It should be pointed out, however, that this discussion is not designed to make a refrigeration expert out of anyone.

Using Figure 6.92, the maintenance operator can gain a complete understanding of the refrigeration cycle of such a mechanical refrigeration system. For example, from Figure 6.92 it can be seen that the pumping action of the compressor (1) draws vapor from the evaporator (2). This action reduces the pressure in the evaporator, causing the liquid particles to evaporate. As the liquid particles evaporate, the evaporator is cooled. Both the liquid and vapor refrigerant tend to extract heat from the warmer objects in the insulated refrigerator cabinet or room.

The ability of the liquid to absorb heat as it vaporizes is very high in comparison to that of the vapor. As the liquid refrigerant is vaporized, the low-pressure vapor is drawn into the suction line by the suction action of the compressor (1). The evaporation of the liquid refrigerant would soon remove the entire refrigerant from the evaporator if it were not replaced. The replacement of the liquid refrigerant is usually controlled by a metering device or expansion valve (3). This device acts as a restrictor to the flow of the liquid refrigerant in the liquid line. Its function is to change the high-pressure, subcooled liquid refrigerant to low-pressure, low-temperature liquid particles, which will continue the cycle by absorbing heat.

The refrigerant low-pressure vapor drawn from the evaporator by the compressor through the suction line, in turn, is compressed by the compressor to a high-pressure vapor, which is forced into the condenser (4). In the condenser, the high-pressure vapor condenses to a liquid under high pressure and gives up heat to the condenser. The heat is removed from the condenser by the cooling medium of air or water. The condensed liquid refrigerant is then forced into the liquid receiver (5) and through the liquid line to the expansion valve by pressure created by the compressor, making a complete cycle.

Important Point: Although the receiver is indicated as being part of the refrigeration system in Figure 6.92, it is not a vital component; however, the omission of the receiver requires exactly the proper amount of refrigerant in the system. The refrigerant charge in systems without receivers is to be considered critical, as any variations in quantity affects the operating efficiency of the unit.

CAUTION: The refrigeration cycle of any refrigeration system must be clearly understood by a maintenance operator before repairing the system.

6.10.1.5 Refrigeration Component Drawings

When work is to be accomplished on a refrigeration system, we often need to refer to drawings of various components. Whether these drawings are assembly or detail drawings, many of them will look much like the drawings we presented earlier in this text. The evaporator drawing shown in Figure 6.93 is an example of a component drawing. This particular evaporator is made of copper tubing with aluminum fins. It functions to cool the air in an air conditioning duct. Complete evaporator units are equipped with accessories such as air-moving devices and motors. Other accessories usually furnished with such units include filters, heating coils, electric resistance or gas heaters, air-humidifying devices, and sheet metal enclosures. Usually, enclosures are insulated and equipped with inlet and outlet connections for ducts.

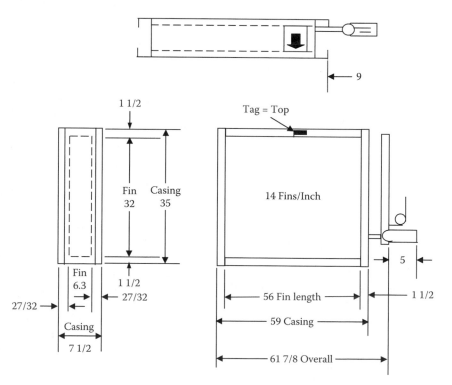

FIGURE 6.93 Evaporator assembly.

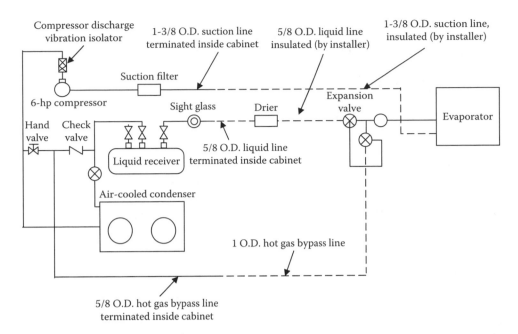

FIGURE 6.94 Refrigerant piping.

6.10.2 AIR CONDITIONING

Air conditioning requires the control of temperature, humidity, purity, and motion of air in an enclosed space, independent of outside conditions. The air conditioning system may be required to heat, cool, humidify, dehumidify, filter, distribute, and deodorize the air.

6.10.2.1 Operation of a Simple Air Conditioning System

In a simple air conditioner, the refrigerant, in a volatile liquid form, is passed through a set of evaporator coils across which air inside the room is passed. The refrigerant evaporates and, in the process, absorbs the heat contained in the air. When the cooled air reaches its saturation point, its moisture content condenses on fins placed over the coils. The water runs down the fins and drains. The cooled and dehumidified air is returned into the room by means of a blower. In the meantime, the vaporized refrigerant passes into a compressor where it is pressurized and forced through condenser coils, which are in contact with outside air. Under these conditions, the refrigerant condenses back into a liquid form and gives off the heat it absorbed inside. This heated air is expelled to the outside, and the liquid recirculates to the evaporator coils to continue the cooling process. In some units, the two sets of coils can reverse functions so in winter the inside coils condense the refrigerant and heat rather than cool the room. Such a unit is known as a *heat pump*. Alternative systems of cooling include the use of chilled water. Water may be cooled by refrigerant at a central location and then run through coils throughout the system.

6.10.2.2 Design of Air Conditioning Systems

The design of air conditioning systems takes many circumstances into consideration. A self-contained unit, described above, serves a space directly. More complex systems, as in tall buildings, use ducts to deliver cooled air. In the *induction system*, air is cooled once at a central plant and then conveyed to individual units, where water is used to adjust the air temperature according to such variables as sunlight exposure and shade. In the *duct–duct system*, warm air and cool air travel through separate ducts and are mixed to reach a desired temperature. A simpler way to control temperature is to regulate the amount of cold air supplied, cutting it off once a desired temperature is reached. This method, known as *variable air volume*, is widely used in both high-rise and low-rise commercial or institutional buildings.

6.10.2.3 Air Conditioning Drawings

Drawings for air conditioning systems usually include the following: the entire system; complete central air conditioning installations; two-, three-, and four-pipe systems; controls; water piping; ductwork; cooling towers; various air conditioning components; and refrigerant piping. Figure 6.94 is an example of a refrigerant piping drawing. The purpose of this drawing is to guide the installer toward making the proper connections in the refrigerant piping. This single-line drawing is, obviously, also helpful in troubleshooting (e.g., finding a leak) the refrigerant system. The drawing clearly depicts the identity of the basic parts of the refrigeration system.

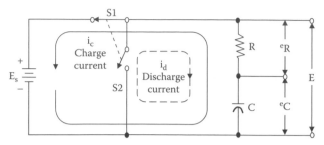

FIGURE 6.95 Schematic of an RC series circuit.

6.11 SCHEMATICS AND SYMBOLS

Because of the complexity of many electrical and mechanical systems, it would be almost impossible to show these systems in full-scale detailed drawing. Instead, we use symbols and connecting lines to represent the parts of a system. Figure 6.95 shows a voltage divider containing resistance and capacitance connected in a circuit by means of a switch. Such a series arrangement is an *RC series circuit*. Note that, unless the reader is an electrician or electronics technician, it is not important to understand this circuit; however, it is important for the reader to understand that Figure 6.95 depicts a *schematic* representation formed by the use of symbols and connecting lines for a technical purpose.

Important Point: A schematic is a line drawing made for a technical purpose that uses symbols and connecting lines to show how a system operates.

6.11.1 How to Use Schematic Diagrams

Learning to read and to use a schematic diagram (any schematic diagram) is a little bit like map reading. In a schematic for an electrical circuit, for example, we need to know which wires connect to which component and where each wire starts and finishes. With a map book, this would be equivalent to knowing the origin and destination points and which roads connect within the highway network. Schematics, however, are a little more complicated, as components must be identified and some are polarity conscious (must be wired up in the circuit the correct way to work). The reader does not need to understand what the circuit does or how it works to read the drawing, but the reader does need to be able to correctly interpret the schematic. Here are some basic rules that will help with reading a simple diagram.

In Figure 6.96, the heavy lines represent wires; for simplicity, we have labeled them as A, B, and C. There are just three components here, and it is easy to see where each wire starts and ends and to which components a wire is connected. As long as the wire labeled A connects to the switch and negative terminal of the battery, wire B

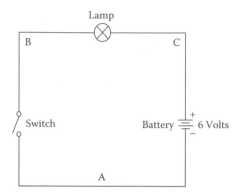

FIGURE 6.96 A single schematic diagram.

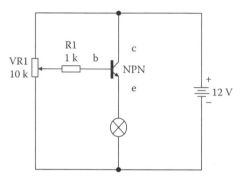

FIGURE 6.97 Schematic of simple lamp dimmer circuit.

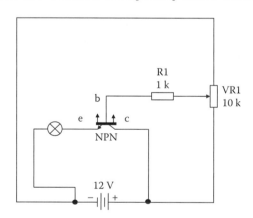

FIGURE 6.98 Schematic of a simple dimmer circuit.

connects to the switch and lamp, and C connects to the lamp and the battery positive terminal, then this circuit should operate.

Before we move on and describe how to read the diagram, it is important to point out that any schematic may be drawn in a number of different ways; for example, in Figure 6.97 and Figure 6.98, we have drawn two electrically equivalent lamp dimmer circuits. They may look very different, but if we mentally label the wires and trace them we will see that in both diagrams each wire starts and finishes at the same components on both diagrams. The components have been labeled and so have the three terminals of the transistor (e.g., NPN is the transistor).

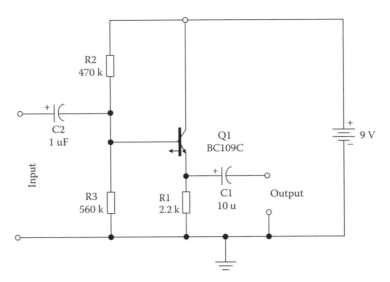

FIGURE 6.99 Schematic of a simple audio circuit.

In Figure 6.98, the two wire junctions are indicated by a dot. A wire connects from the positive battery terminal to the C (transistor collector) terminal, and a wire also runs from the collector terminal to one end of the potentiometer (VR1). The wires could be joined at the transistor collector, the positive battery terminal, or even one end of the potentiometer; it does not matter, as long as both wires exist. Similarly, a wire runs from battery negative to the lamp and also from the lamp to the other end of VR1. The wires could be joined at the negative terminal of the battery, the lamp, or the opposite leg of VR1. In Figure 6.98, we could have drawn the wires from the lamp and bottom terminal of VR1 back to the negative battery terminal and placed the dot there; it would be the same. Looking at Figure 6.98, we see that one wire junction appears at the negative battery terminal, and the other junction is at a similar place on the positive battery terminal.

6.11.2 SCHEMATIC CIRCUIT LAYOUT

Sometimes the way a circuit is wired may compromise its performance. This is particularly true for high-frequency and radio circuits and some high-gain audio circuits. Consider the audio circuit shown in Figure 6.99. (For our purposes here, we have simplified the following explanation.) Although this circuit has a voltage gain of less than one, wires to and from the transistor should be kept as short as possible. This will prevent a long wire from picking up radio interference or hum from a transformer. Moreover, in this circuit the input and output terminals have been labeled, and a common reference point or earth (ground) is indicated. The ground terminal would be connected to the chassis (metal framework of the enclosure) in which the circuit is built. Many schematics contain a chassis or ground point. Generally, it is just to indicate the common reference terminal of the circuit, but in radio

work the ground symbol usually requires a physical connection to a cold water pipe or a length of pipe or earth spike buried in the soil.

6.11.3 SCHEMATIC SYMBOLS

Water and wastewater operators and maintenance operators must be Jacks or Jills of many trades. Simply, a good maintenance operator must be able to do many different kinds of jobs. To become a fully qualified Jack or Jill, the maintenance operator must learn to perform electrical tasks, mechanical tasks, piping tasks, fluid-power tasks, AC&R tasks, hotwork tasks, and many other special tasks. Moreover, maintenance operators must be flexible; they must be able to work on both familiar as well as new equipment and systems. We have all heard seasoned maintenance operators claim that they can fix anything and everything using nothing more than their own intuition (i.e., "seat of the pants" troubleshooting). In the real world, however, maintenance operators must be able to read and understand schematics to troubleshoot systems. By learning this skill, operators will have little difficulty understanding, maintaining, and repairing almost any equipment or unit process in the plant—old or new.

6.11.3.1 Lines on a Schematic

As mentioned, symbols are used instead of pictures on schematics, and a schematic is simply a line diagram. Lines on a schematic show the connections between the symbols (devices) in a system. Each line has meaning; thus, we can say that schematic lines are part of the symbolism employed. The meaning of certain lines, however, depends on the kind of system the schematic portrays; for example, a simple solid line can have totally different meanings. On an electrical diagram, it probably represents wiring. On a

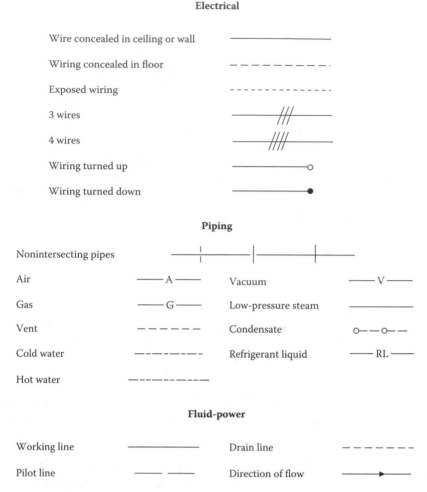

FIGURE 6.100 Examples of lines used in schematics.

fluid-power diagram, it stands for a working line. On a piping diagram, it could mean a low-pressure steam line. Figure 6.100 shows some other common lines used in schematics. A schematic diagram is not necessarily limited to one kind of line. In fact, several kinds of lines may appear on a single schematic. Following applicable ANSI standards, most schematics use only one thickness, but they may use various combinations of solid and broken lines.

Important Point: Not all schematics adhere to standards set by national organizations as an aid in providing uniform drawings. Some designers prefer to use their own line symbols. These symbols are usually identified in a legend.

6.11.3.2 Lines Connect Symbols

If we look at a diagram filled with lines, we may simply have nothing more than a diagram filled with lines. Likewise, if we look at a diagram with assorted symbols, we may simply have a diagram filled with various symbols. Such diagrams may have meaning to someone but prob-

ably have little meaning to most of us. To make a schematic readable (understandable) for a wide audience, we must have a diagram that uses a combination of recognizable lines and symbols. When symbols are combined with lines in schematic form, we must also understand the meaning of the symbols used. The meaning of certain symbols depends on the kind of system the schematic shows; for example, the symbols used in electrical systems differ from those used in piping and fluid-power systems.

The bottom line: To understand and properly use a schematic diagram, we must understand the meaning of both the lines and the symbols used.

6.11.4 SCHEMATIC DIAGRAM: AN EXAMPLE

Figure 6.101 shows a schematic diagram used in electronics and communications (ANSI, 1982). The layout of this schematic involves the same principles and procedures (except for lesser detail) suggested for more complex schematics. Although less complex than most schematics, Figure 6.101 serves our intended purpose: to provide a

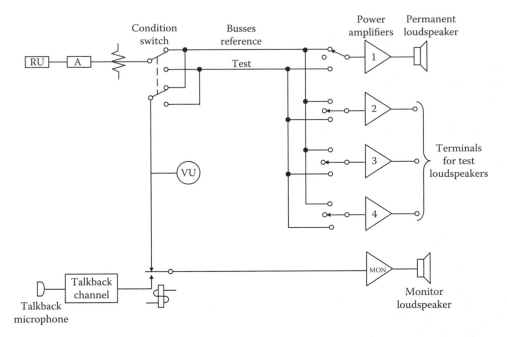

FIGURE 6.101 Single-line diagram. (From ANSI, *ANSI Y14.15: Dimensioning and Tolerancing*, ANSI, New York, 1982.)

simplified schematic diagram for basic explanation and easier of understanding of a few key points—essential to understanding schematics and on how to use them.

6.11.4.1 A Schematic by Any Other Name Is a Line Diagram

The schematic (or line) diagram is intended to describe the basic functions of a circuit or system. As such, the individual lines connecting the symbols may represent single conductors or multiple conductors. The emphasis is on the function of each stage of a device and the composition of the stage. The various parts or symbols used in a schematic (or line) diagram are typically arranged to provide a pleasing balance between blank areas and lines (see Figure 6.101). Sufficient blank spaces are provided adjacent to symbols for insertion of reference designations and notes.

It is standard practice to arrange schematic or line diagrams so the signal or transmission path from input to output proceeds from left to right (see Figure 6.101) and from top to bottom for a diagram in successive layers.

Supplementary circuits, such as a power supply and an oscillator circuit, are usually shown below the main circuit.

Stages of an electronic device, such as shown in Figure 6.101, are groups of components, usually associated with a transistor or other semiconductor, which together perform one function of the device. *Connecting lines* (for conductors) are drawn horizontally or vertically, for the most part, minimizing bends and crossovers. Typically, long interconnecting lines are avoided; instead, *interrupted paths* are used in place of long, awkward, interconnecting lines or where a diagram occupies more than one sheet. When parallel connecting lines are drawn close together, the spacing between lines is not less than .06 in. after reduction. As a further visual aid, parallel lines are grouped with consideration of function and with double spacing between groups.

Crossovers are usually necessary in schematic diagrams. The looped crossovers shown in Figure 6.102A have been used for several years to avoid confusion; however, this method is not approved by ANSI. A simpler practice recognized by ANSI is shown in Figure 6.102B.

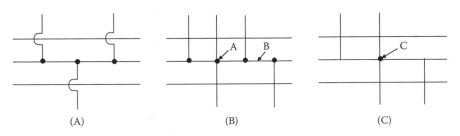

(A)   (B)   (C)

FIGURE 6.102 Crossovers.

Connection of more than three lines at one point, shown at A, is not recommended and can usually be avoided by moving or staggering one of more lines, as at B. ANSI Y14.15 recommends crossovers as shown in Figure 6.102C. In this system, it is understood that termination of a line signifies a connection. If more than three lines come together, as at C, the dot symbol becomes necessary. *Interrupted paths*, either for a single line or groups of lines, may be used where desirable for overall simplification of a diagram.

6.11.5 SCHEMATICS AND TROUBLESHOOTING

As mentioned, one of the primary purposes of schematic diagrams is to assist the maintenance operator in troubleshooting system, component, or unit process faults. A basic schematic can be the troubleshooter's best friend, but experience has shown that many mistakes and false starts can be avoided by taking a step-by-step approach to troubleshooting. Seasoned water/wastewater maintenance operators usually develop a standard troubleshooting protocol or step-by-step procedure to assist them in their troubleshooting activities. No single protocol is the same, and each troubleshooter proceeds based on intuition and experience (not on "seat of the pants" solutions). The simple 15-step protocol described below, however, has worked well (along with an accurate system schematic) for those of us who have used it (note that several steps may occur at the same time):

1. Recognize that a problem exists; figure out what it is designed to do and how it should work.
2. Review all available data.
3. Find the part of the schematic that shows the troubled area, and study it in detail.
4. Evaluate the current plant operation.
5. Decide what additional information is needed.
6. Collect the additional data.
7. Test the process by making modifications and observing the results.
8. Develop an initial opinion as to the cause of the problem and potential solutions.
9. Fine tune your opinion.
10. Develop alternative actions to be taken.
11. Prioritize alternatives (e.g., based on its chances of success, how much it will cost).
12. Confirm your opinion.
13. Implement the alternative actions (this step may be repeated several times).
14. Observe the results of the alternative actions implemented; that is, observe the impact on effluent quality, the impact on individual unit process performance, changes (trends) in the results of process control tests and calculations, and the impact on operational costs.
15. During project completion, evaluate other, more permanent long-term solutions to the problem, such as chemical addition, improved preventive maintenance, or design changes. Continue to monitor results. Document the actions taken and the results produced for use in future problems.

6.12 ELECTRICAL SCHEMATICS

A good deal of the water/wastewater treatment process equipment in use today runs by electricity. Plants must keep their electrical equipment working. When a machine fails or a system stops working, the plant maintenance operator must find the problem and solve it quickly. No maintenance operator can be expected to remember all of the details of the electrical equipment in the plant. This information must be stored in diagram or drawings, in a format that can be readily understood by trained and qualified maintenance operators. Electrical schematics store the information in a user-friendly form. In this section, the basics of electrical schematics and wiring diagrams are discussed. Typical symbols and circuits are used as examples.

When describing electrical systems, three kinds of drawings are typically used: pictorial, wiring, and schematic drawings. *Pictorial drawings* show an object or system much as it would appear in a photograph—as if we were viewing the actual object. Several sides of the object are visible in the one pictorial view. Pictorial drawings are quite easy to understand. They can be used when making or servicing simple objects but are usually not adequate for complicated parts or systems, such as electrical components and systems. A *wiring diagram* shows the connections of an installation or its component devices or parts. It may cover internal or external connections, or both, and contains such detail as is needed to make or trace connections that are involved. The wiring diagram usually shows the general physical arrangement of the component devices or parts. A *schematic diagram* uses symbols instead of pictures for the working parts of the circuit. These symbols are used in an effort to make the diagrams easier to draw and easier to understand. In this respect, schematic symbols aid the maintenance operator in the same way that shorthand aids the stenographer.

Important Point: A schematic diagram emphasizes the flow in a system. It shows how a circuit functions rather than how each part actually looks. Stated differently, a schematic represents the *electrical*, not the physical, situation in a circuit.

6.12.1 ELECTRICAL SYMBOLS

Electrical and electronic circuits are indicated by very simple drawings, called *schematic symbols*, which are standardized throughout the world, with minor variations. Some of

Single wire	———————
Wiring concealed in floor	– – – – – – –
Exposed wiring	- - - - - - - - - -
Wires crossing but not connected	(symbols)
Wires connected (dot required)	(symbol)

FIGURE 6.103 Symbols for wires.

these symbols look like the components they represent. Some look like key parts of the components they represent. Maintenance operators must know these symbols so they can read the diagrams and keep the plant equipment in working order. The more schematics are used, the easier it becomes to remember what these symbols mean.

6.12.1.1 Schematic Lines

In electrical and electronic schematics, lines symbolize (or stand for) wires connecting various components. Different kinds of lines have different meanings in schematic diagrams. Figure 6.103 shows examples of some lines and their meanings; other lines are usually identified by a diagram legend. To understand any schematic diagram, we must observe how the lines intersect. These intersections show that two or more wires are connected or that the wires pass over or under each other without connecting. Figure 6.103 shows some of the connections and crossings of lines. When wires intersect in a connection, a dot is used to indicate this. If it is clear that the wires connect, the dot is not used.

Important Point: Wires and how they intersect are important. Maintenance operators must be able to tell the difference between wires that connect and those that do not to be able to properly read the schematic and determine the flow of current in a circuit.

6.12.1.2 Power Supplies: Electrical Systems

Most water and wastewater treatment facilities receive electrical power from the transmission lines of a utility company. On a schematic, the entry of power lines into the electrical system can be shown in several ways. Electrical power supply lines to a motor are shown in Figure 6.104. Another source of electrical power is a battery. A battery consists of two or more cells, and each cell is a unit that produces electricity by chemical means. The cells can be connected together to produce the necessary volt-

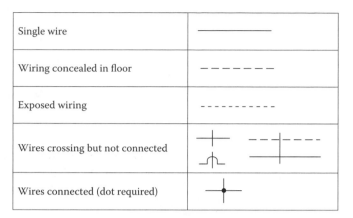

FIGURE 6.104 Electrical power supply lines.

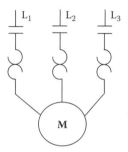

FIGURE 6.105 Schematic symbol for a battery.

age and current; for example, a 12-volt storage battery might consist of six 2-volt cells.

On a schematic, a battery power supply is represented by a symbol. Consider the symbol in Figure 6.105. The symbol is rather simple and straightforward but is very important. By convention, the shorter line in the symbol for a battery represents the negative terminal. It is important to remember this, because it is sometimes necessary to note the direction of current flow, which is from negative to positive, when examining a schematic. The battery symbol shown in Figure 6.105 has a single cell, so only one short and one long line are used. The number of lines used to represent a battery vary (and they are not necessarily equivalent to the number of cells), but they are always in pairs, with long and short lines alternating. In the circuit shown in Figure 6.106, the current would flow in a *counterclockwise* direction—that is, in the opposite direction that the hands of a clock move. If the long and short lines of the battery symbol (symbol shown in Figure 6.106) were reversed, the current in the circuit shown in Figure 6.106 would flow *clockwise*—that is, in the direction that the hands of a clock move.

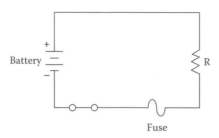

FIGURE 6.106 Schematic of a simple fused circuit.

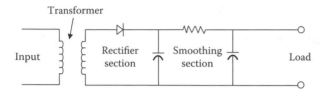

FIGURE 6.107 Basic power supply.

Important Point: Current flows from the negative (–) terminal of the battery (see Figure 6.107) through the switch, fuse, and resistor (R) to the positive (+) battery terminal, and it continues flowing through the battery from the positive (+) terminal to the negative (–) terminal. As long as the pathway is unbroken, it is a closed circuit, and current will flow; however, if the path is broken (e.g., switch is in open position), it is an open circuit, and no current flows.

Important Point: In Figure 6.106, a fuse is placed directly into the circuit. A fuse will open the circuit whenever a dangerous large current starts to flow; a short-circuit condition occurs due to an accidental connection between two points in a circuit that offer very little resistance. A fuse will permit currents smaller than the fuse value to flow but will melt and therefore break or open the circuit if a larger current flows.

In water and wastewater treatment operations, maintenance operators are most likely to maintain or troubleshoot circuits connected to an outside power source. Occasionally, however, they may also work on some circuits that are battery powered. In fact, in work on electronics systems, more work may be performed on battery-power supplied systems than on outside sources. Electronic power supply systems are discussed in the following section.

6.12.1.3 Power Supplies: Electronics

In electronics, power supplies perform two important functions: (1) they provide electrical power when no other source is available, and (2) they convert available power into power that can be used by electronic circuits. Power supplies provide the DC supply voltage required for an amplifier, oscillator, or other electronic device.

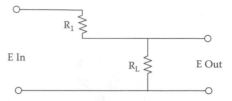

FIGURE 6.108 Schematic of basic DC-to-DC power supply.

Important Point: Conversion from one type of power supply (AC to DC, for example) does not improve the quality of the input power, so power conditioners are added for smoothing.

A simple schematic diagram (see Figure 6.107) best demonstrates the actual makeup of an electronic power supply. As shown in the figure, a basic power supply consists of four sections: a transformer, a rectifier, a smoothing section, and a load. The transformer converts the 120-V line AC to a lower AC voltage. The rectifier section is used to convert the AC input to DC. Unfortunately, the DC produced is not smooth DC but instead is pulsating DC. The smoothing, or conditioning, section functions to convert the pulsating DC to a pure DC with as little AC ripple as possible. The smoothed DC is then applied to the load.

Electronic components must have electrical power to operate, obviously. This electrical power is usually direct current (in electronic equipment). Components typically use a single low voltage, usually 5 volts, and single polarity, either positive (+5 V) or negative (–5 V). Circuits often require several voltages and both polarities, typically +5 V, –5 V, +12 V, and –12 V. One of the simplest power supplies, a DC-to-DC power supply, is shown in schematic form in Figure 6.108. In this simple circuit, if the current drawn by the load does not change, the voltage across the resistor (often referred to as a *dropping resistor*) will be as steady as the source voltage.

In an AC-to-DC power supply, a *rectifier* changes the 60-hertz AC input voltage to fluctuating, or pulsating, DC output voltage. The diode allows current in only one direction, for one polarity of applied voltage. Thus, current flows in the output circuit only during the half-cycles of the AC input voltage that turn the diode on. Figure 6.109 shows a schematic of a typical AC-to-DC power supply. Notice that the input is 120-V AC, which is typical of the voltage supplied to most households and businesses from a utility line. The 120-V AC is fed to the primary of a transformer. The transformer reduces the voltage of the AC output from the secondary. The transformer output (secondary output) is the input to the rectifier, which delivers a pulsating DC output. The rectified output is the input to the filter, which smoothes out the pulses from the rectifier. The filter output is the input to the voltage regulator,

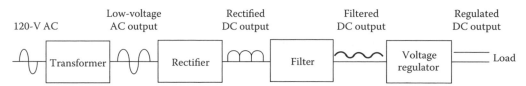

FIGURE 6.109 Schematic of an AC-to-DC power supply.

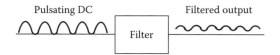

FIGURE 6.110 Ripple in filter output.

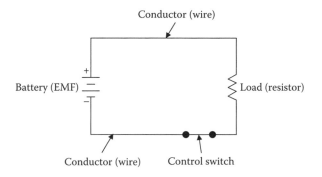

FIGURE 6.111 Schematic of a simple closed circuit.

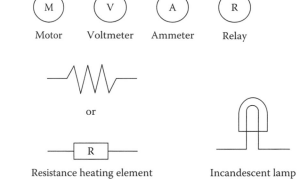

FIGURE 6.112 Symbols for electrical loads.

which maintains a constant voltage output, even if the power drawn by the load changes. The voltage regulator output is then fed to the load.

Power smoothers, or conditioners, are built into or added to power supplies to regulate and stabilize the power supply. They include filters and voltage regulators. A *filter* is used in both DC output and AC output power supplies. In DC output power supplies, filters help smooth out the rectifier output pulses, as shown in Figure 6.110. In AC output power supplies, filters are used to shape waveforms (i.e., they remove the undesired parts of the waveform).

6.12.1.4 Electrical Loads

As mentioned, an electric circuit, which provides a complete path for electric current, includes an energy (voltage) source, such as a battery or utility line; a conductor (wire); a means of control (switch); and a load. As shown in the schematic representation in Figure 6.111, the energy source is a battery. The battery is connected to the circuit by conductors (wire). The circuit includes a switch for control. The circuit also consists of a load (resistive component). The *load* that dissipates battery-stored energy could be a lamp, a motor, heater, a resistor, or some other device (or devices) that does useful work, such as an electric toaster, a power drill, radio, or soldering iron. Figure 6.112 shows some symbols for common loads and

other components in electrical circuits. The maintenance operator should become familiar with these symbols (and others not shown here), because they are widely used.

Important Point: Every complete electrical circuit has at least one load.

6.12.1.5 Switches

In Figure 6.107 and Figure 6.112, the schematics show simple circuits with switches. A *switch* is a device for making or breaking the electrical connection at one point in a wire. A switch allows starting, stopping, or changing the direction of current flow in a circuit. Figure 6.113 shows some common switches and their symbols.

6.12.1.6 Inductors (Coils)

Simply put, an *inductor* is a coil of wire, usually many turns (of wire) around a piece of soft iron (magnetic core). In some cases, the wire is wound around a nonconducting material. Inductors are used as ballasts in fluorescent lamps and for magnets and solenoids. When electric current flows through a coil, it creates a magnetic field (an electromagnetic field). The magnetism causes certain effects required in electric circuits (e.g., in an alarm circuit, the magnetic field in a coil can cause the alarm bell to ring). It is not important to understand these effects to be able to read schematic diagrams. It is, however, important to recognize the symbols for inductors (or coils). These symbols are shown in Figure 6.114.

Type of switch	Abbreviation	Symbol
Single-pole, single-throw	SPST	Toggle switch Knife switch
Double-pole, single-throw	DPST	
Single-pole, double-throw	SPDT	
Double-pole, double-throw	DPDT	
Normally open	NO	
Normally closed	NC	
Two-position	NC-NO	

FIGURE 6.113 Types of switches.

Relay coil	─(R)─
Fixed coil	
Solenoid	
Tapped coil	
Variable coil	or

FIGURE 6.114 Symbols for coils and inductors.

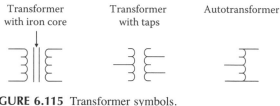

Transformer   Transformer   Autotransformer
with iron core   with taps

FIGURE 6.115 Transformer symbols.

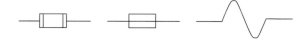

FIGURE 6.116 Fuse symbols.

6.12.1.7 Transformers

Transformers are used to increase or decrease AC voltages and currents in circuits. The operation of transformers is based on the principal of *mutual inductance*. A transformer usually consists of two coils of wire wound on the same core. The primary coil is the input coil of the transformer, and the secondary coil is the output coil. Mutual induction causes voltage to be induced in the secondary coil. If the output voltage of a transformer is greater than the input voltage, it is a *step-up* transformer. If the output voltage of a transformer is less than the input voltage it is a *step-down* transformer. Figure 6.115 shows some of the basic symbols that are used to designate transformers on schematic diagrams.

6.12.1.8 Fuses

A *fuse* is a device that automatically opens a circuit when the current rises above a certain limit. When the current becomes too high, part of the fuse melts. Melting opens the electrical path, stopping the flow of electricity. To restore the flow of electricity, the fuse must be replaced. Figure 6.116 shows some of the basic symbols that are used to designate fuses on schematic diagrams.

6.12.1.9 Circuit Breakers

A *circuit breaker* is an electric device (similar to a switch) that, like a fuse, interrupts an electric current in a circuit when the current becomes too high. The advantage of a circuit breaker is that it can be reset after it has been tripped; a fuse must be replaced after it has been used once. When a current supplies enough energy to operate a trigger device in a breaker, a pair of contacts conducting the current are separated by preloaded springs or some similar mechanism. Generally, a circuit breaker registers the current either by the heating effect of the current or by the magnetism it creates in passing through a small coil. Figure 6.117 shows some of the basic symbols that are used to represent circuit breakers on schematic diagrams.

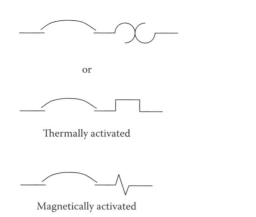

FIGURE 6.117 Circuit breaker symbols.

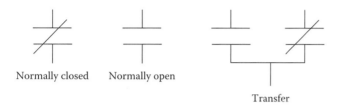

Normally closed   Normally open

Transfer

FIGURE 6.118 Electrical contacts symbols.

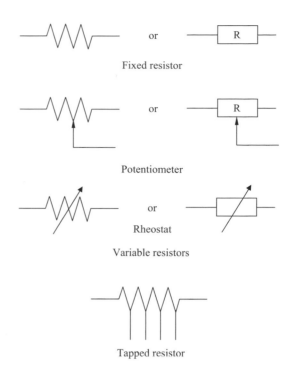

Fixed resistor

Potentiometer

Rheostat

Variable resistors

Tapped resistor

FIGURE 6.119 Resistor symbols.

6.12.1.10 Electrical Contacts

Electrical *contacts* (usually wires) join two conductors in an electrical circuit. *Normally closed* (NC) contacts allow current to flow when the switching device is at rest. *Normally open* (NO) contacts prevent current from flowing when the switching device is at rest. Figure 6.118 shows some of the basic symbols that are used to designate contacts on schematic diagrams.

6.12.1.11 Resistors

Electricity travels through a conductor (wire) easily and efficiently, with almost no other energy released as it passes. On the other hand, electricity cannot travel through a resistor easily. When electricity is forced through a resistor, often the energy in the electricity is changed into another form of energy, such as light or heat. The reason a light bulb glows is because electricity is forced through the tungsten filament, which is a resistor. Resistors are commonly used for controlling the current flowing in a circuit. A *fixed resistor* provides a constant amount of resistance in a circuit. A *variable resistor* (also called a potentiometer) can be adjusted to provide varying amounts of resistance, such as in a dimmer switch for lighting systems. A resistor also acts as a load in a circuit in that there is always a voltage drop across it. Figure 6.119 shows some of the basic symbols that are used to designate resistors on schematic diagrams. A summary of basic electrical symbols that are used to designate electrical components or devices on schematic diagrams is provided in Figure 6.120.

6.12.2 READING PLANT SCHEMATICS

The information provided in the preceding sections on electrical schematic symbols and their functions should help in reading simple schematic diagrams. Many of the schematics that are used in water and wastewater treatment operations are of simple motor circuits, such as the one shown in Figure 6.121, which is for a reversing motor starter.

Important Point: The *reversing starter* has two starters of equal size for a given horsepower motor application. The reversing of a three-phase, squirrel-cage induction motor is accomplished by interchanging any two line connections to the motor. The concern is to properly connect the two starters to the motor so the line feed from one starter is different from the other. Both mechanical and electrical interlocks are used to prevent both starters from closing their line contacts at the same time. Only one set of overloads is required, as the same load current is available for both directions of rotation.

From the schematic shown in Figure 6.121, it can be seen that the motor is connected to the power source of the plant by the three power lines (line leads): L_1, L_2, and L_3. The circuits for forward and reverse drive are also shown. For forward drive, lead L_1 is connected to terminal T_1 (known as a T lead) on the motor. Likewise, L_2 is connected to T_2 and L_3 to T_3. When the three normally open F contacts (F for forward) are closed, these connections are made, current flows between the power source

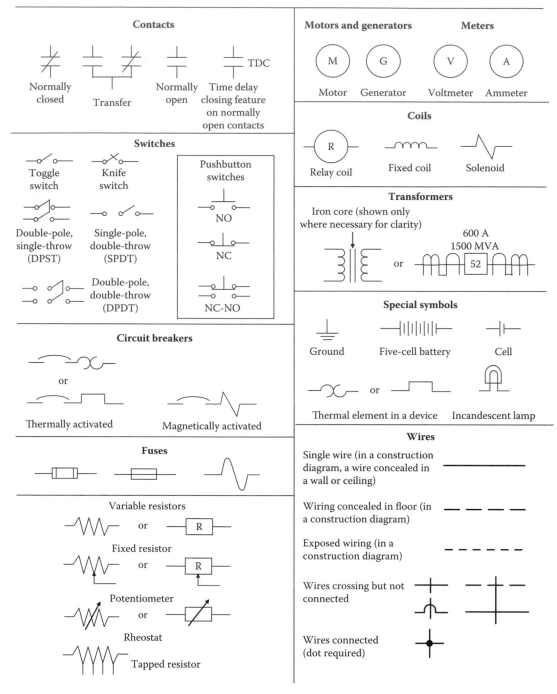

FIGURE 6.120 Summary of electrical symbols.

and the motor, and the motor rotor turns in the forward direction. In the reverse drive condition, the three leads (L_1, L_2, L_3) connect to a set of R contacts. The R contacts reverse the connections of terminals T_1 and T_3 which reverses the rotation of the motor rotor. To reverse the motor, the three normally open R contacts must close and the F contacts must be open. The three fuses located on lines L_1, L_2, and L_3 protect the circuit from overloads.

Moreover, three thermal overload cutouts protect the motor from damage.

Important Point: In actual operation, the circuit shown in Figure 6.121 utilizes a separate control to open or close the forward and reverse contacts. It has a mechanical interlock to make sure the R contacts stay open when the F contacts are closed and *vice versa*.

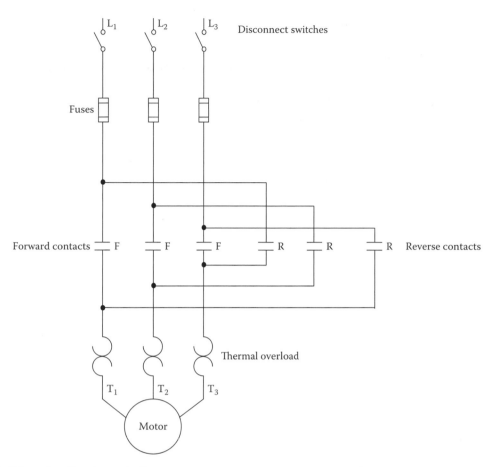

FIGURE 6.121 Schematic of motor controller.

6.13 GENERAL PIPING SYSTEMS AND SYSTEM SCHEMATICS

It would be difficult to imagine any modern water or wastewater treatment process without pipes. Pipes convey all the fluids—that is, the liquids, gases, and semisolids (sludge or biosolids)—either processed or used in plant operations. In addition to conveying raw water or wastewater influent into the plant for treatment, piping systems also bring in water for drinking, for flushing toilets, and for removing wastes. They also carry steam or hot water for heating, refrigerant for cooling, pneumatic and hydraulic fluids (gases and liquids flow under pressure) for equipment operation, and gas for auxiliary uses (e.g., for incineration of biosolids, methane off-gases for heating).

With the exception of in ground distribution and interceptor lines, in water/wastewater operations almost all pipes are visible; however, they may be arranged in complex ways that look confusing to the untrained eye. Notwithstanding the complexity of a piping network, the maintenance operator can trace through a piping system, no matter how complex the system, by reading a piping schematic of the system. In this section, piping systems, piping system schematic diagrams, and schematic symbols typical of water and wastewater operations are described. In addition, we describe schematic diagrams used for hydraulic/pneumatic systems and AC&R systems. We also describe how piping system symbols are used to represent various connections and fittings used in piping arrangements.

6.13.1 PIPING SYSTEMS

Pipe is used for conveying fluids (liquid and gases)—water, steam, wastewater, petroleum, off-gases, and chemicals—and for structural elements such as columns and handrails. Pipe can carry semisolid material (e.g., sludge or biosolids) if it is processed fine enough and mixed with liquid. The choice of the type of pipe is determined by the purpose for which it is to be used. A *pipe* is defined as an enclosed, stationary device that conducts a fluid or a semisolid from one place to another in a controlled way. A *piping system* is a set of pipes and control devices that work together to deliver a fluid where it is needed, in the right amount, and at the proper rate.

Pipes are made of several kinds of solid materials. They can be made of wood, glass, porcelain, cast iron, lead, aluminum, stainless steel, brass, copper, plastic, clay, concrete, and many other materials. Cast iron, steel, wrought iron, brass, copper, plastic, and lead pipes are most commonly used for conveying water and wastewater.

When conveying a substance in a piping system, the flow of the substance must be controlled. The substance must be directed to the place where it is needed. The amount of substance that flows and how quickly it flows must be regulated. The flow of a substance in a piping system is controlled, adjusted, and regulated by *valves*. In this section, the symbols that represent valves on various kinds of piping schematics are described, as are basic *joints* and other fittings used in piping systems.

6.13.2 PIPING SYMBOLS: GENERAL

To read piping schematics correctly, maintenance operators must identify and understand the symbols used. It is not necessary to memorize them, but maintenance operators should keep a table of basic symbols handy and refer to it whenever the need arises. In the following sections, we describe most of the common symbols used in general piping systems.

6.13.3 PIPING JOINTS

The joints between pipes, fittings, and valves may be screwed (or threaded), flanged, or welded; for nonferrous materials, joints may be soldered. A *joint* is the connection between two elements in a piping system. The five major types of joints are screwed (or threaded), welded, flanged, bell-and-spigot, and soldered.

6.13.3.1 Screwed Joints

Screwed joints are threaded together; that is, screwed joints can be made up tightly by simply screwing the threads together. Figure 6.122 shows the schematic symbol for a screwed joint. Screwed joints are usually used with pipe ranging from 1/4 in. to about 6 in. in diameter.

Important Point: In most screwed joints, the threads are on the inside of the fitting and on the outside of the pipe.

6.13.3.2 Welded Joints

Piping construction employing *welded joints* is almost universal practice today. This kind of joint is used when the coupling must be permanent, particularly for higher pressure and temperature conditions. Figure 6.123 shows the schematic symbol for welded joints.

Important Point: Welded joints may be either socket welded or butt welded.

FIGURE 6.122 Screwed joint symbol.

FIGURE 6.123 Welded joint symbol.

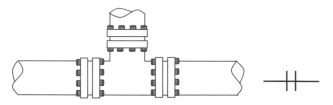

FIGURE 6.124 Flanged joint.

FIGURE 6.125 Symbol for bell-and-spigot joint.

FIGURE 6.126 Symbol for soldered joint.

6.13.3.3 Flanged Joints

Flanged joints are made by bolting two flanges together with a gasket between the flange faces. Flanges may be attached to the pipe, fitting, or appliance by means of a screwed, joint, by welding, by lapping the pipe, or by being cast integrally with the pipe, fitting, or appliance. A flanged joint is shown in Figure 6.124, along with the symbol used on schematics.

6.13.3.4 Bell-and-Spigot Joints

Cast-iron pipes for handling wastewater, in particular, usually fit together in a special way. When each fitting and section of pipe is cast, one end is made large enough to fit loosely around the opposite end of another fitting or pipe. When this type of fitting is connected, it forms a joint called a *bell-and-spigot joint*. Figure 6.125 shows a bell-and-spigot joint and the symbol used on schematics.

6.13.3.5 Soldered Joints

Nonferrous fittings, such as copper piping and fittings, are often joined by *soldering* them with a torch or soldering iron and then melting solder on the joint. The solder flows into the narrow space between the two mating parts and seals the joint. Figure 6.126 shows the schematic symbol for a soldered joint.

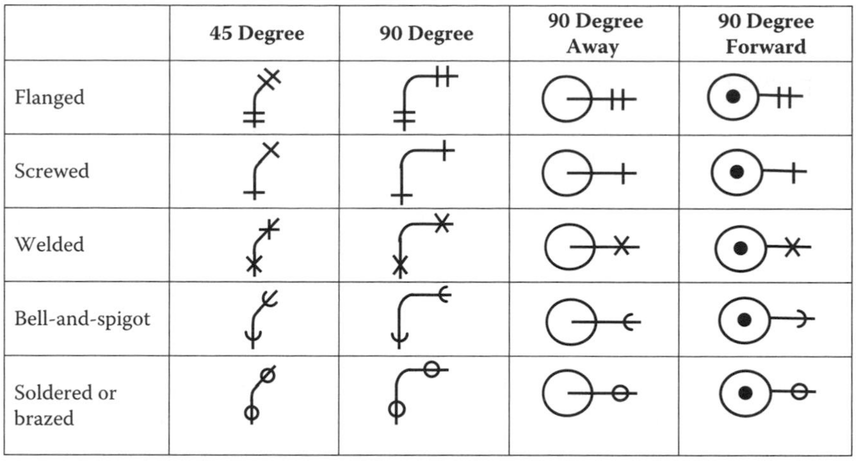

	45 Degree	90 Degree	90 Degree Away	90 Degree Forward
Flanged				
Screwed				
Welded				
Bell-and-spigot				
Soldered or brazed				

FIGURE 6.127 Symbols for elbow fittings.

6.13.3.6 Symbols for Joints and Fittings

Figure 6.127 shows most of the common schematic piping joint and fitting symbols (for elbows only) used today; however, it does not show all the symbols used on all piping schematics.

6.13.4 VALVES

Any water or wastewater operation will have many valves that are important components of different piping systems. Simply as a matter of routine, a maintenance operator must be able to identify and locate valves to inspect them, adjust them, and repair or replace them. For this reason, the maintenance operator should be familiar with schematic diagrams that include valves and with schematic symbols used to designate valves.

6.13.4.1 Valves: Definition and Function

A *valve* is defined as any device by which the flow of fluid may be started, stopped, or regulated by a movable part that opens or obstructs passage. As applied in fluid power systems, valves are used for controlling the flow, the pressure, and the direction of the fluid flow through a piping system. The fluid may be a liquid, a gas, or some loose material in bulk (such as a biosolids slurry). Designs of valves vary, but all valves have two features in common:

- A passageway through which fluid can flow
- Some kind of movable (usually machined) part that opens and closes the passageway.

Important Point: It is all but impossible, obviously, to operate a practical fluid power system without some means of controlling the volume and pressure of the fluid and directing the flow of fluid to the operating units. This is accomplished by the incorporation of different types of valves.

Whatever type of valve is used in a system, it must be accurate in the control of fluid flow and pressure and the sequence of operation. Leakage between the valve element and the valve seat is reduced to a negligible quantity by precision-machined surfaces, resulting in carefully controlled clearances. This is, of course, one of the very important reasons for minimizing contamination in fluid power systems. Contamination causes valves to stick, plugs small orifices, and causes abrasions of the valve seating surfaces, which results in leakage between the valve element and valve seat when the valve is in the closed position. Any of these can result in inefficient operation or complete stoppage of the equipment. Valves may be controlled manually, electrically, pneumatically, mechanically, hydraulically, or by combinations of two or more of these methods. Factors that determine the method of control include the purpose of the valve, the design and purpose of the system, the location of the valve within the system, and the availability of the source of power.

Valves are made from bronze, cast iron, steel, Monel®, stainless steel, and other metals. They are also made from plastic and glass. Special valve trim is used where seating and sealing materials are different from the basic material of construction; valve trim usually refers to those internal parts of a valve controlling the flow and in physical contact with the line fluid. Valves are made in a full range of sizes, with matching pipe and tubing sizes. Actual valve size is based upon the internationally agreed-upon definition of nominal size. *Nominal size* (DN) is a numerical designation of size that is common to all components in a piping system other than components designated by outside diameters. It is a convenient number for reference purposes and is only loosely related to manufacturing dimensions.

Valves are made for service at the same pressures and temperatures to which the piping and tubing are subject. Valve pressures are based on the internationally agreed-upon definition of nominal pressure. *Nominal pressure*

TABLE 6.4
Valve Special Features

High temperature	Valves are usually able to operate continuously on services above 250°C.
Cryogenic	Valves will operate continuously on services in the range of –50°C to 196°C.
Bellows sealed	Valves are glandless designs having a metal bellows for stem sealing.
Actuated	Valves may be operated by a gear box, pneumatic or hydraulic cylinder (including diaphragm actuator), or electric motor and gear box.
Fire-tested design	Valves have passed a fire test procedure specified in an appropriate inspection standard.

(PN) is a pressure that is conventionally accepted or used for reference purposes. All equipment of the same nominal size designated by the same nominal pressure number must have the same mating dimensions appropriate to the type of end connections. The permissible working pressure depends on materials, design, and working temperature and should be selected from the relevant pressure/temperature tables. The pressure rating of many valves is designated under the ANSI classification system. The equivalent class ratings to PN ratings are based on international agreement. Usually, valve end connections are classified as flanged, threaded, or other. Valves are also covered by various codes and standards, as are the other components of piping and tubing systems. Many valve manufacturers offer valves with special features. Table 6.4 lists a few of these special features; however, this is not an exhaustive list. For more details of other features, the various manufacturers should be consulted.

6.13.4.2 Valve Construction

Figure 6.128 shows the basic construction and principle of operation of a common valve type. Fluid flows into the valve through the inlet. The fluid flows through passages in the body and past the opened element that closes the valve. It then flows out of the valve through the outlet or discharge. If the closing element is in the closed position, the passageway is blocked. Fluid flow is stopped at that point. The closing element keeps the flow blocked until the valve is opened again. Some valves are opened automatically, and others are controlled by manually operated handwheels. Other valves, such as check valves, operate in response to pressure or the direction of flow.

6.13.4.3 Types of Valves

The types of valves covered in this text include the following

- Ball valve
- Cock valve
- Gate valve
- Globe valve
- Check valve

Each of these valves is designed to control the flow, pressure, or direction of fluid flow or for some other special application. With a few exceptions, these valves take their names from the type of internal element that controls the passageway. The exception is the check valve.

6.13.4.3.1 Ball Valve

Ball valves, as the name implies, are stop valves that use a ball to stop or start a flow of fluid. The ball performs the same function as the disk in other valves. As the valve handle is turned to open the valve, the ball rotates to a

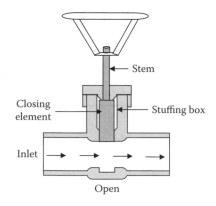
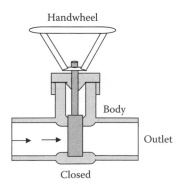

FIGURE 6.128 Basic valve operation.

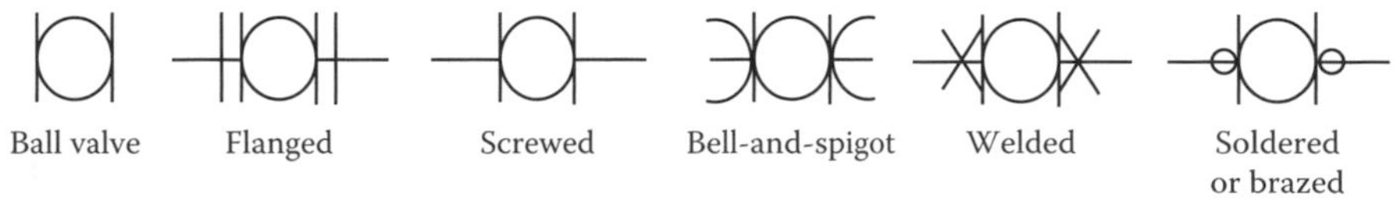

FIGURE 6.129 Symbols for ball valves.

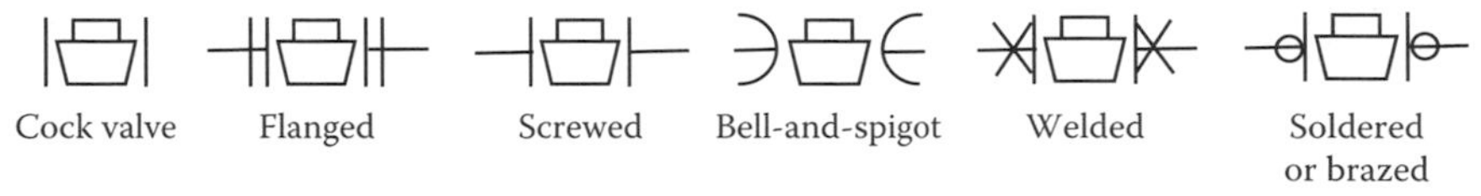

FIGURE 6.130 Symbols for cock valves.

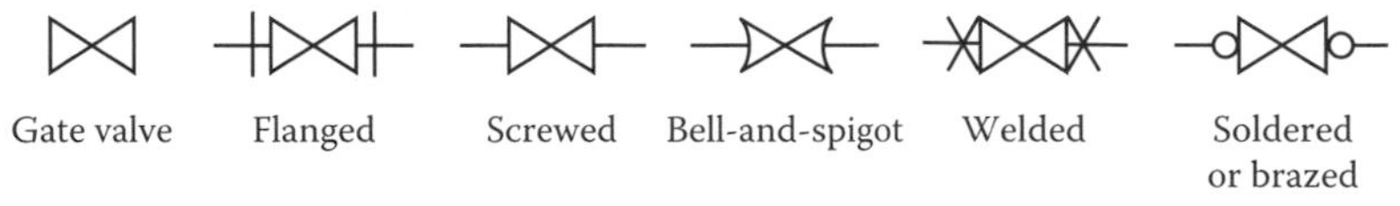

FIGURE 6.131 Symbols for gate valves.

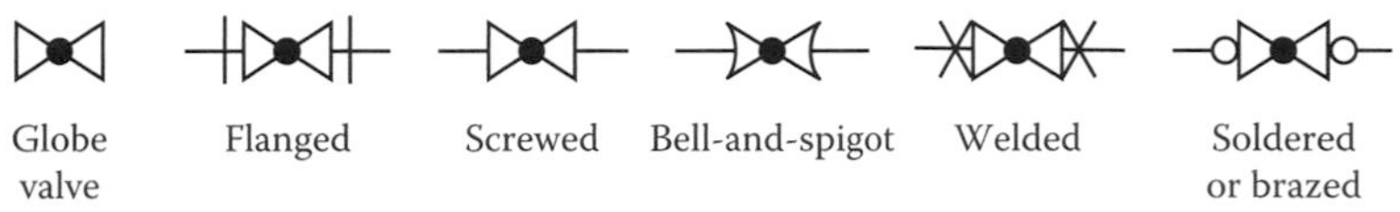

FIGURE 6.132 Symbols for globe valves.

point where part or all of the hole through the ball is in line with the valve body inlet and outlet, allowing fluid to flow through the valve. When the ball is rotated so the hole is perpendicular to the flow openings of the valve body, the flow of fluid stops. Most ball valves are the quick-acting type. They require only a 90° turn to completely open or close the valve. Others are operated by planetary gears. This type of gearing allows the use of a relatively small handwheel and operating force to operate a fairly large valve. The gearing does, however, increase the operating time for the valve. Some ball valves also contain a swing check located within the ball to give the valve a check valve feature. The two main advantages of using ball valves are that: (1) the fluid can flow through it in either direction, as desired, and (2) when closed, pressure in the line helps to keep it closed. Symbols that represent the ball valve in schematic diagrams are shown in Figure 6.129.

6.13.4.3.2 Cock Valves

The *cock* valve, like the gate valve, has only two positions—on and off. The difference is in the speed of operation. The gate valve, for example, opens and closes gradually. The cock valve opens and closes quickly and is used when the flow must be started or stopped quickly. The schematic symbols for a cock valve appear in Figure 6.130.

6.13.4.3.3 Gate Valves

Gate valves are used when a straight-line flow of fluid and minimum flow restriction are necessary; they are the most common type of valve found in a water distribution sys-

tem. Gate valves are so named because the part that either stops or allows flow through the valve acts somewhat like a gate. The gate is usually wedge shaped. When the valve is wide open the gate is fully drawn up into the valve bonnet. This leaves an opening for flow through the valve the same size as the pipe in which the valve is installed. For these reasons, the pressure loss (pressure drop) through these types of valves is about equal to the loss in a piece of pipe of the same length. Gate valves are not suitable for throttling (controlling the flow by means of intermediate steps between fully open and fully closed) purposes. The control of flow is difficult because of the design of the valve, and the flow of fluid slapping against a partially open gate can cause extensive damage to the valve. The schematic symbols for a gate valve appear in Figure 6.131.

Important Point: Gate vales are well suited to service on equipment in distant locations, where they may remain in the open or closed position for a long time. Generally, gate valves are not installed where they will have to be operated frequently because they require too much time to transition from fully open to closed (AWWA, 1998).

6.13.4.3.4 Globe Valves

Probably the most common valve type in existence, the *globe* valve is commonly used for water faucets and other household plumbing. As illustrated in Figure 6.132, the valves have a circular disk—the "globe"—that presses against the valve seat to close the valve. The disk is the

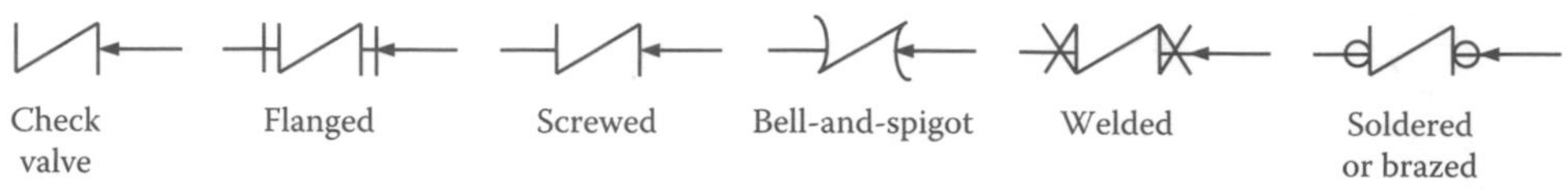

FIGURE 6.133 Symbols for check valves.

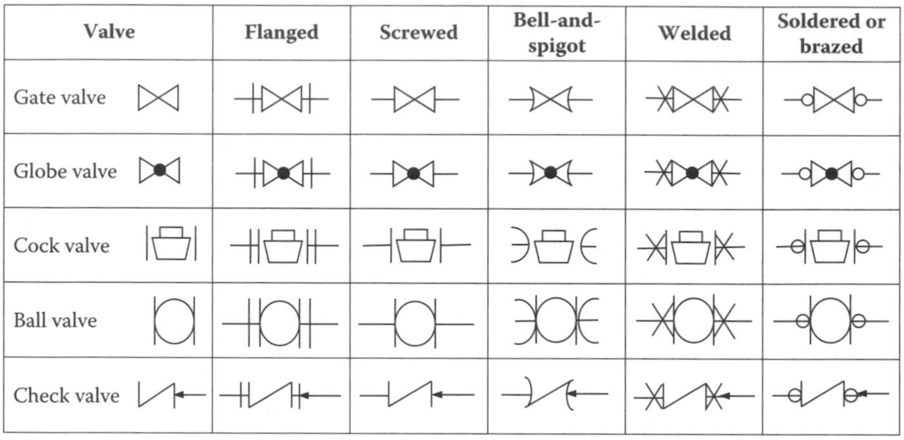

Valve	Flanged	Screwed	Bell-and-spigot	Welded	Soldered or brazed
Gate valve					
Globe valve					
Cock valve					
Ball valve					
Check valve					

FIGURE 6.134 Standard symbols for valves.

part of the globe valve that controls flow. The disk is attached to the valve stem. Fluid flow through a globe valve is at right angles to the direction of flow in the conduits. Globe valves seat very tightly and can be adjusted with fewer turns of the wheel than gate valves; thus, they are preferred for applications that call for frequent opening and closing. On the other hand, globe valves create high head loss when fully open, so they are not suited in systems where head loss is critical. The schematic symbols that represent the globe valve are also shown in Figure 6.132.

Important Point: The globe valve should never be jammed in the open position. After a valve is fully opened, the handwheel should be turned toward the closed position approximately one half turn. Unless this is done, the valve is likely to seize in the open position, making it difficult, if not impossible, to close the valve. Another reason for not leaving a globe valve in the fully open position is that it is sometimes difficult to determine if the valve is open or closed (Globe Valve, 1998).

6.13.4.3.5 Check Valves

Check valves are usually self-acting and designed to allow the flow of fluid in one direction only. They are commonly used at the discharge of a pump to prevent backflow when the power is turned off. When the direction of flow is moving in the proper direction, the valve remains open. When the direction of flow reverses, the valve closes automatically from the fluid pressure against it. Several types of check valves are used in water/wastewater operations, including:

- Slanting disk check valves
- Cushioned swing check valves
- Rubber flapper swing check valves
- Double door check valves
- Ball check valves
- Foot valves
- Backflow prevention devices

In each case, pressure from the flow in the proper direction pushes the valve element to an open position. Flow in the reverse direction pushes the valve element to a closed position. Symbols that represent check valves are shown in Figure 6.133, and Figure 6.134 shows the standard symbols for valves discussed in this text.

Important Point: Check valves are also commonly referred to as *non-return* or *reflux valves*.

6.14 HYDRAULIC AND PNEUMATIC SYSTEM SCHEMATIC SYMBOLS

In water and wastewater operations, hydraulic and pneumatic systems are very common. As a maintenance operator, training and the skill to solve various problems that may occur in hydraulic and pneumatics systems in the plant are paramount. The purpose of this section is to provide the basic operating principles of hydraulic and pneumatic systems so maintenance operators will be able to diagnose and fix hydraulic and pneumatic equipment. To fix hydraulic and pneumatic equipment, maintenance operators must be familiar with the symbols used in the diagrams of these systems.

Lines and line functions			
Line, working	 ———————	Line, pilot	 —— — ——
Line, drain	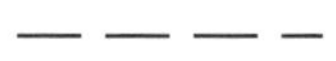 — — — —	Connector (dot 5× thickness of line)	●
Line, flexible		Lines, joining	
Lines, crossing	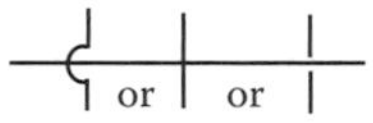 or | or |	Direction of flow	→
Line to reservoir, above fluid level and below fluid level		Line to vented manifold	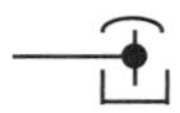
Pumps			
Pump, fixed displacement		Pump, variable displacement	
Motors			
Motor, hydraulic, fixed displacement		Motor, pneumatic, variable displacement	
Valves			
Valve, check	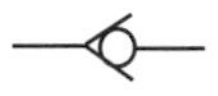	Valve, manual shutoff	
Valve, maximum pressure (relief) normally closed	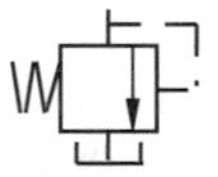	Valve, basic symbol, single flow path	
Valve, basic symbol, multiple flow paths	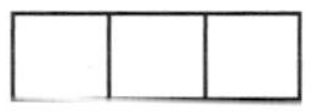	Valve, single flow path, normally closed	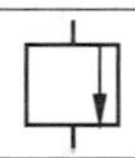
Valve, single flow path, normally open	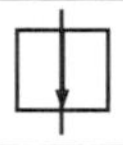	Valve, multiple flow path, closed position	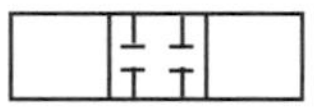

FIGURE 6.135 Symbols for hydraulic and pneumatic components.

6.14.1 FLUID POWER SYSTEMS

Fluid power is the generation, control, and application of smooth, effective power of pumped or compressed fluids (either liquids or gases) to provide force and motion to mechanisms. This force and motion may be in the form of pushing, pulling, rotating, regulating, or driving. Fluid power includes *hydraulics*, which involves liquids, and *pneumatics*, which involves gases. Liquids and gases are similar in many respects. They do not, however, behave in the same way under pressure. Moreover, even the terms used for some of the basic components of both types of fluid power systems are different—and so are some of the symbols. In any hydraulic or pneumatic fluid power system, the basic components consist of the following:

• A *reservoir* stores the fluid in a hydraulic system; in a pneumatic system, this reservoir is referred to as a *receiver*.

• The *pump* in a hydraulic system provides the pressure that results in work; in a pneumatic system, it is referred to as a *compressor*.

• In both systems, the *actuator* reacts to the pressure of the fluid and does the work.

• *Valves* direct and adjust the flow of fluid in both systems.

• The lines that circulate the fluid are called *piping* in a hydraulic system and *tubing* in a pneumatic system.

6.14.2 SYMBOLS USED FOR HYDRAULIC AND PNEUMATIC COMPONENTS

A schematic diagram explains how a hydraulic or pneumatic system operates. It is made up of symbols. Keep in mind that some of these symbols are combinations of other, more basic symbols. To understand the operation of a hydraulic or pneumatic system, the maintenance operator must be familiar with the symbols shown in Figure 6.135.

Refrigerant discharge	——— RD ———
Refrigerant suction	– – – RD – – –
Brine supply	——— B ———
Brine return	– – – BR – – –
Condenser water flow	——— C ———
Condenser water return	– – – CR – – –
Chilled water supply	——— CH ———
Chilled water return	– – – CHR – – –
Fill line	——— FILL ———
Humidification line	– · – · H · – · – ·
Drain	——— D ———

FIGURE 6.136 Symbols for refrigeration lines.

6.14.3 AC&R System Schematic Symbols

Major components of AC&R systems and system operation were explained earlier. This section describes the symbols for the components described earlier. An understanding of AC&R schematic symbols enables the maintenance operator to more effectively and efficiently troubleshoot AC&R systems.

Important Point: In practice, refrigeration piping diagrams may be either orthographic or isometric projections.

Air distribution systems can be either double-line or single-line schematics. For our purposes, we refer to diagrams for both systems as *schematics*. The importance rests not in the type of drawing but, instead, in recognition of the symbols used in both systems.

6.14.4 Schematic Symbols Used in Refrigeration Systems

Many of the components used in refrigeration systems include components described in other sections of this text. These components include electrical, piping, hydraulic, and pneumatic equipment. Refrigeration piping, however, conveys both gases and liquids; therefore, the symbols used sometimes differ from symbols used in simpler piping systems. In the following sections, we describe both these differences and the symbols used.

6.14.4.1 Refrigeration Piping Symbols

In refrigeration systems, three different kinds of piping are used, depending on whether the refrigerant is a gas, a liquid, or both; for example, the discharge line going from the compressor to the condenser carries hot gas at high pressure. The liquid line going from the condenser to the receiver and from the receiver to the expansion valve carries lower temperature liquid at the same high pressure as the discharge side. The suction line going from the evaporator to the compressor intake valve conveys relatively cool refrigerant vapor at low pressures. Figure 6.136 shows symbols for refrigeration lines used on refrigeration schematic drawings.

6.14.4.2 Refrigeration Fittings Symbols

Refrigeration fittings are connected by brazing, threading, flanging, or welding. Figure 6.137 shows the symbols for screwed or threaded connections. Symbols for fittings with other connections have the same body shapes.

Bushing	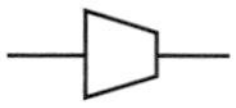	Tee	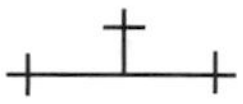
Connection, bottom	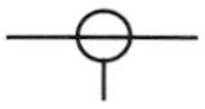	Cap	
Coupling (joint)		Connection, top	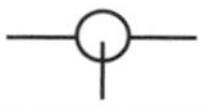
Elbow, 90°		Cross	
Elbow, turned down	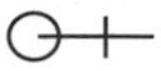	Elbow, 45°	
Elbow, reducing		Reducer, concentric	

FIGURE 6.137 Symbols for refrigeration fittings.

Blueprint Reading   **263**

Air line		Safety valve	
Pressure-reducing		Cock valve	
Quick-opening	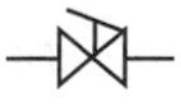	Control, two-way	
Solenoid		Control, three-way	

FIGURE 6.138 Symbols for refrigeration valves.

Air eliminator		Expansion joint	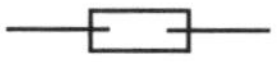
Refrigerant filter and strainer	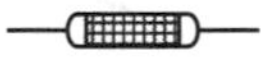	Vibration absorber	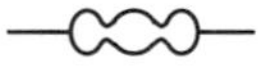
Oil separator		Line filter	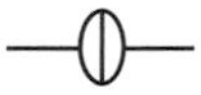
Drier	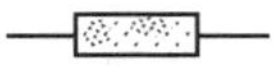	Heat exchanger	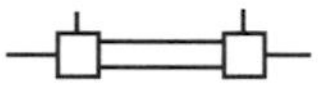

FIGURE 6.139 Symbols for refrigeration accessories.

Compressor, centrifugal		Compressor, rotary	
Evaporator, finned type, natural convection	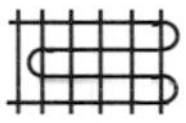	Condenser, air-cooled, forced air	
Condenser, water-cooled shell and tube	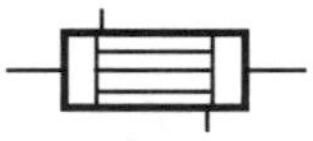	Condenser, evaporative	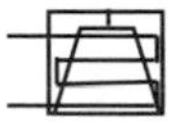
Cooling tower		Receiver, horizontal	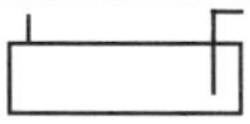
Spray pond		Receiver, vertical	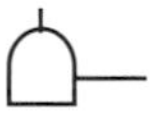
Electric motor (number indicates horsepower)		Engine (letter indicates fuel)	

FIGURE 6.140 Symbols for refrigeration.

6.14.4.3 Refrigeration Valve Symbols

Like refrigeration fittings, valves used in refrigeration systems are brazed, threaded, flanged, or welded. Figure 6.138 shows the symbols for various valves.

6.14.4.4 Refrigeration Accessory Symbols

Specialized piping accessories are used in refrigeration systems. These accessories include expansion joints, driers, strainers, heat exchangers, filters, and other such devices. Symbols for refrigeration accessories are shown in Figure 6.139.

6.14.4.5 Refrigeration Component Symbols

In addition to piping, fittings, accessories, and valves, various major components make up a refrigeration system. These components include a compressor, an evaporator, condensers, refrigerant receivers and accumulators, and heat exchangers. Symbols for these components are shown in Figure 6.140.

Duct		Exhaust duct section	
Splitter damper		Turning vanes	
Supply outlet, ceiling diffuser		Direction of flow	
Supply duct section		Access door	

FIGURE 6.141 Symbols for ducting.

6.14.4.6 Schematic Symbols Used in AC&R Air Distribution System

A major subsystem of an AC&R system is air distribution. Air is distributed to various spaces that require cooling or heating (i.e., if the system is equipped with both heating and cooling equipment). The air is delivered through ducting systems. Figure 6.141 shows standard symbols for ducting schematics.

CHAPTER REVIEW QUESTIONS

The symbols below represent various components used in blueprints or schematics for electrical, welding, hydraulic, pneumatic, piping, and AC&R systems. Identify the meaning of each symbol. Use the space alongside each symbol to describe what the symbol represents.

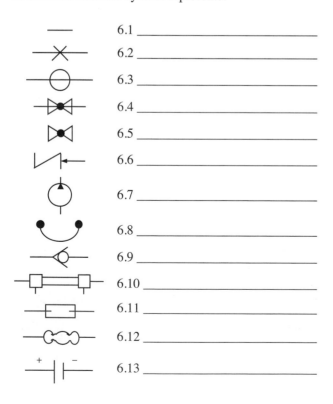

6.1 _____

6.2 _____

6.3 _____

6.4 _____

6.5 _____

6.6 _____

6.7 _____

6.8 _____

6.9 _____

6.10 _____

6.11 _____

6.12 _____

6.13 _____

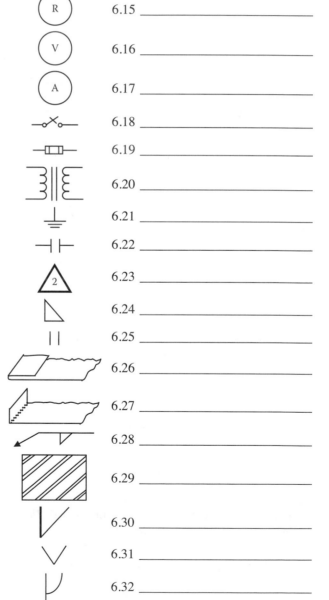

6.14 _____

6.15 _____

6.16 _____

6.17 _____

6.18 _____

6.19 _____

6.20 _____

6.21 _____

6.22 _____

6.23 _____

6.24 _____

6.25 _____

6.26 _____

6.27 _____

6.28 _____

6.29 _____

6.30 _____

6.31 _____

6.32 _____

6.33 If the scale box says 4, it means the object is drawn at _____.

6.34 The _____ is the basic identification assigned to a drawing.

6.35 When two lines cross, the sum of any two adjacent angles is always _____ degrees.

6.36 Ruled line and zig-zag line for long breaks indicate _____.

6.37 This type of line represents the outline of an object: _____.

6.38 A _____ displays three sides of an object.

6.39 The number and selection of views are governed by the _____ and _____ of the object.

6.40 _____ are the maximum and minimum sizes indicated by a toleranced dimension.

6.41 Location dimensions are usually made from either a _____ or a _____.

6.42 A pump casing is called the _____.

6.43 _____ are used to eliminate the raw edge and also to stiffen the material.

6.44 When drawing a pattern on sheet metal, you should provide some extra metal for making a _____.

6.45 A _____ is set to open and bleed off some of the air if the pressure becomes to high.

6.46 A _____ is easier to compress than a _____.

6.47 What kind of joint joins the edges of two metals without overlapping? _____

6.48 An _____ drawing shows the physical locations of the electric lines in a plant building.

6.49 The _____ shows how electrical power is distributed in all buildings.

6.50 Another name for a schematic diagram is a _____ diagram.

REFERENCES AND SUGGESTED READING

ANSI. 1973a. *Line Conventions and Lettering*. New York: American National Standards Institute.

ANSI. 1973b. *Dimension and Tolerance*. New York: American National Standards Institute.

ANSI. 1980. *ANSI Y14.1: Blueprint Sheet Size*. New York: American National Standards Institute.

ANSI. 1982. *Dimensioning and Tolerancing*. New York: American National Standards Institute.

ANSI/AWS. 1991. *A2.4-93 Welding: An American National Standard*. New York: American National Standards Institute/American Welding Society.

AWS. 1996. *Materials and Applications. Part I. Welding Handbook*, Vol. 3, 8th ed. Miami, FL: American Welding Society.

AWWA. 1998. *Water Transmission and Distribution*, 2nd ed. Denver, CO: American Water Works Association.

Brown, W.C. 1989. *Blueprint Reading for Industry*. South Holland, IL: The Goodheart–Wilcox Company.

Cheremisinoff, N.P. and Cheremisinoff, P.N. 1989. *Pumps, Compressors, Fans: Pocket Handbook*. Boca Raton, FL: CRC Press.

Globe Valve 1998. Integrated Publishing (www.tpub.com).

Olivo, C.T. and Olivo, T.P. 1999. *Basic Blueprint Reading and Sketchings*, 7th ed. Albany, NY: Delmar Publishers.

Spellman, F.R. 2000. *Spellman's Standard Handbook for Wastewater Operators*, Vol. 3. Boca Raton, FL: CRC Press.

Spellman, F.R. and Drinan, J. 2000a. *Pumping*. Boca Raton, FL: CRC Press.

Spellman, F.R. and Drinan, J. 2000b. *Electricity*. Boca Raton, FL: CRC Press.

Spellman, F.R. and Drinan, J. 2001. *Piping and Valves*. Boca Raton, FL: CRC Press.

Valves. 1998. Integrated Publishing (www.tpub.com).

7 Water Hydraulics

Beginning students of water hydraulics and its principles often come to the subject matter with certain misgivings. For example, water and wastewater operators quickly learn on the job that their primary operational/maintenance concerns involve a daily routine of monitoring, sampling, laboratory testing, operation, and process maintenance. How does water hydraulics relate to daily operations? The hydraulic functions of the treatment process have already been designed into the plant. Why learn water hydraulics at all?

Simply put, while having hydraulic control of the plant is obviously essential to the treatment process, maintaining and ensuring continued hydraulic control is also essential. No water/wastewater facility (and/or distribution collection system) can operate without proper hydraulic control. The operator must know what hydraulic control is and what it entails to know how to ensure proper hydraulic control. Moreover, in order to understand the basics of piping and pumping systems, water and wastewater maintenance operators must have a fundamental knowledge of basic water hydraulics.

Spellman and Drinan (2001)

Note: The practice and study of water hydraulics are not new. Even in medieval times, water hydraulics was not new, as "Medieval Europe had inherited a highly developed range of Roman hydraulic components" (Magnusson, 2001). The basic conveyance technology, based on low-pressure systems of pipe and channels, had already been established. In studying "modern" water hydraulics, it is important to remember that the science of water hydraulics is the direct result of two immediate and enduring problems: "the acquisition of freshwater and access to continuous strip of land with a suitable gradient between the source and the destination" (Magnusson, 2001).

7.1 WHAT IS WATER HYDRAULICS?

The word "hydraulic" is derived from the Greek words *hydro* (meaning water) and *aulis* (meaning pipe). Originally, the term referred only to the study of water at rest and in motion (flow of water in pipes or channels). Today, it is taken to mean the flow of any liquid in a system.

What is a liquid? In terms of hydraulics, a liquid can be either oil or water. In fluid power systems used in modern industrial equipment, the hydraulic liquid of choice is oil. Some common examples of hydraulic fluid power systems include automobile braking and power steering systems, hydraulic elevators, and hydraulic jacks or lifts. Probably the most familiar hydraulic fluid power systems in water/wastewater operations are those in dump trucks, front-end loaders, graders, and earth-moving and excavations equipment. In this text, we are concerned with liquid water.

Many find the study of water hydraulics difficult and puzzling (especially the licensure examination questions), but we know it is not mysterious or difficult. It is the function or output of practical applications of the basic principles of water physics. Because water/wastewater treatment is based on the principles of water hydraulics, concise, real-world training is necessary for operators who must operate the plant and for those sitting for state licensure/certification examinations.

7.2 BASIC CONCEPTS

Air pressure (at sea level) = 14.7 pounds per square inch (psi)

This relationship is important because our study of hydraulics begins with air. A blanket of air many miles thick surrounds the Earth. The weight of this blanket on a given square inch of the Earth's surface will vary according to the thickness of the atmospheric blanket above that point. As shown above, at sea level, the pressure exerted is 14.7 pounds per square inch (psi). On a mountain top, air pressure decreases because the blanket is not as thick.

$$1 \text{ ft}^3 \text{ of water} = 62.4 \text{ lb}$$

This relationship is also important; note that both cubic feet and pounds are used to describe a volume of water. A defined relationship exists between these two methods of measurement. The specific weight of water is defined relative to a cubic foot. One cubic foot of water weighs 62.4 lb (see Figure 7.1). This relationship is true only at a temperature of 4°C and at a pressure of 1 atmosphere, conditions referred to as *standard temperature and pressure* (STP). Note that 1 atmosphere = 14.7 lb/in.^{2} at sea level and 1 ft^3 of water contains 7.48 gal.

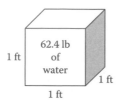

FIGURE 7.1 One cubic foot of water weighs 62.4 lb.

The weight varies so little that, for practical purposes, this weight is used for temperatures ranging from 0 to 100°C. One cubic inch of water weighs 0.0362 lb. Water 1 ft deep will exert a pressure of 0.43 lb/in.² on the bottom area (12 in. × 0.0362 lb/in.³). A column of water 2 ft high exerts 0.86 psi (2 ft × 0.43 psi/ft); one 10 ft high exerts 4.3 psi (10 ft × 0.43 psi/ft); and one 55 ft high exerts 23.65 psi (55 ft × 0.43 psi/ft). A column of water 2.31 feet high will exert 1.0 psi (2.31 ft × 0.43 psi/ft). To produce a pressure of 50 psi requires a 115.5-ft water column (50 psi × 2.31 ft/psi).

Remember the important points being made here:

1. 1 ft³ of water = 62.4 lb (see Figure 7.1).
2. A column of water 2.31 ft high will exert 1.0 psi.

Another relationship is also important:

$$1 \text{ gal water} = 8.34 \text{ lb}$$

At standard temperature and pressure, 1 ft³ of water contains 7.48 gal. With these two relationships, we can determine the weight of 1 gal of water. This is accomplished by:

Weight of 1 gal of water = 62.4 lb ÷ 7.48 gal = 8.34 lb/gal

Thus,

$$1 \text{ gal water} = 8.34 \text{ lb}$$

Note: Further, this information allows cubic feet to be converted to gallons by simply multiplying the number of cubic feet by 7.48 gal/ft³.

■ **EXAMPLE 7.1**

Problem: Find the number of gallons in a reservoir that has a volume of 855.5 ft³.

Solution:

$$855.5 \text{ ft}^3 \times 7.48 \text{ gal/ft}^3 = 6399 \text{ gal (rounded)}$$

Note: As mentioned in Chapter 4, the term *head* is used to designate water pressure in terms of the height of a column of water in feet; for example, a 10-foot column of water exerts 4.3 psi. This can be referred to as 4.3-psi pressure or 10 feet of head.

7.2.1 Stevin's Law

Stevin's law deals with water at rest. Specifically, it states: "The pressure at any point in a fluid at rest depends on the distance measured vertically to the free surface and the density of the fluid." Stated as a formula, this becomes

$$p = w \times h \tag{7.1}$$

where:

p = pressure in pounds per square foot (lb/ft² or psf).
w = density in pounds per cubic foot (lb/ft³).
h = vertical distance in feet.

■ **EXAMPLE 7.2**

Problem: What is the pressure at a point 18 ft below the surface of a reservoir?

Solution: To calculate this, we must know that the density of the water (w) is 62.4 lb/ft³.

$$p = w \times h$$

$$p = 62.4 \text{ lb/ft}^3 \times 18 \text{ ft} = 1123 \text{ lb/ft}^2 \text{ (psf)}$$

Water/wastewater operators generally measure pressure in pounds per square inch rather than pounds per square foot; to convert, divide by 144 in.²/ft² (12 in. × 12 in. = 144 in.²):

$$p = \frac{1123 \text{ psf}}{144 \text{ in.}^2/\text{ft}^2} = 7.8 \text{ lb/in.}^2 \text{ or psi (rounded)}$$

7.3 DENSITY AND SPECIFIC GRAVITY

Table 7.1 shows the relationships among temperature, specific weight, and density of water. When we say that iron is heavier than aluminum, we say that iron has a greater density than aluminum. In practice, what we are really saying is that a given volume of iron is heavier than the same volume of aluminum.

Note: What is density? Density is the *mass per unit volume* of a substance.

Suppose you have a tub of lard and a large box of cold cereal, each having a mass of 600 g. The density of the cereal would be much less than the density of the lard because the cereal occupies a much larger volume than the lard occupies. The density of an object can be calculated by using the formula:

$$\text{Density} = \frac{\text{Mass}}{\text{Volume}} \tag{7.2}$$

TABLE 7.1
Water Properties (Temperature, Specific Weight, and Density)

Temperature (°F)	Specific Weight (lb/ft³)	Density (slugs/ft³)	Temperature (°F)	Specific Weight (lb/ft³)	Density (slugs/ft³)
32	62.4	1.94	130	61.5	1.91
40	62.4	1.94	140	61.4	1.91
50	62.4	1.94	150	61.2	1.90
60	62.4	1.94	160	61.0	1.90
70	62.3	1.94	170	60.8	1.89
80	62.2	1.93	180	60.6	1.88
90	62.1	1.93	190	60.4	1.88
100	62.0	1.93	200	60.1	1.87
110	61.9	1.92	210	59.8	1.86
120	61.7	1.92			

In water and wastewater treatment, perhaps the most common measures of density are pounds per cubic foot (lb/ft³) and pounds per gallon (lb/gal):

- 1 ft³ of water weighs 62.4 lb; density = 62.4 lb/ft³.
- 1 gal of water weighs 8.34 lb; density = 8.34 lb/gal.

The density of a dry material, such as cereal, lime, soda, or sand, is usually expressed in pounds per cubic foot. The density of a liquid, such as liquid alum, liquid chlorine, or water, can be expressed either as pounds per cubic foot or as pounds per gallon. The density of a gas, such as chlorine gas, methane, carbon dioxide, or air, is usually expressed in pounds per cubic foot.

As shown in Table 7.1, the density of a substance such as water changes slightly as the temperature of the substance changes. This occurs because substances usually increase in volume (size) by expanding as they become warmer. Because of this expansion with warming, the same weight is spread over a larger volume, so the density is lower when a substance is warm than when it is cold.

Note: What is specific gravity? Specific gravity is the weight (or density) of a substance compared to the weight (or density) of an equal volume of water. The specific gravity of water is 1.

This relationship is easily seen when a cubic foot of water, which weighs 62.4 lb, is compared to a cubic foot of aluminum, which weighs 178 lb. Aluminum is 2.8 times heavier than water.

It is not that difficult to find the specific gravity of a piece of metal. All you have to do is weigh the metal in air, then weigh it under water. The loss of weight is the weight of an equal volume of water. To find the specific gravity, divide the weight of the metal by its loss of weight in water.

$$\text{Specific gravity} = \frac{\text{Weight of substance}}{\text{Weight of equal volume of water}} \qquad (7.3)$$

■ **EXAMPLE 7.3**

Problem: Suppose a piece of metal weighs 150 lb in air and 85 lb under water. What is the specific gravity?

Solution:

150 lb – 85 lb = 65 lb loss of weight in water

$$\text{Specific gravity} = \frac{150}{65} = 2.3$$

Note: In a calculation of specific gravity, it is *essential* that the densities be expressed in the same units.

As stated earlier, the specific gravity of water is 1, which is the standard, the reference that all other liquid or solid substances are compared. Specifically, any object that has a specific gravity greater than 1 will sink in water (e.g., rocks, steel, iron, grit, floc, sludge). Substances with specific gravities of less than 1 will float (e.g., wood, scum, gasoline). Considering the total weight and volume of a ship, its specific gravity is less than 1; therefore, it can float.

The most common use of specific gravity in water/wastewater treatment operations is in gallon-to-pound conversions. In many cases, the liquids being handled have a specific gravity of 1 or very nearly 1 (between 0.98 and 1.02), so 1 may be used in the calculations without introducing significant error. For calculations involving a liquid with a specific gravity of less than 0.98 or greater than 1.02, however, the conversions from gallons to pounds must consider specific gravity. The technique is illustrated in the following example.

■ **EXAMPLE 7.4**

Problem: A basin contains 1455 gal of a liquid. If the specific gravity of the liquid is 0.94, how many pounds of liquid are in the basin?

Solution: Normally, for a conversion from gallons to pounds, we would use the factor 8.34 lb/gal (the density of water) if the specific gravity of the substance is between 0.98 and 1.02. In this instance, however, the substance has a specific gravity outside this range, so the 8.34 factor must be adjusted by multiplying 8.34 lb/gal by the specific gravity to obtain the adjusted factor:

$$8.34 \text{ lb/gal} \times 0.94 = 7.84 \text{ lb/gal (rounded)}$$

Then convert 1455 gal to pounds using the corrected factor:

$$1455 \text{ gal} \times 7.84 \text{ lb/gal} = 11,407 \text{ lb (rounded)}$$

7.4 FORCE AND PRESSURE

Water exerts force and pressure against the walls of its container, whether it is stored in a tank or flowing in a pipeline. Force and pressure are different, although they are closely related. *Force* is the push or pull influence that causes motion. In the English system, force and weight are often used in the same way. The weight of 1 ft³ of water is 62.4 lb. The force exerted on the bottom of a 1-ft cube is 62.4 lb (see Figure 7.1). If we stack two 1-ft cubes on top of one another, the force on the bottom will be 124.8 lb. *Pressure* is the force per unit of area. In equation form, this can be expressed as:

$$P = \frac{F}{A} \qquad (7.4)$$

where:

    P = pressure.
    F = force.
    A = area over which the force is distributed.

Earlier we pointed out that pounds per square inch (lb/in.² or psi) or pounds per square foot (lb/ft²) are common expressions of pressure. The pressure on the bottom of the cube is 62.4 lb/ft² (see Figure 7.1). It is normal to express pressure in pounds per square inch. This is easily accomplished by determining the weight of 1 in.² of a 1-ft cube. If we have a cube that is 12 in. on each side, the number of square inches on the bottom surface of the cube is 12 × 12 = 144 in². Dividing the weight by the number of square inches determines the weight on each square inch:

$$\text{psi} = \frac{62.4 \text{ lb/ft}}{144 \text{ in.}^2} = 0.433 \text{ psi/ft}$$

This is the weight of a column of water 1 in. square and 1 ft tall. If the column of water were 2 ft tall, the pressure would be 2 ft × 0.433 psi/ft = 0.866.

Note: 1 foot of water = 0.433 psi.

With this information, feet of head can be converted to psi by multiplying the feet of head by 0.433 psi/ft.

■ **EXAMPLE 7.5**

Problem: A tank is mounted at a height of 90 ft. Find the pressure at the bottom of the tank.

Solution:

$$90 \text{ ft} \times 0.433 \text{ psi/ft} = 39 \text{ psi (rounded)}$$

Note: To convert psi to feet, divide the psi by 0.433 psi/ft.

■ **EXAMPLE 7.6**

Problem: Find the height of water in a tank if the pressure at the bottom of the tank is 22 psi.

Solution:

$$\text{Height in feet} = \frac{22 \text{ psi}}{0.433 \text{ psi/ft}} = 51 \text{ ft (rounded)}$$

Important Point: One of the problems encountered in a hydraulic system is storing the liquid. Unlike air, which is readily compressible and is capable of being stored in large quantities in relatively small containers, a liquid such as water cannot be compressed. It is not possible to store a large amount of water in a small tank, as 62.4 lb of water occupies a volume of 1 ft³, regardless of the pressure applied to it.

7.4.1 HYDROSTATIC PRESSURE

Figure 7.2 shows a number of differently shaped, connected, open containers of water. Note that the water level is the same in each container, regardless of the shape or size of the container. This occurs because pressure is developed within a liquid by the weight of the liquid above. If the water level in any one container is momentarily higher than that in any of the other containers, the higher pressure at the bottom of this container would cause some water to flow into the container having the lower liquid level. In addition, the pressure of the water at any level (such as line T) is the same in each of the containers. Pressure increases because of the weight of the water. The farther down from the surface, the more pressure is created. This illustrates that the weight, not the volume, of water contained in a vessel determines the pressure at the bottom of the vessel.

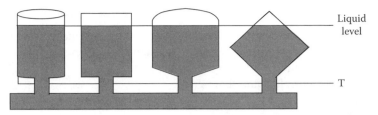

FIGURE 7.2 Hydrostatic pressure.

Nathanson (1997) pointed out some very important principles that always apply for hydrostatic pressure.

1. The pressure depends only on the depth of water above the point in question (not on the water surface area).
2. The pressure increases in direct proportion to the depth.
3. The pressure in a continuous volume of water is the same at all points that are at the same depth.
4. The pressure at any point in the water acts in all directions at the same depth.

7.4.2 Effects of Water under Pressure

Hauser (1996) pointed out that water under pressure and in motion can exert tremendous forces inside a pipeline. One of these forces, called *hydraulic shock* or *water hammer*, is the momentary increase in pressure that occurs due to a sudden change of direction or velocity of the water. When a rapidly closing valve suddenly stops water from flowing in a pipeline, pressure energy is transferred to the valve and pipe wall. Shock waves are set up within the system. Waves of pressure move in horizontal yo-yo fashion—back and forth—against any solid obstacles in the system. Neither the water nor the pipe will compress to absorb the shock, which may result in damage to pipes and valves and shaking of loose fittings.

Another effect of water under pressure is *thrust*, which is the force that water exerts on a pipeline as it rounds a bend. As shown in Figure 7.3, thrust usually acts perpendicular (90°) to the inside surface it pushes against. It affects not only bends in a pipe but also reducers, dead ends, and tees. Uncontrolled, thrust can cause movement in the fitting or pipeline, which will lead to separation of the pipe coupling away from both sections of pipeline or at some other nearby coupling upstream or downstream of the fitting.

Two types of devices commonly used to control thrust in larger pipelines are thrust blocks and thrust anchors. A *thrust block* is a mass of concrete cast in place onto the pipe and around the outside bend of the turn. An example is shown in Figure 7.4. Thrust blocks are used for pipes with tees or elbows that turn left or right or slant upward. The thrust is transferred to the soil through the larger bearing surface of the block. A *thrust anchor* is a massive block

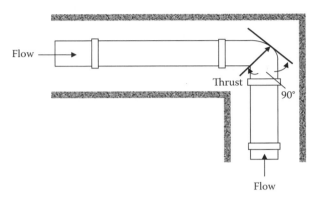

FIGURE 7.3 Direction of thrust in a pipe in a trench (viewed from above).

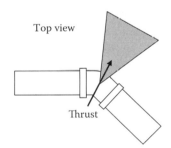

FIGURE 7.4 Thrust block.

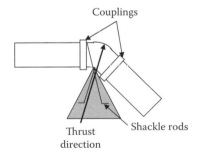

FIGURE 7.5 Thrust anchor.

of concrete, often a cube, cast in place below the fitting to be anchored (see Figure 7.5). As shown in Figure 7.5, imbedded steel shackle rods anchor the fitting to the concrete block, effectively resisting upward thrusts. The size and shape of a thrust control device depend on pipe size, type of fitting, water pressure, water hammer, and soil type.

7.5 HEAD

Head is defined as the vertical distance the water/wastewater must be lifted from the supply tank to the discharge or as the height a column of water would rise due to the pressure at its base. A perfect vacuum plus atmospheric pressure of 14.7 psi would lift the water 34 ft. When the top of the sealed tube is open to the atmosphere and the reservoir is enclosed, the pressure in the reservoir is increased; the water will rise in the tube. Because atmospheric pressure is essentially universal, we usually ignore the first 14.7 psi of actual pressure measurements and measure only the difference between the water pressure and the atmospheric pressure; we call this *gauge pressure*. Consider water in an open reservoir subjected to 14.7 psi of atmospheric pressure; subtracting this 14.7 psi leaves a gauge pressure of 0 psi, indicating that the water would rise 0 feet above the reservoir surface. If the gauge pressure in a water main were 120 psi, the water would rise in a tube connected to the main:

$$120 \text{ psi} \times 2.31 \text{ ft/psi} = 277 \text{ ft (rounded)}$$

The *total head* includes the vertical distance the liquid must be lifted (static head), the loss to friction (friction head), and the energy required to maintain the desired velocity (velocity head).

$$\text{Total head} = \begin{bmatrix} \text{Static head} + \text{friction head} \\ + \text{velocity head} \end{bmatrix} \quad (7.5)$$

7.5.1 STATIC HEAD

Static head is the actual vertical distance the liquid must be lifted:

$$\text{Static head} = \begin{bmatrix} \text{Discharge elevation} \\ - \text{supply elevation} \end{bmatrix} \quad (7.6)$$

■ EXAMPLE 7.7

Problem: The supply tank is located at elevation 118 ft. The discharge point is at elevation 215 ft. What is the static head in feet?

Solution:

$$\text{Static head (ft)} = 215 \text{ ft} - 118 \text{ ft} = 97 \text{ ft}$$

7.5.2 FRICTION HEAD

Friction head is the equivalent distance of the energy that must be supplied to overcome friction. Engineering references include tables showing the equivalent vertical distance for various sizes and types of pipes, fittings, and valves. The total friction head is the sum of the equivalent vertical distances for each component.

$$\text{Friction head (ft)} = \text{Energy losses due to friction} \quad (7.7)$$

7.5.3 VELOCITY HEAD

Velocity head is the equivalent distance of the energy consumed to achieve and maintain the desired velocity in the system:

$$\text{Velocity head (ft)} = \text{Energy losses to maintain velocity} \quad (7.8)$$

7.5.4 TOTAL DYNAMIC HEAD (TOTAL SYSTEM HEAD)

$$\text{Total head} = \begin{bmatrix} \text{Static head} + \text{friction head} \\ + \text{velocity head} \end{bmatrix} \quad (7.9)$$

7.5.5 PRESSURE AND HEAD

The pressure exerted by water/wastewater is directly proportional to its depth or head in the pipe, tank, or channel. If the pressure is known, the equivalent head can be calculated:

$$\text{Head (ft)} = \text{Pressure (psi)} \times 2.31 \text{ (ft/psi)} \quad (7.10)$$

■ EXAMPLE 7.8

Problem: The pressure gauge on the discharge line from the influent pump reads 72.3 psi. What is the equivalent head in feet?

Solution:

$$\text{Head (ft)} = 72.3 \times 2.31 \text{ ft/psi} = 167 \text{ ft}$$

7.5.6 HEAD AND PRESSURE

If the head is known, the equivalent pressure can be calculated by:

$$\text{Pressure (psi)} = \frac{\text{Head (ft)}}{2.31 \text{ ft/psi}} \quad (7.11)$$

■ EXAMPLE 7.9

Problem: A tank is 22 ft deep. What is the pressure in psi at the bottom of the tank when it is filled with water?

Solution:

$$\text{Pressure (psi)} = \frac{22 \text{ ft}}{2.31 \text{ ft/psi}} = 9.52 \text{ psi (rounded)}$$

7.6 FLOW AND DISCHARGE RATES: WATER IN MOTION

The study of fluid flow is much more complicated than that of fluids at rest, but it is important to have an understanding of these principles because the water in a waterworks and distribution system and in a wastewater treatment plant and collection system is nearly always in motion. *Discharge (or flow)* is the quantity of water passing a given point in a pipe or channel during a given period. Stated another way for open channels, the flow rate through an open channel is directly related to the velocity of the liquid and the cross-sectional area of the liquid in the channel:

$$Q = A \times V \tag{7.12}$$

where:

- Q = flow, or discharge in cubic feet per second (cfs).
- A = cross-sectional area of the pipe or channel (ft²).
- V = water velocity in feet per second (fps or ft/sec).

■ EXAMPLE 7.10

Problem: A channel is 6 ft wide and the water depth is 3 ft. The velocity in the channel is 4 fps. What is the discharge or flow rate in cubic feet per second?

Solution:

Flow (cfs) = 6 ft × 3 ft × 4 ft/sec = 72 cfs

Discharge or flow can be recorded as gal/day (gpd), as gal/min (gpm), or as cubic feet per second (cfs). Flows treated by many waterworks or wastewater treatment plants are large and are often referred to in million gallons per day (MGD). The discharge or flow rate can be converted from cfs to other units such as gpm or MGD by using appropriate conversion factors.

■ EXAMPLE 7.11

Problem: A 12-in.-diameter pipe has water flowing through it at 10 fps. What is the discharge in (a) cfs, (b) gpm, and (c) MGD?

Solution: Before we can use the basic formula, we must determine the area (*A*) of the pipe. The formula for the area of a circle is:

$$\text{Area } (A) = \pi \times \frac{D^2}{4} = \pi \times r^2 \tag{7.13}$$

Note: π is the constant value 3.14159, or simply 3.14.

where:

- D = diameter of the circle in feet.
- r = radius of the circle in feet.

Therefore, the area of the pipe is:

$$A = \pi \times \frac{D^2}{4} = 3.14 \times \frac{1 \text{ ft}^2}{4} = 0.785 \text{ ft}^2$$

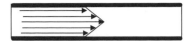

FIGURE 7.6 Laminar (streamline) flow.

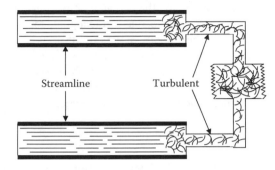

FIGURE 7.7 Turbulent flow.

(a) Now we can determine the discharge in cfs:

$$Q = V \times A$$

$$Q = 10 \text{ ft/sec} \times 0.785 \text{ ft}^2 = 7.85 \text{ ft}^3/\text{sec (cfs)}$$

(b) We need to know that 1 cfs is 449 gpm, so 7.85 cfs × 449 gpm/cfs = 3525 gpm (rounded).

(c) 1 million gallons per day is 1.55 cfs, so:

$$\frac{7.85 \text{ cfs}}{1.55 \text{ cfs/MGD}} = 5.06 \text{ MGD}$$

Note: Flow may be *laminar* (i.e., streamline; see Figure 7.6) or *turbulent* (see Figure 7.7). Laminar flow occurs at extremely low velocities. The water moves in straight parallel lines, called *streamlines* or *laminae*, which slide upon each other as they travel, rather than mixing up. Normal pipe flow is turbulent flow, which occurs because of friction encountered on the inside of the pipe. The outside layers of flow are thrown into the inner layers; the result is that all of the layers mix and are moving in different directions, and at different velocities; however, the direction of flow is forward.

Note: Flow may be steady or unsteady. For our purposes, we consider steady-state flow only; that is, most of the hydraulic calculations in this manual assume steady-state flow.

7.6.1 Area and Velocity

The *law of continuity* states that the discharge at each point in a pipe or channel is the same as the discharge at any other point (if water does not leave or enter the pipe or channel). That is, under the assumption of steady-state flow, the flow that enters the pipe or channel is the same flow that exits the pipe or channel. In equation form, this becomes:

$$Q_1 = Q_2 \quad \text{or} \quad A_1V_1 = A_2V_2 \qquad (7.14)$$

Note: With regard to the area/velocity relationship, Equation 7.14 also makes clear that, for a given flow rate, the velocity of the liquid varies indirectly with changes in cross-sectional area of the channel or pipe. This principle provides the basis for many of the flow measurement devices used in open channels (weirs, flumes, and nozzles).

■ EXAMPLE 7.12

Problem: A pipe 12 in. in diameter is connected to a 6-in.-diameter pipe. The velocity of the water in the 12-in. pipe is 3 fps. What is the velocity in the 6-in. pipe?

Solution: Using the equation $A_1V_1 = A_2V_2$, we need to determine the area of each pipe:

$$A = \pi \times \frac{D^2}{4}$$

$$A \,(\text{12-in. pipe}) = 3.14 \times \frac{(1 \text{ ft})^2}{4} = 0.785 \text{ ft}^2$$

$$A \,(\text{6-in. pipe}) = 3.14 \times \frac{(0.5 \text{ ft})^2}{4} = 0.196 \text{ ft}^2$$

The continuity equation now becomes:

$$0.785 \text{ ft}^2 \times 3 \text{ ft/sec} = 0.196 \text{ ft}^2 \times V_2$$

Solving for V_2:

$$V_2 = \frac{0.785 \text{ ft}^2 \times 3 \text{ ft/sec}}{0.196 \text{ ft}^2} = 12 \text{ ft/sec (fps)}$$

7.6.2 Pressure and Velocity

In a closed pipe flowing full (under pressure), the pressure is indirectly related to the velocity of the liquid. This principle, when combined with the principle discussed in the previous section, forms the basis for several flow measurement devices (Venturi meters and rotameters), as well as the injector used for dissolving chlorine into water and chlorine, sulfur dioxide, or other chemicals into wastewater:

$$\text{Velocity}_1 \times \text{Pressure}_1 = \text{Velocity}_2 \times \text{Pressure}_2 \qquad (7.15)$$

or

$$V_1P_1 = V_2P_2$$

7.7 PIEZOMETRIC SURFACE AND BERNOULLI'S THEOREM

They will take your hand and lead you to the pearls of the desert, those secret wells swallowed by oyster crags of wadi, underground caverns that bubble rusty salt water you would sell your own mothers to drink.

Holman (1998)

To keep the systems in your plant operating properly and efficiently, you must understand the basics of hydraulics—the laws of force, motion, and others. As stated previously, most applications of hydraulics in water/wastewater treatment systems involve water in motion—in pipes under pressure or in open channels under the force of gravity. The volume of water flowing past any given point in the pipe or channel per unit time is called the *flow rate* or *discharge rate*—or just *flow*. The *continuity of flow* and the *continuity equation* have already been discussed (see Equation 7.15). Along with the continuity of flow principle and continuity equation, the law of conservation of energy, piezometric surface, and Bernoulli's theorem (or principle) are also important to our study of water hydraulics.

7.7.1 Conservation of Energy

Many of the principles of physics are important to the study of hydraulics. When applied to problems involving the flow of water, few of the principles of physical science are more important and useful to us than the *law of conservation of energy*. Simply, the law of conservation of energy states that energy can be neither created nor destroyed, but it can be converted from one form to another. In a given closed system, the total energy is constant.

7.7.2 Energy Head

In hydraulic systems, two types of energy (kinetic and potential) and three forms of mechanical energy (potential energy due to elevation, potential energy due to pressure, and kinetic energy due to velocity) exist. Energy is measured in units of foot-pounds (ft-lb). It is convenient to express hydraulic energy in terms of *energy head* in feet of water. This is equivalent to foot-pounds per pound of water (ft-lb/lb = ft).

7.7.3 Piezometric Surface

As mentioned earlier, we have seen that when a vertical tube, open at the top, is inserted into a vessel of water, the water will rise in the tube to the water level in the tank. The water level to which the water rises in a tube is the *piezometric surface*. That is, the piezometric surface is an imaginary surface that coincides with the level of the water to which water in a system would rise in a *piezometer* (an instrument used to measure pressure).

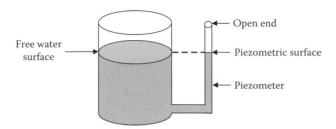

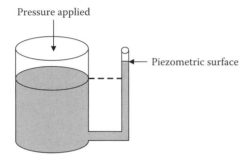

FIGURE 7.8 A container not under pressure where the piezo-metric surface is the same as the free water surface in the vessel.

FIGURE 7.9 A container under pressure where the piezometric surface is above the level of the water in the tank.

The surface of water that is in contact with the atmosphere is known as *free water surface*. Many important hydraulic measurements are based on the difference in height between the free water surface and some point in the water system. The piezometric surface is used to locate this free water surface in a vessel where it cannot be observed directly.

To understand how a piezometer actually measures pressure, consider the following example. If a clear, see-through pipe is connected to the side of a clear glass or plastic vessel, the water will rise in the pipe to indicate the level of the water in the vessel. Such a see-through pipe—a piezometer—allows us to see the level of the top of the water in the pipe; this is the piezometric surface.

In practice, a piezometer is connected to the side of a tank or pipeline. If the water-containing vessel is not under pressure (as is the case in Figure 7.8), the piezometric surface will be the same as the free water surface in the vessel, just as when a drinking straw (the piezometer) is left standing in a glass of water.

When a tank and pipeline system is pressed, as they often are, the pressure will cause the piezometric surface to rise above the level of the water in the tank. The greater the pressure, the higher the piezometric surface (see Figure 7.9). An increased pressure in a water pipeline system is usually obtained by elevating the water tank.

Note: In practice, piezometers are not installed on water towers, because water towers are hundreds of feet high, or on pipelines. Instead, pressure gauges are used that record pressure in feet of water or in psi.

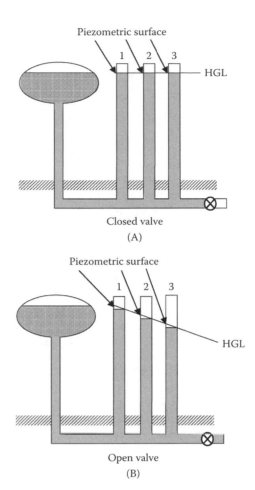

FIGURE 7.10 Head loss and piezometric surface changes when water is flowing.

Water only rises to the water level of the main body of water when it is at rest (static or standing water). The situation is quite different when water is flowing. Consider, for example, an elevated storage tank feeding a distribution system pipeline. When the system is at rest, with all of the valves closed, all of the piezometric surfaces are the same height as the free water surface in storage. On the other hand, when the valves are opened and the water begins to flow, the piezometric surface changes. This is an important point because, as water continues to flow down a pipeline, less and less pressure is exerted. This happens because some pressure is lost (used up) to keep the water moving over the interior surface of the pipe (friction). The pressure that is lost is called *head loss*.

7.7.4 HEAD LOSS

Head loss is best explained by example. Figure 7.10 shows an elevated storage tank feeding a distribution system pipeline. When the valve is closed (Figure 7.10A), all the piezometric surfaces are the same height as the free water surface in storage. When the valve opens and water begins to flow (Figure 7.10B), the piezometric surfaces *drop*. The farther along the pipeline, the lower the piezometric surface, because some of the pressure is used up keeping the water moving over the rough interior surface of the pipe. Thus, pressure is lost and is no longer available to push water up in a piezometer; this, again, is the head loss.

7.7.5 HYDRAULIC GRADE LINE (HGL)

When the valve shown in Figure 7.10 is opened, flow begins with a corresponding energy loss due to friction. The pressures along the pipeline can measure this loss. In Figure 7.10B, the difference in pressure heads between sections 1, 2, and 3 can be seen in the piezometer tubes attached to the pipe. A line connecting the water surface in the tank with the water levels at sections 1, 2, and 3 shows the pattern of continuous pressure loss along the pipeline. This is the *hydraulic grade line* (HGL), or *hydraulic gradient*, of the system.

Note: It is important to point out that in a static water system the HGL is always horizontal. The HGL is a very useful graphical aid when analyzing pipe flow problems.

Note: During the early design phase of a treatment plant, it is important to establish the hydraulic grade line across the plant because both the proper selection of the plant site elevation and the suitability of the site depend on this consideration. Typically, most conventional water treatment plants require 16 to 17 ft of head loss across the plant.

Key Point: Changes in the piezometric surface occur when water is flowing.

7.7.6 BERNOULLI'S THEOREM

Nathanson (1997) noted that Swiss physicist and mathematician Samuel Bernoulli developed the calculation for the total energy relationship from point to point in a steady-state fluid system in the 1700s. Before discussing Bernoulli's energy equation, it is important to understand the basic principle behind Bernoulli's equation. Water (and any other hydraulic fluid) in a hydraulic system possesses two types of energy—kinetic and potential. *Kinetic energy* is present when the water is in motion. The faster the water moves, the more kinetic energy is used. *Potential energy* is a result of the water pressure. The *total energy* of the water is the sum of the kinetic and potential energy. Bernoulli's principle states that the total energy of the water (fluid) always remains constant; therefore, when the water flow in a system increases, the pressure must decrease. When water starts to flow in a hydraulic system, the pressure drops. When the flow stops, the pressure rises again. The pressure gauges shown in Figure 7.11 illustrate this balance more clearly.

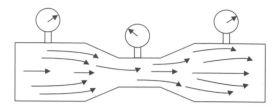

FIGURE 7.11 Bernoulli's principle.

Note: This discussion of Bernoulli's equation ignores friction losses from point to point in a fluid system employing steady-state flow.

7.7.7 BERNOULLI'S EQUATION

In a hydraulic system, total energy head is equal to the sum of three individual energy heads. This can be expressed as:

$$\text{Total head} = \begin{bmatrix} \text{Elevation head} + \text{pressure head} \\ + \text{velocity head} \end{bmatrix}$$

where elevation head is the pressure due to the elevation of the water, pressure head is the height of a column of water that a given hydrostatic pressure in a system could support, and velocity head is the energy present due to the velocity of the water. This can be expressed mathematically as:

$$E = z + \frac{P}{w} + \frac{V^2}{2g} \qquad (7.16)$$

where:

E = total energy head.
z = height of the water above a reference plane (ft).
P = pressure (psi).
w = unit weight of water (62.4 lb/ft^3).
V = flow velocity (ft/sec).
g = acceleration due to gravity (32.2 ft/sec^2).

Consider the constriction in section of pipe shown in Figure 7.11. We know, based on the law of energy conservation, that the total energy head at section A (E_1) must equal the total energy head at section B (E_2). Using Equation 7.16, we get Bernoulli's equation:

$$z_A = \frac{P_A}{w} + \frac{V_A^2}{2g} = z_a + \frac{P_B}{w} + \frac{V_B^2}{2g} \qquad (7.17)$$

The pipeline system shown in Figure 7.11 is horizontal; therefore, we can simplify Bernoulli's equation because $z_A = z_B$. Because they are equal, the elevation heads cancel out from both sides, leaving:

$$\frac{P_A}{w} + \frac{V_A^2}{2g} + \frac{P_B}{w} + \frac{V_B^2}{2g} \qquad (7.18)$$

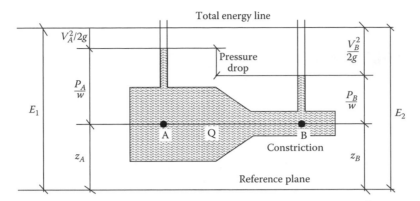

FIGURE 7.12 The law of conservation: Because the velocity and kinetic energy of the water flowing in the constricted section must increase, the potential energy will decrease. This is observed as a pressure drop in the constriction. (Adapted from Nathanson, J.A., *Basic Environmental Technology: Water Supply, Waste Management, and Pollution Control*, 2nd ed., Prentice Hall, Upper Saddle River, NJ, 1997, p. 29.)

As water passes through the constricted section of the pipe (section B), we know from continuity of flow that the velocity at section B must be greater than the velocity at section A, because of the smaller flow area at section B. This means that the velocity head in the system increases as the water flows into the constricted section; however, the total energy must remain constant. For this to occur, the pressure head, and therefore the pressure, must drop. In effect, pressure energy is converted into kinetic energy in the constriction.

The fact that the pressure in the narrower pipe section (constriction) is less than the pressure in the bigger section seems to defy common sense; however, it does follow logically from continuity of flow and conservation of energy. The fact that there is a pressure difference allows measurement of flow rate in the closed pipe.

■ EXAMPLE 7.13

Problem: In Figure 7.12, the diameter at section A is 8 in., and at section B it is 4 in. The flow rate through the pipe is 3.0 cfs and the pressure at section A is 100 psi. What is the pressure in the constriction at section B?

Solution: Compute the flow area at each section:

$$A_A = \frac{\pi \times (0.666 \text{ ft})^2}{4} = 0.349 \text{ ft}^2$$

$$A_B = \frac{\pi \times (0.333 \text{ ft})^2}{4} = 0.087 \text{ ft}^2$$

From $Q = A \times V$ or $V = Q/A$, we get:

$$V_A = \frac{3.0 \text{ ft}^3/\text{sec}}{0.349 \text{ ft}^2} = 8.6 \text{ ft/sec}$$

$$V_B = \frac{3.0 \text{ ft}^3/\text{sec}}{0.087 \text{ ft}^2} = 34.5 \text{ ft/sec}$$

And we get:

$$\frac{100 \times 144}{62.4} + \frac{(8.6)^2}{2 \times 32.2} = \frac{p_B \times 144}{62.4} + \frac{(34.5)^2}{2 \times 32.2}$$

Note: The pressures are multiplied by 144 in.²/ft² to convert from psi to lb/ft² to be consistent with the units for w; the energy head terms are in feet of head.

Continuing, we get:

$$231 + 1.15 = 2.3p_B + 18.5$$

and

$$P_B = \frac{232.2 - 18.5}{2.3} = \frac{213.7}{2.3} = 93 \text{ psi}$$

7.8 WELL AND WET-WELL HYDRAULICS

When the source of water for a water distribution system is from a groundwater supply, knowledge of well hydraulics is important to the operator. In this section, basic well hydraulics terms are presented and defined, and they are related pictorially (see Figure 7.13). Also discussed are wet wells, which are important in both water and wastewater operations.

7.8.1 WELL HYDRAULICS

- *Static water level*—The water level in a well when no water is being taken from the groundwater source (i.e., the water level when the pump is off; see Figure 7.13). Static water level is normally measured as the distance from the ground surface to the water surface. This is an important parameter because it is used to measure changes in the water table.

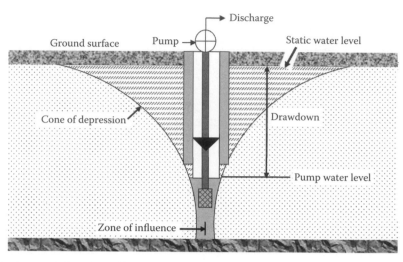

FIGURE 7.13 Hydraulic characteristics of a well.

- *Pumping water level*—The water level when the pump is off. When water is pumped out of a well, the water level usually drops below the level in the surrounding aquifer and eventually stabilizes at a lower level; this is the pumping level (see Figure 7.13).
- *Drawdown*—The difference, or the drop, between the static water level and the pumping water level, measured in feet. Simply, it is the distance the water level drops when pumping begins (see Figure 7.13).
- *Cone of depression*—In unconfined aquifers, water flows in the aquifer from all directions toward the well during pumping. The free water surface in the aquifer then takes the shape of an inverted cone or curved funnel line. The curve of the line extends from the pumping water level to the static water level at the outside edge of the zone (or radius) of influence (see Figure 7.13).

Note: The shape and size of the cone of depression are dependent on the relationship between the pumping rate and the rate at which water can move toward the well. If the rate is high, the cone will be shallow and its growth will stabilize. If the rate is low, the cone will be sharp and continue to grow in size.

- *Zone (or radius) of influence*—The distance between the pump shaft and the outermost area affected by drawdown (see Figure 7.13). The distance depends on the porosity of the soil and other factors. This parameter becomes important in well fields with many pumps. If wells are set too closely together, the zones of influence will overlap, increasing the drawdown in all wells. Obviously, pumps should be spaced apart to prevent this from happening.

Two important parameters not shown in Figure 7.13 are well yield and specific capacity:

1. *Well yield* is the rate of water withdrawal that a well can supply over a long period, or, alternatively, the maximum pumping rate that can be achieved without increasing the drawdown. The yield of small wells is usually measured in gallons per minute (liters per minute) or gallons per hour (liters per hour). For large wells, it may be measured in cubic feet per second (cubic meters per second).
2. *Specific capacity* is the pumping rate per foot of drawdown (gpm/ft), or:

$$\text{Specific capacity} = \text{Well yield} \div \text{drawdown} \quad (7.19)$$

■ EXAMPLE 7.14

Problem: If the well yield is 300 gpm and the drawdown is measured to be 20 ft, what is the specific capacity?

Solution:

Specific capacity = 300 ÷ 20 = 15 gpm per ft of drawdown

Specific capacity is one of the most important concepts in well operation and testing. The calculation should be made frequently in the monitoring of well operation. A sudden drop in specific capacity indicates problems such as pump malfunction, screen plugging, or other problems that can be serious. Such problems should be identified and corrected as soon as possible.

7.8.2 WET-WELL HYDRAULICS

Water pumped from a wet well by a pump set above the water surface exhibits the same phenomena as a groundwater well. In operation, a slight depression of the water

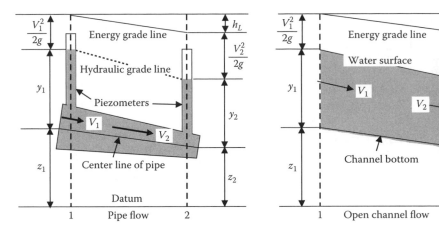

FIGURE 7.14 Comparison of pipe flow and open-channel flow. (Adapted from Metcalf & Eddy, *Wastewater Engineering: Collection and Pumping of Wastewater*, McGraw–Hill, New York, 1981, p. 11.)

surface forms right at the intake line (drawdown), but in this case it is minimal because there is free water at the pump entrance at all times (at least there should be). The most important consideration in wet-well operations is to ensure that the suction line is submerged far enough below the surface so air entrained by the active movement of the water at this section is not able to enter the pump. Because water or wastewater flow is not always constant or at the same level, variable speed pumps are commonly used in wet-well operations, or several pumps are installed for single or combined operation. In many cases, pumping is accomplished in an on/off mode. Control of pump operation is in response to water level in the well. Level control devices such as mercury switches are used to sense high or low levels in the well and to transmit the signal to pumps for action.

7.9 FRICTION HEAD LOSS

Materials or substances capable of flowing cannot flow freely. Nothing flows without encountering some type of resistance. Consider electricity, the flow of free electrons in a conductor. Whatever type of conductor used (e.g., copper, aluminum, silver) offers some resistance. In hydraulics, the flow of water/wastewater is analogous to the flow of electricity. Within a pipe or open channel, for example, flowing water, like electron flow in a conductor, encounters resistance; however, resistance to the flow of water is generally termed *friction loss* (or more appropriately, head loss).

7.9.1 FLOW IN PIPELINES

The problem of waste/wastewater flow in pipelines—the prediction of flow rate through pipes of given characteristics, the calculation of energy conversions therein, and so forth—is encountered in many applications of water/wastewater operations and practice. Although the subject of pipe flow embraces only those problems in which pipes

flow completely full (as in water lines), we also address pipes that flow partially full (wastewater lines, normally treated as open channels) in this section.

Also discussed is the solution of practical pipe flow problems resulting from application of the energy principle, the equation of continuity, and the principle and equation of water resistance. Resistance to flow in pipes occurs due to not only long reaches of pipe but also pipe fittings, such as bends and valves, which dissipate energy by producing relatively large-scale turbulence.

To gain an understanding of what friction head loss is all about, it is necessary to review a few terms presented earlier in the text and to introduce some new terms pertinent to the subject (Lindeburg, 1986):

- *Laminar and turbulent flow*—Laminar flow is ideal flow; that is, water particles move along straight, parallel paths in layers or streamlines. Moreover, laminar flow has no turbulence and no friction loss. This is not typical of normal pipe flow because the water velocity is too great, but it is typical of groundwater flow. Turbulent flow (characterized as normal for a typical water system) occurs when water particles move in a haphazard fashion and continually cross each other in all directions, resulting in pressure losses along the length of pipe.
- *Hydraulic grade line (HGL)*—Recall that the hydraulic grade line (shown in Figure 7.14) is a line connecting two points to which the liquid would rise at various places along any pipe or open channel if piezometers were inserted in the liquid. It is a measure of the pressure head available at these various points.

Note: When water flows in an open channel, the HGL coincides with the profile of the water surface.

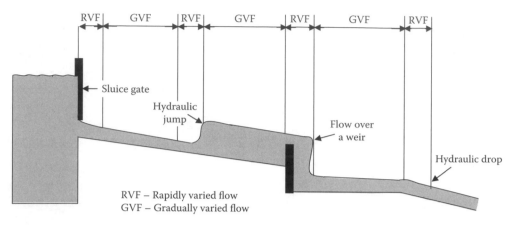

FIGURE 7.15 Varied flow.

- *Energy grade line*—The total energy of flow in any section with reference to some datum (i.e., a reference line, surface, or point) is the sum of the elevation head (z), the pressure head (y), and the velocity head ($V^2/2g$). Figure 7.14 shows the energy grade line or energy gradient, which represents the energy from section to section. In the absence of frictional losses, the energy grade line remains horizontal, although the relative distribution of energy may vary among the elevation, pressure, and velocity heads. In all real systems, however, losses of energy occur because of resistance to flow, and the resulting energy grade line is sloped (i.e., the energy grade line is the slope of the specific energy line).
- *Specific energy (E)*—Sometimes called *specific head*, the specific energy is the sum of the pressure head (y) and the velocity head ($V^2/2g$). The specific energy concept is especially useful in analyzing flow in open channels.
- *Steady flow*—Steady flow occurs when the discharge or rate of flow at any cross-section is constant.
- *Uniform and nonuniform flow*—Uniform flow occurs when the depth, cross-sectional area, and other elements of flow are substantially constant from section to section. Nonuniform flow occurs when the slope, cross-sectional area, and velocity change from section to section. The flow through a Venturi section used for measuring flow is a good example.
- *Varied flow*—Flow in a channel is considered varied if the depth of flow changes along the length of the channel. The flow may be gradually varied or rapidly varied (i.e., when the depth of flow changes abruptly) as shown in Figure 7.15.
- *Slope*—Slope (gradient) is the head loss per foot of channel.

7.9.2 Major Head Loss

Major head loss consists of pressure decreases along the length of pipe caused by friction created as water encounters the surfaces of the pipe. It typically accounts for most of the pressure drop in a pressurized or dynamic water system. The components that contribute to major head loss are roughness, length, diameter, and velocity:

- *Roughness*—Even when new, the interior surfaces of pipes are rough. The roughness varies, of course, depending on pipe material, corrosion (tuberculation and pitting), and age. Because normal flow in a water pipe is turbulent, the turbulence increases with pipe roughness, which, in turn, causes pressure to drop over the length of the pipe.
- *Pipe length*—With every foot of pipe length, friction losses occur. The longer the pipe, the greater the head loss. Friction loss due to pipe length must be factored into head loss calculations.
- *Pipe diameter*—Generally, small-diameter pipes have more head loss than large-diameter pipes. In large-diameter pipes, less of the water actually touches interior surfaces of the pipe (encountering less friction) than in a small-diameter pipe.
- *Water velocity*—Turbulence in a water pipe is directly proportional to the speed (or velocity) of the flow; thus, the velocity head also contributes to head loss.

Note: For pipe with a constant diameter, when flow increases, head loss increases.

7.9.3 Calculating Major Head Loss

Darcy, Weisbach, and others developed the first practical equation used to determine pipe friction in about 1850. The equation or formula now known as the *Darcy–Weisbach* equation for circular pipes is:

$$h_f = f \frac{LV^2}{D^2 g} \qquad (7.20)$$

$$h_f = \frac{8 f L Q^2}{\pi^2 g D^5} \qquad (7.21)$$

where:

h_f = head loss (ft).
f = coefficient of friction.
L = length of pipe (ft).
V = mean velocity (ft/sec).
D = diameter of pipe (ft).
g = acceleration due to gravity (32.2 ft/sec²).
Q = flow rate, (ft³/sec).

The Darcy–Weisbach formula was meant to apply to the flow of any fluid, and into this friction factor was incorporated the degree of roughness and an element known as the *Reynold's number*, which is based on the viscosity of the fluid and the degree of turbulence of flow. The Darcy–Weisbach formula is used primarily for head loss calculations in pipes. For open channels, the *Manning* equation was developed during the latter part of the 19th century. Later, this equation was used for both open channels and closed conduits.

In the early 1900s, a more practical equation, the *Hazen–Williams* equation, was developed for use in making calculations related to water pipes and wastewater force mains:

$$Q = 0.435 \times C D^{2.63} \times S^{0.54} \qquad (7.22)$$

where:

Q = flow rate (ft³/sec).
C = coefficient of roughness (C decreases with roughness).
D = hydraulic radius R (ft).
S = slope of energy grade line (ft/ft).

7.9.4 C FACTOR

The C factor, as used in the Hazen–Williams formula, designates the coefficient of roughness. C does not vary appreciably with velocity, and by comparing pipe types and ages it includes only the concept of roughness, ignoring fluid viscosity and Reynold's number. Based on experience (experimentation), accepted tables of C factors have been established for pipe (see Table 7.2). Generally, the C factor decreases by one with each year of pipe age. Flow for a newly designed system is often calculated with a C factor of 100, based on averaging it over the life of the pipe system.

Note: A high C factor means a smooth pipe; a low C factor means a rough pipe.

TABLE 7.2
C Factors

Type of Pipe	C Factor
Asbestos cement	140
Brass	140
Brick sewer	100
Cast iron	
10 years old	110
20 years old	90
Ductile iron (cement-lined)	140
Concrete or concrete-lined	
Smooth, steel forms	140
Wooden forms	120
Rough	110
Copper	140
Fire hose (rubber-lined)	135
Galvanized iron	120
Glass	140
Lead	130
Masonry conduit	130
Plastic	150
Steel	
Coal-tar-enamel-lined	150
New unlined	140
Riveted	110
Tin	130
Vitrified	120
Wood stave	120

Source: Adapted from Lindeburg, M.R., *Civil Engineering Reference Manual*, 4th ed., Professional Publications, San Carlos, CA, 1986.

Note: An alternative to calculating the Hazen–Williams formula, called an *alignment chart*, has become quite popular for fieldwork. The alignment chart can be used with reasonable accuracy.

7.9.5 SLOPE

Slope is defined as the head loss per foot. In open channels, where the water flows by gravity, slope is the amount of incline of the pipe and is calculated as feet of drop per foot of pipe length (ft/ft). Slope is designed to be just enough to overcome frictional losses, so the velocity remains constant, the water keeps flowing, and solids will not settle in the conduit. In piped systems, where pressure loss for every foot of pipe is experienced, slope is not provided by slanting the pipe but instead by adding pressure to overcome friction.

7.9.6 MINOR HEAD LOSS

In addition to the head loss caused by friction between the fluid and the pipe wall, losses also are caused by turbulence created by obstructions (i.e., valves and fittings of all types) in the line, changes in direction, and changes in flow area.

Note: In practice, if minor head loss is less than 5% of the total head loss, it is usually ignored.

7.10 BASIC PIPING HYDRAULICS

Water, regardless of the source, is conveyed to the water-works for treatment and distributed to the users. Conveyance from the source to the point of treatment occurs by aqueducts, pipelines, or open channels, but the treated water is normally distributed in pressurized closed conduits. After use, whatever the purpose, the water becomes wastewater, which must be disposed of somehow but almost always ends up being conveyed back to a treatment facility before being outfalled to some water body, to begin the cycle again. We call this an *urban water cycle*, because it provides a human-generated imitation of the natural water cycle. Unlike the natural water cycle, however, without pipes the cycle would be nonexistent or, at the very least, short circuited.

For use as water mains in a distribution system, pipes must be strong and durable to resist applied forces and corrosion. The pipe is subjected to internal pressure from the water and to external pressure from the weight of the backfill (soil) and vehicles above it. The pipe may also have to withstand water hammer. Damage due to corrosion or rusting may also occur internally because of the water quality or externally because of the nature of the soil conditions.

Pipes used in a wastewater system must be strong and durable to resist the abrasive and corrosive properties of the wastewater. Like water pipes, wastewater pipes must also be able to withstand stresses caused by the soil backfill material and the effect of vehicles passing above the pipeline. Joints between wastewater collection/interceptor pipe sections should be flexible but tight enough to prevent excessive leakage, either of sewage out of the pipe or groundwater into the pipe. Of course, pipes must be constructed to withstand the expected conditions of exposure, and pipe configuration systems for water distribution or wastewater collection and interceptor systems must be properly designed and installed in terms of water hydraulics. Because the water and wastewater operator should have a basic knowledge of water hydraulics related to commonly used standard piping configurations, piping basics are briefly discussed in this section.

7.10.1 Piping Networks

It would be far less costly and make for more efficient operation if municipal water and wastewater systems were built with individual, single-pipe networks extending from the treatment plant to the user's residence or from the user's sink or bathtub drain to the local wastewater treat-

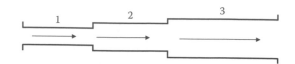

FIGURE 7.16 Pipes in series.

ment plant. Unfortunately, this ideal single-pipe scenario is not practical for real-world applications. Instead of a single piping system, a network of pipes is laid under the streets. Each of these piping networks is composed of different materials that vary (sometimes considerably) in diameter, length, and age. These networks range in complexity to varying degrees, and each of these joined-together pipes contributes energy losses to the system.

Water and wastewater flow networks may consist of pipes arranged in series, parallel, or some complicated combination. In any case, an evaluation of friction losses for the flows is based on energy conservation principles applied to the flow junction points. Methods of computation depend on the particular piping configuration. In general, however, they involve establishing a sufficient number of simultaneous equations or employing a friction loss formula where the friction coefficient depends only on the roughness of the pipe (e.g., Hazen–Williams equation, Equation 7.22).

Note: Demonstrating the procedure for making these complex computations is beyond the scope of this text. Here, we present only an operator's need-to-know aspects of complex or compound piping systems.

When two pipes of different sizes or roughnesses are connected in series (see Figure 7.16), head loss for a given discharge, or discharge for a given head loss, may be calculated by applying the appropriate equation between the bonding points, taking into account all losses in the interval; thus, head losses are cumulative. Series pipes may be treated as a single pipe of constant diameter to simplify the calculation of friction losses. The approach involves determining an equivalent length of a constant-diameter pipe that has the same friction loss and discharge characteristics as the actual series pipe system. In addition, application of the continuity equation to the solution allows the head loss to be expressed in terms of only one pipe size.

Note: In addition to the head loss caused by friction between the water and the pipe wall, minor losses are caused by obstructions in the line, changes in directions, and changes in flow area. In practice, the method of equivalent length is often used to determine these losses. The method of equivalent length uses a table to convert each valve or fitting into an equivalent length of straight pipe.

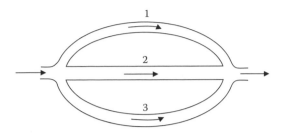

FIGURE 7.17 Pipes in parallel.

When making calculations involving pipes in series, remember these two important basic operational tenets:

1. The same flow passes through all pipes connected in series.
2. The total head loss is the sum of the head losses of all of the component pipes.

In some operations involving series networks where the flow is given and the total head loss is unknown, we can use the Hazen–Williams formula to solve for the slope and the head loss of each pipe as if they were separate pipes. Adding up the head losses to get the total head loss is then a simple matter.

Other series network calculations may not be as simple to solve using the Hazen–Williams equation; for example, one problem we may be faced with is what diameter to use with varying sized pipes connected together in a series combination. Moreover, head loss is applied to both pipes (or other multiples), and it is not known how much loss originates from each one; thus, determining slope would be difficult—but not impossible.

In such cases, the equivalent pipe theory, as mentioned earlier, can be used. Again, one single "equivalent pipe" is created that will carry the correct flow. This is practical because the head loss through it is the same as that in the actual system. The equivalent pipe can have any *C* factor and diameter, just as long as those same dimensions are maintained all the way through to the end. Keep in mind that the equivalent pipe must have the correct length so it will allow the correct flow through that yields the correct head loss (the given head loss) (Lindeburg, 1986).

Two or more pipes connected (as in Figure 7.17) so flow is first divided among the pipes and then rejoined comprise a parallel pipe system. A parallel pipe system is a common method for increasing the capacity of an existing line. Flows in pipes arranged in parallel are determined by applying energy conservation principles—specifically, energy losses through all pipes connecting common junction points must be equal. Each leg of the parallel network is treated as a series piping system and converted to a single equivalent length pipe. The friction losses through the equivalent length parallel pipes are then considered equal, and the respective flows are determined by proportional distribution.

Note: Computations used to determine friction losses in parallel combinations may be accomplished using a simultaneous solution approach for a parallel system that has only two branches; however, if the parallel system has three or more branches, a modified procedure using the Hazen–Williams loss formula is easier.

7.11 OPEN-CHANNEL FLOW

Water is transported over long distances through aqueducts to locations where it is to be used or treated. Selection of an aqueduct type rests on such factors as topography, head availability, climate, construction practices, economics, and water quality protection. Along with pipes and tunnels, aqueducts may also include or be solely composed of open channels (Lindeburg, 1986). In this section, we deal with water passage in open channels, which allow part of the water to be exposed to the atmosphere. This type of channel—an open-flow channel—includes natural waterways, canals, culverts, flumes, and pipes flowing under the influence of gravity.

7.11.1 CHARACTERISTICS OF OPEN-CHANNEL FLOW

McGhee (1991) pointed out that basic hydraulic principles apply in open-channel flow (with water depth constant) although there is no pressure to act as the driving force. Velocity head is the only natural energy this water possesses, and at normal water velocities it is a small value ($V^2/2g$). Several parameters can be (and often are) used to describe open-channel flow; however, we begin our discussion by addressing several characteristics of open-channel flow, including whether it is laminar or turbulent, uniform or varied, or subcritical, critical, or supercritical.

7.11.1.1 Laminar and Turbulent Flow

Laminar and *turbulent* flows in open channels are analogous to those in closed pressurized conduits (e.g., pipes). It is important to point out, however, that flow in open channels is usually turbulent. In addition, laminar flow essentially never occurs in open channels in either water or wastewater unit processes or structures.

7.11.1.2 Uniform and Varied Flow

Flow can be a function of time and location. If the flow quantity is invariant, it is said to be steady. *Uniform* flow is flow in which the depth, width, and velocity remain constant along a channel; that is, if the flow cross-section does not depend on the location along the channel, the flow is said to be uniform. *Varied* or *nonuniform* flow involves a change in these variables, with a change in one producing a change in the others. Most circumstances of open-channel flow in water and wastewater systems involve varied flow. The concept of uniform flow is valuable, however, in that it defines a limit that the varied flow may be considered to be approaching in many cases.

Note: Uniform channel construction does not ensure uniform flow.

7.11.1.3 Critical Flow

Critical flow (i.e., flow at the critical depth and velocity) defines a state of flow between two flow regimes. Critical flow coincides with minimum specific energy for a given discharge and maximum discharge for a given specific energy. Critical flow occurs in flow measurement devices at or near free discharges and establishes controls in open-channel flow. Critical flow occurs frequently in water/wastewater systems and is very important in their operation and design.

Note: Critical flow minimizes the specific energy and maximizes discharge.

7.11.2 Parameters Used in Open-Channel Flow

The three primary parameters used in open-channel flow are *hydraulic radius*, *hydraulic depth*, and *slope (S)*.

7.11.2.1 Hydraulic Radius

The *hydraulic radius* is the ratio of area in flow to wetted perimeter:

$$r_H = \frac{A}{P} \qquad (7.23)$$

where:

r_H = hydraulic radius.
A = cross-sectional area of the water.
P = wetted perimeter.

Why is hydraulic radius important? Good question.

Probably the best way to answer this question is by illustration. Consider, for example, that in open channels it is of primary importance to maintain the proper velocity. This is the case, of course, because if velocity is not maintained then flow stops (theoretically). To maintain velocity at a constant level, the channel slope must be adequate to overcome friction losses. As with other flows, calculation of head loss at a given flow is necessary, and the Hazen–Williams equation is useful ($Q = .435 \times C \times D^{2.63} \times S^{.54}$). Keep in mind that the concept of slope has not changed. The difference? We are now measuring, or calculating for, the physical slope of a channel (ft/ft), equivalent to head loss.

The preceding seems logical and makes sense—but, there is a problem. The problem is with the diameter. In conduits that are not circular (e.g., grit chambers, contact basins, streams, and rivers) or in pipes only partially full

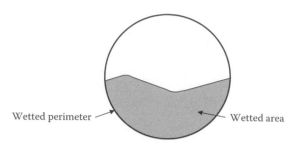

FIGURE 7.18 Hydraulic radius.

(e.g., drains, wastewater gravity mains, sewers) where the cross-sectional area of the water is not circular, there is no diameter. Without a diameter, what do we do? Another good question.

Because we do not have a diameter in situations where the cross-sectional area of the water is not circular, we must use another parameter to designate the size of the cross-section and the amount of it that contacts the sides of the conduit. This is where the hydraulic radius (r_H) comes in. The hydraulic radius is a measure of the efficiency with which the conduit can transmit water. Its value depends on pipe size, and amount of fullness. We use the hydraulic radius to measure how much of the water is in contact with the sides of the channel or how much of the water is not in contact with the sides (see Figure 7.18).

Note: For a circular channel flowing either full or half full, the hydraulic radius is ($D/4$). Hydraulic radii of other channel shapes are easily calculated from the basic definition.

7.11.2.2 Hydraulic Depth

The *hydraulic depth* is the ratio of area in flow to the width of the channel at the fluid surface (note that other names for hydraulic depth are *hydraulic mean depth* and *hydraulic radius*):

$$d_H = \frac{A}{w} \qquad (7.24)$$

where:

d_H = hydraulic depth.
A = area in flow.
w = width of the channel at the fluid surface.

7.11.2.3 Slope (S)

The *slope (S)* in open-channel equations is the slope of the energy line. If the flow is uniform, the slope of the energy line will be parallel to the water surface and channel bottom. In general, the slope can be calculated from the Bernoulli equation as the energy loss per unit length of channel:

$$S = \frac{D_h}{D_l} \qquad (7.25)$$

TABLE 7.3
Manning Roughness Coefficient (*n*)

Type of Conduit	*n*	Type of Conduit	*n*
		Pipe	
Cast iron, coated	0.012–0.014	Cast iron, uncoated	0.013–0.015
Wrought iron, galvanized	0.015–0.017	Wrought iron, black	0.012–0.015
Steel, riveted and spiral	0.015–0.017	Corrugated	0.021–0.026
Wood stave	0.012–0.013	Cement surface	0.010–0.013
Concrete	0.012–0.017	Vitrified	0.013–0.015
Clay, drainage tile	0.012–0.014		
		Lined channels	
Metal, smooth semicircular	0.011–0.015	Metal, corrugated	0.023–0.025
Wood, planed	0.010–0.015	Wood, unplaned	0.011–0.015
Cement lined	0.010–0.013	Concrete	0.014–0.016
Cement rubble	0.017–0.030	Grass	0.020
		Unlined channels	
Earth, straight and uniform	0.017–0.025	Earth, dredged	0.025–0.033
Earth, winding	0.023–0.030	Earth, stony	0.025–0.040
Rock, smooth and uniform	0.025–0.035	Rock, jagged and irregular	0.035–0.045

7.11.3 Open-Channel Flow Calculations

As mentioned, the calculation for head loss at a given flow is typically accomplished by using the Hazen–Williams equation. In addition, in open-channel flow problems, although the concept of slope has not changed, a problem again rises with the diameter. In pipes only partially full where the cross-sectional area of the water is not circular, we have no diameter to work with, and the hydraulic radius is used for these noncircular areas. In the original version of the Hazen–Williams equation, the hydraulic radius was incorporated. Moreover, similar versions developed by Chezy (pronounced "Shay-zee"), Manning, and others incorporated the hydraulic radius. For use in open channels, Manning's formula has become the most commonly used:

$$Q = \frac{1.5}{n} A \times R^{.66} \times s^2 \qquad (7.26)$$

where:

Q = channel discharge capacity (ft³/sec).
1.5 = constant.
n = channel roughness coefficient.
A = cross-sectional flow area (ft²).
R = hydraulic radius of the channel (ft).
S = slope of the channel bottom, dimensionless.

The hydraulic radius of a channel is defined as the ratio of the flow area to the wetted perimeter (*P*). In formula form, *R* = *A/P*. The new component, *n* (the roughness coefficient), depends on the material and age for a pipe or lined channel and on topographic features for a natural streambed. It approximates roughness in open channels and can range from a value of 0.01 for a smooth

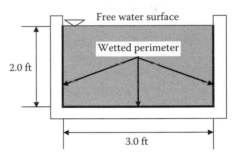

FIGURE 7.19 Illustration for Example 7.15.

clay pipe to 0.1 for a small natural stream. The value of *n* commonly assumed for concrete pipes or lined channels is 0.013. The *n* values decrease as the channels become smoother (see Table 7.3). The following example illustrates the application of Manning's formula for a channel with a rectangular cross-section.

■ EXAMPLE 7.15

Problem: A rectangular drainage channel is 3 ft wide and is lined with concrete, as illustrated in Figure 7.19. The bottom of the channel drops in elevation at a rate of 0.5 per 100 ft. What is the discharge in the channel when the depth of water is 2 ft?

Solution: Assume that *n* = 0.013. Referring to Figure 7.19, we see that the cross-sectional flow area (*A*) = 3 ft × 2 ft = 6 ft², and the wetted perimeter (*P*) = 2 ft + 3 ft + 2 ft = 7 ft. The hydraulic radius (*R*) = *A/P* = 6 ft²/7 ft = 0.86 ft. The slope (*S*) = 0.5/100 = 0.005. Applying Manning's formula, we get:

$$Q = \frac{2.0}{0.013} \times 6 \times 0.86^{.66} \times 0.005^{.5}$$

$$= 59 \text{ cfs}$$

To this point, we have set the stage for explaining (in the simplest possible way) what open-channel flow is—what it is all about. Thus, now that we have explained the necessary foundational material and important concepts, we are ready to explain open-channel flow in a manner in which it will be easily understood.

We stated that, when water flows in a pipe or channel with a free surface exposed to the atmosphere, it is referred to as *open-channel flow*. We also know that gravity provides the motive force, the constant push, while friction resists the motion and causes energy expenditure. River and stream flows are open-channel flows. Flows in sanitary sewers and stormwater drains are open-channel flows, except in force mains where the water is pumped under pressure.

The key to solving routine stormwater and sanitary sewer problems is a condition known as *steady uniform flow*; that is, we assume steady uniform flow. Steady flow, of course, means that the discharge is constant with time. Uniform flow means that the slope of the water surface and the cross-sectional flow area are also constant. It is common practice to call a length of channel, pipeline, or stream that has a relatively constant slope and cross-section a *reach* (Nathanson, 1997).

The slope of the water surface under steady uniform flow conditions is the same as the slope of the channel bottom. The hydraulic grade line (HGL) lies along the water surface and, as in pressure flow in pipes, the HGL slopes downward in the direction of flow. Energy loss is evident as the water surface elevation drops. Figure 7.20 illustrates a typical profile view of uniform steady flow. The slope of the water surface represents the rate of energy loss.

Note: Rate of energy loss (see Figure 7.20) may be expressed as the ratio of the drop in elevation of the surface in the reach to the length of the reach.

Figure 7.21 shows typical cross-sections of open-channel flow. In Figure 7.21A, the pipe is only partially filled with water and there is a free surface at atmospheric pressure. This is still open-channel flow, although the pipe is a closed underground conduit. Remember, the important point is that gravity and not a pump is moving the water.

7.12 FLOW MEASUREMENT

Although it is clear that maintaining water and wastewater flow is at the heart of any treatment process, clearly it is the measurement of flow that is essential to ensuring the

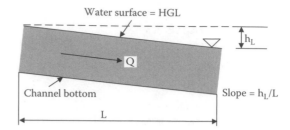

FIGURE 7.20 Steady uniform open-channel flow, where the slope of the water surface (or HGL) is equal to the slope of the channel bottom.

proper operation of a water and wastewater treatment system. Few knowledgeable operators would argue with this statement. Hauser (1996) asks: "Why measure flow?" Then she explains: "The most vital activities in the operation of water and wastewater treatment plants are dependent on a knowledge of how much water is being processed." In this statement, Hauser makes clear that flow measurement is not only important but also routine in water/wastewater operations. Routine, yes, but also the most important variable measured in a treatment plant. Hauser also discussed several reasons for measuring flow in a treatment plant. The American Water Works Association (1995) listed several additional reasons to measure flow:

- The flow rate through the treatment processes must be controlled so it matches distribution system use.
- It is important to determine the proper feed rate of chemicals added in the processes.
- The detention times through the treatment processes must be calculated. This is particularly applicable to surface water plants that must meet $C \times T$ values required by the surface water treatment rule.
- Flow measurement allows operators to maintain a record of water furnished to the distribution system for periodic comparison with the total water metered to customers. This provides a measure of water accounted for or, conversely (as pointed out earlier by Hauser), the amount of water wasted, leaked, or otherwise not paid for—that is, lost water.
- Flow measurement allows operators to determine the efficiency of pumps. Pumps that are not delivering their designed flow rate are probably not operating at maximum efficiency, so power is being wasted.
- For well systems, it is very important to maintain records of the volume of water pumped and the hours of operation for each well. The periodic computation of well pumping rates can identify problems such as worn pump impellers and blocked well screens.

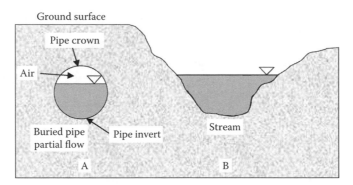

FIGURE 7.21 Open-channel flow, whether in a surface stream or in an underground pipe. (Adapted from Nathanson, J.A., *Basic Environmental Technology: Water Supply, Waste Management, and Pollution Control*, Prentice Hall, Upper Saddle River, NJ, 1997, p. 35.)

- Reports that must be furnished to the state by most water systems must include records of raw and finished water pumpage.
- Wastewater generated by a treatment system must also be measured and recorded.
- Individual meters are often required for the proper operation of individual pieces of equipment; for example, the makeup water to a fluoride saturator is always metered to assist in tracking the fluoride feed rate.

Note: Simply put, measurement of flow is essential for operation, process control, and recordkeeping of water and wastewater treatment plants.

All of the uses just discussed create the need, obviously, for a number of flow measuring devices, often with different capabilities. In this section, we discuss many of the major flow measuring devices currently used in water/wastewater operations.

7.12.1 FLOW MEASUREMENT: THE OLD-FASHIONED WAY

An approximate but very simple method to determine open-channel flow has been used for many years. The procedure involves measuring the velocity of a floating object moving in a straight uniform reach of the channel or stream. If the cross-sectional dimensions of the channel are known and the depth of flow is measured, then flow area can be computed. From the relationship $Q = A \times V$, the discharge Q can be estimated. In preliminary fieldwork, this simple procedure is useful in obtaining a ballpark estimate for the flow rate but is not suitable for routine measurements.

■ EXAMPLE 7.16

Problem: A floating object is placed on the surface of water flowing in a drainage ditch and is observed to travel a distance of 20 m downstream in 30 sec. The ditch is 2 m wide, and the average depth of flow is estimated to be 0.5 m. Estimate the discharge under these conditions.

Solution: The flow velocity is computed as distance over time, or:

$$V = D/T = 20 \text{ m}/30 \text{ sec} = 0.67 \text{ m/sec}$$

The channel area is $A = 2 \text{ m} \times 0.5 \text{ m} = 1.0 \text{ m}^2$. The discharge $(Q) = A \times V = 1.0 \text{ m}^2 \times 0.66 \text{ m}^2 = 0.66 \text{ m}^3/\text{sec}$.

7.12.2 BASIS OF TRADITIONAL FLOW MEASUREMENT

Flow measurement can be based on flow rate or flow amount. *Flow rate* is measured in gallons per minute (gpm), million gallons per day (MGD), or cubic feet per second (cfs). Water/wastewater operations require flow rate meters to determine process variables within the treatment plant, in wastewater collection, and in potable water distribution. Typically, the flow rate meters used are pressure differential meters, magnetic meters, and ultrasonic meters. Flow rate meters are designed for metering flow in closed pipe or open-channel flow.

Flow amount is measured in either gallons (gal) or cubic feet (ft³). Typically, a totalizer, which sums up the gallons or cubic feet that pass through the meter, is used. Most service meters are of this type. They are used in private, commercial, and industrial activities where the total amount of flow measured is used for customer billing. In wastewater treatment, where sampling operations are important, automatic composite sampling units—flow proportioned to grab a sample every so many gallons—are used. Totalizer meters can be of the velocity (propeller or turbine), positive displacement, or compound type. In addition, weirs and flumes are used extensively for measuring flow in wastewater treatment plants because they are not affected (to a degree) by dirty water or floating solids.

7.12.3 FLOW MEASURING DEVICES

In recent decades, flow measurement technology has evolved rapidly from the old-fashioned way of measuring flow discussed earlier to the use of simple practical measuring devices to much more sophisticated devices. Physical phenomena discovered centuries ago have been the starting point for many of the viable flowmeter designs used today. Moreover, the recent technology explosion has enabled flowmeters to handle many more applications than could have only been imagined centuries ago. Before selecting a particular type of flow measurement device, Kawamura (2000) recommends consideration of several questions:

1. Is liquid or gas flow being measured?
2. Is the flow occurring in a pipe or in an open-channel?
3. What is the magnitude of the flow rate?
4. What is the range of flow variation?
5. Is the liquid being measured clean, or does it contain suspended solids or air bubbles?
6. What is the accuracy requirement?
7. What is the allowable head loss by the flow meter?
8. Is the flow corrosive?
9. What types of flowmeters are available to the region?
10. What types of post-installation service is available to the area?

7.12.3.1 Differential Pressure Flowmeters

For many years, *differential pressure* flowmeters have been the most widely applied flow measuring device for water flow in pipes that require accurate measurement at reasonable cost. The differential pressure type of flowmeter makes up the largest segment of the total flow measurement devices currently being used. Differential pressure-producing meters currently on the market include the Venturi, Dall type, Hershel Venturi, universal Venturi, and Venturi inserts.

The differential pressure-producing device has a flow restriction in the line that causes a differential pressure, or head, to be developed between the two measurement locations. Differential pressure flowmeters are also referred to as *head meters*, and, of all the head meters, the orifice flowmeter is the most widely applied device. The advantages of differential pressure flowmeters include:

- Simple construction
- Relatively low cost
- No moving parts
- External transmitting instruments
- Low maintenance
- Wide application of flowing fluid, suitable for measuring both gas and liquid flow
- Ease of instrument and range selection
- Extensive product experience and performance database

Disadvantages include:

- Flow rate being a nonlinear function of the differential pressure
- Low flow rate range with normal instrumentation

7.12.3.1.1 Operating Principle

Differential pressure flowmeters operate on the principle of measuring pressure at two points in the flow, which provides an indication of the rate of flow that is passing by. The difference in pressures between the two measurement locations of the flowmeter is the result of the change in flow velocities. Simply, there is a set relationship between the flow rate and volume, so the meter instrumentation automatically translates the differential pressure into a volume of flow. The volume of flow rate through the cross-sectional area is given by:

$$Q = A \times v$$

where:

Q = the volumetric flow rate.
A = flow in the cross-sectional area.
v = the average fluid velocity.

Differential pressure flowmeters operate on the principle of developing a differential pressure across a restriction that can be related to the fluid flow rate.

Note: Optimum measurement accuracy is maintained when the flowmeter is calibrated, the flowmeter is installed in accordance with standards and codes of practice, and the transmitting instruments are periodically calibrated.

7.12.3.2 Types of Differential Pressure Flowmeters

The most commonly used differential pressure flowmeter types used in water/wastewater treatment are:

1. Orifice
2. Venturi
3. Nozzle
4. Pitot–static tube

7.12.3.2.1 Orifice

The most commonly applied, *orifice* is a thin, concentric, and flat metal plate with an opening in the plate (see Figure 7.22), installed perpendicular to the flowing stream in a circular conduit or pipe. Typically, a sharp-edged hole is bored in the center of the orifice plate. As the flowing water passes through the orifice, the restriction causes an increase in velocity. A concurrent decrease in pressure occurs as potential energy (static pressure) is converted into kinetic energy (velocity). As the water leaves the

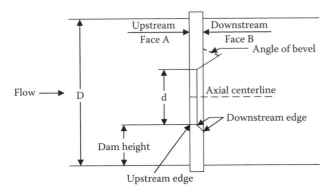

FIGURE 7.22 Orifice plate.

Concentric   Eccentric   Segmental

FIGURE 7.23 Orifice plate.

orifice, its velocity decreases and its pressure increases as kinetic energy is converted back into potential energy according to the laws of conservation of energy; however, some permanent pressure loss due to friction always occurs, and the loss is a function of the ratio of the diameter of the orifice bore (*d*) to the pipe diameter (*D*).

For dirty water applications (e.g., wastewater), the performance of a concentric orifice plate will eventually be impaired due to dirt buildup at the plate, so eccentric or segmental orifice plates (see Figure 7.23) can be used. Measurements are typically less accurate than those obtained from the concentric orifice plate, so eccentric or segmental orifices are rarely applied in current practice.

The orifice differential pressure flowmeter is the lowest cost differential flowmeter, is easy to install, and has no moving parts; however, it also has high permanent head loss (ranging from 40 to 90%), higher pumping costs, and an accuracy of ±2 % for a flow range of 4:1, and it is affected by wear or damage.

Note: Orifice meters are not recommended for permanent installation to measure wastewater flow; sol-

ids in the water easily catch on the orifice, throwing off accuracy. For installation, it is necessary to have ten diameters of straight pipe ahead of the orifice meter, to create a smooth flow pattern, and five diameters of straight pipe on the discharge side.

7.12.3.2.2 Venturi

A *Venturi* is a restriction with a relatively long passage with smooth entry and exit (see Figure 7.24). It features long life expectancy, simplicity of construction, and relatively high-pressure recovery (i.e., produces less permanent pressure loss than a similar sized orifice), but it is more expensive, is not linear with flow rate, and is the largest and heaviest differential pressure flowmeter. It is often used in wastewater flows because the smooth entry allows foreign material to be swept through instead of building up as would occur in front of an orifice. The accuracy of this type flowmeter is ±1% for a flow range of 10:1. The head loss across a Venturi flowmeter is relatively small, ranging from 3 to 10% of the differential, depending on the ratio of the throat diameter to the inlet diameter (i.e., beta ratio).

7.12.3.2.3 Nozzle

Flow nozzles (flow tubes) have a smooth entry and sharp exit. For the same differential pressure, the permanent pressure loss of a nozzle is of the same order as that of an orifice, but it can handle wastewater and abrasive fluids better than an orifice can. Note that, for the same line size and flow rate, the differential pressure at the nozzle is lower (head loss ranges from 10 to 20% of the differential) than the differential pressure for an orifice; hence, the total pressure loss is lower than that of an orifice. Nozzles are primarily used in steam service because of their rigidity, which makes them dimensionally more stable at high temperatures and velocities than orifices.

Note: A useful characteristic of nozzles it that they reach a critical flow condition—that is, a point at which further reduction in downstream pressure does not produce a greater velocity through the nozzle. When operated in this mode, nozzles are very predictable and repeatable.

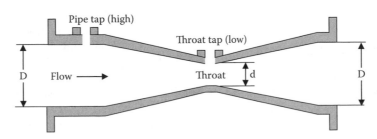

FIGURE 7.24 Venturi tube.

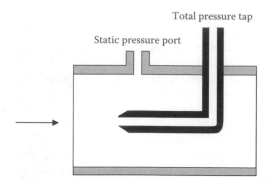

FIGURE 7.25 Pitot tube.

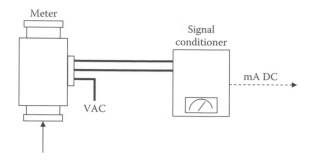

FIGURE 7.27 Magnetic flowmeter system.

7.12.3.2.4 Pitot Tube

A *Pitot tube* is a point velocity-measuring device (see Figure 7.25) with an impact port. As fluid hits the port, the velocity is reduced to zero, and kinetic energy (velocity) is converted to potential energy (pressure head). The pressure at the impact port is the sum of the static pressure and the velocity head. The pressure at the impact port is also known as the *stagnation pressure* or *total pressure*. The pressure difference between the impact pressure and the static pressure measured at the same point is the velocity head. The flow rate is the product of the measured velocity and the cross-sectional area at the point of measurement. Note that the Pitot tube has negligible permanent pressure drop in the line, but the impact port must be located in the pipe where the measured velocity is equal to the average velocity of the flowing water through the cross-section.

7.12.3.3 Magnetic Flowmeters

Magnetic flowmeters are relatively new to the water/wastewater industry (USEPA, 1991). They are volumetric flow devices designed to measure the flow of electrically conductive liquids in a closed pipe. They measure the flow rate based on the voltage created between two electrodes (in accordance with Faraday's law of electromagnetic induction) as the water passes through an electromagnetic field (see Figure 7.26). Induced voltage is proportional to

flow rate. Voltage depends on magnetic field strength (constant), distance between electrodes (constant), and velocity of flowing water (variable). Properties of the magnetic flowmeter include: (1) minimal head loss (no obstruction with line size meter); (2) no effect on flow profile; (3) suitable for sizes ranging between 0.1 and 120 in.; (4) accuracy rating of from 0.5 to 2% of flow rate; and (5) an ability to measure forward or reverse flow. The advantages of magnetic flowmeters include:

- Obstruction less flow
- Minimal head loss
- Wide range of sizes
- Bidirectional flow measurement
- Negligible effect of variations in density, viscosity, pressure, and temperature
- Suitability for wastewater
- No moving parts

Disadvantages include:

- The metered liquid must be conductive (but you would not use this type of meter on clean fluids anyway).
- They are bulky and expensive in smaller sizes, and they may require periodic calibration to correct drifting of the signal.

The combination of a magnetic flowmeter and transmitter is considered to be a system. A typical system, illustrated in Figure 7.27, shows a transmitter mounted remote from the magnetic flowmeter. Some systems are available with transmitters mounted integral to the magnetic flowmeter. Each device is individually calibrated during the manufacturing process, and the accuracy statement of the magnetic flowmeter includes both pieces of equipment. One is not sold or used without the other. It is also interesting to note that since 1983 almost every manufacturer now offers the microprocessor-based transmitter.

With regard to a minimum piping, straight-run requirement, magnetic flowmeters are quite forgiving of piping configuration. The downstream side of the magnetic

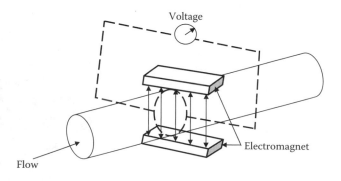

FIGURE 7.26 Magnetic flowmeter.

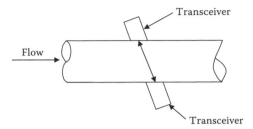

FIGURE 7.28 Time-of-flight ultrasonic flowmeter.

flowmeter is much less critical than the upstream side. Essentially, all that is required of the downstream side is that sufficient backpressure be provided to keep the magnetic flowmeter full of liquid during flow measurement. Two diameters downstream should be acceptable (Mills, 1991).

Note: Magnetic flowmeters are designed to measure conductive liquids only. If air or gas is mixed with the liquid, the output becomes unpredictable.

7.12.3.4 Ultrasonic Flowmeters

Ultrasonic flowmeters use an electronic transducer to send a beam of ultrasonic sound waves through the water to another transducer on the opposite side of the unit. The velocity of the sound beam varies with the liquid flow rate, so the beam can be electronically translated to indicate flow volume. The accuracy is ±1% for a flow velocity ranging from 1 to 25 ft/sec, but the meter reading is greatly affected by a change in the fluid composition Two types of ultrasonic flowmeters are in general use for closed-pipe flow measurements. The first (time-of-flight or transit time) usually uses pulse transmission and is intended for use with clean liquids; the second (Doppler) usually utilizes continuous wave transmission and is intended for use with dirty liquids.

7.12.3.7.1 Time-of-Flight Ultrasonic Flowmeters

Time-of-flight flowmeters make use of the difference in time required for a sonic pulse to travel a fixed distance, first in the direction of flow and then against the flow (Brown, 1991). This is accomplished by positioning opposing transceivers on a diagonal path across a meter spool, as shown in Figure 7.28. Each transmits and receives ultrasonic pulses with flow and against flow. The fluid velocity is directly proportional to the time difference of pulse travel. The time-of-flight ultrasonic flowmeter operates with minimal head loss and has an accuracy range of 1 to 2.5% full scale. These flowmeters can be mounted as integral spool piece transducers or as externally mountable clamp-ons. They can measure flow accurately when properly installed and applied. The advantages of time-of-flight ultrasonic flowmeters include:

- No obstruction or interruption of flow
- Minimal head loss
- Clamp on
- Can be portable
- No moving parts
- Linear over wide range
- Wide range of pipe sizes
- Bidirectional flow measurement

Disadvantages include:

- Sensitive to solids or bubble content
- Interference with sound pulses
- Sensitive to flow disturbances
- Critical alignment of transducers
- Requirement for pipe walls to freely pass ultrasonic pulses (clamp-on type)

7.12.3.7.2 Doppler Ultrasonic Flowmeters

Doppler ultrasonic flowmeters make use of the Doppler frequency shift caused by sound scattered or reflected from moving particles in the flow path. Doppler meters are not considered to be as accurate as time-of-flight flowmeters; however, they are very convenient to use and generally more popular and less expensive than time-of-flight flowmeters. In operation, a propagated ultrasonic beam is interrupted by particles in moving fluid and reflected toward a receiver. The difference of propagated and reflected frequencies is directly proportional to fluid flow rate. Doppler ultrasonic flowmeters feature minimal head loss with an accuracy of 2 to 5% full scale. They can be the integral spool piece transducer type or externally mountable clamp-ons. The advantages of the Doppler ultrasonic flowmeter include:

- No obstruction or interruption of flow
- Minimal head loss
- Clamp-on
- Can be portable
- No moving parts
- Linear over wide range
- Wide range of pipe sizes
- Low installation and operating costs
- Bidirectional flow measurement

The disadvantages include:

- Minimum concentration and size of solids or bubbles required for reliable operation (see Figure 7.29)
- Minimum speed required to maintain suspension
- Limited to sonically conductive pipe (clamp-on type)

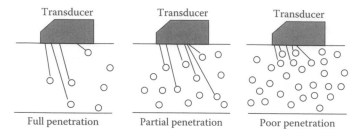

FIGURE 7.29 Particle concentration effect—the more particles, the greater the error.

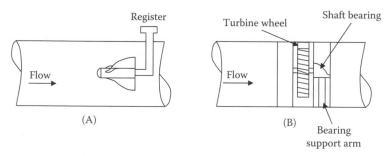

FIGURE 7.30 (A) Propeller meter; (B) turbine meter.

7.12.3.5 Velocity Flowmeters

Velocity or *turbine* flowmeters use a propeller or turbine to measure the velocity of the flow passing the device (see Figure 7.30). The velocity is then translated into a volumetric amount by the meter register. Sizes exist from a variety of manufacturers to cover the flow range from 0.001 gpm to over 25,000 gpm for liquid service. End connections are available to meet the various piping systems. The flowmeters are typically manufactured of stainless steel but are also available in a wide variety of materials, including plastic. Velocity meters are applicable to all clean fluids. Velocity meters are particularly well suited for measuring intermediate flow rates on clean water (Oliver, 1991). The advantages of the velocity flowmeter include:

- Accuracy
- Composed of corrosion-resistant materials
- Long-term stability
- Liquid or gas operation
- Wide operating range
- Low pressure drop
- Wide temperature and pressure limits
- High shock capability
- Wide variety of electronics available

As shown in Figure 7.30, a turbine flowmeter consists of a rotor mounted on a bearing and shaft in a housing. The fluid to be measured is passed through the housing, causing the rotor to spin with a rotational speed proportional to the velocity of the flowing fluid within the meter. A device to measure the speed of the rotor is employed to make the actual flow measurement. The sensor can be a mechanically gear-driven shaft connected to a meter or an electronic sensor that detects the passage of each rotor blade generating a pulse. The rotational speed of the sensor shaft and the frequency of the pulse are proportional to the volumetric flow rate through the meter.

7.12.3.6 Positive-Displacement Flowmeters

Positive-displacement flowmeters are most commonly used for customer metering and have long been used to measure liquid products. These meters are very reliable and accurate for low flow rates because they measure the exact quantity of water passing through them. Positive-displacement flowmeters are frequently used for measuring small flows in a treatment plant because of their accuracy. Repair or replacement is easy because they are so common in the distribution system (Barnes, 1991).

In essence, a positive-displacement flowmeter is a hydraulic motor with high volumetric efficiency that absorbs a small amount of energy from the flowing stream. This energy is used to overcome internal friction in driving the flowmeter and its accessories and is reflected as a pressure drop across the flowmeter. Pressure drop is considered unavoidable and must be minimized. It is the pressure drop across the internals of a positive-displacement flowmeter that actually creates a hydraulically unbalanced rotor, which causes rotation. A positive-displacement flowmeter continuously divides the flowing stream into known volumetric segments, isolates the segments momentarily, and returns them to the flowing stream while counting the number of displacements.

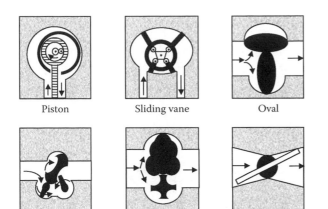

FIGURE 7.31 Six common positive-displacement meter principles.

A positive-displacement flowmeter can be broken down into three basic components: the external housing, the measuring unit, and the counter drive train. The external housing is the pressure vessel that contains the product being measured. The measuring unit is a precision metering element made up of a measuring chamber and a displacement mechanism. The most common displacement mechanisms include the oscillating piston, sliding vane, oval gear, trirotor, birotor, and nutating disc types (see Figure 7.31). The counter drive train is used to transmit the internal motion of the measuring unit into a usable output signal. Many positive-displacement flowmeters use a mechanical gear train that requires a rotary shaft seal or packing gland where the shaft penetrates the external housing.

The positive-displacement flowmeter can offer excellent accuracy, repeatability, and reliability in many applications. It has satisfied many needs in the past and should play a vital role in serving future needs as required.

7.12.4 OPEN-CHANNEL FLOW MEASUREMENT

The majority of industrial liquid flows are carried in closed conduits that flow completely full and under pressure; however, this is not the case for high-volume flows of liquids in waterworks, sanitary, and stormwater systems that are commonly carried in open channels. Low system heads and high volumetric flow rates characterize flow in open channels (Grant, 1991). The most commonly used method of measuring the rate of flow in open-channel flow configurations is that of hydraulic structures. In this method, flow in an open channel is measured by inserting a hydraulic structure into the channel, which changes the level of liquid in or near the structure. By selecting the shape and dimensions of the hydraulic structure, the rate of flow through or over the restriction will be related to the liquid level in a known manner. Thus, the flow rate through the open channel can be derived from a single measurement of the liquid level in or near the structure.

The hydraulic structures used in measuring flow in open channels are known as primary measuring devices and may be divided into two broad categories—weirs and flumes, which are covered in the following subsections.

7.12.4.1 Weirs

The *weir* is a widely used device to measure open-channel flow. As can be seen in Figure 7.32, a weir is simply a dam or obstruction placed in the channel so water backs up behind it and then flows over it. The sharp crest or edge allows the water to spring clear of the weir plate and to fall freely in the form of a nappe. As Nathanson (1997) pointed out, when the nappe discharges freely into the air, a relationship exists between the height or depth of water flowing over the weir crest and the flow rate. This height, the vertical distance between the crest and the water surface, is the head on the weir; it can be measured directly with a meter or yardstick or automatically by float-operated recording devices. Two common weirs, rectangular and triangular, are shown in Figure 7.33.

Rectangular weirs are commonly used for large flows (see Figure 7.33A). The formula to compute the rectangular weir is:

$$Q = 3.33 \times L \times h^{1.5} \qquad (7.27)$$

where:

Q = flow.
L = width of weir.
h = head on weir (measured from edge of weir in contact with the water, up to the water surface).

■ EXAMPLE 7.17

Problem: A weir 4-ft high extends 15 ft across a rectangular channel with a flow of 80 cfs. What is the depth just upstream from the weir?

Solution:

$$Q = 3.33 \times L \times h^{1.5}$$

$$80 = 3.33 \times 15 \times h^{1.5}$$

$$h = 1.4 \text{ ft (w/calculator: 1.6 INV } y^{x1.5} = 1.36, \text{ or 1.4)}$$

4 ft height of weir + 1.4 ft head of water = 5.4 ft depth

Triangular weirs, also called V-notch weirs, can have notch angles ranging from 22.5° to 90°, but right-angle notches are the most common (see Figure 7.33B). The formula used to make V-notch (90°) weir calculations is:

$$Q = 2.5 \times h^{2.5} \qquad (7.28)$$

where:

Q = flow.
h = head on weir (measured from bottom of notch to water surface).

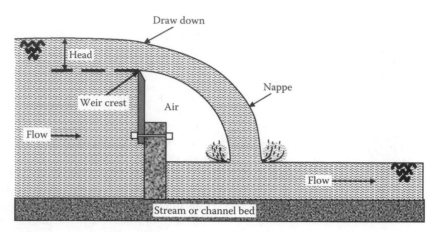

FIGURE 7.32 Side view of a weir.

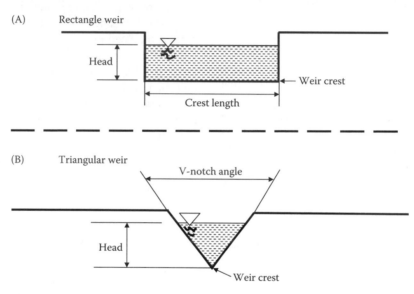

FIGURE 7.33 (A) Rectangular weir; (B) triangular V-notch weir.

■ EXAMPLE 7.18

Problem: What should be the minimum weir height for measuring a flow of 1200 gpm with a 90° V-notch weir, if the flow is moving at 4 ft/sec in a 2.5-ft-wide rectangular channel?

Solution:

$$\frac{1200 \text{ gpm}}{60 \text{ sec/min} \times 7.48 \text{ gal/ft}^3} = 2.67 \text{ cfs}$$

$$Q = A \times V$$

$$2.67 = 2.5 \times d \times 4$$

$$d = 0.27 \text{ ft}$$

$$Q = 2.5 \times h^{2.5}$$

$$2.67 = 2.5 \times h^{2.5}$$

$h = 1.03$ (w/calculator: 1.06 INV $y^{x2.5}$=1.026, or 1.03)

0.27 ft (original depth) + 1.03 (head on weir) = 1.3 ft

It is important to point out that weirs, aside from being operated within their flow limits, must also be operated within the available system head. In addition, the operation of the weir is sensitive to the approach velocity of the water, often necessitating a stilling basin or pond upstream of the weir. Weirs are not suitable for water that carries excessive solid materials or silt, which deposit in the approach channel behind the weir and destroy the conditions required for accurate discharge measurements.

Note: Accurate flow rate measurements with a weir cannot be expected unless the proper conditions and dimensions are maintained.

7.12.4.2 Flumes

A *flume* is a specially shaped constricted section in an open channel (similar to the Venturi tube in a pressure conduit). The special shape of the flume (see Figure 7.34) restricts the channel area and changes the channel slope,

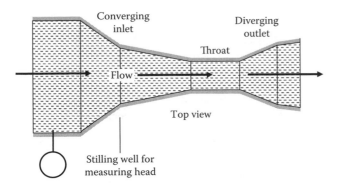

FIGURE 7.34 Parshall flume

resulting in an increased velocity and a change in the level of the liquid flowing through the flume. The flume restricts the flow and then expands it in a definite fashion. The flow rate through the flume may be determined by measuring the head on the flume at a single point, usually at some distance downstream from the inlet.

Flumes can be categorized as belonging to one the three general families, depending on the state of flow induced: subcritical, critical, or supercritical. Typically, flumes that induce a critical or supercritical state of flow are most commonly used. When critical or supercritical flow occurs in a channel, one head measurement can indicate the discharge rate if it is made far enough upstream so the flow depth is not affected by the drawdown of the water surface as it achieves or passes through a critical state of flow. For critical or supercritical states of flow, a definitive head–discharge relationship can be established and measured, based on a single head reading. Thus, most commonly encountered flumes are designed to pass the flow from subcritical through critical or near the point of measurement.

The most common flume used for a permanent wastewater flowmetering installation is called the *Parshall flume*, shown in Figure 7.34. Formulae for flow through Parshall flumes differ, depending on throat width. The formula below can be used for widths of 1 to 8 ft and applies to a medium range of flows:

$$Q = 4 \times W \times H_a^{1.52} \times W^{.026} \qquad (7.29)$$

where:

Q = flow.
W = width of throat.
H_a = depth in stilling well upstream.

Note: Parshall flumes are low-maintenance items.

CHAPTER REVIEW QUESTIONS

7.1   What should be the minimum weir height for measuring a flow of 900 gpm with a 90° V-notch weir, if the flow is now moving at 3 ft/sec in a 2-ft-wide rectangular channel?

7.2   A 90° V-notch weir is to be installed in a 30-in.-diameter sewer to measure 600 gpm. What head should be expected?

7.3   For dirty water operations, a _____ or _____ orifice plate should be used.

7.4   A _____ has a smooth entry and sharp exit.

7.5   _____ send a beam of ultrasonic sound waves through the water to another transducer on the opposite side of the unit.

7.6   Find the number of gallons in a storage tank that has a volume of 660 ft³.

7.7   Suppose a rock weighs 160 lb in air and 125 lb under water. What is the specific gravity?

7.8   A 110-ft-diameter cylindrical tank contains 1.6 MG water. What is the water depth?

7.9   The pressure in a pipeline is 6400 psf. What is the head on the pipe?

7.10   The pressure on a surface is 35 psig. If the surface area is 1.6 ft², what is the force (lb) exerted on the surface?

7.11   Bernoulli's principle states that the total energy of a hydraulic fluid is _____
_____.

7.12   What is *pressure head*?

7.13   What is a *hydraulic grade line*?

7.14   A flow of 1500 gpm takes place in a 12-in. pipe. Calculate the velocity head.

7.15   Water flows at 5.00 mL/sec in a 4-in. line under a pressure of 110 psi. What is the pressure head (ft of water)?

7.16   In Question 7.15, what is the velocity head in the line?

7.17   What is velocity head in a 6-in. pipe connected to a 1-ft pipe, if the flow in the larger pipe is 1.46 cfs?

7.18   What is *velocity head*?

7.19   What is *suction lift*?

7.20   Explain *energy grade line*.

REFERENCES AND SUGGESTED READING

AWWA. 1995. *Basic Science Concepts and Applications: Principles and Practices of Water Supply Operations*, 2nd ed. Denver, CO: American Water Works Association.

Barnes, R.G. 1991. Positive displacement flowmeters for liquid measurement. In *Flow Measurement*, Spitzer, D.W., Ed. Research Triangle Park, NC: Instrument Society of America.

Brown, A.E. 1991. Ultrasonic flowmeters. In *Flow Measurement*, Spitzer, D.W., Ed. Research Triangle Park, NC: Instrument Society of America.

Grant, D.M. 1991. Open-channel flow measurement. In *Flow Measurement*, Spitzer, D.W., Ed. Research Triangle Park, NC: Instrument Society of America.

Husain, Z.D. and Sergesketter, M.J. 1991. Differential pressure flowmeters. In *Flow Measurement*, Spitzer, D.W., Ed. Research Triangle Park, NC: Instrument Society of America.

Hauser, B.A. 1993. *Hydraulics for Operators*. Boca Raton, FL: Lewis Publishers.

Hauser, B.A. 1996. *Practical Hydraulics Handbook*, 2nd ed. Boca Raton, FL: Lewis Publishers.

Holman, S. 1998. *A Stolen Tongue*. New York: Anchor Press.

Kawamura, S. 2000. *Integrated Design and Operation of Water Treatment Facilities*, 2nd ed. New York: John Wiley & Sons.

Lindeburg, M.R. 1986. *Civil Engineering Reference Manual*, 4th ed. San Carlos, CA: Professional Publications.

Magnusson, R.J. 2001. *Water Technology in the Middle Ages*. Baltimore, MD: The John Hopkins University Press.

McGhee, T.J. 1991. *Water Supply and Sewerage*, 2nd ed. New York: McGraw–Hill.

Metcalf & Eddy. 1981. *Wastewater Engineering: Collection and Pumping of Wastewater*. New York: McGraw–Hill.

Mills, R.C. 1991. Magnetic flowmeters. In *Flow Measurement*, Spitzer, D.W., Ed. Research Triangle Park, NC: Instrument Society of America.

Nathanson, J.A. 1997. *Basic Environmental Technology: Water Supply Waste Management, and Pollution Control*, 2nd ed. Upper Saddle River, NJ: Prentice Hall.

Oliver, P.D. 1991. Turbine flowmeters. In *Flow Measurement*, Spitzer, D.W., Ed. Research Triangle Park, NC: Instrument Society of America.

Spellman, F.R., 2007. *The Science of Water*, 2nd ed. Boca Raton, FL: CRC Press.

Spellman, F.R. and Drinan, J. 2001. *Water Hydraulics*. Boca Raton, FL: CRC Press.

USEPA. 1991. *Flow Instrumentation: A Practical Workshop on Making Them Work*. Sacramento, CA: Water and Wastewater Instrumentation Testing Association.

Viessman, Jr., W. and Hammer, M.J. 1998. *Water Supply and Pollution Control*, 6th ed. Menlo Park, CA: Addison–Wesley.

8 Fundamentals of Electricity

When Gladstone was British Prime Minister, he visited Michael Faraday's laboratory and asked if some esoteric substance called "electricity" would ever have practical significance. "One day, sir, you will tax it," was his answer.

Quoted in *Science* (1994)

Electricity. What is it? Water and wastewater operators generally have little difficulty in recognizing electrical equipment. Electrical equipment is everywhere and is easy to spot; for example, typical plant sites are outfitted with equipment to:

- Generate electricity (a generator or emergency generator)
- Store electricity (batteries)
- Change electricity from one form to another (transformers)
- Transport or transmit and distribute electricity throughout the plant site (wiring distribution systems)
- Measure electricity (meters)
- Convert electricity into other forms of energy (mechanical energy, heat energy, light energy, chemical energy, or radio energy)
- Protect other electrical equipment (fuses, circuit breakers, or relays)
- Operate and control other electrical equipment (motor controllers)
- Convert some condition or occurrence into an electric signal (sensors)
- Convert some measured variable to a representative electrical signal (transducers or transmitters)

Recognizing electrical equipment is easy because we use so much of it. If we ask typical operators where such equipment is located in their plant site, they know, because they probably operate these devices or their ancillaries. If we asked these same operators what a particular electrical device does, they could probably tell us. If we were to ask if their plant electrical equipment was important to plant operations, the chorus would resound "absolutely."

Here is another question that does not always result in such an assured answer. If we asked these same oper-

ators to explain to us in very basic terms how electricity works to make their plant equipment operate, the answers we would receive probably would be varied, jumbled, disjointed—and probably not all that accurate. Even on a more basic level, how many operators would be able to accurately answer the question: What is electricity?

Probably very few. Why do so many operators in both water and wastewater know so little about electricity? Part of the answer resides in the fact that operators are expected to know so much (and they do) but are given so little opportunity to be properly trained. We all know that experience is a great trainer. As an example, let us look at what an operator assigned to change the bearings on a 5-hp three-phase motor would need to know to accomplish this task. (Remember that it is not uncommon for water/wastewater operators to maintain as well as operate plant equipment.) The operator would have to know:

- How to deenergize the equipment (i.e., proper lockout/tagout procedures)
- After the equipment has been deenergized, how to properly disassemble the motor coupling from the device it operates (e.g., a motor coupling from a pump shaft) and the proper tools to use
- When the equipment has been uncoupled, how to properly disassemble the motor end-bells (preferably without damaging the rotor shaft)
- When the equipment has been disassembled, how to recognize if the bearings are really in need of replacement (although when they have been removed from the end-bells, the bearings are typically replaced)

Questions for which the operator would need to know how to find answers include the following:

- If the bearings are in need of replacement, how are they to be removed without causing damage to the rotor shaft?
- When they have been removed, what bearings should be used to replace the old bearings?
- When the proper bearings are identified and obtained, how are they to be installed properly?
- When the bearings are replaced properly, how is the motor to be reassembled properly?

- When the motor is correctly put back together, how is it properly aligned to the pump and then reconnected?
- What is the test procedure to ensure that the motor has been restored properly to full operational status?

Every one of these procedures is important, and, obviously, errors at any point in the process could cause damage (maybe more damage than occurred in the first place). Here is another question to go along with our earlier one: Does the operator need to know electricity to perform the sequence of tasks described above? The short answer is no, not exactly. Fully competent operators (who received most of their training via on-the-job experience) are usually qualified to perform a bearing change-out on most plant motors with little difficulty. The long answer is yes. Consider the motor mechanic who tunes your automobile engine, then ask yourself whether or not it is important for the mechanic to have some understanding of internal combustion engines. We think it is important. You probably do, too. We also think it is important for the operator who changes bearings on an electrical motor to have some understanding of how the electric motor operates.

Here is another issue to consider. Have you taken an operator's state licensure examination? If you have, then you know that, typically, state licensure examinations test the examinee's knowledge of basic electricity. (This is especially the case for water operators.) Some states definitely consider operator knowledge of electricity to be important. For reasons of licensure and of job competence, water/wastewater operators should have some basic electrical knowledge, but how and where can operators quickly and easily acquire such knowledge? In this chapter, we provide the how and the where—here and now.

8.1 NATURE OF ELECTRICITY

The word "electricity" is derived from the Greek word *electron* for "amber," which is a translucent (semitransparent) yellowish fossilized mineral resin. The ancient Greeks used the words "electric force" to refer to the mysterious forces of attraction and repulsion exhibited by amber when it was rubbed with a cloth. They did not understand the nature of this force and could not answer the question: "What is electricity?" The fact is this question remains unanswered. Today, we often attempt to answer this question by describing the effect and not the force; that is, the standard answer given is that electricity is "the force that moves electrons," which is about the same as defining a sail as "the force that moves a sailboat."

At the present time, little more is known than the ancient Greeks knew about the fundamental nature of electricity, but we have made tremendous strides in harnessing and using it. As with many other unknown (or unexplainable) phenomena, elaborate theories concerning the nature and behavior of electricity have been advanced and have gained wide acceptance because of their apparent truth—and because they work.

Scientists have determined that electricity seems to behave in a constant and predictable manner in given situations or when subjected to given conditions. Scientists such as Faraday, Ohm, Lenz, and Kirchhoff have described the predictable characteristics of electricity and electric current in the form of certain rules. These rules are often referred to as *laws*. Thus, although electricity itself has never been clearly defined, its predictable nature and ease of use have made it one of the most widely used power sources in modern times.

The bottom line: We can still learn about electricity by studying the rules, or laws, applying to the behavior of electricity and by understanding the methods of producing, controlling, and using it. Thus, this learning about electricity can be accomplished without ever having determined its fundamental identity.

You are probably scratching your head in puzzlement, and a question almost certainly running through your brain at this exact moment is: "This is a text about basic electricity and the authors can't even explain what electricity is?" That is correct. We cannot. The point is, no one can definitively define electricity. Electricity is one of those subject areas where the old saying "We don't know what we don't know about it" fits perfectly.

Again, a few theories about electricity have so far stood the test of extensive analysis and much time (relatively speaking, of course). One of the oldest and most generally accepted theories concerning electric current flow (or electricity) is known as the *electron theory*, which states that electricity or current flow is the result of the flow of free electrons in a conductor. Thus, electricity is the flow of free electrons or simply electron flow. In this text this is how we define electricity; that is, *electricity is the flow of free electrons*.

Electrons are extremely tiny particles of matter. To gain an understanding of electrons and exactly what is meant by electron flow, it is necessary to briefly discuss the structure of matter.

8.2 THE STRUCTURE OF MATTER

Matter is anything that has mass and occupies space. To study the fundamental structure or composition of any type of matter, it must be reduced to its fundamental components. All matter is made of *molecules*, or combinations of *atoms* (a term which, in Greek, means "not able to be divided"). These are bound together to produce a given substance, such as salt, glass, or water. If we were to keep dividing water into smaller and smaller drops, we

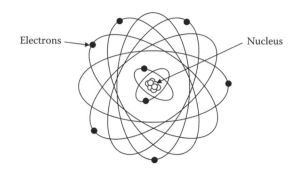

FIGURE 8.1 Electrons and nucleus of an atom.

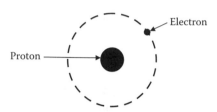

FIGURE 8.2 One proton and one electron = electrically neutral.

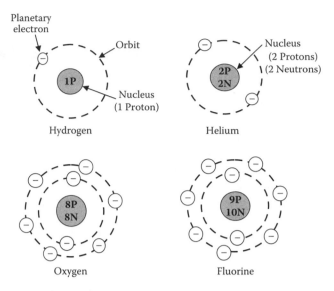

FIGURE 8.3 Atomic structure of elements.

would eventually arrive at the smallest particle that was still water. That particle is the molecule, which is defined as the smallest bit of a substance that retains the characteristics of that substance.

Note: Molecules are made up of atoms, which are bound together to produce a given substance.

Atoms are composed, in various combinations, of sub-atomic particles of *electrons*, *protons*, and *neutrons*. These particles differ in weight (a proton is much heavier than the electron) and charge. We are not concerned with the weights of particles in this text, but the charge is extremely important in electricity. The electron is the fundamental negative (–) charge of electricity. Electrons revolve about the nucleus or center of the atom in paths of concentric *orbits*, or shells (see Figure 8.1). The proton is the fundamental positive (+) charge of electricity. Protons are found in the nucleus. The number of protons within the nucleus of any particular atom specifies the atomic number of that atom; for example, the helium atom has 2 protons in its nucleus so the atomic number is 2. The neutron, which is the fundamental neutral charge of electricity, is also found in the nucleus.

Most of the weight of the atom is in the protons and neutrons of the nucleus. Whirling around the nucleus are one or more negatively charged electrons. Normally, there is one proton for each electron in the entire atom so the net positive charge of the nucleus is balanced by the net negative charge of the electrons rotating around the nucleus (see Figure 8.2).

Note: Most batteries are marked with the symbols + and – or even with the abbreviations POS (positive) and NEG (negative). The concept of a positive or

negative polarity and its importance in electricity will become clear later; however, for the moment, it is important to remember that an electron has a negative charge and a proton has a positive charge.

We stated earlier that in an atom the number of protons is usually the same as the number of electrons. This is an important point because this relationship determines the kind of element in question. The atom is the smallest particle that makes up an element, and an element retains its characteristics when subdivided into atoms. Figure 8.3 shows a simplified drawing of several atoms of various materials based on the concept of electrons orbiting around the nucleus. Hydrogen, for example, has a nucleus consisting of 1 proton around which rotates 1 electron. The helium atom has a nucleus containing 2 protons and 2 neutrons with 2 electrons encircling the nucleus. Both of these elements are electrically neutral (or balanced) because each has an equal number of electrons and protons. Because the negative (–) charge of each electron is equal in magnitude to the positive (+) charge of each proton, the two opposite charges cancel.

A balanced (neutral or stable) atom has a certain amount of energy that is equal to the sum of the energies of its electrons. Electrons, in turn, have different energies, called *energy levels*. The energy level of an electron is proportional to its distance from the nucleus; therefore, the energy levels of electrons in shells farther from the nucleus are higher than those of electrons in shells closer to the nucleus.

When an electric force is applied to a conducting medium, such as copper wire, electrons in the outer orbits of the copper atoms are forced out of orbit (i.e., the electrons are liberated or freed) and are impelled along the wire. This electrical force, which forces electrons out of

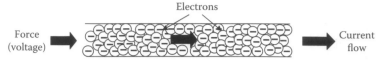

FIGURE 8.4 Electron flow in a copper wire.

orbit, can be produced in a number of ways, such as by moving a conductor through a magnetic field; by friction, such as when a glass rod is rubbed with cloth (silk); or by chemical action, such as in a battery.

When the electrons are forced from their orbits, they are called *free electrons*. Some of the electrons of certain metallic atoms are so loosely bound to the nucleus that they are relatively free to move from atom to atom. These free electrons constitute the flow of an electric current in electrical conductors.

Note: When an electric force is applied to a copper wire, free electrons are displaced from the copper atoms and move along the wire, producing electric current as shown in Figure 8.4.

If the internal energy of an atom is raised above its normal state, the atom is said to be *excited*. Excitation may be produced by causing the atoms to collide with particles that are impelled by an electric force, as shown in Figure 8.4. In effect, what occurs is that energy is transferred from the electric source to the atom. The excess energy absorbed by an atom may become sufficient to cause loosely bound outer electrons (as shown in Figure 8.4) to leave the atom against the force that acts to hold them within.

Note: An atom that has lost or gained one or more electrons is said to be *ionized*. If the atom loses electrons, it becomes positively charged and is referred to as a *positive ion*; conversely, if the atom gains electrons, it becomes negatively charged and is referred to as a *negative ion*.

8.3 CONDUCTORS, SEMICONDUCTORS, AND INSULATORS

Electric current moves easily through some materials but with greater difficulty through others. Substances that permit the free movement of a large number of electrons are called *conductors*. The most widely used electrical conductor is copper because of its high conductivity (which refers to how good a conductor the material is) and cost effectiveness.

Electrical energy is transferred through a copper or other metal conductor by means of the movement of free electrons that migrate from atom to atom inside the conductor (see Figure 8.4). Each electron moves a very short distance to the neighboring atom, where it replaces one or more electrons by forcing them out of their orbits. The

TABLE 8.1
Electrical Conductors

Silver	Brass
Copper	Iron
Gold	Tin
Aluminum	Mercury
Zinc	

replaced electrons repeat the process in other nearby atoms until the movement is transmitted throughout the entire length of the conductor. A good conductor is said to have a low opposition, or resistance, to the electron (current) flow. Table 8.1 lists some of the metals commonly used as electric conductors. The best conductors appear at the top of the list, with the poorer ones shown last.

Note: If lots of electrons flow through a material with only a small force (voltage) applied, we call that material a *conductor*.

Note: The movement of each electron (e.g., in copper wire) takes a very small amount of time; it is almost instantaneous. This is an important point to keep in mind later in the chapter, when events in an electrical circuit seem to occur simultaneously.

Although it is true that electron motion is known to exist to some extent in all matter, some substances, such as rubber, glass, and dry wood, have very few free electrons. In these materials, large amounts of energy must be expended to break the electrons loose from the influence of the nucleus. Substances containing very few free electrons are called *insulators*. Insulators are important in electrical work because they prevent the current from being diverted from the wires. Table 8.2 lists some materials that are often used as insulators in electrical circuits. The list is in decreasing order of ability to withstand high voltages without conducting.

Note: If the voltage is large enough, even the best insulators will break down and allow their electrons to flow.

A material that is neither a good conductor nor a good insulator is called a *semiconductor*. Silicon and germanium are substances that fall into this category. Because of their peculiar crystalline structure, these materials may under certain conditions act as conductors; under other

TABLE 8.2
Common Insulators

Rubber
Mica
Wax or paraffin
Porcelain
Bakelite
Plastics
Glass
Fiberglass
Dry wood
Air

conditions, as insulators. As the temperature is raised, however, a limited number of electrons become available for conduction.

8.4 STATIC ELECTRICITY

Electricity at rest is often referred to as *static electricity.* More specifically, when two bodies of matter have unequal charges and are near one another, an electric force exists between them because of their unequal charges. Because they are not in contact, their charges cannot equalize. Such an electric force, where current cannot flow, is considered static electricity; however, static electricity, or electricity at rest, will flow if given the opportunity. An example of this phenomenon is often experienced when one walks across a dry carpet and then touches a doorknob—a slight shock is usually felt and a spark at the fingertips is likely noticed. In the workplace, static electricity is prevented from building up by properly bonding equipment to ground or earth.

8.4.1 CHARGED BODIES

To fully grasp a clear understanding of static electricity, it is necessary to know one of the fundamental laws of electricity and its significance. The fundamental law of charged bodies states that *like charges repel each other and unlike charges attract each other.* A positive charge and negative charge, being opposite or unlike, tend to move toward each other—they are attracted to each other. In contrast, like bodies tend to repel each other. Electrons repel each other because of their like negative charges, and protons repel each other because of their like positive charges. Figure 8.5 demonstrates the law of charged bodies.

It is important to point out another significant part of the fundamental law of charged bodies: *The force of attraction or repulsion existing between two magnetic poles decreases rapidly as the poles are separated from each other.* More specifically, the force of attraction or repulsion varies directly as the product of the separate pole strengths and inversely as the square of the distance separating the magnetic poles, provided the poles are small enough to be considered as points.

Let's look at an example. If we increase the distance between two north poles from 2 feet to 4 feet, the force of repulsion between them is decreased to one fourth of its original value. If either pole strength is doubled, with the distance remaining the same, the force between the poles will be doubled.

8.4.2 COULOMB'S LAW

Simply put, Coulomb's law points out that the amount of attracting or repelling force that acts between two electrically charged bodies in free space depends on two things:

- Their charges
- The distance between them

Specifically, Coulomb's law states: *Charged bodies attract or repel each other with a force that is directly proportional to the product of their charges and is inversely proportional to the square of the distance between them.*

Note: The magnitude of electric charge a body possesses is determined by the number of electrons compared with the number of protons within the body. The symbol for the magnitude of electric charge is Q, expressed in units of coulombs (C). A charge of 1 positive coulomb means a body contains a charge of 6.25×10^{18}. A charge of 1 negative coulomb ($-Q$) means a body contains a charge of 6.25×10^{18} more electrons than protons.

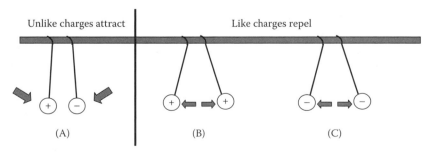

FIGURE 8.5 Reaction between two charged bodies: (A) opposite charges attract; (B) and (C) like charges repel each other.

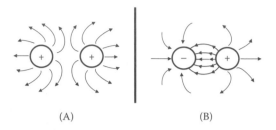

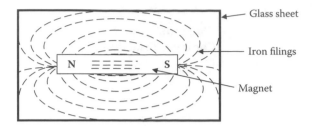

FIGURE 8.6 Electrostatic lines of force: (A) repulsion of like-charged bodies and their associated fields; (B) attraction between unlike-charged bodies and their associated fields.

FIGURE 8.7 Magnetic field around a bar magnet. If the glass sheet is tapped gently, the filings will move into a definite pattern that describes the field of force around the magnet.

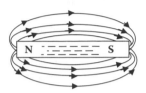

FIGURE 8.8 Magnetic field of force around a bar magnet, indicated by lines of force.

8.4.3 ELECTROSTATIC FIELDS

The fundamental characteristic of an electric charge is its ability to exert force. The space between and around charged bodies in which their influence is felt is an *electric field of force*. The electric field is always terminated on material objects and extends between positive and negative charges. This region of force can consist of air, glass, paper, or a vacuum. This region of force is referred to as an *electrostatic field*.

When two objects of opposite polarity are brought near each other, the electrostatic field is concentrated in the area between them. Lines referred to as *electrostatic lines of force* generally represent the field. These lines are imaginary and are used merely to represent the direction and strength of the field. To avoid confusion, the positive lines of force are always shown as a leaving charge; for a negative charge, they are shown as entering. Figure 8.6 illustrates the use of lines to represent the field about charged bodies.

Note: A charged object will retain its charge temporarily if no immediate transfer of electrons to or from it is occurring. In this condition, the charge is said to be *at rest*. Remember that electricity at rest is *static* electricity.

8.5 MAGNETISM

Most electrical equipment depends directly or indirectly on *magnetism*, which is defined as a phenomenon associated with magnetic fields; that is, it has the power to attract such substances as iron, steel, nickel, or cobalt (metals that are known as magnetic materials). Correspondingly, a substance is said to be a magnet if it has the property of magnetism; for example, a piece of iron can be magnetized and thus becomes a magnet.

For our discussion here, we will assume that the piece of iron is a flat bar 6 in. long × 1 in. wide × 0.5 in. thick (see Figure 8.7). When magnetized (i.e., becomes a bar magnet), this piece of iron will have two points opposite each other, which most readily attract other pieces of iron. The points of maximum attraction (one on each end) are

called the *magnetic poles* of the magnet: the north (N) pole and the south (S) pole. Just as like electric charges repel each other and opposite charges attract each other, like magnetic poles repel each other and unlike poles attract each other. Although invisible to the naked eye, the force can be shown to exist by sprinkling small iron filings on a glass covering a bar magnet, as shown in Figure 8.7.

Figure 8.8 shows how the field looks without iron filings; it is shown as lines of force in the field, repelled away from the north pole of the magnet and attracted to its south pole. These lines of force are *magnetic flux* or *flux lines*; the symbol for magnetic flux is the Greek lowercase letter ϕ (phi).

Note: A magnetic circuit is a complete path through which magnetic lines of force may be established under the influence of a magnetizing force. Most magnetic circuits are composed largely of magnetic materials to contain the magnetic flux. These circuits are similar to the electric circuit (an important point), which is a complete path through which current is caused to flow under the influence of an electromotive force.

Three types or groups of magnets exist:

- *Natural magnets* are found in the natural state in the form of a mineral (an iron compound) called *magnetite*.
- *Permanent magnets* (artificial magnets) are hardened steel or some alloy such as Alnico™ bars that have been permanently magnetized. The permanent magnet most people are familiar with is the horseshoe magnet (see Figure 8.9).

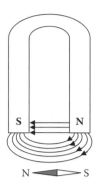

FIGURE 8.9 Horseshoe magnet.

- *Electromagnets* (artificial magnets) are composed of soft-iron cores around which are wound coils of insulated wire. When an electric current flows through the coil, the core becomes magnetized. When the current ceases to flow, the core then loses most of the magnetism.

8.5.1 MAGNETIC MATERIALS

Natural magnets are no longer used in electrical circuitry because more powerful and more conveniently shaped permanent magnets can be produced artificially. Commercial magnets are made from special steels and alloys that are *magnetic materials*, which are materials that are attracted or repelled by a magnet and can be magnetized themselves. Iron, steel, and alloy bars are the most common magnetic materials. These materials can be magnetized by inserting the material (in bar form) into a coil of insulated wired and passing a heavy direct current through the coil. The same material may also be magnetized if it is stroked with a bar magnet. It will then have the same magnetic property that the magnet used to induce the magnetism has—namely, there will be two poles of attraction, one at either end. This process produces a permanent magnet by induction; that is, the magnetism is induced in the bar by the influence of the stroking magnet.

Note: Permanent magnets are made of hard magnetic materials (hard steel or alloys) that retain their magnetism when the magnetizing field is removed. A temporary magnet is one that has *no* ability to retain a magnetized state when the magnetizing field is removed.

Even though they are classified as permanent magnets, it is important to point out that hardened steel and certain alloys are relatively difficult to magnetize and are said to have a low permeability because the magnetic lines of force do not easily permeate or distribute themselves readily through the steel.

Note: *Permeability* refers to the ability of a magnetic material to concentrate magnetic flux. Any material that is easily magnetized has high permeability. A measure of permeability for different materials in comparison with air or vacuum is called *relative permeability*, symbolized by μ (mu).

When hard steel and other alloys are magnetized, they retain a large part of their magnetic strength and are considered to be *permanent magnets*. Conversely, materials that are relatively easy to magnetize, such as soft iron and annealed silicon steel, are said to have a high permeability. Such materials retain only a small part of their magnetism after the magnetizing force is removed and are considered to be *temporary magnets*. The magnetism that remains in a temporary magnet after the magnetizing force is removed is *residual magnetism*.

Early magnetic studies classified magnetic materials merely as being magnetic or nonmagnetic; that is, they were classified based on the strong magnetic properties of iron. Because weak magnetic materials can be important in some applications, current studies classify materials into the following three groups:

- *Paramagnetic materials* include aluminum, platinum, manganese, and chromium—materials that become only slightly magnetized even when under the influence of a strong magnetic field. This slight magnetization is in the same direction as the magnetizing field. Relative permeability is slightly more than 1 (i.e., they are considered nonmagnetic materials).
- *Diamagnetic materials* include bismuth, antimony, copper, zinc, mercury, gold, and silver—materials that can also be slightly magnetized when under the influence of a very strong field. Relative permeability is less than 1 (i.e., they are considered nonmagnetic materials).
- *Ferromagnetic materials* include iron, steel, nickel, cobalt, and commercial alloys—materials that comprise the most important group for applications of electricity and electronics. Ferromagnetic materials are easy to magnetize and have high permeability, ranging from 50 to 3000.

8.5.2 MAGNETIC EARTH

The Earth is a huge magnet, and surrounding Earth is a magnetic field produced by the Earth's magnetism. Most people would have no problem understanding or at least accepting this statement, but they might not accept or understand this statement that the Earth's north magnetic pole is actually its south magnetic pole and the south magnetic pole is actually the Earth's north magnetic pole. In terms of a magnet, though, this is true.

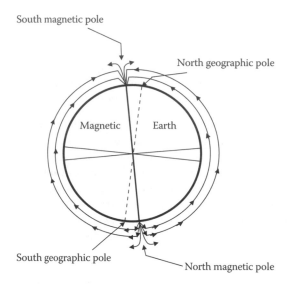

FIGURE 8.10 Magnetic poles of the Earth.

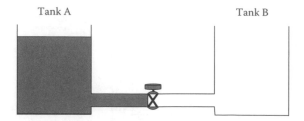

FIGURE 8.11 Water analogy of electric difference of potential.

Note: The basic unit of potential difference is the *volt* (V). The symbol for potential difference is *V*, indicating the ability to do the work of forcing electrons (current flow) to move. Because the volt unit is used, potential difference is called *voltage*.

8.6.1 THE WATER ANALOGY

When teaching the concepts of basic electricity, especially with regard to difference of potential (voltage), current, and resistance relationships in a simple electrical circuit, it has been common practice to use what is referred to as the *water analogy*. We use the water analogy later to explain (in simple, straightforward fashion) voltage, current, and resistance and their relationships in more detail, but for now we will use the analogy to explain the basic concept of electricity: difference of potential, or voltage. Because a difference in potential causes current flow (against resistance), it is important that this concept be understood first before the concept of current flow and resistance are explained.

Consider the two water tanks connected by a pipe and valve shown in Figure 8.11. At first, the valve is closed and all the water is in Tank A; thus, the water pressure across the valve is at maximum. When the valve is opened, the water flows through the pipe from A to B until the water level becomes the same in both tanks. The water then stops flowing in the pipe, because a difference in water pressure (difference in potential) no longer exists between the two tanks. Just as the flow of water through the pipe in Figure 8.11 is directly proportional to the difference in water level in the two tanks, current flow through an electric circuit is directly proportional to the difference in potential across the circuit.

Important Point: A fundamental law of electricity is that the current is directly proportional to the applied voltage; that is, if the voltage is increased, the current is increased. If the voltage is decreased, the current is decreased.

8.6.2 PRINCIPAL METHODS FOR PRODUCING A VOLTAGE

Electromotive force, or voltage, can be produced in many ways. Some of these methods are much more widely used than others. The following is a list of the six most common methods for producing electromotive force:

Figure 8.10 shows the magnetic polarities of the Earth. The geographic poles are also shown at each end of the axis of rotation of the Earth. Clearly, the magnetic axis does not coincide with the geographic axis; therefore, the magnetic and geographic poles are not the same on the surface of the Earth.

Recall that magnetic lines of force are assumed to emanate from the north pole of a magnet and to enter the south pole as closed loops. Because the Earth is a magnet, lines of force emanate from its north magnetic pole and enter the south magnetic pole as closed loops. A compass needle aligns itself in such a way that the Earth's lines of force enter at its south pole and leave at its north pole. Because the north pole of the needle is defined as the end that points in a northerly direction, it follows that the magnetic pole near the north geographic pole is in reality a south magnetic pole and *vice versa*.

8.6 DIFFERENCE IN POTENTIAL

Because of the force of its electrostatic field, an electric charge has the ability to do the work of moving another charge by attraction or repulsion. The force that causes free electrons to move in a conductor as an electric current may be referred to as follows:

- Electromotive force (emf)
- Voltage
- Difference in potential

When a difference in potential exists between two charged bodies that are connected by a wire (conductor), electrons (current) will flow along the conductor. This flow is from the negatively charged body to the positively charged body until the two charges are equalized and the potential difference no longer exists.

- *Friction*—Voltage is produced by rubbing two materials together.
- *Pressure* (piezoelectricity)—Voltage is produced by squeezing crystals of certain substances.
- *Heat* (thermoelectricity)—Voltage is produced by heating the joint (junction) where two unlike metals are joined.
- *Light* (photoelectricity)—Voltage is produced by light striking photosensitive (light sensitive) substances.
- *Chemical action*—Voltage is produced by chemical reaction in a battery cell.
- *Magnetism*—Voltage is produced in a conductor when the conductor moves through a magnetic field or a magnetic field moves through the conductor in such a manner as to cut the magnetic lines of force of the field.

In the study of basic electricity, we are most concerned with magnetism and chemistry as the means to produce voltage. Friction has little practical application, although we discussed it earlier with regard to static electricity. Pressure, heat, and light do have useful applications, but we do not need to consider them in this text. Magnetism and chemistry, on the other hand, are the principal sources of voltage and arc discussed at length in this text.

8.7 CURRENT

The movement or the flow of electrons is called *current*. To produce current, the electrons must be moved by a potential difference.

Note: Terms such as *current*, *current flow*, *electron flow*, and *electron current* may be used to describe the same phenomenon. Current is represented by the letter *I*. The basic unit in which current is measured is the ampere, or amp (A). One ampere of current is defined as the movement of 1 coulomb past any point of a conductor during 1 second of time.

Electron flow, or current, in an electric circuit moves from a region of less negative potential to a region of more positive potential. Recall that we used the water analogy to help us understand potential difference. We can also use the water analogy to help us understand current flow through a simple electric circuit.

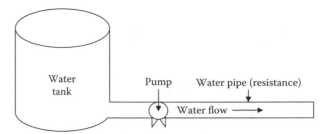

FIGURE 8.12 Water analogy: current flow.

Consider Figure 8.12, which shows a water tank connected via a pipe to a pump with a discharge pipe. If the water tank contains an amount of water above the level of the pipe opening to the pump, the water exerts pressure (a difference in potential) against the pump. When sufficient water is available for pumping with the pump, water flows through the pipe against the resistance of the pump and pipe. The analogy should be clear—in an electric circuit, if a difference of potential exists, current will flow in the circuit. Another simple way of looking at this analogy is to consider Figure 8.13, where the water tank has been replaced with a generator, the pipe with a conductor (wire), and water flow with the flow of electric current. Again, the key point illustrated by Figure 8.12 and Figure 8.13 is that to produce current the electrons must be moved by a potential difference.

Electric current is generally classified into two general types:

- Direct current (DC)
- Alternating current (AC)

Direct current moves through a conductor or circuit in one direction only; alternating current periodically reverses direction.

8.8 RESISTANCE

Earlier we pointed out that free electrons, or electric current, could move easily through a good conductor, such as copper, but that an insulator, such as glass, was an obstacle to current flow. In the water analogy shown in Figure 8.12 and the simple electric circuit shown in Figure 8.13, either the pipe or the conductor represents resistance. Every material offers some resistance, or opposition, to the flow of electric current through it. Good conductors,

FIGURE 8.13 Simple electric circuit with current flow.

such as copper, silver, and aluminum, offer very little resistance. Poor conductors, or insulators, such as glass, wood, and paper, offer a high resistance to current flow.

Note: The amount of current that flows in a given circuit depends on two factors: voltage and resistance.

Note: The letter R represents resistance. The basic unit in which resistance is measured is the ohm (Ω). One ohm is the resistance of a circuit element, or circuit, that permits a steady current of 1 ampere (1 coulomb per second) to flow when a steady electromotive force (emf) of 1 volt is applied to the circuit. Manufactured circuit parts containing definite amounts of resistance are called *resistors*.

The size and type of material of the wires in an electric circuit are chosen to keep the electrical resistance as low as possible. In this way, current can flow easily through the conductors, just as water flows through the pipe between the tanks in Figure 8.11. If the water pressure remains constant, the flow of water in the pipe will depend on how far the valve is opened. The smaller the opening, the greater the opposition (resistance) to the flow, and the smaller will be the rate of flow in gallons per second.

In the electric circuit shown in Figure 8.13, the larger the diameter of the wire, the lower will be its electrical resistance (opposition) to the flow of current through it. In the water analogy, pipe friction opposes the flow of water between the tanks. This friction is similar to electrical resistance. The resistance of the pipe to the flow of water through it depends on: (1) the length of the pipe, (2) the diameter of the pipe, and (3) the nature of the inside walls (rough or smooth). Similarly, the electrical resistance of the conductors depends on: (1) the length of the wires, (2) the diameter of the wires, and (3) the material of the wires (e.g., copper, silver).

It is important to note that temperature also affects the resistance of electrical conductors to some extent. In most conductors (e.g., copper, aluminum), the resistance increases with temperature. Carbon is an exception. In carbon, the resistance decreases as temperature increases.

Important Note: Electricity is a phenomenon that is frequently explained in terms of opposites. The term that is exactly the opposite of resistance is *conductance*. Conductance (G) is the ability of a material to pass electrons. The unit of conductance is the *mho*, which is ohm spelled backwards. The relationship that exists between resistance and conductance is the reciprocal. A reciprocal of a number is obtained by dividing the number into one. If the resistance of a material is known, dividing its value into one will give its conductance. Similarly, if the conductance is known, dividing its value into one will give its resistance.

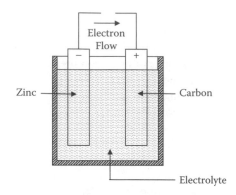

FIGURE 8.14 Simple voltaic cell.

8.9 BATTERY-SUPPLIED ELECTRICITY

Battery-supplied direct current electricity has many applications and is widely used in water and wastewater treatment operations. Applications include providing electrical energy to plant vehicles and emergency diesel generators, material-handling equipment (forklifts), portable electric/electronics equipment, hazard warning signal lights, and flashlights, as well as serving as backup emergency power for light packs and as standby power supplies or uninterruptible power supplies (UPS) for computer systems. In some instances, they are used as the only source of power, while in others they are used as a secondary or standby power supply.

8.9.1 THE VOLTAIC CELL

The simplest cell (a device that transforms chemical energy into electrical energy) is known as a *voltaic* (or galvanic) *cell* (see Figure 8.14). It consists of a piece of carbon (C) and a piece of zinc (Zn) suspended in a jar that contains a solution of water (H_2O) and sulfuric acid (H_2SO_4).

Note: A simple cell consists of two strips, or electrodes, placed in a container that holds the electrolyte. A battery is formed when two or more cells are connected.

The electrodes are the conductors by which the current leaves or returns to the electrolyte. In the simple cell described above, they are carbon and zinc strips placed in the electrolyte. Zinc contains an abundance of negatively charged atoms, and carbon has an abundance of positively charged atoms. When the plates of these materials are immersed in an electrolyte, chemical action between the two begins. In a dry cell (see Figure 8.15), the electrodes are the carbon rod in the center and the zinc container in which the cell is assembled.

The electrolyte is the solution that acts upon the electrodes that are placed in it. The electrolyte may be a salt, an acid, or an alkaline solution. In the simple voltaic cell

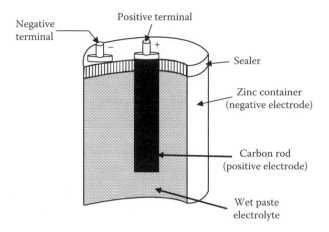

FIGURE 8.15 Dry cell (cross-sectional view).

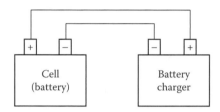

FIGURE 8.16 Hookup for charging a secondary cell with a battery charger.

and in the automobile storage battery, the electrolyte is in a liquid form; in the dry cell (see Figure 8.15), the electrolyte is a moist paste.

8.9.2 PRIMARY AND SECONDARY CELLS

Primary cells are normally those that cannot be recharged or returned to good condition after their voltage drops too low. Dry cells in flashlights and transistor radios are examples of primary cells. Some primary cells can be recharged. A *secondary cell* is one in which the electrodes and the electrolyte are altered by the chemical action that takes place when the cell delivers current. These cells are rechargeable. During recharging, the chemicals that provide electric energy are restored to their original condition. Recharging is accomplished by forcing an electric current through them in the direction opposite to that of discharge. Connecting, as shown in Figure 8.16, recharges a cell. Some battery chargers have a voltmeter and an ammeter that indicate the charging voltage and current. The automobile storage battery is the most common example of the secondary cell.

8.9.3 BATTERY

A *battery* consists of two or more cells placed in a common container. The cells are connected in series, in parallel, or in some combination of series and parallel, depending on the amount of voltage and current required of the battery.

8.9.4 BATTERY OPERATION

The chemical reaction within a battery provides the voltage. This occurs when a conductor is connected externally to the electrodes of a cell, causing electrons to flow under the influence of a difference in potential across the electrodes from the zinc (negative) through the external conductor to the carbon (positive), returning within the solution to the zinc. After a short period, the zinc will begin to waste away because of the acid.

The voltage across the electrodes depends on the materials from which the electrodes are made and the composition of the solution. The difference of potential between the carbon and zinc electrodes in a dilute solution of sulfuric acid and water is about 1.5 volts.

The current that a primary cell may deliver depends on the resistance of the entire circuit, including that of the cell itself. The internal resistance of the primary cell depends on the size of the electrodes, the distance between them in the solution, and the resistance of the solution. The larger the electrodes and the closer together they are in solution (without touching), the lower the internal resistance of the primary cell and the more current it is capable of supplying to the load.

Note: When current flows through a cell, the zinc gradually dissolves in the solution and the acid is neutralized.

8.9.5 COMBINING CELLS

In many operations, battery-powered devices may require more electrical energy than one cell can provide. Various devices may require either a higher voltage or more current, and in some cases both. Under such conditions, it is necessary to combine, or interconnect, a sufficient number of cells to meet the higher requirements. Cells connected in series provide a higher voltage, whereas cells connected in parallel provide a higher current capacity. To provide adequate power when both voltage and current requirements are greater than the capacity of one cell, a combination series–parallel network of cells must be devised.

When cells are connected in series, the total voltage across the battery of cells is equal to the sum of the voltage of each of the individual cells. In Figure 8.17, the four 1.5-V cells in series provide a total battery voltage of 6 V. When cells are placed in series, the positive terminal of one cell is connected to the negative terminal of the other cell. The positive electrode of the first cell and negative electrode of the last cell then serve as the power takeoff terminals of the battery. The current flowing through such a battery of series cells is the same as from one cell because the same current flows through all the series cells.

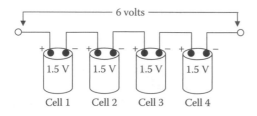

FIGURE 8.17 Cells in series.

To obtain a greater current, a battery has cells connected in parallel as shown in Figure 8.18. In this parallel connection, all the positive electrodes are connected to one line, and all negative electrodes are connected to the other. Any point on the positive side can serve as the positive terminal of the battery and any point on the negative side can be the negative terminal. The total voltage output of a battery of three parallel cells is the same as that for a single cell (Figure 8.18), but the available current is three times that of one cell; that is, the current capacity has been increased. Identical cells in parallel all supply equal parts of the current to the load; for example, of three different parallel cells producing a load current of 210 milliamperes (mA), each cell contributes 70 mA.

Figure 8.19 depicts a schematic of a series–parallel battery network supplying power to a load requiring both voltage and current greater than what one cell can provide. To provide the required increased voltage, groups of three 1.5-V cells are connected in series. To provide the required increased amperage, four series groups are connected in parallel.

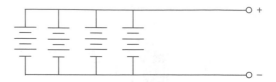

FIGURE 8.19 Series–parallel connected cells.

8.9.6 Types of Batteries

In the past 25 years, several different types of batteries have been developed. In this text, we briefly discuss five types: the dry cell, lead–acid battery, alkaline cell, nickel–cadmium, and mercury cell.

8.9.6.1 Dry Cell

The dry cell, or carbon–zinc cell, is so called because its electrolyte is not in a liquid state (the electrolyte is a moist paste). The carbon, in the form of a rod that is placed in the center of the cell, is the positive terminal. The case of the cell is made of zinc, which is the negative terminal (see Figure 8.15). Between the carbon electrode and the zinc case is the electrolyte of a moist chemical paste-like mixture. The cell is sealed to prevent the liquid in the paste from evaporating. The voltage of a cell of this type is about 1.5 V.

8.9.6.2 Lead–Acid Battery

The *lead–acid battery* is a secondary cell, commonly termed a *storage battery*, that stores chemical energy until it is released as electrical energy.

Note: The lead–acid battery differs from the primary cell type of battery mainly in that it may be recharged,

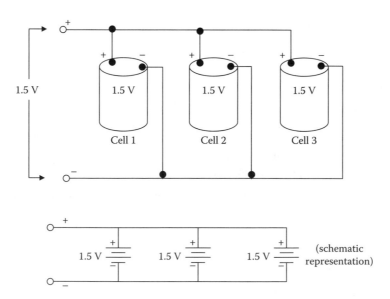

FIGURE 8.18 Cells in parallel.

whereas most primary cells are not normally recharged. In addition, the term *storage battery* is somewhat deceiving because this battery does not store electrical energy but is a source of chemical energy that produces electrical energy.

As the name implies, the lead–acid battery consists of a number of lead–acid cells immersed in a dilute solution of sulfuric acid. Each cell has two groups of lead plates; one set is the positive terminal and the other is the negative terminal. Active materials within the battery (lead plates and sulfuric acid electrolyte) react chemically to produce a flow of direct current whenever current consuming devices are connected to the battery terminal posts. This current is produced by chemical reaction between the active material of the plates (electrodes) and the electrolyte (sulfuric acid). This type of cell produces slightly more than 2 V. Most automobile batteries contain six cells connected in series so the output voltage from the battery is slightly more than 12 V. In addition to being rechargeable, the main advantage of the lead–acid storage battery over the dry cell battery is that the storage battery can supply current for a much longer time than the average dry cell.

Safety Note: Whenever a lead–acid storage battery is charging, the chemical action produces dangerous hydrogen gas; thus, the charging operation should take place only in a well-ventilated area.

8.9.6.3 Alkaline Cell

The *alkaline cell* is a secondary cell that gets its name from its alkaline electrolyte—potassium hydroxide. Another type battery, sometimes referred to as the *alkaline battery*, has a negative electrode of zinc and a positive electrode of manganese dioxide. It generates 1.5 V.

8.9.6.4 Nickel–Cadmium Cell

The *nickel–cadmium cell*, or Ni–Cad cell, is the only dry battery that is a true storage battery with a reversible chemical reaction, allowing recharging many times. In the secondary nickel–cadmium dry cell, the electrolyte is potassium hydroxide, the negative electrode is nickel hydroxide, and the positive electrode is cadmium oxide. The operating voltage is 1.25 V. Because of its rugged characteristics (stands up well to shock, vibration, and temperature changes) and availability in a variety of shapes and sizes, it is ideally suited for use in powering portable communication equipment.

8.9.6.5 Mercury Cell

The *mercury cell* was developed for space exploration activities, for which small transceivers and miniaturized equipment required a small power source. In addition to reduced size, the mercury cell has a good shelf life and is very rugged; it also produces a constant output voltage under different load conditions. Two types of mercury cells are available. One is a flat cell that is shaped like a button, and the other is a cylindrical cell that looks like a standard flashlight cell. The advantage of the button-type cell is that several of them can be stacked inside one container to form a battery. A mercury cell produces 1.35 V.

8.9.7 Battery Characteristics

Batteries are generally classified by their various characteristics. Parameters such as internal resistance, specific gravity, capacity, and shelf life are used to classify batteries by type. Regarding internal resistance, it is important to keep in mind that a battery is a DC voltage generator. As such, the battery has internal resistance. In a chemical cell, the resistance of the electrolyte between the electrodes is responsible for most of the internal resistance of the cell. Because any current in the battery must flow through the internal resistance, this resistance is in series with the generated voltage. With no current, the voltage drop across the resistance is zero, so fully generated voltage develops across the output terminals. This is the open-circuit voltage, or no-load voltage. If a load resistance is connected across the battery, the load resistance is in series with internal resistance. When current flows in this circuit, the internal voltage drop decreases the terminal voltage of the battery.

The ratio of the weight of a certain volume of liquid to the weight of the same volume of water is the *specific gravity* of the liquid. Pure sulfuric acid has a specific gravity of 1.835, as it weighs 1.835 times as much as water per unit volume. The specific gravity of a mixture of sulfuric acid and water varies with the strength of the solution from 1.000 to 1.830.

The specific gravity of the electrolyte solution in a lead–acid cell ranges from 1.210 to 1.300 for new, fully charged batteries. The higher the specific gravity, the less internal resistance of the cell and the higher the possible load current. As the cell discharges, the water formed in the process dilutes the acid and the specific gravity gradually decreases to about 1.150, at which time the cell is considered to be fully discharged.

The specific gravity of the electrolyte is measured with a hydrometer, which has a compressible rubber bulb at the top, a glass barrel, and a rubber hose at the bottom of the barrel. When taking readings from a hydrometer, the decimal point is usually omitted; for example, a specific gravity of 1.260 is read simply as "twelve-sixty." A hydrometer reading of 1210 to 1300 indicates full charge, about 1250 is half charge, and 1150 to 1200 is complete discharge.

The capacity of a battery is measured in ampere-hours (Ah).

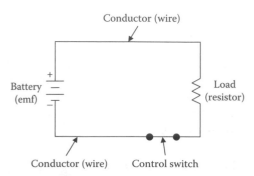

FIGURE 8.20 Simple closed circuit.

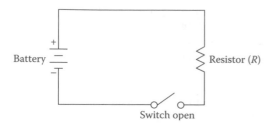

FIGURE 8.21 Open circuit.

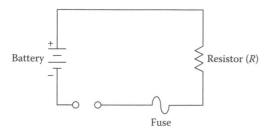

FIGURE 8.22 Simple fused circuit.

Note: The ampere-hour capacity is equal to the product of the current in amperes and the time in hours during which the battery is supplying this current. The ampere-hour capacity varies inversely with the discharge current. The size of a cell is determined generally by its ampere-hour capacity.

The capacity of a storage battery determines how long it will operate at a given discharge rate and depends on many factors, the most important being:

- Area of the plates in contact with the electrolyte
- Quantity and specific gravity of the electrolyte
- Type of separators
- General condition of the battery (e.g., degree of sulfating, buckled plates, warped separators, sediment in bottom of cells)
- Final limiting voltage

The shelf life of a cell is that period of time during which the cell can be stored without losing more than approximately 10% of its original capacity. The loss of capacity of a stored cell is due primarily to the drying out of its electrolyte in a wet cell and to chemical actions, which change the materials within the cell. Keeping it in a cool, dry place can extend the shelf life.

8.10 SIMPLE ELECTRICAL CIRCUIT

An electric circuit includes (Figure 8.20):

- An energy source—a source of electromotive force (emf), or voltage, such as a battery or generator
- A conductor (wire)
- A load
- A means of control

The energy source could be a battery, as shown in Figure 8.20, or some other means for producing a voltage. The load that dissipates the energy could be a lamp, a resistor, or some other device (or devices) that does useful work,

such as an electric toaster, a power drill, radio, or soldering iron. Conductors are wires that offer low resistance to current; they connect all the loads in the circuit to the voltage source. No electrical device dissipates energy unless current flows through it. Because conductors, or wires, are not perfect conductors, they heat up (dissipate energy), so they are actually part of the load. For simplicity, however, we usually think of the connecting wiring as having no resistance, as it would be tedious to assign a very low resistance value to the wires every time we wanted to solve a problem. Control devices might be switches, variable resistors, circuit breakers, fuses, or relays.

A complete pathway for current flow, or closed circuit (Figure 8.20), is an unbroken path for current from the emf, through a load, and back to the source. A circuit is open (see Figure 8.21) if a break in the circuit (e.g., open switch) does not provide a complete path for current.

Important Point: Current flows from the negative (−) terminal of the battery, shown in Figure 8.20 and Figure 8.21, through the load to the positive (+) battery terminal and then continues through the battery from the positive (+) terminal to the negative (−) terminal. As long as this pathway is unbroken, it is a closed circuit and current will flow; however, if the path is broken at *any* point, it becomes an open circuit and no current flows.

To protect a circuit, a fuse is placed directly in the circuit (see Figure 8.22). A fuse will open the circuit whenever a dangerous large current begins to flow (i.e., when a short circuit condition occurs caused by an accidental connection between two points in a circuit offering very little resistance). A fuse will permit currents smaller than the fuse value to flow but will melt and therefore break or open the circuit if a larger current flows.

FIGURE 8.23 Schematic symbol for a battery.

8.10.1 SCHEMATIC REPRESENTATIONS

The simple circuits shown in Figure 8.20, Figure 8.21, and Figure 8.22 are displayed in schematic form. A *schematic diagram* (usually shortened to "schematic") is a simplified drawing that represents the electrical, not the physical, situation in a circuit. The symbols used in schematic diagrams are the electrician's shorthand; they make the diagrams easier to draw and easier to understand. Consider the symbol used to represent a battery power supply (see Figure 8.23). The symbol is rather simple and straightforward but is very important; for example, by convention, the shorter line in the symbol for a battery represents the negative terminal. It is important to remember this, because it is sometimes necessary to note the direction of current flow, which is from negative to positive, when examining a schematic. The battery symbol shown in Figure 8.23 has a single cell, so only one short and one long line are used. The number of lines used to represent a battery vary (and they are not necessarily equivalent to the number of cells), but they are always in pairs, with long and short lines alternating. In the circuit shown in Figure 8.22, the current would flow in a *counterclockwise* direction—that is, opposite the direction that the hands of a clock move. If the long and short lines of the battery symbol (symbol shown in Figure 8.23) were reversed, the current in the circuit shown in Figure 8.22 would flow *clockwise*; that is, in the direction that the hands of a clock move.

Note: In studies of electricity and electronics, many circuits are analyzed that consist mainly of specially designed resistive components. As previously stated, these components are called *resistors*. Throughout our remaining analysis of the basic circuit, the resistive component will be a physical resistor; however, the resistive component could be any one of several electrical devices.

Keep in mind that the simple circuits shown in the figures to this point illustrate only a few of the many symbols used in schematics to represent circuit components. Other symbols will be introduced, as we need them. It is also important to keep in mind that a closed loop of wire (conductor) is not necessarily a circuit. A source of voltage must be included to make it an electric circuit. In any electric circuit where electrons move around a closed loop, current, voltage, and resistance are present. The physical pathway for current flow is actually the circuit. By knowing any two of the three quantities, such as voltage and current, the third (resistance) may be determined.

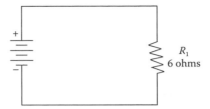

FIGURE 8.24 Determining current in a simple circuit.

This is done mathematically using *Ohm's law*, which is the foundation on which electrical theory is based.

8.11 OHM'S LAW

Simply put, Ohm's law defines the relationship between current, voltage, and resistance in electric circuits. Ohm's law can be expressed mathematically in three ways:

1. The *current* (*I*) in a circuit is equal to the voltage applied to the circuit divided by the resistance of the circuit. Stated another way, the current in a circuit is *directly* proportional to the applied voltage and *inversely* proportional to the circuit resistance. Ohm's law may be expressed as:

$$E = \frac{I}{R} \tag{8.1}$$

where:

E = voltage in volts.
I = current in amps.
R = resistance in ohms.

2. The *resistance* (*R*) of a circuit is equal to the voltage applied to the circuit divided by the current in the circuit:

$$R = \frac{E}{I} \tag{8.2}$$

3. The applied *voltage* (*E*) to a circuit is equal to the product of the current and the resistance of the circuit:

$$E = I \times R = IR \tag{8.3}$$

If any two of the quantities in Equations 8.1 through 8.3 are known, the third may be easily found. Let us look at an example.

■ EXAMPLE 8.1

Problem: Figure 8.24 shows a circuit containing a resistance (*R*) of 6 ohms and a source of voltage (*E*) of 3 volts. How much current (*I*) flows in the circuit?

Solution:

$$I = \frac{E}{R} = \frac{3}{6} = 0.5 \text{ amp}$$

To observe the effect of source voltage on circuit current, we use the circuit shown in Figure 8.24 but double the voltage to 6 volts. Notice that, as the source of voltage doubles, the circuit current also doubles.

■ EXAMPLE 8.2

Problem: Given that E = 6 volts and R = 6 ohms, what is I?

Solution:

$$I = \frac{E}{R} = \frac{6}{6} = 1 \text{ amp}$$

Key Point: Circuit current is directly proportional to applied voltage and will change by the same factor that the voltage changes.

To verify that current is inversely proportional to resistance, assume that the resistor in Figure 8.24 has a value of 12 ohms.

■ EXAMPLE 8.3

Problem: Given that E = 3 volts and R = 12 ohms, what is I?

Solution:

$$I = \frac{E}{R} = \frac{3}{12} = 0.25 \text{ amps}$$

Comparing the current of 0.25 amp for the 12-ohm resistor to the 0.5-amp current obtained with the 6-ohm resistor shows that doubling the resistance reduces the current to one half the original value. The point here is that *circuit current is inversely proportional to circuit resistance*.

Recall that, if we know any two quantities (E, I, or R), we can calculate the third. In many circuit applications, current is known and either the voltage or the resistance will be the unknown quantity. To solve a problem in which voltage (E) and resistance (R) are known, the basic formula for Ohm's law must be transposed to solve for I. The Ohm's law equations can be memorized and practiced effectively by using an Ohm's law circle (see Figure 8.25).

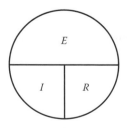

FIGURE 8.25 Ohm's law circle.

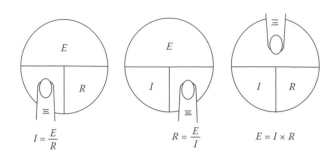

$$I = \frac{E}{R} \qquad\qquad R = \frac{E}{I} \qquad\qquad E = I \times R$$

FIGURE 8.26 Putting the Ohm's law circle to work.

To find the equation for E, I, or R when two quantities are known, cover the unknown third quantity with your finger, ruler, or piece of paper as shown in Figure 8.26.

■ EXAMPLE 8.4

Problem: Find I when E = 120 volts and R = 40 ohms.

Solution: Place finger on I as shown in the figure below.

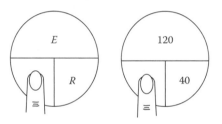

Use Equation 8.1 to find the unknown I:

$$I = \frac{E}{R} = \frac{120}{40} = 3 \text{ amps}$$

■ EXAMPLE 8.5

Problem: Find R when E = 220 volts and I = 10 amps.

Solution: Place finger on R as shown in the figure below.

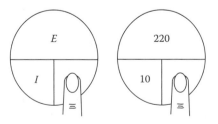

Use Equation 8.2 to find the unknown R:

$$R = \frac{E}{I} = \frac{220}{10} = 22 \text{ ohms}$$

■ EXAMPLE 8.6

Problem: Find *E* when *I* = 2.5 amps and *R* = 25 ohms.

Solution: Place finger on *E* as shown in the figure below.

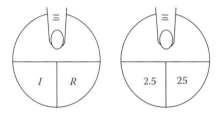

Use Equation 8.3 to find the unknown *E*:

$$E = IR = 2.5 \times 25 = 62.5 \text{ volts}$$

Note: In the previous examples, we have demonstrated how the Ohm's law circle can help solve simple voltage, current, and amperage problems. Beginning students are cautioned, however, not to rely entirely on the use of this circle when transposing simple formulae but rather to use it to supplement their knowledge of the algebraic method. Algebra is a basic tool in the solution of electrical problems, and the importance of knowing how to use it should not be underemphasized, and its use should not be bypassed after the operator has learned a shortcut method such as the one indicated in this circle.

■ EXAMPLE 8.7

Problem: An electric light bulb draws 0.5 amp when operating on a 120-volt DC circuit. What is the resistance of the bulb?

Solution: The first step in solving a circuit problem is to sketch a schematic diagram of the circuit itself, labeling each of the parts and showing the known values (see Figure 8.27). Because *I* and *E* are known, we can use Equation 8.2 to solve for *R*:

$$R = \frac{E}{I} = \frac{120}{0.5} = 240 \text{ ohms}$$

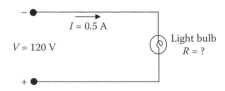

FIGURE 8.27 Simple circuit.

8.12 ELECTRICAL POWER

Power, whether electrical or mechanical, pertains to the rate at which work is being done, so the power consumption in a plant is related to current flow. A large electric motor or air dryer consumes more power (and draws more current) in a given length of time than, for example, an indicating light on a motor controller. Work is done whenever a force causes motion. If a mechanical force is used to lift or move a weight, work is done; however, force exerted *without* causing motion, such as the force of a compressed spring acting between two fixed objects, does not constitute work.

Key Point: Power is the rate at which work is done.

8.12.1 ELECTRICAL POWER CALCULATIONS

The electric power (*P*) used in any part of a circuit is equal to the current (*I*) in that part multiplied by the voltage across that part of the circuit. In equation form:

$$P = EI \tag{8.4}$$

where:

P = power (watts, W).
E = voltage (volts, V).
I = current (amps, A).

If we know the current (*I*) and the resistance (*R*) but not the voltage, we can find the power (*P*) by using Ohm's law for voltage, so by substituting Equation 8.3:

$$E = IR$$

into Equation 8.4, we obtain:

$$P = IR \times I = I^2R \tag{8.5}$$

In the same manner, if we know the voltage and the resistance but not the current, we can find the *P* by using Ohm's law for current, so by substituting Equation 8.1:

$$I = \frac{E}{R}$$

into Equation 8.4, we obtain:

$$P = E \times \frac{E}{R} = \frac{E^2}{R} \tag{8.6}$$

Key Point: If we know any two quantities, we can calculate the third.

■ **EXAMPLE 8.8**

Problem: The current through a 200-ohm resistor to be used in a circuit is 0.25 amp. Find the power rating of the resistor.

Solution: Because the current (*I*) and resistance (*R*) are known, use Equation 8.5 to find *P*:

$$P = I^2R = (0.25)^2 \times 200 = 0.0625 \times 200 = 12.5 \text{ W}$$

Important Point: The power rating of any resistor used in a circuit should be twice the wattage calculated by the power equation to prevent the resistor from burning out; thus, the resistor used in Example 8.8 should have a power rating of 25 W.

■ **EXAMPLE 8.9**

Problem: How many kilowatts of power are delivered to a circuit by a 220-V generator supplying 30 A to the circuit?

Solution: Because the voltage (*E*) and current (*I*) are given, use Equation 8.4 to find *P*:

$$P = EI = 220 \times 30 = 6600 \text{ W} = 6.6 \text{ kW}$$

■ **EXAMPLE 8.10**

Problem: If the voltage across a 30,000-ohm resistor is 450 volts, what is the power dissipated in the resistor?

Solution: Because the resistance (*R*) and voltage (*E*) are known, use Equation 8.6 to find *P*:

$$P = \frac{E^2}{R} = \frac{(450)^2}{30,000} = \frac{202,500}{30,000} = 6.75 \text{ W}$$

In this section, *P* was expressed in terms of various pairs of the other three basic quantities *E*, *I*, and *R*. In practice, we should be able to express any one of the three basic quantities, as well as *P*, in terms of any two of the others. Figure 8.28 provides a summary of the 12 basic formulae we should know. The four quantities *E*, *I*, *R*, and *P* are at the center of the figure. Adjacent to each quantity are three segments. Note that in each segment, the basic quantity is expressed in terms of two other basic quantities, and no two segments are alike.

8.13 ELECTRICAL ENERGY (KILOWATT-HOURS)

Energy (the mechanical definition) is defined as the ability to do work (energy and time are essentially the same and are expressed in identical units). Energy is expended when work is done, because it takes energy to maintain a force when that force acts through a distance. The total energy expended to do a certain amount of work is equal to the working force multiplied by the distance through which

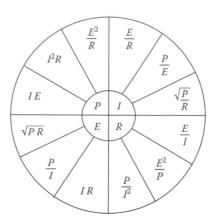

FIGURE 8.28 Ohm's law circle—summary of basic formulae.

the force moves to do the work. In electricity, total energy expended is equal to the *rate* at which work is done, multiplied by the length of time the rate is measured. Essentially, energy (*W*) is equal to power (*P*) times time (*t*).

The kilowatt-hour (kWh) is a unit commonly used for large amounts of electric energy or work. The amount of kilowatt-hours is calculated as the product of the power in kilowatts (kW) and the time in hours (hr) during which the power is used.

$$\text{kWh} = \text{kW} \times \text{hr} \tag{8.7}$$

■ **EXAMPLE 8.11**

Problem: How much energy is delivered in 4 hours by a generator supplying 12 kW?

Solution:

$$\text{kWh} = \text{kW} \times \text{hr} = 12 \times 4 = 48$$

The energy delivered is 48 kWh.

8.14 SERIES DC CIRCUIT CHARACTERISTICS

As previously mentioned, an electric circuit is made up of a voltage source, the necessary connecting conductors, and the effective load. If the circuit is arranged so the electrons have only *one* possible path, the circuit is a *series circuit*. A series circuit, then, is defined as a circuit that contains only one path for current flow. Figure 8.29 shows a series circuit having several loads (resistors).

Key Point: A series circuit is a circuit having only one path for the current to flow along.

8.14.1 SERIES CIRCUIT RESISTANCE

To follow its electrical path, the current in a series circuit must flow through resistors inserted in the circuit (see Figure 8.30); thus, each additional resistor offers added

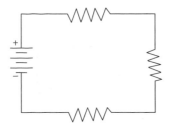

FIGURE 8.29 Series circuit.

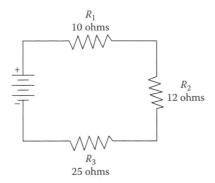

FIGURE 8.30 Solving for total resistance in a series circuit.

resistance. In a series circuit, *the total circuit resistance (R_T) is equal to the sum of the individual resistances*, or:

$$R_T = R_1 + R_2 + R_3 + \ldots + R_n \qquad (8.8)$$

where:

R_T = total resistance (ohms).
R_1, R_2, R_3 = resistance in series (ohms).
R_n = any number of additional resistors in the series.

■ **EXAMPLE 8.12**

Problem: Three resistors of 10 ohms, 12 ohms, and 25 ohms are connected in series across a battery whose emf is 110 volts (Figure 8.30). What is the total resistance?

Solution:

Given:
R_1 = 10 ohms
R_2 = 12 ohms
R_3 = 25 ohms

$$R_T = R_1 + R_2 + R_3$$
$$R_T = 10 + 12 + 25 = 47 \text{ ohms}$$

Equation 8.8 can be transposed to solve for the value of an unknown resistance; for example, transposition can be used in some circuit applications where the total resistance is known but the value of a circuit resistor has to be determined.

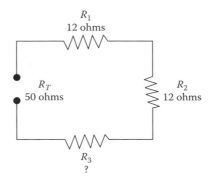

FIGURE 8.31 Calculating the value of one resistance in a series circuit.

■ **EXAMPLE 8.13**

Problem: The total resistance of a circuit containing three resistors is 50 ohms (see Figure 8.31). Two of the circuit resistors are 12 ohms each. Calculate the value of the third resistor (R_3).

Solution:

Given:
R_T = 50 ohms
R_1 = 12 ohms
R_2 = 12 ohms

$$R_T = R_1 + R_2 + R_3$$
$$R_3 = R_T - R_1 - R_2$$
$$R_3 = 50 - 12 - 12 = 26 \text{ ohms}$$

Key Point: When resistances are connected in series, the total resistance in the circuit is equal to the sum of the resistances of all the parts of the circuit.

8.14.2 SERIES CIRCUIT CURRENT

Because there is but one path for current in a series circuit, the same current (I) must flow through each part of the circuit. Thus, to determine the current throughout a series circuit, only the current through one of the parts must be known. The fact that the same current flows through each part of a series circuit can be verified by inserting ammeters into the circuit at various points as shown in Figure 8.32. As indicated in Figure 8.32, each meter indicates the same value of current.

Key Point: In a series circuit, the same current flows in every part of the circuit. *Do not* add the currents in each part of the circuit to obtain I.

8.14.3 SERIES CIRCUIT VOLTAGE

The *voltage* drop across the resistor in the basic circuit is the total voltage across the circuit and is equal to the

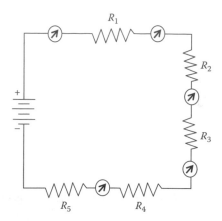

FIGURE 8.32 Current in a series circuit.

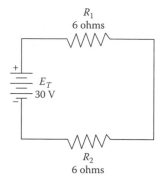

FIGURE 8.33 Calculating total resistance in a series circuit.

applied voltage. The total voltage across a series circuit is also equal to the applied voltage but consists of the sum of two or more individual voltage drops. This statement can be proven by an examination of the circuit shown in Figure 8.33. In this circuit, a source potential (E_T) of 30 volts is impressed across a series circuit consisting of two 6-ohm resistors. The total resistance of the circuit is equal to the sum of the two individual resistances, or 12 ohms. Using Ohm's law, the circuit current may be calculated as follows:

$$I = \frac{E_T}{R_T} = \frac{30}{12} = 2.5 \text{ amps}$$

Because we know that the value of the resistors is 6 ohms each, and the current through the resistors is 2.5 amps, we can calculate the voltage drops across the resistors. The voltage (E_1) across R_1 is, therefore:

$$E_1 = IR_1$$
$$E_1 = 2.5 \text{ amps} \times 6 \text{ ohms}$$
$$E_1 = 15 \text{ volts}$$

Because R_2 is the same ohmic value as R_1 and carries the same current, the voltage drop across R_2 is also equal to 15 volts. Adding these two 15-volt drops together gives a

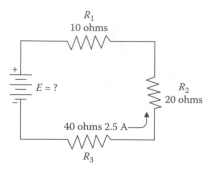

FIGURE 8.34 Solving for applied voltage in a series circuit.

total drop of 30 volts, exactly equal to the applied voltage. For a series circuit then,

$$E_T = E_1 + E_2 + E_3 + \ldots + E_n \qquad (8.9)$$

where:

    E_T = total voltage (volts).
    E_1 = voltage across resistance R_1 (volts).
    E_2 = voltage across resistance R_2 (volts).
    E_3 = voltage across resistance R_3 (volts).

■ EXAMPLE 8.14

Problem: A series circuit consists of three resistors having values of 10 ohms, 20 ohms, and 40 ohms respectively. Find the applied voltage if the current through the 20-ohm resistor is 2.5 amps.

Solution: To solve this problem, a circuit diagram is first drawn and labeled as shown in Figure 8.34.

    Given:
        R_1 = 10 ohms
        R_2 = 20 ohms
        R_3 = 40 ohms
        I = 2.5 amps

Because the circuit involved is a series circuit, the same 2.5 amps of current flow through each resistor. Using Ohm's law, the voltage drops across each of the three resistors can be calculated:

$$E_1 = 25 \text{ volts}$$
$$E_2 = 50 \text{ volts}$$
$$E_3 = 100 \text{ volts}$$

When the individual drops are known, they can be added to find the total or applied voltage by using Equation 8.9:

$$E_T = E_1 + E_2 + E_3$$
$$E_T = 25 \text{ volts} + 50 \text{ volts} + 100 \text{ volts}$$
$$E_T = 175 \text{ volts}$$

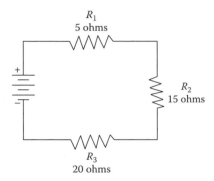

FIGURE 8.35 Solving for total power in a series circuit.

Key Point 1: The total voltage (E_T) across a series circuit is equal to the sum of the voltages across each resistance of the circuit.

Key Point 2: The voltage drops that occur in a series circuit are in direct proportions to the resistance across which they appear. This is the result of having the same current flow through each resistor. Thus, the larger the resistor, the larger will be the voltage drop across it.

8.14.4 Series Circuit Power

Each resistor in a series circuit consumes *power*. This power is dissipated in the form of heat. Because this power must come from the source, the total power must be equal in amount to the power consumed by the circuit resistances. In a series circuit, the total power is equal to the sum of the powers dissipated by the individual resistors. Total power (P_T) is thus equal to:

$$P_T = P_1 + P_2 + P_3 \ldots P_n \qquad (8.10)$$

where:

P_T = total power (watts).
P_1 = power used in first part (watts).
P_2 = power used in second part (watts).
P_3 = power used in third part (watts).
P_n = power used in nth part (watts).

■ EXAMPLE 8.15

Problem: A series circuit consists of three resistors having values of 5 ohms, 15 ohms, and 20 ohms. Find the total power dissipation when 120 volts is applied to the circuit (see Figure 8.35).

Solution:

Given:

$R_1 = 5$ ohms
$R_2 = 15$ ohms
$R_3 = 20$ ohms
$E = 120$ volts

The total resistance is found first.

$$R_T = R_1 + R_2 + R_3$$
$$R_T = 5 + 15 + 20 = 40 \text{ ohms}$$

Using total resistance and the applied voltage, we can calculate the circuit current:

$$I = \frac{E_T}{R_T} = \frac{120}{40} = 3 \text{ amps}$$

Using the power formula, we can calculate the individual power dissipations:

For resistor R_1:

$$P_1 = I^2R_1 = (3)^2 \times 5 = 45 \text{ watts}$$

For resistor R_2:

$$P_2 = I^2R_2 = (3)^2 \times 15 = 135 \text{ watts}$$

For resistor R_3:

$$P_3 = I^2R_3 = (3)^2 \times 20 = 180 \text{ watts}$$

To obtain total power:

$$P_T = P_1 + P_2 + P_3$$
$$P_T = 45 + 135 + 180 = 360 \text{ watts}$$

To check our answer, the total power delivered by the source can be calculated:

$$P = EI = 3 \text{ amps} \times 120 \text{ volts} = 360 \text{ watts}$$

Thus, the total power is equal to the sum of the individual power dissipations.

Key Point: We found that Ohm's law can be used for total values in a series circuit as well as for individual parts of the circuit. Similarly, the formula for power may be used for total values:

$$P_T = IE_T \qquad (8.11)$$

8.14.5 Summary of the Rules for Series DC Circuits

To this point, we have covered many of the important factors governing the operation of basic series circuits. In essence, what we have really done is to lay a strong foundation for the more advanced circuit theory that follows. Some of the important factors governing the operation of a series circuit include:

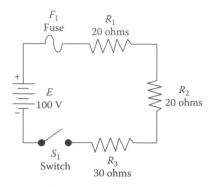

FIGURE 8.36 Solving for various values in a series circuit.

- The same current flows through each part of a series circuit.
- The total resistance of a series circuit is equal to the sum of the individual resistances.
- The total voltage across a series circuit is equal to the sum of the individual voltage drops.
- The voltage drop across a resistor in a series circuit is proportional to the size of the resistor.
- The total power dissipated in a series circuit is equal to the sum of the individual dissipations.

8.14.6 GENERAL SERIES CIRCUIT ANALYSIS

Now that we have discussed the pieces involved in solving the puzzle that is series circuit analysis, we can move on to the next step in the process: solving series circuit analysis in total.

■ **EXAMPLE 8.16**

Problem: Three resistors of 20 ohms, 20 ohms, and 30 ohms are connected across a battery supply rated at 100-volt terminal voltage. Completely solve the circuit shown in Figure 8.36.

Note: To solve the circuit, the total resistance must be found first, then the circuit current can be calculated. When the current is known, the voltage drops and power dissipations can be calculated.

Solution: The total resistance is:

$R_T = R_1 + R_2 + R_3$

$R_T = 20 \text{ ohms} + 20 \text{ ohms} + 30 \text{ ohms} = 70 \text{ ohms}$

By Ohm's law the current is:

$$I = \frac{E}{R_T} = \frac{100}{70} = 1.43 \text{ amps}$$

The voltage (E_1) across R_1 is:

$E_1 = IR_1 = 1.43 \text{ amps} \times 20 \text{ ohms} = 28.6 \text{ volts}$

The voltage (E_2) across R_2 is:

$E_2 = IR_2 = 1.43 \text{ amps} \times 20 \text{ ohms} = 28.6 \text{ volts}$

The voltage (E_3) across R_3 is:

$E_3 = IR_2 = 1.43 \text{ amps} \times 30 \text{ ohms} = 42.9 \text{ volts}$

The power dissipated by R_1 is:

$P_1 = IE_1 = 1.43 \text{ amps} \times 28.6 \text{ volts} = 40.9 \text{ watts}$

The power dissipated by R_2 is:

$P_2 = IE_2 = 1.43 \text{ amps} \times 28.6 \text{ volts} = 40.9 \text{ watts}$

The power dissipated by R_3 is:

$P_3 = IE_3 = 1.43 \text{ amps} \times 42.9 \text{ volts} = 61.3 \text{ watts}$

The total power dissipated is:

$P_T = E_T I = 100 \text{ volts} \times 1.43 \text{ amps} = 143 \text{ watts}$

Note: Keep in mind when applying Ohm's law to a series circuit to consider whether the values used are component values or total values. When the information available enables the use of Ohm's law to find total resistance, total voltage, and total current, then total values must be inserted into the formula.

To find total resistance:

$$R_T = \frac{E_T}{I_T}$$

To find total voltage:

$$E_T = I_T \times R_T$$

To find total current:

$$I_T = \frac{E_T}{R_T}$$

8.14.7 KIRCHHOFF'S VOLTAGE LAW

Kirchhoff's voltage law states that the voltage applied to a closed circuit equals the sum of the voltage drops in that circuit. It should be obvious that this fact was used in the study of series circuits to this point. It was expressed as follows:

Voltage applied = Sum of voltage drops

$$E_A = E_1 + E_2 + E_3$$

where E_A is the applied voltage, and E_1, E_2, and E_3 are voltage drops.

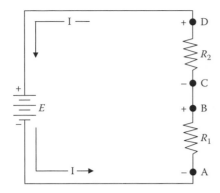

FIGURE 8.37 Polarity of voltage drops.

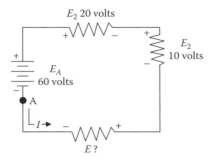

FIGURE 8.38 Determining unknown voltage in a series circuit.

Another way of stating Kirchhoff's law is that the algebraic sum of the instantaneous emf values and voltage drops around any closed circuit is zero. Through the use of Kirchhoff's law, circuit problems can be solved that would be difficult and often impossible with only knowledge of Ohm's law. When Kirchhoff's law is properly applied, an equation can be set up for a closed loop and the unknown circuit values may be calculated.

8.14.8 Polarity of Voltage Drops

When a voltage drop occurs across a resistance, one end must be more positive or more negative than the other end. The polarity of the voltage drop is determined by the direction of current flow. In the circuit shown in Figure 8.37, the current is flowing in a counterclockwise direction due to the arrangement of the battery source (E). Notice that the end of resistor R_1 into which the current flows is marked negative (–). The end of R_1 at which the current leaves is marked positive (+). These polarity markings are used to show that the end of R_1 into which the current flows is at a higher negative potential than is the end of the resistor from which the current leaves. Point A is thus more negative than point B.

Point C, which is at the same potential as point B, is labeled negative. This indicates that point C, although positive with respect to point A, is more negative than point D. To say that a point is positive (or negative), without stating what it is positive (or negative) with respect to, has no meaning.

Kirchhoff's voltage law can be written as an equation as shown below:

$$E_a + E_b + E_c + \dots E_n = 0 \qquad (8.12)$$

where E_a, E_b, etc. are the voltage drops and emf values around any closed circuit loop.

■ EXAMPLE 8.17

Problem: Three resistors are connected across a 60-volt source. What is the voltage across the third resistor if the voltage drops across the other two resistors are 10 volts and 20 volts?

Solution: First, draw a diagram like the one shown in Figure 8.38. Next, assume a direction of current as shown. Using this current, place the polarity markings at each end of each resistor and on the terminals of the source. Starting at point A, trace around the circuit in the direction of current flow, recording the voltage and polarity of each component. Starting at point A, these voltages would be as follows:

Basic formula:

$$E_a + E_b + E_c \dots E_n = 0$$

From the circuit:

$$(+E_?) + (+E_2) + (+E_3) - (E_A) = 0$$

Substituting values from the circuit:

$$E_? + 10 + 20 - 60 = 0$$

$$E_? - 30 = 0$$

$$E_? = 30 \text{ volts}$$

Thus, the unknown voltage ($E_?$) is found to be 30 volts.

Using the same idea as above, a problem can be solved in which the current is the unknown quantity.

8.14.9 Series Aiding and Opposing Sources

Sources of voltage that cause current to flow in the same direction are considered to be *series aiding,* and their voltages add. Sources of voltage that would tend to force current in opposite directions are said to be *series opposing,* and the effective source voltage is the difference between the opposing voltages. When two opposing sources are inserted into a circuit, current flow would be in a direction determined by the larger source. Examples of series aiding and opposing sources are shown in Figure 8.39.

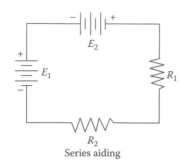

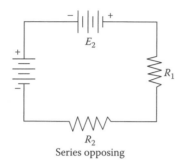

FIGURE 8.39 Aiding and opposing sources.

8.14.10 KIRCHHOFF'S LAW AND MULTIPLE-SOURCE SOLUTIONS

Kirchhoff's law can be used to solve multiple-source circuit problems. When applying this method, the exact same procedure is used for multiple-source circuits as was used for single-source circuits. This is demonstrated by the following example.

■ EXAMPLE 8.18

Problem: Find the amount of current in the circuit shown in Figure 8.40.

Solution: Start at point A.

Basic equation:

$$E_a + E_b + E_c + \ldots E_n = 0$$

From the circuit:

$$E_{b2} + E_1 - E_{b1} + E_{b3} + E_2 = 0$$

$$40 + 40I - 140 + 20 + 20I = 0$$

Combining like terms:

$$60I - 80 = 0$$

$$60I = 80$$

$$I = 80/60 = 1.33 \text{ amps}$$

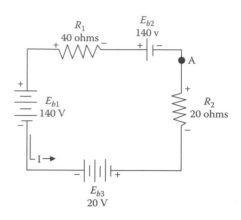

FIGURE 8.40 Solving for circuit current in a multiple-source circuit.

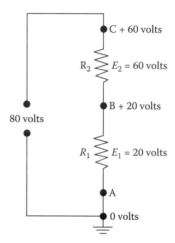

FIGURE 8.41 Use of ground symbols.

8.15 GROUND

The term *ground* is used to denote a common electrical point of zero potential. The reference point of a circuit is always considered to be at zero potential. The earth (ground) is said to be at zero potential. In Figure 8.41, point A is the zero reference or ground and is symbolized as such. Point C is 60 volts positive and point B is 20 volts positive with respect to ground. The common ground for much electrical/electronics equipment is the metal chassis. The value of ground is noted when considering its contribution to economy, simplification of schematics, and ease of measurement. When completing each electrical circuit, common points of a circuit at zero potential are connected directly to the metal chassis, thereby eliminating a large amount of connecting wire. An example of a grounded circuit is illustrated in Figure 8.42.

Note: Most voltage measurements used to check proper circuit operation in electronic equipment are taken with respect to ground. One meter lead is attached to ground, and the other meter lead is moved to various test points.

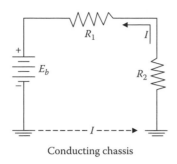

FIGURE 8.42 Ground used as a conductor.

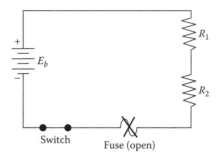

FIGURE 8.43 Open circuit with blown fuse.

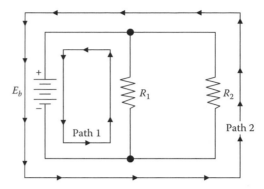

FIGURE 8.44 Basic parallel circuit.

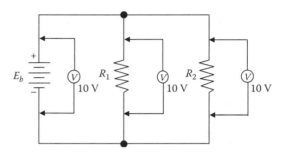

FIGURE 8.45 Voltage comparison in a parallel circuit.

8.16 OPEN AND SHORT CIRCUITS

A circuit is *open* if a break in the circuit does not provide a complete path for current. Figure 8.43 shows an open circuit, where the fuse is blown. To protect a circuit, a fuse is placed directly into the circuit. A fuse will open the circuit whenever a dangerously large current begins to flow. A fuse will permit currents smaller than the fuse value to flow but will melt and therefore break or open the circuit if a larger current flows. A dangerously large current will flow when a *short circuit* occurs. A short circuit is usually caused by an accidental connection between two points in a circuit that offers very little resistance and passes an abnormal amount of current. A short circuit often occurs because of improper wiring or broken insulation.

8.17 PARALLEL DC CIRCUITS

The principles we applied to solving simple series circuit calculations for determining the reactions of such quantities as voltage, current, and resistance can be used in parallel circuits.

8.17.1 PARALLEL CIRCUIT CHARACTERISTICS

A *parallel circuit* is defined as one having two or more components connected across the same voltage source (see Figure 8.44). Recall that a series circuit has only one path for current flow. As additional loads (resistors, etc.) are added to the circuit, the total resistance increases and

the total current decreases. This is *not* the case in a parallel circuit. In a parallel circuit, each load (or branch) is connected directly across the voltage source. In Figure 8.44, commencing at the voltage source (E_b) and tracing counterclockwise around the circuit, two complete and separate paths can be identified in which current can flow. One path is traced from the source through resistance R_1 and back to the source, the other from the source through resistance R_2 and back to the source.

8.17.2 VOLTAGE IN PARALLEL CIRCUITS

Recall that in a series circuit the source voltage divides proportionately across each resistor in the circuit. In a parallel circuit (see Figure 8.44), the same voltage is present across all of the resistors of a parallel group. This voltage is equal to the applied voltage (E_b) and can be expressed in equation form as:

$$E_b = E_{R1} = E_{R2} = E_{Rn} \qquad (8.13)$$

We can verify Equation 8.13 by taking voltage measurements across the resistors of a parallel circuit, as illustrated in Figure 8.45. Notice that each voltmeter indicates the same amount of voltage; that is, the voltage across each resistor is the same as the applied voltage.

Key Point: In a parallel circuit the voltage remains the same throughout the circuit.

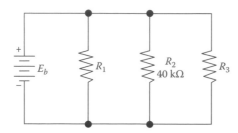

FIGURE 8.46 Illustration for Example 8.19.

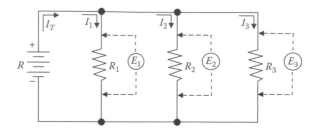

FIGURE 8.47 Parallel circuit.

■ **EXAMPLE 8.19**

Problem: Assume that the current through a resistor of a parallel circuit is known to be 4 milliamperes (mA) and the value of the resistor is 40,000 ohms. Determine the potential (voltage) across the resistor. The circuit is shown in Figure 8.46.

Solution:

Given:

$$R_2 = 40 \text{ k}\Omega$$
$$I_{R2} = 4 \text{ mA}$$

Find E_{R2} and E_b.

Select the appropriate equation:

$$E = IR$$

Substitute known values:

$$E_{R2} = I_{R2} \times R_2$$
$$E_{R2} = 4 \text{ mA} \times 40{,}000 \text{ ohms}$$

Using power of tens:

$$E_{R2} = (4 \times 10^{-3}) \times (40 \times 10^3)$$
$$E_{R2} = 4 \times 40 = 160 \text{ volts}$$

Therefore:

$$E_b = 160 \text{ volts}$$

8.17.3 Current in Parallel Circuits

In a series circuit, a single current flows. Its value is determined in part by the total resistance of the circuit; however, the source current in a parallel circuit divides among the available paths in relation to the value of the resistors in the circuit. Ohm's law remains unchanged. For a given voltage, current varies inversely with resistance.

Important Point: Ohm's law states that the *current in a circuit is inversely proportional to the circuit resistance.* This fact, important as a basic building block of electrical theory, is also important in the following explanation of current flow in parallel circuits.

The behavior of current in a parallel circuit is best illustrated by example (see Figure 8.47). The resistors R_1, R_2, and R_3 are in parallel with each other and with the battery. Each parallel path is then a branch with its own individual current. When the total current (I_T) leaves the voltage source (E), part I_1 of current I_T will flow through R_1, part I_2 will flow through R_2, and I_3 through R_3. The branch currents I_1, I_2, and I_3 can be different; however, if a voltmeter (used for measuring the voltage of a circuit) is connected across R_1, R_2, and R_3, then the respective voltages E_1, E_2, and E_3 will be equal. Therefore,

$$E = E_1 = E_2 = E_3 \qquad (8.14)$$

The total current (I_T) is equal to the sum of all branch currents:

$$I_T = I_1 = I_2 = I_3 \qquad (8.15)$$

This formula applies for any number of parallel branches, whether the resistances are equal or unequal.

By Ohm's law, each branch current equals the applied voltage divided by the resistance between the two points where the voltage is applied. Hence, for each branch we have the following equations:

$$\text{Branch 1: } I_1 = \frac{E_1}{R_1} = \frac{E}{R_1}$$

$$\text{Branch 2: } I_2 = \frac{E_2}{R_2} = \frac{E}{R_2} \qquad (8.16)$$

$$\text{Branch 3: } I_3 = \frac{E_3}{R_3} = \frac{V}{R_3}$$

With the same applied voltage, any branch that has less resistance allows more current through it than a branch with higher resistance.

■ **EXAMPLE 8.20**

Problem: Two resistors, each drawing 2 amps, and a third resistor drawing 1 amp are connected in parallel across a 100-volt line (see Figure 8.48). What is the total current?

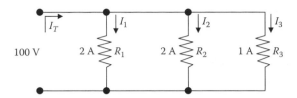

FIGURE 8.48 Illustration for Example 8.20.

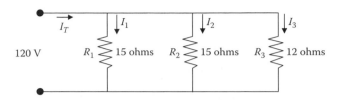

FIGURE 8.50 Illustration for Example 8.22.

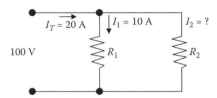

FIGURE 8.49 Illustration for Example 8.21.

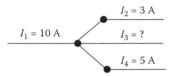

FIGURE 8.51 Illustration for Example 8.23.

Now find total current, using Equation 8.16:

$$I_T = I_1 + I_2 + I_3$$
$$I_T = 8 + 8 + 10 = 26 \text{ amps}$$

Solution: The formula for total current is:

$$I_T = I_1 = I_2 = I_3$$

Thus,

$$I_T = 2 + 2 + 1 = 5 \text{ amps}$$

The total current, then, is 5 amps.

■ EXAMPLE 8.21

Problem: Two branches, R_1 and R_2, across a 100-volt power line draw total line current of 20 amps (Figure 8.49). Branch R_1 takes 10 amps. What is the current (I_2) in branch R_2?

Solution: Beginning with Equation 8.15, transpose to find I_2 and then substitute given values:

$$I_T = I_1 + I_2$$
$$I_2 = I_T - I_1 = 20 - 10 = 10 \text{ amps}$$

The current in branch R_2, then, is 10 amps.

■ EXAMPLE 8.22

Problem: A parallel circuit consists of two 15-ohm and one 12-ohm resistor across a 120-volt line (see Figure 8.50). What current will flow in each branch of the circuit and what is the total current drawn by all the resistors?

Solution: There is 120-volt potential across each resistor. Using Equation 8.16, apply Ohm's law to each resistor.

$$I_1 = \frac{V}{R_1} = \frac{120}{15} = 8 \text{ amps}$$

$$I_2 = \frac{V}{R_2} = \frac{120}{15} = 8 \text{ amps}$$

$$I_3 = \frac{V}{R_3} = \frac{120}{12} = 10 \text{ amps}$$

8.17.4 PARALLEL CIRCUITS AND KIRCHHOFF'S CURRENT LAW

The division of current in a parallel network follows a definite pattern. This pattern is described by *Kirchhoff's current law*, which is stated as follows:

> The algebraic sum of the currents entering and leaving any junction of conductors is equal to zero.

This can be stated mathematically as:

$$I_a + I_b + \dots + I_n = 0 \qquad (8.17)$$

where $I_a, I_b, \dots I_n$ are the currents entering and leaving the junction. Currents entering the junction are assumed to be positive, and currents leaving the junction are considered negative. When solving a problem using Equation 8.17, the currents must be placed in the equation with the proper polarity.

■ EXAMPLE 8.23

Problem: Solve for the value of I_3 in Figure 8.51.

Solution: First, give the currents the appropriate signs:

$I_1 = +10$ amps
$I_2 = -3$ amps
$I_3 = ?$ amps
$I_4 = -5$ amps

Then, place these currents into Equation 8.17:

$$I_a + I_b + \dots + I_n = 0$$

with the proper signs as follows:

$$I_1 + I_2 + I_3 + I_4 = 0$$

$$(+10) + (-3) + (I_3) + (-5) = 0$$

Combining like terms, we obtain:

$$I_3 + 2 = 0$$

$$I_3 = -2 \text{ amps}$$

thus, I_3 has a value of 2 amps, and the negative sign shows it to be a current leaving the junction.

8.17.5 PARALLEL CIRCUIT RESISTANCE

Unlike series circuits, where total resistance (R_T) is the sum of the individual resistances, in a parallel circuit the total resistance is *not* the sum of the individual resistances. In a parallel circuit, we can use Ohm's law to find total resistance:

$$R = \frac{E}{I}$$

or

$$R_T = \frac{E_S}{I_T}$$

where R_T is the total resistance of all of the parallel branches across the voltage source E_S, and I_T is the sum of all the branch currents.

■ EXAMPLE 8.24

Problem: Given that $E_S = 120$ volts and $I_T = 26$ amps, what is the total resistance of the circuit shown in Figure 8.52?

Solution: In Figure 8.52, the line voltage is 120 volts and the total line current is 26 amps; therefore,

$$R_T = \frac{E_S}{I_T} = \frac{120}{26} = 4.62 \text{ ohms}$$

Important Point: Notice that R_T is smaller than any of the three resistances in Figure 8.52. This fact may surprise you; it may seem strange that the total circuit resistance is *less* than that of the smallest resistor (R_3, 12 ohms). If we refer back to the water analogy we have used previously, it makes sense. Consider water pressure and water pipes, and assume that we can keep the water pressure constant. A small pipe offers more resistance to the flow of water than a larger pipe, but if we add another pipe in parallel, one of even smaller diameter, the total resistance to water flow is decreased. In an electrical circuit, even a

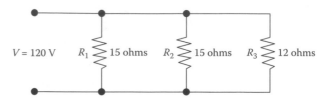

FIGURE 8.52 Illustration for Example 8.24.

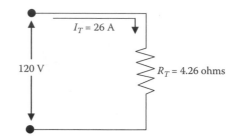

FIGURE 8.53 Circuit equivalent to that of Figure 8.52.

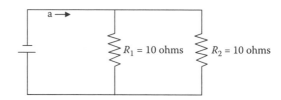

FIGURE 8.54 Two equal resistors connected in parallel.

larger resistor in another parallel branch provides an additional path for current flow, so the total resistance is less. Remember, if we add one more branch to a parallel circuit, the total resistance decreases and the total current increases.

Back to Example 8.24 and Figure 8.52. What we essentially demonstrated in working this particular problem is that the total load connected to the 120-V line is the same as the single equivalent resistance of 4.62 ohms connected across the line. It is probably more accurate to call this total resistance the *equivalent resistance*, but by convention R_T (total resistance) is generally used, although they are often used interchangeably. The equivalent resistance is illustrated in the equivalent circuit shown in Figure 8.53.

Other methods are used to determine the equivalent resistance of parallel circuits. The most appropriate method for a particular circuit depends on the number and value of the resistors; for example, consider the parallel circuit shown in Figure 8.54. For this circuit, the following simple equation is used:

$$R_{eq} = \frac{R}{N}$$

where:

R_{eq} = equivalent parallel resistance.
R = ohmic value of one resistor.
N = number of resistors.

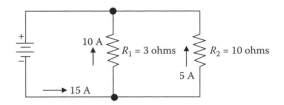

FIGURE 8.55 Illustration for Example 8.26.

Thus,

$$R_{eq} = \frac{10 \text{ ohms}}{2} = 5 \text{ ohms}$$

Note: Equation 8.18 is valid for any number of equal value parallel resistors.

Key Point: When two equal value resistors are connected in parallel, they present a total resistance equivalent to a single resistor of one half the value of either of the original resistors.

■ EXAMPLE 8.25

Problem: Five 50-ohm resistors are connected in parallel. What is the equivalent circuit resistance?

Solution: Using Equation 8.18:

$$R_{eq} = \frac{R}{N} = \frac{50}{5} = 10 \text{ ohms}$$

What about parallel circuits containing resistance of unequal value? How is equivalent resistance determined? Example 8.26 demonstrates how this is accomplished.

■ EXAMPLE 8.26

Problem: Refer to Figure 8.55.

Solution:

Given:
$R_1 = 3$ ohms
$R_2 = 6$ ohms
$E_a = 30$ volts

We know that:
$I_1 = 10$ amps
$I_2 = 5$ amps
$I_t = 15$ amps

Now determine R_{eq}:

$$R_{eq} = \frac{E_a}{I_t} = \frac{30}{15} = 2 \text{ ohms}$$

Key Point: In Example 8.26, the equivalent resistance of 2 ohms is less than the value of either branch resistor. Remember, in parallel circuits the equivalent resistance will always be smaller than the resistance of any branch.

8.17.6 RECIPROCAL METHOD

When circuits are encountered in which resistors of unequal value are connected in parallel, the equivalent resistance may be computed by using the *reciprocal method*.

Note: A *reciprocal* is an inverted fraction; the reciprocal of the fraction 3/4, for example, is 4/3. We consider a whole number to be a fraction with 1 as the denominator, so the reciprocal of a whole number is that number divided into 1; for example, the reciprocal of R_T is $1/R_T$.

The equivalent resistance in parallel is given by the formula:

$$\frac{1}{R_T} = \frac{1}{R_1} + \frac{1}{R_2} + \frac{1}{R_3} + \cdots + \frac{1}{R_n} \tag{8.19}$$

where R_T is the total resistance in parallel, and R_1, R_2, R_3, and R_n are the branch resistances.

■ EXAMPLE 8.27

Problem: Find the total resistance of a 2-ohm, a 4-ohm, and an 8-ohm resistor in parallel (Figure 8.56).

Solution: Write the formula for the three resistors in parallel:

$$\frac{1}{R_T} = \frac{1}{R_1} + \frac{1}{R_2} + \frac{1}{R_3}$$

Substitute the resistance values:

$$\frac{1}{R_T} = \frac{1}{2} + \frac{1}{4} + \frac{1}{8}$$

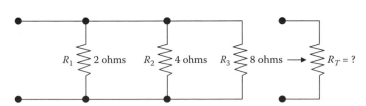

FIGURE 8.56 Illustration for Example 8.27.

Add the fractions:

$$\frac{1}{R_T} = \frac{4}{8} + \frac{2}{8} + \frac{1}{8} = \frac{7}{8}$$

Invert both sides of the equation to solve for R_T:

$$\frac{1}{R_T} = \frac{8}{7} = 1.14 \text{ ohms}$$

Note: When resistances are connected in parallel, the total resistance is always less than the smallest resistance of any single branch.

8.17.7 PRODUCT OVER THE SUM METHOD

When any two unequal resistors are in parallel, it is often easier to calculate the total resistance by multiplying the two resistances and then dividing the product by the sum of the resistances:

$$R_T = \frac{R_1 \times R_2}{R_1 + R_2}$$

where R_T is the total resistance in parallel, and R_1 and R_2 are the two resistors in parallel.

■ EXAMPLE 8.28

Problem: What is the equivalent resistance of a 20-ohm and a 30-ohm resistor connected in parallel?

Solution:

Given:

$R_1 = 20$ ohms
$R_2 = 30$ ohms

Determine the equivalent resistance:

$$R_T = \frac{R_1 \times R_2}{R_1 + R_2} = \frac{20 \times 30}{20 + 30} = 12 \text{ ohms}$$

8.17.8 REDUCTION TO AN EQUIVALENT CIRCUIT

In the study of basic electricity, it is often necessary to resolve a complex circuit into a simpler form. Any complex circuit consisting of resistances can be reduced to a basic equivalent circuit containing the source and total resistance. This process is called *reduction to an equivalent circuit*. An example of circuit reduction was demonstrated in Example 8.78 and is illustrated in Figure 8.57. The circuit shown in Figure 8.57A is reduced to the simple circuit shown in Figure 8.57B.

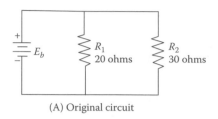

(A) Original circuit

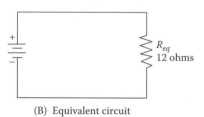

(B) Equivalent circuit

FIGURE 8.57 Parallel circuit with equivalent circuit.

8.17.9 POWER IN PARALLEL CIRCUITS

As in the series circuit, the total *power* consumed in a parallel circuit is equal to the sum of the power consumed in the individual resistors.

Note: Because power dissipation in resistors consists of a heat loss, power dissipations are additive regardless of how the resistors are connected in the circuit.

$$P_T = P_1 + P_2 + P_3 + \dots + P_n \qquad (8.21)$$

where P_T is the total power, and P_1, P_2, P_3, ... P_n are the branch powers.

Total power can also be calculated by the equation

$$P_T = EI_T \qquad (8.22)$$

where P_T is the total power, E is the voltage source across all parallel branches, and I_T is the total current. The power dissipated in each branch is equal to EI and equal to V^2/R.

Note: In both parallel and series arrangements, the sum of the individual values of power dissipated in the circuit equal the total power generated by the source. The circuit arrangements cannot change the fact that all power in the circuit comes from the source.

8.17.10 RULES FOR SOLVING PARALLEL DC CIRCUITS

Problems involving the determination of resistance, voltage, current, and power in a parallel circuit are solved as simply as in a series circuit. The procedure is the same: (1) draw a circuit diagram, (2) state the values given and the values to be found, (3) state the applicable equations, and (4) substitute the given values and solve for the unknown. Along with this problem-solving procedure, it is also important to remember and apply the rules for solving parallel DC circuits:

- The same voltage exists across each branch of a parallel circuit and is equal to the source voltage.
- The current through a branch of a parallel network is inversely proportional to the amount of resistance of the branch.
- The total current of a parallel circuit is equal to the sum of the currents of the individual branches of the circuit.
- The total resistance of a parallel circuit is equal to the reciprocal of the sum of the reciprocals of the individual resistances of the circuit.
- The total power consumed in a parallel circuit is equal to the sum of the power consumption of the individual resistances.

8.18 SERIES–PARALLEL CIRCUITS

To this point, we have discussed series and parallel DC circuits; however, operators will seldom encounter a circuit that consists solely of either type of circuit. Most circuits consist of both series and parallel elements. A circuit of this type is referred to as a *series–parallel circuit* or as a *combination circuit*. Analyzing a series–parallel (combination) circuit is simply a matter of applying the laws and rules discussed up to this point.

8.18.1 SOLVING A SERIES–PARALLEL CIRCUIT

At least three resistors are required to form a series–parallel circuit: two parallel resistors connected in series with at least one other resistor. In a circuit of this type, the current (I_T) divides after it flows through R_1, and part of it flows through R_2 and part flows through R_3. Then, the current joins at the junction of the two resistors and flows back to the positive terminal of the voltage source (E) and through the voltage source to the positive terminal.

When solving for values in a series–parallel circuit (current, voltage, and resistance), follow the rules that apply to a series circuit for the series part of the circuit, and follow the rules that apply to a parallel circuit for the parallel part of the circuit. Solving series–parallel circuits is simplified if all parallel and series groups are first reduced to single equivalent resistances and the circuits are redrawn in simplified form. Recall that the redrawn circuit is called an *equivalent circuit*.

Note: No general formulae are available for solving series–parallel circuits because so many different forms of these circuits exist.

Note: The total current in the series–parallel circuit depends on the effective resistance of the parallel portion and on the other resistances.

8.19 CONDUCTORS

Earlier we mentioned that electric current moves easily through some materials but with greater difficulty through others. Three good electrical conductors are copper, silver, and aluminum (generally, we can say that most metals are good conductors). Today, copper is the material of choice for electrical conductors. Under special conditions, certain gases are also used as conductors; for example, neon gas, mercury vapor, and sodium vapor are used in various kinds of lamps.

The function of the wire conductor is to connect a source of applied voltage to a load resistance with a minimum *IR* voltage drop in the conductor so most of the applied voltage can produce current in the load resistance. Ideally, a conductor must have a very low resistance; a typical value for a conductor such as copper is less than 1 ohm per 10 feet.

Because all electrical circuits utilize conductors of one type or another, in this section we discuss the basic features and electrical characteristics of the most common types of conductors. Moreover, because conductor splices and connections (and insulation of such connections) are also an essential part of any electric circuit, they are also discussed.

8.19.1 UNIT SIZE OF CONDUCTORS

A standard (or unit size) of a conductor has been established to compare the resistance and size of one conductor with another. The unit of linear measurement used (with regard to the diameter of a piece of wire) is the *mil* (0.001 of an inch). A convenient unit of wire length is the foot. Thus, the standard unit of size in most cases is the mil-foot; that is, a wire will have unit size if it has diameter of 1 mil and a length of 1 foot. The resistance in ohms of a unit conductor or a given substance is called the *resistivity* (or *specific resistance*) of the substance. As a further convenience, gauge numbers are also used to compare the diameter of wires. The Browne and Sharpe (B&S) gauge was used in the past; now the most commonly used gauge is the American Wire Gauge (AWG).

8.19.2 SQUARE MIL

Figure 8.58 shows a square mil. The *square mil* is a convenient unit of cross-sectional area for square or rectangular conductors. As shown in Figure 8.58, a square mil is the area of a square, the sides of which are 1 mil. To obtain the cross-sectional area in square mils of a square conductor, square one side measured in mils. To obtain

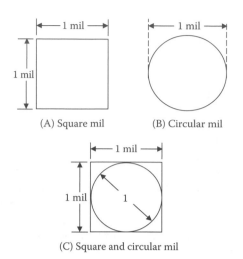

FIGURE 8.58 (A) Square mil; (B) circular mil; (C) comparison of circular to square mil.

the cross-sectional area in square mils of a rectangular conductor, multiply the length of one side by that of the other, each length being expressed in mils.

■ **EXAMPLE 8.29**

Problem: Find the cross-sectional area of a large rectangular conductor 5/8 inch thick and 5 inches wide.

Solution: The thickness may be expressed in mils as 0.625 × 1000 = 625 mils and the width as 5 × 1000 = 5000 mils. The cross-sectional area is 625 × 5000, or 3,125,000 square mils.

8.19.3 CIRCULAR MIL

The *circular mil* is the standard unit of wire cross-sectional area used in most wire tables. To avoid the use of decimals (because most wires used to conduct electricity may be only a small fraction of an inch), it is convenient to express these diameters in mils. As an example, the diameter of a wire is expressed as 25 mils instead of 0.025 inch. A circular mil is the area of a circle having a diameter of 1 mil, as shown in Figure 8.58B. The area in circular mils of a round conductor is obtained by squaring the diameter measured in mils. Thus, a wire having a diameter of 25 mils has an area of $(25)^2$ or 625 circular mils. By way of comparison, the basic formula for the area of a circle is:

$$A = \pi r^2 \qquad (8.22)$$

In this example, the area in square inches is:

$$A = \pi r^2 = 3.14 \times (0.0125)^2 = 0.00049 \text{ in.}^2$$

If D is the diameter of a wire in mils, the area in square mils can be determined using:

$$A = \pi \times (D/2)^2 \qquad (8.23)$$

which translates to:

$$A = 3.14/4 \times D^2 = 0.785D^2 \text{ square mils}$$

Thus, a wire 1 mil in diameter has an area of:

$$A = 0.785 \times (1)^2 = 0.785 \text{ square mils}$$

which is equivalent to 1 circular mil. The cross-sectional area of a wire in circular mils is therefore determined as:

$$A = \frac{0.785D^2}{0.785} = D^2 \text{ circular mils}$$

where D is the diameter in mils; therefore, the constant $\pi/4$ is eliminated from the calculation.

It should be noted that in comparing square and round conductors that the circular mil is a smaller unit of area than the square mil; therefore, there are more circular mils than square mils in any given area. The comparison is shown in Figure 8.58C. The area of a circular mil is equal to 0.785 of a square mil.

Important Point: To determine the circular-mil area when the square-mil area is given, divide the area in square mils by 0.785. Conversely, to determine the square-mil area when the circular-mil area is given, multiply the area in circular mils by 0.785.

■ **EXAMPLE 8.30**

Problem: A No. 12 wire has a diameter of 80.81 mils. What are (1) its area in circular mils and (2) its area in square mils?

Solution:

1. $A = D^2 = (80.81)^2 = 6530$ circular mils
2. $A = 0.785 \times 6530 = 5126$ square mils

■ **EXAMPLE 8.31**

Problem: A rectangular conductor is 1.5 inches wide and 0.25 inch thick. (1) What is its area in square mils? (2) What size of round conductor in circular mils is necessary to carry the same current as the rectangular bar?

Solution:

1. 1.5 in. = 1.5 × 1000 = 1500 mils, and 0.25 in. = 0.25 × 1000 = 250 mils; thus,

$$A = 1500 \times 250 = 375,000 \text{ square mils}$$

2. To carry the same current, the cross-sectional area of the rectangular bar and the cross-

FIGURE 8.59 Circular-mil-foot.

sectional area of the round conductor must be equal. There are more circular mils than square mils in this area; therefore,

$$A = \frac{375,000}{0.785} = 477,700 \text{ circular mils}$$

Note: Many electric cables are composed of stranded wires. The strands are usually single wires twisted together in sufficient numbers to make up the necessary cross-sectional area of the cable. The total area in circular mils is determined by multiplying the area of one strand in circular mils by the number of strands in the cable.

8.19.4 CIRCULAR-MIL-FOOT

As shown in Figure 8.59, a *circular-mil-foot* is actually a unit of volume. More specifically, it is a unit conductor 1 foot in length and having a cross-sectional area of 1 circular mil. The circular-mil-foot is useful in making comparisons between wires that are made of different metals because it is considered a unit conductor. Because it is considered a unit conductor, the circular-mil-foot is useful in making comparisons between wires that are made of different metals; for example, a basis of comparison of the resistivity of various substances may be the resistance of a circular-mil-foot of each of the substances.

Note: It is sometimes more convenient to employ a different unit of volume when working with certain substances. Accordingly, unit volume may also be taken as the centimeter cube. The inch cube may also be used. The unit of volume employed is given in tables of specific resistances.

8.19.5 RESISTIVITY

All materials differ in their atomic structure and therefore in their ability to resist the flow of an electric current. The measure of the ability of a specific material to resist the flow of electricity is called its *resistivity* or *specific resistance*—the resistance in ohms offered by unit volume (the circular-mil-foot) of a substance to the flow of electric current. Resistivity is the reciprocal of conductivity (i.e., the ease with which current flows in a conductor). A substance that has a high resistivity will have a low conductivity, and *vice versa*.

TABLE 8.3
Resistivity (Specific Resistance)

Substance	Specific Resistance at 20° (CM × ft/ohm)
Silver	9.8
Copper (drawn)	10.37
Gold	14.7
Aluminum	17.02
Tungsten	33.2
Brass	42.1
Steel (soft)	95.8
Nichrome	660.0

The resistance of a given length for any conductor depends on the resistivity of the material, the length of the wire, and the cross-sectional area of the wire according to the equation

$$R = \rho \frac{L}{A} \tag{8.24}$$

where:

R = resistance of the conductor (ohms).
ρ = specific resistance or resistivity (circular mil ohm/ft).
L = length of the wire (ft).
A = cross-sectional area of the wire (circular mil).

The factor ρ (Greek letter rho) permits different materials to be compared for resistance according to their nature without regard to different lengths or areas. Higher values of ρ mean more resistance.

Key Point: The resistivity of a substance is the resistance of a unit volume of that substance.

Many tables of resistivity are based on the resistance in ohms of a volume of the substance 1 foot long and 1 circular mil in cross-sectional area. The temperature at which the resistance measurement is made is also specified. If we know the metal of which the conductor is made, the resistivity of the metal may be obtained from a table. Table 8.3 provides the resistivity, or specific resistance, of some common substances.

Note: Because silver, copper, gold, and aluminum have the lowest values of resistivity, they are the best conductors. Tungsten and iron have a much higher resistivity.

■ EXAMPLE 8.32

Problem: What is the resistance of 1000 feet of copper wire having a cross-sectional area of 10,400 circular mils (No. 10 wire) when the wire temperature is 20°C?

Solution: The resistivity (specific resistance) from Table 8.3 is 10.37. Substituting the known values in Equation 8.24, we can determine the resistance (R) as:

$$R = \rho \frac{L}{A} = 10.37 \times \frac{1000}{10,400} = 1 \text{ ohm (approx.)}$$

8.19.6 WIRE MEASUREMENT

Wires are manufactured in sizes numbered according to a table known as the American Wire Gauge (AWG). Table 8.4 lists the standard wire sizes that correspond to the AWG. The gauge numbers specify the size of round wire in terms of its diameter and cross-sectional area. Note the following:

1. As the gauge numbers increase from 1 to 40, the diameter and circular area decrease. Higher gauge numbers mean smaller wire sizes; thus, No. 12 is a smaller wire than No. 4.
2. The circular area doubles for every three gauge sizes; for example, No. 12 wire has about twice the area of No. 15 wire.
3. The higher the gauge number and the smaller the wire, the greater the resistance of the wire for any given length; therefore, for 1000 ft of wire, No. 12 has a resistance of 1.62 ohms and No. 4 has a resistance of 0.253 ohms.

8.19.7 FACTORS GOVERNING THE SELECTION OF WIRE SIZE

Several factors must be considered when selecting the size of wire to be used for transmitting and distributing electric power. These factors include allowable power loss in the line, the permissible voltage drop in the line, the current-carrying capacity of the line, and the ambient temperatures at which the wire is to be used:

- *Allowable power loss (I^2R) in the line*—This loss represents electrical energy converted into heat. The use of large conductors will reduce the resistance and therefore the I^2R loss; however, large conductors are heavier and require more substantial supports, so they are more expensive initially than small ones.
- *Permissible voltage drop (IR drop) in the line*—If the source maintains a constant voltage at the input to the line, any variation in the load on the line will cause a variation in line current and a consequent variation in the IR drop in the line. A wide variation in the IR drop in the line causes poor voltage regulation at the load.
- *The current-carrying capacity of the line*—When current is drawn through the line, heat is

TABLE 8.4
Copper Wire Table

Gauge No.	Diameter	Circular mils	Ohms/1000 ft at 25°C
1	289.0	83,700.0	.126
2	258.0	66,400.0	.159
3	229.0	52,600.0	.201
4	204.0	41,700.0	.253
5	182.0	33,100.0	.319
6	162.0	26,300.0	.403
7	144.0	20,800.0	.508
8	128.0	16,500.0	.641
9	114.0	13,100.0	.808
10	102.0	10,400.0	1.02
11	91.0	8230.0	1.28
12	81.0	6530.0	1.62
13	72.0	5180.0	2.04
14	64.0	4110.0	2.58
15	57.0	3260.0	3.25
16	51.0	2580.0	4.09
17	45.0	2050.0	5.16
18	40.0	1620.0	6.51
19	36.0	1290.0	8.21
20	32.0	1020.0	10.4
21	28.5	810.0	13.1
22	25.3	642.0	16.5
23	22.6	509.0	20.8
24	20.1	404.0	26.4
25	17.9	320.0	33.0
26	15.9	254.0	41.6
27	14.2	202.0	52.5
28	12.6	160.0	66.2
29	11.3	127.0	83.4
30	10.0	101.0	105.0
31	8.9	79.7	133.0
32	8.0	63.2	167.0
33	7.1	50.1	211.0
34	6.3	39.8	266.0
35	5.6	31.5	335.0
36	5.0	25.0	423.0
37	4.5	19.8	533.0
38	4.0	15.7	673.0
39	3.5	12.5	848.0
40	3.1	9.9	1070.0

generated. The temperature of the line will rise until the heat that is radiated, or otherwise dissipated, is equal to the heat generated by the passage of current through the line. If the conductor is insulated, the heat generated in the conductor is not so readily removed, as it would be if the conductor were not insulated.

- *Conductors installed where ambient temperature is relatively high*—When installed in such surroundings, the heat generated by external sources constitutes an appreciable part of the

TABLE 8.5
Copper vs. Aluminum

Characteristic	Copper	Aluminum
Tensile strength (lb/in.²)	55,000	25,000
Tensile strength for the same conductivity (lb)	55,000	40,000
Weight for same conductivity (lb)	100	48
Cross-section for the same conductivity (CM)	100	160
Specific resistance (Ω/mil-ft)	10.6	17

total conductor heating. Due allowance must be made for the influence of external heating on the allowable conductor current, and each case has its own specific limitations.

8.19.8 COPPER VS. OTHER METAL CONDUCTORS

If it were not cost prohibitive, silver, the best conductor of electron flow (electricity), would be the conductor of choice in electrical systems. Instead, silver is used only in special circuits where a substance with high conductivity is required. The two most generally used conductors are copper and aluminum. Each has characteristics that make its use advantageous under certain circumstances. Likewise, each has certain disadvantages, or limitations.

Copper has a higher conductivity. It is more ductile (can be drawn out into wire), has relatively high tensile strength, and can be easily soldered. It is more expensive and heavier than aluminum. Aluminum has only about 60% of the conductivity of copper, but its lightness makes possible long spans, and its relatively large diameter for a given conductivity reduces corona (i.e., the discharge of electricity from the wire when it has a high potential). The discharge is greater when smaller diameter wire is used than when larger diameter wire is used. However, aluminum conductors are not easily soldered, and the relatively large size of aluminum for a given conductance does not permit the economical use of an insulation covering. A comparison of some of the characteristics of copper and aluminum is given in Table 8.5.

Note: Recent practice involves using copper wiring (instead of aluminum wiring) in housing and some industrial applications because aluminum connections are not as easily made as with copper. In addition, over the years, many fires have been caused by improperly connected aluminum wiring, as poor connections are high resistance connections that can result in excessive heat generation.

8.19.9 TEMPERATURE COEFFICIENT

The resistance of pure metals—such as silver, copper, and aluminum—increases as the temperature increases. The *temperature coefficient* of resistance (α, Greek letter

TABLE 8.6
Properties of Conducting Materials (Approximate)

Material	Temperature Coefficient (Ω/°C)
Aluminum	0.004
Carbon	–0.0003
Constantan	0 (average)
Copper	0.004
Gold	0.004
Iron	0.006
Nichrome	0.0002
Nickel	0.005
Silver	0.004
Tungsten	0.005

alpha), indicates how much the resistance changes for a change in temperature. A positive value for α means that R increases with temperature; a negative α means that R decreases; and a zero α means that R is constant, not varying with changes in temperature. Typical values of α are listed in Table 8.6. The amount of increase in the resistance of a 1-ohm sample of the copper conductor per degree rise in temperature (i.e., the temperature coefficient of resistance) is approximately 0.004. For pure metals, the temperature coefficient of resistance ranges between 0.004 and 0.006 ohm. Thus, a copper wire having a resistance of 50 ohms at an initial temperature of 0°C will have an increase in resistance of 50 × 0.004, or 0.2 ohm (approximate) for the entire length of wire for each degree of temperature rise above 0°C. At 20°C, the increase in resistance is approximately 20 × 0.2, or 4 ohms. The total resistance at 20°C is 50 + 4, or 54 ohms.

Note: As shown in Table 7.6, carbon has a negative temperature coefficient. In general, α is negative for all semiconductors such as germanium and silicon. A negative value for α means less resistance at higher temperatures; therefore, the resistance of semiconductor diodes and transistors can be reduced considerably when they become hot with normal load current. Observe, also, that constantan has a value of 0 for α (Table 8.6); thus, it can be used for precision wire-wound resistors, which do not change resistance when the temperature increases.

8.19.10 Conductor Insulation

Electric current must be contained; it must be channeled from the power source to a useful load safely. To accomplish this, electric current must be forced to flow only where it is needed. Moreover, current-carrying conductors must not be allowed (generally) to come in contact with one another, their supporting hardware, or personnel working near them. To accomplish this, conductors are coated or wrapped with various materials. These materials have such a high resistance that they are, for all practical purposes, nonconductors. They are generally referred to as *insulators* or *insulating materials*. A wide variety of insulated conductors is available to meet the requirements of any job; however, only the necessary minimum of insulation is applied for any particular type of cable designed to do a specific job. This is because insulation is expensive and has a stiffening effect, and it is required to meet a great variety of physical and electrical conditions.

Two fundamental but distinctly different properties of insulation materials (e.g., rubber, glass, asbestos, and plastics) are insulation resistance and dielectric strength:

- *Insulation resistance* is the resistance to current leakage through and over the surface of insulation materials.
- *Dielectric strength* is the ability of the insulator to withstand potential difference and is usually expressed in terms of the voltage at which the insulation fails because of the electrostatic stress.

Various types of materials are used to provide insulation for electric conductors, including rubber, plastics, varnished cloth, paper, silk, cotton, and enamel.

8.19.11 Conductors, Splices, and Terminal Connections

When conductors join each other, or connect to a load, *splices* or *terminals* must be used. It is important that they be properly made, because any electric circuit is only as good as its weakest connection. The basic requirement of any splice or connection is that it be both mechanically and electrically as strong as the conductor or device with which it is used. High-quality workmanship and materials must be employed to ensure lasting electrical contact, physical strength, and insulation (if required).

Important Point: Conductor splices and connections are an essential part of any electric circuit.

8.19.12 Soldering Operations

Soldering operations are a vital part of electrical or electronics maintenance procedures. Soldering is a manual skill that must be learned by all personnel who work in the field of electricity. Obviously, practice is required to develop proficiency in the techniques of soldering. When performing a soldering operation, both the solder and the material to be soldered (e.g., electric wire or terminal lugs) must be heated to a temperature that allows the solder to flow. If either is heated inadequately, cold solder joints result (i.e., high-resistance connections are created). Such joints do not provide either the physical strength or the electrical conductivity required. Moreover, in soldering operations it is necessary to select a solder that will flow at a temperature low enough to avoid damage to the part being soldered or to any other part or material in the immediate vicinity.

8.19.13 Solderless Connections

Generally, terminal lugs and splicers that do not require solder are more widely used (because they are easier to mount correctly) than those that do require solder. Solderless connectors—made in a wide variety of sizes and shapes—are attached to their conductors by means of several different devices, but the principle of each is essentially the same. They are all crimped (squeezed) tightly onto their conductors. They offer adequate electrical contact plus great mechanical strength.

8.19.14 Insulation Tape

The carpenter has saws, the dentist has pliers, the plumber has wrenches, and the electrician has insulating tape. Accordingly, one of the first things the rookie maintenance operator (a rookie who is also learning proper and safe techniques for performing electrical work) learns is the value of electrical insulating tape. Normally, the use of electrical insulating tape comes into play as the final step in completing a splice or joint to place insulation over the bare wire at the connection point.

Typically, the insulating tape should be of the same basic substance as the original insulation, usually a rubber-splicing compound. When using rubber (latex) tape as the splicing compound where the original insulation was rubber, it should be applied to the splice with a light tension so each layer presses tightly against the one underneath it. In addition to the rubber tape application (which restores the insulation to its original form), restoring with friction tape is also often necessary.

In recent years, plastic electrical tape has come into wide use. It has certain advantages over rubber and friction tape; for example, it will withstand higher voltages for a given thickness. Single thin layers of some commercially available plastic tape will stand several thousand volts without breaking down.

Important Point: Be advised that, although the use of plastic electrical tape has become almost universal in industrial applications, to ensure an extra margin of safety it must be applied in more layers because it is thinner than rubber or friction tape.

8.20 ELECTROMAGNETISM

Earlier, we discussed the fundamental theories concerning simple magnets and magnetism. Those discussions dealt mainly with forms of magnetism that were not related directly to electricity—permanent magnets for example. Further, only brief mention was made of those forms of magnetism having direct relation to electricity—such as producing electricity with magnetism.

In medicine, anatomy and physiology are so closely related that the medical student cannot study one at length without involving the other. A similar relationship holds for the electrical field; that is, magnetism and basic electricity are so closely related that one cannot be studied at length without involving the other. This close fundamental relationship is continually borne out in subsequent sections of this chapter, such as in the study of generators, transformers, and motors. To be proficient in electricity, the operator must become familiar with such general relationships that exist between magnetism and electricity as follows:

- Electric current flow will always produce some form of magnetism.
- Magnetism is by far the most commonly used means for producing or using electricity.
- The peculiar behavior of electricity under certain conditions is caused by magnetic influences.

8.20.1 MAGNETIC FIELD AROUND A SINGLE CONDUCTOR

In 1819, Hans Christian Oersted, a Danish scientist, discovered that a field of magnetic force exists around a single wire conductor carrying an electric current. Figure 8.60 shows a wire being passed through a piece of cardboard and connected through a switch to a dry cell. With the switch open (no current flowing), we can sprinkle iron filings on the cardboard and then tap it gently, and the filings will fall back haphazardly. Now, if we close the switch, current will begin to flow in the wire. If we tap the cardboard again, the magnetic effect of the current in the wire will cause the filings to fall back into a definite pattern of concentric circles with the wire as the center of the circles. Every section of the wire has this field of force around it in a plane perpendicular to the wire, as shown in Figure 8.61.

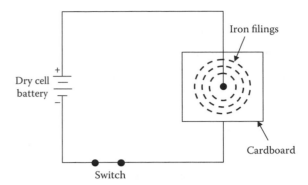

FIGURE 8.60 Circular pattern of magnetic force exists around a wire carrying an electric current.

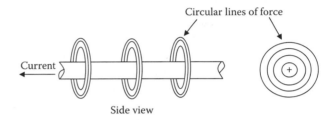

FIGURE 8.61 Circular fields of force around a wire carrying a current are in planes perpendicular to the wire.

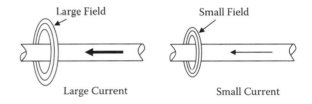

FIGURE 8.62 The strength of the magnetic field around a wire carrying a current depends on the amount of current.

The ability of the magnetic field to attract bits of iron (as demonstrated in Figure 8.60) depends on the number of lines of force present. The strength of the magnetic field around a wire carrying a current depends on the current, because it is the current that produces the field. The greater the current, the greater the strength of the field. A large current will produce many lines of force extending far from the wire, while a small current will produce only a few lines close to the wire, as shown in Figure 8.62.

8.20.2 POLARITY OF A SINGLE CONDUCTOR

The relationship between the direction of the magnetic lines of force around a conductor and the direction of current flow along the conductor may be determined by means of the left-hand rule for a conductor. If the conductor is grasped in the left hand with the thumb extended in the direction of electron flow (– to +), the fingers will point in the direction of the magnetic lines of force. The

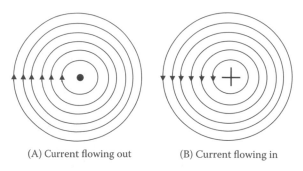

(A) Current flowing out   (B) Current flowing in

FIGURE 8.63 Magnetic field around current-carrying conductor.

north pole of a compass would point this same direction if the compass were placed in the magnetic field.

Important Note: Arrows are generally used in electric diagrams to denote the direction of current flow along the length of wire. Where cross-sections of wire are shown, a special view of the arrow is used. A cross-sectional view of a conductor that is carrying current toward the observer is illustrated in Figure 8.63A. The direction of current is indicated by a dot, which represents the head of the arrow. A conductor that is carrying current away from the observer is illustrated in Figure 8.63B. A cross, which represents the tail of the arrow, indicates the direction of current.

8.20.3 FIELD AROUND TWO PARALLEL CONDUCTORS

When two parallel conductors carry current in the same direction, the magnetic fields tend to encircle both conductors, drawing them together with a force of attraction, as shown in Figure 8.64A. Two parallel conductors carrying currents in opposite directions are shown in Figure 8.64B. The field around one conductor is opposite in direction to the field around the other conductor. The resulting lines of force are crowded together in the space between the wires and tend to push the wires apart. Thus, two parallel adjacent conductors carrying currents in the same direction attract each other, and two parallel conductors carrying currents in opposite directions repel each other.

8.20.4 MAGNETIC FIELD OF A COIL

The magnetic field around a current-carrying wire exists at all points along its length. Bending the current-carrying wire into the form of a single loop has two results. First, the magnetic field consists of more dense concentric circles in a plane perpendicular to the wire, although the total number of lines is the same as for the straight conductor. Second, all of the lines inside the loop are in the same direction. When this straight wire is wound around a core, it becomes a coil and the magnetic field assumes a different shape. When current is passed through the coiled conductor, the magnetic field of each turn of wire links with

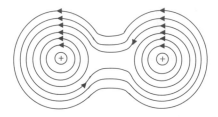

(A) Currents flowing in the same direction

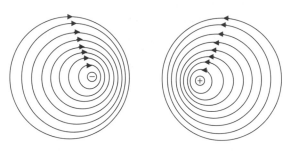

(B) Currents flowing in opposite directions

FIGURE 8.64 Magnetic field around two parallel conductors.

the fields of adjacent turns. The combined influence of all the turns produces a two-pole field similar to that of a simple bar magnet. One end of the coil will be a north pole and the other end will be a south pole.

8.20.5 POLARITY OF AN ELECTROMAGNETIC COIL

It was shown that the direction of the magnetic field around a straight conductor depends on the direction of current flow through that conductor; thus, a reversal of current flow through a conductor causes a reversal in the direction of the magnetic field that is produced. It follows that a reversal of the current flow through a coil also causes a reversal of its two-pole field. This is true because that field is the product of the linkage between the individual turns of wire on the coil. If the field of each turn is reversed, it follows that the total field (coil field) is also reversed. When the direction of electron flow through a coil is known, its polarity may be determined by use of the left-hand rule for coils. This rule is stated as follows: Grasping the coil in the left hand, with the fingers wrapped around in the direction of electron flow, the thumb will point toward the north pole (see Figure 8.65).

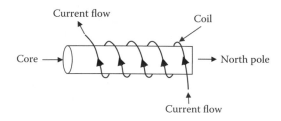

FIGURE 8.65 Current-carrying coil.

8.20.6 Strength of an Electromagnetic Field

The strength, or intensity, of the magnetic field of a coil depends on a number of factors:

- The number of turns of conductor
- The amount of current flow through the coil
- The ratio of the coil's length to its width
- The type of material in the core

8.20.7 Magnetic Units

The law of current flow in the electric circuit is similar to the law for the establishing of flux in the magnetic circuit. The *magnetic flux* (ϕ, phi) is similar to current in the Ohm's law formula and is the total number of lines of force existing in the magnetic circuit. The Maxwell is the unit of flux—that is, 1 line of force is equal to 1 Maxwell.

Note: The Maxwell is often referred to as simply a *line of force*, *line of induction*, or *line*.

The *strength* of a magnetic field in a coil of wire depends on how much current flows in the turns of the coil—the more current, the stronger the magnetic field. In addition, the more turns, the more concentrated are the lines of force. The *force* that produces the flux in the magnetic circuit (comparable to electromotive force in Ohm's law) is known as *magnetomotive force* (mmf). The practical unit of magnetomotive force is the ampere-turn (At). In equation form,

$$F = \text{ampere-turns} = NI \qquad (8.25)$$

where:

F = magnetomotive force (At).
N = number of turns.
I = current (A).

■ EXAMPLE 8.33

Problem: Calculate the ampere-turns for a coil with 2000 turns and a 5-Ma current.

Solution: Use Equation 8.25 and substitute $N = 2000$ and $I = 5 \times 10^{-3}$ amp:

$$NI = 2000 \times (5 \times 10^{-3}) = 10 \text{ At}$$

The unit of *intensity* of magnetizing force per unit of length is designated as H and is sometimes expressed as Gilberts per centimeter of length. Expressed as an equation:

$$H = \frac{NI}{L} \qquad (8.26)$$

where:

H = magnetic field intensity (ampere-turns per meter, At/m).
NI = ampere-turns (At).
L = length between poles of the coil (m).

Note: Equation 8.26 is for a solenoid. H is the intensity of an air core. For an iron core, H is the intensity through the entire core, and L is the length or distance between poles of the iron core.

8.20.8 Properties of Magnetic Materials

In this section, we discuss two important properties of magnetic materials: *permeability* and *hysteresis*.

8.20.8.1 Permeability

When the core of an electromagnet is made of annealed sheet steel, it produces a stronger magnet than if a cast-iron core is used, because the magnetizing force of the coil more readily acts upon annealed sheet steel than the hard cast iron. Simply put, soft sheet steel is said to have greater permeability because of the greater ease with which magnetic lines are established in it. Recall that permeability is the relative ease with which a substance conducts magnetic lines of force. The permeability of air is arbitrarily set at 1. The permeability of other substances is the ratio of their ability to conduct magnetic lines compared to that of air. The permeability of nonmagnetic materials, such as aluminum, copper, wood, and brass, is essentially unity, or the same as for air.

Note: The permeability of magnetic materials varies with the degree of magnetization, being smaller for high values of flux density.

Note: *Reluctance,* which is analogous to resistance, is the opposition to the production of flux in a material and is inversely proportional to permeability. Iron has high permeability and, therefore, low reluctance. Air has low permeability and hence high reluctance.

8.20.8.2 Hysteresis

When the current in a coil of wire reverses thousands of times per second, a considerable loss of energy can occur. This loss of energy is caused by hysteresis. Hysteresis means "a lagging behind"; that is, the magnetic flux in an iron core lags behind the increases or decreases of the magnetizing force. The simplest method of illustrating the property of hysteresis is by graphical means, such as the hysteresis loop shown in Figure 8.66. The hysteresis loop is a series of curves showing the characteristics of a magnetic material. Opposite directions of current result are in

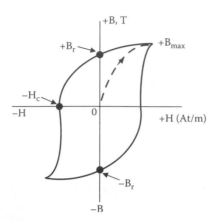

FIGURE 8.66 Hysteresis loop.

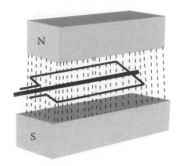

FIGURE 8.67 Loop rotating in magnetic field produces an AC voltage.

the opposite directions of +H and −H for field intensity. Similarly, opposite polarities are shown for flux density as +B and −B. The current starts at the center 0 (zero) when the material is unmagnetized. Positive H values increase B to saturation at $+B_{max}$. H then decreases to zero, but B drops to the value of B_r because of hysteresis. The current that produced the original magnetization now is reversed so H becomes negative. B drops to zero and continues to $-B_{max}$. As the −H values decrease, B is reduced to $-B_r$ when H is zero. Now with a positive swing of current, H becomes positive, producing saturation at $+B_{max}$ again. The hysteresis loop is now completed. The curve does not return to zero at the center because of hysteresis.

8.20.9 ELECTROMAGNETS

An *electromagnet* is composed of a coil of wire wound around a core that is normally soft iron because of its high permeability and low hysteresis. When direct current flows through the coil, the core will become magnetized with the same polarity that the coil would have without the core. If the current is reversed, the polarities of the coil and core are reversed.

The electromagnet is of great importance in electricity simply because the magnetism can be turned on or turned off at will. The starter solenoid (an electromagnet) in automobiles and powerboats is a good example. In an automobile or boat, an electromagnet is part of a relay that connects the battery to the induction coil, which generates the very high voltage required to start the engine. The starter solenoid isolates this high voltage from the ignition switch. When no current flows in the coil it is an *air core*, but when the coil is energized a movable soft-iron core does two things. First, the magnetic flux is increased because the soft-iron core is more permeable than the air core. Second, the flux is more highly concentrated. All of this concentration of magnetic lines of force in the soft-iron core results in a very good magnet when current flows in the coil; however, soft iron loses its magnetism quickly

when the current is shut off. The effect of the soft iron is, of course, the same whether it is movable, as in some solenoids, or permanently installed in the coil. An electromagnet then consists of a coil and a core; it becomes a magnet when current flows through the coil. Because the action of magnetic forces can be controlled, the electromagnet is very useful in many circuit applications.

8.21 AC THEORY

Because voltage is induced in a conductor when lines of force are cut, the amount of the induced emf depends on the number of lines cut in a unit time. To induce an emf of 1 volt, a conductor must cut 100,000,000 lines of force per second. To obtain this great number of cuttings, the conductor is formed into a loop and rotated on an axis at great speed (see Figure 8.67). The two sides of the loop become individual conductors in series, each side of the loop cutting lines of force and inducing twice the voltage that a single conductor would induce. In commercial generators, the number of cuttings and the resulting emf are increased by: (1) increasing the number of lines of force by using more magnets or stronger electromagnets, (2) using more conductors or loops, and (3) rotating the loops faster.

How an alternating-current (AC) generator operates to produce an AC voltage and current is a basic concept today, taught in elementary and middle school science classes. Of course, we accept technological advances as commonplace today. We surf the Internet, watch cable television, use our cell phones, take space flight as a given—and consider the production of electricity that makes all these technologies possible to be our right. These technologies are bottom shelf to us today; they are available to us so we use them.

In the groundbreaking years of electric technology development, the geniuses of the science of electricity (including George Simon Ohm) achieved their technological breakthroughs in faltering steps. We tend to forget that those first faltering steps of scientific achievement in the field of electricity were performed with crude and, for the most part, homemade apparatus. Indeed, the innovators of electricity had to fabricate nearly all of the laboratory equipment used in their experiments. At the time, the only convenient source of electrical energy available to these early scientists was the voltaic cell, invented some years earlier. Because cells and batteries were the only sources of power available, some of the early electrical devices were designed to operate from direct current (DC). For this reason, initially direct current was used extensively; however, when the use of electricity became widespread, certain disadvantages in the use of direct current became apparent. In a DC system, the supply voltage must be generated at the level required by the load. To operate a 240-volt lamp, for example, the generator must deliver 240 volts. A 120-volt lamp could not be operated from this generator by any convenient means. A resistor could be placed in series with the 120-volt lamp to drop the extra 120 volts, but the resistor would waste an amount of power equal to that consumed by the lamp.

Another disadvantage of DC systems is the large amount of power lost due to the resistance of the transmission wires used to carry current from the generating station to the consumer. This loss could be greatly reduced by operating the transmission line at very high voltage and low current. This is not a practical solution in a DC system, however, because the load would also have to operate at high voltage. Because of the difficulties encountered with direct current, practically all modern power distribution systems use alternating current, including water/wastewater treatment plants.

Unlike DC voltage, AC voltage can be stepped up or down by a device called a *transformer*. Transformers permit the transmission lines to be operated at high voltage and low current for maximum efficiency. At the consumer end, the voltage is stepped down to whatever value the load requires by using a transformer. Due to its inherent advantages and versatility, alternating current has replaced direct current in all but a few commercial power distribution systems.

8.21.1 Basic AC Generator

As shown in Figure 8.67, an AC voltage and current can be produced when a conductor loop rotates through a magnetic field and cuts lines of force to generate an induced AC voltage across its terminals. This describes the basic principle of operation of an alternating current generator, or alternator. An alternator converts mechanical energy into electrical energy. It does this by utilizing the principle of electromagnetic induction. The basic components of an alternator are an armature, about which many turns of conductor are wound and which rotates in a magnetic field, and some means of delivering the resulting alternating current to an external circuit.

8.21.2 Cycle

An AC voltage is one that continually changes in magnitude and periodically reverses in polarity (see Figure 8.68). The zero axis is a horizontal line across the center.

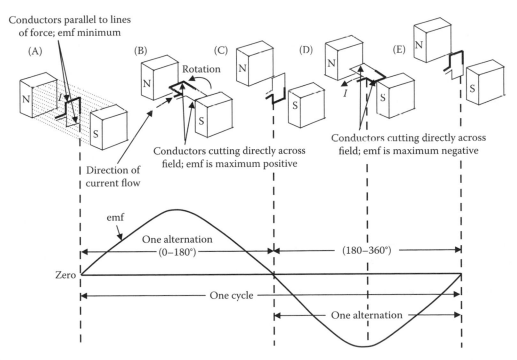

FIGURE 8.68 Basic AC sine wave and AC generator.

The vertical variations on the voltage wave show the changes in magnitude. The voltages above the horizontal axis have positive (+) polarity, and voltages below the horizontal axis have negative (–) polarity.

Figure 8.68 shows a suspended loop of wire (conductor or armature) being rotated (moved) in a counterclockwise direction through the magnetic field between the poles of a permanent magnet. For ease of explanation, the loop has been divided into a thick and thin half. Notice that in part A, the thick half is moving along (parallel to) the lines of force; consequently, it is cutting none of these lines. The same is true of the thin half, moving in the opposite direction. Because the conductors are not cutting any lines of force, no emf is induced. As the loop rotates toward the position shown in part B, it cuts more and more lines of force per second because it is cutting more directly across the field (lines of force) as it approaches the position shown in part B. At position B, the induced voltage is greatest because the conductor is cutting directly across the field.

As the loop continues to be rotated toward the position shown in part C, it cuts fewer and fewer lines of force per second. The induced voltage decreases from its peak value. Eventually, the loop is once again moving in a plane parallel to the magnetic field, and no voltage (zero voltage) is induced. The loop has now been rotated through half a circle (one alternation, or 180°). The sine curve shown in the lower part of Figure 8.68 shows the induced voltage at every instant of rotation of the loop. Notice that this curve contains 360°, or two alternations. Two alternations represent one complete circle of rotation.

Important Point: Two complete alternations in a period are called a *cycle*.

In Figure 8.68, if the loop is rotated at a steady rate and if the strength of the magnetic field is uniform, the number of cycles per second (cps), or hertz, and the voltage will remain at fixed values. Continuous rotation will produce a series of sine-wave voltage cycles, or, in other words, an AC voltage. In this way, mechanical energy is converted into electrical energy.

8.21.3 FREQUENCY, PERIOD, AND WAVELENGTH

The *frequency* of an alternating voltage or current is the number of complete cycles occurring in each second of time. It is indicated by the symbol f and is expressed in hertz (Hz). One cycle per second equals 1 hertz; thus, 60 cycles per second (cps) equals 60 Hz. A frequency of 2 Hz (Figure 8.69A) is twice the frequency of 1 Hz (Figure 8.69B). The amount of time for the completion of 1 cycle is the *period*. It is indicated by the symbol T for time and is expressed in seconds (sec). Frequency and period are reciprocals of each other:

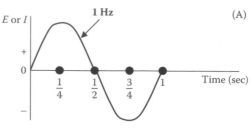

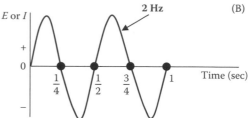

FIGURE 8.69 Comparison of frequencies.

$$f = \frac{1}{T} \qquad (8.27)$$

$$T = \frac{1}{f} \qquad (8.28)$$

Important Point: The higher the frequency, the shorter the period.

The angle of 360° represents the time for 1 cycle or the period T; therefore, we can show the horizontal axis of the sine wave in units of either electrical degrees or seconds (see Figure 8.70).

The *wavelength* is the length of one complete wave or cycle. It depends on the frequency of the periodic variation and its velocity of transmission. It is indicated by the symbol λ (lambda). Expressed as a formula:

$$\lambda = \frac{\text{Velocity}}{\text{Frequency}} \qquad (8.29)$$

8.21.4 CHARACTERISTIC VALUES OF AC VOLTAGE AND CURRENT

Because an AC sine wave voltage or current has many instantaneous values throughout the cycle, it is convenient to specify magnitudes for comparing one wave with another. The peak, average, or root-mean-square (RMS) value can be specified (see Figure 8.71). These values apply to current or voltage.

8.21.5 PEAK AMPLITUDE

One of the most frequently measured characteristics of a sine wave is its amplitude. Unlike DC measurement, the amount of alternating current or voltage present in a circuit

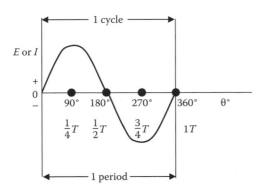

FIGURE 8.70 Relationship between electrical degrees and time.

can be measured in various ways. In one method of measurement, the maximum amplitude of either the positive or the negative alternation is measured. The value of current or voltage obtained is called the *peak voltage* or the *peak current*. To measure the peak value of current or voltage, an oscilloscope must be used. The peak value is illustrated in Figure 8.71.

8.21.6 PEAK-TO-PEAK AMPLITUDE

A second method of indicating the amplitude of a sine wave consists of determining the total voltage or current between the positive and negative peaks. This value of current or voltage is the *peak-to-peak value* (see Figure 8.71). Because both alternations of a pure sine wave are identical, the peak-to-peak value is twice the peak value. Peak-to-peak voltage is usually measured with an oscilloscope, although some voltmeters have a special scale calibrated in peak-to-peak volts.

8.21.7 INSTANTANEOUS AMPLITUDE

The *instantaneous value* of a sine wave of voltage for any angle of rotation is expressed by the formula:

$$e = E_m \times \sin\theta \qquad (8.30)$$

where:

e = the instantaneous voltage.
E_m = the maximum or peak voltage.
$\sin\theta$ = the sine of angle at which e is desired.

Similarly the equation for the instantaneous value of a sine wave of current is:

$$i = I_m \times \sin\theta \qquad (8.31)$$

where:

i = the instantaneous current.
I_m = the maximum or peak current.
$\sin\theta$ = the sine of the angle at which i is desired.

Note: The instantaneous value of voltage constantly changes as the armature of an alternator moves through a complete rotation. Because current varies directly with voltage, according to Ohm's law, the instantaneous changes in current also result in a sine wave whose positive and negative peaks and intermediate values can be plotted exactly as we plotted the voltage sine wave. Because instantaneous values are not useful in solving most AC problems, an effective value is used instead.

8.21.8 EFFECTIVE OR RMS VALUE

The *effective value* of an AC voltage or current of sine waveform is defined in terms of an equivalent heating effect of a direct current. Heating effect is independent of the direction of current flow.

Important Point: Because all instantaneous values of induced voltage are somewhere between 0 and E_m (maximum or peak voltage), the effective value of a sine wave voltage or current must be greater than 0 and less than E_m.

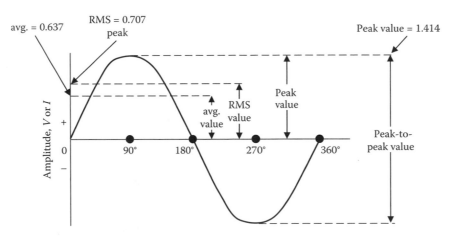

FIGURE 8.71 Amplitude values for AC sine wave.

The AC of a sine waveform having a maximum value of 14.14 amps produces the same amount of heat in a circuit having a resistance of 1 ohm as a direct current of 10 amps. For this reason, we can work out a constant value for converting any peak value to a corresponding effective value. In the simple equation below, x represents this constant (solve for x to three decimal places):

$$14.14x = 10$$

$$x = 0.707$$

The effective value is also called the *root-mean-square* (RMS) value because it is the square root of the average of the squared values between zero and maximum. The effective value of an AC current is stated in terms of an equivalent DC current. The phenomenon used for standard comparisons is the heating effect of the current.

Important Point: Anytime an AC voltage or current is stated without any qualifications, it is assumed to be an effective value.

In many instances, it is necessary to convert from effective to peak or *vice versa* using a standard equation. Figure 8.71 shows that the peak value of a sine wave is 1.414 times the effective value; therefore, the equation we use is:

$$E_m = E \times 1.414 \qquad (8.32)$$

where:

E_m = maximum or peak voltage.
E = effective or RMS voltage.

and

$$I_m = I \times 1.414 \qquad (8.33)$$

where:

I_m = maximum or peak current.
I = effective or RMS current.

Occasionally, it is necessary to convert a peak value of current or voltage to an effective value. This is accomplished by using the following equations:

$$E = E_m \times 0.707 \qquad (8.34)$$

where:

E = effective voltage.
E_m = maximum or peak voltage.

$$I = I_m \times 0.707 \qquad (8.35)$$

where:

I = effective current.
I_m = maximum or peak current.

TABLE 8.7
AC Sine Wave Conversion Table

Multiply the Value:	by	To Obtain:
Peak	2	Peak-to-peak
Peak-to-peak	0.5	Peak
Peak	0.637	Average
Average	1.637	Peak
Peak	0.707	RMS (effective)
RMS (effective)	1.414	Peak
Average	1.110	RMS (effective)
RMS (effective)	0.901	Average

8.21.9 AVERAGE VALUE

Because the positive alternation is identical to the negative alternation, the *average value* of a complete cycle of a sine wave is zero. In certain types of circuits however, it is necessary to compute the average value of one alternation. Figure 8.71 shows that the average value of a sine wave is 0.637 times peak value:

$$\text{Average value} = 0.637 \times \text{peak value} \qquad (8.36)$$

or

$$E_{avg} = E_m \times 0.637$$

where:

E_{avg} = average voltage of one alternation.
E_m = maximum or peak voltage.

Similarly,

$$I_{avg} = I_m \times 0.637 \qquad (8.37)$$

where:

I_{avg} = average current in one alternation.
I_m = maximum or peak current.

Table 8.7 lists values of sine wave amplitude used for the conversion of AC sine wave voltage and current.

8.21.10 RESISTANCE IN AC CIRCUITS

If a sine wave of voltage is applied to a resistance, the resulting current will also be a sine wave. This follows Ohm's law, which states that the current is directly proportional to the applied voltage. Figure 8.72 shows a sine wave of voltage and the resulting sine wave of current superimposed on the same time axis. Notice that as the voltage increases in a positive direction the current increases along with it. When the voltage reverses direction, the current reverses direction. At all times, the voltage and current pass through the same relative parts of their respective cycles at the same time. When two waves, such as those shown in Figure 8.72, are precisely in step with one another, they are said to be *in phase*. To be in phase, the two waves reach their maximum and minimum points

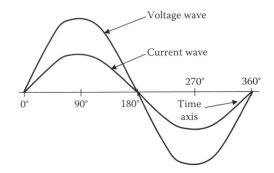

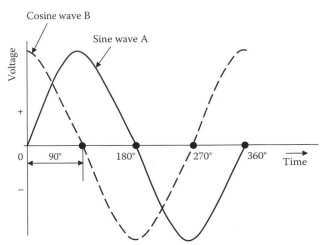

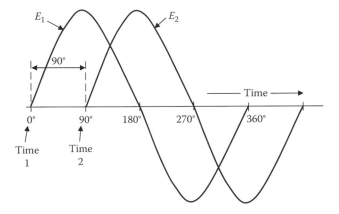

FIGURE 8.72 Voltage and current waves in phase.

FIGURE 8.73 Voltage waves 90° out of phase.

FIGURE 8.74 Wave B leads wave A by a phase angle of 90°.

at the same time and in the same direction. In some circuits, several sine waves can be in phase with each other; thus, it is possible to have two or more voltage drops in phase with each other and in phase with the circuit current.

Note: It is important to remember that Ohm's law for DC circuits is applicable to AC circuits with resistance only.

Voltage waves are not always in phase. Figure 8.73 shows a voltage wave (E_1) considered to start at 0° (time 1). As voltage wave E_1 reaches its positive peak, a second voltage wave (E_2) begins to rise (time 2). Because these waves do not pass through their maximum and minimum points at the same instant of time, a phase difference exists between the two waves. The two waves are said to be *out of phase*. For the two waves in Figure 8.73, this phase difference is 90°.

8.21.11 PHASE RELATIONSHIPS

In the preceding section, we discussed the important concepts of *in phase* and *phase difference*. Another important phase concept is *phase angle*. The phase angle between two waveforms of the same frequency is the angular difference at a given instant of time. As an example, the phase angle between waves B and A (see Figure 8.74) is 90°. Take the instant of time at 90°. The horizontal axis is shown

in angular units of time. Wave B begins at maximum value and reduces to 0 value at 90°, whereas wave A begins at 0 and increases to maximum value at 90°. Wave B reaches its maximum value 90° ahead of wave A, so wave B leads wave A by 90° (and wave A lags wave B by 90°). This 90° phase angle between waves B and A is maintained throughout the complete cycle and all successive cycles. At any instant of time, wave B has the value that wave A will have 90° later. Wave B is a cosine wave because it is displaced 90° from wave A, which is a sine wave.

Important Point: The amount by which one wave leads or lags another is measured in degrees.

To compare phase angles or phases of alternating voltages or currents, it is more convenient to use vector diagrams corresponding to the voltage and current waveforms. A *vector* is a straight line used to denote the magnitude and direction of a given quantity. The length of the line drawn to scale denotes magnitude, and the direction is indicated by the arrow at one end of the line, together with the angle that the vector makes with a horizontal reference vector.

Note: In electricity, because different directions really represent time expressed as a phase relationship, an electrical vector is called a *phasor*. In an AC circuit containing only resistance, the voltage and current occur at the same time, or are in phase. To indicate this condition by means of phasors, all that is necessary is to draw the phasors for the voltage and current in the same direction. The length of the phasor indicates the value of each.

A vector, or phasor, diagram is shown in Figure 8.75, where vector V_B is vertical to show the phase angle of 90° with respect to vector V_A, which is the reference. Because lead angles are shown in the counterclockwise direction from the reference vector, V_B leads V_A by 90°.

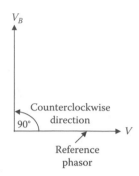

FIGURE 8.75 Phasor diagram.

8.22 INDUCTANCE

To this point, we have learned the following key points about magnetic fields:

- A field of force exists around a wire carrying a current.
- This field has the form of concentric circles around the wire in planes perpendicular to the wire, with the wire at the center of the circles.
- The strength of the field depends on the current. Large currents produce large fields; small currents produce small fields.
- When lines of force cut across a conductor, a voltage is induced in the conductor.

Moreover, to this point we have studied circuits that have been resistive (i.e., resistors presented the only opposition to current flow). Two other phenomena—*inductance* and *capacitance*—exist in DC circuits to some extent, but they are major players in AC circuits. Both inductance and capacitance present a kind of opposition to current flow that is called reactance. (Other than this brief introduction to capacitance and reactance, we do not discuss these two electrical properties in detail in this text; instead, our focus is on the basics, covering only those electrical properties important to water/wastewater operators.)

Inductance is the characteristic of an electrical circuit that makes itself evident by opposing the starting, stopping, or changing of current flow. A simple analogy can be used to explain inductance. We are all familiar with how difficult it is to push a heavy load (such as a cart full of heavy items). It takes more work to start the load moving than it does to keep it moving. This is because the load possesses the property of *inertia*. Inertia is the characteristic of mass that opposes a change in velocity; it can hinder us in some ways and help us in others. Inductance exhibits the same effect on current in an electric circuit as inertia does on velocity of a mechanical object. The effects of inductance are sometimes desirable and sometimes undesirable.

FIGURE 8.76 Schematic symbol for an inductor.

Important Point: Simply put, inductance is the characteristic of an electrical conductor that opposes a change in current flow.

Inductance is the property of an electric circuit that opposes any change in the current passing through that circuit, so, if the current increases, a self-induced voltage opposes this change and delays the increase. On the other hand, if the current decreases, a self-induced voltage tends to aid (or prolong) the current flow, delaying the decrease. Thus, current can neither increase nor decrease as quickly in an inductive circuit as it can in a purely resistive circuit.

In AC circuits, this effect becomes very important because it affects the phase relationships between voltage and current. Earlier, we learned that voltages (or currents) could be out of phase if they are induced in separate armatures of an alternator. In that case, the voltage and current generated by each armature were in phase. When inductance is a factor in a circuit, the voltage and current generated by the same armature are out of phase. We will examine these phase relationships later. Our objective here is to understand the nature and effects of inductance in an electric circuit.

The unit for measuring inductance (*L*) is the *henry* (named for the American physicist Joseph Henry), which is abbreviated as h. Figure 8.76 shows the schematic symbol for an inductor. An inductor has an inductance of 1 henry if an emf of 1 volt is induced in the inductor when the current through the inductor is changing at the rate of 1 ampere per second. The relationships among the induced voltage, inductance, and rate of change of current with respect to time can be stated mathematically as follows:

$$E = L \frac{\Delta I}{\Delta t} \qquad (8.38)$$

where:

    E = the induced emf (volts).
    L = the inductance (henrys).
    Δ*I* = is the change in amperes occurring in Δ*t* seconds.

Note: The symbol Δ (delta) means "change in."

The henry is a large unit of inductance and is used with relatively large inductors. The unit employed with small inductors is the millihenry (mh). For still smaller inductors, the unit of inductance is the microhenry (μh).

8.22.1 Self-Inductance

As previously, explained, current flow in a conductor always produces a magnetic field surrounding, or linking with, the conductor. When the current changes, the magnetic field changes, and an emf is induced in the conductor. This emf is referred to as a *self-induced emf* because it is induced in the conductor carrying the current.

Note: Even a perfectly straight length of conductor has some inductance.

The direction of the induced emf has a definite relation to the direction in which the field that induces the emf varies. When the current in a circuit is increasing, the flux linking with the circuit is increasing. This flux cuts across the conductor and induces an emf in the conductor in such a direction as to oppose the increase in current and flux. This emf is sometimes referred to as *counterelectromotive force* (cemf). The two terms are used synonymously throughout this manual. Likewise, when the current is decreasing, an emf is induced in the opposite direction and opposes the decrease in current.

Important Point: The effects just described are summarized by *Lenz's law*, which states that the induced emf in any circuit is always in a direction opposed to the effect that produced it.

Shaping a conductor so the electromagnetic field around each portion of the conductor cuts across some other portion of the same conductor increases the inductance. This is shown in its simplest form in Figure 8.77A. A loop of conductor is looped so two portions of the conductor lie adjacent and parallel to one another. These portions are labeled conductor 1 and conductor 2. When the switch is closed, electron flow through the conductor establishes a typical concentric field around all portions of the conductor. The field is shown in a single plane (for simplicity) that is perpendicular to both conductors. Although the field originates simultaneously in both conductors, it is considered as originating in conductor 1, and its effect on conductor 2 will be noted. With increasing current, the field expands outward, cutting across a portion of conductor 2. The dashed arrow shows the resultant induced emf in conductor 2. Note that it is in opposition to the battery current and voltage, according to Lenz's law. In Figure 8.77B, the same section of conductor 2 is shown but with the switch open and the flux collapsing.

Important Point: In Figure 8.77, the important point to note is that the voltage of self-induction opposes both changes in current. It delays the initial buildup of current by opposing the battery voltage and delays the breakdown of current by exerting an induced voltage in the same direction in which the battery voltage acted.

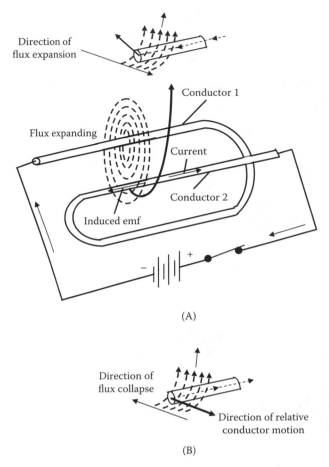

FIGURE 8.77 Self-inductance.

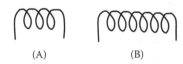

FIGURE 8.78 (A) Few turns, low inductance; (B) more turns, higher inductance.

Four major factors affect the self-inductance of a conductor, or circuit:

1. *Number of turns*—Inductance depends on the number of wire turns. Wind more turns to increase inductance; take turns off to decrease the inductance. Figure 8.78 compares the inductance of two coils made with different numbers of turns.
2. *Spacing between turns*—Inductance depends on the spacing between turns, or the length of the inductor. Figure 8.79 shows two inductors with the same number of turns. The turns of the first inductor have a wide spacing. The turns of the second inductor are close together. The second coil, though shorter, has a larger inductance value because of its close spacing between turns.

FIGURE 8.79 (A) Wide spacing between turns, low inductance; (B) close spacing between turns, higher inductance.

FIGURE 8.80 (A) Small diameter, low inductance; (B) larger diameter, higher inductance.

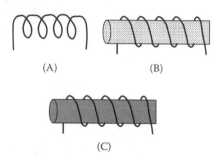

FIGURE 8.81 (A) Air core, low inductance; (B) powdered iron core, higher inductance; (C) soft iron core, highest inductance

3. *Coil diameter*—Coil diameter, or cross-sectional area, is highlighted in Figure 8.80. The larger diameter inductor has more inductance. Both coils shown have the same number of turns, and the spacing between turns is the same. The first inductor has a small diameter, and the second one has a larger diameter. The second inductor has more inductance than the first one.

4. *Type of core material*—Permeability, as pointed out earlier, is a measure of how easily a magnetic field goes through a material. Permeability also tells us how much stronger the magnetic field will be with the material inside the coil.

Figure 8.81 shows three identical coils. One has an air core, one has a powdered-iron core in the center, and the other has a soft-iron core. This figure illustrates the effects of core material on inductance. The inductance of a coil is affected by the magnitude of current when the core is a magnetic material. When the core is air, the inductance is independent of the current.

Key Point: The inductance of a coil increases very rapidly as the number of turns is increased. It also increases as the coil is made shorter, the cross-sectional area is made larger, or the permeability of the core is increased.

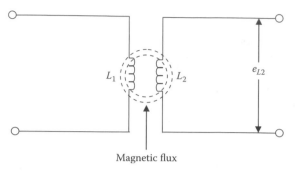

FIGURE 8.82 Mutual inductance between L_1 and L_2.

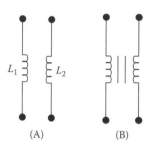

FIGURE 8.83 (A) Schematic symbol for two coils (air core) with mutual inductance; (B) two coils (iron core) with mutual inductance.

8.22.2 MUTUAL INDUCTANCE

When the current in a conductor or coil changes, the varying flux can cut across any other conductor or coil located nearby, thus inducing voltages in both. A varying current in L_1, therefore, induces voltage across L_1 and across L_2 (Figure 8.82; see Figure 8.83 for the schematic symbol for two coils with mutual inductance). When the induced voltage e_{L2} produces current in L_2, its varying magnetic field induces voltage in L_1; hence, the two coils L_1 and L_2 have *mutual inductance* because current change in one coil can induce voltage in the other. The unit of mutual inductance is the henry, and the symbol is L_M. Two coils have L_M of 1 henry when a current change of 1 A/sec in one coil induces 1 E in the other coil.

Factors affecting the mutual inductance of two adjacent coils include:

• Physical dimensions of the two coils
• Number of turns in each coil
• Distance between the two coils
• Relative positions of the axes of the two coils
• Permeability of the cores

Important Point: The amount of mutual inductance depends on the relative position of the two coils. If the coils are separated a considerable distance, the amount of flux common to both coils is small and the mutual inductance is low. Conversely, if the coils are close together so nearly all the flow of one coil links the turns of the other, mutual inductance is high. The mutual inductance can be increased greatly by mounting the coils on a common iron core.

8.22.3 Calculation of Total Inductance

(In the study of advanced electrical theory, it is necessary to know the effect of mutual inductance when solving for total inductance in both series and parallel circuits. For our purposes here, however, we will not attempt to make these calculations; instead, we discuss the basic total inductance calculations with which the maintenance operator should be familiar.)

If inductors in series are located far enough apart, or well shielded to make the effects of mutual inductance negligible, the total inductance is calculated in the same manner as for resistances in series; we merely add them:

$$L_t = L_1 + L_2 + L_3 + \dots \quad (8.39)$$

■ **EXAMPLE 8.34**

Problem: If a series circuit contains three inductors with values of 40 µh, 50 µh, and 20 µh, what is the total inductance?

Solution:

$$L_t = 40\ \mu h + 50\ \mu h + 20\ \mu h = 110\ \mu h$$

In a parallel circuit containing inductors (without mutual inductance), the total inductance is calculated in the same manner as for resistances in parallel:

$$\frac{1}{L_t} = \frac{1}{L_1} + \frac{1}{L_2} + \frac{1}{L_3} + \dots \quad (8.40)$$

■ **EXAMPLE 8.35**

Problem: A circuit contains three totally shielded inductors in parallel. The values of the three inductances are 4 mh, 5 mh, and 10 mh. What is the total inductance?

Solution:

$$\frac{1}{L_t} = \frac{1}{4} + \frac{1}{5} + \frac{1}{10} = 0.25 + 0.2 + 0.1 = 0.55$$

$$L_t = \frac{1}{0.55} = 1.8\ mh$$

8.23 PRACTICAL ELECTRICAL APPLICATIONS

As mentioned, water and wastewater operators normally have little difficulty recognizing electrical equipment within their plant site. This is the case, of course, because there are few places within the plant site (at least in the majority of plant sites) where electricity is not performing some important function. Simply stated, whether this important function is powering lighting, heating, air conditioning, pump motors, mechanized bar screens, control systems, communications equipment, or computerized systems, it would be difficult for the modern operator to imagine plant operations without the use of electrical power.

Up to this point, our goal has been to concentrate on the fundamentals of electricity and electric circuits. Along with satisfying our goal, we also understand that having a basic knowledge of electrical theory is a great accomplishment; however, knowledge of basic theory (of any type) that is not put to practical use is analogous to understanding the operation of an internal combustion engine without ever having the opportunity to work on one.

In short, for the water/wastewater operator, having an understanding of basic electrical fundamentals assists in the successful passing of various certification examinations—no doubt an important career enhancing achievement. Certification, however, is just one critical element required of operators. Operators must also be qualified to operate the plant and its associated machinery—much of which is electrical equipment. To this end, we have incorporated (along with the required theory) pertinent information on electrical applications most important to operators in their daily tasks of operating plant electrical equipment as it was intended to be operated and as it should be operated—that is, operated *with understanding*.

8.23.1 Electrical Power Generation

Most water and wastewater treatment plants do not generate their own general-service electrical power. Instead, and as with most other industrial users, treatment plants typically purchase electrical power from a local electrical utility company, and any on-site generation is provided only for standby power. Additionally, in smaller communities, it is not unusual for a pumping station to be located in residential areas where an associated emergency generator has been installed as an integral part of the station. Portable generators have also seen increased use in recent years. Generators used for standby power will normally be the gasoline, diesel, or gas-turbine types, because these can be started immediately.

Note: Most electrical utility companies operate with reliability in excess of 90%; however, even a momentary loss of electrical power to plant operations cannot, in many instances, be tolerated. For this reason, standby electrical generators are important in maintaining plant operations. One of the main reasons why standby power is installed in plants is to provide emergency lighting for safe egress in

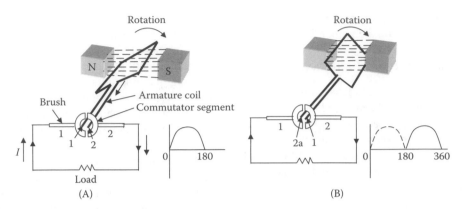

FIGURE 8.84 Basic operation of a DC generator.

the event of a failure of the utility supply. Exit lights, stairway lighting, and a portion of the corridor lighting systems are typically connected to the standby generator.

Fire and safety equipment often is connected to the standby power system to assist in emergency operations. Fire pumps, communications systems, fire detection and alarm systems, and security systems remain operational in the event of a disruption of the normal electrical supply. Also, critical mechanical equipment, such as pumps, is normally connected to the standby power system to protect the facility from damage if the equipment is taken out of service during an electrical outage.

Generators can be designed to supply small amounts of power or many thousands of kilowatts of power. In addition, generators may be designed to supply either direct current or alternating current.

8.23.1.1 DC Generators

A *DC generator* is a rotating machine that converts mechanical energy into electrical energy. This conversion is accomplished by rotating an armature, which carries conductors, in a magnetic field, thus inducing an emf in the conductors. As stated previously, for an emf to be induced in the conductors a relative motion must always exist between the conductors and the magnetic field in such a manner that conductors cut through the field. In most DC generators, the armature is the rotating member and the field is the stationary member. A mechanical force is applied to the shaft of the rotating member to cause the relative motion. Thus, when mechanical energy is put into the machine in the form of a mechanical force or twist on the shaft, causing the shaft to turn at a certain speed, electrical energy in the form of voltage and current is delivered to the external load circuit.

Important Point: Mechanical power must be applied to the shaft constantly as long as the generator is supplying electrical energy to the external load circuit.

To gain a basic understanding of the operation of a DC generator, consider the following explanation. A simple DC generator consists of an armature coil with a single turn of wire (see Figure 8.84). (The armature coils used in large DC machines are usually wound in their final shape before being put on the armature. The sides of the preformed coil are placed in the slots of the laminated armature core.) This armature coil cuts across the magnetic field to produce voltage. If a complete path is present, current will move through the circuit in the direction shown by the arrows (Figure 8.84A). In this position of the coil, commutator segment 1 is in contact with brush 1, while commutator segment 2 is in contact with brush 2. As the armature rotates a half turn in a clockwise direction, the contacts between the commutator segments and the brushes are reversed (Figure 8.84B). At this moment, segment 1 is in contact with brush 2 and segment 2 is in contact with brush 1. Because of this commutator action, that side of the armature coil that is in contact with either of the brushes is always cutting across the magnetic field in the same direction. Thus, brushes 1 and 2 have constant polarity, and a pulsating DC current is delivered to the external load circuit.

Note: In DC generators, voltage induced in individual conductors is AC. It is converted to DC (rectified) by the commutator that rotates in contact with carbon brushes so current generated is in one direction, or direct current.

The various types of DC generators take their names from the type of field excitation used; that is, they are classified according to the manner in which the field windings are connected to the armature circuit. For example, when the field of the generator is excited (or supplied) from a separate DC source (such as a battery) other than its own armature, it is referred to as a *separately excited DC generator* (Figure 8.85). The field windings of a *shunt generator* (self-excited) are connected in series with a rheostat, across the armature in shunt with the load, as

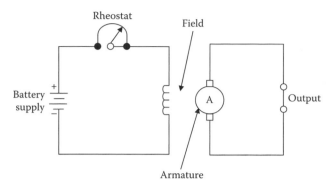

FIGURE 8.85 Separately excited DC generator.

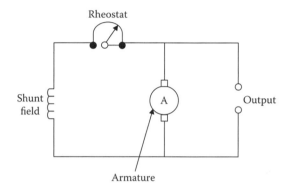

FIGURE 8.86 DC shunt generator.

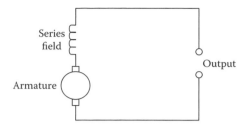

FIGURE 8.87 DC series generator.

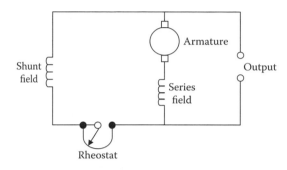

FIGURE 8.88 DC compound generator.

shown in Figure 8.86. The shunt generator is widely used in industry. The field windings of a *series generator* (self-excited) are connected in series with the armature and load, as shown in Figure 8.87. Series generators are seldom used. *Compound generators* (self-excited) contain both series and shunt field windings, as shown in Figure 8.88. Compound generators are widely used in industry.

Note: As central generating stations increased in size along with number and power distribution distances, DC generating systems, because of the high power losses in long DC transmission lines, were replaced by AC generating systems to reduce power transmission costs.

8.23.1.2 AC Generators

Most electric power utilized today is generated by *alternating-current generators* (also called *alternators*). They are made in many different sizes, depending on their intended use. Regardless of size, however, all generators operate on the same basic principle—a magnetic field cutting through conductors or conductors passing through a magnetic field. In one type of conductor, the output voltage is generated; in a second type of conductor, direct current is passed through to obtain an electromagnetic field of fixed polarity. The conductors in which the electromagnetic field originates are always referred to as the *field windings*. In addition to the armature and field, there must also be motion between the two. To provide this, AC generators are built with two major assemblies: the *stator* and the *rotor*. The rotor rotates inside the stator.

The revolving-field AC generator (see Figure 8.89) is the most widely used type. In this type of generator, direct current from a separate source is passed through windings on the rotor by means of sliprings and brushes. (Sliprings and brushes are adequate for the DC field supply because the power level in the field is much smaller than in the armature circuit.) This maintains a rotating electromagnetic field of fixed polarity. The rotating magnetic field, following the rotor, extends outward and cuts through the armature windings imbedded in the surrounding stator. As the rotor turns, AC voltages are induced in the windings because magnetic fields of first one polarity and then the other cut through them. Because the output power is taken from stationary windings, the output may be connected through fixed output terminals (T1 and T2 in Figure 8.89). This is advantageous in that there are no sliding contacts, and the entire output circuit is continuously insulated.

Important Point: In AC generators, frequency and electromagnetic wave cycles per second depend on how fast the rotor turns and the number of electromagnetic field poles. The voltage generated depends on the rotor speed, number of coils in the armature, and strength of the magnetic field.

8.23.1.3 Motors

At least 60% of the electrical power fed to a typical waterworks or wastewater treatment plant is consumed by *electric motors*. One thing is certain: Electric motors perform

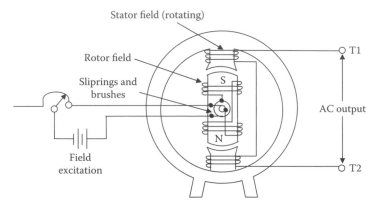

FIGURE 8.89 Essential parts of a rotating field AC generator.

an almost endless variety of tasks in water and wastewater treatment. An electric motor changes electrical energy to mechanical energy to do the work. (Recall that a generator does just the opposite; that is, a generator changes mechanical energy to electrical energy.)

Previously, we pointed out that when a current passes through a wire a magnetic field is produced around the wire. If this magnetic field passes through a stationary magnetic field, the fields either repel or attract, depending on their relative polarity. If both are positive or negative, they repel. If they are opposite polarity, they attract. Applying this basic information to motor design, an electromagnetic coil, the *armature*, rotates on a *shaft*. The armature and shaft assembly is called the *rotor*. The rotor is assembled between the poles of a permanent magnet, and each end of the rotor coil (armature) is connected to a *commutator* also mounted on the shaft. A commutator is composed of copper segments insulated from the shaft and from each other by an insulting material. As like poles of the electromagnet in the rotating armature pass the stationary permanent magnet poles, they are repelled, continuing the motion. As the opposite poles near each other, they attract, continuing the motion.

8.23.1.4 DC Motors

The construction of a DC motor is essentially the same as that of a DC generator; however, it is important to remember that the DC generator converts mechanical energy into electrical energy and back into mechanical energy. A DC generator may be made to function as a motor by applying a suitable source of DC voltage across the normal output electrical terminals. DC motors vary, depending on the way in which the field coils are connected. Each has characteristics that are advantageous under given load conditions.

The field coils of *shunt motors* (Figure 8.90) are connected in parallel with the armature circuit. This type of motor, with constant potential applied, develops variable torque at an essentially constant speed, even under chang-

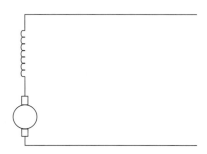

FIGURE 8.90 DC shunt motor.

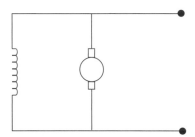

FIGURE 8.91 DC series motor.

ing load conditions. Such loads are found in machine-shop equipment such as lathes, shapes, drills, milling machines, and so forth.

The field coils of *series motors* (Figure 8.91) are connected in series with the armature circuit. This type of motor, with constant potential applied, develops variable torque, but its speed varies widely under changing load conditions. The speed is low under heavy loads but becomes excessively high under light loads. Series motors are commonly used to drive electric hoists, winches, cranes, and certain types of vehicles (e.g., electric trucks). In addition, series motors are used extensively to start internal combustion engines.

Compound motors (see Figure 8.92) have one set of field coils in parallel with the armature circuit and another set of field coils in series with the armature circuit. This type of motor is a compromise between shunt and series motors. It develops an increased starting torque over that

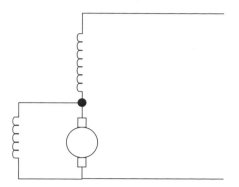

FIGURE 8.92 DC compound motor.

of the shunt motor and has less variation in speed than the series motor.

The speed of a DC motor is variable. It is increased or decreased by a rheostat connected in series with the field or in parallel with the rotor. Interchanging either the rotor or field winding connections reverses direction.

8.23.1.5 AC Motors

Alternating-current voltage can be easily transformed from low voltages to high voltages or *vice versa* and can be moved over a much greater distance without too much loss in efficiency. Most of the power generating systems today, therefore, produce alternating current; thus, it logically follows that a great majority of the electrical motors utilized today are designed to operate on alternating current. AC motors, however, offer several other advantages in addition to the wide availability of AC power. In general, AC motors are less expensive than DC motors. Most types of AC motors do not employ brushes and commutators. This eliminates many problems of maintenance and wear and eliminates dangerous sparking. AC motors are manufactured in many different sizes, shapes, and ratings for use on a great number of jobs. They are designed for use with either polyphase or single-phase power systems. This chapter cannot possibly cover all aspects of AC motors; consequently, it deals primarily with the operating principles of the two most common types—the *induction motor* and the *synchronous motor*.

8.23.1.5.1 Induction Motors

The induction motor is the most commonly used type of AC motor because of its simple, rugged construction and good operating characteristics. It consists of two parts: the *stator* (stationary part) and the *rotor* (rotating part). The most important type of polyphase induction motor is the *three-phase motor*.

Important Note: A three-phase (3-θ) system is a combination of three single-phase (1-θ) systems. In a 3-θ balanced system, the power comes from an AC generator that produces three separate but equal voltages, each of which is out of phase with the other voltages by 120°. Although 1-θ circuits are widely used in electrical systems, most generation and distribution of AC current is 3-θ.

The driving torque of both DC and AC motors is derived from the reaction of current-carrying conductors in a magnetic field. In the DC motor, the magnetic field is stationary and the armature, with its current-carrying conductors, rotates. The current is supplied to the armature through a commutator and brushes. In induction motors, the rotor currents are supplied by electromagnet induction. The stator windings, connected to the AC supply, contain two or more out-of-time-phase currents, which produce corresponding mmfs. These mmfs establish a rotating magnetic field across the air gap. This magnetic field rotates continuously at constant speed regardless of the load on the motor. The stator winding corresponds to the armature winding of a DC motor or to the primary winding of a transformer. The rotor is not connected electrically to the power supply.

The induction motor derives its name from the fact that mutual induction (or transformer action) takes place between the stator and the rotor under operating conditions. The magnetic revolving field produced by the stator cuts across the rotor conductors, inducing a voltage in the conductors. This induced voltage causes rotor current to flow; hence, motor torque is developed by the interaction of the rotor current and the magnetic revolving field.

8.23.1.5.2 Synchronous Motors

Like induction motors, *synchronous motors* have stator windings that produce a rotating magnetic field. Unlike the induction motor, however, the synchronous motor requires a separate source of DC from the field. It also requires special starting components. These include a salient-pole field with starting grid winding. The rotor of the conventional type synchronous motor is essentially the same as that of the salient-pole AC generator. The stator windings of induction and synchronous motors are essentially the same.

In operation, the synchronous motor rotor locks into step with the rotating magnetic field and rotates at the same speed. If the rotor is pulled out of step with the rotating stator field, no torque is developed and the motor stops. Because a synchronous motor develops torque only when running at synchronous speed, it is not self-starting and hence requires some device to bring the rotor to synchronous speed; for example, a synchronous motor may be started rotating with a DC motor on a common shaft. After the motor is brought to synchronous speed, AC current is applied to the stator windings. The DC starting motor now acts as a DC generator, which supplies DC field excitation for the rotor. The load then can be coupled to the motor.

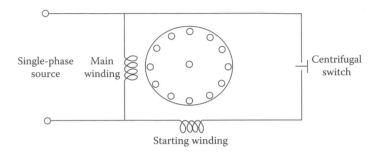

FIGURE 8.93 Split-phase motor.

8.23.1.5.3 Single-Phase Motors

Single-phase ($1\text{-}\theta$) motors are so called because their field windings are connected directly to a single-phase source. These motors are used extensively in fractional horsepower sizes in commercial and domestic applications. The advantages of using single-phase motors are that they are less expensive to manufacture than other types, and they eliminate the need for three-phase AC lines. Single-phase motors are used in fans, refrigerators, portable drills, grinders, and so forth.

A single-phase induction motor with only one stator winding and a cage rotor is like a three-phase induction motor with a cage rotor except that the single-phase motor has no magnetic revolving field at start and hence no starting torque. If the rotor is brought up to speed by external means, however, the induced currents in the rotor will cooperate with the stator currents to produce a revolving field, which causes the rotor to continue to run in the direction in which it was started.

Several methods are used to provide the single-phase induction motor with starting torque. These methods identify the motor as a *split-phase*, *capacitor*, *shaded-pole*, or *repulsion-start* induction motor. Another class of single-phase motors is the *AC series* (universal) type. Only the more commonly used types of single-phase motors are described here.

8.23.1.5.3.1 Split-Phase Motors

The split-phase motor (see Figure 8.93), has a stator composed of slotted lamination that contain a starting winding and a running winding.

Note: If two stator windings of unequal impedance are spaced 90 electrical degrees apart but connected in parallel to a single-phase source, the field produced will appear to rotate. This is the principle of *phase splitting*.

The starting winding has fewer turns and utilizes smaller wire than the running winding; hence, it offers higher resistance and less reactance. The main winding occupies the lower half of the slots, and the starting winding occupies the upper half. When the same voltage is applied to both windings, the current in the main winding lags behind the current in the starting winding. The angle θ between the main and starting windings is enough phase difference to provide a weak rotating magnetic field to produce a starting torque. When the motor reaches a predetermined speed, usually 75% of synchronous speed, a centrifugal switch mounted on the motor shaft opens, thereby disconnecting the starting winding. Because of the low starting torque, fractional-horsepower, split-phase motors are used in a variety of equipment such as washers, oil burners, ventilating fans, and woodworking machines. Interchanging the starting winding leads can reverse the direction of rotation of the split-phase motor.

8.23.1.5.3.2 Capacitor Motors

The capacitor motor is a modified form of split-phase motor having a capacitor in series with the starting winding. The capacitor motor operates with an auxiliary winding and series capacitor permanently connected to the line (see Figure 8.94). The capacitance in series may be of one value for starting and another value for running. As the

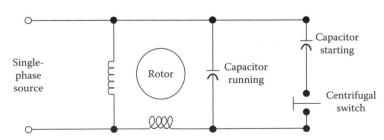

FIGURE 8.94 Capacitor motor.

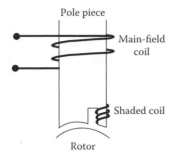

FIGURE 8.95 Shaded pole.

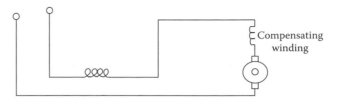

FIGURE 8.96 AC series motor.

motor approaches synchronous speed, the centrifugal switch disconnects one section of the capacitor. If the starting winding is cut out after the motor has increased in speed, the motor is referred to as a *capacitor-start motor*. If the starting winding and capacitor are designed to be left in the circuit continuously, the motor is a *capacitor-run motor*. Capacitor motors are used to drive grinders, drill presses, refrigerator compressors, and other loads that require relatively high starting torque. Interchanging the starting winding leads may reverse the direction of rotation of the capacitor motor.

8.23.1.5.3.3   Shaded-Pole Motor

A shaded-pole motor employs a salient-pole stator and a cage rotor. The projecting poles on the stator resemble those of DC machines except that the entire magnetic circuit is laminated and a portion of each pole is split to accommodate a short-circuited coil called a *shading coil* (see Figure 8.95). The coil is usually a single band or strap of copper. The effect of the coil is to produce a small sweeping motion of the field flux from one side of the pole piece to the other as the field pulsates. This slight shift in the magnetic field produces a small starting torque; thus, shaded-pole motors are self-starting. This motor is generally manufactured in very small sizes, up to 1/20 horsepower, for driving small fans, small appliances, and clocks.

In operation, during that part of the cycle when the main pole flux is increasing, the shading coil is cut by the flux, and the resulting induced emf and current in the shading coil tend to prevent the flux from rising readily through it. Thus, the greater portion of the flux rises in that portion of the pole that is not near the shading coil. When the flux reaches its maximum value, the rate of change of flux is zero, and the voltage and current in the shading coil are zero. At this time, the flux is distributed more uniformly over the entire pole face. Then as the main flux decreases toward zero, the induced voltage and current in the shading coil reverse their polarity, and the resulting mmf tends to prevent the flux from collapsing through the iron in the region of the shading coil. The result is that the main flux first rises in the unshaded portion of the pole and later in the shaded portion. This action is equivalent to a sweeping movement of the field across the pole face in the direction of the shaded pole. This moving field cuts the rotor conductors, and the force exerted on them causes the rotor to turn in the direction of the sweeping field. The shaded-pole method of starting is used in very small motors, up to about 1/25 horsepower, for driving small fans, small appliances, and clocks.

8.23.1.5.3.4   Repulsion-Start Motor

Like a DC motor, the *repulsion-start motor* has a form-wound rotor with commutator and brushes. The stator is laminated and contains a distributed single-phase winding. In its simplest form, the stator resembles that of the single-phase motor. In addition, the motor has a centrifugal device, which removes the brushes from the commutator and places a short-circuiting ring around the commutator. This action occurs at about 75% of synchronous speed; thereafter, the motor operates with the characteristics of the single-phase induction motor. This type of motor is made in sizes ranging from 1/2 to 15 horsepower and is used in applications requiring a high starting torque.

8.23.1.5.3.5   Series Motor

The AC series motor operates on either AC or DC circuits. When an ordinary DC series motor is connected to an AC supply, the current drawn by the motor is low due to the high series-field impedance. The result is low running torque. To reduce the field reactance to a minimum, AC series motors are built with as few turns as possible. Armature reaction is overcome by using compensating windings (see Figure 8.96) in the pole pieces. As with DC series motors, in an AC series motor the speed increases to a high value with a decrease in load. The torque is high for high armature currents so the motor has a good starting torque. AC series motors operate more efficiently at low frequencies. Fractional-horsepower AC series motors are called *universal motors*. They do not have compensating windings. They are used extensively to operate fans and portable tools, such as drills, grinders, and saws.

8.23.2   Transformers

A *transformer* is an electric control device (with no moving parts) that raises or lowers voltage or current in an electric distribution system. The basic transformer consists of two

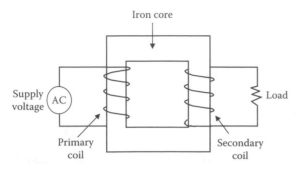

FIGURE 8.97 Basic transformer.

coils electrically insulated from each other and wound upon a common core (see Figure 8.97). Magnetic coupling is used to transfer electric energy from one coil to another. The coil that receives energy from an AC source is called the *primary*. The coil that delivers energy to an AC load is the *secondary*. The core of transformers used at low frequencies is generally made of magnetic material, usually laminated sheet steel. Cores of transformers used at higher frequencies are made of powdered iron and ceramics or nonmagnetic materials. Some coils are simply wound on nonmagnetic hollow forms such as cardboard or plastic so the core material is actually air.

In operation, an alternating current will flow when an AC voltage is applied to the primary coil of a transformer. This current produces a field of force that changes as the current changes. The changing magnetic field is carried by the magnetic core to the secondary coil, where it cuts across the turns of that coil. In this way, an AC voltage in one coil is transferred to another coil, even though there is no electrical connection between them. The primary voltage and the number of turns on the primary determine the number of lines of force available in the primary—each turn producing a given number of lines. If there are many turns on the secondary, each line of force will cut many turns of wire and induce a high voltage. If the secondary contains only a few turns, there will be few cuttings and low induced voltage. The secondary voltage, then, depends on the number of secondary turns as compared with the number of primary turns. If the secondary has twice as many turns as the primary, the secondary voltage will be twice as large as the primary voltage. If the secondary has half as many turns as the primary, the secondary voltage will be one half as large as the primary voltage.

Important Point: The voltage on the coils of a transformer is directly proportional to the number of turns on the coils.

A voltage ratio of 1:4 means that for each volt on the primary there are 4 volts on the secondary. This is called a *step-up transformer*. A step-up transformer receives a low voltage on the primary and delivers a high voltage from the secondary. A voltage ratio of 4:1 means that for 4 volts on the primary there is only 1 volt on the secondary. This is called a *step-down transformer*. A step-down transformer receives a high voltage on the primary and delivers a low voltage from the secondary.

8.23.3 Power Distribution System Protection

Interruptions are very rare in a power distribution system that has been properly designed. Still, protective devices are necessary because of the load diversity. Most installations are quite complex. In addition, externally caused variations might overload them or endanger personnel. Figure 8.98 shows the general relationship between protective devices and various components of a complete system. Each part of the circuit has its own protective device or devices that protect not only the load but also the wiring and control devices themselves. These disconnect and protective devices are described in the following sections.

8.23.3.1 Fuses

The passage of an electric current produces heat. The larger the current, the more heat is produced. To prevent large currents from accidentally flowing through expensive apparatus and burning it up, a *fuse* is placed directly into the circuit, as shown in Figure 8.98, and all current must flow through the fuse.

Key Point: A fuse is a thin strip of easily melted material. It protects a circuit from large currents by melting quickly, thereby breaking the circuit.

The fuse will permit currents smaller than the fuse value to flow but will melt and therefore break the circuit if a larger, dangerous current ever appears; for example, a dangerously large current will flow when a short circuit occurs. A short circuit is usually caused by an accidental connection between two points in a circuit offering very little resistance to the flow of electrons. If the resistance is small, nothing will stop the flow of the current, and the current will increase enormously. The resulting heat generated might cause a fire. If the circuit is protected by a fuse, however, the heat caused by the short-circuit current will melt the fuse wire, thus breaking the circuit and reducing the current to zero.

The number of amps of current that can flow through a fuse before it melts and breaks the circuit determines the rating of a fuse; for example, we have 10-, 15-, 20-, and 30-amp fuses. We must be sure that any fuse inserted in a circuit is rated low enough to melt, or blow, before the apparatus is damaged. In a plant building wired to carry a current of 10 amps, for example, it is best to use a fuse no larger than 10 amps so a current larger than 10 amps can never flow.

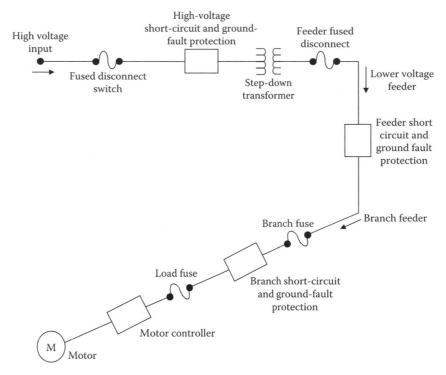

FIGURE 8.98 Motor power distribution system.

Some equipment, such as an electric motor, requires more current during starting than for normal running; in such cases, fast-time or medium-time fuses that provide running protection might blow during the initial period when high starting current is required. *Delayed-action fuses* are used to handle these situations.

8.23.3.2 Circuit Breakers

Circuit breakers are protective devices that open automatically at a preset ampere rating to interrupt an overload or short circuit. Unlike fuses, they do not require replacement when they are activated. They are simply reset to restore power after the overload has been cleared.

Key Point: A circuit breaker is designed to break the circuit and stop the current flow when the current exceeds a predetermined value.

Circuit breakers are made in both plug-in and bolt-on designs. Plug-in breakers are used in load centers. Bolt-on breakers are used in panelboards and exclusively for high interrupting current applications.

Circuit breakers are rated according to current and voltage, as well as short-circuit interrupting current. A single handle opens or closes contacts between two or more conductors. Breakers are single pole but can be ganged single-pole units to form double- or triple-pole devices opened with a single handle.

Several types of circuit breakers are commonly used. They may be thermal or magnetic, or a combination of the two. Thermal breakers are tripped when the temperature rises because of heat created by the overcurrent condition. Bimetallic strips provide the time delay for overload protection. Magnetic breakers operate on the principle that a sudden current rise creates enough magnetic field to turn an armature, tripping the breaker and opening the circuit. Magnetic breakers provide the instantaneous action required for short-circuit protection. They are also used in circumstances where ambient temperature might adversely affect the action of a thermal breaker. Thermal–magnetic breakers combine features of both types of breakers.

An important feature of the circuit breaker is its *arc chutes*, which enable the breaker to extinguish very hot arcs harmlessly. Some circuit breakers must be reset by hand, but others reset themselves automatically. If the overload condition still exists when the circuit breaker is reset, the circuit breaker will trip again to prevent damage to the circuit.

8.23.3.3 Control Devices

Control devices are electrical accessories (switches and relays) that govern the power delivered to any electrical load. In its simplest form, the control applies voltage to or removes it from a single load. In more complex control systems, the initial switch may set into action other control devices (relays) that govern motor speeds, servomechanisms, temperatures, and numerous other types of equipment. In fact, all electrical systems and equipment are controlled in some manner by one or more controls. A controller is a device or group of devices that serves to govern, in some predetermined manner, the device to which it is connected. In large electrical systems, it is necessary to have a variety of controls for operation of the equipment. These controls range from simple pushbuttons to heavy-duty contactors that are designed to control the operation of large motors. The pushbutton is manually operated and a contactor is electrically operated.

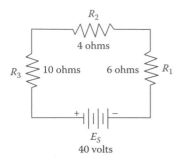

FIGURE 8.99 Illustration for Questions 8.19 through 8.23.

CHAPTER REVIEW QUESTIONS

8.1 What is another name for an AC generator?
8.2 What is electromagnetic induction?
8.3 An alternator converts _____ energy into _____ energy.
8.4 A step-up transformer _____ the voltage and _____ the current. A step-down transformer _____ the voltage and _____ the current.
8.5 What is the purpose of a fuse?
8.6 An electrical circuit with a conductance of 5 mho would have a resistance of _____.
8.7 Electrons move about the nucleus of an atom in paths that are called _____.
8.8 The nucleus of an atom consists of particles called _____ and _____.
8.9 What three factors affect the resistance in a circuit?
8.10 What is the difference between direct and alternating current?
8.11 What are the points of maximum attraction on a magnet?
8.12 What is a magnetic field?
8.13 What are the three general groups of magnets?
8.14 What method of producing a voltage is used in batteries?
8.15 A _____ consists of _____ or more cells connected in series or parallel.
8.16 Explain the difference between a series and a parallel circuit.
8.17 The sum of all voltages in a series circuit is equal to _____.
8.18 For any total voltage rise in a circuit, there must be an equal total _____.

Refer to Figure 8.99 for Questions 8.19 through 8.23

8.19 Is the direction of current flow clockwise or counterclockwise?
8.20 $I_t = ?$
8.21 E dropped across $R_1 = ?$
8.22 What is the power absorbed by R_2?

8.23 $P_T = ?$
8.24 The equivalent resistance (R_T) of parallel branches is _____ than the smallest branch resistance since all the branches must take _____ current from the source than any one branch.
8.25 The resistance in ohms of a unit conductor or a given substance is called the _____.
8.26 The _____ is the standard unit of wire cross-sectional area used in most wire tables.
8.27 Which is smaller, circular mil or square mil?
8.28 Resistivity is the reciprocal of _____.
8.29 No. 8 wire is _____ than No. 4 wire.
8.30 For every three gauge sizes, the circular area of a wire _____.
8.31 What is the biggest advantage of using plastic insulation tape over other types of tape?
8.32 What is the relationship between flux density and magnetic field strength?
8.33 Permeability depends on what two factors?
8.34 Lines of force flow from _____.
8.35 If a conductor is rotated faster in the magnetic field of an alternator, what will happen to the frequency of the voltage?
8.36 If the peak AC voltage across a resistor is 200 volts, what is the RMS voltage?
8.37 What is electromagnetic induction?
8.38 When an AC voltage is impressed across a coil, the resulting current is an _____ current. This changing current produces changing fields of force that _____ the wires of the coil. These cuttings induce a _____ emf in the coil.
8.39 The inductance of a coil is a measure of its ability to produce a _____ emf when the current through it is changing.
8.40 What characteristics of electric current results in self-inductance?
8.41 An increase in the cross-sectional area of a coil will _____ inductance.
8.42 An increase in the permeability of a coil will _____ inductance.

REFERENCES AND SUGGESTED READING

Kirschner, M.W., Marincola, E., and Olmsted Teisberg, E. 1994. The role of biomedical research in health care reform. *Science*, 266(5182), 49–51.

Spellman, F.R. and Drinan, J. 2001. *Electricity*. Boca Raton, FL: CRC Press.

9 Hydraulic Machines: Pumps

A hydraulic machine, or pump, is a device that raises, compresses, or transfers fluids.

9.1 INTRODUCTION

Garay (1990) pointed out that "few engineered artifacts are as essential as pumps in the development of the culture which our Western civilization enjoys." This statement is germane to any discussion about pumps simply because humans have always needed to move water from one place to another against the forces of nature. As the demand for potable water increases, the need to pump water from distant locations to where it is most needed is also growing.

Initially, humans relied on one of the primary forces of nature—gravity—to move water from one place to another. Gravity only works, of course, if the water is moving downhill on a sloping grade. It was soon discovered that the pressure built up by accumulating water behind the water source (e.g., behind a barricade, levy, or dam) moved the water farther. But, when pressure is dissipated by various losses (e.g., friction loss) or when water in low-lying areas must be moved to higher areas, the energy required to move that water must be created. Simply, some type of pump is needed.

In 287 B.C., Archimedes, a Greek mathematician and physicist, invented the screw pump (see Figure 9.1). It is believed that around 100 A.D. the Roman emperor Nero developed the piston pump. The piston pump displaces volume after volume of water with each stroke. The piston pump has two basic problems: (1) its size limits capacity, and (2) its energy consumption is high. It was not until the 19th century that pumping technology took a leap forward from its rudimentary beginnings. The first fully functional centrifugal pumps were developed in the 1800s. Centrifugal pumps can move great quantities of water with much smaller units compared to earlier versions of pumps.

The pump is a type of hydraulic machine. Pumps convert mechanical energy into fluid energy. Whether water is being moved from groundwater or a surface water body, from one unit treatment process to another, or to a storage tank for eventual final delivery through various sizes and types of pipes to the customer, pumps are the usual source of energy necessary for the conveyance of water. Again, the only exception may be, of course, where the source of energy is supplied entirely by gravity. Waterworks and wastewater maintenance operators must therefore be familiar with pumps, pump characteristics, pump operation, and maintenance.

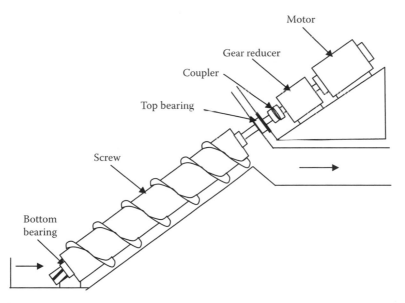

FIGURE 9.1 Archimedes's screw pump.

There are three general requirements of pump and motor combinations. These requirements are (1) reliability, (2) adequacy, and (3) economy. Reliability is generally obtained by installing in duplicate the very best equipment available and by the use of an auxiliary power source. Adequacy is obtained by securing liberal sizes of pumping equipment. Economics can be achieved by taking into account the life and depreciation, first cost, standby charges, interest and operating costs.

Texas Utilities Association (1988)

Over the past several years, it has become more evident that many waterworks and wastewater facilities have been unable to meet their optimum supply or treatment requirements for one of three reasons:

1. Untrained operations and maintenance staff
2. Poor plant maintenance
3. Improper plant design

9.2 BASIC PUMPING CALCULATIONS

Calculations, calculations, calculations, and more calculations! Indeed, we cannot get away from them—not in water/wastewater treatment, collection and distribution operations, licensure certification examinations, nor in real life. Basic calculations are a fact of life that the water/wastewater maintenance operator must accept and should learn well enough to use as required to operate a water/wastewater facility correctly. The following sections address the basic calculations used frequently in water hydraulic and pumping applications. The basic calculations that water and wastewater maintenance operators may be required to know for operational and certification purposes are also discussed. In addition, calculations for pump specific speed, suction specific speed, and affinity, among other advanced calculations, are also covered in this section, although at a higher technical level.

9.2.1 Velocity of a Fluid through a Pipeline

The speed or velocity of a fluid flowing through a channel or pipeline is related to the cross-sectional area of the pipeline and the quantity of water moving through the line; for example, if the diameter of a pipeline is reduced, then the velocity of the water in the line must increase to allow the same amount of water to pass through the line.

$$\text{Velocity } (V) \text{ (fps)} = \frac{\text{Flow } (Q) \text{ (cfs)}}{\text{Cross-sectionial area } (A) \text{ (ft}^2)} \quad (9.1)$$

■ EXAMPLE 9.1

Problem: If the flow through a 2-ft-diameter pipe is 9 MGD, the velocity is:

$$\text{Velocity } (V) \text{ (fps)} = \frac{9 \text{ MGD} \times 1.55 \text{ cfs/MGD}}{0.785 \times 2 \text{ ft} \times 2 \text{ ft}}$$

$$= \frac{14 \text{ cfs}}{3.14 \text{ ft}^2} = 4.5 \text{ fps}$$

■ EXAMPLE 9.2

Problem: If the same 9-MGD flow used in Example 9.1 is transferred to a pipe with a 1-ft diameter, what would the velocity be?

Solution:

$$\text{Velocity } (V) \text{ (fps)} = \frac{9 \text{ MGD} \times 1.55 \text{ cfs/MGD}}{0.785 \times 1 \text{ ft} \times 1 \text{ ft}}$$

$$= \frac{14 \text{ cfs}}{0.785 \text{ ft}^2} = 17.8 \text{ fps}$$

Based on these sample problems, you can see that if the cross-sectional area is decreased, the velocity of the flow must be increased. Mathematically, we can say that the velocity and cross-sectional area are inversely proportional when the amount of flow (Q) is constant:

$$\text{Area}_1 \times \text{Velocity}_1 = \text{Area}_2 \times \text{Velocity}_2 \quad (9.2)$$

Note: The concept just explained is extremely important in the operation of a centrifugal pump and will be discussed later.

9.2.2 Pressure–Velocity Relationship

A relationship similar to that of velocity and cross-sectional area exists for velocity and pressure. As the velocity of flow in a full pipe increases, the pressure of the liquid decreases. This relationship is:

$$\text{Pressure}_1 \times \text{Velocity}_1 = \text{Pressure}_2 \times \text{Velocity}_2 \quad (9.3)$$

■ EXAMPLE 9.3

Problem: If the flow in a pipe has a velocity of 3 fps and a pressure of 4 psi and the velocity of the flow increases to 4 fps, what will the pressure be?

Solution:

$$\text{Pressure}_1 \times \text{Velocity}_1 = \text{Pressure}_2 \times \text{Velocity}_2$$

$$4 \text{ psi} \times 3 \text{ fps} = \text{Pressure}_2 \times 4 \text{ fps}$$

Rearranging:

$$\text{Pressure}_2 = \frac{4 \text{ psi} \times 3 \text{ fps}}{4 \text{ fps}} = \frac{12 \text{ psi}}{4} = 3 \text{ psi}$$

Again, this is another important hydraulics principle that is very important to the operation of a centrifugal pump.

9.2.3 STATIC HEAD

Pressure at a given point originates from the height, or depth of water above it. It is this pressure, or *head*, which gives the water energy, and causes it to flow. By definition, *static head* is the vertical distance the liquid travels from the supply tank to the discharge point. This relationship is shown as:

$$\text{Static head (ft)} = \begin{bmatrix} \text{Discharge level (ft)} \\ -\text{supply level (ft)} \end{bmatrix} \quad (9.4)$$

In many cases, it is desirable to separate the static head into two separate parts: (1) the portion that occurs before the pump (suction head or suction lift), and (2) the portion that occurs after the pump (discharge head). When this is done, the center (or datum) of the pump becomes the reference point.

9.2.3.1 Static Suction Head

Static suction head refers to when the supply is located above the pump datum:

$$\begin{matrix} \text{Total velocity} \\ \text{head (ft)} \end{matrix} = \begin{bmatrix} \text{Velocity head discharge (} \\ \text{velocity head suction (ff} \end{bmatrix} \quad (9.5)$$

9.2.3.2 Static Suction Lift

Static suction lift refers to when the supply is located below the pump datum:

$$\text{Static suction lift (ft)} = \begin{bmatrix} \text{Pump level (ft)} \\ -\text{supply level (ft)} \end{bmatrix} \quad (9.6)$$

9.2.3.3 Static Discharge Head

$$\text{Static discharge head (ft)} = \begin{bmatrix} \text{Discharge level (ft)} \\ -\text{pump level (ft)} \end{bmatrix} \quad (9.7)$$

If the total static head is to be determined after calculating the static suction head or lift and static discharge head individually, two different calculations can be used, depending on suction head or suction lift.

For suction head:

$$\text{Total static head (ft)} = \begin{bmatrix} \text{Static discharge head (ft)} \\ -\text{static suction lift (ft)} \end{bmatrix} \quad (9.8)$$

For suction lift:

$$\text{Total static head (ft)} = \begin{bmatrix} \text{Static discharge head (ft)} \\ +\text{static suction lift (ft)} \end{bmatrix} \quad (9.9)$$

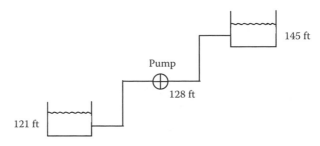

FIGURE 9.2 Illustration for Example 9.4.

■ EXAMPLE 9.4

Problem: Refer to Figure 9.2.

Solution:

$$\text{Static suction lift (ft)} = \text{Pump level (ft)} - \text{supply level (ft)}$$

$$= 128 \text{ ft} - 121 \text{ ft} = 7 \text{ ft}$$

$$\text{Total static head (ft)} = \begin{bmatrix} \text{Static discharge head (ft)} \\ +\text{static suction lift (ft)} \end{bmatrix}$$

$$= 17 \text{ ft} + 7 \text{ ft} = 24 \text{ ft}$$

or

$$\text{Static head (ft)} = \text{Discharge level (ft)} - \text{supply level (ft)}$$

$$= 145 \text{ ft} - 119 \text{ ft}$$

$$= 24 \text{ ft}$$

■ EXAMPLE 9.5

Problem: See Figure 9.3.

Solution:

$$\text{Static suction head (ft)} = \text{Supply level (ft)} - \text{pump level (ft)}$$

$$= 124 \text{ ft} - 117 \text{ ft}$$

$$= 7 \text{ ft}$$

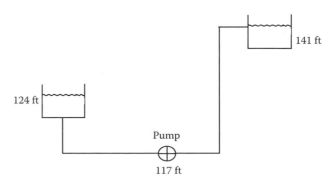

FIGURE 9.3 Illustration for Example 9.5.

$$\text{Static discharge head (ft)} = \begin{bmatrix} \text{Discharge level (ft)} \\ -\text{pump level (ft)} \end{bmatrix}$$

$$= 141\ \text{ft} - 117\ \text{ft} = 24\ \text{ft}$$

$$\text{Total static head (ft)} = \begin{bmatrix} \text{Static discharge head (ft)} \\ -\text{static suction head (ft)} \end{bmatrix}$$

$$= 24\ \text{ft} - 7\ \text{ft} = 17\ \text{ft}$$

or

$$\text{Total static head (ft)} = \begin{bmatrix} \text{Discharge level (ft)} \\ -\text{supply level (ft)} \end{bmatrix}$$

$$= 141\ \text{ft} - 124\ \text{ft} = 17\ \text{ft}$$

9.2.4 Friction Head

Various formulae calculate friction losses. Hazen–Williams wrote one of the most common for smooth steel pipe. Usually we do not need to calculate the friction losses, because handbooks such as the *Hydraulic Institute Pipe Friction Manual* tabulated these long ago. This important manual also shows velocities in different pipe diameters at varying flows, as well as the resistance coefficient (K) for valves and fittings (Wahren, 1997). *Friction head* (in feet) is the amount of energy used to overcome resistance to the flow of liquids through a system. It is affected by the length and diameter of the pipe, the roughness of the pipe, and the velocity head. It is also affected by the physical construction of the piping system. The number of and types of elbows, valves, T's, etc. will greatly influence the friction head for the system. These must be converted to their equivalent length of pipe and included in the calculation.

$$\text{Friction head (ft)} = \begin{bmatrix} \text{Roughness factor } (f) \times \dfrac{\text{length}}{\text{diameter}} \\ \times \dfrac{\text{velocity}^2}{2g} \end{bmatrix} \quad (9.10)$$

The *roughness factor* (f) varies with length and diameter as well as the condition of the pipe and the material from which it is constructed; it normally ranges from .01 to .04.

Important Point: For centrifugal pumps, good engineering practice is to try to keep velocities in the suction pipe to 3 ft/sec or less. Discharge velocities higher than 11 ft/sec may cause turbulent flow or erosion in the pump casing.

■ EXAMPLE 9.6

Problem: What is the friction head in a system that uses 150 ft of 6-in.-diameter pipe when the velocity is 3 fps? The valving of the system is equivalent to an additional 75 ft of pipe. Reference material indicates a roughness factor (f) of 0.025 for this particular pipe and flow rate.

Solution:

$$\text{Friction head (ft)} = \begin{bmatrix} \text{Roughness factor } (f) \times \dfrac{\text{length}}{\text{diameter}} \\ \times \dfrac{\text{velocity}^2}{2g} \end{bmatrix}$$

$$= 0.025 \times \frac{(150\ \text{ft} + 75\ \text{ft})}{0.5\ \text{ft}} \times \frac{(3\ \text{fps})^2}{2 \times 32\ \text{fps}}$$

$$= 0.025 \times \frac{225\ \text{ft}}{0.5\ \text{ft}} \times \frac{9\ \text{ft}^2/\text{sec}^2}{64\ \text{fps}}$$

$$= 0.025 \times 450 \times 0.140\ \text{ft} = 1.58\ \text{ft}$$

It is also possible to compute friction head using tables. Friction head can also be determined on both the suction side of the pump and the discharge side of the pump. In each case, it is necessary to determine:

1. The length of pipe
2. The diameter of the pipe
3. Velocity
4. Pipe equivalent of valves, elbows, T's, etc.

9.2.5 Velocity Head

Velocity head is the amount of head or energy required to maintain a stated velocity in the suction and discharge lines. The design of most pumps makes the total velocity head for the pumping system zero.

Note: Velocity head only changes from one point to another on a pipeline if the diameter of the pipe changes.

Velocity head and total velocity are determined by:

$$\text{Velocity head (ft)} = \frac{\text{Velocity}^2}{2g} \quad (9.11)$$

$$\begin{array}{l} \text{Total velocity} \\ \text{head (ft)} \end{array} = \begin{bmatrix} \text{Velocity head discharge (ft)} \\ -\text{velocity head suction (ft)} \end{bmatrix} \quad (9.12)$$

■ EXAMPLE 9.7

Problem: What is the velocity head for a system that has a velocity of 5 fps?

Solution:

$$\text{Velocity head (ft)} = \frac{\text{Velocity}^2}{2 \times \text{acceleration due to gravity}}$$

$$= \frac{(5\ \text{fps})^2}{2 \times 32\ \text{fps}} = \frac{25\ \text{ft}^2/\text{sec}^2}{64\ \text{fps}} = 0.39\ \text{fps}$$

Note: There is no velocity head in a static system, as the water is not moving.

9.2.6 TOTAL HEAD

Total head is the sum of the static, friction, and velocity heads.

$$\text{Total head (ft)} = \begin{bmatrix} \text{Static head (ft)} + \text{friction head (ft)} \\ + \text{velocity head (ft)} \end{bmatrix} \quad (9.13)$$

9.2.7 CONVERSION OF PRESSURE HEAD

Pressure is directly related to the head. If liquid in a container subjected to a given pressure is released into a vertical tube, the water will rise 2.31 ft for every pound per square inch of pressure. To convert pressure to head in feet:

$$\text{Head (ft)} = \text{Pressure (psi)} \times 2.31 \text{ ft/psi} \quad (9.14)$$

This calculation can be very useful in cases where liquid is moved through another line that is under pressure. Because the liquid must overcome the pressure in the line it is entering, the pump must supply this additional head.

■ EXAMPLE 9.8

Problem: A pump is discharging to a pipe that is full of liquid under a pressure of 20 psi. The pump and piping system has a total head of 97 ft. How much additional head must the pump supply to overcome the line pressure?

Solution:

$$\text{Head (ft)} = \text{Pressure (psi)} \times 2.31 \text{ ft/psi}$$

$$= 20 \text{ psi} \times 2.31 \text{ ft/psi} = 46 \text{ ft}$$

Note: The pump must supply and additional head of 46 ft to overcome the internal pressure of the line.

9.2.8 HORSEPOWER

The unit of work is the foot pound (ft-lb), which is the amount of work required to lift a 1-lb object 1 ft off the ground. For practical purposes, we consider the amount of work being done. It is more valuable, obviously, to be able to work faster; that is, for economic reasons, we consider the rate at which work is being done (i.e., power or ft-lb/sec). At some point, the horse was determined to be the ideal work animal; it could move 550 lb 1 foot in 1 sec, which is considered to be equivalent to 1 horsepower.

$$550 \text{ ft-lb/sec} = 1 \text{ horsepower (hp)}$$

or

$$33,000 \text{ ft-lb/min} = 1 \text{ horsepower (hp)}$$

A pump performs work while it pushes a certain amount of water at a given pressure. The two basic terms for horsepower are (1) *hydraulic horsepower* and (2) *brake horsepower*.

9.2.8.1 Hydraulic (Water) Horsepower (WHP)

A pump has power because it does work. A pump lifts water (which has weight) a given distance in a specific amount of time (ft-lb/min). One hydraulic (water) horsepower (WHP) provides the necessary power to lift the water to the required height; it equals the following:

- 550 ft-lb/sec
- 33,000 ft-lb/min
- 2545 British thermal units per hour (Btu/hr)
- 0.746 kW
- 1.014 metric hp

To calculate the hydraulic horsepower (WHP) using flow in gpm and head in feet, use the following formula for centrifugal pumps:

$$\text{WHP} = \frac{\text{Flow (gpm)} \times \text{head (ft)} \times \text{specific gravity}}{3960} \quad (9.15)$$

Note: 3960 is derived by dividing 33,000 ft-lb by 8.34 lb/gal.

9.2.8.2 Brake Horsepower (BHP)

A water pump does not operate alone. It is driven by a motor, and electrical energy drives the motor. Brake horsepower (BHP) is the horsepower applied to the pump. The BHP of a pump equals its hydraulic horsepower divided by the efficiency of the pump. Note that neither the pump nor its prime mover (motor) is 100% efficient. Both of these units experience friction losses, and more horsepower will have to be applied to the pump to achieve the required amount of horsepower to move the water, and even more horsepower must be applied to the motor to get the job done (Hauser, 1993). The formula for BHP is:

$$\text{BHP} = \frac{\text{Flow (gpm)} \times \text{head (ft)} \times \text{specific gravity}}{3960 \times \text{efficiency}} \quad (9.16)$$

Important Points: (1) *Water horsepower* is the power necessary to lift the water to the required height, (2) *brake horsepower* is the horsepower applied to the pump, (3) *motor horsepower* is the horsepower applied to the motor, and (4) *efficiency* is the power produced by the unit divided by the power used in operating the unit.

9.2.9 SPECIFIC SPEED

The capacity of flow rate of a centrifugal pump is governed by the impeller thickness (Lindeburg, 1986). For a given impeller diameter, the deeper the vanes, the greater the capacity of the pump. Each desired flow rate or desired discharge head will have one optimum impeller design. The impeller that is best for developing a high discharge pressure will have different proportions from an impeller designed to produce a high flow rate. The quantitative index of this optimization is the *specific speed* (N_s)—the higher the specific speed of a pump, the higher its efficiency.

The specific speed of an impeller is its speed when pumping 1 gpm of water at a differential head of 1 ft. The following formula is used to determine specific speed (where H is at the best efficiency point):

$$N_S = \frac{\text{rpm} \times Q^{0.5}}{H^{0.75}} \qquad (9.17)$$

where:

rpm = revolutions per minute.
Q = flow (gpm).
H = head (ft).

Pump specific speeds vary between pumps. Although no absolute rule sets the specific speed for different kinds of centrifugal pumps, the following rule of thumb for N_s can be used:

Volute, diffuser, and vertical turbine = 500–5000
Mixed flow = 5000–10,000
Propeller pumps = 9000–15,000

9.2.9.1 Suction Specific Speed

Suction specific speed (N_{ss}), another impeller design characteristic, is an index of the suction characteristics of the impeller (i.e., the suction capacities of the pump) (Wahren, 1997). For practical purposes, N_{ss} ranges from about 3000 to 15,000. The limit for the use of suction specific speed impellers in water is approximately 11,000. The following equation is used for N_{ss}:

$$N_{ss} = \frac{\text{rpm} \times Q^{0.5}}{\text{NPSHR}^{0.75}} \qquad (9.18)$$

where:

rpm = revolutions per minute.
Q = flow (gpm).
NPSHR = net positive suction head required.

Ideally, N_{ss} should be approximately 7900 for single-suction pumps and 11,200 for double-suction pumps. (The value of Q in Equation 9.21 should be halved for double suction pumps.)

9.2.10 Affinity Laws—Centrifugal Pumps

Most parameters (impeller diameter, speed, and flow rate) determining the performance of a pump can vary. If the impeller diameter is held constant and the speed varied, the following ratios are maintained with no change of

efficiency (because of inexact results, some deviations may occur in the calculations):

$$Q_2/Q_1 = D_2/D_1 \qquad (9.19)$$
$$H_2/H_1 = (D_2/D_1)^2 \qquad (9.20)$$
$$BHP_2/BHP_1 = (D_2/D_1)^3 \qquad (9.21)$$

where:

Q = flow.
H_1 = head before change.
H_2 = head after change.
BHP = brake horsepower.
D_1 = impeller diameter before change.
D_2 = impeller diameter after change.

The relationships for speed (N) changes are as follows:

$$Q_2/Q_1 = N_2/N_1 \qquad (9.22)$$
$$H_2/H_1 = (N_2/N_1)^2 \qquad (9.23)$$
$$BHP_2/BHP_1 = (N_2/N_1)^3 \qquad (9.24)$$

where:

N_1 = initial rpm.
N_2 = changed rpm.

■ EXAMPLE 9.9

Problem: Change an 8-in.-diameter impeller for a 9-in.-diameter impeller, and find the new flow (Q), head (H), and brake horsepower (BHP) when the 8-in.-diameter data are:

Q_1 = 340 gpm
H_1 = 110 ft
BHP_1 = 10

Solution: Data for the 9-in.-diameter impeller would be as follows:

Q_2 = 340 × 9/8 = 383 gpm
H_2 = 110 × (9/8)² = 139 ft
BHP_2 = 10 × (9/8)³ = 14

9.2.11 Net Positive Suction Head (NPSH)

Earlier we referred to the *net positive suction head required* (NPSHR); also important in pumping technology is *net positive suction head* (NPSH) (Lindeburg, 1986; Wahren, 1997). NPSH is different from both suction head and suction pressure. This important point tends to be confusing for those first introduced to the term and to pumping technology in general. When an impeller in a centrifugal pump spins, the motion creates a partial vacuum in the impeller eye. The NPSH is the height of the column of liquid that will fill this partial vacuum without allowing the vapor pressure of the liquid to drop below its flash point; that is, this is the NPSH required (NPSHR) for the pump to function properly.

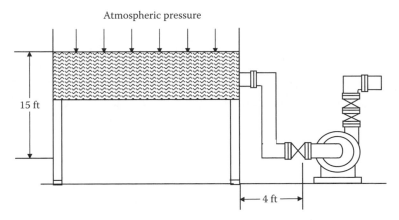

FIGURE 9.4 Open atmospheric tank.

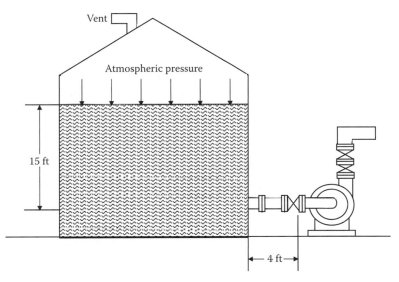

FIGURE 9.5 Roofed water storage tank.

The Hydraulic Institute (1994) defined NPSH as "the total suction head in feet of liquid absolute determined at the suction nozzle and referred to datum less the vapor pressure of the liquid in feet absolute." This defines the *NPSH available* (NPSHA) for the pump. (Note that NPSHA is the actual water energy at the inlet.) The important point here is that a pump will run satisfactorily if the NPSHA equals or exceeds the NPSHR. Most authorities recommend that the NPSHA be at least 2 ft absolute or 10% larger than the NPSHR, whichever number is larger.

Note: With regard to NPSHR, contrary to popular belief water is not sucked into a pump. A positive head (normally atmospheric pressure) must push the water into the impeller (i.e., flood the impeller). NPSHR is the minimum water energy required at the inlet by the pump for satisfactory operation. The pump manufacturer usually specifies NPSHR.

It is important to point out that if NPSHA is less than NPSHR the water will cavitate. *Cavitation* is the vapor-

ization of fluid within the casing or suction line. If the water pressure is less than the vapor pressure, pockets of vapor will form. As vapor pockets reach the surface of the impeller, the local high water pressure will collapse the pockets, causing noise, vibration, and possible structural damage to the pump.

9.2.11.1 Calculating NPSHA

In the following two examples, we demonstrate how to calculate NPSH for two real-world situations: (1) determining NPSHA for an open-top water tank or a municipal water storage tank with a roof and correctly sized vent, and (2) determining the NPSHA for a suction lift from an open reservoir.

9.2.11.1.1 NPSHA: Atmospheric Tank

The following calculation may be used for an open-top water tank or a municipal water storage tank with a roof and correctly sized vent, as shown in Figure 9.4 and Figure 9.5. The formula for calculating NPSHA is:

$$\text{NPSHA} = P_a + h - P_v - h_e - h_f \qquad (9.25)$$

where:

P_a = atmospheric pressure in absolute or pressure of gases against surface of water.

h = weight of liquid column from surface of water to center of pump suction nozzle in feet absolute.

P_v = vapor pressure in absolute of water at given temperature.

h_e = entrance losses in feet absolute.

h_f = friction losses in suction line in feet absolute.

■ **EXAMPLE 9.10**

Problem: Given the following, find the NPSHA.

Liquid = water
Temperature (t) = 60°F
Specific gravity = 1.0
P_a = 14.7 psia (34 ft)
h = 15 ft
P_v = 0.256 psia (0.6 ft)
h_e = 0.4 ft
h_f = 2 ft

Solution:

NPSHA = 34 ft + 15 ft − 0.6 ft − 0.4 ft − 2 ft = 46 ft

9.2.11.1.2   NPSHA: Suction Lift from Open Reservoir

See Figure 9.6.

■ **EXAMPLE 9.11**

Problem: Find the NPSHA, where:

Liquid = water
Temperature (t) = 60°F
Specific gravity = 1.0
P_a = 14.7 psia (34 ft)
h = −20 ft
P_v = 0.256 psia (0.6 ft)
Q = 120 gpm
h_e = 0.4 ft
h_f = 2 ft

Solution:

NPSHA = 34 ft + (−20 ft) − 0.6 ft − 0.4 ft − 2 ft = 11 ft

9.2.12   Pumps in Series and Parallel

Parallel operation occurs when two pumps discharge into a common header. This type of connection is advantageous when the system demand varies greatly. An advantage of operating pumps in parallel is that when two pumps are

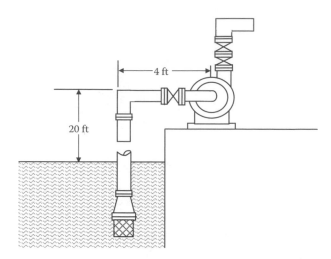

FIGURE 9.6 Suction lift from open reservoir.

online, one can be shut down during low demand. This allows the remaining pump to operate close to its optimum efficiency. *Series* operation is achieved by having one pump discharge into the suction of the next. This arrangement is used primarily to increase the discharge head, although a small increase in capacity also results.

9.3   CENTRIFUGAL PUMPS

The *centrifugal pump* (and its modifications) is the most widely used type of pumping equipment in water/wastewater operations. This type of pump is capable of moving high volumes of water/wastewater (or other liquids) in a relatively efficient manner. The centrifugal pump is very dependable, has relatively low maintenance requirements, and can be constructed out of a wide variety of construction materials. It is considered one of the most dependable systems available for water transfer.

9.3.1   Description

The centrifugal pump consists of a rotating element (*impeller*) sealed in a casing (*volute*). The rotating element is connected to a drive unit (*motor/engine*) that supplies the energy to spin the rotating element. As the impeller spins inside the volute casing, an area of low pressure is created in the center of the impeller. This low pressure allows the atmospheric pressure on the liquid in the supply tank to force the liquid up to the impeller. Because the pump will not operate if no low-pressure zone is created at the center of the impeller, it is important that the casing be sealed to prevent air from entering the casing. To ensure that the casing is airtight, the pump employs some type of seal (*mechanical or conventional packing*) assembly at the point where the shaft enters the casing. This seal also includes lubrication, provided by water, grease, or oil, to prevent excessive wear.

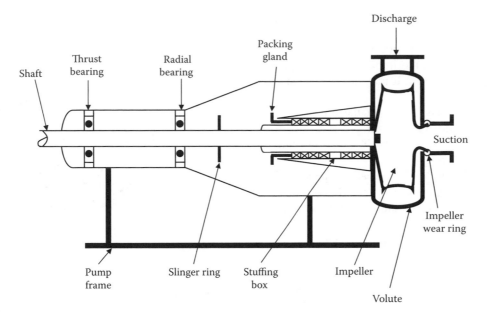

FIGURE 9.7 Major components of a centrifugal pump.

From a hydraulic standpoint, note the energy changes that occur in the moving water. As water enters the casing, the spinning action of the impeller imparts (transfers) energy to the water. This energy is transferred to the water in the form of increased speed or velocity. The liquid is thrown outward by the impeller into the volute casing where the design of the casing allows the velocity of the liquid to be reduced which, in turn, converts the velocity energy (*velocity head*) to pressure energy (*pressure head*). The process by which this change occurs is described later. The liquid then travels out of the pump through the pump discharge. The major components of the centrifugal pump are shown in Figure 9.7.

Key Point: A centrifugal pump is a pumping mechanism whose rapidly spinning impeller imparts a high velocity to the water that enters, then converts that velocity to pressure upon exit.

9.3.2 TERMINOLOGY

To understand centrifugal pumps and their operation, we must understand the terminology associated with centrifugal pumps:

- *Base plate*—The foundation under a pump. It usually extends far enough to support the drive unit. The base plate is often referred to as the *pump frame*.
- *Bearings*—Devices used to reduce friction and allow the shaft to rotate easily. Bearings may be sleeve, roller, or ball.
  - *Thrust bearing*—In a single suction pump, this is the bearing located nearest the motor,

farthest from the impeller. It takes up the major thrust of the shaft, which is opposite from the discharge direction.
  - *Radial (line) bearing*—In a single suction pump, this is the one closest to the pump. It rides free in its own section and takes up and down stresses.

Note: In most cases, where the pump and motor are constructed on a common shaft with no coupling, the bearings will be part of the motor assembly.

- *Casing*—The housing surrounding the rotating element of the pump. In the majority of centrifugal pumps, this casing can also be called the *volute*.
  - *Split casing*—A pump casing that is manufactured in two pieces fastened together by means of bolts. Split casing pumps may be vertically split (perpendicular to the shaft direction) or horizontally split (parallel to the shaft direction).
- *Coupling*—Device to join the pump shaft to the motor shaft. A pump and motor constructed on a common shaft is referred to as a *close-coupled arrangement*.
- *Extended shaft*—For a pump constructed on one shaft that must be connected to the motor by a coupling.
- *Frame*—The housing that supports the pump bearing assemblies. In an end suction pump, it may also be the support for the pump casing and the rotating element.

- *Impeller*—The rotating element in the pump that actually transfers the energy from the drive unit to the liquid. Depending on the pump application, the impeller may be open, semi-open, or closed. It may also be single or double suction.
- *Impeller eye*—The center of the impeller, the area that is subject to lower pressures due to the rapid movement of the liquid to the outer edge of the casing.
- *Priming*—Filling the casing and impeller with liquid. If this area is not completely full of liquid, the centrifugal pump will not pump efficiently.
- *Seals*—Devices used to stop the leakage of air into the inside of the casing around the shaft.
  - *Gland*—Also known as the *packing gland*, this is a metal assembly that is designed to apply even pressure to the packing to compress it tightly around the shaft.
  - *Lantern ring*—Also known as the *seal cage*, this is positioned between the rings of packing in the stuffing box to allow the introduction of a lubricant (water, oil, or grease) onto the surface of the shaft to reduce the friction between the packing and the rotating shaft.
  - *Mechanical seal*—This is a device consisting of a stationary element, a rotating element, and a spring to supply force to hold the two elements together. Mechanical seals may be either single or double units.
  - *Packing*—This is material placed around the pump shaft to seal the shaft opening in the casing and to prevent air leakage into the casing.
  - *Stuffing box*—The assembly located around the shaft at the rear of the casing, this holds the packing and lantern ring.
- *Shaft*—The rigid steel rod that transmits the energy from the motor to the pump impeller. Shafts may be either vertical or horizontal.
- *Shaft sleeve*—A piece of metal tubing placed over the shaft to protect the shaft as it passes through the packing or seal area. In some cases, the sleeve may also help to position the impeller on the shaft.
- *Shroud*—The metal plate that is used to either support the impeller vanes (open or semi-open impeller) or to enclose the vanes of the impeller (closed impeller).
- *Shut-off head*—The head or pressure at which the centrifugal pump will stop discharging. It is also the pressure developed by the pump when it is operated against a closed discharge valve. This is also known as a *cut-off head*.

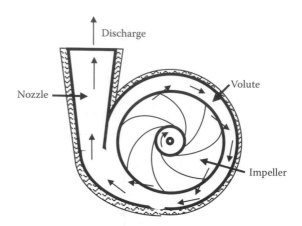

FIGURE 9.8 Cross-sectional diagram showing the features of a centrifugal pump.

- *Slinger ring*—A device to prevent pumped liquids from traveling along the shaft and entering the bearing assembly. A slinger ring is also called a *deflector*.
- *Wearing ring*—Device that is installed on stationary or moving parts within the pump casing to protect the casing and the impeller from wear due to the movement of liquid through points of small clearances.
  - *Impeller ring*—A wearing ring installed directly on the impeller.
  - *Casing ring*—A wearing ring installed in the casing of the pump; a casing ring is also known as the *suction head ring*.
  - *Stuffing box cover ring*—A wearing ring installed at the impeller in an end suction pump to maintain the impeller clearances and to prevent casing wear.

9.3.3 Pump Theory

The *volute-cased centrifugal pump* (see Figure 9.8) provides the pumping action necessary to transfer liquids from one point to another. First, the drive unit (usually an electric motor) supplies energy to the pump impeller to make it spin. This energy is then transferred to the water by the impeller. The vanes of the impeller spin the liquid toward the outer edge of the impeller at a high rate of speed or velocity. This action is very similar to that which would occur when a bucket full of water with a small hole in the bottom is attached to a rope and spun. When the bucket is sitting still, the water in it will drain out slowly; however, when the bucket is spinning, the water will be forced through the hole at a much higher rate of speed.

Centrifugal pumps may be single stage with a single impeller, or they may be multiple stage with several impellers through which the fluid flows in series. Each impeller in the series increases the pressure of the fluid at the pump discharge. Pumps may have thirty or more stages in extreme cases. In centrifugal pumps, a correlation of pump capacity, head, and speed at optimum efficiency is used to classify the pump impellers with respect to their specific geometry. This correlation is called *specific speed*, and is an important parameter for analyzing pump performance (Garay, 1990).

The volute of the pump is designed to convert velocity energy to pressure energy. As a given volume of water moves from one cross-sectional area to another with the volute casing, the velocity or speed of the water changes proportionately. The volute casing has a cross-sectional area that is extremely small at the point in the case that is farthest from the discharge (see Figure 9.8). This area increases continuously to the discharge. As this area increases, the velocity of the water passing through it decreases as it moves around the volute casing to the discharge point.

As the velocity of the water decreases, the velocity head decreases and the energy is converted to pressure head. A direct relationship exists between the velocity of the water and the pressure it exerts; therefore, as the velocity of the water decreases, the excess energy is converted to additional pressure (pressure head). This pressure head supplies the energy to move the water through the discharge piping.

9.3.4 PUMP CHARACTERISTICS

The centrifugal pump operates on the principle of an energy transfer and therefore has certain characteristics that make it unique. The type and size of the impeller limit the amount of energy that can be transferred to the water, the characteristics of the material being pumped, and the total head of the system through which the water is moving. For any one centrifugal pump, a definite relationship exists between these factors along with head (capacity), efficiency, and brake horsepower.

9.3.4.1 Head (Capacity)

As might be expected, the capacity of a centrifugal pump is directly related to the total head of the system. If the total head on the system is increased, the volume of the discharge will be reduced proportionately. As the head of the system increases, the capacity of the pump decreases proportionately until the discharge stops. The head at which the discharge no longer occurs is known as the *cut-off head*. As pointed out earlier, the total head includes a certain amount of energy to overcome the friction of the system. This friction head can be greatly affected by the size and configuration of the piping and the condition of the valving of the system. If the control valves on the system are closed partially, the friction head can increase dramatically. When this happens, the total head increases and the capacity or volume discharged by the pump decreases. In many cases, this method is employed to reduce the discharge of a centrifugal pump. It should be noted, however, that this does increase the load on the pump and drive system, causing additional energy requirements and additional wear.

The total closure of the discharge control valve increases the friction head to the point where all of the energy supplied by the pump is consumed in the friction head and is not converted to pressure head; consequently, the pump exceeds its cut-off head, and the pump discharge is reduced to zero. Again, it is important to note that, although the operation of a centrifugal pump against a closed discharge may not be hazardous (as with other types of pumps), it should be avoided because of the excessive load placed on the drive unit and pump. Our experience has shown that on occasion the pump can produce pressure higher than the pump discharge piping can withstand. Whenever this occurs, the discharge piping may be severely damaged by the operation of the pump against a closed or plugged discharge.

9.3.4.2 Efficiency

Every centrifugal pump will operate with varying degrees of efficiency over its entire capacity and head ranges. The important factor in selecting a centrifugal pump is to select a unit that will perform near its maximum efficiency in the expected application.

9.3.4.3 Brake Horsepower Requirements

In addition to the head capacity and efficiency factors, most pump literature includes a graph showing the amount of energy in horsepower that must be supplied to the pump to obtain optimal performance.

9.3.5 ADVANTAGES AND DISADVANTAGES OF THE CENTRIFUGAL PUMP

The primary reason why centrifugal pumps have become one of the most widely used types of pumps are the several advantages it offers, including:

- *Construction*—The pump consists of a single rotating element and simple casing, which can be constructed using a wide assortment of materials. If the fluids to be pumped are highly corrosive, the pump parts that are exposed to the fluid can be constructed of lead or other material that is not likely to corrode. If the fluid being pumped is highly abrasive, the internal parts can be made of abrasion-resistant material or coated with a protective material. Also, the simple design of a centrifugal pump allows the pump to be constructed in a variety of sizes and configurations. No other pump currently available offers the range of capacities or applications that the centrifugal pump does.

- *Operation*—"Simple and quiet" best describes the operation of a centrifugal pump. An operator-in-training with a minimum amount of experience may be capable of operating facilities that use centrifugal-type pumps. Even when improperly operated, the rugged construction of the centrifugal pump allows it to operate (in most cases) without incurring major damage.
- *Maintenance*—The amount of wear on the moving parts of a centrifugal pump is reduced and its operating life is extended because its moving parts are not required to be constructed to very close tolerances.
- *Pressure is self-limited*—Because of the nature of its pumping action, the centrifugal pump will not exceed a predetermined maximum pressure. Thus, if the discharge valve is suddenly closed, the pump cannot generate additional pressure that might result in damage to the system or could potentially result in a hazardous working condition. The power supplied to the impeller will only generate a specified amount of head (pressure). If a major portion of this head or pressure is consumed in overcoming friction or is lost as heat energy, the pump will have a decreased capacity.
- *Adaptable to high-speed drive systems*—Centrifugal pumps can make use of high-speed, high-efficiency motors. In situations where the pump is selected to match a specific operating condition, which remains relatively constant, the pump drive unit can be used without the need for expensive speed reducers.
- *Small space requirements*—For most pumping capacities, the amount of space required for installation of the centrifugal-type pump is much less than that of any other type of pump.
- *Fewer moving parts*—Rotary rather than reciprocating motion employed in centrifugal pumps reduces space and maintenance requirements due to the fewer number of moving parts required.

Although the centrifugal pump is one of the most widely used pumps, it does have a few disadvantages:

- *Additional equipment is needed for priming*—The centrifugal pump can be installed in a manner that will make it self-priming, but it is not capable of drawing water to the pump impeller unless the pump casing and impeller are filled with water. This can cause problems, because if the water in the casing drains out the pump ceases pumping until it is refilled; therefore, it is normally necessary to start a centrifugal pump with the discharge valve closed. The valve is then gradually opened to its proper operating level. Starting the pump against a closed discharge valve is not hazardous provided the valve does not remain closed for extended periods.
- *Air leaks affect pump performance*—Air leaks on the suction side of the pump can cause reduced pumping capacity in several ways. If the leak is not serious enough to result in a total loss of prime, the pump may operate at a reduced head or capacity due to air mixing with the water. This causes the water to be lighter than normal and reduces the efficiency of the energy transfer process.
- *Narrow range of efficiency*—Centrifugal pump efficiency is directly related to the head capacity of the pump. The highest performance efficiency is available for only a very small section of the head-capacity range. When the pump is operated outside of this optimum range, the efficiency may be greatly reduced.
- *Pump may run backwards*—If a centrifugal pump is stopped without closing the discharge line, it may run backwards, because the pump does not have any built-in mechanism to prevent flow from moving through the pump in the opposite direction (i.e., from discharge side to suction). If the discharge valve is not closed or the system does not contain the proper check valves, the flow that was pumped from the supply tank to the discharge point will immediately flow back to the supply tank when the pump shuts off. This results in increased power consumption due to the frequent start-up of the pump to transfer the same liquid from supply to discharge.

Note: It is sometimes difficult to tell whether a centrifugal pump is running forward or backward because it appears and sounds like it is operating normally when operating in reverse.

- *Pump speed is difficult to adjust*—Centrifugal pump speed cannot usually be adjusted without the use of additional equipment, such as speed-reducing or speed-increasing gears or special drive units. Because the speed of the pump is directly related to the discharge capacity of the pump, the primary method available to adjust the output of the pump other than a valve on the discharge line is to adjust the speed of the impeller. Unlike some other types of pumps, the delivery of the centrifugal pump cannot be adjusted by changing some operating parameter of the pump.

9.3.6 Centrifugal Pump Applications

The centrifugal pump is probably the most widely used pump available at this time because of its simplicity of design and wide-ranging diversity (it can be adjusted to suit a multitude of applications). Proper selection of the pump components (e.g., impeller, casing) and construction materials can produce a centrifugal pump capable of transporting not only water but also other materials ranging from material or chemical slurries to air (centrifugal blowers). To attempt to list all of the various applications for the centrifugal pump would exceed the limitations of this handbook; therefore, our discussion of pump applications is limited to those that frequently occur in water/wastewater operations.

- *Large-volume pumping*—In water/wastewater operations, the primary use of centrifugal pumps is large-volume pumping. Generally, in large-volume pumping, low-speed, moderate-head, vertically shafted pumps are used. Centrifugal pumps are well suited for water/wastewater system operations because they can be used in conditions where high volumes are required and a change in flow is not a problem. As the discharge pressure on a centrifugal pump is increased, the quantity of water/wastewater pumped is reduced. Also, centrifugal pumps can be operated for short periods with the discharge valve closed.
- *Non-clog pumping*—These specially designed centrifugal pumps utilize closed impellers with, at most, two to three vanes. They are usually designed to pass solids or trash up to 3 inches in diameter.
- *Dry-pit pump*—Depending on the application, the dry-pit pump may be either a large-volume pump or a non-clog pump. It is located in a dry pit that shares a common wall with the wet well. This pump is normally placed in such a position as to ensure that the liquid level in the wet well is sufficient to maintain the prime of the pump.
- *Wet-pit or submersible pump*—This type of pump is usually a non-clog pump that can be submerged, with its motor, directly in the wet well. In a few instances, the pump may be submerged in the wet well while the motor remains above the water level. In these cases, the pump is connected to the motor by an extended shaft.
- *Underground pump stations*—Utilizing a wet-well/dry-well design, these pumps are located in an underground facility. Wastes are collected in a separate wet well, then pumped upward and discharged into another collector line or manhole. This system normally uses a non-clog type pump and is designed to add sufficient head to water/wastewater flow to allow gravity to move the flow to the plant or the next pump station.
- *Recycle or recirculation pumps*—Because the liquids being transferred by the recycle or recirculation pump normally do not contain any large solids, the use of the nonclog type of centrifugal pump is not always required. A standard centrifugal pump may be used to recycle trickling filter effluent, return activated sludge, or digester supernatant.
- *Service water pumps*—The wastewater plant effluent may be used for many purposes, such as to clean tanks, water lawns, provide water to operate the chlorination system, and to backwash filters. Because the plant effluent used for these purposes is normally clean, the centrifugal pumps used in this case closely parallel the units used for potable water. In many cases, the double-suction, closed-impeller, or turbine type of pump is used.

9.3.7 Pump Control Systems

Pump operations usually control only one variable: flow, pressure or level. All pump control systems have a measuring device that compares a measured value with a desired one. This information relays to a control element that makes the changes. ...The user may obtain control with manually operated valves or sophisticated microprocessors. Economics dictate the accuracy and complication of a control system.

Wahren (1997)

Most centrifugal pumps require some form of pump control system. The only exception to this practice is when the plant pumping facilities are designed to operate continuously at a constant rate of discharge. The typical pump control system includes a sensor to determine when the pump should be turned on or off and the electrical/electronics controls to actually start and stop the pump. The control systems currently available for the centrifugal pump range from a very simple on/off float control to an extremely complex system capable of controlling several pumps in sequence. In the following sections, we briefly describe the operation of various types of control devices used with centrifugal pumps.

9.3.7.1 Float Control

Currently, the *float control system* is the simplest of the centrifugal pump controls (see Figure 9.9). In the float control system, the float rides on the surface of the water in the well, storage tank, or clear well and is attached to the pump controls by a rod with two collars. One collar activates the pump when the liquid level in the well or

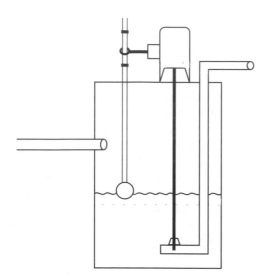

FIGURE 9.9 Float system for pump motor control.

tank reaches a preset level, and a second collar shuts the pump off when the level in the well reaches a minimum level. This type of control system is simple to operate and relatively inexpensive to install and maintain. The system has several disadvantages; for example, it operates at one discharge rate, which can result in: (1) extreme variations in the hydraulic loading on succeeding units, and (2) long periods of not operating due to low flow periods or maintenance activities.

9.3.7.2 Pneumatic Controls

Pneumatic control systems (also called a *bubbler tube control system*) are relatively simple systems that can be used to control one or more pumps. The system consists of an air compressor; a tube extending into the well, clear well, or storage tank or basin; and pressure-sensitive switches with varying on/off set points and a pressure-relief valve (see Figure 9.10). The system works on the basic principle of measuring the depth of the water in the well or tank by determining the air pressure necessary to just release a bubble from the bottom of the tube (see Figure 9.10)—hence the name *bubbler tube*. The air pressure required to force a bubble out of the tube is determined by the liquid pressure, which is directly related to the depth of the liquid (1 psi = 2.31 ft). By installing a pressure switch on the air line to activate the pump starter at a given pressure, the level of the water can be controlled by activating one or more pumps.

Installation of additional pressure switches with slightly different pressure settings allows several pumps to be activated in sequence. As an example, the first pressure switch can be adjusted to activate a pump when the level in the well or tank is 3.8 ft (1.6 psi) and shut off at 1.7 ft (0.74 psi). If the flow into the pump well or tank varies greatly and additional pumps are available to ensure that the level in the well or tank does not exceed the design capacity, additional pressure switches may be installed. These additional pressure switches are set to activate a second pump when the level in the well or tank reaches a preset level (e.g., 4.5 ft or 1.95 psi) and to cut off when the well or tank level is reduced to a preset level (e.g., 2.7 ft or 1.2 psi). If the capacity of the first pump is less than the rate of flow into the well or tank, the level will continue to rise. When the preset level is reached, the second pump will be activated. If necessary, a third pump can be added to the system that will activate at a higher preset well or tank depth (e.g., 4.6 ft or 1.99 psi) and to cut off a preset depth (e.g., 3.0 ft or 1.3 psi).

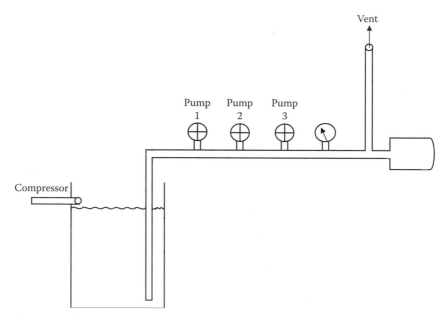

FIGURE 9.10 Pneumatic system for pump motor control.

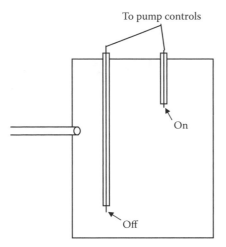

FIGURE 9.11 Electrode system for pump motor control.

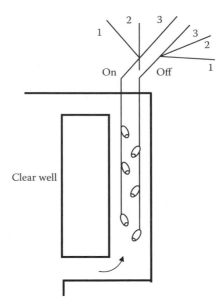

FIGURE 9.12 Electrical contacts for pump motor control.

The pneumatic control system is relatively simple and has minimal operation and maintenance requirements. The major operational problem involved with this control system is clogging of the bubbler tube. If, for some reason, the tube becomes clogged, the pressure on the system can increase and may activate all pumps to run even when the level in the well or tank is low. This can result in excessive power consumption, which, in turn, may damage the pumps.

9.3.7.3 Electrode Control Systems

The *electrode control system* uses a probe or electrode to control the pump on/off cycle. A relatively simple control system, it consists of two electrodes extending into the clear well, storage tank, or basin. One electrode is designed to activate the pump starter when it is submerged in the water; the second electrode extends deeper into the well or tank and is designed to open the pump circuit when the water drops below the electrode (see Figure 9.11). The major maintenance requirement of this system is keeping the electrodes clean.

Important Point: Because the electrode control system uses two separate electrodes, the unit may be locked into an on cycle or an off cycle, depending on which electrode is compromised.

9.3.7.4 Other Control Systems

Several other systems that use electrical energy are available for control of the centrifugal pump. These include a *tube-like device* that has several electrical contacts mounted inside (see Figure 9.12). As the water level rises in the clear well, storage tank, or basin, the water rises in the tube, making contact with the electrical contacts and activating the motor starter. Again, this system can be used to activate several pumps in series by installation of sev-

eral sets of contact points. As the water level drops in the well or tank, the level in the tube drops below a second contact that deactivates the motor and stops the pumping. Another control system uses a *mercury switch* (or a similar type of switch) enclosed in a protective capsule. Again, two units are required per pump. One switch activates the pump when the liquid level rises, and the second switch shuts the pump off when the level reaches the desired minimum depth.

9.3.8 ELECTRONIC CONTROL SYSTEMS

Several centrifugal pump control systems are available that use electronic systems for the control of pump operation. A brief description of some of these systems is provided in the sections that follow.

9.3.8.1 Flow Equalization System

In any multiple pump operation, the flow delivered by each pump will vary due to the basic hydraulic design of the system. To obtain equal loads on each pump when two or more are in operation, the flow equalization system electronically monitors the delivery of each pump and adjusts the speed of the pumps to obtain similar discharge rates for each pump.

9.3.8.2 Sonar or Other Transmission Types of Controllers

A *sonar* or *low-level radiation system* can be used to control centrifugal pumps. These types of controllers use a transmitter and receiver to locate the level of the water in a tank, clear well, or basin. When the level reaches a predetermined set point, the pump is activated; when the level is reduced to a predetermined set point, the pump is shut off. Basically, the system is very similar to a radar unit. The transmitter sends out a beam that travels to the liquid, bounces off the surface, and returns to the receiver. The time required for this is directly proportional to the distance from the liquid to the instrument. The electronic components of the system can be adjusted to activate the pump when the time interval corresponds to a specific depth in the well or tank. The electronic system can also be set to shut off the pump when the time interval corresponds to a preset minimum depth.

9.3.8.3 Motor Controllers

Several types of controllers are available that protect the motor not only from overloads but also from short-circuit conditions. Many motor controllers also function to adjust motor speed to increase or decrease the discharge rate for a centrifugal pump. This type of control may use one of the previously described controls to start and stop the pump and, in some cases, adjust the speed of the unit. As the depth of the water in a well or tank increases, the sensor automatically increases the speed of the motor in predetermined steps to the maximum design speed. If the level continues to increase, the sensor may be designed to activate an additional pump.

9.3.8.4 Protective Instrumentation

Protective instrumentation of some type is normally employed in pump or motor installation. (Note that the information provided in this section applies not only to the centrifugal pump but also many other types of pumps.) Protective instrumentation for centrifugal pumps (or most other types of pumps) is dependent on pump size, application, and the amount of operator supervision; that is, pumps under 500 hp often only come with pressure gauges and temperature indicators. These gauges or transducers may be mounted locally (on the pump itself) or remotely (in suction and discharge lines immediately upstream and downstream of the suction and discharge nozzles). If transducers are employed, readings are typically displayed and taken (or automatically recorded) at a remote operating panel or control center.

9.3.8.5 Temperature Detectors

Resistance temperature devices (RTDs) and *thermocouples* (Figure 9.13) (Grimes, 1976) are commonly used as temperature detectors on the pump prime movers (motors) to indicate temperature problems. In some cases, dial thermometers, armored glass-stem thermometers, or bimetallic-actuated temperature indicators are used. Whichever device is employed, it typically monitors temperature variances that may indicate a possible source of trouble. On electric motors greater than 250 hp, RTD elements are used to

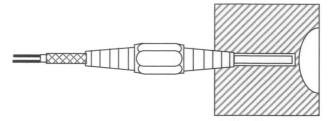

FIGURE 9.13 Thermocouple installation in journal bearing.

monitor temperatures in stator winding coils. Two RTDs per phase are standard. One RTD element is usually installed in the shoe of the loaded area employed on journal bearings in pumps and motors. Normally, tilted-pad thrust bearings have an RTD element on the active, as well as the inactive, side. RTDs are used when remote indication, recording, or automatic logging of temperature readings is required. Because of their smaller size, RTDs provide more flexibility in locating the measuring device near the measuring point. When dial thermometers are installed, they monitor oil thrown from bearings. Sometimes temperature detectors also monitor bearings with water-cooled jackets to warn against water supply failure. Pumps with heavy wall casing may also have casing temperature monitors.

9.3.8.6 Vibration Monitors

Vibration sensors are available to measure either bearing vibration or shaft vibration direction directly. Direct measurement of shaft vibration is desirable for machines with stiff bearing supports where bearing-cap measurements will be only a fraction of the shaft vibration. Wahren (1997) noted that pumps and motors 1000 hp and larger may have the following vibration monitoring equipment:

*   Seismic pickup with double set points installed on the pump outboard housing
*   Proximators with *x–y* vibration probes complete with interconnecting coaxial cables at each radial and thrust journal bearing
*   Key phasor with proximator and interconnecting coaxial cables

9.3.8.7 Supervisory Instrumentation

Supervisory instruments are used to monitor the routine operation of pumps, their prime movers, and their accessories to sustain a desired level of reliability and performance. Generally, these instruments are not used for accurate performance tests or for automatic control, although they may share connections or functions. Supervisory instruments consist of annunciators and alarms that provide operators with warnings of abnormal conditions that, unless corrected, will cause pump failure. Annunciators that are used for both alarm and pre-alarm have visible and audible signals.

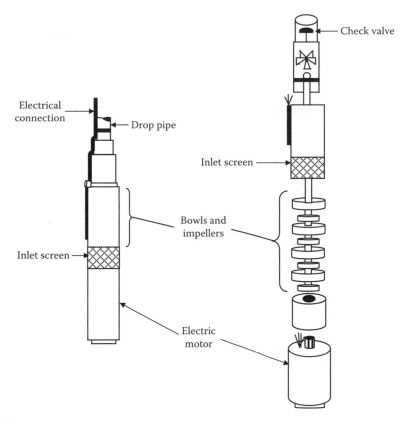

FIGURE 9.14 Submersible pump.

9.3.9 CENTRIFUGAL PUMP MODIFICATIONS

The centrifugal pump can be modified to meet the needs of several different applications. If it is necessary to produce higher discharge heads, the pump may be modified to include several additional impellers. If the material being pumped contains a large amount of material that could clog the pump, the pump construction may be modified to remove a major portion of the impeller from direct contact with the material being pumped. Although numerous modifications of the centrifugal pump are available, the scope of this text covers only those that have found wide application in the water distribution and wastewater collection and treatment fields. Modifications presented in this section include:

- Submersible pumps
- Recessed impeller or vortex pumps
- Turbine pumps

9.3.9.1 Submersible Pumps

The *submersible pump*, as the name suggests, is placed directly in the wet well or groundwater well. It uses a waterproof electric motor located below the static level of the well to drive a series of impellers. In some cases, only the pump is submerged, while in other cases the entire pump–motor assembly is submerged. Figure 9.14 illustrates this system.

9.3.9.1.1 Description
The submersible pump may be either a close-coupled centrifugal pump or an extended-shaft centrifugal pump. If the system is a close-coupled system, then both motor and pump are submerged in the liquid being pumped. Seals prevent water and wastewater from entering the inside of the motor, thus protecting the electric motor in a close-coupled pump from shorts and motor burnout. In the extended shaft system, the pump is submerged and the motor is mounted above the pump wet well. In this situation, an extended shaft assembly must connect the pump and motor.

9.3.9.1.2 Applications
The submersible pump has wide applications in the water/wastewater treatment industry. It generally can be substituted in any application of other types of centrifugal pumps; however, it has found its widest application in distribution or collector system pump stations.

9.3.9.1.3 Advantages
In addition to the advantages discussed earlier for a conventional centrifugal pump, the submersible pump has additional advantages:

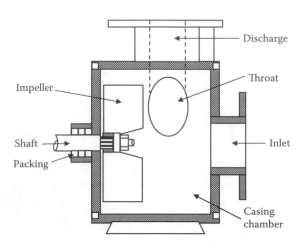

FIGURE 9.15 Schematic of a recessed impeller or vortex pump.

- Because it is located below the surface of the liquid, it is not as likely that the pump will lose its prime, develop air leaks on the suction side of the pump, or require initial priming.
- The pump or the entire assembly is located in the well, so costs associated with the construction and operation of this system are reduced. It is not necessary to construct a dry well or a large structure to hold the pumping equipment and necessary controls.

9.3.9.1.4 Disadvantages

The major disadvantage associated with the submersible pump is the lack of access to the pump or pump and motor. The performance of any maintenance requires either drainage of the wet well or extensive lift equipment to remove the equipment from the wet well, or both. This may be a major factor in determining if a pump receives the attention it requires. Also, in most cases, all major maintenance on close-coupled submersible pumps must be performed by outside contractors due to the need to reseal the motor to prevent leakage.

9.3.9.2 Recessed Impeller or Vortex Pumps

The *recessed impeller* or *vortex pump* uses an impeller that is either partially or wholly recessed into the rear of the casing (Figure 9.15). The spinning action of the impeller creates a vortex or whirlpool. This whirlpool increases the velocity of the material being pumped. As in other centrifugal pumps, this increased velocity is then converted to increased pressure or head.

9.3.9.2.1 Applications

The recessed impeller or vortex pump is used widely in applications where the liquid being pumped contains large amounts of solids or debris and slurries that could clog or damage the impeller of the pump. It has found increasing use as a sludge pump in facilities that withdraw sludge continuously from their primary clarifiers.

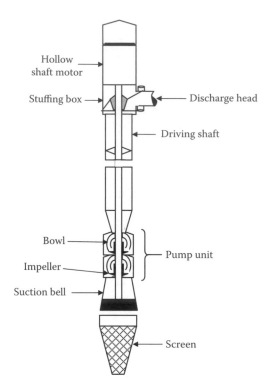

FIGURE 9.16 Vertical turbine pump.

9.3.9.2.2 Advantages

The major advantage of this modification is the increased ability to handle materials that would normally clog or damage the pump impeller. Because the majority of the flow does not come in direct contact with the impeller, the potential for problems is reduced.

9.3.9.2.3 Disadvantages

Because of the reduced direct contact between the liquid and the impeller, the energy transfer is less efficient. This results in somewhat higher power costs and limits application of the pump to low to moderate capacities. Objects that might have clogged a conventional type of centrifugal pump are able to pass through the pump. Although this is very beneficial in reducing pump maintenance requirements, it has, in some situations, allowed material to be passed into a less accessible location where it becomes an obstruction. To be effective, the piping and valving must be designed to pass objects of a size equal to that which the pump will discharge.

9.3.9.3 Turbine Pumps

The turbine pump consists of a motor, drive shaft, a discharge pipe of varying lengths, and one or more impeller–bowl assemblies (see Figure 9.16). The pump is normally a vertical assembly where water enters at the bottom, passes axially through the impeller–bowl assembly where the energy transfer occurs, and then moves upward through additional impeller–bowl assemblies to the discharge pipe. The length of this discharge pipe will vary with the distance from the wet well to the desired point of discharge.

9.3.9.3.1 Application

Due to the construction of the turbine pump, the major application has traditionally been to pump relatively clean water. The line shaft turbine pump has been used extensively for drinking water pumping, especially in those situations where water is withdrawn from deep wells. The main wastewater plant application has been pumping plant effluent back into the plant for use as service water.

9.3.9.3.2 Advantages

The turbine pump has a major advantage in the amount of head it is capable of producing. By installing additional impeller–bowl assemblies, the pump is capable of even greater production. Moreover, the turbine pump has simple construction and a low noise level and is adaptable to several drive types—motor, engine, or turbine.

9.3.9.3.3 Disadvantages

High initial cost and high repair costs are two of the major disadvantages of turbine pumps. In addition, the presence of large amounts of solids within the liquid being pumped can seriously increase the amount of maintenance the pump requires; consequently, the unit has not found widespread use in any situation other than service water pumping.

9.4 POSITIVE-DISPLACEMENT PUMPS

Positive-displacement pumps force or displace water through the pumping mechanism. Most have a reciprocating element that draws water into the pump chamber on one stroke and pushes it out on the other. Unlike centrifugal pumps that are meant for low-pressure, high-flow applications, positive-displacement pumps can achieve greater pressures but are slower moving, low-flow pumps. Other positive-displacement pumps include the piston pump, diaphragm pump, and peristaltic pumps, which are the focus of our discussion. In the water/wastewater industry, positive-displacement pumps are most often found as chemical feed pumps. It is important to remember that positive-displacement pumps cannot be operated against a closed discharge valve. As the name indicates, something must be displaced with each stroke of the pump. Closing the discharge valve can cause rupturing of the discharge pipe, the pump head, the valve, or some other component.

9.4.1 Piston Pump or Reciprocating Pump

The *piston* or *reciprocating pump* is one type of positive-displacement pump. This pump works just like the piston in an automobile engine—on the intake stroke, the intake valve opens, filling the cylinder with liquid. As the piston reverses direction, the intake valve is pushed closed and

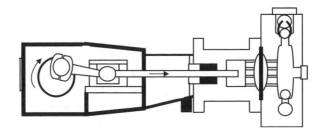

FIGURE 9.17 Diaphragm pump.

the discharge valve is pushed open; the liquid is pushed into the discharge pipe. With the next reversal of the piston, the discharge valve is pulled closed and the intake valve pulled open, and the cycle repeats. A piston pump is usually equipped with an electric motor and a gear-and-cam system that drives a plunger connected to the piston. Just like an automobile engine piston, the piston must have packing rings to prevent leakage and must be lubricated to reduce friction. Because the piston is in contact with the liquid being pumped, only good-grade lubricants can be used for pumping materials that will be added to drinking water. The valves must be replaced periodically as well.

9.4.2 Diaphragm Pump

A *diaphragm pump* is composed of a chamber used to pump the fluid, a diaphragm that is operated by either electric or mechanical means, and two valve assemblies—a suction and a discharge valve assembly (see Figure 9.17). A diaphragm pump is a variation of the piston pump in which the plunger is isolated from the liquid being pumped by a rubber or synthetic diaphragm. As the diaphragm is moved back and forth by the plunger, liquid is pulled into and pushed out of the pump. This arrangement provides better protection against leakage of the liquid being pumped and allows the use of lubricants that otherwise would not be permitted. Care must be taken to ensure that diaphragms are replaced before they rupture. Diaphragm pumps are appropriate for discharge pressures up to about 125 psi but do not work well if they must lift liquids more than about 4 feet. Diaphragm pumps are frequently used for chemical feed pumps. By adjusting the frequency of the plunger motion and the length of the stroke, extremely accurate flow rates can be metered. The pump may be driven hydraulically by an electric motor or by an electronic driver in which the plunger is operated by a solenoid. Electronically driven metering pumps are extremely reliable (few moving parts) and inexpensive.

9.4.3 Peristaltic Pumps

Peristaltic pumps (sometimes called *tubing pumps*) use a series of rollers to compress plastic tubing to move the liquid through the tubing. A rotary gear turns the rollers at a constant speed to meter the flow. Peristaltic pumps are mainly used as chemical feed pumps. The flow rate is adjusted by changing the speed the roller-gear rotates (to push the waves faster) or by changing the size of the tubing (so there is more liquid in each wave). As long as the appropriate type of tubing is used, peristaltic pumps can operate at discharge pressures up to 100 psi. Note that the tubing must be resistant to deterioration from the chemical being pumped. The principle item of maintenance is the periodic replacement of the tubing in the pump head. There are no check valves or diaphragms in this type of pump.

CHAPTER REVIEW QUESTIONS

9.1   Applications in which chemicals must be metered under high pressure require high-powered _____ pumps.

9.2   _____ materials are materials that resist any flow-producing force.

9.3   What type of pump is usually used for pumping high-viscosity materials?

9.4   High-powered positive-displacement pumps are used to pump chemicals that are under _____ pressure.

9.5   _____ viscosity materials are thick.

9.6   When the _____ of a pump impeller is above the level of the pumped fluid, the condition is called suction lift.

9.7   When a pump is not running, conditions are referred to as _____; when a pump is running the conditions are _____.

9.8   With the _____, the difference in elevation between the suction and discharge liquid levels is called static head.

9.9   Velocity head is expressed mathematically as _____.

9.10   The sum of total static head, head loss, and dynamic head is called _____.

9.11   What are the three basic types of controls used for centrifugal pumps?

9.12   The liquid used to rate pump capacity is _____.

9.13   Because of the reduced amount of air pressure at high altitudes, less _____ is available for the pump.

9.14   With the pump shut off, the difference between the suction and discharge liquid levels is called _____.

9.15   _____ and _____ are the largest contributing factors to the reduction of pressure at a pump impeller.

9.16   The operation of a centrifugal pump is based on _____.

9.17   The casing of a pump encloses the pump impeller, the shaft, and the _____.

9.18   The _____ is the part of the pump that supplies energy to the fluid.

9.19   If wearing rings are used only on the volute case, we must replace the _____ and _____ at the same time.

9.20   Which part of the end-suction pump directs water flow into and out of the pump?

9.21   What is the function of the pump's impeller?

9.22   The _____ pump has no bearings.

9.23   _____ split casings split perpendicular to the pump shaft.

9.24   Name three types of impellers.

9.25   A _____ casing adds a guiding vane to the fluid passage.

REFERENCES AND SUGGESTED READING

AWWA. 1995. *Basic Science Concepts and Applications: Principles and Practices of Water Supply Operations*, 2nd ed. Denver, CO: American Water Works Association.

Garay, P.N. 1990. *Pump Application Desk Book*. Lilburn, GA: The Fairmont Press.

Grimes, A.S. 1976. Supervisory and monitoring instrumentation. In *Pump Handbook*, Karassik, I.J. et al., Eds. New York: McGraw-Hill.

Hauser, B.A. 1993. *Hydraulics for Operators*. Boca Raton, FL: Lewis Publishers.

Hauser, B.A. 1996. *Practical Hydraulics Handbook*, 2nd ed. Boca Raton, FL: Lewis Publishers.

Hydraulic Institute. 1990. *The Hydraulic Institute Engineering Data Book*, 2nd ed. Cleveland, OH: Hydraulic Institute.

Hydraulic Institute. 1994. *Hydraulic Institute Complete Pump Standards*, 4th ed. Cleveland, OH: Hydraulic Institute.

Lindeburg, M.R. 1986. *Civil Engineering Reference Manual*, 4th ed. San Carlos, CA: Professional Publications.

OCDDS. 1986. *Basic Maintenance Training Course*. North Syracuse, NY: Onondaga County Department of Drainage and Sanitation.

Spellman, F.R. 1997. *The Science of Water*. Lancaster, PA: Technomic.

Spellman, F.R. 2000. *The Handbook for Waterworks Operator Certification*. Vol. 2. *Intermediate Level*. Lancaster, PA: Technomic.

TUA. 1988. *Manual of Water Utility Operations*, 8th ed. Austin: Texas Water Utilities Association.

U.S. Navy. 1963. *Class A Engineman Training Program*. Fleet Training Center.

VPISU. 2007. *Water Treatment Operators Short Course*. Blacksburg, VA: Virginia Polytechnic Institute and State University.

Wahren, U. 1997. *Practical Introduction to Pumping Technology*. Houston, TX: Gulf Publishing.

10 Water/Wastewater Conveyance

The design considerations for the piping system are the function of the specifics of the system. However, all piping systems have a few common issues: The pipe strength must be able to resist internal pressure, handling, and earth and traffic loads; the pipe characteristics must enable the pipe to withstand corrosion and abrasion and expansion and contraction of the pipeline (if the line is exposed to atmospheric conditions); engineers must select the appropriate pipe support, bedding, and backfill conditions; the design must account for the potential for pipe failure at the connection point to the basins due to subsidence of a massive structure; and the composition of the pipe must not give rise to any adverse effects on the health of consumers.

Kawamura (1999)

In the United States, water and wastewater conveyance systems are quite extensive. Consider, for example, wastewater conveyance. In the year 2000, the United States operated 21,264 collection and conveyance systems that included both sanitary and combined sewer systems. Publicly owned sewer systems in the country account for 724,000 miles of sewer pipe, and privately owned sewer pipe represents an additional 500,000 miles. Most of our nation's conveyance systems are beginning to show signs of aging, with some systems dating back more than 100 years (USEPA, 2006).

10.1 DELIVERING THE LIFEBLOOD OF CIVILIZATION

Conveyance or piping systems resemble veins, arteries, and capillaries. According to Nayyar (2000), "They carry the lifeblood of modern civilization. In a modern city they transport water from the sources of water supply to the points of distribution [and] convey waste from residential and commercial buildings and other civic facilities to the treatment facility or the point of discharge." Water and wastewater operators must be familiar with piping, piping systems, and the many components that make piping systems function. Operators are directly concerned with various forms of piping, tubing, hoses, and the fittings that connect these components to create workable systems.

This chapter covers important, practical information about the piping systems that are a vital part of plant operation, essential to the success of the total activity. To prevent major system trouble, skilled operators are called upon to perform the important function of preventive maintenance to avoid major breakdowns and must be able to make needed repairs when breakdowns do occur. A comprehensive knowledge of piping systems and accoutrements is essential to maintaining plant operations.

10.2 CONVEYANCE SYSTEMS

With regard to early conveyance systems, the prevailing practice in medieval England was the use of closed pipes. This practice was contrary to the Romans, who generally employed open channels in their long-distance aqueducts and used pipes mainly to distribute water within cities. The English preferred to lay long runs of pipes from the water source to the final destination. The Italians, on the other hand, where antique aqueduct arches are still visible, seem to have had more of a tendency to follow the Roman tradition of long-distance channel conduits. At least some of the channel aqueducts seem to have fed local distribution systems of lead or earthenware pipes (Magnusson, 2001).

With today's water and wastewater conveyance, not that much has changed from the past. Our goals today remain the same: (1) Convey water from source to treatment facility to user, and (2) convey wastewater from user to treatment to the environment. In water and wastewater operations, the term *conveyance* or *piping system* refers to a complete network of pipes, valves, and other components. For water and wastewater operations in particular, the piping system is all inclusive; it includes both the network of pipes, valves, and other components that bring the flow (water or wastewater) to the treatment facility as well as piping, valves, and other components that distribute treated water to the end-user or treated wastewater to outfall. In short, all piping systems are designed to perform a specific function.

Probably the best way to illustrate the importance of a piping system is to describe many of their applications used in water and wastewater operations. In the modern water and wastewater treatment plant, piping systems are critical to successful operation. In water and wastewater operations, fluids and gases are used extensively in processing operations; they usually are conveyed through pipes. Piping carries water and wastewater into the plant for treatment, fuel oil to heating units, steam to steam services, lubricants to machinery, compressed air to pneumatic service outlets for air-powered tools, etc., and chemicals to

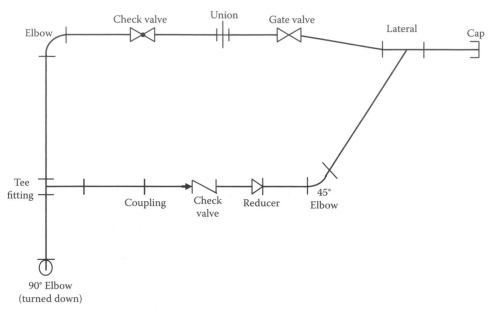

FIGURE 10.1 Various components in a single-line piping diagram.

unit processes. For water treatment alone, as Kawamura (1999) pointed out, there are "six basic piping systems: (1) raw water and finished waste distribution mains; (2) plant yard piping that connects the unit processes; (3) plant utility, including the fire hydrant lines; (4) chemical lines; (5) sewer lines; and (6) miscellaneous piping, such as drainage and irrigation lines." In addition to raw water, treated water, wastewater influent, and treated wastewater effluent, the materials conveyed through piping systems include oils, chemicals, liquefied gases, acids, paints, sludge, and many other others.

Important Point: Because of the wide variety of materials that piping systems can convey, the components of piping systems are themselves made of different materials and are furnished in many sizes to accommodate the requirements of numerous applications; for example, pipes and fittings can be made of stainless steel, many different types of plastic, brass, lead, glass, steel, and cast iron.

Any waterworks or wastewater treatment plant has many piping systems, not just the systems that convey water and wastewater. Along with those mentioned earlier, keep in mind that plant piping systems also include those that provide hot and cold water for plant personnel use. Another system heats the plant, while still another may be used for air conditioning.

Water and wastewater operators have many responsibilities and basic skills. The typical plant operator is skilled in heating, ventilating, and air conditioning (HVAC) systems, chemical feed systems, mechanical equipment operation and repair, and piping system maintenance activities; however, only the fluid transfer systems themselves are important to us in this text. The units that the piping system serves or supplies (such as pumps, unit processes, and machines) are discussed in other chapters of the text.

For water and wastewater operators, a familiar example of a piping system is the network of sodium hypochlorite pipes in treatment plants that use this chemical for disinfection and other purposes. The whole group of components—pipes, fittings, and valves—working together for one purpose makes up a *system*. This particular system has a definite purpose: to carry sodium hypochlorite and distribute it, conveying it to the point of application.

Note: This chapter is concerned only with the piping system used to circulate the chemical, not with the hypochlorination equipment itself. Our concern begins where the chemical outlet is connected to the storage tank and continues to the point where the pipe is connected to the point of application. The piping, fittings, and valves of the hypochlorination pipeline (and others) are important to us. Gate, needle, pressure-relief, air-and-vacuum relief, diaphragm, pinch butterfly, check, rotary and globe valves, traps, expansion joints, plugs, elbows, tee fittings, couplings, reducers, laterals, caps, and other fittings help ensure the effective flow of fluids through the lines. Tracing a piping system through a plant site will reveal many of them (see Figure 10.1). They are important because they are directly related to the operation of the system. Piping system maintenance is concerned with keeping the system functioning properly, and to function properly piping systems must be kept closed and leakproof.

Important Point: Figure 10.1 shows a single-line diagram that is similar to an electrical schematic. It uses symbols for all the diagram components. A double-line diagram (not shown here) is a pictorial view of the pipe, joints, valves, and other major components similar to an electrical wiring diagram vs. an electrical schematic.

10.3 DEFINITIONS

Key terms related to water/wastewater conveyance are defined in this section.

Absolute pressure—Gauge pressure plus atmospheric pressure.

Alloy—A substance composed of two or more metals.

Anneal—To heat and then cool a metal to make it softer and less brittle.

Annealing—Process of heating and then cooling a metal, usually to make it softer and less brittle.

Asbestos—Fibrous mineral form of magnesium silicate.

Backsiphonage—A condition in which the pressure in the distribution system is less than atmospheric pressure, which allows contamination to enter a water system through a cross-connection.

Bellows—A device that uses a bellow to measure pressure.

Bimetallic—Made of two different types of metal.

Bourdon tube—A semicircular tube of elliptical cross-section, used to sense pressure changes.

Brazing—Soldering with a nonferrous alloy that melts at a lower temperature than that of the metals being joined; also known as *hard soldering*.

Butterfly valve—A valve in which a disk rotates on a shaft as the valve opens and closes. In the full open position, the disk is parallel to the axis of the pipe.

Carcass—The reinforcement layers of a hose, between the inner tube and the outer cover.

Cast iron—A generic term for the family of high carbon–silicon–iron casting alloys including gray, white, malleable, and ductile iron.

Check valve—A valve designed to open in the direction of normal flow and close with reversal of flow. An approved check valve has substantial construction and suitable materials, is positive in closing, and permits no leakage in a direction opposite to normal flow.

Condensate—Steam that condenses into water in a piping system.

Diaphragm valve—Valve in which the closing element is a thin, flexible disk; often used in low-pressure systems.

Differential pressure—The difference between the inlet and outlet pressures in a piping system.

Double-line diagram—Pictorial view of the pipes, joints, valves, and other major components similar to an electrical wiring diagram.

Ductile—A term applied to a metal that can be fashioned into a new form without breaking.

Expansion joint—Absorbs thermal expansion and contraction in piping systems.

Extruding—Process of shaping a metal or plastic by forcing it through a die.

Ferrous—A term applied to a metal that contains iron.

Ferrule—A short bushing used to make a tight connection.

Filter—An accessory fitting used to remove solids from a fluid stream.

Fluids—Any substance that flows.

Flux—Used in soldering to prevent the formation of oxides during the soldering operation and to increase the wetting action so solder can flow more freely.

Friable—Readily crumbled by hand.

Gate valve—A valve in which the closing element consists of a disk that slides across an opening to stop the flow of water.

Gauge pressure—The amount by which the total absolute pressure exceeds the ambient atmospheric pressure.

Globe valve—A valve having a round, ball-like shell and horizontal disk.

Joint—A connection between two lengths of pipe or between a length of pipe and a fitting.

Laminar—Flow arranged in or consisting of thin layers.

Mandrel—A central core or spindle around which material may be shaped.

Metallurgy—The science and study of metals.

Neoprene—A synthetic material that is highly resistant to oil, flame, various chemicals, and weathering.

Nominal pipe size—The thickness given in the product material specifications or standard to which manufacturing tolerances are applied.

Nonferrous—A term applied to a material that does not contain iron.

Piping systems—A complete network of pipes, valves, and other components.

Ply—One of several thin sheets or layers of material.

Pressure-regulating valve—A valve with a horizontal disk for automatically reducing water pressures in a main to a preset value.

Prestressed concrete—Concrete that has been compressed with wires or rods to reduce or eliminate cracking and tensile forces.

PVC—Polyvinyl chloride plastic pipe.

Schedule—Approximate value of the expression $1000P/S$, where P is the service pressure and S is the allowable stress, both expressed in pounds per square inch.

Single-line diagram—Uses symbols for all the diagram components.

Soldering—A form of brazing in which nonferrous filler metals having melting temperatures below 800°F (427°C) are used. The filler material is called *solder* and is distributed between surfaces by *capillary action*.

Solenoid—An electrically energized coil of wire surrounding a movable iron case.

Stainless steel—Alloy steel having unusual corrosion-resisting properties, usually imparted by nickel and chromium.

Strainer—An accessory fitting used to remove large particles of foreign matter from a fluid.

Throttle—Controlling flow through a valve in intermediate steps between fully open and fully closed.

Tinning—Covering metal to be soldered with a thin coat of solder to work properly. Overheating or failure to keep the metal clean causes the point to become covered with oxide. The process of replacing this coat of oxide is called *tinning*.

Trap—An accessory fitting used to remove condensate from steam lines.

Vacuum breaker—A mechanical device that allows air into the piping system, thereby preventing backflow that could otherwise be caused by the siphoning action created by a partial vacuum.

Viscosity—The thickness or resistance to flow of a liquid.

Vitrified clay—Clay that has been treated in a kiln to produce a glazed, watertight surface.

Water hammer—The concussion of moving water against the sides of pipe, caused by a sudden change in the rate of flow or stoppage of flow in the line.

10.4 FLUIDS VS. LIQUIDS

We use the term *fluids* throughout this text to describe the substances being conveyed through various piping systems from one part of the plant to another. We normally think of pipes conveying some type of liquid substance, which most of us take to have the same meaning as fluid; however, a subtle difference exists between the two terms. The dictionary's definition of *fluid* is any substance that flows—which can mean a liquid or gas (e.g., air, oxygen, nitrogen). Some fluids carried by piping systems include thick viscous mixtures such as sludge in a semifluid state. Although sludge and other such materials might seem more solid (at times) than liquid, they do flow, and are considered fluids. In addition to carrying liquids such as oil, hydraulic fluids, and chemicals, piping systems carry compressed air and steam, which also are considered fluids because they flow.

Important Point: Fluids travel through a piping system at various pressures, temperature, and speeds.

10.5 MAINTAINING FLUID FLOW IN PIPING SYSTEMS

The primary purpose of any piping system is to maintain free and smooth flow of fluids through the system. Another purpose is to ensure that the fluids being conveyed are kept in good condition (i.e., free of contamination). Piping systems are purposely designed to ensure free and smooth flow of fluids throughout the system, but additional system components are often included to ensure that fluid quality is maintained. Piping system filters are one example, and strainers and traps are two others.

It is extremely important to maintain free and smooth flow and fluid quality in piping systems, especially those that feed vital pieces of equipment or machinery. Consider the internal combustion engine, for example. Impurities such as dirt and metal particles can damage internal components and cause excessive wear and eventual breakdown. To help prevent such wear, the oil is run continuously through a filter designed to trap and filter out the impurities.

Other piping systems require the same type of protection that the internal combustion engine does, which is why most piping systems include filters, strainers, and traps. These filtering components may prevent damage to valves, fittings, the pipe itself, and to downstream equipment/machinery. Chemicals, various types of waste products, paint, and pressurized steam are good examples of potentially damaging fluids. Filters and strainers play an important role in piping systems—protecting both the piping system and the equipment that the piping system serves.

10.5.1 SCALING

Because sodium and calcium hypochlorite are widely used in water/wastewater treatment operations, problems common in piping systems feeding this chemical are of special concern. In this section, we discuss *scaling* problems that can occur in piping systems that convey hypochlorite solution. To maintain the chlorine in solution (used primarily as a disinfectant), sodium hydroxide (caustic) is used to raise the pH of the hypochlorite; the excess caustic raises the shelf life. A high pH caustic solution raises the pH of the dilution water to over pH 9.0 after it is diluted. The calcium in the dilution water reacts with dissolved CO_2 and forms calcium carbonate. Experience has shown that 2-inch pipes have turned into 3/4-inch pipes due to scale buildup. The scale deposition is greatest in areas of turbulence such as pumps, valves, rotameters, and backpressure devices.

If lime (calcium oxide) is added (for alkalinity), plant water used as dilution water will have higher calcium levels and will generate more scale. Although it is true that softened water will not generate scale, it is also true that it is expensive in large quantities. Many facilities use softened water on hypochlorite mist odor scrubbers only.

Scaling also often occurs in solution rotameters, making flow readings impossible and freezing the flow indicator in place. Various valves can freeze up, and pressure-sustaining valves can freeze and become plugged. Various small diffuser holes fill with scale. To slow the rate of scaling, many facilities purchase water from local suppliers to dilute hypochlorite for the return activated sludge and miscellaneous uses. Some facilities have experimented with the system by not adding lime to it, but manganese dioxide (black deposits) has developed on the rotameters glass, making viewing the float impossible. In many instances, moving the point of hypochlorite addition downstream of the rotameter seems to solve the problem.

If remedial steps are not taken, scaling from hypochlorite solutions can cause problems; for example, scale buildup can reduce the inside diameter of a pipe so much that the actual supply of hypochlorite solution required to properly disinfect water or wastewater is reduced. As a result, the water sent to the customer or outfalled to the receiving body may not be properly disinfected. Because of the scale buildup, the treatment system itself will not function as designed and could result in a hazardous situation in which the reduced pipe size increases the pressure level to the point of catastrophic failure. Scaling, corrosion, or other clogging problems in certain piping systems are far from an ideal situation.

■ EXAMPLE 10.1

The scale problem can be illustrated by use of an example. Assume that we have a piping system designed to provide chemical feed to a critical plant unit process. The motive force for the chemical being conveyed is provided by a positive-displacement pump at a given volume of solution at 70 psi through clean pipe. After clogging occurs, the pump continues trying to force the same volume of chemical through the system at 70 psi, but the pressure drops to 25 psi because friction has caused the pressure drop. The reduction of the inside diameter of the pipe increased the friction between the chemical solution and the inside wall of the pipe.

Important Point: A basic principle in fluid mechanics states that fluid flowing through a pipe is affected by friction—the greater the friction, the greater the loss of pressure.

Important Point: Another principle or rule states that the amount of friction increases as the square of the velocity. (Note that speed and velocity are not the same, but common practice refers to the "velocity" of a fluid.) In short, if the velocity of the fluid doubles, the friction is increased four times what it was before. If the velocity is multiplied by 5, the friction is multiplied by 25, and so on.

In Example 10.1, the pressure dropped from 70 psi to 25 psi because the solution had to run faster to move through the pipe. Because the velocity of the solution pushed by the pump had to increase to levels above what it was when the pipe was clean, the friction increased at a higher rate than before. The friction loss was the reason why a pressure of 25 psi reached the far end of the piping system. The equipment designed to operate at a pressure of 70 psi could not work on the 25 psi of pressure being supplied.

Important Point: After reading the previous example, you might ask: "Why couldn't the pump be slowed down so the chemical solution could pass more slowly through the system, thus avoiding the effect of increased friction?" Lower pressure results when the pump speed is reduced. This causes other problems as well. Pumps that run at a speed other than that for which they are designed do so with a reduction in efficiency.

What is the solution to our pressure loss problem in Example 10.1? Actually, we can solve this problem two possible ways—either replace the piping or clean it. Replacing the piping or cleaning it sounds simple and straightforward, but it can be complicated. If the pipe is relatively short, no more than 20 to a few hundred feet in length, then we may decide to replace the pipe. What would we do if the pipe were 3 to 5 miles or more in length? Cleaning this length of pipe probably makes more sense than replacing its entire length. Each situation is different, requiring remedial choices based on practicality and expense.

10.5.2 PIPING SYSTEM MAINTENANCE

Maintaining a piping system can be an involved process; however, good maintenance practices can extend the life of piping system components and rehabilitation can further prolong their life. The performance of a piping system depends on the ability of the pipe to resist unfavorable conditions and to operate at or near the capacity and efficiency that is was designed for. This performance can be checked in several ways: flow measurement, fire flow tests, loss-of-head tests, pressure tests, simultaneous flow and pressure tests, tests for leakage, and chemical and bacteriological water tests. These tests are an important part of system maintenance. They should be scheduled as part of the regular operation of the system (AWWA, 1996).

Most piping systems are designed with various protective features, included minimizing wear and catastrophic failure, as well as the amount of maintenance required. Such protective features include pressure-relief valves, blow-off valves, and clean-out plugs:

- *Pressure-relief valves*—A valve that opens automatically when the fluid pressure reaches a preset limit to relieve the stress on a piping system
- *Blow-off valve*—A valve that can be opened to blow out any foreign material in a pipe
- *Clean-out plug*—A threaded plug that can be removed to allow access to the inside of the pipe for cleaning

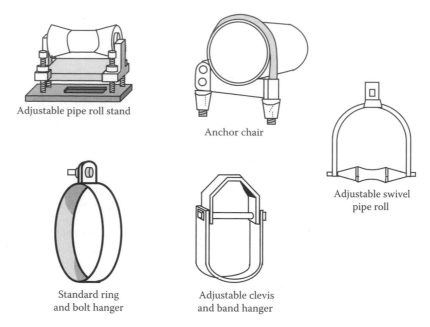

Adjustable pipe roll stand

Anchor chair

Adjustable swivel
pipe roll

Standard ring
and bolt hanger

Adjustable clevis
and band hanger

FIGURE 10.2 Pipe hangers and supports.

Important Point: Use caution when removing a clean-out plug from a piping system. Before removing the plug, pressure must be cut off and system bled of residual pressure.

Many piping systems (including water distribution networks and wastewater lines and interceptors) can be cleaned either by running chemical solvents through the lines or by using mechanical clean-out devices.

10.6 PIPING SYSTEM ACCESSORIES

Depending on the complexity of the piping system, the number of valves included in a system can range from no more than one in a small, simple system to a large number in very complex systems such as water distributions systems. *Valves* are necessary for both the operation of a piping system and for control of the system and system components. In water/wastewater treatment, this control function is used to control various unit processes, pumps, and other equipment. Valves also function as protective devices; for example, valves used to protect a piping system may be designed to open automatically to vent fluid out of the pipe when the pressure in the lines becomes too high. In lines that carry liquids, relief valves preset to open at a given pressure are commonly used.

Important Point: Not all valves function as safety valves; for example, hand-operated gate and globe valves function primarily as control valves.

The size and type of valve are selected depending on its intended use. Most valves require periodic inspection to ensure that they are operating properly.

Along with valves, piping systems typically include accessories such as pressure and temperature gauges, filters, strainers, and pipe hangers, and supports:

- *Pressure gauges* indicate the pressure in the piping system.
- *Temperature gauges* indicate the temperature in the piping system.
- *Filters* and *strainers* are installed in piping systems to help keep fluids clean and free from impurities.
- *Pipe hangers and supports* support piping to keep the lines straight and prevent sagging, especially in long runs. Various types of pipe hangers and supports are shown in Figure 10.2.

10.7 PIPING SYSTEMS: TEMPERATURE EFFECTS AND INSULATION

Most materials, especially metals, expand as the temperature increases and contract as the temperature decreases. This can be a significant problem in piping systems. To combat this problem and to allow for expansion and contraction in piping systems, expansion joints must be installed in the line between sections of rigid pipe. An *expansion joint* absorbs thermal expansion or terminal movement; as the pipe sections expand or contract with the temperature, the expansion joint expands or compresses accordingly, eliminating stress on the pipes.

Piping system temperature requirements also have an impact on how pipes are insulated; for example, we do not have to wander too far in most plant sites to find pipes covered with layers of piping insulation. Piping insulation amounts to wrapping the pipe in a envelope of insulating material. The thickness of the insulation depends on the application. Under normal circumstances, heat passes from a hot or warm surface to a cold or cooler one. Insulation helps prevent hot fluid from cooling as it passes through the system. For systems conveying cold fluid, insulation helps keep the fluid cold. Materials used for insulation vary, and they are selected according to the requirements of application. Various types of insulating materials are also used to protect underground piping against rusting and corrosion caused by exposure to water and chemicals in the soil.

10.8 METALLIC PIPING

Pipe materials that are used to transport water may also be used to collect wastewater. It is more usual, however, to employ less expensive materials because wastewater lines rarely are required to withstand any internal pressure. Iron and steel pipe are used to convey wastewater only under unusual loading conditions or for force mains (interceptor lines) in which the wastewater flow is pressurized (McGhee, 1991).

10.8.1 PIPING MATERIALS

Materials selected for piping applications must be chosen keeping in mind the requirements for the intended service; for example, the piping material selected must be suitable for the flow medium and the given operating conditions of temperature and pressure during the intended design life of the product. For long-term service capability, the mechanical strength of the material must be appropriate; for example, the piping material must be able to resist operational variables such as thermal or mechanical cycling. Extremes in application temperatures must also be considered with respect to material capabilities.

Environmental factors must also be considered. The operating environment surrounding the pipe or piping components affects pipe durability and life span. Corrosion, erosion, or a combination of the two can result in degradation of material properties or loss of effective load-carrying cross-section. The nature of the substance contained by the piping is an important factor, as well.

Knowledge of the basic characteristics of the metals and nonmetals used for piping provides clues to the uses of the piping materials in water/wastewater treatment operations. Such knowledge is especially helpful to operators, as it makes their jobs much easier and more interesting. In this section, metallic piping is discussed. Piping joints, how to join or connect sections of metallic piping, and how to maintain metallic pipe are also discussed.

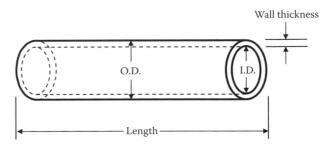

FIGURE 10.3 Pipe terminology.

10.8.2 PIPING: THE BASICS

Earlier, we pointed out that piping includes pipe, flanges, fittings, bolting, gaskets, valves, and the pressure-containing portions of other piping components.

Important Point: According to Nayyar (2000), "A *pipe* is a tube with round cross-section conforming to the dimensional requirements of ASME B36.10M (Welded and Seamless Wrought Steel Pipe) and ASME B36.19M (Stainless Steel Pipe)."

Piping also includes pipe hangers and supports and other accessories necessary to prevent overpressurization and overstressing of the pressure-containing components. From a system viewpoint, a pipe is one element or a part of piping. Accordingly, when joined with fittings, valves, and other mechanical devices or equipment, pipe sections are referred to as *piping*.

10.8.3 PIPE SIZES

With time and technological advancements (development of stronger and corrosion-resistant piping materials), pipe sizes have become standardized and are usually expressed in inches or fractions of inches. As a rule, the size of a pipe is given in terms of its outside or inside diameter. Figure 10.3 shows the terminology that applies to a section of pipe. Pipes are designated by diameter. The principal dimensions are as follows:

- *Wall thickness*
- *Length*
- *Outside diameter* (O.D.), which is used to designate pipe sizes greater than 12 inches in diameter
- *Inside diameter* (I.D.), which is used to designate pipe sizes less than 12 inches in diameter

Important Point: Another important pipe consideration not listed above or shown in Figure 10.3 is *weight per foot*, which varies according to the pipe material and the wall thickness of the pipe.

In the continuing effort to standardize pipe size and wall thickness of pipe, the designation *nominal pipe size* (NPS) replaced the iron pipe size designation, and the term *schedule* (SCH) was introduced to specify the nominal wall thickness of pipe. The NPS diameter (approximate dimensionless designator of pipe size) is generally somewhat different from its actual diameter; for example, the pipe we refer to as a "3-in.-diameter pipe" has an actual O.D. of 3.5 in., while the actual O.D. of a "12-in. pipe" may be .075 in. greater (i.e., 12.750 in.) than the nominal diameter. On the other hand, a pipe 14 in. or greater in diameter has an actual O.D. equal to the nominal size. The inside diameter will depend on the pipe wall thickness specified by the schedule number.

Important Point: Keep in mind that whether the O.D. is small or large, the dimensions must be within certain tolerances in order to accommodate various fittings.

10.8.3.1 Pipe Wall Thickness

Original pipe wall thickness designations of STD (standard), XS (extra-strong), and XXS (double extra-strong) are still in use today; however, because this system allows no variation in wall thickness and because pipe requirements became more numerous, greater variation was needed. As a result, today the pipe wall thickness, or *schedule*, is expressed in numbers (e.g., 5, 5S, 10, 10S, 20, 20S, 30, 40, 40S, 60, 80, 80S, 100, 120, 140, 160). (Note that you may often hear piping referred to either in terms of its diameter or schedule number.) The most common schedule numbers are 40, 80, 120, and 160. The outside diameter of each pipe size is standardized; therefore, a particular nominal pipe size will have a different inside diameter depending on the schedule number specified; for example, a Schedule 40 pipe with a 3-in. nominal diameter (actual O.D. of 3.500 in.) has a wall thickness of 0.216 in. The same pipe in a Schedule 80 (XS) would have a wall thickness of 0.300 in.

Important Point: A schedule number indicates the approximate value of the expression $1000P/S$, where P is the service pressure and S is the allowable stress, both expressed in pounds per square inch (psi). The higher the schedule number, the thicker the pipe is.

Important Point: Schedule numbers followed by the letter S are per ASME B36.19M, and they are primarily intended for use with stainless steel pipe (ASME, 1996).

10.8.3.2 Piping Classification

The usual practice is to classify pipe in accordance with the pressure/temperature rating system used for classifying flanges; however, because of the increasing variety and complexity of requirements for piping, a number of engineering societies and standards groups have devised codes, standards, and specifications that meet most applications. By consulting such codes (e.g., ASTM, manufacturers' specifications, NFPA, AWWA, and others), a designer can determine exactly what piping specification should be used for any application.

Important Point: Because pipelines often carry hazardous materials and fluids under high pressures, following a code helps ensure the safety of personnel, equipment, and the piping system itself.

1. *ASTM ratings.* The American Society for Testing and Materials (ASTM) publishes standards (codes) and specifications that are used to determine the minimum pipe size and wall thickness for a given application.
2. *Manufacturers' ratings.* Pipe manufacturers, because of their proprietary pipe, fitting, or joint designs, often assign a pressure/temperature rating that may form the basis of the design of the piping system. (In addition, the manufacturer may impose limitations that must be adhered to.)

Important Point: Under no circumstances should the manufacturer's rating be exceeded.

3. *NFPA ratings.* Certain piping systems fall within the jurisdiction of the National Fire Protection Association (NFPA). These pipes are required to be designed and tested to certain required pressures (usually rated for 175 psi, 200 psi, or as specified).
4. *AWWA ratings.* The American Water Works Association (AWWA) publishes standards and specifications that are used to design and install water pipelines and distribution system piping. The ratings used may be in accordance with the flange ratings of AWWA, or the rating could be based on the rating of the joints used in the piping.
5. *Other ratings.* Sometimes a piping system may not fall within the above rating systems. In this case, the designer may assign a specific rating to the piping system. This is a common practice in classifying or rating piping for main steam or hot reheat piping of power plants whose design pressure and design temperature may exceed the pressure/temperature rating of ASME B16.5. In assigning a specific rating to such piping, the rating must be equal to or higher than the design conditions.

Important Point: The rating of all pressure-containing components in the piping system must meet or exceed the specific rating assigned by the designer (Nayyar, 2000).

When piping systems are subjected to full-vacuum conditions or are submerged in water, they experience both the internal pressure of the flow medium and external pressure. In such instances, piping must be rated for both internal and external pressures at the given temperature. Moreover, if a piping system is designed to handle more than one flow medium during its different modes of operation, it must be assigned a dual rating for two different flow media.

10.8.4 Types of Piping Systems

Piping systems consist of two main categories: process lines and service lines. Process lines convey the flow medium used in a manufacturing process or a treatment process (such as fluid flow in water or wastewater treatment); for example, a major unit process operation in wastewater treatment is sludge digestion. The sludge is converted from bulky, odorous, raw sludge to a relatively inert material that can be rapidly dewatered with the absence of obnoxious odors. Because sludge digestion is a unit process operation, the pipes used in the system are called *process lines*. *Service lines* (or utility lines) carry water, steam, compressed air, air conditioning fluids, and gas. Normally, all or part of the general service system of a plant is composed of service lines. Service lines cool and heat the plant, provide water where it is needed, and carry the air that drives air equipment and tools.

10.8.4.1 Code for Identification of Pipelines

Under guidelines provided by the American National Standards Institute (ANSI A13.1), a code has been established for the identification of pipelines. This code involves the use of nameplates (tags), legends, and colors. The code states that the contents of a piping system must be identified by lettered legend giving the name of the contents. In addition, the code requires that information relating to temperature and pressure be included. Stencils, tape, or markers can be used to accomplish the marking. To identify the characteristic hazards of the contents, color should be used, but its use must be in combination with legends.

Important Point: Not all plants follow the same code recommendations, which can be confusing for those not familiar with the system used. Standard piping color codes are often used in water and wastewater treatment operations. Plant maintenance operators must be familiar with the pipe codes used in their plants.

10.8.5 Metallic Piping Materials

In the not too distant past, it was not (relatively speaking) that difficult to design certain pipe delivery systems; for example, several hundred years ago (and even more recently, in some cases), when it was desirable to convey water from a source to point of use, the designer was faced with only two issues. First, a source of freshwater had to be found. Next, if the source was found and it was determined suitable for whatever the need, a means of conveying the water to point of use was necessary.

When designing early water conveyance systems, gravity was the key player. This point is clear when we consider that, before the advent of the pump, before a motive force to power the pump and the energy required to provide power to the motive force were developed, gravity was the means by which water was conveyed from one location to another (with the exception of physically carrying the water). Early gravity conveyance systems employed clay pipe, wood pipe, natural gullies or troughs, aqueducts fashioned from stone, and any other means available to convey the water. Some of these earlier pipe or conveyance materials are still in use today. Along with the advent of modern technology (electricity, the electric motor, the pump, and various machines and processes) and the need to convey fluids other than water came the need to develop piping materials that could carry a wide variety of fluids.

Modern waterworks have a number of piping systems made up of various materials. One of the principal materials used in piping systems is metal. Metal pipes may be made of cast iron, stainless steel, brass, copper, or various alloys. Waterworks and wastewater maintenance operators who work with metal piping must be knowledgeable about the characteristics of individual metals as well as the various considerations common to all piping systems. These considerations include the effect of temperature changes, impurities in the line, shifting of pipe supports, corrosion, and water hammer.

In this section, we present information about pipes made of cast iron, steel, copper, and other metals. We also discuss the behavior of fluids in a piping system, and the methods of connection sections of pipe.

10.8.6 Characteristics of Metallic Materials

Metallurgy (the science and study of metals) deals with the extraction of metals from ores and with combining, treating, and processing of metals into useful materials. Different metals have different characteristics, making them usable in a wide variety of applications. Metals are divided into two types: *ferrous*, which includes iron and iron-base alloys (a metal made up of two or more metals that dissolve into each other when melted together), and *nonferrous*, which includes other metals and alloys.

Important Point: Mixing a metal and a nonmetal, such as steel, which is a mixture of iron (a metal) and carbon (a nonmetal), can also form an alloy.

A *ferrous* metal is one that contains iron (Fe). Iron is one of the most common of metals but is rarely found in nature in its pure form. Comprising about 6% of the

Earth's crust, iron ore is actually in the form of iron oxides (Fe_2O_3 or Fe_3O_4). Coke and limestone are used in the reduction of iron ore in a blast furnace, where oxygen is removed from the ore, leaving a mixture of iron and carbon and small amounts of other impurities. The end product removed from the furnace is called *pig iron*—an impure form of iron. Sometimes the liquid pig iron is cast from the blast furnace and used directly for metal castings; however, the iron is more often remelted in a furnace, to further refine it and adjust its composition (Babcock & Wilcox, 1972).

Important Point: Piping is commonly made of wrought iron, cast iron, or steel. The difference among them lies largely in the amount of carbon that each contains.

Remelted pig iron is known as *cast iron* (meaning the iron possesses carbon in excess of 2% weight). Cast iron is inferior to steel in malleability, strength, toughness, and ductility (i.e., it is hard and brittle). Cast iron, however, has better fluidity in the molten state and can be cast satisfactorily into complicated shapes.

Steel is an alloy of iron having not more than 2.0% by weight carbon. The most common method of producing steel is to refine pig iron by oxidation of impurities and excess carbon, both of which have a higher affinity for oxygen than iron. *Stainless steel* is an alloy of steel and chromium.

Important Point: When piping is made of stainless steel, an "S" after the schedule number identifies it as such.

Various heat treatments can be used to manipulate specific properties of steel, such as hardness and ductility (meaning it can be fashioned into a new form without breaking). One of the most common heat treatments employed in steel processing is *annealing*. Annealing (sometimes referred to as *stress relieving*) consists of heating the metal and permitting it to cool gradually to make it softer and less brittle.

Important Point: Steel is one of the most important basic production materials of modern industry.

Nonferrous metals, unlike ferrous metals, do not contain iron. A common example of a nonferrous metal used in piping is brass. Other examples of nonferrous materials used in pipe include polyethylene, polybutylene, polyurethane, and polyvinyl chloride (PVC). Pipes of these materials are commonly used in low-pressure applications for transporting coarse solids (Snoek and Carney, 1981).

In addition to the more commonly used ferrous and nonferrous metals, special pipe materials for special applications are also gaining wider use in industry—even though they are more expensive. Probably one of the most commonly used materials that falls into this category is aluminum pipe. Aluminum pipe has the advantage of being lightweight and corrosion resistant with relatively good strength characteristics.

Important Point: Although aluminum is relatively strong, it is important to note that its strength decreases as temperature increases.

Lead is another special pipe material used for certain applications, especially where a high degree of resistance to corrosive materials is desired. Tantalum, titanium, and zirconium piping materials are also highly resistant to corrosives.

Piping systems convey many types of water including service water, city water, treated or processed water, and distilled water. Service water, used for flushing and cooling purposes, is untreated water that is usually strained but is otherwise raw water taken directly from a source (e.g., lake, river, or deep well). City water is treated potable water. Treated water has been processed to remove various minerals that could cause deterioration or sludge in piping. Distilled water is specially purified.

Important Point: Piping materials selection for use in water treatment and distribution operations should be based on commonly accepted piping standards such as those provided by the American Society for Testing and Materials (ASTM), American Water Works Association (AWWA), American National Standards Institute (ANSI), the American Society of Mechanical Engineers (ASME), and the American Petroleum Industry (API).

10.8.6.1 Cast-Iron Pipe

According to the AWWA (1996), "There are more miles of [cast-iron pipe] in use today than of any other type. There are many water systems having cast-iron mains that are over 100 years old and still function well in daily use." Cast-iron pipe has the advantages of strength, long service life, and being reasonably maintenance free. The disadvantages include its being subject to electrolysis and attack from acid and alkali soil and its heaviness (Gagliardi and Liberatore, 2000).

10.8.6.2 Ductile-Iron Pipe

Ductile-iron pipe resembles cast-iron pipe in appearance and has many of the same characteristics. It differs from cast-iron pipe in that the graphite in the metal is of a spheroidal or nodular form; that is, it is in a ball-shape form rather than a flake form. Ductile-iron pipe is strong and durable, has high flexural strength and good corrosion resistance, is lighter weight than cast iron, has greater carrying capacity for same external diameter, and is easily tapped. Ductile-iron pipe, however, is subject to general corrosion when installed unprotected in a corrosive environment (Gigliarda and Liberatore, 2000).

10.8.6.3 Steel Pipe

Steel pipe is sometimes used for large feeder mains in water distribution systems. It is frequently used where the pressure is particularly high or where very large diameter pipe is required. Steel pipe has high tensile strength and offers lower costs; it is relatively easy to install; and it has good hydraulic characteristics when lined. It can also be adapted to locations where some movement may occur. Steel pipe, however, is subject to electrolysis and external corrosion in acid or alkali soil, and it has poor corrosion resistance unless properly lined, coated, and wrapped.

Note: The materials of which street wastewater (sewer) pipes are most commonly constructed are vitrified clay pipe, plastic, concrete, and ductile-iron pipe; however, metallic ductile-iron pipe is most commonly used in wastewater collection—primarily for force mains (e.g., interceptor lines) and for piping in and around buildings. Ductile-iron pipe is generally not used for gravity sewer applications, however.

10.8.7 Maintenance Characteristics of Metallic Piping

The maintenance required for metallic piping is determined in part by characteristics of the metal (i.e., expansion, flexibility, and support) but also includes the kind of maintenance common to nonmetallic piping systems. The major considerations include:

- Expansion and flexibility
- Pipe support systems
- Valve selection
- Isolation
- Preventing backflow
- Water hammer
- Air binding
- Corrosion effects

10.8.7.1 Expansion and Flexibility

Because of thermal expansion, water and wastewater systems (which are rigid and laid out in specified lengths) must have adequate flexibility. In water and wastewater systems without adequate flexibility, thermal expansion may lead to failure of piping or anchors; moreover, it may also lead to joint leakage and excessive loads on appurtences. The thermal expansion of piping can be controlled by properly locating anchors, guides, and snubbers. Where expansion cannot be controlled, flexibility is provided by use of bends, loops, or expansion joints (Gigliardi and Liberatore, 2000).

Important Point: Metals expand or contract according to temperature variations. Over a long run (length of pipe), the effects can cause considerable strain on the lines, and damage or failure may result.

10.8.7.2 Pipe Support Systems

Pipe supports are normally used to carry dead weight and thermal expansion loads. Pipe supports may loosen in time, so they require periodic inspection. Along with normal expansion and contraction, vibration (water hammer or fluids traveling at high speeds and pressures) can cause the supports to loosen.

10.8.7.3 Valve Selection

Proper valve selection and routine preventive maintenance are critical to the operation and maintenance of any piping system. In water/wastewater piping systems, valves are generally used for isolating a section of a water main/wastewater collection line, draining the water/wastewater line, throttling liquid flow, regulating water/wastewater storage levels, controlling water hammer, bleeding off air, or preventing backflow.

10.8.7.4 Isolation

Various valves are used in piping systems to provide for isolation; for example, gate valves are used to isolate specific areas (valve closed) of the system during repair work or to reroute water/wastewater flow (valve open) throughout the distribution or collection system. Service stop valves are commonly used to shut off service lines to individual homes or industries. Butterfly valves are also used for isolation purposes.

10.8.7.5 Preventing Backflow

Backflow, or reversed flow, could result in contaminated or polluted water entering the potable water system. Water distribution systems have numerous places where unsafe water may be drawn into the potable water mains if a temporary vacuum should occur in the system. In addition, contaminated water from a higher pressure source can be forced through a water system connection that is not properly controlled. A typical backflow condition from a recirculated system is illustrated in Figure 10.4.

Important Point: Valves, air gaps, reduced-pressure-zone backflow preventers, vacuum breakers, and barometric loops are often used as backflow-prevention devices, depending on the situation.

10.8.7.6 Water Hammer

In water/wastewater operations specifically involving flow through piping, we often hear the term *water hammer*,

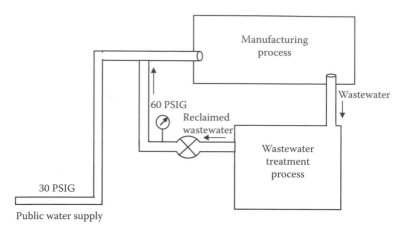

FIGURE 10.4 Backflow from recirculated system

which is actually a misnomer in that it implies only water and the connotation of a hammering noise; however, it has become a generic term for pressure wave effects in liquids. By definition, *water hammer* (often called *surging*) is a pressure (acoustic) wave phenomenon created by relatively sudden changes in the liquid velocity. In pipelines, sudden changes in the flow (velocity) can occur as a result of: (1) pump and valve operation in pipelines, (2) vapor pocket collapse, or (3) even the impact of water following the rapid expulsion of air out of a vent or a partially open valve (Marine, 1999). Water hammer can damage or destroy piping, valves, fittings, and equipment.

Important Point: When water hammer occurs, there is little the maintenance operator can do except to repair any damage that results.

10.8.7.7 Air Binding

Air enters a piping system from several sources, such as the release of air from the water, air carried in through vortices into the pump suction, air leaking in through joints that may be under negative pressure, and air present in the piping system before it is filled. The problem with air entry or air binding due to the accumulation of air in the piping is that the effective cross-sectional area for water/wastewater flow in the piping is reduced. This flow reduction can, in turn, lead to an increase in pumping costs through the resulting extra head loss.

10.8.7.8 Corrosion Effects

All metallic pipes are subject to corrosion. Many materials react chemically with metal piping to produce rust, scale, and other oxides. With regard to water treatment processes, when raw water is taken from wells, rivers, or lakes, the water solution is an extremely dilute liquid of mineral salts and gases. The dissolved mineral salts are a result of water flowing over and through the earth layers. The dissolved

gases are atmospheric oxygen and carbon dioxide, picked up by water–atmosphere contact. Wastewater picks up corrosive materials mainly from industrial processes or from chemicals added to the wastewater during treatment.

Important Point: Materials such as acids, caustic solutions, and similar solutions are typical causes of pipe corrosion.

Several types of corrosion should be considered in water or wastewater distribution and collection piping systems (AWWA, 1996):

1. *Internal corrosion*, which is caused by aggressive water flowing through the pipes
2. *External corrosion*, which is caused by chemical and electrical conditions in the soil
3. *Bimetallic corrosion*, which is caused when components made of dissimilar metals are connected
4. *Stray-current corrosion*, which is caused by uncontrolled DC electrical currents flowing in the soil

10.8.8 Joining Metallic Pipe

According to Crocker (2000), pipe joint design and selection can have a major impact on the initial cost, long-range operating cost, and performance of the piping system. When determining the type of joint to be used in connecting pipe, certain considerations must be made; for example, initial considerations include material costs, installation labor costs, and degree of leakage integrity required. The operator is also concerned with periodic maintenance requirements and specific performance requirements.

Metallic piping can be joined or connected in a number of ways. The method used depends on: (1) the nature of the metal sections (ferrous, nonferrous) being joined, (2) the kind of liquid or gas to be carried by the system,

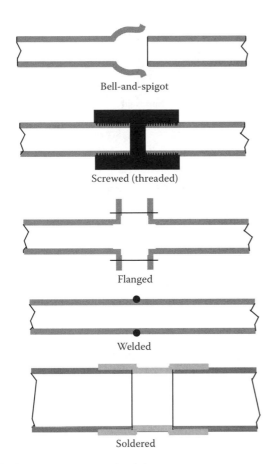

FIGURE 10.5 Common pipe joints.

(3) pressure and temperature in the line, and (4) access requirements.

A *joint* is defined simply as the connection between elements in a piping system. Five major types of joints, each for a special purpose, are used for joining metal pipe (see Figure 10.5):

1. Bell-and-spigot joints
2. Screwed or threaded joints
3. Flanged joints
4. Welded joints
5. Soldered joints

10.8.8.1 Bell-and-Spigot Joints

The bell-and-spigot joint has been around since its development in the late 1780s. The joint is used for connecting lengths of cast-iron water and wastewater pipe (gravity flow only). The *bell* is the enlarged section at one end of the pipe; the plain end is the *spigot* (see Figure 10.5). The spigot end is placed into the bell, and the joint is sealed. The joint sealing compound is typically made up with lead and oakum. Lead and oakum are the prevailing joint sealers for sanitary systems. Bell-and-spigot joints are usually reserved for sanitary sewer systems; they are no longer used in water systems.

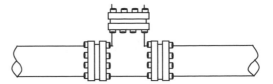

FIGURE 10.6 Flanged assembly.

Important Point: Bell-and-spigot joints are not used for ductile-iron pipe.

10.8.8.2 Screwed or Threaded Joints

Screwed or threaded joints (see Figure 10.6) are commonly used to join sections of smaller diameter, low-pressure pipe; they are used in low-cost, noncritical applications such as domestic water, industrial cooling, and fire protection systems. Diameters of ferrous or nonferrous pipe joined by threading range from 1/8 in. up to 8 in. Most couplings have threads on the inside surface. The advantages of this type of connection are its relative simplicity, ease of installation (where disassembly and reassembly are necessary to accommodate maintenance needs or process changes), and high leakage integrity at low pressure and temperature where vibration is not encountered. Screwed construction is commonly used with galvanized pipe and fittings for domestic water and drainage applications.

Important Point: Maintenance supervisors must ensure that screwed or threaded joints are used within the limitations imposed by the rules and requirements of the applicable code.

10.8.8.3 Flanged Joints

As shown in Figure 10.6, flanged joints consist of two machined surfaces that are tightly bolted together with a gasket between them. The flange is a rim or ring at the end of the fitting, which mates with another section. Flanges are joined by being either bolted together or welded together. Some flanges have raised faces and others have plain faces, as shown in Figure 10.7. Steel flanges generally have raised faces, and iron flanges usually have plain or flat faces.

Important Point: A flange with a raised face should never be joined to one with a plain face.

Flanged joints are used extensively in water and wastewater piping systems because of their ease of assembly and disassembly; however, they are expensive. Contributing to the higher cost are the material costs of the flanges themselves and the labor costs for attaching the flanges to the pipe and then bolting the flanges to each other (Crocker, 2000). Flanged joints are not normally

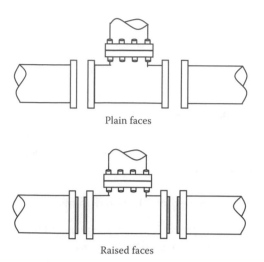

Plain faces

Raised faces

FIGURE 10.7 Flange faces.

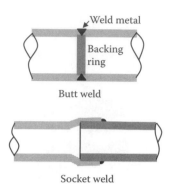

Butt weld

Socket weld

FIGURE 10.8 Two kinds of welding pipe joints.

used for buried pipe because of their lack of flexibility to compensate for ground movement. Instead, flanged joints are primarily used in exposed locations where rigidity, self-restraint, and tightness are required (e.g., inside treatment plants and pumping stations).

10.8.8.4 Welded Joints

For applications involving high pressures and temperatures, *welded joints* are preferred. Welding of joints is the process whereby metal sections to be joined are heated to such a high temperature that they melt and blend together. The advantage of welded joints is obvious: The pieces joined become one continuous piece. When a joint is properly welded, the joint is as strong as the piping itself. The two basic welded joints are (see Figure 10.8):

1. *Butt-welded joints*, in which the sections to be welded are placed end to end; this is the most common method of joining pipe used in large industrial piping systems.
2. *Socket-welded joints*, in which one pipe fits inside the other, the weld being made on the outside of the lap; this type of joint is used in applications where leakage integrity and structural strength are important.

10.8.8.5 Soldered and Brazed Joints

Soldered and brazed joints are most often used to join copper and copper-alloy (nonferrous metals) piping systems, although brazing of steel and aluminum pipe and tubing is possible. The main difference between *brazing* and welding is the temperatures used with each process. Brazing is accomplished at far lower temperatures. Brazing, in turn, requires higher temperatures than *soldering*. In both brazing and soldering, the joint is cleaned (using

emery cloth) and then coated with flux that prevents oxides from forming. The clean, hot joint draws solder or brazing rod (via capillary action) into the joint to form the connection. The parent metal does not melt in brazed or soldered construction.

10.9 NONMETALLIC PIPING

Although metal piping is in wide use today, nonmetallic piping (especially clay and cement) is of equal importance. New processes to make them more useful in meeting today's requirements have modified these older materials. Relatively speaking, using metallic piping is a new practice. Originally, all piping was made from clay or wood, and stone soon followed. Open stone channels or aqueducts were used to transport water over long distances. After nearly 2000 years of service, some of these open channels are still in use today.

Common practice today is to use metal piping, although nonmetallic piping is of equal importance and has many applications in water/wastewater operations. Many of the same materials that have been used for centuries (clay, for example) are still used today, but now many new piping materials are available, and the choice depends on the requirements of the planned application. The development of new technological processes has enabled the modification of older materials for new applications in modern facilities and has brought about the use of new materials for old applications, as well.

In this section, we discuss nonmetallic piping materials: what they are and where they are most commonly used. We also describe how to join sections of nonmetallic piping and how to maintain them.

10.9.1 NONMETALLIC PIPING MATERIALS

Nonmetallic piping materials used in water/wastewater applications include clay (wastewater), concrete (water/wastewater), asbestos–cement pipe (water/wastewater), and plastic (water/wastewater). Other nonmetallic piping materials include glass (chemical porcelain pipe) and wood (continuous-strip wooden pipes for carrying water and waste chemicals are used in some areas, especially in the western part of the United States); however, these materials are not discussed in this text because of their limited application in water and wastewater operations.

Important Point: As with the use of metallic piping, non-metallic piping must be used in accordance with specifications established and codified by a number of engineering societies and standards organizations. These codes were devised to help ensure personnel safety and protection of equipment.

10.9.1.1 Clay Pipe

Clay pipes are used to carry or collect industrial wastes, wastewater, and stormwater (they are not typically used to carry potable water). Clay pipes typically range in size from 4 to 36 inches in diameter and are available in more than one grade and strength. Clay pipe is used in nonpressurized systems. When used in drainpipe applications, for example, liquid flow is solely dependent on gravity; that is, the clay pipe is used as an open-channel pipe, whether partially or completely filled. Clay pipe is manufactured in two forms:

* Vitrified (glass-like)
* Unglazed (nonglassy)

Important Point: *Vitrified clay pipe* is extremely corrosion proof. It is ideal for many industrial waste and wastewater applications.

Important Point: McGhee (1991) recommended that wyes and tees (see Figure 10.9) should be used for joining various sections of wastewater piping. Failure to provide wyes and tees in common wastewater lines invites builders to break the pipe to make new connections. Obviously, this practice should be avoided, because such breaks are seldom properly sealed and can be a major source of infiltration.

Both vitrified and unglazed clay pipe is made and joined with the same type of bell-and-spigot joint described earlier. The bell-and-spigot shape is shown in Figure 10.10. In joining sections of clay pipe, both ends of the pipe must first be thoroughly cleaned. The small (spigot) end of the pipe must be centered properly and then seated securely in the large (bell) end. The bell is then packed with fibrous material (usually jute) for solid joints, which is tamped down until about 30% of the space is filled. The joint is then filled with sealing compound. In flexible joint applications, the sealing elements are made from natural or synthetic rubber or from a plastic-type material.

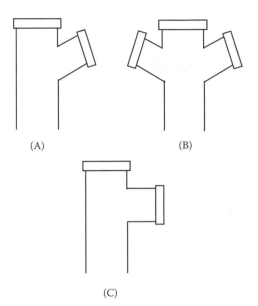

FIGURE 10.9 Section of bell-and-spigot fillings for clay pipe: (A) wye, (B) double wye, and (C) tee.

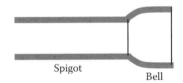

Spigot   Bell

FIGURE 10.10 Bell-and-spigot ends of clay pipe sections.

Drainage and wastewater collection lines designed for gravity flow are laid downgrade at an angle, with the bell ends of the pipe pointing upgrade. The pipe is normally placed in a trench with strong support members (along its small dimension and not on the bell end). Vitrified clay pipe can be placed directly into a trench and covered with soil; however, unglazed clay pipe must be protected against the effects of soil contaminants and ground moisture.

10.9.1.2 Concrete Pipe

Concrete is another common pipe material and is sometimes used for sanitary sewers in locations where grades, temperatures, and wastewater characteristics prevent corrosion (ACPA, 1987). The pipe provides both high tensile and compressive strength and corrosion resistance. Concrete pipe is generally found in three basic forms: (1) nonreinforced concrete pipe; (2) reinforced concrete cylinder and noncylinder pipe; and (3) reinforced and prestressed concrete pressure pipe. With the exception of reinforced and prestressed pressure pipe, most concrete pipe is limited to low-pressure applications. Moreover, almost all concrete piping is used for conveying industrial wastes, wastewater, and stormwater, although some is used for water service connections. Rubber gaskets are used to join sections of nonreinforced concrete pipe; however, for circular concrete sewer and culvert pipe, flexible, watertight, rubber joints are used to join pipe sections.

The general advantages of concrete pipe include the following:

- It is relatively inexpensive to manufacture.
- It can withstand relatively high internal pressure or external loads.
- It has high resistance to corrosion (internal and external).
- Generally, when installed properly, it has a very long, trouble-free life.
- Bedding requirements during installation are minimal.

Disadvantages of concrete pipe include:

- It is very heavy and thus expensive to ship long distances.
- Its weight makes special handling equipment necessary.
- Exact pipes and fittings must be laid out in advance for installation (AWWA, 1996).

10.9.1.2.1 Nonreinforced Concrete Pipe

Nonreinforced concrete pipe, or ordinary concrete pipe, is manufactured in diameters from 4 to 24 inches. As in vitrified clay pipe, nonreinforced concrete pipe is made with bell-and-spigot ends. Nonreinforced concrete pipe is normally used for small wastewater (sewer) lines and culverts.

10.9.1.2.2 Reinforced Concrete Pipe

All concrete pipe made in sizes larger that 24 inches is reinforced; however, reinforced pipe can also be obtained in sizes as small as 12 inches. Reinforced concrete pipe is used for water conveyance (cylinder pipe), to carry wastewater, stormwater, and industrial wastes. It is also used in culverts. It is manufactured by wrapping high-tensile strength wire or rods about a steel cylinder that has been lined with cement mortar. Joints are either bell-and-spigot or tongue-and-groove in sizes up to 30 inches; they are exclusively tongue-and-grove above that size.

10.9.1.2.3 Reinforced and Prestressed Concrete Pipe

When concrete piping is to be used for heavy-load, high-pressure applications (up to 600 psi), it is strengthened by reinforcement and prestressing. Prestressed concrete pipe is reinforced by steel wire, steel rods, or bars embedded lengthwise in the pipe wall. If wire is used, it is wound tightly to prestress the core and is covered with an outer coating of concrete. Prestressing is accomplished by manufacturing the pipe with a permanent built-in compression force.

10.9.1.2.4 Asbestos–Cement (A-C) Pipe

Asbestos–cement pipe is composed of a mixture of Portland cement and asbestos fiber built up on a rotating steel mandrel and then compacted with steel pressure rollers. This pipe has been used for over 70 years in the United States. Because it has a very smooth inner surface, it has excellent hydraulic characteristics (McGhee, 1991). Before beginning our brief discussion of asbestos–cement (A-C) pipe, however, it is necessary to discuss the safety and health implications involved with performing maintenance activities on A-C pipe. Prior to 1971, asbestos was known as the "material of a thousand uses" (Coastal Video Communications, 1994). It was used primarily for fireproofing, as well as for insulation (e.g., on furnaces, ducts, boilers, and hot water pipes), soundproofing, and a host of other applications. Other applications included its use in the conveyance of water and wastewater. Although still used in some industrial applications and in many water/wastewater piping applications, asbestos-containing materials (ACMs), including asbestos–cement pipe, are not as widely used as they were before 1971.

Asbestos-containing materials lost favor with regulators and users primarily because of the health risks involved. Asbestos has been found to cause chronic and often fatal lung diseases, including asbestosis and certain forms of lung cancer. Although debatable, some evidence suggests that asbestos fibers in water may cause intestinal cancers, as well. Although asbestos fibers can be found in some natural waters (Bales et al., 1984) and can be leached from asbestos–cement pipe by very aggressive waters (i.e., those that dissolve the cement itself) (Webber et al., 1989), it is also true that the danger from asbestos exposure is not so much due to the danger of specific products (A-C pipe, for example) as it is to the overall exposure of people involved in the mining, production, installation, and the ultimate removal and disposal of asbestos products (AWWA, 1996).

In water and wastewater operations, the ultimate removal and disposal of asbestos–cement pipe pose a problem for operators. Consider, for example, an underground wastewater line break that must be repaired. After locating exactly where the line break is (which can sometimes be difficult to accomplish because A-C pipe is not as easily located as conventional pipe), the work crew must first excavate the soil covering the line break, being careful not to cause further damage, as A-C pipe is relatively fragile. When the soil has been removed and the line break exposed, the damaged pipe section must be removed. In some instances, it may be more economical or practical to remove only the damaged portion of the pipe, install a replacement portion, and then girdle it with a clamping mechanism (sometimes referred to as a *saddle clamp*).

To this point in the described repair operation, the likelihood of exposure to asbestos by the personnel is small because, to be harmful, ACMs must release fibers that can be inhaled. The asbestos in undamaged A-C pipe is not friable (nonfriable asbestos); that is, it cannot be readily reduced to powder form by hand pressure when it is dry. Thus, it poses little or no hazard in this condition. If, however, the maintenance crew making the pipe repair must cut, grind, or sand the A-C pipe section under repair, the nonfriable asbestos will become separated from its bond. This type of repair activity is capable of releasing friable airborne fibers—and herein lies the hazard of working with A-C pipe.

To guard against the hazard of exposure to asbestos fibers, A-C pipe repairs must be accomplished in a safe manner. Operators must avoid any contact with ACMs that could disturb its position or arrangement, that could disturb its matrix or render it friable, or that could generate any visible debris.

Important Point: The presence of visibly damaged, degraded, or friable ACMs is always an indicator that surface debris or dust could be contaminated with asbestos. OSHA standards require that we assume that such dust or debris contains asbestos fibers (Coastal Video Communications, 1994).

In the A-C pipe repair operation described above, repairs to the A-C pipe require that prescribed U.S. Environmental Protection Agency (USEPA), Occupational Safety and Health Administration (OSHA), state, and local guidelines be followed. General USEPA/OSHA guidelines, at a minimum, require that only trained personnel perform repairs on the A-C pipe. The following safe work practice is provided for those who must work with ACMs (Spellman, 2001).

10.9.1.2.4.1 Safe Work Practices for A-C Pipe

1. When repairs/modifications are conducted that require cutting, sanding, or grinding on cement pipe containing asbestos, USEPA-trained asbestos workers or supervisors are to be called to the work site *immediately.*
2. Excavation personnel will unearth buried pipe to the point necessary to make repairs or modifications. The immediate work area will then be cleared of personnel as directed by the asbestos-trained supervisor.
3. The on-scene supervisor will direct the asbestos-trained workers as required to accomplish the work task.
4. The work area will be barricaded 20 feet in all directions to prevent unauthorized personnel from entering.
5. Asbestos-trained personnel will wear *all* required personal protective equipment (PPE). Required PPE includes Tyvek® totally enclosed suits, half-face respirators equipped with HEPA filters, rubber boots, goggles, gloves, and hardhats.

6. Supervisor will perform the required air sampling before entry.
7. Air sampling must be conducted using the NIOSH 7400 protocol.
8. A portable decontamination station will be set up as directed by supervisor.
9. Workers will enter the restricted area *only* when directed by the supervisors and, using wet methods only, will either perform pipe cutting using a rotary cutter assembly or inspect the broken area to be covered with a repair saddle device.
10. After performing the required repair or modifications, workers will encapsulate bitter ends and fragmented sections.
11. After encapsulation, the supervisor can authorize entry into restricted area for other personnel.
12. Broken ACM pipe pieces must be properly disposed of following USEPA, state, and local guidelines.

Important Point: Although exposure to asbestos fibers is dangerous, it is important to note that studies by USEPA, AWWA, and other groups have concluded that the asbestos in water mains does not generally constitute a health threat to the public (AWWA, 1996).

Because A-C pipe is strong and corrosion resistant, it is widely used for carrying water and wastewater. Standard sizes range from 3 to 36 inches. Though highly resistant to corrosion, A-C pipe should not be used for carrying highly acid solutions or unusually soft water, unless its inner and outer surface walls are specially treated. A-C pipe is preferred for use in many outlying areas because of its light weight, which offers greater ease of handling. An asbestos–cement sleeve joins A-C pipe. The inner diameter of the sleeve is larger than the outer diameter of the pipe. The ends of the pipes fit snugly into the sleeve and are sealed with a natural or synthetic rubber seal or gasket, which acts as an expansion joint.

10.9.1.3 Plastic Pipe

Plastic pipe has been used in the United States for about 60 years, and its use is becoming increasingly common. In fact, because of its particular advantages, plastic pipe is replacing both metallic and nonmetallic piping. The advantages of plastic piping include:

- High internal and external corrosion resistance
- Rarely requires insulating or painting
- Light weight
- Ease of joining
- Freedom from rot and rust
- Resistance to burning

- Lower cost
- Long service life
- Easy to maintain

Several types of plastic pipe are available; still, where plastic pipe is commonly used in water and wastewater service, polyvinyl chloride (PVC) is the most common plastic pipe for municipal water distribution systems. PVC is a polymer extruded (shaped by forcing through a die) under heat and pressure into a thermoplastic that is nearly inert when exposed to most acids, fuels, and corrosives. PVC is commonly used to carry cold drinking water, because PVC is nontoxic and will not affect the taste of the water or cause odor. The limitations of PVC pipe include its limited temperature range (approximately 150° to 250° F) and low pressure capability (usually 75 to 100 psi). Joining sections of plastic pipe is accomplished by welding (solvent, fusion, fillet), threading, and flanges.

Important Point: The strength of plastic piping decreases as the temperature of the materials it carries increases.

10.10 TUBING

Piping by another name might be tubing? A logical question might be "When is a pipe a tube, or a tube a pipe?" Does it really matter whether we refer to piping and tubing by two distinct, separate, and different names? It depends, of course, on the differences between the two. When we think of piping and tubing, we think of tubular, which infers cylindrical products that are hollow. Does this description help us determine the difference between piping and tubing? No, not really. We need more—a better, more concise description or delineation.

Maybe *size* will work. When we normally think of pipe, we think in terms of either metallic or nonmetallic cylindrical products that are hollow and range in nominal size from about 0.5 inch (or less) to several feet in diameter. On the other hand, when we think of tubing, we think of cylindrical, hollow products that are relatively smaller in diameter to that of many piping materials.

Maybe *application* will work. When we normally think of pipe, we think of any number of possible applications from conveying raw petroleum from field to refinery, to the conveyance of raw water from source to treatment facility, to wastewater discharge point to treatment to outfall, and several others. On the other hand, when we think in terms of tubing applications and products conveyed, the conveyance of compressed air, gases (including liquefied gas), steam, water, lubricating oil, fuel oil, chemicals, fluids in hydraulic systems, and waste products comes to mind.

On the surface, it is apparent that when we attempt to classify or differentiate piping and tubing, our effort is best characterized as somewhat arbitrary, capricious,

vague, or ambiguous. It appears that piping by any other name is just piping. In reality, however, piping is not tubing and in the end (so to speak) the difference may come down to the end use.

The bottom line: It is important to differentiate between *piping* and *tubing*, because they are different. They are different in physical characteristics and methods of installation, as well as in their advantages and disadvantages. In this section, these differences will become clear.

10.10.1 Tubing vs. Piping: The Difference

Lohmeir and Avery (2000) pointed out that piping and tubing are considered separate and distinct products, even though they are geometrically quite similar. Moreover, the classification of pipe or tube is determined by end use. As mentioned, many of the differences between piping and tubing are related to physical characteristics and methods of installation, as well as the advantages and disadvantages.

Simply, *tubing* refers to tubular materials (products) made to either an inside diameter (I.D.) or outside diameter (O.D.), expressed in even inches or fractions. Tubing walls are generally much thinner than those of piping; thus, wall thickness in tubing is of particular importance.

Important Point: Wall thickness tolerance in tubing is held so closely that wall thickness is usually given in thousandths of an inch rather than as a fraction of an inch. Sometimes a gauge number is used to indicate the thickness according to a given system.

Tubing of various diameters has different wall thicknesses; for example, the wall thickness of a commercial type of 8-in. pipe is 0.406 in. Light-wall 8-in. copper tubing, by contrast, has a wall thickness of 0.050 in. When we compare these figures, it is clear that tubing has much thinner walls than piping of the same general diameter (Basavaraju, 2000).

Important Point: It is important to note that the range between *thick* and *thin* is narrower for tubing than it is for piping.

The list of tubing applications is a lengthy one. Some tubing types can be used not only as conduits for electrical wire but also to convey waste products, compressed air, hydraulic fluids, gases, fuel oil, chemicals, lubricating oil, stream, waters, and other fluids (i.e., both gaseous and liquid).

Tubing is made from both metals and plastics. One of the primary reasons why tubing is employed for industrial applications is the fact that some tubing materials are extremely resistant to deterioration by corrosive chemicals. Metal tubing is designed to be somewhat flexible but

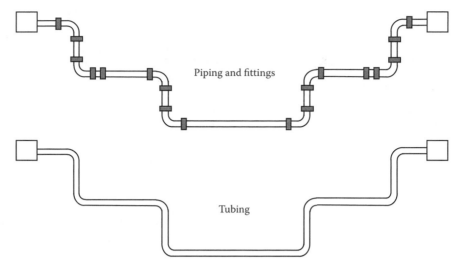

FIGURE 10.11 Tubing eliminates fittings.

also strong. Metallic materials such as copper, aluminum, steel, and stainless steel are used in applications where fluids are carried under high pressure; some types of tubing (e.g., stainless steel) can accommodate very high pressures (>5000 psi). As the diameter of the tubing increases, the wall thickness increases accordingly (slightly).

Ranging in size from 1/32 to 12 in. in diameter, the smaller sizes are most commonly used. Standard copper tubing ranges from 1/32 to 10 in. in diameter, steel from 3/15 to 10 3/4 in., and aluminum from 1/8 to 12 in.; special alloy tubing is available up to 8 in. in diameter.

Typically, in terms of initial cost, metal tubing materials are more expensive than iron piping; however, high initial cost vs. the ability to accommodate a particular application as designed (as desired) is a consideration that cannot be overlooked or underemphasized. Consider, for example, an air compressor. Typically, while in operation, air compressors are mechanical devices that not only produce a lot of noise but also vibrate. Installing a standard rigid metal piping system onto such a device might not be practical. Installing tubing that is flexible onto the same device, however, may have no detrimental impact on operation whatsoever. An even more telling example is the internal combustion engine. A lawnmower engine, like the air compressor, also vibrates and is used in less than static conditions (i.e., the lawnmower is typically exposed to all kinds of various dynamic stresses). Obviously, we would not want the fuel lines (tubing) in such a device to be made with rigid pipe; instead, we would want the fuel lines to be durable but also somewhat flexible. Thus, flexible metal tubing is called for in this application because it will hold up.

Simply put, initial cost can be important; however, considerations such as maintenance requirements, durability, length of life, and ease of installation often favor the use of metallic tubing over the use of metallic pipe.

Although it is true that most metallic tubing materials have relatively thin walls, it is also true that most are quite strong. Small tubing material with thin walls (i.e., soft materials up to approximately 1 in. O.D.) can be bent quite easily by hand. Tubing with larger diameters requires special bending tools. The big advantage of flexible tubing should be obvious: Tubing can be run from one point to another with fewer fittings than if piping was used.

Note: Figure 10.11 shows how the use of tubing can eliminate several pipefittings.

The advantages of the tubing type of arrangement shown in Figure 10.11 include the following:

- It eliminates 18 potential sources of leaks.
- The cost of the 18 90° elbow fittings required for the piping installation is eliminated.
- The time required to cut, gasket, and flange the separate sections of pipe is saved (obviously, it takes little time to bend tubing into the desired configuration).
- The tubing configuration is much lighter in weight than the separate lengths of pipe and the pipe flanges would have been.

For the configuration shown in Figure 10.11, the weight is considerably less for the copper tubing than the piping arrangement. Moreover, the single length of tubing bent to follow the same general conveyance route is much easier to install.

It may seem apparent to some readers that many of the weight and handling advantages of tubing compared to piping can be eliminated or at least matched simply by reducing the wall thickness of the piping. It is important to remember, however, that piping has a thick wall because

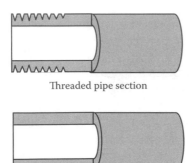

Threaded pipe section

Pipe section without threads

FIGURE 10.12 Pipe wall thickness is important when threading is required.

it often has to be threaded to make connections. If the wall thickness of iron pipe, for example, was made comparable to the thickness of copper tubing and then threaded at connection points, its mechanical integrity would be reduced. The point is that piping must have sufficient wall thickness left after threading not only to provide a tight fit but also to handle the fluid pressure. On the other hand, copper tubing is typically designed for brazed and soldered connections, rather than threaded ones; thus, its wall thickness can be made uniformly thin. This advantage of tubing over iron piping is illustrated in Figure 10.12.

Important Point: The lighter weight of tubing offers greater ease of handling, as well as lower shipping costs.

10.10.2 ADVANTAGES OF TUBING

To this point, with regard to design requirements, reliability, and maintenance activities of using tubing instead of piping, we have pointed out several advantages of tubing. These advantages can be classified as *mechanical* or *chemical* advantages.

10.10.2.1 Mechanical Advantages of Tubing

Probably the major mechanical advantage of using tubing is its relatively small diameter and its flexibility, which makes it user friendly in tight spaces where piping would be difficult to install and to maintain (e.g., for the tightening, repair, or replacement of fittings). Another mechanical advantage of tubing important to water and wastewater maintenance operators is the ability of tubing to absorb shock from *water hammer*. Water hammer can occur whenever fluid flow is started or stopped. In water/wastewater operations, certain fluid flow lines have a frequent on/off cycle. In a conventional piping system, this may produce vibration, which is transmitted along the rigid conduit, shaking joints, valves, and other fittings. The resulting damage usually results in leaks, which, of course, necessitates repairs. In addition, the piping supports can

also be damaged. When tubing, with its built in flexibility, is used in place of conventional iron piping, however, the conduit absorbs most of the vibration and shock. The result is far less wear and tear on the fittings and other appurtenances.

As mentioned, sections of tubing are typically connected by means of soldering, brazing, or welding rather than by threaded joints, although steel tubing is sometimes joined by threading. In addition to the advantages in cost and time savings, not using threaded joints precludes other problems; for example, any time piping is threaded it is weakened, although threading is commonly used for most piping systems and usually presents no problem.

Another advantage of tubing over iron piping is the difference in inner-wall surfaces between the two. Specifically, tubing generally has a smoother inner-wall surface than does iron piping. This smoother inner-wall characteristic aids in reducing turbulent flow (wasted energy and decreased pressure) in tubing. Flow in the smoother walled tubing is more laminar; that is, it has less turbulence. Laminar flow is characterized as flow in layers—very thin layers. (Somewhat structurally analogous to this liquid laminar flow phenomenon is wood-type products such as kitchen cabinets, many of which are constructed of laminated materials.)

This might be a good time to address laminar flow inside of a section of tubing. First, we need to discuss both laminar and turbulent flow to clarify the distinct difference between them. Simply, in *laminar* flow, streamlines remain parallel to one another and no mixing occurs between adjacent layers. In *turbulent* flow, mixing occurs across the pipe. The distinction between the two regimes lies in the fact that the shear stress in laminar flow results from viscosity, while that in turbulent flow results from momentum exchanges occurring as a result of motion of fluid particles from one layer to another (McGhee, 1991). Normally, flow inside tubing is laminar; however, if the tubing wall has any irregularities (dents, scratches, or bumps), the fluid will be forced across the otherwise smooth surface at different velocities which causes turbulence. Iron piping has more irregularities along its inner walls which cause turbulence in the fluid flowing along the conduit. Ultimately, this turbulence can reduce the delivery rate of the piping system considerably.

10.10.2.2 Chemical Advantages of Tubing

The major chemical advantage of tubing as compared to piping comes from the corrosion-resistant properties of the metals used to make the tubing. Against some corrosive fluids, most tubing materials do very well. Some metals perform better than others, though, depending on the metal and the corrosive nature of the fluid. It is important to also point out that tubing must be compatible with the fluid it is conveying. When conveying a liquid stream from one point to another, the last thing we want is to add contamination from the tubing. Many tubing conveyance systems are designed for use in food-processing operations. If we were conveying raw milk to or from a unit process, for example, we certainly would not want to contaminate the milk. To avoid such contamination, where conditions of particular sanitation are necessary, stainless steel, aluminum, or appropriate plastic tubing must be used.

10.10.3 CONNECTING TUBING

The skill required to properly connect metal or nonmetallic tubing can be learned by just about anyone; however, significant practice and experience are required to ensure that the tubing is properly connected. Moreover, certain tools are required for connecting sections of tubing. The tools used to make either a soldered connection or a compression connection (where joint sections are pressed together) include:

- Hacksaw
- Tube cutter
- Scraper
- Flat file
- Burring tool
- Flaring tool
- Presetting tool for flareless fittings
- Assorted wrenches
- Hammer
- Tube bender

10.10.3.1 Cutting Tubing

No matter what type of connection is being made (soldered or compressed), it is important to cut the tubing cleanly and squarely. This can be accomplished using a tubing cutter. Use of a tubing cutter is recommended because it provides a much smoother cut than that made with a hacksaw. A typical tubing cutter has a pair of rollers on one side and a cutting wheel on the other. The tubing cutter is turned all the way around the tubing, making a clean cut.

Important Point: When cutting stainless steel tubing, cut the tubing as rapidly and safely as possible with as few strokes as possible. This is necessary because as stainless steel is cut, it hardens, especially when cut with a hacksaw.

After making the tubing cut, the rough edge of the cut must be smoothed with a burring tool to remove the small metal chads, burrs, or whiskers. If a hacksaw is used to cut the tubing, be sure that the rough cut is filed until it is straight and square to the length of tubing.

10.10.3.2 Soldering Tubing

Soldering is a form of brazing in which nonferrous filler metals having melting temperatures below 800°F (427°C) are used. The filler metal is called *solder* (usually a tin–lead alloy, which has a low melting point) and is distributed between surfaces by capillary action. Whether soldering two sections of tubing together or connecting tubing to a fitting, such as an elbow, the soldering operation is the same. Using emery cloth or a wire brush, the two pieces to be soldered must first be cleaned (turned to bright metal). Clean, oxide-free surfaces are necessary to make sound soldered joints. Uniform capillary action is possible only when surfaces are completely free of foreign substances such as dirt, oil, grease, and oxide.

Important Point: During the cleaning process care must be taken to avoid getting the prepared adjoining surfaces too smooth. Surfaces that are too smooth will prevent the filler metal (solder) from effectively wetting the joining areas.

The next step is to ensure that both the tubing outside and the fitting inside are covered with soldering flux and fitted together. When joining two tubing ends, use a sleeve. The purpose of flux is to prevent or inhibit the formation of oxide during the soldering process. The two ends are fitted into the sleeve from opposite sides. Make sure the fit is snug.

Next, heat the joint. Heat the tubing next to the fitting first, then the fitting itself. When the flux begins to spread, solder should be added (this is known as *tinning*). The heat will suck the solder into the space between the tubing and the sleeve. The next step is to heat the fitting, on and off, and apply more solder until the joint is fully penetrated (Giachino and Weeks, 1985).

Important Point: During the soldering operation, it is important to ensure that the heat is applied evenly around the tubing. A continuous line of solder will appear where the fitting and tubing meet at each end of the sleeve. Also, be sure that the joined parts are held so they will not move. After soldering the connection, wash the connection with hot water to prevent future corrosion.

The heat source normally used to solder is heated using an oxyacetylene torch or some other high-temperature heat source. Important soldering points to remember include:

1. Always use the recommended flux when soldering.
2. Make sure parts to be soldered are clean and their surfaces fit closely together.
3. During the soldering process, do not allow the parts to move while the solder is in a liquid state.

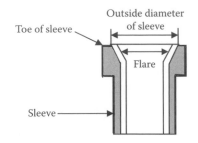

FIGURE 10.13 Flared tubing end.

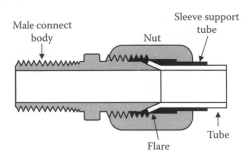

FIGURE 10.14 Flared fitting.

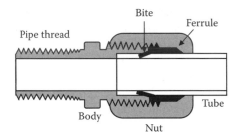

FIGURE 10.15 Flareless fitting.

4. Be sure the soldering heat is adequate for the soldering job to be done, including the types of metal and the fluxes.
5. Wash the solder work in hot water to stop later corrosive action.

10.10.3.3 Connecting Flared/Nonflared Joints

In addition to being connected by brazing or soldering, tubing can also be connected by either *flared* or *nonflared* joints. Flaring is accomplished by evenly spreading the end of the tube outward, as shown in Figure 10.13. The accuracy of the angle of flare is important; it must match the angle of the fitting being connected. The flaring tool is inserted into the squared end of the tubing and then hammered or impacted into the tube a short distance, spreading the tubing end as required.

Figure 10.14 shows the resulting flared connection. The flared section is inserted into the fitting in such a way that the flared edge of the tube rests against the angled face of the male connector body—a sleeve supports the tubing. The nut is tightened firmly on the male connector

body, making a firm joint that will not leak, even if the tubing ruptures because of excess pressure. Figure 10.15 shows a flareless fitting. As shown, the plain tube end is inserted into the body of the fitting. Notice the two threaded outer sections with a ferrule or bushing located between them. As the threaded members are tightened, the ferrule bites into the tubing, making a tight connection.

10.10.4 Bending Tubing

A type of tool typically used in water/wastewater maintenance applications for bending tubing is the *hand bender*, which is nothing more than a specifically sized, spring-type apparatus. Spring-type benders come in several different sizes (the size that fits the particular sized tubing to be bent is the one to use). The spring-type tubing bender is slipped over the tubing section to be bent, then the spring and tubing are carefully bent by hand to conform to the angle of bend desired. When using any type of tubing bender, it is important to obtain the desired bend without damaging (flattening, kinking, or wrinkling) the tubing. As mentioned, any distortion of the smooth, inner wall of a tubing section causes turbulence in the flow, which lowers the pressure. Figure 10.16 shows three different kinds of incorrect bends and one correct bend. From the figure, it should be apparent how the incorrect bends constrict the flow, causing turbulence and lower pressure.

10.10.5 Types of Tubing

Common types of metal tubing in industrial service include:

- *Copper* is seamless, fully annealed, and furnished in coils or in straight lengths. In water treatment applications, copper tubing has replaced lead and galvanized iron in service line installations because it is flexible, easy to install, corrosion resistant in most soils, and able to withstand high pressure. It is not sufficiently soluble in most water to be a health hazard, but corrosive water may dissolve enough copper to cause green stains on plumbing fixtures. Copper water service tubing is usually connected by either flare or compression fittings. Copper plumbing is usually connected with solder joints (AWWA, 1996).

Important Point: *Annealing* is the process of reheating a metal and then letting it cool slowly. In the production of tubing, annealing is performed to make the tubing softer and less brittle.

- *Aluminum* is seamless, annealed, and suitable for bending and flaring.
- *Steel* is seamless and fully annealed and is also available as a welded type, suitable for bending and flaring.

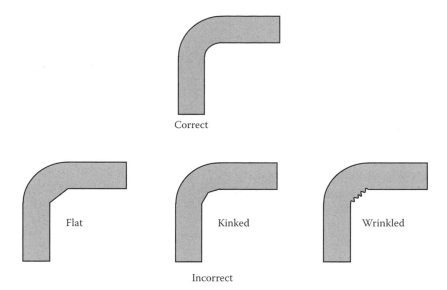

Correct

Flat   Kinked   Wrinkled

Incorrect

FIGURE 10.16 Correct and incorrect tubing bends.

- *Stainless steel* is seamless and fully annealed and is also available as a welded type, suitable for bending and flaring.
- *Special alloy* is made for carrying corrosive materials.

Like metal piping, metal tubing is made in both welded and seamless styles. *Welded tubing* begins as flat strips of metal that is then rolled and formed into tubing. The seam is then welded. *Seamless tubing* is formed as a long, hot metal ingot and then shaped into a cylindrical shape. The cylinder is then extruded (passed through a die), producing tubing in the larger sizes and wall thicknesses. If smaller tubing (with thinner walls and closer tolerances) is desired, the extruded tubing is reworked by drawing it through another die.

10.10.6 TYPICAL TUBING APPLICATIONS

In a typical water/wastewater operation, tubing is used in unit processes and machinery. Heavy-duty tubing is used for carrying gas, oxygen, steam, and oil in many underground services, interior plumbing, and heating and cooling systems throughout the plant site. Steel tubing is used in high-pressure hydraulic systems. Stainless steel tubing is used in many chemical systems. In addition, in many plants, aluminum tubing is used as raceways or containers for electrical wires. Plastics have become very important as nonmetallic tubing materials. The four most common types of plastic tubing are Plexiglas® (acrylic), polycarbonate, vinyl, and polyethylene (PE). For plant operations, plastic tubing usage is most prevalent where it meets corrosion resistance demands and the temperatures are within its working range (primarily in chemical processes). Plastic tubing is connected either by fusing with solvent cement

or by heating. Reducing the plastic ends of the tubing to a soft, molten state and then pressing them together produces a fused joint. In the solvent-cement method, the ends of the tubing are coated with a solvent that dissolves the plastic. The tube ends are firmly pressed together, and as the plastic hardens they are securely joined. For heat fusion, the tubes are held against a hot plate. When molten, the ends are joined and the operation is complete.

10.11 INDUSTRIAL HOSES

Earlier we described the uses and merits of piping and tubing. This section describes industrial hoses, which are classified as a slightly different tubular product. Their basic function is the same, however, and that is to carry fluids (liquids and gases) from one point to another. The outstanding feature of industrial hose is its flexibility, which allows it to be used in application where vibrations would make the use of rigid pipe impossible. Most water/wastewater treatment plants use industrial hoses to convey steam, water, air, and hydraulic fluids over short distances. It is important to point out that each application must be analyzed individually, and an industrial hose must be selected that is compatible with the system specification.

In this section, we address industrial hoses—what they are, how they are classified and constructed, and the ways in which sections of hose are connected to one another and to piping or tubing. We will also read about the maintenance requirements of industrial hoses and what to look for when we make routine inspections or checks for specific problems.

Industrial hoses, piping, and tubing all are used to convey a variety of materials under a variety of circumstances. Beyond this similar ability to convey a variety of materials, however, differences do exist among industrial hoses and piping and tubing; for example, in their construction and in their advantages, industrial hoses are different from piping and tubing. As mentioned, the outstanding advantage of hose is its flexibility; its ability to bend means that hose can meet the requirements of numerous applications that cannot be met by rigid piping and some tubing systems. Two examples of hose offering such flexibility are camel hose, which is used in wastewater collection systems to clean out interceptor lines or to remove liquid from excavations where broken lines are in need of repair, and hose that supplies hydraulic fluids used on many forklifts. Clearly, rigid piping would be impractical to use in both situations.

Industrial hose is not only flexible but also has a dampening effect on vibration. Certain tools used in water and wastewater maintenance activities must vibrate to do their jobs. Probably the most well known such tool is the power hammer, or jackhammer. Obviously, the built-in rigidity of piping and tubing would not allow vibrating tools to stand up very long under such conditions. Other commonly used tools and machines in water/wastewater operations have pneumatically or hydraulically driven components. Many of these devices are equipped with moving members that require the air or oil supply to move with them. In such circumstances, of course, rigid piping could not be used.

It is important to note that the flexibility of industrial hose is not the only consideration that must be taken into account when selecting hose over either piping or tubing; that is, hose must be selected according to the potential damaging conditions of an application. These conditions include the effects of pressure, temperature, and corrosion.

Hose applications include lightweight ventilating hose (commonly called *elephant trunk*) used to supply fresh air to maintenance operators working in manholes, vaults, or other tight places. Also, in water and wastewater treatment plants, hoses are used to carry water, steam, corrosive chemicals and gases, and hydraulic fluids under high pressure. To meet such service requirements, hoses are manufactured from a number of different materials.

10.11.1 HOSE NOMENCLATURE

To gain a fuller understanding of industrial hoses and their applications, it is important to be familiar with the nomenclature or terminology normally associated with industrial hoses. Accordingly, in this section, we explain hose terminology with which water/wastewater operators should be familiar. Figure 10.17 provides a cutaway view of a high-pressure air hose of the kind that supplies portable air hammers and drills and other pneumatic tools commonly used in water/wastewater maintenance operations. This hose is the most common type of reinforced nonmetallic hose in general use. Many of the terms given here

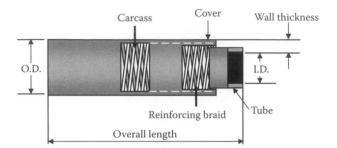

FIGURE 10.17 Common hose nomenclature.

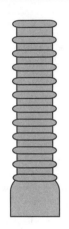

FIGURE 10.18 Expanded-end hose.

have already been mentioned. The I.D., which designates the hose size, refers to the inside diameter throughout the length of the hose body, unless the hose has enlarged ends. The O.D. is the diameter of the outside wall of the hose.

Important Point: If the ends of an industrial hose are enlarged, as shown in Figure 10.18, the letters E.E. are used (meaning *expanded* or *enlarged end*). Some hoses have enlarged ends to fit a fixed end of piping tightly (e.g., an automobile engine).

As shown in Figure 10.17, the *tube* is the inner section (i.e., the core) of the hose through which the fluid flows. Surrounding the tube is the *reinforcement material*, which provides resistance to pressure—either from the inside or outside. Notice that the hose shown in Figure 10.17 has two layers of reinforcement *braid* (fashioned from high-strength, synthetic cord). The hose is said to be *mandrel braided*, because a spindle or core (the mandrel) is inserted into the tube before the reinforcing materials are put on. The mandrel provides a firm foundation over which the cords are evenly and tightly braided. The *cover* of the hose is an outer, protective covering. The hose in Figure 10.17 has a cover of tough, abrasion-resistant material.

The *overall length* is the true length of a straight piece of hose. Hose that is not too flexible is formed or molded in a curve (e.g., automobile hose used in heating systems). As shown in Figure 10.19, the *arm* is the section of a

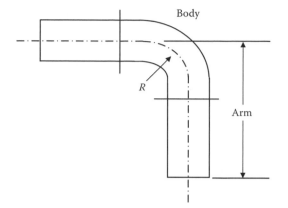

FIGURE 10.19 Bend radius.

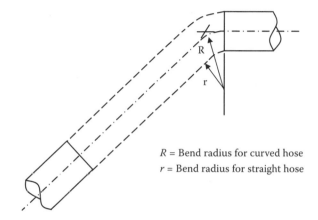

R = Bend radius for curved hose
r = Bend radius for straight hose

FIGURE 10.20 Bend radius: measurement.

curved hose that extends from the end of the hose to the nearest centerline intersection. The *body* is the middle section or sections of the curved hose. Figure 10.20 shows the *bend radius* (i.e., the radius of the bend measured to the centerline) of the curved hose, which is designated as radius *R*. In a straight hose, bent on the job, the radius of the bend is measured to the surface of the hose (i.e., radius *r* in Figure 10.20).

Important Point: Much of the nomenclature used above does not apply to nonmetallic hose that is not reinforced; however, nonreinforced nonmetallic hose is not very common in water/wastewater treatment plant operations.

10.11.2 FACTORS GOVERNING HOSE SELECTION

The amount of *pressure* to which a hose will be exposed is one of the important factors governing hose selection. Typically, the pressures encountered fall into one of three general groups:

- <250 psi (low-pressure applications)
- 250 to 3000 psi (medium-pressure applications)
- 3000 to 6000+ psi (high-pressure applications)

Important Point: Note that some manufacturers have their own distinct systems for rating hose pressure; we cannot assume that a hose rated as a low-pressure hose will automatically be useful at 100 or 200 psi. It may, in fact, be built for pressures not to exceed 50 psi, for example; therefore, whenever we replace a particular hose, we must be sure to use the same type of hose with the same pressure rating as the original hose. In high-pressure applications, this precaution is of particular importance.

In addition to the pressure rating of a hose, we must also consider, for some applications, the *vacuum rating* of a hose which refers to suction hose applications in which the pressure outside the hose is greater than the pressure inside the hose. It is important, obviously, to know the degree of vacuum that can be created before a hose begins to collapse. A drinking straw, for example, collapses rather easily if too much vacuum is applied; thus, it has a low vacuum rating. In contrast, the lower automobile radiator hose (which also works under vacuum) has a relatively higher vacuum rating.

10.11.3 STANDARDS, CODES, AND SIZES

Just as they have for piping and tubing, authoritative standards organizations have devised standards and codes for hoses. Standards and codes are safety measures designed to protect personnel and equipment; for example, specifications are provided for working pressures, sizes, and material requirements. The working pressure of a hose, for example, is typically limited to one fourth, or 25%, of the amount of pressure required to burst the hose. Let's look at an example. If we have a hose that has a maximum rated working pressure of 200 psi, it should not rupture until 800 psi has been reached, and possibly not even then. Thus, the importance of using hoses that meet specified standards or codes is quite evident.

10.11.3.1 Hose Size

The parameter typically used to designate hose size is its inside diameter (I.D.). With regard to the classification of hose, ordinarily a *dash numbering* system is used. Current practice by most manufacturers is to use the dash system to identify both hose and fittings. To determine the size of a hose, we simply convert the size in 16ths; for example, a hose size of 1/2 in. (a hose with a 1/2-in. I.D.) is the same as 8/16 in. The numerator of the fraction (the top number, or 8 in this case) is the dash size of the hose. In the same way, a size of 1-1/2 in. can be converted to 24/16 in. and so is identified as a -24 (pronounced "dash 24") hose. By using the dash system, we can match a hose line to tubing or piping section and be sure the I.D. of both will be the same. This means, of course, that the nonturbulent flow of fluid will not be interrupted. Based on I.D., hoses range in size from 3/16 to as large as 24 in.

10.11.3.2 Hose Classifications

Hose is classified in a number of ways; for example, hose can be classified by type of service (hydraulic, pneumatic, corrosion-resistant), by material, by pressure, or by type of construction. Hose may also be classified by type. The three types include *nonmetallic*, *metallic*, and *reinforced nonmetallic*. Generally, terminology is the same for each type.

10.11.3.3 Nonmetallic Hose

Relatively speaking, the use of hose is not a recent development. Hoses, in fact, have been used for one application or another for hundreds of years. Approximately 100 years ago, after new developments in the processing of rubber, hoses were usually made by layering rubber around mandrels; the mandrel was later removed, leaving a flexible rubber hose. These flexible hoses tended to collapse easily. Even so, they were an improvement over the earlier types. Manufacturers later added layers of rubberized canvas. This improvement gave hoses more strength and gave them the ability to handle higher pressures. Later, after the development of synthetic materials, manufacturers had more rugged and more corrosion-resistant rubber-type materials to work with. Today, neoprene, nitrile rubber, and butyl rubber are commonly used in hose. Current manufacturing practice is not to make hoses from a single material; instead, various materials form layers in the hose, reinforcing it in various ways for strength and resistance to pressure. Hose manufactured today usually has a rubber-type inner tube or a synthetic (e.g., plastic) lining surrounded by a *carcass* (usually braided) and cover. The type of carcass braiding used is determined by the requirements of the application. To reinforce a hose, two types of braiding are used: *vertical braiding* and *horizontal braiding*. Vertical braiding strengthens the hose against pressure applied at right angles to the centerline of the hose. Horizontal braiding strengthens the hose along its length, giving it greater resistance to expansion and contraction. Descriptions of the types of nonmetallic hose follow, with references to their general applications.

10.11.3.3.1 Vertical-Braided Hose

Vertical-braided hose has an inner tube of seamless rubber (see Figure 10.21). The reinforcing wrapping (carcass) around the tube is made of one or more layers of braided yarn. This type of hose is usually made in lengths of up

FIGURE 10.22 Wrapped hose.

to 100 ft with an I.D. of up to 1.5 in. Considered a small hose, it is used in low-pressure applications to carry fuel oil; acetylene gas and oxygen for welding; water for lawns, gardens, and other household uses; and paint for spraying.

10.11.3.3.2 Horizontal-Braided Hose

Horizontal-braided hose is mandrel built; it is used to make hose with an I.D. of up to 3 in. Used in high-pressure applications, the seamless rubber tube is reinforced by one or more layers of braided fibers or wire. This hose is used to carry propane and butane gas and steam and for various hydraulic applications that require high working pressures.

10.11.3.3.3 Reinforced Horizontal Braided-Wire Hose

In this type of hose, the carcasses around the seamless tube are made up of two or more layers of fiber braid with steel wire reinforcement between them. The I.D. may be up to 4 in. Mechanically very strong, this hose is used in applications with high working pressures or strong suction (vacuum) forces, such as in chemical transfer and petroleum applications.

10.11.3.3.4 Wrapped Hose

Made in diameters up to 24 in., wrapped hose is primarily used for pressure service rather than suction. The hose is constructed of mandrels and to close tolerances (Figure 10.22). It also has a smooth bore, which encourages laminar flow and avoids turbulence. Several plies (layers) of woven cotton or synthetic fabric make up the reinforcement. To achieve its resistance to corrosive fluids, the tube is made from a number of synthetic rubbers. It is also used in sandblasting applications.

10.11.3.3.5 Wire-Reinforced Hose

In this type of hose, wires wound in a spiral around the tube, or inside the carcass, in addition to a number of layers of wrapped fabrics, provide the reinforcement (see Figure 10.23). With I.D.s of 16 to 24 in. common, this type of hose is used in oil suction and discharge situations

FIGURE 10.21 Vertical-braided hose.

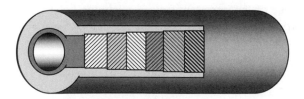

FIGURE 10.23 Wire-reinforced hose.

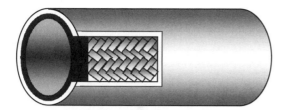

FIGURE 10.24 Wire-woven hose.

that require special hose ends, maximum suction (without collapsing), or special flexing characteristics (must be able to bend in a small radius without collapsing).

10.11.3.3.6 Wire-Woven Hose

Wire-woven hose (see Figure 10.24) features cords interwoven with wire running spirally around the tube; it is highly flexible, low in weight, and resistant to collapse even under suction conditions. This kind of hose is well suited for such negative pressure applications.

10.11.3.3.7 Other Types of Nonmetallic Hose

Hoses are also made of other nonmetallic materials, many of which are nonreinforced; for example, materials such as Teflon®, Dacron®, polyethylene, and nylon are being used for the manufacture of hose. Dacron remains flexible at very low temperatures, even as low as −200°C (up to −350°F), nearly the temperature of liquid nitrogen. Consequently, these hoses are used to carry liquefied gas in cryogenic applications. Where corrosive fluids and fluids up to 230°C (up to 450°F) must be carried, Teflon is often used. Teflon can also be used at temperatures as low as −55°C (−65°F). Usually sheathed in a flexible, braided metal covering, Teflon hoses are well protected against abrasion, and they also have added resistance to pressure.

Nylon hoses (small-diameter) are commonly used as air hoses, supplying compressed air to small pneumatic tools. The large plastic hoses (up to 24 in.) we use to ventilate manholes are made of such neoprene-coated materials as nylon fabric, glass fabric, and cotton duck. The cotton duck variety is for light-duty applications. The glass fabric type is used with portable heaters and for other applications involving hot air and fumes.

Various hoses made from natural latex, silicone rubber, and pure gum are available. The pure gum hose will safely carry acids, chemicals, and gases. Small hoses of natural latex, which can be sterilized, are used in hospitals for pharmaceuticals, blood, and intravenous solutions, as well as in food-handling operations and laboratories. Silicone rubber hose is used in situations where extreme temperatures and chemical reactions are possible. It is also used as in aircraft starters, where it provides compressed air in very large volumes. Silicone rubber hose works successfully over a temperature range from −57°C (−70°F) to 232°C (450°F).

10.11.3.4 Metallic Hose

The construction of a braided, flexible all-metal hose includes a tube of corrugated bronze. The tube is covered with the woven metallic braid to protect against abrasion and to provide increased resistance to pressure. Metal hose is also available in steel, aluminum, Monel®, stainless steel, and other corrosion-resistant metals in diameters up to 3 in. and in lengths of 24 in. In addition to providing protection against abrasion and resistance to pressure, flexible metal hose also dampens vibration. A plant air compressor, for example, produces a considerable amount of vibration. The use of flexible hoses in such machines increases their portability and dampens vibrations. Other considerations such as constant bending at high temperatures and pressures can be extremely damaging to most other types of hoses.

Other common uses for metallic hoses include serving as steam lines, lubricating lines, gas and oil lines, and exhaust hose for diesel engines. The corrugated type, for example, is used for high-temperature, high-pressure leak-proof service. Another type of construction is the *interlocked* flexible metal hose, used mainly for low-pressure applications. The standard shop oil can has a flexible hose for its spout. Other metal hose, with a liner of flexible, corrosion-resistant material, is available in diameters of up to 24 in.

Another type of metallic hose is used in *ductwork*. This type of hose is usually made of aluminum, galvanized steel, and stainless steel and is used to protect against corrosive fumes, as well as gases at extreme hot or cold temperatures. This hose is also fire resistant because it usually does not burn.

10.11.4 HOSE COUPLINGS

The methods of connecting or coupling hoses vary. Hose couplings may be either permanent or reusable. They can also be manufactured for the obvious advantage of quick connect or quick disconnect. Probably the best example of the need for quick connect is the fire hose—quick-disconnect couplings permit rapid connection between separate lengths of hose and between hose ends and hydrants or nozzles. Another good example of when the feature of quick connect and quick disconnect is user friendly is in plant or mobile compressed-air systems where a single line may have a number of uses. Changes involve disconnecting one section and connecting another. In plant shops, for example, compressed air from a single source is used to power pneumatic tools, cleaning units, paint sprayers, and so on. Each unit has a hose that is equipped for rapidly connecting and disconnecting at the fixed airline.

Caution: Before connections are broken, unless quick-acting, self-closing connectors are used, pressure must be released first.

FIGURE 10.25 Low-pressure hose coupling.

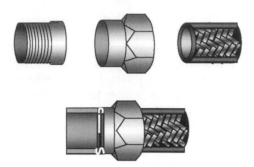

FIGURE 10.26 Coupling installation for all-metal hose.

For general low-pressure applications, a coupling such as that shown in Figure 10.25 is used. To place this coupling on the hose by hand, first cut the hose to the proper length, then oil the inside of the hose and the outside of the coupling stem. Force the hose over the stem into the protective cap until it seats against the bottom of the cap. No brazing is involved, and the coupling can be used repeatedly. After the coupling has been inserted in the hose, a yoke is placed over it in such a way that its arms are positioned along opposite sides of the hose behind the fitting. The arms are then tightly strapped or banded.

Caution: Where the pressure demands are greater, such a coupling can be blown out of the tube. Hose couplings designed to meet high-pressure applications must be used.

A variation of this type uses a clamp that is put over the inner end of the fitting and is then tightly bolted, thus holding the hose firmly. In other cases, a plain clamp is used. Each size clamp is designed for a hose of a specified size (diameter). The clamp slides snugly over the hose and is then crimped tight by means of a special hand tool or air-powered tool.

Coupling for all-metal hose, as described earlier, involves two brazing operations, as shown in Figure 10.26. The sleeve is slipped over hose end and brazed to it, and then the nipple is brazed to the sleeve.

Important Point: For large hoses of rugged wall construction, it is not possible to insert push-on fittings by hand. Special bench tools are required.

Quick-connect, quick-disconnect hose couplings provide flexibility in many plant process lines where a number of different fluids or dry chemicals from a single source are to be either blended or routed to different vats or other containers. Quick-connect couplings can be used to pump out excavations, manholes, and so forth. They would not be used, however, where highly corrosive materials are involved.

10.11.5 HOSE MAINTENANCE

All types of equipment and machinery require proper care and maintenance, including hoses. Depending on the hose type and its application, some require more frequent checking than others. The maintenance procedures required for most hoses are typical and are outlined here as an example. To maintain a hose, we should:

1. Examine for cracks in the cover caused by weather, heat, oil, or usage.
2. Look for a restricted bore because of tube swelling or foreign objects.
3. Look for cover blisters, which permit material pockets to form between the carcass and cover.
4. Look for leaking materials, which is usually caused by improper couplings or faulty fastenings of couplings.
5. Look for corrosion damage to couplings.
6. Look for kinked or otherwise damaged hose.

Caution: Because any of the faults listed above can result in a dangerous hose failure, regular inspection is necessary. At the first sign of weakness or failure, replace the hose. System pressure and temperature gauges should be checked regularly. Do not allow the system to operate above design conditions—especially when hose is a component of the system.

10.12   PIPE AND TUBE FITTINGS

The term *piping* refers to the overall network of pipes or tubing, fittings, flanges, valves, and other components that comprise a conduit system used to convey fluids. Whether a piping system is used to simply convey fluids from one point to another or to process and condition the fluid, piping components serve an important role in the composition and operation of the system. A system used solely to convey fluids may consist of relatively few components, such as valves and fittings, whereas a complex chemical processing system may consist of a variety of components used to measure, control, condition, and convey the fluids. In this section, the characteristics and functions of various piping and tubing fittings are described (Geiger, 2000).

10.12.1 FITTINGS

The primary function of *fittings* is to connect sections of piping and tubing and to change direction of flow. Whether used in piping or tubing, fittings are similar in shape and type, even though pipefittings are usually heavier than tubing fittings. Several methods can be used to connect fittings to piping and tubing systems; however, most tubing is threadless because it does not have the wall thickness required to for threading. Most pipes, on the other hand, because they have heavier walls, are threaded.

With regard to changing direction of flow, the simplest way would be simply to bend the conduit, which, of course, is not always practical or possible. When piping is bent, it is usually accomplished by the manufacturer in the production process (in larger shops equipped with their own pipe-bending machines), but not by the maintenance operator on the job. Tube bending, on the other hand, is a common practice. Generally, a tubing line requires fewer fittings than a pipeline; however, in actual practice many tube fittings are used.

Important Point: Recall that improperly made bends can restrict fluid flow by changing the shape of the pipe and weakening the pipe wall.

Fittings are basically made from the same materials (and in the same broad ranges of sizes) as piping and tubing, including bronze, steel, cast iron, glass, and plastic. Various established standards are in place to ensure that fittings are made from the proper materials and are able to withstand the pressures required; they are also made to specific tolerances, so they will properly match the piping or tubing that they join. A fitting stamped "200 lb," for example, is suitable (and safe) for use up to 200 psi.

10.12.2 FUNCTIONS OF FITTINGS

Fittings in piping and tubing systems have five main functions:

- Changing the direction of flow
- Providing branch connections
- Changing the sizes of lines
- Closing lines
- Connecting lines

10.12.2.1 Changing the Direction of Flow

Usually, a 45° or 90° *elbow* (or "ell") fitting is used to change the direction of flow. Elbows are among the most commonly used fittings in piping and are occasionally used in tubing systems. Two types of 90° elbows are shown in Figure 10.27. From the figure, it is apparent that the *long-radius* fitting (the most preferred elbow) has the more gradual curve of the two. This type of elbow is used in applications where the rate of flow is critical, and space presents no problem. Flow loss caused by turbulence is minimized by the gradual curve. The *short-radius* elbow (see Figure 10.27) should not be used in a system made up of long lines that has many changes in direction. Because of the greater frictional loss in the short-radius

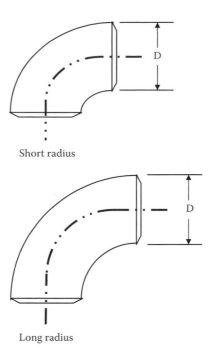

FIGURE 10.27 Short- and long-radius elbows.

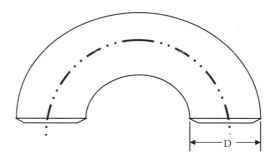

FIGURE 10.28 Long-radius return bend.

elbow, heavier more expensive pumping equipment may be required. Figure 10.28 shows a *return bend* fitting that carries fluid through a 180° ("hairpin") turn. This type of fitting is used for piping in heat exchangers and heater coils. Note that tubing, which can be bent into this form, does not require any fittings in this kind of application.

10.12.2.2 Providing Branch Connections

Because they are often more than single lines running from one point to another point, piping and tubing systems usually have a number of intersections. In fact, many complex piping and tubing systems resemble the layout of a town or city.

10.12.2.3 Changing the Sizes of Lines

For certain applications, it is important to reduce the volume of fluid flow or to increase flow pressure in a piping or tubing system. To accomplish this a *reducer* (which reduces a line to a smaller pipe size) is commonly used.

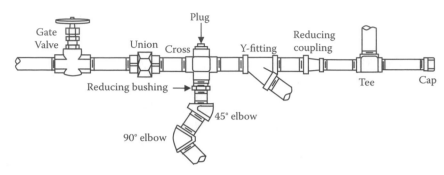

FIGURE 10.29 Diagram of a hypothetical shortened piping system.

Important Point: Reducing is also sometimes accomplished by means of a bushing inserted into the fitting.

10.12.2.4 Sealing Lines

Pipe *caps* are used to seal or close off the end of a pipe or tube (similar to corking a bottle). Usually, caps are used in a part of the system that has been dismantled. To seal off openings in fittings, *plugs* are used. Plugs also provide a means of access into the piping or tubing system, in case the line becomes clogged.

10.12.2.5 Connecting Lines

To connect two lengths of piping or tubing together, a coupling or union is used. A *coupling* is simply a threaded sleeve. A *union* is a three-piece device that includes a threaded end, an internally threaded bottom end, and a ring. A union does not change the direction of flow, close off the pipe, or provide for a branch line. Unions make it easy to connect or disconnect pipes without disturbing the position of the pipes. Figure 10.29 is a diagram of a shortened piping system that illustrates how some fittings are used in a piping system. (Figure 10.29 is only for illustrative purposes; it is unlikely that such a system with so many fittings would actually be used.)

10.12.3 TYPES OF CONNECTIONS

Pipe connections may be screwed, flanged, or welded. Each method is widely used, and each has its own advantages and disadvantages.

10.12.3.1 Screwed Fittings

Screwed fittings are joined to the pipe by means of threads. The main advantage of using threaded pipefittings is that they can be easily replaced. The actual threading of a section of replacement pipe can be accomplished on the job. The threading process itself, however, which cuts right into the pipe material, may weaken the pipe in the joint area. The weakest links in a piping system are the connection points. Because threaded joints can be potential

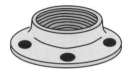

FIGURE 10.30 Flanged fitting.

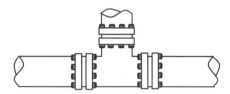

FIGURE 10.31 Flanged joint.

problem areas, especially where higher pressures are involved, the threads must be properly cut to ensure that the weakest link is not further compromised. Typically, the method used to ensure a good seal in a threaded fitting is to coat the threads with a paste dope. Another method is to wind the threads with Teflon® tape.

10.12.3.2 Flanged Connections

Figure 10.30 shows a flanged fitting. *Flanged* fittings are forged or cast-iron pipe. The *flange* is a rim at the end of the fitting, which mates with another section. Pipe sections are also made with flanged ends. Flanges are joined by being either bolted or welded together. The flange faces may be ground and lapped to provide smooth, flat mating surfaces. Obviously, a tight joint must be provided to prevent leakage of fluid and pressure. Figure 10.31 shows a typical example of a flanged joint. The mating parts are bolted together with a gasket inserted between their faces to ensure a tight seal. The procedure requires proper alignment of clean parts and tightening of bolts.

Important Point: Some flanges have raised faces and others have plain faces. Like faces must be matched—a flange with a raised face should never be joined to one with a plain face.

10.12.3.3 Welded Connections

Currently, because of improvements in piping technology and welding techniques and equipment, the practice of using welded joints is increasing. When properly welded, a piping system forms a continuous system that combines piping, valves, flanges, and other fittings. Along with providing a long leakproof and maintenance-free life, the smooth joints simplify applying insulation and they take up less room.

10.12.4 Tubing Fittings and Connections

Tubing is connected by brazed or welded flange fittings, compression fittings, and flare fittings. The *welded flange* connection is a reliable means of connecting tubing components. The flange welded to the tube end fits against the end of the fitting. The locknut of the flange is then tightened securely onto the fitting. The *compression fitting* connection uses a ferrule that pinches the tube as the locknut is tightened on the body of the fitting. The *flare fitting* connection uses tubing flared on one end of the tubing that matches the angle of the fitting. The flared end of the tube is butted against the fitting, and a locknut is screwed tightly onto the fitting, sealing the tube connection properly.

Other fittings used for flanged connections include *expansion joints* and *vibration dampeners*. Expansion joints function to compensate for slight changes in the length of pipe by allowing joined sections of rigid pipe to expand and contract with changes in temperature. They also allow pipe motion, either along the length of the pipe or to the side, as the pipe shifts around slightly after installation. Finally, expansion joints also help dampen vibration and noise carried along the pipe from distant equipment (e.g., pumps). One type of expansion joint has a leakproof tube that extends through the bore and forms the outside surfaces of the flanges. Natural or synthetic rubber compounds are normally used, depending on the application. Other types of expansion joints include metal corrugated types, slip-joint types, and spiral-wound types. In addition, high-temperature lines are usually made up with a large bend or loop to allow for expansion. Vibration dampeners absorb vibrations that, unless reduced, could shorten the life of the pipe and the service life of the operating equipment. They also eliminate line humming and hammering (water hammer) carried by the pipes.

10.13 VALVES

Any water or wastewater operation will have many valves that require attention. A maintenance operator must be able to identify and locate different valves to inspect them, adjust them, or repair or replace them. For this reason, the operator should be familiar with all valves, especially those that are vital parts of a piping system. A *valve* is defined as any device by which the flow of fluid may be started, stopped, or regulated by a movable part that opens or obstructs passage. As applied in fluid power systems, valves are used for controlling the flow, the pressure, and the direction of the fluid flow through a piping system. The fluid may be a liquid, a gas, or some loose material in bulk (such as a biosolids slurry). Designs of valves vary, but all valves have two features in common: a passageway through which fluid can flow and some kind of movable (usually machined) part that opens and closes the passageway (*Valves*, 1998).

Important Point: It is all but impossible, obviously, to operate a practical fluid power system without some means of controlling the volume and pressure of the fluid and directing the flow of fluid to the operating units. This is accomplished by the incorporation of various types of valves.

Whatever type of valve is used in a system, it must be accurate in the control of fluid flow and pressure and the sequence of operation. Leakage between the valve element and the valve seat is reduced to a negligible quantity by precision-machined surfaces, resulting in carefully controlled clearances. This is, of course, one of the very important reasons for minimizing contamination in fluid power systems. Contamination causes valves to stick, plugs small orifices, and causes abrasions of the valve seating surfaces, resulting in leakage between the valve element and valve seat when the valve is in the closed position. Any of these can result in inefficient operation or complete stoppage of the equipment. Valves may be controlled manually, electrically, pneumatically, mechanically, or hydraulically, or by combinations of two or more of these methods. Factors that determine the method of control include the purpose of the valve, the design and purpose of the system, the location of the valve within the system, and the availability of the source of power.

Valves are made from bronze, cast iron, steel, Monel®, stainless steel, and other metals. They are also made from plastic and glass (see Table 10.1). Special valve trim is used where seating and sealing materials are different from the basic material of construction (see Table 10.2). (*Valve trim* usually refers to the internal parts of a valve controlling the flow and in physical contact with the line fluid.) Valves are made in a full range of sizes that match pipe and tubing sizes. Actual valve size is based on the internationally agreed-upon definition of *nominal size* (DN), which is a numerical designation of size that is common to all components in a piping system other than components designated by outside diameters. It is a convenient number for reference purposes and is only loosely related to manufacturing dimensions.

TABLE 10.1
Valves: Materials of Construction

Cast iron	Gray cast iron; also referred to as flake graphite iron
Ductile iron	May be malleable iron or spheroidal graphite (nodular) cast iron
Carbon steel	May be as steel forgings, or steel castings, according to the method of manufacture; may also be manufactured by fabrication using wrought steels
Stainless steel	May be in the form of forgings, castings, or wrought steels for fabrication
Copper alloy	May be gunmetal, bronze, or brass; aluminum bronze may also be used
High-duty alloys	Usually nickel or nickel molybdenum alloys manufactured under various trade names
Other metals	Pure metals having extreme corrosion resistance, such as titanium, or aluminum
Nonmetals	Typically plastic materials such as PVC or polypropylene

TABLE 10.2
Valve Trim

Metal seating	Commonly used in gate and globe valves, particularly in the latter for control applications where seatings may additionally be coated with hard metal.
Soft seating	Commonly used in ball, butterfly, and diaphragm valves; seatings may be made from a wide variety of elastomers and polymers including fluorocarbons.
Lined	Usually made in cast iron with an internal lining of elastomer of polymer material. Inorganic materials such as glass, together with metals such as titanium, are also used for lining. Lining thickness will depend on the design and type of material used. In many cases, the valve lining will also form the seating trim.

Valves are made for service at the same pressures and temperatures that piping and tubing is subject to. Valve pressures are based on the internationally agreed definition of *nominal pressure* (PN), which is a pressure that is conventionally accepted or used for reference purposes. All equipment of the same nominal size (DN) designated by the same nominal pressure (PN) number must have the same mating dimensions appropriate to the type of end connections. The permissible working pressure depends on materials, design, and working temperature and should be selected from the (relevant) pressure/temperature tables. The pressure ratings of many valves are designated under the ANSI class system. The equivalent class rating to PN ratings is based on international agreement. Valves are also covered by various codes and standards, as are the other components of piping and tubing systems.

Usually, valve end connections are classified as *flanged*, *threaded*, or *other* (see Table 10.3). Many valve manufacturers offer valves with special features. Table 10.4 lists a few of these special features; however, this is not an exhaustive list and for more details of other features the manufacturer should be consulted. The various types of valves used in fluid power systems, their classifications, and their applications are discussed in this section.

10.13.1 Valve Construction

Figure 10.32 shows the basic construction and principle of operation of a common valve type. Fluid flows into the valve through the inlet. The fluid flows through passages in the body and past the opened element that closes the valve. It then flows out of the valve through the outlet or discharge. If the closing element is in the closed position, the passageway is blocked. Fluid flow is stopped at that point. The closing element keeps the flow blocked until the valve is opened again. Some valves are opened automatically, whereas manually operated hand wheels control others. Other valves, such as check valves, operate in response to pressure or the direction of flow. To prevent leakage whenever the closing element is positioned in the closed position, a seal is used. In Figure 10.32, the seal consists of a *stuffing box* fitted with packing. The closing element fits against the *seat* in the valve body to keep the valve tightly closed.

10.13.2 Types of Valves

The types of valves covered in this text include:

1. Ball valves
2. Gate valves
3. Globe valves
4. Needle valves
5. Butterfly valves
6. Plug valves
7. Check valves
8. Quick-opening valves
9. Diaphragm valves
10. Regulating valves
11. Relief valves
12. Reducing valves

TABLE 10.3
Valve End Connections

Flanged	Valves are normally supplied with flanges conforming to either BS4505 (equivalent to DIN) or BS 1560 (equivalent to ANSI), according to specifications. Manufacturers may be able to supply valves with flanges to other standards.
Threaded	Valves are normally supplied with threads to BS21 (ISO/7), parallel or taper.
Other	End connections include butt or socket weld ends and wafer valves are designed to fit between pipe flanges.

TABLE 10.4
Valve Special Features

High-temperature	Valves are usually able to operate continuously on services above 250°C.
Cryogenic	Valves will operate continuously on services in the range of –50°C to 196°C.
Bellows-sealed	Valves are glandless designs having a metal bellows for stem sealing.
Actuated	Valves may be operated by a gearbox, pneumatic or hydraulic cylinder (including diaphragm actuator), or electric motor and gearbox.
Fire-tested design	Valves have passed a fire test procedure specified in an appropriate inspection standard.

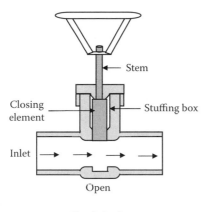

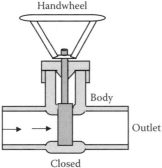

FIGURE 10.32 Basic valve operation.

Each of these valves is designed to control the flow, pressure, and direction of fluid flow or for some other special application. With a few exceptions, these valves take their names from the type of internal element that controls the passageway. The exceptions are the check valve, quick-opening valve, regulating valve, relief valve, and reducing valves.

10.13.2.1 Ball Valves

Ball valves, as the name implies, are stop valves that use a ball to stop or start fluid flow. The ball performs the same function as the disk in other valves. As the valve handle is turned to open the valve, the ball rotates to a point where part or all of the hole through the ball is in line with the valve body inlet and outlet, allowing fluid to flow through the valve. When the ball is rotated so the hole is perpendicular to the flow openings of the valve body, the flow of fluid stops. Most ball valves are the quick-acting type and require only a 90° turn to either completely open or close the valve; however, many are operated by planetary gears. This type of gearing allows the use of a relatively small hand wheel and operating force to operate a large valve. This gearing increases the operating time for the valve. Some ball valves also contain a swing check located within the ball to give the valve a check valve feature. The two main advantages of using ball valves are that: (1) the fluid can flow through it in either direction, as desired, and (2) when closed, pressure in the line helps to keep it closed.

10.13.2.2 Gate Valves

Gate valves are used when a straight-line flow of fluid and minimal flow restriction are necessary; they are the most common type of valve found in a water distribution system. Gate valves are so named because the part that either stops or allows flow through the valve acts somewhat like a gate. The gate is usually wedge shaped. When the valve is wide open, the gate is fully drawn up into the valve bonnet. This leaves an opening for flow through the valve

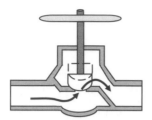

FIGURE 10.33 Globe valve.

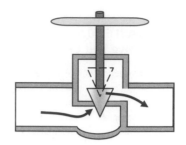

FIGURE 10.34 Common needle valve.

the same size as the pipe in which the valve is installed. For these reasons, the pressure loss (pressure drop) through these types of valves is about equal to the loss in a piece of pipe of the same length. Gate valves are not suitable for throttling purposes (i.e., controlling the flow by means of intermediate steps between fully open and fully closed). The control of flow is difficult because of the design of the valve, and the flow of fluid slapping against a partially open gate can cause extensive damage to the valve.

Important Point: Gate vales are well suited to service on equipment in distant locations, where they may remain in the open or closed position for a long time. Generally, gate valves are not installed where they will have to be operated frequently because they require too much time to switch from being fully open to closed (AWWA, 1996).

10.13.2.3 Globe Valves

Probably the most common valve type in existence, the globe valve is commonly used for water faucets and other household plumbing. As illustrated in Figure 10.33, these valves have a circular disk (the globe) that presses against the valve seat to close the valve. The disk is the part of the globe valve that controls flow. The disk is attached to the valve stem. As shown in Figure 10.33, fluid flow through a globe valve is at right angles to the direction of flow in the conduits. Globe valves seat very tightly and can be adjusted with fewer turns of the wheel than gate valves; thus, they are preferred for applications that call for frequent opening and closing. On the other hand, globe valves create high head loss when fully open; thus, they are not suited in systems where head loss is critical.

Important Point: The globe valve should never be jammed in the open position. After a valve is fully opened, the hand wheel should be turned toward the closed position approximately one half turn. Unless this is done, the valve

is likely to seize in the open position, making it difficult, if not impossible, to close the valve. Another reason for not leaving globe valves in the fully open position is that it is sometimes difficult to determine if the valve is open or closed (*Globe Valves*, 1998).

10.13.2.4 Needle Valves

Although similar in design and operation to the globe valve (a variation of globe valves), the needle valve has a closing element in the shape of a long tapered point, which is at the end of the valve stem. Figure 10.34 shows a cross-sectional view of a needle valve. In Figure 10.34, the long taper of the valve closing element permits a much smaller seating surface area than that of the globe valve; accordingly, the needle valve is more suitable for use as a throttle valve. In fact, needle valves are used for very accurate throttling.

10.13.2.5 Butterfly Valves

Figure 10.35 shows a cross-sectional view of a butterfly valve. The valve itself consists of a body in which a disk (butterfly) rotates on a shaft to open or close the valve. Butterfly valves may be flanged or wafer design, the latter intended for fitting directly between pipeline flanges. In the full open position, the disk is parallel to the axis of the pipe and the flow of fluid. In the closed position, the disk seals against a rubber gasket-type material bonded either on the valve seat of the body or on the edge of the disk. Because the disk of a butterfly valve stays in the fluid path in the open position, the valve creates more turbulence (higher resistance to flow and thus greater pressure loss) than a gate valve. On the other hand, butterfly valves are compact. They can also be used to control flow in either direction. This feature is useful in water treatment plants that periodically backwash to clean filter systems.

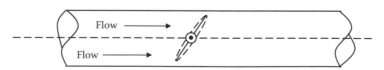

FIGURE 10.35 Cross-section of butterfly valve.

10.13.2.6 Plug Valves

A plug valve (also known as a *cock*, or *petcock*) is similar to a ball valve. Plug valves:

- Offer high-capacity operation (1/4 turn operation).
- Use either a cylindrical or conical plug as the closing member.
- Are directional.
- Offer moderate vacuum service.
- Allow flow throttling with interim positioning.
- Are of a simple construction (O-ring seal).
- Are not necessarily fully on and off.
- Are easily adapted to automatic control.
- Can safely handle gases and liquids.

10.13.2.7 Check Valves

Check valves are usually self-acting and designed to allow the flow of fluid in one direction only. They are commonly used at the discharge of a pump to prevent backflow when the power is turned off. When the direction of flow is moving in the proper direction, the valve remains open. When the direction of flow reverses, the valve closes automatically from the fluid pressure against it. Several types of check valves are used in water/wastewater operations, including:

- Slanting-disk check valves
- Cushioned swing check valves
- Rubber-flapper swing check valves
- Double-door check valves
- Ball check valves
- Foot valves
- Backflow prevention devices

In each case, pressure from the flow in the proper direction pushes the valve element to an open position. Flow in the reverse direction pushes the valve element to a closed position.

Important Point: Check valves are also commonly referred to as *nonreturn* or *reflux valves*.

10.13.2.8 Quick-Opening Valves

Quick-opening valves are nothing more than adaptations of some of the valves already described. Modified to provide a quick on/off action, these valves use a lever device in place of the usual threaded stem and control handle to operate the valve. This type of valve is commonly used in water/ wastewater operations when deluge showers and emergency eyewash stations are installed in work areas where chemicals are loaded or transferred or where chemical systems are maintained. They also control the air supply for some emergency alarm horns around chlorine storage areas, for example. Moreover, they are usually used to cut off the flow of gas to a main or to individual outlets.

10.13.2.9 Diaphragm Valves

Diaphragm valves are glandless valves that use a flexible elastomeric diaphragm (a flexible disk) as the closing member and in addition provide an external seal. They are well suited to service in applications where tight, accurate closure is important. The tight seal is effective whether the fluid is a gas or a liquid. This tight closure feature makes these valves useful in vacuum applications. Diaphragm valves operate similar to globe valves and are usually multi-turn in operation; they are available as weir type and full bore. A common application of diaphragm valves in water or wastewater operations is to control fluid to an elevated tank.

10.13.2.10 Regulating Valves

As their name implies, *regulating* valves regulate either pressure or temperature in a fluid line, keeping them very close to a preset level. If the demands and conditions of a fluid line remained steady at all times, no regulator valve would be needed. In the real world, however, ideal conditions do not occur. *Pressure-regulating valves* regulate fluid pressure levels to meet flow demand variations. Flow variations vary with the number of pieces of equipment in operation and with changes in demand as pumps and other machines operate. In such fluid line systems, demands are constantly changing. Probably the best example of this situation is seen in the operation of a low-pressure air supply system. For shop use, no more than 30 psi air is usually required (depending on required usage, of course). This air is supplied by an air compressor, which normally operates long enough to fill an accumulator with pressurized air at a set pressure level.

When shop air is required, for whatever reason, compressed air is drawn from the connection point in the shop. The shop connection point is usually connected via a pressure reducer (sets the pressure at the desired usage level) that, in turn, is fed from the accumulator, where the compressed air is stored. If the user draws a large enough quantity of compressed air from the system (from the accumulator), a sensing device within the accumulator will send a signal to the air compressor to start, which will produce compressed air to recharge the accumulator. In addition to providing service in air lines, pressure-regulating valves are also used in liquid lines. The operating principle is much the same for both types of service. Simply, the valve is set to monitor the line, and to make needed adjustments in response to a signal from a sensing device.

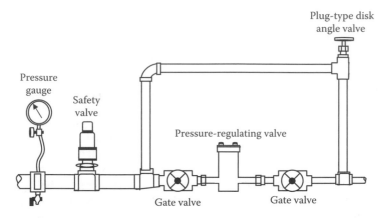

FIGURE 10.36 Pressure-regulating valve system.

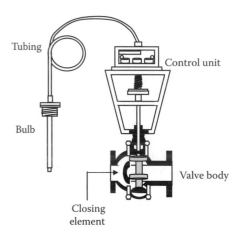

FIGURE 10.37 Temperature-regulating valve assembly.

Temperature-regulating valves (also referred to as *thermostatic control valves*) are closely related to pressure-regulating valves (see Figure 10.36). Their purpose is to monitor the temperature in a line or process solution tank and to regulate it—to raise or lower the temperature as required. In water and wastewater operations, probably the most familiar application for temperature-regulating valves (see Figure 10.37) is in *heat exchangers*. A heat exchanger type of water system utilizes a water-to-coolant heat exchanger for heat dissipation. This is an efficient and effective method to dispose of unwanted heat. Heat exchangers are equipped with temperature-regulating valves that automatically modulate the shop process water, limiting usage to just what is required to achieve the desired coolant temperature.

10.13.2.11 Relief Valves

Some fluid power systems, even when operating normally, may temporarily develop excessive pressure; for example, whenever an unusually strong work resistance is encountered, dangerously high pressure may develop. *Relief* valves are used to control this excess pressure. Such valves are automatic valves; they start to open at a preset pressure but require a 20% overpressure to open wide. As the pressure increases, the valve continues to open farther until it has reached its maximum travel. As the pressure drops, it starts to close and finally shuts off at about the set pressure. Main system relief valves are generally installed between the pump or pressure source and the first system isolation valve. The valve must be large enough to allow the full output of the hydraulic pump to be delivered back to the reservoir.

Important Point: Relief valves do not maintain flow or pressure at a given amount but prevent pressure from rising above a specific level when the system is temporarily overloaded.

10.13.2.12 Reducing Valves

Pressure-reducing valves provide a steady pressure into a system that operates at a lower pressure than the supply system. In practice, they are very much like pressure-regulating valves. A pressure-reducing valve reduces pressure by throttling the fluid flow. A reducing valve can normally be set for any desired downstream pressure within the design limits of the valve. Once the valve is set, the reduced pressure will be maintained regardless of changes in supply pressure (as long as the supply pressure is at least as high as the reduced pressure desired) and regardless of the system load, providing the load does not exceed the design capacity of the reducer.

10.13.3 Valve Operators

In many modern water and wastewater operations, devices called *operators* or *actuators* mechanically operate many valves. These devices may be operated by air, electricity, or fluid—that is, by pneumatic, magnetic, or hydraulic operators.

10.13.3.1 Pneumatic and Hydraulic Valve Operators

Pneumatic and hydraulic valve operators are similar in appearance and work in much the same way. Hydraulic cylinders using either plant water pressure or hydraulic fluid are frequently used to operate valves in treatment plants and pumping stations (AWWA, 1996). In a typical pneumatic ball-valve actuator, the cylinder assembly is attached to the ball-valve stem close to the pipe. A piston inside the cylinder can move in either direction. The piston rod is linked to the valve stem, opening or closing the valve, depending on the direction in which the piston is traveling. As a fail-safe feature, some of these valves are spring loaded. In case of hydraulic or air pressure failure, the valve operator returns the valve to the safe position.

Note: According to Casada (2000), valve operators and positioners usually require more maintenance than the valves themselves.

10.13.3.2 Magnetic Valve Operators

Magnetic valve operators use electric *solenoids*. A solenoid is a coil of magnetic wire that is roughly in the shape of a doughnut. When a bar of iron is inserted as a plunger mechanism inside an energized coil, it moves along the coil because of the magnetic field that is created. If the plunger (the iron bar) is fitted with a spring, it returns to its starting point when the electric current is turned off. Solenoids are used as operators for many different types of valves used in water/wastewater operations; for example, in a direct-operating valve, the solenoid plunger is used in place of a valve stem and hand wheel. The plunger is connected directly to the disk of a globe valve. As the solenoid coil is energized or deenergized, the plunger rises or falls, operating or closing the valve.

10.13.4 VALVE MAINTENANCE

As with any other mechanical device, effective valve maintenance begins with its correct operation. As an example of incorrect operation, consider the standard household water faucet. As the faucet washers age, they harden and deteriorate. The valve becomes more difficult to operate properly, and eventually the valve begins to leak. A common practice is simply to apply as much force as possible to the faucet handle. Doing so, however, damages the valve stem and the body of the valve body. Good maintenance includes preventive maintenance, which, in turn, includes inspection of valves, correct lubrication of all moving parts, and the replacement of seals or stem packing.

10.14 PIPING SYSTEMS: PROTECTIVE DEVICES

Piping systems must be protected from the harmful effects of undesirable impurities (solid particles) entering the fluid stream. Because of the considerable variety of materials carried by piping systems, an equal range of choices in protective devices is available. Such protective devices include *strainers*, *filters*, and *traps*. In this section, we describe the design and function of strainers, filters, and traps. The major maintenance considerations of these protective devices also are explained.

10.14.1 APPLICATIONS

Filters, strainers, and traps are normally thought of in terms of specific components used in specific systems; however, it is important to keep in mind that the basic principles apply in many systems. Although the examples used in this chapter include applications found in water/wastewater treatment, collection, and distribution systems, the applications are also found in almost every plant—hot and cold water lines, lubricating lines, pneumatic and hydraulic lines, and steam lines. With regard to steam lines, it is important to point out that, in our discussion of traps, their primary application is in steam systems, where they remove unwanted air and condensate from lines.

Important Point: A very large percentage (estimated to be >70%) of all plant facilities in the United States make use of steam in some applications.

Other system applications of piping protective devices include the conveyance of hot and chilled water for heating and air conditioning and lines that convey fluids for various processes. Any foreign contamination in any of these lines can cause problems. Piping systems can become clogged, thereby causing greatly increased friction and lowering the line pressure. Foreign contaminants (dirt and other particles) can also damage valves, seals, and pumping components.

Important Point: Foreign particles in a high-pressure line can damage a valve by clogging the valve so it cannot close tightly. In addition, foreign particles may wear away the closely machined valve parts.

10.14.2 STRAINERS

Strainers, usually wire mesh screens, are used in piping systems to protect equipment sensitive from contamination that may be carried by the fluid. Strainers can be used in pipelines conveying air, gas, oil, steam, water, wastewater, and nearly any other fluid conveyed by pipes. Generally, strainers are installed ahead of valves, pumps, regulators,

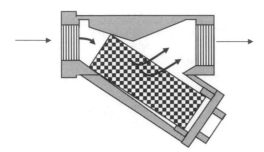

FIGURE 10.38 A common strainer.

and traps to protect them against the damaging effects of corrosive products that may become dislodged and conveyed throughout the piping system (Geiger, 2000).

A common strainer is shown in Figure 10.38. This type of strainer is generally used upstream of traps, control valves, and instruments. This strainer resembles a lateral branch fitting with the strainer element installed in the branch. The end of the lateral branch is removable to permit servicing of the strainer. In operation, the fluid passes through the strainer screen, which catches most of the contaminants. The fluid then passes back into the line. Contaminants in the fluid are caught in two ways—either they do not make it through the strainer screen or they do not make the sharp turn that the fluid must take as it leaves the unit. The bottom of the unit serves as a sump where the solids collect. A blowout connection may be provided in the end cap to flush the strainer. The blowout plug can be removed, and pressure in the line can be used to blow the fixture clean.

Important Point: Before removing the blowout plug, the valve system must first be locked out or tagged out.

10.14.3 FILTERS

The purpose of any filter is to reduce or remove impurities or contaminants from a fluid (liquid or gas) to an acceptable or predetermined level. This is accomplished by passing the fluid through some kind of porous barrier. Filter cartridges have replaceable elements made of paper, wire cloth, nylon cloth, or fine-mesh nylon cloth between layers of coarse wire. These materials filter out unwanted contaminants that collect on the entry side of the filter element. When clogged, the element is replaced. Most filters operate in two ways: (1) they cause the fluid to make sharp changes in direction as it passes through (this is important, because the larger particles are too heavy to change direction quickly), or (2) they contain some kind of barrier that will not let larger contaminants pass.

10.14.4 TRAPS

Traps used in steam processes are automatic valves that release condensate (condensed steam) from a steam space

FIGURE 10.39 A thermostatic trap (shown in the open position).

while preventing the loss of live steam. Condensate is undesirable because water produces rust, and water plus steam leads to water hammer. In addition, steam traps remove air and noncondensate from the steam space. The operation of a trap depends on what is called *differential pressure* (or delta-P) (measured in psi). Differential pressure is the difference between the inlet and outlet pressures. A trap will not operate correctly at a differential pressure higher than the one for which it was designed.

Many types of steam traps are available because of their many different types of applications. Each type of trap has a range of applications for which it is best suited; for example, thermostatic and float-and-thermostatic are two general types of traps. *Thermostatic traps* have a corrugated bellows-operating element that is filled with an alcohol mixture that has a boiling point lower than that of water (see Figure 10.39). The bellows contracts when it comes into contact with condensate and expands when stream is present. If a heavy condensate load occurs, the bellows will remain in the contracted state, allowing condensate to flow continuously. As steam builds up, the bellows closes. Thus, at times the trap acts as a continuous-flow type of valve, but at other times it acts intermittently as it opens and closes to condensate and steam, or it may remain totally closed (Bandes and Gorelick, 2000).

Important Point: The thermostatic trap is designed to operate at a definite temperature drop of a certain number of degrees below the saturated temperature for the existing steam pressure.

A *float-and-thermostatic trap* is shown in Figure 10.40. It consists of a ball float and a thermostatic bellows element. As condensate flows through the body, the float rises and falls, opening the valve according to the flow rate. The thermostatic element discharges air from the

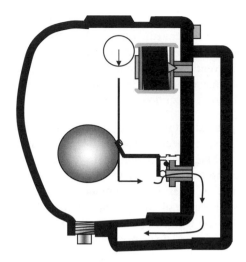

FIGURE 10.40 A float-and-thermostatic trap.

steam lines. These traps are suitable for heavy, and light loads and high and low pressures, but they are not recommended where water hammer is a possibility.

10.14.4.1 Trap Maintenance and Testing

Because they operate under constantly varying pressure and temperature conditions, traps used in steam systems require maintenance. Just as significant, because of these varying conditions, traps can fail. When they do fail, most traps fail in the open mode, which may require the boiler to work harder to perform a task which, in turn, can create high backpressure to the condensate system. This reduces the discharge capacities of some traps, which may be operating beyond their rating and thus causing system inefficiency.

Important Point: Although it is true that most traps operate with backpressure, it is also true that they do so only at a percentage of their rating, affecting everything down the line of the failed trap. Steam quality and product can be affected.

A closed trap produces condensate back up into the steam space. The equipment cannot produce the intended heat. Consider, as an example, a four-coil dryer with only three of the coils operating. In this setup, it will take longer for the dryer to dry a product, which has a negative effect on production.

10.14.4.1.1 Trap Maintenance

Excluding design problems, two of the most common causes of trap failure are oversizing and dirt. Oversizing causes traps to work too hard. In some cases, this can result in the blowing of live stream. Certain types of traps, for example, can lose their prime due to an abrupt change in pressure. This will cause the valve to open. Traps tend to accumulate dirt (sludge) that hinders tight closing. The

moving parts of the traps are subject to wear. Because the moving parts of traps operate in a mixture of steam and water, or sometimes in a mixture of compressed air and water, they are difficult to lubricate.

Important Point: Dirt (sludge) is generally produced from pipe scale or from overtreating of chemicals in a boiler.

Trap maintenance includes periodic cleaning, removing dirt that interferes with valve action, adjusting the mechanical linkage between moving parts and valves, and reseating the valves when necessary. If these steps are not taken, the trap will not operate properly.

10.14.4.2 Trap Testing

Important Point: A word of caution is advised before testing any steam trap: Inspectors should be familiar with the particular functions and types of traps and should know the various pressures within the system. This can help to ensure inspector safety, help avoid misdiagnosis, and allow proper interpretation of trap conditions.

The three main categories of online trap inspection are visual, thermal, and acoustic. *Visual inspection* depends on a release valve situated downstream of certain traps. A maintenance operator opens these valves and looks to see if the trap is discharging condensate or steam. *Thermal inspection* relies on upstream/downstream temperature variations in a trap. It includes pyrometry, infrared, heat bands (which are wrapped around a trap and change color as temperature increases), and heat sticks (which melt at various temperatures). *Acoustic techniques* require a maintenance operator to listen to and detect stream trap operations and malfunction. This approach utilizes various forms of listening devices such as medical stethoscopes, screwdrivers, mechanical stethoscopes, and ultrasonic detection instruments.

Important Point: A simple trap test—just listening to the trap action—tells us how the trap is opening and closing. Moreover, if the trap has a bypass line around it, leaky valves will be apparent when the main line to the trap is cut off, forcing all the fluid through the bypass.

10.15 PIPING ANCILLARIES

Earlier, we described various devices associated with process piping systems designed to protect the system. In this section, we discuss some of the most widely used ancillaries (or accessories) designed to improve the operation and control the system. These include pressure and temperature gauges, vacuum breakers, accumulators, receivers, and heat exchangers. It is important for us to know how these ancillary devices work, how to care for them, and, more importantly, how to use them.

10.15.1 Gauges

To properly operate a system, any system, the operator must know certain things. As an example, to operate a plant air compressor the operator must know: (1) how to operate it, (2) how to maintain it, (3) how to monitor its operation, and, in many cases, (4) how to repair it. In short, the operator must know system parameters and how to monitor them. Simply, *operating parameters* refer to the physical indications of system operation. The term *parameter* refers to the limits or restrictions of a system. Let's consider, again, an air compressor. Obviously, it is important to know how the air compressor operates or at least how to start and place the compressor online properly; however, it is also just as important to determine if the compressor is operating as per design. Experienced operators know that to ensure that an air compressor is operating correctly (i.e., as per design) they must monitor its operation. Again, they do this by monitoring the operation of the air compressor by observing certain parameters.

Before starting any machine or system, however, we must first perform a prestart check to ensure that it has the proper level of lubricating oil, etc. Then, after starting the compressor, we need to determine (observe) if the compressor is actually operating (which usually is not difficult to discern considering that most air compressor systems make a lot of noise while in operation). Once the compressor is in operation, our next move is to double check the system line-up to ensure that various valves in the system are correctly positioned (opened or closed). We might even go to a remote plant compressed-air service outlet to make sure that the system is producing compressed air. (Keep in mind that some compressed air systems have a supply of compressed air stored in an air receiver; thus, when an air outlet is opened, air pressure might be present even if the compressor is not functioning properly.) On the other hand, instead of using a remote outlet to test for compressed air supply, all we need do is look at the air-pressure gauge for the compressor. This gauge should indicate that the compressor is producing compressed air.

Gauges are the main devices that provide us with the parameter indications that we need to evaluate equipment or system operation. With regard to the air compressor, the parameter we are most concerned with now is air pressure (gauge pressure). Not only is correct pressure generation by the compressor important, but maintaining the correct pressure in system pipes, tubes, and hoses is also essential. Keeping air pressure at the proper level is necessary mainly for four reasons:

1. Safe operation
2. Efficient, economic conveyance of air through the entire system, without waste of energy
3. Delivery of compressed air to all outlet points in the system (the places where the air is to be used) at the required pressure
4. Prevention of too much or too little pressure (either condition can damage the system and become hazardous to personnel)

We pointed out that before starting the air compressor, certain prestart checks must be made. This is important for all machinery, equipment, and systems. In the case of our air compressor example, we want to ensure that proper lubricating oil pressure is maintained. This is important, of course, because pressure failure in the lubricating line that serves the compressor can mean inadequate lubrication of bearings and, in turn, expensive mechanical repairs.

10.15.2 Pressure Gauges

As mentioned, many pressure-measuring instruments are called *gauges*. Generally, pressure gauges are located at key points in piping systems. Usually expressed in pounds per square inch (psi), there is a difference between *gauge pressure* (psig) and *absolute pressure* (psia). Gauge pressure refers to the pressure level indicated by the gauge; however, even when the gauge reads 0, it is subject to ambient atmospheric pressure (i.e., 14.7 psi at sea level). When a gauge reads 50 psi, that is 50 pounds *gauge pressure* (psig). The true pressure is the 50 pounds shown plus the 14.7 pounds of atmospheric pressure acting on the gauge. The total pressure is called the *absolute pressure*, which is the gauge pressure plus atmospheric pressure (50 psi + 14.7 psi = 64.7). It is written 64.7 psia.

Important Point: Pressure in any fluid pushes equally in all directions. The total force on any surface is the psi multiplied by the area in square inches; for example, a fluid under a pressure of 10 psi, pushing against an area of 5 in.^{2}, produces a total force against that surface of 50 lb (10 × 5).

10.15.2.1 Spring-Operated Pressure Gauges

Pressure, by definition, must operate against a surface. Thus, the most common method of measuring pressure in a piping system is to have the fluid press against some type of surface—a flexible surface that moves slightly. This movable surface, in turn, is linked mechanically to a gear–lever mechanism that moves the indicator arrow to indicate the pressure on the dial (i.e., a pressure gauge). The surface that the pressure acts against may be a disk or diaphragm, the inner surface of a coiled tube, a set of bellows, or the end of a plunger. No matter the element type, if the mechanism is fitted with a spring that resists the pressure and returns the element (i.e., the indicator pointer) back to the zero position when the spring drops to zero, it is called a *spring-loaded* gauge.

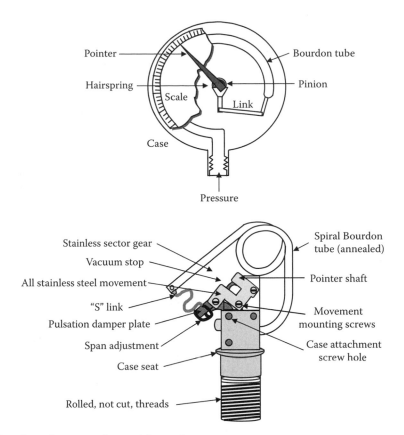

FIGURE 10.41 (Top) Bourdon tube gauge; (bottom) internal components.

10.15.2.2 Bourdon-Tube Gauges

Many pressure gauges in use today use a coiled tube as a measuring element called a *Bourdon tube* (named for its inventor, Eugene Bourdon, a French engineer). The Bourdon tube is a device that senses pressure and converts the pressure to displacement. Under pressure, the fluid fills the tube (see Figure 10.41). Because the Bourdon-tube displacement is a function of the pressure applied, it may be mechanically amplified and indicated by a pointer; thus, the pointer position indirectly indicates pressure.

Important Point: The Bourdon-tube gauge is available in various tube shapes: helical, C-shaped or curved, and spiral. The size, shape, and material of the tube depend on the pressure range and the type of gauge desired.

10.15.2.3 Bellows Gauge

Figure 10.42 shows how a simplified bellows gauge works. The bellows itself is a convoluted unit that expands and contracts axially with changes in pressure. The pressure to be measured can be applied to either the outside or the inside of the bellows; in practice, most bellows measuring devices have the pressure applied to the outside of the bellows. When pressure is released, the spring returns the bellows and the pointer to the zero position.

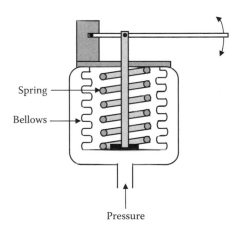

FIGURE 10.42 Bellows gauge.

10.15.2.4 Plunger Gauge

Most of us are familiar with the simple tire-pressure gauge. This device is a type of plunger gauge. Figure 10.43 shows a plunger gauge used in industrial hydraulic systems. The bellows gauge is a spring-loaded gauge, where pressure from the line acts on the bottom of a cylindrical plunger in the center of the gauge and moves it upward. At full pressure, the plunger extends above the gauge, indicating the measured pressure. As the pressure drops,

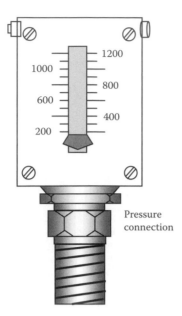

FIGURE 10.43 Plunger gauge.

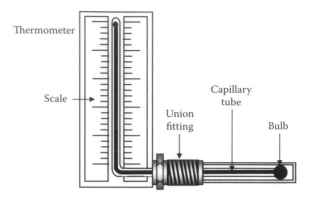

FIGURE 10.44 Industrial thermometer.

the spring contracts to pull the plunger downward, back into the body (the zero reading indication).

Note: Spring-loaded gauges are not extremely accurate, but they are entirely adequate when more precise readings are not necessary.

10.15.3 TEMPERATURE GAUGES

As mentioned, ensuring that system pressures are properly maintained in equipment and piping systems is critical to safe and proper operation. Likewise, ensuring that the temperature of fluids in industrial equipment and piping systems is correct is just as critical. Various temperature-measuring devices are available for measuring the temperature of fluids in industrial systems.

Temperature has been defined in a variety of ways. One approach defines temperature as the measure of heat (thermal energy) associated with the movement (kinetic energy) of the molecules of a substance. This definition is based on the theory that molecules of all matter are in continuous motion, which is sensed as heat. For our purposes, we define temperature as the degree of hotness or coldness of a substance measured on a definite scale. Temperature is measured when a measuring instrument (e.g., a thermometer) is brought into contact with the medium being measured. All temperature-measuring instruments use some change in a material to indicate temperature. Some of the effects that are used to monitor the temperature are changes in physical properties as well as altered physical dimensions (e.g., the change in the length of a material in the form of expansion and contraction).

Several temperature scales have been developed to provide a standard for indicating the temperatures of substances. The most commonly used scales include the Fahrenheit, Celsius, Kelvin, and Rankine temperature scales. The Celsius scale is also called the *centigrade scale*. The Fahrenheit (°F) and Celsius (°C) scales are based on the freezing point and boiling point of water. The freezing point of a substance is the temperature at which changes its physical state from a liquid to a solid. The boiling point is the temperature at which a substance changes from a liquid state to a gaseous state.

Note: Thermometers are classified as mechanical temperature-sensing devices because they produce some type of mechanical action or movement in response to temperature changes. The many types of thermometers include liquid-, gas-, and vapor-filled systems and bimetallic thermometers.

Figure 10.44 shows an *industrial-type thermometer* that is commonly used for measuring the temperature of fluids in industrial piping systems. This type of measuring instrument is nothing more than a rugged version of the familiar mercury thermometer. The bulb and capillary tube are contained inside a protective metal tube called a *well*. The thermometer is attached to the piping system (such as a vat, tank, or other component) by a union fitting.

Another common type of temperature gauge is the *bimetallic gauge* shown in Figure 10.45. If two materials with different linear coefficients of expansion (i.e., how much a material expands with heat) are bonded together, their rates of expansion will be different as the temperature changes. This will cause the entire assembly to bend in an arc. When the temperature is raised, an arc is formed around the material with the smaller expansion coefficient. The amount of arc is reflected in the movement of the pointer on the gauge. Because two dissimilar materials form the assembly, it is known as a *bimetallic element*, which is also commonly used in thermostats.

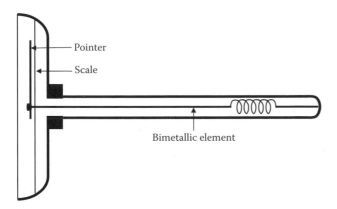

FIGURE 10.45 Bimetallic gauge.

10.15.4 VACUUM BREAKERS

Another common ancillary device found in pipelines is a *vacuum breaker* (see Figure 10.46). Simply, a vacuum breaker is a mechanical device that allows air into the piping system, thereby preventing backflow that could otherwise be caused by the siphoning action created by a partial vacuum. In other words, a vacuum breaker is designed to admit air into the line whenever a vacuum develops. A vacuum, obviously, is the absence of air. Vacuum in a pipeline can be a serious problem; for example, it can cause fluids to run in the wrong direction, possibly mixing contaminants with purer solutions. In water systems, backsiphonage can occur when a partial vacuum pulls nonpotable liquids back into the supply lines (AWWA, 1996). In addition, a vacuum can cause the collapse of tubing or equipment.

As illustrated in Figure 10.46, this particular type of vacuum breaker uses a ball that usually is held against a seat by a spring. The ball is contained in a retainer tube mounted inside the piping system or inside the component being protected. If a vacuum develops, the ball is forced (sucked) down into the retainer tube, where it works against the spring. Air then flows into the system to fill the vacuum. In water systems, when air enters the line between a cross-connection and the source of the vacuum, the vacuum will be broken and backsiphonage is prevented (AWWA, 1996). The spring then returns the ball to its usual position, which acts to once again seal the system.

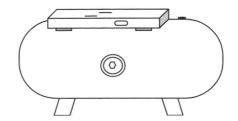

FIGURE 10.47 Air receiver.

10.15.5 ACCUMULATORS

As mentioned, in a plant compressed-air system, a means of storing and delivering air as needed is usually provided. An air receiver normally accomplishes this. In a hydraulic system, an accumulator provides the functions provided by an air receiver for an air system; that is, the accumulator (usually a dome-shaped or cylindrical chamber or tank attached to a hydraulic line) in a hydraulic system works to help store and deliver energy as required. Moreover, accumulators work to help keep pressure in the line smoothed out. If, for example, pressure in the line rises suddenly, the accumulator absorbs the rise, preventing shock to the piping. If pressure in the line drops, the accumulator acts to bring it up to normal.

Important Point: The primary function of an accumulator in a hydraulic system is to supplement pump flow.

10.15.6 AIR RECEIVERS

As shown in Figure 10.47, an air receiver is a tank or cylindrical-type vessel used for a number of purposes. Most important is their ability to store compressed air. Much like accumulators, they cushion shock from sudden pressure rises in an air line. In this way, the air receiver serves to absorb the shock of valve closure and load starts, stops, and reversals. There is no liquid in an air receiver. The air compresses as pressure rises. As pressure drops, the air expands to maintain pressure in the line.

Important Point: OSHA standard, 29 CFR 1910.169(a), requires air receivers to be drained. Specifically, the standard states, "A drain pipe and valve shall be installed at the lowest point of every air receiver to provide for the removal of accumulated oil and water" (OSHA, 1978). This is an item that should be taken seriously, not only

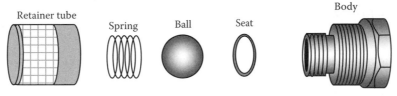

FIGURE 10.46 Vacuum breaker components.

for safety reasons but also because it is a compliance item that OSHA inspectors often check.

10.15.7 HEAT EXCHANGERS

Operating on the principle that heat flows from a warmer body to a cooler one, *heat exchangers* are devices used for adding or removing heat and cold from a liquid or gas. The purpose may be to cool one body or to warm the other; nonetheless, whether used to warm or to cool, the principle remains the same. Various designs are used in heat exchangers. The simplest form consists of a tube or possibly a large coil of tubing placed inside a larger cylinder. In an oil lubrication system, the purpose of a heat exchanger is to cool the hot oil. A heat exchanger system can also be used to heat up a process fluid circulating through part of the heat exchanger while steam circulates through its other section.

Final Note: In this section, we have discussed the major ancillary or accessory equipment used in many piping systems. It is important to point out that other accessories are commonly used in piping systems (e.g., rotary pressure joints, actuators, intensifiers, pneumatic pressure line accessories); however, discussion of these accessories is beyond the scope of this text.

CHAPTER REVIEW QUESTIONS

10.1 What is an expansion joint?

10.2 A _____ is defined as any substance or material that flows.

10.3 Compressed air is considered to be a _____.

10.4 Sections or lengths of pipe are _____ with fittings.

10.5 The _____ of fluids through a pipe is controlled by valves.

10.6 Friction causes _____ in a piping system.

10.7 As friction _____ in a piping system, the output pressure decreases.

10.8 Relief valves are designed to open _____.

10.9 _____ is used to help keep the fluids carried in piping systems hot or cold.

10.10 The major problems in piping systems are caused by _____ and corrosion.

10.11 If the speed of fluid in a pipe doubles, the friction is _____.

10.12 The most important factor in keeping a piping system operating efficiently is _____.

10.13 Pipe sizes above _____ in. are usually designated by outside diameter.

10.14 The difference in _____ numbers represents the difference in the wall _____ of pipes.

10.15 When pipe wall thickness _____, the I.D. decreases.

10.16 A _____ metal contains iron.

10.17 As temperature _____, the viscosity of a liquid decreases.

10.18 Another name for rust is _____.

10.19 Sections of _____ water pipe are usually connected with a bell-and-spigot joint.

10.20 A ferrous metal always contains _____.

10.21 Asbestos–cement pipe has the advantage of being highly resistant to _____.

10.22 As temperature increases, the strength of plastic pipe _____.

10.23 Name four basic nonmetallic piping materials.

10.24 Vitrified clay pipe is the most _____ pipe available for carrying industrial wastes.

10.25 Cast-iron pipe can be lined with _____ to increase its resistance to corrosion.

10.26 A joint made so the sections of tubing are _____ together is called a compression joint.

10.27 Incorrect tube bends can cause _____ flow and _____ pressure.

10.28 High-pressure hydraulic systems use _____ tubing.

10.29 One process used to join plastic tubing is called _____ welding.

10.30 Compared to pipe, tubing is more _____.

10.31 _____ tubing is most likely used in food-processing applications.

10.32 Before tubing can be bent or flared, it should be _____.

10.33 Plastic tubing is usually joined by _____.

10.34 The materials used most commonly for tubing are _____ and _____.

10.35 Smooth fluid flow is called _____ flow.

10.36 The _____ hose is the most common type of hose in general use.

10.37 The type of hose construction most suitable for maximum suction conditions is _____.

10.38 _____ is the nonmetallic hose best suited for use at extremely low temperatures.

10.39 Each size of hose clamp is designed for hose of a specific _____.

10.40 _____ is the outstanding advantage of hose.

10.41 Applied to hose, the letters _____ stand for enlarged end.

10.42 Hose is _____ in order to provide strength and greater resistance to _____.

10.43 Dacron hose remains _____ at extremely low temperatures.

10.44 The _____ fitting allows for a certain amount of pipe movement.

10.45 The _____ fitting helps reduce the effects of water hammer.

10.46 A flange that has a plain face should be joined to a flange that has a _____ face.

10.47 Improperly made _____ restrict fluid flow in a pipeline.

10.48 The designation 200 lb refers to the _____ at which a fitting can safely be used.

10.49 Used to close off an unused outlet in a fitting with a _____.

10.50 A _____ connects two or more pipes of different diameters.

REFERENCES AND SUGGESTED READING

ACPA. 1987. *Concrete Pipe Design Manual*. Vienna, VA: American Concrete Pipe Association.

ASME. 1996. *ASME B36.10M: Welded and Seamless Wrought Steel Pipe*. New York: American Society of Mechanical Engineers.

AWWA. 1996. *Water Transmission and Distribution*, 2nd ed. Denver, CO: American Water Works Association.

Babcock & Wilcox. 1972. *Steam: Its Generation and Use*. Cambridge, Ontario: The Babcock & Wilcox Company.

Bales, R.C., Newkirk, D.D., and Hayward, S.B. 1984. Chrysoltile asbestos in California surface waters from upstream rivers through water treatment. *J. Am. Water Works Assoc.*, 76, 66.

Bandes, A. and Gorelick, B. 2000. *Inspect Steam Traps for Efficient System*. Terre Haute, IN: TWI Press.

Basavaraju, C. 2000. Simplified analysis of shrinkage in pipe to pipe butt welds. *Nucl. Eng. Des.*, 197, 239–247.

Casada, D., 2000. *Valve Replacement Savings*. Oak Ridge, TN: Oak Ridge Laboratories.

Coastal Video Communications. 1994. *Asbestos Awareness: Controlling Exposure*. Virginia Beach, VA: Coastal Video Communications.

Crocker, Jr., S. 2000. Hierarchy of design documents. In *Piping Handbook*, 7th ed., Nayyar, M.L., Ed. New York: McGraw-Hill.

Gagliardi, M.G. and Liberatore, L.J., 2000. Water piping systems. In *Piping Handbook*, 7th ed., Nayyar, M.L., Ed. New York: McGraw-Hill.

Geiger, E.L. 2000a. Piping components. In *Piping Handbook*, 7th ed., Nayyar, M.L., Ed. New York: McGraw-Hill.

Geiger, E.L. 2000b. Tube properties. In *Piping Handbook*, 7th ed., Nayyar, M.L., Ed. New York: McGraw-Hill.

Giachino. J.W. and Weeks, W. 1985. *Welding Skills*. Homewood, IL: American Technical Publishers.

Globe Valves. 1998. Integrated Publishing (http/tpub.com/fluid/ch2c.htm).

Kawamura, S. 1999. *Integrated Design and Operation of Water Treatment Facilities*, 2nd ed. New York: John Wiley & Sons.

Lohmeir, A. and Avery, D.R. 2000. Manufacture of metallic pipe. In *Piping Handbook*, 7th ed., Nayyar, M.L., Ed. New York: McGraw-Hill.

Magnusson, R.J. 2001. *Technology in the Middle Ages*. Baltimore, MD: The Johns Hopkins University.

Marine, C.S. 1999. Hydraulic transient design for pipeline systems. In *Water Distribution Systems Handbook*, Mays, L.W., Ed. New York: McGraw-Hill.

McGhee, T.J. 1991. *Water Supply and Sewerage*, 6th ed. New York: McGraw-Hill.

Nayyar, M.L. 2000. Introduction to piping. In *Piping Handbook*, 7th ed. Nayyar, M.L., Ed. New York: McGraw-Hill.

OSHA. 1978. 29 CFR 1910.169: *Drain on Air Receivers*. Washington, D.C.: Occupational Safety and Health Administration.

Snoek, P.E. and Carney, J.C. 1981. Pipeline material selection for transport of abrasive tailings. In *Proc. of the 6th Internal Technical Conference on Slurry Transportation*, Las Vegas, NV.

Spellman, F.R., 1996. *Safe Work Practices for Wastewater Treatment Plants*. Boca Raton, FL: CRC Press.

USEPA 2006. *Emerging Technologies for Conveyance Systems*, EPA 832-R-06-004. Washington, D.C.: U.S. Environmental Protection Agency.

Valves. 1998. Integrated Publishing (http://pub.com/fluid/ch2c.htm).

Webber, J.S., Covey, J.R., and King, M.V. 1989. Asbestos in drinking water supplied through grossly deteriorated pipe. *J. Am Water Works Assoc.*, 81, 80.

Part III

Characteristics of Water

11 Basic Water Chemistry

Chemical testing can be divided into two types. The first type measures a bulk physical property of the sample, such as volume, temperature, melting point, or mass. These measurements are normally performed with an instrument, and one simply has to calibrate the instrument to perform the test. Most analyses, however, are of the second type, in which a chemical property of the sample is determined that generates information about how much of what is present.

Smith (1993)

As mentioned, although no one has seen a water molecule, we have determined through x-rays that atoms in water are elaborately meshed. Moreover, although it is true that we do not know as much as we need to know about water (our growing knowledge of water is a work in progress), we have determined many things about water. A large amount of our current knowledge comes from studies of water chemistry.

Water chemistry is important because several factors about water to be treated and then distributed or returned to the environment are determined through simple chemical analysis. Probably the most important determination that the water practitioner makes about water is its hardness.

Why chemistry? "I'm not a chemist," you say. But, when you add chlorine to water to make it safe to drink or safe to discharge into a receiving body (usually a river or lake), you *are* a chemist. Chemistry is the study of substances and the changes they undergo. Water specialists and those interested in the study of water must possess a fundamental knowledge of chemistry. Before beginning our discussion of water chemistry, it is important for the reader to have some basic understanding of chemistry concepts and chemical terms. The following section presents a review of chemistry terms, definitions, and concepts.

11.1 CHEMISTRY CONCEPTS AND DEFINITIONS

Chemistry, like the other sciences, has its own language; thus, to understand chemistry, it is necessary to understand the following concepts and key terms.

11.1.1 CONCEPTS

11.1.1.1 Miscible and Solubility

Substances that are *miscible* are capable of being mixed in all proportions. Simply, when two or more substances disperse themselves uniformly in all proportions when brought into contact, they are said to be completely soluble in one another, or completely miscible. The precise chemistry definition is a "homogenous molecular dispersion of two or more substances" (Jost, 1992). Examples include the following:

- All gases are completely miscible.
- Water and alcohol are completely miscible.
- Water and mercury (in its liquid form) are immiscible liquids.

Between the two extremes of miscibility is a range of *solubility*; that is, various substances mix with one another up to a certain proportion. In many environmental situations, a rather small amount of a contaminant may be soluble in water in contrast to the complete miscibility of water and alcohol. The amounts are measured in parts per million (ppm).

11.1.1.2 Suspension, Sediment, Particles, and Solids

Often water carries *solids* or *particles* in *suspension*. These dispersed particles are much larger than molecules and may be comprised of millions of molecules. The particles may be suspended in flowing conditions and initially under quiescent conditions, but eventually gravity causes settling of the particles. The resultant accumulation by settling is referred to as *sediment* or *biosolids* (sludge) or *residual solids* in wastewater treatment vessels. Between this extreme of readily falling out by gravity and permanent dispersal as a solution at the molecular level are intermediate types of dispersion or suspension. Particles can be so finely milled or of such small intrinsic size as to remain in suspension almost indefinitely and in some respects similarly to solutions.

11.1.1.3 Emulsion

Emulsions represent a special case of a suspension. As you know, oil and water do not mix. Oil and other hydrocarbons derived from petroleum generally float on water with negligible solubility in water. In many instances, oils may be dispersed as fine oil droplets (an emulsion) in water and not readily separated by floating because of size or the addition of dispersal-promoting additives. Oil and, in particular, emulsions can prove detrimental to many treatment technologies and must be treated in the early steps of a multistep treatment train.

11.1.1.4 Ion

An ion is an electrically charged particle; for example, sodium chloride or table salt forms charged particles on dissolution in water. Sodium is positively charged (a cation), and chloride is negatively charged (an anion). Many salts similarly form cations and anions on dissolution in water.

11.1.1.5 Mass Concentration

Concentration is often expressed in terms of parts per million (ppm) or mg/L. Sometimes parts per thousand (ppt) and parts per billion (ppb) are also used:

$$ppm = \frac{\text{Mass of substance}}{\text{Mass of solutions}} \quad (11.1)$$

Because 1 kg of solution with water as a solvent has a volume of approximately 1 liter:

$$1 \text{ ppm} \approx 1 \text{ mg/L}$$

11.1.2 Definitions

Chemistry—The science that deals with the composition and changes in composition of substances. Water is an example of this composition; it is composed of two gases: hydrogen and oxygen. Water also changes form from liquid to solid to gas but does not necessarily change composition.

Matter—Anything that has weight (mass) and occupies space. Kinds of matter include elements, compounds, and mixtures.

Solids—Substances that maintain definite size and shape. Solids in water fall into one of these categories:

- *Dissolved solids*, which are in solution and pass through a filter. The solution consisting of the dissolved components and water forms a single phase (a homogenous solution).
- *Colloidal solids* (sols), which are uniformly dispersed in solution. They form a solid phase that is distinct from the water phase.

- *Suspended solids*, which are also a separate phase from the solution. Some suspended solids are classified as settleable solids, which are determined by placing a sample in a cylinder and measuring the amount of solids that have settled after a set amount of time. The size of solids increases moving from dissolved solids to suspended solids.

Liquids—Having a definite volume but not shape, liquids will fill containers to certain levels and form free level surfaces.

Gases—Having neither definite volume nor shape, gases completely fill any container in which they are placed.

Element—The simplest form of chemical matter. Each element has chemical and physical characteristics different from all other kinds of matter.

Compound—A substance of two or more chemical elements chemically combined. Examples include water (H_2O), which is a compound formed by hydrogen and oxygen, and carbon dioxide (CO_2), which is composed of carbon and oxygen.

Mixture—A physical, not chemical, intermingling of two or more substances. Sand and salt stirred together form a mixture.

Atom—The smallest particle of an element that can unite chemically with other elements. All the atoms of an element are the same in chemical behavior, although they may differ slightly in weight. Most atoms can combine chemically with other atoms to form molecules.

Molecule—The smallest particle of matter or a compound that possesses the same composition and characteristics as the rest of the substance. A molecule may consist of a single atom, two or more atoms of the same kind, or two or more atoms of different kinds.

Radical—Two or more atoms that unite in a solution and behave chemically as if a single atom.

Solvent—The component of a solution that does the dissolving.

Solute—The component of a solution that is dissolved by the solvent.

Ion—An atom or group of atoms that carries a positive or negative electric charge as a result of having lost or gained one or more electrons.

Ionization—The formation of ions by splitting of molecules or electrolytes in solution. Water molecules are in continuous motion, even at lower temperatures. When two water molecules collide, a hydrogen ion is transferred from one molecule to the other. The water molecule that loses the hydrogen ion becomes a negatively charged hydroxide ion. The water molecule that gains the hydrogen ion becomes a positively charged hydronium ion. This process is commonly referred to as the *self-ionization of water*.

Cation—A positively charged ion.

Anion—A negatively charged ion.

Organic—Chemical substances of animal or vegetable origin made of carbon structure.

Inorganic—Chemical substances of mineral origin.

Solids—As it pertains to water, suspended and dissolved material in water.

Dissolved solids—The material in water that will pass through a glass-fiber filter and remain in an evaporating dish after evaporation of the water.

Suspended solids—The quantity of material deposited when a quantity of water, sewage, or other liquid is filtered through a glass-fiber filter.

Total solids—The solids in water, sewage, or other liquids. Total solids include suspended solids (largely removable by a filter) and filterable solids (those that pass through the filter).

Saturated solution—The physical state in which a solution will no longer dissolve more of the dissolving substance (solute).

Colloidal—Any substance in a certain state of fine division in which the particles are less than 1 micron in diameter.

Turbidity—A condition in water caused by the presence of suspended matter. Turbidity results in the scattering and absorption of light rays.

Precipitate—A solid substance that can be dissolved but is separated from the solution because of a chemical reaction or change in conditions such as pH or temperature.

11.2 CHEMISTRY FUNDAMENTALS

Whenever water and wastewater practitioners add a substance to another substance (e.g., adding sugar to a cup of tea or adding chlorine to water to make it safe to drink), they perform chemistry. Water and wastewater operators (as well as many others) are chemists, because they are working with chemical substances, and it is important for operators to know and understand how those substances react.

11.2.1 MATTER

Going through a day without coming in contact with many kinds of matter is impossible. Paper, coffee, gasoline, chlorine, rocks, animals, plants, water, air—all the materials of which the world is made—are all different forms or kinds of matter. Earlier matter was defined as anything that has mass (weight) and occupies space; matter is distinguishable from empty space by its presence. Thus, obviously, the statement about the impossibility of going through a day without coming into contact with matter is correct; in fact, avoiding some form of matter is virtually impossible. Not all matter is the same, even though we narrowly classify all matter into three groups: solids, liquids, and gases. These three groups are the *physical states of matter* and are distinguishable from one another by means of two general features: shape and volume.

Important Point: *Mass* is closely related to the concept of *weight*. On Earth, the weight of matter is a measure of the force with which it is pulled by gravity toward the center of the Earth. As we leave Earth's surface, the gravitational pull decreases, eventually becoming virtually insignificant, while the weight of matter accordingly reduces to zero. Yet, the matter still possesses the same amount of mass. Hence, the mass and weight of matter are proportional to each other.

Important Point: Because matter occupies space, a given form of matter is also associated with a definite volume. Space should not be confused with air, as air is itself a form of matter. *Volume* refers to the actual amount of space that a given form of matter occupies.

Solids have a definite, rigid shape; their particles are closely packed together and stick firmly to each other. A solid does not change its shape to fit a container. Put a solid on the ground and it will keep its shape and volume—it will never spontaneously assume a different shape. Solids also possess a definite volume at a given temperature and pressure.

Liquids maintain a constant volume but change shape to fit the shape of their container; they do not possess a characteristic shape. The particles of the liquid move freely over one another but still stick together enough to maintain a constant volume. Consider a glass of water. The liquid water takes the shape of the glass up to the level it occupies. If we pour the water into a drinking glass, the water takes the shape of the glass; if we pour it into a bowl, the water takes the shape of the bowl. Thus, if space is available, a liquid assumes whatever shape its container possesses. Like solids, liquids possess a definite volume at a given temperature and pressure, and they tend to maintain this volume when they are exposed to a change in either of these conditions.

Gases have no definite fixed shape, and their volume can be expanded or compressed to fill different sizes of containers. A gas or mixture of gases such as air can be put into a balloon and will take the shape of the balloon. Particles of gases do not stick together at all and move about freely, filling containers of any shape and size. A gas is identified by its lack of a characteristic volume. When confined to a container with nonrigid, flexible walls, for example, the volume that a confined gas occupies depends on its temperature and pressure. When confined to a container with rigid walls, however, the volume of the gas is forced to remain constant.

Internal linkages among its units, including between one atom and another, maintain the constant composition associated with a given substance. These linkages are called *chemical bonds*. When a particular process occurs that involves the making and breaking of these bonds, we say that a *chemical reaction* or a *chemical change* has occurred. Let's take a closer look at both chemical and physical changes of matter.

Chemical changes occur when new substances are formed that have entirely different properties and characteristics. When wood burns or iron rusts, a chemical change has occurred; the linkages—the chemical bonds—are broken. *Physical changes* occur when matter changes its physical properties such as size, shape, and density, as well as when it changes its state (e.g., from gas to liquid to solid). When ice melts or when a glass window breaks into pieces, a physical change has occurred.

11.2.2 THE CONTENT OF MATTER: THE ELEMENTS

Matter is composed of pure basic substances. Earth is made up of the fundamental substances of which all matter is composed. These substances that resist attempts to decompose them into simpler forms of matter are called *elements*. To date, more than 100 elements are known to exist. They range from simple, lightweight elements to very complex, heavyweight elements. Some of these elements exist in nature in pure form; others are combined.

The smallest unit of an element is the *atom*. The simplest atom possible consists of a nucleus having a single proton with a single electron traveling around it. This is an atom of hydrogen, which has an atomic weight of one because of the single proton. The *atomic weight* of an element is equal to the total number of protons and neutrons in the nucleus of an atom of an element.

To better understand the basic atomic structure and related chemical principles it is useful to compare the atom to our solar system. In our solar system, the sun is the center of everything, whereas the *nucleus* is the center in the atom. The sun has planets orbiting around it, whereas the atom has *electrons* orbiting about the nucleus. It is interesting to note that astrophysicists, who would likely find this analogy overly simplistic, are concerned primarily with activity within the nucleus. This is not the case, however, with chemists, who deal principally with the activity of the planetary electrons. Chemical reactions between atoms or molecules involve only electrons, with no changes in the nuclei.

The nucleus is made up of positive electrically charge *protons* and *neutrons*, which are neutral (no charge). The negatively charged *electrons* orbiting the nucleus balance the positive charge in the nucleus. An electron has negligible mass (less than 0.02% of the mass of a proton), which makes it practical to consider the weight of the atom as the weight of the nucleus.

Atoms are identified by name, atomic number, and atomic weight. The *atomic number* or *proton number* is the number of protons in the nucleus of an atom. It is equal to the positive charge on the nucleus. In a neutral atom, it is also equal to the number of electrons surrounding the nucleus. As mentioned, the atomic weight of an atom depends on the number of protons and neutrons in the nucleus, the electrons having negligible mass. Atoms (elements) have received their names and symbols in interesting ways. The discoverer of the element usually proposes a name for it. Some elements get their symbols from languages other than English. The following is a list of common elements with their common names and the names from which the symbol is derived.

Chlorine	Cl
Copper	Cu (Latin *cuprum*)
Hydrogen	H
Iron	Fe (Latin *ferrum*)
Nitrogen	N
Oxygen	O
Phosphorus	P
Sodium	Na (Latin *natrium*)
Sulfur	S

As shown above, a capital letter or a capital letter and a small letter designate each element. These are called *chemical symbols*. As is apparent from the above list, most of the time the symbol is easily recognized as an abbreviation of the atom name, such as O for oxygen.

Typically, we do not find most of the elements as single atoms. They are more often found in combinations of atoms called *molecules*. Basically, a molecule is the least common denominator of what makes a substance what it is. A system of formulae has been devised to show how atoms are combined into molecules. When a chemist writes the symbol for an element, it stands for one atom of the element. A subscript following the symbol indicates the number of atoms in the molecule. O_2 is the chemical formula for an oxygen molecule. It shows that oxygen occurs in molecules consisting of two oxygen atoms. As you know, a molecule of water contains two hydrogen atoms and one oxygen atom, so the formula is H_2O.

Important Point: H_2O, the chemical formula of the water molecule, was defined in 1860 by the Italian scientist Stanisloa Cannizzarro.

Some elements have similar chemical properties; for example, a chemical such as bromine (atomic number 35) has chemical properties that are similar to the chemical properties of the element chlorine (atomic number 17), with which most water operators are familiar, and iodine (atomic number 53).

In 1865, English chemist John Newlands arranged some of the known elements in increasing order of atomic weights. Newlands arranged the lightest element known at the time at the top of his list and the heaviest element at the bottom. Newlands was surprised when he observed that starting from a given element, every eighth element repeated the properties of the given element.

Later, in 1869, Mendeleev, a Russian chemist, published a table of the 63 known elements. In his table, Mendeleev, like Newlands, arranged the elements in an increasing order of atomic weights. He also grouped them in eight vertical columns so the elements with similar chemical properties would be found in one column. It is interesting to note that Mendeleev left blanks in his table. He correctly hypothesized that undiscovered elements existed that would fill in the blanks when they were discovered. Because he knew the chemical properties of the elements above and below the blanks in his table, he was able to predict quite accurately the properties of some of the undiscovered elements.

Today our modern form of the periodic table is based on work done by the English scientist Henry Moseley, who was killed during World War I. Following the work of Ernest Rutherford (a New Zealand physicist) and Niels Bohr (a Danish physicist), Moseley used x-ray methods to determine the number of protons in the nucleus of an atom.

The atomic number, or number of protons, of an atom is related to its atomic structure. In turn, atomic structure governs chemical properties. The atomic number of an element is more directly related to its chemical properties than is its atomic weight. It is more logical to arrange the periodic table according to atomic numbers than atomic weights. By demonstrating the atomic numbers of elements, Moseley helped chemists to make a better periodic table.

In the periodic table, each box or section contains the atomic number, symbol, and atomic weight of an element. The numbers down the left side of the box show the arrangement, or configuration, of the electrons in the various shells around the nucleus. For example, the element carbon has an atomic number of 6, its symbol is C, and its atomic weight is 12.01.

In the periodic table, a horizontal row of boxes is called a *period* or *series*. Hydrogen is all by itself because of its special chemical properties. Helium is the only element in the first period. The second period contains lithium, beryllium, boron, carbon, nitrogen, oxygen, fluorine, and neon. Other elements may be identified by looking at the table. A vertical column is called a *group* or *family*. Elements in a group have similar chemical properties. The periodic table is useful because knowing where an element is located in the table gives us a general idea of its chemical properties.

TABLE 11.1
Elements and Their Symbols

Element	Symbol
Aluminum*	Al
Arsenic	As
Barium	Ba
Cadmium	Ca
Carbon*	C
Calcium	Ca
Chlorine*	Cl
Chromium	Cr
Cobalt	Co
Copper	Cu
Fluoride*	F
Helium	He
Hydrogen*	H
Iodine	I
Iron*	Fe
Lead	Pb
Magnesium*	Mg
Manganese*	Mn
Mercury	Hg
Nitrogen*	N
Nickel	Ni
Oxygen*	O
Phosphorus	P
Potassium	K
Silver	Ag
Sodium*	Na
Sulfur*	S
Zinc	Zn

* Indicates elements familiar to water/wastewater treatment operators.

As mentioned, for convenience, elements have a specific name and symbol but are often identified by chemical symbol only. The symbols of the elements consist of either one or two letters, with the first letter capitalized. Table 11.1 lists the elements important to the water practitioner (about a third of the over 100 known elements). Those elements most closely associated with water treatment are marked with an asterisk (*).

11.2.3 Compound Substances

If we take a pure substance such as calcium carbonate (limestone) and heat it, the calcium carbonate ultimately crumbles to a white powder; however, careful examination of the process shows that carbon dioxide also evolves from the calcium carbonate. Substances such as calcium carbonate that can be broken down into two or more simpler substances are called *compound substances* or simply *compounds*. Heating is a common way of decomposing compounds, but other forms of energy are often used as well.

Chemical elements that make up compounds such as calcium carbonate combine with each other in definite proportions. When atoms of two or more elements are bonded together to form a compound, the resulting particle is called a *molecule*.

Important Point: This law simply means that only a certain number of atoms or radicals of one element will combine with a certain number of atoms or radicals of a different element to form a chemical compound.

Water (H_2O) is a compound. As stated, compounds are chemical substances made up of two or more elements bonded together. Unlike elements, compounds can be separated into simpler substances by chemical changes. Most forms of matter in nature are composed of combinations of the over 100 pure elements.

If we have a particle of a compound—for example, a crystal of salt (sodium chloride)—and subdivide, subdivide, and subdivide until we get the smallest unit of sodium chloride possible, we would have a molecule. As stated, a molecule (or least common denominator) is the smallest particle of a compound that still has the characteristics of that compound.

Important Point: Because the weights of atoms and molecules are relative and the units are extremely small, the chemist works with units identified as moles. A mole (symbol, mol) is defined as the amount of a substance that contains as many elementary entities (atoms, molecules, and so on) as there are atoms in 12 g of the isotope carbon-12.

Important Point: An *isotope* of an element is an atom having the same structure as the element—the same electrons orbiting the nucleus and the same protons in the nucleus—but having more or fewer neutrons.

One mole of an element that exists as single atoms weighs as many grams as its atomic number (so 1 mole of carbon weighs 12 g), and it contains 6.022045×10^{23} atoms, which is the *Avogadro's number*.

As stated previously, symbols are used to identify elements and are a shorthand method for writing the names of the elements. This shorthand method is also used for writing the names of compounds. Symbols used in this manner show the kinds and numbers of different elements in the compound. These shorthand representations of chemical compounds are called chemical *formulae*; for example, the formula for table salt (sodium chloride) is NaCl. The formula shows that one atom of sodium combines with one atom of chlorine to form sodium chloride. Let's look at a more complex formula for the compound sodium carbonate (soda ash): Na_2CO_3. The formula shows that this compound is made up of three elements: sodium, carbon, and oxygen. In addition, each molecule has two atoms of sodium, one atom of carbon, and three atoms of oxygen.

When depicting chemical reactions, chemical *equations* are used. The following equation shows a chemical reaction with which most water/wastewater operators are familiar—chlorine gas added to water. It shows the molecules that react together and the resulting product molecules:

$$Cl_2 + H_2O \rightarrow HOCl + HCl$$

As stated previously, a chemical equation tells what elements and compounds are present before and after a chemical reaction. Sulfuric acid poured over zinc will cause the release of hydrogen and the formation of zinc sulfate. This is shown by the following equation:

$$Zn + H_2SO_4 \rightarrow ZnSO_4 + H_2$$

One atom (also one molecule) of zinc unites with one molecule of sulfuric acid to give one molecule of zinc sulfate and one molecule (two atoms) of hydrogen. Notice the same number of atoms of each element on each side of the arrow, even though the atoms are combined differently.

Let's look at another example. When hydrogen gas is burned in air, the oxygen from the air unites with the hydrogen and forms water. The water is the product of burning hydrogen. This can be expressed as an equation:

$$2H_2 + O_2 \rightarrow 2H_2O$$

This equation indicates that two molecules of hydrogen unite with one molecule of oxygen to form two molecules of water.

11.3 WATER SOLUTIONS

A *solution* is a condition in which one or more substances are uniformly and evenly mixed or dissolved. A solution has two components, a *solvent* and a *solute*. The solvent is the component that does the dissolving. The solute is the component that is dissolved. In water solutions, water is the solvent. Water can dissolve many other substances; in fact, given enough time, not too many solids, liquids, or gases exist that water cannot dissolve. When water dissolves substances, it creates solutions with many impurities. Generally, a solution is usually transparent and not cloudy; however, a solution may have some color when the solute remains uniformly distributed throughout the solution and does not settle with time.

When molecules dissolve in water, the atoms making up the molecules come apart (dissociate) in the water. This dissociation in water is called *ionization*. When the atoms in the molecules come apart, they do so as charged atoms (both negatively and positively charged) called *ions*. The positively charged ions are called *cations* and the negatively charged ions are called *anions*. A good example of the ionization process is when calcium carbonate ionizes:

$$CaCO_3 \leftrightarrow Ca^{2+} + CO_3^{2-}$$

Calcium   Calcium ion   Carbonate ion
carbonate   (cation)   (anion)

Another good example is the ionization that occurs when table salt (sodium chloride) dissolves in water:

$$NaCl \leftrightarrow Na^+ + Cl^-$$

Sodium   Sodium ion   Chloride ion
chloride   (cation)   (anion)

The symbols of some of the ions commonly found in water are provided below:

Hydrogen H^+
Sodium Na^+
Potassium K^+
Chloride Cl^-
Bromide Br^-
Iodide I^-
Bicarbonate HCO^{3-}

Water dissolves polar substances better than nonpolar substances. This makes sense when we consider that water is a polar substance. Polar substances such as mineral acids, bases, and salts are easily dissolved in water, while nonpolar substances such as oils, fats, and many organic compounds do not dissolve easily in water.

Water dissolves polar substances better than nonpolar substances, but only to a point. Polar substances dissolve in water up to a point, but only so much solute will dissolve at a given temperature, for example. When that limit is reached, the resulting solution is saturated. When a solution becomes saturated, no more solute can be dissolved. For solids dissolved in water, if the temperature of the solution is increased, the amount of solids (solutes) required to reach saturation increases.

11.4 WATER CONSTITUENTS

Natural water can contain a number of substances (what we may call *impurities*) or constituents in water treatment operations. The concentrations of various substances in water in dissolved, colloidal, or suspended form are typically low but vary considerably. A hardness value of up to 400 ppm of calcium carbonate, for example, is sometimes tolerated in public supplies, whereas 1 ppm of dissolved iron would be unacceptable. When a particular constituent can affect the health of the water user or the environment, it is considered a *contaminant* or *pollutant*.

These contaminants, of course, are what the water operator removes from or tries to prevent from entering the water supply. In this section, we discuss some of the more common constituents of water.

11.4.1 SOLIDS

Other than gases, all contaminants of water contribute to the solids content. Natural water carries many dissolved and undissolved solids. The undissolved solids are nonpolar substances and consist of relatively large particles of materials such as silt, that will not dissolve. Classified by their size and state, by their chemical characteristics, and by their size distribution, solids can be dispersed in water in both suspended and dissolved forms.

The sizes of solids in water can be classified as *suspended*, *settleable*, *colloidal*, or *dissolved*. *Total solids* are the suspended and dissolved solids that remain behind when the water is removed by evaporation. Solids are also characterized as being *volatile* or *nonvolatile*.

The distribution of solids is determined by computing the percentage of filterable solids by size range. Solids typically include inorganic solids such as silt and clay from riverbanks and organic matter such as plant fibers and microorganisms from natural or manmade sources.

Important Point: Though not technically accurate from a chemical point of view because some finely suspended material can actually pass through the filter, *suspended solids* are defined as those that can be filtered out in the suspended solids laboratory test. The material that passes through the filter is defined as *dissolved solids*.

As mentioned, colloidal solids are extremely fine suspended solids (particles) less than 1 micron in diameter; they are so small (though they still can make water cloudy) that they will not settle even if allowed to sit quietly for days or weeks.

11.4.2 TURBIDITY

Simply, turbidity refers to how clear the water is. The clarity of water is one of the first characteristics people notice. Turbidity in water is caused by the presence of suspended matter, which results in the scattering and absorption of light rays. The greater the amount of *total suspended solids* (TSS) in the water, the murkier it appears and the higher the measured turbidity. Thus, in plain English, turbidity is a measure of the light-transmitting properties of water. Natural water that is very clear (low turbidity) allows us to see images at considerable depths. High-turbidity water, on the other hand, appears cloudy. Keep in mind that water of low turbidity is not necessarily without dissolved solids. Dissolved solids do not cause light to be scattered or absorbed; thus, the water looks clear. High turbidity causes problems for the waterworks operator, as components that cause high turbidity can cause taste and odor problems and will reduce the effectiveness of disinfection.

11.4.3 COLOR

Color in water can be caused by a number of contaminants such as iron, which changes in the presence of oxygen to yellow or red sediments. The color of water can be deceiving. In the first place, color is considered an aesthetic quality of water with no direct health impact. Second, many of the colors associated with water are not true colors but the result of colloidal suspension and are referred to as the *apparent color*. This apparent color can often be attributed to iron and to dissolved tannin extracted from decaying plant material. *True color* is the result of dissolved chemicals (most often organics) that cannot be seen. True color is distinguished from apparent color by filtering the sample.

11.4.4 DISSOLVED OXYGEN

Although water molecules contain an oxygen atom, this oxygen is not what is needed by aquatic organisms living in our natural waters. A small amount of oxygen, up to about ten molecules of oxygen per million molecules of water, is actually dissolved in water. This dissolved oxygen is breathed by fish and zooplankton and is needed by them to survive. Other gases can also be dissolved in water. In addition to oxygen, carbon dioxide, hydrogen sulfide, and nitrogen are examples of gases that dissolve in water. Gases dissolved in water are important; for example, carbon dioxide is important because of the role it plays in pH and alkalinity. Carbon dioxide is released into the water by microorganisms and consumed by aquatic plants. Dissolved oxygen (DO) in water, however, is of the most importance to us here, not only because it is important to most aquatic organisms but also because dissolved oxygen is an important indicator of water quality.

Like terrestrial life, aquatic organisms need oxygen to live. As water moves past their breathing apparatus, microscopic bubbles of oxygen gas in the water—dissolved oxygen (DO)—are transferred from the water to their blood. Like any other gas diffusion process, the transfer is efficient only above certain concentrations. In other words, oxygen can be present in the water but at too low a concentration to sustain aquatic life. Oxygen also is needed by virtually all algae and macrophytes and for many chemical reactions that are important to water body functioning.

Rapidly moving water, such as in a mountain stream or large river, tends to contain a lot of dissolved oxygen, while stagnant water contains little. Bacteria in water can consume oxygen as organic matter decays; thus, excess organic material in our lakes and rivers can cause an

TABLE 11.2
Common Metals Found in Water

Metal	Health Hazard
Barium	Circulatory system effects and increased blood pressure
Cadmium	Concentration in the liver, kidneys, pancreas, and thyroid
Copper	Nervous system damage, kidney effects; toxic to humans
Lead	Same as copper
Mercury	Central nervous system (CNS) disorders
Nickel	CNS disorders
Selenium	CNS disorders
Silver	Gray skin
Zinc	Taste effects; not a health hazard

oxygen-deficient situation to occur. Aquatic life can have a difficult time surviving in stagnant water that has a lot of rotting, organic material in it, especially in the summer, when dissolved oxygen levels are at a seasonal low.

Important Point: As mentioned, solutions can become saturated with solute. This is the case with water and oxygen. As with other solutes, the amount of oxygen that can be dissolved at saturation depends on the temperature of the water. In the case of oxygen, the effect is just the opposite of other solutes. The higher the temperature, the lower the saturation level; the lower the temperature, the higher the saturation level.

11.4.5 METALS

Metals are elements present in chemical compounds as positive ions or in the form of cations (+ ions) in solution. Metals with a density over 5 kg/dm³ are known as *heavy metals*. Metals are one of the constituents or impurities often carried by water. Although most of the metals are not harmful at normal levels, a few metals can cause taste and odor problems in drinking water. In addition, some metals may be toxic to humans, animals and microorganisms. Most metals enter water as part of compounds that ionize to release the metal as positive ions. Table 11.2 lists some metals commonly found in water and their potential health hazards.

Important Point: Metals may be found in various chemical and physical forms. These forms, or *species*, can be particles or simple organic compounds, organic complexes, or colloids. The dominating form is determined largely by the chemical composition of the water, the matrix, and in particular the pH.

11.4.6 ORGANIC MATTER

Organic matter or compounds are those that contain the element carbon and are derived from material that was once alive (i.e., plants and animals). Organic compounds include fats, dyes, soaps, rubber products, plastics, wood, fuels, cotton, proteins, and carbohydrates. Organic compounds in water are usually large, nonpolar molecules that do not dissolve well in water. They often provide large amounts of energy to animals and microorganisms.

Important Point: *Natural organic matter* (NOM) is used to describe the complex mixture of organic material, such as humic and hydrophilic acids present in all drinking water sources. NOM can cause major problems in the treatment of water as it reacts with chlorine to form *disinfection byproducts* (DBPs). Many of the DBPs formed by the reaction of NOM with disinfectants are reported to be toxic and carcinogenic to humans if ingested over an extended period. The removal of NOM and subsequent reduction in DBPs are major goals in the treatment of any water source.

11.4.7 INORGANIC MATTER

Inorganic matter or compounds are carbon free, not derived from living matter, and easily dissolved in water; inorganic matter is of mineral origin. The inorganics include acids, bases, oxides, and salts. Several inorganic components are important in establishing and controlling water quality. Two important inorganic constituents in water are nitrogen and phosphorus.

11.4.8 ACIDS

Lemon juice, vinegar, and sour milk are acidic or contain acid. The common acids used in waterworks operations are hydrochloric acid (HCl), sulfuric acid (H_2SO_4), nitric acid (HNO_3), and carbonic acid (H_2CO_3). Note that in each of these acids, hydrogen (H) is one of the elements.

Important Point: An acid is a substance that produces hydrogen ions (H^+) when dissolved in water. Hydrogen ions are hydrogen atoms stripped of their electrons. A single hydrogen ion is nothing more than the nucleus of a hydrogen atom.

The relative strengths of acids in water (listed in descending order of strength) are shown in Table 11.3.

Note: Acids and bases become solvated; they loosely bond to water molecules.

11.4.9 BASES

A *base* is a substance that produces hydroxide ions (OH⁻) when dissolved in water. Lye, or common soap, (bitter things), contains bases. The bases used in waterworks operations are calcium hydroxide, $Ca(OH)_2$; sodium hydroxide, NaOH; and potassium hydroxide, KOH. Note that the hydroxyl group (OH) is found in all bases. In addition,

TABLE 11.3
Relative Strengths of Acids in Water

Acid	Formula
Perchloric acid	$HClO_4$
Sulfuric acid	$H2SO_4$
Hydrochloric acid	HCl
Nitric acid	HNO_3
Phosphoric acid	H_3PO_4
Nitrous acid	HNO_2
Hydrofluoric acid	HF
Acetic acid	CH_3COOH
Carbonic acid	$H2CO_3$
Hydrocyanic acid	HCN
Boric acid	H_3BO_3

note that bases contain metallic substances, such as sodium (Na), calcium (Ca), magnesium (Mg), and potassium (K). These bases contain the elements that produce the alkalinity in water.

11.4.10 SALTS

When acids and bases chemically interact, they neutralize each other. The compound (other than water) that forms from the neutralization of acids and bases is called a *salt*. Salts constitute, by far, the largest group of inorganic compounds. A common salt used in waterworks operations, copper sulfate is utilized to kill algae in water.

11.4.11 pH

pH is a measure of the hydrogen ion (H^+) concentration. Solutions range from very acidic (having a high concentration of H^+ ions) to very basic (having a high concentration of OH⁻ ions). The pH scale ranges from 0 to 14, with 7 being the neutral value (see Figure 4.1). The pH of water is important to the chemical reactions that take place within water, and pH values that are too high or low can inhibit the growth of microorganisms. High pH values are considered basic, and low pH values are considered acidic. Stated another way, low pH values indicate a high H^+ concentration, and high pH values indicate a low H^+ concentration. Because of this inverse logarithmic relationship, each pH unit represents a tenfold difference in H^+ concentration.

Natural water varies in pH depending on its source. Pure water has a neutral pH, with equal H^+ and OH⁻. Adding an acid to water causes additional positive ions to be released, so the H^+ ion concentration goes up and the pH value goes down:

$$HCl \leftrightarrow H^+ + Cl^-$$

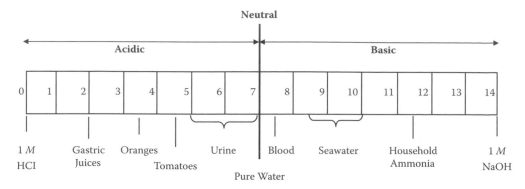

FIGURE 11.1 pH scale.

To control water coagulation and corrosion, the waterworks operator must test for the hydrogen ion concentration of the water to determine the pH of the water. In a coagulation test, as more alum (acid) is added, the pH value lowers. If more lime (alkali) is added, the pH value rises. This relationship should be remembered—if a good floc is formed, the pH should then be determined and maintained at that pH value until the raw water changes.

Pollution can change the pH of water, which in turn can harm animals and plants living in the water. Water coming out of an abandoned coal mine, for example, can have a pH of 2, which is very acidic and would definitely affect any fish crazy enough to try to live in that water. By using the logarithm scale, this mine-drainage water would be 100,000 times more acidic than neutral water—so stay out of abandoned mines.

Important Point: Seawater is slightly more basic (the pH value is higher) than most natural freshwater. Neutral water (such as distilled water) has a pH of 7, which is in the middle of being acidic and alkaline. Seawater happens to be slightly alkaline (basic), with a pH of about 8. Most natural water has a pH range of 6 to 8, although acid rain can have a pH as low as 4.

11.5   COMMON WATER MEASUREMENTS

Water and wastewater practitioners and regulators, such as waterworks operators and the U.S. Environmental Protection Agency (USEPA), along with their scientific counterparts at the U.S. Geological Survey (USGS), have been analyzing water for decades. Millions of measurements and analyses have been made. Some measurements are taken almost every time water is sampled and investigated, no matter where in the United States the water is being studied. Even these simple measurements can sometimes reveal something important about the water and the environment around it.

The USGS (2006) has noted that the results of a single measurement of the properties of water are actually less important than looking at how those properties vary over time. Suppose we take the pH of the river running through our town and find that it is 5.5. We might say, "Wow, the water is acidic!" But, a pH of 5.5 might be normal for that particular river. It is similar to how an adult's normal body temperature is about 97.5 degrees, but a youngster's normal temperature is *really* normal—right on the 98.6 mark. As with our temperatures, if the pH of a river begins to change, then we might suspect that something is going on somewhere that is affecting the water and possibly the water quality. For this reason, changes in water measurements are more important than the actual measured values.

Up to this point, we have addressed the important constituents and parameters of turbidity, dissolved oxygen, pH, and others, but there are others that we should address. In the following sections, the parameters of alkalinity, water temperature, specific conductance, and hardness are discussed.

11.5.1   ALKALINITY

Alkalinity is defined as the capacity of water to accept protons; it can also be defined as a measure of the ability of water to neutralize an acid. Bicarbonates, carbonates, and hydrogen ions cause alkalinity and create hydrogen compounds in a raw or treated water supply. Bicarbonates are the major components because of carbon dioxide action on basic materials of soil; borates, silicates, and phosphates may be minor components. The alkalinity of raw water may also contain salts formed from organic acids such as humic acids.

Alkalinity in water acts as a buffer that tends to stabilize and prevent fluctuations in pH. In fact, alkalinity is closely related to pH, but the two must not be confused. Total alkalinity is a measure of the amount of alkaline materials in the water. The alkaline materials act as buffers to changes in the pH. If the alkalinity is too low (below 80 ppm), the pH can fluctuate rapidly because of insufficient buffer. High alkalinity (above 200 ppm) results in the water being too buffered. Thus, having significant alkalinity in water is usually beneficial,

TABLE 11.4
Water Hardness

Classification	mg/L CaCo₃
Soft	0–75
Moderately hard	75–150
Hard	150–300
Very hard	Over 300

because it tends to prevent quick changes in pH that interfere with the effectiveness of common water treatment processes. Low alkalinity also contributes to the corrosive tendencies of water.

Note: When alkalinity is below 80 mg/L, it is considered to be low.

11.5.2 WATER TEMPERATURE

Water temperature is important not only to fishermen but also to industries and even fish and algae. A lot of water is used for cooling purposes in power plants that generate electricity. These plants need to cool the water to begin with and then generally release warmer water back to the environment. The temperature of the released water can affect downstream habitats. Temperature can also affect the ability of water to hold oxygen as well as the ability of organisms to resist certain pollutants.

11.5.3 SPECIFIC CONDUCTANCE

Specific conductance is a measure of the ability of water to conduct an electrical current. It is highly dependent on the amount of dissolved solids (such as salt) in the water. Pure water, such as distilled water, will have a very low specific conductance, and seawater will have a high specific conductance. Rainwater often dissolves airborne gasses and airborne dust while it is in the air and thus often has a higher specific conductance than distilled water. Specific conductance is an important water quality measurement because it gives a good idea of the amount of dissolved material in the water. When electrical wires are attached to a battery and light bulb and the wires are put into a beaker of distilled water, the light will not light. But, the bulb does light up when the beaker contains saline (saltwater). In saline water, the salt has dissolved and released free electrons, so the water will conduct an electric current.

11.5.4 HARDNESS

Hardness may be considered a physical or chemical characteristic or parameter of water. Hardness represents the total concentration of calcium and magnesium ions,

reported as calcium carbonate. Simply, the amount of dissolved calcium and magnesium in water determines its hardness. Hardness causes soaps and detergents to be less effective and contributes to scale formation in pipes and boilers. Hardness is not considered a health hazard; however, water that contains hardness must often be softened by lime precipitation or ion exchange. Hardwater can even shorten the life of fabrics and clothes. Low hardness contributes to the corrosive tendencies of water. Hardness and alkalinity often occur together, because some compounds can contribute both alkalinity and hardness ions. Hardness is generally classified as shown in Table 11.4.

11.5.5 ODOR CONTROL (WASTEWATER TREATMENT)

There is an old saying in wastewater treatment: "Odor is not a problem until the neighbors complain" (Spellman, 1997). Experience has shown that when treatment plant odor is apparent it is not long before the neighbors do complain. Thus, odor control is an important factor affecting the performance of any wastewater treatment plant, especially with regard to public relations.

According to Metcalf & Eddy (1991), in wastewater operations, "The principal sources of odors are from (1) septic wastewater containing hydrogen sulfide and odorous compounds, (2) industrial wastes discharged into the collection system, (3) screenings and unwanted grit, (4) septage handling facilities, (5) scum on primary settling tanks, (6) organically overloaded treatment processes, (7) [biosolids]-thickening tanks, (8) waste gas-burning operations where lower-than optimum temperatures are used, (9) [biosolids]-conditioning and dewatering faculties, (10) [biosolids] incineration, (11) digested [biosolids] in drying beds or [biosolids]-holding basins, and (12) [biosolids]-composting operations."

Odor control can be accomplished by chemical or physical means. Physical means include utilizing buffer zones between the process operation and the public, making operation changes, controlling discharges to collection systems, containments, dilution, fresh air, adsorption, activated carbon, and scrubbing towers, among other means. Odor control by chemical means involves scrubbing with various chemicals, chemical oxidation, and chemical precipitation methods. For *scrubbing with chemicals*, odorous gases are passed through specially designed scrubbing towers to remove odors. The commonly used chemical scrubbing solutions are chlorine and potassium permanganate. When hydrogen sulfide concentrations are high, sodium hydroxide is often used. In *chemical oxidation* applications, the oxidants chlorine, ozone, hydrogen peroxide, and potassium permanganate are used to oxidize the odor compounds. *Chemical precipitation* works to precipitate sulfides from odor compounds using iron and other metallic salts.

11.6 WATER TREATMENT CHEMICALS

To operate a water treatment process correctly and safely, water operators need to know the types of chemical used in the processes, the purpose of each, and the safety precautions required for the use of each. This section briefly discusses chemicals used in:

- Disinfection (also used in wastewater treatment)
- Coagulation
- Taste and odor removal
- Water softening
- Recarbonation
- Ion exchange softening
- Scale and corrosion control

11.6.1 DISINFECTION

In water practice, disinfection is often accomplished using chemicals. The purpose of disinfection is to selectively destroy disease-causing organisms. Chemicals commonly used in disinfection include chlorine and its compounds (most widely used), ozone, bromide, iodine, and hydrogen peroxide. Many factors must be considered when choosing the type of chemical to be used for disinfection, such as contact time, intensity and nature of the physical agent, temperature, and type and number of organisms.

11.6.2 COAGULATION

Chemical coagulation conditions water for further treatment by the removal of:

- Turbidity, color, and bacteria
- Iron and manganese
- Tastes, odors, and organic pollutants

In water treatment, normal sedimentation processes do not always settle out particles efficiently. This is especially the case when attempting to remove particles less than 50 μm in diameter.

In some instances, it is possible to agglomerate (to make or form into a rounded mass) particles into masses or groups. These rounded masses are of increased size and therefore increased settling velocities. For colloidal-sized particles, however, agglomeration is difficult because the colloidal particles are difficult to clarify without special treatment.

Chemical coagulation is usually accomplished by adding metallic salts such as aluminum sulfate (alum) or ferric chloride. Alum is the most commonly used coagulant in water treatment and is most effective between pH ranges of 5.0 and 7.5. Sometimes polymer is added to alum to help form small floc together for faster settling. Ferric chloride, effective down to a pH of 4.5, is sometimes used.

In addition to pH, a variety of other factors influence the chemical coagulation process, including:

- Temperature
- Influent quality
- Alkalinity
- Type and amount of coagulant used
- Type and length of flocculation
- Type and length of mixing

11.6.3 TASTE AND ODOR REMOVAL

Although odor can be a problem with wastewater treatment, the taste and odor parameters are primarily associated with potable water. Either organic or inorganic materials may produce tastes and odors in water. The perceptions of taste and odor are closely related and often confused by water practitioners as well as by consumers; thus, it is difficult to precisely measure either one. Experience has shown that a substance that produces an odor in water almost invariably imparts a perception of taste as well; however, taste is generally attributed to mineral substances present in the water, but most of these minerals do not cause odors.

Along with the impact that minerals can have on the taste of water, other substances or practices can also affect both water tastes and odors (e.g., metals, salts from the soil, constituents of wastewater, end products generated from biological reactions). When water has a distinct taste but no odor, the taste might be the result of inorganic substances. Anyone who has tasted alkaline water has experienced its biting bitterness. Salts not only give water a salty taste but also contribute to a bitter taste. In addition to natural sources, water can take on a distinctive color or taste, or both, from human contamination of the water. Organic materials can produce both taste and odor in water. Petroleum-based products are probably the prime contributors to both of these problems in water.

Biological degradation or decomposition of organics in surface waters also contributes to both taste and odor problems in water. Algae are another problem. Certain species of algae produce oily substances that may result in both altered taste and an odor. Synergy can also work to produce taste and odor problems in water. Mixing water and chlorine is one example.

With regard to chemically treating water for odor and taste problems, oxidants such as chlorine, chlorine dioxide, ozone, and potassium permanganate can be used. These chemicals are especially effective when water is associated with an earthy or musty odor caused by the nonvolatile metabolic products of actinomycetes and blue–green algae. Tastes and odors associated with dissolved gases and some volatile organic materials are normally removed by oxygen in aeration processes.

11.6.4 Water Softening

The reduction of hardness, or softening, is a process commonly practiced in water treatment. Chemical precipitation and ion exchange are the two softening processes most commonly used. Softening of hardwater is desired (for domestic users) to reduce the amount of soap used, increase the life of water heaters, and reduce encrustation of pipes (cementing together the individual filter media grains).

11.6.4.1 Chemical Precipitation

In chemical precipitation, it is necessary to adjust pH. To precipitate the two ions most commonly associated with hardness in water, calcium (Ca^{2+}) and magnesium (Mg^{2+}), the pH must be raised to about 9.4 for calcium and about 10.6 for magnesium. To raise the pH to the required levels, lime is added.

Chemical precipitation is accomplished by converting calcium hardness to calcium carbonate and magnesium hardness to magnesium hydroxide. This is normally accomplished by using the lime–soda ash or caustic soda processes. The lime–soda ash process reduces the total mineral content of the water, removes suspended solids, removes iron and manganese, and reduces color and bacterial numbers. The process, however, has a few disadvantages. McGhee (1991) pointed out, for example, that the process produces large quantities of sludge, requires careful operation, and, as stated earlier, if the pH is not properly adjusted, may create operational problems downstream of the process.

In the caustic soda process, the caustic soda reacts with the alkalinity to produce carbonate ions for reduction with calcium. The process works to precipitate calcium carbonate in a fluidized bed of sand grains, steel grit, marble chips, or some other similar dense material. As particles grow in size by deposition of $CaCO_3$, they migrate to the bottom of the fluidized bed from which they are removed. This process has the advantages of requiring short detention times (about 8 seconds) and producing no sludge.

11.6.4.2 Ion Exchange Softening

Hardness can be removed by ion exchange. In water softening, ion exchange replaces calcium and magnesium with a nonhardness cation, usually sodium. Calcium and magnesium in solution are removed by interchange with sodium within a solids interface (matrix) through which the flow is passed. Similar to the filter, the ion exchanger contains a bed of granular material, a flow distributor, and an effluent vessel that collects the product. The exchange media include greensand (a sand or sediment given a dark greenish color by grains of glauconite), aluminum silicates, synthetic siliceous gels, bentonite clay, sulfonated coal, and synthetic organic resins and are generally in particle form, usually ranging up to a diameter of 0.5 mm. Modern applications more often employ artificial organic resins. These clear, BB-sized resins are spherical and have the advantage of providing a greater number of exchange sites. Each of these resin spheres contains sodium ions, which are released into the water in exchange for calcium and magnesium. As long as exchange sites are available, the reaction is virtually instantaneous and complete.

When all of the exchange sites have been utilized, hardness begins to appear in the influent, known as *breakthrough*. Breakthrough requires the regeneration of the medium by bringing it into contact with a concentrated solution of sodium chloride.

Ion exchange used in water softening has both advantages and disadvantages. One of its major advantages is that it produces a softer water than does chemical precipitation. Additionally, ion exchange does not produce the large quantity of sludge encountered in the lime–soda process. One disadvantage is that, although it does not produce sludge, ion exchange does produce concentrated brine. Moreover, the water must be free of turbidity and particulate matter or the resin might function as a filter and become plugged.

11.6.5 Recarbonation

Recarbonation (stabilization) is the adjustment of the ionic condition of water so it will neither corrode pipes nor deposit calcium carbonate, which can produce an encrusting film. During or after the lime–soda ash softening process, this recarbonation is accomplished through the reintroduction of carbon dioxide into the water. Lime softening of hardwater supersaturates the water with calcium carbonate, and it may have a pH of greater than 10. Because of this, pressurized carbon dioxide is bubbled into the water to lower the pH and remove calcium carbonate. The high pH can also create a bitter taste in drinking water, but recarbonation removes this bitterness.

11.6.6 Scale and Corrosion Control

Controlling scale and corrosion is important in water systems. Carbonate and noncarbonate hardness constituents in water cause *scale*, a chalky-white deposit frequently found on teakettle bottoms. When controlled, this scale can be beneficial as it forms a protective coating inside tanks and pipelines. A problem arises when scale is not controlled. Excessive scaling reduces the capacity of pipelines and heat transfer efficiency in boilers. *Corrosion* is the oxidation of unprotected metal surfaces. A concern in water treatment is the corrosion of iron and its alloys (i.e., the formation of rust). Several factors contribute to the corrosion of iron and steel. Alkalinity, pH, DO, and carbon dioxide can all cause corrosion. Along with the corrosion potential of these chemicals, their corrosive tendencies are significantly increased when water temperature and flow are increased.

11.7 DRINKING WATER PARAMETERS: CHEMICAL

Water, in any of its forms, also ... [has] scant respect for the laws of chemistry.

Most materials act either as acids or bases, settling on either side of a natural reactive divided. Not water. It is one of the few substances that can behave both as an acid and as a base, so that under certain conditions it is capable of reacting chemically with itself. Or with anything else.

Molecules of water are off balance and hard to satisfy. They reach out to interfere with every other molecule they meet, pushing its atoms apart, surrounding them, and putting them into solution. Water is the ultimate solvent, wetting everything, setting other elements free from the rocks, making them available for life. Nothing is safe. There isn't a container strong enough to hold it.

Watson (1988)

Water chemical parameters are categorized into two basic groups: *inorganic* and *organic*. Both groups enter water from natural causes or pollution.

Note: The solvent capabilities of water are directly related to its chemical parameters.

In this section, we do not look at each organic or inorganic chemical individually; instead, we look at general chemical parameter categories such as dissolved oxygen organics (BOD and COD), dissolved oxygen (DO), synthetic organic compounds (SOCs), volatile organic compounds (VOCs), total dissolved solids (TDS), fluorides, metals, and nutrients—the major chemical parameters of concern.

11.7.1 ORGANICS

Natural organics contain carbon and consist of biodegradable organic matter such as wastes from biological material processing, human sewage, and animal feces. Microbes aerobically break down the complex organic molecules into simpler, more stable end products. Microbial degradation yields end products such as carbon dioxide, water, phosphate, and nitrate. Organic particles in water may harbor harmful bacteria and pathogens. Infection by microorganisms may occur if the water is used for primary contact or as a raw drinking water source. Treated drinking water will not present the same health risks. In a potable drinking water plant, all organics should be removed in the water before disinfection.

Organic chemicals also contain carbon; they are substances that come directly from, or are manufactured from, plant or animal matter. Plastics provide a good example of organic chemicals that are made from petroleum, which originally came from plant and animal matter. Some organic chemicals (such as those discussed above) released by decaying vegetation occur naturally and by themselves tend not to pose health problems when they get in our drinking water; however, more serious problems are caused by the more than 100,000 different manufactured or synthetic organic compounds in commercial use today. These include paints, herbicides, synthetic fertilizers, pesticides, fuels, plastics, dyes, preservatives, flavorings, and pharmaceuticals, to name a few.

Many organic materials are soluble in water and are toxic, and many of them are found in public water supplies. According to Tchobanoglous and Schroeder (1987), the presence of organic matter in water is troublesome. Organic matter causes: (1) color formation, (2) taste and odor problems, (3) oxygen depletion in streams, (4) interference with water treatment processes, and (5) the formation of halogenated compounds when chlorine is added to disinfect water.

Remember, organics in natural water systems may come from natural sources or may result from human activities. Generally, the principle source of organic matter in water is from natural sources including decaying leaves, weeds, and trees; the amount of these materials present in natural waters is usually low. *Anthropogenic* (manmade) sources of organic substances come from pesticides and other synthetic organic compounds.

Again, many organic compounds are soluble in water, and surface waters are more prone to contamination by natural organic compounds than are groundwaters. In water, dissolved organics are usually divided into two categories: *biodegradable* and *nonbiodegradable*. Material that is biodegradable (capable of breaking down) consists of organics that can be used for food (nutrients) by naturally occurring microorganisms within a reasonable length of time. Alcohols, acids, starches, fats, proteins, esters, and aldehydes are the main constituents of biodegradable materials. They may result from domestic or industrial wastewater discharges, or they may be end products of the initial microbial decomposition of plant or animal tissue. Biodegradable organics in surface waters cause problems mainly associated with the effects that result from the action of microorganisms. As the microbes metabolize organic material, they consume oxygen.

When this process occurs in water, the oxygen consumed is dissolved oxygen (DO). If the oxygen is not continually replaced in the water by artificial means, the DO level will decrease as the organics are decomposed by the microbes. This need for oxygen is the *biochemical oxygen demand* (BOD), which is the amount of dissolved oxygen demanded by bacteria to break down the organic materials during the stabilization action of the decomposable organic matter under aerobic conditions over a 5-day incubation period at 20°C (68°F). This bioassay test measures the oxygen consumed by living organisms using the organic matter contained in the sample and dissolved oxygen in the liquid. The organics are broken down into simpler compounds, and the microbes use the energy released for growth and reproduction. A BOD test is not required for monitoring drinking water.

Note: The more organic material in the water, the higher the BOD exerted by the microbes will be. Some biodegradable organics can cause color, taste, and odor problems.

Nonbiodegradable organics are resistant to biological degradation. The constituents of woody plants are a good example. These constituents, including tannin and lignic acids, phenols, and cellulose, are found in natural water systems and are considered *refractory* (resistant to biodegradation). Some polysaccharides with exceptionally strong bonds and benzene (associated, for example, with the refining of petroleum) with its ringed structure are essentially nonbiodegradable.

Certain nonbiodegradable chemicals can react with oxygen dissolved in water. The *chemical oxygen demand* (COD) is a more complete and accurate measurement of the total depletion of dissolved oxygen in water. *Standard Methods* (Greenberg et al., 1999) defines COD as a test that provides a measure of the oxygen equivalent of that portion of the organic matter in a sample that is susceptible to oxidation by a strong chemical oxidant. The procedure is detailed in *Standard Methods*.

Note: COD is not normally used for monitoring water supplies but is often used for evaluating contaminated raw water.

11.7.2 SYNTHETIC ORGANIC COMPOUNDS

Synthetic organic compounds (SOCs) are manmade, and because they do not occur naturally in the environment they are often toxic to humans. More than 50,000 SOCs are in commercial production, including common pesticides, carbon tetrachloride, chloride, dioxin, xylene, phenols, aldicarb, and thousands of other synthetic chemicals. Unfortunately, even though they are so prevalent, few data have been collected on these toxic substances. Determining definitively just how dangerous many of the SOCs are is rather difficult.

11.7.3 VOLATILE ORGANIC COMPOUNDS

Volatile organic compounds (VOCs) are a particularly dangerous type of organic chemical. VOCs are absorbed through the skin during contact with water—as in the shower or bath. Hot-water exposure allows these chemicals to evaporate rapidly, and they are harmful if inhaled. VOCs can be in any tap water, regardless of where in the country one lives and the water supply source.

11.7.4 TOTAL DISSOLVED SOLIDS

Earlier we pointed out that solids in water occur either in solution or in suspension and are distinguished by passing the water sample through a glass-fiber filter. By definition, the *suspended solids* are retained on top of the filter, and the *dissolved solids* pass through the filter with the water. When the filtered portion of the water sample is placed in a small dish and then evaporated, the solids in the water remain as residue in the evaporating dish. This material represents the *total dissolved solids* (TDS). Dissolved solids may be organic or inorganic. Water may come into contact with these substances within the soil, on surfaces, and in the atmosphere. The organic dissolved constituents of water can be the decay products of vegetation, organic chemicals, or organic gases. Removing these dissolved minerals, gases, and organic constituents is desirable, because they may cause physiological effects and produce aesthetically displeasing color, taste, and odors.

Note: In water distribution systems, a high TDS means high conductivity with consequent higher ionization during corrosion control efforts; however, high TDS also suggests a greater likelihood of a protective coating, a positive factor in corrosion control.

11.7.5 FLUORIDES

According to Phyllis J. Mullenix, Ph.D., water fluoridation is not the safe public health measure we have been led to believe. Concerns about uncontrolled dosage, accumulation in the body over time, and effects beyond those on the teeth (e.g., brain as well as bones) have not been resolved for fluoride. The health of citizens requires that all the facts be considered, not just those that are politically expedient (Mullenix, 1997).

Most medical authorities would take issue with Mullenix's view on the efficacy of fluoride in reducing tooth decay. Most authorities seem to hold that a moderate amount of fluoride ions (F^-) in drinking water contributes to good dental health. Fluoride is seldom found in appreciable quantities of surface waters and appears in groundwater in only a few geographical regions, although it is sometimes found in a few types of igneous or sedimentary rocks. Fluoride is toxic to humans in large quantities (the key words are "large quantities" or, in Mullenix's view, "uncontrolled dosages") and is also toxic to some animals.

Fluoride used in small concentrations (about 1.0 mg/L in drinking water) can be beneficial. Experience has shown that drinking water containing a proper amount of fluoride can reduce tooth decay by 65% in children between the ages of 12 and 15. When the concentration of fluorides in untreated natural water supplies is excessive, however, either alternative water supplies must be used or treatment to reduce the fluoride concentration must be applied, because excessive amounts of fluoride can cause mottled or discolored teeth, a condition known as *dental fluorosis*.

11.7.6 HEAVY METALS

Heavy metals are elements with atomic weights between 63.5 and 200.5, and a specific gravity greater than 4.0. Living organisms require trace amounts of some heavy metals, including cobalt, copper, iron, manganese, molybdenum, vanadium, strontium, and zinc. Excessive levels of essential metals, however, can be detrimental to the organism. Nonessential heavy metals of particular concern to surface water systems are cadmium, chromium, mercury, lead, arsenic, and antimony.

Heavy metals in water are classified as either *nontoxic* or *toxic*. Only those metals that are harmful in relatively small amounts are labeled toxic; other metals fall into the nontoxic group. In natural waters (other than in groundwaters), sources of metals include dissolution from natural deposits and discharges of domestic, agricultural, or industrial wastes.

All heavy metals exist in surface waters in colloidal, particulate, and dissolved phases, although dissolved concentrations are generally low. The colloidal and particulate metal may be found in: (1) hydroxides, oxides, silicates, or sulfides, or (2) adsorbed to clay, silica, or organic matter. The soluble forms are generally ions or unionized organometallic chelates or complexes. The solubility of trace metals in surface waters is predominately controlled by water pH, the type and concentration of liquids on which the metal could adsorb, the oxidation state of the mineral components, and the redox environment of the system. The behavior of metals in natural waters is a function of the substrate sediment composition, the suspended sediment composition, and the water chemistry. Sediment composed of fine sand and silt will generally have higher levels of adsorbed metal than will quartz, feldspar, and detrital carbonate-rich sediment.

The water chemistry of the system controls the rate of adsorption and desorption of metals to and from sediment. Adsorption removes the metal from the water column and stores the metal in the substrate. Desorption returns the metal to the water column, where recirculation and bioassimilation may take place. Metals may be desorbed from the sediment if the water experiences increases in salinity, decreases in redox potential, or decreases in pH.

Although heavy metals such as iron (Fe) and manganese (Mn) do not cause health problems, they do impart a noticeable bitter taste to drinking water, even at very low concentrations. These metals usually occur in groundwater in solution, and these and others may cause brown or black stains on laundry and on plumbing fixtures.

11.7.7 NUTRIENTS

Elements in water (such as carbon, nitrogen, phosphorus, sulfur, calcium, iron, potassium, manganese, cobalt, and boron, all essential to the growth and reproduction of plants and animals) are called *nutrients* (or biostimulants). The two nutrients that concern us in this text are nitrogen and phosphorus. Nitrogen (N_2), an extremely stable gas, is the primary component of the Earth's atmosphere (78%). The nitrogen cycle is composed of four processes. Three of the processes—fixation, ammonification, and nitrification—convert gaseous nitrogen into usable chemical forms. Denitrification, the fourth process, converts fixed nitrogen back to the unusable gaseous nitrogen state.

Nitrogen occurs in many forms in the environment and takes part in many biochemical reactions. Major sources of nitrogen include runoff from animal feedlots, fertilizer runoff from agricultural fields, municipal wastewater discharges, and certain bacteria and blue–green algae that obtain nitrogen directly from the atmosphere. Certain forms of acid rain can also contribute nitrogen to surface waters.

Nitrogen in water is commonly found in the form of *nitrate* (NO_3), which indicates that the water may be contaminated with sewage. Nitrates can also enter the groundwater from chemical fertilizers used in agricultural areas. Excessive nitrate concentrations in drinking water pose an immediate health threat to infants, both human and animal, and can cause death. The bacteria commonly found in the intestinal tract of infants can convert nitrate to highly toxic nitrites (NO_2). Nitrite can replace oxygen in the bloodstream, resulting in oxygen starvation, which causes a bluish discoloration of the infant ("blue baby" syndrome).

Note: Lakes and reservoirs usually have less than 2 mg/L of nitrate measured as nitrogen. Higher nitrate levels are found in groundwater ranging up to 20 mg/L, but much higher values are detected in shallow aquifers polluted by sewage or excessive use of fertilizers.

Phosphorus (P) is an essential nutrient that contributes to the growth of algae and the eutrophication of lakes, although its presence in drinking water has little effect on health. In aquatic environments, phosphorus is found in the form of phosphate and is a limiting nutrient. If all phosphorus is used, plant growth ceases, no matter the amount of nitrogen available. Many bodies of freshwater currently experience influxes of nitrogen and phosphorus from outside sources. The increasing concentration of available phosphorus allows plants to assimilate more nitrogen before the phosphorus is depleted. If sufficient phosphorus is available, high concentrations of nitrates will lead to phytoplankton (algae) and macrophyte (aquatic plant) production.

Major sources of phosphorus include phosphates in detergents, fertilizer and feedlot runoff, and municipal wastewater discharges. The 1976 USEPA *Quality Criteria for Water* recommended a phosphorus criterion of 0.10 µg/L (elemental) phosphorus for marine and estuarine waters but offered no freshwater criterion.

The biological, physical, and chemical condition of our water is of enormous concern to us all, because we must live in such intimate contact with water. When these parameters shift and change, the changes affect us, often in ways science cannot yet define for us. Water pollution is an external element that can and does significantly affect our water. But, what exactly is water pollution? We quickly learn that water pollution does not always occur directly from the source to the water. Controlling what goes into our water is difficult, because the hydrologic cycle carries water (and whatever it picks up along the way) through all of our environment's media, affecting the biological, physical, and chemical condition of the water we must drink to live.

CHAPTER REVIEW QUESTIONS

11.1 The chemical symbol for sodium is _____.
11.2 The chemical symbol for sulfuric acid is _____.
11.3 Neutrality on the pH scale is _____.
11.4 Is NaOH a salt or a base?
11.5 Chemistry is the study of substances and the _____ they undergo.
11.6 The three stages of matter are _____, _____, and _____.
11.7 A basic substance that cannot be broken down any further without changing the nature of the substance is _____.
11.8 A combination of two or more elements is a _____.
11.9 A table of the basic elements is called the _____ table.
11.10 When a substance is mixed into water to form a solution, the water is called the _____, and the substance is called the _____.
11.11 Define ion.
11.12 A solid that is less than 1 µm in size is called a _____.
11.13 What is the property of water that causes light to be scattered and absorbed?
11.14 What is true color?
11.15 What is the main problem with metals found in water?
11.16 Compounds derived form material that once was alive are called _____ chemicals.
11.17 The pH range is from _____ to _____.
11.18 What is alkalinity?
11.19 The two ions that cause hardness are _____ and _____.
11.20 What type of substance produces hydroxide ions (OH⁻) in water?

REFERENCES AND SUGGESTED READING

Greenberg, A.E. et al., Eds. 1999. *Standard Methods for Examination of Water and Wastewater*, 20th ed. Washington, D.C.: American Public Health Association.

Jost, N.J. 1992. Surface and ground water pollution control technology. In *Fundamentals of Environmental Science and Technology*, Knowles, P.-C., Ed. Rockville, MD: Government Institutes, Inc.

McGhee, T.J. 1991. *Water Supply and Sewerage*, 6th ed. New York: McGraw-Hill.

Metcalf & Eddy. 1991. *Wastewater Engineering: Treatment, Disposal, Reuse*, 3rd ed. New York: McGraw-Hill.

Mullenix, P. J. 1997. Letter to the Operations and Environmental Committee, City of Calgary, Canada.

Smith, R.K. 1993. *Water and Wastewater Laboratory Techniques*. Alexandria, VA: Water Environment Federation.

Spellman, F.R. 1997. *Wastewater Biosolids to Compost*. Boca Raton, FL: CRC Press.

Tchobanoglous, G. and Schroeder E.D. 1987. *Water Quality*. Reading, MA: Addison-Wesley.

USEPA. 1976. *Quality Criteria for Water*. Washington, D.C.: U.S. Environmental Protection Agency.

USGS. 2006. *Water Science for Schools: Water Measurements*. Washington, D.C.: U.S. Geological Survey.

Watson, L. 1988. *The Water Planet: A Celebration of the Wonder of Water*. New York: Crown Publishers.

12 Water Microbiology

Scientists picture the primordial Earth as a planet washed by a hot sea and bathed in an atmosphere containing water vapor, ammonia, methane, and hydrogen. Testing this theory, Stanley Miller at the University of Chicago duplicated these conditions in the laboratory. He distilled seawater in a special apparatus and passed the vapor with ammonia, methane, and hydrogen through an electrical discharge at frequent intervals; the condensed "rain" was returned to the boiling seawater. Within a week, the seawater had turned red. Analysis showed that it contained amino acids, which are the building blocks of protein substances. Whether this is what really happened early in the Earth's history is not important; the experiment demonstrated that the basic ingredients of life could have been made in some such fashion, setting the stage for life to come into existence in the sea. The saline fluids in most living things may be an inheritance from such early beginnings.

Kemmer (1979)

12.1 INTRODUCTION

Microorganisms are significant in water and wastewater because of their roles in disease transmission, and they are the primary agents of biological treatment. Thus, water and wastewater practitioners must have considerable knowledge of the microbiological characteristics of water and wastewater. Simply put, waterworks operators cannot fully comprehend the principles of effective water treatment without knowing the fundamental factors concerning microorganisms and their relationships to one another, their effect on the treatment process, and their impact on consumers, animals, and the environment.

Water/wastewater operators must know the principal groups of microorganisms found in water supplies (surface and groundwater) and wastewater, as well as those that must be treated (pathogenic organisms) and removed or controlled for biological treatment processes; they must be able to identify the organisms used as indicators of pollution and know their significance; and they must know the methods used to enumerate the indicator organisms. This chapter provides microbiology fundamentals specifically targeting the needs of water and wastewater specialists.

Note: To have microbiological activity, the body of water or wastewater must possess the appropriate environmental conditions. The majority of waste-

water treatment processes, for example, are designed to operate using an aerobic process. The conditions required for aerobic operation include: (1) sufficient free, elemental oxygen; (2) sufficient organic matter (food); (3) sufficient water; (4) enough nitrogen and phosphorus (nutrients) to permit oxidation of the available carbon materials; (5) proper pH (6.5 to 9.0); and (6) lack of toxic materials.

12.2 MICROBIOLOGY: WHAT IS IT?

Biology is generally defined as the study of living organisms (i.e., the study of life). *Microbiology* is a branch of biology that deals with the study of microorganisms so small in size that they must be studied under a microscope. Microorganisms of interest to the water and wastewater operator include bacteria, protozoa, viruses, algae, and others.

Note: The science and study of bacteria are known as *bacteriology* (discussed later).

As mentioned, the primary concern of waterworks operators is how to control microorganisms that cause waterborne diseases (waterborne pathogens) to protect the consumer, both human and animal. Wastewater operators have the same microbiological concerns as water operators, but instead of directly purifying water for consumer consumption their focus is on removing harmful pathogens from the wastestream before outfalling it to the environment. In summary (Spellman, 1996):

- *Water operators* are concerned with water supply and water purification through a treatment process. When treating water, the primary concern is producing potable water that is safe to drink (free of pathogens) with no accompanying offensive characteristics such as foul taste and odor. The treatment operator must possess a wide range of knowledge to be able to correctly examine water for pathogenic microorganisms and to determine the type of treatment necessary to ensure that the water quality of the end product, potable water, meets regulatory requirements.

- *Wastewater operators* are also concerned with water quality; however, they are not as concerned as water specialists with total removal or reduction of most microorganisms. The wastewater treatment process actually benefits from the presence of microorganisms that act to degrade organic compounds and thus stabilize organic matter in the wastestream. Wastewater operators must be trained to operate the treatment process in a manner that controls the growth of microorganisms and puts them to work. Moreover, to fully understand wastewater treatment, it is necessary to determine which microorganisms are present and how they function to break down components in the wastewater stream. The operator must then ensure that, before outfalling or dumping treated effluent into the receiving body, the microorganisms that worked so hard to degrade organic waste products, especially pathogenic microorganisms, are not sent from the plant with effluent as viable organisms.

12.3 WATER/WASTEWATER MICROORGANISMS

As mentioned, microorganisms of interest to water and wastewater operators include bacteria, protozoa, rotifers, viruses, algae, fungi, and nematodes. These organisms are the most diverse group of living organisms on Earth and occupy important niches in the ecosystem. Their simplicity and minimal survival requirements allow them to exist in diverse situations. Because microorganisms are a major health issue, water treatment specialists are concerned about controlling the *waterborne pathogens* (e.g., bacteria, virus, protozoa) that cause *waterborne diseases*. The focus of wastewater operators, on the other hand, is on the millions of organisms that arrive at the plant with the influent. The majority of these organisms are nonpathogenic and beneficial to plant operations. From a microbiological standpoint, the mix of microorganism species depends on the characteristics of the influent, environmental conditions, process design, and mode of plant operation. This mix may also include pathogenic organisms, including those responsible for diseases such as typhoid, tetanus, hepatitis, dysentery, and gastroenteritis, among others.

To understand how to minimize or maximize the growth of microorganisms and control pathogens, one must study the structure and characteristics of the microorganisms. In the sections that follow, we will look at each of the major groups of microorganisms (those important to water/wastewater operators) with regard to their size, shape, types, nutritional needs, and control.

Note: Koren (1991) pointed out that, in a water environment, water is not the medium for growth of micro-

organisms but is instead a means of transmission of the pathogen (that is, it serves as a conduit, thus the name *waterborne*). Individuals drink the water carrying the pathogens and thus begins an outbreak of disease. When the topic of waterborne disease is brought up, many might mistakenly assume that waterborne diseases are at home in water. Nothing could be further from the truth. A water-filled environment is not one in which pathogenic organisms would choose to live—that is, if it had such a choice. The point is that microorganisms do not normally grow, reproduce, languish, and thrive in watery surroundings. Pathogenic microorganisms temporarily residing in water are simply biding their time, going with the flow, waiting for their opportunity to meet up with their unsuspecting hosts. To some degree, when the pathogenic microorganism finds a host, it is finally home and may have found its final resting place (Spellman, 1997).

12.4 KEY TERMS

Algae, simple—Plants, many microscopic, containing chlorophyll. Freshwater algae are diverse in shape, color, size, and habitat. They are the basic link in the conversion of inorganic constituents in water into organic constituents.

Algal bloom—Sudden spurts of algal growth which can affect water quality adversely and indicate potentially hazardous changes in local water chemistry.

Anaerobic—Able to live and grow in the absence of free oxygen.

Autotrophic organisms—Produces food from inorganic substances.

Bacteria—Single-celled microorganisms that have rigid cell walls. They may be aerobic, anaerobic, or facultative. They can cause disease, but some are important in pollution control.

Biogeochemical cycle—The chemical interactions between the atmosphere, hydrosphere, and biosphere.

Coliform organism—Microorganisms found in the intestinal tract of humans and animals. Their presence in water indicates fecal pollution and potentially adverse contamination by pathogens.

Denitrification—The anaerobic biological reduction of nitrate to nitrogen gas.

Fungi—Simple plants lacking in ability to produce energy through photosynthesis.

Heterotrophic organism—Organisms that are dependent on organic matter for foods.

Prokaryotic cell—The simple cell type, characterized by a lack of nuclear membrane and the absence of mitochondria.

Virus—The smallest form of microorganisms capable of causing disease.

12.5 MICROORGANISM CLASSIFICATION AND DIFFERENTIATION

The microorganisms we are concerned with are tiny organisms made up a large and diverse group of free-living forms; they exist as single cells, cell bunches, or clusters. Found in abundance almost anywhere on Earth, the vast majority of microorganisms are not harmful. Many microorganisms, or microbes, occur as single cells (unicellular), others are multicellular, and still others—viruses—do not have a true cellular appearance. A single microbial cell, for the most part, exhibits the characteristic features common to other biological systems, such as metabolism, reproduction, and growth.

12.5.1 CLASSIFICATION

For centuries, scientists classified the forms of life visible to the naked eye as either animal or plant. The Swedish naturalist Carolus Linnaeus organized much of the current knowledge about living things in 1735. The importance of organizing or classifying organisms cannot be overstated, for without a classification scheme it would be difficult to establish criteria for identifying organisms and to arrange similar organisms into groups. Probably the most important reason for classifying organisms is to make things less confusing (Wistriech and Lechtman, 1980). Linnaeus was quite innovative in his classification of organisms. One of his innovations still with us today is the *binomial system of nomenclature*. Under the binomial system, all organisms are generally described by a two-word scientific name: *genus* and *species*. Genus and species are groups that are part of a hierarchy of groups of increasing size, based on their taxonomy:

Kingdom
  Phylum
    Class
      Order
        Family
          Genus
            Species

Using this system, a fruit fly might be classified as:

Animalia
  Arthropoda
    Insecta
      Diptera
        Drosophilidae
          Drosophila
            melanogaster

This means that this organism is the species *melanogaster* in the genus *Drosophila* in the family Drosophilidae in the order Diptera in the class Insecta in the phylum Arthropoda in the kingdom Animalia.

To further illustrate how the hierarchical system is exemplified by the classification system, the standard classification of the mayfly is provided below:

Kingdom—Animalia
  Phylum—Arthropoda
    Class—Insecta
      Order—Ephemeroptera
        Family—Ephemeridae
          Genus—*Hexagenia*
            Species—*limbata*

Utilizing this hierarchy and Linnaeus' binomial system of nomenclature, the scientific name of any organism includes both the generic and specific names. To uniquely name a species, it is necessary to supply both the genus and the species; for our examples, those would be *Drosophila melanogaster* for the fruit fly and *Hexagenia limbota* for the mayfly. The first letter of the generic name is usually capitalized; hence, for example, *E. coli* indicates that *coli* is the species and *Escherichia* (abbreviated to *E.*) is the genus. The largest, most inclusive category is the kingdom. The genus and species names are always in Latin, so they are usually printed in italics. Some organisms also have English common names. Microbe names of particular interest in water/wastewater treatment include:

- *Escherichia coli* (a coliform bacterium)
- *Salmonella typhi* (the typhoid bacillus)
- *Giardia lamblia* (a protozoan)
- *Shigella* spp.
- *Vibrio cholerae*
- *Campylobacter*
- *Leptospira* spp.
- *Entamoeba histolytica*
- *Cryptosporidium*

Note: *Escherichia coli* is commonly known as simply *E. coli*, and *Giardia lamblia* is usually referred to by only its genus name, *Giardia*.

Generally, we use a simplified system of microorganism classification in water science by breaking down the classification into the kingdoms of Animal, Plant, and Protista. As a general rule, the Animal and Plant kingdoms contain all of the multicell organisms, and the Protista kingdom includes all single-cell organisms. Along with a microorganism classification based on the Animal, Plant, and Protista kingdoms, microorganisms can be further classified as being *eucaryotic* or *prokaryotic* (see Table 12.1).

Note: A eucaryotic organism is characterized by a cellular organization that includes a well-defined nuclear membrane. The prokaryotes have a structural

TABLE 12.1
Simplified Classification of Microorganisms

Kingdom	Members	Cell Classification
Animal	Rotifers	
	Crustaceans	
	Worms and larvae	
Plant	Ferns	Eucaryotic
	Mosses	
Protista	Protozoa	
	Algae	
	Fungi	
	Bacteria	Prokaryotic
	Lower algae forms	

organization that sets them off from all other organisms. They are simple cells characterized by a nucleus *lacking* a limiting membrane, an endoplasmic reticulum, chloroplasts, and mitochondria.

· They are remarkably adaptable and exist abundantly in the soil, sea, and freshwater.

12.5.2 DIFFERENTIATION

Differentiation among the higher forms of life is based almost entirely on morphological (form or structure) differences; however, differentiation (even among the higher forms) is not as easily accomplished as we might expect, because normal variations among individuals of the same species occur frequently. Because of this variation, even within a species, securing accurate classifications when dealing with single-celled microscopic forms that present virtually no visible structural differences becomes extremely difficult. Under these circumstances, it is necessary to consider physiological, cultural, and chemical differences, as well as structure and form. Differentiation among the smaller groups of bacteria is based almost entirely on chemical differences.

12.6 THE CELL

The structural and fundamental unit of both plants and animals, no matter how complex, is the cell. Since the 19th century, scientists have known that all living things, whether animal or plant, are made up of cells. A typical cell is an entity, isolated from other cells by a membrane or cell wall. The cell membrane contains protoplasm and the nucleus (see Figure 12.1). The protoplasm within the cell is a living mass of viscous, transparent material. Within the protoplasm is a dense spherical mass called the *nucleus* or *nuclear material*. In a typical mature plant cell, the cell wall is rigid and is composed of nonliving material, while in the typical animal cell the wall is an elastic, living membrane. Cells exist in a very great variety of sizes and shapes, and their functions also vary widely.

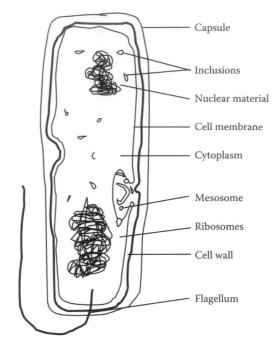

FIGURE 12.1 Bacterial cell.

The size of cells ranges from bacteria too small to be seen with the light microscope to the largest known single cell, the ostrich egg. Microbial cells also have an extensive size range, some being larger than human cells (Kordon, 1992).

Note: The nucleus cannot always be observed in bacteria.

12.6.1 STRUCTURE OF THE BACTERIAL CELL

The structural form and various components of the bacterial cell are probably best understood by referring to the simplified diagram of the rod-form bacterium shown in Figure 12.1. When studying Figure 12.1, keep in mind that cells of different species may differ greatly, both in structure and chemical composition; for this reason, no typical bacterium exists. Figure 12.1 shows a generalized bacterium and should be referred to for the discussion that follows; however, not all bacteria have all of the features shown in the figure, and some bacteria have structures not shown in the figure.

12.6.1.1 Capsules

Bacterial *capsules* are organized accumulations of gelatinous materials on cell walls, in contrast to *slime layers* (a water secretion that adheres loosely to the cell wall and commonly diffuses into the cell), which are unorganized accumulations of similar material. The capsule is usually thick enough to be seen under the ordinary light microscope (macrocapsule), whereas thinner capsules (microcapsules) can be detected only by electron microscopy (Singleton and Sainsbury, 1994). The production of capsules is determined largely by genetics as well as environmental conditions and depends on the presence or absence of capsule-degrading enzymes and other growth factors. Varying in composition, capsules are mainly composed of water; the organic contents are made up of complex polysaccharides, nitrogen-containing substances, and polypeptides. Capsules confer several advantages when bacteria grow in their normal habitat; for example, they help to: (1) prevent desiccation, (2) resist phagocytosis by host phagocytic cells, (3) prevent infection by bacteriophages, and (4) aid bacterial attachment to tissue surfaces in plant and animal hosts or to surfaces of solids objects in aquatic environments. Capsule formation often correlates with pathogenicity.

12.6.1.2 Flagella

Many bacteria are motile, and this ability to move independently is usually attributed to a special structure, the *flagella* (singular: flagellum). Depending on species, a cell may have a single flagellum (*monotrichous* bacteria, *trichous* meaning "hair"), one flagellum at each end (*amphitrichous* bacteria, *amphi* meaning "on both sides"), a tuft of flagella at one or both ends (*lophotrichous* bacteria, *lopho* meaning "tuft"), or flagella that arise all over the cell surface (*peritrichous* bacteria, *peri* meaning "around").

A flagellum is a threadlike appendage extending outward from the plasma membrane and cell wall. Flagella are slender, rigid, locomotor structures, about 20 nm across and up to 15 to 20 μm long. Flagellation patterns are very useful in identifying bacteria and can be seen by light microscopy, but only after being stained with special techniques designed to increase their thickness. The detailed structure of flagella can be seen only in the electron microscope.

Bacterial cells benefit from flagella in several ways. They can increase the concentration of nutrients or decrease the concentration of toxic materials near the bacterial surfaces by causing a change in the flow rate of fluids. They can also disperse flagellated organisms to areas where colony formation can take place. The main benefit of flagella to organisms is their increased ability to flee from areas that might be harmful.

12.6.1.3 Cell Wall

The main structural component of most prokaryotes is the rigid *cell wall*. Functions of the cell wall include: (1) providing protection for the delicate protoplast from osmotic lysis (bursting); (2) determining the shape of a cell; (3) acting as a permeability layer that excludes large molecules and various antibiotics and plays an active role in regulating the intake of ions by the cell; and (4) providing a solid support for flagella. Cell walls of different species may differ greatly in structure, thickness, and composition. The cell wall accounts for about 20 to 40% of the dry weight of a bacterium.

12.6.1.4 Plasma Membrane (Cytoplasmic Membrane)

Surrounded externally by the cell wall and composed of a lipoprotein complex, the *plasma membrane* or cell membrane is the critical barrier separating the inside from outside the cell. About 7 to 8 nm thick and comprising 10 to 20% of the dry weight of a bacterium, the plasma membrane controls the passage of all material into and out of the cell. The inner and outer faces of the plasma membrane are embedded with water-loving (hydrophilic) lipids, whereas the interior is hydrophobic. Control of material into the cell is accomplished by screening, as well as by electric charge. The plasma membrane is the site of the surface charge of the bacteria.

In addition to serving as an osmotic barrier that passively regulates the passage of material into and out of the cell, the plasma membrane participates in the entire active transport of various substances into the bacterial cell. Inside the membrane, many highly reactive chemical groups guide the incoming material to the proper points for further reaction. This active transport system provides bacteria with certain advantages, including the ability to maintain a fairly constant intercellular ionic state in the presence of varying external ionic concentrations. In addition to participating in the uptake of nutrients, the cell membrane transport system participates in waste excretion and protein secretions.

12.6.1.5 Cytoplasm

Within a cell and bounded by the cell membrane is a complicated mixture of substances and structures called the *cytoplasm*. The cytoplasm is a water-based fluid containing ribosomes, ions, enzymes, nutrients, storage granules (under certain circumstances), waste products, and various molecules involved in synthesis, energy metabolism, and cell maintenance.

12.6.1.6 Mesosome

A common intracellular structure found in bacterial cytoplasm is the *mesosome*. Mesosomes are invaginations of the plasma membrane in the shape of tubules, vesicles, or lamellae. Their exact function is unknown. Currently, many bacteriologists believe that mesosomes are artifacts generated during the fixation of bacteria for electron microscopy.

12.6.1.7 Nucleoid (Nuclear Body or Region)

The *nuclear region* of the prokaryotic cell is primitive and a striking contrast to that of the eucaryotic cell. Prokaryotic cells lack a distinct nucleus, the function of the nucleus being carried out by a single, long, double strand of DNA that is efficiently packaged to fit within the nucleoid. The nucleoid is attached to the plasma membrane. A cell can have more than one nucleoid when cell division occurs after the genetic material has been duplicated.

12.6.1.8 Ribosomes

The bacterial cytoplasm is often packed with ribosomes. *Ribosomes* are minute, rounded bodies made of RNA and are loosely attached to the plasma membrane. Ribosomes are estimated to account for about 40% of the dry weight of a bacterium; a single cell may have as many as 10,000 ribosomes. Ribosomes serve as sites for protein synthesis and are part of the translation process.

12.6.1.9 Inclusions

Inclusions (or storage granules) are often seen within bacterial cells. Some inclusion bodies are not bound by a membrane and lie free in the cytoplasm. A single-layered membrane about 2 to 4 nm thick encloses other inclusion bodies. Many bacteria produce polymers that are stored as granules in the cytoplasm.

12.7 BACTERIA

The simplest wholly contained life systems are *bacteria* or *prokaryotes*, which are the most diverse group of microorganisms. They are among the most common microorganisms in water. They are primitive, unicellular (single celled) organisms possessing no well-defined nucleus and presenting a variety of shapes and nutritional needs. Bacteria contain about 85% water and 15% ash or mineral matter. The ash is largely composed of sulfur, potassium, sodium, calcium, and chlorides, with small amounts of iron, silicon, and magnesium. Bacteria reproduce by binary fission.

Note: *Binary fission* occurs when one organism splits or divides into two or more new organisms.

Bacteria, once considered the smallest living organism (although now it is known that smaller forms of matter exhibit many of the characteristics of life), range in size from 0.5 to 2 μm in diameter and about 1 to 10 μm long.

Note: A *micron* is a metric unit of measurement equal to 1/1000 of a millimeter. To visualize the size of bacteria, consider that about 1000 average bacteria lying side by side would reach across the head of a straight pin.

Bacteria are categorized into three general groups based on their physical form or shape (though almost every variation has been found; see Table 12.2). The simplest form is the sphere. Spherical-shaped bacteria are called *cocci* (meaning "berries"). They are not necessarily perfectly round but may be somewhat elongated, flattened on one side, or oval. Rod-shaped bacteria are called *bacilli*. Spiral-shaped bacteria, called *spirilla*, have one or

TABLE 12.2
Forms of Bacteria

Form	Technical Name		Example
	Singular	Plural	
Sphere	Coccus	Cocci	*Streptococcus*
Rod	Bacillus	Bacilli	*Bacillus typhosis*
Curved or spiral	Spirillum	Spirilla	*Spirillum cholera*

more twists and are never straight (see Figure 12.2). Such formations are usually characteristic of a particular genus or species. Within these three groups are many different arrangements. Some exist as single cells; others as pairs, as packets of four or eight, as chains, or as clumps.

Most bacteria require organic food to survive and multiply. Plant and animal material that gets into the water provides the food source for bacteria. Bacteria convert the food to energy and use the energy to make new cells. Some bacteria can use inorganics (e.g., minerals such as iron) as an energy source and can multiply even when organics (pollution) are not available.

12.7.1 Bacterial Growth Factors

Several factors affect the rate at which bacteria grow, including temperature, pH, and oxygen levels. The warmer the environment, the faster the rate of growth. Generally, for each increase of 10°C, the growth rate doubles. Heat can also be used to kill bacteria. Most bacteria grow best at neutral pH. Extreme acidic or basic conditions generally inhibit growth, although some bacteria may require acidic conditions and some alkaline conditions for growth.

Bacteria are aerobic, anaerobic, or facultative. If *aerobic*, they require free oxygen in the aquatic environment. *Anaerobic* bacteria exist and multiply in environments that lack dissolved oxygen. *Facultative* bacteria (e.g., iron bacteria) can switch from aerobic to anaerobic growth or grow in an anaerobic or aerobic environment.

Under optimum conditions, bacteria grow and reproduce very rapidly. As noted previously, bacteria reproduce by *binary fission*. An important point to consider with regard to bacterial reproduction is the rate at which the process can take place. The total time required for an organism to reproduce and the offspring to reach maturity is the *generation time*. Bacteria growing under optimal conditions can double their number about every 20 to 30 minutes. Obviously, this generation time is very short compared with that of higher plants and animals. Bacteria continue to grow at this rapid rate as long as nutrients hold out—even the smallest contamination can result in a sizable growth in a very short time.

Note: Even though wastewater can contain bacteria counts in the millions per milliliter, in wastewater treatment under controlled conditions bacteria can

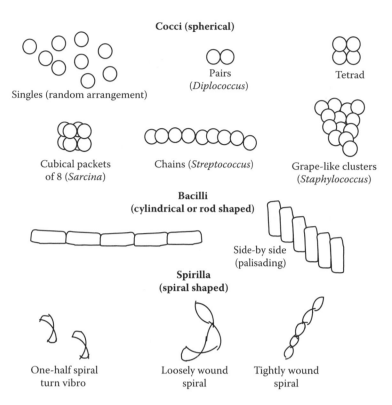

FIGURE 12.2 Bacterial shapes and arrangements.

help to destroy and identify pollutants. In such a process, bacteria stabilize organic matter (e.g., activated sludge processes) and thereby assist the treatment process in producing effluent that does not impose an excessive oxygen demand on the receiving body. Coliform bacteria can be used as an indicator of pollution by human or animal wastes.

12.7.2 DESTRUCTION OF BACTERIA

In water and wastewater treatment, the destruction of bacteria is usually called *disinfection*. Disinfection does not mean that all microbial forms are killed. That would be *sterilization*. Instead, disinfection reduces the number of disease-causing organisms to an acceptable number. Growing bacteria are generally easy to control by disinfection; however, some bacteria form survival structures known as *spores*, which are much more difficult to destroy.

Note: Inhibiting the growth of microorganisms is termed *antisepsis*, whereas destroying them is called *disinfection*.

12.7.3 WATERBORNE BACTERIA

All surface waters contain bacteria. Waterborne bacteria, as we have said, are responsible for infectious epidemic diseases. Bacterial numbers increase significantly during storm events when streams are high. Heavy rainstorms increase stream contamination by washing material from the ground surface into the stream. After the initial washing occurs, few impurities are left to be washed into the stream, which may then carry relatively "clean" water. A river of good quality shows its highest bacterial numbers during rainy periods; however, a much-polluted stream may show the highest numbers during low flows because of the constant influx of pollutants. Water and wastewater operators are primarily concerned with bacterial pathogens responsible for disease. These pathogens enter potential drinking water supplies through fecal contamination and are ingested by humans if the water is not properly treated and disinfected.

Note: Regulations require that owners of all public water supplies collect water samples and deliver them to a certified laboratory for bacteriological examination at least monthly. The number of samples required is usually in accordance with federal standards, which generally require that one sample per month be collected for each 1000 persons served by the waterworks.

12.8 PROTOZOA

Protozoa (or "first animals") are a large group of eucaryotic organisms of more than 50,000 known species belonging to the kingdom Protista. They have adapted a form of

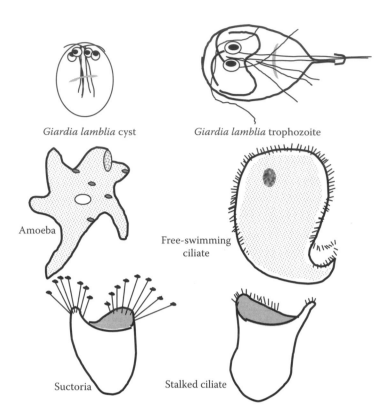

Giardia lamblia cyst   *Giardia lamblia* trophozoite

Amoeba

Free-swimming
ciliate

Suctoria   Stalked ciliate

FIGURE 12.3 Protozoa.

cell to serve as the entire body; in fact, protozoa are one-celled, animal-like organisms with complex cellular structures. In the microbial world, protozoa are giants, many times larger than bacteria. They range in size from 4 to 500 μm. The largest ones can almost be seen by the naked eye. They can exist as solitary or independent organisms, such as the stalked ciliates (e.g., *Vorticella* sp.) (Figure 12.3), or they can colonize (e.g., the sedentary *Carchesium* sp.). Protozoa get their name because they employ the same type of feeding strategy as animals; that is, they are *heterotrophic*, meaning that they obtain cellular energy from organic substances such as proteins. Most are harmless, but some are parasitic. Some forms have two life stages: *active trophozoites* (capable of feeding) and *dormant cysts*.

The major groups of protozoa are based on their method of locomotion (motility). The Mastigophora are motile by means of one or more *flagella*, the whip-like projection that propels the free-swimming organisms (*Giardia lamblia* is a flagellated protozoan). The Ciliophora move by means of shortened modified flagella called *cilia*, which are short hair-like structures that beat rapidly and propel them through the water. The Sarcodina rely on *amoeboid movement*, which is a streaming or gliding action; the shape of the amoebae changes as they stretch and then contract to move from place to place. The Sporozoa, in contrast, are nonmotile; they are simply swept along, riding the current of the water.

Protozoa consume organics to survive; their favorite food is bacteria. Protozoa are mostly aerobic or facultative with regard to their oxygen requirements. Toxic materials, pH, and temperature affect protozoan rates of growth in the same way as they affect bacteria.

Most protozoan life cycles alternate between an active growth phase (*trophozoites*) and a resting stage (*cysts*). Cysts are extremely resistant structures that protect the organism from destruction when it encounters harsh environmental conditions—including chlorination.

Note: Those protozoa not completely resistant to chlorination require higher disinfectant concentrations and longer contact time for disinfection than normally used in water treatment.

The protozoa and associated waterborne diseases of most concern to waterworks operators include:

- *Entamoeba histolytica*—amoebic dysentery
- *Giardia lamblia*—giardiasis
- *Cryptosporidium*—cryptosporidiosis

In wastewater treatment, protozoa are a critical part of the purification process and can be used to indicate the condition of treatment processes. Protozoa normally associated with wastewater include amoebae, flagellates, free-swimming ciliates, and stalked ciliates.

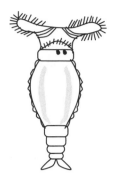

FIGURE 12.4 *Philodina*, a common rotifer.

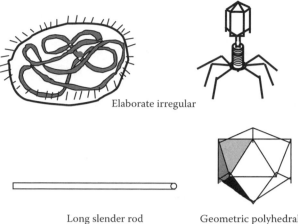

Elaborate irregular

Long slender rod   Geometric polyhedral

FIGURE 12.5 Virus shapes.

Amoebae are associated with poor wastewater treatment of a young biosolids mass (see Figure 12.3). They move through wastewater by a streaming or gliding motion, accomplished by moving liquids stored within the cell wall. They are normally associated with an effluent high in biochemical oxygen demand (BOD) and suspended solids.

Flagellates (flagellated protozoa) have a single, long, hair-like or whip-like projection (flagellum) that is used to propel the free-swimming organisms through wastewater and to attract food (see Figure 12.3). Flagellated protozoa are normally associated with poor treatment and a young biosolids. When the predominate organism is the flagellated protozoa, the plant effluent will contain large amounts of BODs and suspended solids.

The *free-swimming ciliated protozoan* uses its tiny, hair-like projections (cilia) to move itself through the wastewater and to attract food (see Figure 12.3). This type of protozoan is normally associated with a moderate biosolids age and effluent quality. When the free-swimming ciliated protozoan is the predominate organisms, the plant effluent will normally be turbid and contain a high amount of suspended solids.

The *stalked ciliated protozoan* attaches itself to the wastewater solids and uses its cilia to attract food (see Figure 12.3). The stalked ciliated protozoan is normally associated with a plant effluent that is very clear and contains low amounts of both BODs and suspended solids.

Rotifers make up a well-defined group of the smallest, simplest multicellular microorganisms and are found in nearly all aquatic habitats (see Figure 12.4). Rotifers are a higher life form associated with cleaner waters. Normally found in well-operated wastewater treatment plants, they can be used to indicate the performance of certain types of treatment processes.

12.9 MICROSCOPIC CRUSTACEANS

Because they are important members of freshwater zooplankton, microscope *crustaceans* are of interest to water and wastewater operators. These microscopic organisms are characterized by a rigid shell structure. They are multicellular animals that are strict aerobes, and as primary producers they feed on bacteria and algae. They are important as a source of food for fish. Additionally, microscopic crustaceans have been used to clarify algae-laden effluents from oxidations ponds. *Cyclops* and *Daphnia* are two microscopic crustaceans of interest to water and wastewater operators.

12.10 VIRUSES

Viruses are very different from the other microorganisms. Consider their size relationship, for example. Relative to size, if protozoa are the Goliaths of microorganisms, then viruses are the Davids. Stated more specifically and accurately, viruses are intercellular parasitic particles that are the smallest living infectious materials known—the midgets of the microbial world. Viruses are very simple life forms consisting of a central molecule of genetic material surrounded by a protein shell called a *capsid* and sometimes by a second layer called an *envelope*. Viruses occur in many shapes, including long slender rods, elaborate irregular shapes, and geometric polyhedrals (see Figure 12.5).

Viruses contain no mechanisms by which to obtain energy or reproduce on their own, thus viruses must have a host to survive. After they invade the cells of their specific host (animal, plant, insect, fish, or even bacteria), they take over the cellular machinery of the host and force it to make more viruses. In the process, the host cell is destroyed and hundreds of new viruses are released into the environment. The viruses of most concern to the waterworks operator are the pathogens that cause hepatitis, viral gastroenteritis, and poliomyelitis.

Smaller and different from bacteria, viruses are prevalent in water contaminated with sewage. Detecting viruses in water supplies is a major problem because of the complexity of the procedures involved, although

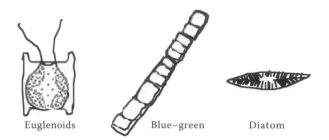

Euglenoids   Blue–green   Diatom

FIGURE 12.6 Algae.

experience has shown that the normal coliform index can be used as a rough guide for viruses as well as for bacteria. More attention must be paid to viruses, however, when surface water supplies have been used for sewage disposal. Viruses are difficult to destroy by normal disinfection practices, as they require increased disinfectant concentration and contact time for effective destruction.

Note: Viruses that infect bacterial cells cannot infect and replicate within cells of other organisms. It is possible to utilize this specificity to identify bacteria, a procedure called *phage typing*.

12.11 ALGAE

You do not have to be a water/wastewater operator to understand that algae can be a nuisance. Many ponds and lakes in the United States are currently undergoing *eutrophication*, which is enrichment of an environment with inorganic substances (e.g., phosphorus and nitrogen), causing excessive algae growth and premature aging of the water body. When eutrophication occurs, especially when filamentous algae such as *Caldophora* breaks loose in a pond or lake and washes ashore, algae makes its stinking, noxious presence known.

Algae are a form of aquatic plants and are classified by color (e.g., green algae, blue–green algae, golden–brown algae). Algae come in many shapes and sizes (see Figure 12.6). Although they are not pathogenic, algae do cause problems with water/wastewater treatment plant operations. They grow easily on the walls of troughs and basins, and heavy growth can plug intakes and screens. Additionally, some algae release chemicals that give off undesirable tastes and odors. Although algae are usually classified by their color, they are also commonly classified based on their cellular properties or characteristics. Several characteristics are used to classify algae, including: (1) cellular organization and cell wall structure; (2) the nature of the chlorophylls; (3) the type of motility, if any; (4) the carbon polymers that are produced and stored; and (5) the reproductive structures and methods.

Many algae (in mass) are easily seen by the naked eye, but others are microscopic. They occur in fresh and polluted water, as well as in saltwater. Because they are

plants, they are capable of using energy from the sun in photosynthesis. They usually grow near the surface of the water because light cannot penetrate very far through the water. Algae are controlled in raw waters with chlorine and potassium permanganate. Algal blooms in raw water reservoirs are often controlled with copper sulfate.

Note: By producing oxygen, which is utilized by other organisms, including animals, algae play an important role in the balance of nature.

12.12 FUNGI

Fungi are of relatively minor importance in water/wastewater operations (except for biosolids composting, where they are critical). Fungi, like bacteria, are extremely diverse. They are multicellular, autotrophic, photosynthetic protists. They grow as filamentous, mold-like forms or as yeast-like (single-celled) organisms. They feed on organic material.

Note: Aquatic fungi grow as *parasites* on living plants or animals and as *saprophytes* on those that are dead.

12.13 NEMATODES AND FLATWORMS

Along with inhabiting organic mud, worms also inhabit biological slimes; they have been found in activated sludge and in trickling filter slimes (wastewater treatment processes). Microscopic in size, they range in length from 0.5 to 3 mm and in diameter from 0.01 to 0.05 mm. Most species have a similar appearance. They have a body that is covered by cuticle and is cylindrical, nonsegmented, and tapered at both ends.

These organisms continuously enter wastewater treatment systems, primarily through attachment to soils that reach the plant through inflow and infiltration (I&I). They are present in large, often highly variable numbers, but as strict aerobes they are found only in aerobic treatment processes where they metabolize solid organic matter.

When nematodes are firmly established in the treatment process, they can promote microfloral activity and decomposition. They feed on bacteria in both the activated sludge and trickling filter systems. Their activities in these systems enhance oxygen penetration due to their tunneling through floc particles and biofilm. In activated sludge processes, they are present in relatively small numbers because the liquefied environment is not a suitable habitat for crawling, which they prefer over the free-swimming mode. In trickling filters where the fine stationary substratum is suitable to permit crawling and mating, nematodes are quite abundant.

Along with preferring the trickling filter habitat, nematodes play a beneficial role in this habitat; for example, they break loose portions of the biological slime coating the filter bed. This action prevents excessive slime growth and filter clogging. They also aid in keeping slime porous and accessible to oxygen by tunneling through the slime. In the activated sludge process, nematodes play important roles as agents of improved oxygen diffusion. They accomplish this by tunneling through floc particles.

Nematodes also act as indicators of operational conditions in the process, such as low dissolved oxygen levels (anoxic conditions) and the presence of toxic wastes. Environmental conditions have an impact on the growth of nematodes; for example, in anoxic conditions their swimming and growth are impaired. These changes in activity are effective indicators of changes in wastewater strength and composition. Temperature fluctuations also directly affect their growth and survival, as their population decreases when temperatures increase.

Aquatic flatworms (improperly named because they are not all flat) feed primarily on algae. Because of their aversion to light, they are found in the lower depths of pools. Two varieties of flatworms are seen in wastewater treatment processes. *Microtubellarians* are more round than flat and average about 0.5 to 5 mm in size, whereas *macrotubellarians* (planarians) are more flat than round and average about 5 to 20 mm in body size. Flatworms are very hardy and can survive in wide variations in humidity and temperature. As inhabitants of sewage sludge, they play an important role in sludge stabilization and serve as bioindicators or parameters of process problems; for example, their inactivity or sluggishness might indicate a low dissolved oxygen level or the presence of toxic wastes.

Surface waters grossly polluted with organic matter (especially domestic sewage) have a fauna capable of thriving in very low concentrations of oxygen. A few species of tubificid worms dominate this environment. Pennak (1989) reported that the bottoms of severely polluted streams can be literally covered with a "writhing" mass of these tubificids.

The *tubifex* (commonly known as *sludge worms*) are small, slender, reddish worms that normally range in length from 25 to about 50 mm. They are burrowers; their posterior end protrudes to obtain nutrients (see Figure 12.7). When found in streams, tubifex are indicators of pollution.

12.14 WATER TREATMENT AND MICROBIOLOGICAL PROCESSES

The primary goal in water treatment is to protect the consumer of potable drinking water from disease. Drinking water safety is a worldwide concern. Drinking nontreated or improperly treated water is a major cause of illness in developing countries. Water contains several biological (as well as chemical) contaminants that must

FIGURE 12.7 Tubificid worms.

be removed efficiently to produce safe drinking water that is also aesthetically pleasing to the consumer. The finished water must be free of microbial pathogens and parasites, turbidity, color, and odor. To achieve this goal, raw surface water or groundwater is subjected to a series of treatment processes that will be described in detail later. Disinfection alone is sufficient if the raw water originates from a protected source. More commonly, several processes are used to treat water; for example, disinfection may be combined with coagulation, flocculation, and filtration.

As mentioned, several unit processes are used in the water treatment process to produce microbiologically (and chemically) safe drinking water. The extent of treatment depends on the source of raw water; surface waters generally require more treatment than groundwaters. With the exception of disinfection, the other unit processes in the treatment train do not specifically address the destruction or removal of pathogens.

Water treatment unit processes include: (1) storage of raw water, (2) prechlorination, (3) coagulation–flocculation, (4) water softening, (5) filtration, and (6) disinfection. Filtration and disinfection are the primary means of removing contaminants and pathogens from drinking water supplies. In each of these unit processes, the reduction or destruction of pathogens is variable and influenced by a number of factors such as sunlight, sedimentation, and temperature.

These water treatment unit processes are important and are described in detail later in this text. For the moment, because of relatively recent events involving pathogenic protozoa causing adverse reactions, including death, among consumers in various locations throughout the United States (and elsewhere), it is important to turn our attention to the pathogenic protozoa. One thing is certain— these pathogenic protozoa have the full attention of water treatment operators everywhere.

12.14.1 PATHOGENIC PROTOZOA

As mentioned, certain types of protozoa can cause disease. Of particular interest to the drinking water practitioner are *Entamoeba histolytica* (amoebic dysentery and amoebic hepatitis), *Giardia lamblia* (giardiasis), *Cryptosporidium* (cryptosporidiosis), and the emerging *Cyclospora* (cyclosporasis). Sewage contamination transports eggs, cysts, and oocysts of parasitic protozoa and helminths (tapeworms, hookworms, etc.) into raw water supplies, leaving water treatment (in particular, filtration) and disinfection as the means by which to diminish the danger of contaminated water for the consumer.

To prevent the occurrence of *Giardia* and *Cryptosporidium* in surface water supplies and to address increasing problems with waterborne diseases, the U.S. Environmental Protection Agency (USEPA) implemented its Surface Water Treatment Rule (SWTR) in 1989. The rule requires both filtration and disinfection of all surface water supplies as the primary means of controlling *Giardia* and enteric viruses. Since implementation of SWTR, USEPA has also recognized that *Cryptosporidium* species are agents of waterborne disease. In its 1996 series of surface water regulations, the USEPA included *Cryptosporidium*.

To test the need for and effectiveness of the Surface Water Treatment Rule, LeChevallier et al. (1991) conducted a study on the occurrence and distribution of *Giardia* and *Cryptosporidium* organisms in raw water supplies to 66 surface water filter plants. These plants were located in 14 states and a Canadian province. A combined immunofluorescence test indicated that cysts and oocysts were widely dispersed in the aquatic environment. *Giardia* was detected in more than 80% of the samples. *Cryptosporidium* was found in 85% of the sample locations. Taking into account several variables, *Giardia* or *Cryptosporidium* were detected in 97% of the raw water samples. After evaluating their data, the researchers concluded that the Surface Water Treatment Rule might have to be upgraded (subsequently, it has been) to require additional treatment.

12.14.2 GIARDIA

Giardia (gee-ar-dee-ah) *lamblia* (also known as the hiker's or traveler's scourge or disease) is a microscopic parasite that can infect warm-blooded animals and humans. Although *Giardia* was discovered in the 19th century, not until 1981 did the World Health Organization (WHO) classify *Giardia* as a pathogen. An outer shell called a *cyst* allows *Giardia* to survive outside the body for long periods of time. If viable cysts are ingested, *Giardia* can cause the illness known as *giardiasis*, an intestinal illness that can cause nausea, anorexia, fever, and severe diarrhea. The symptoms last only for several days, and the body can naturally rid itself of the parasite in 1 to 2 months; however, for individuals with weakened immune systems, the body often cannot rid itself of the parasite without medical treatment.

In the United States, *Giardia* is the most commonly identified pathogen in waterborne disease outbreaks. Contamination of a water supply by *Giardia* can occur in two ways: (1) by the activity of animals in the watershed area of the water supply, or (2) by the introduction of sewage into the water supply. Wild and domestic animals are major contributors to the contamination of water supplies. Studies have also shown that, unlike many other pathogens, *Giardia* is not host specific. In short, *Giardia* cysts excreted by animals can infect and cause illness in humans. Additionally, in several major outbreaks of waterborne diseases, the *Giardia* cyst source was sewage-contaminated water supplies.

Treating the water supply, however, can effectively control waterborne *Giardia*. Chlorine and ozone are examples of two disinfectants known to effectively kill *Giardia* cysts. Filtration of the water can also effectively trap and remove the parasite from the water supply. The combination of disinfection and filtration is the most effective water treatment process available today for prevention of *Giardia* contamination.

In drinking water, *Giardia* is regulated under the Surface Water Treatment Rule. Although the SWTR does not establish a maximum contaminant level (MCL) for *Giardia*, it does specify treatment requirements to achieve at least 99.9% (3 log) removal or inactivation of *Giardia*. This regulation requires that all drinking water systems using surface water or groundwater under the influence of surface water must disinfect and filter the water. The Enhanced Surface Water Treatment Rule (ESWTR), which includes *Cryptosporidium* and further regulates *Giardia*, was established in December 1996.

12.14.2.1 Giardiasis

Giardiasis is recognized as one of the most frequently occurring waterborne diseases in the United States. *Giardia lamblia* cysts have been discovered in places as widely scattered as Estes Park, Colorado (near the Continental Divide); Missoula, Montana; Wilkes-Barre, Scranton, and Hazleton, Pennsylvania; and Pittsfield and Lawrence, Massachusetts, just to name a few (CDC, 1995).

Giardiasis is characterized by intestinal symptoms that usually last a week or more and may be accompanied by one or more of the following: diarrhea, abdominal cramps, bloating, flatulence, fatigue, and weight loss. Although vomiting and fever are commonly listed as relatively frequent symptoms, people involved in waterborne outbreaks in the United States have not commonly reported them. Although most *Giardia* infections persist only for 1 or 2 months, some people experience a more chronic phase that can follow the acute phase or it may become manifest without an antecedent acute illness. Loose stools and increased abdominal gassiness with cramping, flatulence, and burping characterize the chronic phase. Fever is not common, but malaise, fatigue, and depression may ensue; for a small number of people, the persistence of infection is associated with the development of marked malabsorption and weight loss (Weller, 1985). Similarly, lactose (milk) intolerance can become a problem for some people. This can develop coincidentally with the infection or be aggravated by it, causing an increase in intestinal symptoms after ingestion of milk products.

Some people may have several of these symptoms without evidence of diarrhea or have only sporadic episodes of diarrhea every three or four days. Still others may have no symptoms at all. The problem, then, may not be one of determining whether or not someone is infected with the parasite but how harmoniously the host and the parasite can live together. When such harmony does not exist or is lost, it then becomes a problem of how to get rid of the parasite, either spontaneously or by treatment.

Note: Three prescription drugs are available in the United States to treat giardiasis: quinacrine, metronidazole, and furazolidone. In a recent review of drug trials in which the efficacies of these drugs were compared, quinacrine produced a cure in 93% of patients, metronidazole cured 92%, and furazolidone cured about 84% of patients (Davidson, 1984).

Giardiasis occurs worldwide. In the United States, *Giardia* is the parasite most commonly identified in stool specimens submitted to state laboratories for parasitologic examination. During a 3-year period, approximately 4% of 1 million stool specimens submitted to state laboratories tested positive for *Giardia* (CDC, 1979). Other surveys have demonstrated *Giardia* prevalence rates ranging from 1 to 20%, depending on the location and ages of persons studied. Giardiasis ranks among the top 20 infectious diseases causing the greatest morbidity in Africa, Asia, and Latin America; it has been estimated that about 2 million infections occur per year in these regions (Walsh, 1981). People who are at highest risk for acquiring *Giardia* infection in the United States may be placed into five major categories:

1. People in cities whose drinking water originates from streams or rivers and whose water treatment process does not include filtration, or where filtration is ineffective because of malfunctioning equipment
2. Hikers, campers, and those who enjoy the outdoors
3. International travelers

4. Children who attend daycare centers, daycare center staff, and parents and siblings of children infected in daycare centers.
5. Homosexual men

People in categories 1, 2, and 3 have in common the same general source of infection; that is, they acquire *Giardia* from fecally contaminated drinking water. The city resident usually becomes infected because the municipal water treatment process does not include the filter necessary to physically remove the parasite from the water. The number of people in the United States at risk (i.e., the number who receive municipal drinking water from unfiltered surface water) is estimated to be 20 million. International travelers may also acquire the parasite from improperly treated municipal waters in cities or villages in other parts of the world, particularly in developing countries. In Eurasia, only travelers to Leningrad appear to be at increased risk. In prospective studies, 88% of U.S. and 35% of Finnish travelers to Leningrad who had negative stool tests for *Giardia* on departure to the Soviet Union developed symptoms of giardiasis and had positive tests for *Giardia* after they returned home (Brodsky et al., 1974). With the exception of visitors to Leningrad, however, *Giardia* has not been implicated as a major cause of traveler's diarrhea, as it has been detected in fewer than 2% of travelers developing diarrhea. Hikers and campers, however, risk infection every time they drink untreated raw water from a stream or river. Persons in categories 4 and 5 become exposed through more direct contact with feces or an infected person by exposure to the soiled diapers of an infected child in cases associated with daycare centers or through direct or indirect anal–oral sexual practices in the case of homosexual men.

Although community waterborne outbreaks of giardiasis have received the greatest publicity in the United States, about half of the *Giardia* cases discussed with the staff of the Centers for Disease Control over a 3-year period had a daycare exposure as the most likely source of infection. Numerous outbreaks of *Giardia* in daycare centers have been reported. Infection rates for children in daycare center outbreaks range from 21 to 44% in the United States and from 8 to 27% in Canada (Black et al., 1981). The highest infection rates are usually observed in children who wear diapers (1 to 3 years of age). In a study of 18 randomly selected daycare centers in Atlanta, 10% of diapered children were found to be infected. Transmission from this age group to older children, daycare staff, and household contacts is also common. About 20% of parents caring for an infected child become infected.

Local health officials and managers or water utility companies need to realize that sources of *Giardia* infection other than municipal drinking water exist. Armed with this knowledge, they are less likely to make a quick (and sometimes wrong) assumption that a cluster of recently

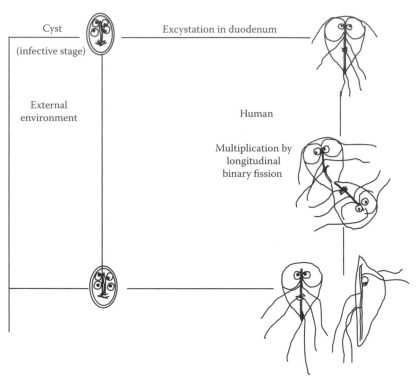

FIGURE 12.8 Life cycle of *Giardia lamblia*.

diagnosed cases in a city is related to municipal drinking water. Of course, drinking water must not be ruled out as a source of infection when a larger than expected number of cases is recognized in a community, but the possibility that the cases are associated with a daycare center outbreak, drinking untreated stream water, or international travel should also be entertained.

To understand the finer aspects of *Giardia* transmission and strategies for control, the drinking water practitioner must become familiar with several aspects of the biology of the parasite. Two forms of the parasite exist: a *trophozoite* and a *cyst*, both of which are much larger than bacteria (see Figure 12.8). Trophozoites live in the upper small intestine, where they attach to the intestinal wall by means of a disc-shaped suction pad on their ventral surface. Trophozoites actively feed and reproduce at this location. At some time during the trophozoite's life, it releases its hold on the bowel wall and floats in the fecal stream through the intestine. As it makes this journey, it undergoes a morphologic transformation into the egg-like cyst. The cyst, about 6 to 9 nm in diameter and 8 to 12 μm in length, has a thick exterior wall that protects the parasite against the harsh elements that it will encounter outside the body. This cyst form of parasite is infectious to other people or animals. Most people become infected either directly (by hand-to-mouth transfer of cysts from the feces of an infected individual) or indirectly (by drinking feces-contaminated water). Less common modes of transmission included ingestion of fecally contaminated food and hand-

to-mouth transfer of cysts after touching a fecally contaminated surface. After the cyst is swallowed, the trophozoite is liberated through the action of stomach acid and digestive enzymes and becomes established in the small intestine.

Although infection after ingestion of only one *Giardia* cyst is theoretically possible, the minimum number of cysts shown to infect a human under experimental conditions is ten (Rendtorff, 1954). Trophozoites divide by binary fission about every 12 hours. What this means in practical terms is that, if a person swallowed only a single cyst, reproduction at this rate would result in more than 1 million parasites 10 days later—1 billion parasites by day 15.

The exact mechanism by which *Giardia* causes illness is not yet well understood, but it apparently is not necessarily related to the number of organisms present. Nearly all of the symptoms, however, are related to dysfunction of the gastrointestinal tract. The parasite rarely invades other parts of the body, such as the gall bladder or pancreatic ducts. Intestinal infection does not result in permanent damage.

Note: *Giardia* has an incubation period of 1 to 8 weeks.

Data reported by the CDC indicate that *Giardia* is the most frequently identified cause of diarrheal outbreaks associated with drinking water in the United States. The remainder of this section is devoted specifically to waterborne transmissions of *Giardia*. *Giardia* cysts have been detected in 16% of potable water supplies (lakes, reservoirs, rivers, springs, groundwater) in the United States at an average concentration of 3 cysts per 100 L (Rose et al., 1983). Waterborne epidemics of giardiasis are a relatively frequent occurrence. In 1983, for example, *Giardia* was identified as the cause of diarrhea in 68% of waterborne outbreaks in which the causal agent was identified. From 1965 to 1982, more than 50 waterborne outbreaks were reported (CDC, 1984). In 1984, about 250,000 people in Pennsylvania were advised to boil drinking water for 6 months because of *Giardia*-contaminated water.

Many of the municipal waterborne outbreaks of *Giardia* have been subjected to intense study to determine their cause. Several general conclusions can be made from data obtained in those studies. Waterborne transmission of *Giardia* in the United States usually occurs in mountainous regions where community drinking water obtained from clear running streams is chlorinated but not filtered before distribution. Although mountain streams appear to be clean, fecal contamination upstream by human residents or visitors, as well as by *Giardia*-infected animals such as beavers, has been well documented. Water obtained from deep wells is an unlikely source of *Giardia* because of the natural filtration of water as it percolates through the soil to reach underground cisterns. Wells that pose the greatest risk of fecal contamination are poorly constructed or improperly located ones. A few outbreaks have occurred in towns that included filtration in the water treatment process but the filtration was not effective in removing *Giardia* cysts because of defects in filter construction, poor maintenance of the filter media, or inadequate pretreatment of the water before filtration. Occasional outbreaks have also occurred because of accidental cross-connections between water and sewage systems.

Important Point: From these data, we can conclude that two major ingredients are necessary for waterborne outbreak: *Giardia* cysts must be present in untreated source water, and the water purification process must fail to either kill or remove *Giardia* cysts from the water.

Although beavers are often blamed for contaminating water with *Giardia* cysts, the fact that they are responsible for introducing the parasite into new areas seem unlikely. Far more likely is that they are also victims: *Giardia* cysts may be carried in untreated human sewage discharged into the water by small-town sewage disposal plants or originate from cabin toilets that drain directly into streams and rivers. Backpackers, campers, and sports enthusiasts may also deposit *Giardia*-contaminated feces in the environment which are subsequently washed into streams by rain. In support of this concept is a growing amount of data indicating a higher *Giardia* infection rate in beavers living downstream from U.S. national forest campgrounds when compared with beavers living in more remote areas that have a near-zero rate of infection.

Although beavers may be unwitting victims of the *Giardia* story, they still play an important part in the contamination scheme because they can (and probably do) serve as amplifying hosts. An *amplifying host* is one that is easy to infect, serves as a good habitat for reproduction of the parasite, and, in the case of *Giardia*, returns millions of cysts to the water for every one ingested. Beavers are especially important in this regard, because they tend to defecate in or very near the water, which ensures that most of the *Giardia* cysts excreted are returned to the water.

The microbial quality of water resources and the management of the microbially laden wastes generated by the burgeoning animal agriculture industry are critical local, regional, and national problems. Animal wastes from cattle, hogs, sheep, horses, poultry, and other livestock and commercial animals can contain high concentrations of microorganism, such as *Giardia,* that are pathogenic to humans.

The contribution of other animals to waterborne outbreaks of *Giardia* is less clear. Muskrats (another semiaquatic animal) have been found in several parts of the United States to have high infection rates (30 to 40%) (Frost et al., 1984). Recent studies have shown that muskrats can be infected with *Giardia* cysts from humans and beavers. Occasional *Giardia* infections have been reported in coyotes, deer, elk, cattle, dogs, and cats (but not in horses and sheep) encountered in mountainous regions of the United States. Naturally occurring *Giardia* infections have not been found in most other wild animals (bear, nutria, rabbit, squirrel, badger, marmot, skunk, ferret, porcupine, mink, raccoon, river otter, bobcat, lynx, moose, and bighorn sheep) (Frost et al., 1984).

Scientific knowledge about what is required to kill or remove *Giardia* cysts from a contaminated water supply has increased considerably. We know, for example, that cysts can survive in cold water (4°C) for at least 2 months and that they are killed instantaneously by boiling water (100°C) (Frost et al., 1984). We do not know how long the cysts will remain viable at other water temperatures (e.g., 0°C or in a canteen at 15 to 20°C), nor do we know how long the parasite will survive on various environment surfaces, such as under a pine tree, in the sun, on a diaper-changing table, or in carpets in a daycare center.

The effect of chemical disinfection (chlorination, for example) on the viability of *Giardia* cysts is an even more complex issue. The number of waterborne outbreaks of *Giardia* that have occurred in communities where chlorination was employed as a disinfectant process demonstrates that the amount of chlorine used routinely for municipal water treatment is not effective against *Giardia* cysts. These observations have been confirmed in the laboratory under experimental conditions (Jarroll et al., 1979). This does not mean that chlorine does not work at all. It does work under certain favorable conditions. Without getting too technical, gaining some appreciation of the problem can be achieved by understanding a few of the variables that influence the efficacy of chlorine as a disinfectant:

- *Water pH*—At pH values above 7.5, the disinfectant capability of chlorine is greatly reduced.
- *Water temperature*—The warmer the water, the higher the efficacy. Chlorine does not work in ice-cold water from mountain streams.
- *Organic content of the water*—Mud, decayed vegetation, or other suspended organic debris in water chemically combines with chlorine, making it unavailable as a disinfectant.
- *Chlorine contact time*—The longer that *Giardia* cysts are exposed to chlorine, the more likely it is that the chemical will kill them.
- *Chlorine concentration*—The higher the chlorine concentration, the more likely it is that chlorine will kill *Giardia* cysts. Most water treatment facilities try to add enough chlorine to give a free (unbound) chlorine residual at the customer tap of 0.5 mg per liter of water.

These five variables are so closely interrelated that improving one can often compensate for another; for example, if chlorine efficacy is expected to be low because water is obtained from an icy stream, the chlorine contact time or chlorine concentration, or both, could be increased. In the case of *Giardia*-contaminated water, producing safe drinking water with a chlorine concentration of 1 mg per liter and contact time as short as 10 minutes might be possible if all the other variables are optimal—a pH of 7.0, water temperature of 25°C, and total organic content of the water close to zero. On the other hand, if all of these variables are unfavorable—pH of 7.9, water temperature of 5°C, and high organic content—chlorine concentrations in excess of 8 mg per liter within several hours of contact time may not be consistently effective. Because water conditions and water treatment plant operations (especially those related to water retention time and, therefore, to chlorine contact time) vary considerably in different parts of the United States, neither the USEPA nor the CDC has been able to identify a chlorine concentration that would be safe yet effective against *Giardia* cysts under all water conditions. For this reason, the use of chlorine as a preventive measure against waterborne giardiasis generally has been used under outbreak conditions when the amount of chlorine and contact time have been tailored to fit specific water conditions and the existing operational design of the water utility.

In an outbreak, for example, the local health department and water utility may issue an advisory to boil water, may increase the chlorine residual at the consumer's tap from 0.5 mg/L to 1 or 2 mg/L, and, if the physical layout and operation of the water treatment facility permit, increase the chlorine contact time. These are emergency procedures intended to reduce the risk of transmission until a filtration device can be installed or repaired or until an alternative source of safe water (a well, for example) can be made operational.

The long-term solution to the problem of municipal waterborne outbreaks of giardiasis involves improvements in and more widespread use of filters in the municipal water treatment process. The sand filters most commonly used in municipal water treatment today cost millions of dollars to install, which makes them unattractive for many small communities. The pore sizes in these filters are not sufficiently small to remove *Giardia* (6 to 9 μm by 8 to 12 μm). For the sand filter to remove *Giardia* cysts from the water effectively, the water must receive some additional treatment before it reaches the filter. The flow of water through the filter bed must also be carefully regulated.

An ideal prefilter treatment for muddy water would include sedimentation (a holding pond where large suspended particles are allowed to settle out by the action of gravity) followed by flocculation or coagulation (the addition of chemicals such as alum or ammonium to cause microscopic particles to clump together). The sand filter easily removes the large particles resulting from the flocculation–coagulation process, including *Giardia* cysts bound to other microparticulates. Chlorine is then added to kill the bacteria and viruses that may escape the filtration process. If the water comes from a relatively clear source, chlorine may be added to the water before it reaches the filter.

The successful operation of a complete waterworks operation is a complex process that requires considerable training. Troubleshooting breakdowns or recognizing the potential problems in the system before they occur often requires the skills of an engineer. Unfortunately, most small water utilities with water treatment facilities that include filtration cannot afford the services of a full-time engineer. Filter operation or maintenance problems in such systems may not be detected until a *Giardia* outbreak is recognized in the community. The bottom line is that, although filtration is the best protection against waterborne giardiasis that water treatment technology has to offer for municipal water systems, it is not infallible. For municipal water filtration facilities to work properly, they must be properly constructed, operated, and maintained.

Whenever possible, persons outdoors should carry drinking water of known purity with them. When this is not practical and when water from streams, lakes, ponds, or other outdoor sources must be used, time should be taken to properly disinfect the water before drinking it.

12.14.3 CRYPTOSPORIDIUM

Ernest E. Tyzzer first described the protozoan parasite *Cryptosporidium* in 1907. Tyzzer frequently found a parasite in the gastric glands of laboratory mice. Tyzzer

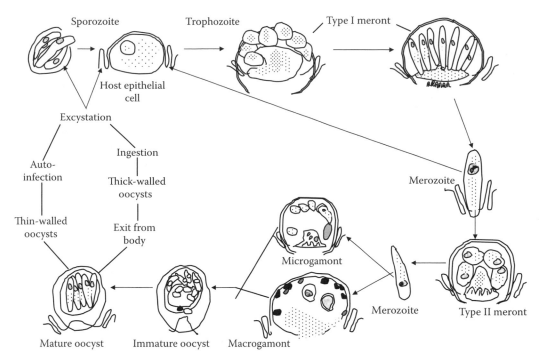

FIGURE 12.9 Life cycle of *Cryptosporidium parvum*.

identified the parasite as a sporozoan but of uncertain taxonomic status, and he named it *Cryptosporidium muris*. Later, in 1910, after more detailed study, he proposed *Cryptosporidium* as a new genus and *muris* as the type of species. Amazingly, except for developmental stages, Tyzzer's original description of the life cycle (see Figure 12.9) was later confirmed by electron microscopy. In 1912, Tyzzer described another new species, *Cryptosporidium parvum* (Tyzzer, 1912).

For almost 50 years, Tyzzer's discovery of the genus *Cryptosporidium* remained (like himself) relatively obscure because it appeared to be of no medical or economic importance. Slight rumblings of the importance of the genus began to be felt in the medical community when Slavin (1955) wrote about a new species, *Cryptosporidium melagridis*, which was associated with illness and death in turkeys. Interest remained slight even when *Cryptosporidium* was found to be associated with bovine diarrhea (Panciera et al., 1971).

Not until 1982 did worldwide interest focus on the study of organisms in the genus *Cryptosporidium*. At that time, the medical community and other interested parties were beginning a full-scale, frantic effort to find out as much as possible about acquired immune deficiency syndrome (AIDS), and the CDC reported that 21 AIDS-infected males from six large cities in the United States had severe protracted diarrhea caused by *Cryptosporidium*. It was in 1993, though, that *Cryptosporidium*—"the pernicious parasite"—made itself and Milwaukee famous (Mayo Foundation, 1996).

Note: The *Cryptosporidium* outbreak in Milwaukee caused the deaths of 100 people—the largest U.S. episode of waterborne disease in the 70 years since health officials began tracking such outbreaks.

The massive waterborne outbreak in Milwaukee (more than 400,000 persons developed acute and often prolonged diarrhea or other gastrointestinal symptoms) increased interest in *Cryptosporidium* at an exponential level. The Milwaukee incident spurred both public interest and the interest of public health agencies, agricultural agencies and groups, environmental agencies and groups, and suppliers of drinking water. This increase in interest level and concern spurred new studies of *Cryptosporidium*, with an emphasis on developing methods for recovery, detection, prevention, and treatment (Fayer et al., 1997).

The USEPA is particularly interested in this pathogen. In its reexamination of regulations on water treatment and disinfection, the USEPA issued a maximum contaminant level goal (MCLG) and contaminant candidate list (CCL) for *Cryptosporidium*. Its similarity to *Giardia lamblia* and the need for an efficient conventional water treatment capable of eliminating viruses forced the USEPA to regulate surface water supplies in particular. The Enhanced Surface Water Treatment Rule (ESWTR) included regulations ranging from watershed protection to specialized operation of treatment plants (certification of operators and state overview) and effective chlorination. Protection against *Cryptosporidium* includes control of waterborne pathogens such as *Giardia* and viruses (De Zuane, 1997).

TABLE 12.3
Valid Named Species
of *Cryptosporidium*

Species	Host
C. baileyi	Chicken
C. felis	Domestic cat
C. meleagridis	Turkey
C. murishouse	House mouse
C. nasorium	Fish
C. parvum	House mouse
C. serpentis	Corn snake
C. wrairi	Guinea pig

Source: Fayer, R., Ed., *Cryptosporidium and Cryptosporidiosis,* CRC Press, Boca Raton, FL, 1997. With permission.

12.14.3.1 The Basics of *Cryptosporidium*

Cryptosporidium (crip-toe-spor-ID-ee-um) is one of several single-celled protozoan genera in the phylum Apircomplexa (all referred to as coccidian). *Cryptosporidium* along with other genera in the phylum Apircomplexa develop in the gastrointestinal tract of vertebrates through all of their life cycle; in short, they live in the intestines of animals and people. This microscopic pathogen causes a disease called *cryptosporidiosis* (crip-toe-spor-id-ee-O-sis). The dormant (inactive) form of *Cryptosporidium* is called an *oocyst* (O-o-sist) and is excreted in the feces (stool) of infected humans and animals. The tough-walled oocysts survive under a wide range of environmental conditions.

Several species of *Cryptosporidium* were incorrectly named after the host in which they were found, and subsequent studies have invalidated many species. Now, eight valid species of *Cryptosporidium* (see Table 12.3) have been named. Upton (1997) reported that *C. muris* infects the gastric glands of laboratory rodents and several other mammalian species but (even though several texts state otherwise) is not known to infect humans. *C. parvum,* however, infects the small intestine of an unusually wide range of mammals, including humans, and is the zoonotic species responsible for human cryptosporidiosis. In most mammals *C. parvum* is predominately a parasite of neonate (newborn) animals. Upton pointed out that, even though exceptions occur, older animals generally develop poor infections, even when unexposed previously to the parasite. Humans are the one host that can be seriously infected at any time in their lives, and only previous exposure to the parasite results in either full or partial immunity to challenge infections.

Oocysts are present in most surface bodies of water across the United States, many of which supply public drinking water. Oocysts are more prevalent in surface waters when heavy rains increase runoff of wild and domestic animal wastes from the land or when sewage treatment plants are overloaded or break down. Only laboratories with specialized capabilities can detect the presence of *Cryptosporidium* oocysts in water. Unfortunately, current sampling and detection methods are unreliable. Recovering oocysts trapped on the material used to filter water samples is difficult. When a sample has been obtained, however, determining whether the oocyst is alive and if it is *C. parvum* and thus can infect humans is easily accomplished by looking at the sample under a microscope.

The number of oocysts detected in raw (untreated) water varies with location, sampling time, and laboratory methods. Water treatment plants remove most, but not always all, oocysts. Low numbers of oocysts are sufficient to cause cryptosporidiosis, but the low numbers of oocysts sometimes present in drinking water are not considered cause for alarm in the public.

Protecting water supplies from *Cryptosporidium* demands multiple barriers. Why? Because *Cryptosporidium* oocysts have tough walls that can withstand many environmental stresses and are resistant to chemical disinfectants such as chlorine that are traditionally used in municipal drinking water systems.

Physical removal of particles, including oocysts, from water by filtration is an important step in the water treatment process. Typically, water pumped from rivers or lakes into a treatment plant is mixed with coagulants, which help settle out particles suspended in the water. If sand filtration is used, even more particles are removed. Finally, the clarified water is disinfected and piped to customers. Filtration is the only conventional method now in use in the United States for controlling *Cryptosporidium.*

Ozone is a strong disinfectant that kills protozoa if sufficient doses and contact times are used, but ozone leaves no residual for killing microorganisms in the distribution system, as does chlorine. The high costs of new filtration or ozone treatment plants must be weighed against the benefits of additional treatment. Even well-operated water treatment plants cannot ensure that drinking water will be completely free of *Cryptosporidium* oocysts. Water treatment methods alone cannot solve the problem; watershed protection and monitoring of water quality are critical. As mentioned earlier, watershed protection is another barrier to *Cryptosporidium* in drinking water. Land use controls such as septic system regulations and best management practices to control runoff can help keep human and animal wastes out of water.

Under the Surface Water Treatment Rule of 1989, public water systems must filter surface water sources unless water quality and disinfection requirements are met and a watershed control program is maintained. This rule, however, did not address *Cryptosporidium.* The USEPA has now set standards for turbidity (cloudiness) and coliform bacteria (which indicate that pathogens are probably present) in drinking water. Frequent monitoring must occur to provide officials with early warning of potential problems to enable them to take steps to protect public health. Unfortunately, no water quality indicators can reliably predict the occurrence of cryptosporidiosis. More accurate and rapid assays of oocysts will make it possible to notify residents promptly if their water supply is contaminated with *Cryptosporidium* and thus avert outbreaks.

The bottom line: The collaborative efforts of water utilities, government agencies, healthcare providers, and individuals are needed to prevent outbreaks of cryptosporidiosis.

12.14.3.2 Cryptosporidiosis

Juranek (1995) wrote in the journal *Clinical Infectious Diseases*:

> *Cryptosporidium parvum* is an important emerging pathogen in the U.S. and a cause of severe, life-threatening disease in patients with AIDS. No safe and effective form of specific treatment for cryptosporidiosis has been identified to date. The parasite is transmitted by ingestion of oocysts excreted in the feces of infected humans or animals. The infection can therefore be transmitted from person to person, through ingestion of contaminated water (drinking water and water used for recreational purposes) or food, from animal to person, or by contact with fecally contaminated environmental surfaces. Outbreaks associated with all of these modes of transmission have been documented. Patients with human immunodeficiency virus infection should be made more aware of the many ways that *Cryptosporidium* species are transmitted, and they should be given guidance on how to reduce their risk of exposure.

Since the Milwaukee outbreak, concern about the safety of drinking water in the United States has increased, and new attention has been focused on determining and reducing the risk of acquiring cryptosporidiosis from community and municipal water supplies. Cryptosporidiosis is spread by putting something in the mouth that has been contaminated with the stool of an infected person or animal. In this way, people swallow the *Cryptosporidium* parasite. As mentioned earlier, a person can become infected by drinking contaminated water or eating raw or undercooked food contaminated with *Cryptosporidium* oocysts, by direct contact with the droppings of infected animals or stools of infected humans, or by hand-to-mouth transfer of oocysts from surfaces that may have become contaminated with microscopic amounts of stool from an infected person or animal.

The symptoms may appear 2 to 10 days after infection by the parasite. Although some persons may not have symptoms, others have watery diarrhea, headache, abdominal cramps, nausea, vomiting, and low-grade fever. These symptoms may lead to weight loss and dehydration. In otherwise healthy persons, these symptoms usually last 1 to 2 weeks, at which time the immune system is able to defeat the infection. In persons with suppressed immune systems, such as persons who have AIDS or who recently have had an organ or bone marrow transplant, the infection may continue and become life threatening.

Currently, no safe and effective cure for cryptosporidiosis exists. People with normal immune systems improve without taking antibiotic or antiparasitic medications. The treatment recommended for this diarrheal illness is to drink plenty of fluids and to get extra rest. Physicians may prescribe medication to slow the diarrhea during recovery.

The best way to prevent cryptosporidiosis is to:

- Avoid water or food that may be contaminated.
- Wash hands after using the toilet and before handling food.
- Be sure, if you work in a daycare center, to wash your hands thoroughly with plenty of soap and warm water after every diaper change, even if you wear gloves when changing diapers.

During community-wide outbreaks caused by contaminated drinking water, drinking water practitioners should inform the public to boil drinking water for 1 minute to kill the *Cryptosporidium* parasite.

12.14.4 CYCLOSPORA

Cyclospora organisms, which until recently were considered blue–green algae, were discovered at the turn of the 19th century. The first human cases of *Cyclospora* infection were reported in the 1970s. In the early 1980s, *Cyclospora* was recognized as a pathogen in patients with AIDS. We now know that *Cyclospora* is endemic in many parts of the world and appears to be an important cause of traveler's diarrhea. *Cyclospora* are two to three times larger than *Cryptosporidium* but otherwise have similar features. *Cyclospora* diarrheal illness in patients with healthy immune systems can be cured by a week of therapy with timethoprim–sulfamethoxazole (TMP–SMX).

So, what exactly is *Cyclospora?* In 1998, the CDC described *Cyclospora cayetanensis* as a unicellular parasite previously known as a cyanobacterium-like (blue–green algae-like) or coccidian-like body. The disease is known as *cyclosporasis. Cyclospora* infects the small intestine and causes an illness characterized by diarrhea with frequent stools. Other symptoms can include loss of appetite, bloating, gas, stomach cramps, nausea, vomiting, fatigue, muscle ache, and fever. Some individuals infected with *Cyclospora* may not show symptoms. Since the first known cases of illness caused by *Cyclospora* infection were reported in the medical journals in the 1970s, cases of cyclosporasis have been reported with increasing frequency from around the world (in part because of the availability of better techniques for detecting the parasite in stool specimens).

Huang et al. (1995) detailed what they believe is the first known outbreak of diarrheal illness associated with *Cyclospora* in the United States. The outbreak, which occurred in 1990, consisted of 21 cases of illness among physicians and others working at a Chicago hospital. Contaminated tap water from a physicians' dormitory at the hospital was the probable source of the organisms. The tap water probably picked up the organism while in a storage tank at the top of the dormitory after the failure of a water pump.

The transmission of *Cyclospora* is not a straightforward process. When infected persons excrete the oocyst state of *Cyclospora* in their feces, the oocysts are not infectious and may require from days to weeks to become so (i.e., to sporulate). Thus, transmission of *Cyclospora* directly from an infected person to someone else is unlikely; however, indirect transmission can occur if an infected person contaminates the environment and oocysts have sufficient time, under appropriate conditions, to become infectious. For example, *Cyclospora* may be transmitted by ingestion of water or food contaminated with oocysts. Outbreaks linked to contaminated water, as well as outbreaks linked to various types of fresh produce, have been reported in recent years (Herwaldt et al., 1997). The various modes of transmission and sources of infection are not yet fully understood nor is it known whether animals can be infected and serve as sources of infection for humans.

Note: *Cyclospora* organisms have not yet been grown in tissue cultures or laboratory animal models.

Persons of all ages are at risk for infection. Persons living or traveling in developing countries may be at increased risk, but infection can be acquired worldwide, including in the United States. In some countries of the world, infection appears to be seasonal. Based on currently available information, avoiding water or food that may be contaminated with stool is the best way to prevent infection. Reinfection can occur.

Note: De Zuane (1997) reported that pathogenic parasites are not easily removed or eliminated by conventional treatment and disinfection unit processes. This is particularly true for *Giardia lamblia*, *Cryptosporidium*, and *Cyclospora*. Filtration facilities can be adjusted with regard to depth, prechlorination, filtration rate, and backwashing to become more effective in the removal of cysts. The pretreatment of protected watershed raw water is a major factor in the elimination of pathogenic protozoa.

12.14.5 HELMINTHS

Helminths are parasitic worms that grow and multiply in sewage (biological slimes) and wet soil (mud). They multiply in wastewater treatment plants. Strict aerobes, they have been found in activated sludge and particularly in trickling filters and therefore appear in large concentrations in treated domestic liquid waste. Either they enter the skin or the worm is ingested in one of its many life-cycle phases. Generally, they are not a problem in drinking water supplies in the United States because both their egg and larval forms are large enough to be trapped during conventional water treatment. In addition, most helminths are not waterborne, so chances of infection are minimal (WHO, 1996).

12.15 WASTEWATER TREATMENT AND BIOLOGICAL PROCESSES

Uncontrolled bacteria in industrial water systems produce an endless variety of problems, including disease, equipment damage, and product damage. Unlike the microbiological problems that can occur in water systems, in wastewater treatment microbiology can be applied as a beneficial science for the destruction of pollutants in wastewater (Kemmer, 1979).

It should be noted that all of the biological processes used for the treatment of wastewater (in particular) are derived or modeled from processes occurring naturally in nature. The processes discussed in the following are typical examples. It also should be noted that "by controlling the environment of microorganisms, the decomposition of wastes is speeded up. Regardless of the type of waste, the biological treatment process consists of controlling the environment required for optimum growth of the microorganism involved" (Metcalf & Eddy, 2003).

12.15.1 AEROBIC PROCESS

In *aerobic treatment processes*, organisms use free, elemental oxygen and organic matter together with nutrients (nitrogen, phosphorus) and trace metals (e.g., iron) to produce more organisms and stable dissolved and suspended solids and carbon dioxide (see Figure 12.10).

12.15.2 ANAEROBIC PROCESS

The *anaerobic treatment process* consists of two steps, occurs completely in the absence of oxygen, and produces a useable byproduct, methane gas. In the first step of the process, facultative microorganisms use the organic matter as food to produce more organisms, volatile (organic) acids, carbon dioxide, hydrogen sulfide, and other gases and some stable solids (see Figure 12.11).

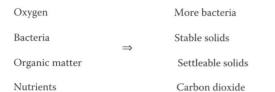

Oxygen		More bacteria
Bacteria	$\Rightarrow$	Stable solids
Organic matter		Settleable solids
Nutrients		Carbon dioxide

FIGURE 12.10 Aerobic decomposition.

Facultative bacteria		More bacteria
		Volatile solids
Organic matter	$\Rightarrow$	Settleable solids
Nutrients		Hydrogen sulfide

FIGURE 12.11 Anaerobic decomposition—first step.

Anaerobic bacteria		More bacteria
		Stable solids
Volatile acids	$\Rightarrow$	Settleable solids
Nutrients		Methane

FIGURE 12.12 Anaerobic decomposition—second step.

In the second step, anaerobic microorganisms use the volatile acids as their food source. The process produces more organisms, stable solids, and methane gas that can be used to provide energy for various treatment system components (see Figure 12.12).

12.15.3 ANOXIC PROCESS

In the *anoxic treatment process* (anoxic means "without oxygen"), microorganisms use the fixed oxygen in nitrate compounds as a source of energy. The process produces more organisms and removes nitrogen from the wastewater by converting it to nitrogen gas that is released into the air (see Figure 12.13).

12.15.4 PHOTOSYNTHESIS

Green algae use carbon dioxide and nutrients in the presence of sunlight and chlorophyll to produce more algae and oxygen (see Figure 12.14).

12.15.5 GROWTH CYCLES

All organisms follow a basic growth cycle that can be shown as a growth curve. This curve occurs when the environmental conditions required for the particular organism are reached. The environmental conditions (i.e., oxygen availability, pH, temperature, presence or absence of nutrients, presence or absence of toxic materials) that determine when a particular group of organ-

Nitrate oxygen		More bacteria
Bacteria	$\Rightarrow$	Stable solids
Organic matter		Settleable solids
Nutrients		Nitrogen

FIGURE 12.13 Anoxic decomposition.

Sun		
Algae	$\Rightarrow$	More algae
Carbon dioxide		Oxygen
Nutrients		

FIGURE 12.14 Photosynthesis.

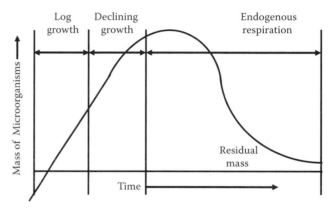

FIGURE 12.15 Microorganism growth curve.

isms will predominate. Obviously, this information can be very useful in operating a biological treatment process (see Figure 12.15).

12.15.6 BIOGEOCHEMICAL CYCLES

Several chemicals are essential to life and follow predictable cycles through nature. In these natural cycles, or *biogeochemical cycles*, the chemicals are converted from one form to another as they progress through the environment. The water/wastewater operator should be aware of those cycles dealing with the nutrients (e.g., carbon, nitrogen, and sulfur) because they have a major impact on the performance of the plant and may require changes in operation at various times of the year to keep them functioning properly; this is especially the case in wastewater treatment. The microbiology of each cycle deals with the biotransformation and subsequent biological removal of these nutrients in wastewater treatment plants.

Note: Smith categorizes biogeochemical cycles into two types, the *gaseous* and the *sedimentary*. Gaseous cycles include the carbon and nitrogen cycles. The main sink of nutrients in the gaseous cycle is the

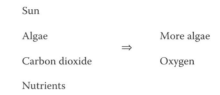

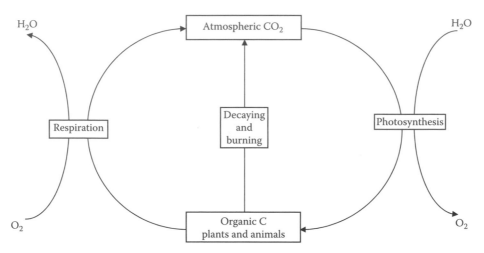

FIGURE 12.16 Carbon cycle.

atmosphere and the ocean. Sedimentary cycles include the sulfur cycle. The main sink for sedimentary cycles is soil and rocks of the Earth's crust (Smith, 1974).

12.15.7 CARBON CYCLE

Carbon, which is an essential ingredient of all living things, is the basic building block of the large organic molecules necessary for life. Carbon is cycled into food chains from the atmosphere, as shown in Figure 12.16. From Figure 12.16, it can be seen that green plants obtain carbon dioxide (CO_2) from the air and through photosynthesis, which was described by Asimov (1989) as the "most important chemical process on Earth." Photosynthesis produces the food and oxygen that all organisms require. Part of the carbon produced remains in living matter; the other part is released as CO_2 in cellular respiration. The carbon dioxide released by cellular respiration in all living organisms is returned to the atmosphere (Miller, 1988).

Some carbon is contained in buried dead and animal and plant materials. Much of these buried animal and plant materials were transformed into fossil fuels. Fossil fuels—coal, oil, and natural gas—contain large amounts of carbon. When fossil fuels are burned, stored carbon combines with oxygen in the air to form carbon dioxide, which enters the atmosphere. In the atmosphere, carbon dioxide acts as a beneficial heat screen, as it does not allow the radiation of Earth's heat into space. This balance is important. The problem is that as more carbon dioxide from burning is released into the atmosphere the balance can and is being altered. Odum (1983) warned that increases in the consumption of fossil fuels "coupled with the decrease in 'removal capacity' of the green belt is beginning to exceed the delicate balance." Massive increases of carbon dioxide into the atmosphere tend to increase the possibility of global warming. The conse-

quences of global warming "would be catastrophic ... and the resulting climatic change would be irreversible" (Abrahamson, 1988).

12.15.8 NITROGEN CYCLE

Nitrogen is an essential element required by all organisms. In animals, nitrogen is a component of crucial organic molecules such as proteins and DNA and constitutes 1 to 3% dry weight of cells. Our atmosphere contains 78% by volume of nitrogen, yet it is not a common element. Although nitrogen is an essential ingredient for plant growth, it is chemically very inactive, and it must be *fixed* before the vast majority of the biomass can incorporate it. Special nitrogen-fixing bacteria found in soil and water fix nitrogen; thus, microorganisms play a major role in nitrogen cycling in the environment. These microorganisms (bacteria) have the ability to take nitrogen gas from the air and convert it to nitrate via a process known as *nitrogen fixation*. Some of these bacteria occur as free-living organisms in the soil. Others live in a *symbiotic relationship* with plants. A symbiotic relationship is a close relationship between two organisms of different species and one where both partners benefit from the association. An example of a symbiotic relationship, related to nitrogen, can be seen, for example, in the roots of peas. These roots have small swellings along their length. These contain millions of symbiotic bacteria that have the ability to take nitrogen gas obtained from the atmosphere and convert it to nitrates that can be used by the plant. The plant is plowed back into the soil after the growing season to improve the nitrogen content. Price (1984) described the nitrogen cycle as an example "of a largely complete chemical cycle in ecosystems with little leaching out of the system." Simply, the nitrogen cycle provides various bridges between the atmospheric reservoirs and the biological communities (see Figure 12.17).

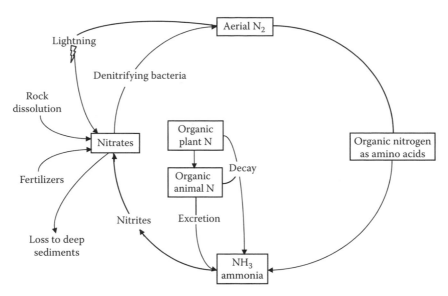

FIGURE 12.17 Nitrogen cycle.

Atmospheric nitrogen is fixed by either natural or industrial means; for example, nitrogen is fixed by lightning or by soil bacteria that convert it to ammonia, then to nitrite, and finally to nitrates, which plants can use. Nitrifying bacteria make nitrogen from animal wastes. Denitrifying bacteria convert nitrates back to nitrogen and release it as nitrogen gas.

The logical question now is "What does all of this have to do with water?" The best way to answer this question is to ask another question. Have you ever dived into a slow-moving stream and had the noxious misfortune to surface right in the middle of an algal bloom? When this happens to you, the first thought that runs through your mind is, "Where is my nose plug?" Why? Because of the horrendous stench, disablement of the olfactory sense is a necessity.

If too much nitrate, for example, enters the water supply as runoff from fertilizers it produces an overabundance of algae known as an *algal bloom*. If this runoff from fertilizer gets into a body of water, algae may grow so profusely that they form a blanket over the surface. This usually happens in summer, when the light levels and warm temperatures favor rapid growth.

In the voluminous and authoritative text *Wastewater Engineering: Treatment, Disposal, and Reuse* (Metcalf & Eddy, 2003), it is noted that nitrogen is found in wastewater in the form of urea. During wastewater treatment, the urea is transformed into ammonia nitrogen. Because ammonia exerts a BOD and chlorine demand, high quantities of ammonia in wastewater effluents are undesirable. The process of nitrification is utilized to convert ammonia to nitrates. *Nitrification* is a biological process that involves the addition of oxygen to the wastewater. If further treatment is necessary, another biological process called *denitrification* is used. In this process, nitrate is converted into nitrogen gas, which is lost to the atmosphere, as can be seen in Figure 12.17. From the wastewater operator's point of view, nitrogen and phosphorus are both considered limiting factors for productivity. Phosphorus discharged into streams contributes to pollution. Of the two, nitrogen is more difficult to control but is found in smaller quantities in wastewater.

12.15.9 Sulfur Cycle

Sulfur, like nitrogen, is characteristic of organic compounds. The *sulfur cycle* is both sedimentary and gaseous (see Figure 12.18). The principal forms of sulfur that are of special significance in water quality management are organic sulfur, hydrogen sulfide, elemental sulfur, and sulfate (Tchobanoglous and Schroeder, 1985). Bacteria play a major role in the conversion of sulfur from one form to another. In an anaerobic environment, bacteria break down organic matter, thereby producing hydrogen sulfide with its characteristic rotten-egg odor. The bacterium *Beggiatoa* converts hydrogen sulfide into elemental sulfur and sulfates. Other sulfates are contributed by the dissolving of rocks and some sulfur dioxide. Sulfur is incorporated by plants as proteins. Organisms then consume some of these plants. Many heterotrophic anaerobic bacteria liberate sulfur from proteins, as hydrogen sulfide.

12.15.10 Phosphorus Cycle

Phosphorus is another chemical element that is common in the structure of living organisms (see Figure 12.19); however, the phosphorus cycle is different from the hydrologic, carbon, and nitrogen cycles because phosphorus is found in sedimentary rock. These massive deposits are gradually eroding to provide phosphorus to ecosystems.

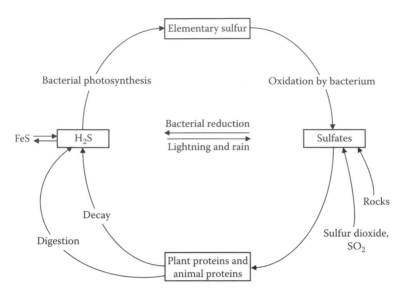

FIGURE 12.18 Sulfur cycle.

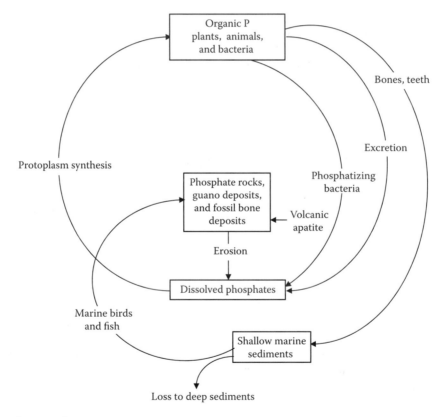

FIGURE 12.19 Phosphorus cycle.

A large amount of eroded phosphorus ends up in deep sediments in the oceans and in lesser amounts in shallow sediments. Part of the phosphorus comes to land when marine animals surface. Decomposing plant or animal tissue and animal droppings return organic forms of phosphorus to the water and soil. Fish-eating birds, for example, play a role in the recovery of phosphorus. Guano deposits (bird excreta) on the Peruvian coast are an example.

Humans have hastened the rate of phosphorus loss through mining and the production of fertilizers, which are washed away and the phosphorus is lost. Odum (1983) suggested, however, that there was no immediate cause for concern, as the known reserves of phosphate are quite large.

Phosphorus has become very important in water quality studies, because it is often found to be a limiting factor. Control of phosphorus compounds that enter surface waters and contribute to the growth of algal blooms is of considerable interest and has generated much study (Metcalf & Eddy, 2003). Upon entering a stream, phosphorus acts as a fertilizer, promoting the growth of undesirable algae populations or algae blooms. As the organic matter decays, dissolved oxygen levels decrease and fish and other aquatic species die.

Although it is true that phosphorus discharged into streams is a contributing factor to stream pollution, it is also true that phosphorus is not the lone factor. Odum (1975) warned against what he called the *one-factor control hypothesis* (i.e., the one-problem/one-solution syndrome). He noted that environmentalists in the past have focused on one or two items, such as phosphorus contamination, and have failed to understand that the strategy for pollution control must involve reducing the input of all enriching and toxic materials.

CHAPTER REVIEW QUESTIONS

12.1 The three major groups of microorganisms that cause disease in water are _____, _____, and _____.
12.2 When does a river of good quality show its highest bacterial numbers?
12.3 Are coliform organisms pathogenic?
12.4 How do bacteria reproduce?
12.5 The three common shapes of bacteria are _____, _____, and _____.
12.6 Three waterborne diseases caused by bacteria are _____, _____, and _____.
12.7 Two protozoa-caused waterborne diseases are _____ and _____.
12.8 When a protozoon is in a resting phase, it is called a _____.
12.9 For a virus to live it must have a _____.
12.10 What problems do algae cause in drinking water?

REFERENCES AND SUGGESTED READING

Abrahamson, D.E., Ed. 1988. *The Challenge of Global Warming.* Washington, D.C.: Island Press.
Asimov, I. 1989. *How Did We Find Out About Photosynthesis?* New York: Walker & Company.
Bingham, A.K., Jarroll, E.L, Meyer, E.A., and Radulescu, S. 1979. Introduction to *Giardia* excystation and the effect of temperature on cyst viability compared by eoxin exclusion and *in vitro* excystation. In *Waterborne Transmission of Giardiasis*, EPA-600/9-79-001, Jakubowski, J. and Hoff H.C., Eds. Washington, D.C.: U.S. Environmental Protection Agency.
Black, R.E., Dykes, A.C., Anderson, K.E., Wells, J.G., Sinclair, S.P., Gary, G.W., Hatch, M.H., and Ginagaros, E.J. 1981. Handwashing to prevent diarrhea in day-care centers. *Am. J. Epidemiol.*, 113, 445–451.

Brodsky, R.E., Spencer, H.C., and Schultz, M.G. 1974. Giardiasis in American travelers to the Soviet Union. *J. Infect. Dis.*, 130, 319–323.
CDC. 1979. *Intestinal Parasite Surveillance, Annual Summary 1978.* Atlanta, GA: Centers for Disease Control.
CDC. 1983. *Water-Related Disease Outbreaks Surveillance, Annual Summary 1983.* Atlanta, GA: Centers for Disease Control.
CDC. 1995. *Giardiasis.* Atlanta, GA: Centers for Disease Control.
Davidson, R.A. 1984. Issues in clinical parasitology: the treatment of giardiasis. *Am. J. Gastrol. Enterol.*, 79(4), 256–261.
De Zuane, J. 1997. *Handbook of Drinking Water Quality.* New York: John Wiley & Sons.
Fayer, R., Speer, C.A., and Dudley, J.P. 1997. The general biology of *Cryptosporidium*. In *Cryptosporidium and Cryptosporidiosis*, Fayer, R., Ed. Boca Raton, FL: CRC Press.
Frost, F., Plan, B., and Liechty, B. 1984. *Giardia* prevalence in commercially trapped mammals. *J. Environ. Health*, 42, 245–249.
Herwaldt, F.L. et al. 1997. An outbreak in 1996 of cyclosporasis associated with imported raspberries. *N. Engl. J. Med.*, 336, 1548–1556.
Huang, P., Weber, J.T., Sosin, D.M. et al. 1995. *Cyclospora. Ann. Intern. Med.*, 123, 401–414.
Jarroll, Jr., E.L., Gingham, A.K, and Meyer, E.A. 1979. *Giardia* cyst destruction: effectiveness of six small-quantity water disinfection methods. *Am J. Trop. Med. Hygiene*, 29, 8–11.
Juranek, D.D. 1995. Cryptosporidiosis: sources of infection and guidelines for prevention. *Clin. Infect. Dis.*, 21, S37–S61.
Kemmer, F.N. 1979. *Water: The Universal Solvent*, 2nd ed. Oak Brook, IL: Nalco Chemical Co.
Kordon, C. 1992. *The Language of the Cell.* New York: McGraw-Hill.
Koren, H. 1991. *Handbook of Environmental Health and Safety: Principles and Practices.* Chelsea, MI: Lewis Publishers.
LeChevallier, M.W., Norton, W.D., and Less, R.G. 1991. Occurrences of *Giardia* and *Cryptosporidium* spp. in surface water supplies. *Appl. Environ. Microbiol.*, 57, 2610–2616.
Mayo Foundation. 1996. *The "Bug" That Made Milwaukee Famous.* Rochester, MN: Mayo Foundation.
Metcalf & Eddy. 2003. *Wastewater Engineering: Treatment, Disposal, and Reuse.* New York: McGraw-Hill.
Miller, G.T. 1988. *Environmental Science: An Introduction.* Belmont, CA: Wadsworth.
Odum, E.P. 1975. *Ecology: The Link Between the Natural and the Social Sciences.* New York: Holt, Rinehart and Winston.
Odum, E.P. 1983. *Basic Ecology.* Philadelphia, PA: Saunders.
Panciera, R.J., Thomassen, R.W., and Garner, R.M. 1971. Cryptosporidial infection in a calf. *Vet. Pathol.*, 8, 479.
Pennak, R.W. 1989. *Fresh-Water Invertebrates of the United States*, 3rd ed. New York: John Wiley & Sons.
Price, P.W. 1984. *Insect Ecology.* New York: John Wiley & Sons.

Rendtorff, R.C. 1954. The experimental transmission of human intestinal protozoan parasites II *Giardia lamblia* cysts given in capsules, *Am. J. Hygiene*, 59, 209–220.

Rose, J.B., Gerb, C.P., and Jakubowski, W. 1983. Survey of potable water supplies for *Cryptosporidium* and *Giardia. Environ. Sci. Technol.*, 25, 1393–1399.

Singleton, P. and Sainsbury, D. 1994. *Dictionary of Microbiology and Molecular Biology*, 2nd ed. New York: John Wiley & Sons.

Slavin, D. 1955. *Cryptosporidium melagridis. J. Comp. Pathol.*, 65, 262.

Smith, R.L. 1974. *Ecology and Field Biology.* New York: Harper & Row.

Spellman, F.R. 1996. *Stream Ecology and Self-Purification: An Introduction for Wastewater and Water Specialists.* Boca Raton, FL: CRC Press.

Spellman, F.R. 1997. *Microbiology for Water/Wastewater Operators.* Boca Raton, FL: CRC Press.

Tchobanoglous, G. and Schroeder, E.D. 1985. *Water Quality.* Reading, MA: Addison-Wesley.

Tyzzer, E.E. 1912. *Cryptosporidium parvum* sp.: a coccidium found in the small intestine of the common mouse. *Arch. Protistenkd.*, 26, 394.

Upton, S.J. 1997. *Basic Biology of Cryptosporidium.* Manhattan: Kansas State University.

Walsh, J.D. and Warren, K.S. 1979. Selective primary health care: an interim strategy for disease control in developing countries, *N. Engl. J. Med.*, 301, 974–976.

Weller, P.F. 1985. Intestinal protozoa: giardiasis. *Sci. Am. Med.*, 12(4), 554–558.

WHO. 1990. *Guidelines for Drinking Water Quality*, Vol. 2, 2nd ed. Geneva: World Health Organization.

Wistriech, G.A. and Lechtman, M.D. 1980. *Microbiology*, 3rd ed. New York: Macmillan.

13 Water Ecology

Streams are arteries of earth, beginning in capillary creeks, brooks, and rivulets. No matter the source, they move in only one direction—downhill, as the heavy hand of gravity tugs and drags the stream toward the sea. During its inexorable flow downward, now and then there is an abrupt change in geology. Boulders are mowed down by "slumping" (gravity) from their in-place points high up on canyon walls.

As stream flow grinds, chisels, and sculpts the landscape, the effort is increased by momentum, augmented by turbulence provided by rapids, cataracts, and waterfalls. These falling waters always hypnotize us, just as fire gazing or wave watching does.

Before emptying into the sea, streams often pause, forming lakes. When one stares into a healthy lake, its phantom blue–green eye stares right back, but only for a moment—relatively speaking, of course, because all lakes are ephemeral, doomed. Eventually the phantom blue–green eye is close lidded by the moist verdant green of landfill.

For water that escapes the temporary bounds of a lake, most of it evaporates or moves on to the gigantic sink—the sea—where the cycles continues, forever more it is hoped.

13.1 INTRODUCTION

The "control of nature" is a phrase conceived in arrogance, born of the Neanderthal age of biology and the convenience of man.

Rachel Carson (1962)

What is ecology? Why is ecology important? Why study ecology? These are all simple, straightforward questions; however, providing simple, straightforward answers is not that easy. Notwithstanding the inherent difficulty with explaining any complex science in simple, straightforward terms, that is the purpose, the goal, the mission of this text and this chapter. In short, the task of this chapter is to outline basic information that explains the functions and values of ecology and its interrelationships with other sciences, including the direct impact of ecology on our lives. In doing so, the author hopes not only to dispel the common misconception that ecology is too difficult for the average person to understand but also to instill the concept of ecology as an asset that can be learned as well as cherished.

13.2 WHAT IS ECOLOGY?

Ecology can be defined in various and numerous ways. Ecology, or ecological science, is commonly defined in the literature as the scientific study of the distribution and abundance of living organisms and how the distribution and abundance are affected by interactions between the organisms and their environment. The term *ecology* was coined in 1866 by the German biologist Haeckel and it loosely means "the study of the household [of nature]". Odum (1983) suggested that the term is derived from the Greek *oikos*, meaning "home." Ecology, then, means the study of organisms at home. It means the study of an organism in its home. Ecology is the study of the relation of an organism or a group of organisms to their environment. In a broader sense, ecology is the study of the relation of organisms or groups to their environment.

Important Point: No ecosystem can be studied in isolation. If we were to describe ourselves, our histories, and what made us the way we are, we could not leave the world around us out of our description! So it is with streams: They are directly tied in with the world around them. They take their chemistry from the rocks and dirt beneath them as well as for a great distance around them (Spellman, 1996).

Charles Darwin explained ecology in a famous passage in *The Origin of Species* (Darwin, 1998), a passage that helped establish the science of ecology. According to Darwin, a "web of complex relations" binds all living things in any region. Adding or subtracting even a single species causes waves of change that race through the web, "onwards in ever-increasing circles of complexity." The simple act of adding cats to an English village would reduce the number of field mice. The reduced number of mice would benefit bumblebees, whose nests and honeycombs the mice often devour. Increasing the number of bumblebees would benefit the heartsease and red clover crops, which are fertilized almost exclusively by bumblebees. So adding cats to the village could end by adding flowers. For Darwin, the whole of the Galapagos archipelago argues this fundamental lesson. The island volcanoes are much more diverse in their ecology than their biology. The contrast suggests that, in the struggle for existence, species are shaped at least as much by the local flora and fauna as by the local soil and climate. "Why else would the plants and animals differ radically among islands that have the same geological nature, the same height, and climate?" (Darwin, 1998).

Probably the best way to understand ecology—to get a really good feel for it or to get to the heart of what ecology is all about—is to read the following by Rachel Carson (1962):

> We poison the caddis flies in a stream and the salmon runs dwindle and die. We poison the gnats in a lake and the poison travels from link to link of the food chain and soon the birds of the lake margins become victims. We spray our elms and the following springs are silent of robin song, not because we sprayed the robins directly but because the poison traveled, step by step, through the now familiar elm leaf–earthworm–robin cycle. These are matters of record, observable, part of the visible world around us. They reflect the web of life—or death—that scientists know as ecology.

As Carson pointed out, what we do to any part of our environment has an impact on other parts. In other words, there is an interrelationship between the parts that make up our environment. Probably the best way to state this interrelationship is to define ecology definitively—that is, to define it as it is used in this text: "Ecology is the science that deals with the specific interactions that exist between organisms and their living and nonliving environment" (Tomera, 1989).

When environment was mentioned in the proceeding and as it is discussed throughout this text, the environment includes everything important to the organism in its surroundings. The organism's environment can be divided into four parts:

1. Habitat and distribution (its place to live)
2. Other organisms (e.g., friendly or hostile?)
3. Food
4. Weather (e.g., light, moisture, temperature, soil)

The four major subdivisions of ecology are:

1. Behavioral ecology
2. Population ecology (autecology)
3. Community ecology (synecology)
4. Ecosystem ecology

Behavioral ecology is the study of the ecological and evolutionary basis for animal behavior. *Population ecology* (or autecology) is the study of an individual organism or a species. It emphasizes life history, adaptations, and behavior. It is the study of communities, ecosystems, and biosphere. An example of autecology would be when biologists study exclusively the ecology of the salmon. *Community ecology* (or synecology), on the other hand, is the

study of groups of organisms associated together as a unit and deals with the environmental problems caused by mankind; for example, the effect of discharging phosphorus-laden effluent into a stream involves several organisms. The activities of human beings are having a major impact on many natural areas. As a result, it is important to realize that the study of ecology must involve people. *Ecosystem ecology* is the study of how energy flow and matter interact with biotic elements of ecosystems (Odum, 1971).

Important Point: Ecology is generally categorized according to complexity, the primary kinds of organism under study (e.g., plant, animal, insect ecology), the biomes principally studied (e.g., forest, desert, benthic, grassland), the climatic or geographic area (e.g., artic or tropics), and the spatial scale (macro or micro) under consideration.

13.3 WHY IS ECOLOGY IMPORTANT?

Ecology, in its true sense, is a holistic discipline that does not dictate what is right or wrong. Instead, ecology is important to life on Earth simply because it makes us aware, to a certain degree, of what life on Earth is all about. Ecology shows us that each living organism has an ongoing and continual relationship with every other element that makes up our environment. Simply, ecology is all about interrelationships, intraspecific and interspecific, and on how important it is to maintain these relationships to ensure our very survival.

At this point in this discussion, literally countless examples could be used to point out the importance of ecology and the interrelationships involved; however, an excerpt from Peter Marshall's *Mr. Jones: Meet the Master* is provided here to demonstrate the importance of ecology as well as to point out that an ecological principle can be a double-edged sword, depending on one's point of view (ecological problems, as with pollution, can be a judgment call; that is, they are a matter of opinion).

The Keeper of the Spring

This is the story of the keeper of the spring. He lived high in the Alps above an Austrian town and had been hired by the town council to clear debris from the mountain springs that fed the stream that flowed through the town. The man did his work well and the village prospered. Graceful swans floated in the stream. The surrounding countryside was irrigated. Several mills used the water for power. Restaurants flourished for townspeople and for a growing number of tourists.

Years went by. One evening at the town council meeting someone questioned the money being paid to the keeper of the spring. No one seemed to know who he was or even if he was still on the job high up in the mountains. Before the evening was over, the council decided to dispense with the old man's services.

Weeks went by and nothing seemed to change. Then autumn came. The trees began to shed their leaves. Branches broke and fell into the pools high up in the mountains. Down below the villagers began to notice the water becoming darker. A foul odor appeared. The swans disappeared. Also, the tourists. Soon disease spread through the town.

When the town council reassembled, they realized that they had made a costly error. They found the old keeper of the spring and hired him back again. Within a few weeks, the stream cleared up and life returned to the village as they had known it before (Marshall, 1950).

After reviewing Marshall's parable about restoration of the spring, the average person might think, "Gee, all is well with the town again." Because of their swans, irrigation, hydropower, and pretty views, residents seem to be pleased that the stream has been restored to its "normal" state. The trained ecologist, however, would take a different view of this same stream. The ecologist would go beyond the hype (as portrayed in the popular media, including literature) about what a healthy stream is. The trained ecologist would know that a perfectly clean stream, clear of all terrestrial plant debris (woody debris and leaves) would not be conducive to ensuring diverse, productive populations of invertebrates and fish, would not preserve natural sediment and water regimes, and would not ensure overall stream health (Dolloff and Webster, 2000).

13.4 WHY STUDY ECOLOGY?

Does anyone really need to be an ecologist or a student of ecology to appreciate the following words of Will Carleton (1845–1912) from his classic poem "Autumn Days"?

Sweet and smiling are thy ways,
Beauteous, gold Autumn days.

Moreover, does anyone need to study ecology to observe, to relish, to feel, to sense the real thing—nature's annual color palette in full kaleidoscopic display, when the "yellow, mellow, ripened days are sheltered in a golden coating?" It is those clear and sunny days and cool and crisp nights of autumn that provide an almost irresistible lure to those of us (ecologist and nonecologist alike) who enjoy the outdoors. To take in the splendor and delight of autumn's color display, many head for the hills, the mountains, countryside, lakes, streams, and recreation areas of our national forests. The more adventurous ride horseback or backpack through nature's glory and solitude on trails winding deep into forest tranquility—just being out of doors in these golden days rivals any thrill in life. Even those of us fettered to the chains of city life are often exposed to city streets with columns of life ablaze in color.

No, one need not study ecology to witness, appreciate, and understand the enchantment of autumn's annual color display—summer extinguished in a blaze of color. It is a different story, however, for those involved in trying to understand all of the complicated actions—and even more complicated interactions involving pigments, sunlight, moisture, chemicals, temperatures, site, hormones, length of daylight, genetic traits, and so on that make for a perfect autumn color display (USDA, 1999). This is the work of the ecologist—to probe deeper and deeper into the basics of nature, constantly seeking answers. To find the answers, the ecologist must be a synthesis scientist; that is, ecologists must be generalists well versed in botany, zoology, physiology, genetics, and other disciplines such as geology, physics, and chemistry.

Earlier, we used Marshall's parable to make the point that a clean stream and other downstream water bodies can be a good thing, depending on one's point of view—pollution is a judgment call. It was also pointed out that this view may not be shared by a trained ecologist, especially a stream ecologist. The ecologist knows, for example, that terrestrial plant debris is not only a good thing but is also absolutely necessary. Why? Consider the following explanation (Spellman, 1996).

A stream has two possible sources of primary energy: (1) instream photosynthesis by algae, mosses, and higher aquatic plants, and (2) imported organic matter from streamside vegetation (e.g., leaves and other parts of vegetation). Simply put, a significant portion of the food that is eaten by aquatic organisms grows right in the stream, such as algae, diatoms, nymphs and larvae, and fish. This food that originates from within the stream is *autochthonous* (Benfield, 1996).

Most food in a stream, however, comes from outside the stream. This is especially the case in small, heavily wooded streams, where there is normally insufficient light to support substantial instream photosynthesis so energy pathways are supported largely by imported energy. A large portion of this imported energy is provided by leaves. Worms drown in floods and get washed in. Leafhoppers and caterpillars fall from trees. Adult mayflies and other insects mate above the stream, lay their eggs in it, and then die in it. All of this food from outside the stream is allochthonous.

Little brook, sing a song of leaf that sailed along.

Down the golden braided center of your current swift and strong.

James Whitcomb Riley ("The Brook Song")

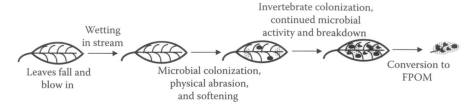

FIGURE 13.1 Leaf processing in streams.

13.4.1 LEAF PROCESSING IN STREAMS

Autumn leaves entering streams are nutrition poor because trees absorb most of the sugars and amino acids (nutrients) that were present in the green leaves (Suberkoop et al., 1976). Leaves falling into streams may be transported short distances but usually are caught by structures in the streambed to form leaf packs. These leaf packs are then processed in place by components of the stream communities in a series of well documented steps (Figure 13.1) (Peterson and Cummins, 1974).

Within 24 to 48 hours of entering a stream, much of the remaining nutrients in leaves leaches into the water. After leaching, these leaves are composed mostly of structural materials such as nondigestible cellulose and lignin. Within a few days, fungi (especially Hyphomycetes), protozoa, and bacteria process the leaves by microbial processing (Figure 13.1) (Barlocher and Kendrick, 1975). Two weeks later, microbial conditioning leads to structural softening of the leaf and, among some species, fragmentation. Reduction in particle size from whole leaves (coarse particulate organic matter, or CPOM) to fine particulate organic matter (FPOM) is accomplished primarily through the feeding activities of a variety of aquatic invertebrates collectively known as *shredders* (Cummins, 1974; Cummins and Klug, 1979). Shredders (stoneflies, for example) help to produce fragments shredded from leaves but not ingested and fecal pellets, which reduce the particle size of organic matter. The particles are then collected (by mayflies, for example) and serve as a food resource for a variety of micro- and macroconsumers. Collectors eat what they want and send even smaller fragments downstream. These tiny fragments may be filtered out of the water by true fly larvae (i.e., *filterers*). Leaves may also be fragmented by a combination of microbial activity and physical factors such as current and abrasion (Benfield et al., 1977; Paul et al., 1978).

Leaf-pack processing by all the elements mentioned above (i.e., leaf species, microbial activity, physical and chemical features of the stream) is important; however, the most important point is that these integrated ecosystem processes convert whole leaves into fine particles that are then distributed downstream and used as an energy source by various consumers.

The bottom line on allochthonous material in a stream: Insects that have fallen into a stream are ready to eat and may join leaves, exuviae (castoff coverings, such as crab shells or the skins of snakes), copepods, dead and dying animals, rotifers, bacteria, dislodged algae, and immature insects in their float (drift) downstream to a waiting hungry mouth.

Note: Anytime the author, a stream ecologist, thinks or writes about streams and allochthonous inputs, haunting wet refrains mingle with his thoughts. For example, consider the following by Henry David Thoreau (1862) in *A Winter Walk*:

> When every stream in its penthouse
> Goes gurgling on its way,
> And in his gallery the mouse
> Nibbleth the meadow of hay.

Another important reason to study and learn ecology can be garnered from another simple stream ecology example:

Family Picnic Hosts Insect Intruders

On one of their late August holiday outings, a family of 18 picnickers from a couple of small rural towns visited a local stream that coursed its way alongside or through their towns. This annual outing was looked upon with great anticipation for it was that one time each year when aunts, uncles, and cousins came together as a one big family. The streamside setting was perfect for such an outing, but historically, until quite recently, the stream had been posted "DANGER—NO SWIMMING, CAMPING, or FISHING!"

Because the picnic area was such a popular location for picnickers, swimmers, and fishermen over the years, several complaints about the polluted stream were filed with the County Health Department. The Health Department finally took action to restore the stream to a relatively clean condition: sanitation workers removed debris and old tires and plugged or diverted end-of-pipe industrial outfalls upstream of the picnic area. After a couple years of continuous stream clean-up and the stream's own natural self-purification process, the stream was given a clean bill of health by the Health Department.

When the stream had been declared clean and fit for use by swimmer and fisher, the postings were removed, and it did not take long for the word to get out. Local folks and others alike made certain, at first opportunity, to flock to the restored picnic and swimming and fishing site alongside the stream.

During most visits to the restored picnic-stream area, visitors, campers, fishermen, and others were pleased with their cleaned-up surroundings. In late summer, however, when the family of 18 and several others visited the restored picnic area, they found themselves swarmed by thousands of speedy dragonflies and damselflies, especially near the bank of the stream. Soon they found the insects too much to deal with so they stayed clear of the stream. To themselves and to anyone who would listen, the same complaint was heard over and over again: "What happened to our nice clean stream? With all those nasty bugs the stream may as well be polluted again." So, when August arrived with its hordes of dragonfly-type insects, the pic-nickers, campers, swimmers, and fishermen avoided the place and did not return until the insects departed and they thought the stream was clean again.

One local family, though, did not avoid the picnic area in August; on the contrary, August became one of their favorite times to visit, camp, swim, take in nature, and fish, as they usually had most of the site to themselves. The family was led by a local university professor of ecology who knew the truth about the area and the dragonflies and other insects. She knew that dragonflies and damselflies are macroinvertebrate indicator organisms; they only inhabit, grow, and thrive in and around streams that are clean and healthy—when dragonflies and damselflies are around, they indicate nonpolluted water. Further, the ecology professor knew that dragonflies are valued as predators, friends, and allies in the continual war against flies and in controlling populations of harmful insects such as mosquitoes. Dragonflies, on swiftest wings (25 to 35 mph), take the wrigglers in the water and the adults that are hovering over streams and ponds laying their eggs.

The ecology professor's husband, an amateur poet, also understood the significance of the presence of the indicator insects and had no problem sharing the same area with them. He viewed the winged insects with the eye of a poet and was aware that poets through the years have lavished their attention on dragonflies and paid them delightful tributes. James Whitcomb Riley (1849–1916) wrote:

> Till the dragon fly, in light gauzy armor
> burnished bright,
> Came tilting down the waters in a wild,
> bewildered flight.

Alfred, Lord Tennyson, drew inspiration for one of his most beautiful poems from the two stages of dragonfly life, but perhaps James Russell Lowell's (1819–1891) poem "The Fountain of Youth" gives us the perfect description of these insects:

> In summer-noon flushes
> When all the wood hushes,
> Blue dragon-flies knitting
> To and fro in the sun,
> With sidelong jerk flitting,
> Sink down on the rushes.
> And, motionless sitting,
> Hear it bubble and run,
> Hear its low inward singing
> With level wings swinging
> On green tasseled rushes,
> To dream in the sun.

13.5 HISTORY OF ECOLOGY

The chronological development of most sciences is clear and direct. Listing the progressive stages in the development of biology, math, chemistry, and physics is a relatively easy, straightforward process. The science of ecology is different. Having only gained prominence in the latter part of the 20th century, ecology is generally spoken of as a new science; however, ecological thinking at some level has been around for a long time, and the principles of ecology have developed gradually and more like a multiple-stemmed bush than a tree with a single trunk (Smith, 1996).

Smith and Smith (2006) suggest that one can argue that ecology can be traced back to Aristotle or perhaps his friend and associate, Theophrastus, both of whom had interest in the relations between organisms and the environment and in many species of animals. Theophrastus described interrelationships between animals and between animals and their environment as early as the 4th century B.C.E. (Ramalay, 1940).

Modern ecology has its early roots in plant geography (i.e., plant ecology, which developed earlier than animal ecology) and natural history. The early plant geographers (ecologists) included Carl Ludwig Willdenow (1765–1812) and Friedrich Alexander von Humboldt (1769–1859). Willdenow was one of the first phytogeographers, and he was also a mentor to von Humboldt. Willdenow, for whom the perennial vine Willdenow's spikemoss (*Selaginella willdenowil*) is named, developed the notion, among many others, that plant distribution patterns changed over time. von Humboldt, considered by many to be the father of ecology, further developed many of the Willdenow's notions, including the notion that barriers to plant dispersion were not absolute.

Another scientist who is considered a founder of plant ecology was Johannes E.B. Warming (1841–1924). Warming studied the tropical vegetation of Brazil and is best known for working on the relations between living plants and their surroundings. He is also recognized for his flagship text on plant ecology, *Plantesamfund* (1895). He also wrote *A Handbook of Systematic Botany* (1878).

Meanwhile, other naturalists were assuming important roles in the development of ecology. First and foremost among the naturalists was Charles Darwin. While working on his *The Origin of Species*, Darwin came across the writings of Thomas Malthus (1766–1834), who advanced the principle that populations grow in a geometric fashion, doubling at regular intervals until they outstrip the food supply—ultimately resulting in death and thus restraining population growth (Smith and Smith, 2006). In his autobiography written in 1876, Darwin wrote:

> In October 1838, that is, fifteen months after I had begun my systematic inquiry, I happened to read for amusement Malthus on *Population* and being well prepared to appreciate the struggle for existence which everywhere goes on from long-continued observation of the habits of animals and plants it at once struck me that under these circumstances favorable variations would tend to be preserved, and unfavorable ones to be destroyed. The results of this would be the formation of a new species. Here, then, I had at last got a theory by which to work.

During the period when Darwin was formulating his theories regarding the origin of species, Gregor Mendel (1822–1884) was studying the transmission of characteristics from one generation of pea plants to another. The work of Mendel and Darwin provided the foundation for *population genetics*, the study of evolution and adaptation.

Time marched on and the work of chemists such as Antoine-Laurent Lavoisier (who lost his head during the French Revolution) and Horace B. de Saussere, as well as the Austrian geologist Eduard Suess, who proposed the term *biosphere* in 1875, all set the foundations of the advanced work that followed.

Important Point: The Russian geologist Vladimir Vernadsky detailed the idea of biosphere in 1926.

Several strides forward in animal ecology, independent of plant ecology, were made during the 19th century that enabled the 20th-century scientists R. Hesse, Charles Elton, Charles Adams, and Victor Shelford to refine the discipline. Many early plant ecologists were "concerned with observing the patterns of organisms in nature, attempting to understand how patterns were formed and maintained by interactions with the physical environment" (Smith and Smith, 2006). Instead of looking for patterns, Frederic E. Clements (1874–1945) sought a system of organizing nature. Conducting his studies on vegetation in Nebraska, he postulated that the plant community behaves as a complex organism that grows and develops

through stages, resembling the development of an individual organism to a mature (climax) stage. Clements's theory of vegetation was roundly criticized by Arthur Tansley, a British ecologist, and others.

In 1935, Tansley coined the term *ecosystem*—the interactive system established between the group of living creatures (*biocoenosis*) and the environment in which they live (*biotype*). Tansley's ecosystem concept was adopted by the well-known and influential biology educator Eugene P. Odum. Along with his brother, Howard Odum, E.P. Odum wrote a textbook that educated multiple generations of biologists and ecologists in North American (including the author of this text). Odum is often called the "father of modern ecosystem ecology."

A new direction in ecology was given a boost in 1913 when Victor Shelford stressed the interrelationship of plants and animals. He conducted early studies on succession in the Indiana dunes and on experimental *physiological ecology*. Because of his work, ecology became a science of communities. His *Animal Communities in Temperate America*, written in 1913, was one of the first books to treat ecology as a separate science. E.P. Odum was one of Shelford's students.

Human ecology began to be studied in the 1920s. At about the same time, the study of populations split into the two fields of *population ecology* and *evolutionary ecology*. Closely associated with population ecology and evolutionary ecology is *community ecology*. At the same time, *physiological ecology* arose. Later, natural history observations spawned *behavioral ecology* (Smith and Smith, 2006).

The history of ecology has been tied to advances in biology, physics, and chemistry that have spawned new areas of study in ecology, such as landscape, conservation, restoration, and global ecology. Ecology has been rife with conflicts and opposing camps. The first major split in ecology was between plant ecology and animal ecology (Smith, 1996), which even led to a controversy over the use of the term *ecology*. Botanists dropped the initial "o" from *oecology*, the spelling in use at the time, and zoologists refused to use the term at all because of its perceived affiliation with botany. Other historical schisms involved organismal and individualist ecology, holism vs. reductionism, and theoretical vs. applied ecology.

To illustrate one way in which the ecosystem classification is used, a real-world example from the U.S. Department of Agriculture is provided below (USDA, 2007).

13.5.1 Example Ecosystem: Agroecosystem Model

What are the basic components of agroecosystems? Just as for natural ecosystems, they can be thought of as including the processes of primary producing, consumption, and decomposition interacting with abiotic environmental components and resulting in energy flow and nutrient cycling. Economic, social, and environmental factors must be added to this primary concept because of the human element that is so closely involved with agroecosystem creation and maintenance.

13.5.1.2 Agroecosystem Characteristics

Agricultural ecosystems, or agroecosystems, have been described by Odum (1984) as domesticated ecosystems. He states that they are in many ways intermediate between natural ecosystems (such as grasslands and forests) and fabricated ecosystems (cities). Agroecosystems are solar powered (as are natural systems) but differ from natural systems in that:

1. Auxiliary energy sources are used to enhance productivity; these sources are processed fuels along with animal and human labor.
2. Species diversity is reduced by human management to maximize yields of specific foodstuffs (plant or animal).
3. Dominant plant and animal species are under artificial rather than natural selection.
4. Control is external and goal oriented rather than internal via subsystem feedback as in natural ecosystems.

Agroecosystems do not exist without human intervention; therefore, creation of these ecosystems (and maintenance of them as well) is necessarily concerned with the human economic goals of production, productivity, and conservation. Agroecosystems are controlled, by definition, by management of ecological processes.

Crossley et al. (1984) addressed the possible use of agroecosystems as a unifying and in many ways clarifying concept for proper management of managed landscape units. All ecosystems are open; that is, they exchange biotic and abiotic elements with other ecosystems. Agroecosystems are extremely open—with major exports of primary and secondary production (plant and animal production) as well as increased opportunity for loss of nutrient elements. Because modern agroecosystems are entirely dependent on human intervention, they would not persist but for that intervention. It is for this reason that they are sometimes considered to be artificial systems as opposed to natural systems that do not require intervention to persist.

Definitions of agroecosystems often include the entire support base of energy and material subsidies, seeds, and chemicals, as well as a sociopolitical–economic matrix in which management decisions are made. Although this may be logical, Crossley (1984) preferred to designate an individual field as the agroecosystem because it is consistent with designating an individual forest catchment or lake as an ecosystem. He envisioned the *farm system* as consisting of a set of agroecosystems—fields with similar or different crops—together with support mechanisms and socioeconomic factors contributing to their management. Agroecosystems retain most if not all of the functional properties of natural ecosystems—nutrient conservation mechanisms, energy storage and use patterns, and regulation of biotic diversity.

13.5.1.3 Ecosystem Pattern and Process

Throughout the United States, the landscape consists of patches of natural ecosystems scattered (or imbedded) in a matrix of various agroecosystems and fabricated ecosystems. In fact, about three quarters of the land area of the United States is occupied by agroecosystems (USDA, 1982). The pattern created by this interspersion incorporates elements of the variability of structure and separation of functions among the various ecosystems. Pattern variables quantify the structure and relationships between systems; the process component implies functional relationships between and within the biotic and abiotic ecosystem components. Within agroecosystems, processes include:

- Enhanced productivity of producers through fertilization
- Improved productivity through selective breeding
- Management of pests by means of various control methods
- Management of various aspects of the hydrologic cycle
- Landforming

13.6 LEVELS OF ORGANIZATION

Odum (1983) suggested that the best way to delimit modern ecology is to consider the concept of *levels of organization*. Levels of organization can be simplified as shown in Figure 13.2. In this relationship, organs form an organism, organisms of a particular species form a population, and populations occupying a particular area form a community. Communities, interacting with nonliving or abiotic factors, separate into a natural unit to create a stable system known as the *ecosystem* (the major ecological unit), and the part of Earth in which the ecosystem operates is known as the *biosphere*. Tomera (1989) pointed

Organs → Organism → Population → Communities → Ecosystem → Biosphere

FIGURE 13.2 Levels of organization.

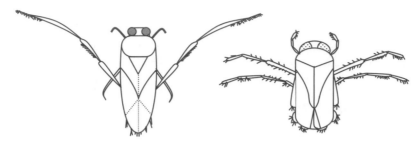

FIGURE 13.3 *Notonecta* (left) and *Corixa* (right). (Adapted from Odum, E.P., *Basic Ecology*, Saunders, Philadelphia, PA, 1983, p. 402.)

out that every community is influenced by a particular set of abiotic factors. Inorganic substances such as oxygen and carbon dioxide and some organic substances represent the abiotic part of the ecosystem.

The physical and biological environment in which an organism lives is referred to as its *habitat*; for example, the habitat of two common aquatic insects, the backswimmer (*Notonecta*) and the water boatman (*Corixa*), is the littoral zone of ponds and lakes (shallow, vegetation-choked areas) (Figure 13.3) (Odum, 1983). Within each level of organization of a particular habitat, each organism has a special role. The role the organism plays in the environment is referred to as its *niche*. A niche might be that the organism is food for some other organism or it is a predator of other organisms. Odum referred to an organism's niche as its "profession" (Odum, 1971). In other words, each organism has a job or role to fulfill in its environment. Although two different species might occupy the same habitat, niche separation based on food habits differentiates two species (Odum, 1983). Comparing the niches of the water backswimmer and the water boatman reveals such niche separation. The backswimmer is an active predator, while the water boatman feeds largely on decaying vegetation (McCafferty, 1981).

13.7 ECOSYSTEMS

An *ecosystem* is an area that includes all organisms therein and their physical environment. The ecosystem is the major ecological unit in nature. Living organisms and their nonliving environment are inseparably interrelated and interact upon each other to create a self-regulating and self-maintaining system. To create a self-regulating and self-maintaining system, ecosystems are homeostatic; that is, they resist any change through natural controls. These natural controls are important in ecology. This is especially the case because it is people through their complex activities who tend to disrupt natural controls.

As stated earlier, the ecosystem encompasses both the living and nonliving factors in a particular environment. The living or biotic part of the ecosystem is formed by two components: *autotrophic* and *heterotrophic*. The autotrophic (self-nourishing) component does not require food from its environment but can manufacture food from inorganic substances; for example, some autotrophic components (plants) manufacture needed energy through photosynthesis. Heterotrophic components, on the other hand, depend on autotrophic components for food.

The nonliving or abiotic part of the ecosystem is formed by three components: inorganic substances, organic compounds (link biotic and abiotic parts), and climate regime. Figure 13.4 is a simplified diagram of a few of the living and nonliving components of an ecosystem found in a freshwater pond.

An ecosystem is a cyclic mechanism in which biotic and abiotic materials are constantly exchanged through *biogeochemical cycles*, where *bio* refers to living organisms; *geo* to water, air, rocks, or solids; and *chemical* to the chemical composition of the Earth. Biogeochemical cycles are driven by energy, directly or indirectly from the sun.

Figure 13.4 depicts an ecosystem where biotic and abiotic materials are constantly exchanged. Producers construct organic substances through photosynthesis and chemosynthesis. Consumers and decomposers use organic matter as their food and convert it into abiotic components; that is, they dissipate energy fixed by producers through food chains. The abiotic part of the pond in Figure 13.4 is formed of inorganic and organic compounds dissolved in sediments such as carbon, oxygen, nitrogen, sulfur, calcium, hydrogen, and humic acids. Producers such as rooted plants and phytoplankton represent the biotic part. Fish, crustaceans, and insect larvae make up the consumers. Mayfly nymphs represent detrivores, which feed on organic detritus. Decomposers make up the final biotic part. They include aquatic bacteria and fungi, which are distributed throughout the pond.

As stated earlier, an ecosystem is a cyclic mechanism. From a functional viewpoint, an ecosystem can be analyzed in terms of several factors. The factors important in this study include the biogeochemical cycles and energy and food chains.

13.8 ENERGY FLOW IN THE ECOSYSTEM

Simply defined, *energy* is the ability or capacity to do work. For an ecosystem to exist, it must have energy. All activities of living organisms involve work, which is the expenditure of energy. This means the degradation of a

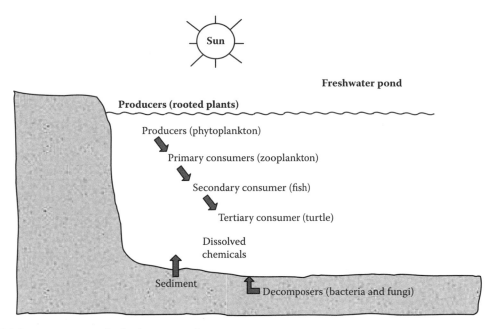

FIGURE 13.4 Major components of a freshwater pond ecosystem.

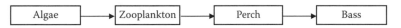

FIGURE 13.5 Aquatic food chain.

higher state of energy to a lower state. Two laws govern the flow of energy through an ecosystem: *the first and second laws of thermodynamics*. The first law, sometimes referred to as the *conservation law*, states that energy may not be created or destroyed. The second law states that no energy transformation is 100% efficient; that is, in every energy transformation, some energy is dissipated as heat. The term *entropy* is used as a measure of the nonavailability of energy to a system. Entropy increases with an increase in dissipation. Because of entropy, the input of energy in any system is higher than the output or work done; thus, the resultant efficiency is less than 100%.

The interaction of energy and materials in the ecosystem is important. Energy drives the biogeochemical cycles. Note that energy does not cycle as nutrients do in biogeochemical cycles. For example, when food passes from one organism to another, energy contained in the food is reduced systematically until all the energy in the system is dissipated as heat. Price (1984) referred to this process as "a unidirectional flow of energy through the system, with no possibility for recycling of energy." When water or nutrients are recycled, energy is required. The energy expended in this recycling is not recyclable.

As mentioned, the principal source of energy for any ecosystem is sunlight. Green plants, through the process of photosynthesis, transform the energy of the sun into carbohydrates, which are consumed by animals. This transfer of energy, again, is unidirectional—from producers to consumers. Often this transfer of energy to different organisms is referred to as a *food chain*. Figure 13.5 shows a simple aquatic food chain.

All organisms, alive or dead, are potential sources of food for other organisms. All organisms that share the same general type of food in a food chain are said to be at the same *trophic level* (nourishment or feeding level). Because green plants use sunlight to produce food for animals, they are the *producers*, or the first trophic level. The herbivores, which eat plants directly, are the *primary consumers*, or the second trophic level. The carnivores are flesh-eating consumers; they include several trophic levels from the third on up. At each transfer, a large amount of energy (about 80 to 90%) is lost as heat and waste; thus, nature normally limits food chains to four or five links. In aquatic ecosystems, however, food chains are commonly longer than those on land. The aquatic food chain is longer because several predatory fish may be feeding on the plant consumers. Even so, the built-in inefficiency of the energy transfer process prevents development of extremely long food chains.

Only a few simple food chains are found in nature. Most simple food chains are interlocked. This interlocking of food chains forms a *food web*. Most ecosystems support a complex food web. A food web involves animals that do not feed on one trophic level; for example, humans feed on both plants and animals. An organism in a food web may occupy one or more trophic levels. Trophic level is determined by an organism's role in its particular community, not by its species. Food chains and webs help to explain how energy moves through an ecosystem.

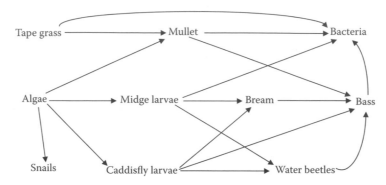

FIGURE 13.6 Aquatic food web.

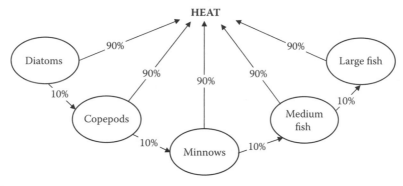

FIGURE 13.7 Simple food chain.

An important trophic level of the food web is comprised of the *decomposers*, which feed on dead plants or animals and play an important role in recycling nutrients in the ecosystem. Simply, there is no waste in ecosystems. All organisms, dead or alive, are potential sources of food for other organisms. An example of an aquatic food web is shown in Figure 13.6.

13.9 FOOD CHAIN EFFICIENCY

Earlier, we pointed out that energy from the sun is captured (via photosynthesis) by green plants and used to make food. Most of this energy is used to carry on the life activities of the plant. The rest of the energy is passed on as food to the next level of the food chain. Nature limits the amount of energy that is accessible to organisms within each food chain. Not all food energy is transferred from one trophic level to the next. Only about 10% (the 10% rule) of the amount of energy is actually transferred through a food chain. If we apply the 10% rule to the diatoms–copepods–minnows–medium fish–large fish food chain shown in Figure 13.7, we can predict that 1000 grams of diatoms produce 100 grams of copepods, which will produce 10 grams of minnows, which will produce 1 gram of medium fish, which, in turn, will produce 0.1 gram of large fish. Thus, only about 10% of the chemical energy available at each trophic level is transferred and stored in usable form at

the next level. The other 90% is lost to the environment as low-quality heat in accordance with the second law of thermodynamics.

13.10 ECOLOGICAL PYRAMIDS

In the food chain, from the producer to the final consumer, it is clear that a particular community in nature often consists of several small organisms associated with a smaller and smaller number of larger organisms. A grassy field, for example, has a larger number of grasses and other small plants, a smaller number of herbivores such as rabbits, and an even smaller number of carnivores such as the fox. The practical significance of this is that we must have several more producers than consumers.

This pound-for-pound relationship, which requires more producers than consumers, can be demonstrated graphically by building an *ecological pyramid*. In an ecological pyramid, separate levels represent the number of organisms at various trophic levels in a food chain, or bars are placed one above the other with a base formed by producers and the apex formed by the final consumer. The pyramid shape is formed due to a great amount of energy loss at each trophic level. The same is true if the corresponding biomass or energy substitutes numbers. Ecologists generally use three types of ecological pyramids: *energy*, *biomass*, and *numbers*. Obviously, differences exist among them, but here are some generalizations:

1. Energy pyramids must always be larger at the base than at the top (because of the second law of thermodynamics and the dissipation of energy as it moves from one trophic level to another).

2. Likewise, biomass pyramids (in which biomass is used as an indicator of production) are usually pyramid shaped. This is particularly true of terrestrial systems and aquatic ones dominated by large plants (marshes), in which consumption by heterotroph is low and organic matter accumulates with time. Biomass pyramids can sometimes be inverted. This is common in aquatic ecosystems, where the primary producers are microscopic planktonic organisms that multiply very rapidly, have very short life spans, and are subject to heavy grazing by herbivores. At any single point in time, the amount of biomass in primary producers is less than that in larger, long-lived animals that consume primary producers.

3. Numbers pyramids can have various shapes (and not be pyramids at all) depending on the sizes of the organisms that make up the trophic levels. In forests, the primary producers are large trees and the herbivore level usually consists of insects, so the base of the pyramid is smaller than the herbivore level above it. In grasslands, the number of primary producers (grasses) is much larger than that of the herbivores above (large grazing animals).

13.11 PRODUCTIVITY

As mentioned earlier, the flow of energy through an ecosystem begins with the fixation of sunlight by plants through photosynthesis. When evaluating an ecosystem, the measurement of photosynthesis is important. Ecosystems may be classified into highly productive or less productive; therefore, the study of ecosystems must involve some measure of the productivity of that ecosystem. Primary production is the rate at which the primary producers of the ecosystem capture and store a given amount of energy, in a specified time interval. In simpler terms, primary productivity is a measure of the rate at which photosynthesis occurs. Four successive steps in the production process are:

1. *Gross primary productivity*—The total rate of photosynthesis in an ecosystem during a specified interval

2. *Net primary productivity*—The rate of energy storage in plant tissues in excess of the rate of aerobic respiration by primary producers

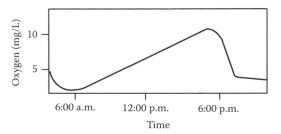

FIGURE 13.8 Diurnal oxygen curve for an aquatic ecosystem.

3. *Net community productivity*—The rate of storage of organic matter not used

4. *Secondary productivity*—The rate of energy storage at consumer levels

When attempting to comprehend the significance of the term *productivity* as it relates to ecosystems, it is wise to consider an example. Consider the productivity of an agricultural ecosystem such as a wheat field. Often, its productivity is expressed as the number of bushels produced per acre. This is an example of the harvest method for measuring productivity. For a natural ecosystem, several 1-m² plots are marked off, and the entire area is harvested and weighed to give an estimate of productivity as grams of biomass per square meter per given time interval. From this method, a measure of net primary production (net yield) can be measured.

Productivity, in both the natural and the cultured ecosystem, may vary considerably, not only between types of ecosystems but also within the same ecosystem. Several factors influence year-to-year productivity within an ecosystem. Such factors as temperature, availability of nutrients, fire, animal grazing, and human cultivation activities are directly or indirectly related to the productivity of a particular ecosystem.

Productivity can be measured in several different ways in the aquatic ecosystem; for example, the production of oxygen may be used to determine productivity. Oxygen content may be measured in several ways. One way is to measure it in the water every few hours for a period of 24 hours. During daylight, when photosynthesis is occurring, the oxygen concentration should rise. At night, the oxygen level should drop. The oxygen level can be measured by using a simple *x–y* graph. The oxygen level can be plotted on the *y*-axis with time plotted on the *x*-axis, as shown in Figure 13.8.

Another method of measuring oxygen production in aquatic ecosystems is to use light and dark bottles. Biochemical oxygen demand (BOD) bottles (300 mL) are filled with water to a particular height. One of the bottles is tested for the initial dissolved oxygen (DO), and then the other two bottles (one clear, one dark) are suspended in the water at the depth from which they were taken. After a 12-hour period, the bottles are collected and the

DO values for each bottle recorded. When the oxygen production is known, the productivity in terms of grams per meter per day can be calculated. In the aquatic ecosystem, pollution can have a profound impact on the productivity of the system.

13.12 POPULATION ECOLOGY

Webster's Third New International Dictionary defines *population* as "the total number or amount of things especially within a given area; the organisms inhabiting a particular area or biotype; and a group of interbreeding biotypes that represents the level of organization at which speciation begins." The concept of population is interpreted differently in various sciences. In *human demography*, a population is a set of humans in a given area. In *genetics*, a population is a group of interbreeding individuals of the same species which is isolated from other groups. In *population ecology*, a population is a group of individuals of the same species inhabiting the same area.

If we want to study the organisms in a slow-moving stream or stream pond, we have two options. We can study each fish, aquatic plant, crustacean, and insect, one by one, in which case we would be studying individuals. It would be relatively simple to do this if the subject were trout, but it would be difficult to separate and study each aquatic plant. The second option would be to study all of the trout, all of the insects of each specific kind, or all of a certain aquatic plant type in the stream or pond at the time of the study. When ecologists study a group of the same kind of individuals in a given location at a given time, they are investigating a *population*. When attempting to determine the population of a particular species, it is important to remember that time is a factor. Time is important because populations change, whether it be at various times of the day, in different seasons of the year, or from year to year.

Population density may change dramatically. As an example, a dam being closed off in a river midway through spawning season, with no provision allowed for fish movement upstream (a fish ladder), would drastically decrease the density of spawning salmon upstream. Along with the swift and sometimes unpredictable consequences of change, it can be difficult to draw exact boundaries between various populations. The population density or level of a species depends on *natality, mortality, immigration*, and *emigration*. Changes in population density are the result of both births and deaths. The birth rate of a population is *natality* and the death rate is *mortality*. In aquatic populations, two factors besides natality and mortality can affect density. In a run of returning salmon to their spawning grounds, for example, the density could vary as more salmon migrated in or as others left the run for their own spawning grounds. The arrival of new salmon to a population from other places is termed *immi-*

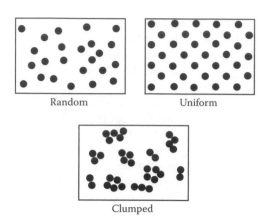

FIGURE 13.9 Basic patterns of distribution. (Adapted from Odum, E.P., *Fundamental of Ecology,* Saunders, Philadelphia, PA, 1971, p. 205.)

gration (ingress). The departure of salmon from a population is *emigration* (egress). Thus, natality and immigration increase population density, whereas mortality and emigration decrease it. The net increase in population is the difference between these two sets of factors.

Each organism occupies only those areas that can provide for its requirements, resulting in an irregular distribution. How a particular population is distributed within a given area has considerable influence on density. As shown in Figure 13.9, organisms in nature may be distributed in three ways. In a *random distribution*, there is an equal probability of an organism occupying any point in space, and each individual is independent of the others. In a *regular* or *uniform distribution*, in turn, organisms are spaced more evenly; they are not distributed by chance. Animals compete with each other and effectively defend a specific territory, excluding other individuals of the same species. In regular or uniform distribution, the competition between individuals can be quite severe and antagonistic to the point where the spacing generated is quite even. The most common distribution is the *contagious* or *clumped distribution*, where organisms are found in groups; this may reflect the heterogeneity of the habitat. Organisms that exhibit a contagious or clumped distribution may develop social hierarchies to live together more effectively. Animals within the same species have evolved many symbolic aggressive displays that carry meanings that not only are mutually understood but also prevent injury or death within the same species.

The size of animal populations is constantly changing due to natality, mortality, emigration, and immigration. The population size will increase if the natality and immigration rates are high, and it will decrease if the mortality and emigration rates are high. Each population has an upper limit on size, often referred to as the *carrying capacity*. Carrying capacity is the optimum number of individuals of a species that can survive in a specific area over time. Stated differently, the carrying capacity is the maximum number of species that can be supported in a bioregion. A pond may be able to support only a dozen frogs depending on the food resources for the frogs in the pond. If there were 30 frogs in the same pond, at least half of them would probably die because the pond environment would not have enough food for them to live. Carrying capacity is based on the quantity of food supplies, the physical space available, the degree of predation, and several other environmental factors.

The carrying capacity can be of two types: ultimate and environmental. The *ultimate carrying capacity* is the theoretical maximum density; that is, it is the maximum number of individuals of a species in a place that can support itself without rendering the place uninhabitable. The *environmental carrying capacity* is the actual maximum population density that a species maintains in an area. Ultimate carrying capacity is always higher than environmental. Ecologists have concluded that a major factor that affects population stability or persistence is *species diversity*. Species diversity is a measure of the number of species and their relative abundance.

If the stress on an ecosystem is small, the ecosystem can usually adapt quite easily. Moreover, even when severe stress occurs, ecosystems have a way of adapting. Severe environmental change to an ecosystem can result from such natural occurrences as fires, earthquakes, and floods and from people-induced changes such as land clearing, surface mining, and pollution. One of the most important applications of species diversity is in the evaluation of pollution. Stress of any kind will reduce the species diversity of an ecosystem to a significant degree. In the case of domestic sewage pollution, for example, the stress is caused by a lack of dissolved oxygen (DO) for aquatic organisms.

Ecosystems can and do change; for example, if a fire devastates a forest, it will grow back eventually because of *ecological succession*. Ecological succession is the observed process of change (a normal occurrence in nature) in the species structure of an ecological community over time. Succession usually occurs in an orderly, predictable manner. It involves the entire system. The science of ecology has developed to such a point that ecologists are now able to predict several years in advance what will occur in a given ecosystem. Scientists know, for example, that if a burned-out forest region receives light, water, nutrients, and an influx or immigration of animals and seeds, it will eventually develop into another forest through a sequence of steps or stages. Ecologists recognize two types of ecological succession: primary and secondary. The particular type that takes place depends on the condition at a particular site at the beginning of the process.

Primary succession, sometimes referred to as *bare-rock succession*, occurs on surfaces such as hardened volcanic lava, bare rock, and sand dunes, where no soil exists

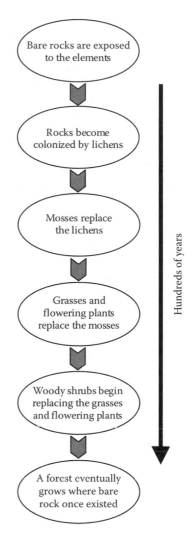

FIGURE 13.10 Bare-rock succession. (Adapted from Tomera, A.N., *Understanding Basic Ecological Concepts*, J. Weston Walch, Publisher, Portland, ME, 1989, p. 67.)

and where nothing has ever grown before (Figure 13.10). Obviously, to grow, plants require soil; thus, soil must form on the bare rock before succession can begin. Usually this soil formation process results from weathering. Atmospheric exposure—weathering, wind, rain, and frost—forms tiny cracks and holes in rock surfaces. Water collects in the rock fissures and slowly dissolves the minerals out of the surface of the rock. A pioneer soil layer is formed from the dissolved minerals and supports such plants as lichens. Lichens gradually cover the rock surface and secrete carbonic acid, which dissolves additional minerals from the rock. Eventually, mosses replace the lichens. Organisms known as *decomposers* move in and feed on dead lichen and moss. A few small animals such as mites and spiders arrive next. The result is a *pioneer community*, which is defined as the first successful integration of plants, animals, and decomposers into a bare-rock community.

After several years, the pioneer community builds up enough organic matter in its soil to be able to support rooted plants such as herbs and shrubs. Eventually, the pioneer community is crowded out and is replaced by a different environment. This, in turn, works to thicken the upper soil layers. The progression continues through several other stages until a mature or climax ecosystem is developed, several decades later. In bare-rock succession, each stage in the complex succession pattern dooms the stage that existed before it. *Secondary succession* is the most common type of succession. Secondary succession occurs in an area where the natural vegetation has been removed or destroyed but the soil is not destroyed; for example, succession that occurs in abandoned farm fields, known as *old-field succession*, illustrates secondary succession. An example of secondary succession can be seen in the Piedmont region of North Carolina. Early settlers of the area cleared away the native oak–hickory forests and cultivated the land. In the ensuing years, the soil became depleted of nutrients, thus reducing the fertility of the soil. As a result, farming ceased in the region a few generations later, and the fields were abandoned. Some 150 to 200 years after abandonment, the climax oak–hickory forest was restored.

In a stream ecosystem, growth is enhanced by biotic and abiotic factors, including:

- Ability to produce offspring
- Ability to adapt to new environments
- Ability to migrate to new territories
- Ability to compete with species for food and space to live
- Ability to blend into the environment so as not to be eaten
- Ability to find food
- Ability to defend itself from enemies
- Favorable light
- Favorable temperature
- Favorable dissolved oxygen (DO) content
- Sufficient water level

The biotic and abiotic factors in an aquatic ecosystem that reduce growth include:

- Predators
- Disease
- Parasites
- Pollution
- Competition for space and food
- Unfavorable stream conditions (e.g., low water levels)
- Lack of food

With regard to the stability of a freshwater ecosystem, the higher the species diversity, the greater the inertia and resilience of the ecosystem. When the species diversity is high within a stream ecosystem, a population within the stream can be out of control because of an imbalance between growth and reduction factors, but the ecosystem will remain stable at the same time. With regard to the instability of a freshwater ecosystem, recall that imbalance occurs when growth and reduction factors are out of balance; for example, when sewage is accidentally dumped into a stream, the stream ecosystem, via a self-purification process, responds and returns to normal. This process can be described as follows:

- Raw sewage is dumped into the stream.
- Available oxygen decreases as the detritus food chain breaks down the sewage.
- Some fish die at the pollution site and downstream.
- Sewage is broken down, washes out to sea, and is finally broken down in the ocean.
- Oxygen levels return to normal.
- Fish populations that were deleted are restored as fish about the spill reproduce and the young occupy the real estate formerly occupied by the dead fish.
- Populations all return to normal.

A shift in balance in the ecosystem of a stream (or in any ecosystem) similar to the one just described is a common occurrence. In this particular case, the stream responded (on its own) to the imbalance the sewage caused and through the self-purification process returned to normal. Recall that succession is the method by which an ecosystem either forms itself or heals itself; thus, we can say that a type of succession occurred in our polluted stream example because, in the end, it healed itself. More importantly, this healing process is a good thing; otherwise, long ago there would have been few streams on Earth suitable for much more than the dumping of garbage.

In summary, through research and observation, ecologists have found that the succession patterns in different ecosystems usually display common characteristics. First, succession brings about changes in the plant and animal members present. Second, organic matter increases from stage to stage. Finally, as each stage progresses, the tendency is toward greater stability or persistence. Remember, succession is usually predictable—that is, unless humans interfere.

13.13 STREAM GENESIS AND STRUCTURE

Consider the following: Early in the spring, on a snow- and ice-covered high alpine meadow, the water cycle continues. The main component of the cycle—water—has been held in reserve, literally frozen over the long dark winter months. Now, though, because of the longer, warmer spring days, the sun is higher, more direct, and of longer duration, and the frozen masses of water respond to the increased warmth. The melt begins with a single drop, then two, then more. As the snow and ice melt, the drops of water join a chorus that continues apparently unending; they fall from ice-bound lips to the bare rock and soil terrain below.

The terrain on which the snow melt falls is not like glacial till, which is an unconsolidated, heterogeneous mixture of clay, sand, gravel, and boulders dug out, ground out, and exposed by the force of a huge, slow, inexorably moving glacier. Instead, this soil and rock ground is exposed to the falling drops of snow melt because of a combination of wind and the tiny, enduring force exerted by drops of water as over seasons after season they collide with the thin soil cover, exposing the intimate bones of the Earth.

Gradually, the individual drops increase to a small rush—they join to form a splashing, rebounding, helter-skelter cascade, many separate rivulets that trickle then run their way down the face of the granite mountain. At an indented ledge halfway down the mountain slope, a pool forms whose beauty, clarity, and sweet iciness provides the visitor with an incomprehensible, incomparable gift—a blessing from Earth.

The mountain pool fills slowly, tranquil under the blue sky, reflecting the pines, snow, and sky around and above it, an open invitation to lie down and drink and to peer into the glass-clear, deep phantom blue–green eye, so clear that it seems possible to reach down over 50 feet and touch the very bowels of the mountain. The pool has no transition from shallow margin to depth; it is simply deep and pure. As the pool fills with more melt water, we wish to freeze time, to hold this place and this pool in its perfect state forever; it is such a rarity to us in our modern world. However, this cannot be, as Mother Nature calls, prodding, urging. For a brief instant, the water laps in the breeze against the outermost edge of the ridge, then a trickle flows over the rim. The giant hand of gravity reaches out and tips the overflowing melt onward and it continues the downward journey, following the path of least resistance to its next destination, several thousand feet below.

When the overflow, still high in altitude but with its rock-strewn bed bent downward, toward the sea, meets the angled, broken rocks below, it bounces, bursts, and mists its way against steep, V-shaped walls that form a small valley, carved out over time by water and the forces of the earth. Within the valley confines, the melt water has grown from drops to rivulets to a small mass of flowing water. It flows through what is at first a narrow opening, gaining strength, speed, and power as the V-shaped valley widens to form a U shape. The journey continues as the water mass picks up speed and tumbles over massive boulders, and then slows again.

At a larger but shallower pool, waters from higher elevations have joined the main body—from the hillsides, crevices, springs, rills, and mountain creeks. At the influent poolsides, all appears peaceful, quiet, and restful, but not far away, at the effluent end of the pool, gravity takes control again. The overflow is flung over the jagged lip and cascades downward several hundred feet, where the waterfall again brings its load to a violent, mist-filled meeting.

The water separates and joins repeatedly, forming a deep, furious, wild stream that calms gradually as it continues to flow over lands that are less steep. The waters widen into pools overhung by vegetation, surrounded by tall trees. The pure, crystalline waters have become progressively discolored on their downward journey, stained brown–black with humic acid and literally filled with suspended sediments; the once pure stream is now muddy.

The mass divides and flows in different directions over different landscapes. Small streams divert and flow into open country. Different soils work to retain or speed the waters, and in some places the waters spread out into shallow swamps, bogs, marshes, fens, or mires. Other streams pause long enough to fill deep depressions in the land and form lakes. For a time, the water remains and pauses in its journey to the sea, but this is only a short-term pause, as lakes are only a short-term resting place in the water cycle. The water will eventually move on, by evaporation or seepage into groundwater. Other portions of the water mass stay with the main flow, and the speed of flow changes to form a river, which braids its way through the landscape, heading for the sea. As it changes speed and slows, the river bottom changes from rock and stone to silt and clay. Plants begin to grow, stems thicken, and leaves broaden. The river is now full of life and the nutrients needed to sustain life. As the river courses onward, though, it meets its destiny when the flowing rich mass slows at last and finally spills into the sea.

Freshwater systems are divided into two broad categories: running waters (*lotic systems*) and standing waters (*lentic systems*). We concentrate here on lotic systems, although many of the principles described herein apply to other freshwater surface bodies as well, which are known by common names. Some examples include seeps, springs, brooks, branches, creeks, streams, and rivers. Again, because it is the best term to use in freshwater ecology, it is the stream we are concerned with here. Although there is no standard scientific definition of a stream, it is usually distinguished subjectively as follows: A stream is of intermediate size that can be waded from one side to the other.

Physical processes involved in the formation of a stream are important to the ecology of the stream, because stream channel and flow characteristics directly influence the functioning of the stream ecosystem and the biota found therein. Thus, in this section, we discuss the pathways of water flow contributing to stream flow; namely, we discuss precipitation inputs as they contribute to flow. We also discuss stream flow discharge, transport of material, characteristics of stream channels, stream profile, sinuosity, the floodplain, pool–riffle sequences, and depositional features—all of which directly or indirectly impact the ecology of the stream.

13.13.1 Water Flow in a Stream

Most elementary students learn early in their education process that water on Earth flows downhill—from land to the sea; however, they may or may not be told that water flows downhill toward the sea by various routes. The route (or pathway) that we are primarily concerned with here is the surface water route taken by surface water runoff. Surface runoff is dependent on various factors; for example, climate, vegetation, topography, geology, soil characteristics, and land use all determine how much surface runoff occurs compared with other pathways.

The primary source (input) of water to total surface runoff is, of course, precipitation. This is the case even though a substantial portion of all precipitation input returns directly to the atmosphere by *evapotranspiration*, which is a combination process, as the name suggests, whereby water in plant tissue and in the soil evaporates and transpires to water vapor in the atmosphere. A substantial portion of precipitation input returns directly to the atmosphere by evapotranspiration. It is important to point out that when precipitation occurs some rainwater is intercepted by vegetation, from which it evaporates, never reaching the ground or being absorbed by plants. A large portion of the rainwater that reaches the ground, lakes, and streams also evaporates directly back to the atmosphere.

Although plants display a special adaptation to minimize transpiration, plants still lose water to the atmosphere during the exchange of gases necessary for photosynthesis. Notwithstanding the large percentage of precipitation that evaporates, rain or melt water that reaches the ground surface follows several pathways to reach a stream channel or groundwater.

Soil can absorb rainfall to its *infiltration capacity* (i.e., to its maximum rate). During a rain event, this capacity decreases. Any rainfall in excess of infiltration capacity accumulates on the surface. When this surface water exceeds the depression storage capacity of the surface, it moves as an irregular sheet of overland flow. In arid areas, overland flow is likely because of the low permeability of the soil. Overland flow is also likely when the surface is frozen or when human activities have rendered the land surface less permeable. In humid areas, where infiltration capacities are high, overland flow is rare.

In rain events, where the infiltration capacity of the soil is not exceeded, rain penetrates the soil and eventually

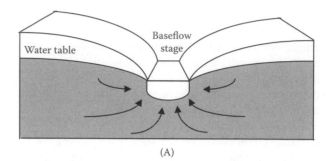

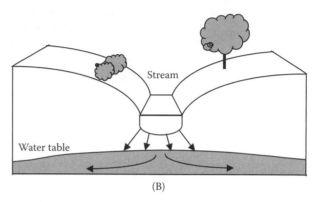

FIGURE 13.11 (A) Cross-section of a gaining stream; (B) cross-section of a losing stream.

reaches the groundwater—from which it discharges to the stream slowly and over a long period. This phenomenon helps to explain why stream flow through a dry-weather region remains constant; the flow is continuously augmented by groundwater. This type of stream is known as a *perennial stream*, as opposed to an *intermittent* one, because the flow continues during periods of no rainfall.

When a stream courses through a humid region, it is fed water via the water table, which slopes toward the stream channel. Discharge from the water table into the stream accounts for flow during periods without precipitation and explains why this flow increases, even without tributary input, as one proceeds downstream. Such streams are called *gaining* or *effluent*, as opposed to *losing* or *influent streams*, which lose water into the ground (see Figure 13.11). The same stream can shift between gaining and losing conditions along its course because of changes in underlying strata and local climate.

13.13.2 Stream-Water Discharge

The current velocity (speed) of water (driven by gravitational energy) in a channel varies considerably within the cross-section of a stream due to friction with the bottom and sides, with sediment, and the atmosphere, as well as to sinuosity (bending or curving) and obstructions. Highest velocities, obviously, are found where friction is least, generally at or near the surface and near the center of the channel. In deeper streams, current velocity is greatest just below the surface due to friction with the atmosphere; in shallower streams, current velocity is greatest at the surface due to friction with the bed. Velocity decreases as a function of depth, approaching zero at the substrate surface.

13.13.3 TRANSPORT OF MATERIAL

Water flowing in a channel may exhibit *laminar flow* (parallel layers of water shear over one another vertically) or *turbulent flow* (complex mixing). In streams, laminar flow is uncommon, except at boundaries where flow is very low and in groundwater. The flow in streams is generally turbulent. Turbulence exerts a shearing force, referred to as the *bed load,* that causes particles to move along the streambed by pushing, rolling, and skipping . This same shear causes turbulent eddies that entrain particles in suspension, referred to as the *suspended load* (particle size under 0.06 mm).

Entrainment is the incorporation of particles when stream velocity exceeds the *entraining velocity* for a particular particle size. The entrained particles in suspension (suspended load) also include fine sediment, primarily clays, silts, and fine sands that require only low velocities and minor turbulence to remain in suspension. These are referred to as the *wash load* (particle size under 0.002 mm). Thus, the suspended load includes the wash load and coarser materials (at lower flows). Together, the suspended load and bed load constitute the *solids load*. It is important to note that in bedrock streams the bed load will be a lower fraction than in alluvial streams where channels are composed of easily transported material.

A substantial amount of material is also transported as the *dissolved load*. Solutes are generally derived from chemical weathering of bedrock and soils, and their contribution is greatest in subsurface flows and in regions of limestone geology. The relative amount of material transported as solute rather than solid load depends on basin characteristics, lithology (i.e., the physical character of rock), and the hydrologic pathways. In areas of very high runoff, the contribution of solutes approaches or exceeds sediment load, whereas in dry regions, sediments make up as much as 90% of the total load.

Deposition occurs when *stream competence*—which refers to the largest particles that a stream can move, which in turn depends on the critical erosion competent of velocity—falls below a given velocity. Simply stated: The size of the particle that can be eroded and transported is a function of current velocity.

Sand particles are the most easily eroded. The greater the mass of larger particles (e.g., coarse gravel) the higher the initial current velocities must be for movement. Smaller particles (silts and clays) require even greater initial velocities, however, because of their cohesiveness and because they present smaller, streamlined surfaces to

the flow. Once in transport, particles will continue in motion at somewhat slower velocities than initially required to initiate movement and will settle at still lower velocities.

Particle movement is determined by size, flow conditions, and mode of entrainment. Particles over 0.02 mm (medium-coarse sand size) tend to move by rolling or sliding along the channel bed as *traction load*. When sand particles fall out of the flow, they move by *saltation* or repeated bouncing. Particles under 0.06 mm (silt) move as suspended load and particles under 0.002 (clay) as wash load. Unless the supply of sediments becomes depleted the concentration and amount of transported solids increase. Discharge is usually too low, throughout most of the year, to scrape or scour, shape channels, or move significant quantities of sediment in all but sand-bed streams, which can experience change more rapidly. During extreme events, the greatest scour occurs and the amount of material removed increases dramatically.

Sediment inflow into streams can be both increased and decreased because of human activities. Poor agricultural practices and deforestation greatly increase erosion. Fabricated structures such as dams and channel diversions, on the other hand, can greatly reduce sediment inflow.

13.13.4 CHARACTERISTICS OF STREAM CHANNELS

Flowing waters (rivers and streams) determine their own channels, and these channels exhibit relationships attesting to the operation of physical laws—laws that are not, as of yet, fully understood. The development of stream channels and entire drainage networks and the existence of various regular patterns in the shape of channels indicate that streams are in a state of dynamic equilibrium between erosion (sediment loading) and deposition (sediment deposit) and are governed by common hydraulic processes. Because channel geometry is four dimensional, with a long profile, cross-section, depth, and slope, and because these mutually adjust over a time scale as short as years and as long as centuries or more, cause-and-effect relationships are difficult to establish. Other variables that are presumed to interact as the stream achieves its graded state include width and depth, velocity, size of sediment load, bed roughness, and the degree of braiding (sinuosity).

13.13.5 STREAM PROFILES

Mainly because of gravity, most streams exhibit a downstream decrease in gradient along their length. Beginning at the headwaters, the steep gradient becomes less so as one proceeds downstream, resulting in a concave longitudinal profile. Although diverse geography provides for almost unlimited variation, a lengthy stream that originates in a mountainous area typically comes into existence as a series of springs and rivulets; these coalesce into a fast-flowing, turbulent mountain stream, and the addition of tributaries results in a large and smoothly flowing river that winds through the lowlands to the sea.

When studying a stream system of any length, it becomes readily apparent (almost from the start) that what we are studying is a body of flowing water that varies considerably from place to place along its length. As an example, increases in discharge cause corresponding changes in the width, depth, and velocity of the stream. In addition to physical changes that occur from location to location along the course of a stream, a legion of biological variables correlate with stream size and distance downstream. The most apparent and striking changes are in steepness of slope and in the transition from a shallow stream with large boulders and a stony substrate to a deep stream with a sandy substrate. The particle size of bed material is also variable along the course of a stream. The particle size usually shifts from an abundance of coarser material upstream to mainly finer material in downstream areas.

13.13.6 Sinuosity

Unless forced by humans in the form of heavily regulated and channelized streams, straight channels are uncommon. Stream flow creates distinctive landforms composed of straight (usually in appearance only), meandering, and braided channels; channel networks; and floodplains. Simply put, flowing water will follow a sinuous course. The most commonly used measure is the *sinuosity index* (SI). Sinuosity equals 1 in straight channels and more than 1 in sinuous channels. *Meandering* is the natural tendency for alluvial channels and is usually defined as an arbitrarily extreme level of sinuosity, typically a SI greater than 1.5. Many variables affect the degree of sinuosity.

Even in many natural channel sections of a stream course that appear straight, meandering occurs in the line of maximum water or channel depth (known as the *thalweg*). Keep in mind that a stream has to meander; that is how they renew themselves. By meandering, they wash plants and soil from the land into their waters, and these serve as nutrients for the plants in the rivers. If rivers are not allowed to meander, if they are *channelized*, the amount of life they can support will gradually decrease. That means fewer fish, as well as fewer bald eagles, herons, and other fishing birds (Spellman, 1996).

Meander flow follows predictable patterns and causes regular regions of erosion and deposition (Figure 13.12). The streamlines of maximum velocity and the deepest part of the channel lie close to the outer side of each bend and cross over near the point of inflection between the banks. A huge elevation of water at the outside of a bend causes a helical flow of water toward the opposite bank. In addition, a separation of surface flow causes a back eddy. The

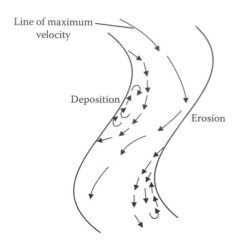

FIGURE 13.12 A meandering reach.

result is zones of erosion and deposition and explains why point bars develop in a downstream direction in depositional zones.

13.13.7 Bars, Riffles, and Pools

Implicit in the morphology and formation of meanders are *bars*, *riffles*, and *pools*. Bars develop by deposition in slower, less competent flow on either side of the sinuous mainstream. Onward moving water, depleted of bed load, regains competence and shears a pool in the meander, reloading the stream for the next bar. Alternating bars migrate to form riffles (see Figure 13.13). As stream flow continues along its course, a pool–riffle sequence is formed. The riffle is a mound or hillock, and the pool is a depression.

13.13.8 The Floodplain

A stream channel influences the shape of the valley floor through which it courses. The self-formed, self-adjusted flat area near the stream is the *floodplain*, which loosely describes the valley floor prone to periodic inundation during overbank discharges. What is not commonly known is that valley flooding is a regular and natural behavior of the stream. The aquatic community of a stream has several unique characteristics. The aquatic community operates under the same ecologic principles as terrestrial ecosystems, but the physical structure of the community is more isolated and exhibits limiting factors that are very different than the limiting factors of a terrestrial ecosystem. Certain materials and conditions are necessary for the growth and reproduction of organisms. If, for example, a farmer plants wheat in a field containing too little nitrogen, it will stop growing when it has used up the available nitrogen, even if the requirements of wheat for oxygen, water, potassium, and other nutrients are met. In this particular case, nitrogen is said to be the *limiting factor*. A

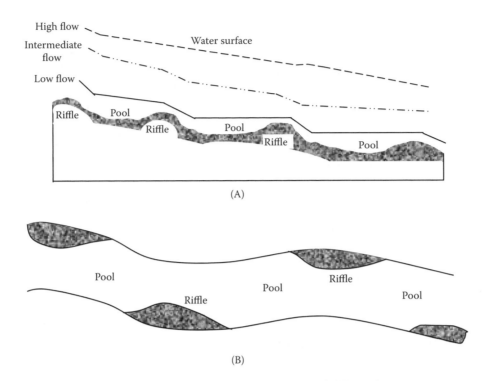

FIGURE 13.13 (A) Longitudinal profile of a riffle-pool sequence; (B) plain view of riffle-pool sequence.

limiting factor is a condition or a substance (the resource in shortest supply) that limits the presence and success of an organism or a group of organisms in an area.

Even the smallest mountain stream provides an astonishing number of different places for aquatic organisms to live, or *habitats*. If it is a rocky stream, every rock of the substrate provides several different habitats. On the side facing upriver, organisms with special adaptations, such as being able to cling to rock, do well. On the side that faces downriver, a certain degree of shelter is provided from current, but organisms can still hunt for food. The top of a rock, if it contacts air, is a good place for organisms that cannot breathe underwater and need to surface now and then. Underneath the rock is a popular place for organisms that hide to prevent predation. Normal stream life can be compared to that of a balanced aquarium (ASTM, 1969); that is, nature continuously strives to provide clean, healthy, normal streams. This is accomplished by maintaining the flora and fauna of the stream in a balanced state. Nature balances stream life by maintaining both the number and the type of species present in any one part of the stream. Such balance prevents an overabundance of one species compared to another. Nature structures the stream environment so plant and animal life is dependent on the existence of others within the stream.

As mentioned, lotic (washed) habitats are characterized by continuously running water or current flow. These running water bodies typically have three zones: riffle, run, and pool. The *riffle zone* contains faster flowing, well-oxygenated water, with coarse sediments. In the riffle zone,

the velocity of current is great enough to keep the bottom clear of silt and sludge, thus providing a firm bottom for organisms. This zone contains specialized organisms adapted to living in running water; for example, organisms adapted to living in fast streams or rapids (e.g., trout) have streamlined bodies that aid in their respiration and in obtaining food (Smith, 1996). Stream organisms that live under rocks to avoid the strong current have flat or streamlined bodies. Others have hooks or suckers to cling or attach to a firm substrate to avoid being washed away by the strong current. The *run zone* (or intermediate zone) is the slow-moving, relatively shallow part of the stream with moderately low velocities and little or no surface turbulence. The *pool zone* of the stream is usually a deeper water region where the velocity of the water is reduced, and silt and other settling solids provide a soft bottom (more homogeneous sediments), which is unfavorable for sensitive bottom dwellers. Decomposition of some of these solids leads to a reduced amount of dissolved oxygen (DO). Some stream organisms spend some of their time in the rapids part of the stream, and at other times they can be found in the pool zone. Trout, for example, typically spend about the same amount of time in the rapid zone pursuing food as they do in the pool zone pursuing shelter.

Organisms are sometimes classified based on their mode of life:

- *Benthos* (mud dwellers)—The term originates from the Greek word for bottom and broadly includes aquatic organisms living on the bottom or on submerged vegetation. They live under and on rocks and in the sediments. A shallow sandy bottom has sponges, snails, earthworms, and some insects. A deep, muddy bottom will support clams, crayfish, and nymphs of damselflies, dragonflies, and mayflies. A firm, shallow, rocky bottom has nymphs of mayflies and stoneflies and larvae of water beetles.

- *Periphytons or aufwuchs*—The first term usually refers to microfloral growth upon substrata (i.e., benthic-attached algae). The second term, *aufwuchs* (pronounced OWF-vooks; German for "growth upon") refers to the fuzzy, sort of furry-looking, slimy green coating that attaches or clings to stems and leaves of rooted plants or other objects projecting above the bottom without penetrating the surface. It consists not only of algae such as Chlorophyta but also diatoms, protozoa, bacteria, and fungi.
- *Plankton* (drifters)—These are small, mostly microscopic plants and animals that are suspended in the water column; movement depends on water currents. They usually float in the direction of current. Two types of plankton have been identified: (1) *Phytoplankton* are assemblages of small plants (algae) and have limited locomotion abilities; they are subject to movement and distribution by water movements. (2) *Zooplankton* are animals that are suspended in water and have limited means of locomotion. Examples of zooplankton include crustaceans, protozoa, and rotifers.
- *Nektons or pelagic organisms* (capable of living in open waters)—Nektons are distinct from other plankton in that they are capable of swimming independently of turbulence. They are swimmers that can navigate against the current. Examples of nektons include fish, snakes, diving beetles, newts, turtles, birds, and large crayfish.
- *Neustons*—They are organisms that float or rest on the surface of the water (never break water tension). Some varieties can spread out their legs so the surface tension of the water is not broken (e.g., water striders) (see Figure 13.14).
- *Madricoles*—Organisms that live on rock faces in waterfalls or seepages.

In a stream, the rocky substrate is the home for many organisms; thus, we need to know something about the particles that make up the substrate. Namely, we need to know how to measure the particles so we can classify them by size.

Substrate particles are measured with a metric ruler in centimeters (cm). Because rocks can be long and narrow, we measure them twice: first the width then the

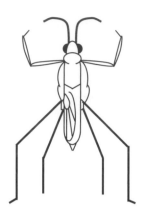

FIGURE 13.14 Water strider. (Adapted from APHA, *Standard Methods for the Examination of Water and Wastewater*, 15th ed. Copyright © 1981 by the American Public Health Association, the American Water Works Association, and the Water Pollution Control Federation.)

length. By adding the width to the length and dividing by two, we obtain the average size of the rock.

It is important to randomly select the rocks we wish to measure; otherwise, we would tend to select larger rocks, more colorful rocks, or those with unusual shapes. Instead, we should just reach down and collect those rocks in front of us and within easy reach. We then measure each rock. Upon completion of measurement, each rock should be classified. Ecologists have developed a standard scale (Wentworth scale) for size categories of substrate rock and other mineral materials:

Boulder	>256 mm
Cobble	64–256 mm
Pebble	16–64 mm
Gravel	2–16 mm
Sand	0.0625—2 mm
Silt	0.0039–0.0625 mm
Clay	<0.0039 mm

Organisms that live in, on, or under rocks or in small spaces occupy what is known as a *microhabitat*. Some organisms make their own microhabitats; many caddisflies build cases about themselves to use as shelter.

Rocks are not the only physical features of streams where aquatic organisms can be found. Fallen logs and branches (commonly referred to as *large woody debris*, or LWD) provide an excellent place for some aquatic organisms to burrow into and for others to attach themselves to, as they might to a rock. They also create areas where small detritus such as leaf litter can pile up underwater. These piles of leaf litter are excellent shelters for many organisms, including large, fiercely predaceous larvae of dobsonflies.

Another important aquatic organism habitat is found in the matter, or *drift*, that floats along downstream. Drift

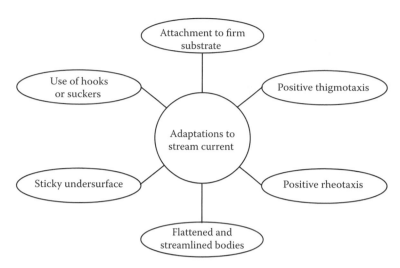

FIGURE 13.15 Adaptations to stream current.

is important because it is the main source of food for many fish. It may include insects such as mayflies (Ephemeroptera), some true flies (Diptera), and some stoneflies (Plecoptera) and caddisflies (Trichoptera). In addition, dead or dying insects and other small organisms, terrestrial insects that fall from the trees, leaves, and other matter are common components of drift. Among the crustaceans, amphipods (small crustaceans) and isopods (small crustaceans including sow bugs and gribbles) also have been reported in the drift.

13.13.9 ADAPTATIONS TO STREAM CURRENT

Current in streams is the outstanding feature of streams and the major factor limiting the distribution of organisms. The current is determined by the steepness of the bottom gradient, the roughness of the streambed, and the depth and width of the streambed. The current in streams has promoted many special adaptations by stream organisms. Odum (1971) listed these adaptations as follows (Figure 13.15):

1. *Attachment to a firm substrate*—Attachment is to stones, logs, leaves, and other underwater objects such as discarded tires, bottles, pipes, etc. Organisms in this group are primarily composed of the primary producer plants and animals, such as green algae, diatoms, aquatic mosses, caddisfly larvae, and freshwater sponges.
2. *Use of hooks or suckers*—These organisms have the unusual ability to remain attached and withstand even the strongest rapids. The larvae of *Simulium* and *Blepharocera* are examples.
3. *A sticky undersurface*—Snails and flatworms are examples of organisms that are able to use their sticky undersurfaces to adhere to underwater surfaces.

4. *Flattened and streamlined bodies*—All macroconsumers have streamlined bodies; that is, the body is broad in front and tapers posteriorly to offer minimum resistance to the current. All nektons such as fish, amphibians, and insect larvae exhibit this adaptation. Some organisms have flattened bodies, which enable them to stay under rocks and in narrow places. Examples are water penny, a beetle larva, mayfly, and stone fly nymphs.
5. *Positive rheotaxis* (*rheo*, "current"; *taxis*, "arrangement")—An inherent behavioral trait of stream animals (especially those capable of swimming) is to orient themselves upstream and swim against the current.
6. *Positive thigmotaxis* (*thigmo*, "touch or contact")—Another inherent behavior pattern for many stream animals is to cling close to a surface or keep the body in close contact with the surface. This is the reason why stonefly nymphs (when removed from one environment and placed into another) will attempt to cling to just about anything, including each other.

It would take an entire text to describe the great number of adaptations made by aquatic organisms to their surroundings in streams. For our purposes here, we cover only those special adaptations that are germane to our discussion. The important thing to remember is that an aquatic organism can adapt to its environment in several basic ways.

13.13.10 TYPES OF ADAPTIVE CHANGES

Adaptive changes are classed as genotypic, phenotypic, behavioral, or ontogenic:

- *Genotypic changes* tend to be great enough to separate closely related animals into species, such as mutations or recombination of genes. A salmonid is an example that has evolved a subterminal mouth (i.e., below the snout) to eat from the benthos.
- *Phenotypic changes* are the changes that an organism might make during its lifetime to better utilize its environment (e.g., a fish that changes sex from female to male because of an absence of males).
- *Behavioral changes* have little to do with the body structure or type; for example, a fish might spend more time under an overhang to hide from predators.
- *Ontogenetic change* takes place as an organism grows and matures (e.g., a Coho salmon that inhabits streams when young and migrates to the sea when older, changing its body chemistry to allow it to tolerate saltwater).

13.13.11 SPECIFIC ADAPTATIONS

Specific adaptations observed in aquatic organisms include mouths, shape, color, aestivation, and schooling:

- *Mouths*—Aquatic organisms such as fish change mouth shape (morphology) depending on the food the fish eats. The arrangement of the jawbones and even other head bones; the length and width of gill rakers; the number, shape, and location of teeth; and the barbels all change to allow fish to eat just about anything found in a stream.
- *Shape*—Changes in shape allow fish to do different things in the water. Some organisms have body shapes that push them down in the water, against the substrate, and allow them to hold their place against even strong current (e.g., chubs, catfish, dace, sculpins). Other organisms (e.g., bass, perch, pike, trout, sunfish) have evolved an arrangement and shape of fins that allow them to lurk without moving so they can lunge suddenly to catch their prey.
- *Color*—Color may change within hours for camouflage or in a matter of days, or it may be genetically predetermined. Fish tend to turn dark in clear water and pale in muddy water.
- *Aestivation*—Aestivation helps fishes survive in arid desert climates, where streams may dry up from time to time. It is the ability of some fishes to burrow into the mud and wait out the dry period.
- *Schooling*—Schooling serves as protection for many fish, particularly those that are subject to predation.

13.14 BENTHIC LIFE

The benthic habitat is found in the streambed, or benthos. As mentioned, the streambed is comprised of various physical and organic materials where erosion and deposition are a continuous characteristic. Erosion and deposition may occur simultaneously and alternately at different locations in the same streambed. Where channels are exceptionally deep and taper slowly to meet the relatively flattened streambed, habitats may form on the slopes of the channel. These habitats are referred to as *littoral habitats*. Shallow channels may dry up periodically in accordance with weather changes. The streambed is then exposed to open air and may take on the characteristics of a wetland.

Silt and organic materials settle and accumulate in the streambed of slowly flowing streams. These materials decay and become the primary food resource for the invertebrates inhabiting the streambed. Productivity in this habitat depends on the breakdown of these organic materials by herbivores. Bottom-dwelling organisms do not use all of the organic materials; a substantial amount becomes part of the streambed in the form of peat.

In faster moving streams, organic materials do not accumulate so easily. Primary production occurs in a different type of habitat found in the riffle regions with shoals and rocky regions for organisms to adhere to. Plants that can root themselves into the streambed dominate these regions. By plants, we are referring mostly to forms of algae, often microscopic and filamentous, that can cover rocks and debris that have settled into the streambed during summer months.

Note: The green, slippery slime on rocks in a streambed is representative of this type of algae.

Although the filamentous algae seem well anchored, strong currents can easily lift the algae from the streambed and carry them downstream, where they become a food resource for low-level consumers. One factor that greatly influences the productivity of a stream is the width of the channel; a direct relationship exists between stream width and richness of bottom organisms. Bottom-dwelling organisms are very important to the ecosystem as they provide food for other, larger benthic organisms by consuming detritus.

13.15 BENTHIC PLANTS AND ANIMALS

Vegetation is not common in the streambed of slow-moving streams; however, vegetation may anchor along the banks. Algae (mainly green and blue–green) as well as common types of water moss attach themselves to rocks in fast-moving streams. Mosses and liverworts often climb up the sides of the channel onto the banks, as well. Some plants similar to the reeds of wetlands with long stems and narrow leaves are able to maintain roots and withstand the current. Aquatic insects and invertebrates dominate slow-moving streams. Most aquatic insects are in their larval and nymph forms such as the blackfly, caddisfly, and stonefly. Adult water beetles and waterbugs are also abundant. Insect larvae and nymphs are the primary food source for many fish species, including American eel and brown bullhead catfish. Representatives of crustaceans, rotifers, and nematodes (flatworms) are sometimes present. The abundance of leeches, worms, and mollusks (especially freshwater mussels) varies with stream conditions but generally favors low phosphate conditions. Larger animals found in slow-moving streams and rivers include newts, tadpoles, and frogs. As mentioned, the important characteristic of all life in streams is adaptability to withstand currents.

13.16 BENTHIC MACROINVERTEBRATES

The emphasis on aquatic insect studies, which have expanded exponentially in the last several decades, has been largely an ecological one. Freshwater macroinvertebrates are ubiquitous; even polluted waters contain some representatives of this diverse and ecologically important group of organisms. Benthic macroinvertebrates are aquatic organisms without backbones that spend at least a part of their life cycle on the stream bottom. Examples include aquatic insects, such as stoneflies, mayflies, caddisflies, midges, and beetles, as well as crayfish, worms, clams, and snails. Most hatch from eggs and mature from larvae to adults. The majority of the insects spend their larval phase on the river bottom, and, after a few weeks to several years, emerge as winged adults. The aquatic beetles, true bugs, and other groups remain in the water as adults. Macroinvertebrates typically collected from the stream substrate are either aquatic larvae or adults.

In practice, stream ecologists observe indicator organisms and their responses to determine the quality of the stream environment. A number of methods can be used to determine water quality based on biologic characteristics. A wide variety of indicator organisms (biotic groups) can be used for biomonitoring. Those used most often include algae, bacteria, fish, and macroinvertebrates.

Notwithstanding their popularity, in this text we discuss benthic macroinvertebrates because they offer a number of advantages:

- They are ubiquitous, so they are affected by perturbations in many different habitats.
- They are species rich, so the large number of species produces a range of responses.
- They are sedentary, so they stay put, which allows determination of the spatial extent of a perturbation.

- They are long lived, thus we can follow temporal changes in abundance and age structure.
- They integrate conditions temporally, so like any biotic group they provide evidence of conditions over long periods.

In addition, benthic macroinvertebrates are preferred as bioindicators because they are easily collected and handled by samplers; they require no special culture protocols. They are visible to the naked eye and samplers can easily distinguish their characteristics. They have a variety of fascinating adaptations to stream life. Certain benthic macroinvertebrates have very special tolerances and thus are excellent specific indicators of water quality. Useful benthic macroinvertebrate data are easy to collect without expensive equipment. The data obtained by macroinvertebrate sampling can serve to indicate the need for additional data collection, possibly including water analysis and fish sampling.

In short, we base the focus of this discussion on benthic macroinvertebrates (with regard to water quality in streams and lakes) simply because some cannot survive in polluted water while others can survive or even thrive in polluted water. In a healthy stream, the benthic community includes a variety of pollution-sensitive macroinvertebrates. In an unhealthy stream or lake, only a few types of nonsensitive macroinvertebrates may be present; thus, the presence or absence of certain benthic macroinvertebrates is an excellent indicator of water quality.

Moreover, it may be difficult to identify stream or lake pollution with water analysis, which can only provide information for the time of sampling (a snapshot of time). Even the presence of fish may not provide information about a polluted stream because fish can move away to avoid polluted water and then return when conditions improve. In contrast, most benthic macroinvertebrates cannot move to avoid pollution; thus, a macroinvertebrate sample may provide information about pollution that is not present at the time of sample collection.

Obviously, before we can use benthic macroinvertebrates to gauge water quality in a stream (or for any other reason), we must be familiar with the macroinvertebrates that are commonly used as bioindicators. Samplers must be aware of basic insect structures before they can classify the macroinvertebrates they collect. Structures that should be stressed include head, eyes (compound and simple), antennae, mouth (no emphasis on parts), segments, thorax, legs and leg parts, gills, abdomen, etc. Samplers also should be familiar with insect metamorphosis—both complete and incomplete—as most of the macroinvertebrates collected are larval or nymph stages.

Note: Information on basic insect structures is beyond the scope of this text, so we highly recommend the standard guide to aquatic insects of North

America, *An Introduction to the Aquatic Insects of North America*, 3rd ed., edited by R.W. Merritt and K.W. Cummins (Kendall/Hunt Publishing, 1996).

13.16.1 Identification of Benthic Macroinvertebrates

Before identifying and describing the key benthic macroinvertebrates significant to water and wastewater operators, it is important first to provide foundational information. We characterize benthic macroinvertebrates using two important descriptive classifications: trophic groups and mode of existence. In addition, we discuss their relationship in the food web—that is, what, or whom, they eat:

- *Trophic groups*. Of the trophic groups (i.e., feeding groups) that Merritt and Cummins (1996) identified for aquatic insects, only five are likely to be found in a stream using typical collection and sorting methods:

  Shredders—These have strong, sharp mouthparts that allow them to shred and chew coarse organic material such as leaves, algae, and rooted aquatic plants. These organisms play an important role in breaking down leaves or larger pieces of organic material to a size that can be used by other macroinvertebrates. Shredders include certain stonefly and caddisfly larvae, sowbugs, scuds, and others.

  Collectors—These gather the very finest suspended matter in the water. To do this, they often sieve the water through rows of tiny hairs. These sieves of hairs may be displayed in fans on their heads (blackfly larvae) or on their forelegs (some mayflies). Some caddisflies and midges spin nets and catch food in them as the water flows through.

  Scrapers—These scrape the algae and diatoms off surfaces of rocks and debris, using their mouthparts. Many of these organisms are flattened to hold onto surfaces while feeding. Scrapers include water pennies, limpets and snails, netwinged midge larvae, and certain mayfly larvae, among others.

  Piercers—These herbivores pierce plant tissues or cells and suck the fluids out. Some caddisflies do this.

  Predators—These eat other living creatures. Some of these are *engulfers*; that is, they eat their prey completely or in parts. This is very common in stoneflies and dragonflies, as well as caddisflies. Others are *piercers*, which are similar to the herbivorous piercers except that they eat live animal tissues.

- *Mode of existence* (habit, locomotion, attachment, concealment):

  Skaters are adapted for skating on the surface where they feed as scavengers on organisms trapped in the surface film (e.g., water striders).

  Planktonic insects inhabit the open water limnetic zone of standing (lentic) waters, such as lakes, bogs, and ponds. Representatives may float and swim about in the open water but usually exhibit a diurnal vertical migration pattern (e.g., phantom midges) or float at the surface to obtain oxygen and food, diving when alarmed (e.g., mosquitoes).

  Divers are adapted for swimming by rowing with their hind legs in lentic habitats and lotic pools. Representatives come to the surface to obtain oxygen but dive and swim when feeding or alarmed; they may cling to or crawl on submerged objects such as vascular plants (e.g., water boatmen, predaceous diving beetle).

  Swimmers are adapted for fishlike swimming in lotic or lentic habitats. Individuals usually cling to submerged objects, such as rocks (lotic riffles) or vascular plants (lentic), between short bursts of swimming (e.g., mayflies).

  Clingers have behavioral adaptations (e.g., fixed retreat construction) and morphological adaptations (e.g., long, curved tarsal claws, dorsoventral flattening, ventral gills arranged as a sucker) for attachment to surfaces in stream riffles and wave-swept rocky littoral zones of lakes (e.g., mayflies and caddisflies).

  Sprawlers inhabit the surface of floating leaves of vascular hydrophytes or fine sediments and usually have modifications for staying on top of the substrate and maintaining the respiratory surfaces free of silt (e.g., mayflies, dobsonflies, damselflies).

  Climbers are adapted for living on vascular hydrophytes or detrital debris (e.g., overhanging branches, roots, and vegetation along streams and submerged brush in lakes) with modifications for moving vertically on stem-type surfaces (e.g., dragonflies and damselflies).

  Burrowers inhabit the fine sediments of streams (pools) and lakes. Some construct discrete burrows, which may have sand grain tubes extending above the surface of the substrate or the individuals; they may ingest their way through the sediments (e.g., mayflies and midges).

13.16.2 Macroinvertebrates and the Food Web

In a stream or lake, the two possible sources of primary energy are (1) photosynthesis by algae, mosses, and higher aquatic plants, and (2) imported organic matter from streamside or lakeside vegetation (e.g., leaves and other parts of vegetation). Simply put, a significant portion of the food that is eaten grows right in the stream or lake—for example, algae, diatoms, nymphs and larvae, and fish. A food that originates from within the stream is *autochthonous*. Most food in a stream, however, comes from outside the stream—especially in small, heavily wooded streams, which normally have insufficient light to support substantial instream photosynthesis so energy pathways are supported largely by imported energy. Leaves provide a large portion of this imported energy. Worms drown in floods and are washed in. Leafhoppers and caterpillars fall from trees. Adult mayflies and other insects mate above the stream, lay their eggs in it, and then die in it. All of this food from outside the stream is *allochthonous*.

13.16.3 Units of Organization

Macroinvertebrates, like all other organisms, are classified and named. Macroinvertebrates are classified and named using a *taxonomic hierarchy*. The taxonomic hierarchy for the caddisfly (a macroinvertebrate insect commonly found in streams) is shown below:

> Kingdom—Animalia (animals)
> Phylum—Arthropoda ("jointed legs")
> Class—Insecta (insect)
> Order—Trichoptera (caddisfly)
> Family—Hydropsychidae (net-spinning caddis)
> Genus and species—*Hydropsyche morosa*

13.17 TYPICAL BENTHIC MACROINVERTEBRATES IN RUNNING WATERS

As mentioned, the macroinvertebrates are the best-studied and most diverse animals in streams. We therefore devote our discussion to the various macroinvertebrate groups. Although it is true that non-insect macroinvertebrates, such as Oligochaeta (worms), Hirudinea (leeches), and Acari (water mites), are frequently encountered groups in lotic environments, the insects are among the most conspicuous inhabitants of streams. In most cases, it is the larval stages of the insects that are aquatic, whereas the adults are terrestrial. Typically, the larval stage is much extended, and the adult lifespan is short. Lotic insects are found among many different orders, and brief accounts of their biology are presented in the following sections.

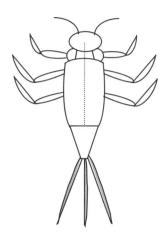

FIGURE 13.16 Mayfly (Emphemeroptera order).

13.17.1 Insect Macroinvertebrates

The most important insect groups found in streams are Ephemeroptera (mayflies), Plecoptera (stoneflies), Trichoptera (caddisflies), Diptera (true flies), Coleoptera (beetles), Hemiptera (true bugs), Megaloptera (alderflies and dobsonflies), and Odonata (dragonflies and damselflies). The identification of these different orders is usually easy, and many keys and specialized references are available to help in the identification of these species (such as Merritt and Cummins, 1996). In contrast, specialist taxonomists are often required to identify some genera and species, particularly the Diptera. As mentioned, insect macroinvertebrates are ubiquitous in streams and are often represented by many species. Although the macroinvertebrates discussed below are aquatic species, a majority of the species can be found in streams.

13.17.1.1 Mayflies (Order: Ephemeroptera)

Streams and rivers are generally inhabited by many species of mayflies, and, in fact, most species are restricted to streams. For the experienced freshwater ecologist who encounters a mayfly nymph, recognition is obtained through trained observation: abdomen with leaf-like or feather-like gills, legs with a single tarsal claw, generally (but not always) having three cerci ("tails"—two cerci and between them usually a terminal filament) (see Figure 13.16). The experienced ecologist knows that mayflies are hemimetabolous insects (i.e., larvae or nymphs resemble wingless adults) that go through many postembryonic molts, often in a range of between 20 and 30. For some species, body length increases about 15% for each instar.

Mayfly nymphs are mainly grazers or collector–gatherers feeding on algae and fine detritus, although a few genera are predatory. Some members filter particles from the water using hair-fringed legs or maxillary palps. Shredders are rare among mayflies. In general, mayfly nymphs tend to live primarily in unpolluted streams, where, with densities of up to 10,000/m², they contribute substantially to secondary producers.

Adult mayflies resemble nymphs but usually possess two pairs of long, lacy wings folded upright; adults usually have only two cerci. The adult lifespan is short, ranging from a few hours to a few days, rarely up to two weeks, and the adults do not feed. Mayflies are unique among insects in having two winged stages, the subimago and the imago. The emergence of adults tends to be synchronous, thus ensuring the survival of enough adults to continue the species.

13.17.1.2 Stoneflies (Order: Plecoptera)

Although many freshwater ecologists would maintain that the stonefly is a well-studied group of insects, this is not exactly the case. Despite their importance, less than 5 to 10% of stonefly species are well known with respect to life history, trophic interactions, growth, development, spatial distribution, and nymphal behavior. Notwithstanding our lack of extensive knowledge with regard to stoneflies, enough is known to provide an accurate characterization of these aquatic insects. We know, for example, that stonefly larvae are characteristic inhabitants of cool, clean streams (most nymphs occur under stones in well-aerated streams). They are sensitive to organic pollution or, more precisely, to low oxygen concentrations accompanying organic breakdown processes, but stoneflies seem rather tolerant to acidic conditions. Lack of extensive gills at least partly explains their relative intolerance of low oxygen levels.

Stoneflies are drab-colored, small- to medium-sized (1/6 to 2-1/4 in., or 4 to 60 mm), rather flattened insects. Stoneflies have long, slender, many-segmented antennae and two long, narrow, antenna-like structures (cerci) on the tip of the abdomen (see Figure 13.17). The cerci may be long or short. At rest, the wings are held flat over the abdomen, giving a "square-shouldered" look compared to the roof-like position of most caddisflies and vertical position of the mayflies. Stoneflies have two pairs of wings. The hind wings are slightly shorter than the forewings and much wider, having a large anal lobe that is folded fanwise when the wings are at rest. This fanlike folding of the wings gives rise to the name of the order: *pleco* ("folded or plaited") and *ptera* ("wings"). The aquatic nymphs are generally very similar to mayfly nymphs except that they have only two cerci at the tip of the abdomen. The stoneflies have chewing mouthparts. They may be found any where in a nonpolluted stream that food is available. Many adults, however, do not feed and have reduced or vestigial mouthparts.

Stoneflies have a specific niche in high-quality streams where they are very important as a fish food source at specific times of the year (winter to spring, especially) and of the day. They complement other important food sources, such as caddisflies, mayflies, and midges.

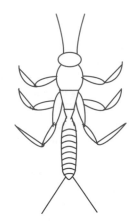

FIGURE 13.17 Stonefly (Plecoptera order).

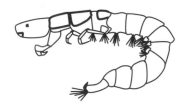

FIGURE 13.18 Caddis (*Hydropsyche*) larvae.

13.16.1.3 Caddisflies (Order: Trichoptera)

Trichoptera (*trichos*, "hair"; *ptera*, "wings") represents one of the most diverse insect orders living in the stream environment, and caddisflies have nearly a worldwide distribution, with the exception of Antarctica. Caddisflies may be categorized broadly into free-living (roving and net spinning) and case-building species.

Caddisflies are described as medium-sized insects with bristle-like and often long antennae. They have membranous hairy wings (which explains the use of *trichos* in the name), which are held tent-like over the body when at rest; most are weak fliers. They have greatly reduced mouthparts and five tarsi. The larvae are mostly caterpillar like and have a strongly sclerotized (hardened) head with very short antennae and biting mouthparts. They have well-developed legs with a single tarsi. The abdomen usually has ten segments; in case-bearing species, the first segment bears three papillae, one dorsally and the other two laterally, which help hold the insect centrally in its case and allows a good flow of water to pass the cuticle and gills. The last or anal segment bears a pair of grappling hooks.

In addition to being aquatic insects, caddisflies are superb architects. Most caddisfly larvae (see Figure 13.18) live in self-designed, self-built houses called *cases*. They spin out silk, and either live in silk nets or use the silk to stick together bits of whatever is lying on the stream bottom. These houses are so specialized that we can usually identify a caddisfly larva to its genus if we can see its house (case). With nearly 1400 species of caddisfly species in North America (north of Mexico), this is a good thing!

FIGURE 13.19 Midge larvae.

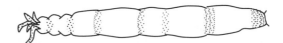

FIGURE 13.20 Cranefly larvae.

Caddisflies are closely related to butterflies and moths (Lepidoptera order). They live in most stream habitats, and that is why they are so diverse. Each species has particular adaptations that allow it to survive in its environment. Mostly herbivorous, most caddisflies feed on decaying plant tissue and algae. Their favorite algae are diatoms, which they scrape off rocks. Some of them, though, are predacious.

Caddisfly larvae can take a year or two to change into adults. They change into *pupae* (the inactive stage in the metamorphosis of many insects, following the larval stage and preceding the adult form) while still inside the cases built for their metamorphosis. It is interesting to note that caddisflies, unlike stoneflies and mayflies, go through a complete metamorphosis. Caddisflies remain as pupae for 2 to 3 weeks, then emerge as adults. When they split open their cases and leave, they must swim to the surface of the water to escape it. The winged adults fly evening and night, and some are known to feed on plant nectar. Most of them will live less than a month; like many other winged stream insects, their adult lives are brief compared to the time they spend in the water as larvae.

Caddisflies are sometimes grouped into five main groups according to the kinds of cases they build: (1) free-living forms that do not make cases, (2) saddle-case makers, (3) purse-case makers, (4) net-spinners and retreat-makers, and (5) tube-case makers. Caddisflies demonstrate their architectural talents in the cases they design and build; for example, a caddisfly might build a perfect, four-sided box case of bits of leaves and bark or tiny bits of twigs. It may make a clumsy dome of large pebbles. Another might construct rounded tubes out of twigs or very small pebbles. In our experience in gathering caddisflies, we have come to appreciate not only their architectural ability but also their flare in the selection of construction materials. We have found many caddisfly cases constructed of silk that was emitted through an opening at the tip of the labium and was combined with bits of ordinary rock mixed with sparkling quartz and red garnet, green peridot, and bright fool's gold.

In addition to the protection their cases provide them, the cases provide another advantage. The cases actually help caddisflies breathe. They move their bodies up and down and back and forth inside their cases, and this produces a current that brings them fresh oxygen. The less oxygen there is in the water, the faster they have to move. It has been found that caddisflies inside their cases get more oxygen than those that are outside of their cases—and

this is why stream ecologists think that caddisflies can often be found even in still waters, where dissolved oxygen is low, in contrast to stoneflies and mayflies.

13.17.1.4 True Flies (Order: Diptera)

True or two- (*di-*) winged (*ptera*) flies include not only the flies that we are most familiar with, such as fruitflies and houseflies, but also midges (see Figure 13.19), mosquitoes, craneflies (see Figure 13.20), and others. Houseflies and fruitflies live only on land, and we do not concern ourselves with them. Some, however, spend nearly their entire lives in water, and they contribute to the ecology of streams.

True flies are in the order Diptera, one of the most diverse orders of the class Insecta, with about 120,000 species worldwide. Dipteran larvae occur almost everywhere except Antarctica and deserts where there is no running water. They may live in a variety of places within a stream: buried in sediments, attached to rocks, beneath stones, in saturated wood or moss, or in silken tubes attached to the stream bottom. Some even live below the stream bottom.

True fly larvae may eat almost anything, depending on their species. Those with brushes on their heads use them to strain food out of the water that passes through. Others may eat algae, detritus, plants, and even other fly larvae.

The longest part of the true fly's life cycle, like that of mayflies, stoneflies, and caddisflies, is the larval stage. It may remain an underwater larva anywhere from a few hours to 5 years. The colder the environment, the longer it takes to mature. It pupates and emerges and then becomes a winged adult. The adult may live 4 months—or it may live for only a few days. While reproducing, it will often eat plant nectar for the energy it needs to make its eggs. Mating sometimes takes place in aerial swarms. The eggs are deposited back in the stream; some females will crawl along the stream bottom, losing their wings in the process, to search for the perfect place to put their eggs. Once they lay them, they die.

Diptera serve an important role in cleaning water and breaking down decaying material, and they are a vital food source for many of the animals living in and around streams, as they play pivotal roles in the processing of food energy. The true flies most familiar to us, however, are the midges, mosquitoes, and the craneflies, because they are pests. Some midge flies and mosquitoes bite; the cranefly does not bite but looks like a giant mosquito.

Like mayflies, stoneflies, and caddisflies, true flies are mostly in larval form. Just as for caddisflies, we can find their pupae, because they are holometabolous insects; that is, they go through complete metamorphosis. Most of them are free living and travel around. Although none of the true fly larvae has the six jointed legs that we see on other insects in the stream, they sometimes have strange little almost-legs (prolegs) to move around with. Others may move somewhat like worms do, and some—the ones who live in waterfalls and rapids—have a row of six suction discs that they use to move much like a caterpillar does. Many use silk pads and hooks at the ends of their abdomens to hold them fast to smooth rock surfaces.

13.17.1.5 Beetles (Order: Coleoptera)

Of the more than 1 million described species of insect, at least one third are beetles, making Coleoptera not only the largest order of insects but also the most diverse order of living organisms. Even though this is the most speciose order of terrestrial insects, surprisingly their diversity is not so apparent in running waters. Coleoptera belongs to the infraclass Neoptera, division Endpterygota. Members of this order have an anterior pair of wings (the *elytra*) that are hard and leathery and not used in flight; the membranous hindwings, which are used for flight, are concealed under the elytra when the organisms are at rest. Only 10% of the 350,000 described species of beetles are aquatic.

Beetles are holometabolous. Eggs of the aquatic coleopterans hatch in 1 or 2 weeks, with diapause occurring rarely. Larvae undergo from three to eight molts. The pupal phase of all coleopternas is technically terrestrial, making this life stage of beetles the only one that has not successfully invaded the aquatic habitat. A few species have diapausing prepupae but most complete transformation to adults in 2 to 3 weeks. Terrestrial adults of aquatic beetles are typically short lived and sometimes nonfeeding, like those of the other orders of aquatic insects. The larvae of Coleoptera are morphologically and behaviorally different from the adults, and their diversity is high.

Aquatic species occur in two major suborders, the Adephaga and the Polyphaga. Both larvae and adults of six beetle families are aquatic: Dytiscidae (predaceous diving beetles), Elmidae (riffle beetles), Gyrinidae (whirligig beetles), Halipidae (crawling water beetles), Hydrophilidae (water scavenger beetles), and Noteridae (burrowing water beetles). Five families—Chrysomelidae (leaf beetles), Limnichidae (marsh-loving beetles), Psephenidae (water pennies), Ptilodactylidae (toe-winged beetles), and Scirtidae (marsh beetles)—have aquatic larvae and terrestrial adults, as do most of the other orders of aquatic insects; adult limnichids, however, readily submerge when disturbed. Three families have species that are terrestrial as larvae and aquatic as

FIGURE 13.21 Riffle beetle larvae.

FIGURE 13.22 Riffle beetle adult.

adults, a highly unusual combination among insects: Curculionidae (weevils), Dryopidae (long-toed water beetles), and Hydraenidae (moss beetles). Because they provide a greater understanding of the condition of a freshwater body (i.e., they are useful indicators of water quality), we focus our discussion here on the riffle beetle, water penny, and whirligig beetle.

Riffle beetle larvae (most commonly found in running waters, hence their name) are up to 3/4 in. long (see Figure 13.21). Their body is not only long but also hard, stiff, and segmented. They have six long segmented legs on the upper middle section of the body; the back end has two tiny hooks and short hairs. Larvae may take 3 years to mature before they leave the water to form pupae; adults return to the stream. Riffle beetle adults are considered better indicators of water quality than larvae because they have been subjected to water quality conditions over a longer period. They walk very slowly under the water (on the stream bottom), and do not swim on the surface. They have small oval-shaped bodies (see Figure 13.22) and are typically about 1/4 in. in length. Both adults and larvae of most species feed on fine detritus with associated microorganisms scraped from the substrate, although others may be xylophagous—that is, wood eating (e.g., *Lara*, Elmidae). Predators do not seem to include riffle beetles in their diet, except perhaps for their eggs, which are sometimes attacked by flatworms.

The adult *water penny* is inconspicuous and often found clinging tightly in a sucker-like fashion to the undersides of submerged rocks, where they feed on attached algae. The body is broad, slightly oval, and flat in shape, ranging from 4 to 6 mm (about 1/4 in.) in length. The body is covered with segmented plates and looks like a tiny round leaf (see Figure 13.23). It has six tiny jointed legs (underneath). The color ranges from light brown to almost black. There are 14 water penny species in the United States. They live predominately in clean, fast-moving

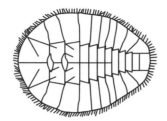

FIGURE 13.23 Water penny larva.

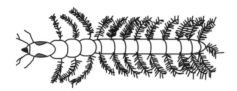

FIGURE 13.24 Whirligig beetle larva.

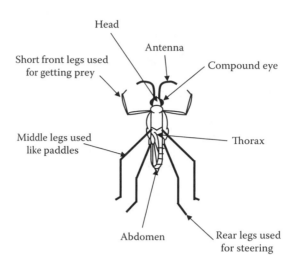

FIGURE 13.25 Water strider.

streams. Aquatic larvae live a year or more; the terrestrial adults survive on land for only a few days. Water pennies scrape algae and plants from surfaces.

Whirligig beetles are common inhabitants of streams and normally are found on the surface of quiet pools. The body of a whirligig beetle has pincher-like mouthparts and six segmented legs on the middle of the body; the legs end in tiny claws. Many filaments extend from the sides of the abdomen. They have four hooks at the end of the body and no tail (see Figure 13.24).

Note: When disturbed, whirligig beetles swim erratically or dive while emitting defensive secretions.

As larvae, they are benthic predators, whereas the adults live on the water surface, attacking dead and living organisms trapped in the surface film. They occur on the surface in aggregations of up to thousands of individuals. Unlike the mating swarms of mayflies, these aggregations serve primarily to confuse predators. Whirligig beetles have other interesting defensive adaptations—for example, the Johnston's organ at the base of the antennae enables them to echolocate using surface-wave signals; their compound eyes are divided into two pairs, one above and one below the water surface, enabling them to detect both aerial and aquatic predators; and they produce noxious chemicals that are highly effective at deterring predatory fish.

13.17.1.6 Water Strider ("Jesus Bugs") (Order: Hemiptera)

It is fascinating to sit on a log at the edge of a stream pool and watch the drama that unfolds among the small water animals. Among the star performers in small streams are the water bugs. These are aquatic members of that large group of insects known as the "true bugs," most of which

live on land. Moreover, unlike many other types of water insects, they do not have gills but get their oxygen directly from the air. Most conspicuous and commonly known are the *water striders* or *water skaters*. These ride the top of the water, with only their feet making dimples in the surface film. Like all insects, water striders have a three-part body (head, thorax, and abdomen), six jointed legs, and two antennae. It has a long, dark, narrow body (see Figure 13.25). The underside of the body is covered with water-repellent hair. Some water striders have wings, others do not. Most water striders are over 5 mm (0.2 in.) long. Water striders eat small insects that fall on the surface of the water and larvae. Water striders are very sensitive to motion and vibrations in the surface of the water. It uses this ability to locate prey. It pushes its mouth into its prey, paralyzes it, and sucks the insect dry. Predators of the water strider, such as birds, fish, water beetles, backswimmers, dragonflies, and spiders, take advantage of the fact that water striders cannot detect motion above or below the surface of the water.

13.17.1.7 Alderflies and Dobsonflies (Order: Megaloptera)

Larvae of all species of Megaloptera ("large wing") are aquatic and attain the largest size of all aquatic insects. Megaloptera is a medium-sized order with less than 5000 species worldwide. Most species are terrestrial; in North America, 64 aquatic species occur. In running waters, alderflies (Sialidae family) and dobsonflies (Corydalidae family; sometimes called *hellgrammites* or *toe biters*) are particularly important, as they are voracious predators, having large mandibles with sharp teeth.

Alderfly brownish-colored larvae possess a single tail filament with distinct hairs. The body is thick skinned with six to eight filaments on each side of the abdomen; gills

FIGURE 13.26 Alderfly larva.

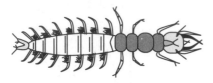

FIGURE 13.27 Dobsonfly larva.

are located near the base of each filament. The mature body size varies from 0.5 to 1.25 in. (see Figure 13.26). Larvae are aggressive predators, feeding on other adult aquatic macroinvertebrates (they swallow their prey without chewing); as secondary consumers, other larger predators eat them. Female alderflies deposit eggs on vegetation that overhangs water; when the larvae hatch, they fall directly into the quiet but moving water. Adult alderflies are dark with long wings folded back over the body; they live only a few days.

Dobsonfly larvae are extremely ugly (thus, they are rather easy to identify) and can be rather large and anywhere from 25 to 90 mm (1 to 3.5 in.) in length. The body is stout, with eight pairs of appendages on the abdomen. Brush-like gills at the base of each appendage look like hairy armpits (see Figure 13.27). The elongated body has spiracles (spines), three pairs of walking legs near the upper body, and one pair of hooked legs at the rear. The head bears four segmented antennae, small compound eyes, and strong mouth parts (large chewing pinchers). Coloration varies from yellowish, brown, gray, to black, often mottled. Dobsonfly larvae, commonly known as *hellgrammites*, are customarily found along stream banks under and between stones. As indicated by the mouthparts, they are predators and feed on all kinds of aquatic organisms.

13.17.1.8 Dragonflies and Damselflies (Order: Odonata)

The order Odonata, which includes dragonflies (suborder Anisoptera) and damselflies (suborder Zygoptera), is a small order of conspicuous, hemimetabolous insects (lacking a pupal stage) representing about 5000 named species and 23 families worldwide. *Odonata* is a Greek word meaning "toothed one." It refers to the serrated teeth located on the insect's chewing mouthparts (mandibles). Characteristics of dragonfly and damselfly larvae include:

- Large eyes
- Three pairs of long segmented legs on the upper middle section (thorax) of the body

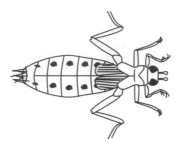

FIGURE 13.28 Dragonfly nymph.

- Large scoop-like lower lip that covers the bottom of the mouth
- No gills on its sides or underneath the abdomen

Note: Dragonflies and damselflies are unable to fold their four elongated wings back over the abdomen when at rest.

Dragonflies and damselflies are medium to large insects with two pairs of long equal-sized wings. The body is long and slender, with short antennae. Immature stages are aquatic, and development occurs in three stages (egg, nymph, adult).

Dragonflies are also known as *darning needles* (at one time, children were warned to keep quiet or the dragonfly's darning needles would sew the child's mouth shut). In their nymphal stage, dragonflies are grotesque creatures, robust and stoutly elongated. They do not have long tails (see Figure 13.28). They are commonly gray, greenish, or brown to black in color. They are medium to large aquatic insects, ranging in size from 15 to 45 mm; the legs are short and used for perching. They are often found on submerged vegetation and at the bottom of streams in the shallows. They are rarely found in polluted waters. Their food consists of other aquatic insects, annelids, small crustacea, and mollusks. Transformation occurs when the nymph crawls out of the water, usually onto vegetation. There it splits its skin and emerges prepared for flight. The adult dragonfly is a strong flier, capable of great speed (>60 mph) and maneuverability. (They can fly backward, stop on a dime, zip 20 feet straight up, and slip sideways in the blink of an eye!) When at rest the wings remain open and out to the sides of the body. A dragonfly's freely movable head has large, hemispherical eyes (nearly 30,000 facets each), which the insects use to locate prey with their excellent vision. Dragonflies eat small insects, mainly mosquitoes (large numbers of mosquitoes) while in flight. Depending on the species, dragonflies lay hundreds of eggs by dropping them into the water and leaving them to hatch or by inserting eggs singly into a slit in the stem of a submerged plant. The complete metamorphosis (egg, nymph, mature nymph, and adult) can take 2 to 3 years. Nymphs are often covered by algal growth.

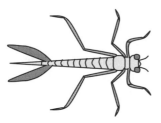

FIGURE 13.29 Damselfly nymph.

Note: Adult dragonflies are sometimes referred to as *mosquito hawks* because they eat such a large number of mosquitoes that they catch while they are flying.

Damselflies are smaller and more slender than dragonflies. They have three long, oar-shaped feathery tails, which are actually gills, and long slender legs (see Figure 13.29). They are gray, greenish, or brown to black in color. Their habits are similar to those of dragonfly nymphs, and they emerge from the water as adults in the same manner. The adult damselflies are slow and seem uncertain in flight. Wings are commonly black or clear, and the body is often brilliantly colored. When at rest, they perch on vegetation with their wings closed upright. Damselflies mature in 1 to 4 years. Adults live for a few weeks or months. Unlike the dragonflies, adult damselflies rest with their wings held vertically over their backs. They mostly feed on live insect larvae.

Note: Relatives of the dragonflies and damselflies are some of the most ancient of the flying insects. Fossils have been found of giant dragonflies with wingspans up to 720 mm (28.4 in.) that lived long before the dinosaurs!

13.17.2 Non-Insect Macroinvertebrates

Non-insect macroinvertebrates are also important to our discussion of stream and freshwater ecology because many of them are used as bioindicators of stream quality. Three frequently encountered groups in running water systems are Oligochaeta (worms), Hirudinea (leeches), and Gastropoda (lung-breathing snails). They are by no means restricted to running-water conditions, and the great majority of them occupy slow-flowing marginal habitats where the sedimentation of fine organic materials takes place.

13.17.2.1 Oligochaeta (Family: Tuificidae, Genus: *Tubifex*)

Tubifex worms (commonly known as sludge worms) are unique in the fact that they build tubes. Sometimes we might find as many as 8000 individuals per square meter. They attach themselves within the tube and wave their posterior end in the water to circulate the water and make more oxygen available to their body surface. These worms are commonly red, because their blood contains hemoglobin. *Tubifex* worms may be very abundant in situations when other macroinvertebrates are absent; they can survive in very low oxygen levels and can live with no oxygen at all for short periods. They are commonly found in polluted streams and feed on sewage or detritus.

13.17.2.2 Hirudinea (Leeches)

Despite the many different families of leeches, they all have common characteristics. They are soft-bodied worm-like creatures that are flattened when extended. Their bodies are dull in color, ranging from black to brown and reddish to yellow, often with a brilliant pattern of stripes or diamonds on the upper body. Their size varies among species but generally ranges from 5 mm to 45 cm when extended. Leeches are very good swimmers, but they typically move in an inchworm fashion. They are carnivorous and feed on other organisms ranging from snails to warm-blooded animals. Leeches are found in warm, protected shallows under rocks and other debris.

13.17.2.3 Gastropoda (Lung-Breathing Snail)

Lung-breathing snails (pulmonates) may be found in streams that are clean; however, their dominance may indicate that dissolved oxygen levels are low. These snails are different from *right-handed snails* because they do not breathe underwater by use of gills but instead have a lung-like sac called a *pulmonary cavity*, which they fill with air at the surface of the water. When the snail takes in air from the surface, it makes a clicking sound. The air taken in allows the snail to remain under water for long periods, sometimes hours.

Lung-breathing snails have two characteristics that help us to identify them. First, they have no operculum or hard cover over the opening to its body cavity. Second, snails are either right-handed or left-handed, and the lung-breathing snails are left-handed. We can tell the difference by holding the shell so its tip is upward and the opening toward us. If the opening is to the left of the axis of the shell, the snail is considered to be *sinistral*—that is, it is left-handed. If the opening is to the right of the axis of the shell, the snail is *dextral*—that is, it is right-handed, and it breathes with gills. Snails are animals of the substrate and are often found creeping along on all types of submerged surfaces in water from 10 cm to 2 m deep.

Before the Industrial Revolution of the 1800s, metropolitan areas were small and sparsely populated; thus, river and stream systems within or next to early communities received insignificant quantities of discarded waste. Early on, these river and stream systems were able to compensate for the small amount of wastes they received;

when wounded (polluted), nature has a way of fighting back. In the case of rivers and streams, nature provides their flowing waters with the ability to restore themselves through their own self-purification process. It was only when humans gathered in great numbers to form great cities that the stream systems were not always able to recover from receiving great quantities of refuse and other wastes. What exactly is it that we are doing to rivers and streams? We are upsetting the delicate balance between pollution and the purification process; that is, we are unbalancing the aquarium.

SUMMARY OF KEY TERMS

Abiotic factor is the nonliving part of the environment composed of sunlight, soil, mineral elements, moisture, temperature, topography, minerals, humidity, tide, wave action, wind, and elevation.

Important Point: Every community is influenced by a particular set of abiotic factors. Whereas it is true that the abiotic factors affect the community members, it is also true that the living (biotic factors) may influence the abiotic factors; for example, the amount of water lost through the leaves of plants may add to the moisture content of the air. Also, the foliage of a forest reduces the amount of sunlight that penetrates the lower regions of the forest. The air temperature is therefore much lower than in non-shaded areas (Tomera, 1989).

Autotrophs (green plants) fix energy of the sun and manufacture food from simple, inorganic substances.

Biogeochemical cycles are cyclic mechanisms in all ecosystems by which biotic and abiotic materials are constantly exchanged.

Biotic factor (community) is the living part of the environment composed of organisms that share the same area; they are mutually sustaining, interdependent, and constantly fixing, utilizing, and dissipating energy.

Community, in an ecological sense, includes all the populations occupying a given area.

Consumers and decomposers dissipate energy fixed by the producers through food chains or webs. The available energy decreases by 80 to 90% during transfer from one trophic level to another.

Ecology is the study of the interrelationship of an organism or a group of organisms and their environment.

Ecosystem is the community and the nonliving environment functioning together as an ecological system.

Environment is everything that is important to an organism in its surroundings.

Heterotrophs (animals) use food stored by the autotroph, rearrange it, and finally decompose complex materials into simple inorganic compounds. Het-

erotrophs may be carnivorous (meat-eaters), herbivorous (plant-eaters), or omnivorous (plant- and meat-eaters).

Homeostasis is a natural occurrence during which an individual population or an entire ecosystem regulates itself against negative factors and maintains an overall stable condition.

Niche is the role that an organism plays in its natural ecosystem, including its activities, resource use, and interaction with other organisms.

Pollution is an adverse alteration to the environment by a pollutant.

CHAPTER REVIEW QUESTIONS

13.1   The major ecological unit is _____.

13.2   Those organisms residing within or on the bottom sediment are _____.

13.3   Organisms attached to plants or rocks are referred to as _____.

13.4   Small plants and animals that move about with the current are _____.

13.5   Free-swimming organisms belong to which group of aquatic organisms?

13.6   Organism that live on the surface of the water are _____.

13.7   Movement of new individuals into a natural area is referred to as _____.

13.8   What fixes energy of the sun and makes food from simple inorganic substances?

13.9   The freshwater habitat that is characterized by normally clean water is _____.

13.10   The amount of oxygen dissolved in water and available for organisms is the _____.

REFERENCES AND SUGGESTED READING

ASTM. 1969. *Manual on Water*. Philadelphia, PA: American Society for Testing and Materials.

Barlocher, R. and Kendrick, L. 1975. Leaf conditioning by microorganisms. *Oecologia*, 20, 359–362.

Benfield, E.F. 1996. Leaf breakdown in streams ecosystems. In *Methods in Stream Ecology*, Hauer, F.R. and Lambertic, G.A., Eds., pp. 579–590. San Diego, CA: Academic Press.

Benfield, E.F., Jones, D.R., and Patterson, M.F. 1977. Leaf pack processing in a pastureland stream. *Oikos*, 29, 99–103.

Benjamin, C.L., Garman, G.R., and Funston, J.H. 1997. *Human Biology*. New York: McGraw-Hill.

Carson, R. 1962. *Silent Spring*. Boston: Houghton Mifflin.

Clements, E.S. 1960. *Adventures in Ecology*. New York: Pageant Press.

Crossley, Jr., D.A. et al. 1984. The positive interactions in agroecosystems. In *Agricultural Ecosystems*, Lowrance, R., Stinner, B.R., and House, G.J., Eds. New York: John Wiley & Sons.

Cummins, K.W. 1974. Structure and function of stream ecosystems. *Bioscience*, 24, 631–641.

Cummins, K.W. and Klug, M.J. 1979. Feeding ecology of stream invertebrates. *Annu. Rev. Ecol. Syst.*, 10, 631–641.

Darwin, C. 1998. *The Origin of Species*, Suriano, G., Ed. New York: Grammercy.

Dolloff, C.A. and Webster, J.R. 2000. Particulate organic contributions from forests to streams: debris isn't so bad. In *Riparian Management in Forests of the Continental Eastern United States*, Verry, E.S., Hornbeck, J.W., and Dolloff C.A., Eds. Boca Raton, FL: CRC Press.

Evans, F.C. 1956. Ecosystem as the basic unit in ecology. *Science*, 23, 1127–1128.

Krebs, C.H. 1972. *Ecology. The Experimental Analysis of Distribution and Abundance*. New York: Harper & Row.

Lindeman, R.L. 1942. The trophic-dynamic aspect of ecology. *Ecology*, 23, 399–418.

Margulis, L. and Sagan, D. 1997. *Microcosmos: Four Billion Years of Evolution from Our Microbial Ancestors*. Berkeley: University of California Press.

Marshall, P. 1950. *Mr. Jones, Meet the Master*. Grand Rapids, MI: Fleming H. Revel Co.

McCafferty, P.W. 1981. *Aquatic Entomology*. Boston: Jones & Bartlett Publishers.

Merrit, R.W. and Cummins, K.W. 1996. *An Introduction to the Aquatic Insects of North America*, 3rd ed. Dubuque, IA: Kendall/Hunt Publishing.

Odum, E.P. 1952. *Fundamentals of Ecology*, 1st ed. Philadelphia, PA: Saunders.

Odum, E.P. 1971. *Fundamentals of Ecology*, 3rd ed. Philadelphia, PA: Saunders.

Odum, E.P. 1983. *Basic Ecology*. Philadelphia, PA: Saunders.

Odum, E.P. 1984. Properties of agroecosystems. In *Agricultural Ecosystems*, Lowrance, R., Stinner, B.R., and House, G.J., Eds. New York: John Wiley & Sons.

Odum, E.P. and Barrett, G.W. 2005. *Fundamentals of Ecology*, 5th ed. Belmont, CA: Thomson Brooks/Cole.

Paul, Jr., R.W., Benfield, E.F., and Cairns, Jr., J. 1978. Effects of thermal discharge on leaf decomposition in a river ecosystem. *Verhandlugen der Internationalen Vereinigung fur Thoeretsche and Angewandte Limnologie*, 20, 1759–1766.

Peterson, R.C. and Cummins, K.W. 1974. Leaf processing in woodland streams. *Freshwater Biol.*, 4, 345–368.

Porteous, A. 1992. *Dictionary of Environmental Science and Technology*. New York: John Wiley & Sons.

Price, P.W. 1984. *Insect Ecology*. New York: John Wiley & Sons.

Ramalay, F. 1940. The growth of a science. *Univ. Colorado Stud.*, 26, 3–14.

Smith, R.L. 1996. *Ecology and Field Biology*. New York: HarperCollins.

Smith, T.M. and Smith, R.L. 2006. *Elements of Ecology*, 6th ed. San Francisco, CA: Pearson, Benjamin Cummings.

Spellman, F.R. 1996. *Stream Ecology and Self-Purification*. Lancaster, PA: Technomic.

Suberkoop, K., Godshalk, G.L., and Klug, M.J. 1976. Changes in the chemical composition of leaves during processing in a woodland stream. *Ecology*, 57, 720–727.

Tansley, A.G. 1935. The use and abuse of vegetational concepts and terms. *Ecology*, 16, 284–307.

Tomera, A.N. 1989. *Understanding Basic Ecological Concepts*. Portland, ME: J. Weston Walch, Publisher.

USDA. 1982. *Agricultural Statistics 1982*. Washington, D.C.: U.S. Department of Agriculture.

USDA. 1999. *Autumn Colors—How Leaves Change Color*. Washington, D.C.: U.S. Department of Agriculture (http://www.na.fs.fed.us/spfo/pubs/misc/autumn/autumn_colors.htm).

USDA, 2007. *Agricultural Ecosystems and Agricultural Ecology*. Washington, D.C.: U.S. Department of Agriculture (http://nrcs.usda.gov/technical/ECS/agecol/ecosystem.html).

USFWS. 2007. *Ecosystem Conservation*. Washington, D.C.: U.S. Fish & Wildlife Service (http://www.fws.gov/ecosystems/).

14 Water Quality

Are we to wait until all frogs "croak"?

The earliest chorus of frogs—those high-pitched rhapsodies of spring peepers, those "jug-o-rum" calls of bullfrogs, those banjo-like bass harmonies of green frogs, those long and guttural cadences of leopard frogs, their singing, a prelude to the splendid song of birds beside an otherwise still pond on an early spring evening—heralds one of nature's most dramatic events: metamorphosis. This metamorphosis begins with masses of eggs that soon hatch into gill-breathing, herbivorous, fishlike tadpole larvae. As they feed and grow, warmed by the spring sun, almost imperceptibly a remarkable transformation begins. Hind legs appear and gradually lengthen. Tails shorten. Larval teeth vanish, and lungs replace gills. Eyes develop lids. Forelegs emerge. In a matter of weeks, the aquatic, vegetarian tadpole (should it escape the many perils of the pond) will complete its metamorphosis into an adult, carnivorous frog.

This springtime metamorphosis is special. This anticipated event (especially for the frog) marks the end of winter, the rebirth of life, a rekindling of hope (especially for mankind). This yearly miracle of change sums up in a few months each spring what occurred over 3000 million years ago, when the frog evolved from its ancient predecessor. Today, however, something is different, strange, and wrong with this striking and miraculous event.

In the first place, where are all the frogs? Where have they gone? Why has their population decreased so dramatically in recent years?

The second problem is that this natural metamorphosis process (perhaps a reenactment of some Paleozoic drama whereby, over countless generations, the first amphibian types equipped themselves for life on land) now demonstrates aberrations of the worst kind, of monstrous proportions and with dire results for frog populations in certain areas. USEPA has received reports about deformed frogs in certain sections of the United States, particularly Minnesota, as well as in Canada and parts of Europe.

Most of the deformities have been in the rear legs and appear to be developmental. The question is why? Researchers have noted that neurological abnormalities have also been found. Again, the question is why?

Researchers have pointed the finger of blame at parasites, pesticides and other chemicals, ultraviolet radiation, acid rain, and metals. Something is going on. What is it? We do not know!

The next question, then, is what are we going to do about it? Are we to wait until all the frogs croak before we act—before we find the source, the cause, the polluter—before we see this same reaction occurring in other species … maybe our own?

The final question is obvious: When frogs are forced by mutation into something else, is this evolution by gunpoint?

Are we holding the gun?

14.1 INTRODUCTION

The quality of water, whether it is used for drinking, irrigation, or recreational purposes, is significant for health in both developing and developed countries worldwide. The first problem with water is rather obvious: A source of water must be found. Second, when accessible water is found it must be suitable for human consumption. Meeting the water needs of those who populate Earth is an ongoing challenge. New approaches to meeting these water needs will not be easy to implement. Economic and institutional structures still encourage the wasting of water and destruction of ecosystems (Gleick, 2001). Again, finding a water source is the first problem. Finding a source of water that is safe to drink is the other problem.

Water quality is important, as it can have a major impact on health through outbreaks of waterborne disease and by contributing to the background rates of disease. Accordingly, water quality standards are important to protect public health.

In this text, *water quality* refers to those characteristics or range of characteristics that make water appealing and useful. Keep in mind that *useful* also means nonharmful or nondisruptive to either ecology or the human condition within the very broad spectrum of possible uses of water. For example, the absence of odor, turbidity, or color is a desirable immediate quality; however, various imperceptible qualities are also important—that is, chemical qualities of the water. The fact is the presence of materials such as toxic metals (e.g., mercury and lead), excessive nitrogen and phosphorus, or dissolved organic material may not be readily perceived by the senses but may exert substantial negative impacts on the health of a stream or humans. The ultimate impact of these imperceptible chemical qualities of water on the

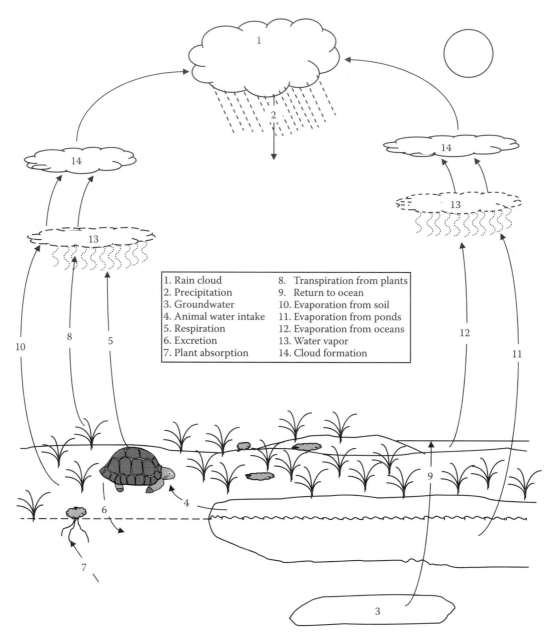

1. Rain cloud	8. Transpiration from plants
2. Precipitation	9. Return to ocean
3. Groundwater	10. Evaporation from soil
4. Animal water intake	11. Evaporation from ponds
5. Respiration	12. Evaporation from oceans
6. Excretion	13. Water vapor
7. Plant absorption	14. Cloud formation

FIGURE 14.1 Hydrologic (water) cycle.

user may be nothing more than a loss of aesthetic values. On the other hand, water containing chemicals could lead to a reduction in biological health or to an outright degradation of human health.

Simply stated, the importance of water quality cannot be overstated.

With regard to water/wastewater treatment operations, water quality management begins with a basic understanding of how water moves through the environment, is exposed to pollutants, and transports and deposits pollutants. The hydrologic (water) cycle depicted in Figure 14.1 illustrates the general links among the atmosphere, soil, surface waters, groundwaters, and plants.

14.2 THE WATER CYCLE

The water cycle describes how water moves through the environment and identifies the links among groundwater, surface water, and the atmosphere (see Figure 14.1). Water is taken from the Earth's surface to the atmosphere by evaporation from the surface of lakes, rivers, streams, and oceans. This evaporation process occurs when the sun heats water. The heat of the sun energizes surface molecules, allowing them to break free of the attractive force binding them together. They then evaporate and rise as invisible vapor in the atmosphere. Water vapor is also emitted from plant leaves by a process called *transpiration*. Every day, an actively growing plant transpires five to ten times as much water as it can hold at once. As water vapor rises, it cools and eventually condenses, usually on tiny particles of dust in the air. When it condenses, it becomes a liquid again or turns directly into a solid (ice, hail, or snow). These water particles then collect and form clouds. The atmospheric water formed in clouds eventually falls to the ground as precipitation. The precipitation can contain contaminants from air pollution. The precipitation may fall directly onto surface waters, be intercepted by plants or structures, or fall onto the ground. Most precipitation falls in coastal areas or in high elevations. Some of the water that falls in high elevations becomes runoff water, the water that runs over the ground (sometimes collecting nutrients from the soil) to lower elevations to form streams, lakes, and fertile valleys. The water we see is known as *surface water*. Surface water can be broken down into five categories: oceans, lakes, rivers and streams, estuaries, and wetlands.

Because the amount of rain and snow remains almost constant, but population and usage per person are both increasing rapidly, water is in short supply. In the United States alone, water usage is four times greater today than it was in 1900. In the home, this increased use is directly related to increases in the number of bathrooms, garbage disposals, home laundries, and lawn sprinklers. In industry, usage has increased 13 times since 1900.

Over 170,000 small-scale suppliers provide drinking water to approximately 200 million Americans by at least 60,000 community water supply systems, as well as to nonresidential locations, such as schools, factories, and campgrounds. The rest of Americans are served by private wells. The majority of the drinking water used in the United States is supplied from groundwater. Untreated water drawn from groundwater and surface waters and used as a drinking water supply can contain contaminants that pose a threat to human health.

Note: The U.S. Environmental Protection Agency has reported that individual American households use approximately 146,000 gallons of freshwater annually and that Americans drink 1 billion glasses of tap water each day.

Obviously, with a limited amount of drinking water available for use, water that is available must be reused or we will be faced with an inadequate supply to meet the needs of all users. Water use/reuse is complicated by water pollution. Pollution is relative and difficult to define; for example, floods and animals (dead or alive) are polluters, but their effects are local and tend to be temporary. Today, water is polluted in many ways, and pollution exists in many forms. Pollution may be apparent as excess aquatic weeds, oil slicks, a decline in sport fish populations, or an increase in carp, sludge worms, and other forms of life that readily tolerate pollution. Maintaining water quality is important because water pollution is detrimental not only to health but also to recreation, commercial fishing, aesthetics, and private, industrial, and municipal water supplies.

At this point the reader might be asking: With all the recent publicity about pollution and the enactment of new environmental regulations, hasn't water quality in the United States improved recently? Answer: With the recent pace of achieving fishable or swimmable waters under the Clean Water Act (CWA), one might think so. The 1994 *National Water Quality Inventory Report to Congress* indicated that 63% of the nation's lakes, rivers, and estuaries met designated uses, which was only a slight increase over that reported in 1992.

The main culprit is *nonpoint source pollution*, which is the leading cause of impairment for rivers, lakes, and estuaries. Impaired sources are those that do not fully support designated uses, such as fish suitable for consumption, drinking water supply, groundwater recharge, aquatic life support, or recreation. According to Fortner and Schechter (1996), the five leading sources of water quality impairment in rivers are agriculture, municipal wastewater treatment plants, habitat and hydrologic modification, resource extraction, and urban runoff and storm sewers.

The health of rivers and streams is directly linked to the integrity of habitat along the river corridor and in adjacent wetlands. Stream quality will deteriorate if activities damage vegetation along riverbanks and in nearby wetlands. Trees, shrubs, and grasses filter pollutants from runoff and reduce soil erosion. Removal of vegetation also eliminates shade that moderates stream temperature. Stream temperature, in turn, affects the availability of dissolved oxygen in the water column for fish and other aquatic organisms. Lakes, reservoirs, and ponds may receive water-carrying pollutants from rivers and streams, melting snow, runoff, or groundwater. Lakes may also receive pollution directly the air.

Thus, in attempting to answer the original question—Has water quality in the United States improved recently?—the best answer probably is that we are holding our own in controlling water pollution, but we need to make more progress. This understates an important point; that is, when it comes to water quality, we need to make more progress on a continuing basis.

14.3 WATER QUALITY STANDARDS

The effort to regulate drinking water and wastewater effluent has increased since the early 1900s. Beginning with an effort to control the discharge of wastewater into the environment, preliminary regulatory efforts focused on protecting public health. The goal of this early wastewater treatment program was to remove suspended and floatable material, treat biodegradable organics, and eliminate pathogenic organisms. Thus, regulatory efforts were

TABLE 14.1
Minimum National Standards for Secondary Treatment

Characteristic of Discharge	Unit of Measure	Average 30-Day	Average 7-Day
BOD_5	mg/L	30	45
Suspended solids	mg/L	30	45
Concentration	pH units	6.0–9.0	—

Source: 40 CFR Part 133, *Secondary Treatment Regulations*, U.S. Environmental Protection Agency, Washington, D.C., 1988.

pointed toward constructing wastewater treatment plants in an effort to alleviate these problems. Another problem then soon developed: progress. Progress in the sense that time marched on and with it so did a proliferation of growing cities, where it became increasingly difficult to find land required for wastewater treatment and disposal. Wastewater professionals soon recognized the need to develop methods of treatment that would accelerate the natural purification of water under controlled conditions in treatment facilities of comparatively smaller size.

Regulatory influence on water quality improvements in both wastewater and drinking water took a giant step forward in the 1970s. The Water Pollution Control Act Amendments of 1972 (Clean Water Act), established national water pollution control goals. At about the same time, the Safe Drinking Water Act (SDWA) passed by Congress in 1974 initiated a new era in the field of drinking water supply to the public.

14.3.1 CLEAN WATER ACT

In 1972, Congress adopted the Clean Water Act (CWA), which established a framework for achieving its national objective "to restore and maintain the chemical, physical, and biological integrity of the nation's waters." Congress decreed that, where attainable, water quality "provides for the protection and propagation of fish, shellfish, and wildlife and provides for recreation in and on the water." These goals are referred to as the "fishable and swimmable" goals of the Act.

Before CWA, no specific national water pollution control goals or objectives existed. Current standards require that municipal wastewater be given secondary treatment (to be discussed in detail later) and that most effluents meet the conditions shown in Table 14.1. The goal, via secondary treatment (i.e., the biological treatment component of a municipal treatment plant), was set so the principal components of municipal wastewater—suspended solids, biodegradable material, and pathogens—could be reduced to acceptable levels. Industrial dischargers are required to treat their wastewater to the level obtainable by the *best available technology* (BAT) for wastewater treatment in that particular type of industry.

Moreover, a National Pollutant Discharge Elimination System (NPDES) program was established based on uniform technological minimums with which each point source discharger must comply. Under NPDES, each municipality and industry discharging effluent into streams is assigned discharge permits. These permits reflect the secondary treatment and BAT standards. Water quality standards are the benchmark against which monitoring data are compared to assess the health of waters to develop *total maximum daily loads* in impaired waters. They are also used to calculate water-quality-based discharge limits in permits issued under NPDES.

14.3.2 SAFE DRINKING WATER ACT

The Safe Drinking Water Act (SDWA) of 1974 mandated the USEPA to establish drinking water standards for all public water systems serving 25 or more people or having 15 or more connections. Pursuant to this mandate, USEPA established maximum contaminant levels for drinking water delivered through public water distribution systems. The *maximum contaminant levels* (MCLs) of inorganics, organic chemicals, turbidity, and microbiological contaminants are shown in Table 14.2. USEPA's primary regulations are mandatory and must be complied with by all public water systems to which they apply. If analysis of the water produced by a water system indicates that an MCL for a contaminant is being exceeded, the system must take steps to stop providing the water to the public or initiate treatment to reduce the contaminant concentration to below the MCL.

The USEPA has also issued guidelines to states with regard to secondary drinking water standards. These appear in Table 14.3. These guidelines apply to drinking water contaminants that may adversely affect the aesthetic qualities of the water (i.e., those qualities that make water appealing and useful), such as odor and appearance. These qualities have no known adverse health effects, so secondary regulations are not mandatory; however, most drinking water systems comply with the limits. They have learned through experience that the odor and appearance of drinking water are not problems until customers complain, and one thing is certain—they will complain.

14.4 WATER QUALITY CHARACTERISTICS OF WATER AND WASTEWATER

In this section, individual pollutants and stressors that affect water quality are described. Knowledge of the parameters and characteristics most commonly associated with water and wastewater treatment processes is essential to the water/wastewater operator. Water and wastewater practitioners are encouraged to take a holistic approach to managing water quality problems. It is important to point out that when this text refers to *water quality* the definition used is predicated on the intended use of the water. Many

TABLE 14.2
USEPA Primary Drinking Water Standards

1. Inorganic Contaminant Levels

Contaminants	Level (mg/L)
Arsenic	0.05
Barium	1.00
Cadmium	0.010
Chromium	0.05
Lead	0.05
Mercury	0.002
Nitrate	10.00
Selenium	0.01
Silver	0.05

2. Organic Contaminant Levels

Chemical	Maximum Contaminant Level (MCL) (mg/L)
Chlorinated hydrocarbons	
Endrin	0.0002
Lindane	0.004
Mexthoxychlor	0.1
Toxaphene	0.005
Chlorophenoxy	
2,4-D	0.1
2,4,5-TP (Silvex®)	0.01

3. Maximum Levels of Turbidity

Reading Basis	Maximum Contaminant Level (MCL) Turbidity Units (TUs)
Turbidity reading (monthly average)	1 TU or up to 5 TUs if the water supplier can demonstrate to the state that the higher turbidity does not interfere with disinfection, prevent maintenance of an effective disinfectant agent throughout the distribution system, or interfere with microbiological determinations
Turbidity reading (based on average of two consecutive days)	5 TUs

4. Microbiological Contaminants

Test Method	Monthly Basis	Individual Sample Basis — Fewer Than 20 Samples/Month	Individual Sample Basis — More Than 20 Samples/Month
Membrane filter technique	1/100 mL average daily	*Number of coliform bacteria not to exceed:*	
		4/100 mL in more than 1 sample	4/100 mL in more than 5% of samples
Fermentation		*Coliform bacteria shall not be present in:*	
10-mL standard portions	More than 10% of the portions	3 or more portions in more than 1 sample	3 or more portions in more than 5% of samples
100-mL standard portions	More than 60% of the portions	5 portions in more than 1 sample	5 portions in more than 20% of samples

Source: Adapted from USEPA, *National Interim Primary Drinking Water Regulations*, U.S. Environmental Protection Agency, Washington, D.C., 1975.

parameters have evolved that qualitatively reflect the impact of various contaminants (impurities) on selected water uses; the following sections provide a brief discussion of these parameters.

14.4.1 Physical Characteristics of Water and Wastewater

The physical characteristics of water and wastewater are germane to the discussion at hand. These represent a category of parameters or characteristics that can be used to describe water quality, including those that are apparent to the senses of smell, taste, sight, and touch. Solids, turbidity, color, taste and odor, and temperature also fall into this category.

14.4.1.1 Solids

Other than gases, all contaminants of water contribute to the solids content. Classified by their size and state, by their chemical characteristics, and by their size distribution, solids can be dispersed in water in both suspended

TABLE 14.3
Secondary Maximum Contaminant Levels

Contaminant	Level	Adverse Effect
Chloride	250 mg/L	Taste
Color	15 cu	Appearance
Copper	1 mg/L	Taste and odor
Corrosivity	Noncorrosive	Taste and odor
Fluoride	2 mg/L	Dental fluorosis
Foaming agents	0.5 mg/L	Appearance
Iron	0.3 mg/L	Appearance
Manganese	0.05 mg/L	Laundry discoloration
Odor	3 TON	Unappealing to drink
pH	6.5–8.5	Corrosion or scaling
Sulfate	250 mg/L	Laxative effect
Total dissolved solids	500 mg/L	Taste, corrosion
Zinc	5 mg/L	Taste, appearance

Note: Cu, color unit; TON, threshold odor number.

Source: Adapted from McGhee, T.J., *Water Supply and Sewerage*, 6th ed., McGraw-Hill, New York, 1991.

and dissolved forms. With regard to size, solids in water and wastewater can be classified as suspended, settleable, colloidal, or dissolved. Solids are also characterized as being *volatile* or *nonvolatile*. The distribution of solids is determined by computing the percentage of filterable solids by size range. Solids typically include inorganic solids such as silt, sand, gravel, and clay from riverbanks and organic matter such as plant fibers and microorganisms from natural or manmade sources. We use the term *siltation* to describe the suspension and deposition of small sediment particles in water bodies. In flowing water, many of these contaminants result from the erosive action of water flowing over surfaces.

Sedimentation and siltation can severely alter aquatic communities. Sedimentation may clog and abrade fish gills, suffocate eggs and aquatic insect larvae on the bottom, and fill in the pore space between bottom cobbles where fish lay eggs. Suspended silt and sediment interfere with recreational activities and aesthetic enjoyment of streams and lakes by reducing water clarity and filling in lakes. Sediment may also carry other pollutants into surface waters. Nutrients and toxic chemicals may attach to sediment particles on land and ride the particles into surface waters, where the pollutants may settle with the sediment or detach and become soluble in the water column.

Suspended solids are a measure of the weight of relatively insoluble materials in the ambient water. These materials enter the water column as soil particles from land surfaces or sand, silt, and clay from stream-bank erosion of channel scour. Suspended solids can include both organic (detritus and biosolids) and inorganic (sand or finer colloids) constituents.

In water, suspended material is objectionable because it provides adsorption sites for biological and chemical agents. These adsorption sites provide a protective barrier for attached microorganisms against the chemical action of chlorine. In addition, suspended solids in water may be degraded biologically, resulting in objectionable byproducts. Thus, the removal of these solids is of great concern in the production of clean, safe drinking water and wastewater effluent.

In water treatment, the most effective means of removing solids from water is by *filtration*. It should be pointed out, however, that not all solids, such as colloids and other dissolved solids, can be removed by filtration. In wastewater treatment, the level of suspended solids is an important water quality parameter and is used to measure the quality of the wastewater influent, to monitor performance of several processes, and to measure the quality of effluent. Wastewater is normally 99.9% water and 0.1% solids. If a wastewater sample is evaporated, the solids remaining are the *total solids*. As shown in Table 14.1, USEPA has set a maximum suspended solids standard of 30 mg/L for most treated wastewater discharges.

14.4.1.2 Turbidity

One of the first things we notice about water is its clarity. The clarity of water is usually measured by its *turbidity*. Turbidity is a measure of the extent to which light is either absorbed or scattered by suspended material in water. Both the size and surface characteristics of the suspended material influence absorption and scattering. Although algal blooms can make waters turbid, in surface water most turbidity is related to the smaller inorganic components of the suspended solids burden, primarily the clay particles. Microorganisms and vegetable material may also contribute to turbidity. Wastewaters from industry and households usually contain a wide variety of turbidity-producing materials. Detergents, soaps, and various emulsifying agents contribute to turbidity.

In water treatment, turbidity is useful in defining drinking water quality. In wastewater treatment, turbidity measurements are particularly important whenever ultraviolet (UV) irradiation is used in the disinfection process. For UV irradiation to be effective in disinfecting wastewater effluent, the UV light must be able to penetrate the stream flow. Obviously, stream flow that is turbid works to reduce the effectiveness of irradiation (penetration of light).

The colloidal material associated with turbidity provides absorption sites for microorganisms and chemicals that may be harmful or cause undesirable tastes and odors. Moreover, the adsorptive characteristics of many colloids provide protection from disinfection processes for microorganisms. Turbidity in running waters also interferes with light penetration and photosynthetic reactions.

14.4.1.3 Color

Color is another physical characteristic by which the quality of water can be judged. Pure water is colorless. Water takes on color when foreign substances such as organic matter from soils, vegetation, minerals, and aquatic organisms are present. Color can also be contributed to water by municipal and industrial wastes. Color in water is classified as either *true color* or *apparent color*. True color is the color of water that is partly due to dissolved solids that remain after removal of suspended matter. Color contributed by suspended matter is said to have apparent color. In water treatment, true color is the most difficult to remove.

Note: Water also has an *intrinsic color*, and this color has a unique origin. Intrinsic color is easy to discern, as can be seen in Crater Lake, Oregon, which is known for its intense blue color. The appearance of the lake varies from turquoise to deep navy blue, depending on whether the sky is hazy or clear. Pure water and ice have a pale blue color.

The obvious problem with colored water is that it is not acceptable to the public; that is, given a choice, the public prefers clear, uncolored water. Another problem with colored water is the effect it has on laundering, papermaking, manufacturing, textiles, and food processing. The color of water has a profound impact on its marketability for both domestic and industrial use.

In water treatment, color is not usually considered unsafe or unsanitary but is a treatment problem in that it exerts a chlorine demand, which reduces the effectiveness of chlorine as a disinfectant. In wastewater treatment, color is not necessarily a problem but instead is an indicator of the *condition* of the wastewater. Condition refers to the age of the wastewater, which, along with odor, provides a qualitative indication of its age. Early in the flow, wastewater is a light brownish-gray color. The color of wastewater containing dissolved oxygen (DO) is nor-

TABLE 14.4
Significance of Color in Wastewater Influent

Color	Problem Indicated
Gray	None
Red	Blood or other industrial wastes
Green, yellow	Industrial wastes (e.g., paints) not pretreated
Red or other soil color	Surface runoff into influent; industrial flows
Black	Septic conditions or industrial flows

mally gray. Black-colored wastewater accompanied by foul odors and containing little or no DO is said to be *septic*. Table 14.4 provides wastewater color information. As the travel time in the collection system increases (flow becomes increasingly more septic) and more anaerobic conditions develop, the color of the wastewater changes from gray to dark gray and ultimately to black.

14.4.1.4 Taste and Odor

Taste and *odor* are used jointly in the vernacular of water science. The term *odor* is used in wastewater; taste, obviously, is not a consideration for wastewater. Domestic sewage should have a musty odor. Bubbling gas or a foul odor may indicate industrial wastes, anaerobic (septic) conditions, and operational problems. Refer to Table 14.5 for typical wastewater odors, possible problems, and solutions.

In wastewater, odors are of major concern, especially to those who reside in close proximity to a wastewater treatment plant. These odors are generated by gases produced by decomposition of organic matter or by substances added to the wastewater. Because these substances are volatile, they are readily released to the atmosphere at any point where the wastestream is exposed, particularly if there is turbulence at the surface.

Most people would argue that all wastewater is the same; it has a disagreeable odor. It is difficult to argue the disagreeable odor point; however, one wastewater operator

TABLE 14.5
Odors in Wastewater Treatment Plant

Odor	Location	Problem	Possible Solution
Earthy, musty	Primary and secondary units	No problem (normal)	None required
Hydrogen sulfide	Influent	Septic (rotten-egg odor)	Aerate, chlorinate, ozonizate
	Primary clarifier	Septic sludge	Remove sludge
	Activated sludge	Septic sludge	Remove sludge
	Trickling filters	Septic conditions	More air, less BOD
	Secondary clarifiers	Septic conditions	Remove sludge
	Chlorine contact tank	Septic conditions	Remove sludge
	General plant	Septic conditions	Good housekeeping
Chlorine	Chlorine contact tank	Improper chlorine dosage	Adjust chlorine dosage controls
Industrial odors	General plant	Inadequate pretreatment	Enforce sewer use regulations

told us that wastewater "smells great—smells just like money to me—money in the bank," she said. This was an operator's view. We also received another opinion of odor problems resulting from wastewater operations. This particular opinion, given by an odor control manager, was quite different. His opinion was "that odor control is a never-ending problem" and that to combat this difficult problem odors must be contained. In most urban plants, it has become necessary to physically cover all source areas such as treatment basins, clarifiers, aeration basins, and contact tanks to prevent odors from leaving the processes. These contained spaces must then be positively vented to wet-chemical scrubbers to prevent the buildup of a toxic concentration of gas.

As mentioned, in drinking water, taste and odor are not normally a problem until the consumer complains. The problem is, of course, that most consumers find taste and odor in water aesthetically displeasing. Also, taste and odor do not directly present a health hazard, but they can cause the customer to seek water that tastes and smells good but may not be safe to drink. Many consumers consider water to be tasteless and odorless; thus, when consumers discover that their drinking water has a taste or odor, or both, they immediately assume the drinking water is contaminated.

Water contaminants are attributable to contact with nature or human use. Taste and odor in water are caused by a variety of substances such as minerals, metals, and salts from the soil; constituents of wastewater; and end products produced in biological reactions. When water has a taste but no accompanying odor, the cause is usually inorganic contamination. Water that tastes bitter is usually alkaline, whereas salty water is commonly the result of metallic salts. When water has both taste and odor, however, the likely cause is organic materials. The list of possible organic contaminants is too long to record here, but petroleum-based products lead the list of offenders. Taste- and odor-producing liquids and gases in water are produced by biological decomposition of organics. A prime example of one of these is hydrogen sulfide, known best for its characteristic rotten-egg taste and odor. Certain species of algae also secrete an oily substance that may produce both taste and odor. When certain substances combine (such as organics and chlorine), the synergistic effect produces taste and odor.

In water treatment, a common method used to remove taste and odor is to oxidize the materials that cause the problem. Oxidants such as potassium permanganate and chlorine are used. Another common treatment method is to feed powdered activated carbon before the filter. The activated carbon has numerous small openings that absorb the components that cause the odor and tastes. These contained spaces must then be positively vented to wet-chemical scrubbers to prevent the buildup of toxic concentrations of gas.

14.4.1.5 Temperature

Heat is added to surface and groundwater in many ways. Some of these are natural, some artificial. Heat is added by natural means to Yellowstone Lake in Wyoming. This lake, one of the world's largest freshwater lakes, resides in a calderas, situated at more than 7700 feet (the largest high-altitude lake in North America). When one attempts to swim in Yellowstone Lake without benefit of a wetsuit, the bitter cold of the water literally takes one's breath away. If it were not for the hydrothermal discharges that occur in Yellowstone, though, the water would be even colder. With regard to human-heated water, this most commonly occurs whenever a raw water source is used for cooling water in industrial operations. The influent to industrial facilities is at normal ambient temperature. When it is used to cool machinery and industrial processes and then is discharged back to the receiving body, it is often heated. The problem with heat or temperature increases in surface waters is that they affect the solubility of oxygen in water, the rate of bacterial activity, and the rate at which gases are transferred to and from the water.

Note: It is important to point out that in the examination of water or wastewater, temperature is not normally used to evaluate either; however, temperature is one of the most important parameters in natural surface water systems. Surface waters are subject to great temperature variations.

Water temperature does determine, in part, how efficiently certain water treatment processes operate; for example, temperature has an effect on the rate at which chemicals dissolve and react. When water is cold, more chemicals are required for efficient coagulation and flocculation to take place. When water temperature is high, the result may be a higher chlorine demand because of the increased reactivity, as well as an increased level of algae and other organic matter in raw water. Temperature also has a pronounced effect on the solubility of gases in water.

Ambient temperature (temperature of the surrounding atmosphere) has the most profound and universal effect on the temperature of shallow, natural water systems. When water is used by industry to dissipate process waste heat, the discharge locations into surface waters may experience localized temperature changes that are quite dramatic. Other sources of increased temperatures in running water systems result because of clear-cutting practices in forests (where protective canopies are removed) and from irrigation flows returned to a body of running water.

In wastewater treatment, the temperature of wastewater varies greatly, depending on the type of operations being conducted at a particular installation. Wastewater is generally warmer than that of the water supply, because of the addition of warm water from industrial activities

TABLE 14.6
Chemical Constituents in Water

Calcium	Fluorine
Magnesium	Nitrate
Sodium	Silica
Potassium	Silica
Potassium	Total dissolved solids (TDS)
Iron	Hardness
Manganese	Color
Bicarbonate	pH
Carbonate	Turbidity
Sulfate	Temperature
Chloride	

and households. Wide variation in the wastewater temperature indicates heated or cooled discharges, often of substantial volume. They have any number of sources. Decreased temperatures after a snowmelt or rain event, for example, may indicate serious infiltration. In the treatment process itself, temperature not only influences the metabolic activities of the microbial population but also has a profound effect on such factors as gas transfer rates and the settling characteristics of the biological solids.

14.4.2 CHEMICAL CHARACTERISTICS OF WATER

Another category used to define or describe water quality is its chemical characteristics. The most important chemical characteristics are *total dissolved solids* (TDS), alkalinity, hardness, fluoride, metals, organics, and nutrients. Chemical impurities can be natural or manmade (industrial), or they can be added to raw water sources by enemy forces. Some chemical impurities cause water to behave as either an acid or a base. Because either condition has an important bearing on the water treatment process, the pH value must be determined. Generally, the pH influences the corrosiveness of the water, chemical dosages necessary for proper disinfection, and the ability to detect contaminants. The principal contaminants found in water are shown in Table 14.6. These chemical constituents are important because each one affects water use in some manner; each one either restricts or enhances specific uses. The pH of water is very important. As pH rises, for example, the equilibrium (between bicarbonate and carbonate) increasingly favors the formation of carbonate, which often results in the precipitation of carbonate salts. If you have ever had flow in a pipe system interrupted or a heat-transfer problem in your water heater system, most likely carbonate salts formed a difficult-to-dissolve scale within the system. It should be pointed out that not all carbonate salts have a negative effect on their surroundings. Consider, for example, the case of blue marl lakes; they owe their unusually clear, attractive appearance to carbonate salts.

Water has been called the *universal solvent*. This is, of course, a fitting description. The solvent capabilities of water are directly related to its chemical characteristics or parameters. As mentioned, in water quality management, total dissolved solids, alkalinity, hardness, fluorides, metals, organics, and nutrients are the major chemical parameters of concern.

14.4.2.1 Total Dissolved Solids

Because of the solvent properties of water, minerals dissolve from rocks and soil as water passes over and through them to produce total dissolved solids (TDS), which are any minerals, salts, metals, cations, or anions dissolved in water. TDS constitute a part of total solids (TS) in water and are the material remaining in water after filtration. Dissolved solids may be organic or inorganic. Water may be exposed to these substances within the soil, on surfaces, and in the atmosphere. The organic dissolved constituents of water come from the decay products of vegetation, from organic chemicals, and from organic gases.

Dissolved solids can be removed from water by distillation, electrodialysis, reverse osmosis, or ion exchange. It is desirable to remove these dissolved minerals, gases, and organic constituents because they may cause psychological effects and produce aesthetically displeasing color, taste, and odors. Although it is desirable to remove many of these dissolved substances from water, it is not prudent to remove them all. Consider the flat taste of pure, distilled water. Further, water has an equilibrium state with respect to dissolved constituents; thus, if water is out of equilibrium or undersaturated, it will aggressively dissolve materials it comes into contact with. Because of this problem, substances that are readily dissolvable are sometimes added to pure water to reduce its tendency to dissolve plumbing.

14.4.2.2 Alkalinity

Another important characteristic of water is its *alkalinity*, which is a measure of the ability of water to neutralize acid, or an expression of its buffering capacity. The major chemical constituents of alkalinity in natural water supplies are the bicarbonate, carbonate, and hydroxyl ions. These compounds are primarily the carbonates and bicarbonates of sodium, potassium, magnesium, and calcium. These constituents originate from carbon dioxide (from the atmosphere and as a byproduct of microbial decomposition of organic material) and minerals (primarily from chemical compounds dissolved from rocks and soil). Highly alkaline waters are unpalatable; however, this condition has little known significance for human health. The principal problem with alkaline water is the reactions that occur between alkalinity and certain substances in the water. Alkalinity is important for fish and aquatic life because it protects or buffers against rapid pH changes.

TABLE 14.7
Classifications of Hardness

Range of Hardness (mg/L [ppm] as CaCO$_3$)	Descriptive Classification
1–50	Soft
51–150	Moderately hard
151–300	Hard
Above 300	Very hard

Moreover, the resultant precipitate can foul water system appurtenances. In addition, alkalinity levels affect the efficiency of certain water treatment processes, especially the coagulation process.

14.4.2.3 Hardness

Hardness is due to the presence of multivalent metal ions, which come from minerals dissolved in water. Hardness is based on the ability of these ions to react with soap to form a precipitate or soap scum. In freshwater, the primary ions are calcium and magnesium; however, iron and manganese may also contribute. Hardness is classified as *carbonate* hardness or *noncarbonate* hardness. Carbonate hardness is equal to alkalinity but a noncarbonate fraction may include nitrates and chlorides. Hardness is either temporary or permanent. Carbonate hardness (temporary hardness) can be removed by boiling. Noncarbonate hardness cannot be removed by boiling and is classified as permanent.

Hardness values are expressed as an equivalent amount or equivalent weight of calcium carbonate (the *equivalent weight* of a substance is its atomic or molecular weight divided by n). Water with a hardness of less than 50 ppm is soft. Above 200 ppm, domestic supplies are usually blended to reduce the hardness value. Table 14.7 shows the U.S. Geological Survey classifications.

The impact of hardness can be measured in economic terms. Soap consumption, for example, represents an economic loss to the hard-water user, as it is necessary to use more soap to get a lather. Also, when a lather is finally built up, the water has been softened by the soap. The precipitate formed by the hardness and the soap (soap curd) adheres to just about anything (tubs, sinks, dishwashers) and may stain clothing, dishes, and other items. The residues of the hardness–soap precipitate may penetrate into the pores of the skin, causing the skin to feel rough and uncomfortable. Today these problems have been largely reduced by the development of synthetic soaps and detergents that do not react with hardness; however, hardness still leads to other problems, such as scaling. Scaling occurs when carbonate hard water is heated and calcium carbonate and magnesium

hydroxide are precipitated out of solution, forming a rock-hard scale that clogs hot water pipes and reduces the efficiency of boilers, water heaters, and heat exchangers. Hardness, especially with the presence of magnesium sulfates, can also lead to the development of a laxative effect in new consumers.

The use of hard water does offer some advantages, though, in that: (1) hard water aids in the growth of teeth and bones, (2) hard water reduces the toxicity of poisoning by lead oxide from lead pipelines, and (3) soft waters are suspected to be associated with cardiovascular diseases (Rowe and Abdel-Magid, 1995).

14.4.2.4 Fluoride

We purposely fluoridate a range of everyday products, notably toothpaste and drinking water, because for decades we have believed that fluoride in small doses has no adverse effects on health that offset its proven benefits in preventing dental decay. The jury is still out, however, on the real benefits of fluoride, even in small amounts. Fluoride is seldom found in appreciable quantities in surface waters and appears in groundwater in only a few geographical regions; for example, fluoride is sometimes found in a few types of igneous or sedimentary rocks. Fluoride is toxic to humans in large quantities. Fluoride is also toxic to some animals. Certain plants used for fodder have the ability to store and concentrate fluoride. When animals consume this forage, they ingest an enormous overdose of fluoride. Animals' teeth become mottled, they lose weight, give less milk, grow spurs on their bones, and become so crippled they must be destroyed (Koren, 1991).

Fluoride used in small concentrations (about 1.0 mg/L in drinking water) can be beneficial. Experience has shown that drinking water containing a proper amount of fluoride can reduce tooth decay by 65% in children between ages 12 to 15. When large concentrations are used (>2.0 mg/L), discoloration of teeth may result. Adult teeth are not affected by fluoride. USEPA has set the upper limits for fluoride based on ambient temperatures. Because people drink more water in warmer climates, fluoride concentrations should be lower in these areas.

Note: How does fluoridization of a drinking water supply actually work to reduce tooth decay? Fluoride combines chemically with tooth enamel when permanent teeth are forming. The result is teeth that are harder, stronger, and more resistant to decay.

14.4.2.5 Metals

Although iron and manganese are most commonly found in groundwaters, surface waters may also contain significant amounts at times. Metal ions are dissolved in groundwater and surface water when the water is exposed to rock or soil containing the metals, usually in the form of metal salts. Metals can also enter with discharges from sewage treatment plants, industrial plants, and other sources. The metals most often found in the highest concentrations in natural waters are calcium and magnesium. These are usually associated with a carbonate anion and come from the dissolution of limestone rock. As mentioned under the discussion of hardness, the higher the concentration of these metal ions, the harder the water; however, in some waters other metals can contribute to hardness. Calcium and magnesium are nontoxic and normally absorbed by living organisms more readily than the other metals; therefore, if the water is hard, the toxicity of a given concentration of a toxic metal is reduced. Conversely, in soft, acidic water, the same concentrations of metals may be more toxic. In natural water systems, other nontoxic metals are generally found in very small quantities. Most of these metals cause taste problems well before they reach toxic levels. Fortunately, toxic metals are present in only minute quantities in most natural water systems; however, even in small quantities, toxic metals in drinking water are harmful to humans and other organisms. Arsenic, barium, cadmium, chromium, lead, mercury, and silver are toxic metals that may be dissolved in water. Arsenic, cadmium, lead, and mercury, all cumulative toxins, are particularly hazardous. These particular metals are concentrated by the food chain, thereby posing the greatest danger to organisms near the top of the chain.

14.4.2.6 Organics

Organic chemicals in water primarily emanate from synthetic compounds that contain carbon, such as polychlorinated biphenyls (PCBs), dioxin, and dichlorodiphenyltrichloroethane (DDT), all of which are toxic organic chemicals. These synthesized compounds often persist and accumulate in the environment because they do not readily breakdown in natural ecosystems. Many of these compounds can cause cancer in people and birth defects in other predators near the top of the food chain, such as birds and fish. The presence of organic matter in water is troublesome for the following reasons: (1) color formation, (2) taste and odor problems, (3) oxygen depletion in streams, (4) interference with water treatment processes, and (5) the formation of halogenated compounds when chlorine is added to disinfect water (Tchobanoglous and Schroeder, 1985).

Generally, the source of organic matter in water is from decaying leaves, weeds, and trees; the amount of these materials present in natural waters is usually low. The general category of "organics" in natural waters includes organic matter whose origins could be from both natural sources and human activities. It is important

to distinguish natural organic compounds from organic compounds that are solely manmade (anthropogenic), such as pesticides and other synthetic organic compounds.

Many organic compounds are soluble in water, and surface waters are more prone to contamination by natural organic compounds than are groundwaters. In water, dissolved organics are usually divided into two categories: *biodegradable* and *nonbiodegradable*. Biodegradable material consists of organics that can be utilized for nutrients (food) by naturally occurring microorganisms within a reasonable length of time. These materials usually consist of alcohols, acids, starches, fats, proteins, esters, and aldehydes. They may result from domestic or industrial wastewater discharges, or they may be end products of the initial microbial decomposition of plant or animal tissue. The principle problem associated with biodegradable organics is the effect resulting from the action of microorganisms. Moreover, some biodegradable organics can cause color, taste, and odor problems.

Oxidation and *reduction* play an important accompanying role in microbial utilization of dissolved organics. In oxidation, oxygen is added or hydrogen is deleted from elements of the organic molecule. Reduction occurs when hydrogen is added to or oxygen is deleted from elements of the organic molecule. The oxidation process is by far more efficient and is predominant when oxygen is available. In *oxygen-present* (aerobic) environments, the end products of microbial decomposition of organics are stable and acceptable compounds. On the other hand, *oxygen-absent* (anaerobic) decomposition results in unstable and objectionable end products.

The quantity of oxygen-consuming organics in water is usually determined by measuring the *biochemical oxygen demand* (BOD), which is the amount of dissolved oxygen required by aerobic decomposers to break down the organic materials in a given volume of water over a 5-day incubation period at 20°C (68°F).

Nonbiodegradable organics are resistant to biological degradation; for example, constituents of woody plants such as tannin and lignic acids, phenols, and cellulose are found in natural water systems and are considered refractory (resistant to biodegradation). In addition, some polysaccharides with exceptionally strong bonds and benzene with its ringed structure are essentially nonbiodegradable. An example is benzene associated with the refining of petroleum. Some organics are toxic to organisms and thus are nonbiodegradable. These include the organic pesticides and compounds that have combined with chlorine. Pesticides and herbicides have found widespread use in agriculture, forestry (silviculture), and mosquito control. Surface streams are contaminated via runoff and wash off by rainfall. These toxic substances are harmful to some fish, shellfish, predatory birds, and mammals. Some compounds are toxic to humans.

14.4.2.7 Nutrients

Nutrients (biostimulants) are essential building blocks for healthy aquatic communities, but excess nutrients (especially nitrogen and phosphorus compounds) overstimulate the growth of aquatic weeds and algae. Excessive growth of these organisms, in turn, can clog navigable waters, interfere with swimming and boating, outcompete native submerged aquatic vegetation, and, with excessive decomposition, lead to oxygen depletion. Oxygen concentrations can fluctuate daily during algal blooms, rising during the day as algae perform photosynthesis and falling at night as algae continue to respire and consume oxygen. Beneficial bacteria also consume oxygen as they decompose the abundant organic food supply in dying algae cells.

Plants require large amounts of the nutrients carbon, nitrogen, and phosphorus; otherwise, their growth will be *limited*. Carbon is readily available from a number of natural sources, including alkalinity, decaying products of organic matter, and dissolved carbon dioxide from the atmosphere. Because carbon is readily available, it is seldom the *limiting nutrient*. This is an important point because it suggests that identifying and reducing the supply of a particular nutrient can control algal growth. In most cases, nitrogen and phosphorus are essential growth factors and are the limiting factors in aquatic plant growth. Freshwater systems are most often limited by phosphorus.

Nitrogen gas (N_2), which is extremely stable, is the primary component of the Earth's atmosphere. Major sources of nitrogen include runoff from animal feedlots, fertilizer runoff from agricultural fields, municipal wastewater discharges, and certain bacteria and blue–green algae that can obtain nitrogen directly from the atmosphere. In addition, certain forms of acid rain can also contribute nitrogen to surface waters. Nitrogen in water is commonly found in the form of nitrate (NO_3). Nitrate in drinking water can lead to serious problems. Specifically, nitrate poisoning in infant humans, as well as animals, can cause serious problems and even death. Bacteria commonly found in the intestinal tract of infants can convert nitrate to highly toxic nitrites (NO_2). Nitrite can replace oxygen in the bloodstream and result in oxygen starvation, which causes a bluish discoloration of the infant known as blue-baby syndrome.

In aquatic environments, phosphorus is found in the form of phosphate. Major sources of phosphorus include phosphates in detergents, fertilizer and feedlot runoff, and municipal wastewater discharges.

14.4.3 CHEMICAL CHARACTERISTICS OF WASTEWATER

The chemical characteristics of wastewater consist of three parts: (1) organic matter, (2) inorganic matter, and (3) gases. In wastewater of medium strength, about 75% of the suspended solids and 40% of the filterable solids are organic in nature (Metcalf & Eddy, 2003). The organic substances of interest in this discussion include proteins, oil and grease, carbohydrates, and detergents (surfactants).

14.4.3.1 Organic Substances

Proteins are nitrogenous organic substances of high molecular weight found in the animal kingdom and to a lesser extent in the plant kingdom. The amount present varies, from a small percentage found in tomatoes and other watery fruits and in the fatty tissues of meat to a high percentage in lean meats and beans. All raw foodstuffs, plant and animal, contain proteins. Proteins consist wholly or partially of very large numbers of amino acids. They also contain carbon, hydrogen, oxygen, sulfur, phosphorus, and a fairly high and constant proportion of nitrogen. The molecular weight of proteins is quite high.

Proteinaceous materials constitute a large part of the wastewater biosolids; biosolids particles that do not consist of pure protein will be covered with a layer of protein that will govern their chemical and physical behavior (Coakley, 1975). Moreover, the protein content ranges from 15 to 30% of the organic matter present for digested biosolids and 28 to 50% in the case of activated biosolids. Proteins and urea are the chief sources of nitrogen in wastewater. When proteins are present in large quantities, microorganisms decompose and produce end products that have objectionable foul odors. During this decomposition process, proteins are hydrolyzed to amino acids and then further degraded to ammonia, hydrogen sulfide, and simple organic compounds.

Oils and *grease* are another major component of foodstuffs. They are also usually related to spills or other releases of petroleum products. Minor oil and grease problems can result from wet weather runoff from highways or the improper disposal in storm drains of motor oil. They are insoluble in water but dissolve in organic solvents such as petroleum, chloroform, and ether. Fats, oils, waxes, and other related constituents found in wastewater are commonly grouped under the category of grease. Fats and oils contributed to domestic wastewater include butter, lard, margarine, and vegetable fats and oils. Fats, which are compounds of alcohol and glycerol, are among the more stable of organic compounds and are not easily decomposed by bacteria; however, they can be broken down by mineral acids, resulting in the formation of fatty acid and glycerin. When these glycerides of fatty acids are liquid at ordinary temperature, they are considered oils and those that are solids are fats.

The grease content of wastewater can cause many problems in wastewater treatment unit processes; for example, high grease content can cause clogging of filters, nozzles, and sand beds (Gilcreas et al., 1975). Moreover, grease can coat the walls of sedimentation tanks and decompose, thus increasing the amount of scum. Additionally, if grease is not removed before discharge of the effluent, it can interfere with the biological processes in the surface waters and create unsightly floating matter and films (Rowe and Abdel-Magid, 1995). In the treatment process, grease can coat trickling filters and interfere with the activated sludge process which, in turn, interferes with the transfer of oxygen from the liquid to the interior of living cells (Sawyer et al., 1994).

Carbohydrates, which are widely distributed in nature and found in wastewater, are organic substances that include starch, cellulose, sugars, and wood fibers; they contain carbon, hydrogen, and oxygen. Sugars are soluble but starches are insoluble in water. The primary function of carbohydrates in higher animals is to serve as a source of energy. In lower organisms (e.g., bacteria), carbohydrates are utilized to synthesize fats and proteins as well as energy. In the absence of oxygen, the end products of decomposition of carbohydrates are organic acids and alcohols, as well as gases such as carbon dioxide and hydrogen sulfide. The formation of large quantities of organic acids can affect the treatment process by overtaxing the buffering capacity of the wastewater, resulting in a drop in pH and a cessation of biological activity.

Detergents (surfactants) are large organic molecules that are slightly soluble in water and cause foaming in wastewater treatment plants and in the surface waters into which the effluent is discharged. Probably the most serious effect detergents can have on wastewater treatment processes is in their tendency to reduce the oxygen uptake in biological processes. According to Rowe and Abdel-Magid (1995), "Detergents affect wastewater treatment processes because they (1) lower the surface, or interfacial, tension of water and increase its ability to wet surfaces with which they come in contact; (2) emulsify grease and oil, deflocculate colloids; (3) induce flotation of solids and give rise to foams; and (4) may kill useful bacteria and other living organisms." Since the development and increasing use of synthetic detergents, many of these problems have been reduced or eliminated.

14.4.3.2 Inorganic Substances

Several inorganic components common to both wastewater and natural waters are important in establishing and controlling water quality. Inorganic load in water is the result of discharges of treated and untreated wastewater, various geologic formations, and inorganic substances left in the water after evaporation. Natural waters dissolve rocks and minerals with which they come into contact. As mentioned, many of the inorganic constituents found in natural waters are also found in wastewater. Many of these constituents are added via human activities. These inorganic constituents include pH, chlorides, alkalinity, nitrogen, phosphorus, sulfur, toxic inorganic compounds, and heavy metals.

When the pH of a water or wastewater is considered, we are simply referring to the hydrogen ion concentration. Acidity, the concentration of hydrogen ions, drives many chemical reactions in living organisms. A pH value of 7 represents a neutral condition. A low pH value (less than 5) indicates acidic conditions; a high pH (greater than 9) indicates alkaline conditions. Many biological processes, such as reproduction, cannot function in acidic or alkaline waters. Acidic conditions also aggravate toxic contamination problems because sediments release toxicants in acidic waters.

Many of the important properties of wastewater are due to the presence of weak acids and bases and their salts. The wastewater treatment process is made up of several different unit processes (these are discussed later). It can be safely stated that one of the most important unit processes in the overall wastewater treatment process is *disinfection*. pH has an effect on disinfection. This is particularly the case with regard to disinfection using chlorine. With increases in pH, the contact time required for disinfection using chlorine increases. Common sources of acidity include mine drainage, runoff from mine tailings, and atmospheric deposition.

In the form of the Cl^- ion, *chloride* is a major inorganic constituent in water and wastewater. Sources of chlorides in natural waters include: (1) leaching of chloride from rocks and soils; (2) saltwater intrusion in coastal areas; (3) agricultural, industrial, domestic, and human wastewater; and (4) infiltration of groundwater into sewers adjacent to salt water. The salty taste produced by the chloride concentration in potable water is variable and depends on the chemical composition of the water. In wastewater, the chloride concentration is higher than in raw water because sodium chloride (salt) is a common part of the diet and passes unchanged through the digestive system. Because conventional methods of waste treatment do not remove chloride to any significant extent, higher than usual chloride concentrations can be taken as an indication that the body of water is being used for waste disposal (Metcalf & Eddy, 2003).

As mentioned earlier, *alkalinity* is a measure of the buffering capacity of water and in wastewater helps to resist changes in pH caused by the addition of acids. Alkalinity is caused by chemical compounds dissolved from soil and geologic formations and is mainly due to the presence of hydroxyl and bicarbonate ions. These compounds are primarily the carbonates and bicarbonates of calcium, potassium, magnesium, and sodium. Wastewater is usually alkaline. Alkalinity is important in wastewater treatment because anaerobic digestion requires sufficient alkalinity to ensure that the pH will not drop below 6.2; if alkalinity does drop below this level, the methane bacteria cannot function. For the digestion process to operate successfully, the alkalinity must range from about 1000 to 5000 mg/L as calcium carbonate. Alkalinity in wastewater is also important when chemical treatment is used, in biological nutrient removal, and whenever ammonia is removed by air stripping.

In domestic wastewater, "nitrogen compounds result from the biological decomposition of proteins and from urea discharged in body waste" (Peavy et al., 1987). In wastewater treatment, biological treatment cannot proceed unless nitrogen, in some form, is present. Nitrogen must be present in the form of organic nitrogen (N), ammonia (NH_3), nitrite (NO_2), or nitrate (NO_3). Organic nitrogen includes such natural constituents as peptides, proteins, urea, nucleic acids, and numerous synthetic organic materials. Ammonia is present naturally in wastewaters. It is produced primarily by deaeration of organic nitrogen-containing compounds and by hydrolysis of urea. Nitrite, an intermediate oxidation state of nitrogen, can enter a water system through use as a corrosion inhibitor in industrial applications. Nitrate is derived from the oxidation of ammonia.

Nitrogen data are essential for evaluating the treatability of wastewater by biological processes. If nitrogen is not present in sufficient amounts, it may be necessary to add it to the waste to make it treatable. When the treatment process is complete, it is important to determine how much nitrogen is in the effluent. This is important because the discharge of nitrogen into receiving waters may stimulate algal and aquatic plant growth. These, of course, exert a high oxygen demand at nighttime, which adversely affects aquatic life and has a negative impact on the beneficial use of water resources.

Phosphorus (P) is a macronutrient that is necessary to all living cells and is a ubiquitous constituent of wastewater. It is primarily present in the form of phosphates, the salts of phosphoric acid. Municipal wastewaters may contain 10 to 20 mg/L phosphorus as P, much of which comes from phosphate builders in detergents. Noxious algal blooms that occur in surface waters have generated much interest in controlling the amount of phosphorus compounds that enter surface waters in domestic and industrial waste discharges and natural runoff. This is particularly the case in the United States, because approximately 15% of the population contributes wastewater effluents to lakes, resulting in *eutrophication* of these water bodies. Eutrophication leads to significant changes in water quality. Reducing phosphorus inputs to receiving waters can control this problem.

Sulfur (S) is required for the synthesis of proteins and is released in their degradation. The sulfate ion occurs naturally in most water supplies and is present in wastewater as well. Sulfate is reduced biologically to sulfide, which in turn can combine with hydrogen to form hydrogen sulfide (H_2S). H_2S is toxic to animals and plants. H_2S in interceptor systems can cause severe corrosion to pipes and appurtenances. Moreover, in certain concentrations, H_2S is a deadly toxin.

Toxic inorganic compounds such as copper, lead, silver, arsenic, boron, and chromium are classified as priority pollutants and are toxic to microorganisms. They must be taken into consideration in the design and operation of a biological treatment process. When introduced into a treatment process, these contaminants can kill off the microorganisms required for treatment and thus halt the treatment process.

Heavy metals are major toxicants found in industrial wastewaters; they may adversely affect the biological treatment of wastewater. Mercury, lead, cadmium, zinc, chromium, and plutonium are among the so-called heavy metals—those with a high atomic mass. (It should be noted that the classification of heavy metals is rather loose and is taken by some to include arsenic, beryllium, and selenium, which are not really metals and are better termed *toxic metals*.) The presence of any of these metals in excessive quantities will interfere with many beneficial uses of water because of their toxicity. Urban runoff is a major source of lead and zinc in many water bodies. The lead comes from the exhaust of automobiles using leaded gasoline, while zinc comes from tire wear.

Note: Lead is a toxic metal that is harmful to human health; there is *no* safe level for lead exposure. It is estimated that up to 20% of the total lead exposure in children can be attributed to a waterborne route (i.e., consumption of contaminated water).

14.4.4 Biological Characteristics of Water and Wastewater

Specialists or practitioners who work in the water or wastewater treatment field must have not only a general understanding of the microbiological principles presented in Chapter 12 but also some knowledge of the biological characteristics of water and wastewater. This knowledge begins with an understanding that water may serve as a medium in which thousands of biological species spend part, if not all, of their life cycles. It is important to understand that, to some extent, all members of the biological community are water quality parameters, because their presence or absence may indicate in general terms the characteristics of a given body of water.

The presence or absence of certain biological organisms is of primary importance to the water/wastewater specialist. These are, of course, *pathogens*. Pathogens are organisms that are capable of infecting or transmitting diseases in humans and animals. It should be pointed out that these organisms are not native to aquatic systems and usually require an animal host for growth and reproduction. They can, however, be transported by natural water systems. These waterborne pathogens include species of bacteria, viruses, protozoa, and parasitic worms (helminths). In the following sections, a brief review of each of these species is provided.

14.4.4.1 Bacteria

The word *bacteria* (singular: *bacterium*) comes from the Greek word for "rod" or" staff," a shape characteristic of many bacteria. Recall that bacteria are single-celled, microscopic organisms that multiply by splitting in two (binary fission). To multiply, they need carbon dioxide from the air if they are autotrophs and from organic compounds (dead vegetation, meat, sewage) if they are heterotrophs. Their energy comes either from sunlight, if they are photosynthetic, or from chemical reaction, if they are chemosynthetic. Bacteria are present in air, water, earth, rotting vegetation, and the intestines of animals. Human and animal wastes are the primary sources of bacteria in water. These sources of bacterial contamination include runoff from feedlots, pastures, dog runs, and other land areas where animal wastes are deposited. Additional sources include seepage or discharge from septic tanks and sewage treatment facilities. Bacteria from these sources can enter wells that are either open at the land surface or do not have watertight casings or caps. Gastrointestinal disorders are common symptoms of most diseases transmitted by waterborne pathogenic bacteria. In wastewater treatment processes, bacteria are fundamental, especially in the degradation of organic matter, which takes place in trickling filters, activated biosolids processes, and biosolids digestion.

14.4.4.2 Viruses

A *virus* is an entity that carries the information required for its replication but does not possess the machinery for such replication (Sterritt and Lester, 1988). Thus, they are obligate parasites that require a host in which to live. They are the smallest biological structures known, so they can only be seen with the aid of an electron microscope. Waterborne viral infections are usually indicated by disorders with the nervous system rather than of the gastrointestinal tract. Viruses that are excreted by human beings may become a major health hazard to public health. Waterborne viral pathogens are known to cause poliomyelitis and infectious hepatitis. Testing for viruses in water is difficult because: (1) they are small, (2) they are of low concentrations in natural waters, (3) there are numerous varieties, (4) they are unstable, and (5) limited identification methods are available. Because of these testing problems and the uncertainty of viral disinfection, direct recycling of wastewater and the practice of land application of wastewater is a cause of concern (Peavy et al., 1987).

14.4.4.3 Protozoa

Protozoa (singular: protozoan) are mobile, single-celled, complete, self-contained organisms that can be free living or parasitic, pathogenic or nonpathogenic, microscopic or macroscopic. Protozoa range in size from two to several hundred microns in length. They are highly adaptable and widely distributed in natural waters, although only a few are parasitic. Most protozoa are harmless; only a few cause illness in humans—*Entamoeba histolytica* (amebiasis) being an exception. Because aquatic protozoa form cysts during adverse environmental conditions, they are difficult to deactivate by disinfection and must undergo filtration to be removed.

14.4.4.4 Worms (Helminths)

Worms are the normal inhabitants in organic mud and organic slime. They have aerobic requirements but can metabolize solid organic matter not readily degraded by other microorganisms. Water contamination may result from human and animal waste that contains worms. Worms pose hazards primarily to those persons who come into direct contact with untreated water; thus, swimmers in surface water polluted by sewage or stormwater runoff from cattle feedlots and sewage plant operators are at particular risk.

CHAPTER REVIEW QUESTIONS

14.1   What are the characteristics or range of characteristics that make water appealing and useful?
14.2   What is the process by which water vapor is emitted by leaves?
14.3   Water we see is known as _____.
14.4   The leading cause of impairment for rivers, lakes, and estuaries is _____.
14.5   All contaminants of water contribute to the _____.
14.6   The clarity of water is usually measured by its _____.
14.7   Water has been called the _____.
14.8   A measure of the ability of water to neutralize acid is _____.
14.9   A pH value of 7 represents a _____.
14.10 There is no safe level for _____ exposure.

REFERENCES AND SUGGESTED READING

Coakley, P. 1975. Developments in our knowledge of sludge dewatering behavior. In *Proceedings of the 8th Public Health Engineering Conference*, Department of Civil Engineering, University of Technology, Loughborough.

Fortner, B. and Schechter, D. 1996. U.S. water quality shows little improvement over 1992 inventory. *Water Environ. Technol.*, 8, 15–16.

Gilcreas, F.W., Sanderson, W.W., and Elmer, R.P. 1953. Two methods for the determination of grease in sewage. *Sewage Indust. Wastes*, 25, 1379.

Gleick, P.H. 2001. Freshwater forum. *U.S. Water News*, 18(6).

Koren, H. 1991. *Handbook of Environmental Health and Safety: Principles and Practices.* Chelsea, MI: Lewis Publishers.

McGhee, T.J. 1991. *Water Supply and Sewerage*, 6th ed. New York: McGraw-Hill.

Metcalf & Eddy. 2003. *Wastewater Engineering: Treatment, Disposal, and Reuse*, 4th ed. New York: McGraw-Hill.

Peavy, H.S., Rowe, D.R., and Tchobanoglous, G. 1987. *Environmental Engineering*. New York: McGraw-Hill.

Rowe, D.R. and Abdel-Magid, I.M. 1995. *Handbook of Wastewater Reclamation and Reuse*. Boca Raton, FL: Lewis Publishers.

Sawyer, C.N., McCarty, A.L., and Parking, G.F. 1994. *Chemistry for Environmental Engineering*, New York: McGraw-Hill.

Sterritt, R.M. and Lester, J.M. 1988. *Microbiology for Environmental and Public Health Engineers*. London: E. & F.N. Spoon.

Tchobanoglous, G. and Schroeder, E.D. 1985. *Water Quality*. Reading, MA: Addison-Wesley.

USEPA. 2007. *Protecting America's Public Health*. Washington, D.C.: U.S. Environmental Protection Agency (www.epa.gov/safewater/publicoutreach.html.)

15 Biomonitoring, Monitoring, Sampling, and Testing

In January, we take our nets to a no-name stream in the foothills of the Blue Ridge Mountains of Virginia to do a special kind of macroinvertebrate monitoring—looking for winter stoneflies. Winter stoneflies have an unusual life cycle. Soon after hatching in early spring, the larvae bury themselves in the streambed. They spend the summer lying dormant in the mud, thereby avoiding problems such as overheated streams, low oxygen concentrations, fluctuating flows, and heavy predation. In later November, they emerge, grow quickly for a couple of months, and then lay their eggs in January.

January monitoring of winter stoneflies aids in interpreting the results of spring and fall macroinvertebrate surveys. In spring and fall, a thorough benthic survey is conducted based on Protocol II of the USEPA's *Rapid Bioassessment Protocols for Use in Streams and Rivers*. Some sites on various rural streams have poor diversity and sensitive families. Is the lack of macroinvertebrate diversity because of specific warm-weather conditions, high water temperature, low oxygen, or fluctuating flows, or is some toxic contamination present? In the January screening, if winter stoneflies are plentiful, seasonal conditions could probably be blamed for the earlier results; if winter stoneflies are absent, the site probably suffers from toxic contamination (based on our rural location, probably emanating from nonpoint sources) that is present all year. Though different genera of winter stoneflies are found in our region (southwestern Virginia), *Allocapnia* is sought because it is present even in the smallest streams.

15.1 WHAT IS BIOMONITORING?

The life in, and physical characteristics of, a stream ecosystem provide insight into the historical and current status of its quality. The assessment of a water body ecosystem based on organisms living in it is called *biomonitoring*. The assessment of the system based on its physical characteristics is called a *habitat assessment*. Biomonitoring and habitat assessments are two tools that stream ecologists use to assess the water quality of a stream.

Biological monitoring involves the use of various organisms, such as periphytons, fish, and macroinvertebrates (combinations of which are referred to as *assemblages*), to assess environmental condition. Biological observation is more representative as it reveals cumulative effects as opposed to chemical observation, which is representative only at the actual time of sampling.

Again, the presence of various assemblages of organisms is used in conducting biological assessments and biosurveys. In selecting the appropriate assemblages for a particular biomonitoring situation, the advantages of using each must be considered along with the objectives of the program. Some of the advantages of using periphytons (algae), benthic macroinvertebrates, and fish in a biomonitoring program are presented in this section.

Important Point: *Periphytons* are a complex matrix of benthic attached algae, cyanobacteria, heterotrophic microbes, and detritus that is attached to submerged surfaces in most aquatic ecosystems.

15.1.1 ADVANTAGES OF USING PERIPHYTONS

1. Algae generally have rapid reproduction rates and very short life cycles, making them valuable indicators of short-term impacts.
2. As primary producers, algae are most directly affected by physical and chemical factors.
3. Sampling is simple and inexpensive, requires few people, and creates minimal impact to resident biota.
4. Relatively standard methods exist for the evaluation of functional and nontaxonomic structural (biomass, chlorophyll measurements) characteristics of algal communities.
5. Algal assemblages are sensitive to some pollutants that may not visibly affect other aquatic assemblages or may only affect other organisms at higher concentrations (e.g., herbicides) (APHA, 1971; Carins and Dickson, 1971; Karr, 1981; Patrick, 1973; Rodgers et al., 1979; USEPA, 1983; Weitzel, 1979).

15.1.2 ADVANTAGES OF USING FISH

1. Fish are good indicators of long-term (several years) effects and broad habitat conditions because they are relatively long lived and mobile (Karr et al., 1986).
2. Fish assemblages include a range of species representing various trophic levels (omnivores, herbivores, insectivores, planktivores, piscivores).

They tend to integrate effects of lower trophic levels; thus, fish assemblage structure is reflective of integrated environmental health.

3. Fish are at the top of the aquatic food web and are consumed by humans, making them an important factor in contamination assessment.

4. Fish are relatively easy to collect and identify to the species level. Most specimens can be sorted and identified in the field by experienced fishery professionals and subsequently released unharmed.

5. Environmental requirements of most fish are comparatively well known. Life history information is extensive for many species, and information on fish distributions is commonly available.

6. Aquatic life uses (water quality standards) are typically characterized in terms of fisheries (coldwater, coolwater, warmwater, sport, forage). Monitoring fish allows direct evaluation of fishability and fish propagation and recognizes the importance of fish to anglers and commercial fishermen.

7. Fish account for nearly half of the endangered vertebrate species and subspecies in the United States (Warren and Burr, 1994).

15.1.3 ADVANTAGES OF USING MACROINVERTEBRATES

As discussed earlier in Chapter 13, benthic macroinvertebrates are the larger organisms such as aquatic insects, insect larvae, and crustaceans that live in the bottom portions of a waterway for part their life cycle. They are ideal for use in biomonitoring, as they are ubiquitous, relatively sedentary, and long lived. They provide a cross-section of the situation, as some species are extremely sensitive to pollution while others are more tolerant. Just as for toxicity testing, however, biomonitoring does not tell us *why* animals are present or absent. Benthic macroinvertebrates are excellent indicators for several reasons:

1. Biological communities reflect overall ecological integrity (i.e., chemical, physical, and biological integrity); therefore, biosurvey results directly assess the status of a water body relative to the primary goal of the Clean Water Act (CWA).

2. Biological communities integrate the effects of different stressors and thus provide a broad measure of their aggregate impact.

3. Because they are ubiquitous, communities integrate the stressors over time and provide an ecological measure of fluctuating environmental conditions.

4. Routine monitoring of biological communities can be relatively inexpensive, because they are easy to collect and identify.

5. The status of biological comminutes is of direct interest to the public as a measure of a particular environment.

6. Where criteria for specific ambient impacts do not exist (e.g., nonpoint sources that degrade habitats), biological communities may be the only practical means of evaluation.

Benthic macroinvertebrates act as continuous monitors of the water they live in. Unlike chemical monitoring, which provides information about water quality at the time of measurement (a snapshot), biological monitoring can provide information about past or episodic pollution (a videotape). This concept is analogous to miners who took canaries into deep mines with them to test for air quality. If the canary died, the miners knew the air was bad and they had to leave the mine. Biomonitoring a water body ecosystem uses the same theoretical approach. Aquatic macroinvertebrates are subject to pollutants in the water body; consequently, the health of the organisms reflects the quality of the water they live in. If the pollution levels reach a critical concentration, certain organisms will migrate away, fail to reproduce, or die, eventually leading to the disappearance of those species at the polluted site. Normally, these organisms will return if conditions improve in the system (Bly and Smith, 1994).

Biomonitoring (and the related concept bioassessment) surveys are conducted before and after an anticipated impact to determine the effect of the activity on the water body habitat. Moreover, surveys are performed periodically to monitor water body habitats and watch for unanticipated impacts. Finally, biomonitoring surveys are designed to reference conditions or to set biocriteria for determining that an impact has occurred; that is, they establish monitoring thresholds that signal future impacts or necessary regulatory actions (Camann, 1996).

Note: The primary justification for bioassessment and monitoring is that degradation of water body habitats affects the biota using those habitats; therefore, the living organisms themselves provide the most direct means of assessing real environmental impacts. Although the focus of this text is on macroinvertebrate protocols, the periphyton and fish protocols are discussed briefly before our in-depth coverage of the macroinvertebrate protocols.

15.2 PERIPHYTON PROTOCOLS

Benthic algae (periphyton or phytobenthos) are primary producers and an important foundation of many stream food webs (Bahls, 1993; Stevenson, 1996). These organisms also stabilize substrata and serve as habitat for many other organisms. Because benthic algal assemblages are attached to substrate, their characteristics are affected by physical, chemical, and biological disturbances that occur in the stream reach during the time in which assemblage developed.

Diatoms in particular are useful ecological indicators because they are found in abundance in most lotic ecosystems. Diatoms and many other algae can be identified to species by experienced algologists. The great numbers of species provide multiple, sensitive indicators of environmental change and the specific conditions of their habitat. Diatom species are differentially adapted to a wide range of ecological conditions.

Periphyton indices of biotic integrity have been developed and tested in several regions (Hill, 1997; KDEP, 1993). Because the ecological tolerances for many species are known, changes in community composition can be used to diagnose the environmental stressors affecting ecological health as well as to assess biotic integrity (Stevenson, 1998; Stevenson and Pan, 1999). Periphyton protocols may be used by themselves, but they are most effective when used with one or more of the other assemblages and protocols. They should be used with habitat and benthic macroinvertebrate assessments particularly because of the close relation between periphyton and these elements of stream ecosystems.

Currently, few states have developed protocols for periphyton assessment. Montana, Kentucky, and Oklahoma have developed periphyton bioassessment programs, and other states are exploring the possibility of developing periphyton programs. Algae have been widely used to monitor water quality in rivers of Europe, where many different approaches have been used for sampling and data analysis.

15.3 FISH PROTOCOLS

Monitoring fish assemblages is an integral component of many water quality management programs, and its importance is reflected in the Aquatic Life Use Support designations of many states. Assessments of the fish assemblage must measure the overall structure and function of the ichthyofaunal community to adequately evaluate biological integrity and protect surface water resource quality. Fish bioassessment data quality and comparability are ensured by utilizing qualified fisheries professionals and consistent methods.

In the fish protocol, the principal evaluation mechanism utilizes the technical framework of the Index of Biotic Integrity (IBI), a fish assemblage assessment approach developed by Karr (1981). The IBI incorporates the zoogeographic, ecosystem, community, and population aspects of the fish assemblage into a single, ecologically based index. Calculation and interpretation of the IBI involves a sequence of activities, including fish sample

collection, data tabulation, and regional modification and calibration of metrics and expectation values.

The fish protocol involves careful, standardized field collection; species identification and enumeration; and analyses using aggregated biological attributes or quantification of numbers of key species. The role of experienced fisheries scientists in the adaptation and application of the bioassessment protocol and the taxonomic identification of fishes cannot be overemphasized. The fish bioassessment protocols survey yields an objective discrete measure of the condition of the fish assemblage. Although the fish survey can usually be completed in the field by qualified fish biologists, difficult species identifications will require laboratory confirmation. Data provided by the fish bioassessment protocols can serve to assess use attainment, to develop biological criteria, to prioritize sites for further evaluation, to provide a reproducible impact assessment, and to evaluate status and trends of the fish assemblage.

Fish collection procedures must focus on a multihabitat approach by sampling habitats in relative proportion to their local representation (as determined during site reconnaissance). Each sample reach should contain riffle, run, and pool habitat, when available. Whenever possible, the reach should be sampled sufficiently upstream of any bridge or road crossing to minimize the hydrological effects on overall habitat quality. The ability to wade in the water and accessibility may ultimately govern the exact placement of the sample reach. A habitat assessment is performed and physicochemical parameters measured concurrently with fish sampling to document and characterize available habitat specifics within the sample reach.

15.4 MACROINVERTEBRATE PROTOCOLS

Benthic macroinvertebrates, by indicating the extent of oxygenation of a stream, may be regarded as indicators of the intensity of pollution from organic waste. The responses of aquatic organisms in water bodies to large quantities of organic wastes are well documented. They occur in a predictable cyclical manner; for example, upstream from the discharge point, a stream can support a wide variety of algae, fish, and other organisms, but in the section of the water body where oxygen levels are low (below 5 ppm), only a few types of worms survive. As stream flow courses downstream, oxygen levels recover, and those species that can tolerate low rates of oxygen (such as gar, catfish, and carp) begin to appear. In a stream, eventually, at some point further downstream, a clean water zone reestablishes itself, and a more diverse and desirable community of organisms returns. Due to this characteristic pattern of varying levels of dissolved oxygen (in response to the dumping of large amounts of biodegradable organic material), a stream goes through an *oxygen sag curve* cycle. Its state can be determined using the biotic index as an indicator of oxygen content.

15.4.1 The Biotic Index

The biotic index is a systematic survey of macroinvertebrates organisms. Because the diversity of species in a stream is often a good indicator of the presence of pollution, the biotic index can be used to evaluate stream quality. Observation of the types of species present or missing is used as an indicator of stream pollution. The biotic index, which reflects the types, species, and numbers of biological organisms present in a stream, is commonly used as an auxiliary to biochemical oxygen demand (BOD) determination when monitoring stream pollution. The biotic index is based on two principles:

1. A large dumping of organic waste into a stream tends to restrict the variety of organisms at a certain point in the stream.
2. As the degree of pollution in a stream increases, key organisms tend to disappear in a predictable order. The disappearance of particular organisms, then, can represent the water quality of the stream.

Several different forms of the biotic index are in use. In Great Britain, for example, the Trent Biotic Index (TBI), the Chandler score, the Biological Monitoring Working Party (BMWP) score, and the Lincoln Quality Index (LQI) are widely used. Most of these instruments use a biotic index that ranges from 0 to 10. The most polluted stream, which therefore contains the smallest variety of organisms, is at the lowest end of the scale (0); clean streams are at the highest end (10). A stream with a biotic index of greater than 5 will support game fish; a stream with a biotic index of less than 4 will not support game fish.

As mentioned, because they are easy to sample, macroinvertebrates have been prominent in biological monitoring. Macroinvertebrates are a diverse group. They demonstrate tolerances that vary among species; thus, discrete differences can be observed between tolerant and sensitive indicators. Macroinvertebrates can be easily identified using identification keys that are portable and easily used in field settings. Current knowledge of macroinvertebrate tolerances and responses to stream pollution is well documented. In the United States, for example, the U.S. Environmental Protection Agency (USEPA) required states to incorporate narrative biological criteria into their water quality standards by 1993. The National Park Service (NPS) has collected macroinvertebrate samples from American streams since 1984. Through their sampling effort, the NPS has been able to derive quantitative biological standards (Huff, 1993).

The biotic index provides a valuable measure of pollution. This is especially the case for species that are very sensitive to lack of oxygen. An example of an organism

TABLE 15.1
BMWP Score System (Modified)

Family	Common Name Example	Score
Heptageniidae	Mayflies	10
Leuctridae	Stoneflies	9–10
Aeshnidae	Dragonflies	8
Polycentropidae	Caddisflies	7
Hydrometridae	Water strider	6–7
Gyrinidae	Whirligig beetle	5
Chironomidae	Mosquitoes	2
Oligochaeta	Worms	1

that is commonly used in biological monitoring is the stonefly. Stonefly larvae live underwater and survive best in well-aerated, unpolluted waters with clean gravel bottoms. When stream-water quality deteriorates due to organic pollution, stonefly larvae cannot survive. The degradation of stonefly larvae has an exponential effect on other insects and fish that feed off the larvae; when the stonefly larvae disappears, so in turn do many insects and fish (O'Toole, 1986).

Table 15.1 shows a modified version of the BMWP biotic index, which takes into account the sensitivities of various macroinvertebrate species represented by diverse populations and that are excellent indicators of pollution. These aquatic macroinvertebrates are organisms that are large enough to be seen by the unaided eye. Moreover, most aquatic macroinvertebrates species live for at least a year, and they are sensitive to stream-water quality both on a short-term basis and over the long term. Mayflies, stoneflies, and caddisflies are aquatic macroinvertebrates that are considered clean-water organisms; they are generally the first to disappear from a stream if water quality declines, so they are given a high score in the index. On the other hand, tubificid worms (which are tolerant to pollution) are given a low score.

As shown in Table 15.1, a score from 1 to 10 is given for each family present. A site score is calculated by adding the individual family scores. The site score or total score is then divided by the number of families recorded to derive the Average Score Per Taxon (ASPT). High ASPT scores are obtained when such taxa as stoneflies, mayflies, and caddisflies are found in the stream. A low ASPT score is obtained for streams that are heavily polluted and are dominated by tubificid worms and other pollution-tolerant organism. From Table 15.1, it can be seen that those organisms having high scores, especially mayflies and stoneflies, are the most sensitive, while others, such as dragonflies and caddisflies, are very sensitive to any pollution (deoxygenation) of their aquatic environment.

As noted earlier, the benthic macroinvertebrate biotic index employs the use of certain benthic macroinvertebrates

TABLE 15.2
Sample Groupings of Macroinvertebrates

Group One (Sensitive)	Group Two (Somewhat Sensitive)	Group Three (Tolerant)
Stonefly larvae	Alderfly larvae	Aquatic worms
Caddisfly larvae	Damselfly larvae	Midgefly larvae
Water penny larvae	Cranefly larvae	Blackfly larvae
Riffle beetle adults	Beetle adults	Leeches
Mayfly larvae	Dragonfly larvae	Snails
Gilled snails	Sowbugs	

to determine, or gauge, the water quality (relative health) of a water body such as a stream or river. Benthic macroinvertebrates are classified into three groups based on their sensitivity to pollution. The number of taxa in each of these groups is tallied and assigned a score. The scores are then summed to yield a value that can be used as an estimate of the quality of the water body life.

15.4.2 Metrics within the Benthic Macroinvertebrates

Table 15.2 provides a sample index of macroinvertebrates and their sensitivity to pollution. The three groups based on their sensitivity to pollution are:

- Group One—indicators of good water quality
- Group Two—indicators of moderate water quality
- Group Three—indicators of poor water quality

In summary, it can be said that unpolluted streams normally support a wide variety of macroinvertebrates and other aquatic organisms with relatively few of any one kind. Any significant change in the normal population usually indicates pollution.

15.5 BIOLOGICAL SAMPLING IN STREAMS

A few years ago, we were preparing to perform benthic macroinvertebrate sampling protocols in a wadeable section of one of the countless reaches of the Yellowstone River in Wyoming. It was autumn, windy and cold. Before we stepped into the slow-moving frigid waters, we stood for a moment at the bank and took in the surroundings. The pallet of autumn is austere in Yellowstone. The coniferous forests west of the Mississippi lack the bronzes, the coppers, the peach-tinted yellows, and the livid scarlets that set the mixed stands of the East aflame. All we could see in the line of trees was the quaking aspen and its gold. This autumnal gold, which provides the closest thing to eastern autumn in the west, is mined from the narrow, rounded crowns of *Populus tremuloides*. The aspen trunks

stand stark white and antithetical against the darkness of the firs and pines, the shiny pale gold leaves sensitive to the slightest rumor of wind. Agitated by the slightest hint of breeze, the gleaming upper surfaces bounced the sun into our eyes. Each tree scintillated, like a show of gold coins in free fall. The bright, metallic flash of the aspens seemed, in all their glittering motion, to be making a valiant dying attempt to fill the spectrum of fall.

Because they were bright and glorious, we did not care that they could not approach the colors of an eastern autumn, although nothing is comparable to leaf fall in autumn along the Appalachian Trail. This spirited display of gold against dark green lightened our hearts and eased the task that was before us: entering the bone-chilling water. With the aspens gleaming gold against the pines and firs, it simply did not seem to matter. Notwithstanding being able to witness such glories of nature, one should not be deceived about biological sampling. Conducting biological sampling in a body of water is not only the nuts and bolts of biological sampling but also very hard and important work.

15.5.1 Biological Sampling Planning

When planning a biological sampling outing, it is important to determine the precise objectives. One important consideration is to determine whether sampling will be accomplished at a single point or at isolated points. Additionally, frequency of sampling must be determined. That is, will sampling be accomplished at hourly, daily, weekly, monthly, or even longer intervals? Whatever sampling frequency of sampling is chosen, the entire process will probably continue over a protracted period (i.e., preparing for biological sampling in the field might take several months from the initial planning stages to the time when actual sampling occurs). An experienced freshwater ecologist should be centrally involved in all aspects of planning.

In its *Monitoring Water Quality: Intensive Stream Bioassay* (08/18/2000), the USEPA recommends that the following issues should be considered when planning a sampling program:

- Availability of reference conditions for the chosen area
- Appropriate dates for sampling in each season
- Appropriate sampling gear
- Availability of laboratory facilities
- Sample storage
- Data management
- Appropriate taxonomic keys, metrics, or measurement for macroinvertebrate analysis
- Habitat assessment consistency
- Availability of a USGS topographical map
- Familiarity with safety procedures

When the initial objectives (issues) have been determined and the plan devised, then the sampler can move to other important aspects of the sampling procedure. Along with the items just mentioned, it is imperative for the sampler to understand what biological sampling is all about.

Sampling is one of the most basic and important aspects of water quality management (Tchobanoglous and Schroeder, 1985). Biological sampling allows for rapid and general water quality classification. Rapid classification is possible because quick and easy cross-checking between stream biota and a standard stream biotic index is possible. Biological sampling is typically used for general water quality classification in the field because sophisticated laboratory apparatus is usually not available. Additionally, stream communities often show a great deal of variation in basic water quality parameters such as dissolved oxygen, BOD, suspended solids, and coliform bacteria. This occurrence can be observed in eutrophic lakes that may vary from oxygen saturation to less than 0.5 mg/L in a single day, and the concentration of suspended solids may double immediate after a heavy rain. Moreover, the sampling method chosen must take into account the differences in the habits and habitats of the aquatic organisms.

The first step toward accurate measurement of the water quality of a stream is to make sure that the sampling targets those organisms (i.e., macroinvertebrates) that are most likely to provide the information being sought. Second, it is essential that representative samples be collected. Laboratory analysis is meaningless if the sample collected was not representative of the aquatic environment being analyzed. As a rule, samples should be taken at many locations, as often as possible. If, for example, we are studying the effects of sewage discharge into a stream, we should take at least six samples upstream of the discharge, six samples at the discharge, and at least six samples at several points below the discharge for two to three days (the six–six sampling rule). If these samples show wide variability, then the number of samples should be increased. On the other hand, if the initial samples exhibit little variation, then a reduction in the number of samples may be appropriate (Kittrell, 1969).

When planning the biological sampling protocol (using biotic indices as the standards), remember that when the sampling is to be conducted in a stream, findings are based on the presence or absence of certain organisms; thus, the absence of these organisms must be a function of pollution and not of some other ecological problem. The preferred aquatic group for biological monitoring in streams, macroinvertebrates are usually retained by 30-mesh sieves (pond nets).

15.5.2 SAMPLING STATIONS

After determining the number of samples to be taken, sampling stations (locations) must be determined. Several factors determine where the sampling stations should be set up. These factors include stream habitat types, the position of the wastewater effluent outfalls, the stream characteristics, stream developments (dams, bridges, navigation locks, and other manmade structures), the self-purification characteristics of the stream, and the nature of the objectives of the study (Velz, 1970). The stream habitat types used in this discussion are macroinvertebrate assemblages in stream ecosystems. Combinations of habitats are sampled in a multi-habitat approach to benthic sampling (Barbour et al., 1997):

1. *Cobble (hard substrate)*—Cobble is prevalent in the riffles (and runs) that are common features throughout most mountain and piedmont streams. In many high-gradient streams, this habitat type will be dominant; however, riffles are not a common feature of most coastal or other low-gradient streams. Sample shallow areas with coarse substrates (mixed gravel, cobble or larger) by holding the bottom of the dip net against the substrate and dislodging organisms by kicking the substrate 0.5 m upstream of the net (this is where the designated kicker, the sampling partner, comes into play).
2. *Snags*—Snags and other woody debris that have been submerged for a relatively long period (not recent deadfall) provide excellent colonization habitat. Sample submerged woody debris by jabbing medium-sized snag material (sticks and branches). The snag habitat may be kicked first to help to dislodge organisms but only after placing the net downstream of the snag. Accumulated woody material in pool areas is considered snag habitat. Large logs should be avoided because they are generally difficult to sample adequately.
3. *Vegetated banks*—When lower banks are submerged and have roots and emergent plants associated with them, they are sampled in a fashion similar to snags. Submerged areas of

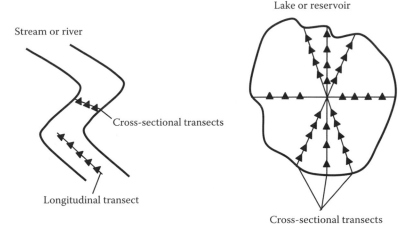

Stream or river

Cross-sectional transects

Longitudinal transect

Lake or reservoir

Cross-sectional transects

FIGURE 15.1 Transect sampling.

undercut banks are good habitats to sample. Sample banks with protruding roots and plants by jabbing into the habitat. Bank habitat can be kicked first to help dislodge organisms, but only after placing the net downstream.

4. *Submerged macrophytes*—Submerged macrophytes are seasonal in their occurrence and may not be a common feature of many streams, particularly those that are high gradient. Sample aquatic plants that are rooted on the bottom of the stream in deep water by drawing the net through the vegetation from the bottom to the surface of the water (maximum of 0.5 m each jab). In shallow water, sample by bumping or jabbing the net along the bottom in the rooted area, avoiding sediments where possible.

5. *Sand* (and other fine sediment)—Usually the least productive macroinvertebrate habitat in streams, this habitat may be the most prevalent in some streams. Sample banks of unvegetated or soft soil by bumping the net along the surface of the substrate rather than dragging the net through soft substrate; this reduces the amount of debris in the sample.

It is usually impossible to go out and count each and every macroinvertebrate present in a waterway. This would be comparable to counting different sizes of grains of sand on the beach. Thus, in a biological sampling program (based on our experience), the most common sampling methods are the *transect* and the *grid*. Transect sampling involves taking samples along a straight line either at uniform or at random intervals (see Figure 15.1). The transect approach samples a cross-section of a lake or stream or a longitudinal section of a river or stream. The transect sampling method allows for a more complete analysis by including variations in habitat.

In grid sampling, an imaginary grid system is placed over the study area. The grids may be numbered, and random numbers are generated to determine which grids should be sampled (see Figure 15.2). This type of sampling method allows for quantitative analysis because the grids are all of a certain size; for example, to sample a stream for benthic macroinvertebrates, grids that are 0.25 m^2 may be used. The weight or number of benthic macroinvertebrates per square meter can then be determined.

Random sampling requires that each possible sampling location have an equal chance of being selected. Numbering all sampling locations and then using a computer, calculator, or a random numbers table to collect a series of random numbers can accomplish this. An illustration of how to put the random numbers to work is provided in the following example. Given a pond that has 300 grid units, find 8 random sampling locations using the following sequence of random numbers taken from a standard random numbers table: 101, 209, 007, 018, 099, 100, 017, 069, 096, 033, 041, 011. The first eight numbers of the sequence could be selected and only those grids would sampled to obtain a random sample.

15.5.3 SAMPLING FREQUENCY AND NOTES

(The sampling procedures that follow have been suggested by USEPA, 1997.) After establishing the sampling methodology and the sampling locations, the frequency of sampling must be determined. The more samples collected, the more reliable the data will be. A frequency of once a week or once a month will be adequate for most aquatic studies. Usually, the sampling period covers an entire year so yearly variations may be included. The details of sample collection will depend on the type of problem that is being solved and will vary with each study. When a sample is collected, it must be carefully identified with the following information:

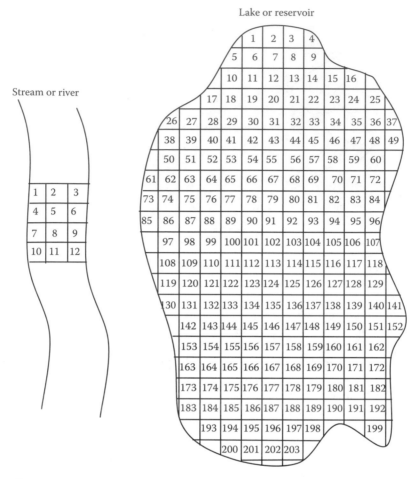

FIGURE 15.2 Grid sampling.

1. Location (name of water body and place of study; longitude and latitude)
2. Date and time
3. Site (sampling location)
4. Name of collector
5. Weather (temperature, precipitation, humidity, wind, etc.)
6. Miscellaneous (any other important information, such as observations)
7. Field notebook

With regard to the last item, on each sampling day notes on field conditions should be taken; for example, miscellaneous observations and weather conditions can be entered. Additionally, notes that describe the condition of the water are also helpful (e.g., color, turbidity, odor, algae). All unusual findings and conditions should also be entered.

15.5.4 MACROINVERTEBRATE SAMPLING EQUIPMENT

In addition to the appropriate sampling equipment, also assemble the following equipment:

1. Jars (two), at least quart size, plastic, wide-mouth with tight cap; one should be empty and the other filled about two thirds with 70% ethyl alcohol
2. Hand lens, magnifying glass, or field microscope
3. Fine-point forceps
4. Heavy-duty rubber gloves
5. Plastic sugar scoop or ice-cream scoop
6. Kink net (rocky-bottom stream) or dip net (muddy-bottom stream)
7. Buckets (two) (see Figure 15.3)
8. String or twine (50 yards) and a tape measure
9. Stakes (four)

FIGURE 15.3 Sieve bucket.

10. Orange to measure velocity (a stick, an apple, or a fish float may also be used in place of an orange)
11. Reference maps indicating general information pertinent to the sampling area, including the surrounding roadways, as well as a hand-drawn station map
12. Station ID tags
13. Spray bottle
14. Pencils (at least two)

15.5.5 Macroinvertebrate Sampling in Rocky-Bottom Streams

Rocky-bottom streams are defined as those with bottoms made up of gravel, cobbles, and boulders in any combination. They usually have definite riffle areas. As mentioned, riffle areas are fairly well oxygenated and, therefore, are prime habitats for benthic macroinvertebrates. In these streams, we use the *rocky-bottom sampling method*. This method of macroinvertebrate sampling is used in streams that have riffles and gravel/cobble substrates. Three samples are to be collected at each site, and a composite sample is obtained (i.e., one large total sample).

- *Step 1. Locate a site on a map, with its latitude and longitude indicated.*
  A. Samples will be taken in three different spots within a 100-yard stream site. These spots may be three separate riffles, one large riffle with different current velocities, or, if no riffles are present, three run areas with gravel or cobble substrate. Combinations are also possible (if, for example, the site has only one small riffle and several run areas). Mark off the 100-yard stream site. If possible, it should begin at least 50 yards upstream of any manmade modification of the channel, such as a bridge, dam, or pipeline crossing. Avoid walking in the stream, because this might dislodge macroinvertebrates and affect later sampling results.
  B. Sketch the 100-yard sampling area. Indicate the location of the three sampling spots on the sketch. Mark the most downstream site as Site 1, the middle site as Site 2, and the upstream site as Site 3.
- *Step 2. Get into place.*
  A. Always approach sampling locations from the downstream end and sample the site farthest downstream first (Site 1). This prevents biasing of the second and third collections with dislodged sediment of macroinvertebrates. Always use a clean kick-seine, relatively free of mud and

debris from previous uses. Fill a bucket about one third full with stream water, and fill the spray bottle.
  B. Select a 3×3-ft riffle area to sample at Site 1. One member of the team, the net holder, should position the net at the downstream end of this sampling area. Hold the net handles at a 45° angle to the surface of the water. Be sure the bottom of the net fits tightly against the streambed so no macroinvertebrates escape under the net. Rocks from the sampling area can be used to anchor the net against the stream bottom. Do not allow any water to flow over the net.
- *Step 3. Dislodge the macroinvertebrates.*
  A. Pick up any large rocks in the 3×3-ft sampling area and rub them thoroughly over the partially filled bucket so any macroinvertebrates clinging to the rocks will be dislodged into the bucket. Place each cleaned rock outside of the sampling area. After sampling is completed, rocks can be returned to the stretch of stream where they were found.
  B. The member of the team designated as the kicker should thoroughly stir up the sampling areas with his feet, starting at the upstream edge of the 3×3-ft sampling area and working downstream, moving toward the net. All dislodged organisms will be carried by the stream flow into the net. Be sure to disturb the first few inches of stream sediment to dislodge burrowing organisms. As a general guide, disturb the sampling area for about 3 minutes, or until the area is thoroughly worked over.
  C. Any large rocks used to anchor the net should be thoroughly rubbed into the bucket as above.
- *Step 4. Remove the net.*
  A. Remove the net without allowing any of the organisms it contains to wash away. While the net holder grabs the top of the net handles, the kicker grabs the bottom of the net handles and the bottom edge of the net. Remove the net from the stream with a forward scooping motion.
  B. Roll the kick net into a cylinder shape and place it vertically in the partially filled bucket. Pour or spray water down the net to flush its contents into the bucket. If necessary, pick debris and organisms from the net by hand. Release back into the stream any fish, amphibians, or reptiles caught in the net.
- *Step 5. Collect the second and third samples.*

When all of the organisms have been removed from the net, repeat the steps above at Sites 2 and 3. Put the samples from all three sites into the same bucket. Combining the debris and organisms from all three sites into the same bucket is called *compositing*.

Note: If the bucket is nearly full of water after washing the net clean, let the debris and organisms settle to the bottom. Then, cup the net over the bucket and pour the water through the net into a second bucket. Inspect the water in the second bucket to be sure no organisms came through.

- *Step 6. Preserve the sample.*
  A. After collecting and compositing samples from all three sites, it is time to preserve the sample. All team members should leave the stream and return to a relatively flat section of the stream bank with their equipment. The next step will be to remove large pieces of debris (leaves, twigs, and rocks) from the sample. Carefully remove the debris one piece at a time. While holding the material over the bucket, use the forceps, spray bottle, and your hands to pick, rub, and rinse the leaves, twigs, and rocks to remove any attached organisms. Use a magnifying lens and forceps to find and remove small organisms clinging to the debris. When satisfied that the material is clean, discard it back into the stream.
  B. The water will have to be drained before transferring material to the jar. This process requires two team members. Place the kick net over the second bucket, which has not yet been used and should be completely empty. One team member should push the center of the net into the second bucket, creating a small indentation or depression. Hold the sides of the net closely over the mouth of the bucket. The second person can now carefully pour the remaining contents of the first bucket onto a small area of the net to drain the water and concentrate the organisms. Use care when pouring so organisms are not lost over the side of the net (see Figure 15.4). Use the spray bottle, forceps, sugar scoop, and gloved hands to transfer all material from the first bucket onto the net. When the first bucket is empty, use your hands and the sugar scoop to transfer the material from the net into the empty jar. The second bucket captures the water and any organisms that might have fallen through the netting during pouring.

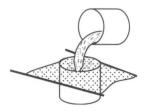

FIGURE 15.4 Pouring sample water through the net.

As a final check, repeat the process above, but this time pour the second bucket over the net into the first bucket. Transfer any organisms on the net into the jar.
  C. Now fill the jar (so all material is submerged) with the alcohol from the second jar. Put the lid tightly back onto the jar and gently turn the jar upside down two or three times to distribute the alcohol and remove air bubbles.
  D. Complete the sampling station ID tag. Be sure to use a pencil, not a pen, because the ink will run in the alcohol! The tag includes the station number, stream, location (e.g., upstream from a road crossing), date, time, and names of the members of the collecting team. Place the ID tag into the sample container, writing side facing out, so the identification can be seen clearly.

15.5.5.1 Rocky-Bottom Habitat Assessment

The habitat assessment (including measuring general characteristics and local land use) for a rocky-bottom stream is conducted in a 100-yard section of stream that includes the riffles from which organisms were collected.

- *Step 1. Delineate the habitat assessment boundaries.*
  A. Begin by identifying the most downstream riffle that was sampled for macroinvertebrates. Using tape measure or twine, mark off a 100-yard section extending 25 yards below the downstream riffle and about 75 yards upstream.
  B. Complete the identifying information of the field data sheet for the habitat assessment site. On the stream sketch, be as detailed as possible, and be sure to note which riffles were sampled.
- *Step 2. Describe the general characteristics and local land use on the field sheet.* For safety reasons as well as to protect the stream habitat, it is best to estimate the following characteristics rather than actually wading into the stream to measure them:

A. Water appearance can be a physical indicator of water pollution.
   1. Clear—colorless, transparent
   2. Milky—cloudy-white or gray, not transparent; might be natural or due to pollution
   3. Foamy—might be natural or due to pollution, generally detergents or nutrients (foam that is several inches high and does not brush apart easily is generally due to pollution)
   4. Turbid—cloudy brown due to suspended silt or organic material
   5. Dark brown—might indicate that acids are being released into the stream due to decaying plants
   6. Oily sheen—multicolored reflection might indicate oil floating in the stream, although some sheens are natural
   7. Orange—might indicate acid drainage
   8. Green—might indicate that excess nutrients are being released into the stream
B. Water odor can be a physical indicator of water pollution.
   1. No odor, or a natural smell
   2. Sewage—might indicate the release of human waste material
   3. Chlorine—might indicate a sewage treatment plant is overchlorinating its effluent
   4. Fishy—might indicate the presence of excessive algal growth or dead fish
   5. Rotten eggs—might indicate sewage pollution (the presence of a natural gas)
C. Water temperature can be particularly important for determining whether the stream is suitable as habitat for some species of fish and macroinvertebrates that have distinct temperature requirements. Temperature also has a direct effect on the amount of dissolved oxygen available to aquatic organisms. Measure temperature by submerging a thermometer for at least 2 minutes in a typical stream run. Repeat once and average the results.
D. The width of the stream channel can be determined by estimating the width of the streambed that is covered by water from bank to bank. If it varies widely along the stream, estimate an average width.
E. Local land use refers to the part of the watershed within 1/4 mile upstream of and adjacent to the site. Note which land uses are present, as well as which ones seem to be having a negative impact on the stream. Base observations on what can be seen, what was passed on the way to the stream, and, if possible, what is noticed when leaving the stream.

- *Step 3. Conduct the habitat assessment.* The following information describes the parameters that will be evaluated for rocky-bottom habitats. Use these definitions when completing the habitat assessment field data sheet. The first two parameters should be assessed directly at the riffles or runs that were used for the macroinvertebrate sampling. The last eight parameters should be assessed in the entire 100-yard section of the stream.
   A. *Attachment sites* for macroinvertebrates are essentially the amount of living space or hard substrates (rocks, snags) available for adequate insects and snails. Many insects begin their life underwater in streams and need to attach themselves to rocks, logs, branches, or other submerged substrates. The greater the variety and number of available living spaces or attachment sites, the greater the variety of insects in the stream. Optimally, cobble should predominate, and boulders and gravel should be common. The availability of suitable living spaces for macroinvertebrates decreases as cobble becomes less abundant and boulders, gravel, or bedrock become more prevalent.
   B. *Embeddedness* refers to the extent to which rocks (gravel, cobble, and boulders) are surrounded by, covered with, or sunken into the silt, sand, or mud of the stream bottom. Generally, as rocks become embedded, fewer living spaces are available to macroinvertebrates and fish for shelter, spawning, and egg incubation.

Note: To estimate the percent of embeddedness, observe the amount of silt or finer sediments overlaying and surrounding the rocks. If kicking does not dislodge the rocks or cobbles, they might be greatly embedded.

   C. *Shelter* for fish includes the relative quantity and variety of natural structures in the stream, such as fallen trees, logs, and branches; cobble and large rock; and undercut banks that are available to fish for hiding, sleeping, or feeding. A wide variety of submerged structures in the stream provide fish with many living spaces; the more living spaces in a stream, the more types of fish the stream can support.

D. *Channel alteration* is a measure of large-scale changes in the shape of the stream channel. Many streams in urban and agricultural areas have been straightened, deepened (e.g., dredged), or diverted into concrete channels, often for flood control purposes. Such streams have far fewer natural habitats for fish, macroinvertebrates, and plants than do naturally meandering streams. Channel alteration is present when the stream runs through a concrete channel; when artificial embankments, riprap, and other forms of artificial bank stabilization or structures are present; when the stream is very straight for significant distances; when dams, bridges, and flow-altering structures such as combined sewer overflow (CSO) are present; when the stream is of uniform depth due to dredging; and when other such changes have occurred. Signs that indicate the occurrence of dredging include straightened, deepened, and otherwise uniform stream channels, as well as the removal of streamside vegetation to provide dredging equipment access to the stream.

E. *Sediment deposition* is a measure of the amount of sediment that has been deposited in the stream channel and the changes to the stream bottom that have occurred as a result of the deposition. High levels of sediment deposition create an unstable and continually changing environment that is unsuitable for many aquatic organisms. Sediments are naturally deposited in areas where the stream flow is reduced, such as in pools and bends, or where flow is obstructed. These deposits can lead to the formation of islands, shoals, or point bars (sediments that build up in the stream, usually at the beginning of a meander) or can result in the complete filling of pools. To determine whether these sediment deposits are new, look for vegetation growing on them; new sediments will not yet have been colonized by vegetation.

F. *Stream velocity and depth* combinations are important to the maintenance of healthy aquatic communities. Fast water increases the amount of dissolved oxygen in the water, keeps pools from being filled with sediment, and helps food items such as leaves, twigs, and algae move more quickly through the aquatic system. Slow water provides spawning areas for fish and shelters

macroinvertebrates that might be washed downstream in higher stream velocities. Similarly, shallow water tends to be more easily aerated (i.e., it holds more oxygen), but deeper water stays cooler longer. Thus, the best stream habitat includes all of the following velocity and depth combinations and can maintain a wide variety of organisms:

- Slow (<1 ft/sec), shallow (<1.5 ft)
- Slow, deep
- Fast, deep
- Fast, shallow

Measure stream velocity by marking off a 10-foot section of stream run and measuring the time it takes an orange, stick, or other floating biodegradable object to float the 10 feet. Repeat five times, in the same 10-foot section, and determine the average time. Divide the distance (10 feet) by the average time (seconds) to determine the velocity in feet per second. Measure the stream depth by using a stick of known length and taking readings at various points within the stream site, including riffles, runs, and pools. Compare velocity and depth at various points within the 100-yard site to see how many of the combinations are present.

G. *Channel flow status* is the percent of the existing channel that is filled with water. The flow status changes as the channel enlarges or as flow decreases because of dams and other obstructions, diversions for irrigation, or drought. When water does not cover much of the streambed, the living area for aquatic organisms is limited.

Note: For the following parameters, evaluate the conditions of the left and right stream banks separately. Define the left and right banks by standing at the downstream end of the study stretch and look upstream. Each bank is evaluated on a scale of 0 to 10.

H. *Bank vegetation protection* measures the amount of the stream bank that is covered by natural (i.e., growing wild and not obviously planted) vegetation. The root system of plants growing on stream banks helps hold soil in place, reducing erosion. Vegetation on banks provides shade for fish and macroinvertebrates and serves as a food source by dropping leaves and other organic matter into the stream. Ideally, a variety of vegetation should be present, including trees, shrubs, and grasses. Vegetation disruption can occur when the grasses and plants on the stream banks are mowed or grazed, or when the trees and shrubs are cut back or cleared.

I. *Condition of banks* measures erosion potential and whether the stream banks are eroded. Steep banks are more likely to collapse and suffer from erosion than are gently sloping banks; therefore, they are considered to have erosion potential. Signs of erosion include crumbling, unvegetated banks, exposed tree roots, and exposed soil.

J. The *riparian vegetative zone* is defined as the width of natural vegetation from the edge of the stream bank. The riparian vegetative zone is a buffer zone to pollutants entering a stream from runoff. It also controls erosion and provides stream habitat and nutrient input into the stream.

Note: A wide, relatively undisturbed riparian vegetative zone reflects a healthy stream system; narrow, far less useful riparian zones occur when roads, parking lots, fields, lawns, and other artificially cultivated areas; bare soil; rock; or buildings are near the stream bank. The presence of old fields (i.e., previously developed agricultural fields allowed to revert to natural conditions) should rate higher than fields in continuous or periodic use. In arid areas, the riparian vegetative zone can be measured by observing the width of the area dominated by riparian or water-loving plants, such as willows, marsh grasses, and cottonwood trees.

15.5.6 MACROINVERTEBRATE SAMPLING IN MUDDY-BOTTOM STREAMS

In muddy-bottom streams, as in rocky-bottom streams, the goal is to sample the most productive habitat available and look for the widest variety of organisms. The most productive habitat is the one that harbors a diverse population of pollution-sensitive macroinvertebrates. Samples should sample by using a D-frame net (see Figure 15.5) to jab at the habitat and scoop up the organisms that are dislodged. The idea is to collect a total sample that consists of 20 jabs taken from a variety of habitats. Use the following method of macroinvertebrate sampling in streams that have muddy-bottom substrates.

- *Step 1. Determine which habitats are present.* Muddy-bottom streams usually have four habitats: vegetated banks margins, snags and logs,

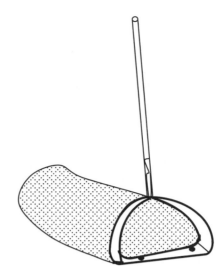

FIGURE 15.5 D-frame aquatic net.

aquatic vegetation beds and decaying organic matter, and silt/sand/gravel substrate. It is generally best to concentrate sampling efforts on the most productive habitat available, yet sample other principal habitats if they are present. This ensures that you will secure as wide a variety of organisms as possible. Not all habitats are present in all streams or are present in significant amounts. If the sampling areas have not been preselected, determine which of the following habitats are present.

Note: Avoid standing in the stream while making habitat determinations.

A. Vegetated bank margins consist of overhanging bank vegetation and submerged root mats attached to the banks. The bank margins may also contain submerged, decomposing leaf packs trapped in root wads or lining the stream banks. This is generally a highly productive habitat in a muddy stream, and it is often the most abundant type of habitat.

B. Snags and logs consist of submerged wood (primarily dead trees), branches, logs, roots, cypress knees, and leaf packs that are lodged between rocks or logs. This is also a very productive muddy-bottom stream habitat.

C. Aquatic vegetation beds and decaying organic matter consist of beds of submerged, green, leafy plants that are attached to the stream bottom. This habitat can be as productive as vegetated bank margins and snags and logs.

D. Silt/sand/gravel substrate includes sandy, silty, or muddy stream bottoms; rocks along the stream bottom; or wetted gravel bars. This habitat may also contain algae-covered rocks (*aufwuchs*). This is the least productive of the four muddy-bottom stream habitats, and it is always present in one form or another (e.g., silt, sand, mud, or gravel might predominate).

- *Step 2. Determine how many times to jab in each habitat type.* The sampler's goal is to jab 20 times. The D-frame net (see Figure 15.5) is 1 foot wide, and a jab should be approximately 1 foot in length. Thus, 20 jabs equal 20 square feet of combined habitat.
  A. If all four habitats are present in plentiful amounts, jab the vegetated banks 10 times and divide the remaining 10 jabs among the remaining 3 habitats.
  B. If three habitats are present in plentiful amounts, and one is absent, jab the silt/sand/gravel substrate, the least productive habitat, 5 times and divide the remaining 15 jabs between the other two more productive habitats.
  C. If only two habitats are present in plentiful amounts, the silt/sand/gravel substrate will most likely be one of those habitats. Jab the silt/sand/gravel substrate 5 times and the more productive habitat 15 times.
  D. If some habitats are plentiful and others are sparse, sample the sparse habitats to the extent possible, even if it is possible to take only one or two jabs. Take the remaining jabs from the plentiful habitats. This rule also applies if a habitat cannot be reached because of unsafe stream conditions. Jab 20 times total.

Note: Because the sampler might need to make an educated guess to decide how many jabs to take in each habitat type, it is critical that the sampler note, on the field data sheet, how many jabs were taken in each habitat. This information can be used to help characterize the findings.

- *Step 3. Get into place.* Outside and downstream of the first sampling location (first habitat), rinse the dip net and check to make sure it does not contain any macroinvertebrates or debris from the last time it was used. Fill a bucket approximately one third with clean stream water. Also, fill the spray bottle with clean stream water. This bottle will be used to wash the net between jabs and after sampling is completed.

Note: This method of sampling requires only one person to disturb the stream habitats. While one person is sampling, a second person should stand outside the sampling area, holding the bucket and spray bottle. After every few jabs, the sampler should hand the net to the second person, who can then rinse the contents of the net into the bucket.

- *Step 4. Dislodge the macroinvertebrates.* Approach the first sample site from downstream, and sample while walking upstream. Sample in the four habitat types as follows:
  A. Sample vegetated bank margins by jabbing vigorously, with an upward motion, brushing the net against vegetation and roots along the bank. The entire jab motion should occur underwater.
  B. To sample snags and logs, hold the net with one hand under the section of submerged wood being sampled. With the other hand (which should be gloved), rub about 1 square foot of area on the snag or log. Scoop organisms, bark, twigs, or other organic matter dislodged into the net. Each combination of log rubbing and net scooping is one jab.
  C. To sample aquatic vegetation beds, jab vigorously, with an upward motion, against or through the plant bed. The entire jab motion should occur underwater.
  D. To sample a silt/sand/gravel substrate, place the net with one edge against the stream bottom and push it forward about a foot (in an upstream direction) to dislodge the first few inches of silt, sand, gravel, or rocks. To avoid gathering a net full of mud, periodically sweep the mesh bottom of the net back and forth in the water, making sure that water does not run over the top of the net. This will allow fine silt to rinse out of the net. When 20 jabs have been completed, rinse the net thoroughly in the bucket. If necessary, pick any clinging organisms from the net by hand, and put them in the bucket.
- *Step 5. Preserve the sample.*
  A. Look through the material in the bucket, and immediately return any fish, amphibians, or reptiles to the stream. Carefully remove large pieces of debris (leaves, twigs, and rocks) from the sample. While holding the material over the bucket, use the forceps, spray bottle, and your hands to pick, rub, and rinse the leaves, twigs, and rocks to remove any attached organisms. Use the

magnifying lens and forceps to find and remove small organisms clinging to the debris. When satisfied that the material is clean, discard it back into the stream.

B. Drain the water before transferring material to the jar. This process will require two people. One person should place the net into the second bucket, like a sieve (this bucket, which has not yet been used, should be completely empty) and hold it securely. The second person can now carefully pour the remaining contents of the first bucket onto the center of the net to drain the water and concentrate the organisms. Use care when pouring so organisms are not lost over the side of the net. Use the spray bottle, forceps, sugar scoop, and gloved hands to remove all the material from the first bucket onto the net. When satisfied that the first bucket is empty, use your hands and the sugar scoop to transfer all the material from the net into the empty jar. The contents of the net can also be emptied directly into the jar by turning the net inside out into the jar. The second bucket captures the water and any organisms that might have fallen through the netting. As a final check, repeat the process above, but this time pour the second bucket over the net into the first bucket. Transfer any organisms on the net into the jar.

C. Fill the jar (so all material is submerged) with alcohol. Put the lid tightly back onto the jar and gently turn the jar upside down two or three times to distribute the alcohol and remove air bubbles.

D. Complete the sampling station ID tag (see Figure 15.6). Be sure to use a pencil, not a pen, because the ink will run in the alcohol. The tag should include the station number, the stream, location (e.g., upstream from a road crossing), date, time, and names of the members of the collecting crew. Place the ID tag into the sample container, writing side facing out, so the identification can be seen clearly.

Note: To prevent samples from being mixed up, samplers should place the ID tag *inside* the sample jar.

15.5.6.1 Muddy-Bottom Stream Habitat Assessment

The muddy-bottom stream habitat assessment (which includes measuring general characteristics and local land

Station # _____

Stream _____

Location _____

Date/Time _____

Team Members: _____

_____

FIGURE 15.6 Station ID tag.

use) is conducted in a 100-yard section of the stream that includes the habitat areas from which organisms were collected.

Note: Reference made previously, and in the following sections, about a field data sheet (habitat assessment field data sheet) assume that the sampling team is using standard forms provided by the USEPA, the U.S. Geological Survey (USGS), or state water control authorities; generic forms put together by the sampling team can also be used. The source of the form and exact type of form are not important, but some type of data recording field sheet should be employed to record pertinent data.

- *Step 1. Delineate habitat assessment boundaries.*
  A. Begin by identifying the most downstream point that was sampled for macroinvertebrates. Using a tape measure or twine, mark off a 100-yard section extending 25 yards below the downstream sampling point and about 75 yards upstream.
  B. Complete the identifying information on the field data sheet for the habitat assessment site. On the stream sketch, be as detailed as possible, and be sure to note which habitats were sampled.

- *Step 2. Record general characteristics and local land use on the data field sheet.* For safety reasons as well as to protect the stream habitat, it is best to estimate these characteristics rather than to actually wade into the stream to measure them. For instructions on completing these sections of the field data sheet, see the rocky-bottom habitat assessment instructions.

- *Step 3. Conduct the habitat assessment.* The following information describes the parameters to be evaluated for muddy-bottom habitats. Use these definitions when completing the habitat assessment field data sheet.
  - A. *Shelter* for fish and attachment sites for macroinvertebrates are essentially the amount of living space and shelter (rocks, snags, and undercut banks) available for fish, insects, and snails. Many insects attach themselves to rocks, logs, branches, or other submerged substrates. Fish can hide or feed in these areas. The greater the variety and number of available shelter sites or attachment sites, the greater the variety of fish and insects in the stream.

Note: Many of the attachment sites result from debris falling into the stream from the surrounding vegetation. When debris first falls into the water, it is termed *new fall*, and it has not yet been broken down by microbes (conditioned) for macroinvertebrate colonization. Leaf material or debris that is conditioned is called *old fall*. Leaves that have been in the stream for some time lose their color, turn brown or dull yellow, become soft and supple with age, and might be slimy to the touch. Woody debris becomes blackened or dark in color; smooth bark becomes coarse and partially disintegrated, creating holes and crevices. It might also be slimy to the touch.

  - B. *Pool substrate characterization* evaluates the type and condition of bottom substrates found in pools. Pools with firmer sediment types (e.g., gravel, sand) and rooted aquatic plants support a wider variety of organisms than do pools with substrates dominated by mud or bedrock and no plants. Also, a pool with one uniform substrate type will support far fewer types of organisms than will a pool with a wide variety of substrate types.
  - C. *Pool variability* rates the overall mixture of pool types found in the stream according to size and depth. The four basic types of pools are large-shallow, large-deep, small-shallow, and small-deep. A stream with many pool types will support a wide variety of aquatic species. Rivers with low sinuosity (few bends) and monotonous pool characteristics do not have sufficient quantities and types of habitats to support a divers aquatic community.
  - D. *Channel alteration*—see the rocky-bottom habitat assessment instructions.
  - E. *Sediment deposition*—see the rocky-bottom habitat assessment instructions.
  - F. *Channel sinuosity* evaluates the sinuosity or meandering of the stream. Streams that meander provide a variety of habitats (such as pools and runs) and stream velocities and reduce the energy from current surges during storm events. Straight stream segments are characterized by even stream depth and unvarying velocity, and they are prone to flooding. To evaluate this parameter, imagine how much longer the stream would be if it were straightened out.
  - G. *Channel flow status*—see the rocky-bottom habitat assessment instructions.
  - H. *Bank vegetation protection*—see the rocky-bottom habitat assessment instructions.
  - I. *Condition of banks*—see the rocky-bottom habitat assessment instructions.
  - J. *Riparian vegetative zone width*—see the rocky-bottom habitat assessment instructions.

Note: Whenever stream sampling is to be conducted, it is a good idea to have a reference collection on hand. A reference collection is a sample of locally found macroinvertebrates that have been identified, labeled, and preserved in alcohol. The program advisor, along with a professional biologist/entomologist, should assemble the reference collection, properly identify all samples, preserve them in vials, and label them. This collection may then be used as a training tool and, in the field, as an aid in macroinvertebrate identification.

15.5.7 Post-Sampling Routine

After completing the stream characterization and habitat assessment, make sure that all of the field data sheets have been completed properly and that the information is legible. Be sure to include the identifying name of the site and the sampling date on each sheet. This information will function as a quality control element. Before leaving the stream location, make sure that all sampling equipment and devices have been collected and rinsed properly. Double-check to make sure that sample jars are tightly closed and correctly identified. All samples, field sheets, and equipment should be returned to the team leader at this point. Keep a copy of the field data sheets for comparison with future monitoring trips and for personal records. The next step is to prepare for macroinvertebrate laboratory work. This step includes setting up a laboratory for processing samples into subsamples and identifying macroinvertebrates to the family level. A professional biologist/entomologist/freshwater ecologist or professional advisor should supervise the identification procedure. (The actual laboratory procedures that follow the sampling and collecting phase are beyond the scope of this text.)

15.5.8 SAMPLING DEVICES

In addition to the sampling equipment mentioned previously, it may be desirable to employ, depending on stream conditions, the use of other sampling devices. Additional sampling devices commonly used, and discussed in the following sections, include dissolved oxygen and temperature monitors, sampling nets (including the D-frame aquatic net), sediment samplers (dredges), plankton samplers, and Secchi disks. The methods described below are approved by the USEPA. Coverage that is more detailed is available in APHA (1998).

15.5.8.1 Dissolved Oxygen and Temperature Monitor

As mentioned, the dissolved oxygen (DO) content of a stream sample can provide the investigator with vital information, as DO content reflects the ability of a stream to maintain aquatic life.

15.5.8.1.1 The Winkler DO with Azide Modification Method

The Winkler DO with azide modification method is commonly used to measure DO content. The Winkler method is best suited for clean waters. It can be used in the field but is better suited for laboratory work where greater accuracy may be achieved. The Winkler method adds a divalent manganese solution followed by a strong alkali to a 300-mL BOD bottle of stream-water sample. Any DO rapidly oxidizes an equivalent amount of divalent manganese to basic hydroxides of higher balance states. When the solution is acidified in the presence of iodide, oxidized manganese again reverts to the divalent state, and iodine, equivalent to the original DO content of the sample, is liberated. The amount of iodine is then determined by titration with a standard, usually thiosulfate, solution.

Fortunately for the field biologist, this is the age of miniaturized electronic circuit components and devices; thus, it is not too difficult to obtain portable electronic measuring devices for DO and temperature that are of quality construction and have better than moderate accuracy. These modern electronic devices are usually suitable for laboratory and field use. The device may be subjected to severe abuse in the field; therefore, the instrument must be durable, accurate, and easy to use. Several quality DO monitors are available commercially.

When using a DO monitor, it is important to calibrate (standardize) the meter prior to use. Calibration procedures can be found in APHA (1998) or in the manufacturer's instructions for the meter. The meter can be cali-

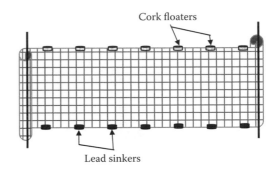

FIGURE 15.7 Two-person seine net.

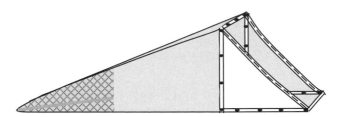

FIGURE 15.8 Surber sampler.

brated by determining the air temperature and the DO at saturation for that temperature and then adjusting the meter so it reads the saturation value. After calibration, the monitor is ready for use. As mentioned, all recorded measurements, including water temperatures and DO readings, should be entered in a field notebook.

15.5.8.2 Sampling Nets

A variety of sampling nets is available for use in the field. The two-person seine net shown in Figure 15.7 is 20 ft long by 4 ft deep with an 1/8-inch mesh and is utilized to collect various organisms. Two people, each holding one end, walk upstream, and small organisms are gathered in the net. Dip nets are used to collect organisms in shallow streams. The Surber sampler collects macroinvertebrates stirred up from the bottom (see Figure 15.8) and can be used to obtain a quantitative sample (number of organisms per square feet). It is designed for sampling riffle areas in streams and rivers up to a depth of about 450 mm (18 in.). It consists of two folding stainless steel frames set at right angles to each other. The frame is placed on the bottom, with the net extending downstream. Using a hand or a rake, all sediment enclosed by the frame is dislodged. All organisms are caught in the net and transferred to another vessel for counting. The D-frame aquatic dip net (see Figure 15.5) is ideal for sweeping over vegetation or for use in shallow streams.

15.5.8.3 Sediment Samplers (Dredges)

A sediment sampler or dredge is designed to obtain a sample of the bottom material in a slow-moving stream and the organisms in it. The simple homemade dredge

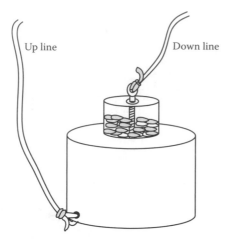

FIGURE 15.9 Home-made dredge.

FIGURE 15.10 Plankton net.

shown in Figure 15.9 works well in water too deep to sample effectively with handheld tools. The homemade dredge is fashioned from a #3 coffee can and a smaller can with a tight-fitting plastic lid (peanut cans work well). To use the homemade dredge, first invert it under water so the can fills with water and no air is trapped. Then, lower the dredge as quickly as possible with the "down" line. The idea is to bury the open end of the coffee can in the bottom. Then, quickly pull the "up" line to bring the can to the surface with a minimum loss of material. Dump the contents into a sieve or observation pan to sort. It works best in bottoms composed of sediment, mud, sand, and small gravel. The bottom sampling dredge can be used to for a number of different analyses. Because the bottom sediments represent a good area in which to find macroinvertebrates and benthic algae, the communities of organisms living on or in the bottom can be easily studied quantitatively and qualitatively. A chemical analysis of the bottom sediment can be conducted to determine what chemicals are available to organisms living in the bottom habitat.

15.5.8.4 Plankton Sampler

(More detailed information on plankton sampling can be found in AWRI, 2000.) *Plankton* (meaning to drift) are distributed through the stream and, in particular, in pool areas. They are found at all depths and are comprised of plant (phytoplankton) and animal (zooplankton) forms. Plankton show a distribution pattern that can be associated with the time of day and seasons. The three fundamental sizes of plankton are *nanoplankton*, *microplankton*, and *macroplankton*. The smallest, nanoplankton, range in size from 5 to 60 μm (millionth of a meter). Because of their small size, most nanoplankton will pass through the pores of a standard sampling net. Special fine mesh nets can be used to capture the larger nanoplankton. Most planktonic organism fall into the microplankton (or net plankton) category, where the sizes range from the largest nano-

plankton to about 2 mm (thousandths of a meter). Nets of various sizes and shapes are used to collect microplankton. The nets collect the organism by filtering water through fine meshed cloth. The third group of plankton, macroplankton, are visible to the naked eye. The largest can be several meters long.

The plankton net or sampler (see Figure 15.10) is a device that makes it possible to collect phytoplankton and zooplankton samples. For quantitative comparisons of different samples, some nets have a flowmeter to determine the amount of water passing through the collecting net. The plankton net or sampler provides a means of obtaining samples of plankton from various depths so distribution patterns can be studied. The net can be towed to sample plankton at a single depth (horizontal tow) or can be lowered into the water to sample the water column (vertical tow). Another possibility is oblique tows where the net is lowered to a predetermined depth and raised at a constant rate as the vessel moves forward.

After towing and removal from the stream, the sides of the net are rinsed to dislodge the collected plankton. If a quantitative sample is desired, a certain quantity of water is collected. If the plankton density is low, then the sample may be concentrated using a low-speed centrifuge or some other filtering device. A definite volume of the sample is studied under the compound microscope for counting and identification of plankton.

15.5.8.5 Secchi Disk

For determining water turbidity or degree of visibility in a stream, a Secchi disk is often used (Figure 15.11). The Secchi disk originated with Father Pietro Secchi, who was

FIGURE 15.11 Secchi disk.

an astrophysicist and scientific advisor to the Pope. When the head of the Papal Navy asked him to measure transparency in the Mediterranean Sea, Secchi used some white disks to measure the clarity of water in April of 1865. Various sizes of disks have been used since that time, but the most frequently used disk is an 8-inch-diameter metal disk painted in alternating black and white quadrants. The disk shown in Figure 15.11 is 20 cm in diameter; it is lowered into the stream using the calibrated line. To use the Secchi disk properly, it should be lowered into the stream water until no longer visible. At the point where it is no longer visible, a measurement of the depth is taken. This depth is called the *Secchi disk transparency light extinction coefficient*. The best results are usually obtained after early morning and before late afternoon.

15.5.8.6 Miscellaneous Sampling Equipment

Several other sampling tools and devices are available for use in sampling a stream; for example, consider the standard sand/mud sieve. Generally made of heavy-duty galvanized 1/8-inch mesh screen supported by a water-sealed 24 × 15 × 3-inch wood frame, this device is useful for collecting burrowing organisms found in soft bottom sediments. Moreover, no stream sampling kit would be complete without a collecting tray, collecting jars of assorted sizes, heavy-duty plastic bags, small pipets, large 2-ounce pipets, fine mesh straining nets, and black china marking pencils. In addition, depending on the quantity of material to be sampled, it is prudent to include several 3- and 5-gallon collection buckets in the stream sampling field kit.

15.5.9 THE BOTTOM LINE ON BIOLOGICAL SAMPLING

This discussion has stressed the practice of biological monitoring, employing the use of biotic indices as key measuring tools. We emphasized biotic indices not only for their simplicity of use but also for the relative accuracy they provide, although their development and use can sometimes be derailed. The failure of a monitoring protocol to assess environmental condition accurately or to protect running waters usually stems from conceptual, sam-

pling, or analytical pitfalls. Biotic indices can be combined with other tools for measuring the condition of ecological systems in ways that enhance or hinder their effectiveness. The point is that, like any other tool, these tools can be misused; however, the fact that biotic indices can be misused does not mean that the approach itself is useless. Thus, to ensure that the biotic indices approach is not useless, it is important for the practicing freshwater ecologist and water sampler to remember a few key guidelines:

1. Sampling everything is not the goal. As Botkin (1990) noted, biological systems are complex and unstable in space and time, and samplers often feel compelled to study all components of this variation. Complex sampling programs proliferate, but every study need not explore everything. Freshwater samplers and monitors should avoid the temptation to sample all the unique habitats and phenomena that make freshwater monitoring so interesting. Emphasis should be placed on the central components of a clearly defined research agenda (a sampling and monitoring protocol) to detect and measure the influence of human activities on the ecological system of a water body.

2. With regard to the influence of human activities on the ecological system of a water body, we must consider protecting biological conditions to be a central responsibility of water resource management. One thing is certain—until biological monitoring is seen as essential to tracking attainment of that goal and biological criteria become enforceable standards mandated by the Clean Water Act, the diversity of life in the nation's freshwater systems will continue to decline.

Biomonitoring is only one of several tools available to the water practitioner. No matter the tool employed, all results depend on proper biomonitoring techniques. Biological monitoring must be designed to obtain accurate results, and current approaches must be strengthened. In addition, "the way it's always been done" must be reexamined, and efforts must be undertaken to do what works to keep freshwater systems alive. We can afford nothing less.

15.6 DRINKING WATER QUALITY MONITORING

When we speak of water quality monitoring, we refer to monitoring practices based on three criteria: (1) to ensure to the extent possible that the water is not a danger to public health, (2) to ensure that the water provided at the tap is as aesthetically pleasing as possible, and (3) to ensure compliance with applicable regulations. To meet these goals, all public systems must monitor water quality to some extent. The degree of monitoring employed is dependent on local needs and requirements, and on the type of water system; small water systems using good-quality water from deep wells may have to conduct only occasional monitoring, but systems using surface water sources must test water quality frequently (AWWA 1995).

Drinking water must be monitored to provide adequate control of the entire water drawing, treatment, and conveyance system. *Adequate control* is defined as monitoring employed to assess the current level of water quality, so action can be taken to maintain the required level (whatever that might be). We define *water quality monitoring* as the sampling and analysis of water constituents and conditions. When we monitor, we collect data. As a monitoring program is developed, deciding the reasons for collecting the information is important. The reasons are defined by establishing a set of objectives that includes a description of who will collect the information.

It may come as a surprise to know that today the majority of people collecting data are not water and wastewater operators; instead, many are volunteers. These volunteers have a stake in their local stream, lake, or other water body, and in many cases are proving they can successfully carry out a water quality monitoring program.

15.6.1 Is the Water Good or Bad?

(Much of the information presented in the following sections is based on our personal experience and on USEPA, 1997.) To answer the question of whether the water is good or bad we must consider two factors. First, we return to the basic principles of water quality monitoring—sampling and analyzing water constituents and conditions. These constituents include:

1. Introduced pollutants, such as pesticides, metals, and oil
2. Constituents found naturally in water that can nevertheless be affected by human sources, such as dissolved oxygen, bacteria, and nutrients

The magnitude of their effects is influenced by properties such as pH and temperature; for example, temperature influences the quantity of dissolved oxygen that water is able to contain, and pH affects the toxicity of ammonia.

The second factor to be considered is that the only valid way to answer this question is to conduct tests, the results of which must then be compared to some form of water quality standards. If simply assigning a "good" and "bad" value to each test factor were possible, the meters and measuring devices in water quality test kits would be much easier to make. Instead of fine graduations, they could simply have a "good" and a "bad" zone.

TABLE 15.4
Total Residual Chlorine Effects

Total Residual Chlorine (mg/L)	Effect
0.06	Toxic to striped bass larvae
0.31	Toxic to white perch larvae
0.5–1.0	Typical drinking water residual
1.0–3.0	Recommended for swimming pools

Water quality—the difference between good and bad water—must be interpreted according to the intended use of the water; for example, the perfect balance of water chemistry that provides a sparkling clear, sanitary swimming pool would not be acceptable for drinking water and would be a deadly environment for many biota (Table 15.4). In another example, widely different levels of fecal coliform bacteria are considered acceptable, depending on the intended use of the water.

State and local water quality practitioners as well as volunteers have been monitoring water quality conditions for many years. In fact, until the past decade or so (until biological monitoring protocols were developed and began to take hold), water quality monitoring was generally considered the primary way to identify water pollution problems. Today, professional water quality practitioners and volunteer program coordinators alike are moving toward approaches that combine chemical, physical, and biological monitoring methods to achieve the best picture of water quality conditions. Water quality monitoring can be used for many purposes:

1. *To identify whether waters are meeting designated uses.* All states have established specific criteria (limits on pollutants) identifying what concentrations of chemical pollutants are allowable in their waters. When chemical pollutants exceed maximum or minimum allowable concentrations, waters may no longer be able to support the beneficial uses—such as fishing, swimming, and drinking—for which they have been designated (see Table 15.5). Designated or intended uses and the specific criteria that protect them (along with antidegradation statements prohibiting waters from deteriorating below existing or anticipated uses) together form water quality standards. State water quality professionals assess water quality by comparing the concentrations of chemical pollutants found in streams to the criteria in the state's standards and so judge whether or not streams are meeting their designated uses. Water quality monitoring, however, might be inadequate for determining whether aquatic life needs are being met in a stream. Although some

TABLE 15.5
Fecal Coliform Bacteria per 100 mL of Water

Desirable	Permissible	Type of Water Use
0	0	Potable and well water (for drinking)
<200	<1000	Primary contact water (for swimming)
<1000	<5000	Secondary contact water (boating and fishing)

constituents (such as dissolved oxygen and temperature) are important to maintaining healthy fish and aquatic insect populations, other factors (such as the physical structure of the stream and the condition of the habitat) play an equal or greater role. Biological monitoring methods are generally better suited to determining whether aquatic life is supported.

2. *To identify specific pollutants and sources of pollution.* Water quality monitoring helps link sources of pollution to water body quality problems because it identifies specific problem pollutants. Because certain activities tend to generate certain pollutants (bacteria and nutrients are more likely to come from an animal feedlot than an automotive repair shop), a tentative link to what would warrant further investigation or monitoring can be formed.

3. *To determine trends.* Chemical constituents that are properly monitored (i.e., using consistent time of day and on a regular basis using consistent methods) can be analyzed for trends over time.

4. *To screen for impairment.* Finding excessive levels of one or more chemical constituents can serve as an early warning for potential pollution problems.

15.6.2 STATE WATER QUALITY STANDARDS PROGRAMS

Each state has a program to set standards for the protection of each body of water within its boundaries. Standards for each body of water are developed that:

1. Depend on the water's designated use
2. Are based on USEPA national water quality criteria and other scientific research into the effects of specific pollutants on different types of aquatic life and on human health
3. May include limits based on the biological diversity of the body of water (the presence of food and prey species)

State water quality standards set limits on pollutants and establish water quality levels that must be maintained for each type of water body, based on its designated use. Resources for this type of information include:

1. USEPA Water Quality Criteria Program
2. U.S. Fish and Wildlife Service Habitat Suitability Index Models (for specific species of local interest)

Monitoring test results can be plotted against these standards to provide a focused, relevant, required assessment of water quality.

15.6.3 DESIGNING A WATER QUALITY MONITORING PROGRAM

The first step in designing a water quality-monitoring program is to determine the purpose for the monitoring. This aids in selection of parameters to monitor. This decision should be based on such factors as:

1. Types of water quality problems and pollution sources that are likely to be encountered (see Table 15.6)
2. Cost of available monitoring equipment
3. Precision and accuracy of available monitoring equipment
4. Capabilities of monitors

Note: We discuss the parameters most commonly monitored by drinking water practitioners in streams (i.e., we assume, for illustration and discussion purposes, that our water source is a surface water stream) in detail in this section. These parameters include dissolved oxygen (DO), biochemical oxygen demand (BOD), temperature, pH, turbidity, total orthophosphate, nitrates, total solids, conductivity, total alkalinity, fecal bacteria, apparent color, odor, and hardness. When monitoring water supplies under the Safe Drinking Water Act (SDWA) or the National Pollutant Discharge Elimination System (NPDES), utilities must follow test procedures approved by the USEPA for these purposes. Additional testing requirements under these and other federal programs are published as amendments in the *Federal Register*.

into the water, facing upstream. Collect a water sample 8 to 12 inches beneath the surface, or midway between the surface and the bottom if the stream reach is shallow.

5. Turn the bottle underwater into the current and away from you. In slow-moving stream reaches, push the bottle underneath the surface and away from you in the upstream direction.
6. Leave a 1-inch air space (except for DO and BOD samples). Do not fill the bottle completely (so the sample can be shaken just before analysis). Recap the bottle carefully, remembering not to touch the inside.
7. Fill in the bottle number and site number on the appropriate field data sheet. This is important because it tells the lab specialist which bottle goes with which site.
8. If the samples are to be analyzed in the lab, place them in the cooler for transport to the lab.

15.7.4 Sample Preservation and Storage

Samples can change very rapidly; however, no single preservation method will serve for all samples and constituents. If analysis must be delayed, follow the instructions for sample preservation and storage listed in *Standard Methods* (APHA, 1998) or those specified by the laboratory that will eventually process the samples (see Table 15.7). In general, handle the sample in a way that does not cause changes in biological activity, physical alterations, or chemical reactions. Cool the sample to reduce biological and chemical reactions. Store in darkness to suspend photosynthesis. Fill the sample container completely to prevent the loss of dissolved gases. Metal cations such as iron and lead and suspended particles may adsorb onto container surfaces during storage.

15.7.5 Standardization of Methods

References used for sampling and testing must correspond to those listed in the most current federal regulations. For the majority of tests, to compare the results of either different water quality monitors or the same monitors over the course of time requires some form of standardization of the methods. The American Public Health Association (APHA) recognized this requirement when, in 1899, it appointed a committee to draw up standard procedures for the analysis of water. The report (originally published in 1905) constituted the first edition of what is now known as *Standard Methods for the Examination of Water and Wastewater* or *Standard Methods*. This book is now in its 20th edition (APHA, 1998) and serves as the primary reference for water testing methods, and as the basis for most USEPA-approved methods.

15.8 TEST METHODS FOR DRINKING WATER AND WASTEWATER

The material presented in this section is based on personal experience and adaptations from *Standard Methods* (APHA, 1998), *Federal Register*, and *The Monitor's Handbook* (LaMotte, 1992). Descriptions of general methods to help you understand how each works in specific test kits follow. Always use the specific instructions included with the equipment and individual test kits. Most water analyses are conducted by either titrimetric analyses or colorimetric analyses. Both methods are simple to use and provide accurate results.

15.8.1 Titrimetric Methods

Titrimetric analyses are based on adding a solution of known strength (the *titrant*, which must have an exact known concentration) to a specific volume of a treated sample in the presence of an indicator. The indicator produces a color change indicating that the reaction is complete. Titrants are generally added by a titrator (microburet) or a precise glass pipet.

15.8.2 Colorimetric Methods

Colorimetric standards are prepared as a series of solutions with increasing known concentrations of the constituent to be analyzed. Two basic types of colorimetric tests are commonly used:

1. The pH is a measure of the concentration of hydrogen ions (the acidity of a solution) determined by the reaction of an indicator that varies in color, depending on the hydrogen ion levels in the water.
2. Tests that determine a concentration of an element or compound are based on sBeer's law. Simply, this law states that the higher the concentration of a substance, the darker the color produced in the test reaction and therefore the more light absorbed. Assuming a constant viewpath, the absorption increases exponentially with concentration.

15.8.3 Visual Methods

The Octet Comparator uses standards that are mounted in a plastic comparator block. It employs eight permanent translucent color standards and built-in filters to eliminate optical distortion. The sample is compared using either of two viewing windows. Two devices that can be used with the comparator are a B color reader, which neutralizes color or turbidity in water samples, and an axial mirror that intensifies faint colors of low concentrations for easy distinction.

TABLE 15.7
Recommended Sample Storage and Preservation Techniques

Test Factor	Container Type	Preservation	Maximum Storage Time Recommended (Required)
Alkalinity	Plastic Glass	Refrigerate	24 hr (14 days)
BOD	Plastic Glass	Refrigerate	6 hr (48 hr)
Conductivity	Plastic Glass	Refrigerate	28 days (28 days)
Hardness	Plastic Glass	Lower pH to <2	6 months (6 months)
Nitrate	Plastic Glass	Analyze ASAP or refrigerate	48 hr (48 hr)
Nitrite	Plastic Glass	Analyze ASAP or refrigerate	None (48 hr)
Odor	Glass	Analyze ASAP	6 hr (not specified)
Dissolved oxygen			
Electrode	Glass	Immediately analyze	0.25 hr (0.25 hr)
Winkler	Glass	Fix immediately	8 hr (8 hr)
pH	Plastic Glass	Immediately analyze	0.25 hr (0.25 hr)
Phosphate	Glass	Refrigerate	48 hr (not specified)
Salinity	Glass	Immediately analyze or use wax seal	6 months (not specified)
Temperature	Plastic Glass	Immediately analyze	0.25 hr
Turbidity	Plastic Glass	Analyze same day or store in dark up to 24 hr; refrigerate	24 hr (48 hr)

15.8.4 ELECTRONIC METHODS

Although the human eye is capable of differentiating color intensity, the interpretation of colors is quite subjective. Electronic colorimeters consist of a light source that passes through a sample and is measured on a photodetector with an analog or digital readout. Besides electronic colorimeters, specific electronic instruments are manufactured for lab and field determination of many water quality factors, including pH, total dissolved solids, conductivity, dissolved oxygen, temperature, and turbidity.

15.8.5 DISSOLVED OXYGEN TESTING

In this section and the sections that follow, we discuss several water quality factors that are routinely monitored in drinking water operations. We do not discuss the actual test procedures used to analyze each water quality factor; instead, we refer you the latest edition of *Standard Methods* (APHA, 1998) for the correct procedures to use when conducting these tests.

A stream system used as a source of water produces and consumes oxygen. It gains oxygen from the atmosphere and from plants through photosynthesis. Because of the churning of running water, it dissolves more oxygen than does still water, such as in a reservoir behind a dam. Respiration by aquatic animals, decomposition, and various chemical reactions consume oxygen.

Oxygen is actually poorly soluble in water. Its solubility is related to pressure and temperature. In water supply systems, dissolved oxygen in raw water is considered the necessary element to support life of many aquatic organisms. From the drinking water practitioner's point of view, DO is an important indicator of the water treatment process and an important factor in corrosivity.

Wastewater from sewage treatment plants often contains organic materials that are decomposed by microorganisms, which use oxygen in the process. The amount of oxygen consumed by these organism in breaking down the waste is known as the *biochemical oxygen demand* (BOD). We include a discussion of BOD and how to monitor it later. Other sources of oxygen-consuming waste

TABLE 15.8
Maximum DO Concentrations vs. Temperature Variations

Temperature (°C)	DO (mg/L)	Temperature (°C)	DO (mg/L)
0	14.60	23	8.56
1	14.19	24	8.40
2	13.81	25	8.24
3	13.44	26	8.09
4	13.09	27	7.95
5	12.75	28	7.81
6	12.43	29	7.67
7	12.12	30	7.54
8	11.83	31	7.41
9	11.55	32	7.28
10	11.27	33	7.16
11	11.01	34	7.05
12	10.76	35	6.93
13	10.52	36	6.82
14	10.29	37	6.71
15	10.07	38	6.61
16	9.85	39	6.51
17	9.65	40	6.41
18	9.45	41	6.31
19	9.26	42	6.22
20	9.07	43	6.13
21	8.90	44	6.04
22	8.72	45	5.95

include stormwater runoff from farmland or urban streets, feedlots, and failing septic systems.

Oxygen is measured in its dissolved form as dissolved oxygen. If more oxygen is consumed than produced, DO levels decline and some sensitive animals may move away, weaken, or die. DO levels fluctuate over a 24-hour period and seasonally. They vary with water temperature and altitude. Cold water holds more oxygen than warm water (see Table 15.8), and water holds less oxygen at higher altitudes. Thermal discharges (such as water used to cool machinery in a manufacturing plant or a power plant) raise the temperature of water and lower its oxygen content. Aquatic animals are most vulnerable to lowered DO levels in the early morning on hot summer days when stream flows are low, water temperatures are high, and aquatic plants have not been producing oxygen since sunset.

15.8.5.1 Sampling and Equipment Considerations

In contrast to lakes, where DO levels are most likely to vary vertically in the water column, changes in DO in rivers and streams move horizontally along the course of the waterway. This is especially true in smaller, shallow streams. In larger, deeper rivers, some vertical stratification of dissolved oxygen might occur. The DO levels in and below riffle areas, waterfalls, or dam spillways are typically higher than those in pools and slower moving stretches. If we want to measure the effect of a dam, sampling for DO behind the dam, immediately below the spillway, and upstream of the dam would be important. Because DO levels are critical to fish, a good place to sample is in the pools that fish tend to favor or in the spawning areas they use.

An hourly time profile of DO levels at a sampling site is a valuable set of data, because it shows the change in DO levels from the low point (just before sunrise) to the high point (sometime near midday); however, this might not be practical for a volunteer monitoring program. Note the time of the DO sampling to help judge when in the daily cycle the data were collected.

Dissolved oxygen samples are collected using a special BOD bottle: a glass bottle with a "turtleneck" and a ground stopper. You can fill the bottle directly in the stream if the stream is wadeable or boatable, or you can use a sampler dropped from a bridge or boat into water deep enough to submerse it. Samplers can be made or purchased.

Dissolved oxygen is measured either as the amount of oxygen in a liter of water (milligrams per liter, or mg/L) or as percent saturation, which is the amount of oxygen in a liter of water relative to the total amount of oxygen that the water can hold at that temperature. Dissolved oxygen is determined using some variation of the Winkler method or by using a meter and probe.

15.8.5.2 Winkler Method (Azide Modification)

The Winkler method (azide modification) involves filling a sample bottle completely with water (no air is left to bias the test). The dissolved oxygen is then fixed using a series of reagents that form a titrated acid compound. Titration involves the drop-by-drop addition of a reagent that neutralizes the acid compound, causing a change in the color of the solution. The point at which the color changes is the *endpoint* and is equivalent to the amount of oxygen dissolved in the sample. The sample is usually fixed and titrated in the field at the sample site, but preparing the sample in the field and delivering it to a lab for titration is possible. The azide modification method is best suited for relatively clean waters; otherwise, substances such as color, organics, suspended solids, sulfide, chlorine, and ferrous and ferric iron can interfere with test results. If fresh azide is used, nitrite will not interfere with the test.

In testing, iodine is released in proportion to the amount of DO present in the sample. By using sodium thiosulfate with starch as the indicator, the sample can be titrated to determine the amount of DO present. The chemicals used include:

1. Manganese sulfate solution
2. Alkaline azide–iodide solution
3. Sulfuric acid (concentrated)
4. Starch indicator
5. Sodium thiosulfate solution (0.025 *N*), phenylarsine solution (0.025 *N*), or potassium biniodate solution (0.025 *N*)
6. Distilled or deionized water

The equipment used includes:

1. Buret, graduated to 0.1 mL
2. Buret stand
3. 300-mL BOD bottles
4. 500-mL Erlenmeyer flasks
5. 1.0-mL pipets with elongated tips
6. Pipet bulb
7. 250-mL graduated cylinder
8. Laboratory-grade water rinse bottle
9. Magnetic stirrer and stir bars (optional)

15.8.5.2.1 Procedure

The procedure for the Winkler method is:

1. Collect sample in a 300-mL BOD bottle.
2. Add 1 mL manganous sulfate solution at the surface of the liquid.
3. Add 1 mL alkaline iodide–azide solution at the surface of the liquid.
4. Stopper bottle, and mix by inverting the bottle.
5. Allow the floc to settle halfway in the bottle, remix, and allow to settle again.

6. Add 1 mL concentrated sulfuric acid at the surface of the liquid.
7. Restopper bottle, rinse top with laboratory-grade water, and mix until precipitate is dissolved (the liquid in the bottle should appear clear and have an amber color).
8. Measure 201 mL from the BOD bottle into an Erlenmeyer flask.
9. Titrate with 0.025 *N* PAO or thiosulfate to a pale yellow color, and note the amount of titrant.
10. Add 1 mL of starch indicator solution.
11. Titrate until blue color first disappears.
12. Record the total amount of titrant.

15.8.5.2.2 Calculation

To calculate the DO concentration when the modified Winkler titration method is used:

$$\text{DO (mg/L)} = \frac{[\text{Buret}_{\text{Final}} \text{ (mL)} - \text{buret}_{\text{Start}} \text{ (mL)}] \times N \times 8000}{\text{Sample volume (mL)}} \quad (15.1)$$

Note: Using a 200-mL sample and a 0.025 *N* (*N* is the normality of the solution used to titrate the sample) titrant reduces this calculation to:

$$\text{DO (mg/L)} = \text{Titrant used (mL)}$$

■ EXAMPLE 15.1

Problem: The operator titrates a 200-mL DO sample. The buret reading at the start of the titration was 0.0 mL. At the end of the titration, the buret read 7.1 mL. The concentration of the titrating solution was 0.025 *N*. What is the DO concentration in mg/L?

Solution:

$$\text{DO (mg/L)} = \frac{(7.1 \text{ mL} - 0.0 \text{ mL}) \times 0.025 \times 8000}{200 \text{ mL}} = 7.1$$

Dissolved oxygen field kits using the Winkler method are relatively inexpensive, especially compared to a meter and probe. Field kits run between \$35 and \$200, and each kit comes with enough reagents to run 50 to 100 DO tests. Replacement reagents are inexpensive and can be purchased already measured out for each test in plastic pillows. Reagents can also be purchased in larger quantities in bottles and measured out with a volumetric scoop. The advantage of the pillows is that they have a longer shelf life and are much less prone to contamination or spillage. Buying larger quantities in bottles has the advantage of considerably lower cost per test.

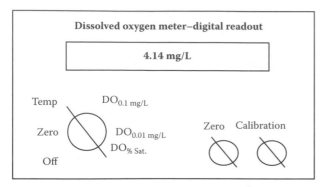

FIGURE 15.13 Dissolved oxygen meter.

The major factor in the expense for the kits is the method of titration used. Eyedropper or syringe-type titration is less precise than digital titration, because a larger drop of titrant is allowed to pass through the dropper opening; on a microscale, the drop size (and thus volume of titrant) can vary from drop to drop. A digital titrator or a buret (a long glass tube with a tapered tip like a pipet) permits much more precision and uniformity with regard to the titrant it allows to pass.

If a high degree of accuracy and precision in DO results is required, a digital titrator should be used. A kit that uses an eyedropper-type or syringe-type titrator is suitable for most other purposes. The lower cost of this type of DO field kit might be attractive if several teams of samplers and testers at multiple sites at the same time are required.

15.8.5.3 Meter and Probe

A *dissolved oxygen meter* is an electronic device that converts signals from a probe placed in the water into units of DO in milligrams per liter. Most meters and probes also measure temperature. The probe is filled with a salt solution and has a selectively permeable membrane that allows DO to pass from the stream water into the salt solution. The DO that has diffused into the salt solution changes the electric potential of the salt solution, and this change is sent by electric cable to the meter, which converts the signal to milligrams per liter on a scale that the user can read.

15.8.5.3.1 Methodology

If samples are to be collected for analysis in the laboratory, a special APHA sampler or the equivalent must be used. If the sample is exposed to or mixed with air during collection, test results can change dramatically; therefore, the sampling device must allow collection of a sample that is not mixed with atmospheric air and allows for at least 3× bottle overflow (see Figure 15.13). Again, because the DO level in a sample can change quickly, only grab samples should be used for dissolved oxygen testing. Samples must be tested immediately (within 15 minutes) after collection.

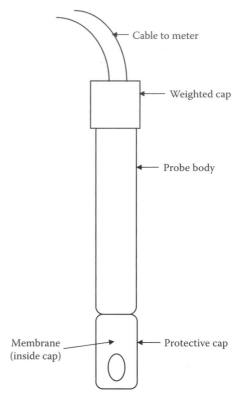

FIGURE 15.14 dissolved oxygen-field probe.

Note: Samples collected for analysis using the modified Winkler titration method may be preserved for up to 8 hours by adding 0.7 mL of concentrated sulfuric acid or by adding all the chemicals required by the procedure. Samples collected from the aeration tank of the activated sludge process must be preserved using a solution of copper sulfate–sulfamic acid to inhibit biological activity.

The advantage of using the DO oxygen meter method is that the meter can be used to determine DO concentration directly (see Figure 15.13). In the field, a direct reading can be obtained using a probe (see Figure 15.14) or by collection of samples for testing in the laboratory using a laboratory probe (see Figure 15.15).

Note: The field probe can be used for laboratory work by placing a stirrer in the bottom of the sample bottle, but the laboratory probe should never be used in any situation where the entire probe might be submerged.

The probe used in the determination of DO consists of two electrodes, a membrane, and a membrane filling solution. Oxygen passes through the membrane into the filling solution and causes a change in the electrical current passing between the two electrodes. The change is measured and displayed as the concentration of DO. For accuracy, the probe membrane must be in proper operating

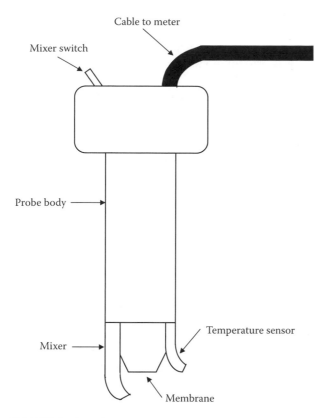

FIGURE 15.15 Dissolved oxygen-lab probe.

condition, and the meter must be calibrated before use. The only chemical used in the DO meter method during normal operation is the electrode filling solution, whereas in the Winkler DO method chemicals are required for meter calibration.

Calibration prior to use is important. Both the meter and the probe must be calibrated to ensure accurate results. The frequency of calibration is dependent on the frequency of use; for example, if the meter is used once a day, then calibration should be performed before use. Three methods are available for calibration: *saturated water*, *saturated air*, and the *Winkler method*. It is important to note that if the Winkler method is not used for routine calibration method, periodic checks using this method are recommended.

15.8.5.3.2 Procedure

It is important to keep in mind that the operating procedures of the meter and probe supplier should always be followed. Normally, the manufacturer's recommended procedure will include the following generalized steps:

1. Turn DO meter on, and allow 15 minutes for it to warm up.
2. Turn meter switch to 0, and adjust as needed.
3. Calibrate meter using the saturated air, saturated water, or Winkler azide procedure for calibration.

4. Collect sample in 300-mL bottle, or place field electrode directly in stream.
5. Place laboratory electrode in BOD bottle without trapping air against membrane; turn on stirrer.
6. Turn meter switch to temperature mode, and measure the temperature.
7. Turn meter switch to DO mode, and allow 10 seconds for the meter reading to stabilize.
8. Read DO mg/L from meter, and record the results.

No calculation is necessary using this method because results are read directly from the meter.

Dissolved oxygen meters are expensive compared to field kits that use the titration method. Meter and probe combinations run between $500 and $1200, including a long cable to connect the probe to the meter. The advantage of a meter and probe is that DO and temperature can be quickly read at any point where the probe is inserted into the stream. DO levels can be measured at a certain point on a continuous basis. The results are read directly as milligrams per liter, unlike the titration methods, in which the final titration result might have to be converted to milligrams per liter. DO meters, however, are more fragile than field kits, and repairs to a damaged meter can be costly. The meter and probe must be carefully maintained and must be calibrated before each sample run and, if many tests are done, between sampling. Because of the expense, a small water/wastewater facility might only have one meter and probe, which means that only one team of samplers can sample DO, and they must test all the sites. With field kits, on the other hand, several teams can sample simultaneously.

15.8.6 Biochemical Oxygen Demand Testing

Biochemical oxygen demand (BOD) measures the amount of oxygen consumed by microorganisms in decomposing organic matter in stream water. BOD also measures the chemical oxidation of inorganic matter (the extraction of oxygen from water via chemical reaction). A test is used to measure the amount of oxygen consumed by these organisms during a specified period of time (usually 5 days at 20°C). The rate of oxygen consumption in a stream is affected by a number of variables: temperature, pH, the presence of certain kinds of microorganisms, and the type of organic and inorganic material in the water. BOD directly affects the amount of dissolved oxygen in water bodies. The greater the BOD, the more rapidly oxygen is depleted in the water body, leaving less oxygen available to higher forms of aquatic life. The consequences of high BOD are the same as those for low dissolved oxygen: aquatic organisms become stressed, suffocate, and die. Most river waters used as water supplies have a BOD less than 7 mg/L; therefore, dilution is not necessary.

Sources of BOD include leaves and woody debris; dead plants and animals; animal manure; effluents from pulp and paper mills, wastewater treatment plants, feedlots, and food-processing plants; failing septic systems; and urban stormwater runoff.

Note: To evaluate the potential for raw water to be used as a drinking water supply, it is usually sampled, analyzed, and tested for biochemical oxygen demand when turbid, polluted water is the only source available.

15.8.6.1 Sampling Considerations

Biochemical oxygen demand is affected by the same factors that affect dissolved oxygen. Aeration of stream water—by rapids and waterfalls, for example—will accelerate the decomposition of organic and inorganic material. BOD levels at a sampling site with slower, deeper waters might be higher for a given column of organic and inorganic material than the levels for a similar site in high aerated waters. Chlorine can also affect BOD measurement by inhibiting or killing the microorganisms that decompose the organic and inorganic matter in a sample. If sampling in chlorinated waters (such as those below the effluent from a sewage treatment plant), it is necessary to neutralize the chlorine with sodium thiosulfate (APHA, 1998).

Biochemical oxygen demand measurement requires taking two samples at each site. One is tested immediately for dissolved oxygen, and the second is incubated in the dark at 20°C for 5 days and then tested for dissolved oxygen remaining. The difference in oxygen levels between the first test and the second test (in milligrams per liter) is the amount of BOD. This represents the amount of oxygen consumed by microorganisms and used to break down the organic matter present in the sample bottle during the incubation period. Because of the 5-day incubation, the tests are conducted in a laboratory.

Sometimes by the end of the 5-day incubation period, the dissolved oxygen level is 0. This is especially true for rivers and streams with a lot of organic pollution. It is not possible to know when the 0 point was reached, so determining the BOD level is also impossible. In this case, diluting the original sample by a factor that results in a final dissolved oxygen level of at least 2 mg/L is necessary. Special dilution water should be used for the dilutions (APHA, 1998).

Some experimentation is necessary to determine the appropriate dilution factor for a particular sampling site. The result is the difference in dissolved oxygen between the first measurement and the second, after multiplying the second result by the dilution factor. *Standard Methods* (APHA, 1998) prescribes all phases of procedures and calculations for BOD determination. A BOD test is not required for monitoring water supplies.

15.8.6.2 BOD Sampling, Analysis, and Testing

The approved biochemical oxygen demand sampling and analysis procedure measures the DO depletion (biological oxidation of organic matter in the sample) over a 5-day period under controlled conditions (20°C in the dark). The test is performed using a specified incubation time and temperature. Test results are used to determine plant loadings, plant efficiency, and compliance with NPDES effluent limitations. The duration of the test (5 days) makes it difficult to use the data effectively for process control.

The standard BOD test does not differentiate between oxygen used to oxidize organic matter and that used to oxidize organic and ammonia nitrogen to more stable forms. Because many biological treatment plants now control treatment processes to achieve oxidation of the nitrogen compounds, it is possible that BOD testing for plant effluent and some process samples may produce BOD test results based on both carbon and nitrogen oxidation. To avoid this situation, a nitrification inhibitor can be added. When this is done, the tests measure *carbonaceous BOD* (CBOD). A second uninhibited BOD should also be run whenever CBOD is determined.

When taking a BOD sample, no special sampling container is required. Either a grab or composite sample can be used. BOD samples can be preserved by refrigeration at or below 4°C (not frozen); composite samples must be refrigerated during collection. Maximum holding time for preserved samples is 48 hours.

Using the incubation of dissolved approved test method, a sample is mixed with dilution water in several different concentrations (dilutions). The dilution water contains nutrients and materials to provide an optimum environment. Chemicals used include dissolved oxygen, ferric chloride, magnesium sulfate, calcium chloride, phosphate buffer, and ammonium chloride.

Note: Remember that all chemicals can be dangerous if not used properly and in accordance with the recommended procedures. Review appropriate sections of the individual chemical Material Safety Data Sheet (MSDS) to determine proper methods for handling and for safety precautions that should be taken.

Sometimes it is necessary to add healthy organisms to the sample (i.e., seed the sample). The DO of the dilution and the dilution water is determined. If seed material is used, a series of dilutions of seed material must also be prepared. The dilutions and dilution blanks are incubated in the dark for 5 days at 20°C ± 1°C. At the end of 5 days, the DO of each dilution and the dilution blanks are determined. For the test results to be valid, certain criteria must be met:

TABLE 15.9
BOD_5 Test Procedure

1. Fill two bottles with BOD dilution water; insert stoppers.
2. Place sample in two BOD bottles; fill with dilution water; insert stoppers.
3. Test for dissolved oxygen (DO).
4. Incubate for 5 days.
5. Test for DO.
6. Add 1 mL $MnSO_4$ below surface.
7. Add 1 mL alkaline KI below surface.
8. Add 1 mL H_2SO_4.
9. Transfer 203 mL to flask.
10. Titrate with PAO or thiosulfate.

1. Dilution water blank DO change must be ≤ 0.2 mg/L.
2. Initial DO must be > 7.0 mg/L but ≤ 9.0 mg/L (or saturation at 20°C and test elevation).
3. Sample dilution DO depletion must be ≥ 2.0 mg/L.
4. Sample dilution residual DO must be ≥ 1.0 mg/L.
5. Sample dilution initial DO must be ≥ 7.0 mg/L.
6. Seed correction should be ≥ 0.6 but ≤ 1.0 mg/L.

The BOD test procedure consists of 10 steps (for unchlorinated water) as shown in Table 15.9.

Note: BOD is calculated individually for all sample dilutions meeting the criteria. The reported result is the average of the BOD of each valid sample dilution.

15.8.6.3 BOD_5 Calculation

Unlike the direct reading instrument used in the DO analysis, BOD_5 results require calculation. Several criteria are used when selecting which BOD_5 dilutions should be used for calculating test results. Consult a laboratory testing reference manual, such as *Standard Methods* (APHA, 1998), for this information. Currently, two basic calculations are used for BOD_5. The first is used for samples that have not been seeded. The second must be used whenever BOD_5 samples are seeded. An example calculation procedure for unseeded samples is presented first.

$$BOD_5 = \frac{\left[DO_{Start}\ (mg/L) - DO_{Final}\ (mg/L)\right] \times 300\ mL}{Sample\ volume\ (mL)} \quad (15.2)$$

■ EXAMPLE 15.2

Problem: The BOD_5 test is completed. Bottle 1 of the test had a DO of 7.1 mg/L at the start of the test. After 5 days, bottle 1 had a DO of 2.9 mg/L. Bottle 1 contained 120 mg/L of sample. Determine the BOD_5.

Solution:

$$BOD_5\ (unseeded) = \frac{(7.1\ mg/L - 2.9\ mg/L) \times 300\ mL}{120\ mL}$$
$$= 10.5\ mg/L$$

If the BOD_5 sample has been exposed to conditions that could reduce the number of healthy, active organisms, the sample must be seeded with organisms. Seeding requires use of a correction factor to remove the BOD_5 contribution of the seed material:

$$Seed\ correction = \frac{Seed\ material\ BOD_5 \times seed\ in\ dilution\ (mL)}{300\ mL} \quad (15.3)$$

$$BOD_5 = \frac{\left[\begin{array}{c}\left((DO_{Start}\ (mg/L) - DO_{Final}\ (mg/L))\right) \\ - seed\ correction\end{array}\right] \times 300}{Sample\ volume\ (mL)} \quad (15.4)$$

■ EXAMPLE 15.3

Problem: Using the data below, determine the BOD_5:

BOD_5 of seed material = 90 mg/L
Dilution #1:
   Seed material = 3 mL
   Sample = 100 mL
   Start DO = 7.6 mg/L
   Final DO = 2.7 mg/L

$$Seed\ correction = \frac{90\ mg/L \times 3\ mL}{300\ mL} = 0.90\ mg/L$$

$$BOD_5\ (seeded) = \frac{[(7.6\ mg/L - 2.7\ mg/L) - 0.90] \times 300}{100\ mL}$$
$$= 12\ mg/L$$

15.8.7 Temperature Measurement

As mentioned, an ideal water supply should have, at all times, an almost constant temperature or one with minimum variation. Knowing the temperature of the water supply is important because the rates of biological and chemical processes depend on it. Temperature affects the oxygen content of the water (oxygen levels become lower as temperature increases), the rate of photosynthesis by aquatic plants, the metabolic rates of aquatic organisms, and the sensitivity of organisms to toxic wastes, parasites, and diseases. Causes of temperature change include weather, removal of shading stream-bank vegetation, impoundments (a body of water confined by a barrier, such as a dam), and discharge of cooling water, urban stormwater, and groundwater into the stream.

15.8.7.1 Sampling and Equipment Considerations

Temperature—for example, in a stream—varies with width and depth, and the temperature of well-sunned portions of a stream can be significantly higher than the shaded portion of the water on a sunny day. In a small stream, the temperature will be relatively constant as long as the stream is uniformly in sun or shade. In a large stream, temperature can vary considerably with width and depth, regardless of shade. If it is safe to do so, temperature measurements should be collected at varying depths and across the surface of the stream to obtain vertical and horizontal temperature profiles. This can be done at each site at least once to determine the necessity of collecting a profile during each sampling visit. Temperature should be measured at the same place every time.

Temperature is measured in the stream with a thermometer or a meter. Alcohol-filled thermometers are preferred over mercury-filled because they are less hazardous if broken. Armored thermometers for field use can withstand more abuse than unprotected glass thermometers and are worth the additional expense. Meters for other tests, such as pH (acidity) or dissolved oxygen, also measure temperature and can be used instead of a thermometer.

15.8.8 Hardness Measurement

Hardness refers primarily to the amount of calcium and magnesium in the water. Calcium and magnesium enter water mainly by leaching of rocks. Calcium is an important component of aquatic plant cell walls and the shells and bones of many aquatic organisms. Magnesium is an essential nutrient for plants and is a component of the chlorophyll molecule. Hardness test kits express test results in ppm of $CaCO_3$, but these results can be converted directly to calcium or magnesium concentrations:

$$\text{Calcium hardness (ppm } CaCO_3) \times 0.40 = \text{ppm Ca} \quad (15.5)$$

$$\text{Magnesium hardness (ppm } CaCO_3) \times 0.24 = \text{ppm Mg} \quad (15.6)$$

Note: Because of less contact with soil minerals and more contact with rain, surface raw water is usually softer than groundwater.

As a rule, when hardness is greater than 150 mg/L, softening treatment may be required for public water systems. Hardness determination via testing is required to ensure efficiency of treatment.

Note: Keep in mind that, when measuring calcium hardness, the concentration of calcium is routinely measured separately from total hardness. Its concentration in waters can range from 0 to several thousand mg/L as $CaCO_3$. Likewise, when measuring magnesium hardness, magnesium is routinely determined by subtracting calcium hardness from total harness. Natural water usually contains less magnesium than calcium. The lime dosage for water softening is partly based on the concentration of magnesium hardness in the water.

In the hardness test, the sample must be carefully measured, and then a buffer is added to the sample to correct pH for the test and an indicator to signal the titration endpoint. The indicator reagent is normally blue in a sample of pure water, but if calcium or magnesium ions are present in the sample the indicator combines with them to form a red complex. The titrant in this test is ethylenediaminetetraacetic acid (EDTA), used with its salts in the titration method; it is a *chelant* that pulls the calcium and magnesium ions away from the red-colored complex. The EDTA is added dropwise to the sample until all the calcium and magnesium ions have been chelated away from the complex and the indicator returns to its normal blue color. The amount of EDTA required to cause the color change is a direct indication of the amount of calcium and magnesium ions in the sample.

Some hardness kits include an additional indicator that is specific for calcium. This type of kit will provide three readings: total hardness, calcium hardness, and magnesium hardness. For more information, consult the latest edition of *Standard Methods* (APHA, 1998).

15.8.9 pH Measurement

pH is defined as the negative log of the hydrogen ion concentration of the solution. This is a measure of the ionized hydrogen in solution. Simply, it is the relative acidity or basicity of the solution. The chemical and physical properties and the reactivity of almost every component in water are dependent on pH. It relates to corrosivity, contaminant solubility, and conductance of the water and has a secondary maximum contaminant level (MCL) range set at 6.5 to 8.5.

15.8.9.1 Analytical and Equipment Considerations

The pH can be analyzed in the field or in the lab. If analyzed in the lab, it must be measured within 2 hours of the sample collection, because the pH will change due to carbon dioxide in the air dissolving in the water, moving the pH toward 7. If your program requires a high degree of accuracy and precision in pH results, the pH should be measured with a laboratory-quality pH meter and electrode. Meters of this quality range in cost from around $250 to $1000. Color comparators and pH "pocket pals" are suitable for most other purposes. The cost of either of these is in the $50 range. The lower cost of the alternatives might be attractive if multiple samplers are used to sample several sites at the same time.

15.8.9.2 pH Meters

A pH meter measures the electric potential (millivolts) across an electrode when immersed in water. This electric potential is a function of the hydrogen ion activity in the sample; therefore, pH meters can display results in either millivolts (mV) or pH units. A pH meter consists of a *potentiometer*, which measures electric potential where it meets the water sample; a *reference electrode*, which provides a constant electric potential; and a *temperature-compensating device,* which adjusts the readings according to the temperature of the sample (because pH varies with temperature). The reference and glass electrodes are frequently combined into a single probe called a *combination electrode.* A wide variety of meters is available, but the most important part of the pH meter is the electrode; thus, purchasing a good, reliable electrode and following the manufacturer's instructions for proper maintenance are important. Infrequently used or improperly maintained electrodes are subject to corrosion, which makes them highly inaccurate.

15.8.9.3 pH "Pocket Pals" and Color Comparators

pH "pocket pals" are electronic handheld "pens" that are dipped in the water to obtain a digital readout of the pH. They can be calibrated to only one pH buffer. (Lab meters, on the other hand, can be calibrated to two or more buffer solutions and thus are more accurate over a wide range of pH measurements.) Color comparators involve adding a reagent to the sample that colors the sample water. The intensity of the color is proportional to the pH of the sample and is matched against a standard color chart. The color chart equates particular colors to associated pH values, which can be determined by matching the colors from the chart to the color of the sample. For instructions on how to collect and analyze samples, refer to *Standard Methods* (APHA, 1998).

15.8.10 TURBIDITY MEASUREMENT

Turbidity is a measure of water clarity—how much the material suspended in water decreases the passage of light through the water. Turbidity consists of suspended particles in the water and may be caused by a number of materials, organic and inorganic. These particles are typically in the size range of 0.004 mm (clay) to 1.0 mm (sand). The occurrence of turbid source waters may be permanent or temporary. It can affect the color of the water. Higher turbidity increases water temperatures,

because suspended particles absorb more heat. This in turn reduces the concentration of dissolved oxygen because warm water holds less DO than cold. Higher turbidity also reduces the amount of light penetrating the water, which reduces photosynthesis and the production of DO. Suspended materials can clog fish gills, reducing resistance to disease in fish, lowering growth rates, and affecting egg and larval development. As the particles settle, they can blanket the stream bottom (especially in slower waters) and smother fish eggs and benthic macroinvertebrates

Turbidity also affects treatment plant operations; for example, turbidity hinders disinfection by shielding microbes, some of them pathogens, from the disinfectant. Obviously, this is the most important significance of turbidity monitoring; testing for turbidity provides an indication of the effectiveness of filtration of water supplies. It is important to note that turbidity removal is the principal reason for chemical addition, settling, coagulation, settling, and filtration in potable water treatment. Sources of turbidity include:

1. Soil erosion
2. Waste discharge
3. Urban runoff
4. Eroding stream banks
5. Large numbers of bottom feeders (such as carp), which stir up bottom sediments
6. Excessive algae growth

15.8.10.1 Sampling and Equipment Considerations

Turbidity can be useful as an indicator of the effects of runoff from construction, agricultural practices, logging activity, discharges, and other sources. Turbidity often increases sharply during rainfall, especially in developed watersheds, which typically have relatively high proportions of impervious surfaces. The flow of stormwater runoff from impervious surfaces rapidly increases stream velocity, which increases the erosion rates of stream banks and channels. Turbidity can also rise sharply during dry weather if earth-disturbing activities occur in or near a stream without erosion control practices in place.

Regular monitoring of turbidity can help detect trends that might indicate increasing erosion in developing watersheds; however, turbidity is closely related to stream flow and velocity and should be correlated with these factors. Comparisons of the change in turbidity over time, therefore, should be made at the same point at the same flow.

Keep in mind that turbidity is not a measurement of the amount of suspended solids present or the rate of sedimentation of a stream because it measures only the amount of light that is scattered by suspended particles. Measurement of total solids is a more direct measurement of the amount of material suspended and dissolved in water.

Turbidity is generally measured by using a turbidity meter or *turbidimeter*. The turbidimeter is a modern *nephelometer*. Early nephelometers were comprised of a box containing a light bulb that directed light at a sample. The amount of light scattered at right angles by the turbidity particles was measured and registered as nephelometric turbidity units (NTU). Today's turbidimeters use a photoelectric cell to register the scattered light on an analog or digital scale, and the instrument is calibrated with permanent turbidity standards composed of the colloidal substance formazin. Meters can measure turbidity over a wide range—from 0 to 1000 NTUs. A clear mountain stream might have a turbidity of around 1 NTU, whereas a large river such as the Mississippi might have a dry-weather turbidity of 10 NTUs. Because these values can jump into hundreds of NTUs during runoff events, the turbidity meter should be reliable over the range in which you will be working. Meters of this quality cost about $800. Many meters in this price range are designed for field or lab use.

An operator may also take samples to a lab for analysis. Another approach, discussed previously, is to measure transparency (an integrated measure of light scattering and absorption) instead of turbidity. Water clarity and transparency can be measured using a *Secchi disk* (see Figure 15.11) or transparency tube. The Secchi disk can only be used in deep, slow-moving rivers; the transparency tube (a comparatively new development) is gaining acceptance but is not yet in wide use.

15.8.10.2　Using a Secchi Disk

A Secchi disk is a black and white disk that is lowered by hand into the water to the depth at which it vanishes from sight (see Figure 15.11). The distance to vanishing is then recorded—the clearer the water, the greater the distance. Secchi disks are simple to use and inexpensive. For river monitoring, they have limited use, because in most cases the river bottom will be visible and the disk will not reach a vanishing point. Deeper, slower moving rivers are the most appropriate places for Secchi disk measurement, although the current might require that the disk be weighted so it does not sway and make measurement difficult. Secchi disks cost about $50 but can be homemade.

The line attached to the Secchi disk must be marked in waterproof ink according to units designated by the sampling program. Many programs require samplers to measure to the nearest 1/10 meter. Meter intervals can be tagged (e.g., with duct tape) for ease of use. To measure water clarity with a Secchi disk:

1. Check to make sure that the Secchi disk is securely attached to the measured line.
2. Lean over the side of the boat and lower the Secchi disk into the water, keeping your back to the sun to block glare.
3. Lower the disk until it disappears from view. Lower it one third of a meter and then slowly raise the disk until it just reappears. Move the disk up and down until you find the exact vanishing point.
4. Attach a clothespin to the line at the point where the line enters the water. Record the measurement on your data sheet. Repeating the measurement provides you with a quality control check.

The key to consistent results is to train samplers to follow standard sampling procedures and, if possible, to have the same individual take the reading at the same site throughout the season.

15.8.10.3　Transparency Tube

Pioneered by Australia's Department of Conservation, the *transparency tube* is a clear, narrow plastic tube marked in units with a dark pattern painted on the bottom. Water is poured into the tube until the pattern painted on the bottom disappears. Some U.S. volunteer monitoring programs, such as the Tennessee Valley Authority (TWA) Clean Water Initiative and the Minnesota Pollution Control Agency (MPCA), are testing the transparency tube in streams and rivers. MPCA uses tubes marked in centimeters and has found that the tube readings relate fairly well to lab measurements of turbidity and total suspended solids, although it does not recommend the transparency tube for applications where very precise measurement is required or in highly colored waters. The TVA and MPCA suggest the following sampling techniques:

1. Collect the sample in a bottle or bucket in midstream and at mid-depth if possible. Avoid stagnant water and sample as far from the shoreline as is safe. Avoid collecting sediment from the bottom of the stream.
2. Face upstream as you fill the bottle or bucket.
3. Take readings in open but shaded conditions. Avoid direct sunlight by turning your back to the sun.
4. Carefully stir or swish the water in the bucket or bottle until it is homogeneous, taking care not to produce air bubbles (these scatter light and affect the measurement). Pour the water slowly in the tube while looking down the tube. Measure the depth of the water column in the tube at the point where the symbol just disappears.

15.8.11　Orthophosphate Measurement

Earlier we discussed the nutrients phosphorus and nitrogen. Both phosphorus and nitrogen are essential nutrients for the plants and animals that make up the aquatic food web. Because phosphorus is the nutrient in short supply in most freshwater systems, even a modest increase in phosphorus can (under the right conditions) set off a whole chain of undesirable events in a stream, including accelerated plant growth, algae blooms, low dissolved oxygen, and the death of certain fish, invertebrates, and other aquatic animals. Phosphorus comes from many sources, both natural and human. These include soil and rocks, wastewater treatment plants, runoff from fertilized lawns and cropland, failing septic systems, runoff from animal manure storage areas, disturbed land areas, drained wetlands, water treatment, and commercial cleaning preparations.

15.8.11.1 Forms of Phosphorus

Phosphorus has a complicated story. Pure, elemental phosphorus (P) is rare. In nature, phosphorus usually exists as part as part of a phosphate molecule (PO_4). Phosphorus in aquatic systems occurs as organic phosphate and inorganic phosphate. Organic phosphate consists of a phosphate molecule associated with a carbon-based molecule, as in plant or animal tissue. Phosphate that is not associated with organic material is inorganic, the form required by plants. Animals can use either organic or inorganic phosphate. Both organic and inorganic phosphate can be dissolved in the water or suspended (attached to particles in the water column).

15.8.11.2 The Phosphorus Cycle

Phosphorus cycles through the environment, changing form as it does so. Aquatic plants take in dissolved inorganic phosphorus, and it becomes part of their tissues. Animals get the organic phosphorus they need by eating aquatic plants, other animals, or decomposing plant and animal material. In water bodies, as plants and animals excrete wastes or die, the organic phosphorus they contain sinks to the bottom, where bacterial decomposition converts it back to inorganic phosphorus, both dissolved and attached to particles. This inorganic phosphorus gets back into the water column when animals, human activity, interactions, or water currents stir up the bottom. Plants then take it up and the cycle begins again. In a stream system, the phosphorus cycle tends to move phosphorus downstream as the current carries decomposing plant and animal tissue and dissolved phosphorus. It becomes stationary only when it is taken up by plants or is bound to particles that settles to the bottom of ponds.

In the field of water quality chemistry, phosphorus is described by several terms. Some of these terms are chemistry based (referring to chemically based compounds), and others are methods based (they describe what is measured by a particular method). The term *orthophosphate* is a chemistry-based term that refers to the phosphate

molecule all by itself. More specifically, orthophosphate is simple phosphate, or reactive phosphate—that is, Na_3PO_4 (sodium phosphate, tribasic) and NaH_2PO_4, (sodium phosphate, monobasic). Orthophosphate is the only form of phosphate that can be directly tested for in the laboratory and is the form that bacteria use directly for metabolic processes. *Reactive phosphorus* is a corresponding method-based term that describes what is actually being measured when the test for orthophosphate is being performed. Because the lab procedure is not quite perfect, mostly orthophosphate is obtained along with a small fraction of some other forms. More complex inorganic phosphate compounds are referred to as *condensed phosphates* or *polyphosphates*. The method-based term for these forms in *acid hydrolyzable*.

15.8.11.3 Testing Phosphorus

Testing phosphorus is challenging because it involves measuring very low concentrations, down to 0.01 mg/L or even lower. Even such very low concentrations of phosphorus can have a dramatic impact on streams. Less sensitive methods should be used only to identify serious problem areas. Although many tests for phosphorus exist, only four are likely to be performed by most samplers. The *total orthophosphate test* is largely a measure of orthophosphate. Because the sample is not filtered, the procedure measures both dissolved and suspended orthophosphate. The USEPA-approved method for measuring phosphorus is known as the *ascorbic acid method*. Briefly, a reagent (either liquid or powder) containing ascorbic acid and ammonium molybdate reacts with orthophosphate in the sample to form a blue compound. The intensity of the blue color is directly proportional to the amount of orthophosphate in the water.

The *total phosphate test* measures all the forms of phosphorus in the sample (orthophosphate, condensed phosphate, and organic phosphate) by first digesting (heating and acidifying) the sample to convert all the other forms to orthophosphate. The orthophosphate is then measured by the ascorbic acid method. Because the sample is not filtered, the procedure measures both dissolved and suspended orthophosphate. The *dissolved phosphorus test* measures that fraction of the total phosphorus that is in solution in the water (as opposed to being attached to suspended particles). It is determined by first filtering the sample, then analyzing the filtered sample for total phosphorus. Insoluble phosphorus is calculated by subtracting the dissolved phosphorus result from the total phosphorus result.

All of these tests have one thing in common—they all depend on measuring orthophosphate. The total orthophosphate test measures the orthophosphate that is already present in the sample. The others measure that which is already present and that which is formed when the other forms of phosphorus are converted to orthophosphate by digestion. Monitoring phosphorus involves two basic steps:

1. Collect a water sample.
2. Analyze it in the field or lab for one of the types of phosphorus described above.

15.8.11.4 Sampling and Equipment Considerations

Sample containers made of either some form of plastic or Pyrex® glass are acceptable to the USEPA. Because phosphorus molecules have a tendency to absorb (attach) to the inside surface of sample containers, containers that are to be reused must be acid-washed to remove absorbed phosphorus. The container must be able to withstand repeated contact with hydrochloric acid. Plastic containers, either high-density polyethylene or polypropylene, might be preferable to glass from a practical standpoint because they are better able to resist breakage. Some programs use disposable, sterile, plastic Whirl-Pak® bags. The size of the container depends on the sample amount required for the phosphorus analysis method chosen and the amount required for other analyses to be performed.

All containers that will hold water samples or come into contact with reagents used in the orthophosphate test must be dedicated. They should not be used for other tests, to eliminate the possibility that reagents containing phosphorus will contaminate the labware. All labware should be acid-washed.

The only form of phosphorus this text recommends for field analysis is total orthophosphate, utilizing the ascorbic acid method on an untreated sample. Analysis of any of the other forms requires adding potentially hazardous reagents, heating the sample to boiling, and using too much time and too much equipment to be practical. In addition, analysis for other forms of phosphorus is prone to errors and inaccuracies in field situations. Pretreatment and analysis for these other forms should be handled in a laboratory.

15.8.11.5 Ascorbic Acid Method for Determining Orthophosphate

In the ascorbic acid method, a combined liquid or prepackaged powder reagent consisting of sulfuric acid, potassium antimonyl tartrate, ammonium molybdate, and ascorbic acid (or comparable compounds) is added to either 50 or 25 mL of the water sample. This colors the sample blue in direct proportion to the amount of orthophosphate in the sample. Absorbence or transmittance is then measured after 10 minutes but before 30 minutes, using a color comparator with a scale in milligrams per liter that increases with the increase in color hue, or an electronic meter that measures the amount of light absorbed or trans-

mitted at a wavelength of 700 to 880 nanometers (again, depending on manufacturer's directions).

A color comparator may be useful for identifying heavily polluted sites with high concentrations (greater than 0.1 mg/L); however, matching the color of a treated sample to a comparator can be very subjective, especially at low concentrations, and lead to variable results. A field spectrophotometer or colorimeter with a 2.5-cm light path and an infrared photocell (set for a wavelength of 700 to 880 nm) is recommended for accurate determination of low concentrations (between 0.2 and 0.02 mg/L). Use of a meter requires that a prepared known standard concentration be analyzed ahead of time to convert the absorbence readings of a stream sample to milligrams per liter or that the meter reads directly in milligrams per liter.

For information on how to prepare standard concentrations and on how to collect and analyze samples, refer to *Standard Methods* (APHA, 1998) and USEPA's *Methods for Chemical Analysis of Water and Wastes* (USEPA, 1991, Method 365.2).

15.8.12 Nitrates Measurement

As mentioned, *nitrates* are a form of nitrogen found in several different forms in terrestrial and aquatic ecosystems. These forms of nitrogen include ammonia (NH_3), nitrates (NO_3), and nitrites (NO_2). Nitrates are essential plant nutrients, but excess amounts can cause significant water quality problems. Together with phosphorus, nitrates in excess amounts can accelerate eutrophication, causing dramatic increases in aquatic plant growth and changes in the types of plants and animals that live in the stream. This, in turn, affects dissolved oxygen, temperature, and other indicators. Excess nitrates can cause hypoxia (low levels of dissolved oxygen) and can become toxic to warm-blooded animals at higher concentrations (10 mg/L or higher) under certain conditions. The natural level of ammonia or nitrate in surface water is typically low (less than 1 mg/L); in the effluent of wastewater treatment plants, it can range up to 30 mg/L. Conventional potable water treatment plants cannot remove nitrate. High concentrations must be prevented by controlling the input at the source. Sources of nitrates include wastewater treatment plants, runoff from fertilized lawns and cropland, failing on-site septic systems, runoff from animal manure storage areas, and industrial discharges that contain corrosion inhibitors.

15.8.12.1 Sampling and Equipment Considerations

Nitrates from land sources end up in rivers and streams more quickly than other nutrients such as phosphorus; they dissolve in water more readily than phosphorus, which has an attraction for soil particles. As a result, nitrates serve as a better indicator of the possibility of a source of sewage or manure pollution during dry weather. Water that is polluted with nitrogen-rich organic matter might show low nitrates. Decomposition of the organic matter lowers the dissolved oxygen level, which in turn slows the rate at which ammonia is oxidized to nitrite (NO_2) and then to nitrate (NO_3). Under such circumstances, monitoring for nitrites or ammonia (considerably more toxic to aquatic life than nitrate) might also be necessary. For appropriate nitrite methods, see *Standard Methods*, Sections 4500-NH3 and 4500-NH2 (APHA, 1998). Water samples to be tested for nitrate should be collected in glass or polyethylene containers that have been prepared by using Method B (described previously). Two methods are typically used for nitrate testing: the cadmium reduction method and the nitrate electrode. The more commonly used cadmium reduction method produces a color reaction measured by comparison to a color wheel or by use of a spectrophotometer. A few programs also use a nitrate electrode, which can measure in the range of 0 to 100 mg/L nitrate. A newer colorimetric immunoassay technique for nitrate screening is also now available.

15.8.12.2 Cadmium Reduction Method

In the *cadmium reduction method*, nitrate is reduced to nitrite by passing the sample through a column packed with activated cadmium. The sample is then measured quantitatively for nitrite. More specifically, the cadmium reduction method is a colorimetric method that involves contact of the nitrate in the sample with cadmium particles which causes the nitrates to be converted to nitrites. The nitrites then react with another reagent to form a red color, the intensity of which is in proportion to the original amount of nitrate. The color is measured either by comparison to a color wheel with a scale in milligrams per liter that increases with the increase in color hue or by use of an electronic spectrophotometer that measures the amount of light absorbed by the treated sample at a 543-nanometer wavelength. The absorbence value converts to the equivalent concentration of nitrate against a standard curve. Methods for making standard solutions and standard curves are presented in *Standard Methods* (APHA, 1998).

Before each sampling run, the sampling/monitoring supervisor should create this curve. The curve is developed by making a set of standard concentrations of nitrate, reacting them, and developing the corresponding color, then plotting the absorbence value for each concentration against concentration. A standard curve could also be generated for the color wheel. Use of the color wheel is appropriate only if nitrate concentrations are greater than 1 mg/L. For concentrations below 1 mg/L, use a spectro-

photometer. Matching the color of a treated sample at low concentrations to a color wheel (or cubes) can be very subjective and can lead to variable results. Color comparators, however, can be effectively used to identify sites with high nitrates.

This method requires that the samples being treated are clear. If a sample is turbid, filter it through a 0.45-μm filter. Be sure to test to make sure the filter is nitrate free. If copper, iron, or others metals are present in concentrations above several milligrams per liter, the reaction with the cadmium will slow down and the reaction time must be increased.

The reagents used for this method are often prepackaged for different ranges, depending on the expected concentration of nitrate in the stream. Manufacturers, for example, provide reagents for the following ranges: low (0 to 0.40 mg/L), medium (0 to 15 mg/L), and high (0 to 30 mg/L). Determining the appropriate range for the stream being monitored is important.

15.8.12.3 Nitrate Electrode Method

A *nitrate electrode* (used with a meter) is similar in function to a dissolved oxygen meter. It consists of a probe with a sensor that measures nitrate activity in the water; this activity affects the electric potential of a solution in the probe. This change is then transmitted to the meter, which converts the electric signal to a scale that is read in millivolts; the millivolts are then converted to mg/L of nitrate by plotting them against a standard curve. The accuracy of the electrode can be affected by high concentrations of chloride or bicarbonate ions in the sample water. Fluctuating pH levels can also affect the meter reading.

Nitrate electrodes and meters are expensive compared to field kits that employ the cadmium reduction method. (The expense is comparable, however, if a spectrophotometer is used rather than a color wheel.) Meter and probe combinations can run between $700 and $1200, including a long cable to connect the probe to the meter. A pH meter that displays readings in millivolts can be used with a nitrate probe, and no separate nitrate meter is needed. Results are read directly as milligrams per liter.

Although nitrate electrodes and spectrophotometers can be used in the field, they have certain disadvantages. These devices are more fragile than the color comparators and therefore are more at risk of breaking in the field. They must be carefully maintained and must be calibrated before each sample run and, if many tests are being run, between samplings. This means that samples are best tested in the lab. Note that samples to be tested with a nitrate electrode should be at room temperature, whereas color comparators can be used in the field with samples at any temperature.

15.8.13 Solids Measurement

Solids in water are defined as any matter that remains as residue upon evaporation and drying at 103°C. They are separated into two classes: *suspended solids* and *dissolved solids*:

$$\text{Total solids} = \frac{\text{Suspended solids}}{\text{(nonfilterable residue)}} + \frac{\text{dissolved solids}}{\text{(filterable residue)}}$$

As shown above, *total solids* are dissolved solids plus suspended and settleable solids in water. In natural freshwater bodies, dissolved solids consist of calcium, chlorides, nitrate, phosphorus, iron, sulfur, and other ions—particles that will pass through a filter with pores of around 2 µm (0.002 cm) in size. Suspended solids include silt and clay particles, plankton, algae, fine organic debris, and other particulate matter. These are particles that will not pass through a 2-µm filter.

The concentration of total dissolved solids affects the water balance in the cells of aquatic organisms. An organism placed in water with a very low level of solids (distilled water, for example) swells because water tends to move into its cells, which have a higher concentration of solids. An organism placed in water with a high concentration of solids shrinks somewhat, because the water in its cells tends to move out. This in turn affects the organism's ability to maintain the proper cell density, making keeping its position in the water column difficult. It might float up or sink down to a depth to which it is not adapted, and it might not survive.

Higher concentrations of suspended solids can serve as carriers of toxics, which readily cling to suspended particles. This is particularly a concern where pesticides are being used on irrigated crops. Where solids are high, pesticide concentrations may increase well beyond those of the original application as the irrigation water travels down irrigation ditches. Higher levels of solids can also clog irrigation devices, and levels might become so high that irrigated plant roots will lose water rather than gaining it.

A high concentration of total solids will make drinking water unpalatable and might have an adverse effect on people who are not used to drinking such water. Levels of total solids that are too high or too low can also reduce the efficiency of wastewater treatment plants, as well as the operation of industrial processes that use raw water.

Total solids affect water clarity. Higher solids decrease the passage of light through water, thereby slowing photosynthesis by aquatic plants. Water heats up more rapidly and holds more heat; this, in turn, might adversely affect aquatic life adapted to a lower temperature regime.

Sources of total solids include industrial discharges, sewage, fertilizers, road runoff, and soil erosion. Total solids are measured in milligrams per liter (mg/L).

15.8.13.1 Sampling and Equipment Considerations

When conducting solids testing, many things can affect the accuracy of the test or result in wide variations in results for a single sample, including:

1. Drying temperature
2. Length of drying time
3. Condition of desiccator and desiccant
4. lack of consistency in test procedures for non-representative samples
5. Failure to achieve constant weight prior to calculating results

Several precautions can be taken to improve the reliability of test results:

1. Use extreme care when measuring samples, weighing materials, and drying or cooling samples.
2. Check and regulate oven and furnace temperatures frequently to maintain the desired range.
3. Use an indicator drying agent in the desiccator that changes color when it is no longer good; change or regenerate the desiccant when necessary.
4. Keep the desiccator cover greased with the appropriate type of grease to seal the desiccator and prevent moisture from entering the desiccator as the test glassware cools.
5. Check ceramic glassware for cracks and the glass-fiber filter for possible holes; a hole in the filter will cause solids to pass through and give inaccurate results.
6. Follow manufacturer's recommendation for care and operation of analytical balances.

Total solids are important to measure in areas where discharges from sewage treatment plants, industrial plants, or extensive crop irrigation may occur. In particular, streams and rivers in arid regions where water is scarce and evaporation is high tend to have higher concentrations of solids and are more readily affected by human introduction of solids from land-use activities.

Total solids measurements can be useful as an indicator of the effects of runoff from construction, agricultural practices, logging activities, sewage treatment plant discharges, and other sources. As with turbidity, concentrations often increase sharply during rainfall, especially in developed watersheds. They can also rise sharply during dry weather if earth-disturbing activities occur in or near the stream without erosion control practices in place. Regular monitoring of total solids can help detect trends that might indicate increasing erosion in developing watersheds. Total solids are closely related to stream flow and velocity and should be correlated with these factors. Any change in total solids over time should be measured at the same site at the same flow.

Total solids are measured by weighing the amount of solids present in a known volume of sample; this is accomplished by weighing a beaker, filling it with a known volume, evaporating the water in an oven to completely dry the residue, and weighing the beaker with the residue. The total solids concentration is equal to the difference between the weight of the beaker with the residue and the weight of the beaker without it. Because the residue is so light in weight, the lab should have a balance that is sensitive to weights in the range of 0.0001 gram. Balances of this type are called *analytical* or *Mettler* balances, and they are expensive (around $3000). The beakers must be kept in a desiccator, a sealed glass container that contains material that absorbs moisture and ensures that the weighing is not biased by water condensing on the beaker. Some desiccants change color to indicate moisture content. Measurement of total solids cannot be done in the field. Samples must be collected using clean glass or plastic bottles, or Whirl-Pak® bags and taken to a laboratory where the test can be run.

15.8.13.2 Total Suspended Solids

As mentioned, the term *solids* means any material suspended or dissolved in water and wastewater. Although normal domestic wastewater contains a very small amount of solids (usually less than 0.1%), most treatment processes are designed specifically to remove or convert solids to a form that can be removed or discharged without causing environmental harm. When sampling for total suspended solids (TSS), samples may be either grab or composite and can be collected in either glass or plastic containers. TSS samples can be preserved by refrigeration at or below 4°C (not frozen); however, composite samples must be refrigerated during collection. The maximum holding time for preserved samples is 7 days.

15.8.13.2.1 *Test Procedure*

To conduct a TSS test procedure, a well-mixed measured sample is poured into a filtration apparatus and, with the aid of a vacuum pump or aspirator, is drawn through a preweighted glass-fiber filter. After filtration, the filter is dried at 103 to 105°C, cooled, and then reweighed. The increase in weight of the filter and solids compared to the filter alone represents the total suspended solids. An example of the specific test procedure used for total suspended solids is given below:

1. Select a sample volume that will yield between 10 and 200 mg of residue with a filtration time of 10 minutes or less.

Note: If filtration time exceeds 10 minutes, increase filter area or decrease volume to reduce filtration time.

Note: For nonhomogenous samples or samples with very high solids concentrations (i.e., raw wastewater or mixed liquor), use a larger filter to ensure that a representative sample volume can be filtered.

2. Place the preweighed glass-fiber filter on the filtration assembly in a filter flask.
3. Mix the sample well, and measure the selected volume of sample.
4. Apply suction to the filter flask, and wet the filter with a small amount of laboratory-grade water to seal it.
5. Pour the selected sample volume into the filtration apparatus.
6. Draw the sample through the filter.
7. Rinse the measuring device into the filtration apparatus with three successive 10-mL portions of laboratory-grade water. Allow complete drainage between rinsings.
8. Continue suction for 3 minutes after filtration of the final rinse is completed.
9. Remove the filter from the filtration assembly (membrane filter funnel or clean Gooch crucible). If using the large disks and membrane filter assembly, then transfer the filter to a support (aluminum pan or evaporating dish) for drying.
10. Place the filter with solids and support (pan, dish, or crucible) in a drying oven.
11. Dry the filter and solids to a constant weight at 103 to 105°C (at least 1 hour).
12. Cool to room temperature in a desiccator.
13. Weigh the filter and support, and record the constant weight in the test record.

15.8.13.2.2 *TSS Calculations*

To determine the total suspended solids concentration in mg/L, we use the following equations:

1. To determine weight of dry solids in grams:

$$\text{Dry solids (g)} = \left| \begin{array}{l} \text{Wt. of dry solids and filter (g)} \\ - \text{wt. of dry filter (g)} \end{array} \right| \quad (15.7)$$

2. To determine weight of dry solids in milligrams (mg):

$$\text{Dry solids (mg)} = \left[\begin{array}{l} \text{Wt. of dry solids and filter (mg)} \\ - \text{wt. of dry filter (mg)} \end{array} \right] \quad (15.8)$$

3. To determine the TSS concentration in mg/L:

$$\text{TSS (mg/L)} = \frac{\text{Dry solids (mg)} \times 1000 \text{ mL}}{\text{Sample (mL)}} \quad (15.9)$$

■ **EXAMPLE 15.4**

Problem: Using the data provided below, calculate the total suspended solids (TSS):

> Sample volume = 250 mL
> Weight of dry solids and filter = 2.305 g
> Weight of dry filter = 2.297 g

Solution:

> Dry solids (g) = 2.305 g − 2.297 g = 0.008 g
> Dry solids (mg) = 0.008 g × 1000 mg/g = 8 mg

$$TSS\ (mg/L) = \frac{8.0 \times 1000\ mL/L}{250\ mL} = 32.0\ mg/L$$

15.8.13.3 Volatile Suspended Solids Testing

When the total suspended solids are ignited at $550 \pm 50°C$, the volatile (organic) suspended solids of the sample are converted to water vapor and carbon dioxide and are released to the atmosphere. The solids that remain after the ignition (ash) are the inorganic or fixed solids. In addition to the equipment and supplies required for the total suspended solids test, we need the following:

1. Muffle furnace ($550 \pm 50°C$)
2. Ceramic dishes
3. Furnace tongs
4. Insulated gloves

15.8.13.3.1 Test Procedure

An example of the test procedure used for volatile suspended solids is given below:

1. Place the weighed filter with the solids and support from the total suspended solids test in the muffle furnace.
2. Ignite the filter, solids, and support at $550 \pm 50°C$ for 15 to 20 minutes.
3. Remove the ignited solids, filter, and support from the furnace, and partially air cool.
4. Cool to room temperature in a desiccator.
5. Weigh the ignited solids, filter, and support on an analytical balance.
6. Record the weight of the ignited solids, filter, and support.

15.8.13.3.2 Total Volatile Suspended Solids Calculations

To calculate total volatile suspended solids (TVSS) requires the following information:

1. Weight of dry solids, filter, and support (g)
2. Weight of ignited solids, filter, and support (g)

$$TVSS\ (mg/L) = \frac{(A - C) \times 1000\ mg/g \times 1000\ mL/L}{Sample\ volume\ (mL)} \quad (15.10)$$

where

> A = Weight of dried solids, filter, and support.
> C = Weight of ignited solids, filter, and support.

■ **EXAMPLE 15.5**

Problem: Using the data provided below, calculate the total volatile suspended solids:

> Weight of dried solids, filter, and support = 1.6530 g
> Weight of ignited solids, filter, and support = 1.6330 g
> Sample volume = 100 mL

Solution:

$$TVSS = \frac{(1.6530\ g - 1.6330\ g) \times 1000\ mg/g \times 1000\ mL/L}{100\ mL}$$
$$= \frac{0.02 \times 1{,}000{,}000\ mg/L}{100} = 200\ mg/L$$

Note: Total fixed suspended solids (TFSS) is the difference between the total volatile suspended solids (TVSS) and the total suspended solids (TSS) concentrations:

$$TFSS\ (mg/L) = TTS - TVSS \quad (15.11)$$

■ **EXAMPLE 15.6**

Problem: Using the data provided below, calculate the total fixed suspended solids:

> Total suspended solids = 202 mg/L
> Total volatile suspended solids = 200 mg/L

Solution:

> TFSS (mg/L) = 202 mg/L − 200 mg/L = 2 mg/L

15.8.14 CONDUCTIVITY TESTING

Conductivity is a measure of the capacity of water to pass an electrical current. Conductivity in water is affected by the presence of inorganic dissolved solids such as chloride, nitrate, sulfate, and phosphate anions (ions that carry a negative charge), or sodium, magnesium, calcium, iron, and aluminum cations (ions that carry a positive charge). Organic compounds such as oil, phenol, alcohol, and sugar do not conduct electrical current very well and therefore have a low conductivity when in water. Conductivity is also affected by temperature: the warmer the water, the higher the conductivity.

Conductivity in streams and rivers is affected primarily by the geology of the area through which the water flows. Streams that run through areas with granite bedrock tend to have lower conductivity because granite is composed of more inert materials that do not ionize (dissolve into ionic components) when washed into the water. On the other hand, streams that run through areas with clay soils tend to have higher conductivity because of the presence of materials that ionize when washed into the water. Groundwater inflows can have the same effects, depending on the bedrock through which they flow.

Discharges to streams can change the conductivity depending on their make-up. A failing sewage system would raise the conductivity because of the presence of chloride, phosphate, and nitrate; an oil spill would lower conductivity.

The basic unit of measurement of conductivity is the mho or siemens. Conductivity is measured in micromho per centimeter (μmho/cm) or microsiemens per centimeter (μS/cm). Distilled water has conductivity in the range of 0.5 to 3 μmho/cm. The conductivity of rivers in the United States generally ranges from 50 to 1500 μmho/cm. Studies of inland freshwaters indicated that streams supporting good mixed fisheries have a range between 150 and 500 μmho/cm. Conductivity outside this range could indicate that the water is not suitable for certain species of fish or macroinvertebrates. Industrial waters can range as high as 10,000 μmho/cm.

15.8.14.1 Sampling, Testing, and Equipment Considerations

Conductivity is useful as a general measure of source water quality. Each stream tends to have a relatively constant range of conductivity that, once established, can be used as a baseline for comparison with regular conductivity measurements. Significant changes in conductivity could indicate that a discharge or some other source of pollution has entered a stream. The conductivity test is not routine in potable water treatment, but when performed on source water it is a good indicator of contamination. Conductivity readings can also be used to indicate wastewater contamination or saltwater intrusion.

Note: Distilled water used for potable water analyses at public water supply facilities must have a conductivity of no more than 1 μmho/cm.

Conductivity is measured with a probe and a meter. Voltage is applied between two electrodes in a probe immersed in the sample water. The drop of voltage caused by the resistance of the water is used to calculate the conductivity per centimeter. The meter converts the probe measurement to micromhos per centimeter (μmho/cm) displays the result for the user.

Note: Some conductivity meters can also be used to test for total dissolved solids and salinity. The total dissolved solids concentration in milligrams per liter (mg/L) can also be calculated by multiplying the conductivity result by a factor between 0.55 and 0.9, which is empirically determined; see *Standard Methods*, Method #2510 (APHA, 1998).

Suitable conductivity meters cost about $350. Meters in this price range should also measure temperature and automatically compensate for temperature in the conductivity reading. Conductivity can be measured in the field or the lab. In most cases, collecting samples in the field and taking them to a lab for testing is probably a better approach. In this way, several teams can collect samples simultaneously. If testing in the field is important, meters designed for field use can be obtained for around the same cost mentioned above. If samples will be collected in the field for later measurement, the sample bottle should be a glass or polyethylene bottle that has been washed in phosphate-free detergent and rinsed thoroughly with both tap and distilled water. Factory-prepared Whirl-Pak® bags may be used.

15.8.15 TOTAL ALKALINITY

Alkalinity is defined as the ability of water to resist a change in pH when acid is added; it relates to the pH buffering capacity of the water. Almost all natural waters have some alkalinity. Alkaline compounds in the water, such as bicarbonates (baking soda is one type), carbonates, and hydroxides, remove H^+ ions and lower the acidity of the water (which means increased pH). They usually do this by combining with the H^+ ions to make new compounds. Without this acid-neutralizing capacity, any acid added to a stream would cause an immediate change in the pH. Measuring alkalinity is important in determining the ability of a stream to neutralize acidic pollution from rainfall or wastewater—one of the best measures of the sensitivity of the stream to acid inputs. Alkalinity in streams is influenced by rocks and soils, salts, certain plant activities, and certain industrial wastewater discharges.

Total alkalinity is determined by measuring the amount of acid (e.g., sulfuric acid) required to bring the sample to a pH of 4.2. At this pH all of the alkaline compounds in the sample are used up. The result is reported as milligrams per liter of calcium carbonate (mg/L $CaCO_3$).

Testing for alkalinity in potable water treatment is most important with regard to its relation to coagulant addition; that is, we must know that enough natural alkalinity exists in the water to buffer chemical acid addition so floc formation will be optimal and turbidity removal can proceed. In water softening, proper chemical dosage will depend on the type and amount of alkalinity in the water. For corrosion control, the presence of adequate alkalinity in a water supply neutralizes any acid tendencies, and prevents it from becoming corrosive.

15.8.15.1 Analytical and Equipment Considerations

For total alkalinity, a double endpoint titration using a pH meter (or pH "pocket pal") and a digital titrator or buret is recommended. This can be done in the field or in the lab. If alkalinity must be analyzed in the field, a digital titrator should be used instead of a buret, because burets are fragile and more difficult to set up. The alkalinity method described below was developed by the Acid Rain Monitoring Project of the University of Massachusetts Water Resources Research Center (River Watch Network, 1992).

15.8.15.1.1 Burets, Titrators, and Digital Titrators for Measuring Alkalinity

The total alkalinity analysis involves titration. In this test, titration is the addition of small, precise quantities of sulfuric acid (the reagent) to the sample, until the sample reaches a certain pH (known as an *endpoint*). The amount of acid used corresponds to the total alkalinity of the sample. Alkalinity can be measured using a buret, titrator, or digital titrator (described below):

1. A buret is a long, graduated glass tube with a tapered tip like a pipet and a valve that opens to allow the reagent to drop out of the tube. The amount of reagent used is calculated by subtracting the original volume in the buret from the column left after the endpoint has been reached. Alkalinity is calculated based on the amount used.
2. Titrators forcefully expel the reagent by using a manual or mechanical plunger. The amount of reagent used is calculated by subtracting the original volume in the titrator from the volume left after the endpoint has been reached. Alkalinity is then calculated based on the amount used or is read directly from the titrator.
3. Digital titrators have counters that display numbers. A plunger is forced into a cartridge containing the reagent by turning a knob on the titrator. As the knob turns, the counter changes in proportion to the amount of reagent used. Alkalinity is then calculated based on the amount used. Digital titrators cost approximately $100.

Digital titrators and burets allow for much more precision and uniformity in the amount of titrant that is used.

15.8.16 Fecal Coliform Bacteria Testing

Much of the information in this section is from USEPA (1985, 1986). Fecal coliform bacteria are non-disease-causing organisms found in the intestinal tract of all warm-blooded animals. Each discharge of body wastes contains large amounts of these organisms. The presence of fecal coliform bacteria in a stream or lake indicates the presence of human or animal wastes. The number of fecal coliform bacteria present is a good indicator of the amount of pollution present in the water. USEPA's 2001 Total Coliform Rule 816-F-01-035:

1. Is intended to improve public health protection by reducing fecal pathogens to minimal levels through control of total coliform bacteria, including fecal coliforms and *Escherichia coli*
2. Establishes a maximum contaminant level (MCL) based on the presence or absence of total coliforms, modifies monitoring requirements including testing for fecal coliforms or *E. coli*, requires use of a sample siting plan, and requires sanitary surveys for systems collecting fewer than five samples per month
3. Applies to all public water systems
4. Has resulted in reduction in risk of illness from disease-causing organisms associated with sewage or animal wastes; disease symptoms may include diarrhea, cramps, nausea, and possibly jaundice, and associated headaches and fatigue

Fecal coliforms are used as indicators of possible sewage contamination because they are commonly found in human and animal feces. Although they are not generally harmful themselves, they indicate the possible presence of pathogenic (disease-causing) bacteria and protozoa that also live in human and animal digestive systems. Their presence in streams suggests that pathogenic microorganisms might also be present and that swimming in or eating shellfish from the waters might present a health risk. Because testing directly for the presence of a large variety of pathogens is difficult, time consuming, and expensive, water is usually tested for coliforms and fecal streptococci instead. Sources of fecal contamination to surface waters include wastewater treatment plants, on-site septic systems, domestic and wild animal manure, and storm runoff. In addition to the possible health risks associated with the presence of elevated levels of fecal bacteria, they can also cause cloudy water, unpleasant odors, and an increased oxygen demand.

Note: In addition to the most commonly tested fecal bacteria indicators—total coliforms, fecal coliforms, and *E. coli*—fecal streptococci and enterococci are also commonly used as bacteria indicators. The focus of this presentation is on total coliforms are total coliforms and fecal coliforms.

Fecal coliforms are widespread in nature. All members of the total coliform group can occur in human feces, but some can also be present in animal manure, soil, and submerged wood, and in other places outside the human body. The usefulness of total coliforms as an indicator of fecal contamination depends on the extent to which the bacteria species found are fecal and human in origin. For recreational waters, total coliforms are no longer recommended as an indicator. For drinking water, total coliforms are still the standard test, because their presence indicates contamination of a water supply by an outside source.

Fecal coliforms, a subset of total coliform bacteria, are more fecal specific in origin; however, even this group contains a genus, *Klebsiella*, with species that are not necessarily fecal in origin. *Klebsiella* are commonly associated with textile and pulp- and paper-mill wastes. If these sources discharge to a local stream, consideration should be given to monitoring more fecal and human-specific bacteria. For recreational waters, this group was the primary bacteria indicator until relatively recently, when USEPA began recommending *E. coli* and enterococci as better indicators of health risk from water contact. Fecal coliforms are still being used in many states as indicator bacteria.

15.8.16.1 USEPA's Total Coliform Rule

Under USEPA's Total Coliform Rule, sampling requirements are specified as follows.

15.8.16.1.1 Routine Sampling Requirements

1. Total coliform samples must be collected at sites that are representative of water quality throughout the distribution system according to a written sample siting plan subject to state review and revision.
2. Samples must be collected at regular time intervals throughout the month, but groundwater systems serving 4900 persons or fewer may collect them on the same day.
3. Monthly sampling requirements are based on population served (see Table 15.10 for the minimum sampling frequency).
4. A reduced monitoring frequency may be available for systems serving 1000 persons or fewer and using only groundwater if a sanitary survey within the past 5 years shows the system is free of sanitary defects (the frequency may be no less than 1 sample per quarter for community and 1 sample per year for non-community systems).
5. Each total coliform-positive routine sample must be tested for the presence of fecal coliforms or *E. coli*.

TABLE 15.10
Public Water System Routine Monitoring Frequencies

Population	Minimum Samples/Month
25–1000[a]	1
1001–2500	2
2501–3300	3
3301–4100	4
4101–4900	5
4901–5800	6
5801–6700	7
6701–7600	8
7601–8500	9
8501–12,900	10
12,901–17,200	15
17,201–21,500	20
21,501–25,000	25
25,001–33,000	30
33,001–41,000	40
41001–50,000	50
50,001–59,000	60
59,001–70,000	70
70,000–83,000	80
83,001–96,000	90
96,001–130,000	100
130,000–220,000	120
220,001–320,000	150
320,001–450,000	180
450,001–600,000	210
600,001–780,000	240
780,001–970,000	270
970,001–1,230,000	330
1,520,001–1,850,000	360
1,850,001–2,270,000	390
2,270,001–3,020,000	420
3,020,001–3,960,000	450
≥3,960,001	480

[a] Includes PWSs that have at least 15 service connections but serve <25 people.

15.8.16.1.2 Repeat Sampling Requirements

1. Within 24 hours of learning of a total coliform-positive *routine* sample result, at least 3 *repeat* samples must be collected and analyzed for total coliforms.
2. One *repeat* sample must be collected from the same tap as the original sample.
3. One *repeat* sample must be collected within five service connections upstream.
4. One *repeat* sample must be collected within five service connections downstream.
5. Systems collecting one *routine* sample a month or fewer must collect a fourth *repeat* sample.

If any *repeat* sample is total coliform positive:

6. The system must analyze that total coliform-positive culture for fecal coliforms or *E. coli*.
7. The system must collect another set of *repeat* samples, as before, unless the MCL has been violated and the system has notified the state.

15.8.16.1.3 Additional Routine Sample Requirements

A positive *routine* or *repeat* total coliform result requires a minimum of five *routine* samples to be collected the following month that the system provides water to the public unless waived by the state.

15.8.16.1.4 Other Total Coliform Rule Provisions

1. Systems collecting fewer than 5 *routine* samples per month must have a sanitary survey every 5 years (or every 10 years if it is a non-community water system using protected and disinfected groundwater).
2. Systems using surface water or groundwater under the direct influence of surface water (GWUDI) and meeting filtration avoidance criteria must collect and have analyzed one coliform sample each day the turbidity of the source water exceeds 1 NTU. This sample must be collected from a tap near the first service connection.

15.8.16.1.5 Compliance

Compliance is based on the presence or absence of total coliforms. Moreover, compliance is determined each calendar month the system serves water to the public (or each calendar month that sampling occurs for systems on reduced monitoring). The results of *routine* and *repeat* samples are used to calculate compliance. With regard to violations, a monthly MCL violation is triggered if a system collecting fewer than 40 samples per month has greater than 1 *routine/repeat* sample per month that is total coliform positive. In addition, a system collecting at least 40 samples per month for which more than 5% of the *routine/repeat* samples are total coliform positive is technically in violation of the Total Coliform Rule. An acute MCL violation is triggered if any public water system has any fecal coliform- or *E. coli*-positive *repeat* sample or has a fecal coliform- or *E. coli*-positive *routine* sample followed by a total coliform-positive *repeat* sample.

The Total Coliform Rule also has requirements for public notification and reporting; for example, for a monthly MCL violation, the violation must be reported to the state no later than the end of the next business day after the system learns of the violation. The public must be notified within 14 days. For an acute MCL violation, the violation must be reported to the state no later than the end of the next business day after the system learns of the violation. The public must be notified within 72

hours. Systems with *routine* or *repeat* samples that are fecal coliform or *E. coli* positive must notify the state by the end of the day they are notified of the result or by the end of the next business day if the state office is already closed.

15.8.16.2 Sampling and Equipment Considerations

Bacteria can be difficult to sample and analyze, for many reasons. Natural bacteria levels in streams can vary significantly; bacteria conditions are strongly correlated with rainfall, making the comparison of wet and dry weather bacteria data a problem. Many analytical methods have a low level of precision, yet can be quite complex to accomplish, and absolutely sterile conditions are essential to maintain while collecting and handling samples. The primary equipment decision to make when sampling for bacteria is what type and size of sample container to use. Once that decision has been made, the same straightforward collection procedure is used, regardless of the type of bacteria being monitored.

When monitoring bacteria, it is critical that all containers and surfaces with which the sample will come into contact are sterile. Containers made of either some form of plastic or Pyrex® glass are acceptable to the USEPA; however, if the containers are to be reused, they must be sturdy enough to survive sterilization using heat and pressure. The containers can be sterilized by using an autoclave, a machine that sterilizes with pressurized steam. When an autoclave is used, the container material must be able to withstand high temperatures and pressure. Plastic containers, either high-density polyethylene or polypropylene, might be preferable to glass from a practical standpoint because they will better resist breakage. In any case, be sure to check the manufacturer's specifications to see whether the container can withstand 15 minutes in an autoclave at a temperature of 121°C without melting. (Extreme caution is advised when working with an autoclave.) Disposable, sterile, plastic Whirl-Pak® bags are used by a number of programs. The size of the container depends on the sample amount required for the bacteria analysis method chosen and the amount required for other analyses. The two basic methods for analyzing water samples for bacteria in common use are the *membrane filtration* and *multiple-tube fermentation* methods (described later).

Given the complexity of the analysis procedures and the equipment required, field analysis of bacteria is not recommended. Bacteria can be analyzed at a well-equipped lab or sent to a state-certified lab for analysis. If a bacteria sample is sent to a private lab, be sure the lab is certified by the state for bacteria analysis. Consider state water quality labs, university and college labs, private labs, wastewater treatment plant labs, and hospitals.

On the other hand, if the treatment plant has a modern lab with the proper equipment and properly trained technicians, the fecal coliform testing procedures described in the following section will be helpful. *A note of caution:* If you decide to analyze samples in your own lab, be sure to carry out a quality assurance/quality control program.

15.8.16.3 Fecal Coliform Testing

Federal regulations cite two approved methods for the determination of fecal coliform in water: (1) multiple-tube fermentation or most probable number (MPN) procedure, and (2) membrane filter (MF) procedure.

Note: Because the MF procedure can yield low or highly variable results for chlorinated wastewater, USEPA requires verification of results using the MPN procedure to resolve any controversies; however, do not attempt to perform the fecal coliform test using the summary information provided in this handbook. Instead, refer to the appropriate reference cited in the federal regulations for a complete discussion of these procedures.

15.8.16.3.1 Basic Equipment and Techniques

Whenever microbiological testing of water samples is performed, certain general considerations and techniques will be required. Because these are basically the same for each test procedure, they are reviewed here prior to discussion of the two methods.

1. *Reagents and media*—All reagents and media utilized in performing microbiological tests on water samples must meet the standards specified in the reference cited in federal regulations.
2. *Reagent-grade water*—Deionized water that is tested annually and found to be free of dissolved metals and bactericidal or inhibitory compounds is preferred for use in preparing culture media and test reagents, although distilled water may be used.
3. *Chemicals*—All chemicals used in fecal coliform monitoring must be ACS reagent grade or the equivalent.
4. *Media*—To ensure uniformity in the test procedures, the use of dehydrated media is recommended. Sterilized, prepared media in sealed test tubes, ampoules, or dehydrated media pads are also acceptable for use in this test.
5. *Glassware and disposable supplies*—All glassware, equipment, and supplies used in microbiological testing should meet the standards specified in the references cited in federal regulations.

15.8.16.3.2 Sterilization

All glassware used for bacteriological testing must be thoroughly cleaned using a suitable detergent and hot water. The glassware should be rinsed with hot water to remove all traces of residual from the detergent and, finally, should be rinsed with distilled water. Laboratories should use a detergent certified to meet bacteriological standards or, at a minimum, rinse all glassware after washing with two tap water rinses followed by five distilled water rinses. For sterilization of equipment, a hot-air sterilizer or autoclave can be used. When using the hot-air sterilizer, all equipment should be wrapped in high-quality (Kraft) paper or placed in containers prior to hot air sterilization. All glassware, except those in metal containers, should be sterilized for a minimum of 60 minutes at 170°C. Sterilization of glassware in metal containers should require a minimum of 2 hours. Hot-air sterilization cannot be used for liquids. An autoclave can be used to sterilize sample bottles, dilution water, culture media, and glassware at 121°C for 15 minutes.

15.8.16.3.3 Sterile Dilution Water Preparation

The dilution water used for making sample serial dilutions is prepared by adding 1.25 mL of stock buffer solution and 5.0 mL of magnesium chloride solution to 1000 mL of distilled or deionized water. The stock solutions of each chemical should be prepared as outlined in the reference cited by the federal regulations. The dilution water is then dispensed in sufficient quantities to produce 9 or 99 mL in each dilution bottle following sterilization. If the membrane filter procedure is used, additional 60- to 100-mL portions of dilution water should be prepared and sterilized to provide rinse water required by the procedure.

15.8.16.3.4 Serial Dilution Procedure

At times, the density of the organisms in a sample makes it difficult to accurately determine the actual number of organisms in the sample. When this occurs, the sample size may have to be reduced to as small as one millionth of a milliliter. To obtain such small volumes, a technique known as *serial dilutions* has been developed.

15.8.16.3.5 Bacteriological Sampling

To obtain valid test results that can be utilized in the evaluation of process efficiency of water quality, proper technique, equipment, and sample preservation are critical. These factors are especially critical in bacteriological sampling:

1. *Sample dechlorination*—When samples of chlorinated effluents are to be collected and tested, the sample must be dechlorinated. Prior to sterilization, place enough sodium thiosulfate solution (10%) in a clean sample container to produce a concentration of 100 mg/L in the

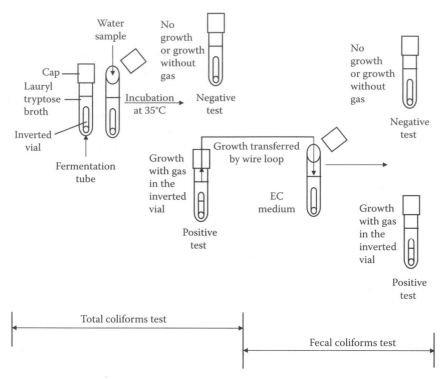

FIGURE 15.16 Multiple-tube fermentation technique.

sample (for a 120-mL sample bottle, 0.1 mL is usually sufficient). Sterilize the sample container as previously described.

2. *Sample procedure:*
   A. Keep the sample bottle unopened after sterilization until the sample is to be collected.
   B. Remove the bottle stopper and hood or cap as one unit. Do not touch or contaminate the cap or the neck of the bottle.
   C. Submerge the sample bottle in the water to be sampled.
   D. Fill the sample bottle approximately three quarters full, but not less than 100 mL.
   E. Aseptically replace the stopper or cap on the bottle.
   F. Record the date, time, and location of sampling, as well as the sampler's name and any other descriptive information pertaining to the sample.

3. *Sample preservation and storage*—Examination of bacteriological water samples should be performed immediately after collection. If testing cannot be imitated within 1 hour of sampling, the sample should be iced or refrigerated at 4°C or less. The maximum recommended holding time for fecal coliform samples from wastewater is 6 hours. The storage temperature and holding time should be recorded as part of the test data.

15.8.16.3.6 Multiple-Tube Fermentation Technique

The multiple-tube fermentation technique for fecal coliform testing is useful in determining the fecal coliform density in most water, solid, or semisolid samples. Wastewater testing normally requires use of the presumptive and confirming test procedures. It is recognized as the method of choice for any samples that may be controversial (enforcement related). The technique is based on the most probable number (MPN) of bacteria present in a sample that produces gas in a series of fermentation tubes with various volumes of diluted sample. The MPN is obtained from charts based on statistical studies of known concentrations of bacteria.

The technique utilizes a two-step incubation procedure (see Figure 15.16). The sample dilutions are first incubated in lauryl (sulfonate) tryptose broth for 24 to 48 hours (presumptive test). Positive samples are then transferred to EC broth and incubated for an additional 24 hours (confirming test). Positive samples from this second incubation are used to statistically determine the MPN from the appropriate reference chart. A single-medium, 24-hour procedure is also acceptable. In this procedure, sample dilutions are inoculated in A-1 medium, incubated for 3 hr at 35°C, and then incubated the remaining 20 hours at 44.5°C. Positive samples from these inoculations are then used to statistically determine the MPN value from the appropriate chart.

15.8.16.3.7 Fecal Coliform MPN Presumptive Test Procedure

The procedure for the fecal coliform MPN presumptive test is described below:

1. Prepare dilutions and inoculate five fermentation tubes for each dilution.
2. Cap all tubes, and transfer to incubator.
3. Incubate 24 ± 2 hr at $35 \pm 0.5°C$.
4. Examine tubes for gas:
   A. Gas present (positive test)—transfer.
   B. No gas—continue incubation.
5. Incubate for a total time of 48 ± 3 hr at $35 \pm 0.5°C$.
6. Examine tubes for gas:
   A. Gas present (positive test)—transfer.
   B. No gas (negative test).

Note: Keep in mind that the fecal coliform MPN confirming procedure of fecal coliform procedure using the A-1 broth test is used to determine the MPN/100 mL. The MPN procedure for fecal coliform determinations requires a minimum of three dilutions with five tubes per dilution.

15.8.16.3.8 Calculation of Most Probable Number (MPN)/100 mL

Calculation of the MPN test results requires selection of a valid series of three consecutive dilutions. The number of positive tubes in each of the three selected dilution inoculations is used to determine the MPN/100 mL. When selecting the dilution inoculations to be used in the calculation, each dilution is expressed as a ratio of positive tubes per tubes inoculated in the dilution (e.g., three positive/five inoculated, or 3/5). Several rules should be followed when determining the most valid series of dilutions. In the following examples, four dilutions were used for the test.

1. Using the confirming test data, select the highest dilution showing all positive results (no lower dilution showing less than all positive) and the next two higher dilutions.
2. If a series shows all negative values with the exception of one dilution, select the series that places the only positive dilution in the middle of the selected series.
3. If a series shows a positive result in a dilution higher than the selected series (using rule 1), it should be incorporated into the highest dilution of the selected series. After selecting the valid series, the MPN/100 mL is determined by locating the selected series on the MPN reference chart. If the selected dilution series matches the dilution series of the reference chart, the MPN value from the chart is the reported value for the test. If the dilution series used for the test does not match the dilution series of the chart, the test result must be calculated.

$$MPN/100 \text{ mL} = MPN_{chart} \times \frac{\text{Sample vol. in 1st dilution}_{chart}}{\text{Sample vol. in 1st dilution}_{sample}} \quad (15.12)$$

■ EXAMPLE 15.7

Problem: Using the results given below, calculate the MPN/100 mL of the example:

Sample in Each Serial Dilution (mL)	Positive Tubes (Inoculated)
10.0	5/5
1.0	5/5
0.1	3/5
0.01	1/5
0.001	1/5

Solution:

1. Select the highest dilution (tube with the lowest amount of sample) with all positive tubes (1.0-mL dilution). Select the next two higher dilutions (0.1 mL and 0.01 mL). In this case, the selected series will be 5–3–1.
2. Include any positive results in dilutions higher than the selected series (0.001-mL dilution 1/5). This changes the selected series to 5–3–2.
3. Find the three columns of Table 15.11 that have this series (5–3–2).
4. Read the MPN value from the next column (140).
5. In Table 15.11, the dilution series begins with 10 mL. For this test, this series begins with 1.0 mL.

$$MPN/100 \text{ mL} = 140 \text{ MPN}/100 \text{ mL} \times \frac{10 \text{ mL}}{1 \text{ mL}}$$

$$= 1400 \text{ MPN}/100 \text{ mL}$$

15.8.16.3.9 Membrane Filtration Technique

The membrane filtration technique can be useful for determining the fecal coliform density in wastewater effluents, except for primary treated wastewater that has not been chlorinated or wastewater containing toxic metals or phenols. Chlorinated secondary or tertiary effluents may be tested using this method, but results are subject to verification by MPN technique. The membrane filter technique

TABLE 15.11
MPN Reference Chart

Sample Volume (mL)			MPN/	Sample Volume (mL)			MPN/
10	1.0	0.1	100 mL	10	1.0	0.1	100 mL
0	0	0	0	4	3	0	27
0	0	1	2	4	3	1	33
0	1	0	2	4	4	0	34
0	2	0	4	5	0	0	23
1	0	0	2	5	0	1	31
1	0	1	4	5	0	2	43
1	1	0	4	5	1	0	33
1	1	1	6	5	1	1	46
1	2	0	6	5	1	2	63
2	0	0	5	5	2	0	49
2	0	1	7	5	2	1	70
2	1	0	7	5	2	2	94
2	1	1	9	5	3	0	79
2	2	0	9	5	3	1	110
2	3	0	12	5	3	2	140
3	0	0	8	5	3	3	180
3	0	1	11	5	4	0	130
3	1	0	11	5	4	1	170
3	1	1	14	5	4	2	220
3	2	0	14	5	4	3	280
3	2	1	17	5	4	4	350
4	0	0	13	5	5	0	240
4	0	1	17	5	5	1	350
4	1	0	17	5	5	2	540
4	1	1	21	5	5	3	920
4	1	2	26	5	5	4	1600
4	2	0	22	5	5	5	≥2400
4	2	1	26				

utilizes a specially designed filter pad with uniformly sized pores (openings) that are small enough to prevent bacteria from entering the filter (see Figure 15.17). Another unique characteristic of the filter allows liquids, such as the media, placed under the filter to pass upward through the filter to provide nourishment required for bacterial growth.

Note: In the membrane filter method, the number of colonies grown estimates the number of coliforms.

15.8.16.3.10 Membrane Filter Procedure

The procedure for the membrane filter method is described below:

1. Sample filtration:
   A. Select a filter, and aseptically separate it from the sterile package.
   B. Place the filter on the support plate with the grid side up.
   C. Place the funnel assembly on the support; secure as needed (see Figure 15.16).

D. Pour 100 mL of sample or serial dilution onto the filter; apply vacuum.

Note: The sample size and necessary serial dilution should produce a growth of 20 to 60 fecal coliform colonies on at least one filter. The selected dilutions must also be capable of showing permit excursions.

E. Allow all of the liquid to pass through the filter.
F. Rinse the funnel and filter with three portions (20 to 30 mL) of sterile, buffered dilution water. (Allow each portion to pass through the filter before the next addition).

Note: Filtration units should be sterile at the start of each filtration series and should be sterilized again if the series is interrupted for 30 minutes or more. A rapid interim sterilization can be accomplished by a 2-minute exposure to ultraviolet (UV) light, flowing steam, or boiling water.

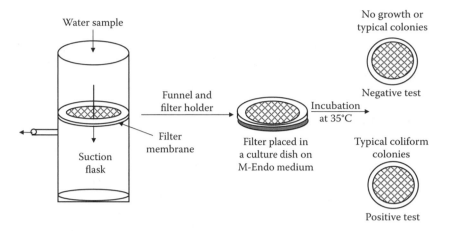

Water sample

Funnel and
filter holder

Filter
membrane

Suction
flask

Filter placed in
a culture dish on
M-Endo medium

Incubation
at 35°C

No growth or
typical colonies

Negative test

Typical coliform
colonies

Positive test

FIGURE 15.17 Membrane filter technique.

2. Incubation
   A. Place absorbent pad into culture dish using sterile forceps.
   B. Add 1.8 to 2.0 mL M-FC media to the absorbent pad.
   C. Discard any media not absorbed by the pad.
   D. Filter sample through sterile filter.
   E. Remove filter from assembly, and place on absorbent pad (grid up).
   F. Cover culture dish.
   G. Seal culture dishes in a weighted plastic bag.
   H. Incubate filters in a water bath for 24 hours at 44.5 ± 0.2°C.

15.8.16.3.11   Colony Counting

Upon completion of the incubation period, the surface of the filter will have growths of both fecal coliform and nonfecal coliform bacterial colonies. The fecal coliform will appear blue in color, while nonfecal coliform colonies will appear gray or cream colored. When counting the colonies, the entire surface of the filter should be scanned using a 10× to 15× binocular, wide-field dissecting microscope. The desired range of colonies, for the most valid fecal coliform determination is 20 to 60 colonies per filter. If multiple sample dilutions are used for the test, counts for each filter should be recorded on the laboratory data sheet.

1. *Too many colonies*—Filters that show a growth over the entire surface of the filter with no individually identifiable colonies should be recorded as confluent growth. Filters that show a very high number of colonies (greater than 200) should be recorded as TNTC (too numerous to count).
2. *Not enough colonies*—If no single filter meets the desired minimum colony count (20 colonies), the sum of the individual filter counts and the respective sample volumes can be used in the formula to calculate the colonies/100 mL.

Note: In each of these cases, adjustments in sample dilution volumes should be made to ensure that future tests meet the criteria for obtaining a valid test result.

15.8.16.3.12   Calculations

The fecal coliform density can be calculated using the following formula:

$$\text{Colonies/100 mL} = \frac{\text{Colonies counted}}{\text{Sample vol. (mL)}} \times 100 \text{ mL} \quad (15.13)$$

■ EXAMPLE 15.8

Problem: Using the data shown below, calculate the colonies per 100 mL for the influent and effluent samples noted.

Sample Location	Influent Sample Dilutions			Effluent Sample Dilutions		
Sample (mL)	1.0	0.1	0.01	10	1.0	0.1
Colonies counted	97	48	16	10	5	3

Solution:

- *Step 1. Influent sample.* Select the influent sample filter that has a colony count in the desired range (20 to 60). Because one filter meets this criterion, the remaining influent filters that did not meet the criterion are discarded.

$$\text{Colonies/100 mL} = \frac{48 \text{ colonies}}{0.1 \text{ mL}} \times 100 \text{ mL}$$

$$= 48,000 \text{ colonies/100 mL}$$

- *Step 2. Effluent sample.* Because none of the filters for the effluent sample meets the minimum test requirement, the colonies per 100 mL must be determined by totaling the colonies on each filter and the sample volumes used for each filter.

$$Total\ colonies = 10 + 5 + 3 = 18\ colonies$$

$$Total\ sample = 10.0\ ml + 1.0\ ml + 0.1\ mL$$
$$= 11.1\ mL$$

$$Colonies/100\ mL = \frac{18\ colonies}{11.1\ mL} \times 100$$
$$= 162\ colonies/100\ mL$$

Note: The USEPA criterion for fecal coliform bacteria in bathing waters is a logarithmic mean of 200 per 100 mL, based on the minimum of 5 samples taken over a 30-day period, with not more than 10% of the total samples exceeding 400 per 100 mL. Because shellfish may be eaten without being cooked, the strictest coliform criterion applies to shellfish cultivation and harvesting. The USEPA criterion states that the mean fecal coliform concentration should not exceed 14 per 100 mL, with not more than 10% of the samples exceeding 43 per 100 mL.

15.8.16.3.13 Interferences

Large amounts of turbidity, algae, or suspended solids may interfere with this technique blocking the filtration of the sample through the membrane filter. Dilution of these samples to prevent this problem may make the test inappropriate for samples with low fecal coliform densities because the sample volumes after dilution may be too small to give representative results. The presence of large amounts of non-coliform group bacteria in the samples may also prohibit the use of this method.

Note: Many NPDES discharge permits require fecal coliform testing. Results for fecal coliform testing must be reported as a geometric mean (average) of all the test results obtained during a reporting period. A geometric mean, unlike an arithmetic mean or average, dampens the effect of very high or low values that otherwise might cause a nonrepresentative result.

15.8.17 Apparent Color Testing/Analysis

Color in water often originates from organic sources: decomposition of leaves, and other forest debris such as bark, pine needles, etc. Tannins and lignins, organic compounds, dissolve in water. Some organics bond to iron to produce soluble color compounds. Biodegrading algae from recent bloom may cause significant color. Though less likely a source of color in water, possible inorganic sources of color are salts of iron, copper, and potassium permanganate added in excess at the treatment plant.

Note: Noticeable color is an objectionable characteristic that makes the water psychologically unacceptable to the consumer.

Recall that *true color* is dissolved. It is measured colorimetrically and compared against a USEPA color standard. *Apparent color* may be caused by suspended material (turbidity) in the water. It is important to point out that even though it may also be objectionable in the water, supply, it is not meant to be measured in the color analysis or test. Probably the most common cause of apparent color is particulate oxidized iron.

By using established color standards, people in different areas can compare test results. Over the years, several attempts have been made to standardize the method of describing the apparent color of water using comparisons to color standards. *Standard Methods* (APHA, 1998) recognizes visual comparison methods as reliable methods for analyzing water from the distribution system. The Forel-Ule Color Scale consists of a dozen shades ranging from deep blue to khaki green that are typical of offshore and coastal bay waters. Another visual comparison method is the Borger Color System, which provides an inexpensive, portable color reference for shades typically found in natural waters; it can also be used for its original purpose—describing the colors of insects and larvae found in streams of lakes. The Borger Color System also allows the recording of the color of algae and bacteria on streambeds. To ensure reliable and accurate descriptions of apparent color, use a system of color comparisons that is reproducible and comparable to the systems used by other groups.

Note: Do not leave color standard charts and comparators in direct sunlight.

Measured levels of color in water can serve as indicators for a number of conditions; for example, transparent water with a low accumulation of dissolved minerals and particulate matter usually appears blue and indicates low productivity. Yellow to brown color normally indicates that the water contains dissolved organic materials, humic substances from soil, peat, or decaying plant material. Deeper yellow to reddish colors indicate the presence of some algae and dinoflagellates. A variety of yellows, reds, browns, and grays is indicative of soil runoff.

Note: Color by itself has no health significance in drinking waters; however, a secondary MCL is set at 15 color units, and it is recommended that community supplies provide water that has less color.

TABLE 15.12
Descriptions of Odors

Nature of Odor	Description	Examples
Aromatic	Spicy	Camphor, cloves, lavender
Balsamic	Flowery	Geranium, violet, vanilla
Chemical	Industrial wastes or chlorinous treatments	Chlorine
Hydrocarbon	Oil refinery wastes	
Medicinal	Phenol and iodine	
Sulfur	Hydrogen sulfide	
Disagreeable	Fishy	Dead algae
Pigpen	Algae	
Septic	Stale sewage	
Earthy	Damp earth	
Peaty	Peat	
Grassy	Crushed grass	
Musty	Decomposing straw; moldy	
Vegetable	Root vegetables	Damp cellar

Source: Adapted from APHA, *Standard Methods for the Examination of Water and Wastewater*, 20th ed., American Public Health Association, Washington, D.C., 1998.

When treating for color in water, alum and ferric coagulation is often effective. It removes apparent color and often much of the true color. Oxidation of color-causing compounds to a noncolored version is sometimes effective. Activated carbon treatment may adsorb some of the organics causing color. For apparent color problems, filtration is usually effective in trapping the colored particles.

15.8.18 ODOR ANALYSIS OF WATER

Odor is expected in wastewater—the fact is, any water containing waste, especially human waste, has a detectable (expected) odor associated with it. Odor in a raw water source (for potable water) is caused by a number of constituents; for example, chemicals that may come from municipal and industrial waste discharges or natural sources such as decomposing vegetable matter or microbial activity may cause odor problems. Odor affects the acceptability of drinking water, the aesthetics of recreation water, and the taste of aquatic foodstuffs.

The human nose can accurately detect a wide variety of smells and serves as the best odor-detection and testing device currently available. To measure odor, collect a sample in a large-mouthed jar. After waving off the air above the water sample with your hand, smell the sample. Use the list of odors provided Table 15.12, which is a system of qualitative descriptions that can help monitors describe and record odors they detect. Record all observations (APHA, 1998).

When treating for odor in water, removal depends on the source of the odor. Some organic substances that cause odor can be removed with powdered activated carbon. If the odor is of gaseous origin, scrubbing (aeration) may remove it. Some odor-causing chemicals can be oxidized to odorless chemicals with chlorine, potassium permanganate, or other oxidizers. Settling may remove some material which, when later dissolved in the water, may have potential odor-causing capacity. Unfortunately, the test for odor in water is subjective and is not very accurate, but no scientific means of measurement exists.

To test odor in water intended for potable water use, a sample is generally heated to 60°C. Odor is observed and recorded. A threshold odor number (TON) is assigned. TON is found by using the following equation:

$$\text{TON} = \frac{\text{Total volume of water sample}}{\text{Lowest sample volume with odor}} \quad (15.14)$$

15.8.19 CHLORINE RESIDUAL TESTING/ANALYSIS

Chlorination is the most widely used means of disinfecting water in the United States. When chlorine gas is dissolved into pure water, it forms hypochlorous acid, hypochlorite ion, and hydrogen chloride (hydrochloric acid). The total concentration of HOCl and OCl ion is known as *free chlorine residual*. Current federal regulations cite six approved methods for determination of total residual chlorine (TRC):

1. DPD–spectrophotometry
2. Titrimetric–amperometric direct
3. Titrimetric–iodometric direct
4. Titrimetric–iodometric back
   A. Starch iodine endpoint, iodine titrant
   B. Starch iodine endpoint, iodate titrant
5. Amperometric endpoint

6. DPD–FAS titration
7. Chlorine electrode

All of these test procedures are approved methods and, unless prohibited by the plant's NPDES discharge permit, can be used for effluent testing. Based on current most popular method usage in the United States, our discussion is limited to

1. DPD–spectrophotometry
2. DPD–FAS titration
3. Titrimetric–amperometric direct

Note: Treatment facilities required to meet nondetectable total residual chlorine limitations must use one of the test methods specified in the plant's NPDES discharge permit.

For information on any of the other approved methods, refer to the appropriate reference cited in the federal regulations.

15.8.19.1 DPD–Spectrophotometry

Diethyl-*p*-phenylenediamine (DPD) reacts with chlorine to form a red color. The intensity of the color is directly proportional to the amount of chlorine present. This color intensity is measured using a colorimeter or spectrophotometer. This meter reading can be converted to a chlorine concentration using a graph developed by measuring the color intensity produced by solutions with precisely known concentrations of chlorine. In some cases, spectrophotometers or colorimeters are equipped with scales that display chlorine concentration directly. In these cases, it is not necessary to prepare a standard reference curve. If the direct reading colorimeter is not used, chemicals that are required include:

1. Potassium dichromate solution (0.100 *N*)
2. Potassium iodine crystals
3. Standard ferrous ammonium sulfate solution (0.00282 *N*)
4. Concentrated phosphoric acid
5. Sulfuric acid solution (1 + 5)
6. Barium diphenylamine sulfonate (0.1%)

If an indicator is not used, the DPD indicator and phosphate buffer (DPD prepared indicator and buffer + indicator together) are required.

Conducting the test requires a direct readout colorimeter designed to meet the test specifications or a spectrophotometer (wavelength of 515 nm and light path of at least 1 cm) or a filter photometer with a filter having maximum transmission in the wavelength range of 490 to 520 nm and a light path of at least 1 cm. In addition, for direct readout colorimeter procedures, a sample test

vial is required. When the direct readout colorimeter procedure is not used, the equipment required for testing includes:

1. 250-mL Erlenmeyer flask
2. 10-mL measuring pipets
3. 15-mL test tubes
4. 1-mL pipets (graduated to 0.1 mL)
5. Sample cuvettes with 1-cm light path

Note: A cuvette is a small, often tubular laboratory vessel, often made of glass.

15.8.19.1.1 Procedure

Note: For direct readout colorimeters, follow the procedure supplied by the manufacturer.

1. Prepare a standard curve for TRC concentrations from 0.05 to 4.0 mg/L—chlorine vs. percent transmittance.

Note: Instructions on how to prepare the TRC concentration curve or a standard curve are normally included in the spectrophotometer manufacturer's operating instructions.

2. Calibrate colorimeter in accordance with the manufacturer's instructions using a laboratory-grade water blank.
3. Add one prepared indicator packet (or tablet) of the appropriate size to match sample volume to a clean test tube or cuvette, or
   A. Pipet 0.5 mL phosphate buffer solution.
   B. Pipet 0.5 mL DPD indicator solution.
   C. Add 0.1 g Kl (potassium iodide) crystals to a clean tube or cuvette.
4. Add 10 mL of sample to the cuvette.
5. Stopper the cuvette, and swirl to mix the contents well.
6. Let stand for 2 minutes.
7. Verify the wavelength of the spectrophotometer or colorimeter, and check and set the 0%*T* using the laboratory-grade water blank.
8. Place the cuvette in the instrument, read %*T*, and record reading.
9. Determine mg/L TRC from the standard curve.

Note: Calculations are not required in this test because TRC (mg/L) is read directly from the meter or from the graph.

15.8.19.2 DPD–FAS Titration

The amount of ferrous ammonium sulfate (FAS) solution required to just remove the red color from a total residual chlorine sample that has been treated with DPD indicator can be used to determine the concentration of chlorine in the sample. This is known as a *titrimetric test procedure*. The chemicals used in the test procedure include:

1. DPD prepared indicator (buffer and indicator together)
2. Potassium dichromate solution (0.100 N)
3. Potassium iodide crystals
4. Standard ferrous ammonium sulfate solution (0.00282 N)
5. Concentrated phosphoric acid
6. Sulfuric acid solution (1 + 5)
7. Barium diphenylamine sulfonate (0.1%)

Note: DPD indicator and phosphate buffer are not required if prepared indicator is used.

The equipment required for this test procedure includes the following:

1. 250-mL graduated cylinder
2. 5-mL measuring pipets
3. 500-mL Erlenmeyer flask
4. 50-mL buret (graduate to 0.1 mL)
5. Magnetic stirrer and stir bars

15.8.19.2.1 Procedure

1. Add the contents of a prepared indicator packet (or tablet) to the Erlenmeyer flask, or
   A. Pipet 5 mL phosphate buffer solution into an Erlenmeyer flask.
   B. Pipet 5 mL DPD indicator solution into the flask.
   C. Add 1 g Kl crystals to the flask.
2. Add 100 mL of sample to the flask.
3. Swirl the flask to mix contents.
4. Let the flask stand for 2 minutes.
5. Titrate with FAS until the red color first disappears.
6. Record the amount of titrant.

The calculation required in this procedure is:

$$TRC \text{ (mg/L)} = mL \text{ of FAS used} \qquad (15.15)$$

15.8.19.3 Titrimetric–Amperometric Direct Titration

In this test procedure, phenylarsine oxide is added to a treated sample to determine when the test reaction has been completed. The volume of phenylarsine oxide (PA) used can then be used to calculate the TRC. The chemicals required include:

1. Phenylarsine oxide solution (0.00564 N)
2. Potassium dichromate solution (0.00564 N)

3. Potassium iodide solution (5%)
4. Acetate buffer solution (pH 4.0)
5. Standard arsenite solution (0.1 N)

Equipment required includes:

1. 250-mL graduated cylinder
2. 5-mL measuring pipets
3. Amperometric titrator

15.8.19.3.1 Procedure

1. Prepare amperometric titrator according to manufacturer.
2. Add 200-mL sample.
3. Place container on titrator stand and turn on mixer.
4. Add 1 g Kl crystals or 1 mL Kl solution.
5. Pipet 1 mL of pH 4 (acetate) buffer into the container.
6. Titrate with 0.0056 N PAO.

When conducting the test procedure, as the downscale endpoint is neared, slow titrant addition to 0.1-mL increments, and note titrant volume used after increment. When no needle movement is noted, the endpoint has been reached. Subtract the final increment from the buret reading to determine the final titrant volume. For this procedure, the only calculation normally required is:

$$TRC \text{ (mg/L)} = mL \text{ PAO used} \qquad (15.16)$$

15.8.20 Fluorides

It has long been accepted that a moderate amount of fluoride ions (F^-) in drinking water contributes to good dental health; it has been added to many community water supplies throughout the United States to prevent dental caries in children's teeth. Fluoride is seldom found in appreciable quantities of surface waters and appears in groundwater in only a few geographical regions. Fluorides are used to make ceramics and glass. Fluoride is toxic to humans in large quantities and to some animals. The chemicals added to potable water in treatment plants include:

- NaF (sodium fluoride, solid)
- Na_2SiF_6 (sodium silicofluoride, solid)
- H_2SiF_6 (hyrofluosilicic acid; most widely used)

Analysis of the fluoride content of water can be performed using the colorimetric method. In this test, fluoride ion reacts with zirconium ion and produces zirconium fluoride, which bleaches an organic red dye in direct proportion to its concentration. This can be compared to standards and read colorimetrically.

CHAPTER REVIEW QUESTIONS

15.1   Explain (in simple terms) the methods involved in analyzing for total coliforms in an effluent sample.

15.2   How soon after the sample is collected must the pH be tested?

15.3   What is a grab sample?

15.4   When is it necessary to use a grab sample?

15.5   What is a composite sample?

15.6   List three rules for sample collection.

15.7   What is the acceptable preservation method for suspended solids samples?

15.8   Most solids test methods are based on changes in weight. What can cause changes in weight during the testing procedure?

REFERENCES AND SUGGESTED READING

AWRI. 2000. *Plankton Sampling*. Allendale, MI: Robert B. Annis Water Resource Institute, Grand Valley State University.

APHA. 1971. *Standard Methods for the Examination of Water and Wastewater*, 17th ed. Washington, D.C.: American Public Health Association.

APHA. 1998. *Standard Methods for the Examination of Water and Wastewater*, 20th ed. Washington, D.C.: American Public Health Association, pp. 4–129.

AWWA. 1995. *Water Treatment*, 2nd ed. Denver, CO: American Water Works Association.

Bahls, L.L. 1993. *Periphyton Bioassessment Methods for Montana Streams*. Helena, MT: Montana Water Quality Bureau, Department of Health and Environmental Science.

Barbour, M.T., Gerritsen, J., Snyder, B.D., and Stibling, J.B. 1997. *Revision to Rapid Bioassessment Protocols for Use in Streams and Rivers: Periphytons, Benthic Macroinvertebrates, and Fish*. Washington, D.C.: U.S. Environmental Protection Agency.

Bly, T.D., and Smith, G.F. 1994. *Biomonitoring Our Streams: What's It All About?* Nashville, TN: U.S. Geological Survey.

Botkin, D.B. 1990. *Discordant Harmonies*. New York: Oxford University Press.

Camann, M. 1996. *Freshwater Aquatic Invertebrates: Biomonitoring*. Arcata, CA: Humboldt State University (http://www.humboldt.edu).

Carins, Jr., J. and Dickson, K.L. 1971. A simple method for the biological assessment of the effects of waste discharges on aquatic bottom-dwelling organisms. *J. Water Pollut. Control Fed.*, 43, 755–772.

Hill, B.H. 1997. The use of periphyton assemblage data in an index of biotic integrity. *Bull. N. Am. Benthol. Soc.*, 14, 158.

Huff, W.R. 1993. Biological indices define water quality standard. *Water Environ. Technol.*, 5, 21–22.

Karr, J.R. 1981. Assessment of biotic integrity using fish communities. *Fisheries*, 66, 21–27.

Karr, J.R. et al. 1986. *Assessing Biological Integrity in Running Waters: A Method and Its Rationale*, Special Publ. 5. Champaign, IL: Illinois Natural History Survey.

KDEP. 1993. *Methods for Assessing Biological Integrity of Surface Waters*. Frankfort: Kentucky Department of Environmental Protection.

Kittrell, F.W. 1969. *A Practical Guide to Water Quality Studies of Streams*. Washington, D.C.: U.S. Department of Interior.

LaMotte Co. 1992. *The Monitor's Handbook: A Reference Guide for Natural Water Monitoring*. Chestertown, MD: LaMotte Co.

O'Toole, C., Ed. 1986. *The Encyclopedia of Insects*. New York: Facts on File.

Patrick, R. 1973. Use of algae, especially diatoms, in the assessment of water quality. In *Biological Methods for the Assessment of Water Quality*, Carins, J. and Dickson, K.L., Eds., Special Tech. Publ. 528. Philadelphia, PA: American Society for Testing and Materials.

River Watch Network. 1992. *Total Alkalinity and pH Field and Laboratory Procedures* (based on University of Massachusetts Acid Rain Monitoring Project), July.

Rodgers, Jr., J.H., Dickson, K.L., and Cairns, Jr., J. 1979. A review and analysis of some methods used to measure functional aspects of periphyton. In *Methods and Measurements of Periphyton Communities: A Review*, Weitzel, R.L., Ed., Special Tech. Publ. 690. Philadelphia, PA: American Society for Testing and Materials.

Stevenson, R.J. 1996. An introduction to algal ecology in freshwater benthic habitats. In *Algal Ecology: Freshwater Benthic Ecosystems*, Stevenson, R.J., Bothwell, M., and Lowe, R.L., Eds., pp. 3–30. San Diego, CA: Academic Press.

Stevenson, R.J. 1998. Diatom indicators of stream and wetland stressors in a risk management framework. *Environ. Monitor. Assess.*, 51, 107–108.

Stevenson, R.J. and Pan, Y. 1999. Assessing ecological conditions in rivers and streams with diatoms. In *The Diatoms: Application to the Environmental and Earth Sciences*, Stoermer, E.F. and Smol, J.P., Eds., pp. 11–40. Cambridge, U.K.: Cambridge University Press.

Tchobanoglous, G. and Schroeder, E.D. 1985. *Water Quality*. Reading, MA: Addison-Wesley.

USEPA. 1983. *Technical Support Manual: Waterbody Survey and Assessments for Conducting Use Attainability Analyses*. Washington, D.C.: U.S. Environmental Protection Agency.

USEPA. 1985. *Test Methods for Escherichia coli and Enterococci in Water by the Membrane Filter Procedure*, Method #1103.1, EPA 600/4-85-076. Cincinnati, OH: Office of Research and Development, U.S. Environmental Protection Agency.

USEPA. 1986. *Bacteriological Ambient Water Quality Criteria for Marine and Fresh Recreational Waters*, EPA 440/5-84-002. Cincinnati, OH: Office of Research and Development, U.S. Environmental Protection Agency.

USEPA. 1991. *Methods for Chemical Analysis of Water and Wastes*, 2nd ed. Washington, D.C.: U.S. Environmental Protection Agency

USEPA. 1997. *Volunteer Stream Monitoring: A Methods Manual*, EPA 08-18-2000. Washington, DC: U.S. Environmental Protection Agency, pp. 1-35.

USEPA. 2000. *Monitoring Water Quality: Intensive Stream Bioassay.* Washington, D.C.: U.S. Environmental Protection Agency.

Velz, C.J. 1970. *Applied Stream Sanitation.* New York: Wiley Interscience.

Warren, Jr., M.L. and Burr, B.M. 1994. Status of freshwater fishes of the U.S.: overview of an imperiled fauna. *Fisheries*, 19(1), 6–18.

Weitzel, R.L. 1979. Periphyton measurements and applications. In *Methods and Measurements of Periphyton Communities: A Review*, Weitzel R.L., Ed., Special Tech. Publ. 690. Philadelphia, PA: American Society for Testing and Material.

Part IV

Water and Water Treatment

16 Potable Water Source

Because of huge volume and flow conditions, the quality of natural water cannot be modified significantly within the body of water. Accordingly, humans must augment Nature's natural processes with physical, chemical, and biological treatment procedures. Essentially, this quality control approach is directed toward the treated water withdrawn from a source for a specific use.

16.1 INTRODUCTION

Before presenting a discussion of potential potable water supplies available to us at the current time, it is important that we define portable water:

Potable water is water fit for human consumption and domestic use, which is sanitary and normally free of minerals, organic substances, and toxic agents in excess or reasonable amounts for domestic usage in the area served and normally adequate in quantity for the minimum health requirements of the persons served.

With regard to a potential potable water supply, the key words are *quality* and *quantity*. Obviously, if we have a water supply that is unfit for human consumption, we have a quality problem. If we do not have an adequate supply of quality water, we have a quantity problem. In this chapter, we discuss the surface water and groundwater hydrology and the mechanical components associated with collection and conveyance of water from its source to the public water supply system for treatment. We also discuss development of well supplies. To better comprehend the material presented in this chapter, we have provided the following list of key terms and their definitions.

16.1.1 KEY TERMS AND DEFINITIONS

Annular space—The space between the casing and the wall of the hole.

Aquifer—A porous, water-bearing geologic formation.

Caisson—Large pipe placed in a vertical position.

Cone of depression—As the water in a well is drawn down, the water near the well drains or flows into it. The water will drain farther back from the top of the water table into the well as drawdown increases.

Confined aquifer—An aquifer surrounded by formations of less permeable or impermeable material.

Contamination—The introduction into water of toxic materials, bacteria, or other deleterious agents that make the water unfit for its intended use.

Drainage basin—An area from which surface runoff or groundwater recharge is carried into a single drainage system; it is also called a *catchment area*, *watershed*, and *drainage area*.

Drawdown—The distance or difference between the static level and the pumping level. When the drawdown for any particular capacity well and rate pump bowls is determined, the pumping level is known for that capacity. The pump bowls are located below the pumping level so they will always be underwater. When the drawdown is fixed or remains steady, the well is then furnishing the same amount of water as is being pumped.

Groundwater—Subsurface water occupying a saturated geological formation from which wells and springs are fed.

Hydrology—The applied science pertaining to properties, distribution, and behavior or water.

Impermeable—A material or substance through which water will not pass.

Overland flow—The movement of water on and just under the Earth's surface.

Permeable—A material or substance through which water can pass.

Porosity—The ratio of pore space to total volume; that portion of a cubic foot of soil that is air space and could therefore contain moisture.

Precipitation—The process by which atmospheric moisture is discharged onto the Earth's crust; precipitation takes the form of rain, snow, hail, and sleet.

Pumping level—The level at which the water stands when the pump is operating.

Raw water—The untreated water to be used after treatment for drinking water.

Radius of influence—Distance from the well to the edge of the cone of depression; the radius of a circle around the well from which water flows into the well.

Recharge area—An area from which precipitation flows into underground water sources.

Specific yield—The geologist's method for determining the capacity of a given well and the production of a given water-bearing formation; it is expressed as gallons per minute per foot of drawdown.

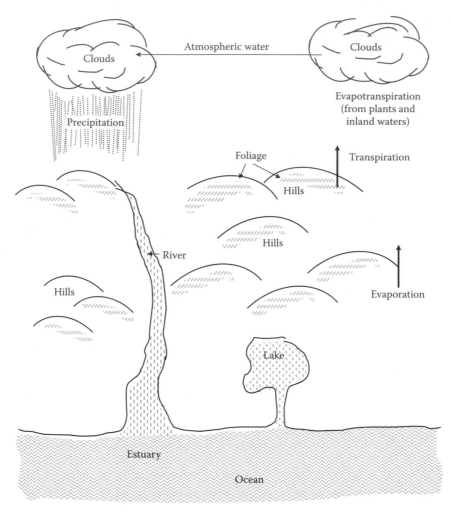

FIGURE 16.1 Natural water cycle.

Spring—A surface feature where, without the aid of humans, water issues from rock or soil onto the land or into a body of water, the place of issuance being relatively restricted in size.

Static level—The height to which the water will rise in the well when the pump is not operating.

Surface runoff—The amount of rainfall that passes over the surface of the Earth.

Surface water—The water on the surface of the Earth, as distinguished from water that is underground (groundwater).

Unconfined aquifer—An aquifer that sits on an impervious layer but is open on the top to local infiltration; the recharge for an unconfined aquifer is local. It is also called a *water table aquifer.*

Water rights—The rights, acquired under the law, to use the water accruing in surface or groundwater for a specified purpose in a given manner and usually within the limits of a given time period.

Water table—Average depth or elevation of the groundwater over a selected area; the upper surface of the zone of saturation, except where that surface is formed by an impermeable body.

Watershed—A drainage basin from which surface water is obtained.

16.2 HYDROLOGIC CYCLE

To gain a better understanding of the *hydrologic cycle (water cycle)*, it is important to review it again (see Figure 16.1). The hydrologic cycle is a cycle without beginning or end. As stated earlier, it transports the Earth's water from one location to another. As can be seen in Figure 16.1, it consists of precipitation, surface runoff, infiltration, percolation, and evapotranspiration. In the hydrologic cycle, water from streams, lakes, and oceans is evaporated by the sun and from the Earth, in addition to transpiration from plants, and furnishes the atmosphere with moisture. Masses of warm air laden with moisture either are forced to cooler upper regions or encounter cool air masses, where the masses condense and form clouds. This condensed moisture falls to Earth in the form of rain, snow, and sleet. Part of the precipitation runs off to streams and lakes. Part enters the ground to supply vegetation and rises through the plants to transpire from the leaves; part seeps or percolates deeply into the ground to supply wells, springs, and the base flow (dry-weather flow) of streams. The cycle constantly repeats itself; a cycle without end.

Note: How long it takes for water that falls from the clouds to return to the atmosphere varies tremendously. After a short summer shower, most of the rainfall on land can evaporate into the atmosphere in only a matter of minutes. A drop of rain falling on the ocean may take as long as 37,000 years before it returns to the atmosphere, and some water has been in the ground or caught in glaciers for millions of years.

16.3 SOURCES OF WATER

Approximately 40 million cubic miles of water cover or reside within the Earth. The oceans contain about 97% of all water on Earth. The other 3% is freshwater: (1) snow and ice on the surface of the Earth, which represents about 2.25% of the water; (2) usable groundwater, which accounts for approximately 0.3%; and (3) surface freshwater, which is less than 0.5%. In the United States, for example, average rainfall is approximately 2.6 feet (a volume of 5900 km^3). Of this amount, approximately 71% evaporates (about 4200 cm^3), and 29% goes to stream flow (about 1700 cubic km^3).

Beneficial freshwater uses include manufacturing, food production, domestic and public needs, recreation, hydroelectric power production, and flood control. Stream flow withdrawn annually is about 7.5% (440 km^3). Irrigation and industry use almost half of this amount (3.4%, or 200 km^3 per year). Municipalities use only about 0.6% (35 km^3 per year) of this amount. Historically, in the United States, water usage has been increasing (as might be expected); for example, in 1975, 40 billion gallons of freshwater were used. In 1990, the total increased to 455 billion gallons. Projected usage in 2002 was about 725 billion gallons.

The primary sources of freshwater include:

- Captured and stored rainfall in cisterns and water jars
- Groundwater from springs, artesian wells, and drilled or dug wells
- Surface water from lakes, rivers, and streams
- Desalinized seawater or brackish groundwater
- Reclaimed wastewater

Current federal drinking water regulations actually define three distinct and separate sources of freshwater: (1) surface water, (2) groundwater, and (3) groundwater under the direct influence of surface water (GWUDI). This last classification is the result of the Surface Water Treatment Rule (SWTR). The definition of what conditions constitute GWUDI, while specific, is not obvious. This classification is discussed in detail later.

16.4 SURFACE WATER

Surface waters are not uniformly distributed over the Earth's surface. In the United States, for example, only about 4% of the landmass is covered by rivers, lakes, and streams. The volumes of these freshwater sources depend on geographic, landscape, and temporal variations, as well as on the impact of human activities. *Surface water* is water that is open to the atmosphere and results from *overland flow* (i.e., *runoff* that has not yet reached a definite stream channel). Put a different way, surface water is the result of *surface runoff*. For the most part, however, *surface* (as used in the context of this text) refers to water flowing in streams and rivers, as well as water stored in natural or artificial lakes, manmade impoundments such as lakes made by damming a stream or river, springs that are affected by a change in level or quantity, shallow wells that are affected by precipitation, wells drilled next to or in a stream or river, rain catchments, and muskeg and tundra ponds.

16.4.1 ADVANTAGES AND DISADVANTAGES OF SURFACE WATER

The biggest advantage of using a surface water supply as a water source is that these sources are readily located; finding surface water sources does not demand sophisticated training or equipment. Many surface water sources have been used for decades and even centuries (in the United States, for example), and considerable data are available on the quantity and quality of the existing water supply. Surface water is also generally softer (i.e., not mineral laden), which makes its treatment much simpler.

The most significant disadvantage of using surface water as a water source is *pollution*. Surface waters are easily contaminated (polluted) with microorganisms that cause waterborne diseases and chemicals that enter the river or stream from surface runoff and upstream discharges. Another problem with many surface water sources is *turbidity,* which fluctuates with the amount of precipitation; increases in turbidity increase treatment costs and operator time. Surface water temperatures can be a problem because they fluctuate with ambient temperature, making consistent water quality production at a waterworks plant difficult. Drawing water from a surface water supply might present other problems; for example, intake structures may clog or become damaged from winter ice, or the source may be so shallow that it completely freezes in the winter. *Water rights* are another issue, in that removing surface water from a stream, lake, or spring requires a legal right. The lingering, seemingly unanswerable, question is who owns the water?

Using surface water as a source means that the purveyor is obligated to meet the requirements of the Surface Water Treatment Rule (SWTR) and Interim Enhanced Surface Water Treatment Rule (IESWTR), which applies only to large public water systems (PWSs) serving more than 10,000 people. The IESWTR tightened controls on disinfection byproducts and turbidity and regulates *Cryptosporidium*.

16.4.2 SURFACE WATER HYDROLOGY

To properly manage and operate water systems, it is important to have a basic understanding of the movement of water and the factors that affect water quality and quantity—in other words, *hydrology*. A discipline of applied science, hydrology includes several components, such as the physical configuration of the watershed, the geology, soils, vegetation, nutrients, energy, wildlife, and the water itself.

The area from which surface water flows is a *drainage basin* or catchment area. With a surface water source, this drainage basin is most often referred to, in nontechnical terms, as a *watershed* (when dealing with groundwater, we call this area a *recharge area*).

Key Point: The area that directly influences the quantity and quality of surface water is the *drainage basin* or *watershed*.

When we trace on a map the course of a major river from its meager beginnings along its seaward path, it is readily apparent that its flow becomes larger and larger. Every tributary adds to its size, and between tributaries the river grows gradually due to overland flow entering it directly (see Figure 16.2). Not only does the river grow, but its entire watershed or drainage basin (basically, the land it drains into) also grows in the sense that it embraces an ever-larger area. The area of the watershed is commonly measured in square miles, sections, or acres. When taking water from a surface water source, knowing the size of the watershed is desirable.

16.4.3 RAW WATER STORAGE

Raw water (i.e., water that has not been treated) is stored for single or multiple uses, such as navigation, flood control, hydroelectric power, agriculture, water supply, pollution abatement, recreation, and flow augmentation. The primary reason for storing water is to meet peak demands or to store water to meet demands when the flow of the source is below demand. Raw water is stored in natural storage sites (such as lakes, muskegs, and tundra ponds) or in manmade storage areas such as dams. Manmade dams are either masonry or embankment. If embankment dams are used, they are typically constructed of local materials with an impermeable clay core. Figure 16.3A–D shows normal flow in Falling Spring, an example raw water source that serves several small communities in southwestern Virginia. Figures 16.3A and B show the original natural upper reaches of Falling Spring. Figure 16.3C shows the human-modified fall lip area (modified to concentrate flow over the lip area to enhance spectacular effect) before the spring flow plunges (Figure 16.3D) toward the crash-landing below (Figure 16.3E and F). The upper, middle, fall, landing, and flow reforming areas are shown in Figure 16.3G.

16.4.4 SURFACE WATER INTAKES

Withdrawing water from a river, lake, or reservoir so it may be conveyed to the first unit process for treatment requires an intake structure. Intakes have no standard design and can range from a simple-pump suction pipe sticking out into the lake or stream to more involved structures costing several thousands of dollars. Typical intakes include submerged intakes, floating intakes, infiltration galleries, spring boxes, and roof catchments. Their primary functions are to supply the highest quality water from the source and to protect piping and pumps from clogging and damage due to wave action, ice formation, flooding, and submerged debris. A poorly conceived or constructed intake can cause many problems. Failure of the intake could result in water system failure.

For a small stream, the most common intake structures are small gravity dams placed across the stream or a submerged intake. In the gravity type of dam, a gravity line or pumps can remove water behind the dam. In the submerged intake type of dam, water is collected in a diversion and carried away by gravity or pumped from a caisson. Another common intake used on small and large streams is an end-suction centrifugal pump or submersible pump placed on a float. The float is secured to the bank, and the water is pumped to a storage area.

Often, the intake structure placed in a stream is an infiltration gallery. The most common infiltration galleries are built by placing well screens or perforated pipe into the streambed. The pipe is covered with clean, graded gravel. When water passes through the gravel, coarse filtration removes a portion of the turbidity and organic material. The water collected by the perforated pipe then flows to a caisson placed next to the stream and is removed from the caisson by gravity or pumping. Intakes used in springs are normally implanted in the water-bearing strata, then covered with clean, washed rock and sealed, usually with clay. The outlet is piped into a spring box. In some locations, a primary source of water is rainwater. Rainwater is collected from the roofs of buildings with a device called a *roof catchment*.

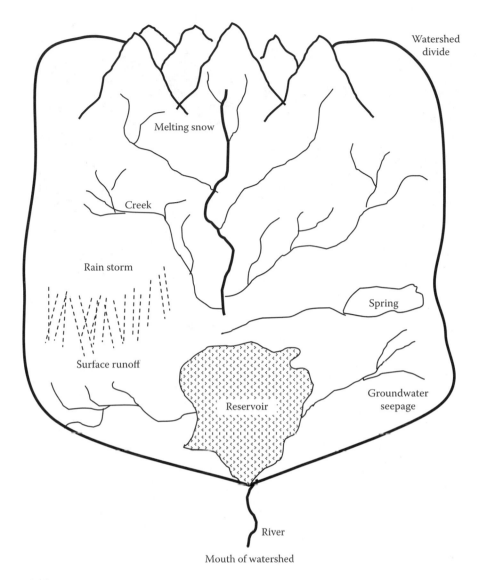

FIGURE 16.2 Watershed.

After determining that a water source provides a suitable quality and quantity of raw water, choosing an intake location includes considering the following:

- Best quality water location
- Dangerous currents
- Sandbar formation
- Wave action
- Ice storms
- Flood factors
- Navigation channels that must be avoided
- Intake accessibility
- Power availability
- Floating or moving objects that pose a hazard
- Distance from pumping station
- Upstream uses that may affect water quality

16.4.5 SURFACE WATER SCREENS

Generally, screening devices are installed to protect intake pumps, valves, and piping. A coarse screen of vertical steel bars, with openings of 1 to 3 inches, is placed in a near-vertical position to exclude large objects. The screen may be equipped with a trash rack rake to remove accumulated debris. A finer screen, one with 3/8-inch openings, removes leaves, twigs, small fish, and other material passing through the bar rack. Traveling screens consist of wire mesh trays that retain solids as the water passes through them. Drive chains and sprockets raise the trays into a head enclosure, where the debris is removed by water sprays. The screen travel pattern is intermittent and controlled by the amount of accumulated material.

(A)

(B)

FIGURE 16.3 Falling Spring, Virginia: (A) Upper reach pool area. (B) Middle reach pool area.

Note: When considering what type of screen should be employed, the most important consideration is ensuring that they can be easily maintained.

16.4.6 SURFACE WATER QUALITY

Surface waters should be of adequate quality to support aquatic life and be aesthetically pleasing; also, waters used as sources of supply should be treatable by conventional processes to provide potable supplies that can meet drink-

ing water standards. Many lakes, reservoirs, and rivers are maintained at a quality suitable for swimming, water skiing, and boating as well as for drinking water. Whether the surface water supply is taken from a river, stream, lake, spring, impoundment, reservoir, or dam, the surface water quality can vary widely, especially in rivers, streams, and small lakes. These water bodies are susceptible not only to waste discharge contamination but also to flash contamination (which can occur almost immediately and not necessarily over time). Lakes are subject to summer/winter

(C)

(D)

FIGURE 16.3 (cont.) Falling Spring, Virginia: (C) Human-modified fall lip area. (D) Plunge to bottom rock.

stratification (turnover) and to algal blooms. Pollution sources include runoff (agricultural, residential, and urban), spills, municipal and industrial wastewater discharges, and recreational users, as well as natural occurrences. Surface water supplies are difficult to protect from contamination and must always be treated.

(E)

(F)

FIGURE 16.3 (cont.) Falling Spring, Virginia: (E) Fall rock floor area. (F) Falling Spring flow reforming area.

16.5 GROUNDWATER

As mentioned, part of the precipitation that falls on land infiltrates the land surface, percolates downward through the soil under the force of gravity, and becomes *ground-* *water*. Groundwater, like surface water, is extremely important to the hydrologic cycle and to our water supplies. Almost half of the people in the United States drink public water from groundwater supplies. Overall, more water exists as groundwater than surface water in the United

(G)

FIGURE 16.3 (cont.) Falling Spring, Virginia: (G) Upper and middle reach pool area, fall overflow lip area, falls, water plunge area, and flow reforming area.

States, including the water in the Great Lakes. Pumping it to the surface is not always economical, though, and in recent years pollution of groundwater supplies from improper waste disposal has become a significant problem.

We find groundwater in saturated layers called *aquifers* under the Earth's surface. Three types of aquifers exist: *unconfined*, *confined*, and *springs*. Aquifers are made up of a combination of solid material such as rock and gravel and open spaces called *pores*. Regardless of the type of aquifer, the groundwater in the aquifer is in a constant state of motion. This motion is caused by gravity or by pumping.

The actual amount of water in an aquifer depends on the amount of space available between the various grains of material that make up the aquifer. The amount of space available is referred to as the *porosity*. The ease of movement through an aquifer is dependent on how well the pores are connected; for example, clay can hold a lot of water and has high porosity, but the pores are not connected, so water moves through the clay with difficulty. The ability of an aquifer to allow water to infiltrate is referred to as *permeability*.

The unconfined aquifer that lies just under the Earth's surface is called the *zone of saturation* (see Figure 16.4). The top of the zone of saturation is the *water table*. An unconfined aquifer is only contained on the bottom and is dependent on local precipitation for recharge. This type of aquifer is often referred to as a *water table aquifer*.

Unconfined aquifers are a primary source of shallow well water (see Figure 16.4). Because these wells are

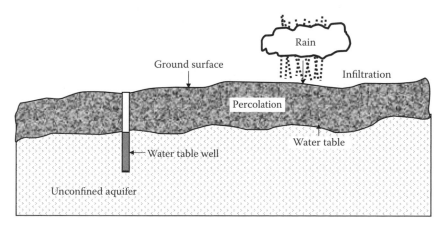

FIGURE 16.4 Unconfined aquifer.

shallow they are not desirable as public drinking water sources. They are subject to local contamination from hazardous and toxic materials, such as fuel and oil, as well as septic tank and agricultural runoff providing increased levels of nitrates and microorganisms. These wells may be classified as groundwater under the direct influence of surface water (GWUDI) and therefore require treatment for control of microorganisms.

A *confined aquifer* is sandwiched between two impermeable layers that block the flow of water. The water in a confined aquifer is under hydrostatic pressure. It does not have a free water table (see Figure 16.5). Confined aquifers are referred to as *artesian* aquifers. Wells drilled into artesian aquifers are *artesian wells* and commonly yield large quantities of high-quality water. An artesian well is any well where the water in the well casing would rise above the saturated strata. Wells in confined aquifers are normally referred to as deep wells and are not generally affected by local hydrological events.

A confined aquifer is recharged by rain or snow in the mountains where the aquifer lies close to the surface of the Earth. Because the recharge area is some distance from areas of possible contamination, the possibility of contamination is usually very low; however, once contaminated, confined aquifers may take centuries to recover.

Groundwater naturally exits the Earth's crust in areas called *springs*. The water in a spring can originate from a water table aquifer or from a confined aquifer. Only water from a confined spring is considered desirable for a public water system.

16.5.1 GROUNDWATER QUALITY

Generally, groundwater is of high chemical, bacteriological, and physical quality. When pumped from an aquifer composed of a mixture of sand and gravel and when not directly influenced by surface water, groundwater is often used without filtration. It can also be used without disinfection if it has a low coliform count; however, as mentioned, groundwater can become contaminated. Septic systems fail, saltwater intrudes, improper disposal of wastes occurs, stockpiled chemicals leach, underground storage tanks leak, hazardous materials spill, fertilizers and pesticides are misplaced, and mines are recklessly abandoned. To understand how an underground aquifer becomes contaminated, we must understand what occurs when pumping is taking place within the well.

When groundwater is removed from its underground source (i.e., from the water-bearing stratum) via a well, water flows toward the center of the well. In a water table aquifer, this movement causes the water table to sag toward the well. This sag is the *cone of depression*. The shape and size of the cone depend on the relationship between the pumping rate and the rate at which water can move toward the well. If the rate is high, the cone is

shallow and its growth stabilizes. The area that is included in the cone of depression is the *cone of influence*, and any contamination in this zone will be drawn into the well.

16.6 GROUNDWATER UNDER THE DIRECT INFLUENCE OF SURFACE WATER

Water under the direct influence of surface water is not classified as a groundwater supply. A supply designated as GWUDI must be treated under the state's surface water rules rather than the groundwater rules. The Surface Water Treatment Rule of the Safe Drinking Water Act requires each site to determine which groundwater supplies are influenced by surface water (i.e., when surface water can infiltrate a groundwater supply and could contaminate it with *Giardia*, viruses, turbidity, and organic material from the surface water source). To determine whether a groundwater supply is under the direct influence of surface water, the U.S. Environmental Protection Agency (USEPA) has developed procedures that focus on significant and relatively rapid shifts in water quality characteristics, including turbidity, temperature, and pH. When these shifts can be closely correlated with rainfall or other surface water conditions, or when certain indicator organisms associated with surface water are found, the source is said to be under the direct influence of surface water.

16.7 SURFACE WATER QUALITY AND TREATMENT REQUIREMENTS

Public water systems (PWSs) must comply with applicable federal and state regulations and must provide the required quantity and quality of water supplies, including proper treatment (where and when required) and competent/qualified waterworks operators. USEPA's regulations require all public water systems using any surface or groundwater under the direct influence of surface water to disinfect it, and they may be required by the state to filter it, unless the water source meets certain requirements and site-specific conditions. Treatment technique requirements are established in lieu of *maximum contaminant levels* (MCLs) for *Giardia*, viruses, heterotrophic plate count bacteria, *Legionella*, and turbidity. Treatment must achieve at least 99.9% removal (3-log removal) or inactivation of *G. lamblia* cysts and 99.9% removal or inactivation of viruses.

Qualified operators (as determined by the state) must operate all systems. To avoid filtration, waterworks must satisfy the following criteria:

1. Fecal coliform concentration must not exceed 20/100 mL or the total coliform concentration must not exceed 100/100 mL before disinfection

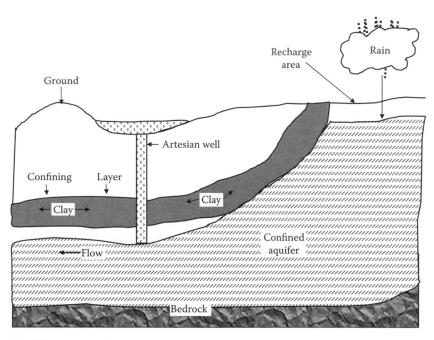

FIGURE 16.5 Confined aquifer.

in more than 10% of the measurements for the previous 6 months, calculated each month.

2. Turbidity levels must be measured every 4 hours by grab samples or continuous monitoring. The turbidity level may not exceed 5 NTU. If the turbidity exceeds 5 NTU, the water supply system must install filtration, unless the state determines that the event is unusual or unpredictable, and the event does not occur more than twice in any one year or five times in any consecutive ten years.

It is important, when considering the choice of a water source, that the source presents minimal risks of contamination from wastewaters and contains a minimum of impurities that may be hazardous to health. Acute (immediate) health effects such as those presented by exposure to *Giardia lamblia* and chronic (those occurring over time) health effects must be guarded against. Maximum contaminant levels must be monitored to ensure that the maximum permissible level of contaminant in water is not exceeded.

Note: Primary maximum contaminant levels (MCLs) are based on health considerations. Secondary MCLs are based on aesthetic considerations (taste, odor, and appearance).

A public water system must also provide water free of pathogens (disease-causing microorganisms, such as bacteria, protozoa, spores, and viruses). Chemical quality must also be monitored to ensure prevention of inorganic and organic contamination.

In 1996, the USEPA finalized the Stage 1 Disinfectants/ Disinfection Byproducts (D/DBP) and Interim Enhanced Surface Water Treatment rules and implemented them in 1998. These amendments tighten controls on DBPs and turbidity and regulate *Cryptosporidium*. Highlights of these changes follow.

16.7.1 STAGE 1 D/DBP RULE

1. The rule tightened the total trihalomethane standard to 0.080 mg/L.
2. The rule set new DBP standards for five haloacetic acids (0.060 mg/L), chlorite (1.0 mg/L), and bromate (0.010 mg/L).
3. The rule established new standards for disinfectant residuals (4.0 mg/L for chlorine, 4.0 mg/L for chloramines, and 0.8 mg/L for chlorine dioxide).
4. The rule requires systems using surface water or groundwater directly influenced by surface water to implement enhanced coagulation or softening to remove DBP precursors unless systems meet alternative criteria
5. The rule applies to all community and non-transient–noncommunity systems that disinfect, including those serving fewer than 10,000 people.
   - *MCLGs*—For maximum contaminant level goals (MCLGs), the USEPA opted to retain the chloroform MCLG at 0 instead of loosening it to 0.3 mg/L as set forth in the Spring 1998 Notice of Data Availability. The USEPA

also loosened the chlorite MCLG from 0.08 mg/L to 0.8 mg/L, loosened the maximum residual disinfectant level goal for chlorine dioxide from 0.3 mg/L to 0.8 mg/L, and set no MCLG for the DBP chloral hydrate (control of which will be covered by the other requirements).

- *Chloroform*—In dropping its plan to loosen the chloroform MCLG, USEPA has backed away (for now) from its first attempt to set a level higher than 0 MCLG for a carcinogenic contaminant, opting for more time to allow the issue to be discussed by stakeholders and the Science Advisory Board. USEPA outlined what it termed a "compelling" case for recognizing a safe (or threshold) exposure level for chloroform, one of the regulated trihalomethanes.

16.7.2 Interim Enhanced Surface Water Treatment (IESWT) Rule

This treatment optimization rule, which only applies to large (those serving more than 10,000 people) public water systems that use surface water or groundwater directly influenced by surface water is the first to directly regulate *Cryptosporidium* (crypto):

1. The rule sets a crypto MCLG of 0.
2. The rule requires systems that filter to remove 99% (2 log) of crypto oocysts.
3. The rule adds crypto control to watershed protection requirements for systems operating under filtration waivers.
4. The rule is particular to the genus *Cryptosporidium*, not the *Cryptosporidium parvum* species.
   - *Turbidity*—The rule requires continuous turbidity monitoring of individual filters and tightens allowable turbidity limits for combined filter effluent, cutting the maximum from 5 NTU to 1 NTU and the average monthly limit from 0.5 NTU to 0.3 NTU.
   - *Benchmarking*—Systems must determine within 15 months of promulgation whether or not they must establish a disinfection benchmark to ensure maintenance of microbial protection as systems comply with new DBP standards. Because the determination is based on whether the public water system exceeds annual average levels of trihalomethanes (THMs) or haloacetic acids, systems that lack such data must have begun collecting it within 3 months of promulgation to have a year's worth by the 15-month deadline.

The rule also requires states to conduct periodic sanitary surveys of all surface water systems regardless of size, and covers all new treated-water reservoirs.

16.7.3 Regulatory Deadlines

Large surface water systems (those serving over 10,000) had to comply with the Stage 1 D/DBP and IESWT rules by December 2001. Smaller surface water systems and all groundwater systems had until December 2003 to comply with the stage 1 D/DBP Rule.

16.8 PUBLIC WATER SYSTEM QUALITY REQUIREMENTS

Many factors affect the use of water, including climate, economic conditions, type of community (i.e., residential, commercial, industrial), integrity of the distribution system (waste pressure, leaks in the system), and water cost. In the United States, the typical per capita usage is approximately 150 gallons per day (gpd) per person. Each residential connection requires approximately 400 gpd per connection. Keep in mind that fire-fighting requirements at a standard fire flow of 500 gpm will use in 1 minute what a family of five normally uses in 24 hours. Water pressure delivered to each service connection should (at a minimum) reach 20 psi under all flow conditions.

16.9 WELL SYSTEMS

The most common method for withdrawing groundwater is to penetrate the aquifer with a vertical well, then pump the water up to the surface. In the past, when someone wanted a well, they simply dug (or hired someone to dig) and hoped (gambled) that they would find water in a quantity suitable for their needs. Today, in most locations in the United States, for example, developing a well supply usually involves a more complicated step-by-step process. Local, state, and federal requirements specify the actual requirements for development of a well supply in the United States. The standard sequence for developing a well supply generally involves a seven-step process:

1. *Application*—Depending on location, filling out and submitting an application (to the applicable authorities) to develop a well supply are usually a standard procedure.
2. *Well site approval*—Once the application has been made, local authorities check various local geological and other records to ensure that the siting of the proposed well coincides with mandated guidelines for approval.
3. *Well drilling*—The well is drilled.
4. *Preliminary engineering report*—After the well is drilled and the results documented, a

preliminary engineering report is made on the suitability of the site to serve as a water source. This procedure involves performing a pump test to determine if the well can supply the required amount of water. The well is generally pumped for at least 6 hours at a rate equal to or greater than the desired yield. A stabilized drawdown should be obtained at that rate and the original static level should be recovered within 24 hours after pumping stops. During this test period, samples are taken and tested for bacteriological and chemical quality.

5. *Submission of documents for review and approval*—The application and test results are submitted to a reviewing authority that determines if the well site meets approval criteria.
6. *Construction permit*—If the site is approved, a construction permit is issued.
7. *Operation permit*—When the well is ready for use, an operation permit is issued.

16.9.1 Well Site Requirements

To protect the groundwater source and provide high-quality safe water, the waterworks industry has developed standards and specifications for wells. The following listing includes industry standards and practices, as well as those items included in example State Department of Environmental Compliance regulations.

Note: Check with your local regulatory authorities to determine well site requirements.

1. Minimum well lot requirements:
   • 50 feet from well to all property lines
   • All-weather access road provided
   • Lot graded to divert surface runoff
   • Recorded well plat and dedication document
2. Minimum well location requirements:
   • At least 50 feet horizontal distance from any actual or potential sources of contamination involving sewage
   • At least 50 feet horizontal distance from any petroleum or chemical storage tank or pipeline or similar source of contamination, except where plastic-type well casing is used the separation distance must be at least 100 feet
3. Vulnerability assessment:
   • Is the wellhead area 1000 ft radius from the well?
   • What is the general land use of the area (residential, industrial, livestock, crops, undeveloped, other)?
   • What are the geologic conditions (sinkholes, surface, subsurface)?

16.9.2 Type of Wells

Water supply wells may be characterized as shallow or deep. In addition, wells are classified as follows:

• Class I, cased and grouted to 100 ft
• Class II A, cased to a minimum of 100 ft and grouted to 20 ft
• Class II B, cased and grouted to 50 ft

Note: During the well development process, mud/silt forced into the aquifer during the drilling process is removed, allowing the well to produce the best-quality water at the highest rate from the aquifer.

16.9.2.1 Shallow Wells

Shallow wells are those that are less than 100 ft deep. Such wells are not particularly desirable for municipal supplies because the aquifers they tap are likely to fluctuate considerably in depth, making the yield somewhat uncertain. Municipal wells in such aquifers cause a reduction in the water table (or phreatic surface) that affects nearby private wells, which are more likely to utilize shallow strata. Such interference with private wells may result in damage suits against the community. Shallow wells may be dug, bored, or driven:

• *Dug wells*—Dug wells are the oldest type of well and date back many centuries; they are dug by hand or by a variety of unspecialized equipment. They range in size from approximately 4 to 15 ft in diameter and are usually about 20 to 40 ft deep. Such wells are usually lined or cased with concrete or brick. Dug wells are prone to failure from drought or heavy pumpage. They are vulnerable to contamination and are not acceptable as a public water supply in many locations.
• *Driven wells*—Driven wells consist of a pipe casing terminating in a point slightly greater in diameter than the casing. The pointed well screen and the lengths of pipe attached to it are pounded down or driven in the same manner as a pile, usually with a drop hammer, to the water-bearing strata. Driven wells are usually 2 to 3 inches in diameter and are used only in unconsolidated materials. This type of shallow well is not acceptable as a public water supply.
• *Bored wells*—Bored wells range from 1 to 36 inches in diameter and are constructed in unconsolidated materials. The boring is accomplished with augers (either hand- or machine-driven) that fill with soil and then are drawn to the surface to be emptied. The casing

may be placed after the well is completed (in materials that are relatively cohesive) but must advance with the well in noncohesive strata. Bored wells are not acceptable as a public water supply.

16.9.2.2 Deep Wells

Deep wells are the usual source of groundwater for municipalities. Deep wells tap thick and extensive aquifers that are not subject to rapid fluctuations in water level (remember that the *piezometric surface* is the height to which water will rise in a tube penetrating a confined aquifer) and that provide a large and uniform yield. Deep wells typically yield water of more constant quality than shallow wells, although the quality is not necessarily better. Deep wells are constructed by a variety of techniques; we discuss two of these techniques below:

- *Jetted wells*—Jetted well construction commonly employs a jetting pipe with a cutting tool. This type of well cannot be constructed in clay or hardpan or where boulders are present. Jetted wells are not acceptable as a public water supply.
- *Drilled wells*—Drilled wells are usually the only type of well allowed for use in most public water supply systems. Several different methods of drilling are available, all of which are capable of drilling wells of extreme depth and diameter. Drilled wells are constructed using a drilling rig that creates a hole into which the casing is placed. Screens are installed at one or more levels when water-bearing formations are encountered.

16.9.3 COMPONENTS OF A WELL

The components that make up a well system include the well itself, the building and the pump, and related piping system. In this section, we focus on the components that make up the well itself. Many of these components are shown in Figure 16.6.

16.9.3.1 Well Casing

A well is a hole in the ground called the *borehole*. To prevent collapse, a casing is placed inside the borehole. The well casing prevents the walls of the hole from collapsing and prevents contaminants (either surface or subsurface) from entering the water source. The casing also provides a column of stored water and housing for the pump mechanisms and pipes. Well casings constructed of steel or plastic material are acceptable. The well casing must extend a minimum of 12 inches above grade.

16.9.3.2 Grout

To protect the aquifer from contamination, the casing is sealed with grout to the borehole near the surface and near the bottom where it passes into the impermeable layer. This sealing process keeps the well from being polluted by surface water and seals out water from water-bearing strata that have undesirable water quality. Sealing also protects the casing from external corrosion and restrains unstable soil and rock formations. Grout consists of cement that is pumped into the annular space (it is completed within 48 hours of well construction); it is pumped under continuous pressure starting at the bottom and progressing upward in one continuous operation.

16.9.3.3 Well Pad

The well pad provides a ground seal around the casing. The pad is constructed of reinforced concrete 6 ft by 6 ft (6 inches thick) with the well head located in the middle. The well pad prevents contaminants from collecting around the well and seeping down into the ground along the casing.

16.9.3.4 Sanitary Seal

To prevent contamination of the well, a sanitary seal is placed at the top of the casing. The type of seal varies depending on the type of pump used. The sanitary seal contains openings for power and control wires, pump support cables, a drawdown gauge, discharge piping, pump shaft, and air vent, while providing a tight seal around them.

16.9.3.5 Well Screen

Screens can be installed at the intake points on the end of a well casing or on the end of the inner casing on gravel packed well. These screens perform two functions: (1) supporting the borehole, and (2) reducing the amount of sand that enters the casing and the pump. They are sized to allow the maximum amount of water while preventing the passage of sand, sediment, and gravel.

16.9.3.6 Casing Vent

The well casing must have a vent to allow air into the casing as the water level drops. The vent terminates 18 inches above the floor with a return bend pointing downward. The opening of the vent must be screened with No. 24 mesh stainless steel to prevent entry of vermin and dust.

16.9.3.7 Drop Pipe

The drop pipe or riser is the line leading from the pump to the well head. It provides adequate support so an aboveground pump does not move and so a submersible pump is not lost down the well. This pipe is either steel or polyvinyl chloride (PVC). Steel is the most desirable.

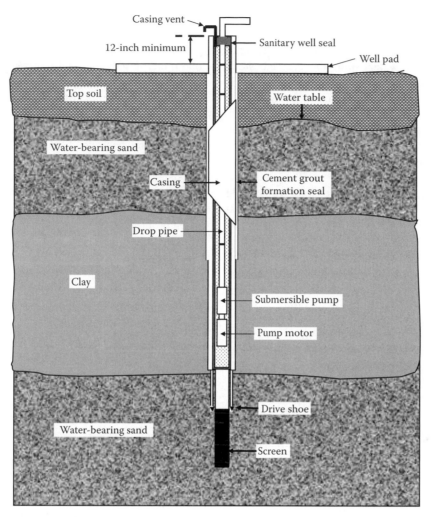

FIGURE 16.6 Components of a well.

16.9.3.8 Miscellaneous Well Components

Miscellaneous well components include:

- *Gauge and air line* are used to measure the water level of the well.
- *Check valve* is located immediately after the well to prevent system water from returning to the well. It must be located above ground and protected from freezing.
- *Flowmeter* is required to monitor the total amount of water withdrawn from the well, including any water blown off.
- *Control switches* control well pump operation.
- *Blow-off* valves are located between the well and storage tank and are used to flush the well of either sediment or turbid or superchlorinated water.
- *Sample taps* include (1) raw water sample taps, which are located before any storage or treatment to permit sampling of the water directly

from the well, and (2) entry point sample taps, located after treatment.
- *Control valves* isolate the well for testing or maintenance or are used to control water flow.

16.9.4 WELL EVALUATION

After a well is developed, conducting a pump test determines if it can supply the required amount of water. The well is generally pumped for at least 6 hours (many states require a 48-hour yield and drawdown test) at a rate equal to or greater than the desired yield. *Yield* is the volume or quantity of water discharged from a well per unit of time (e.g., gpm, ft^3/sec). Regulations usually require that a well produce a minimum of 0.5 gpm per residential connection. *Drawdown* is the difference between the static water level (level of the water in the well when it has not been used for some time and has stabilized) and the pumping water level in a well. Drawdown is measured by using an air line and pressure gauge to monitor the water level during the 48 hours of pumping.

The procedure calls for the air line to be suspended inside the casing down into the water. At the other end are the pressure gauge and a small pump. Air is pumped into the line (displacing the water) until the pressure stops increasing. The highest pressure reading on the gauge is recorded. During the 48 hours of pumping, the yield and drawdown are monitored more frequently during the beginning of the testing period, because the most dramatic changes in flow and water level usually occur then. The original static level should be recovered within 24 hours after pumping stops.

Testing is accomplished on a bacteriological sample for analysis by the most probable number (MPN) method every half hour during the last 10 hours of testing. The results are used to determine if chlorination is required or if chlorination alone will be sufficient to treat the water. Chemical, physical, and radiological samples are collected for analyses at the end of the test period to determine if treatment other than chlorination may be required.

Note: Recovery from the well should be monitored at the same frequency as during the yield and drawdown testing and for at least the first 8 hours, or until 90% of the observed drawdown is obtained.

Specific capacity (often referred to as the *productivity index*) is a test method for determining the relative adequacy of a well; over a period of time, it is a valuable tool for evaluating well production. Specific capacity is expressed as a measure of well yield per unit of drawdown (yield divided by drawdown). When conducting this test, if possible always run the pump for the same length of time and at the same pump rate.

16.9.5 Well Pumps

Pumps are used to move the water out of the well and deliver it to the storage tank or distribution system. The type of pump chosen should provide optimum performance based on the location and operating conditions, required capacity, and total head. Two types of pumps commonly installed in groundwater systems are *lineshaft turbines* and *submersible turbines*. Whichever type of pump is used, they are rated on the basis of pumping capacity expressed in gpm (e.g., 40 gpm), not on horsepower.

16.9.6 Routine Operation and Recordkeeping Requirements

Ensuring the proper operation of a well requires close monitoring; wells should be visited regularly. During routine monitoring visits, check for any unusual sounds in the pump, line, or valves and for any leaks. In addition, as a routine, cycle valves to ensure good working condition. Check motors to make sure they are not overheating. Check the well pump to guard against short cycling. Col-lect a water sample for a visual check for sediment. Also, check chlorine residual and treatment equipment. Measure gallons on the installed meter for one minute to obtain the pump rate in gallons per minute (look for gradual trends or big changes). Check water level in the well at least monthly (perhaps more often in summer or during periods of low rainfall). Finally, from recorded meter readings, determine gallons used and compare with water consumed to determine possible distribution system leaks. Along with meter readings, other records must be accurately and consistently maintained for water supply wells. Such recordkeeping is absolutely imperative. The records (an important resource for troubleshooting) can be useful when problems develop or helpful in identifying potential problems.

The *well log* provides documentation of what materials were found in the borehole and at what depth. The well log includes the depths at which water was found, the casing length and type, the depth at which what type of soils were found, testing procedures, well development techniques, and well production. In general, the following items should be included in the well log:

1. Well location
2. Who drilled the well
3. When the well was completed
4. Well class
5. Total depth to bedrock
6. Hole and casing size
7. Casing material and thickness
8. Screen size and locations
9. Grout depth and type
10. Yield and drawdown (test results)
11. Pump information (type, horsepower, capacity, intake depth, and model number)
12. Geology of the hole
13. A record of yield and drawdown data

Pump data that should be collected and maintained include:

1. Pump brand and model number
2. Rate capacity
3. Date of installation
4. Maintenance performed
5. Date replaced
6. Pressure reading or water level when the pump is set to cut on and off
7. Pumping time (hours per day the pump is running)
8. Output in gallons per minute

A record of water quality should also be maintained, including bacteriological, chemical, physical (inorganics, metals, nitrate/nitrite, VOCs), and radiological reports.

System-specific monthly operation reports should contain information and data from meter readings (total gallons per day and month), chlorine residuals, amount and type of chemicals used, turbidity readings, physical parameters (pH, temperature), pumping rate, total population served, and total number of connections.

A record of water level (static and dynamic levels) should be maintained, in addition to a record of any changes in conditions (such as heavy rainfall, high consumption, leaks, and earthquakes) and a record of specific capacity.

16.9.7 Well Maintenance

Wells do not have an infinite life, and their output is likely to reduce with time as a result of hydrological or mechanical factors. Protecting the well from possible contamination is an important consideration. Locating the well properly (based on knowledge of the local geological conditions and a vulnerability assessment of the area) can minimize potential problems.

During the initial assessment, ensuring that the well is not located in a sinkhole area is important. Locations where unconsolidated or bedrock aquifers could be subject to contamination must be identified. Several other important determinations must also be made: Is the well located on a floodplain? It is located next to a drainfield for septic systems or near a landfill? Are petroleum or gasoline storage tanks nearby? Is any pesticide or plastics manufacturing conducted near the well site?

Along with proper well location, proper well design and construction prevent wells from acting as conduits for the vertical migration of contaminants into the groundwater. Basically, the pollution potential of a well equals how well it was constructed. Contamination can occur during the drilling process, and an unsealed or unfinished well is an avenue for contamination. Any opening in the sanitary seal or break in the casing may cause contamination, as can a reversal of water flow. In routine well maintenance operations, corroded casing or screens are sometimes withdrawn and replaced, but this is difficult and not always successful. Simply constructing a new well may be less expensive.

16.9.7.1 Troubleshooting Well Problems

During operation, various problems may develop; for example, the well may pump sand or mud. When this occurs, the well screen may have collapsed or corroded, causing the slot openings of the screen to become enlarged (allowing debris, sand, and mud to enter). If the well screen is not the problem, the pumping rate should be checked, as it may be too high. In the following, we provide a few other well problems, their probable causes, and the remediation required:

- If the water is white, the pump might be sucking air; reduce the pump rate.
- If water rushes backwards when the pump shuts off, check the valve, as it may be leaking.
- If the well yield has decreased, check the static water level. A downward trend in static water level suggests that the aquifer is becoming depleted, which could be the result of the following:
  - Local overdraft (well spacings are too close)
  - General overdraft (the pumpage exceeds the recharge)
  - Temporary decrease in recharge (dry cycles)
  - Permanent decrease in recharge (less flow in rivers)
  - Decreased specific capacity (if it has dropped 10 to 15%, determine the cause; it may be a result of incrustation)

Note: *Incrustation* occurs when clogging, cementation, or stoppage of a well screen and water-bearing formation occurs. Incrustations on screens and adjacent aquifer materials result from chemical or biological reactions at the air–water interface in the well. The chief encrusting agent is calcium carbonate, which cements the gravel and sand grains together. Incrustation could also be a result of carbonates of magnesium, clays and silts, or iron bacteria. Treatment involves pulling the screen and removing incrusted material, replacing the screen, or treating the screen and water-bearing formation with acids. If severe, treatment may involve rehabilitating the well.

- Pump rate is dropping, but the water level is not—probable cause is pump impairment
- Worn impellers
- Change in hydraulic head against which the pump is working (head may change as a result of corrosion in the pipelines, higher pressure setting, or newly elevated tank)

16.9.8 Well Abandonment

In the past, common practice was simply to walk away and forget about a well when it ran dry. Today, while dry or failing wells are still abandoned, we know that they must be abandoned with care (and not completely forgotten). An abandoned well can become a convenient (and dangerous) receptacle for wastes, thus contaminating the aquifer. An improperly abandoned well could also become a haven for vermin or, worse, a hazard for children. A temporarily abandoned well must be sealed with a water-tight cap or wellhead seal. The well must be maintained so it does not become a source or channel of contamination during temporary abandonment.

When a well is permanently abandoned, all casing and screen materials may be salvaged. The well should be checked from top to bottom to ensure that no obstructions interfere with plugging and sealing operations. Prior to plugging, the well should be thoroughly chlorinated. Bored wells should be completely filled with cement grout. If the well was constructed in an unconsolidated formation, it should be completely filled with cement grout or clay slurry introduced through a pipe that initially extends to the bottom of the well. As the pipe is raised, it should remain submerged in the top layers of grout as the well is filled.

Wells constructed in consolidated rock or that penetrate zones of consolidated rock can be filled with sand or gravel opposite zones of consolidated rock. The sand or gravel fill is terminated 5 feet below the top of the consolidated rock. The remainder of the well is filled with sand–cement grout.

CHAPTER REVIEW QUESTIONS

16.1   When water is withdrawn from a well, a _____ will develop.

16.2   How far should the well casing extend above the ground or well-house floor?

16.3   A well casing should be grouted for at least 10 ft, with the first 20 ft grouted with _____.

16.4   List three sources of drinking water.

16.5   Explain GWUDI.

16.6   What are two advantages of surface water sources?

16.7   Define hydrology.

16.8   The area inside the cone of depression is called the _____.

16.9   A spring is an example of what type of water source?

16.10 Describe the function of the bar screen at a surface water intake.

REFERENCES AND SUGGESTED READING

USEPA. 2007. *Public Drinking Water Systems: Facts and Figures*. Washington, D.C.: U.S. Environmental Protection Agency (http://www.epa.gov/safewater/pws/factoid.html).

17 Watershed Protection

Watershed protection is one of the barriers in the multiple-barrier approach to protecting source water. In fact, watershed protection is the primary barrier, the first line of defense against contamination of drinking water at its source. Ideally, under the general concept of "quality in, quality out," a protected watershed ensures that surface runoff and inflow to the source waters occur within a pristine environment.

17.1 INTRODUCTION

Water regulates population growth, influences world health and living conditions, and determines biodiversity. For thousands of years, people have tried to control the flow and quality of water. Water provided resources and a means of transportation for development in some areas. Even today, the presence or absence of water is critical in determining how we can use land. Yet, despite this long experience in water use and water management, humans often fail to manage water well. Sound water management was pushed aside in favor of rapid, never-ending economic development in many countries. Often, optimism about the applications of technology (e.g., dam building, wastewater treatment, irrigation measures) exceeded concerns for, or even interest in, environmental shortcomings. Pollution was viewed as the inevitable consequence of development, the price that must be paid to achieve economic progress.

Clearly, we now have reached the stage of our development when the need for management of water systems is apparent, beneficial, and absolutely imperative. Land use and activities in the watershed directly impact raw water quality. Effective watershed management improves raw water quality, controls treatment costs, and provides additional health safeguards. Depending on the goals, watershed management can be simple or complex.

This chapter discusses the need for watershed management on a multiple-barrier basis and provides a brief overview of the range of techniques and approaches that can be used to investigate the biophysical, social, and economic forces affecting water and its use. Water utility directors are charged with providing potable water in a quantity and quality to meet the public's demand. They are also charged with providing effective management of the entire water supply system, and such management responsibility includes proper management of the relevant watershed.

Key Point: Integrated water management means putting all of the pieces together, including the social, environmental, and technical aspects.

17.2 CURRENT ISSUES IN WATER MANAGEMENT

(Note that much of the information provided in this section is adapted from Viessman, 1991.) Remarkable consensus exists among worldwide experts over the current issues confronted by waterworks managers and others. These issues include the following:

- *Water availability, requirements, and use*
  Protection of aquatic and wetland habitat
  Management of extreme events (e.g., droughts, floods)
  Excessive extractions from surface and groundwater
  Global climate change
  Safe drinking water supply
  Waterborne commerce
- *Water quality*
  Coastal and ocean water quality
  Lake and reservoir protection and restoration
  Water quality protection, including effective enforcement of legislation
  Management of point and nonpoint source pollution
  Impacts on land, water, and air relationships
  Health risks
- *Water management and institutions*
  Coordination and consistency
  Capturing a regional perspective
  Respective roles of federal and state or provincial agencies
  Respective roles of projects and programs
  Economic development philosophy that should guide planning
  Financing and cost sharing
  Information and education
  Appropriate levels of regulation and deregulation
  Water rights and permits
  Infrastructure
  Population growth

TABLE 17.1
How Land Use Impacts Water Quality

	Sediment	Nutrients	Viruses and Bacteria	Trihalomethane (THM)	Fe, Mn
Urban	✓	✓	✓	✓	✓
Agriculture	✓	✓	✓	✓	✓
Logging	✓	✓	✓	✓	
Industrial	✓	✓	✓	✓	
Septic tanks	✓	✓	✓		
Construction	✓	✓			

- *Water resources planning, including:*
  Consideration of the watershed as an integrated system

  Planning as a foundation for, not a reaction to, decision making

  Establishment of dynamic planning processes incorporating periodic review and redirection

  Sustainability of projects beyond construction and early operation

  A more interactive interface between planners and the public

  Identification of sources of conflict as an integral part of planning

  Fairness, equity, and reciprocity between affected parties

17.3 WHAT IS A WATERSHED?

At the simplest level, we all live in a watershed—the area that drains to a common waterway, such as a lake, estuary, wetland, river or stream, or even an ocean—and our individual actions can directly affect it. More specifically, a watershed is a protected, reserved area, usually distant from the treatment plant, where natural or artificial lakes are used for water storage, natural sedimentation, and seasonal pretreatment, with or without disinfection. A watershed is also defined as a collecting area into which water drains. The area of land encompassed can be tiny or immense. The size of a watershed and the direction of flow of its rivers are determined by landforms. Watersheds are associated with surface water (usually fed by gravity) to distinguish them from groundwater (usually fed by pumping).

Note: The USEPA's watershed approach is an "integrated, holistic strategy for more effectively restoring and protecting aquatic ecosystems and protecting human health (e.g., drinking water supplies and fish consumption)."

17.3.1 WATER QUALITY IMPACT

Generally, in a typical river system, water quality is impacted by about 60% nonpoint pollution, 21% municipal discharge, 18% industrial discharge, and 1% sewer overflows. Of the nonpoint pollution, about 67% is from agriculture, 18% is urban, and 15% comes from other sources. Land use directly impacts water quality. The impact of land use on water quality is clearly evident in Table 17.1. From the point of view of waterworks operators, water quality issues for nutrient contamination can be summarized quite simply:

Nutrients + algae = Taste and odor problems

Nutrients + algae + macrophytes + decay = Trihalomethane (THM) precursors

17.4 WATERSHED PROTECTION AND REGULATIONS

The Clean Water Act (CWA) and Safe Drinking Water Act (SDWA) address source water protection. Implementation of regulatory compliance requirements (with guidance provided by the U.S. Department of Health) is left up to state and local health department officials to implement. Water protection regulations in force today provide not only guidance and regulation for watershed protection but also additional options for those tasked with managing drinking water utilities.

The typical drinking water utility that provides safe drinking water to the consumer has two choices in water pollution control: "Keep it out or take it out." The "keep it out" part pertains to effective watershed management. In contrast, contaminants that have found their way into the water supply must be removed by treatment, which is the "take it out" part. Obviously, utility directors and waterworks managers are concerned with controlling treatment costs. An effective watershed management program can reduce treatment costs by reducing source water contamination. The "take it out" option is much more expensive and time consuming than keeping it out in the first place. Proper watershed management also works to maintain consumer confidence. If the consumer is aware that the water source from the area's watershed is of the highest quality, then, logically, confidence in the quality of the water is high. High-quality water also works directly to reduce public health risks.

17.5 A WATERSHED PROTECTION PLAN

Watershed protection begins with planning. The watershed protection plan consists of several elements, which include the need to:

- Inventory and characterize water sources.
- Identify pollutant sources.
- Assess vulnerability of intake.
- Establish program goals.
- Develop protection strategies.
- Implement the program.
- Monitor and evaluate program effectiveness.

17.6 RESERVOIR MANAGEMENT PRACTICES

To ensure an adequate and safe supply of drinking water for a municipality, watershed management must utilize proper reservoir management practices. These practices include proper lake aeration, harvesting, dredging, and use of algaecide. Water quality improvements from lake aeration include reduced iron, manganese, phosphorus, ammonia, and sulfide content. Lake aeration also reduces the costs of capital and operation for water supply treatment. Algaecide treatment controls algae which in turn reduces taste and odor problems. The drawback of using algaecides is that they are successful for only a brief period.

17.7 WATERSHED MANAGEMENT PRACTICES

Watershed management practices include land acquisitions, land use controls, and best management practices (BMPs). *Land acquisition* refers to the purchase of watershed lands—those land areas that form the watershed for a particular locality. The advantage of having ownership of lands included within a particular watershed are obvious in that the owner (in this case, the local utility) has better control of land use and thus can put protective measures in place to ensure a quality water supply.

Land use controls (those measures deemed necessary to protect the watershed from contamination or destruction) vary from location to location; for example, land use controls may be designed to prohibit mining or other industrial activities from taking place within the watershed, for protection of the water supply. *Best management practices* for watershed management refer specifically to agriculture, logging, urban, and construction practices. The chief problem with best management practices is that they are nonstructural measures. They are often difficult to implement because they require that people must change the way they behave.

In agricultural systems, BMPs may include measures such as conservation tillage and contour plowing; confined animal facility management, such as containing or using waste on-site and keeping animals out of waterways; and appropriate pesticide and herbicide application practices, such as minimizing their use or using alternative chemicals.

Examples of logging BMPs include construction of streamside buffer zones to protect the watercourse. Logging plans should also incorporate water quality and habitat planning.

Urban BMPs revolve around targeted categories such as reduction of impervious areas (reducing tarmac, asphalt coverings, and cement coverings to allow for precipitation infiltration), control of non-stormwater discharges, and proper disposal of residential chemicals. The primary types of BMPs include public education programs, inspections and enforcement, structural controls (e.g., end-of-pipe solutions that seek to treat or remove pollution that has already occurred), and preventive options that are implemented to prevent or reduce the creation of waste within a process.

Examples of construction BMPs include enforcement of stormwater pollution plans and inspections. Types of construction BMPs include erosion and sediment control (e.g., minimize clearing, stage construction, and stabilize stockpiles and finished areas) and chemical control (e.g., proper storage, handling, application, covering, and isolation of materials).

17.8 EIGHT TOOLS OF WATERSHED PROTECTION

Because many water resource managers now follow the guidelines presented in the multiple-barrier model currently used to protect water resource areas, this section discusses the USEPA watershed protection approach—a critical element of the multiple-barrier model. The USEPA approach applies eight tools to protect or restore aquatic resources in an urbanized or developing watershed. The USEPA document (USEPA, 2007) describes the nature and purpose of the eight watershed protection tools, outlines some specific techniques for applying the tools, and highlights some key choices a watershed manager should consider when applying or adapting the tools within a given watershed. The eight tools are:

- Tool 1—Land Use Planning
- Tool 2—Land Conservation
- Tool 3—Aquatic Buffers
- Tool 4—Better Site Design
- Tool 5—Erosion and Sediment Control
- Tool 6—Stormwater Best Management Practices
- Tool 7—Non-Stormwater Discharges
- Tool 8—Watershed Stewardship Programs

17.8.1 Tool 1—Land Use Planning

Because impervious cover has such a strong influence on watershed quality, a watershed manager must critically analyze the degree and location of future development (and impervious cover) that is expected in a watershed; consequently, land use planning ranks as perhaps the single most important watershed protection tool. When preparing a watershed plan, a watershed manager must:

- Predict the impacts of future land use change on water resources.
- Obtain consensus on the most important water resource goals in the watershed.
- Develop a future land use plan that can help meet these goals.
- Select the most acceptable and effective land use planning technique to reduce or shift future impervious cover.
- Select the most appropriate combination of other watershed protection tools to apply to individual subwatersheds.
- Devise an ongoing management structure to adopt and implement the watershed plan.

Land use planning is best conducted at the subwatershed scale, where it is recognized that stream quality is related to land use and consequently impervious cover. One of the goals of land use planning is to shift development toward subwatersheds that can support a particular type of land use or density. The basic goal of the watershed plan is to apply land use planning techniques to redirect development, preserve sensitive areas, and maintain or reduce the impervious cover within a given watershed.

Understanding the concept of impervious cover is essential to understanding one of the biggest ways that urbanization impacts our streams. Impervious cover is defined as the sum total of all hard surfaces within a watershed, including rooftops, parking lots, streets, sidewalks, driveways, and surfaces that are impermeable to infiltration of rainfall into underlying soils and groundwater. Impervious cover changes the natural landscape and is a major influence on aquatic resources because instead of allowing precipitation to permeate the ground, it runs off. Generally the more impervious surfaces in a watershed, the smaller the amount of groundwater recharge, in addition to greater runoff and related erosion in streambeds from greater flow. Further, studies show a direct correlation between the percentage of impervious cover in a watershed and the level of degradation to aquatic organisms. Streams degraded by high percentages of impervious surfaces in their watersheds are often prone to larger and more frequent floods (which cause property damage as well as ecological harm) and lower base flows (which degrade or eliminate fish and other stream life, in addition to reducing the aesthetics of the stream). Impervious surfaces also raise the temperature of runoff, which reduces dissolved oxygen in the stream, harms some gamefish populations, and promotes excess algal growth.

Classification of subwatersheds by the amount of impervious cover is one step toward determining land use planning goals. Although the presence of vegetated streamside buffer zones or wetlands can help counteract impervious cover impacts, a watershed exceeding 10% impervious cover will generally not be able to support a high-quality stream system. In this particular classification system, subwatersheds with impervious cover of less than 10% are classified as sensitive. A subwatershed with 10 to 25% impervious cover is classified as a degraded or impacted system. Any stream with a watershed having greater than 25% impervious cover is classified as a nonsupporting stream and is characterized by eroding banks, poor biological diversity, and high bacterial levels.

17.8.2 Tool 2—Land Conservation

Whereas the first tool emphasizes how much impervious cover is created in a watershed, the second tool focuses on land conservation. Five types of land may have to be conserved in a subwatershed:

1. *Critical habitats*—The essential spaces for plant and animal communities or populations. Examples include tidal wetlands, freshwater wetlands, large forest clumps, springs, spawning areas in streams, habitat for rare or endangered species, potential restoration areas, native vegetation areas, and coves.
2. *Aquatic corridors*—The area where land and water meet. This can include floodplains, stream channels, springs and seeps, small estuarine covers, littoral areas, stream crossings, shorelines, riparian forest, caves, and sinkholes.
3. *Hydrologic reserve areas*—Any undeveloped areas responsible for maintaining the predevelopment hydrologic response of a watershed. The three most common land uses are crops, forest, and pasture. From a hydrologic standpoint, forest is the most desirable land use, followed by pasture, then crops.
4. *Water pollution hazard*—Any land use or activity that is expected to create a relatively high risk of potential water pollution. Examples of water pollution hazards may include septic systems, landfills, hazardous waste generators, above- or below-ground tanks, land application sites, impervious cover, stormwater hotspots, and road and salt storage areas. One way to avoid possible contamination to waterways is to locate such facilities at a designated distance

away from the water body to decrease the chance of contamination.

5. *Cultural areas*—Areas that provide a sense of place in the landscape and are important habitats for people. Examples include historic or archeologic sites, trails, parkland, scenic views, water access, bridges, and recreational areas.

A watershed manager must choose which of these natural and cultural areas to conserve in a subwatershed to sustain the integrity of its aquatic and terrestrial ecosystems and maintain the desired human uses of its waters. Four out of five of these are clearly important because they are key parts of healthy watersheds that should be conserved; water pollution hazards, however, seem out of place on the list. Nevertheless, such areas are intentionally conserved at locations carefully selected to minimize their negative impacts on important water resources.

Although land conservation is most important in sensitive watersheds, it is also a critical tool for other types of resources. Each subwatershed should have its own land conservation strategy based on its management category, inventory of conservation areas, and land ownership patterns. The five conservation areas are not always differentiable. Some of the natural areas may overlap among the conservation areas; for example, a freshwater wetland may serve as a critical habitat, be part of the aquatic corridor, and also comprise part of the hydrologic reserve areas. Most of the critical areas, however, are covered in at least one of these five categories.

17.8.3 Tool 3—Aquatic Buffers

The aquatic corridor, where land and water meet, deserves special protection in the form of buffers. A buffer can be placed along a stream or shoreline or around a natural wetland. A buffer has many uses and benefits. Its primary use is to physically protect and separate a stream, lake, or wetland from a wide variety of water pollutants and habitat impacts that can arise from too much land use too close to the water. For streams, a network of buffers acts as a right-of-way during floods and sustains the integrity of stream ecosystems and habitats. Technically, buffers are one type of land conservation area, but they are especially important because they can:

- Regulate light and temperature conditions, improving the habitat for aquatic plants and animals.
- Be effective in removing sediment, nutrients, and bacteria from stormwater and groundwater.
- Help to stabilize and protect the stream banks.

The basic structure of a stream buffer in an urban setting is broken up into three zones which differ in functions, width, vegetative target, and allowed uses. In the eastern and northwestern United States the streamside zone is often maintained as mature forest, with strict limitations on all other uses (in some arid desert or grassland regions of the country, a forested buffer may not be achievable). The streamside zone also produces the shade and woody debris that are so important to stream quality and biota. The middle zone is typically a forested area 50 to 100 feet wide that is managed to allow some clearing. The outer zone, usually about 25 feet wide, is ideally forest but also can include turf. The three-zone buffer is variable in width and should be increased to allow for protection of special areas such as wetlands and the floodplain.

Once established, the boundaries of buffers must be clearly marked and enforced. A buffer that is well planned, designed, and maintained can help maximize its many potential benefits. Buffers are important because they make up an integral part of the watershed protection strategy and complement other programs and efforts to protect water quality.

17.8.4 Tool 4—Better Site Design

Urban development is often characterized by large amounts of impervious cover. Wide streets, huge cul de sacs, long driveways, and sidewalks lining both sides of the street are all features of site design that create impervious cover and consume natural areas. Many innovative site planning techniques have been shown to sharply reduce the impact of new development. Developers, however, are often unable to apply these techniques in many communities because of outdated local zoning, parking, or subdivision codes. Three categories of better site design that have special merit for watershed protection are:

1. *Residential street and parking lots*—Approximately 65% of total impervious cover in the landscape is habitat for cars in the form of parking lots, roads, and driveways. Much of this impervious cover is often unnecessary and can be minimized at every stage of parking lot and residential street planning and design.
2. *Lot development*—Habitat for people (the area used in house lots) is a necessary part of any community, but many lot development practices result in excessive impervious cover and clearing of natural areas. These impacts can be reduced by changing the shape, orientation, and layout of residential lots. One popular technique, open space design development, minimizes lot sizes within a compact developed portion of a property while leaving the remaining portion open.
3. *Conservation of natural areas*—The goals are to protect all bodies of water and existing vegetation and minimize clearing. These techniques

include the establishment and maintenance of buffers, tree conservation, and providing conservation incentives.

17.8.5 Tool 5—Erosion and Sediment Control

Perhaps the most destructive stage of the development cycle is the relatively short period when vegetation is cleared and a site is graded to create a buildable landscape. The potential impacts to receiving waters are particularity severe at this stage. Trees and topsoil are removed, soils are exposed to erosion, natural topography and drainage patterns are altered, and sensitive areas are often disturbed. Thus, the fifth watershed protection tool seeks to reduce sediment loss during construction and to ensure that conservation areas, buffers, and forests are not cleared or otherwise disturbed during construction. Every community should have an effective erosion and sediment control program to reduce the potentially severe impacts generated by the construction process. Of the numerous techniques available to provide erosion and sediment control, the most effective is to minimize clearing. Some examples include regulations that require exposed soil to be stabilized within 7 to 10 days. Other common methods of erosion and sediment control include sedimentation basins and silt fences. Unfortunately, without proper installation and maintenance, erosion and sediment controls are ineffective.

17.8.6 Tool 6—Stormwater Management Practices

A watershed manager needs to make careful choices about what stormwater management practices should be installed in the subwatershed to compensate for the hydrological changes caused by new and exiting development. Stormwater management practices are used to delay, capture, store, treat, or infiltrate stormwater runoff. A key choice is to determine the primary stormwater objectives for a subwatershed that will govern the selection, design, and location of stormwater management practices at individual sites. Although specific design objectives for stormwater management practices are often unique to each subwatershed, the general goals for stormwater management practices are often the same and include:

- Maintain groundwater quality and recharge.
- Reduce stormwater pollutants loads.
- Protect stream channels.
- Prevent increased overbank flooding.
- Safely convey extreme floods.

Numerous structural stormwater management techniques are available for controlling stormwater quantity and quality. These five practices can be categorized into five broad groups:

- Ponds
- Wetlands
- Infiltration
- Filtering systems
- Grassed channels

Many advances have been made recently in innovative stormwater management designs, but their ability to maintain resource quality in the absence of other watershed protection tools is limited. In fact, stormwater management practices designed or located improperly can sometimes cause more secondary environmental impacts than if they were not installed at all.

17.8.7 Tool 7—Non-Stormwater Discharges

This tool is concerned with how wastewater and other non-stormwater flows are treated and discharged in a watershed. Key program elements consist of inspecting private septic systems, repairing or replacing failing systems, utilizing more advanced on-site septic controls, identifying and eliminating illicit connections from municipal stormwater systems, and spill prevention. Most non-stormwater discharges are strictly governed under the National Pollutant Discharge Elimination System (NPDES) and require a state or federal permit. The three basic categories of non-stormwater discharges are septic systems, sanitary sewers, and other miscellaneous non-stormwater discharges:

1. *Septic systems*, or on-site sewage disposal systems, are used to treat and discharge wastewater from toilets, wash basins, bathtubs, washing machines, and other water consumptive items that can be sources of high pollutant loads. One out of four homes in the country uses a septic system, collectively discharging a trillion gallons of wastewater annually (NSFC, 2007). Because of their widespread use and high-volume discharges, septic systems have the potential to pollute groundwater, lakes, and streams if they are located improperly or if they fail. Even properly functioning septic systems can be substantial sources of nutrient loads in some settings. Unlike other non-stormwater discharges, septic systems are not regulated under NPDES but are approved by local and state health agencies. It is estimated that as many as 20 to 25% of septic systems nationwide are not operating as designed and are failing.
2. *Sanitary sewers* collect wastewater in a central sewer pipe and send it to a municipal treatment plant. Ideally, this allows more efficient collection of wastewater and often results in higher levels of pollutant reduction. The extension of

sanitary sewer lines is not without some risk, however, as it has the potential to induce more development than may have been possible in a watershed previously served only by on-site sewage disposal systems (particularly when soils are limiting). Most communities cannot refuse service to new developments within the water and sewer envelope, so the decision to extend lines into undeveloped areas will allow future developers to tap into the line. In addition, not all sanitary sewer conveyance and treatment systems are capable of achieving high levels of pollutant reduction. Examples include:

- Package treatment plants
- Combined sewer overflows
- Sanitary sewer overflows
- Illicit or illegal connections to the storm drain network

3. Wastewater is not the only non-stormwater discharge possible in a watershed. A planner should also investigate whether *other non-stormwater discharges* are a factor in the subwatershed. Examples include:

- Industrial NPDES discharges
- Urban return flows (discharges caused by activities such as car washing and watering lawns)
- Water diversions
- Runoff from confined animal feeding lots

17.8.8 Tool 8—Watershed Stewardship Programs

When a subwatershed has been developed, communities still need to invest in ongoing watershed stewardship. The goals of watershed stewardship are to increase public awareness about watershed management efforts and to get local landowners' participation in the process to ensure stewardship on their own property and homes. The six basic programs that watershed managers should consider to promote greater watershed stewardship are discussed below:

1. *Watershed advocacy*—Advocacy of watershed protection is important because it can lay the foundations for public support and greater watershed stewardship. One of the most important investments that can be made in a watershed is to seed and support a watered management structure to carry out the long-term stewardship unction. Often, grassroots watershed management organizations are uniquely prepared to handle many critical stewardship programs, given their watershed focus, volunteers, low cost, and ability to reach into communities.

Watershed organizations can be effective advocates for better land management and can develop broad popular support and involvement of watershed protection. Local governments often have an important role to play in watershed advocacy; in many watersheds, local governments create or direct the watershed management structure.

2. *Watershed education*—A basic premise of watershed stewardship is that we must learn two things: that we all live in a watershed and that we understand how to live within it. The design of watershed education programs that create this awareness is of fundamental importance; some of the most effective programs merge leaning, enjoyment of outdoor recreation, and development of personal involvement in watershed stewardship. Four types of watershed education programs are:

- *Watershed awareness*—Raising basic watershed awareness using signs, storm-drain stenciling, streamwalk maps
- *Personal stewardship*—Educating residents about the individual roles they play in the watershed and communicating specific messages about positive and negative behaviors
- *Professional training*—Educating the development community on how to apply the tools of watershed protection
- *Watershed engagement*—Providing opportunities for the public to actively engage in watershed protection and restoration

3. *Watershed maintenance*—Most watershed protection tools require maintenance if they are to function properly over the long run. Some of the most critical watershed maintenance functions include management of conservation areas and buffer networks, as well as maintenance of stormwater practices, septic systems, and sewer networks. Maintenance of the quality of watersheds may require replacing some natural vegetation cover, which can provide another opportunity for public involvement and education.

4. *Pollution prevention*—Some businesses may require special training on how to manage their operations to prevent pollution and thereby protect the watershed. In some cases, local or state government may have a regulatory responsibility to develop pollution prevention programs for certain businesses and industrial categories (e.g., under industrial or municipal NPDES permits).

5. *Watershed indicator monitoring*—An ongoing stewardship responsibility is to monitor key indicators to track the health of the watershed.

Public agencies, as well as private corporations, citizen groups, and even landowners, should serious consider monitoring to provide appropriate indicator data that will answer their own questions. Monitoring water quality can include assessing flow, the quantity and quality of aquatic biota, pollutant levels, and many other characteristics as appropriate to the type of water body and its problems.

6. *Watershed restoration*—The last phase of watershed stewardship is to restore or at least rehabilitate streams that have been degraded by past development. Urban watershed restoration is an emerging art and science seeking to remove pollutants and enhance habitat to restore urban streams. The urban watershed restoration process should include three main themes: stormwater retrofitting, source control through pollution prevention, and stream enhancement.

CHAPTER REVIEW QUESTIONS

17.1   Define watershed.

17.2   Another name for watershed is _____.

17.3   Explain "keep it out or take it out."

17.4   What is the purpose of algaecide in reservoir management?

17.5   What is BMP?

17.6   Land use planning is the single most important tool of watershed protection. *True* or *False*?

17.7   The basic goals of land used planning are to: (a) apply land use planning techniques to redirect development while preserving sensitive areas, and (b) to maintain or reduce the impervious cover within a given watershed. *True* or *False*?

17.8   The most destructive stage of the development cycle is clearing and grading. *True* or *False*?

17.9   Properly functioning septic systems are never a soruce of nutrient loads. *True* or *False*?

17.10 Urban return flows are septic system or sanitary sewer back-ups. *True* or *False*?

REFERENCES AND SUGGESTED READING

NSFC. 2007. *Onsite Wastewater Treatment*. Morgantown, WV: National Small Flows Clearinghouse.

USEPA. 2007. *8 Tools of Watershed Protection*. Washington, D.C.: U.S. Environmental Protection Agency (www.epa.gov/owow/watershed/wacademy/acad2000/protection/index19.html).

Viessman, Jr., W. 1991. Water management issues for the nineties. *Water Res. Bull.*, 26(6), 883–981.

18 Water Treatment Operations

Municipal water treatment operations and associated treatment unit processes are designed to provide reliable, high-quality water service for customers and to preserve and protect the environment for future generations.

Water management officials and treatment plant operators are tasked with exercising responsible financial management, ensuring fair rates and charges, providing responsive customer service, providing a consistent supply of safe potable water for consumption by the user, and promoting environmental responsibility.

18.1 INTRODUCTION

In this chapter the focus is on water treatment operations and the various unit processes currently used to treat raw source water before it is distributed to the user. In addition, the reasons for water treatment and the basic theories associated with individual treatment unit processes are discussed. Water treatment systems are installed to remove those materials that cause disease or create nuisances. At its simplest level, the basic goal of water treatment operations is to protect public health, with a broader goal to provide potable and palatable water. The bottom line is that the water treatment process functions to provide water that is safe to drink and is pleasant in appearance, taste, and odor.

In this text, *water treatment* is defined as any unit process that changes or alters the chemical, physical, or bacteriological quality of water with the purpose of making it safe for human consumption or appealing to the customer. Treatment also is used to protect the water distribution system components from corrosion.

Many water treatment unit processes are commonly used today. The treatment processes used depend on the evaluation of the nature and quality of the particular water to be treated and the desired quality of the finished water. For water treatment unit processes employed to treat raw water, one thing is certain: As new U.S. Environmental Protection Agency (USEPA) regulations take effect, many more processes will come into use in an attempt to produce water that complies with all current regulations, despite source water conditions.

Small water systems tend to use a smaller number of the wide array of unit treatment processes available, in part because they usually rely on groundwater as the source and because their small size makes many sophisticated processes impractical (e.g., too expensive to install, too expensive to operate, too sophisticated for limited operating staff). This chapter concentrates on those individual treatment unit processes usually found in conventional water treatment systems, corrosion control methods, and fluoridation. A summary of basic water treatment processes (many of which are discussed in this chapter) is presented Table 18.1.

18.2 WATERWORKS OPERATORS

Operation of a water treatment system, no matter the size or complexity, requires operators. To perform their functions at the highest knowledge and experience level possible, operators must understand the basic principles and theories behind many complex water treatment concepts and treatment systems. Under new regulations, waterworks operators must be certified or licensed. Actual water treatment protocols and procedures are important; however, without proper implementation, they are nothing more than hollow words occupying space on reams of paper. This is where the waterworks operator comes in. To successfully treat water requires skill, dedication, and vigilance. The waterworks operator must be not only highly trained and skilled but also conscientious—the ultimate user demands nothing less. The role of the waterworks operator can be succinctly stated:

- Waterworks operators provide water that complies with state waterworks regulations, water that is safe to drink and ample in quantity and pressure without interruption.
- Waterworks operators must know their facilities.
- Waterworks operators must be familiar with bacteriology, chemistry, and hydraulics.
- Waterworks operators must stay abreast of technological change and stay current with water supply information.

When operating a waterworks facility, waterworks operator duties include:

- Maintaining distribution system
- Collecting or analyzing water samples
- Operating chemicals feed equipment

TABLE 18.1
Basic Water Treatment Processes

Process	Purpose
Screening	Removes large debris (leaves, sticks, fish) that can foul or damage plant equipment
Chemical pretreatment	Conditions the water for removal of algae and other aquatic nuisances
Presedimentation	Removes gravel, sand, silt, and other gritty materials
Microstraining	Removes algae, aquatic plants, and small debris
Chemical feed and rapid mix	Adds chemicals (e.g., coagulants, pH, adjusters) to water
Coagulation/flocculation	Converts nonsettleable or settable particles
Sedimentation	Removes settleable particles
Softening	Removes hardness-causing chemicals from water
Filtration	Removes particles of solid matter which can include biological contamination and turbidity
Disinfection	Kills disease-causing organisms
Adsorption using granular activated	Removes radon and many organic chemicals such as pesticides, solvents, and trihalomethanes
Aeration	Removes volatile organic compounds (VOCs), radon, H_2S, and other dissolved gases; oxidizes iron and manganese
Corrosion control	Prevents scaling and corrosion
Reverse osmosis, electrodialysis	Removes nearly all inorganic contaminants
Ion exchange	Removes some inorganic contaminants including hardness-causing chemicals
Activated alumina	Removes some inorganic contamination
Oxidation filtration	Removes some inorganic contaminants (e.g., iron, manganese, radium)

Source: Adapted from AWWA, *Introduction to Water Treatment*, Vol. 2, American Water Works Association, Denver, CO, 1984.

- Keeping records
- Operating treatment unit processes
- Performing sanitary surveys of the water supply watershed
- Operating a cross-connection control program

- Bacteriological contamination
- Hydrogen sulfide odors
- Hard water
- Corrosive water
- Iron and manganese

18.3 PURPOSE OF WATER TREATMENT

As mentioned, the purpose of water treatment is to condition, modify, or remove undesirable impurities and to provide water that is safe, palatable, and acceptable to users. This may seem an obvious, expected purpose of treating water, but various regulations also require water treatment. Some regulations state that if the contaminants listed under the various regulations are found in excess of maximum contaminant levels (MCLs), the water must be treated to reduce the levels. If a well or spring source is surface influenced, treatment is required, regardless of the actual presence of contamination. Some impurities affect the aesthetic qualities (taste, odor, color, and hardness) of the water; if they exceed secondary MCLs established by USEPA and the state, the water may have to be treated.

If we assume that the water source used to feed a typical water supply system is groundwater (usually the case in the United States), a number of common groundwater problems may require water treatment. Keep in mind that water that must be treated for any one of these problems may also exhibit several other problems. Among these other problems are:

18.4 STAGES OF WATER TREATMENT

Earlier it was stated that the focus of our discussion in this text is a conventional model of water treatment. Figure 18.1 presents the conventional model discussed throughout this text. Figure 18.1 clearly illustrates that water treatment is made up of various stages or unit processes combined to form one treatment system. Note that a given waterworks may contain all the unit processes discussed in the following or any combination of them. One or more of these stages may be used to treat any one or more of the source water problems listed above. Also note that the model shown in Figure 18.1 does not necessarily apply to very small water systems. In some small systems, water treatment may consist of nothing more than removal of water via pumping from a groundwater source to storage to distribution. In some small water supply operations, disinfection may be added because it is required. Although it is likely that the basic model shown in Figure 18.1 does not reflect the type of treatment process used in most small systems, we use it in this handbook for illustrative and instructive purposes because higher level licensure requires operators, at a minimum, to learn these processes.

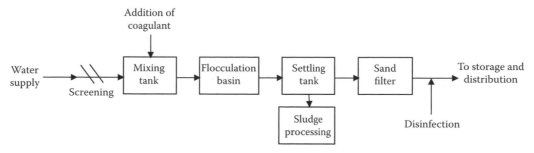

FIGURE 18.1 Conventional water treatment model.

18.5 PRETREATMENT

Simply stated, water pretreatment (also called *preliminary treatment*) is any physical, chemical, or mechanical process used before main water treatment processes. It can include screening, presedimentation, and chemical addition (see Figure 18.1). Pretreatment in water treatment operations usually consists of oxidation or other treatment for the removal of tastes and odors, iron and manganese, trihalomethane precursors, or entrapped gases (such as hydrogen sulfide). Unit processes may include chlorine, potassium permanganate or ozone oxidation, activated carbon addition, aeration, and presedimentation. Pretreatment of surface water supplies accomplishes the removal of certain constituents and materials that interfere with or place an unnecessary burden on conventional water treatment facilities.

Based on our experience and according to the Texas Water Utilities Association (TWUA, 1988), typical pretreatment processes include the following:

- Removal of debris from water from rivers and reservoirs that would clog pumping equipment
- Destratification of reservoirs to prevent anaerobic decomposition that could result in reducing iron and manganese in the soil to a state that would be soluble in water which can cause subsequent removal problems in the treatment plant; the production of hydrogen sulfide and other taste- and odor-producing compounds also results from stratification
- Chemical treatment of reservoirs to control the growth of algae and other aquatic growths that could result in taste and odor problems
- Presedimentation to remove excessively heavy silt loads prior to the treatment processes
- Aeration to remove dissolved odor-causing gases such as hydrogen sulfide and other dissolved gases or volatile constituents and to aid in the oxidation of iron and manganese, although manganese or high concentrations of iron are not removed in detention provided in conventional aeration units

- Chemical oxidation of iron and manganese, sulfides, taste- and odor-producing compounds, and organic precursors that may produce trihalomethanes upon the addition of chlorine
- Adsorption for removal of tastes and odors

Note: An important point to keep in mind is that in small systems using groundwater as a source, pretreatment may be the only treatment process used.

Note: Pretreatment may be incorporated as part of the total treatment process or may be located adjacent to the source before the water is sent to the treatment facility.

18.5.1 AERATION

Aeration is commonly used to treat water that contains trapped gases (such as hydrogen sulfide) that can impart an unpleasant taste and odor to the water. Just allowing the water to rest in a vented tank will (sometimes) drive off much of the gas, but usually some form of forced aeration is needed. Aeration works well (about 85% of the sulfides may be removed) whenever the pH of the water is less than 6.5. Aeration may also be useful in oxidizing iron and manganese, oxidizing humic substances that might form trihalomethanes when chlorinated, eliminating other sources of taste and odor, or imparting oxygen to oxygen-deficient water.

Note: Iron is a naturally occurring mineral found in many water supplies. When the concentration of iron exceeds 0.3 mg/L, red stains will occur on fixtures and clothing. The customer then incurs costs for cleaning and replacement of damaged fixtures and clothing. Manganese, like iron, is a naturally occurring mineral found in many water supplies. When the concentration of manganese exceeds 0.05 mg/L, black stains occur on fixtures and clothing. As with iron, this increases customer costs for cleaning and replacement of damaged fixtures and clothing. Iron and manganese are commonly found together in the same water supply. We discuss iron and manganese later.

18.5.2 Screening

Screening is usually the first major step in the water pre-treatment process (see Figure 18.1). It is defined as the process whereby relatively large and suspended debris is removed from the water before it enters the plant. River water, for example, typically contains suspended and floating debris varying in size from small rocks to logs. Removing these solids is important, not only because these items have no place in potable water but also because this river trash may cause damage to downstream equipment (e.g., clogging and damaging pumps), increase chemical requirements, impede hydraulic flow in open channels or pipes, or hinder the treatment process. The most important criteria used in the selection of a particular screening system for water treatment technology are the screen opening size and flow rate. Other important criteria include costs related to operation and equipment, plant hydraulics, debris handling requirements, and operator qualifications and availability. Large surface water treatment plants may employ a variety of screening devices including rash screens (or trash rakes), traveling water screens, drum screens, bar screens, or passive screens.

18.5.3 Chemical Addition

Much of the procedural information presented in this section applies to both water and wastewater operations. Two of the major chemical pretreatment processes used in treating water for potable use are iron and manganese and hardness removal. Another chemical treatment process that is not necessarily part of the pretreatment process, but is also discussed in this section, is corrosion control. Corrosion prevention is provided by chemical treatment—not only in the treatment process but also in the distribution process. Before discussing each of these treatment methods in detail, however, it is important to describe chemical addition, chemical feeders, and chemical feeder calibration.

When chemicals are used to in the pretreatment process, they must be the proper ones, fed in the correct concentration and introduced to the water at the proper locations. Determining the proper amount of chemical to use is accomplished by testing. The operator must test the raw water periodically to determine if the chemical dosage should be adjusted. For surface supplies, checking must be done more frequently than for groundwater (remember, surface water supplies are subject to change on short notice, while groundwaters generally remain stable). The operator must be aware of the potential for interactions between various chemicals and how to determine the optimum dosage (e.g., adding both chlorine and activated carbon at the same point will minimize the effectiveness of both processes, as the adsorptive power of the carbon will be used to remove the chlorine from the water).

Note: Sometimes using too many chemicals can be worse than not using enough.

Prechlorination (distinguished from chlorination used in disinfection at the end of treatment) is often used as an oxidant to help with the removal of iron and manganese; however, a concern for systems that prechlorinate is the potential for formation of total trihalomethanes (TTHMs), which form as a byproduct of the reaction between chlorine and naturally occurring compounds in raw water. The USEPA TTHM standard does not apply to water systems that serve less than 10,000 people, but operators should be aware of the impact and causes of TTHMs. Chlorine dosage or application point may be changed to reduce problems with TTHMs.

Note: TTHMs such as chloroform are known or suspected to be carcinogenic and are limited by water and state regulations.

Note: To be effective, pretreatment chemicals must be thoroughly mixed with the water. Short-circuiting or plug flows of chemicals that do not come in contact with most of the water will not result in proper treatment.

All chemicals intended for use in drinking water must meet certain standards. Thus, when ordering water treatment chemicals, the operator must be confident that they meet all appropriate standards for drinking water use.

Chemicals are normally fed with dry chemical feeders or solution (metering) pumps. Operators must be familiar with all of the adjustments required to control the rate at which the chemical is fed to the water (wastewater). Some feeders are manually controlled and must be adjusted by the operator when the raw water quality or the flow rate changes; other feeders are paced by a flowmeter to adjust the chemical feed so it matches the water flow rate. Operators must also be familiar with chemical solution and feeder calibration.

As mentioned, a significant part of the waterworks operator's daily functions includes measuring quantities of chemicals and applying them to water at preset rates. Normally accomplished semiautomatically by use of electromechanical–chemical feed devices, waterworks operators must still know what chemicals to add, how much to add to the water (wastewater), and the purpose of the chemical addition.

18.5.3.1 Chemical Solutions

A *water solution* is a homogeneous liquid made of the *solvent* (the substance that dissolves another substance) and the *solute* (the substance that dissolves in the solvent). Water is the solvent (see Figure 18.2). The solute (whatever it may be) may dissolve up to a certain limit. This

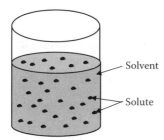

FIGURE 18.2 Solution with two components: solvent and solute.

level is its *solubility*—that is, the solubility of the solute in the particular solvent (water) at a particular temperature and pressure.

Note: Temperature and pressure influence the stability of solutions but not the filtration, because only suspended material can be eliminated by filtration or by sedimentation.

Remember, in chemical solutions, the substance being dissolved is called the *solute*, and the liquid present in the greatest amount in a solution (that does the dissolving) is called the *solvent*. The operator should also be familiar with another term—*concentration*, which is the amount of solute dissolved in a given amount of solvent. Concentration is measured as:

$$\% \text{ Strength} = \frac{\text{Wt. of solute}}{\text{Wt. of solution}} \times 100$$
$$= \frac{\text{Wt. of solute}}{\text{Wt. of solute + solvent}} \times 100$$
(18.1)

■ **EXAMPLE 18.1**

Problem: If 30 lb of chemical is added to 400 lb of water, what is the percent strength (by weight) of the solution?

Solution:

$$\% \text{ Strength} = \frac{30 \text{ lb solute}}{\text{Wt. of solution}} \times 100$$

$$= \frac{30 \text{ lb solute}}{30 \text{ lb solute + 400 lb water}} \times 100$$

$$= \frac{30 \text{ lb solute}}{430 \text{ lb solute/water}} \times 100$$

$$= 7.0 \text{ (rounded)}$$

Important to the process of making accurate computations of chemical strength is a complete understanding of the dimensional units involved; for example, operators should understand exactly what *milligrams per liter* (mg/L) signify:

$$\text{Milligrams per liter (mg/L)} = \frac{\text{Milligrams of solute}}{\text{Liters of solution}} \quad (18.2)$$

Another important dimensional unit commonly used when dealing with chemical solutions is *parts per million* (ppm):

$$\text{Parts per million (ppm)} = \frac{\text{Parts of solute}}{\text{Million parts of solution}} \quad (18.3)$$

Note: "Parts" is usually a weight measurement.

An example is:

$$9 \text{ ppm} = \frac{9 \text{ lb solids}}{1,000,000 \text{ lb solution}}$$

or

$$9 \text{ ppm} = \frac{9 \text{ mg solids}}{1,000,000 \text{ mg solution}}$$

This leads us to two important parameters that operators should commit to memory:

- 1 mg/L = 1 ppm
- 1% = 10,000 mg/L

When working with chemical solutions, it is also necessary to be familiar with two chemical properties that we briefly described earlier: density and specific gravity. *Density* is defined as the weight of a substance per a unit of its volume—for example, pounds per cubic foot or pounds per gallon. *Specific gravity* is defined as the ratio of the density of a substance to a standard density.

$$\text{Density} = \frac{\text{Mass of substance}}{\text{Volume of substance}} \quad (18.4)$$

Here are a few key facts about density:

- Density is measured in units of lb/ft³, lb/gal, or mg/L.
- Density of water = 62.5 lb/ft³ = 8.34 lb/gal.
- Density of concrete = 130 lb/ft³.
- Density of alum (liquid @ 60°F) = 1.33.
- Density of hydrogen peroxide (35%) = 1.132.

$$\text{Specific gravity} = \frac{\text{Density of substance}}{\text{Density of water}} \quad (18.5)$$

Here are a few facts about specific gravity:

- Specific gravity has no units.
- Specific gravity of water = 1.0.
- Specific gravity of concrete = 2.08.
- Specific gravity of alum (liquid @ 60°F) = 1.33.
- Specific gravity of hydrogen peroxide (35%) = 1.132.

18.5.3.2 Chemical Feeders

Simply put, a chemical feeder is a mechanical device for measuring a quantity of chemical and applying it to water at a preset rate.

18.5.3.3 Types of Chemical Feeders

Two types of chemical feeders are commonly used: solution (or liquid) feeders and dry feeders. Liquid feeders apply chemicals in solutions or suspensions, and dry feeders apply chemicals in granular or powdered forms. In a solution feeder, the chemical enters and leaves the feeder in a liquid state; in a dry feeder, the chemical enters and leaves the feeder in a dry state.

18.5.3.3.1 Solution Feeders

Solution feeders are small, positive-displacement metering pumps of three types: (1) reciprocating (piston-plunger or diaphragm type), (2) vacuum type (e.g., gas chlorinator), or (3) gravity feed rotameter (e.g., drip feeder). Positive-displacement pumps are used in high-pressure, low-flow applications; they deliver a specific volume of liquid for each stroke of a piston or rotation of an impeller.

18.5.3.3.2 Dry Feeders

Two types of dry feeders are *volumetric* and *gravimetric*, depending on whether the chemical is measured by volume (volumetric) or weight (gravimetric). Simpler and less expensive than gravimetric pumps, volumetric dry feeders are also less accurate. Gravimetric dry feeders are extremely accurate, deliver high feed rates, and are more expensive than volumetric feeders.

18.5.3.4 Chemical Feeder Calibration

Chemical feeder calibration ensures effective control of the treatment process. Obviously, chemical feed without some type of metering and accounting of chemical used adversely affects the water treatment process. Chemical feeder calibration also optimizes economy of operation; it ensures the optimum use of expensive chemicals. Finally, operators must have accurate knowledge of the capabilities of each individual feeder at specific settings. When a certain dose must be administered, the operator must rely on the feeder to feed the correct amount of chemical. Proper calibration ensures that chemical dos-

ages can be set with confidence. At a minimum, chemical feeders must be calibrated on an annual basis. During operation, when the operator changes chemical strength or chemical purity or makes any adjustment to the feeder, or when the treated water flow changes, the chemical feeder should be calibrated. Ideally, any time maintenance is performed on chemical feed equipment, calibration should be performed.

What factors affect chemical feeder calibration (i.e., feed rate)? For solution feeders, calibration is affected any time solution strength changes, any time a mechanical change is introduced in the pump (change in stroke length or stroke frequency), or whenever flow rate changes. In the dry chemical feeder, calibration is affected any time the chemical purity changes, mechanical damage occurs (e.g., belt change), or the flow rate changes. In the calibration process, standard calibration charts are usually used, or charts are made up to fit the calibration equipment. The calibration chart is also affected by certain factors, including changes in chemical, changes in the flow rate of the water being treated, or mechanical changes in the feeder.

18.5.3.5 Calibration Procedures

When calibrating a positive-displacement pump (liquid feeder), the operator should always refer to the manufacturer's technical manual. Keeping in mind the need to refer to the manufacturer's specific guidelines, we provide general examples here of calibration procedures for simple positive-displacement pump and dry feeder calibration procedures.

18.5.3.5.1 Positive-Displacement Pump

The following equipment is required:

- Graduated cylinder (1000 mL or less)
- Stopwatch
- Calculator
- Graph paper
- Plain paper
- Straight edge

The procedure is as follows:

1. Fill graduated cylinder with solution.
2. Insert pump suction line into graduated cylinder.
3. Run pump 5 minutes at highest setting (100%).
4. Divide the milliliters of liquid withdrawn by 5 minutes to determine pumping rate (mL/min) and record on plain paper.
5. Repeat steps 3 and 4 at 100% setting.
6. Repeat steps 3 and 4 for 20%, 50%, and 70% settings twice.

7. Average the mL/min pumped for each setting.
8. Calculate the weight of chemical pumped for each setting.
9. Calculate the dosage for each setting.
10. Graph dosage vs. setting.

18.5.3.5.2 Dry Feeder

The following equipment is required:

- Weighing pan
- Balance
- Stopwatch
- Plain paper
- Graph paper
- Straight edge
- Calculator

The procedure is as follows:

1. Weigh pan and record.
2. Set feeder at 100% setting.
3. Collect sample for 5 minutes.
4. Calculate weight of sample and record in table.
5. Repeat steps 3 and 4 twice.
6. Repeat steps 3 and 4 for settings of 25%, 50%, and 75%.
7. Calculate the average sample weight per minute for each setting and record in table.
8. Calculate weight per day fed for each setting.
9. Plot weight per day vs. setting on graph paper.

Note: Pounds per day (lb/day) is not normally useful information for setting the feed rate setting on a feeder because process control usually determines a dosage in ppm, mg/L, or grains/gallon. A separate chart may be necessary for another conversion based on the individual treatment facility flow rate.

■ EXAMPLE 18.2

To demonstrate that performing a chemical feed procedure is not necessarily as simple as opening a bag of chemicals and dumping the contents into the feed system, we provide a real-world example below.

Problem: Consider the chlorination dosage rates below:

Setting		Dosage
100%	111/121	0.93 mg/L
70%	78/121	0.66 mg/L
50%	54/121	0.45 mg/L
20%	20/121	0.16 mg/L

Solution: This is not a good dosage setup for a chlorination system. The chlorine residual at the ends of the distribution system should be within 0.5 and 1.0 ppm. At 0.9 ppm, the dosage will probably fall within this range—depending on the chlorine demand of the raw water and detention time in the system. However, the pump is set at its highest setting. We have room to decrease the dosage but no ability to increase the dosage without changing the solution strength in the solution tank. In this example, doubling the solution strength to 1% provides the ideal solution, resulting in the following chart changes:

Setting		Dosage
100%	222/121	1.86 mg/L
70%	154/121	1.32 mg/L
50%	108/121	0.90 mg/L
20%	40/121	0.32 mg/L

This is ideal, because the dosage we want to feed is at the 50% setting for our chlorinator. We can now easily increase or decrease the dosage, whereas the previous setup only allowed the dosage to be decreased.

18.5.3.6 Iron and Manganese Removal

Iron and manganese are frequently found in groundwater and in some surface waters. They do not cause health-related problems but are objectionable because they may cause aesthetic problems. Severe aesthetic problems may cause consumers to avoid an otherwise safe water supply in favor of one of unknown or questionable quality, or they may cause customers to incur unnecessary expense for bottled water. Aesthetic problems associated with iron and manganese include the discoloration of water (iron, reddish water; manganese, brown or black water), staining of plumbing fixtures, a bitter taste, and the growth of microorganisms.

As mentioned, no health concerns are directly associated with iron and manganese, although the growth of iron bacteria slimes may cause indirect health problems. Economic problems include damage to textiles, dye, paper, and food. Iron residue (or tuberculation) in pipes increases pumping head, decreases carrying capacity, may clog pipes, and may corrode through pipes.

Note: Iron and manganese are secondary contaminants. Their secondary maximum contaminant levels (SMCLs) are 0.3 mg/L for iron and 0.05 mg/L for manganese.

Iron and manganese are most likely found in groundwater supplies, industrial waste, and acid mine drainage, and are byproducts of pipeline corrosion. They may accumulate in lake and reservoir sediments, causing possible problems during lake/reservoir turnover. They are not usually found in running waters (e.g., streams, rivers).

18.5.3.7 Iron and Manganese Removal Techniques

Chemical precipitation treatments for iron and manganese removal are called *deferrization* and *demanganization*, respectively. The usual process is *aeration*, where dissolved oxygen in the chemical causes precipitation; chlorine or potassium permanganate may also be required.

18.5.3.7.1 Precipitation

Precipitation (or pH adjustment) of iron or manganese from water in their solid forms can be performed in treatment plants by adjusting the pH of the water through the addition of lime or other chemicals. Some of the precipitate will settle out with time, while the rest is easily removed by sand filters. This process requires the pH of the water to be in the range of 10 to 11.

Note: Although the precipitation or pH adjustment technique for treating water containing iron and manganese is effective, note that the pH level must be adjusted higher (10 to 11) to cause the precipitation, which means that the pH level must also then be lowered (to 8.5 or a bit lower) to use the water for consumption.

18.5.3.7.2 Oxidation

One of the most common methods for removing iron and manganese is the process of oxidation (another chemical process), usually followed by settling and filtration. Air, chlorine, or potassium permanganate can oxidize these minerals. Each oxidant has advantages and disadvantages, as each operates slightly differently:

- *Air*—To be effective as an oxidant, the air must come in contact with as much of the water as possible. Aeration is often accomplished by bubbling diffused air through the water by spraying the water up into the air or by trickling the water over rocks, boards, or plastic packing materials in an aeration tower. The more finely divided the drops of water, the more oxygen comes in contact with the water and the dissolved iron and manganese.
- *Chlorine*—This is one of the most popular oxidants for iron and manganese control because it is also widely used as a disinfectant; controlling iron and manganese by prechlorination can be as simple as adding a new chlorine feed point in a facility already feeding chlorine. It also provides a predisinfecting step that can help control bacterial growth throughout the rest of the treatment system. The downside to using chorine is that when chlorine reacts with the organic materials found in surface water and some groundwaters it forms TTHMs. This pro-

cess also requires that the pH of the water must be in the range of 6.5 to 7; because many groundwaters are more acidic than this, pH adjustment with lime, soda ash, or caustic soda may be necessary when oxidizing with chlorine.

- *Potassium permanganate*—This is the best oxidizing chemical to use for manganese control removal. An extremely strong oxidant, it has the additional benefit of producing manganese dioxide during the oxidation reaction. Manganese dioxide acts as an adsorbent for soluble manganese ions. This attraction for soluble manganese provides removal to extremely low levels.

The oxidized compounds form precipitates that are removed by a filter. Note that sufficient time should be allowed from the addition of the oxidant to the filtration step; otherwise, the oxidation process will be completed after filtration, creating insoluble iron and manganese precipitates in the distribution system.

18.5.3.7.3 Ion Exchange

The ion exchange process is used primarily to soften hard waters, but it will also remove soluble iron and manganese. The water passes through a bed of resin that adsorbs undesirable ions from the water, replacing them with less troublesome ions. When the resin has given up all of its donor ions, it is regenerated with strong salt brine (sodium chloride); the sodium ions from the brine replace the adsorbed ions and restore the ion exchange capabilities.

18.5.3.7.4 Sequestering

Sequestering or stabilization may be used when the water contains mainly low concentration of iron and the volumes required are relatively small. This process does not actually remove the iron or manganese from the water but complexes (binds it chemically) it with other ions in a soluble form that is not likely to come out of solution (i.e., not likely oxidized).

18.5.3.7.5 Aeration

The primary physical process uses air to oxidize the iron and manganese. The water is either pumped up into the air or allowed to fall over an aeration device. The air oxidizes the iron and manganese that is then removed by use of a filter. The addition of lime to raise the pH is often utilized in the process. Although this is referred to as a physical process, removal is accomplished by chemical oxidation.

18.5.3.7.6 Potassium Permanganate Oxidation and Manganese Greensand

The continuous regeneration potassium greensand filter process is another commonly used filtration technique for iron and manganese control. Manganese greensand is a mineral (gluconite) that has been treated with alternating solutions of manganous chloride and potassium permanganate. The result is a sand-like (zeolite) material coated with a layer of manganese dioxide—an adsorbent for soluble iron and manganese. Manganese greensand has the ability to capture (adsorb) soluble iron and manganese that may have escaped oxidation, as well as the capability of physically filtering out the particles of oxidized iron and manganese. Manganese greensand filters are generally set up as pressure filters, totally enclosed tanks containing the greensand. The process of adsorbing soluble iron and manganese uses up the greensand by converting the manganese dioxide coating to manganic oxide, which does not have the adsorption property. The greensand can be regenerated in much the same way as ion exchange resins by washing the sand with potassium permanganate.

18.5.3.8 Hardness Treatment

Hardness in water is caused by the presence of certain positively charged metallic ions in solution in the water. The most common of these hardness-causing ions are calcium and magnesium; others include iron, strontium, and barium. As a general rule, groundwaters are harder than surface waters, so hardness is frequently of concern to the small water system operator. This hardness is derived from contact with soil and rock formations such as limestone. Although rainwater itself will not dissolve many solids, the natural carbon dioxide in the soil enters the water and forms carbonic acid (HCO), which is capable of dissolving minerals. Where soil is thick (contributing more carbon dioxide to the water) and limestone is present, hardness is likely to be a problem. The total amount of hardness in water is expressed as the sum of its calcium carbonate ($CaCO_3$) and its magnesium hardness; however, for practical purposes, hardness is expressed as calcium carbonate. This means that, regardless of the amount of the various components that make up hardness, they can be related to a specific amount of calcium carbonate (e.g., hardness is expressed as "mg/L as $CaCO_3$," or milligrams per liter as calcium carbonate).

Note: The two types of water hardness are temporary hardness and permanent hardness. Temporary hardness is also known as *carbonate hardness* (hardness that can be removed by boiling); permanent hardness is also known as *noncarbonate hardness* (hardness that cannot be removed by boiling).

Hardness is of concern in domestic water consumption because hard water increases soap consumption, leaves a soapy scum in the sink or tub, can cause water heater electrodes to burn out quickly, can cause discoloration of plumbing fixtures and utensils, and is perceived as being less desirable water. In industrial water use, hardness is a concern because it can cause boiler scale and damage to industrial equipment.

TABLE 18.2
Classification of Hardness

Classification	mg/L as $CaCO_3$
Soft	0–75
Moderately hard	75–150
Hard	150–300
Very hard	Over 300

The objection of customers to hardness is often dependent on the amount of hardness they are used to. People familiar with water with a hardness of 20 mg/L might think that a hardness of 100 mg/L is too much. On the other hand, a person who has been using water with a hardness of 200 mg/L might think that 100 mg/L is very soft. Table 18.2 lists the classifications of hardness.

18.5.3.8.1 Hardness Calculation

Recall that hardness is expressed as mg/L as $CaCO_3$. The mg/L of calcium and manganese must be converted to mg/L as $CaCO_3$ before they can be added. The hardness (in mg/L as $CaCO_3$) for any given metallic ion is calculated using the formula:

$$\text{Hardness (mg/L as } CaCO_3) = M \text{ (mg/L)} \times \frac{50}{\text{EW of } M} \quad (18.6)$$

where:

M = metal ion concentration (mg/L).
EW = equivalent weight = gram molecular weight ÷ valence.

18.5.3.8.2 Treatment Methods

Two common methods are used to reduce hardness: *ion exchange* and *cation exchange*:

- *Ion exchange*—The ion exchange process is the process most frequently used for softening water. As a result of charging a resin with sodium ions, the resin exchanges the sodium ions for calcium or magnesium ions. Naturally occurring and synthetic cation exchange resins are available. Natural exchange resins include such substances as aluminum silicate, zeolite clays (zeolites are hydrous silicates found naturally in the cavities of lavas [greensand], glauconite zeolites, or synthetic, porous zeolites), humus, and certain types of sediments. These resins are placed in a pressure vessel. Salt brine is flushed through the resins. The sodium ions in the salt brine attach to the resin. The resin is now said to be charged. Once charged, water is passed through the resin, and the resin exchanges the sodium ions attached to the resin for calcium and magnesium ions, thus removing them from the water. The zeolite clays are most common because they are quite durable, can tolerate extreme ranges in pH, and are chemically stable. They have relatively limited exchange capacities, however, so they should be used only for water with a moderate total hardness. One of the results is that the water may be more corrosive than before. Another concern is that addition of sodium ions to the water may increase the health risk of those with high blood pressure.

- *Cation exchange*—The cation exchange process takes place with little or no intervention from the treatment plant operator. Water containing hardness-causing cations (Ca^{2+}, Mg^{2+}, Fe^{3+}) is passed through a bed of cation exchange resin. The water coming through the bed contains hardness near zero, although it will have elevated sodium content. (The sodium content is not likely to be high enough to be noticeable, but it could be high enough to pose problems to people on highly restricted salt-free diets.) The total lack of hardness in the finished water is likely to make it very corrosive, so normal practice bypasses a portion of the water around the softening process. The treated and untreated waters are blended to produce an effluent with a total hardness around 50 to 75 mg/L as $CaCO_3$.

18.5.3.9 Corrosion

Water operators add chemicals (e.g., lime or sodium hydroxide) to water at the source or at the waterworks to control corrosion. Using chemicals to achieve slightly alkaline chemical balance prevents the water from corroding distribution pipes and consumers' plumbing and keeps substances such as lead from leaching out of plumbing and into the drinking water. For our purposes, we define *corrosion* as the conversion of a metal to a salt or oxide with a loss of desirable properties such as mechanical strength. Corrosion may occur over an entire exposed surface, or may be localized at micro- or macroscopic discontinuities in metal. In all types of corrosion, a gradual decomposition of the material occurs, often due to an electrochemical reaction. Corrosion may be caused by: (1) stray current electrolysis, (2) galvanic corrosion caused by dissimilar metals, or (3) differential concentration cells. Corrosion begins at the surface of a material and moves inward.

The adverse effects of corrosion can be categorized according to health, aesthetics, economic effects, and other effects. The corrosion of toxic metal pipe made from lead creates a serious health hazard. Lead tends to accumulate in the bones of humans and animals. Signs of lead intoxication include gastrointestinal disturbances, fatigue, anemia, and muscular paralysis. Lead is not a natural contaminant in either surface waters or groundwaters, and the MCL of 0.005 mg/L in source waters is rarely exceeded. It is a corrosion byproduct from high lead solder joints in copper and lead piping. Small dosages of lead can lead to developmental problems in children. The USEPA's Lead and Copper Rule addresses the matter of lead in drinking water exceeds specified action levels.

Note: The USEPA's Lead and Copper Rule requires that a treatment facility achieve optimum corrosion control. Because lead and copper contamination generally occurs after water has left the public water system, the best way for the water system operator to find out if customer water is contaminated is to test water that has come from a household faucet.

Cadmium is the only other toxic metal found in samples from plumbing systems. Cadmium is a contaminant found in zinc. Its adverse health effects are best known for being associated with severe bone and kidney disorders in Japan. The proposed maximum contaminant level (PMCL) for cadmium is 0.01 mg/L.

Note: Water systems should try to supply water free of lead and that has no more than 1.3 mg of copper per liter. This is a nonenforceable health goal.

Aesthetic effects that are a result of corrosion of iron are characterized by pitting and are a consequence of the deposition of ferric hydroxide and other products and the solution of iron—*tuberculation*. Tuberculation reduces the hydraulic capacity of the pipe. Corrosion of iron can cause customer complaints of reddish or reddish-brown staining of plumbing fixtures and laundry. Corrosion of copper lines can cause customer complaints of bluish or blue–green stains on plumbing fixtures. Sulfide corrosion of copper and iron lines can cause a blackish color in the water. The byproducts of microbial activity (especially iron bacteria) can cause foul tastes or odors in the water.

The economic effects of corrosion may include the need for water main replacement, especially when tuberculation reduces the flow capacity of the main. Tuberculation increases pipe roughness, causing an increase in pumping costs and reducing distribution system pressure. Tuberculation and corrosion can cause leaks in distribution mains and household plumbing. Corrosion of household plumbing may require extensive treatment, public education, and other actions under the Lead and Copper Rule.

Other effects of corrosion include short service life of household plumbing caused by pitting. A build-up of mineral deposits in a hot water system may eventually restrict hot-water flow. Also, the structural integrity of steel water storage tanks may deteriorate, causing structural failures. Steel ladders in clearwells or water storage tanks may corrode, introducing iron into the finished water. Steel parts in flocculation tanks, sedimentation basins, clarifiers, and filters may also corrode.

18.5.3.9.1 Types of Corrosion

Three types of corrosion occur in water mains: galvanic, tuberculation, and pitting:

- *Galvanic* occurs when two dissimilar metals are in contact and are exposed to a conductive environment; a potential exists between them and current flows. This type of corrosion is the result of an electrochemical reaction when the flow of electric current itself is an essential part of the reaction.
- *Tuberculation* refers to the formation of localized corrosion products scattered over the surface in the form of knob-like mounds. These mounds increase the roughness of the inside of the pipe, increasing resistance to water flow and decreasing the *C* factor of the pipe.
- *Pitting* is localized corrosion that is classified as pitting when the diameter of the cavity at the metal surface is the same or less than the depth.

18.5.3.9.2 Factors Affecting Corrosion

The primary factors affecting corrosion are pH, alkalinity, hardness (calcium), dissolved oxygen, and total dissolved solids. Secondary factors include temperature, velocity of water in pipes, and carbon dioxide (CO_2).

18.5.3.9.3 Determination of Corrosion Problems

To determine if corrosion is taking place in water mains, materials removed from the distribution system should be examined for signs of corrosion damage. A primary indicator of corrosion damage is pitting. (Measure the depth of pits to gauge the extent of damage.) Another common method used to determine if corrosion or scaling is taking place in distribution lines is to insert special steel specimens of a known weight (called *coupons*) in the pipe and examine them for corrosion after a period of time. Detecting evidence of leaks, conducting flow tests and chemical tests for dissolved oxygen and toxic metals, and customer complaints (e.g., red or black water, laundry and fixture stains) can also reveal corrosion problems.

Formulas can also be used to determine corrosion (to an extent). The *Langlier Saturation Index* (LI) and the *Aggressive Index* (AI) are two of the most commonly used indices. The LI determines whether water is corrosive. The AI is used for waters that have low natural pH, are high in dissolved oxygen, are low in total dissolved solids, and have low alkalinity and low hardness. These waters are very aggressive and can be corrosive. Both of these indices are typically used as starting points in determining the adjustments required to produce a film:

- LI approximately 0.5
- AI value of 12 or higher

Note: The LI and AI are based on the dissolving of and precipitation of calcium carbonate; therefore, the respective indices may not actually reflect the corrosive nature of the particular water for a specific pipe material. They can be useful tools, however, in selecting materials or treatment options for corrosion control.

18.5.3.10 Corrosion Control

As mentioned, one method used to reduce the corrosive nature of water is *chemical addition*. Selection of the chemicals depends on the characteristics of the water, where the chemicals can be applied, how they can be applied and mixed with water, and the cost of the chemicals.

If the product of the calcium hardness times the alkalinity of the water is less than 100, treatments may be required. Both lime and carbon dioxide may be required for proper treatment of the water. If the calcium hardness and alkalinity levels are between 100 and 500, either lime or soda ash (Na_2CO_3) will be satisfactory. The decision regarding which chemical to use depends on the cost of the equipment and chemicals. If the product of the calcium hardness times the alkalinity is greater than 500, either lime or caustic (NaOH) may be used. Soda ash will be ruled out because of the expense.

The chemicals chosen for treatment of public drinking water supplies modify the water characteristics, making the water less corrosive to the pipe. Modification of water quality can increase the pH of the water, thus reducing the hydrogen ions available for galvanic corrosion, as well as reducing the solubility of copper, zinc, iron, lead, and calcium and increasing the possibility of forming carbonate protective films.

Calcium carbonate stability is the most effective means of controlling corrosion. Lime, caustic soda, or soda ash is added until the pH and the alkalinity indicate the water is saturated with calcium carbonate. Saturation does *not* always ensure noncorrosiveness. Utilities should exercise caution when applying sodium compounds, because high sodium content in water can be a health concern for some customers. By increasing the alkalinity of the water, the bicarbonate and carbonate available to form a protective carbonate film also increase. By decreasing the dissolved oxygen of the water, the rate of galvanic corrosion is reduced, along with the possibility of iron tuberculation.

Inorganic phosphates used include:

- Zinc phosphates, which can cause algal blooms on open reservoirs
- Sodium silicate, which is used by individual customers, such as apartments, houses, and office buildings
- Sodium polyphosphates (tetrasodium pyrophosphate or sodium hexametaphosphate), which control scale formation in supersaturated waters and are known as *sequestering agents*
- Silicates (SiO_2), which form a film; an initial dosage of 12 to 16 mg/L for about 30 days will adequately coat the pipes, and a 1.0-mg/L concentration should be maintained thereafter

Caution: Great care and caution must be exercised any time corrosion control chemicals are fed into a public drinking water system!

Another corrosion control method is *aeration*. Aeration works to remove carbon dioxide (CO_2); it can be reduced to about 5 mg/L. *Cathodic protection*, often employed to control corrosion, is achieved by applying an outside electric current to the metal to reverse the electromechanical corrosion process. The application of DC current prevents normal electron flow. Cathodic protection uses a sacrificial metal electrode (a magnesium anode) that corrodes instead of the pipe or tank. *Linings*, *coatings*, and *paints* can also be used in corrosion control. Slip-line with a plastic liner, cement mortar, zinc or magnesium, polyethylene, epoxy, and coal tar enamels are some of the materials that can be used.

Caution: Before using any protective coatings, consult the district engineer first!

Several *corrosive-resistant pipe materials* are used to prevent corrosion, including:

- PVC plastic pipe
- Aluminum
- Nickel
- Silicon
- Brass
- Bronze
- Stainless steel
- Reinforced concrete

In addition to internal corrosion problems, waterworks operators must also be concerned with external corrosion problems. The primary culprit involved with external corrosion of distribution system pipe is soil. The measure of corrosivity of the soil is the *soil resistivity*. If the soil resistivity is greater than 5000 ohm/cm, serious corrosion is unlikely. Steel pipe may be used under these conditions.

If soil resistivity is less than 500 ohm/cm, plastic PVC pipe should be used. For intermediate ranges of soil resistivity (500 to 5000 ohm/cm), use ductile iron pipe, linings, and coatings.

Common operating problems associated with corrosion control include:

- $CaCO_3$ not depositing a film is usually a result of poor pH control (out of the normal range of 6.5 to 8.5). This may also cause excessive film deposition.
- Persistence of red water problems are most probably a result of poor flow patterns, insufficient velocity, tuberculation of pipe surface, and the presence of iron bacteria:
  1. *Velocity*—Chemicals must make contact with the pipe surface. Dead ends and low-flow areas should have flushing program; dead ends should be looped.
  2. *Tuberculation*—The best approach is to clean with *pig*. In extreme cases, clean pipe with metal scrapers and install cement-mortar lining.
  3. *Iron bacteria*—Slime prevents film contact with the pipe surface. Slime will grow and the coating will be lost. Pipe cleaning and disinfection programs are needed.

18.6 COAGULATION

The primary purpose in surface-water treatment is chemical clarification by coagulation and mixing, flocculation, sedimentation, and filtration. These unit processes, along with disinfection, work to remove particles, naturally occurring organic matter (NOM; bacteria, algae, zooplankton, and organic compounds), and microbes from water to produce water that is noncorrosive. Specifically, coagulation and flocculation work to destabilize particles and agglomerate dissolved and particulate matter. Sedimentation removes solids and provides 1/2-log *Giardia* and 1-log virus removal. Filtration removes solids and provides 2-log *Giardia* and 1-log virus removal. Finally, disinfection provides microbial inactivation and 1/2-log *Giardia* and 2-log virus removal.

From Figure 18.3, it can be seen that, following screening and the other pretreatment processes, the next unit process in a conventional water treatment system is a mixer where chemicals are added in what is known as *coagulation*. The exception to this unit process configuration occurs in small systems using groundwater, when chlorine or other taste and odor control measures are introduced at the intake and are the extent of treatment.

Materials present in raw water may vary in size, concentration, and type. Dispersed substances in the water may be classified as *suspended*, *colloidal*, or *solution*.

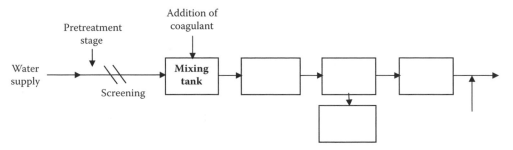

FIGURE 18.3 Coagulation.

Suspended particles may vary in mass and size and are dependent on the flow of water. High flows and velocities can carry larger material. As velocities decrease, the suspended particles settle according to size and mass.

Other material may be in solution; for example, salt dissolves in water. Matter in the colloidal state does not dissolve, but the particles are so small they will not settle out of the water. Color (as in tea-colored swamp water) is mainly due to colloids or extremely fine particles of matter in suspension. Colloidal and solute particles in water are electrically charged. Because most of the charges are alike (negative) and repel each other, the particles stay dispersed and remain in the colloidal or soluble state.

Suspended matter will settle without treatment, if the water is still enough to allow it to settle. The rate of settling of particles can be determined, as this settling follows certain laws of physics; however, much of the suspended matter may be so slow in settling that the normal settling processes become impractical, and if colloidal particles are present settling will not occur. Moreover, water drawn from a raw water source often contains many small unstable (unsticky) particles; therefore, sedimentation alone is usually an impractical way to obtain clear water in most locations, and another method of increasing the settling rate must be used: coagulation, which is designed to convert stable (unsticky) particles to unstable (sticky) particles.

The term *coagulation* refers to the series of chemical and mechanical operations by which coagulants are applied and made effective. These operations are comprised of two distinct phases: (1) rapid mixing to disperse coagulant chemicals by violent agitation into the water being treated, and (2) flocculation to agglomerate small particles into well-defined floc by gentle agitation for a much longer time.

Coagulation results from adding salts of iron or aluminum to the water. The coagulant must be added to the raw water and perfectly distributed into the liquid; such uniformity of chemical treatment is reached through rapid agitation or mixing. Common coagulants (salts) include:

- Alum (aluminum sulfate)
- Sodium aluminate
- Ferric sulfate
- Ferrous sulfate
- Ferric chloride
- Polymers

Coagulation is the reaction between one of these salts and water. The simplest coagulation process occurs between alum and water. Alum, or aluminum sulfate, is produced by a chemical reaction between bauxite ore and sulfuric acid. The normal strength of liquid alum is adjusted to 8.3%, while the strength of dry alum is 17%.

When alum is placed in water, a chemical reaction occurs that produces positively charged aluminum ions. The overall result is the reduction of electrical charges and the formation of a sticky substance—the formation of *floc*, which when properly formed, will settle. These two destabilizing factors are the major contributions that coagulation makes to the removal of turbidity, color, and microorganisms.

Liquid alum is preferred in water treatment because it has several advantages over other coagulants, including:

- Ease of handling
- Lower costs
- Less labor required to unload, store, and convey
- Elimination of dissolving operations
- Less storage space required
- Greater accuracy in measurement and control
- Elimination of the nuisance and unpleasantness of handling dry alum
- Easier maintenance

The formation of floc is the first step of coagulation; for greatest efficiency, rapid, intimate mixing of the raw water and the coagulant must occur. After mixing, the water should be slowly stirred so the very small, newly formed particles can attract and enmesh colloidal particles, holding them together to form larger floc. This slow mixing is the second stage of the process (flocculation), covered later.

A number of factors influence the coagulation process—pH, turbidity, temperature, alkalinity, and the use of polymers. The degree to which these factors influence coagulation depends on the coagulant use. The raw water

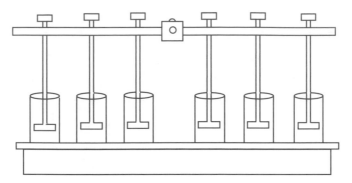

FIGURE 18.4 Variable-speed paddle mixer used in jar testing procedure.

conditions, optimum pH for coagulation, and other factors must be considered before deciding which chemical is to be fed and at what levels.

To determine the correct chemical dosage, a *jar test* or *coagulation test* is performed. Jar tests (widely used for many years by the water treatment industry) simulate full-scale coagulation and flocculation processes to determine optimum chemical dosages. It is important to note that jar testing is only an attempt to achieve a ballpark approximation of correct chemical dosage for the treatment process. The test conditions are intended to reflect the normal operation of a chemical treatment facility. The test can be used to:

- Select the most effective chemical.
- Select the optimum dosage.
- Determine the value of a flocculant aid and the proper dose.

The testing procedure requires a series of samples to be placed in testing jars (see Figure 18.4) and mixed at 100 ppm. Varying amounts of the process chemical or specified amounts of several flocculants are added (one volume/sample container). The mix is continued for 1 minute. Next, the mixing is slowed to 30 rpm to provide gentle agitation, and then the floc is allowed to settle. The flocculation period and settling process arer observed carefully to determine the floc strength, settleability, and clarity of the *supernatant liquor* (the water that remains above the settled floc). Additionally, the supernatant can be tested to determine the efficiency of the chemical addition for removal of total suspended solids (TSS), biochemical oxygen demand (BOD5), and phosphorus.

The equipment required for the jar test includes a six-position, variable-speed paddle mixer (see Figure 18.4); six 2-quart, wide-mouthed jars; an interval timer; and assorted glassware, pipettes, graduates, and so forth. The jar test procedure follows:

1. Place an appropriate volume of water sample in each of the jars (250- to 1000-mL samples

may be used, depending on the size of the equipment being use). Start mixers and set for 100 rpm.
2. Add previously selected amounts of the chemical being evaluated. (Initial tests may use wide variations in chemical volumes to determine the approximate range; this is then narrowed in subsequent tests).
3. Continue mixing for 1 minute.
4. Reduce the mixer speed to a gentle agitation (30 rpm), and continue mixing for 20 minutes. Again, time and mixer speed may be varied to reflect the facility.

Note: During this time, observe the floc formation—that is, how well the floc holds together during the agitation (floc strength).

5. Turn off the mixer and allow solids to settle for 20 to 30 minutes. Observe the settling characteristics, the clarity of the supernatant, the settleability of the solids, the flocculation of the solids, and the compactability of the solids.
6. Perform phosphate tests to determine removals.
7. Select the dose that provided the best treatment based on observations made during the analysis.

Note: After initial ranges and chemical selections are determined, repeat the test using a smaller range of dosages to optimize performance.

18.7 FLOCCULATION

As we see in Figure 18.5, flocculation follows coagulation in the conventional water treatment process. Flocculation is the physical process of slowly mixing the coagulated water to increase the probability of particle collision; unstable particles collide and stick together to form fewer larger flocs. Through experience, we have found that effective mixing reduces the required amount of chemicals and greatly improves the sedimentation process, resulting in longer filter runs and higher quality finished water.

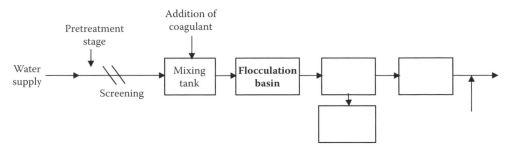

FIGURE 18.5 Flocculation.

The goal of flocculation is to form a uniform, feather-like material similar to snowflakes—a dense, tenacious floc that entraps the fine, suspended, and colloidal particles and carries them down rapidly in the settling basin. Proper flocculation requires from 15 to 45 minutes. The time is based on water chemistry, water temperature, and mixing intensity. Temperature is the key component in determining the amount of time required for floc formation. To increase the speed of floc formation and the strength and weight of the floc, polymers are often added.

18.8 SEDIMENTATION

After raw water and chemicals have been mixed and the floc formed, the water containing the floc (because it has a higher specific gravity than water) flows to the sedimentation or settling basin (see Figure 18.6). *Sedimentation* is also called *clarification*. Sedimentation removes settleable solids by gravity. Water moves slowly though the sedimentation tank/basin with a minimum of turbulence at entry and exit points with minimum short-circuiting. Sludge accumulates at the bottom of the tank/basin. Typical tanks or basins used in sedimentation include conventional rectangular basins, conventional center-feed basins, peripheral-feed basins, and spiral-flow basins.

In conventional treatment plants, the amount of detention time required for settling can vary from 2 to 6 hours. Detention time should be based on the total filter capacity when the filters are passing 2 gpm per square foot of superficial sand area. For plants with higher filter rates, the detention time is based on a filter rate of 3 to 4 gpm

per square foot of sand area. The time requirement is dependent on the weight of the floc, the temperature of the water, and how quiescent (still) the basin is.

A number of conditions affect sedimentation: (1) uniformity of flow of water through the basin, (2) stratification of water due to difference in temperature between water entering and water already in the basin, (3) release of gases that may collect in small bubbles on suspended solids, causing them to rise and float as scum rather than settle as sludge, (4) disintegration of previously formed floc, and (5) size and density of the floc.

18.9 FILTRATION

In the conventional water treatment process, *filtration* usually follows coagulation, flocculation, and sedimentation (see Figure 18.7). At present, filtration is not always used in small water systems; however, recent regulatory requirements under the USEPA Interim Enhanced Surface Water Treatment rules may make water filtering necessary at most water supply systems. Water filtration is a physical process of separating suspended and colloidal particles from water by passing water through a granular material. The process of filtration involves straining, settling, and adsorption. As floc passes into the filter, the spaces between the filter grains become clogged, reducing this opening and increasing removal. Some material is removed merely because it settles on a media grain. One of the most important processes is adsorption of the floc onto the surface of individual filter grains. This helps collect the floc and reduces the size of the openings

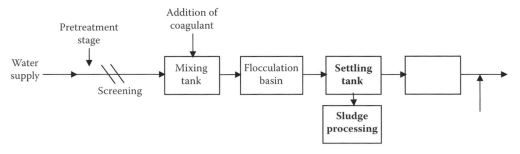

FIGURE 18.6 Sedimentation.

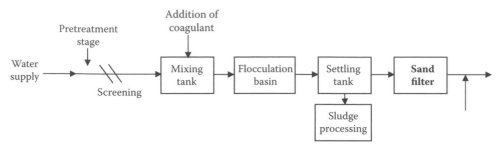

FIGURE 18.7 Filtration.

between the filter media grains. In addition to removing silt and sediment, floc, algae, insect larvae, and any other large elements, filtration also contributes to the removal of bacteria and protozoa such as *Giardia lamblia* and *Cryptosporidium*. Some filtration processes are also used for iron and manganese removal.

18.9.1 Types of Filter Technologies

The Surface Water Treatment Rule (SWTR) specifies four filtration technologies, although SWTR also allows the use of alternative filtration technologies (e.g., cartridge filters). The specified technologies are (1) slow sand filtration/rapid sand filtration, (2) pressure filtration, (3) diatomaceous earth filtration, and (4) direct filtration. Of these, all but rapid sand filtration are commonly employed in small water systems that use filtration. Each type of filtration system has advantages and disadvantages. Regardless of the type of filter, however, filtration involves the processes of *straining* (where particles are captured in the small spaces between filter media grains), *sedimentation* (where the particles land on top of the grains and stay there), and *adsorption* (where a chemical attraction occurs between the particles and the surface of the media grains).

18.9.1.1 Slow Sand Filters

The first slow sand filter was installed in London in 1829, and the technique was used widely throughout Europe, although not in the United States. By 1900, rapid sand filtration began taking over as the dominant filtration technology, although a few slow sand filters are still in operation today. With the advent of the Safe Drinking Water Act (SDWA) and its regulations (especially the Surface Water Treatment Rule) and recognition of the problems associated with *Giardia lamblia* and *Cryptosporidium* in surface water, the water industry is reexamining the use of slow sand filters.

On the plus side, slow sand filtration is well suited for small water systems. It is a proven, effective filtration process with relatively low construction costs and low operating costs (it does not require constant operator attention). It is quite effective for water systems as large

as 5000 people; beyond that, the surface area requirements and manual labor required to recondition the filters make rapid sand filters the more effective choice. The filtration rate is generally in the range of 45 to 150 gallons per day per square foot. Components of a slow sand filter include:

- A covered structure to hold the filter media
- An underdrain system
- Graded rock that is placed around and just above the underdrain
- The filter media, consisting of 30 to 55 inches of sand with a grain size of 0.25 to 0.35 mm
- Inlet and outlet piping to convey the water to and from the filter and the means to drain filtered water to waste

The area above the top of the sand layer is flooded with water to a depth of 3 to 5 feet, and the water is allowed to trickle down through the sand. An overflow device prevents excessive water depth. The filter must have provisions for filling it from the bottom up, and it must be equipped with a loss-of-head gauge, a rate-of-flow control device (such as an orifice or butterfly valve), a weir or effluent pipe that ensures that the water level cannot drop below the sand surface, and filtered waste sample taps.

When the filter is first placed in service, the head loss through the media caused by the resistance of the sand is about 0.2 feet (i.e., a layer of water 0.2 feet deep on top of the filter will provide enough pressure to push the water downward through the filter). As the filter operates, the media become clogged with the material being filtered out of the water, and the head loss increases. When it reaches about 4 to 5 feet, the filter must be cleaned.

For efficient operation of a slow sand filter, the water being filtered should have a turbidity averaging less than 5 TU, with a maximum of 30 TU. Slow sand filters are not backwashed the way conventional filtration units are. One to 2 inches of material must be removed on a periodic basis to keep the filter operating.

18.9.1.2 Rapid Sand Filters

The rapid sand filter, which is similar in some ways to the slow sand filter, is one of the most widely used filtration units. The major difference is in the principle of operation; that is, in the speed or rate at which water passes through the media. In operation, water passes downward through a sand bed that removes the suspended particles. The suspended particles consist of the coagulated matter remaining in the water after sedimentation, as well as a small amount of uncoagulated suspended matter.

Some significant differences exist in construction, control, and operation between slow sand filters and rapid sand filters. Because of the design and construction of the rapid sand filtration, the land area required to filter the same quantity of water is reduced. Components of a rapid sand filter include:

- Structure to house media
- Filter media
- Gravel media support layer
- Underdrain system
- Valves and piping system
- Filter backwash system
- Waste disposal system

Usually 2 to 3 feet deep, the filter media are supported by approximately 1 foot of gravel. The media may be fine sand or a combination of sand, anthracite coal, and coal (dual- or multimedia filter). Water is applied to a rapid sand filter at a rate of 1.5 gallons per minute per square foot of filter media surface. When the rate is between 4 and 6 gpm/ft^2, the filter is referred to as a *high-rate filter*; at a rate over 6 gpm/ft^2, the filter is referred to as a *ultra-high-rate filter*. These rates compare to the slow sand filtration rate of 45 to 150 gallons per day per square foot. High-rate and ultra-high-rate filters must meet additional conditions to assure proper operation.

Generally, raw water turbidity is not that high; however, even if raw water turbidity values exceed 1000 TU, properly operated rapid sand filters can produce filtered water with a turbidity or well under 0.5 TU. The time the filter is in operation between cleanings (filter runs) usually ranges from 12 to 72 hours, depending on the quality of the raw water; the end of the run is indicated by the head loss approaching 6 to 8 feet. Filter *breakthrough* (when filtered material is pulled through the filter into the effluent) can occur if the head loss becomes too great. Operation with head loss too high can also cause *air binding* (which blocks part of the filter with air bubbles), increasing the flow rate through the remaining filter area.

Rapid sand filters have the advantage of lower land requirements, and they have other advantages, as well; for example, rapid sand filters cost less, are less labor intensive to clean, and offer higher efficiency with highly turbid waters. On the downside, the operation and maintenance costs of rapid sand filters are much higher in comparison because of the increased complexity of the filter controls and backwashing system.

When *backwashing* a rapid sand filter, the filter is cleaned by passing treated water backward (upward) through the filter media and agitating the top of the media. The need for backwashing is determined by a combination of filter run time (i.e., the length of time since the last backwashing), effluent turbidity, and head loss through the filter. Depending on the raw water quality, the run time varies from one filtration plant to another (and may even vary from one filter to another in the same plant).

Note: Backwashing usually requires 3 to 7% of the water produced by the plant.

18.9.1.3 Pressure Filter Systems

When raw water is pumped or piped from the source to a gravity filter, the head (pressure) is lost as the water enters the floc basin. When this occurs, pumping the water from the plant clearwell to the reservoir is usually necessary. One way to reduce pumping is to place the plant components into pressure vessels, thus maintaining the head. This type of arrangement is known as a pressure filter system. Pressure filters are also quite popular for iron and manganese removal and for filtration of water from wells. They may be placed directly in the pipeline from the well or pump with little head loss. Most pressure filters operate at a rate of about 3 gpm/ft^2.

Operationally the same, and consisting of components similar to those of a rapid sand filter, the main difference between a rapid sand filtration system and a pressure filtration system is that the entire pressure filter is contained within a pressure vessel. These units are often highly automated and are usually purchased as self-contained units with all necessary piping, controls, and equipment contained in a single unit. They are backwashed in much the same manner as the rapid sand filter.

The major advantage of the pressure filter is its low initial cost. They are usually prefabricated, with standardized designs. A major disadvantage is that the operator is unable to observe the filter in the pressure filter and so is unable to determine the condition of the media. Unless the unit has an automatic shutdown feature on high effluent turbidity, driving filtered material through the filter is possible.

18.9.1.4 Diatomaceous Earth Filters

Diatomaceous earth is a white material made from the skeletal remains of diatoms. The skeletons are microscopic and in most cases porous. Diatomaceous earth is available in various grades, and the grade is selected based on filtration requirements. These diatoms are mixed in water slurry and fed onto a fine screen called a *septum,* usually made of stainless steel, nylon, or plastic. The slurry is fed at a rate of 0.2 lb/ft^2 of filter area. The diatoms collect in a precoat over the septum, forming an extremely fine screen. Diatoms are fed continuously with the raw water, causing the buildup of a filter cake approximately 1/8 to 1/5 inch thick. The openings are so small that the fine particles that cause turbidity are trapped on the screen. Coating the septum with diatoms gives it the ability to filter out very small microscopic material. The fine screen and the buildup of filtered particles cause a high head loss through the filter. When the head loss reaches a maximum level (30 psi on a pressure-type filter or 15 inches of mercury on a vacuum-type filter), the filter cake must be removed by backwashing.

A slurry of diatoms is fed with raw water during filtration in a process called *body feed,* which prevents premature clogging of the septum cake. These diatoms are caught on the septum, increasing the head loss and preventing the cake from clogging too rapidly by the particles being filtered. Body feed increases head loss, but the head loss increases are more gradual than if body feed were not use.

Diatomaceous earth filters are relatively low in cost to construct, but they have high operating costs and can cause frequent operating problems if not properly operated and maintained. They can be used to filter raw surface waters or surface-influenced groundwaters with low turbidity (<5 NTU) and low coliform concentrations (no more than 50 coliforms per 100 mL) and may also be used for iron and manganese removal following oxidation. Filtration rates are between 1.0 and 1.5 gpm/ft^2.

18.9.1.5 Direct Filtration

Direct filtration is a treatment scheme that omits the flocculation and sedimentation steps prior to filtration. Coagulant chemicals are added, and the water is passed directly onto the filter. All solids removal takes place on the filter, which can lead to much shorter filter runs, more frequent backwashing, and a greater percentage of finished water used for backwashing. The lack of a flocculation process and sedimentation basin reduces construction costs but increases the requirement for skilled operators and high-quality instrumentation. Direct filtration must be used only where the water flow rate and raw water quality are fairly consistent and where the incoming turbidity is low.

18.9.1.6 Alternative Filters

A *cartridge filter system* can be employed as an alternative filtering system to reduce turbidity and remove *Giardia.* A cartridge filter is made of a synthetic medium contained in a plastic or metal housing. These systems are normally installed in a series of three or four filters. Each filter contains a medium that is successively smaller than the previous filter. The media sizes typically range from 50

to 5 μm or less. The filter arrangement is dependent on the quality of the water, the capability of the filter, and the quantity of water needed. The USEPA and state agencies have established criteria for the selection and use of cartridge filters. Generally, cartridge filter systems are regulated in the same manner as other filtration systems.

Because of new regulatory requirements and the need to provide more efficient removal of pathogenic protozoa (e.g., *Giardia* and *Cryptosporidium*) from water supplies, *membrane filtration systems* are finding increased application in water treatment systems. A *membrane* is a thin film separating two different phases of a material acting as a selective barrier to the transport of matter operated by some driving force. Simply, a membrane can be regarded as a sieve with very small pores. Membrane filtration processes are typically pressure, electrically, vacuum, or thermally driven. The types of drinking water membrane filtration systems include microfiltration, ultrafiltration, nanofiltration, and reverse osmosis. In a typical membrane filtration process, there is one input and two outputs. Membrane performance is largely a function of the properties of the materials to be separated and can vary throughout operation.

18.9.2 Common Filter Problems

Two common types of filter problems occur: those caused by filter runs that are too long (infrequent backwash) and those caused by inefficient backwash (cleaning). A filter run that is too long can cause *breakthrough* (the pushing of debris removed from the water through the media and into the effluent) and *air binding* (the trapping of air and other dissolved gases in the filter media). Air binding occurs when the rate at which water exits the bottom of the filter exceeds the rate at which the water penetrates the top of the filter. When this happens, a void and partial vacuum occur inside the filter media. The vacuum causes gases to escape from the water and fill the void. When the filter is backwashed, the release of these gases may cause a violent upheaval in the media and destroy the layering of the media bed, gravel, or underdrain. Two solutions to the problems are to (1) check the filtration rates to be sure they are within the design specifications, and (2) remove the top 1 inch of media and replace with new media. This keeps the top of the media from collecting the floc and sealing the entrance into the filter media.

Another common filtration problem is associated with poor backwashing practices: the formation of *mud balls* that get trapped in the filter media. In severe cases, mud balls can completely clog a filter. Poor agitation of the surface of the filter can form a crust on top of the filter; the crust later cracks under the water pressure, causing uneven distribution of water through the filter media. Filter cracking can be corrected by removing the top 1 inch of the filter media, increasing the backwash rate, or checking the effectiveness of the surface wash (if installed). Back-washing at too high a rate can cause the filter media to wash out of the filter over the effluent troughs and may damage the filter underdrain system. Two possible solutions are to (1) check the backwash rate to be sure that it meets the design criteria, and (2) check the surface wash (if installed) for proper operation.

18.9.3 FILTRATION AND COMPLIANCE WITH TURBIDITY REQUIREMENTS (IESWTR)

Under the 1996 Safe Drinking Water Act (SDWA) Amendments, USEPA was directed to supplement the existing Surface Water Treatment Rule (SWTR) with the Interim Enhanced Surface Water Treatment Rule (IESWTR) (USEPA, 1999a) to improve protection against waterborne pathogens. Key provisions established in the IESWTR include (USEPA, 1998):

- A maximum contaminant level goal (MCLG) of 0 for *Cryptosporidium*, with a 2-log (99%) percent) *Cryptosporidium* removal requirement for systems that filter
- Strengthened combined filter effluent turbidity performance standards
- Individual filter turbidity monitoring provisions
- Disinfection benchmark provisions to ensure continued levels of microbial protection while facilities take the necessary steps to comply with new disinfection byproduct standards
- Inclusion of *Cryptosporidium* in the definition of groundwater under the direct influence of surface water (GWUDI) and in the watershed as well as in the watershed control requirements for unfiltered public water systems
- Requirements for covers on new finished water reservoirs
- Sanitary surveys for all surface water systems regardless of size

This section outlines the regulatory, reporting, and record-keeping requirements with which all waterworks operators should be familiar, as well as additional compliance aspects of the IESWTR related to turbidity.

18.9.3.1 IESWTR Regulatory Requirements

The Interim Enhanced Surface Water Treatment Rule contains several key provisions that strengthen combined filter effluent turbidity performance standards and individual filter turbidity monitoring.

18.9.3.1.1 Applicability

Entities potentially regulated by the IESWTR are public water systems that use surface water or groundwater under the direct influence of surface water and serve at least 10,000 people (including industries and state, local, tribal, or federal governments). To determine whether a facility may be regulated by this action, it is necessary to carefully examine the applicability criteria subpart H (systems subject to the Surface Water Treatment Rule) and subpart P (subpart H systems that serve 10,000 or more people) of the final rule.

Note: Systems subject to the turbidity provisions of the IESWTR are a subset of systems subject to the IESWTR, which utilize rapid granular filtration (i.e., conventional filtration treatment and direct filtration) or other filtration processes (excluding slow sand and diatomaceous earth filtration).

18.9.3.1.2 Combined Filter Effluent Monitoring

Under the SWTR, a subpart H system that provides filtration treatment must monitor turbidity in the combined filter effluent. Turbidity measurements must be performed on representative samples of the system's filtered water every 4 hours (or more frequently) that the system serves water to the public. A public water system may substitute continuous turbidity monitoring for grab sample monitoring if it validates the continuous measurement for accuracy on a regular basis using a protocol approved by the state. The turbidity performance provisions of the IESWTR require that all surface water systems that use conventional treatment or direct filtration and serve a population of 10,000 people must meet two distinct filter effluent limits; a maximum limit and a 95% limit. These limits, set forth in the IESWTR, are outlined below for the different types of treatment employed by systems:

1. *Conventional treatment or direct filtration*— For conventional and direct filtration systems (including those systems utilizing inline filtration), the turbidity level of representative samples of a system's filtered water (measured every 4 hours) must be less than or equal to 0.3 NTU in at least 95% of the measurements taken each month. The turbidity level of representative samples of a system's filtered water must not exceed 1 NTU at any time. Conventional filtration is defined as a series of processes including coagulation, flocculation, sedimentation, and filtration resulting in substantial particulate removal. Direct filtration is defined as a series of processes including coagulation and filtration but excluding sedimentation resulting in substantial particle removal.
2. *Other treatment technologies (alternative filtration)*—For other filtration technologies (technologies other than conventional, direct, slow sand, or diatomaceous earth filtration), a system may demonstrate to the state, using pilot plant

studies or other means, that the alternative filtration technology, in combination with disinfection treatment, consistently achieves 99.9% removal or inactivation of *Giardia lamblia* cysts, 99.99% removal or inactivation of viruses, and 99% removal of *Cryptosporidium* oocysts. For a system that makes this demonstration, representative samples of a system's filtered water must be less than or equal to a value determined by the state which the state determines is indicative of 2-log *Cryptosporidium* removal, 3-log *Giardia* removal, and 4-log virus removal in at least 95% of the measurements taken each month, and the turbidity level of representative samples of a system's filtered water must at no time exceed a maximum turbidity value determined by the state. Examples of such technologies include bag or cartridge filtration, microfiltration, and reverse osmosis. USEPA recommends a protocol similar to the *Protocol for Equipment Verification Testing for Physical Removal of Microbiological and Particulate Contaminants* prepared by NSF International with support from USEPA.

3. *Slow sand and diatomaceous earth filtration*— The IESWTR does not contain new turbidity provisions for slow sand or diatomaceous earth (DE) filtration systems. Utilities utilizing either of these filtration processes must continue to meet the requirements for their respective treatment as set forth in the SWTR (1 NTU 95%, 5 NTU maximum).

4. *Systems that utilize lime softening*—Systems that practice lime softening may experience difficulty in meeting the turbidity performance requirements due to residual lime floc carryover inherent in the process. USEPA is allowing such systems to acidify turbidity samples prior to measurement using a protocol approved by states. The chemistry supporting this decision is well documented in environmental chemistry texts. USEPA recommends that acidification protocols lower the pH of samples to <8.3 to ensure an adequate reduction in carbonate ions and corresponding increase in bicarbonate ions. Acid should consist of either hydrochloric acid or sulfuric acid of standard lab grade. Care should be taken when adding acid to samples. Operators should always follow the sampling guidelines as directed by their supervisors and standard protocols. If systems choose to use acidification, USEPA recommends that systems maintain documentation regarding the turbidity with and without acidification as well as pH values and quantity of acid added to the sample.

18.9.3.1.3 Individual Filter Monitoring

In addition to the combined filter effluent monitoring discussed above, those systems that use conventional treatment or direct filtration (including inline filtration) must conduct continuous monitoring of turbidity for each individual filter using an approved method in §141.74(a) and must calibrate turbidimeters using the procedure specified by the manufacturer. Systems must record the results of individual filter monitoring every 15 minutes. If the individual filter is not providing water that contributes to the combined filter effluent (i.e., it is not operating, is filtering to waste, or is recycled), the system does not have to record or monitor the turbidity for that specific filter.

Note: Systems that utilize filtration other than conventional or direct filtration are not required to conduct individual filter monitoring, although USEPA recommends such systems consider individual filter monitoring.

If a failure occurs in the continuous turbidity monitoring equipment, the system must conduct grab sampling every 4 hours in lieu of continuous monitoring, but must return to 15-minute monitoring no more than 5 working days following the failure of the equipment.

18.9.3.2 Reporting and Recordkeeping

Distinct reporting and recordkeeping requirements for the turbidity provisions of the IESWTR for both systems and states include the following:

18.9.3.2.1 System Reporting Requirements

Under the IESWTR, systems are tasked with specific reporting requirements associated with combined filter effluent monitoring and individual filter effluent monitoring.

- *Combined filter effluent reporting*—Turbidity measurements as required by §141.173 must be reported within 10 days after the end of each month the system serves water to the public. Information that must be reported includes:
  1. The total number of filtered water turbidity measurements taken during the month
  2. The number and percentage of filtered water turbidity measurements taken during the month that are less than or equal to the turbidity limits specified in §141.173 (0.3 NTU for conventional and direct filtration and the turbidity limit established by the state for other filtration technologies)
  3. The date and value of any turbidity measurements taken during the month that exceed 1 NTU for systems using conventional filtration

treatment or direct filtration and the maximum limit established by the state for other filtration technologies. This reporting requirement is similar to the reporting requirements currently found under the SWTR.

- *Individual filter requirements*—Systems utilizing conventional and direct filtration must report that they have conducted individual filter monitoring in accordance with the requirements of the IESWTR within 10 days after the end of each month the system serves water to the public. Additionally, systems must report individual filter turbidity measurements within 10 days after the end of each month the system serves water to the public only if measurements demonstrate one of the following:

1. Any individual filter has a measured turbidity level greater than 1.0 NTU in two consecutive measurements taken 15 minutes apart. The system must report the filter number, the turbidity measurement, and the date(s) on which the exceedance occurred. In addition, the system must either produce a filter profile for the filter within 7 days of the exceedance (if the system is not able to identify an obvious reason for the abnormal filter performance) and report that the profile has been produced or report the obvious reason for the exceedance.

2. Any individual filter has a measured turbidity level of greater than 0.5 NTU in two consecutive measurements taken 15 minutes apart at the end of the first 4 hours of continuous filter operation after the filter has been backwashed or otherwise taken offline. The system must report the filter number, the turbidity, and the date(s) on which the exceedance occurred. In addition, the system must either produce a filter profile for the filter within 7 days of the exceedance (if the system is not able to identify an obvious reason for the abnormal filter performance) and report that the profile has been produced or report the obvious reason for the exceedance.

3. Any individual filter has a measured turbidity level of greater than 1.0 NTU in two consecutive measurements taken 15 minutes apart at any time in each of 3 consecutive months. The system must report the filter number, the turbidity measurement, and the date(s) on which the exceedance occurred. In addition, the system must conduct a self-assessment of the filter.

4. Any individual filter has a measured turbidity level of greater than 2.0 NTU in two consecutive measurements taken 15 minutes apart at any time in each of two consecutive measurements taken 15 minutes apart at any time in each of 2 consecutive months. The system must report the filter number, the turbidity measurement, and the date(s) on which the exceedance occurred. In addition, the system must contact the state or a third party approved by the state to conduct a comprehensive performance evaluation.

18.9.3.2.2 State Reporting Requirements

Under §142.15, each state that has primary enforcement responsibility is required to submit quarterly reports to the Administrator of the USEPA on a schedule and in a format prescribed by the Administrator, which includes:

1. New violations by public water systems in the state during the previous quarter with respect to state regulations adopted to incorporate the requirements of national primary drinking water regulations

2. New enforcement actions taken by the state during the previous quarter against public water systems with respect to state regulations adopted to incorporate the requirements of national primary drinking water standards

Any violations or enforcement actions with respect to turbidity would be included in the quarterly report noted above. USEPA has developed a state implementation guidance manual that includes additional information on state reporting requirements.

18.9.3.2.3 System Recordkeeping Requirements

Systems must maintain the results of individual filter monitoring taken under §141.174 for at least 3 years. These records must be readily available for state representatives to review during sanitary surveys on other visits.

18.9.3.2.4 State Recordkeeping Requirements

Records of turbidity measurements must be kept for not less than 1 year. The information retained must be set forth in a form that makes possible comparison with limits specified in §§141.71, 141.73, 141.173, and 141.175. Records of decisions made on a system-by-system and case-by-case basis under provisions of part 141, subpart H or subpart P, must be made in writing and kept by the state (this includes records regarding alternative filtration determinations). USEPA has developed a state implementation guidance manual that includes additional information on state recordkeeping requirements.

18.9.3.3 Additional Compliance Issues

The following section outlines additional compliance issues associated with the IESWTR.

18.9.3.3.1 Schedule

The IESWTR was published on December 16, 1998, and became effective on February 16, 1999. The SDWA requires, within 24 months following the promulgation of a rule, that the primacy agencies adopt any state regulations necessary to implement the rule. Under §14.13, these rules must be at least as stringent as those required by USEPA. Thus, primacy agencies had to promulgate regulations at least as stringent as the IESWTR by December 17, 2000. Beginning December 17, 2001, systems serving at least 10,000 people had to meet the turbidity requirements in §141.173.

18.9.3.3.2 Individual Filter Follow-Up Action

Based on the monitoring results obtained through continuous filter monitoring, a system may have to conduct one of the following follow-up actions due to persistently high turbidity levels at an individual filter:

- Filter profile
- Individual filter self-assessment
- Comprehensive performance evaluation

These requirements are found in §141.175(b)(1)–(4).

18.9.3.3.3 Abnormal Filter Operations— Filter Profile

A filter profile must be produced if no obvious reason for abnormal filter performance can be identified. A filter profile is a graphical representation of individual filter performance based on continuous turbidity measurements or total particle counts vs. time for an entire filter run, from startup to backwash inclusively, that includes assessment of filter performance while another filter is being backwashed. The run length during this assessment should be representative of typical plant filter runs. The profile should include an explanation of the cause of any filter performance spikes during the run. Examples of possible abnormal filter operations that may be obvious to operators include the following:

- Outages or maintenance activities at processes within the treatment train
- Coagulant feed pump or equipment failure
- Filters being run at significantly higher loading rates than approved

It is important to note that, although the reasons for abnormal filter operation may appear obvious, they could be masking other reasons that are more difficult to identify. These may include such situations as:

- Distribution in filter media
- Excessive or insufficient coagulant dosage
- Hydraulic surges due to pump changes or other filters being brought on/offline.

Systems must use their best professional judgment and discretion when determining when to develop a filter profile. Attention at this stage will help systems avoid the other forms of follow-up action described below.

18.9.3.3.4 Individual Filter Self-Assessment

A system must conduct an individual filter self-assessment for any individual filter that has a measured turbidity level of greater than 1.0 NTU in two consecutive measurements taken 15 minutes apart in each of 3 consecutive months. The system must report the filter number, the turbidity measurement, and the date(s) on which the exceedance occurred.

18.9.3.3.5 Comprehensive Performance Evaluation

A system must conduct a comprehensive performance evaluation (CPE) if any individual filter has a measured turbidity level of greater than 2.0 NTU in two consecutive measurements taken 15 minutes apart in 2 consecutive months. The system must report the filter number, the turbidity measurement, and the date(s) on which the exceedance occurred. The system must contact the state or a third party approved by the state to conduct a comprehensive performance evaluation.

Note: USEPA has developed a guidance document called *Handbook: Optimizing Water Treatment Plant Performance Using the Composite Correction Program*, EPA/625/6-91/027 (1998).

18.9.3.3.6 Notification

The IESWTR contains two distinct types of notification: state and public. It is important to understand the differences between each and the requirements of each:

- *State notification*—Systems are required to notify states under §141.31. Systems must report to the state within 48 hours the failure to comply with any national primary drinking water regulation. The system within 10 days of completion of each public notification required pursuant to §141.32 must submit to the state a representative copy of each type of notice distributed, published, posted, or made available to persons served by the system or the media. The water supply system must also submit to the state (within the time stated in the request made by the state) copies of any records required to be maintained under §141.33 or copies of any documents then in existence which the state or the Administrator is entitled to inspect pursuant to the authority of §1445 of the Safe Drinking Water Act or the equivalent provisions of the state law.

- *Public notification*—The IESWTR specifies that the public notification requirements of the Safe Drinking Water Act (SDWA) and the implementation regulations of §141.32 must be followed. These regulations divide public notification requirements into two tiers. These tiers are defined as follows:

Tier 1

- Failure to comply with MCLs
- Failure to comply with prescribed treatment technique
- Failure to comply with a variance or exemption schedule

Tier 2

- Failure to comply with monitoring requirements
- Failure to comply with a testing procedure prescribed by a NPDWR
- Operating under a variance/exemption (this is not considered a violation but public notification is required)

Certain general requirements must be met by all public notices. All notices must provide a clear and readily understandable explanation of the violation, any potential adverse health effects, the population at risk, the steps the system is taking to correct the violation, the necessity of seeking alternate water supplies (if any), and any preventative measures the consumer should take. The notice must be conspicuous and not contain any unduly technical language, unduly small print, or similar obstacles. The notice must include the telephone number of the owner or operator or designee of the public water system as a source of additional information concerning the violation where appropriate. The notice must be bi- or multilingual if appropriate.

18.9.3.3.7 Tier 1 Violations

In addition, the public notification rule requires that when providing notification on potential adverse health effects in Tier 1 public notices and in notices on the granting and continued existence of a variance or exemption, the owner/operator of a public water system must include certain mandatory health effects language. For violations of treatment technique requirements for filtration and disinfection, the mandatory health effects language is:

The USEPA sets drinking water standards and has determined that the presence of microbiological contaminants is a health concern at certain levels of exposure. If water is inadequately treated, microbiological contaminants in

that water cause disease. Disease symptoms may include diarrhea, cramps, nausea, and possibly jaundice, and any associated headaches and fatigue. These symptoms, however, are not just associated with disease-causing organisms in drinking water but also may be caused by a number of factors other than your drinking water. USEPA has set enforceable requirements for treating drinking water to reduce the risk of these adverse health effects. Treatment such as filtering and disinfection of the water removes or destroys microbiological contaminants. Drinking water which is treated to meet UEPA requirements is associated with little to none of this risk and should be considered safe.

Further the owner or operator of a community water system must give a copy of the most recent notice for any Tier 1 violations to all new billing units or hookups prior to or at the time service begins.

The medium for performing public notification and the time period in which notification must be sent vary with the type of violation and are specified in §141.32. For Tier 1 violations, the owner or operator of a public water system must give notice:

By publication in a local daily newspaper as soon as possible but in no case later than 14 days after the violation or failure. If the area does not have a daily newspaper, then notice shall be given by publication in a weekly newspaper of general circulation in the area, and

By either direct mail delivery or hand delivery of the notice, either by itself or with the water bill no later than 45 days after the violation or failure. The primacy agency may waive the requirement if it determines that the owner or operator has corrected the violation with 45 days.

Although the IESWTR does not specify any acute violations, the primacy agency may specify some Tier 1 violations as posing an acute risk to human health; examples might include:

- A waterborne outbreak in an unfiltered supply
- Turbidity of a filtered water exceeding 1.0 NTU at any time
- Failure to maintain a disinfectant residual of at least 0.2 mg/L in the water being delivered to the distribution system.

For these violations or any others defined by the primacy agency as "acute" violations, the system must furnish a copy of the notice to the radio and television stations serving the area as soon as possible but in no case later than 72 hours after the violation. Depending on the circumstances particular to the system, as determined by the primacy agency, the notice may instruct that all water be boiled prior to consumption.

Following the initial notice, the owner or operator must give notice at least once every 3 months by mail delivery (either by itself or with the water bill) or by hand delivery for as long as the violation or failures exist.

There are two variations on these requirements. First, the owner or operator of a community water system in an area not served by a daily or weekly newspaper must give notice within 14 days after the violation by hand delivery or continuous posting of a notice of the violation. The notice must continue for as long as the violation exists. Notice by hand delivery must be repeated at least every 3 months for the duration of the violation. Second, the owner or operator of a noncommunity water system (i.e., one serving a transitory population) may give notice by hand delivery or continuous posting of the notice in conspicuous places in the area served by the system. Notice must be given within 14 days after the violation. If notice is given by posting, then it must continue as long as the violations exist. Notice given by hand delivery must be repeated at least every 3 months for as long as the violation exists.

18.9.3.3.8   Tier 2 Violations

For Tier 2 violations (i.e., violations of §141.74 and §141.174), notice must be given within 3 months after the violation by publication in a daily newspaper of general circulation, or if there is no daily newspaper then in a weekly newspaper. In addition, the owner or operator must give notice by mail (either by itself or with the water bill) or by hand delivery at least once every 3 months for as long as the violation exists. Notice of a variance or exemption must be given every 3 months from the date it is granted for as long as it remains in effect.

If a daily or weekly newspaper does not serve the area, the owner or operator of a community water system must give notice by continuous posting in conspicuous places in the area served by the system. This must continue as long as the violation exists or the variance or exemption remains in effect. Notice by hand delivery must be repeated at least every 3 months for the duration of the violation or the variance or exemption.

For noncommunity water systems, the owner or operator may give notice by hand delivery or continuous posting in conspicuous places, beginning within 3 months of the violation or the variance or exemption. Posting must continue for the duration of the violation or variance or exemption, and notice by hand delivery must be repeated at least every 3 months during this period.

The primacy agency may allow for the owner or operator to provide less frequent notice for minor monitoring violations (as defined, by the primacy agency if USEPA has approved the primacy agency's substitute requirements contained in a program revision application).

18.9.3.4   Variances and Exemptions

As with the SWTR, no variances from the requirements in §141 are permitted for subpart H systems. Under §1416(a) of the SDWA, USEPA or a state may exempt a public water system from any requirements related to an MCL or treatment technique of a national primary drinking water regulation (NPDWR) if it finds that: (1) due to compelling factors (which may include economic factors such as qualifications of the PWS as serving a disadvantaged community), the public water system (PWS) is unable to comply with the requirement or implement measures to develop an alternative source of water supply; (2) the exemption will not result in an unreasonable risk to health; (3) the PWS was in operation on the effective date of the NPDWR, or, for a system that was not in operation by that date, only if no reasonable alternative source of drinking water is available to the new systems; and (4) management or restructuring changes (or both) cannot reasonably result in compliance with the Act or improve the quality of drinking water.

18.10   DISINFECTION

Disinfection is a unit process used in both water and wastewater treatment. Many of the terms, practices, and applications discussed in this section apply to both water and wastewater treatment; however, there are also some differences between the use of disinfection in water and wastewater treatment—mainly in the types of disinfectants used and applications. Thus, in this section we discuss disinfection as it applies to water treatment. Later, we will cover disinfection as it applies to wastewater treatment. Much of the information presented in this section is based on personal experience and on USEPA (1999b).

To comply with the SDWA regulations, the majority of PWSs use some form of water treatment. The 1995 Community Water System Survey reported that, in the United States, 99% of surface water systems provided some treatment to their water, and 99% of these treatment systems utilized disinfection/oxidation as part of the treatment process. Although 45% of groundwater systems provided no treatment, 92% of those groundwater plants providing some form of treatment included disinfection/oxidation as part of the treatment process (USEPA, 1997).

Why such a public health concern regarding groundwater supplies? According to USEPA's Bruce Macler (1996):

> There are legitimate concerns for public health from microbial contamination of groundwater systems. Microorganisms and other evidence of fecal contamination have been detected in a large number of wells tested, even those wells that had been previously judged not vulnerable to such contamination. The scientific community believes

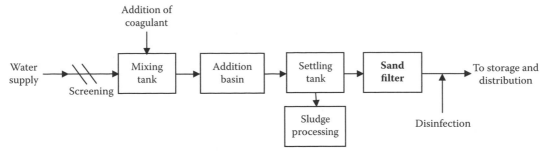

FIGURE 18.8 Disinfection.

that microbial contamination of groundwater is real and widespread. The public health impact from this contamination, while not well quantified, appears to be large. Disease outbreaks have occurred in many groundwater systems. Risk estimates suggest several million illnesses each year. Additional research is underway to better characterize the nature and magnitude of the public health problem.

The most commonly used disinfectants/oxidants (in no particular order) are chlorine, chlorine dioxide, chloramines, ozone, and potassium permanganate.

As mentioned, the process used to control waterborne pathogenic organisms and prevent waterborne disease is disinfection. The goal in proper disinfection in a water system is to destroy all disease-causing organisms. Disinfection should not be confused with sterilization. *Sterilization* is the complete killing of all living organisms. Waterworks operators disinfect by destroying organisms that might be dangerous; they do not attempt to sterilize water.

Disinfectants are also used to achieve other specific objectives in drinking water treatment. These other objectives include nuisance control (e.g., for zebra mussels and Asiatic clams), oxidation of specific compounds (i.e., taste- and odor-causing compounds, iron, and manganese), and use as a coagulant and filtration aid. The goals of this section are to:

- Provide a brief overview of the need for disinfection in water treatment.
- Provide basic information that is common to all disinfectants.
- Discuss other uses for disinfectant chemicals (e.g., as oxidants).
- Describe trends in disinfection byproduct (DBP) formation and the health effects of DBPs found in water treatment.
- Discuss microorganisms of concern in water systems, their associated health impact, and the inactivation mechanisms and efficiencies of various disinfectants.
- Summarize current disinfection practices in the Unites States, including the use of chlorine as a disinfectant and an oxidant.

In water treatment, disinfection is almost always accomplished by adding chlorine or chlorine compounds after all other treatment steps (see Figure 18.8), although in the United States ultraviolet (UV) light and potassium permanganate and ozone processes may be encountered.

The effectiveness of disinfection in a drinking water system is measured by testing for the presence or absence of coliform bacteria. *Coliform bacteria* found in water are generally not pathogenic, although they are good indicators of contamination. Their presence indicates the possibility of contamination, and their absence indicates the possibility that the water is potable—if the source is adequate, the waterworks history is good, and acceptable residual chlorine is present.

Desired characteristics of a disinfectant include the following:

- It must be able to deactivate or destroy any type or number of disease-causing microorganisms that may be in a water supply, in reasonable time, within expected temperature ranges, and despite changes in the character of the water (pH, for example).
- It must be nontoxic.
- It must not add unpleasant taste or odor to the water.
- It must be readily available at a reasonable cost and be safe and easy to handle, transport, store, and apply.
- It must be quick and easy to determine the concentration of the disinfectant in the treated water.
- It should persist within the disinfected water at a high enough concentration to provide residual protection through the distribution.

18.10.1 NEED FOR DISINFECTION IN WATER TREATMENT

Although the epidemiological relation between water and disease had been suggested as early as the 1850s, it was not until the establishment of the germ theory of disease by Pasteur in the mid-1880s that water as a carrier of

TABLE 18.3
Waterborne Diseases from Bacteria

Causative Agent	Disease
Salmonella	
S. typhosa	Typhoid fever
S. paratyphi	Paratyphoid fever
S. schottinulleri	
S. hirschfeldi C.	
Shigella	
S. flexneri	Bacillary dysentery
S. dysenteriae	
S. sonnel	
S. paradysinteriae	
Vibrio	
V. comma	Cholera
V. cholerae	
Pasteurella tularensis	Tularemia
Brucella melitensis	Brucellosis
Leptospira icterchaemorrihagiae	Leptospirosis
Enteropathogenic *Escherichia coli*	Gastroenteritis

TABLE 18.4
Waterborne Human Enteric Viruses

Group	Subgroup
Enterovirus	Poliovirus
	Echovirus
	Coxsackievirus A and B
Reovirus	
Adenovirus	
Hepatitis	

TABLE 18.5
Waterborne Diseases from Parasites

Causative Agent	Disease
Ascario lumbricoides (round worm)	Ascariasis
Cryptosporidium muris, C. parvum	Cryptosporidiosis
Entamoeba histolytica	Amebiasis
Giardia lamblia	Giardiasis
Naegleria gruberi	Amoebic meningoencephalitis
Schistosoma mansoni	Schistosomiasis
Taenis saginata (beef tapeworm)	Taeniasis

disease-producing organisms was understood. In the 1850s, while London was experiencing the "Broad Street well" cholera epidemic, Dr. John Snow conducted his now-famous epidemiological study. Dr. Snow concluded that the well had become contaminated by a visitor with the disease who had arrived in the vicinity. Cholera was one of the first diseases to be recognized as capable of being waterborne. Also, this incident was probably the first reported disease epidemic attributed to direct recy-

cling of nondisinfected water. Now, over 150 years later, the list of potential waterborne disease due to pathogens is considerably larger and includes bacterial, viral, and parasitic microorganisms (see Table 18.3, Table 18.4, and Table 18.5).

A major cause for disease outbreaks in potable water is contamination of the distribution system from cross-connections and backsiphonage with nonpotable water; however, outbreaks resulting from distribution system contamination are usually quickly contained and result in relatively few illnesses compared to contamination of the source water or a breakdown in the treatment system, which typically produce many cases of illnesses per incident. When considering the number of cases, the major causes of disease outbreaks are source water contamination and treatment deficiencies (White, 1992). Historically, about 46% of the outbreaks in the public water systems are found to be related to deficiencies in source water and treatment systems, with 92% of the causes of illness being due to these two particular problems.

All natural waters support biological communities. Because some microorganisms can be responsible for public health problems, biological characteristics of the source water are one of the most important parameters in water treatment. In addition to public health problems, microbiology can also affect the physical and chemical water quality and treatment plant operation.

18.10.2 PATHOGENS OF PRIMARY CONCERN

Table 18.6 shows the attributes of three groups of pathogens of concern in water treatment: bacteria, viruses, and protozoa:

- *Bacteria*—Recall that bacteria are single-celled organisms typically ranging in size from 0.1 to 10 µm. Shape, components, size, and the manner in which they grow can characterize the physical structure of the bacterial cell. Most bacteria can be grouped by shape into four general categories: spheroid, rod, curved rod or spiral, and filamentous. Cocci, or spherical bacteria, are approximately 1 to 3 µm in diameter. Bacilli (rod-shaped bacteria) are variable in size and range from 0.3 to 1.5 µm in width (or diameter) and from 1.0 to 10.0 µm in length. *Vibrios*, or curved rod-shaped bacteria, typically vary in size from 0.6 to 1.0 µm in width (or diameter) and from 2 to 6 µm in length. *Spirilla* (spiral bacteria) can be found in lengths up to 50 µm, whereas filamentous bacteria can occur in lengths in excess of 100 µm.
- *Viruses*—Viruses are microorganisms composed of the genetic material deoxyribonucleic acid (DNA) or ribonucleic acid (RNA) and a

TABLE 18.6
Attributes of the Three Waterborne Pathogens in Water Treatment

Organism	Size (µm)	Mobility	Points of Origin	Resistance to Disinfection
Bacteria	0.1–10	Motile, nonmotile	Humans and animals, water, contamined food	Type specific—bacterial spores typically have the highest resistance, vegetative bacteria have the lowest resistance.
Viruses	0.01–0.1	Nonmotile	Humans and animals, polluted water, contaminated food	Generally more resistant than vegetative bacteria
Protozoa	1–20	Motile, nonmotile	Humans and animals, sewage, decaying vegetation, water	More resistant than viruses or vegetative bacteria

protective protein coat (single-, double-, or partially double-stranded). All viruses are obligate parasites, unable to carry out any form of metabolism, and are completely dependent on host cells for replication. Viruses are typically 0.01 to 0.1 µm in size and are very species specific with respect to infection, typically attacking only one type of host. Although the principal modes of transmission for the hepatitis B virus and poliovirus are through food, personal contact, or exchange of body fluids, these viruses can be transmitted through potable water. Some viruses, such as the retroviruses (including HIV group), appear to be too fragile for water transmission to be a significant danger to public health (Spellman, 2007).

- *Protozoa* Protozoa are single cell eukaryotic microorganisms without cell walls that utilize bacteria and other organisms for food. Most protozoa are free living in nature and can be encountered in water; however, several species are parasitic and live on or in host organisms. Host organisms can vary from primitive organism such as algae to highly complex organisms such as human beings. Several species of protozoa known to utilize human beings as hosts are shown in Table 18.7.

18.10.3 RECENT WATERBORNE OUTBREAKS

Within the past 40 years, several pathogenic agents never before associated with documented waterborne outbreaks have appeared in the United States. Enteropathogenic *Escherichia coli* and *Giardia lamblia* were first identified as the etiological agents responsible for waterborne outbreaks in the 1960s. The first recorded *Cryptosporidium* infection in humans occurred in the mid-1970s. Also occurring during that time was the first recorded outbreak of pneumonia caused by *Legionella pneumophila*. Recently, numerous documented waterborne disease outbreaks have been caused by *E. coli*, *G. lamblia*, *Cryptosporidium*, and *L. pneumophila*.

18.10.3.1 *Escherichia coli*

The first documented case of waterborne disease outbreaks in the United States associated with enteropathogenic *Escherichia coli* occurred in the 1960s. Various serotypes of *E. coli* have been implicated as the etiological agent responsible for disease in newborn infants, usually as the result of cross-contamination in nurseries; also, several well-documented waterborne outbreaks of *E. coli* have occurred in adults. In 1975, the etiologic agent of a large outbreak at Crater Lake National Park was *E. coli* serotype 06:H16 (Craun, 1981).

TABLE 18.7
Human Parasitic Protozoa

Protozoan	Hosts	Disease	Transmission
Acanathamoeba castellannii	Freshwater, sewage, humans, soil	Amoebic meningoencephalitis	Gains entry through abrasions and ulcers and as secondary invaders during other infections
Balantidium coli	Pigs, humans	Balantidiasis (dysentery)	Contaminated water
Cryptosporidium parvum	Animals, humans	Cryptosporidiosis	Person-to-person or animal-to-person contact, ingestion of fecally contaminated water or food, contact with fecally contaminated environmental surfaces
Entamoeba histolytica	Humans	Amoebic dysentery	Contaminated water
Giardia lamblia	Animals, humans	Giardiasis (gastroenteritis)	Contaminated water
Naegleria	Soil, water, humans, decaying vegetation	Primary amoebic meningoencephalitis	Nasal inhalation with subsequent penetration of nasopharynx, swimming in freshwater lakes

18.10.3.2 *Giardia lamblia*

Similar to *Escherichia coli*, *Giardia lamblia* was first identified in the 1960s to be associated with waterborne outbreaks in the United States. Recall that *G. lamblia* is a flagellated protozoan that is responsible for giardiasis, a disease that can range from being mildly to extremely debilitating. *Giardia* is currently one of the most commonly identified pathogens responsible for waterborne disease outbreaks. The life cycle of *Giardia* includes a cyst stage when the organism remains dormant and is extremely resilient (i.e., the cyst can survive some extreme environmental conditions). Once ingested by a warm-blooded animal, the life cycle of *Giardia* continues with excystation.

The cysts are relatively large (8 to 14 μm) and can be removed effectively by filtration using diatomaceous earth, granular media, or membranes. Giardiasis can be acquired by ingesting viable cysts from food or water or by direct contact with fecal material. In addition to humans, wild and domestic animals have been implicated as hosts. Between 1972 and 1981, 50 waterborne outbreaks of giardiasis were reported involving about 20,000 cases (Craun and Jakubowski, 1996). Currently, no simple and reliable method exists to assay *Giardia* cysts in water samples. Microscopic methods for detection and enumeration are tedious and require examiner skill and patience. *Giardia* cysts are relatively resistant to chlorine, especially at higher pH and low temperatures.

18.10.3.3 *Cryptosporidium*

Cryptosporidium is a protozoan similar to *Giardia*. It forms resilient oocysts as part of its life cycle. The oocysts are smaller than *Giardia* cysts, as they are typically about 4 to 6 μm in diameter. These oocysts can survive under adverse conditions until ingested by a warm-blooded animal and then continue with excystation. Due to the increase in the number of outbreaks of cryptosporidiosis, a tremendous amount of research has focused on *Cryptosporidium* within the last 10 years. Medical interest has increased because of its occurrence as a life-threatening infection in individuals with depressed immune systems.

In 1993, the largest documented waterborne disease outbreak in the United States occurred in Milwaukee and was determined to be caused by *Cryptosporidium*. An estimated 403,000 people became ill, 4400 people were hospitalized, and 100 people died. The outbreak was associated with a deterioration in raw water quality and a simultaneous decrease in effectiveness of the coagulation/filtration process, which led to an increase in the turbidity of treated water and inadequate removal of *Cryptosporidium* oocysts.

18.10.3.4 *Legionella pneumophila*

An outbreak of pneumonia occurred in 1976 at the annual convention of the Pennsylvania American Legion. A total of 221 people were affected by the outbreak, and 35 of those afflicted died. The cause of pneumonia was not determined immediately, despite an intense investigation by the Centers for Disease Control and Prevention. Six months after the incident, microbiologists were able to isolate a bacterium from the autopsy lung tissue of one of the Legionnaires. The bacterium responsible for the outbreak was found to be distinct from other known bacterium and was named *Legionella pneumophila* (Witherell et al., 1988). Following the discovery of this organism, other *Legionella*-like organisms were discovered. Legionnaires' disease does not appear to be transferred person to person. Epidemiological studies have shown that the disease enters the body through the respiratory system. *Legionella* can be inhaled in water particles less than 5 μm in size from facilities such as cooling towers, hospital hot water systems, and recreational whirlpools.

18.10.4 MECHANISM OF PATHOGEN INACTIVATION

The three primary mechanisms of pathogen inactivation are:

- Destroy or impair cellular structural organization by attacking major cell constituents, such as destroying the cell wall or impairing the functions of semi-permeable membranes.
- Interfere with energy-yielding metabolism through enzyme substrates in combination with prosthetic groups of enzymes, thus rendering them nonfunctional.
- Interfere with biosynthesis and growth by preventing synthesis of normal proteins, nucleic acids, coenzymes, or the cell wall.

Depending on the disinfectant and microorganism type, combinations of these mechanisms can also be responsible for pathogen inactivation. In water treatment, it is believed that the primary factors controlling disinfection efficiency are: (1) the ability of the disinfectant to oxidize or rupture the cell wall, and (2) the ability of the disinfectant to diffuse into the cell and interfere with cellular activity (Montgomery, 1985). In addition, it is important to point out that disinfection is effective in reducing waterborne diseases because most pathogenic organisms are more sensitive to disinfection than are nonpathogens; however, disinfection is only as effective as the care used in controlling the process and ensuring that all of the water supply is continually treated with the amount of disinfectant required to produce safe water.

TABLE 18.8
The Effects of Various Oxidants on Mortality of the Asiatic Clam

Chemical	Residual (mg/L)	Temperature (°C)	pH	Life Stage	LT_{50} (Days)
Free chlorine	0.5	23	8.0	Adult	8.7
	4.8	21	7.9	Adult	5.9
	4.7	16	7.8	Juvenile	4.8
Potassium	1.1	17	7.6	Juvenile	7.9
Permanganate	4.8	17	7.6	Juvenile	8.6
Chlorine dioxide	1.2	24	6.9	Juvenile	0.7
	4.7	22	6.6	Juvenile	0.6

Source: Adapted from Cameron, G.N. et al., *J. AWWA*, 81(10), 53–62, 1989.

18.10.5 OTHER USES OF DISINFECTANTS IN WATER TREATMENT

Disinfectants are used for more than just disinfection in drinking water treatment. Although inactivation of pathogenic organisms is a primary function, disinfectants are also used as oxidants in drinking water treatment for several other functions:

- Minimization of DBP formation
- Control of nuisance Asiatic clams and zebra mussels
- Oxidation of iron and manganese
- Prevention of regrowth in the distribution system and maintenance of biological stability
- Removal of taste and odors through chemical oxidation
- Improvement of coagulation and filtration efficiency
- Prevention of algal growth in sedimentation basins and filters
- Removal of color

A brief discussion of these additional uses for oxidants follows.

18.10.5.1 Minimization of Disinfection Byproduct Formation

Strong oxidants may play a role in disinfection and DBP control strategies in water treatment. Several strong oxidants, including potassium permanganate and ozone, may be used to control DBP precursors.

Note: Potassium permanganate can be used to oxidize organic precursors at the head of the treatment plant, thus minimizing the formation of byproducts at the downstream disinfection stage of the plant. The use of ozone for oxidation of DBP precursors is currently being studied. Early work has shown

that the effects of ozonation, prior to chlorination, were highly site specific and unpredictable. The key variables that seem to determine the effect of ozone are dose, pH, alkalinity, and the nature of the organic material Ozone has been shown to be effective for DBP precursor reduction at low pHs; however, at higher pHs (i.e., above 7.5), ozone may actually increase the amount of chlorination byproduct precursors.

18.10.5.2 Control of Nuisance Asiatic Clams and Zebra Mussels

The Asiatic clam (*Corbicula fluminea*) was introduced to the United States from Southeast Asia in 1938 and now inhabits almost every major river system south of the 40° latitude. Asiatic clams have been found in the Trinity River in Texas; the Ohio River at Evansville, IN; New River at Narrows and Glen Lyn, VA; and the Catawba River in Rock Hill, SC. This animal has invaded many water utilities. It has clogged source water transmission systems and valves, screens, and meters; damaged centrifugal pumps; and caused taste and odor problems (Belanger et al., 1991; Britton and Morton, 1982; Cameron et al., 1989; Sinclair, 1964).

Cameron et al. (1989) investigated the effectiveness of several oxidants to control the Asiatic clam in both the juvenile and adult phases. As expected, the adult clam was found to be much more resistant to oxidants than the juvenile form. In many cases, the traditional method of control, free chlorination, cannot be used because of the formation of excessive amounts of trihalomethane. Cameron et al. (1989) compared the effectiveness of four oxidants for controlling the juvenile Asiatic clam in terms of the LT_{50} (time required for 50% mortality) (Table 18.8). Monochloramine was found to be the best for controlling the juvenile clams without forming THMs. The effectiveness of monochloramine increased greatly as the temperature increased. Clams can tolerate temperatures between 2 and 35°C.

In a similar study, Belanger et al. (1991) studied the biocidal potential of total residual chlorine, monochloramine, monochloramine plus excess ammonia, bromine, and copper for controlling the Asiatic clam. They found that monochloramine with excess ammonia was the most effective for controlling the clams at 30°C. Chlorination at 0.25 to 0.40 mg/L total residual chlorine at 20 to 25°C controlled clams of all sizes but had minimal effect at 12 to 15°C (as low as zero mortality). As in other studies, the toxicity of all the biocides was highly dependent on temperature and clam size.

The zebra mussel (*Dreissena polymorpha*) is a recent addition to the fauna of the Great Lakes. It was first found in Lake St. Clair in 1988, although it is believed that this native of the Black and Caspian seas was brought over from Europe in ballast water around 1985. The zebra mussel population in the Great Lakes has expanded very rapidly, both in size and geographical distribution (Herbert, 1989; Roberts, 1990). Lang (1994) reported that zebra mussels have been found in the Ohio River, Cumberland River, Arkansas River, Tennessee River, and the Mississippi River south to New Orleans.

Klerks and Fraleigh (1991) evaluated hypochlorite, permanganate, and hydrogen peroxide with iron for their effectiveness in controlling adult zebra mussels. Both continuous and intermittent 28-day static renewal tests were conducted to determine the impact of intermittent dosing. Intermittent treatment proved to be much less effective than continuous dosing. The hydrogen peroxide–iron combination (1 to 5 mg/L with 25% iron) was less effective in controlling the zebra mussel than either permanganate or hypochlorite. Permanganate (0.5 to 2.5 mg $KMnO_4$ per L) was usually less effective than hypochlorite (0.5 to 10 mg Cl_2 per L).

Van Benschoten et al. (1995) developed a kinetic model to predict the rate of mortality of the zebra mussel in response to chlorine. The model shows the relationship between residual chlorine and temperature on the exposure time required to achieve 50 and 95% mortality. Data were collected for chlorine residuals between 0.5 and 3.0 mg Cl_2 per L and temperatures from 0.3 to 24°C. The results show a strong dependence on temperature and required contact times ranging from 2 days to more than a month, depending on environmental factors and mortality required.

Brady et al. (1996) evaluated the ability of chlorine to control the growth of zebra mussel and quagga mussel (*Dreissenda bugensis*). The quagga mussel is a newly identified mollusk within the Great Lakes that is similar in appearance to the zebra mussel. Full-scale chlorination treatment produced a significantly higher mortality for the quagga mussel. The required contact time for 100% mortality for quagga and zebra mussels was 23 days and 37 days, respectively, suggesting that chlorination programs designed to control zebra mussels should also be effective for controlling populations of quagga mussels.

Mastisoff et al. (1996) evaluated chlorine dioxide (ClO_2) to control adult zebra mussels using simple, intermittent, and continuous exposures. A single 30-minute exposure to 20 mg/L chlorine dioxide or higher concentration induced at least 50% mortality, while sodium hypochlorite produced only 26% mortality, and permanganate and hydrogen peroxide were totally ineffective when dosed at 30 mg/L for 30 minutes under the same conditions. These high dosages, even though used only for a short period of time, may not allow application directly in water for certain applications due to byproducts that remain in the water. Continuous exposure to chlorine dioxide for 4 days was effective at concentrations above 0.5 mg/L (LC_{50} = 0.35 mg/L), and 100% mortality was achieved at chlorine dioxide concentrations above 1 mg/L.

These studies all show that the dose required to induce mortality in these nuisance organism is extremely high, both in terms of chemical dose and contact time. The potential impact on DBPs is significant, especially when the water is high in organic content with a high propensity to form THMs and other DBPs.

18.10.5.3 Oxidation of Iron and Manganese

Iron and manganese occur frequently in groundwaters but are less problematic in surface waters. Although not harmful to human health at the low concentrations typically found in water, these compounds can cause staining and taste problems. These compounds are readily treated by oxidation to produce a precipitant that is removed in subsequent sedimentation and filtration processes. Almost all the common oxidants except chloramines will convert ferrous (2+) iron to the ferric (3+) state and manganese (2+) to the (4+) state, which will precipitate as ferric hydroxide and manganese dioxide, respectively (AWWA, 1990). The precise chemical composition of the precipitate will depend on the nature of the water, temperature, and pH. Table 18.9 shows that oxidant doses for iron and manganese control are relatively low. In addition, the reactions are relatively rapid, on the order of seconds, whereas DBP formation occurs over hours. Thus, with proper dosing, residual chlorine during iron and manganese oxidation is therefore relatively low and short lived. These factors reduce the potential for DBP formation as a result of oxidation for iron and manganese removal.

18.10.5.4 Prevention of Regrowth in the Distribution System and Maintenance of Biological Stability

Biodegradable organic compounds and ammonia in treated water can cause microbial growth in the distribution system. *Biological stability* refers to a condition wherein the water quality does not enhance biological growth in the distribution system. Biological stability can be accomplished in several ways:

TABLE 18.9
Oxidant Doses Required for Oxidation of Iron and Manganese

Oxidant	Iron (II) (mg/mg Fe)	Manganese (II) (mg/mg Mn)
Chorine	0.62	0.77
Chorine dioxide	1.21	2.45
Ozone	0.43	0.85
Oxygen	0.14	0.29
Potassium permanganate	0.94	1.92

Source: Adapted from Culp G.L. and Culp, R.L., *New Concepts in Water Purification,* Van Nostrand Reinhold, New York, 1974.

- Removing nutrients from the water prior to distribution
- Maintaining a disinfectant residual in the treated water
- Combining nutrient removal and disinfectant residual maintenance

To maintain biological stability in the distribution system, the Total Coliform Rule (TCR) requires that treated water must have a residual disinfectant of 0.2 mg/L when entering the distribution system. A measurable disinfectant residual must be maintained in the distribution system, or the utility must show through monitoring that the heterotrophic plate count (HPC) remains less than 500/100 mL. A system remains in compliance as long as 95% of samples meet these criteria. Chlorine, monochloramine, and chlorine dioxide are typically used to maintain a disinfectant residual in the distribution system. Filtration can also be used to enhance biological stability by reducing the nutrients in the treated water.

The level of secondary disinfectant residual maintained is low, typically in the range of 0.1 to 0.3 mg/L, depending on the distribution system and water quality; however, because the contact times in the system are quite long, it is possible to generate significant amounts of DBPs in the distribution system, even at low disinfectant does. Distribution system problems associated with the use of combined residual chlorine (chloramines) or no residual have been documented. The use of combined chlorine is characterized by an initial satisfactory phase in which chloramine residuals are easily maintained throughout the system and bacterial counts are very low; however, problems may develop over a period of years, including increased bacterial counts, reduced combined residual chlorine, increased taste and odor complaints, and reduced transmission main carrying capacity. Conversion of the system to free residual chlorine produces an initial increase in consumer complaints of taste and odors resulting from oxidation of accumulated organic material. Also, it is difficult to maintain a free chlorine concentration at the ends of the distribution system (AWWA, 1990).

18.10.5.5 Removal of Taste and Odors through Chemical Oxidation

Taste and odors in drinking water are caused by several culprits, including microorganisms, decaying vegetation, hydrogen sulfide, and specific compounds of municipal, industrial, or agricultural origin. Disinfectants themselves can also create taste and odor problems. In addition to specific taste- and odor-causing compounds, the sanitary impact is often magnified by various combinations of compounds. Recently, significant attention has been given to tastes and odors caused by compounds such as geosmin, 2-methylisoborneol (MIB), and chlorinated inorganic and organic compounds (AWWARF, 1987).

Oxidation is commonly used to remove taste- and odor-causing compounds. Because many of these compounds are very resistant to oxidation, advanced oxidation processes (e.g., ozone/hydrogen peroxide, ozone/UV) and ozone by itself are often used to address taste and odor problems. The ability of various chemicals to control taste and odors can be site specific. Suffet et al. (1986) found that ozone is generally the most effective oxidant for use in taste and odor treatment. They found ozone doses of 2.5 to 2.7 mg/L and 10 minutes of contact time (residual 0.2 mg/L) significantly reduced levels of taste and odors. Lalezary et al. (1986) used chlorine, chlorine dioxide, ozone, and permanganate to treat earthy–musty smelling compounds. In that study, chlorine dioxide was found most effective, although none of the oxidants was able to remove geosmin and MIB by more than 40 to 60%. Potassium permanganate has been used in doses of 0.25 to 20 mg/L.

Prior experiences with taste and odor treatment indicate that oxidant doses are dependent on the source of the water and causative compounds. In general, small doses can be effective for many taste and odor compounds, but some of the difficult-to-treat compounds require strong oxidants such as ozone or advanced oxidation processes or alternative technologies such as granular activated carbon (GAC) adsorption.

18.10.5.6 Improvement of Coagulation and Filtration Efficiency

Oxidants, specifically ozone, have been reported to improve coagulation and filtration efficiency. Other studies, however, have found no improvement in effluent turbidity from oxidation. Prendiville (1986) collected data from a large treatment plant showing that preozonation was more effective than prechlorination to reduce filter effluent turbidities. The cause of the improved coagulation is not clear, but several possibilities have been offered, including (Gurol and Pidatella, 1983; Reckhow et al., 1986):

- Oxidation of organics into more polar forms
- Oxidation of metal ions to yield insoluble complexes such as ferric iron complexes
- Change in the structure and size of suspended particles.

18.10.5.7 Prevention of Algal Growth in Sedimentation Basins and Filters

Prechlorination is often used to minimize operational problems associated with biological growth in water treatment plants (AWWA, 1990). Prechlorination will prevent slime formation on filters, pipes, and tanks and reduce potential taste and odor problems associated with such slimes. Many sedimentation and filtration facilities operate with a small chlorine residual to prevent growth of algae and bacteria in the launders and on the filter surfaces. This practice has increased in recent years as utilities take advantage of additional contact time in the treatment units to meet disinfection requirements under the SWTR.

18.10.5.8 Removal of Color

Free chlorine is used for color removal. A low pH is favored. Humic compounds that have a high potential for DBP formation cause color. The chlorine dosage and kinetics for color removal are best determined through bench studies.

18.10.6 TYPES OF DISINFECTION BYPRODUCTS AND DISINFECTANT RESIDUALS

Table 18.10 provides a list compiled by USEPA of DBPs and disinfection residuals that may be of health concern. The table includes both the disinfectant residuals and

---

TABLE 18.10
Disinfectant Residuals and Disinfection Byproducts

Disinfectant Residuals
Free chlorine
   Hypochlorous acid
   Hypochlorite ion
Chloramines
   Monochloramine
Chlorine dioxide

Inorganic Byproducts
Chlorate ion
Chlorite ion
Bromate ion
Iodate ion
Hydrogen peroxide
Ammonia

Organic Oxidation Byproducts
Aldehydes
Formaldehyde
Acetaldehyde
Glyoxal
Hexanal
Heptanal
Carboxylic acids
   Hexanoic acid
   Heptanoic acid
   Oxalic acid
Assimilable organic carbon

Halogenated Organic Byproducts
Trihalomethanes
   Chloroform
   Bromodichloromethane
   Dibromochloromethane
   Bromoform
Haloacetic acids
   Monochloroacetic acid
   Dichloroacetic acid
   Trichloroacetic acid
   Monobromoacetic acid
   Dibromoacetic acid
Haloacetonitriles
   Dichloroacetonitrile
   Bromochloroacetonitrile
   Dibromoacetonitrile
   Trichloroacetonitrile
Haloketones
   1,1-Dichloropropanone
   1,1,1-Trichloropropanone
Chlorophenols
   2-Chlorophenol
   2,4-Dichlorophenol
   2,4,6-Trichlorophenol
Chloropicrin
Chloral hydrate
Cyanogen chloride
N-organochloramines

TABLE 18.11
Health Information for Disinfectants and Disinfection Byproducts

Contaminant	Cancer Classification
Chloroform	Probable human carcinogen
Bromodichloromethane	Probable human carcinogen
Dibromochloromethane	Possible human carcinogen
Bromoform	Probable human carcinogen
Monochloroacetic acid	—
Dichloroacetic acid	Probable human carcinogen
Trichloroacetic acid	Possible human carcinogen
Dichloroacetonitrile	Possible human carcinogen
Bromochloroacetonitrile	—
Dibromoacetonitrile	Possible human carcinogen
Trichloroacetonitrile	—
1,1-Dichloropropanone	—
1,1,1-Trichloropropanone	—
2-Chlorophenol	Not classifiable
2,4-Dichlorphenol	Not classifiable
2,4,6-Trichlorophenol	Probable human carcinogen
Chloropicrin	—
Chloral hydrate	Possible human carcinogen
Cyanogen chloride	—
Formaldehyde	Probable human carcinogen
Chlorate	—
Chlorite	Not classifiable
Bromate	Probable human carcinogen
Ammonia	Not classifiable
Hypochlorous acid	—
Hypochlorite	—
Monochloramine	—
Chlorine dioxide	Not classifiable

Source: USEPA, *Drinking Water Regulations and Health Advisories*, EPA-822-B-96-002, U.S. Environmental Protection Agency, Washington, D.C., 1996.

the specific byproducts produced by these disinfectants of interest in drinking water treatment. These contaminants of concern are grouped into four distinct categories of *disinfectant residuals*, *inorganic byproducts*, *organic oxidation byproducts*, and *halogenated organic byproducts*.

The production of DBPs depends on the type of disinfectant, the presence of organic material (e.g., TOC), bromide ion, and other environmental factors as discussed in this section. By removing DBP precursors, the formation of DBPs can be reduced. The health effects of DBPs and disinfectants are generally evaluated with epidemiological studies and toxicological studies using laboratory animals. Table 18.11 shows the cancer classifications of both disinfectants and DBPs. The USEPA classification scheme for carcinogenicity weighs both animal studies and epidemiologic studies but places greater weight on evidence of carcinogenicity in humans.

18.10.7 Disinfection Byproduct Formation

Halogenated organic byproducts are formed when natural organic matter (NOM) reacts with free chlorine or free bromine. Free chlorine can be introduced to water directly as a primary or secondary disinfectant, with chlorine dioxide, or with chloramines. Free bromine results from oxidation of the bromide ion in the source water. Factors affecting formation of halogenated DBPs include type and concentration of natural organic matter, oxidant type and dose, time, bromide ion concentration, pH, organic nitrogen concentration, and temperature. Organic nitrogen significantly influences the formation of nitrogen-containing DBPs such as the haloacetonitriles, halopicrins, and cyanogen halides. The parameter TOX represents the concentration of total organic halides in a water sample (calculated as chloride). In general, less than 50% of the TOX content has been identified, despite evidence that several of these unknown halogenated byproducts of water chlorination may be harmful to humans (Reckhow et al., 1990; Singer and Chang, 1989).

Nonhalogenated DBPs are also formed when strong oxidants react with organic compounds found in water. Ozone and peroxone oxidation of organics leads to the production of aldehydes, aldo- and keto-acids, organic acids, and, when bromide ion is present, brominated organics. Many of the oxidation byproducts are biodegradable and appear as biodegradable dissolved organic carbon (BDOC) and assimilable organic carbon (AOC) in treated water.

Bromide ion plays a key role in DBP formation. Ozone or free chlorine oxidizes bromide ion to hypobromate ion/ hypobromous acid, which subsequently forms brominated DBPs. Brominated organic byproducts include compounds such as bromoform, brominated acetic acids and acetonitriles, bromopicrin, and cyanogen bromide. Only about one third of the bromide ions incorporated into byproducts have been identified.

18.10.7.1 DBP Precursors

Numerous researchers have documented that NOM is the principal precursor of organic DBP formation. Chlorine reacts with NOM to produce a variety of DBPs, including THMs and haloacetic acids (HAAs). Ozone reacts with NOM to produce aldehydes, organic acids, and aldo and keto acids; many of these are produced by chlorine as well (Singer et al., 1993; Stevens, 1976). Natural waters contain mixtures of both humic and nonhumic organic substances. NOM can be subdivided into a hydrophobic fraction composed of primarily humic material and a hydrophilic fraction composed of primarily fulvic material. The type and concentration of NOM are often assessed using surrogate measures. Although surrogate parameters have limitations, they are used because they may be measured more easily, rapidly, and inexpensively than the parameter of interest, often allowing online monitoring of the operation and performance of water treatment plants. Surrogates used to assess NOM include:

- Total and dissolved organic carbon (TOC and DOC)
- Specific ultraviolet light absorbence (SUVA), which is the absorbence at a 254-nm wavelength (UV-254) divided by the DOC; SUVA = (UV-254 ÷ DOC) × 100 (L/mg-m)
- THM formation potential (THMFP), a test that measures the quantity of THMs formed with a high dosage of free chlorine and a long reaction time
- TTHM stimulated distribution system (SDS), a test that predicts the total trihalomethane concentration at some selected point in a given distribution system where the conditions of the chlorination test simulate the distribution system at the point desired

On average, about 90% of the TOC is dissolved. DOC is defined as the TOC able to pass through a 0.45-μm filter. UV absorbance is a good technique for assessing the presence of DOC because DOC primarily consists of humic substances, which contain aromatic structures that absorb light in the UV spectrum. Oxidation of DOC reduces the UV absorbance of the water due to oxidation of some of the organic bonds that absorb UV absorbance. Complete mineralization of organic compounds to carbon dioxide usually does not occur under water treatment conditions; therefore, the overall TOC concentration usually is constant.

Concentrations of DBPs vary seasonally and are typically greatest in the summer and early fall for several reasons:

- The rate of DBP formation increases with increasing temperature.
- The nature of organic DBP precursors varies with season.
- Due to warmer temperatures, chlorine demand may be greater during summer months, requiring higher dosages to maintain disinfection.

If the bromide ion is present in source waters, it can be oxidized to hypobromous acid that can react with NOM to form brominated DBPs, such as bromoform. Furthermore, under certain conditions, ozone may react with the hypobromite ion to form bromate ion (Singer 1982).

The ratio of bromide ion to the chlorine dose affects THM formation and bromine substitution of chlorine. Increasing the ratio of bromide ions to chlorine dose shifts the speciation of THMs to produce more brominated

forms. In the Krasener et al. (1989) study, the chlorine dose was roughly proportional to TOC concentration. As TOC was removed through the treatment train, the chlorine dose decreased and TTHM formation declined; however, at the same time, the ratio of bromide ion to chlorine dose increased, thereby shifting TTHM concentrations to the more brominated THMs. Thus, improving the removal of NOM prior to chlorination can shift the speciation of halogenated byproducts toward more brominated forms.

Chloropicrin is produced by the chlorination of humic materials in the presence of nitrate ion. Thibaud et al. (1988) chlorinated humic compounds in the presence of bromide ion to demonstrate the formation of brominated analogs to chloropicrin.

18.10.7.2 Impacts of pH on DBP Formation

The pH of water being chlorinated has an impact on the formation of halogenated byproducts. THM formation increases with increasing pH. Trichloroacetic acid, dichloroacetonitrile, and trichloropropanone formation decrease with increased pH. Overall TOX formation decreases with increasing pH. Based on chlorination studies of humic material in model systems, high pH tends to favor chloroform formation over the formation of trichloroacetic acid and other organic halides. Accordingly, water treatment plants practicing precipitative softening at pH values greater than 9.5 to 10 are likely to have a higher fraction of TOX attributable to THMs than plants treating surface waters by conventional treatment in pH ranges of 6 to 8 (Reckhow and Singer, 1989).

Because the application of chlorine dioxide and chloramines may introduce free chlorine into water, chlorination byproducts that may be formed would be influenced by pH as discussed above. Ozone application to waters containing bromide ion at high pH favors the formation of bromate ion, whereas application at low pH favors the formation of brominated organic byproducts.

The pH also impacts enhanced coagulation (e.g., for ESWTR compliance) and compliance with the Lead and Copper Rule. These issues are addressed in USEPA's *Microbial and Disinfection Byproduct Rules Simultaneous Compliance Guidance Manual* (USEPA, 1999c).

18.10.7.3 Organic Oxidation Byproducts

Organic oxidation byproducts are formed by reactions between NOM and all oxidizing agents added during drinking water treatment. Some of these byproducts are halogenated, as discussed in the previous section; others are not. The types and concentrations of organic oxidation byproducts produced depend on the type and dosage of the oxidant being used, chemical characteristics and concentration of the NOM being oxidized, and other factors such as the pH and temperature.

TABLE 18.12
Inorganic DBPs Produced During Disinfection

Disinfectant	Inorganic Byproduct or Disinfectant Residual Discussed
Chlorine dioxide	Chlorine dioxide, chlorite ion, chlorate ion, bromate ion
Ozone	Bromate ion, hydrogen peroxide
Chloramination	Monochloramine, dichloramine, trichloramine, ammonia

18.10.7.4 Inorganic Byproducts and Disinfectants

Table 18.12 summarizes some of the inorganic DBPs that are produced or remain as residual during disinfection. As discussed earlier, bromide ion reacts with strong oxidants to form bromate ion and other organic DBPs. Chlorine dioxide and chloramines leave residuals that are of concern for health considerations, as well as for taste and odor.

18.10.8 DISINFECTION BYPRODUCT CONTROL

In 1983, the USEPA identified technologies, treatment techniques, and plant modifications that community water systems could use to comply with the maximum contaminant level for TTHMs. The principal treatment modifications involved moving the point of chlorination downstream in the water treatment plant, improving the coagulation process to enhance the removal of DBP precursors, and using chloramines to supplement or replace the use of free chlorine (Singer and Harrington, 1993). Moving the point of chlorination downstream in the treatment train often is very effective in reducing DBP formation, because it allows the NOM precursor concentration to be reduced during treatment prior to chlorine addition. Replacing prechlorination by preoxidation with an alternative disinfectant that produces fewer DBPs is another option for reducing formation of chlorinated byproducts. Other options to control the formation of DBPs include source water quality control, DBP precursor removal, and disinfection strategy selection. An overview of each is provided below.

18.10.8.1 Source Water Quality Control

Source water control strategies involve managing the source water to lower the concentrations of NOM and bromide ion in the source water. Research has shown that algal growth leads to the production of DBP precursors (Oliver and Shindler, 1980); therefore, nutrient and algal management is one method of controlling the DBP formation potential of source waters. Control of bromide ion in source waters may be accomplished by preventing brine or salt water intrusion into the water source.

18.10.8.2 DBP Precursor Removal

Raw water can include DBP precursors in both dissolved and particulate forms. For the dissolved precursors to be removed in conventional treatment, they must be converted to particulate form for subsequent removal during settling and filtering. The THM formation potential generally decreases by about 50% through conventional coagulation and settling, indicating the importance of moving the point of chlorine application after coagulation and settling (and even filtration) to control TOX as well as TTHM formation (Singer and Harrington, 1989). Conventional systems can lower the DBP formation potential of water prior to disinfection by further removing precursors with enhanced coagulation, granular activated carbon (GAC) adsorption, or membrane filtration prior to disinfection. Precursor removal efficiencies are site specific and vary with different source waters and treatment techniques.

Aluminum (alum) and iron (ferric) salts can remove variable amounts of NOM. For alum, the optimal pH for NOM removal is in the range of 5.5 to 6.0. The addition of alum decreases pH and may allow the optimal pH range to be reached without acid addition; however, waters with very low or very high alkalinities may require the addition of base or acid to reach the optimal NOM coagulation pH (Singer, 1992).

Granular activated carbon adsorption can be used following filtration to remove additional NOM. For most applications, empty bed contact times in excess of 20 minutes are required, with regeneration frequencies on the order of 2 to 3 months. These long control times and frequent regeneration requirements make GAC an expensive treatment option. In cases where prechlorination is practiced, the chlorine rapidly degrades GAC. Addition of a disinfectant to the GAC bed can result in specific reactions in which previously absorbed compounds leach into the treated water.

Membrane filtration has been shown effective in removing DBP precursors in some instances. In pilot studies, ultrafiltration (UF) with a molecular weight cutoff (MWCO) of 100,000 Da was ineffective for controlling DBP formation; however, when little or no bromide ion was present in source water, nanofiltration (NF) membranes with MWCOs of 400 to 800 Da effectively controlled DBP formation (Laine et al., 1993).

In waters containing bromide ion, higher bromoform concentrations were observed after chlorination of membrane permeate (compared with raw water). This occurs as a result of filtration removing NOM while concentrating bromide ions in the permeate, thus providing a higher ratio of bromide ions to NOM than in raw water. This reduction in chlorine demand increases the ratio of bromide to chlorine, resulting in higher bromoform concentrations after chlorination of NF membrane permeate

TABLE 18.13
Required Removal of TOC by Enhanced Coagulation for Surface Water Systems Using Conventional Treatment (Percent Reduction)

Source Water TOC (mg/L)	Source Water Alkalinity (mg/L as CaCO₃)		
	0–60	>60–120	>120
>2.0 to 4.0	35.0%	25.0%	15.0%
>4.0 to 8.0	45.0%	35.0%	25.0%
>8.0	50.0%	40.0%	30.0%

(compared with the raw water). TTHMs were lower in chlorinated permeate than chlorinated raw water; however, due to the shift in speciation of THMs toward more brominated forms, bromoform concentrations were actually greater in chlorinated treated water than in chlorinated raw water. Use of spiral-wound NF membranes (200 to 300 Da) more effectively controlled the formation of brominated THMs, but pretreatment of the water was necessary. Significant limitations in the use of membranes are disposal of the waste brine generated, membrane fouling, cost of membrane replacement, and increasing energy cost.

Disinfection byproduct regulations require enhanced coagulation as an initial step for removal of DBP precursors. In addition to meeting maximum contaminant levels (MCLs) and maximum residual disinfectant levels (MRDLs), some water suppliers also must meet treatment requirements to control the organic material (DBP precursors) in the raw water that combines with disinfectant residuals to form DBPs. Systems using conventional treatment are required to control precursors (measured as TOC) by using enhanced coagulation or enhanced softening. A system must remove a specified percentage of TOC (based on raw water quality) prior to the point of continuous disinfection (Table 18.13).

Systems using ozone followed by biologically active filtration or chlorine dioxide that meet specific criteria would be required to meet the TOC removal requirements prior to addition of a residual disinfectant. Systems able to reduce TOC by a specified percentage level have met the DBP treatment technique requirements. If the system does not meet the percent reduction, it must determine its alternative minimum TOC removal level. The primacy agency approves the alternative minimum TOC removal possible for the system on the basis of the relationship between coagulant dose and TOC in the system based on results of bench or pilot-scale testing. Enhanced coagulation is determined in part as the coagulant doses where an incremental addition of 10 mg/L of alum (or an equivalent amount of ferric salt) results in a TOC removal below 0.3 mg/L.

18.10.9 DISINFECTION STRATEGY SELECTION

In addition to improving the raw or predisinfectant water quality, alternative disinfection strategies can be used to control DBPs. These strategies include the following:

- Use an alternative or supplemental disinfectant or oxidant such as chloramines or chlorine dioxide that will produce fewer DBPs.
- Move the point of chlorination to reduce TTHM formation and, where necessary, substitute chloramines, chlorine dioxide, or potassium permanganate for chlorine as a preoxidant.
- Use two different disinfectants or oxidants at various points in the treatment plant to avoid DBP formation at locations where precursors are still present in high quantities.
- Use powdered activated carbon for THM precursor or TTHM reduction seasonally or intermittently.
- Maximize precursor removal.

18.10.10 CT FACTOR

One of the most important factors for determining or predicting the germicidal efficiency of any disinfectant is the CT factor, a version of the Chick–Watson law (1908). The CT factor is defined as the product of the residual disinfectant concentration (C), in mg/L, and the contact time (T), in minutes, that residual disinfectant is in contact with the water. USEPA developed CT values for the inactivation of *Giardia* and viruses under the SWTR. Table 18.14 compares the CT values for virus inactivation using chlorine, chlorine dioxide, ozone, chloramine, and ultraviolet light disinfection under specified conditions. Table 18.15 shows the CT values for inactivation of *Giardia* cyst using chlorine, chloramine, chlorine dioxide, and ozone under specified conditions. The CT values shown in Table 18.14 and Table 18.15 are based on water temperatures of 10°C and pH values in the range of 6 to 9. CT values for chlorine disinfection are based on a free chlorine residual. Note that chlorine is less effective as pH increases from 6 to 9.

TABLE 18.14
CT Values for Inactivation of Viruses

Disinfectant (at 10°C)	Units	Inactivation 2-log	Inactivation 3-log
Chlorine	mg·min/L	4	4
Chloramine	mg·min/L	643	1067
Chlorine dioxide	mg·min/L	4.2	12.8
Ozone	mg·min/L	0.5	0.8
Ultraviolet	mW·S/cm^2	21	36

Source: Data from AWWA, *Guidance Manual for Compliance with the Filtration and Disinfection Requirements for Public Works Systems Using Surface Water Sources,* American Water Works Association, Denver, CO, 1991.

TABLE 18.15
CT Values for Inactivation of *Giardia* Cysts

Disinfectant (@ 10°C)	Inactivation (mg·min/L) 1-log	2-log	3-log
Chlorine	35	69	104
Chloramine	615	1240	1850
Chorine dioxide	7.7	15	23
Ozone	0.48	0.95	1.43

Source: Data from AWWA, *Guidance Manual for Compliance with the Filtration and Disinfection Requirements for Public Works Systems Using Surface Water Sources,* American Water Works Association, Denver, CO, 1991.

In addition, for a given CT value, a low C and a high T is more effective than the reverse (i.e., a high C and a low T). For all disinfectants, as temperature increases, effectiveness increases.

18.10.11 DISINFECTANT RESIDUAL REGULATORY REQUIREMENTS

One of the most important factors for evaluating the merits of alternative disinfectants is their ability to maintain the microbial quality in the water distribution system. Disinfectant residuals may serve to protect the distribution system against regrowth. The Surface Water Treatment Rule requires that filtration and disinfection must be provided to ensure that the total treatment of the system achieves at least a 3-log removal/inactivation of *Giardia* cysts and 4-log removal/inactivation of viruses. In addition, the disinfection process must demonstrate by continuous monitoring and recording that the disinfection residual in the water entering the distribution system is never less than 0.2 mg/L for more than 4 hours (Snead, 1980).

Several of the alternative disinfectants examined in this handbook cannot be used to meet the residual requirements stated in the SWTR; for example, if either ozone or ultraviolet light disinfection is used as the primary disinfectant, a secondary disinfectant such as chlorine or chloramines should be utilized to obtain a residual in the distribution system.

Disinfection byproduct formation continues in the distribution system due to reactions between the residual disinfectant and organics in the water. Koch et al. (1991) found that, with a chlorine dose of 3 to 4 mg/L, THM and haloacetic acid (HAA) concentrations increased rapidly during the first 24 hours in the distribution system. After the initial 48 hours, the subsequent increase in THMs was found to be very small. Chloral hydrate concentrations continued to increase after the initial 24 hours, but at a reduced rate. Haloketones actually decreased in the distribution system.

Nieminski et al. (1993) evaluated DBP formation in the simulated distribution systems of treatment plants in Utah. Finished water chlorine residuals ranged from 0.4 to 2.8 mg/L. Generally, THM values in the distribution system studies increased by 50 to 100% (range of 30 to 200%) of the plant effluent value after a 24-hour contact time. The 24-hour THM concentration was essentially the same as the 7-day THM formation potential. HAA concentrations in the simulated distribution system were about 100% (range of 30 to 200%) of the HAA in the plant effluent. The 7-day HAA formation potential was sometimes higher, or below the distribution system values. If chlorine is used as a secondary disinfectant, one should therefore anticipate a 100% increase in the plant effluent THMs or plan to reach the 7-day THM formation level in the distribution system.

18.10.12 SUMMARY OF CURRENT NATIONAL DISINFECTION PRACTICES

Most water treatment plants disinfect water prior to distribution. The 1995 Community Water Systems Survey reports that 81% of all community water systems provide some form of treatment on all or a portion of their water sources. The survey also found that virtually all surface water systems provide some treatment of their water. Of those systems reporting no treatment, 80% rely on groundwater as their only water source.

The most commonly used disinfectants/oxidants are chlorine, chlorine dioxide, chloramines, ozone, and potassium permanganate. Chlorine is predominately used in surface and groundwater disinfection treatment systems; more than 60% of the treatment systems use chlorine as a disinfectant/oxidant. Potassium permanganate on the other hand, is used by many systems, but its application is primarily for oxidation, rather than for disinfection.

Permanganate will have some beneficial impact on disinfection because it is a strong oxidant that will reduce the chemical demand for the ultimate disinfection chemical. Chloramine is used by some systems and is more frequently used as a post-treatment disinfectant.

TABLE 18.16
Ozone Application in U.S. Water Treatment Plants

Ozone Objective	Number of Plants	Percent
THM control	50	32
Disinfection	63	40
Iron/manganese, taste and odor control	92	58
Total	158	—

The International Ozone Association (IOA, 1997) conducted a survey of ozone facilities in the United States. The survey documented the types of ozone facilities, size, objective of ozone application, and year of operation; the findings are summarized in Table 18.16. The most common use of ozone is for oxidation of iron and manganese and for taste and odor control. Of the 158 ozone facilities, 24 used GAC following ozonation. In addition to the 158 operating ozone facilities, the survey identified 19 facilities under construction and another 30 under design. The capacities of the systems range from less than 25 gpm to exceeding 500 gpm. Nearly half of the operating facilities have a capacity exceeding 1 MGD. Rice et al. (1998) found that, as of May 1998, 264 drinking water plants in the United States were using ozone.

18.10.13 SUMMARY OF METHODS OF DISINFECTION

Methods of disinfection include:

- *Heat*—Possibly the first method of disinfection which is accomplished by boiling water for 5 to 10 minutes; good, obviously, only for household quantities of water when bacteriological quality is questionable
- *Ultraviolet (UV) light*—A practical method of treating large quantities but adsorption of UV light is very rapid so this method is limited to nonturbid waters close to the light source
- *Metal ions*—Silver, copper, mercury
- *Alkalis and acids*
- *pH adjustment*—To under 3.0 or over 11.0
- *Oxidizing agents*—Bromine, ozone, potassium permanganate, and chlorine

The vast majority of drinking water systems in the United States use chlorine for disinfection (Spellman, 2007). Along with meeting the desired characteristics listed above, chlorine has the added advantage of a long history of use and is fairly well understood. Although some small water systems may use other disinfectants, we focus on chlorine in this handbook and provide only a brief overview of other disinfection alternatives.

Note: A recent development in chlorine disinfection is the use of multiple and interactive disinfectants. In these applications, chlorine is combined with a second disinfectant to achieve improved disinfection efficiency and effective DBP control.

Note: As described earlier, the 1995 Community Water System Survey found that all surface water and groundwater systems in the United States use chlorine for disinfection (Spellman, 2007).

18.10.14 CHLORINATION

The addition of chlorine or chlorine compounds to water is called *chlorination*. Chlorination is considered to be the single most important process for preventing the spread of waterborne disease. Chlorine has many attractive features that contribute to its wide use in industry. Five key attributes of chlorine are:

- It damages the cell wall.
- It alters the permeability of the cell (the ability of water to pass in and out through the cell wall).
- It alters the cell protoplasm.
- It inhibits the enzyme activity of the cell so it is unable to use its food to produce energy.
- It inhibits cell reproduction.

Some concerns regarding the use of chlorine that may restrict its use include:

- Chlorine reacts with many naturally occurring organic and inorganic compounds in water to produce undesirable DBPs.
- Hazards associated with using chlorine, specifically chlorine gas, require special treatment and response programs.
- High chlorine doses can cause taste and odor problems.

Chlorine is used in water treatment facilities primarily for disinfection. Because of the oxidizing powers of chlorine, it has been found to serve other useful purposes, as well (White, 1992):

- Taste and odor control
- Prevention of algal growths
- Maintenance of clear filter media
- Removal of iron and manganese
- Destruction of hydrogen sulfide
- Bleaching of certain organic colors
- Maintenance of distribution system water quality by controlling slime growth

- Restoration and preservation of pipeline capacity
- Restoration of well capacity
- Water main sterilization
- Improvement in coagulation

Chlorine is available in a number of different forms:

- As pure elemental gaseous chlorine, a greenish-yellow gas possessing a pungent and irritating odor that is heavier than air, nonflammable, and nonexplosive; when released to the atmosphere, this form is toxic and corrosive
- As solid calcium hypochlorite (in tablets or granules)
- As a liquid sodium hypochlorite solution (in various strengths)

The selection of one form of chlorine over the others for a given water system depends on the amount of water to be treated, configuration of the water system, the local availability of the chemicals, and the skill of the operator.

One of the major advantages of using chlorine is the effective residual that it produces. A residual indicates that disinfection is completed, and the system has an acceptable bacteriological quality. Maintaining a residual in the distribution system provides another line of defense against pathogenic organisms that could enter the distribution system and helps to prevent regrowth of those microorganisms that were injured but not killed during the initial disinfection stage.

18.10.14.1 Chlorine Terminology

Often it is difficult for new waterworks operators to understand the terminology used to describe the various reactions and processes used in chlorination. Common chlorination terms include the following:

- *Chlorine reaction*—Regardless of the form of chlorine used for disinfection, the reaction in water is basically the same. The same amount of disinfection can be expected, provided the same amount of available chlorine is added to the water. The standard units used to express the concentration of chlorine in water are milligrams per liter (mg/L) and parts per million (ppm); these terms indicate the same quantity.
- *Chlorine dose*—The amount of chlorine added to the system. It can be determined by adding the desired residual for the finished water to the chlorine demand of the untreated water. Dosage can be either milligrams per liter (mg/L) or pounds per day. The most common is mg/L.

- *Chlorine demand*—The amount of chlorine used by iron, manganese, turbidity, algae, and microorganisms in the water. Because the reaction between chlorine and microorganisms is not instantaneous, demand is relative to time. For example, the demand 5 minutes after applying chlorine will be less than the demand after 20 minutes. Demand, like dosage, is expressed in mg/L. The chlorine demand is as follows:

$$Cl_2 \text{ demand} = Cl_2 \text{ dose} - Cl_2 \text{ residual} \quad (18.7)$$

- *Residual chlorine*—The amount of chlorine (determined by testing) that remains after the demand is satisfied. Residual, like demand, is based on time. The longer the time after dosage, the lower the residual will be, until all of the demand has been satisfied. Residual, like dosage and demand, is expressed in mg/L. The presence of a *free residual* of at least 0.2 to 0.4 ppm usually provides a high degree of assurance that the disinfection of the water is complete. *Combined residual* is the result of combining free chlorine with nitrogen compounds. Combined residuals are also called *chloramines*. The *total residual chlorine* (TRC) is the mathematical combination of free and combined residuals. Total residual can be determined directly with standard residual chlorine test kits.
- *Chorine contact time*—A key item in predicting the effectiveness of chlorine on microorganisms. It is the interval (usually only a few minutes) between the time when chlorine is added to the water and the time the water passes by the sampling point. Contact time is the "T" in CT. CT is calculated based on the free residual chlorine prior to the first consumer times the contact time in minutes.

$$CT = \text{Concentration} \times \text{contact time}$$
$$= \text{mg/L} \times \text{minutes} \quad (18.8)$$

A certain minimum time period is required for the disinfecting action to be completed. The contact time is usually a fixed condition determined by the rate of flow of the water and the distance from the chlorination point to the first consumer connection. Ideally, the contact time should not be less than 30 minutes, but even more time is needed at lower chlorine doses, in cold weather, or under other conditions.

Pilot studies have shown that specific CT values are necessary for the inactivation of viruses and *Giardia*. The required CT value will vary depending on pH, temperature, and the organisms to be killed. Charts and formulae are available to make this determination. USEPA has set a CT value of 3-log ($CT_{99.9}$) inactivation to ensure that the water is free of *Giardia*. State drinking water regulations provide charts containing CT values for various pH and temperature combinations. Filtration, in combination with disinfection, must provide a 3-log removal/inactivation of *Giardia*. Charts in the USEPA Surface Water Treatment Rule Guidance manual list the required CT values for various filter systems.

Under the 1996 Interim Enhanced Surface Water Treatment Rule (IESWTR), the USEPA requires systems that filter to remove 99% (2 log) of *Cryptosporidium* oocysts. To be sure that the water is free of viruses, a combination of filtration and disinfection that provides a 4-log removal of viruses has been judged the best for drinking water safety—99.99% removal. Viruses are inactivated (killed) more easily than cysts or oocysts.

18.10.14.2 Chlorine Chemistry

The reactions of chlorine with water and the impurities that might be in the water are quite complex, but a basic understanding of these reactions can aid the operator in keeping the disinfection process operating at its highest efficiency. When dissolved in pure water, chlorine reacts with the H^+ ions and the OH^- radicals in the water. Two of the products of this reaction (the actual disinfecting agents) are *hypochlorous acid* (HOCl), and the *hypochlorite radical* (OCl^-). If microorganisms are present in the water, the HOCl and the OCl^- penetrate the microbe cells and react with certain enzymes. This reaction disrupts the metabolism of the organisms and kills them. The chemical equation for hypochlorous acid is as follows:

$$\underset{\text{(chlorine)}}{Cl_2} + \underset{\text{(water)}}{H_2O} \leftrightarrow \underset{\substack{\text{(hypochlorous}\\\text{acid)}}}{HOCl} + \underset{\substack{\text{(hydrochloric}\\\text{acid)}}}{HCl}$$

Note: The symbol ↔ indicates that the reactions are reversible.

Hypochlorous acid (HOCl) is a weak acid, meaning that it dissociates slightly into hydrogen and hypochlorite ions, but it is a strong oxidizing and germicidal agent. Hydrochloric acid (HCl) in the above equation is a strong acid and retains more of the properties of chlorine. HCl tends to lower the pH of the water, especially in swimming pools where the water is recirculated and continually chlorinated. The total hypochlorous acid and hypochlorite ions in water constitute the *free available chlorine*. Hypochlorites act in a manner similar to HCl when added to water, because hypochloric acid is formed.

When chlorine is first added to water containing some impurities, the chlorine immediately reacts with the dissolved inorganic or organic substances and is then unavailable for disinfection. The amount of chlorine used in this initial reaction is the *chlorine demand* of the water. If dissolved ammonia (NH_3) is present in the water, the chlorine will react with it to form compounds called *chloramines*. Only after the chlorine demand is satisfied and the reaction with all the dissolved ammonia is complete is the chlorine actually available in the form of HOCl and OCl^-. The equation for the reaction of hypochlorous acid (HOCl) and ammonia (NH_3) is as follows:

$$\underset{\substack{\text{(hypochlorous}\\\text{acid)}}}{HOCl} + \underset{\text{(ammonia)}}{NH_3} \leftrightarrow \underset{\substack{\text{(mono-}\\\text{chloramine)}}}{NH_2Cl} + \underset{\text{(water)}}{H_2O} \quad (18.10)$$

Note: The chlorine as hypochlorous acid and hypochlorite ions remaining in the water after the above reactions are complete is known as *free available chlorine*, and it is a very active disinfectant.

18.10.14.3 Breakpoint Chlorination

To produce a free chlorine residual, enough chlorine must be added to the water to produce what is referred to as *breakpoint chlorination*, which is the point at which near complete oxidation of nitrogen compounds is reached; any residual beyond breakpoint is mostly free chlorine (see Figure 18.9). When chlorine is added to natural waters, the chlorine begins combining with and oxidizing the chemicals in the water before it begins disinfecting. Although residual chlorine will be detectable in the water, the chlorine will be in the combined form with a weak disinfecting power. As we see in Figure 18.9, adding more chlorine to the water at this point actually decreases the residual chlorine as the additional chlorine destroys the combined chlorine compounds. At this stage, water may have a strong swimming pool or medicinal taste and odor. To avoid this taste and odor, add still more chlorine to produce a free residual chlorine. Free chlorine has the highest disinfecting power. The point at which most of the combined chlorine compounds have been destroyed and the free chlorine starts to form is the *breakpoint*.

The chlorine breakpoint of water can only be determined by experimentation. This simple experiment requires 20 1000-mL breakers and a solution of chlorine. Place the raw water in the beakers and dose with progressively larger amounts of chlorine; for example, we might start with 0 in the first beaker, then 0.5 mg/L, then 1.0 mg/L, and so on. After a period of time, say 20 minutes, test each beaker for total residual chlorine and plot the results.

18.10.14.4 Breakpoint Chlorination Curve

Refer to Figure 18.9. Where the curve starts, no residual exists, even though a dosage was applied. This is the *initial demand*, when microorganisms and interfering agents are

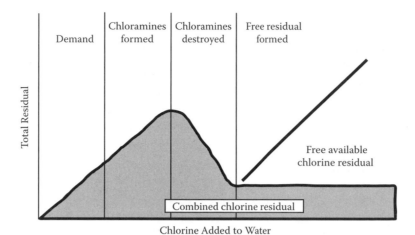

FIGURE 18.9 Breakpoint chlorination curve.

using the result of the chlorine. After the initial demand, the curve slopes upward. Chlorine combining to form chloramines produces this part of the curve. All of the residual measured on this part of the curve is combined residual. At some point, the curve begins to drop back toward zero. This portion of the curve results from a reduction in combined residual, which occurs because enough chlorine has been added to destroy (oxidize) the nitrogen compounds used to form combined residuals. The breakpoint is the point where the downward slope of the curve breaks upward. At this point, all of the nitrogen compounds that could be destroyed have been destroyed. After breakpoint, the curve begins to move upward again, usually at a 45° angle. Only on this part of the curve can free residuals be found. Notice that the breakpoint is not zero. The distance that the breakpoint is above zero is a measure of the remaining combined residual in the water. This combined residual exists because some of the nitrogen compound will not have been oxidized by chlorine. If irreducible combined residual is more than 15% of the total residual, chlorine odor and taste complaints will be high.

18.10.14.5 Gas Chlorination

Gas chlorine is provided in 100-lb to 1-ton containers. Chlorine is placed in the container as a liquid. The liquid boils at room temperature and is reduced to a gas that builds pressure in the cylinder. At room temperature (70°F), a chlorine cylinder will have a pressure of 85 psi; 100- to 150-lb cylinders should be maintained in an upright position and chained to the wall. To prevent a chlorine cylinder from rupturing in a fire, the cylinder valves are equipped with special fusible plugs that melt between 158 and 164°F.

Chlorine gas is 99.9% chlorine. A gas chlorinator meters the gas flow and mixes it with water, which is then injected as a water solution of pure chlorine. As the com-

pressed liquid chlorine is withdrawn from the cylinder, it expands as a gas, withdrawing heat from the cylinder. Care must be taken not to withdraw the chlorine at too fast a rate; if the operator attempts to withdraw more than about 40 lb of chlorine per day from a 150-lb cylinder, it will freeze up.

Note: All chlorine gas feed equipment sold today is vacuum operated. This safety feature ensures that, if a break occurs in one of the components in the chlorinator, the vacuum will be lost, and the chlorinator will shut down without allowing gas to escape.

Chlorine gas is a highly toxic lung irritant, and special facilities are required for storing and housing it. Chlorine gas will expand to 500 times its original compressed liquid volume at room temperature (1 gallon of liquid chlorine will expand to about 67 ft³). Its advantage as a drinking water disinfectant is the convenience afforded by a relatively large quantity of chlorine available for continuous operation for several days or weeks without the need for mixing chemicals. Where water flow rates are highly variable, the chlorination rate can be synchronized with the flow.

Chlorine gas has a very strong, characteristic odor that can be detected by most people at concentrations as low as 3.5 ppm. Highly corrosive in moist air, it is extremely toxic and irritating in concentrated form. Its toxicity ranges from being a throat irritant at 15 ppm to causing rapid death at 1000 ppm. Although chlorine does not burn, it supports combustion, so open flames should never be used around chlorination equipment.

When changing chlorine cylinders, an accidental release of chlorine may occasionally occur. To handle this type of release, an approved (NIOSH-approved) self-contained breathing apparatus (SCBA) must be worn. Special emergency repair kits are available from the Chlorine

Institute for use by emergency response teams to deal with chlorine leaks. Because chlorine gas is 2.5 times heavier than air, exhaust and inlet air ducts should be installed at floor level. A chlorine gas leak can be found by using the fumes from a strong ammonia mist solution. A white cloud develops when ammonia mist and chlorine combine.

18.10.14.6 Hypochlorination

Combining chlorine with calcium or sodium produces hypochlorites. Calcium hypochlorites are sold in powder or tablet forms and can contain chlorine concentrations up to 67%. Sodium hypochlorite is a liquid (bleach, for example) and is found in concentrations up to 16%. Chlorine concentrations of household bleach range from 4.75 to 5.25%. Most small system operators find using these liquid or dry chlorine compounds more convenient and safer than chlorine gas.

The compounds are mixed with water and fed into the water with inexpensive solution feed pumps. These pumps are designed to operate against high system pressures but can also be used to inject chlorine solutions into tanks, although injecting chlorine into the suction side of a pump is not recommended as the chlorine may corrode the pump impeller.

Calcium hypochlorite can be purchased as tablets or granules, with approximately 65% available chlorine (10 lb of calcium hypochlorite granules contain only 6.5 lb of chlorine). Normally, 6.5 lb of calcium hypochlorite will produce a concentration of 50-mg/L chlorine in 10,000 gal of water. Calcium hypochlorite can burn (at 350°F) if combined with oil or grease. When mixing calcium hypochlorite, operators must wear chemical safety goggles, a cartridge breathing apparatus, and rubberized gloves. Always place the powder in the water. Placing the water into the dry powder could cause an explosion.

Sodium hypochlorite is supplied as a clear, greenish-yellow liquid in strengths from 5.25 to 16% available chlorine. Often referred to as "bleach," it is, in fact, used for bleaching. As we stated earlier, common household bleach is a solution of sodium hypochlorite containing 4.75 to 5.25% available chlorine. The amount of sodium hypochlorite required to produce a 50-mg/L chlorine concentration in 10,000 gal of water can be calculated using the solutions equation:

$$C_1V_1 = C_2V_2 \qquad (18.11)$$

where:

    C = the solution concentration (mg/L or %).
    V = the solution volume (e.g., liters, gallons, quarts).
    1.0% = 10,000 mg/L.

In this example, C_1 and V_1 are associated with the sodium hypochlorite, and C_2 and V_2 are associated with the 10,000 gallons of water with a 50-mg/L chlorine concentration. Therefore:

$$C_1 = 5.25\% = \frac{5.25\% \times 10,000 \text{ mg/L}}{1.0\%} = 52,500 \text{ mg/L}$$

V_1 = Unknown volume of sodium hypochlorite

$C_2 = 50$ mg/L

$V_2 = 10,000$ gal

$$C_1 \times V_1 = C_2 \times V_2$$

$$52,500 \text{ mg/L} \times V_1 = 50 \text{ mg/L} \times 10,000 \text{ gal}$$

$$V_1 = \frac{50 \text{ mg/L} \times 10,000 \text{ gal}}{52,500 \text{ mg/L}} = 9.52 \text{ gal}$$

Sodium hypochlorite solutions are introduced to the water in the same manner as calcium hypochlorite solutions. The purchased stock bleach is usually diluted with water to produce a feed solution that is pumped into the water system.

Hypochlorites must be stored properly to maintain their strengths. Calcium hypochlorite must be stored in airtight containers in cool, dry, dark locations. Sodium hypochlorite degrades relatively quickly even when properly stored; it can lose more than half of its strength in 3 to 6 months. Operators should purchase hypochlorites in small quantities to be sure they are used while still strong. Old chemicals should be discarded safely.

The pumping rate of a chemical metering pump is usually manually adjusted by varying the length of the piston or diaphragm stroke. Once the stroke is set, the hypochlorinator feeds accurately at that rate; however, chlorine measurements must be made occasionally at the beginning and end of the well pump cycle to ensure correct dosage. A metering device may be used to vary the hypochlorinator feed rate, synchronized with the water flow rate. Where a well pump is used, the hypochlorinator is connected electrically with the on/off controls of the pump so the chlorine solution is not fed into the pipe when the well is not pumping.

18.10.14.7 Determining Chlorine Dosage

Proper disinfection requires calculation of the amount of chlorine that must be added to the water to produce the required dosage. The type of calculation used depends on the form of chlorine being used. The basic chlorination calculation used is the same one used for all chemical addition calculations—the *pounds formula*:

$$\text{Pounds} = \text{mg/L} \times 8.34 \times \text{MG} \qquad (18.12)$$

where:

    Pounds = pounds of available chlorine required.
    mg/L = desired concentration in milligrams per liter.
    8.24 = conversion factor.
    MG = millions of gallons of water to be treated.

■ EXAMPLE 18.3

Problem: Calculate the number of pounds of gaseous chlorine needed to treat 250,000 gal of water with 1.2 mg/L of chlorine.

Solution:

$$\text{Pounds} = 1.2 \text{ mg/L} \times 8.34 \times 0.25 \text{ MG} = 2.5 \text{ lb}$$

Note: Hypochlorites contain less than 100% available chlorine; thus, we must use more hypochlorite to get the same number of pounds of chlorine into the water.

If we substitute calcium hypochlorite with 65% available chlorine in our example, 2.5 lb of available chlorine is still needed, but more than 2.5 lb of calcium hypochlorite is required to provide that much chlorine. Determine how much of the chemical is needed by dividing the pounds of chlorine required by the decimal form of the percent available chlorine. Because 65% is the same as 0.65, we need to add:

$$\frac{2.5 \text{ lb}}{0.65 \text{ available chlorine}} = 3.85 \text{ lb Ca(OCl)}_2 \quad (18.13)$$

to get that much chlorine.

In practice, because most hypochlorites are fed as solutions, we often need to know how much chlorine solution we should feed. In addition, the practical problems faced in day-to-day operation are never so clearly stated as the practice problems we work; for example, small water systems do not usually deal with water flow in million gallons per day. Real-world problems usually require a lot of intermediate calculations to get everything ready to plug into the pounds formula.

■ EXAMPLE 18.4

Problem: We have raw water with a chlorine demand 2.2 mg/L. We need a final residual of 1.0 mg/L at the entrance to the distribution system. We can use sodium hypochlorite or calcium hypochlorite granules as the source of chlorine. Well output is 65 gallons per minute (gpm), and the chemical feed pump can inject 100 milliliters per minute (mL/min) at the 50% setting. What is the required strength of the chlorine solution we will feed? What volume of 5.20% sodium hypochlorite will be required to produce 1 gal of the chlorine feed solution? How many pounds of 65% calcium hypochlorite will be needed to mix each gallon of solution?

Solution:

- *Step 1. Determine the amount of chlorine to be added to the water (the chlorine dose).* The dose is defined as the chlorine demand of the water, plus the desired residual, or in this case:

$$Q_1 = 65 \text{ gal/min} \qquad Q_t = Q_1 + Q_2$$
$$C_1 = 0.0 \text{ mg/L} \qquad C_t = 3.2 \text{ mg/L}$$

Well pump → → → Treated water

↑

Chlorination metering pump
$Q_2 = 100 \text{ mL/min}$
$C_2 = \text{unknown}$

FIGURE 18.10 Chlorination system problem.

$$\text{Dose} = \text{demand} + \text{residual}$$
$$= 2.2 \text{ mg/L} + 1.0 \text{ mg/L} = 3.2 \text{ mg/L}$$

To obtain a 1.0-mg/L residual when the water enters the distribution system, we must add 3.2 mg/L of chlorine to the water.

- *Step 2. Determine the strength of the chlorine feed solution that would add 3.2 mg/L of chlorine to 65 gpm of water when fed at a rate of 100 mL/min.* To make this calculation, consider the diagram of this chlorination system in Figure 18.10. The well pump is producing water at a flow rate (Q_1) of 65 gpm, with a chlorine concentration (C_1) of 0.0 mg/L. The metering pump will add 100 mL/min (Q_2) of chlorine solution, but we do not know its concentration (C_2) yet. The finished water will have been dosed with a chlorine concentration of 3.2 mg/L (C_1) and will enter the distribution system at a rate (Q_t) of 65 gal + 100 mL per minute, with a free residual chlorine concentration of 1.0 mg/L after the chlorine contact time.

- *Step 3. Convert the metering pump flow (100 mL/min) to gpm so we can calculate Q_1.* To do this, we use standard conversion factors:

$$\frac{100 \text{ mL}}{\text{min}} \times \frac{1 \text{ L}}{1000 \text{ mL}} \times \frac{1 \text{ gal}}{3.785 \text{ L}} = 0.026 \text{ gal/min}$$

So, now we know that Q_2 is 0.026 gpm and that Q_t is 65 gpm + 0.026 gpm = 65.026 gpm.

- *Step 4. Apply the mass balance equation.* The mass balance equation says that the flow rate times the concentration of the output is equal to the flow rate times the concentration of each of the inputs added together:

$$(Q_1 \times C_1) + (Q_2 \times C_2) = (Q_1 \times C_1) \quad (18.14)$$

Substituting the numbers given and those we have calculated so far gives us:

$$(65 \text{ gpm} \times 0.0 \text{ mg/L}) + (0.026 \text{ gpm} \times x \text{ mg/L})$$

$$= (65 + 0.026 \text{ gpm}) \times 3.2 \text{ mg/L}$$

$$0 + 0.026x = 65.026 \times 3.2$$

$$0.026x = 208.1$$

$$x = \frac{208.1}{0.026} = 8004 \text{ mg/L}$$

This is the answer to the first part of the question, the required strength of the chlorine feed solution. Because a 1% solution is equal to 10,000 mg/L, a solution strength of 8004 mg/L is an approximately 0.80% solution.

To determine the required volume of bleach per gallon to produce this 0.80% solution, we go back to the pounds formula:

$$\text{Pounds} = \text{mg/L} \times 8.34 \times \text{MG}$$

$$8004 \text{ mg/L} \times 8.34 \times 0.000001 \text{ MG} = 0.067 \text{ lb Cl}$$

Note: Remember to convert gallons to MG.

Recalling that the bleach, like water, weighs 8.34 lb/gal and contains approximately 5.20% (0.0520) available chlorine,

$$1 \text{ gal bleach} = 8.34 \text{ lb/gal} \times 0.0520$$

$$= 0.43 \text{ lb available chlorine per gal bleach}$$

If 1 gal of bleach contains 0.43 lb of available chlorine, how many gallons of bleach do we need to provide the 0.067 lb of chlorine required for each gallon of chlorine feed solution? To determine the gallons of bleach needed, we use the simple ratio equation:

$$\frac{1 \text{ gal}}{0.43 \text{ lb Cl}_2} = \frac{x \text{ gal}}{0.067 \text{ lb Cl}_2}$$

$$x = \frac{1 \text{ gal} \times 0.067 \text{ lb}}{0.43 \text{ lb}} = 0.16 \text{ gal bleach per gal solution}$$

Or, to determine the required volume of bleach per gallon to produce the 0.80% solution, we can use the solutions equation and calculate it this way:

$$C_1 \times V_1 = C_2 \times V_2$$

$$C_1 = 0.80\%$$

$$V_1 = 1.0 \text{ gal}$$

$$C_2 = 5.20\%$$

$$V_2 = \text{unknown}$$

$$0.80\% \text{ solution} \times 1.0 \text{ gal} = 5.20\% \text{ solution} \times x \text{ gal}$$

$$x = \frac{0.80\% \times 1.0 \text{ gal}}{5.20\%} = 0.15 \text{ gal bleach per gal solution}$$

To summarize, 1 gallon of household bleach:

- Contains 0.85 gal water
- Contains 5.20%, or 52,000 mg/L, available chlorine
- Contains 0.15 gal available chlorine
- Contains 0.43 lb available chlorine
- Weighs 8.34 lb

The third part of the problem requires that we determine the pounds of calcium hypochlorite required for each gallon of feed solution. We know that we need 0.066 lb of chlorine for each gallon of solution and that hypochlorite contains 65% available chlorine; that is, 1.0 lb of hypochlorite contains 0.65 lb of available chlorine. Using the ratio equation,

$$\frac{1 \text{ lb hypochlorite}}{0.65 \text{ lb Cl}_2} = \frac{x \text{ lb hypochlorite}}{0.067 \text{ lb Cl}_2}$$

$$x = \frac{1 \text{ lb} \times 0.067 \text{ lb}}{0.65 \text{ lb}}$$

$$x = \frac{0.067}{0.65} = 0.1 \text{ lb hypochlorite per gal solution}$$

18.10.14.8 Chlorine Generation

On-site generation of chlorine has recently become practical. These generation systems, using only salt and electric power, can be designed to meet disinfection and residual standards and to operate unattended at remote sites. Considerations for chlorine generation include cost, concentration of the brine produced, and availability of the process (AWWA, 1997).

18.10.14.8.1 Chlorine

Chlorine gas can be generated by a number of processes including the electrolysis of alkaline brine or hydrochloric acid, the reaction between sodium chloride and nitric acid, or the oxidation of hydrochloric acid. About 70% of the chlorine produced in the United States is manufactured from the electrolysis of salt brine and caustic solutions in a diaphragm cell (White, 1992). Because chlorine is a stable compound, it is typically produced offsite by a chemical manufacturer. Once produced, chlorine is packaged as a liquefied gas under pressure for delivery to the site in railcars, tanker trucks, or cylinders.

18.10.14.8.2 Sodium Hypochlorite

Dilute sodium hypochlorite solutions (less than 1%) can be generated electrochemically on-site from salt brine solution. Typically, sodium hypochlorite solutions are referred to as *liquid bleach* or *Javelle water*. Generally, the commercial- or industrial-grade solutions produced have hypochlorite strengths of 10 to 16%. The stability

TABLE 18.17
Chlorine Uses and Doses

Application	Typical Dose	Optimal pH	Reaction Time	Effectiveness	Other Considerations
Iron	0.62 mg/mg Fe	7.0	Less than 1 hour	Good	—
Manganese	0.77 mg/mg Mn	7–8	1–3 hour	Slow kinetics	Reaction time increases at lower pH
		9.5	Minutes		
Biological growth	1–2 mg/L	6–8	NA	Good	DBP formation
Taste/odor	Varies	6–8	Varies	Varies	Effectiveness depends on compound
Color removal	Varies	4.0–6.8	Minutes	Good	DBP formation
Zebra mussels	2–5 mg/L	—	Shock level	Good	DBP formation
	0.2–0.5 mg/L	—	Maintenance level		
Asiatic clams	0.3–0.5 mg/L	—	Continuous	Good	DBP formation

Source: Data from White (1992), Connell (1996), and Culp et al. (1986).

of a sodium hypochlorite solution depends on the hypochlorite concentration, the storage temperature, the length of storage (time), the impurities of the solution, and exposure to light. Decomposition of hypochlorite over time can affect the feed rate and dosage, as well as produce undesirable byproducts such as chlorite ions or chlorate (Gordon et al., 1995). Because of the storage problems, many systems are investigating on-site generation of hypochlorite in lieu of its purchase from a manufacturer or vendor.

18.10.14.8.3 Calcium Hypochlorite

To produce calcium hypochlorite, hypochlorous acid is made by adding chlorine monoxide to water and then neutralizing it with lime slurry to create a solution of calcium hypochlorite. Generally, the final product contains up to 70% available chlorine and 4 to 6% lime. Storage of calcium hypochlorite is a major safety consideration. It should never be stored where it is subject to heat or allowed to contact any organic material of an easily oxidized nature (Spellman, 2007).

18.10.14.9 Primary Uses and Points of Application of Chlorine

18.10.14.9.1 Uses

The main usage of chlorine in drinking water treatment is for disinfection; however, chlorine has also found application for a variety of other water treatment objectives, such as the control of nuisance organisms, oxidation of taste and odor compounds, oxidation of iron and manganese, color removal, and as a general treatment aid to filtration and sedimentation processes (White, 1992). Table 18.17 presents a summary of chlorine uses and doses.

18.10.14.9.2 Points of Application

At conventional surface water treatment plants, chlorine is typically added for prechlorination at either the raw

TABLE 18.18
Typical Chlorine Points of Application

Raw water intake
Flash mixer (prior to sedimentation)
Filter influent
Filter clearwell
Distribution system

Sources: Data from Connell (1996), White (1992), and AWWA (1990).

TABLE 18.19
Typical Chlorine Dosages at Water Treatment Plants

Chlorine Compound	Range of Doses (mg/L)
Calcium hypochlorite	0.6–5
Sodium hypochlorite	0.2–2
Chlorine gas	1–16

Source: Data are from USEPA's review of public water systems' initial sampling plans, which were required by the USEPA Information Collection Rule (ICR).

water intake or flash mixer, for intermediate chlorination ahead of the filters, for postchlorination at the filter clearwell, or for rechlorination of the distribution system (Connell, 1996). Table 18.18 summarizes typical points of application.

18.10.14.9.3 Typical Doses

Table 18.19 shows the typical dosages for the various forms of chlorine. The wide range of chlorine gas dosages most likely represents its use as both an oxidant and a disinfectant. Sodium hypochlorite and calcium hypochlorite can also serve as both an oxidant and a disinfectant, but their higher cost may limit their use.

inactivation was observed when oocysts were exposed to free chlorine concentrations ranging from 5 to 80 mg/L at pH 8, a temperature of 22°C, and contact times of 48 to 245 minutes (Gyurek et al., 1996). CT values ranging from 3000 to 4000 mg·min/L were required to achieve 1-log *Cryptosporidium* inactivation at pH 6.0 and temperature of 22°C. During this study, when oocysts were exposed to 80 mg/L of free chlorine for 120 minutes, inactivation greater than 3 logs was produced.

18.10.14.13 Disinfection Byproducts

Halogenated organics are formed when natural organic matter (NOM) reacts with free chlorine or free bromine. Free chlorine is normally introduced into water directly as a primary or secondary disinfectant. Free bromine results from the oxidation by chlorine of the bromide ion in the source water. Factors affecting the formation of these halogenated DBPs include type and concentration of NOM, chlorine form and dose, time, bromide ion concentration, pH, organic nitrogen concentration, and temperature. Organic nitrogen significantly influenced the formation of nitrogen-containing DBPs, including haloacetonitriles (Reckhow et al., 1990), halopicrins, and cyanogen halides. Because most water treatment systems have been required to monitor for total trihalomethanes (TTHMs) in the past, most water treatment operators are probably familiar with some of the requirements of DBP regulations. The key points of the DBP Rule and some of the key changes with which water supply systems are required to comply are summarized below:

- *Chemical limits and testing*—Testing requirements include total trihalomethanes (TTHMs) and five haloacetic acids (HAAs). The maximum contaminant level (MCL) for TTHM is 0.080 mg/L for surface water systems. In addition, a new MCL of 0.060 mg/L has been established for haloacetic acids. New MCLs have been established for bromate (0.010 mg/L and chlorite (1.0 mg/L). Bromate monitoring is required of systems that use ozone. Chlorite monitoring is required only for systems that use chlorine dioxide (i.e., sodium and calcium hypochlorite are not included). Maximum residual disinfectant levels (MRDLs) were established for total chlorine (4.0 mg/L) and chlorine dioxide (0.8 mg/L).
- *Operational requirements*—Analytical requirements for measuring residual chlorine have been changed to require digital equipment (i.e., no color wheels or analog test kits). The test kit must have a detection limit of at least 0.1 mg/L.

- *Monitoring and Reporting*—Individual state requirements will differ, but at a minimum the following requirements are in effect:
  - Surface water system monitoring requirements include four quarterly samples per treatment plant (Source Treatment Unit, or STU) for both TTHMs and HAAs. One of these quarterly samples, or 25% of the total samples, must be collected at the maximum residence time location. The remaining quarterly samples must be collected at representative locations throughout the entire distribution system. Compliance is based on a running annual average computed quarterly.
  - For those surface water systems using conventional filtration or lime softening, a disinfectant/disinfection byproduct (D/DBP) monthly operating report for total organic carbon (TOC) removal must be completed and filed with the state EPA. This monthly report should include TOC, alkalinity, and specific ultraviolet absorption (SUVA) parameters. In addition, a monthly operating report is also required for bromate, chlorite, chlorine dioxide, and residual chlorine.
  - TTHM monitoring results may indicate the possible need for additional treatment to include best available technology for reduction of DBP. This may include granular activated carbon, enhanced coagulation (for surface water systems using conventional filtration), or enhanced softening (for systems using lime softening).
  - Operators are required to develop and implement sample-monitoring plans for disinfectant residuals and disinfection byproducts. The plans must be submitted to and approved by the state EPA. Disinfection residual monitoring compliance for total chlorine, including chloramines, is based on a running annual average, which is computed quarterly, of the monthly average of all samples collected under this rule. Disinfectant residual monitoring compliance for chlorine dioxide is based on consecutive daily samples. Disinfectant residual monitoring is required at the same distribution point and time as total coliform monitoring. In addition, if the operator feeds ozone or chlorine dioxide, then a sample monitoring plan for bromate or chlorite, respectively, must be submitted to and approved by the state EPA.

18.10.14.14 Application Methods

Different application methods are used depending on the form of chlorine used. The following describes the typical application methods for chlorine, sodium hypochlorite, and calcium hypochlorite:

- *Chlorine*—Liquefied chlorine gas is typically evaporated to gaseous chlorine prior to metering. The heat required for evaporation can be provided through either a liquid chlorine evaporator or the ambient heat input to the storage container. When the compressed liquid chlorine has evaporated, the chlorine gas is typically fed under vacuum conditions. Either an injector or a vacuum induction mixer creates the required vacuum. The injector uses water flowing through a Venturi to draw the chlorine gas into a side stream of carrier water to form a concentrated chlorine solution. This solution is then introduced into the process water through a diffuser or it is mixed with a mechanical mixer. A vacuum induction mixer uses the motive forces of the mixer to create a vacuum and draws the chlorine gas directly into the process water at the mixer.
- *Sodium hypochlorite*—Sodium hypochlorite solutions degrade over time; for example, a 12.5% hypochlorite solution will degrade to 10% in 30 days under best-case conditions. Increased temperature, exposure to light, and contact with metals all can increase the rate of sodium hydroxide degradation (White, 1992). The sodium hypochlorite solution is typically fed directly into the process water using a type of metering pump. Similar to theh procedure for a chlorine solution, sodium hypochlorite is mixed with the process water with either a mechanical mixer or an induction mixer. Sodium hypochlorite solution usually is not diluted prior to mixing to reduce scaling problems.
- *Calcium hypochlorite*—Commercial high-level calcium hypochlorite contains at least 70% available chlorine. Under normal storage conditions, calcium hypochlorite loses 3 to 5% of its available chlorine in a year. Calcium hypochlorite comes in powder, granular, and compressed tablet forms. Typically, calcium hypochlorite solution is prepared by mixing powdered or granular calcium hypochlorite with a small flow. The highly chlorinated solution is then flow paced into drinking water flow (USEPA, 1991).

18.10.14.15 Safety and Handling Considerations

Chlorine gas is a strong oxidizer. The U.S. Department of Transportation classifies chlorine as a poisonous gas (Connell, 1996). Fire codes typically regulate the storage and use of chlorine. In addition, facilities storing more than 2500 pounds of chlorine are subject to the following two safety programs:

- Process Safety Management (PSM) standard regulated by the Occupational Safety and Health Administration (OSHA) under 29 CFR 1910.119
- Risk Management Program (RMP) administered by USEPA under Section 112(r) of the Clean Air Act (CAA)

These regulations (as well as local and state codes and regulations) must be considered during the design and operation of chlorination facilities at a water treatment plant.

Sodium hypochlorite solution is a corrosive liquid with an approximate pH of 12. Typical precautions for handling corrosive materials, such as avoiding contact with metals, including stainless steel, should be taken (AWWA, 1990). Sodium hypochlorite solutions may contain chlorate. Chlorate is formed during both the manufacturing and storage of sodium hypochlorite (due to degradation of the product). Chlorate formation can be minimized by reducing the degradation of sodium hypochlorite by limiting storage times, avoiding high temperatures, and reducing the amount of light exposure (Gordon 1995). Spill containment must be provided for sodium hypochlorite storage tanks. Typical spill containment features for this purpose include containment for the entire contents of the largest tank (plus freeboard for rainfall or fire sprinklers), no uncontrolled floor drains, and separate containment areas for each incompatible chemical.

Calcium hypochlorite is an oxidant and as such it should be stored separately from organic materials that can be readily oxidized. It should also be stored away from any source of heat. Improperly stored calcium hypochlorite has been known to cause spontaneous combusting fires.

18.10.14.16 Advantages and Disadvantages of Chlorine Use

The following list presents selected advantages and disadvantages of using chlorine as a disinfection method for drinking water. Because of the wide variation of system size, water quality, and dosages applied, some of these advantages and disadvantages may not apply to a particular system (Masschelein, 1992).

Advantages
- Chlorine oxidizes soluble iron, manganese, and sulfides.
- Chlorine enhances color removal.
- Chlorine enhances taste and odor.
- Chlorine may enhance coagulation and filtration of particulate contaminants.
- Chlorine is an effective biocide.
- Chlorine is the easiest and least disinfection method, regardless of system size.
- Chlorine is the most widely used disinfection method and therefore the best understood.
- Chlorine is available as calcium and sodium hypochlorite (these solutions offer more advantages than chlorine gas for smaller systems because they are easier to use, are safer, and require less equipment compared to chlorine gas).
- Chlorine provides a residual.

Disadvantages
- Chlorine may cause a deterioration in coagulation/filtration of dissolved organic substances.
- Chlorine forms halogen-substituted byproducts.
- Chlorine could cause taste and odor problems in the finished water, depending on the water quality and dosage.
- Chlorine gas is a hazardous corrosive gas.
- Chlorine gas requires special leak containment and scrubber facilities.
- Typically, sodium and calcium hypochlorite are more expensive than chlorine gas.
- Sodium hypochlorite degrades over time and with exposure to light.
- Sodium hypochlorite is a corrosive chemical.
- Calcium hypochlorite must be stored in a cool, dry place because of its reaction with moisture and heat.
- A precipitate may form in a calcium hypochlorite solution because of impurities; therefore, an antiscalant chemical may be needed.
- Higher concentrations of hypochlorite solutions are unstable and will produce chlorate as a byproduct.
- Chlorine is less effective at high pH.
- Chlorine forms oxygenated byproducts that are biodegradable and can enhance subsequent biological growth if a chlorine residual is not maintained.
- Chlorine can cause the release of constituents bound in the distribution system (e.g., arsenic) by changing the redox state.

18.10.14.17 Chlorine Summary Table

Table 18.20 presents a summary of the considerations for the use of chlorine as a disinfectant.

TABLE 18.20
Summary of Chlorine Disinfection

Consideration	Description
Generation	Chlorination may be performed using chlorine gas or other chlorinated compounds that may be in liquid or solid form. Chlorine gas can be generated by any number of processes, including the electrolysis of alkaline brine or hydrochloric acid, the reaction between sodium chloride and nitric acid, or the oxidation of hydrochloric acid. Because chlorine is a stable compound, chlorine gas, sodium hypochlorite, and calcium hypochlorite are typically produced offsite by a chemical manufacturer.
Primary uses	The primary use of chlorination is disinfection. Chlorine also serves as an oxidizing agent for taste and odor control, prevention of algal growths, maintaining clear filter media, removal of iron and manganese, destruction of hydrogen sulfide, color removal, maintaining the water quality at the distribution systems, and improving coagulation.
Inactivation efficiency	The general order of increasing chlorine disinfection difficulty is bacteria, viruses, and then protozoa. Chlorine is an extremely effective disinfectant for inactivating bacteria and highly effective viricide; however, chlorine is less effective against *Giardia* cysts. *Cryptosporidium* oocysts are highly resistant to chlorine.
Byproduct formation	When added to the water, free chlorine reacts with NOM and bromide to form DBPs, primarily THMs, and some haloacetic acids (HAAs), among others.
Point of application	Chlorine is applied during raw water storage, precoagulation/post-raw water storage, presedimentation/postcoagulation, postsedimentation/prefiltration, or postfiltration (disinfection), or in the distribution system.
Special considerations	Because chlorine is such a strong oxidant and extremely corrosive, special storage and handling considerations should be taken into account when planning a water treatment plant. Additionally, health concerns associated with the handling and use of chlorine must be considered.

18.11 ARSENIC REMOVAL FROM DRINKING WATER

(Much of the following information is based on USEPA, 2000.) Operators may be familiar with the controversy created when newly elected President George W. Bush placed the pending arsenic standard on temporary hold. The President prevented implementation of the arsenic standard to give scientists time to review the standard, to take a closer look at the possible detrimental effects on the health and well-being of consumers in certain geographical areas of the United States, and to give economists time to determine the actual cost of implementation.

President Bush's decision caused quite a stir, especially among environmentalists, the media, and others who felt that the arsenic standard should be enacted immediately to protect affected consumers. The President was aware of the emotional and political implications of shelving the arsenic standard and understood the staggering economical implications involved in implementing the new, tougher standard. Many view the President's decision as wrong, but others consider his decision to be the right one, basing their opinion on the old adage, "It is best to make scientific judgments based on good science instead of on 'feel good' science." Whether the reader shares the latter view or not, the point is that arsenic levels in potable water supplies must be reduced to a set level and in the future will have to be reduced to an even lower level. Accordingly, water treatment plants affected by the existing arsenic requirements and the pending tougher arsenic requirements should be familiar with the technologies for removal of arsenic from water supplies. In this section, we describe a number of these technologies.

18.11.1 ARSENIC EXPOSURE

Arsenic (As) is a naturally occurring element present in food, water, and air. Known for centuries to be an effective poison, some animal studies suggest that arsenic may be an essential nutrient at low concentrations. Nonmalignant skin alterations, such as keratosis and hypo- and hyperpigmentation, have been linked to arsenic ingestion, and skin cancers have developed in some patients. Additional studies indicate that arsenic ingestion may result in internal malignancies, including cancers of the kidney, bladder, liver, lung, and other organs. Vascular system effects have also been observed, including peripheral vascular disease, which in its most severe form results in gangrene or blackfoot disease. Other potential effects include neurologic impairment.

The primary route of exposure to arsenic for humans is ingestion. Exposure via inhalation is considered minimal, although in some areas of the world elevated levels of airborne arsenic occur periodically (Hering and Chiu,

1998). Arsenic occurs in two primary forms: organic and inorganic. Organic species of arsenic are predominately found in foodstuffs, such as shellfish, and include such forms as monomethyl arsenic acid (MMAA), dimethyl arsenic acid (DMAA), and arseno-sugars. Inorganic arsenic occurs in two valence states, arsenite and arsenate. In natural surface waters, arsenate is the dominant species.

18.11.2 ARSENIC REMOVAL TECHNOLOGIES

Some of the arsenic removal technologies discussed here are traditional treatment processes that have been tailored to improve removal of arsenic from drinking water. Several treatment techniques are at the experimental stage with regard to arsenic removal, and some have not been demonstrated at full scale. Although some of these processes may be technically feasible, their costs may be prohibitive. Technologies discussed in this section are grouped into four broad categories: *prescriptive processes*, *adsorption processes*, *ion exchange processes*, and *separation* (membrane) *processes*. Each category is presented with at least one treatment technology.

18.11.2.1 Prescriptive Processes

18.11.2.1.1 Coagulation/Filtration

Coagulation/flocculation (C/F) is a treatment process in which the physical or chemical properties of dissolved colloidal or suspended matter are altered such that agglomeration is enhanced to an extent that the resulting particles will settle out of solution by gravity or will be removed by filtration. Coagulants change the surface charge properties of solids to allow agglomeration or enmeshment of particles into a flocculated precipitate. In either case, the final products are larger particles, or floc, which more readily filter or settle under the influence of gravity.

The coagulation/filtration process has traditionally been used to remove solids from drinking water supplies; however, the process is not restricted to the removal of particles. Coagulants render some dissolved species, such as natural organic matter (NOM), inorganics, and hydrophobic synthetic organic compounds (SOCs), insoluble, and the metal hydroxide particles produced by the addition of metal salt coagulants (typically aluminum sulfate, ferric chloride, or ferric sulfate) can adsorb other dissolved species. Major components of a basic coagulation/filtration facility include chemical feed systems, mixing equipment, basins for rapid mix, flocculation, settling, filter media, sludge handling equipment, and filter backwash facilities. Settling may not be necessary in situations where the influent particle concentration is very low. Treatment plants without settling are known as direct filtration plants.

18.11.2.1.2 Iron/Manganese Oxidation

Iron/manganese (Fe/Mn) oxidation is commonly used by facilities treating groundwater. The oxidation process used to remove iron and manganese leads to the formation of hydroxides that remove soluble arsenic by precipitation or adsorption reactions. Arsenic removal during iron precipitation is fairly efficient. Removal of 2 mg/L of iron achieved a 92.5% removal of arsenic from an initial 10-μg/L arsenate concentration by adsorption alone. Even removal of 1 mg/L of iron resulted in the removal of 83% of influent arsenic from a source with 22 μg/L arsenate. Indeed, field studies of iron removal plants have indicated that this treatment can feasibly reach 3 g/L.

The removal efficiencies achieved by iron removal are not as high or as consistent as those realized by activated alumina or ion exchange (Edwards, 1994). Note, however, that arsenic removal during manganese precipitation is relatively ineffective when compared to iron even when removal by both adsorption and coprecipitation are considered. Precipitation of 3 mg/L manganese removed only 69% of arsenate in a 12.5-μg/L arsenate influent concentration.

Oxidation filtration technologies may be effective arsenic removal technologies. Research of oxidation filtration technologies has primarily focused on greensand filtration. As a result, the following discussion focuses on the effectiveness of greensand filtration as an arsenic removal technology. Substantial arsenic removal has been achieved using greensand filtration (Subramanian et al., 1997). The active material in greensand is glauconite, a green, iron-rich, clay-like mineral that has ion exchange properties. Glauconite often occurs in nature as small pellets mixed with other sand particles, giving a green color to the sand. The glauconite sand is treated with $KMnO_4$ until the sand grains are coated with a layer of manganese oxides, particularly manganese dioxide. The mechanisms behind this arsenic removal treatment are multifaceted and include oxidation, ion exchange, and adsorption. Arsenic compounds displace species from the manganese oxide (presumably OH^- and H_2O), becoming bound to the greensand surface—in effect, an exchange of ions occurs. The oxidative nature of the manganese surface converts arsenite to arsenate, and arsenate is adsorbed to the surface. As a result of the transfer of electrons and adsorption of arsenate, reduced manganese is released from the surface.

The effectiveness of greensand filtration for arsenic filtration for arsenic removal is dependent on the influent water quality. Surmanian et al. (1997) demonstrated a strong correlation between influent Fe concentration and arsenic percent removal. Removal increased from 41% to more than 80% as the Fe/As ratio increased from 0 to 20 when treating a tap water with a spiked arsenite concentration of 200 mg/L. The tap water contained 366 mg/L

sulfate and 321 mg/L total dissolved solids (TDS); neither constituent seemed to affect arsenic removal. The authors suggested that the influent manganese concentration may play an important role. Divalent ions, such as calcium, can also compete with arsenic for adsorption sites. Water quality would have to be carefully evaluated for applicability for treatment using greensand. Other researchers have also reported substantial arsenic removal using this technology, including arsenic removals of greater than 90% for treatment of groundwater.

As with other treatment media, greensand must be regenerated when its oxidative and adsorptive capacity has been exhausted. Greensand filters are regenerated using a solution of excess potassium permanganate ($KMnO_4$). Like other treatment media, the regeneration frequency will depend on the influent water quality in terms of constituents that will degrade the filter capacity. Regenerant disposal for greensand filtration has not been addressed in previous research.

18.11.2.1.3 Coagulation-Assisted Microfiltration

Arsenic is removed effectively by the coagulation process. Microfiltration is used as a membrane separation process to remove particulates, turbidity, and microorganisms. In coagulation-assisted microfiltration technology, microfiltration is used in a manner similar to a conventional gravity filter. The advantages of microfiltration over conventional filtration include (Muilenberg, 1997):

- More effective microorganism barrier during coagulation process upsets
- Removal of smaller floc sizes (smaller amounts of coagulants are required)
- Increased total plant capacity

Vickers et al. (1997) reported that microfiltration exhibited excellent arsenic removal capability. This report is corroborated by pilot studies conducted by Clifford et al. (1997), who found that coagulation-assisted microfiltration could reduce arsenic levels below 2 g/L in waters with a pH of between 6 and 7, even when the influent concentration of Fe is approximately 2.5 mg/L. These studies also found that the same level of arsenic removal could be achieved by this treatment process even if source water sulfate and silica levels were high. Further, coagulation-assisted microfiltration can reduce arsenic levels to an even greater extend at a slightly lower pH (approximately 5.5).

Addition of a coagulant did not significantly affect the membrane-cleaning interval, although the solids level to the membrane system increased substantially. With an iron and manganese removal system, it is critical that all of the iron and manganese be fully oxidized before reaching the membrane to prevent fouling (Muilenberg, 1997).

18.11.2.1.4 Enhanced Coagulation

The Disinfectant/Disinfection Byproduct (D/DBP) Rule requires the use of enhanced coagulation treatment for the reduction of disinfection byproduct (DBP) precursors for surface water systems that have sedimentation capabilities. The enhanced process involves modifications to the existing coagulation process such as increasing the coagulant dosage or reducing the pH, or both. Cheng et al. (1994) conducted bench-, pilot-, and demonstration-scale studies to examine arsenate removals during enhanced coagulation. The enhanced coagulation conditions in these studies included an increase in alum and ferric chloride coagulant dosage from 10 to 30 mg/L or a decrease in pH from 7 to 5.5, or both. Results from these studies indicated the following:

- Greater than 90% arsenate removal can be achieved under enhanced coagulation conditions. Arsenate removals greater than 90% were easily attained under all conditions when ferric chloride was used.
- Enhanced coagulation using ferric salts is more effective for arsenic removal than enhanced coagulation using alum. At an influent arsenic concentration of 5 μg/L, ferric chloride achieved 96% arsenate removal with a dosage of 10 mg/L and no acid addition. When alum was used, 90%, arsenate removal could not be achieved without reducing the pH.
- Lowering pH during enhanced coagulation improved arsenic removal by alum coagulation. With ferric coagulation, pH does not have a significant effect between 5.5 and 7.0.

Note: Post-treatment pH adjustment may be required for corrosion control when the process is operated at a low pH.

18.11.2.1.5 Lime Softening

Recall that hardness is predominately caused by calcium and magnesium compounds in solution. Lime softening removes this hardness by creating a shift in the carbonate equilibrium. The addition of lime to water raises the pH. Bicarbonate is converted to carbonate as the pH increases, and as a result calcium is precipitated as calcium carbonate. Soda ash (sodium carbonate) is added if insufficient bicarbonate is present in the water to remove hardness to the desired level. Softening for calcium removal is typically accomplished at a pH range of 9 to 9.5. For magnesium removal, excess lime is added beyond the point of calcium carbonate precipitation. Magnesium hydroxide precipitates at pH levels greater than 10.5. Neutralization is required if the pH of the softened water is excessively high (above 9.5) for potable use. The most common form of pH adjustment in softening plants is recarbonation with carbon dioxide.

Lime softening has been widely used in the United States for reducing hardness in large water treatment systems. Lime softening, excess lime treatment, split lime treatment, and lime–soda softening are all common in municipal water systems. All of these treatment methods are effective in reducing arsenic. Considerable amounts of sludge are produced in a lime softening system, and its disposal is expensive. Large-capacity systems may find it economically feasible to install recalcination equipment to recover and reuse the lime sludge and reduce disposal problems. Construction of a new lime softening plant for the removal of arsenic would not generally be recommended unless hardness must also be reduced.

18.11.2.2 Adsorptive Processes

18.11.2.2.1 Activated Alumina

Activated alumina is a physicochemical process by which ions in the feed water are sorbed to the oxidized activated alumina surface. Activated alumina is considered an adsorption process, although the chemical reactions involved are actually an exchange of ions. Activated alumina is prepared through dehydration of $Al(OH)_3$ at high temperatures and consists of amorphous and gamma alumina oxide (AWWA, 1990; Clifford et al., 1985). Activated alumina is used in packed beds to remove contaminants such as fluoride, arsenic, selenium, silica, and NOM. Feed water is continuously passed through the bed to remove contaminants. The contaminant ions are exchanged with the surface hydroxides on the alumina. When adsorption sites on the activated alumina surface become filled, the bed must be regenerated. Regeneration is accomplished through a sequence of rinsing with regenerant, flushing with water, and neutralizing with acid. The regenerant is a strong base, typically sodium hydroxide; the neutralizer is a strong acid, typically sulfuric acid. Many studies have shown that activated alumina is an effective treatment technique for arsenic removal. Factors such as pH, arsenic oxidation state, competing ions, empty bed contact time, and regeneration have significant effects on the removals achieved with activated alumina. Other factors include spent regenerant disposal, alumina disposal, and secondary water quality.

18.11.2.2.2 Ion Exchange

Ion exchange is a physicochemical process by which an ion on the solid phase is exchanged for an ion in the feed water. This solid phase is typically a synthetic resin that has been chosen to preferentially adsorb the particular contaminant of concern. To accomplish this exchange of ions, feed water is continuously passed through a bed of ion exchange resin beads in a downflow or upflow mode until the resin is exhausted. Exhaustion occurs when all sites on the resin beads have been filled by contaminant ions. At this point, the bed is regenerated by rinsing the ion exchange column with a regenerant—a concentrated solution of ions initially exchanged from the resin. The number of bed volumes that can be treated before exhaustion varies with resin type and influent water quality. Typically from 300 to 60,000 bed volume (BV) can be treated before regeneration is required. In most cases, regeneration of the bed can be accomplished with only 1 to 5 BV of regenerant followed by 2 to 20 BV of rinse water. Important considerations in the applicability of the ion exchange process for removal of a contaminant include water quality parameters such as pH, competing ions, resin type, alkalinity, and influent arsenic concentration. Other factors include the affinity of the resin for the contaminant, spent regenerant and resin disposal requirements, secondary water quality effects, and design operating parameters.

18.11.2.3 Membrane Processes

Membranes are a selective barrier, allowing some constituents to pass while blocking the passage of others. The movement of constituents across a membrane requires a driving force (i.e., a potential difference between the two sides of the membrane). Membrane processes are often classified by the type of driving force, including pressure, concentration, electrical potential, and temperature. The processes discussed here include only pressure-driven and electrical potential-driven types.

Pressure-driven membrane processes are often classified by pore size into four categories: microfiltration (MF), ultrafiltration (UF), nanofiltration (NF), and reverse osmosis (RO). Typical pressure ranges for these processes are given in Table 18.21. NF and UF are high-pressure processes. NF and RO primarily remove constituents through chemical diffusion, and MF and UF primarily remove constituents through physical sieving. An advantage of high-pressure processes is that they tend to remove a broader range of constituents than low-pressure processes; however, the drawback to broader removal is the increase in energy required for high-pressure processes (Aptel and Buckley, 1996). Electrical potential-driven membrane processes can also be used for arsenic removal. These processes include, for the purposes of this document, only electrodialysis reversal (EDR). In terms of achievable contaminant removal, EDR is comparable to RO. The separation process used in EDR, however, is ion exchange.

18.11.2.4 Alternative Technologies

18.11.2.4.1 Iron-Oxide-Coated Sand

Iron-oxide-coated sand is a rare process that has shown some tendency for arsenic removal. Iron-oxide-coated sand consists of sand grains coated with ferric hydroxide which are used in fixed bed reactors to remove various dissolved metal species. The metal ions are exchanged with the surface hydroxides on the iron-oxide-coated sand.

TABLE 18.21
Typical Pressure Ranges for Membrane Processes

Membrane Process	Pressure Range (psi)
Microfiltration (MF)	5–45
Ultrafiltration (UF)	7–100
Nanofiltration (NF)	50–150
Reverse osmosis (RO)	100–150

Iron-oxide-coated sand exhibits selectivity in the adsorption and exchange of ions present in the water. Like other processes, when the bed is exhausted it must be regenerated by a sequence of operations consisting of rinsing with regenerant, flushing with water, and neutralizing with strong acid. Sodium hydroxide is the most common regenerant and sulfuric acid the most common neutralizer. Several studies have shown that iron-oxide-coated sand is effective for arsenic removal, depending on such factors as pH, arsenic oxidation state, competing ions, and regeneration.

18.11.2.4.2 Sulfur-Modified Iron

A patented sulfur-modified iron (SMI) process for arsenic removal (Hydrometrics, 1998) consists of three components: (1) finely divided metallic iron, (2) powdered elemental sulfur or other sulfur compounds, and (3) an oxidizing agent. The powdered iron, powdered sulfur, and oxidizing agent (H_2O_2 in preliminary tests) are thoroughly mixed and then added to the water to be treated. The oxidizing agent serves to convert arsenite to arsenate. The solution is then mixed and settled. Using the sulfur-modified iron process on several water types, high adsorptive capacities were obtained with a final arsenic concentration of 0.050 mg/L. Arsenic removal was influenced by pH. Approximately 20 mg/L arsenic per gram of iron was removed at pH 8, and 50 mg arsenic per gram of iron was removed at pH 7. Arsenic removal seems to be very dependent on the iron-to-arsenic ratio.

Packed-bed column tests demonstrated significant arsenic removal at residence times of 5 to 15 minutes. Significant removal of both arsenate and arsenite was measured. The highest adsorption capacity measured was 11 mg arsenic removed per gram of iron. Flow distribution problems were evident, as several columns became partially plugged, and better arsenic removal was observed with reduced flow rates.

Spent media from the column tests were classified as nonhazardous waste. Projected operating costs for sulfur-modified iron, when the process is operated below a pH of 8, are much lower than alternative arsenic removal technologies such as ferric chloride addition, reverse osmosis, and activated alumina. Cost savings would

TABLE 18.22
Adsorption Tests on Granular Ferric Hydroxide

Parameter	Test 1	Test 2	Test 3	Test 4
Raw Water Parameters				
pH	7.8	7.8	8.2	7.6
Arsenate concentration (μg/L)	100–800	21	16	15–20
Phosphate concentration (μg/L)	0.70	0.22	0.15	0.30
Conductivity (μS/cm)	780	480	200	460
Adsorption capacity for arsenate (g/kg)	8.5	4.5	3.2	Not determined
Adsorber Parameters				
Bed height (m)	0.24	0.16	0.15	0.82
Filter rate (m/hr)	6-10	7.6	5.7	15
Treatment capacity (BV)	34,000	37,000	32,000	85,000
Maximum effluent concentration (μg/L)	10	10	10	7
Arsenate content of GFH (g/kg)	8.5	1.4	0.8	1.7
Mass of spent GFH (dry weight) (g/m^3)	20.5	12	18	8.6

increase proportionally with increased flow rates and increased arsenic concentrations.

Possible treatment systems using sulfur-modified iron include continuous stirred tank reactors, packed bed reactors, fluidized bed reactors, and passive *in situ* reactors. Packed bed and fluidized bed reactors appear to be the most promising for successful arsenic removal in pilot-scale and full-scale treatment systems based on current knowledge of the sulfur-modified iron process.

18.11.2.4.3   Granular Ferric Hydroxide

A removal technique for arsenate developed at the Technical University of Berlin, Department of Water Quality Control, is adsorption on granular ferric hydroxide (GFH) in fixed-bed reactors. This technique combines the advantages of the coagulation/filtration process (efficiency and small residual mass) with the fixed-bed adsorption on activated alumina and sample processing. Demers and Renner (1992) reported that the application of granular ferric hydroxide in test adsorbers showed a high treatment capacity of 30,000 to 40,000 BV with an effluent arsenate concentration never exceeding 10 μm/L. The typical residual mass was in the range of 5 to 25 g/m^3 treated water. The residue was a solid with an arsenate content of 1 to 10 g/kg. Table 18.22 summarizes data from the adsorption tests.

The competition of sulfate on arsenate adsorption was not very strong. Phosphate, however, competed strongly with arsenate, which reduced arsenate removal with GFH. Arsenate adsorption decreased with pH, which is typical for anion adsorption. At high pH values, GFH outperformed alumina. Below a pH of 7.6, the performance is comparable.

A field study reported by Simms et al. (2000) confirmed the efficacy of GFH for arsenic removal. Over the course of this study, a 5.3-MGD GHF facility located in the United Kingdom was found to reliably and consistently reduce average influent arsenic concentrations of 20 g/L to less than 10 g/L for 200,000 BV (over a year of operation) at an empty bed contact time of 3 minutes. Despite insignificant head loss, routine backwashing was conducted on a monthly basis to maintain media condition and to reduce the possibility of bacterial growth. The backwash was not hazardous and could be recycled or disposed to a sanitary sewer. At the time of replacement, arsenic loading on the media was 2.3%. Leachate tests conducted on the spent media found that arsenic did not leach from the media.

The most significant weakness of this technology appears to be its cost. Currently, GFH media cost approximately $4000 per ton; however, if a GFH bed can be used several times longer than an alumina bed, for example, it may prove to be the more cost-effective technology. Indeed, the system profiled in the field study presented here tested activated alumina as well as GFH and found the GFH was sufficiently more efficient that smaller adsorption vessels and smaller quantities of media could be used to achieve the same level of arsenic removal (reducing costs). In addition, unlike activated alumina, GFH does not require preoxidation.

A treatment for leaching arsenic from the media to enable regeneration of GFH seems feasible, but it results in the generation of an alkaline solution with high levels of arsenate which requires further treatment to obtain a solid waste. Thus, direct disposal of spent GFH should be favored.

18.11.2.4.4   Iron Filings

Iron filings and sand may be used to reduce inorganic arsenic species to iron coprecipitates, mixed precipitates, and, in conjunction with sulfates, arsenopyrites. This type of process is essentially a filter technology, much like greensand filtration, wherein the source water is filtered through a bed of sand and iron filings. Unlike some technologies (e.g., ion exchange), sulfate is actually introduced in this process to encourage arsenopyrite precipitation. This arsenic removal method was originally developed as a batch arsenic remediation technology. It appears to be quite effective in this use. Bench-scale tests indicate an average removal efficiency of 81% with much higher removals at lower influent concentrations. This method was tested to arsenic levels of 20,000 ppb, and at 2000 ppb it consistently reduced arsenic levels to less than 50 ppb (the current MCL). Although it is quite effective in this capacity, its use as a drinking water treatment technology appears to be limited. In batch tests, a residence time of approximately 7 days was required to reach the desired arsenic removal. In flowing conditions, even though removals averaged 81% and reached greater than 95% at 2000 ppb arsenic, there is no indication that this technology can reduce arsenic levels below approximately 25 ppb. No data are available that indicate how the technology can reduce arsenic levels below approximately 25 ppb, nor do data exist regarding how the technology performs at normal source water arsenic levels. This technology must be further evaluated before it can be recommended as an approved arsenic removal technology for drinking water.

18.11.2.4.5 *Photooxidation*

Researchers at the Australian Nuclear Science and Technology Organization (ANSTO) have found that, in the presence of light and naturally occurring light-absorbing materials, the oxidation rate of arsenite by oxygen can be increased 10,000-fold. The oxidized arsenic, now arsenate, can then be effectively removed by coprecipitation. ANSTO evaluated both UV lamp reactors and sunlight-assisted photooxidation using acidic, metal-bearing water from an abandoned gold, silver, and lead mine. Air sparging was required for sunlight-assisted oxidation due to the high initial arsenate concentration (12 mg/L). Tests demonstrated that near-complete oxidation of arsenite could be achieved using the photochemical process. Analysis of process waters indicated that 97% of the arsenic in the process stream was present as arsenate. Researchers concluded that arsenite was preferentially oxidized in the presence of excess dissolved Fe (22:1 iron-to-arsenic mole ratio). This is in contrast to conventional plants where dissolved Fe represents an extra chemical oxidant demand that has to be satisfied during oxidation of arsenite (ANSTO, 1999).

Photooxidation of the mine water followed by coprecipitation reduced arsenic concentrations to as low as 17 g/L, which meets the current MCL for arsenic. Initial total arsenic concentrations were unknown, although the arsenite concentration was given as approximately 12 mg/L, which is considerably higher than typical raw water arsenic concentrations. ANSTO reported that residuals from this process are environmentally stable and passed the Toxicity Characteristic Leaching Procedure (TCLP) test necessary to declare waste nonhazardous suitable for landfill disposal. Based on the removals achieved and residuals characteristics, it is expected that photooxidation followed by coprecipitation would be an effective arsenic removal technology; however, this technology is still largely experimental and should be further evaluated before being recommended as an approved arsenic removal technology for drinking water.

18.12 WHO IS ULTIMATELY RESPONSIBLE FOR DRINKING WATER QUALITY?

The Safe Drinking Water Act gives the USEPA the responsibility for setting national drinking water standards that protect the health of the 250 million people who get their water from public water systems. Other people get their water from private wells that are not subject to federal regulations. Since 1974, USEPA has set national standards for over 80 contaminants that may occur in drinking water.

Although USEPA and state governments set and enforce standards, local governments and private water supplies have direct responsibility for the quality of the water that flows to the customer's tap. Water systems test and treat their water, maintain the distribution systems that deliver water to consumers, and report on their water quality to the state. States and USEPA provide technical assistance to water suppliers and can take legal action against systems that fail to provide water that meets state and USEPA standards. USEPA has set standards for more than 180 contaminants that may occur in drinking water and pose a risk to human health. USEPA sets these standards to protect the health of everybody, including vulnerable groups such as children. The contaminants fall into two groups according to the health effects that they cause. Local water suppliers normally alert customers through the local media, direct mail, or other means if an acute or a chronic health effect is possible from compounds in the drinking water. Customers may want to contact them for additional information specific to their area.

Acute effects occur within hours or days of the time that a person consumes a contaminant. People can suffer acute health effects from almost any contaminant if they are exposed to extraordinarily high levels (as in the case of a spill). In drinking water, microbes, such as bacteria and viruses, are the contaminants with the greatest chance of reaching levels high enough to cause acute health effects. Most people can fight off these microbial contaminants the way they fight off germs, and these acute contaminants typically do not have permanent effects. Nonetheless, when high enough levels occur, they can make people ill and can be dangerous or deadly for a person whose immune system is already weak due to HIV/AIDS, chemotherapy, steroid use, or another reason.

Chronic effects occur after people consume a contaminant at levels exceeding USEPA safety standards for many years. The drinking water contaminants that can have chronic effects are chemicals (such as disinfection byproducts, solvents, and pesticides), radionuclides (such as radium), and minerals (such as arsenic). Examples of these chronic effects include cancer, liver or kidney problems, or reproductive difficulties.

CHAPTER REVIEW QUESTIONS

18.1 What two minerals are primarily responsible for causing hard water?

18.2 The power of a substance to resist pH changes is referred to as what?

18.3 What chemical is used as a titrant when analyzing a water sample for carbon dioxide (CO_2)?

18.4 How many pounds of chlorine a day will be used if the dosage is 1.2 mg/L for a flow of 1,600,000 gpd?

18.5 What is the specific capacity of a well having a yield of 60 gpm with a drawdown of 25 ft?

18.6 What is the chlorine demand in mg/L if the chlorine dosage is 1.0 mg/L and the residual chlorine is 0.5 mg/L?

18.7 If the chlorine dosage is 6 mg/L, what must the chlorine residual be if the chlorine demand is 3.3 mg/L?

18.8 A water treatment plant has a daily flow of 3.1 MGD. If the chlorinator setting is 220 lb/day, and the chlorine demand is 6.9 mg/L, what is the chlorine residual?

18.9 A well log is best described as _____.

18.10 The type of well construction not normally permitted for a public water supply is _____.

18.11 Name three water quality tests routinely performed on water samples collected from storage tanks.

18.12 Paint for the interior of a drinking water storage tank must be approved by _____.

18.13 Potable water is water that _____.

18.14 A test of the effluent in the clear well shows that the required dosage of chlorine is 0.6 mg/L. The average daily flow at the treatment plant is 1 MGD. If we are using a hypochlorite solution with 68% available chlorine, how many lb/day hypochlorite will be required?

18.15 A waterworks conveys piped water to the _____.

18.16 The hydrologic cycle describes _____.

18.17 Is *Giardia lamblia* a chronic or an acute health threat?

18.18 Disinfection can be defined as _____.

18.19 As disinfectants, what are as common as household bleach?

18.20 Effective disinfectants must _____.

18.21 How many pounds of chlorine will be used if the dosage is 0.4 mg/L for a flow of 5,300,000 gpd?

18.22 If the chlorine dosage is 10 mg/L, what must the chlorine residual be if the demand is 2.6 mg/L?

18.23 Out of the number of possible interferences with chlorine disinfection, name one.

18.24 Given that the dose of soda ash = 0.8 mg/L and flow = 2.6 MGD, what is the feed rate of soda ash in lb/day?

18.25 A circular clarifier handles a flow of 0.75 MGD. The clarifier has a 20-ft radius and a depth of 10 ft. What is the detention time?

18.26 A water treatment plant operates at a rate of 2 mgd. The dosage of alum is 35 ppm (or mg/L); how many pounds of alum are used in a day?

18.27 Is the following statement true? By weight, more pounds (65%) of hypochlorite material (HTH) than pounds of chlorine are required to get the same number of pounds of available chlorine into the water.

18.28 What is the amount of chlorine present in water after a specified time period?

18.29 For a potable water system to be contaminated by water from a nonpotable system through a cross-connection, two conditions must exist simultaneously. What are they?

18.30 A piping arrangement that could allow a nontoxic substance such as milk, beer, or orange juice to contaminate a potable water system would be classified as a _____ hazard situation.

18.31 When water main breaks and heavy water usage during firefighting withdraw more water from a potable water system than is being supplied to the system, _____ pressure may develop in the potable system.

18.32 What type of pump uses roller and tubing?

18.33 List three watershed management practices.

18.34 A filter plant has three filters, each measuring 10 ft long by 7 ft wide. One filter is out of service, and the other two together are capable of filtering 280 gpm. How many gallons per square foot per minute will each unit filter?

18.35 A filter having an area of 300 ft^2 is ready to be backwashed. Assume a rate of 15 gal/ft^2/min and 8 minutes of backwash is required. What is the amount of water in gallons required for each backwash?

18.36 If water travels 600 ft in 5 minutes, what is the velocity?

18.37 Where should you look to find information about the hazards associated with the various chemicals you come into contact with at your treatment plant?

18.38 What two water unit treatment processes have done the most to eradicate or reduce the level of waterborne disease in the United States?

18.39 If the discharge pressure is lower than the pump is rated for, the pump will _____.

18.40 What is the most effective chlorine compound for killing or inactivating pathogens?

18.41 What is the group of microorganisms that form cysts and thus become resistant to disinfection?

18.42 The Surface Water Treatment Rule contains operational and monitoring requirements to ensure that _____.

18.43 What is the Aggressive Index an indicator of?

18.44 List three factors that influence coagulant dose.

18.45 Flow in a rectangular channel 2.5 ft wide by 1.4 ft deep is 11.2 cfs. What is the average velocity?

18.46 A cylindrical tank is 100 ft high and 20 ft in diameter. How many gallons of water will it contain?

18.47 What is the correct sequence for running a jar test?

18.48 What is the purpose of coagulation and flocculation?

18.49 What is the goal of chemical precipitation?

18.50 To achieve optimum removal of hardness, one should add how much lime and soda ash?

18.51 What percentage of positive samples cannot be exceeded for bacteriological compliance monitoring of a distribution system?

18.52 What is the velocity in ft/min if water travels 1500 ft in 4 minutes? What is the velocity in ft/sec?

18.53 What type of valve in a water distribution system is used to isolate a damaged line?

18.54 The Lead and Copper Rule requires that a treatment facility _____.

18.55 Fluoridation at 0.2 mg/L below optimum cuts effectiveness by _____.

18.56 What is the chemical normally used in a fluoride saturator?

18.57 Long-term consumption of water with a fluoride concentration of 3.0 mg/L or more may cause _____.

18.58 If a raw water source has a fluoride ion concentration of 0.15 mg/L and the optimum concentration for the fluoride ion is 0.9 mg/L, what is the desired fluoride dose?

18.59 What is the purpose of the jar test in water treatment?

18.60 Inorganic phosphate addition normally functions to _____.

18.61 The flow in a 6-in. pipe is 350 gpm. What is the average velocity?

18.62 What is used to oxidize iron and manganese?

18.63 List three factors that affect corrosion.

18.64 Activated carbon removes taste- and odor-producing substances by what method?

18.65 To achieve good coagulation in low-alkalinity waters, an additional source of alkalinity is most effective when added _____.

18.66 To prevent media loss, supplemental backwash air flow and surface sweeps should be turned off _____.

18.67 The most common complaint concerning taste and odor primarily involves _____.

18.68 The concentration of volatile organic compounds (VOCs) is usually greater in groundwater. *True* or *False*?

18.69 VOCs are suspected of being potential carcinogens. *True* or *False*?

18.70 A round tank 30-ft in diameter is filled with water to a depth of 15 ft. How many gallons of water are in the tank?

18.71 What chemical substance would you use to reduce trihalomethane formation?

18.72 What is the cause of brownish-blackish waters?

18.73 Waters that cause bluish-green stains on household fixtures often contain _____.

18.74 What causes water hardness?

18.75 When Fe^{2+} is chemically changed to Fe^{3+}, the iron atom _____.

18.76 What causes alkalinity?

18.77 If the head loss, in feet, at any level in the filter bed exceeds the depth of the water above the same point (static head), a vacuum can result. This situation is referred to as _____.

18.78 Sedimentation is the removal of settleable solids by _____.

18.79 In sedimentation, the _____ zone decreases the velocity of the incoming water and distributes the flow evenly across the basin.

18.80 As water enters a sedimentation basin, _____ flow distribution is important to achieve proper velocity throughout the basin.

18.81 List the three purposes of enhanced coagulation.

18.82 For a water plant that performs bacteriological analysis, what is the maximum hold time from when the sample is collected until analysis is started in the lab?

18.83 When performing a chlorine residual test with an amperometric titrator, what reagent is used as the titrant?

18.84 How many gallons of water fell into a 20-acre reservoir if the water level of the pond rose 2 in. after a storm event?

18.85 To maintain a 0.5-mg/L chlorine residual throughout the distribution system, a chlorine dosage of 1.2 mg/L is required at the clear well. If the average daily flow reaches 2.0 MGD, how many lb/day of chlorine must be added?

18.86 A cylindrical water storage tank is 70 ft tall and 30 ft in diameter. The tank is 40% full. What is the water pressure in PSI at the base of the tank?

18.87 The concentration of volatile organic compounds (VOCs) is usually greater in _____.

18.88 List three methods to remove VOCs.

18.89 The addition of _____ is a method to reduce trihalomethane formation.

18.90 The ability of soil to allow water to pass through it is called its _____.

18.91 The top of an aquifer is called the _____.

18.92 Diseases that are carried by water are referred to as _____ diseases.

18.93 A chemical that combines with suspended particles in water is called a _____.

18.94 When water freezes, its volume becomes _____.

18.95 The most commonly used algaecide is _____.

18.96 Carbonate hardness can be removed by adding _____ to the water.

18.97 To destroy coliforms and pathogens, water must be _____.

18.98 BOD helps the plant operator determine how much _____ will be needed to stabilize the organic matter.

18.99 Bacteria are produced by dividing in half, a process that is called _____.

REFERENCES AND SUGGESTED READING

Aptel, P. and Buckley, C.A. 1996. Categories of membrane operations. In *Water Treatment Membrane Processes*. New York: McGraw-Hill.

AWWA. 1984. *Introduction to Water Treatment*, Vol. 2. Denver, CO: American Water Works Association.

Brady, T.J. et al. 1996. Chlorination effectiveness for zebra and quagga mussels. *J. AWWA*, 88(1), 107–110.

Britton, J.C. and Morton, B.A. 1982. Dissection guide, field and laboratory manual for the introduced bivalve *Corbicula fluminea*. *Malacol. Rev.*, 3(1).

Butterfield, C.T. et al. 1943. Chlorine vs. hypochlorite. *Public Health Rep.*, 58, 1837.

Cameron, G.N., Symons, J.M., Spencer, S.R., and Ja, J.Y. 1989. Minimizing THM formation during control of the Asiatic clam: a comparison of biocides. *J. AWWA*, 81(10), 53–62.

Cheng, R.C. et al. 1994. Enhanced coagulation for arsenic removal, *J. AWWA*, 9, 79–90.

Chick, H. 1908. Investigation of the laws of disinfection. *J. Hygiene*, 8, 92.

Clarke, N.A. et al. 1962. Human enteric viruses in water, source, survival, and removability. In *Proc. of the First International Conference on Water Pollution Research*, September, London.

Clifford, D.A. and Lin, C.C. 1985. *Arsenic (Arsenite) and Arsenic (Arsenate) Removal from Drinking Water in San Ysidro, New Mexico*. Houston, TX: University of Houston.

Clifford, D.A. et al. 1997. *Final Report: Phases 1 and 2 City of Albuquerque Arsenic Study Field Studies on Arsenic Removal in Albuquerque, New Mexico, Using the University of Houston/EPA Mobile Drinking Water Treatment Research Facility.* Houston, TX: University of Houston.

Connell, G.F. 1996. *The Chlorination/Chloramination Handbook*. Denver, CO: American Water Works Association.

Craun, G.F. 1981. Outbreaks of waterborne disease in the United States. *J. AWWA*, 73(7), 360.

Craun, G.F. and Jakubowski, W. 1996. *Status of Waterborne Giardiasis Outbreaks and Monitoring Methods*. Atlanta, GA: American Water Resources Association.

Culp, G.L. and Culp, R.L. 1974. Outbreaks of waterborne disease in the United States. *J. AWWA*, 73(7), 360.

Culp, G.L. et al. 1986. *Handbook of Public Water Systems*. New York: Van Nostrand Reinhold.

Demers, L.D. and Renner, R.C. 1992. *Alternative Disinfection Technologies for Small Drinking Water Systems*, Denver, CO: AWWA/AWWART.

Edwards, M.A. 1994. Chemistry of arsenic removal during coagulation and Fe–Mn oxidation, *J. AWWA*, 86, 64–77.

Finch, G.R. et al. 1994. Ozone and chlorine inactivation of *Cryptosporidium*. In *Proc. of the Water Quality Technology Conference*, Part II, San Francisco, CA.

Gordon, G. et al. 1995. *Minimizing Chlorate Ion Formation in Drinking Water When Hypochlorite Ion Is the Chlorinating Agent*. Denver, CO: AWWA/AWWARF.

Gurol, M.D. and Pidatclla, M.A. 1983. Study on ozone-induced coagulation. In *Proc. of the ASCE Environmental Engineering Division Specialty Conference*, Medicine, A. and Anderson, M., Eds. Boulder, CO: American Society of Civil Engineers.

Gyurek, L.L. et al. 1996. Disinfection of *Cryptosporidium parvum* using single and sequential application of ozone and chlorine species. In *Proc. of the AWWA 1996 Water Quality Technology Conference*, November, Boston, MA.

Hass, C.N. and Englebrecht, R.S. 1980. Physiological alterations of vegetative microorganisms resulting from aqueous chlorination. *J. Water Pollut. Control Fed.*, 52(7).

Herbert, P.D.N. et al. 1989. Ecological and genetic studies on *Dresissmena polymorpha* (Pallas): a new mollusc in the Great Lakes. *Can. J. Fish. Aquat. Sci.*, 46, 187.

Hering, J.G. and Chiu, V.Q. 1998. The chemistry of arsenic: treatment and implications of arsenic speciation and occurrence. In *Proc. of the AWWA Inorganic Contaminants Workshop*, San Antonio, TX.

Hoff, J.C. et al. 1984. Disinfection and the control of waterborne giardiasis. In *Proc. of ASCE Specialty Conference.*

Hydrometrics. 1998. *Second Interim Report on the Sulfur-Modified Iron (SMI) Process for Arsenic Removal.* Helena, MO: Hydrometrics, Inc.

IOA. 1997. *Survey of Water Treatment Plants.* Stanford, CT: International Ozone Association.

Klerks, P.L and Fraleigh, P.C. 1991. Controlling adult zebra mussels with oxidants. *J. AWWA*, 83(12), 92–100.

Koch, B. et al. 1991. Predicting the formation of DBPs by the simulated distribution system. *J. AWWA*, 83(10), 62–70.

Krasner, S.W. 1989. The occurrence of disinfection byproducts in U.S. drinking water. *J. AWWA*, 81(8), 41–53.

Laine, J.M. 1993. Influence of bromide on low-pressure membrane filtration for controlling DBPs in surface waters. *J. AWWA*, 85(6), 87–99.

Lang, C.L. 1994. The impact of the freshwater macrofouling zebra mussel (*Dretssena polymorpha*) on drinking water supplies. In *Proc. of the Water Quality Technology Conference*, Part II, San Francisco, CA.

Lalezary, S. et al. 1986. Oxidation of five earthy–musty taste and odor compounds. *J. AWWA*, 78(3), 62.

Liu, O.C. et al. 1971. Relative resistance of twenty human enteric viruses to free chlorine—viruses and water quality: occurrence and control. In *Proc. of the Thirteenth Water Quality Conference.* Urbana-Champaign, IL: University of Illinois.

Macler, B.A. 1996. Developing the groundwater disinfection rule. *J. AWWA*, 88(3), 47–55.

Masschelein, W.J. 1992. *Unit Processes in Drinking Water Treatment.* New York: Marcel Decker.

Matisoff, G. et al. 1996. Toxicity of chlorine dioxide to adult zebra mussels. *J. AWWA*, 88(8), 93–106.

Montgomery, J.M. 1985. *Water Treatment Principles and Design.* New York: John Wiley & Sons.

Muilenberg, T. 1997. Microfiltration basics: theory and practice. In *Proc. of the Membrane Technology Conference*, February 23–26, New Orleans, LA.

Nieminski, E.C. et al. 1993. The occurrence of DBPs in Utah drinking waters. *J. AWWA*, 85(9), 98–105.

Oliver, B.G. and Shindler, D.B. 1980. Trihalomethanes for chlorination of aquatic algae. *Environ. Sci. Technol.*, 1412, 1502–1505.

Prendiville, P.W. 1986. Ozonation at the 900 cfs Los Angeles water purification plant. *Ozone Sci. Eng.*, 8, 77.

Reckhow, D.A. and Singer, P.C. 1985. Mechanisms of organic halide formation during fulvic acid chlorination and implications with respect to prezonation. In *Water Chlorination: Chemistry, Environmental Impact and Health Effects*, Vol. 5, Jolley, R.L. et al., Eds. Chelsea, MI: Lewis Publishers.

Reckhow, D.A. et al. 1986. Ozone as a coagulant aid. In *Proc. of the AWWA Annual Conference: Ozonation, Recent Advances and Research Needs*, Denver, CO.

Reckhow, D.A. et al. 1990. Chlorination of humic materials: byproduct formation and chemical interpretations. *Environ. Sci. Technol.*, 24(11), 1655.

Rice, R.G. et al. 1998. Ozone Treatment for Small Water Systems. Paper presented at the First International Symposium on Safe Drinking Water in Small Systems, NSF International, Arlington, VA.

Roberts, R. 1990. Zebra mussel invasion threatens U.S. waters. *Science*, 249, 1370.

Sawyer, C.N. 1994. *Chemistry for Environmental Engineering.* New York: McGraw-Hill.

Scarpino, P.V. et al. 1972. A comparative study of the inactivation of viruses in water by chlorine. *Water Res.*, 6, 959.

Simms, J. et al. 2000. Arsenic removal studies and the design of a 20,000 m³ per day plant in U.K. In *Proc. of the AWWA Inorganic Contaminants Workshop*, Albuquerque, NM.

Sinclair, R.M. 1964. Clam pests in Tennessee water supplies. *J AWWA*, 56(5), 592.

Singer, P.C. 1989. Correlations between trihalomethanes and total organic halides formed during water treatment. *J. AWWA*, 81(8), 61–65.

Singer P.C. 1992. Formation and Characterization of Disinfection Byproducts. Paper presented at the First International Conference on the Safety of Water Disinfection: Balancing Chemical and Microbial Risks, Washington, D.C.

Singer, P.C., and Chang, S.D. 1989. Correlations between trihalomethanes and total organic halides formed during water treatment. *J. AWWA*, 81(8), 61–65.

Singer, P.C. and Harrington, G.W. 1993. Coagulation of DBP precursors: theoretical and practical considerations. In *Proc. of the AWWA Water Quality Technology Conference*, Miami, FL.

Snead, M.C. et al. 1980. *Benefits of Maintaining a Chlorine Residual in Water Supply Systems*, EPA600/2-80-010. Washington, D.C.: U.S. Environmental Protection Agency.

Spellman, F.R. 2007. *The Science of Water*, 2nd ed. Boca Raton, FL: CRC Press.

Stevens, A.A. 1976. Chlorination of organics in drinking water. *J. AWWA*, 8(11), 615.

Subramanian, K.D. et al. 1997. Manganese greensand for removal of arsenic in drinking water. *Water Qual. Res.. J. Can.*, 32(3), 551–561.

Suffet, I.H. et al. 1986. Removal of tastes and odors by ozonation. In *Proc. of the AWWA Seminar on Ozonation: Recent Advances and Research Needs*, Denver, CO.

TWUA. 1988. *Manual of Water Utility Operations*, 8th ed. Austin: Texas Water Utilities Association.

Thibaud, H. et al. 1988. Effects of bromide concentration on the production of chloropicrin during chlorination of surface waters: formation of brominated trihalonitromethanes. *Water Res.*, 22(3), 381.

USEPA. 1991. *Manual of Individual and Non-Public Works Supply Systems*, EPA-5709-91-004. Washington, D.C.: U.S. Environmental Protection Agency.

USEPA. 1997. *Community Water System Survey*, Vols. I and II, EPA-815-R-97-001a. Washington, D.C.: U.S. Environmental Protection Agency.

USEPA. 1998. *National Primary Drinking Water Regulations: Interim Enhanced Surface Water Treatment Final Rule.* Washington, D.C.: U.S. Environmental Protection Agency.

USEPA. 1999a. *Turbidity Requirements: IESWTR Guidance Manual: Turbidity Provisions.* Washington, D.C.: U.S. Environmental Protection Agency.

USEPA. 1999b. *Guidance Manual: Alternative Disinfectants and Oxidants*, Chapters 1 and 2. Washington, D.C.: U.S. Environmental Protection Agency.

USEPA. 1999c. *Microbial and Disinfection Byproduct Rules Simultaneous Compliance Guidance Manual*, EPA-815-R-99-015. Washington, D.C.: U.S. Environmental Protection Agency.

USEPA. 2000. *Technologies and Costs for Removal of Arsenic from Drinking Water*, EPA-815-R-00-028. Washington, D.C.: U.S. Environmental Protection Agency (www.epa.gov/safewater).

Van Benschoten, J.E. et al. 1995. Zebra mussel mortality with chorine. *J. AWWA*, 87(5), 101–108.

Vickers, J.C. et al. 1997. Bench scale evaluation of microfiltration for removal of particles and natural organic matter. In *Proc. of the Membrane Technology Conference*, February 23–26, New Orleans, LA.

Watson, H.E. 1908. A note on the variation of the rate of disinfection with change in the concentration of the disinfectant. *J. Hygiene*, 8, 538.

White, G.C. 1992. *Handbook of Chlorination and Alternative Disinfectants*. New York: Van Nostrand Reinhold.

Witherell, L.E. et al. 1988. Investigation of *Legionella pneumophila* in drinking water. *J. AWWA*, 80(2), 88–93.

Part V

Wastewater and Wastewater Treatment

19 Wastewater Treatment Operations

The Code of Federal Regulations (40 CFR, Part 403) established procedures in the late 1970s and early 1980s to help publicly owned treatment works (POTW) control industrial discharges to sewers. These regulations were designed to prevent pass-through and interference at the treatment plants, as well as interference in the collection and transmission systems.

Pass-through occurs when pollutants literally pass through a POTW without being properly treated. These pollutants cause the POTW to have an effluent violation or can increase the magnitude or duration of a violation.

Interference occurs when a pollutant discharge causes a POTW to violate its permit by inhibiting or disrupting treatment processes, treatment operations, or processes related to sludge use or disposal.

19.1 WASTEWATER OPERATORS

Like waterworks operators, wastewater operators are highly trained and artful practitioners and technicians of their trade. Moreover, wastewater operators, again like waterworks operators, are required by the states to be licensed or certified to operate a wastewater treatment plant. When learning wastewater operator skills, there are a number of excellent texts are available to aid in the training process. Many of these texts are listed in Table 19.1.

19.1.1 THE WASTEWATER TREATMENT PROCESS: THE MODEL

Figure 19.1 shows a basic schematic of an example wastewater treatment process providing primary and secondary treatment using the *activated sludge process*. This is the model, the prototype, the paradigm used in this book. Although secondary treatment (which provides BOD removal beyond what is achievable by simple sedimentation) commonly utilizes three different approaches—trickling filter, activated sludge, and oxidation ponds—we focus here, for instructive and illustrative purposes, on the activated sludge process. The purpose of Figure 19.1 is to allow the reader to follow the treatment process step by step as it is presented (and as it is actually configured in the real world) and to assist in demonstrating how all of the various unit processes sequentially follow and tie into

each other. We will begin various sections with reference to Figure 19.1; it is important to follow through these sections in this manner because wastewater treatment is a series of individual steps (unit processes) that treat the wastestream as it makes its way through the entire process. A pictorial presentation of the treatment process, along with pertinent written information, enhances the learning process. It should also be pointed out, however, that, even though the model shown in Figure 19.1 does not include all of the unit processes currently used in wastewater treatment, we do not ignore the other major processes of trickling filters, rotating biological contactors (RBCs), and oxidation ponds.

19.2 WASTEWATER TERMINOLOGY AND DEFINITIONS

Wastewater treatment technology, like many other technical fields, has its own vocabulary. Although some of the terms are unique to this field, many of them are common to other professions, as well. Remember that the science of wastewater treatment is a combination of engineering, biology, mathematics, hydrology, chemistry, physics, and other disciplines; therefore, many of the terms used in engineering, biology, mathematics, hydrology, chemistry, physics, and elsewhere are also used for wastewater treatment. Those terms not listed or defined in the following section will be defined as they appear throughout the text.

Activated sludge—The solids formed when microorganisms are used to treat wastewater using the activated sludge treatment process. It includes organisms, accumulated food materials, and waste products from the aerobic decomposition process.

Advanced waste treatment—Treatment technology to produce an extremely high-quality discharge.

Aerobic—Conditions in which free, elemental oxygen is present; also used to describe organisms, biological activity, or treatment processes that require free oxygen.

Anaerobic—Conditions in which no oxygen (free or combined) is available; also used to describe organisms, biological activity, or treatment processes that function in the absence of oxygen.

TABLE 19.1
Recommended Reference/Study Materials

1. *Advanced Waste Treatment*, 3rd ed., Kerri, K. et al., California State University, Sacramento, 1995.
2. *Aerobic Biological Wastewater Treatment Facilities*, EPA 430/9-77-006, U.S. Environmental Protection Agency, Washington, D.C., 1977.
3. *Anaerobic Sludge Digestion*, EPA-430/9-76-001, U.S. Environmental Protection Agency, Washington, D.C., 1977.
4. *Annual Book of ASTM Standards*. Section 11. *Water and Environmental Technology*, American Society for Testing and Materials (ASTM), Philadelphia, PA, 2001.
5. *Basic Math Concepts for Water and Wastewater Plant Operators*, Price, J.K., Technomic, Lancaster, PA, 1991.
6. *Guidelines Establishing Test Procedures for the Analysis of Pollutants*, 40 CFR 136, *Federal Register*, 60(64), 17160, 1995.
7. *Handbook of Water Analysis*, 2nd ed., HACH Chemical Co., P.O. Box 389, Loveland, CO, 1992.
8. *Industrial Waste Treatment*, Vols. 1 and 2, Kerri, K. et al., California State University, Sacramento, 1994/1995.
9. *Methods for Chemical Analysis of Water and Wastes*, EPA-6000/4-79-020, U.S. Environmental Protection Agency, Environmental Monitoring Systems Laboratory, Cincinnati, OH, revised 1979 and 1983.
10. *O&M of Trickling Filters, RBC, and Related Processes*, Manual of Practice OM-10, Water Pollution Control Federation (now called Water Environment Federation), Alexandria, VA, 1988.
11. *Operation of Wastewater Treatment Plants*, Vols. 1 and 2, 4th ed./5th ed., Kerri, K. et al., California State University, Sacramento, 1992/2001.
12. *Simplified Wastewater Treatment Plant Operations*, Haller, E.J., Technomic, Lancaster, PA, 1995.
13. *Standard Methods for the Examination of Water and Wastewater*, 18th ed., American Public Health Association, American Water Works Association–Water Environment Federation, Washington, D.C., 1992.
14. *The Science of Water*, 2nd ed., Spellman, F.R., CRC Press, Boca Raton, FL, 2007.
15. *Treatment of Metal Wastestreams*, Kerri, K. et. al., California State University, Sacramento, 1993.
16. *Wastewater Treatment Plants: Planning, Design, and Operation*, Qasim, S.R., Technomic, Lancaster, PA, 1994.

Anoxic—Conditions in which no free, elemental oxygen is present, and the only source of oxygen is combined oxygen, such as that found in nitrate compounds; also used to describe biological activity or treatment processes that function only in the presence of combined oxygen.

Average monthly discharge limitation—The highest allowable discharge over a calendar month.

Average weekly discharge limitation—The highest allowable discharge over a calendar week.

Biochemical oxygen demand (BOD₅)—The amount of organic matter that can be biologically oxidized under controlled conditions (5 days at 20°C in the dark).

Biosolids—Solid organic matter recovered from a sewage treatment process and used especially as fertilizer; usually used in plural (*Merriam-Webster's Collegiate Dictionary, 10th ed.*, 1998).

Note: In this text, the term *biosolids* is used in many places to replace the standard term *sludge* (activated sludge being the exception). The author (along with others in the field) views the term *sludge* as an ugly four-letter word that is inappropriate to use when describing biosolids. Biosolids are a product that can be reused; they have some value. Because biosolids have some value, they should not be classified as a waste product, and when we address biosolids for beneficial reuse, they are not.

Buffer—A substance or solution that resists changes in pH.

Carbonaceous biochemical oxygen demand (CBOD₅)—The amount of biochemical oxygen demand that can be attributed to carbonaceous material.

Chemical oxygen demand (COD)—The amount of chemically oxidizable materials present in the wastewater.

Clarifier—A device designed to permit solids to settle or rise and be separated from the flow. Also known as a settling tank or sedimentation basin.

Coliform—A type of bacteria used to indicate possible human or animal contamination of water.

Combined sewer—A collection system that carries both wastewater and stormwater flows.

Comminution—A process to shred solids into smaller, less harmful particles.

Composite sample—A combination of individual samples taken in proportion to flow.

Daily discharge—The discharge of a pollutant measured during a calendar day or any 24-hour period that reasonably represents a calendar day for the purposes of sampling. Limitations expressed as weight are the total mass (weight) discharged over the day. Limitations expressed in other units are average measurements of the day.

Daily maximum discharge—The highest allowable values for a daily discharge.

Detention time—The theoretical time water remains in a tank at a given flow rate.

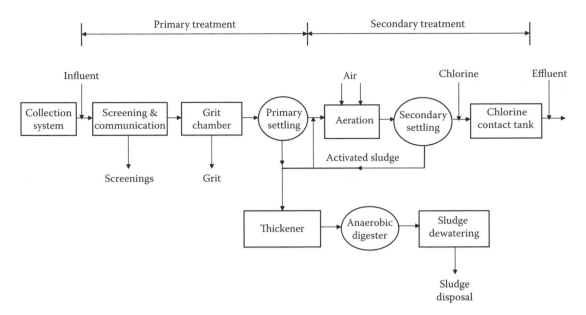

FIGURE 19.1 Schematic of an example wastewater treatment process providing primary and secondary treatment using the activated sludge process.

Dewatering—The removal or separation of a portion of water present in a sludge or slurry.

Discharge monitoring report (DMR)—Monthly report required by a treatment plant's NPDES discharge permit.

Dissolved oxygen (DO)—Free or elemental oxygen dissolved in water.

Effluent—The flow leaving a tank, channel, or treatment process.

Effluent limitation—Any restriction imposed by the regulatory agency on quantities, discharge rates, or concentrations of pollutants that are discharged from point sources into state waters.

Facultative—Organisms that can survive and function either in the presence or absence of free, elemental oxygen.

Fecal coliform—A type of bacteria found in the bodily discharges of warm-blooded animals; used as an indicator organism.

Floc—Solids that join together to form larger particles that will settle better.

Flume—A flow-rate measurement device.

Food-to-microorganism ratio (F/M)—An activated sludge process control calculation based on the amount of food (BOD_5 or COD) available per pound of mixed liquor volatile suspended solids.

Grab sample—An individual sample collected at a randomly selected time.

Grit—Heavy inorganic solids such as sand, gravel, egg shells, or metal filings.

Industrial wastewater—Wastes associated with industrial manufacturing processes.

Infiltration/inflow—Extraneous flows in sewers; defined by Metcalf & Eddy (1991, pp. 29–31) as follows:

• *Infiltration*—Water entering the collection system through cracks, joints, or breaks.

• *Steady inflow*—Water discharged from cellar and foundation drains, cooling water discharges, and drains from springs and swampy areas. This type of inflow is steady and is identified and measured along with infiltration.

• *Direct flow*—Those types of inflow that have a direct stormwater runoff connection to the sanitary sewer and cause an almost immediate increase in wastewater flows. Possible sources are roof leaders, yard and areaway drains, manhole covers, cross connections from storm drains and catch basins, and combined sewers.

• *Total inflow*—The sum of the direct inflow at any point in the system plus any flow discharged from the system upstream through overflows, pumping station bypasses, and the like.

• *Delayed inflow*—Stormwater that may require several days or more to drain through the sewer system. This category can include the discharge of sump pumps from cellar drainage as well as the slowed entry of surface water through manholes in ponded areas.

Influent—The wastewater entering a tank, channel, or treatment process.

Inorganic—Mineral materials such as salt, ferric chloride, iron, sand, or gravel.

License—A certificate issued by the State Board of Waterworks/Wastewater Works Operators authorizing the holder to perform the duties of a wastewater treatment plant operator.

Mean cell residence time (MCRT)—The average length of time a mixed liquor suspended solids particle remains in the activated sludge process; may also be known as sludge retention time.

Milligrams/liter (mg/L)—A measure of concentration equivalent to parts per million (ppm).

Mixed liquor—The combination of return activated sludge and wastewater in the aeration tank.

Mixed liquor suspended solids (MLSS)—The suspended solids concentration of the mixed liquor.

Mixed liquor volatile suspended solids (MLVSS)—The concentration of organic matter in the mixed liquor suspended solids.

Nitrogenous oxygen demand (NOD)—A measure of the amount of oxygen required to biologically oxidize nitrogen compounds under specified conditions of time and temperature.

NPDES permit—National Pollutant Discharge Elimination System permit that authorizes the discharge of treated wastes and specifies the conditions that must be met for discharge.

Nutrients—Substances required to support living organisms; usually refers to nitrogen, phosphorus, iron, and other trace metals.

Organic—Materials that consist of carbon, hydrogen, oxygen, sulfur, and nitrogen. Many organics are biologically degradable. All organic compounds can be converted to carbon dioxide and water when subjected to high temperatures.

Parts per million—An alternative (but numerically equivalent) unit used in chemistry that is equal to milligrams per liter (mg/L). As an analogy think of 1 ppm as being equivalent to a full shot glass in a swimming pool.

Pathogenic—Disease causing; a pathogenic organism is capable of causing illness.

Point source—Any discernible, defined, and discrete conveyance from which pollutants are or may be discharged.

Return activated sludge solids (RASS)—The concentration of suspended solids in the sludge flow being returned from the settling tank to the head of the aeration tank.

Sanitary wastewater—Wastes discharged from residences and from commercial, institutional, and similar facilities that include both sewage and industrial wastes.

Scum—The mixture of floatable solids and water that is removed from the surface of the settling tank.

Septic—A wastewater that has no dissolved oxygen present; generally characterized by a black color and rotten-egg (hydrogen sulfide) odor.

Settleability—A process control test used to evaluate the settling characteristics of the activated sludge. Readings taken at 30 to 60 minutes are used to calculate the settled sludge volume (SSV) and the sludge volume index (SVI).

Settled sludge volume—The volume in percent occupied by an activated sludge sample after 30 to 60 minutes of settling; normally written as SSV with a subscript to indicate the time of the reading used for calculation (e.g., SSV_{60} or SSV_{30}).

Sewage—Wastewater containing human wastes.

Sludge—The mixture of settleable solids and water that is removed from the bottom of the settling tank.

Sludge retention time (SRT)—See mean cell residence time.

Sludge volume index (SVI)—A process control calculation used to evaluate the settling quality of the activated sludge; calculating the SVI requires the SSV_{30} and mixed liquor suspended solids test results.

Storm sewer—A collection system designed to carry only stormwater runoff.

Stormwater—Runoff resulting from rainfall and snowmelt.

Supernatant—In a digester, it is the amber-colored liquid above the sludge.

Waste activated sludge solids (WASS)—The concentration of suspended solids in the sludge being removed from the activated sludge process.

Wastewater—The water supply of the community after it has been soiled by use.

Weir—A device used to measure wastewater flow.

Zoogleal slime—The biological slime that forms on fixed film treatment devices; it contains a wide variety of organisms essential to the treatment process.

19.3 MEASURING PLANT PERFORMANCE

To evaluate how well a plant or treatment unit process is operating, *performance efficiency* or *percent (%) removal* is used. The results can be compared with those listed in the plant's operation and maintenance manual (O&M) to determine if the facility is performing as expected. In this section, sample calculations often used to measure plant performance and efficiency are presented.

19.3.1 PLANT PERFORMANCE AND EFFICIENCY

The calculation used to determine the performance (percent removal) of a digester is different from that used for performance (percent removal) for other processes. Care must be taken to select the correct formula.

$$\% \text{ Removal} = \frac{(\text{Influent conc.} - \text{effluent conc.}) \times 100}{\text{Influent conc.}} \quad (19.1)$$

■ EXAMPLE 19.1

Problem: The influent BOD is 247 mg/L, and the plant effluent BOD is 17 mg/L. What is the percent removal?

Solution:

$$\% \text{ Removal} = \frac{(247 \text{ mg/L} - 17 \text{ mg/L}) \times 100}{247 \text{ mg/L}} = 93\%$$

19.3.2 UNIT PROCESS PERFORMANCE AND EFFICIENCY

Equation 19.1 is used again to determine unit process efficiency. The concentration entering the unit and the concentration leaving the unit (e.g., primary, secondary) are used to determine the unit performance:

$$\% \text{ Removal} = \frac{(\text{Influent conc.} - \text{effluent conc.}) \times 100}{\text{Influent conc.}}$$

■ EXAMPLE 19.2

Problem: The primary influent BOD is 235 mg/L and the primary effluent BOD is 169 mg/L. What is the percent removal?

Solution:

$$\% \text{ Removal} = \frac{(235 \text{ mg/L} - 169 \text{ mg/L}) \times 100}{235 \text{ mg/L}} = 28\%$$

19.3.3 PERCENT VOLATILE MATTER REDUCTION IN SLUDGE

The calculation used to determine *percent volatile matter reduction* is more complicated because of the changes occurring during sludge digestion.

$$\% \text{VM reduction} = \frac{(\% \text{VM}_{in} - \% \text{VM}_{out}) \times 100}{\% \text{VM}_{in} - (\% \text{VM}_{in} \times \% \text{VM}_{out})} \quad (19.2)$$

■ EXAMPLE 19.3

Problem: Using the digester data provided below, determine the percent volatile matter reduction for the digester:

   Raw sludge volatile matter = 74%
   Digested sludge volatile matter = 54%

Solution:

$$\% \text{VM reduction} = \frac{(0.74 - 0.54) \times 100}{0.74 - (0.74 \times 0.54)} = 59\%$$

19.3.4 HYDRAULIC DETENTION TIME

The term *detention time* or *hydraulic detention time* (HDT) refers to the average length of time (theoretical time) a drop of water, wastewater, or suspended particles remains in a tank or channel. It is calculated by dividing the water/wastewater in the tank by the flow rate through the tank. The units of flow rate used in the calculation are dependent on whether the detention time is to be calculated in seconds, minutes, hours, or days. Detention time is used in conjunction with various treatment processes, including sedimentation and coagulation/flocculation. Generally, in practice, detention time is associated with the amount of time required for a tank to empty. The range of detention time varies with the process; for example, in a tank used for sedimentation, detention time is commonly measured in minutes. The calculation methods used to determine detention time are illustrated in the following sections.

19.3.4.1 Detention Time in Days

$$\text{DT (days)} = \frac{\text{Tank vol. (ft}^3) \times 7.48 \text{ gal/ft}^3}{\text{Flow (gal/day)}} \quad (19.3)$$

■ EXAMPLE 19.4

Problem: An anaerobic digester has a volume of 2,400,000 gal. What is the detention time in days when the influent flow rate is 0.07 MGD?

Solution:

$$\text{DT} = \frac{2,400,000 \text{ gal}}{0.07 \text{ MGD} \times 1,000,000 \text{ gal/MG}} = 34 \text{ days}$$

19.3.4.2 Detention Time in Hours

$$\text{DT (hr)} = \frac{\text{Tank vol. (ft}^3) \times 7.48 \text{ gal/ft}^3 \times 24 \text{ hr/day}}{\text{Flow (gal/day)}} \quad (19.4)$$

■ EXAMPLE 19.5

Problem: A settling tank has a volume of 44,000 ft³. What is the detention time in hours when the flow is 4.15 MGD?

Solution:

$$\text{DT} = \frac{44,000 \text{ ft}^3 \times 7.48 \text{ gal/ft}^3 \times 24 \text{ hr/day}}{4.15 \text{ MGD} \times 1,000,000 \text{ gal/MG}} = 1.9 \text{ hr}$$

19.3.4.3 Detention Time in Minutes

$$\text{DT (min)} = \frac{\text{Tank vol. (ft}^3) \times 7.48 \text{ gal/ft}^3 \times 1440 \text{ min/day}}{\text{Flow (gal/day)}} \quad (19.5)$$

■ **EXAMPLE 19.6**

Problem: A grit channel has a volume of 1340 ft³. What is the detention time in minutes when the flow rate is 4.3 MGD?

Solution:

$$DT = \frac{1340 \text{ ft}^3 \times 7.48 \text{ gal/ft}^3 \times 1440 \text{ min/day}}{4,300,000 \text{ gal/day}}$$

$$= 3.36 \text{ min}$$

Note: The tank volume and the flow rate must be in the same dimensions before calculating the hydraulic detention time.

19.4 WASTEWATER SOURCES AND CHARACTERISTICS

Wastewater treatment is designed to use the natural purification processes (self-purification processes of streams and rivers) to the maximum level possible. It is also designed to complete these processes in a controlled environment rather than over many miles of streams or rivers. Moreover, the treatment plant is also designed to remove other contaminants that are not normally subjected to natural processes, in addition to treating the solids that are generated through the treatment unit steps. The typical wastewater treatment plant is designed to achieve many different purposes:

- Protect public health.
- Protect public water supplies.
- Protect aquatic life.
- Preserve the best uses of the waters.
- Protect adjacent lands.

Wastewater treatment is a series of steps. Each of the steps can be accomplished using one or more treatment processes or types of equipment. The major categories of treatment steps are:

1. *Preliminary treatment*—Removes materials that could damage plant equipment or would occupy treatment capacity without being treated
2. *Primary treatment*—Removes settleable and floatable solids (may not be present in all treatment plants)
3. *Secondary treatment*—Removes BOD_5 and dissolved and colloidal suspended organic matter by biological action; organics are converted to stable solids, carbon dioxide, and more organisms
4. *Advanced waste treatment*—Uses physical, chemical, and biological processes to remove

additional BOD_5, solids, and nutrients (not present in all treatment plants)
5. *Disinfection*—Removes microorganisms to eliminate or reduce the possibility of disease when the flow is discharged
6. *Sludge treatment*—Stabilizes the solids that are removed from wastewater during treatment, inactivates pathogenic organisms, and reduces the volume of the sludge by removing water.

The various treatment processes described above are discussed in detail later.

19.4.1 WASTEWATER SOURCES

The principal sources of domestic wastewater in a community are the residential areas and commercial districts. Other important sources include institutional and recreational facilities and stormwater (runoff) and groundwater (infiltration). Each source produces wastewater with specific characteristics. In this section, wastewater sources and the specific characteristics of wastewater are described.

19.4.1.1 Generation of Wastewater

Wastewater is generated by five major sources: human and animal wastes, household wastes, industrial wastes, stormwater runoff, and groundwater infiltration:

- *Human and animal wastes*—Wastes that contain the solid and liquid discharges of humans and animals and are considered by many to be the most dangerous from a human health viewpoint. The primary health hazard is presented by the millions of bacteria, viruses, and other microorganisms (some of which may be pathogenic) present in the wastestream.
- *Household wastes*—Wastes, other than human and animal wastes, discharged from the home. Household wastes usually contain paper, household cleaners, detergents, trash, garbage, and other substances homeowners discharge into the sewer system.
- *Industrial wastes*—Materials discharged from industrial processes into the collection system. Industrial wastes typically contain chemicals, dyes, acids, alkalis, grit, detergents, and highly toxic materials.
- *Stormwater runoff*—Many collection systems are designed to carry both the wastes of the community and stormwater runoff. In this type of system, when a storm event occurs the wastestream can contain large amounts of sand, gravel, and other grit as well as excessive amounts of water.

- *Groundwater infiltration*—Groundwater will enter older, improperly sealed collection systems through cracks or unsealed pipe joints. This can add not only large amounts of water to wastewater flows but also additional grit.

19.4.1.2 Classification of Wastewater

Wastewater can be classified according to the sources of flows: domestic, sanitary, industrial, combined, and stormwater:

- *Domestic (sewage) wastewater*—Mainly contains human and animal wastes, household wastes, small amounts of groundwater infiltration, and small amounts of industrial wastes.
- *Sanitary wastewater*—Consists of domestic wastes and significant amounts of industrial wastes. In many cases, the industrial wastes can be treated without special precautions; however, in some cases the industrial wastes will require special precautions or a pretreatment program to ensure that the wastes do not cause compliance problems for the wastewater treatment plant.
- *Industrial wastewater*—Industrial wastes only; often, the industry will determine that it is safer and more economical to treat its waste independent of domestic waste.
- *Combined wastewater*—The combination of sanitary wastewater and stormwater runoff. All the wastewater and stormwater of the community is transported through one system to the treatment plant.
- *Stormwater*—A separate collection system (no sanitary waste) that carries stormwater runoff including street debris, road salt, and grit.

19.4.2 Wastewater Characteristics

Wastewater contains many different substances that can be used to characterize it. The specific substances and amounts or concentrations of each will vary, depending on the source; thus, it is difficult to precisely characterize wastewater. Instead, wastewater characterization is usually based on and applied to an average domestic wastewater. Wastewater is characterized in terms of its physical, chemical, and biological characteristics.

Note: Keep in mind that other sources and types of wastewater can dramatically change the characteristics.

19.4.2.1 Physical Characteristics

The *physical characteristics* of wastewater are based on color, odor, temperature, and flow:

- *Color*—Fresh wastewater is usually a light brownish-gray color; however, typical wastewater is gray and has a cloudy appearance. The color of the wastewater will change significantly if allowed to go septic (if travel time in the collection system increases). Typical septic wastewater will have a black color.
- *Odor*—Odors in domestic wastewater usually are caused by gases produced by the decomposition of organic matter or by other substances added to the wastewater. Fresh domestic wastewater has a musty odor. If the wastewater is allowed to go septic, this odor will change significantly to a rotten-egg odor associated with the production of hydrogen sulfide (H_2S).
- *Temperature*—The temperature of wastewater is commonly higher than that of the water supply because of the addition of warm water from households and industrial plants; however, significant amounts of infiltration or stormwater flow can cause major temperature fluctuations.
- *Flow*—The actual volume of wastewater is commonly used as a physical characterization of wastewater and is normally expressed in terms of gallons per person per day. Most treatment plants are designed using an expected flow of 100 to 200 gallons per person per day. This figure may have to be revised to reflect the degree of infiltration or storm flow the plant receives. Flow rates will vary throughout the day. This variation, which can be as much as 50 to 200% of the average daily flow, is known as the *diurnal flow variation*.

Note: *Diurnal* means "occurs in a day or each day; daily."

19.4.2.2 Chemical Characteristics

When describing the chemical characteristics of wastewater, the discussion generally includes topics such as organic matter, the measurement of organic matter, inorganic matter, and gases. For the sake of simplicity, in this handbook we specifically describe chemical characteristics in terms of alkalinity, biochemical oxygen demand (BOD), chemical oxygen demand (COD), dissolved gases, nitrogen compounds, pH, phosphorus, solids (organic, inorganic, suspended, and dissolved solids), and water:

- *Alkalinity* is a measure of the capability of the wastewater to neutralize acids. It is measured in terms of bicarbonate, carbonate, and hydroxide alkalinity. Alkalinity is essential to buffer (hold the neutral pH) of the wastewater during the biological treatment processes.

- *Biochemical oxygen demand (BOD)* is a measure of the amount of biodegradable matter in the wastewater. It is normally measured by a 5-day test conducted at 20°C. The BOD_5 for domestic waste is normally in the range of 100 to 300 mg/L.
- *Chemical oxygen demand (COD)* is a measure of the amount of oxidizable matter present in the sample. The COD is normally in the range of 200 to 500 mg/L. The presence of industrial wastes can increase this significantly.
- *Dissolved gases* are gases that are dissolved in wastewater. The specific gases and normal concentrations are based on the composition of the wastewater. Typical domestic wastewater contains oxygen in relatively low concentrations, carbon dioxide, and hydrogen sulfide (if septic conditions exist).
- The type and amount of *nitrogen compounds* present will vary from the raw wastewater to the treated effluent. Nitrogen follows a cycle of oxidation and reduction. Most of the nitrogen in untreated wastewater will be in the forms of organic nitrogen and ammonia nitrogen. Laboratory tests are used to determine both of these forms. The sum of these two forms of nitrogen is also measured and is known as *total Kjeldahl nitrogen* (TKN). Wastewater will normally contain between 20 to 85 mg/L of nitrogen. Organic nitrogen will normally be in the range of 8 to 35 mg/L, and ammonia nitrogen will be in the range of 12 to 50 mg/L.
- *pH* is used to express the acid condition of the wastewater. pH is expressed on a scale of 1 to 14. For proper treatment, wastewater pH should normally be in the range of 6.5 to 9.0 (ideal is 6.5 to 8.0).
- *Phosphorus* is essential to biological activity and must be present in at least minimum quantities or secondary treatment processes will not perform. Excessive amounts can cause stream damage and excessive algal growth. Phosphorus levels will normally be in the range of 6 to 20 mg/L. The removal of phosphate compounds from detergents has had a significant impact on the amounts of phosphorus found in wastewater.
- Most pollutants found in wastewater can be classified as *solids*. Wastewater treatment is generally designed to remove solids or to convert solids to a form that is more stable or can be removed. Solids can be classified by their chemical composition (organic or inorganic) or by their physical characteristics (settleable, floatable, and colloidal). Concentrations of total

solids in wastewater are normally in the range of 350 to 1200 mg/L.
- *Organic solids* consist of carbon, hydrogen, oxygen, and nitrogen and can be converted to carbon dioxide and water by ignition at 550°C; also known as *fixed solids* or *loss on ignition*.
- *Inorganic solids* are mineral solids that are unaffected by ignition; also known as *fixed solids* or *ash*.
- *Suspended solids* will not pass through a glass-fiber filter pad; they can be further classified as total suspended solids (TSS), volatile suspended solids, and fixed suspended solids. They can also be separated into three components based on settling characteristics: settleable solids, floatable solids, and colloidal solids. Total suspended solids in wastewater are normally in the range of 100 to 350 mg/L.
- *Dissolved solids* will pass through a glass-fiber filter pad. They can also be classified as total dissolved solids (TDS), volatile dissolved solids, and fixed dissolved solids. Total dissolved solids are normally in the range of 250 to 850 mg/L.
- *Water* is always the major constituent of wastewater. In most cases, water makes up 99.5 to 99.9% of the wastewater. Even in the strongest wastewater, the total amount of contamination present is less than 0.5% of the total, and in average-strength wastes it is usually less than 0.1%.

19.4.2.3 Biological Characteristics and Processes

(Note that the biological characteristics of water were discussed in detail earlier in this text.) After undergoing the physical aspects of treatment (i.e., screening, grit removal, and sedimentation) in preliminary and primary treatment, wastewater still contains some suspended solids and other solids that are dissolved in the water. In a natural stream, such substances are a source of food for protozoa, fungi, algae, and several varieties of bacteria. In secondary wastewater treatment, these same microscopic organisms (which are one of the main reasons for treating wastewater) are allowed to work as fast as they can to biologically convert the dissolved solids to suspended solids that will physically settle out at the end of secondary treatment.

Raw wastewater influent typically contains millions of organisms. The majority of these organisms are not pathogenic; however, several pathogenic organisms may also be present (these may include the organisms responsible for diseases such as typhoid, tetanus, hepatitis, dysentery, gastroenteritis, and others). Many of the organisms

TABLE 19.2
Typical Domestic Wastewater Characteristics

Characteristic	Typical Characteristic
Color	Gray
Odor	Musty
Dissolved oxygen	>1.0 mg/L
pH	6.5–9.0
TSS	100–350 mg/L
BOD_5	100–300 mg/L
COD	200–500 mg/L
Flow	100–200 gallons per person per day
Total nitrogen	20–85 mg/L
Total phosphorus	6–20 mg/L
Fecal coliform	500,000–3,000,000 MPN/100 mL

found in wastewater are microscopic (microorganisms); they include algae, bacteria, protozoa (such as amoebae, flagellates, free-swimming ciliates, and stalked ciliates), rotifers, and viruses. Table 19.2 provides a summary of typical domestic wastewater characteristics.

19.5 WASTEWATER COLLECTION SYSTEMS

Wastewater collection systems collect and convey wastewater to the treatment plant. The complexity of the system depends on the size of the community and the type of system selected. Methods of collection and conveyance of wastewater include gravity systems, force main systems, vacuum systems, and combinations of all three types of systems.

19.5.1 Gravity Collection System

In a *gravity collection system*, the collection lines are sloped to permit the flow to move through the system with as little pumping as possible. The slope of the lines must keep the wastewater moving at a velocity (speed) of 2 to 4 feet per second (fps); otherwise, at lower velocities, solids will settle out, causing clogged lines, overflows, and offensive odors. To keep collection systems lines at a reasonable depth, wastewater must be lifted (pumped) periodically so it can continue flowing downhill to the treatment plant. Pump stations are installed at selected points within the system for this purpose.

19.5.2 Force Main Collection System

In a typical *force main collection system*, wastewater is collected to central points and pumped under pressure to the treatment plant. The system is normally used for conveying wastewater long distances. The use of the force main system allows the wastewater to flow to the treatment plant at the desired velocity without using sloped lines. It should be noted that the pump station discharge lines in a gravity system are considered to be force mains, as the content of the lines is under pressure.

Note: Extra care must be taken when performing maintenance on force main systems because the content of the collection system is under pressure.

19.5.3 Vacuum System

In a *vacuum collection system*, wastewaters are collected to central points and then drawn toward the treatment plant under vacuum. The system consists of a large amount of mechanical equipment and requires a large amount of maintenance to perform properly. Generally, the vacuum type of collection system is not economically feasible.

19.5.4 Pumping Stations

Pumping stations provide the motive force (energy) to keep the wastewater moving at the desired velocity. They are used in both the force main and gravity systems. They are designed in several different configurations and may use different sources of energy to move the wastewater (i.e., pumps, air pressure, or vacuum). One of the more commonly used types of pumping station designs is the wet-well/dry-well design.

19.5.4.1 Wet-Well/Dry-Well Pumping Stations

The wet-well/dry-well pumping station consists of two separate spaces or sections separated by a common wall. Wastewater is collected in one section (wet well section) and the pumping equipment (and, in many cases, the motors and controllers) are located in a second section known as the dry well. Many different designs for this type of system are available, but in most cases the pumps selected for this system are of a centrifugal design. Among the major considerations when selecting the centrifugal design are that it: (1) allows for the separation of mechanical equipment (pumps, motors, controllers, wiring, etc.) from the potentially corrosive atmosphere (e.g., sulfides) of the wastewater, and (2) this type of design is usually safer for workers because they can monitor, maintain, operate, and repair equipment without entering the pumping station wet well.

Note: Most pumping station wet wells are confined spaces. To ensure safe entry into such spaces compliance with OSHA's 29 CFR 1910.146 (Confined Space Entry Standard) is required.

19.5.4.2 Wet-Well Pumping Stations

Another type of pumping station design is the wet-well type. This type consists of a single compartment that collects the wastewater flow. The pump is submerged in the wastewater with motor controls located in the space or has a weatherproof motor housing located above the wet well. In this type of station, a submersible centrifugal pump is normally used.

19.5.4.3 Pneumatic Pumping Stations

The *pneumatic pumping station* consists of a wet well and a control system that controls the inlet and outlet value operations and provides pressurized air to force or push the wastewater through the system. The exact method of operation depends on the system design. When wastewater in the wet well reaches a predetermined level, an automatic valve is activated that closes the influent line. The tank (i.e., wet well) is then pressurized to a predetermined level. When the pressure reaches the predetermined level, the effluent line valve is opened and the pressure pushes the wastestream out through the discharge line.

19.5.5 Wet-Well Pumping Station Calculations

Calculations normally associated with wet well pumping station design (such as determining design lift or pumping capacity) are usually left up to design and mechanical engineers; however, on occasion, wastewater operators or interceptor technicians may be called upon to make certain basic calculations. Usually these calculations deal with determining either pump capacity without influent (to check the pumping rate of the constant-speed pump) or pump capacity with influent (to check how many gallons per minute the pump is discharging). In this section, we use examples to describe instances on how and where these two calculations are made.

■ EXAMPLE 19.7. Determining Pump Capacity without Influent

Problem: A pumping station wet well is 10 ft by 9 ft. To check the pumping rate of the constant-speed pump, the operator closed the influent valve to the wet well for a 5-min test. The level in the well dropped 2.2 ft. What is the pumping rate in gallons per minute?

Solution: Using the length and width of the well, we can find the area of the water surface:

$$10 \text{ ft} \times 9 \text{ ft} = 90 \text{ ft}^2$$

The water level dropped 2.2 ft. From this information, we can find the volume of water removed by the pump during the test:

$$\text{Area} \times \text{Depth} = \text{Volume} \qquad (19.6)$$

$$90 \text{ ft}^2 \times 2.2 \text{ ft} = 198 \text{ ft}^3$$

One cubic foot of water holds 7.48 gal. We can convert this volume in cubic feet to gallons:

$$198 \text{ ft}^3 \times \frac{7.48 \text{ gal}}{1 \text{ ft}^3} = 1481 \text{ gal}$$

The test was done for 5 min. From this information, a pumping rate can be calculated:

$$\frac{1481 \text{ gal}}{5 \text{ min}} = \frac{296.2}{1 \text{ min}} = 296.2 \text{ gpm}$$

■ EXAMPLE 19.8. Determining Pump Capacity with Influent

Problem: A wet well is 8.2 ft by 9.6 ft. The influent flow to the well, measured upstream, is 365 gpm. If the wet well rises 2.2 in. in 5 min, how many gallons per minute is the pump discharging?

Solution:

$$\text{Influent} = \text{Discharge} + \text{accumulation} \qquad (19.7)$$

$$\frac{365 \text{ gal}}{1 \text{ min}} = \text{Discharge} + \text{accumulation}$$

We want to calculate the discharge. Influent is known, and we have enough information to calculate the accumulation:

$$\text{Volume accumulated} = 8.2 \text{ ft} \times 9.6 \text{ ft} \times 2.2 \text{ in.}$$

$$\times \frac{1 \text{ ft}}{12 \text{ in}} \times \frac{7.48 \text{ gal}}{1 \text{ ft}^3}$$

$$= 108 \text{ gal}$$

$$\text{Accumulation} = \frac{108 \text{ gal}}{5 \text{ min}} = \frac{21.6 \text{ gal}}{1 \text{ min}} = 21.6 \text{ gpm}$$

Using Equation 19.7:

$$\text{Influent} = \text{Discharge} + \text{accumulation}$$

$$365 \text{ gpm} = \text{Discharge} + 21.6$$

Subtracting from both sides:

$$365 \text{ gpm} - 21.6 \text{ gpm} = \text{Discharge} + 21.6 \text{ gpm} - 21.6 \text{ gpm}$$

$$343.4 \text{ gpm} = \text{Discharge}$$

The wet well pump is discharging 343.4 gallons each minute.

19.6 PRELIMINARY TREATMENT

The initial stage in the wastewater treatment process (following collection and influent pumping) is *preliminary treatment*. Raw influent entering the treatment plant may contain many kinds of materials (trash). The purpose of preliminary treatment is to protect plant equipment by removing these materials, which can cause clogs, jams, or excessive wear to plant machinery. In addition, the removal of various materials at the beginning of the treatment process saves valuable space within the treatment plant.

Preliminary treatment may include many different processes; each is designed to remove a specific type of material that poses a potential problem for the treatment process. Processes include wastewater collections—influent pumping, screening, shredding, grit removal, flow measurement, preaeration, chemical addition, and flow equalization; the major processes are shown in Figure 19.1. In this section, we describe and discuss each of these processes and their importance in the treatment process.

Note: As mentioned, not all treatment plants will include all of the processes shown in Figure 19.1. Specific processes have been included to facilitate discussion of major potential problems with each process and its operation; this is information that may be important to the wastewater operator.

19.6.1 SCREENING

The purpose of *screening* is to remove large solids such as rags, cans, rocks, branches, leaves, or roots from the flow before the flow moves on to downstream processes.

Note: Typically, a treatment plant will remove anywhere from 0.5 to 12 ft^3 of screenings for each million gallons of influent received.

A *bar screen* traps debris as wastewater influent passes through. Typically, a bar screen consists of a series of parallel, evenly spaced bars or a perforated screen placed in a channel (see Figure 19.2). The wastestream passes through the screen and the large solids (*screenings*) are trapped on the bars for removal.

Note: The screenings must be removed frequently enough to prevent accumulation that will block the screen and cause the water level in front of the screen to build up.

The bar screen may be coarse (2- to 4-in. openings) or fine (0.75- to 2.0-in. openings). The bar screen may be manually cleaned (bars or screens are placed at an angle of 30° for easier solids removal; see Figure 19.2) or mechanically cleaned (bars are placed at an angle of 45° to 60° to improve mechanical cleaner operation).

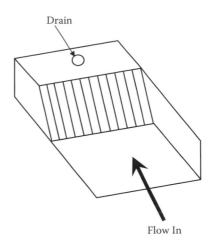

FIGURE 19.2 Basic bar screen.

The screening method employed depends on the design of the plant, the amount of solids expected, and whether the screen is for constant or emergency use only.

19.6.1.1 Manually Cleaned Screens

Manually cleaned screens are cleaned at least once per shift (or often enough to prevent buildup that may cause reduced flow into the plant) using a long tooth rake. Solids are manually pulled to the drain platform and allowed to drain before storage in a covered container. The area around the screen should be cleaned frequently to prevent a buildup of grease or other materials that can cause odors, slippery conditions, and insect and rodent problems. Because screenings may contain organic matter as well as large amounts of grease, they should be stored in a covered container. Screenings can be disposed of by burial in approved landfills or by incineration. Some treatment facilities grind the screenings into small particles, which are then returned to the wastewater flow for further processing and removal later in the process.

19.6.1.1.1 Operational Problems

Manually cleaned screens require a certain amount of operator attention to maintain optimum operation. Failure to clean the screen frequently can lead to septic wastes entering the primary, surge flows after cleaning, or low flows before cleaning. On occasion, when such operational problems occur, it becomes necessary to increase the frequency of the cleaning cycle. Another operational problem is excessive grit in the bar screen channel. Improper design or construction or insufficient cleaning may cause this problem. The corrective action required is either to modify the design or to increase cleaning frequency and flush the channel regularly. Another common problem with manually cleaned bar screens is their tendency to clog frequently. This may be caused by excessive debris in the wastewater or the screen might be too fine for its current application. The operator should locate the source of the excessive debris and eliminate it. If the screen is the problem, a coarser screen may have to be installed. If the bar screen area is filled with obnoxious odors, flies, and other insects, it may be necessary to dispose of screenings more frequently.

19.6.1.2   Mechanically Cleaned Screens

Mechanically cleaned screens use a mechanized rake assembly to collect the solids and move them out of the wastewater flow for discharge to a storage hopper. The screen may be continuously cleaned or cleaned on a time- or flow-controlled cycle. As with the manually cleaned screen, the area surrounding the mechanically operated screen must be cleaned frequently to prevent the buildup of materials that can cause unsafe conditions. As with all mechanical equipment, operator vigilance is required to ensure proper operation and that proper maintenance is performed. Maintenance includes lubricating equipment and maintaining it in accordance with the manufacturer's recommendations or the plant's operations and maintenance (O&M) manual. Screenings from mechanically operated bar screens are disposed of in the same manner as screenings from manually operated screen: landfill disposal, incineration, or ground into smaller particles for return to the wastewater flow.

19.6.1.2.1   Operational Problems

Many of the operational problems associated with mechanically cleaned bar screens are the same as those for manual screens: septic wastes enters the primary, surge flows after cleaning, the bar screen channel has excessive grit, or the screen clogs frequently. Basically the same corrective actions employed for manually operated screens would be applied for these problems in mechanically operated screens. In addition to these problems, however, mechanically operated screens also have other problems, including the cleaner not operating at all or the rake not operating even though the motor is. Obviously, these are mechanical problems that could be caused by jammed cleaning mechanism, broken chain, broken cable, or a broken shear pin. Authorized and fully trained maintenance operators should be called in to handle these types of problems.

19.6.1.3   Screening Safety

The screening area is the first location where the operator is exposed to the wastewater flow. Any toxic, flammable or explosive gases present in the wastewater can be released at this point. Operators who enter enclosed bar screen areas should be equipped with personal air monitors. Adequate ventilation must be provided. It is also important to remember that, due to the grease attached to the screenings, this area of the plant can be extremely slippery. Routine cleaning is required to minimize this problem.

Note: Never override safety devices on mechanical equipment. Overrides can result in dangerous conditions, injuries, and major mechanical failure.

19.6.1.4   Screenings Removal Computations

Operators responsible for screenings disposal are typically required to keep a record of the amount of screenings removed from the wastewater flow. To keep and maintain accurate screening records, the volume of screenings withdrawn must be determined. Two methods are commonly used to calculate the volume of screenings withdrawn.

$$\text{Screenings removed (ft}^3/\text{day)} = \frac{\text{Screenings (ft}^3)}{\text{Days}} \quad (19.8)$$

$$\text{Screenings removed (ft}^3/\text{MG)} = \frac{\text{Screenings (ft}^3)}{\text{Flow (MG)}} \quad (19.9)$$

■ EXAMPLE 19.9

Problem: A total of 65 gal of screenings is removed from the wastewater flow during a 24-hr period. What is the screenings removal reported as ft³/day?

Solution: First, convert gallons screenings to cubic feet:

$$\frac{65 \text{ gal}}{7.48 \text{ gal/ft}^3} = 8.7 \text{ ft}^3 \text{ screenings}$$

Next, calculate screenings removed as ft³/day:

$$\text{Screenings removed (ft}^3/\text{day)} = \frac{8.7 \text{ ft}^3}{1 \text{ day}} = 8.7 \text{ ft}^3/\text{day}$$

■ EXAMPLE 19.10

Problem: During one week, a total of 310 gal of screenings was removed from the wastewater screens. What is the average screening removal in ft³/day?

Solution: First, gallons screenings must be converted to cubic feet off screenings:

$$\frac{310 \text{ gal}}{7.48 \text{ gal/ft}^3} = 41.4 \text{ ft}^3 \text{ screenings}$$

$$\text{Screenings removed (ft}^3/\text{day)} = \frac{41.4 \text{ ft}^3}{7} = 5.9 \text{ ft}^3/\text{day}$$

19.6.2   Shredding

As an alternative to screening, shredding can be used to reduce solids to a size that can enter the plant without causing mechanical problems or clogging. Shredding processes include comminution (*comminute* means to cut up) and barminution devices.

19.6.2.1 Comminution

The *comminutor* is the most common shredding device used in wastewater treatment. In this device all the wastewater flow passes through the grinder assembly. The grinder consists of a screen or slotted basket, a rotating or oscillating cutter, and a stationary cutter. Solids pass through the screen and are chopped or shredded between the two cutters. The comminutor will not remove solids that are too large to fit through the slots, and it will not remove floating objects. These materials must be removed manually. Maintenance requirements for comminutors include aligning, sharpening, and replacing cutters and corrective and preventive maintenance performed in accordance with the plant O&M manual.

19.6.2.1.1 Operational Problems

Common operational problems associated with comminutors include output containing coarse solids. When this occurs it is usually a sign that the cutters are dull or misaligned. If the system does not operate at all, the unit is clogged or jammed, a shear pin or coupling is broken, or electrical power is shut off. If the unit stalls or jams frequently, this usually indicates cutter misalignment, excessive debris in the influent, or dull cutters.

Note: Only qualified maintenance operators should perform maintenance of shredding equipment.

19.6.2.2 Barminution

In barminution, the *barminutor* uses a bar screen to collect solids, which are then shredded and passed through the bar screen for removal at a later process. The cutter alignment and sharpness of each device are critical factors in effective operation. Cutters must be sharpened or replaced and alignment must be checked in accordance with the manufacturer's recommendations. Solids that are not shredded must be removed daily, stored in closed containers, and disposed of by burial or incineration. Barminutor operational problems are similar to those listed above for comminutors. Preventive and corrective maintenance as well as lubrication must be performed by qualified personnel and in accordance with the plant O&M manual. Because of its higher maintenance requirements, the barminutor is less frequently used.

19.6.3 GRIT REMOVAL

The purpose of *grit removal* is to remove the heavy inorganic solids that could cause excessive mechanical wear. Grit is heavier than inorganic solids and includes sand, gravel, clay, egg shells, coffee grounds, metal filings, seeds, and other similar materials. Several processes or devices are used for grit removal. All of the processes are based on the fact that grit is heavier than the organic solids that should be kept in suspension for treatment in subsequent processes. Grit removal may be accomplished in grit chambers or by the centrifugal separation of sludge. Processes use gravity/velocity, aeration, or centrifugal force to separate the solids from the wastewater.

19.6.3.1 Gravity/Velocity-Controlled Grit Removal

Gravity/velocity-controlled grit removal is normally accomplished in a channel or tank where the speed or the velocity of the wastewater is controlled to about 1 foot per second (ideal), so the grit will settle while organic matter remains suspended. As long as the velocity is controlled in the range of 0.7 to 1.4 fps, the grit removal will remain effective. Velocity is controlled by the amount of water flowing through the channel, the depth of the water in the channel, the width of the channel, or the cumulative width of channels in service.

19.6.3.1.1 Process Control Calculations

Velocity of the flow in a channel can be determined either by the float and stopwatch method or by channel dimensions.

■ EXAMPLE 19.11. Velocity by Float and Stopwatch

$$\text{Velocity (ft/sec)} = \frac{\text{Distance traveled (ft)}}{\text{Time required (sec)}} \quad (19.10)$$

Problem: A float requires 25 sec to travel 34 ft in a grit channel. What is the velocity of the flow in the channel?

Solution:

$$\text{Velocity (fps)} = \frac{34 \text{ ft}}{25 \text{ sec}} = 1.4 \text{ fps}$$

■ EXAMPLE 19.12. Velocity by Flow and Channel Dimensions

This calculation can be used for a single channel or tank or for multiple channels or tanks with the same dimensions and equal flow. If the flows through each unit of the unit dimensions are unequal, the velocity for each channel or tank must be computed individually.

$$\text{Velocity (fps)} = \frac{\text{Flow (MGD)} \times 1.55 \text{ cfs/MGD}}{\begin{array}{c}\text{No. of channels in service} \\ \times \text{ channel width (ft)} \\ \times \text{ water depth (ft)}\end{array}} \quad (19.11)$$

Problem: A plant is currently using two grit channels. Each channel is 3 ft wide and has a water depth of 1.2 ft. What is the velocity when the influent flow rate is 3.0 MGD?

Solution:

$$\text{Velocity} = \frac{3.0 \text{ MGD} \times 1.55 \text{ cfs/MGD}}{2 \text{ channels} \times 3 \text{ ft} \times 1.2 \text{ ft}}$$

$$= \frac{4.65 \text{ cfs}}{7.2 \text{ ft}^2} = .65 \text{ fps}$$

Note: The channel dimensions must always be in feet. Convert inches to feet by dividing by 12 inches per foot.

■ EXAMPLE 19.13. Required Settling Time

This calculation can be used to determine the time required for a particle to travel from the surface of the liquid to the bottom at a given settling velocity. To compute the settling time, the settling velocity in feet per second must be provided or determined experimentally in a laboratory.

$$\text{Settling time (sec)} = \frac{\text{Liquid depth (ft)}}{\text{Settling velocity (fps)}} \quad (19.12)$$

Problem: The grit channel of a plant is designed to remove sand that has a settling velocity of 0.085 fps. The channel is currently operating at a depth of 2.2 ft. How many seconds will it take for a sand particle to reach the channel bottom?

Solution:

$$\text{Settling time} = \frac{2.2 \text{ ft}}{0.085 \text{ fps}} = 25.9 \text{ sec}$$

■ EXAMPLE 19.14. Required Channel Length

This calculation can be used to determine the length of channel required to remove an object with a specified settling velocity.

$$\text{Required channel length} = \frac{\begin{array}{c}\text{Channel depth (ft)} \\ \times \text{flow velocity (fps)}\end{array}}{\text{Settling velocity (fps)}} \quad (19.13)$$

Problem: The grit channel of a plant is designed to remove sand that has a settling velocity of 0.070 fps. The channel is currently operating at a depth of 3 ft. The calculated velocity of flow through the channel is 0.80 fps. The channel is 35 ft long. Is the channel long enough to remove the desired sand particle size?

Solution:

$$\text{Required channel length (ft)} = \frac{3 \text{ ft} \times 0.80 \text{ fps}}{0.070 \text{ fps}} = 34.3 \text{ ft}$$

Yes, the channel is long enough to ensure that all of the sand will be removed.

19.6.3.1.2 Cleaning

Gravity-type systems may be manually or mechanically cleaned. Manual cleaning normally requires that the channel be taken out of service, drained, and manually cleaned. Mechanical cleaning systems are operated continuously or on a time cycle. Removal should be frequent enough to prevent grit carryover into the rest of the plant.

Note: Before and during cleaning activities always ventilate the area thoroughly.

19.6.3.1.3 Operational Observations, Problems, and Troubleshooting

As mentioned earlier, gravity/velocity-controlled grit removal normally occurs in a channel or tank where the speed or the velocity of the wastewater is controlled to about 1 fps (ideal), so grit settles while organic matters remains suspended. As long as the velocity is controlled in the range of 0.7 to 1.4 fps, the grit removal remains effective. Velocity is controlled by the amount of water flowing through the channel, the depth of the water in the channel, the width of the channel, or the cumulative width of channels in service. During operation, the operator must pay particular attention to grit characteristics and for evidence of organic solids in the channel, grit carryover into the plant, and mechanical problems, as well as to grit storage and disposal (housekeeping).

19.6.3.2 Aeration

Aerated grit removal systems use aeration to keep the lighter organic solids in suspension while allowing the heavier grit articles to settle out. Aerated grit removal systems may be manually or mechanically cleaned; however, the majority of the systems are mechanically cleaned. During normal operation, adjusting the aeration rate produces the desired separation. This requires monitoring mixing and aeration and sampling fixed suspended solids. Actual grit removal is controlled by the rate of aeration. If the rate is too high, all of the solids remain in suspension. If the rate is too low, both grit and organics will settle out. The operator observes the same kinds of conditions as those listed for the gravity/velocity-controlled system but must also pay close attention to the air distribution system to ensure proper operation.

19.6.3.3 Centrifugal Force

The *cyclone degritter* uses a rapid spinning motion (centrifugal force) to separate the heavy inorganic solids or grit from the light organic solids. This unit process is normally used on primary sludge rather than the entire wastewater flow. This critical control factor for the process is the inlet pressure. If the pressure exceeds the recommendations of the manufacturer, the unit will flood, and grit will carry through with the flow. Grit is separated from flow, washed, and discharged directly to a strange container. Grit removal performance is determined by calculating the percent removal for inorganic (fixed) suspended solids. The operator observes the same kinds of conditions listed for the gravity/velocity-controlled and aerated grit removal systems, with the exception of the air distribution system. Typical problems associated with grit removal include mechanical malfunctions and rotten-egg odor in the grit chamber (hydrogen sulfide formation), which can lead to metal and concrete corrosion problems. Low recovery rate of grit is another typical problem. Bottom scour, overaeration, or not enough detention time normally causes this. When such a problem occurs, the operator must make the required adjustments or repairs to correct the problem.

19.6.3.4 Grit Removal Calculations

Wastewater systems typically average 1 to 15 ft³ of grit per million gallons of flow (sanitary systems, 1 to 4 ft³/million gal; combined wastewater systems, 4 to 15 ft³/million gal), with higher ranges during storm events. Generally, grit is disposed of in sanitary landfills. Because of this practice, for planning purposes, operators must keep accurate records of grit removal. Most often, the data are reported as cubic feet of grit removed per million gallons of flow:

$$\text{Grit removed (ft}^3/\text{MG)} = \frac{\text{Grit volume (ft}^3)}{\text{Flow (MG)}} \quad (19.14)$$

Over a given period, the average grit removal rate at a plant (at least a seasonal average) can be determined and used for planning purposes. Typically, grit removal is calculated as cubic yards, because excavation is normally expressed in terms of cubic yards.

$$\text{Grit (yd}^3) = \frac{\text{Total grit (ft}^3)}{27 \text{ ft}^3/\text{yd}^3} \quad (19.15)$$

■ EXAMPLE 19.15

Problem: A treatment plant removes 10 ft³ of grit in one day. How many cubic feet of grit are removed per million gallons (MG) if the plant flow was 9 MGD?

Solution:

$$\text{Grit removed} = \frac{\text{Grit volume (ft}^3)}{\text{Flow (MG)}}$$

$$= \frac{10 \text{ ft}^3}{9 \text{ MG}} = 1.1 \text{ ft}^3/\text{MG}$$

■ EXAMPLE 19.16

Problem: The total daily grit removed for a plant is 250 gal. If the plant flow is 12.2 MGD, how many cubic feet of grit are removed per MG flow?

Solution: First, convert gallon grit removed to ft³:

$$\frac{250 \text{ gal}}{7.48 \text{ gal/ft}^3} = 33 \text{ ft}^3$$

Next, complete the calculation of ft³/MG:

$$\text{Grit removal (ft}^3/\text{MG)} = \frac{\text{Grit volume (ft}^3)}{\text{Flow (MG)}}$$

$$= \frac{33 \text{ ft}^3}{12.2 \text{ MGD}}$$

$$= 2.7 \text{ ft}^3/\text{MGD}$$

■ EXAMPLE 19.17

Problem: The monthly average grit removal is 2.5 ft³/MG. If the monthly average flow is 2,500,000 gpd, how many cubic yards must be available for grit disposal if the pit is to have a 90-day capacity?

Solution: First, calculate the grit generated each day:

$$\frac{(2.5 \text{ ft}^3)}{\text{MG}} \times (2.5 \text{ MGD}) = 6.25 \text{ ft}^3$$

The ft³ grit generated for 90 days would be:

$$\frac{(6.25 \text{ ft}^3)}{\text{day}} \times (90 \text{ days}) = 562.5 \text{ ft}$$

Convert ft³ grit to yd³ grit:

$$\frac{562.5 \text{ ft}^3}{27 \text{ ft}^3/\text{yd}^3} = 21 \text{ yd}^3$$

19.6.4 PREAERATION

In the *preaeration process* (diffused or mechanical), we aerate wastewater to achieve and maintain an aerobic state (to freshen septic wastes), strip off hydrogen sulfide (to reduce odors and corrosion), agitate solids (to release trapped gases and improve solids separation and settling), and to reduce BOD_5. All of this can be accomplished by aerating the wastewater for 10 to 30 min. To reduce BOD_5, preaeration must be conducted for 45 to 60 min.

19.6.4.1 Operational Observations, Problems, and Troubleshooting

In preaeration grit removal systems, the operator is concerned with maintaining proper operation and must be alert to any possible mechanical problems. In addition, the operator monitors dissolved oxygen levels and the impact of preaeration on influent.

19.6.5 Chemical Addition

Chemical addition to the wastestream is done (either via dry chemical metering or solution feed metering) to improve settling, reduce odors, neutralize acids or bases, reduce corrosion, reduce BOD_5, improve solids and grease removal, reduce loading on the plant, add or remove nutrients, add organisms, or aid subsequent downstream processes. The particular chemical and amount used depends on the desired result. Chemicals must be added at a point where sufficient mixing will occur to obtain maximum benefit. Chemicals typically used in wastewater treatment include chlorine, peroxide, acids and bases, miner salts (e.g., ferric chloride, alum), and bioadditives and enzymes.

19.6.5.1 Operational Observations, Problems, and Troubleshooting

When adding chemicals to the wastestream to remove grit, the operator monitors the process for evidence of mechanical problems and takes proper corrective actions when necessary. The operator also monitors the current chemical feed rate and dosage. The operator ensures that mixing at the point of addition is accomplished in accordance with standard operating procedures and monitors the impact of chemical addition on influent.

19.6.6 Equalization

The purpose of *flow equalization* (whether by surge, diurnal, or complete methods) is to reduce or remove the wide swings in flow rates normally associated with wastewater treatment plant loading; it minimizes the impact of storm flows. The process can be designed to prevent flows above maximum plant design hydraulic capacity, to reduce the magnitude of diurnal flow variations, and to eliminate flow variations. Flow equalization is accomplished using mixing or aeration equipment, pumps, and flow measurement. Normal operation depends on the purpose and requirements of the flow equalization system. Equalized flows allow the plant to perform at optimum levels by providing stable hydraulic and organic loading. The downside to flow equalization is the additional expense associated with construction and operation of the flow equalization facilities.

19.6.6.1 Operational Observations, Problems, and Troubleshooting

During normal operations, the operator must monitor all mechanical systems involved with flow equalization, watch for mechanical problems, and be prepared to take the appropriate corrective action. The operator also monitors dissolved oxygen levels, the impact of equalization on influent, and water levels in equalization basins, in addition to making necessary adjustments.

19.6.7 Aerated Systems

Aerated grit removal systems use aeration to keep the lighter organic solids in suspension while allowing the heavier grit particles to settle out. Aerated grit removal may be manually or mechanically cleaned; however, the majority of the systems are mechanically cleaned. In normal operation, the aeration rate is adjusted to produce the desired separation, which requires observation of mixing and aeration and sampling of fixed suspended solids. Actual grit removal is controlled by the rate of aeration. If the rate is too high, all of the solids remain in suspension. If the rate is too low, both the grit and the organics will settle out.

19.6.8 Cyclone Degritter

The *cyclone degritter* uses a rapid spinning motion (centrifugal force) to separate the heavy inorganic solids or grit from the light organic solids. This unit process is normally used on primary sludge rather than the entire wastewater flow. The critical control factor for the process is the inlet pressure. If the pressure exceeds the recommendations of the manufacturer, the unit will flood and grit will carry through with the flow. Grit is separated from the flow and discharged directly to a storage container. Grit removal performance is determined by calculating the percent removal for inorganic (fixed) suspended solids.

19.6.9 Preliminary Treatment Sampling and Testing

During normal operation of grit removal systems (with the exception of the screening and shredding processes), the plant operator is responsible for sampling and testing as shown in Table 19.3.

19.6.10 Other Preliminary Treatment Process Control Calculations

The desired velocity in sewers in approximately 2 fps at peak flow, because this velocity normally prevents solids from settling from the lines; however, when the flow reaches the grit channel, the velocity should decrease to about 1 fps to permit the heavy inorganic solids to settle.

TABLE 19.3
Sampling and Testing Grit Removal Systems

Process	Location	Test	Frequency
Grit removal (velocity)	Influent	Suspended solids (fixed)	Variable
	Channel	Depth of grit	Variable
	Grit	Total solids (fixed)	Variable
	Effluent	Suspended solids (fixed)	Variable
Grit removal (aerated)	Influent	Suspended solids (fixed)	Variable
	Channel	Dissolved oxygen	Variable
	Grit	Total solids (fixed)	Variable
	Effluent	Suspended solids (fixed)	Variable
Chemical addition	Influent	Jar test	Variable
Preaeration	Influent	Dissolved oxygen	Variable
	Effluent	Dissolved oxygen	Variable
Equalization	Effluent	Dissolved oxygen	Variable

In the example calculations that follow, we describe how the velocity of the flow in a channel can be determined by the float and stopwatch method and by channel dimensions.

■ **EXAMPLE 19.18. Velocity by Float and Stopwatch**

$$\text{Velocity (fps)} = \frac{\text{Distance traveled (ft)}}{\text{Time required (sec)}} \quad (19.16)$$

Problem: A float requires 30 sec to travel 37 ft in a grit channel. What is the velocity of the flow in the channel?

Solution:

$$\text{Velocity} = \frac{37 \text{ ft}}{30 \text{ sec}} = 1.2 \text{ fps}$$

■ **EXAMPLE 19.19. Velocity by Flow and Channel Dimensions**

This calculation can be used for a single channel or tank or for multiple channels or tanks with the same dimensions and equal flow. If the flow through each of the unit dimensions is unequal, the velocity for each channel or tank must be computed individually.

$$\text{Velocity (fps)} = \frac{\text{Flow (MGD)} \times 1.55 \text{ cfs/MGD}}{\substack{\text{No. of channels in service} \\ \times \text{channel width (ft)} \\ \times \text{water depth (ft)}}} \quad (19.17)$$

Problem: A plant is currently using two grit channels. Each channel is 3 ft wide and has a water depth of 1.3 ft. What is the velocity when the influent flow rate is 4.0 MGD?

Solution:

$$\text{Velocity} = \frac{4.0 \text{ MGD} \times 1.55 \text{ cfs/MGD}}{2 \text{ channels} \times 3 \text{ ft} \times 1.3 \text{ ft}}$$

$$= \frac{6.2 \text{ cfs}}{7.8 \text{ ft}^2} = 0.79 \text{ fps}$$

Note: Because 0.79 is within the range of 0.7 to 1.4, the operator of this unit would not make any adjustments.

Note: The channel dimensions must always be in feet. Convert inches to feet by dividing by 12 inches per foot.

■ **EXAMPLE 19.20. Required Settling Time**

This calculation can be used to determine the time required for a particle to travel from the surface of the liquid to the bottom at a given settling velocity. To compute the settling time, settling velocity in fps must be provided or determined by experiment in a laboratory.

$$\text{Settling time (sec)} = \frac{\text{Liquid depth (ft)}}{\text{Settling velocity (fps)}} \quad (19.18)$$

Problem: The grit channel of a plant is designed to remove sand that has a settling velocity of 0.080 fps. The channel is currently operating at a depth of 2.3 ft. How many seconds will it take for a sand particle to reach the channel bottom?

Solution:

$$\text{Settling time} = \frac{2.3 \text{ ft}}{0.080 \text{ fps}} = 28.7 \text{ sec}$$

■ **EXAMPLE 19.21. Required Channel Length**

This calculation can be used to determine the length of channel required to remove an object with a specified settling velocity.

$$\text{Required channel length} = \frac{\text{Channel depth (ft)}}{0.080 \text{ fps}} \times \text{flow velocity (fps)} \quad (19.19)$$

Problem: The grit channel of a plant is designed to remove sand that has a settling velocity of 0.080 fps. The channel is currently operating at a depth of 3 ft. The calculated velocity of flow through the channel is 0.85 fps. The channel is 36 ft long. Is the channel long enough to remove the desired sand particle size?

Solution:

$$\text{Required channel length} = \frac{3 \text{ ft} \times 0.85 \text{ fps}}{0.080 \text{ fps}} = 31.9 \text{ ft}$$

Yes, the channel is long enough to ensure that all of the sand will be removed.

Caution: Before and during cleaning activities, always ventilate the area thoroughly.

19.7 PRIMARY TREATMENT (SEDIMENTATION)

The purpose of primary treatment (primary sedimentation or primary clarification) is to remove settleable organic and floatable solids. Normally, each primary clarification unit can be expected to remove 90 to 95% settleable solids, 40 to 60% total suspended solids, and 25 to 35% BOD_5.

Note: Performance expectations for settling devices used in other areas of plant operation are normally expressed as overall unit performance rather than settling unit performance.

Sedimentation may be used throughout the treatment plant to remove settleable and floatable solids. It is used in primary treatment, secondary treatment, and advanced wastewater treatment processes. In this section, we focus on primary treatment or primary clarification, which utilizes large basins where primary settling is achieved under relatively quiescent conditions (see Figure 19.1). Within these basins, mechanical scrapers collect the primary settled solids into a hopper, from which they are pumped to a sludge processing area. Oil, grease, and other floating materials (scum) are skimmed from the surface. The effluent is discharged over weirs into a collection trough.

19.7.1 PROCESS DESCRIPTION

In primary sedimentation, wastewater enters a settling tank or basin. Velocity is reduced to approximately 1 foot per minute.

Note: Notice that the velocity is based on minutes instead of seconds, as was the case in the grit channels. A grit channel velocity of 1 ft/sec would be 60 ft/min.

Solids that are heavier than water settle to the bottom, while solids that are lighter than water float to the top. Settled solids are removed as sludge, and floating solids are removed as scum. Wastewater leaves the sedimentation tank over an effluent weir and moves on to the next step in treatment. Detention time, temperature, tank design, and condition of the equipment control the efficiency of the process.

19.7.2 OVERVIEW OF PRIMARY TREATMENT

- Primary treatment reduces the organic loading on downstream treatment processes by removing a large amount of settleable, suspended, and floatable materials.
- Primary treatment reduces the velocity of the wastewater through a clarifier to approximately 1 to 2 ft/min so settling and flotation can take place. Slowing the flow enhances removal of suspended solids in wastewater.
- Primary settling tanks remove floated grease and scum, as well as the settled sludge solids, and collect them for pumped transfer to disposal or further treatment.
- Clarifiers may be rectangular or circular. In rectangular clarifiers, wastewater flows from one end to the other, and the settled sludge is moved to a hopper at the one end, either by flights set on parallel chains or by a single bottom scraper set on a traveling bridge. Floating material (mostly grease and oil) is collected by a surface skimmer.
- In circular tanks, the wastewater usually enters at the middle and flows outward. Settled sludge is pushed to a hopper in the middle of the tank bottom, and a surface skimmer removes floating material.
- Factors affecting primary clarifier performance include:
  - Rate of flow through the clarifier
  - Wastewater characteristics (strength, temperature, amount and type of industrial waste, and the density, size, and shapes of particles)
  - Performance of pretreatment processes
  - Nature and amount of any wastes recycled to the primary clarifier

- Key factors in primary clarifier operation include the following concepts:

$$\text{Retention time (hr)} = \frac{\text{Vol. (gal)} \times 24 \text{ hr/day}}{\text{Flow (gpd)}}$$

$$\text{Surface loading rate (gpd/ft}^2) = \frac{Q \text{ (gpd)}}{\text{Surface area (ft}^2)}$$

$$\text{Solids loading rate (lb/day/ft}^2) = \frac{\text{Solids in clarifier (lb/day)}}{\text{Surface area (ft}^2)}$$

$$\text{Weir overflow rate (gpd/linear ft)} = \frac{Q \text{ (gpd)}}{\text{Weir length (linear ft)}}$$

19.7.3 TYPES OF SEDIMENTATION TANKS

Sedimentation equipment includes septic tanks, two-story tanks, and plain settling tanks or clarifiers. All three devices may be used for primary treatment, but plain settling tanks are normally used for secondary or advanced wastewater treatment processes.

19.7.3.1 Septic Tanks

Septic tanks are prefabricated tanks that serve as a combined settling and skimming tank and as an unheated, unmixed anaerobic digester. Septic tanks provide long settling times (6 to 8 hr or more) but do not separate decomposing solids from the wastewater flow. When the tank becomes full, solids will be discharged with the flow. The process is suitable for small facilities (e.g., schools, motels, homes), but, due to the long detention times and lack of control, it is not suitable for larger applications.

19.7.3.2 Two-Story (Imhoff) Tank

The two-story or Imhoff tank is similar to a septic tank with regard to the removal of settleable solids and the anaerobic digestion of solids. The difference is that the two-story tank consists of a settling compartment where sedimentation is accomplished, a lower compartment where settled solids and digestion takes place, and gas vents. Solids removed from the wastewater by settling pass from the settling compartment into the digestion compartment through a slot in the bottom of the settling compartment. The design of the slot prevents solids from returning to the settling compartment. Solids decompose anaerobically in the digestion section. Gases produced as a result of the solids decomposition are released through the gas vents running along each side of the settling compartment.

19.7.3.3 Plain Settling Tanks (Clarifiers)

The plain settling tank or clarifier optimizes the settling process. Sludge is removed from the tank for processing in other downstream treatment units. Flow enters the tank, is slowed and distributed evenly across the width and depth of the unit, passes through the unit, and leaves over the effluent weir. Detention time within the primary settling tank is from 1 to 3 hr (2 hr on average). Sludge removal is accomplished frequently on either a continuous or an intermittent basis. Continuous removal requires additional sludge treatment processes to remove the excess water resulting from the removal of sludge containing less than 2 to 3% solids. Intermittent sludge removal requires the sludge be pumped from the tank on a schedule frequent enough to prevent large clumps of solids rising to the surface but infrequent enough to obtain 4 to 8% solids in the sludge withdrawn.

Scum must be removed from the surface of the settling tank frequently. This is normally a mechanical process but may require manual start-up. The system should be operated frequently enough to prevent excessive buildup and scum carryover but not so frequent as to cause hydraulic overloading of the scum removal system. Settling tanks require housekeeping and maintenance. Baffles (which prevent floatable solids and scum from leaving the tank), scum troughs, scum collectors, effluent troughs, and effluent weirs require frequent cleaning to prevent heavy biological growths and solids accumulations. Mechanical equipment must be lubricated and maintained as specified in the manufacturer's recommendations or in accordance with procedures listed in the plant O&M manual.

Process control sampling and testing are used to evaluate the performance of the settling process. Settleable solids, dissolved oxygen, pH, temperature, total suspended solids, and BOD_5, as well as sludge solids and volatile matter testing, are routinely carried out.

19.7.4 OPERATOR OBSERVATIONS, PROBLEMS, AND TROUBLESHOOTING

Before identifying a primary treatment problem and proceeding with the appropriate troubleshooting effort, the operator must be cognizant of what constitutes normal operation (i.e., is there really a problem or is the system operating as per design?). Several important items of normal operation can have a strong impact on performance. In the following section, we discuss the important operational parameters and normal observations.

19.7.4.1 Primary Clarification: Normal Operation

In primary clarification, wastewater enters a settling tank or basin. Velocity reduces to approximately 1 ft/min.

Note: Notice that the velocity is based on minutes instead of seconds, as was the case in the grit channels. A grit channel velocity of 1 ft/sec would be 60 ft/min.

Solids heavier than water settle to the bottom while solids lighter than water float to the top. Settled solids are removed as sludge and floating solids are removed as scum. Wastewater leaves the sedimentation tank over an effluent weir and moves on to the next step in treatment. Detention time, temperature, tank design, and condition of the equipment control the efficiency of the process.

19.7.4.2 Primary Clarification: Operational Parameters (Normal Observations)

- *Flow distribution*—Normal flow distribution is indicated by flow to each in-service unit being equal and uniform with no indication of short-circuiting. The surface loading rate is within design specifications.
- *Weir condition*—Weirs are level, flow over the weir is uniform, and the weir overflow rate is within design specifications.
- *Scum removal*—The surface is free of scum accumulations; scum removal does not operate continuously.
- *Sludge removal*—No large clumps of sludge appear on the surface, the system operates as designed, the pumping rate is controlled to prevent coning or buildup, and the sludge blanket depth is within desired levels.
- *Performance*—The unit is removing expected levels of BOD_5, TSS, and settleable solids.
- *Unit maintenance*—Mechanical equipment is maintained in accordance with planned schedules; equipment is available for service as required.

To assist the operator in judging primary treatment operation, several process control tests can be used for process evaluation and control. These tests include the following:

- *pH*—Normal, 6.5 to 9.0
- *Dissolved oxygen*—Normal, <1.0 mg/L
- *Temperature*—Varies with climate and season
- *Settleable solids*—Influent, 5 to 15 mL/L; effluent, 0.3 to 5 mL/L
- *BOD_5*—Influent, 150 to 400 mg/L; effluent, 50 to 150 mg/L
- *% Solids*—Normal, 4 to 8%
- *% Volatile matter*—Normal, 40 to 70%
- *Heavy metals*—As required
- *Jar tests*—As required

Note: Testing frequency should be determined on the basis of the process influent and effluent variability and the available resources. All should be performed periodically to provide reference information for evaluation of performance.

19.7.5 Process Control Calculations

As with many other wastewater treatment plant unit processes, process control calculations aid in determining the performance of the sedimentation process. Process control calculations are used in the sedimentation process to determine:

- Percent removal
- Hydraulic detention time
- Surface loading rate (surface settling rate)
- Weir overflow rate (weir loading rate)
- Sludge pumping
- Percent total solids (%TS)

In the following sections we take a closer look at a few of these process control calculations and example problems.

Note: The calculations presented in the following sections allow us to determine values for each function performed. Keep in mind that an optimally operated primary clarifier should have values in an expected range.

19.7.5.1 Percent Removal

The expected ranges of percent removal for a primary clarifier are:

- Settleable solids, 90 to 95%
- Suspended solids, 40 to 60%
- BOD_5, 25 to 35%

19.7.5.2 Detention Time

The primary purpose of primary settling is to remove settleable solids. This is accomplished by slowing the flow down to approximately 1 ft/min. The flow at this velocity will stay in the primary tank from 1.5 to 2.5 hr. The length of time the water stays in the tank is called the *hydraulic detention time*.

19.7.5.3 Surface Overflow Rate (Surface Settling Rate, Surface Loading Rate)

Surface loading rate is the number of gallons of wastewater passing over 1 ft² of tank per day. This can be used to compare actual conditions with design. Plant designs generally use a surface loading rate of 300 to 1200 gpd/ft². Other terms used synonymously with surface loading rate include *surface settling rate* and *surface loading rate*.

$$\text{Surface overflow rate (gpd/ft}^2) = \frac{\text{Flow (gpd)}}{\text{Settling tank area (ft}^2)} \quad (19.20)$$

■ **EXAMPLE 19.22**

Problem: The settling tank is 120 ft in diameter and flow to the unit is 4.5 MGD. What is the surface loading rate in gpd/ft²?

Solution:

$$\text{Surface loading rate} = \frac{4.5 \text{ MGD} \times 1,000,000 \text{ gal/MGD}}{0.785 \times 120 \text{ ft} \times 120 \text{ ft}}$$

$$= 398 \text{ gpd/ft}^2$$

■ **EXAMPLE 19.23**

Problem: A circular clarifier has a diameter of 50 ft. If the primary effluent flow is 2,150,000 gpd, what is the surface overflow rate in gpd/ft²?

Solution:

$$\text{Area} = 0.785 \times 50 \text{ ft} \times 50 \text{ ft}$$

$$\text{Surface overflow rate} = \frac{\text{Flow (gpd)}}{\text{Area (ft}^2)}$$

$$= \frac{2,150,000}{0.785 \times 50 \text{ ft} \times 50 \text{ ft}} = 1096 \text{ gpd/ft}^2$$

19.7.5.4 Weir Overflow Rate (Weir Loading Rate)

The weir overflow rate or weir loading rate is the amount of water leaving the settling tank per linear foot of weir. The result of this calculation can be compared with design. Normally, weir overflow rates of 10,000 to 20,000 gpd/ft are used in the design of a settling tank.

$$\text{Weir overflow rate (gpd/ft)} = \frac{\text{Flow (gpd)}}{\text{Weir length (ft)}} \quad (19.21)$$

■ **EXAMPLE 19.24**

Problem: The circular settling tank is 90 ft in diameter and has a weir along its circumference. The effluent flow rate is 2.55 MGD. What is the weir overflow rate (gpd/ft)?

Solution:

$$\text{Weir overflow (gpd/ft)} = \frac{2.55 \text{ MGD} \times 1,000,000 \text{ gal/MG}}{3.14 \times 90 \text{ ft}}$$

$$= 9023 \text{ gpd/ft}$$

19.7.5.5 Sludge Pumping

Determination of sludge pumping (the quantity of solids and volatile solids removed from the sedimentation tank)

provides accurate information needed for process control of the sedimentation process.

$$\text{Solids pumped (lb/day)} = \begin{bmatrix} \text{Pump rate} \times \text{pump time} \\ \times 8.34 \text{ lb/gal} \times \% \text{ solids} \end{bmatrix} \quad (19.22)$$

$$\text{Volatile solids (lb/day)} = \begin{bmatrix} \text{Pump rate} \times \text{pump time} \\ \times 8.34 \times \% \text{ solids} \\ \times \% \text{ volatile solids} \end{bmatrix} \quad (19.23)$$

■ **EXAMPLE 19.25**

Problem: The sludge pump operates 20 min/hr. The pump delivers 20 gal/min of sludge. Laboratory tests indicate that the sludge is 5.2% solids and 66% volatile matter. How many pounds of volatile matter are transferred from the settling tank to the digester?

Solution:

Pump time = 20 min/hr
Pump rate = 20 gpm
% Solids = 5.2%
% Volatile matter = 66%

$$\text{Volatile solids (lb/day)} = \begin{bmatrix} 20 \text{ gpm} \times (20 \text{ min/hr} \times 24 \text{ hr/day}) \\ \times 8.34 \text{ lb/gal} \times 0.052 \times 0.66 \end{bmatrix}$$

$$= 2748 \text{ lb/day}$$

19.7.5.6 Percent Total Solids (%TS)

■ **EXAMPLE 19.26**

Problem: A settling tank sludge sample is tested for solids. The sample and dish weighed 74.69 g. The dish alone weighed 21.2 g. After drying, the dish with dry solids weighed 22.3 g. What is the percent total solids (%TS) of the sample?

Solution:

Sample + Dish	74.69 g	Dish + Dry solids	22.3 g
Dish alone	−21.20 g	Dish alone	−21.2 g
Sample weight	53.49 g	Dry solids weight	1.1 g

$$\frac{1.1g}{53.49g} \times 100\% = 2\%$$

19.7.5.7 BOD and SS Removal

To calculate the pounds of BOD or suspended solids removed each day, we need to know the mg/L BOD or SS removed and the plant flow. Then, we can use the mg/L to lb/day equation:

$$\text{SS Removed} = \text{mg/L} \times \text{MGD} \times 8.34 \text{ lb/gal} \quad (19.24)$$

■ **EXAMPLE 19.27**

Problem: If 120 mg/L suspended solids are removed by a primary clarifier, how many lb/day suspended solids are removed when the flow is 6,230,000 gpd?

Solution:

$$\text{SS removed} = 120 \text{ mg/L} \times 6.25 \text{ MGD} \times 8.34 \text{ lb/gal}$$
$$= 6255 \text{ lb/day}$$

■ **EXAMPLE 19.28**

Problem: The flow to a secondary clarifier is 1.6 MGD. If the influent BOD concentration is 200 mg/L and the effluent BOD concentration is 70 mg/L, how many pounds of BOD are removed daily?

Solution:

$$\text{BOD removed (lb/day)} = 200 \text{ mg/L} - 70 \text{ mg/L} = 130 \text{ mg/L}$$

After calculating the mg/L BOD removed, calculate the lb/day BOD removed:

$$\text{BOD removed (lb/day)} = \begin{bmatrix} 130 \text{ mg/L} \times 1.6 \text{ MGD} \\ \times 8.34 \text{ lb/gal} \end{bmatrix}$$
$$= 1735 \text{ lb/day}$$

19.7.6 PROBLEM ANALYSIS

In primary treatment (as is also clear in the operation of other unit processes), the primary function of the operator is to identify causes of process malfunctions, develop solutions, and prevent recurrence. In other words, the operator's goal is to perform problem analysis or troubleshooting on unit processes when required and to restore the unit processes to optimal operating condition. Obviously, the immediate goal in problem analysis is to solve the immediate problem. The long-term goal is to ensure that the problem does not pop up again, causing poor performance in the future. In this section, we cover a few indicators or observations of operational problems with the primary treatment process. The observations presented are not all inclusive but highlight the most frequently confronted problems.

19.7.6.1 Causal Factors for Poor Suspended Solids Removal (Primary Clarifier)

- Hydraulic overload
- Sludge buildup in tanks and decreased volume allowing solids to scour out tanks
- Strong recycle flows
- Industrial waste concentrations

- Wind currents
- Temperature currents

19.7.6.2 Causal Factors for Floating Sludge

- Sludge becoming septic in tank
- Damaged or worn collection equipment
- Recycled waste sludge
- Primary sludge pumps malfunctions
- Sludge withdrawal line plugged
- Return of well-nitrified waste activated sludge
- Too few tanks in service
- Damaged or missing baffles

19.7.6.3 Causal Factors for Septic Wastewater or Sludge

- Damaged or worn collection equipment
- Infrequent sludge removal
- Insufficient industrial pretreatment
- Septic sewage from collection system
- Strong recycle flows
- Primary sludge pump malfunction
- Sludge withdrawal line plugged
- Sludge collectors not run often enough
- Septage dumpers

19.7.6.4 Causal Factors for Too Low Primary Sludge Solids Concentrations

- Hydraulic overload
- Overpumping of sludge
- Collection system problem
- Decreased influent solids loading

19.7.6.5 Causal Factors for Too High Primary Sludge Solids Concentrations

- Excessive grit and compacted material
- Primary sludge pump malfunction
- Sludge withdrawal line plugged
- Sludge retention time too long
- Increased influent loadings

19.7.7 EFFLUENT FROM SETTLING TANKS

Upon completion of screening, degritting, and settling in sedimentation basins, large debris, grit, and many settleable materials have been removed from the wastestream. What is left is referred to as *primary effluent*. Usually cloudy and frequently gray in color, primary effluent still contains large amounts of dissolved food and other chemicals (nutrients). These nutrients are treated in the next step in the treatment process (secondary treatment) which is discussed in the next section.

Note: Two of the most important nutrients left to remove are phosphorus and ammonia. Although we want to remove these two nutrients from the wastestream, we do not want to remove two much. Carbonaceous microorganisms in secondary treatment (biological treatment) require both phosphorus and ammonia.

19.8 SECONDARY TREATMENT

The main purpose of *secondary treatment* (sometimes referred to as *biological treatment*) is to provide biochemical oxygen demand (BOD) removal beyond what is achievable by primary treatment. Three commonly used approaches all take advantage of the ability of microorganisms to convert organic wastes (via biological treatment) into stabilized, low-energy compounds. Two of these approaches, the *trickling filter* or its variation, the *rotating biological contactor* (RBC), and the *activated sludge process* sequentially follow normal primary treatment. The third approach, *ponds* (oxidation ponds or lagoons), can provide equivalent results without preliminary treatment. In this section, we present a brief overview of the secondary treatment process followed by a detailed discussion of wastewater treatment ponds (used primarily in smaller treatment plants), trickling filters, and RBCs. We then shift focus to the activated sludge process—the secondary treatment process used primarily in large installations and which is the main focus of this handbook.

Secondary treatment refers to those treatment processes that use biological processes to convert dissolved, suspended, and colloidal organic wastes to more stable solids that can be either removed by settling or discharged to the environment without causing harm. Exactly what is secondary treatment? As defined by the Clean Water Act (CWA), secondary treatment produces an effluent with not more than 30 mg/L BOD_5 and 30 mg/L total suspended solids.

Note: The CWA also states that ponds and trickling filters will be included in the definition of secondary treatment even if they do not meet the effluent quality requirements continuously.

Most secondary treatment processes decompose solids aerobically producing carbon dioxide, stable solids, and more organisms. Because solids are produced, all of the biological processes must include some form of solids removal (e.g., settling tank, filter). Secondary treatment processes can be separated into two large categories: fixed-film systems and suspended growth systems.

Fixed-film systems are processes that use a biological growth (biomass or slime) attached to some form of media. Wastewater passes over or around the media and the slime. When the wastewater and slime are in contact, the organisms remove and oxidize the organic solids. The media may be stone, redwood, synthetic materials, or any other substance that is durable (capable of withstanding weather conditions for many years), provides a large area for slime growth while providing open space for ventilation, and is not toxic to the organisms in the biomass. Fixed-film devices include trickling filters and rotating biological contactors. *Suspended growth systems* are processes that use a biological growth mixed with the wastewater. Typical suspended growth systems consist of various modifications of the activated sludge process.

19.8.1 TREATMENT PONDS

Wastewater treatment can be accomplished using *ponds*. Ponds are relatively easy to build and manage, they accommodate large fluctuations in flow, and they can also provide treatment that approaches the effectiveness of conventional systems (producing a highly purified effluent) at a much lower cost. It is the cost factor that drives many managers to decide on the pond option. The actual degree of treatment provided depends on the type and number of ponds used. Ponds can be used as the sole type of treatment or they can be used in conjunction with other forms of wastewater treatment; that is, other treatment processes can be followed by a pond or a pond can be followed by other treatment processes.

19.8.1.1 Types of Ponds

Ponds can be classified based on their location in the system, by the type of wastes they receive, and by the main biological process occurring in the pond. First, we will take a look at the types of ponds according to their location and the type wastes they receive: *raw sewage stabilization ponds* (see Figure 19.3), *oxidation ponds*, and *polishing ponds*. Then, we will look at ponds classified by the type of processes occurring within the pond: *aerobic ponds*, *anaerobic ponds*, *facultative ponds*, and *aerated ponds*.

19.8.1.2 Ponds Based on Location and Types of Wastes They Receive

19.8.1.2.1 Raw Sewage Stabilization Pond

The raw sewage stabilization pond is the most common type of pond (see Figure 19.3). With the exception of screening and shredding, this type of pond receives no prior treatment. Generally, raw sewage stabilization ponds are designed to provide a minimum of 45 days of detention time and to receive no more than 30 lb of BOD_5 per day per acre. The quality of the discharge is dependent on the time of the year. Summer months produce high BOD_5 removal and excellent suspended solids removals.

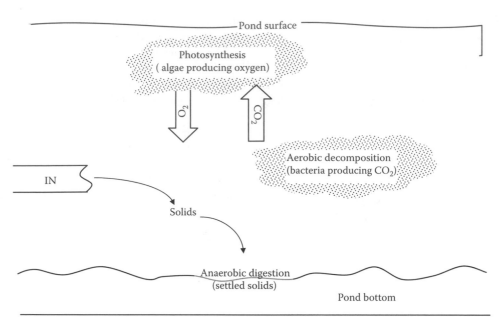

FIGURE 19.3 Stabilization pond processes.

The pond consists of an influent structure, a pond berm or walls, and an effluent structure designed to permit selection of the best quality effluent. Normal operating depth of the pond is 3 to 5 feet. The process occurring in the pond involves bacteria decomposing the organics in the wastewater (aerobically and anaerobically) and algae using the products of the bacterial action to produce oxygen (photosynthesis). Because this type of pond is commonly used in wastewater treatment, the process occurring within the pond is described in greater detail in the following text.

When wastewater enters the stabilization pond, several processes begin to occur. These include settling, aerobic decomposition, anaerobic decomposition, and photosynthesis (see Figure 19.3). Solids in the wastewater will settle to the bottom of the pond. In addition to the solids in the wastewater entering the pond, solids that are produced by the biological activity will also settle to the bottom. Eventually this will reduce the detention time and the performance of the pond. When this occurs (normally 20 to 30 years), the pond will have to be replaced or cleaned.

Bacteria and other microorganisms use the organic matter as a food source. They use oxygen (aerobic decomposition), organic matter, and nutrients to produce carbon dioxide, water, and stable solids, which may settle out, as well as more organisms. The carbon dioxide is an essential component of the photosynthesis process occurring near the surface of the pond. Organisms also use the solids that settle out as food material; however, the oxygen levels at the bottom of the pond are extremely low so the process used is anaerobic decomposition. The organisms use the organic matter to produce gases (hydrogen sulfide, methane, etc.) dissolved in the water, stable solids, and more

organisms. Near the surface of the pond a population of green algae develops that can use the carbon dioxide produced by the bacterial population, nutrients, and sunlight to produce more algae and oxygen, which is dissolved into the water. The dissolved oxygen is then used by organisms in the aerobic decomposition process.

When compared with other wastewater treatment systems involving biological treatment, a stabilization pond treatment system is the simplest to operate and maintain. Operation and maintenance activities include collecting and testing samples for dissolved oxygen and pH, removing weeds and other debris (scum) from the pond, mowing the berms, repairing erosion, and removing burrowing animals.

Note: Dissolved oxygen and pH levels in the pond will vary throughout the day. Normal operation will result in very high DO and pH levels due to the natural processes occurring.

Note: When operating properly, the stabilization pond will exhibit a wide variation in both dissolved oxygen and pH. This is due to photosynthesis occurring in the system.

19.8.1.2.2 Oxidation Pond

An oxidation pond, which is normally designed using the same criteria as the stabilization pond, receives flows that have passed through a stabilization pond or primary settling tank. This type of pond provides biological treatment, additional settling, and some reduction in the number of fecal coliform present.

19.8.1.2.3 Polishing Pond

A polishing pond, which uses the same equipment as a stabilization pond, receives flow from an oxidation pond or from other secondary treatment systems. Polishing ponds remove additional BOD_5, solids, fecal coliform, and some nutrients. They are designed to provide 1 to 3 days of detention time and normally operate at a depth of 5 to 10 ft. Excessive detention time or too shallow a depth will result in algae growth, which increases influent suspended solids concentrations.

19.8.1.3 Ponds Based on the Type of Processes Occurring Within

The type of processes occurring within the pond may also classify ponds. These include the aerobic, anaerobic, facultative, and aerated processes.

19.8.1.3.1 Aerobic Ponds

In aerobic ponds, which are not widely used, oxygen is present throughout the pond. All biological activity is aerobic decomposition.

19.8.1.3.2 Anaerobic Ponds

Anaerobic ponds are normally used to treat high-strength industrial wastes. No oxygen is present in the pond, and all biological activity is anaerobic decomposition.

19.8.1.3.3 Facultative Ponds

The facultative pond is the most common type of pond (based on the processes occurring). Oxygen is present in the upper portions of the pond, and aerobic processes are occurring. No oxygen is present in lower levels of the pond, where the processes occurring are anoxic and anaerobic.

19.8.1.3.4 Aerated Ponds

In the aerated pond, oxygen is provided through the use of mechanical or diffused air systems. When aeration is used, the depth of the pond and the acceptable loading levels may increase. Mechanical or diffused aeration is often used to supplement natural oxygen production or to replace it.

19.8.1.4 Process Control Calculations for Stabilization Ponds

Process control calculations are an important part of wastewater treatment operations, including pond operations. More significantly, process control calculations are an important part of state wastewater licensing examinations—it is simply not possible to master the licensing examinations without being able to perform the required calculations. Thus, as with previous sections (and with sections to follow), whenever possible, example process control problems are provided to enhance the reader's knowledge and skill.

19.8.1.4.1 Determining Pond Area in Acres

$$\text{Area (ac)} = \frac{\text{Area (ft}^2)}{43{,}560 \text{ ft}^2/\text{ac}} \qquad (19.25)$$

19.8.1.4.2 Determining Pond Volume in Ac-Ft

$$\text{Volume (ac-ft)} = \frac{\text{Volume (ft}^3)}{43{,}560 \text{ ft}^3/\text{ac-ft}} \qquad (19.26)$$

19.8.1.4.3 Determining Flow Rate in Ac-Ft/Day

$$\text{Flow (ac-ft/day)} = \text{Flow (MGD)} \times 3.069 \text{ ac-ft/MG} \quad (19.27)$$

Note: Acre-feet (ac-ft) is a unit that can cause confusion, especially for those not familiar with pond or lagoon operations. 1 ac-ft is the volume of a box with a 1-ac top and depth of 1 ft—but the top does not have to be an even number of acres in size to use acre-feet.

19.8.1.4.4 Determining Flow Rate in Ac-In./Day

$$\text{Flow (ac-in./day)} = \text{Flow (MGD)} \times 36.8 \text{ ac-in./MG} \quad (19.28)$$

19.8.1.4.5 Hydraulic Detention Time in Days

$$\text{Detention time (days)} = \frac{\text{Pond volume (ac-ft)}}{\text{Influent flow (ac-ft/day)}} \quad (19.29)$$

Note: Normally, hydraulic detention time ranges from 30 to 120 days for stabilization ponds.

■ EXAMPLE 19.29

Problem: A stabilization pond has a volume of 53.5 ac-ft. What is the detention time in days when the flow is 0.30 MGD?

Solution:

$$\text{Flow} = 0.30 \text{ MGD} \times 3.069 = 0.92 \text{ ac-ft/day}$$

$$\text{Detention time} = \frac{53.5 \text{ ac}}{0.92 \text{ ac-ft/day}} = 58.2 \text{ days}$$

19.8.1.4.6 Hydraulic Loading in In./Day (Overflow Rate)

$$\begin{array}{l} \text{Hydraulic loading} \\ \text{(in./day)} \end{array} = \frac{\text{Influent flow (ac-in./day)}}{\text{Pond area (ac)}} \quad (19.30)$$

$$\begin{array}{l} \text{Population loading} \\ \text{(people/ac/day)} \end{array} = \frac{\begin{array}{c} \text{Population that} \\ \text{system serves (people)} \end{array}}{\text{Pond area (ac)}} \quad (19.31)$$

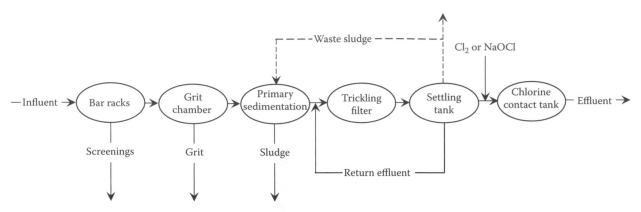

FIGURE 19.4 Simplified flow diagram of trickling filter used for wastewater treatment.

Note: Population loading normally ranges from 50 to 500 people per acre.

19.8.1.4.7 Organic Loading

Organic loading can be expressed as pounds of BOD_5 per acre per day (most common), pounds BOD_5 per acre-foot per day, or people per acre per day.

$$\frac{\text{Organic loading}}{(\text{lb } BOD_5\text{ /ac/day})} = \frac{\begin{array}{c}BOD_5 \text{ (mg/L)}\\ \times \text{influent flow (MGD)}\\ \times 8.34\end{array}}{\text{Pond area (ac)}} \quad (19.32)$$

Note: Normal range is 10 to 50 lb BOD_5 per day per acre.

■ EXAMPLE 19.30

Problem: A wastewater treatment pond has an average width of 380 ft and an average length of 725 ft. The influent flow rate to the pond is 0.12 MGD with a BOD concentration of 160 mg/L. What is the organic loading rate to the pond in pounds per day per acre (lb/day/ac)?

Solution:

$$725 \text{ ft} \times 380 \text{ ft} \times \frac{1 \text{ ac}}{43,560 \text{ ft}^2} = 6.32 \text{ ac}$$

$$0.12 \text{ MGD} \times 160 \text{ mg/L} \times 8.34 \text{ lb/gal} = 160.1 \text{ lb/day}$$

$$\frac{160.1 \text{ lb/day}}{6.32 \text{ ac}} = 25.3 \text{ lb/day/ac}$$

19.8.2 TRICKLING FILTERS

Trickling filters have been used to treat wastewater since the 1890s. It was found that if settled wastewater was passed over rock surfaces, slime grew on the rocks and the water became cleaner. Today we still use this principle, but, in many installations, instead of rocks we use plastic media. In most wastewater treatment systems, the *trickling filter* follows primary treatment and includes a secondary settling tank or clarifier as shown in Figure 19.4. Trickling filters are widely used for the treatment of domestic and industrial wastes. The process is a fixed-film biological treatment method designed to remove BOD_5 and suspended solids.

A trickling filter consists of a rotating distribution arm that sprays and evenly distributes liquid wastewater over a circular bed of fist-sized rocks, other coarse materials, or synthetic media (see Figure 19.5). The spaces between the media allow air to circulate easily so aerobic conditions can be maintained. The spaces also allow wastewater to trickle down through, around, and over the media. A layer of biological slime that absorbs and consumes the wastes trickling through the bed covers the media material. The organisms aerobically decompose the solids, producing more organisms and stable wastes, which either become part of the slime or are discharged back into the wastewater flowing over the media. This slime consists mainly of bacteria, but it may also include algae, protozoa, worms, snails, fungi, and insect larvae. The accumulating slime occasionally sloughs off (*sloughings*) individual media materials (see Figure 19.6); it is collected at the bottom of the filter, along with the treated wastewater, and is passed on to the secondary settling tank where it is removed. The overall performance of the trickling filter is dependent on hydraulic and organic loading, temperature, and recirculation.

19.8.2.1 Trickling Filter Definitions

To clearly understand the correct operation of the trickling filter, the operator must be familiar with certain terms.

Note: The following list of terms applies to the trickling filter process. We assume that other terms related to other units within the treatment system (plant) are already familiar to operators.

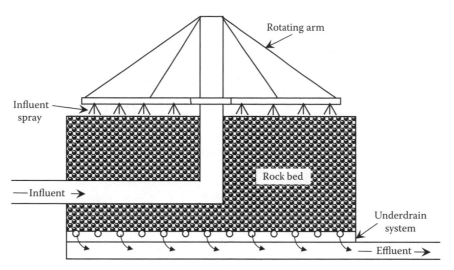

FIGURE 19.5 Schematic of cross-section of a trickling filter.

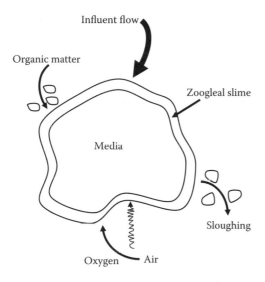

FIGURE 19.6 Filter media showing biological activities that take place on the surface area.

Biological towers—A type of trickling filter that is very deep (10 to 20 ft). Filled with lightweight synthetic media, these towers are also know as *oxidation towers* or *roughing towers* or (because of their extremely high hydraulic loading) *super-rate trickling filters*.

Biomass—The total mass of organisms attached to the media. Similar to the solids inventory in the activated sludge process, it is sometimes referred to as *zoogleal slime*.

Distribution arm—The device most widely used to apply wastewater evenly over the entire surface of the media. In most cases, the force of the wastewater being sprayed through the orifices moves the arm.

Filter underdrain—The open space provided under the media to collect the liquid (wastewater and slough-ings) and to allow air to enter the filter. It has a sloped floor to collect the flow to a central channel for removal.

High-rate trickling filters—A classification (see Table 19.4) in which the organic loading is in the range of 25 to 100 lb of BOD_5 per 1000 ft^3 of media per day. The standard rate filter may also produce a highly nitrified effluent.

Hydraulic loading—The amount of wastewater flow applied to the surface of the trickling filter media. It can be expressed in several ways: flow per square foot of surface per day (gpd/ft^2), flow per acre per day (MGAD), or flow per acre foot per day (MGAFD). The hydraulic loading includes all flow entering the filter.

Media—An inert substance placed in the filter to provide a surface for the microorganism to grow on. The media can be filed stone, crushed stone, slag, plastic, or redwood slats.

Organic loading—The amount of BOD_5 or chemical oxygen demand (COD) applied to a given volume of filter media. It does not include the BOD_5 or COD contributed to any recirculated flow and is commonly expressed as pounds of BOD_5 or COD per 1000 ft^3 of media.

Recirculation—The return of filter effluent back to the head of the trickling filter. It can level flow variations and assist in solving operational problems, such as ponding, filter flies, and odors.

Roughing filters—A classification of trickling filters (see Table 19.4) in which the organic loading is in excess of 200 lb of BOD_5 per 1000 ft^3 of media per day. A roughing filter is used to reduce the loading on other biological treatment processes to produce an industrial discharge that can be safely treated in a municipal treatment facility.

TABLE 19.4
Trickling Filter Classification

Filter Class	Standard	Intermediate	High Rate	Super High Rate	Roughing
Hydraulic loading (gpd/ft²)	25–90	90–230	230–900	350–2100	>900
Organic loading (lb BOD per 1000 ft³)	5–25	15–30	25–300	Up to 300	>300
Sloughing frequency	Seasonal	Varies	Continuous	Continuous	Continuous
Distribution	Rotary	Rotary Fixed	Rotary Fixed	Rotary	Rotary fixed
Recirculation	No	Usually	Always	Usually	Not usually
Media depth (ft)	6–8	6–8	3–8	Up to 40	3–20
Media type	Rock	Rock	Rock	Rock	Rock
	Plastic	Plastic	Plastic	Plastic	—
	Wood	Wood	Wood	Wood	—
Nitrification	Yes	Some	Some	Limited	None
Filter flies	Yes	Variable	Variable	Very few	Not usually
BOD removal	80–85%	50–70%	65–80%	65–85%	40–65%
TSS removal	80–85%	50–70%	65–80%	65–85%	40–65%

Sloughing—The process in which the excess growths break away from the media and washes through the filter to the underdrains with the wastewater. These sloughings must be removed from the flow by settling.

Staging—The practice of operating two or more trickling filters in series. The effluent of one filter is used as the influent of the next. This practice can produce a higher quality effluent by removing additional BOD₅ or COD.

19.8.2.2 Trickling Filter Equipment

The trickling filter distribution system is designed to spread wastewater evenly over the surface of the media. The most common system is the *rotary distributor*, which moves above the surface of the media and sprays the wastewater on the surface. The force of the water leaving the orifices drives the rotary system. The distributor arms usually have small plates below each orifice to spread the wastewater into a fan-shaped distribution system. The second type of distributor is the *fixed-nozzle system*. In this system, the nozzles are fixed in place above the media and are designed to spray the wastewater over a fixed portion of the media. This system is used frequently with deep-bed synthetic media filters.

Note: Trickling filters that use ordinary rock are normally only about 10 ft in depth because of structural problems caused by the weight of rocks, which also requires the construction of beds that are quite wide—in many applications, up to 60 ft in diameter. When synthetic media are used, the bed can be much deeper.

No matter what type of media are selected, the primary consideration is that it must be capable of providing the desired film location for the development of the biomass. Depending on the type of media used and the filter classification, the media may be 3 to 20 ft or more in depth.

The underdrains are designed to support the media, collect the wastewater and sloughings, and carry them out of the filter and to provide ventilation to the filter.

Note: To ensure sufficient airflow to the filter, the underdrains should never be allowed to flow more than 50% full of wastewater.

The effluent channel is designed to carry the flow from the trickling filter to the secondary settling tank. The secondary settling tank provides 2 to 4 hr of detention time to separate the sloughing materials from the treated wastewater. The design, construction, and operation are similar to those for the primary settling tank. Longer detention times are provided because the sloughing materials are lighter and settle more slowly.

Recirculation pumps and piping are designed to recirculate (thus improving the performance of the trickling filter or settling tank) a portion of the effluent back to be mixed with the filter influent. When recirculation is used, pumps and metering devices must be provided.

19.8.2.3 Filter Classifications

Trickling filters are classified by hydraulic and organic loading. Moreover, the expected performance and the construction of the trickling filter are determined by the filter classification. Filter classifications include standard rate, intermediate rate, high rate, super high rate (plastic media), and roughing. Standard-rate, high-rate, and roughing filters are the ones most commonly used. The *standard-rate filter* has a hydraulic loading that varies from 25 to 90 gpd/ft³. It has a seasonal sloughing frequency and does not employ recirculation. It typically has an 80

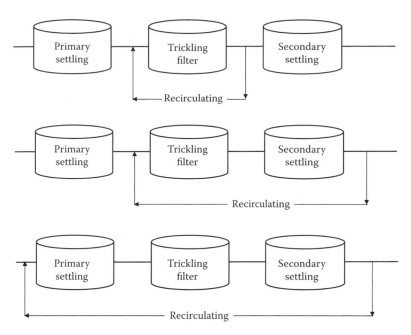

FIGURE 19.7 Common forms of recirculation.

to 85% BOD_5 removal rate and 80 to 85% TSS removal rate. The *high-rate filter* has a hydraulic loading of 230 to 900 gpd/ft³. It has a continuous sloughing frequency and always employs recirculation. It typically has a 65 to 80% BOD_5 removal rate and 65 to 80% TSS removal rate. The *roughing filter* has a hydraulic loading of >900 gpd/ft³. It has a continuous sloughing frequency and does not normally include recirculation. It typically has a 40 to 65% removal rate and 40 to 65% TSS removal rate.

19.8.2.4 Standard Operating Procedures

Standard operating procedures for trickling filters include sampling and testing, observation, recirculation, maintenance, and expectations of performance. Collection of influent and process effluent samples to determine performance and to monitor the process condition of trickling filters is required. Dissolved oxygen, pH, and settleable solids testing should be performed daily. BOD_5 and suspended solids testing should be done as often as practical to determine the percent removal.

The operation and condition of the filter should be observed daily. Items to observe include the distributor movement, uniformity of distribution, evidence of operation or mechanical problems, and the presence of objectionable odors. In addition, normal observation for a settling tank should also be performed.

Recirculation is used to reduce organic loading, improve sloughing, reduce odors, and reduce or eliminate filter fly or ponding problems. The amount of recirculation is dependent on the design of the treatment plant and the operational requirements of the process. Recirculation

flow may be expressed as a specific flow rate (e.g., 2 MGD). In most cases, it is expressed as a ratio (e.g., 3:1, 0.5:1). The recirculation is always listed as the first number and the influent flow is listed as the second number. Because the second number in the ratio is always 1, the 1 is sometimes dropped, and the ratio is written as a single number.

Flows can be recirculated from various points following the filter to various points before the filter. The most common form of recirculation removes flow from the filter effluent or settling tank and returns it to the influent of the trickling filter as shown in Figure 19.7.

Maintenance requirements include lubrication of mechanical equipment, removal of debris from the surface and orifices, and adjustment of flow patterns and maintenance associated with the settling tank.

19.8.2.5 General Process Description

The trickling filter process involves spraying wastewater over solid media such as rock, plastic, or redwood slats (or laths). As the wastewater trickles over the surface of the media, a growth of microorganisms (bacteria, protozoa, fungi, algae, helminthes or worms, and larvae) develops. This growth is visible as a shiny slime very similar to the slime found on rocks in a stream. As the wastewater passes over this slime, the slime adsorbs the organic matter. This organic matter is used for food by the microorganisms. At the same time, air moving through the open spaces in the filter transfers oxygen to the wastewater. This oxygen is then transferred to the slime to keep the outer layer aerobic. As the microorganisms use the food and oxygen, they produce more organisms, carbon dioxide, sulfates, nitrates, and other stable byproducts; these materials are then discarded from the slime back into the wastewater flow and are carried out of the filter:

$$\text{Organics} + \text{organisms} + O_2 = \begin{bmatrix} \text{More organisms} \\ + CO_2 + \text{solid wastes} \end{bmatrix} \quad (19.33)$$

The growth of the microorganisms and the buildup of solid wastes in the slime make it thicker and heavier. When this slime becomes too thick, the wastewater flow breaks off parts of the slime. These must be removed in the final settling tank. In some trickling filters, a portion of the filter effluent is returned to the head of the trickling filter to level out variations in flow and improves operations (recirculation).

19.8.2.6　Overview and Brief Summary of Trickling Filter Process

A trickling filter consists of a bed of coarse media, usually rocks or plastic, covered with microorganisms.

Note: Trickling filters that use ordinary rock are normally only about 10 ft in depth because of structural problems caused by the weight of rocks, which also requires the construction of beds that are quite wide—in many applications, up to 60 ft in diameter. When synthetic media are used, the bed can be much deeper.

- The wastewater is applied to the media at a controlled rate, using a rotating distributor arm or fixed nozzles. Organic material is removed by contact with the microorganisms as the wastewater trickles down through the media openings. The treated wastewater is collected by an underdrain system.

Note: To ensure sufficient air flow to the filter, the underdrains should never be allowed to flow more than 50% full of wastewater.

- The trickling filter is usually built into a tank that contains the media. The filter may be square, rectangular, or circular.
- The trickling filter does not provide any actual filtration. The filter media provide a large amount of surface area that the microorganisms can cling to and grow in a slime that forms on the media as they feed on the organic material in the wastewater.
- The slime growth on the trickling filter media periodically sloughs off and is settled and removed in a secondary clarifier that follows the filter.

- Key factors in trickling filter operation include the following concepts:
  1. *Hydraulic loading rate*:

  $$\frac{\text{gal/day}}{\text{ft}^2} = \frac{\text{Flow (gal/day) (incl. recirculation)}}{\text{Media top surface (ft}^2)}$$

  2. *Organic loading rate*:

  $$\frac{\text{lb/day}}{1000\ \text{ft}^3} = \frac{\text{BOD in filter (lb/day)}}{\text{Media volume (1000 ft}^3)}$$

  3. *Recirculation*:

  $$\text{Recirculation ratio} = \frac{\text{Recirculation flow (MGD)}}{\text{Average influent flow (MGD)}}$$

19.8.2.7　Operator Observations

Trickling filter operation requires routine observation, meter readings, process control sampling and testing, and process control calculations. Comparison of daily results with expected "normal" ranges is the key to identifying problems and appropriate corrective actions.

- *Slime*—The operator checks the thickness of slime to ensure that it is thin and uniform (normal) or thick and heavy (indicates organic overload); the operator is also concerned with ensuring that excessive recirculation is not taking place and checks slime toxicity (if any). The operator is also concerned about the color of the slime. Green slime is normal, dark green/black slime indicates organic overload, and other colors may indicate industrial waste or chemical additive contamination. The operator should check the subsurface growth of the slime to ensure that it is normal (thin and translucent). If the growth is thick and dark, organic overload conditions are indicated. Distribution arm operation is a system function important to slime formation, and it must be checked regularly for proper operation; for example, the distribution of slime should be even and uniform. Striped conditions indicate clogged orifices or nozzles.
- *Flow*—Flow distribution must be checked to ensure uniformity. If nonuniform, the arms are not level or the orifices are plugged. Flow drainage is also important. Drainage should be uniform and rapid. If not, ponding may occur from media breakdown or debris on surface.
- *Distributor*—Movement of the distributor is critical to proper operation of the trickling filter. Movement should be uniform and smooth. Chattering, noisy operation may indicate bearing

failure. The distributor seal must be checked to ensure that there is no leakage.

- *Recirculation*—The operator must check the rate of recirculation to ensure that it is within design specifications. Rates above design specifications indicate hydraulic overloading; rates under design specifications indicate hydraulic underloading.

Note: Recirculation is used to reduce organic loading, improve sloughing, reduce odors, and reduce or eliminate filter fly or ponding problems. The amount of recirculation is dependent on the design of the treatment plant and the operational requirements of the process. Recirculation flow may be expressed as a specific flow rate (e.g., 2 MGD). In most cases, it is expressed as a ratio (e.g., 3:1, 0.5:1). The recirculation is always listed as the first number and the influent flow listed as the second number. Because the second number in the ratio is always 1, the 1 is sometimes dropped, and the ratio is written as a single number.

- *Media*—The operator should check to ensure that the medium is uniform.

19.8.2.8 Process Control Sampling and Testing

To ensure proper operation of the trickling filter, sampling and scheduling are important; however, for samples and the tests derived from the samples to be beneficial, operators must perform a variety of daily or variable tests. Individual tests and sampling may be required daily, weekly, or monthly, depending on seasonal change. The frequency of testing may be lower during normal operations and higher during abnormal conditions. The information gathered through collection and analysis of samples from various points in the trickling filter process is helpful in determining the current status of the process as well as identifying and correcting operational problems. The following routine sampling points and types of tests will permit the operator to identify normal and abnormal operating conditions.

Filter influent tests

- Dissolved oxygen
- pH
- Temperature
- Settleable solids
- BOD_5
- Suspended solids
- Metals

Recirculated flow tests

- Dissolved oxygen
- pH
- Flow rate
- Temperature

Filter effluent tests

- Dissolved oxygen
- pH
- Jar tests

Process effluent tests

- Dissolved oxygen
- pH
- Settleable solids
- BOD_5
- Suspended solids

19.8.2.9 Troubleshooting Operational Problems

(Much of the information included in this section is based on the USEPA's *Field Manual for Performance Evaluation and Troubleshooting at Municipal Wastewater Treatment Facilities*; see Culp and Heim, 1978.) The following sections are not intended to be all inclusive; that is, they do not cover all of the operational problems associated with the trickling filter process. They do, however, provide information on the most common operational problems.

19.8.2.9.1 Ponding

Symptoms:

- Small pools or puddles of water on the surface of the media
- Decreased performance in the removal of BOD and TSS
- Possible odors due to anaerobic conditions in the media
- Poor air flow through the media

Causal factors:

- Inadequate hydraulic loading to keep the media voids flushed clear
- Application of high-strength wastes without sufficient recirculation to provide dilution
- Nonuniform media
- Degradation of the media due to aging or weathering
- Uniform media being too small
- Debris (moss, leaves, sticks) or living organisms (snails) clogging the void spaces

Corrective actions (listed in increasing impact on the quality of the plant effluent):

- Remove all leaves, sticks, and other debris from the media.
- Increase recirculation of dilute, high-strength wastes to improve sloughing to keep voids open.
- Use high-pressure stream of water to agitate and flush the ponded area.

- Rake or fork the ponded area.
- Dose the filter with chlorine solution for 2 to 4 hr. The specific dose of chlorine required will depend on the severity of the ponding problem. When using elemental chlorine, the dose must be sufficient to provide a residual at the orifices of 1 to 50 mg/L. If the filter is severely clogged, the higher residuals may be required to unload the majority of the biomass. If the filter cannot be dosed by elemental chlorine, chlorinated lime or HTH powder may be used. Dosing should be in the range of 8 to 10 lb of chlorine per 1000 ft^2 of media.
- If the filter design permits, the filter media can be flooded for a period of 4 hr. Remember, if the filter is flooded, care must be taken to prevent hydraulic overloads of the final settling tank. The trickling filter should be drained slowly at low flow periods.
- Dry the media. By stopping the flow to the filter, the slime will dry and loosen. When the flow is restarted, the loosened slime will flow out of the filter. The amount of drying time will be dependent on the thickness of the slime and the amount of removal desired. Time may range from a few hours to several days.

Note: Portions of the media can be dried without taking the filter out of service by plugging the orifices that normally service the area.

Note: If these corrective actions do not provide the desired improvement, the media must be carefully inspected. Remove a sample of the media from the affected area. Carefully clean it, inspect for its solidity, and determine its size uniformity (3 to 5 in.). If it is acceptable, the sample must be carefully replaced. If the media appear to be decomposing or are not uniform, then they should be replaced.

19.8.2.9.2 Odors

Frequent offensive odors usually indicate an operational problem. These foul odors occur within the filter periodically and are normally associated with anaerobic conditions. Under normal circumstances, a slight anaerobic slime layer forms due to the inability of oxygen to penetrate all the way to the media; however, under normal operation, the outer slime layers will remain aerobic, and no offensive odors are produced.

Causal factors:

- Excessive organic loading due to poor filter effluent quality (recirculation), poor primary treatment operation, poor control of sludge treatment process that results in high BOD$_5$ recycle flows

- Poor ventilation because of submerged or obstructed underdrains, clogged vent pipes, or clogged void spaces
- Overloaded filter (hydraulically or organically)
- Poor housekeeping

Corrective actions:

- Evaluate the operation of the primary treatment process. Eliminate any short-circuiting. Determine any other actions that can be taken to improve the performance of the primary process.
- Evaluate and adjust control of sludge treatment processes to reduce the BOD$_5$ or recycle flows.
- Increase the recirculation rate to add additional dissolved oxygen to the filter influent. Do not increase the recirculation rate if the flow rate through the underdrains would cause less than 50% open space.
- Maintain aerobic conditions in the filter influent.
- Remove debris from media surfaces.
- Flush underdrains and vent pipes.
- Add a commercially available masking agent to reduce odors and prevent complaints.
- Add chlorine at a 1- to 2-mg/L residual for several hours at low flow to reduce activity and cut down on the oxygen demand. Chlorination only treats symptoms; a permanent solution must be determined and instituted.

19.8.2.9.3 High Clarifier Effluent Suspended Solids or BOD

Symptom:

- The effluent from the trickling filter process-settling unit contains a high concentration of suspended solids.

Causal factors:

- Recirculated flows are too high, causing hydraulic overloading of the settling tank. In multiple unit operations, the flow is not evenly distributed.
- Settling tank baffles or skirts have corroded or broken.
- Sludge collection mechanism is broken or malfunctioning.
- Effluent weirs are not level.
- Short-circuiting is occurring because of temperature variations.
- Withdrawal rate or frequency is not correct.
- Excessive solids loading is occurring due to excessive sloughing.

Corrective actions:

- Check hydraulic loading and adjust recirculated flow if the hydraulic loading is too high.
- Adjust flow to ensure equal distribution.
- Inspect sludge removal equipment; repair broken equipment.
- Monitor sludge blanket depth and sludge solids concentration; adjust withdrawal rate and frequency to maintain aerobic conditions in the settling tank.
- Adjust effluent weir to obtain equal flow over all parts of the weir length.
- Determine temperature in the clarifier at various points and depths throughout the clarifier. If depth temperatures are consistently 1 to 2°F lower than surface readings, a temperature problem exists. Baffles may be installed to help to break up these currents.
- Determine whether high sloughing rates caused by biological activity or temperature changes are creating excessive solids loading. The addition of 1 to 2 mg/L of cationic polymer may be helpful in improving solids capture. Remember, if polymer is added, solids withdrawal must be increased.
- Control high sloughings due to organic overloading, toxic wastes, or wide variations in influent flow at their source.

19.8.2.9.4 Filter Flies

Symptoms:

- The trickling filter and surrounding area become populated with large numbers of very small flying insects (*Psychoda* moths).

Causal factors:

- Poor housekeeping
- Insufficient recirculation
- Intermittent wet and dry conditions
- Warm weather

Corrective actions (note that corrective actions for filter fly problems revolve around disrupting the fly's life cycle—about 7 to 10 days in warm weather):

- Increase the recirculation rate to obtain a hydraulic loading of at least 200 gpd/ft^2. At this rate, filter fly larvae are normally flushed out of the filter.
- Clean filter walls and remove weeds, brush, and shrubbery around the filter. This removes some of the area for fly breeding.
- Dose the filter periodically with low chlorine concentrations (less than 1 mg/L). This normally destroys larvae.

- Dry the filter media for several hours.
- Flood the filter for 24 hr.
- Spray area around the filter with insecticide. Do not use insecticide directly on the media because of the chance of carryover and unknown effects on the slime populations.

19.8.2.9.5 Freezing

Symptoms:

- Decreased air temperature results in visible ice formation and decreased performance.
- Distributed wastes are in a thin film or spray. This is more likely to cause ice formation.

Causal factors:

- Recirculation causes increased temperature drops and losses.
- Strong prevailing winds cause heat losses.
- Intermittent dosing allows water to stand too long, causing freezing.

Corrective actions:

- All corrective actions are based upon a need to reduce heat loss as the wastes move through the filter.
- Reduce recirculation as much as possible to minimize cooling effects.
- Operate two-stage filters in parallel to reduce heat loss.
- Adjust splash plates and orifices to obtain a coarse spray.
- Construct a windbreak or plant evergreens or shrubs in the direction of the prevailing wind.
- If intermittent dosing is used, leave dump gates open.
- Cover pump wet wells and dose tanks to reduce heat losses.
- Cover filter media to reduce heat loss.
- Remove ice before it becomes large enough to cause stoppage of arms.

Note: During periods of cold weather, the filter will show decreased performance; however, the filter should not be shut off for any extended length of time. Freezing of the moisture trapped within the media causes expansion and may cause structural damage.

19.8.2.10 Process Calculations

Several calculations are useful in the operation of a trickling filter, including total flow, hydraulic loading, and organic loading.

19.8.2.10.1 Total Flow

If the recirculated flow rate is given, total flow is:

$$\text{Total flow (MGD)} = \begin{bmatrix} \text{Influent flow (MGD)} \\ + \text{recirculation flow (MGD)} \end{bmatrix}$$

$$\text{Total flow (gpd)} = \begin{bmatrix} \text{Total flow (MGD)} \\ \times 1,000,000 \text{ gal/MG} \end{bmatrix} \quad (19.34)$$

Note: The total flow to the trickling filter includes the influent flow and the recirculated flow. This can be determined using the recirculation ratio:

$$\text{Total flow (MGD)} = \begin{bmatrix} \text{Influent flow} \\ \times (\text{recirculation rate} + 1.0) \end{bmatrix}$$

■ EXAMPLE 19.31

Problem: The trickling filter is currently operating with a recirculation ratio of 1.5. What is the total flow applied to the filter when the influent flow rate is 3.65 MGD?

Solution:

Total flow (MGD) = 3.65 MGD × (1.5 + 1.0) = 9.13 MGD

19.8.2.10.2 Hydraulic Loading

Calculating the hydraulic loading rate is important in accounting for both the primary effluent as well as the recirculated trickling filter effluent. Both of these are combined before being applied to the surface of the filter. The hydraulic loading rate is calculated based on the surface area of the filter.

■ EXAMPLE 19.32

Problem: A trickling filter 90 ft in diameter is operated with a primary effluent of 0.488 MGD and a recirculated effluent flow rate of 0.566 MGD. Calculate the hydraulic loading rate on the filter in units gpd/ft².

Solution: The primary effluent and recirculated trickling filter effluent are applied together across the surface of the filter; therefore,

$$0.488 \text{ MGD} + 0.566 \text{ MGD} = 1.054 \text{ MGD}$$
$$= 1,054,000 \text{ gpd}$$

$$\text{Circular surface area} = 0.785 \times (\text{diameter})^2$$
$$= 0.785 \times (90 \text{ ft})^2 = 6359 \text{ ft}^2$$

$$\frac{1,054,000 \text{ gpd}}{6359 \text{ ft}^2} = 165.7 \text{ gpd/ft}^2$$

19.8.2.10.3 Organic Loading Rate

As mentioned earlier, trickling filters are sometimes classified by the organic loading rate applied. The organic loading rate is expressed as a certain amount of BOD applied to a certain volume of media.

■ EXAMPLE 19.33

Problem: A trickling filter, 50 ft in diameter, receives a primary effluent flow rate of 0.445 MGD. Calculate the organic loading rate in units of pounds of BOD applied per day per 900 ft³ of media volume. The primary effluent BOD concentration is 85 mg/L. The media depth is 9 ft.

Solution:

$$0.445 \text{ MGD} \times 85 \text{ mg/L} \times 8.34 \text{ lb/gal} = 315.5 \text{ lb/day BOD}$$

$$\text{Surface area} = 0.785 \times (50)^2 = 1962.5 \text{ ft}^2$$

$$\text{Area} \times \text{depth} = \text{Volume}$$

$$1962.5 \text{ ft}^2 \times 9 \text{ ft} = 17,662.5$$

Note: To determine the pounds of BOD per 1000 ft³ in a volume of thousands of cubic feet, we must set up the equation as shown below:

$$\frac{315.5 \text{ lb BOD/day}}{17,662.5} \times \frac{1000}{1000}$$

Regrouping the numbers and the units together:

$$\text{BOD} = \frac{315.5 \text{ lb} \times 1000}{17,662.5} \times \frac{\text{lb BOD/day}}{1000 \text{ ft}^3}$$

$$= 17.9 \text{ lb BOD/day/1000 ft}^3$$

19.8.2.10.4 Settling Tanks

In the operation of settling tanks that follow trickling filters, various calculations are routinely made to determine detention time, surface settling rate, hydraulic loading, and sludge pumping.

19.8.3 ROTATING BIOLOGICAL CONTACTORS

The *rotating biological contactor* (RBC) is a biological treatment system (see Figure 19.8) and is a variation of the attached-growth idea provided by the trickling filter. Still relying on microorganisms that grow on the surface of a medium, the RBC is instead a fixed-film biological treatment device, but the basic biological process is similar to that occurring in the trickling filter. An RBC consists of a series of closely spaced (mounted side by side), circular, plastic (synthetic) disks that are typically about 3.5 m in diameter and attached to a rotating horizontal

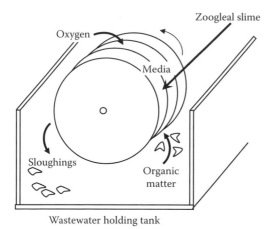

FIGURE 19.8 Rotating biological contactor (RBC) cross-section and treatment system.

19.8.3.1 RBC Equipment

The equipment that makes up a RBC includes the rotating biological contactor (the media, either standard or high density), a center shaft, drive system, tank, baffles, housing or cover, and a settling tank. The *rotating biological contactor* consists of circular sheets of synthetic material (usually plastic) mounted side by side on a shaft. The *sheets* (media) contain large amounts of surface area for growth of the biomass. The *center shaft* provides the support for the disks of media and must be strong enough to support the weight of the media and the biomass; experience has indicated that a major problem is collapse of the support shaft. The *drive system* provides the motive force to rotate the disks and shaft. The drive system may be mechanical or air driven or a combination of each. When the drive system does not provide uniform movement of the RBC, major operational problems can arise.

The *tank* holds the wastewater in which the RBC rotates. It should be large enough to permit variation of the liquid depth and detention time. *Baffles* are required to permit proper adjustment of the loading applied to each stage of the RBC process. Adjustment can be made to increase or decrease the submergence of the RBC. RBC stages are normally enclosed in some type of protective structure (*cover*) to prevent loss of biomass due to severe weather changes (e.g., snow, rain, temperature, wind, sunlight). In many instances, this housing greatly restricts access to the RBC. The *settling tank* is provided to remove the sloughing material created by the biological activity and is similar in design to the primary settling tank. The settling tank provides 2- to 4-hr detention times to permit settling of lighter biological solids.

shaft (see Figure 19.8). Approximately 40% of each disk is submersed in a tank containing the wastewater to be treated. As the RBC rotates, the attached biomass film (zoogleal slime) that grows on the surface of the disk moves into and out of the wastewater. While submerged in the wastewater, the microorganisms absorb organics; when they are rotated out of the wastewater, they are supplied with the oxygen required for aerobic decomposition. As the zoogleal slime reenters the wastewater, excess solids and waste products are stripped off the media as sloughings. These sloughings are transported with the wastewater flow to a settling tank for removal.

Modular RBC units are placed in series (see Figure 19.9) simply because a single contactor is not sufficient to achieve the desired level of treatment; the resulting treatment achieved exceeds conventional secondary treatment. Each individual contactor is called a *stage* and the group is known as a *train*. Most RBC systems consist of two or more trains with three or more stages in each. The key advantage in using RBCs instead of trickling filters is that RBCs are easier to operate under varying load conditions, as it is easier to keep the solid medium wet at all times. Moreover, the level of nitrification that can be achieved by a RBC system is significant, especially when multiple stages are employed.

19.8.3.2 RBC Operation

During normal operation, operator vigilance is required to observe the RBC movement, slime color, and appearance; however, if the unit is covered, observations may be limited to that portion of the media that can be viewed through the access door. Slime color and appearance can indicate process condition; for example:

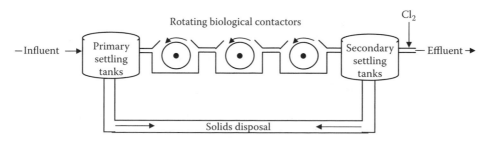

FIGURE 19.9 Rotating biological contactor (RBC) treatment system.

- Gray, shaggy slime growth indicates normal operation.
- Reddish brown, golden shaggy growth indicates nitrification.
- White chalky appearance indicates high sulfur concentrations.
- No slime indicates severe temperature or pH changes.

Sampling and testing should be conducted daily for dissolved oxygen content and pH. BOD_5 and suspended solids testing should also be performed to aid in assessing performance.

19.8.3.3 RBC Expected Performance

The RBC normally produces a high-quality effluent with BOD_5 at 85 to 95% and suspended solids removal at 85 to 95%. The RBC treatment process may also significantly reduce (if designed for this purpose) the levels of organic nitrogen and ammonia nitrogen.

19.8.3.4 Operator Observations

Rotating biological filter operation requires routine observation, process control sampling and testing, troubleshooting, and process control calculations. Comparison of daily results with expected normal ranges is the key to identifying problems and appropriate corrective actions.

Note: If the RBC is covered, observations may be limited to the portion of the media that can be viewed through the access door.

- *Rotation*—The operator routinely checks the operation of the RBC to ensure that smooth, uniform rotation is occurring (normal operation). Erratic, nonuniform rotation indicates a mechanical problem or uneven slime growth. If no movement is observed, mechanical problems or extreme excess of slime growth are indicated.
- *Slime color and appearance*—Gray, shaggy slime growth on the RBC indicates normal operation. Reddish brown or golden brown shaggy growth indicates normal operation during nitrification. Very dark brown, shaggy growth (with worms present) indicates very old slime. White chalky growth indicates high influent sulfur or sulfide levels. No visible slime growth indicates a severe pH or temperature change.

19.8.3.5 RBC Process Control Sampling and Testing

For process control, the RBC process does not require large amounts of sampling and testing to provide the information required. The frequency for performing suggested testing depends on available resources and variability of the process. Frequency may be lower during normal operation and higher during abnormal conditions. The following routine sampling points and types of tests will permit the operator to identify normal and abnormal operating conditions.

RBC train influent tests
- Dissolved oxygen
- pH
- Temperature
- Settleable solids
- BOD_5
- Suspended solids
- Metals

RBC test
- Speed of rotation

RBC train effluent tests
- Dissolved oxygen
- pH
- Jar tests

Process effluent tests
- Dissolved oxygen
- pH
- Settleable solids
- BOD_5
- Suspended solids

19.8.3.6 Troubleshooting Operational Problems

(Much of the information in this section is based on USEPA's *Field Manual for Performance Evaluation and Troubleshooting at Municipal Wastewater Treatment Facilities*; see Culp and Heim, 1978.) Table 19.5 lists several indicators of poor process performance, their causal factors, and corrective actions. The table is not all inclusive, as it does not cover all of the operational problems associated with the rotating biological contactor process. It does, however, provide information on the most common operational problems.

19.8.3.7 RBC Process Control Calculations

Several process control calculations may be useful in the operation of an RBC. These include soluble BOD, total media area, organic loading rate, and hydraulic loading rate. Settling tank calculations and sludge pumping calculations may be helpful for evaluation and control of the settling tank following the RBC.

19.8.3.7.1 RBC Soluble BOD

The soluble BOD_5 concentration of the RBC influent can be determined experimentally in the laboratory, or it can be estimated using the suspended solids concentration and the K factor. The K factor is used to approximate the BOD_5

TABLE 19.5
Symptoms, Causes, and Corrective Actions for Rotating Biological Contactor Problems

Symptom	Cause	Corrective Action
White slime		
White slime on most of the disk area	High hydrogen sulfide in influent	Aerate RBC or plant influent.
	Septic influent	Add sodium nitrate or hydrogen peroxide to influent.
	First stage overloaded	Adjust baffles between stages 1 and 2 to increase fraction of total surface area in first stage.
Excessive sloughing		
Loss of slime	Excessive pH variance	Install pH control equipment.
	Toxic influent	Implement and enforce pretreatment program; equalize flow to acclimate organisms.
RBC rotation		
Uneven RBC rotation	Mechanical problem	Repair mechanical problem.
	Uneven growth	Increase rotational speed; adjust baffles to decrease loading; increase sloughing.
Solids		
Solids accumulating in reactors	Inadequate pretreatment	Identify and correct grit removal problem; identify and correct primary settling problem.
Shaft bearings		
Shaft bearings running hot or failing	Inadequate maintenance	Follow manufacturer's recommendations.
Drive motor		
Drive motor running hot	Inadequate maintenance	Follow manufacturer's recommendations.
	Improper chain drive alignment	Adjust alignment.

(particulate BOD) contributed by the suspended matter. The K factor must be provided or determined experimentally in the laboratory. The K factor for domestic wastes is normally in the range in the range of 0.5 to 0.7.

$$\text{Soluble BOD}_5 = \text{Total BOD}_5 - \left(\frac{K \text{ factor} \times \text{total}}{\text{suspended solids}} \right) \quad (19.35)$$

■ **EXAMPLE 19.34**

Problem: The suspended solids concentration of a wastewater is 250 mg/L. If the normal K value at the plant is 0.6, what is the estimated particulate BOD concentration of the wastewater?

Note: The K value of 0.6 indicates that about 60% of the suspended solids are organic suspended solids (particulate BOD).

Solution:

250 mg/L × 0.6 = 150 mg/L particulate BOD

■ **EXAMPLE 19.35**

Problem: A rotating biological contactor receives a flow of 2.2 MGD with a BOD content of 170 mg/L and suspended solids concentration of 140 mg/L. If the K value is 0.7, how many pounds of soluble BOD enter the RBC daily?

Solution:

Total BOD = Particulate BOD + soluble BOD

170 mg/L = (140 mg/L × 0.7) + x mg/L

170 mg/L = 98 mg/L + x mg/L

170 mg/L − 98 mg/L = x

x = 72 mg/L soluble BOD

Now, we can determine the lb/day soluble BOD:

Soluble BOD (mg/L) × flow (MGD) × 8.34 lb/gal = lb/day

72 mg/L × 2.2 MGD × 8.34 lb/gal = 1321 lb/day

19.8.3.7.2 RBC Total Media Area

Several process control calculations for the RBC use the total surface area of all the stages within the train. As was the case with the soluble BOD calculation, plant design information or information supplied by the unit manufacturer must provide the individual stage areas (or the total train area), because physical determination of this would be extremely difficult.

$$\text{Total area} = \left[\begin{array}{l} \text{1st stage area} + \text{2nd stage area} \\ + \ldots + n\text{th stage area} \end{array} \right] \quad (19.36)$$

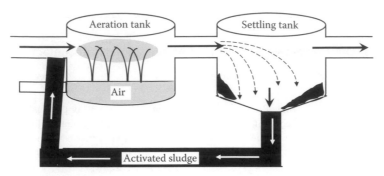

FIGURE 19.10 The activated sludge process.

19.8.3.7.3 RBC Organic Loading Rate

If the soluble BOD concentration is known, the organic loading on a RBC can be determined. Organic loading on an RBC based on soluble BOD concentration can range from 3 to 4 lb/day/1000 ft^2.

■ **EXAMPLE 19.36**

Problem: An RBC has a total media surface area of 102,500 ft^2 and receives a primary effluent flow rate of 0.269 MGD. If the soluble BOD concentration of the RBC influent is 159 mg/L, what is the organic loading rate in lb/1000 ft^2?

Solution:

$$0.269 \text{ MGD} \times 159 \text{ mg/L} \times 8.34 \text{ lb/gal} = 356.7 \text{ lb/day}$$

$$\frac{356.7 \text{ lb/day}}{102,500 \text{ ft}^2} \times \frac{1000 \text{ (number)}}{1000 \text{ (unit)}} = 3.48 \text{ lb/day/1000 ft}^2$$

19.8.3.7.4 RBC Hydraulic Loading Rate

The manufacturer normally specifies the RBC media surface area and the hydraulic loading rate is based on the media surface area, usually in square feet (ft^2). Hydraulic loading on a RBC can range from 1 to 3 gpd/ft^2.

■ **EXAMPLE 19.37**

Problem: An RBC treats a primary effluent flowrate of 0.233 MGD. What is the hydraulic loading rate in gpd/ft^2 if the media surface area is 96,600 ft^2?

Solution:

$$\frac{233,000 \text{ gpd}}{96,600 \text{ ft}^2} = 2.41 \text{ gpd/ft}^2$$

19.9 ACTIVATED SLUDGE

The biological treatment systems discussed to this point—ponds, trickling filters, and rotating biological contactors—have been around for years. The trickling filter, for

example, has been around and successfully used since the late 1800s. The problem with ponds, trickling filters, and RBCs is that they are temperature sensitive, remove less BOD, and cost more to build (particularly trickling filters) than the activated sludge systems that were later developed.

Note: Although trickling filters and other systems cost more to build than activated sludge systems, it is important to point out that activated sludge systems cost more to operate because of the need for energy to run pumps and blowers.

As shown in Figure 19.1, the activated sludge process follows primary settling. The basic components of an activated sludge sewage treatment system include an *aeration tank* and a *secondary basin*, *settling basin*, or *clarifier* (see Figure 19.10). Primary effluent is mixed with settled solids recycled from the secondary clarifier and is then introduced into the aeration tank. Compressed air is injected continuously into the mixture through porous diffusers located at the bottom of the tank, usually along one side.

Wastewater is fed continuously into an aerated tank, where the microorganisms metabolize and biologically flocculate the organics. Microorganisms (activated sludge) are settled from the aerated mixed liquor under quiescent conditions in the final clarifier and are returned to the aeration tank. Left uncontrolled, the number of organisms would eventually become too great; therefore, some must periodically be removed (wasted). A portion of the concentrated solids from the bottom of the settling tank must be removed from the process (waste activated sludge, or WAS). Clear supernatant from the final settling tank is the plant effluent.

19.9.1 ACTIVATED SLUDGE TERMINOLOGY

To better understand the discussion of the activated sludge process presented in the following sections, it is necessary to understand the terms associated with the process. Some of these terms have been used and defined earlier in the text, but we list them here again to refresh your memory. Review these terms and remember them, as they are used throughout the discussion:

Absorption—Taking in or reception of one substance into the body of another by molecular or chemical actions and distribution throughout the absorber.

Activated—To speed up reaction. When applied to sludge, it means that many aerobic bacteria and other microorganisms are in the sludge particles.

Activated sludge—A floc or solid formed by the microorganisms. It includes organisms, accumulated food materials, and waste products from the aerobic decomposition process.

Activated sludge process—A biological wastewater treatment process in which a mixture or influent and activated sludge is agitated and aerated. The activated sludge is subsequently separated from the treated mixed liquor by sedimentation and is returned to the process as needed. The treated wastewater overflows the weir of the settling tank in which separation from the sludge takes place.

Adsorption—The adherence of dissolved, colloidal, or finely divided solids to the surface of solid bodies when they are brought into contact.

Aeration—Mixing air and a liquid by one of the following methods: spraying the liquid in the air, diffusing air into the liquid, or agitating the liquid to promote surface adsorption of air.

Aerobic—Free or dissolved oxygen is present in the aquatic environment. Aerobic organisms must be in the presence of dissolved oxygen to be active.

Bacteria—Single-cell plants that play a vital role in stabilization of organic waste.

Biochemical oxygen demand (BOD)—A measure of the amount of food available to the microorganisms in a particular waste. It is measured by the amount of dissolved oxygen used up during a specific time period (usually 5 days, expressed as BOD_5).

Biodegradable—From *degrade* ("to wear away or break down chemically") and *bio* ("by living organisms"). Put it all together, and you have a substance, usually organic, which can be decomposed by biological action.

Bulking—A problem in activated sludge plants that results in poor settleability of sludge particles.

Coning—A condition that may be established in a sludge hopper during sludge withdrawal when part of the sludge moves toward the outlet while the remainder tends to stay in place; development of a cone or channel of moving liquids surrounded by relatively stationary sludge.

Decomposition—Generally, in waste treatment, refers to the changing of waste matter into simpler, more stable forms that will not harm the receiving stream.

Diffused air aeration—A diffused-air-activated sludge plant takes air, compresses it, then discharges the air below the water surface to the aerator through some type of air diffusion device.

Diffuser—A porous plate or tube through which air is forced and divided into tiny bubbles for distribution in liquids; commonly made of carborundum, aluminum, or silica sand.

Dissolved oxygen—Atmospheric oxygen dissolved in water or wastewater; usually abbreviated as DO.

Note: The typical required DO for a well-operated activated sludge plant is between 2.0 and 2.5 mg/L.

Facultative bacteria—Bacteria that can use molecular (dissolved) oxygen or oxygen obtained from food materials. In other words, facultative bacteria can live under aerobic or anaerobic conditions.

Filamentous bacteria—Organisms that grow in thread or filamentous form.

Food-to-microorganism ratio—A process control calculation used to evaluate the amount of food (BOD or COD) available per pound of mixed liquor volatile suspended solids. This may be written as the F/M ratio:

$$\frac{\text{Food}}{\text{Microorganism}} = \frac{\text{BOD (lb/day)}}{\text{MLVSS (lb)}}$$

$$= \frac{\text{Flow (MGD)} \times \text{BOD (mg/L)} \times 8.34 \text{ lb/gal}}{\text{Vol. (MG)} \times \text{MLVSS (mg/L)} \times 8.34 \text{ lb/gal}}$$

Fungi—Multicellular aerobic organisms.

Gould sludge age—A process control calculation used to evaluate the amount of influent suspended solids available per pound of mixed liquor suspended solids.

Mean cell residence time (MCRT)—The average length of time particles of mixed liquor suspended solids remain in the activated sludge process; may also be referred to as the *sludge retention rate* (STR):

$$\text{MCRT (days)} = \frac{\text{Solids in activated sludge process (lb)}}{\text{Solids removed from process (lb/day)}}$$

Mixed liquor—The contribution of return activated sludge and wastewater (either influent or primary effluent) that flows into the aeration tank.

Mixed liquor suspended solids (MLSS)—The suspended solids concentration of the mixed liquor. Many references use this concentration to represent the amount of organisms in the activated sludge process.

Mixed liquor volatile suspended solids (MLVSS)—The organic matter in the mixed liquor suspended solids; can also be used to represent the amount of organisms in the process.

Nematodes—Microscopic worms that may appear in biological waste treatment systems.

Nutrients—Substances required to support plant organisms. Major nutrients are carbon, hydrogen, oxygen, sulfur, nitrogen, and phosphorus.

Protozoa—Single-cell animals that are easily observed under the microscope at a magnification of 100×. Bacteria and algae are prime sources of food for advanced forms of protozoa.

Return activated sludge (RAS)—The solids returned from the settling tank to the head of the aeration tank.

Rising sludge—Occurs in the secondary clarifiers or activated sludge plant when the sludge settles to the bottom of the clarifier, is compacted, and then rises to the surface in relatively short time.

Rotifiers—Multicellular animals with flexible bodies and cilia near the mouth that are used to attract food. Bacteria and algae are their major source of food.

Secondary treatment—A wastewater treatment process used to convert dissolved or suspended materials into a form that can be removed.

Settleability—A process control test used to evaluate the settling characteristics of the activated sludge. Readings taken at 30 to 60 min are used to calculate the settled sludge volume (SSV) and the sludge volume index (SVI).

Settled sludge volume (SSV)—The volume (mL/L or percent) occupied by an activated sludge sample after 30 or 60 min of settling. Normally written as SSV with a subscript to indicate the time of the reading used for calculation (e.g., SSV_{30} or SSV_{60}).

Shock load—The arrival at a plant of a waste toxic to organisms, in sufficient quantity or strength to cause operating problems, such as odor or sloughing off of the growth of slime on the trickling filter media. Organic overloads also can cause a shock load.

Sludge volume index (SVI)—A process control calculation used to evaluate the settling quality of the activated sludge. Requires the SSV_{30} and mixed liquor suspended solids test results to calculate:

$$SVI (mL/g) = \frac{\text{30-min settled vol. (ml/L)} \times 1000 \text{ mg/g}}{\text{Mixed liquor suspended solids (mg/L)}}$$

Solids—Material in the solid state.
- *Dissolved solids*—Solids present in solution; solids that will pass through a glass-fiber filter.
- *Fixed solids*—Also known as the *inorganic solids*; the solids left after a sample is ignited at 550°C for 15 min.
- *Floatable solids (scum)*—Solids that will float to the surface of still water, sewage, or other liquid; usually composed of grease particles, oils, light plastic material, etc.

- *Nonsettleable solids*—Finely divided suspended solids that will not sink to the bottom in still water, sewage, or other liquid in a reasonable period, usually two hours; also known as *colloidal solids*.
- *Suspended solids*—Solids that will not pass through a glass-fiber filter.
- *Total solids*—Solids in water, sewage, or other liquids, including suspended solids and dissolved solids.
- *Volatile solids*—Organic solids; measured as the solids that are lost on ignition of the dry solids at 550°C.
- *Waste activated sludge (WAS)*—The solids being removed from the activated sludge process.

19.9.2 ACTIVATED SLUDGE PROCESS EQUIPMENT

Equipment requirements for the activated sludge process are more complex than other processes discussed. The equipment includes an *aeration tank*, *aeration system*, *system-settling tank*, *return sludge system*, and *waste sludge system*. These are discussed in the following.

19.9.2.1 Aeration Tank

The *aeration tank* is designed to provide the required detention time (depending on the specific modification) and ensure that the activated sludge and the influent wastewater are thoroughly mixed. Tank design normally attempts to ensure that no dead spots are created.

19.9.2.2 Aeration

Aeration can be mechanical or diffused. Mechanical aeration systems use agitators or mixers to mix air and mixed liquor. Some systems use *sparge rings* to release air directly into the mixer Diffused aeration systems use pressurized air released through diffusers near the bottom of the tank. Efficiency is directly related to the size of the air bubbles produced. Fine bubble systems have a higher efficiency. The diffused air system has a blower to produce large volumes of low pressure air (5 to 10 psi), air lines to carry the air to the aeration tank, and headers to distribute the air to the diffusers, which release the air into the wastewater.

19.9.2.3 Settling Tank

Activated sludge systems are equipped with plain *settling tanks* designed to provide 2 to 4 hr of hydraulic detention time.

19.9.2.4 Return Sludge

The return sludge system includes pumps, a timer or variable speed drive to regulate pump delivery, and a flow measurement device to determine actual flow rates.

19.9.2.5 Waste Sludge

In some cases, the waste activated sludge withdrawal is accomplished by adjusting valves on the return system. When a separate system is used it includes pumps, a timer or variable speed drive, and a flow measurement device.

19.9.3 Overview of Activated Sludge Process

The activated sludge process is a treatment technique in which wastewater and reused biological sludge full of living microorganisms are mixed and aerated. The biological solids are then separated from the treated wastewater in a clarifier and are returned to the aeration process or wasted. The microorganisms are mixed thoroughly with the incoming organic material, and they grow and reproduce by using the organic material as food. As they grow and are mixed with air, the individual organisms cling together (flocculate). Once flocculated, they more readily settle in the secondary clarifiers.

The wastewater being treated flows continuously into an aeration tank where air is injected to mix the wastewater with the returned activated sludge and to supply the oxygen required by the microbes to live and feed on the organics. Aeration can be supplied by injection through air diffusers in the bottom of tank or by mechanical aerators located at the surface. The mixture of activated sludge and wastewater in the aeration tank is called the *mixed liquor*. The mixed liquor flows to a secondary clarifier where the activated sludge is allowed to settle.

The activated sludge is constantly growing, and more is produced than can be returned for use in the aeration basin. Some of this sludge, therefore, must be transferred to a sludge handling system for treatment and disposal. The volume of sludge returned to the aeration basins is normally 40 to 60% of the wastewater flow. The rest is wasted.

19.9.4 Factors Affecting Operation of the Activated Sludge Process

A number of factors affect the performance of an activated sludge system. These include the following:

- Temperature
- Return rates
- Amount of oxygen available
- Amount of organic matter available
- pH
- Waste rates
- Aeration time
- Wastewater toxicity

To obtain the desired level of performance in an activated sludge system, a proper balance must be maintained among the amounts of food (organic matter), organisms (activated sludge), and oxygen (dissolved oxygen). The

majority of problems with the activated sludge process result from an imbalance among these three items.

To fully appreciate and understand the biological process taking place in a normally functioning activated sludge process, the operator must have knowledge of the key players in the process: the organisms. This makes a certain amount of sense when we consider that the heart of the activated sludge process is the mass of settleable solids formed by aerating wastewater containing biological degradable compounds in the presence of microorganisms. Activated sludge consists of organic solids plus bacteria, fungi, protozoa, rotifers, and nematodes.

19.9.5 Growth Curve

To understand the microbiological population and its function in an activated sludge process, the operator must be familiar with the microorganism *growth curve* (refer to Chapter 12, Figure 12.15). In the presence of excess organic matter, the microorganisms multiply at a fast rate. The demand for food and oxygen is at its peak. Most of this is used for the production of new cells. This condition is known as the *log growth phase*. Over time, the amount of food available for the organisms declines. Floc begins to form while the growth rate of bacteria and protozoa begins to decline. This is referred to as the *declining growth phase*. The *endogenous respiration* phase occurs as the food available becomes extremely limited and the organism mass begins to decline. Some of the microorganisms may die and break apart, thus releasing organic matter that can be consumed by the remaining population.

The actual operation of an activated-sludge system is regulated by three factors: (1) the quantity of air supplied to the aeration tank, (2) the rate of activated sludge recirculation, and (3) the amount of excess sludge withdrawn from the system. Sludge wasting is an important operational practice because it allows the operator to establish the desired concentration of MLSS, food-to-microorganism ratio, and sludge age.

Note: Air requirements in an activated sludge basin are governed by: (1) biochemical oxygen demand (BOD) loading and the desired removal effluent, (2) volatile suspended solids concentration in the aerator, and (3) suspended solids concentration of the primary effluent.

19.9.6 Activated Sludge Formation

The formation of activated sludge is dependent on three steps. The first step is the transfer of food from wastewater to organism. Second is the conversion of wastes to a usable form. Third is the flocculation step.

1. *Transfer*—Organic matter (food) is transferred from the water to the organisms. Soluble material

is absorbed directly through the cell wall. Particulate and colloidal matter is adsorbed to the cell wall, where it is broken down into simpler soluble forms, then absorbed through the cell wall.

2. *Conversion*—Food matter is converted to cell matter by synthesis and oxidation into end products such as CO_2, H_2O, NH_3, stable organic waste, and new cells.

3. *Flocculation*—Flocculation is the gathering of fine particles into larger particles. This process begins in the aeration tank and is the basic mechanism for removal of suspended matter in the final clarifier. The concentrated *biofloc* that settles and forms the sludge blanket in the secondary clarifier is known as activated sludge.

19.9.7 Activated Sludge Performance-Controlling Factors

To maintain the working organisms in the activated sludge process, the operator must be sure that a suitable environment is maintained by being aware of the many factors influencing the process and by monitoring them repeatedly. Control, here, is defined as maintaining the proper solids (floc mass) concentration in the aerator for the incoming water (food) flow by adjusting the return and waste sludge pumping rate and regulating the oxygen supply to maintain a satisfactory level of dissolved oxygen in the process.

19.9.7.1 Aeration

The activated sludge process must receive sufficient aeration to keep the activated sludge in suspension and to satisfy the organism oxygen requirements. Insufficient mixing results in dead spots, septic conditions, and a loss of activated sludge.

19.9.7.2 Alkalinity

The activated sludge process requires sufficient alkalinity to ensure that the pH remains in the acceptable range of 6.5 to 9.0. If organic nitrogen and ammonia are being converted to nitrate (nitrification), sufficient alkalinity must be available to support this process, as well.

19.9.7.3 Nutrients

The microorganisms of the activated sludge process require nutrients (nitrogen, phosphorus, iron, and other trace metals) to function. If sufficient nutrients are not available, the process will not perform as expected. The accepted minimum ratio of carbon to nitrogen, phosphorus, and iron is 100 parts carbon to 5 parts nitrogen, 1 part phosphorus, and 0.5 parts iron.

19.9.7.4 pH

The pH of the mixed liquor should be maintained within the range of 6.5 to 9.0 (6.0 to 8.0 is ideal). Gradual fluctuations within this range will normally not upset the process. Rapid fluctuations or fluctuations outside of this range can reduce organism activity.

19.9.7.5 Temperature

As temperature decreases, activity of the organisms will also decrease. Cold temperatures also require longer recovery times for systems that have been upset. Warm temperatures tend to favor denitrification and filamentous growth.

Note: The activity level of bacteria within the activated sludge process increases with rise in temperature.

19.9.7.6 Toxicity

Sufficient concentrations of elements or compounds that enter a treatment plant that have the ability to kill the microorganisms (the activated sludge) are known as *toxic waste* (shock level). Common to this group are cyanides and heavy metals.

Note: A typical example of a toxic substance added by operators is chlorine used for odor control or control of filamentous organisms (prechlorination). Chlorination is for disinfection. Chlorine is a toxicant and should not be allowed to enter the activated sludge process; it is not selective with respect to the type of organisms damaged or killed. It may kill the organisms that should be retained in the process as workers. Chlorine is very effective in disinfecting the plant effluent after treatment by the activated sludge process, however.

19.9.7.7 Hydraulic Loading

Hydraulic loading is the amount of flow entering the treatment process. When compared with the design capacity of the system, it can be used to determine if the process is hydraulically overloaded or underloaded. If more flow is entering the system than it was designed to handle, the system is hydraulically overloaded. If less flow is entering the system than it was designed for, the system is hydraulically underloaded. Generally, the system is more affected by overloading than by underloading. Overloading can be caused by stormwater, infiltration of groundwater, excessive return rates, or many other causes. Underloading normally occurs during periods of drought or in the period following initial startup when the plant has not reached its design capacity. Excess hydraulic flow rates through the treatment plant will reduce the efficiency of the clarifier by allowing activated sludge solids to rise in the clarifier and pass over the effluent weir. This loss of solids in the effluent degrades effluent quality and reduces the amount of activated sludge in the system, in turn, thus reducing process performance.

19.9.7.8 Organic Loading

Organic loading is the amount of organic matter entering the treatment plant. It is usually measured as biochemical oxygen demand (BOD). An organic overload occurs when the amount of BOD entering the system exceeds the design capacity of the system. An organic underload occurs when the amount of BOD entering the system is significantly less than the design capacity of the plant. Organic overloading may occur when the system receives more waste than it was designed to handle. It can also occur when an industry or other contributor discharges more wastes to the system than originally planned. Wastewater treatment plant processes can also cause organic overloads returning high-strength wastes from the sludge treatment processes.

Regardless of the source, an organic overloading of the plant results in increased demand for oxygen. This demand may exceed the air supply available from the blowers. When this occurs, the activated sludge process may become septic. Excessive wasting can also result in a type of organic overload. The food available exceeds the number of activated sludge organisms, resulting in increased oxygen demand and very rapid growth.

Organic underloading may occur when a new treatment plant is initially put into service. The facility may not receive enough waste to allow the plant to operate at its design level. Underloading can also occur when excessive amounts of activated sludge are allowed to remain in the system. When this occurs, the plant will have difficulty in developing and maintaining a good activated sludge.

19.9.8 ACTIVATED SLUDGE MODIFICATIONS

First developed in 1913, the original activated sludge process has been modified over the years to provide better performance for specific operating conditions or with different influent waste characteristics.

1. Conventional activated sludge:
   - Employing the conventional activated sludge modification requires primary treatment.
   - Conventional activated sludge provides excellent treatment; however, a large aeration tank capacity is required, and construction costs are high.
   - In operation, initial oxygen demand is high. The process is also very sensitive to operational problems (e.g., bulking).

2. Step aeration:
   - Step aeration requires primary treatment.
   - It provides excellent treatment.
   - Operation characteristics are similar to conventional.
   - It distributes organic loading by splitting influent flow.
   - It reduces oxygen demand at the head of the system.
   - It reduces solids loading on the settling tank.

3. Complete mix:
   - Complete mix may or may not include primary treatment.
   - It distributes waste, return, and oxygen evenly throughout the tank.
   - Aeration may be more efficient.
   - It maximizes tank use.
   - It permits higher organic loading.

Note: During the complete mix, activated sludge process organisms are in the declining phase on the growth curve.

4. Pure oxygen:
   - Pure oxygen requires primary treatment.
   - It permits higher organic loading.
   - It uses higher solids levels.
   - It operates at higher F/M ratios.
   - It uses covered tanks.
   - It poses a potential safety hazard.
   - Oxygen production is expensive.

5. Contact stabilization:
   - Contact stabilization does not require primary treatment.
   - During operation, organisms collect organic matter (during contact).
   - Solids and activated sludge are separated from flow via settling.
   - Activated sludge and solids are aerated for 3 to 6 hours (stabilization).

Note: Return sludge is aerated before it is mixed with influent flow.

   - The activated sludge oxidizes available organic matter.
   - Although the process is complicated to control, it requires less tank volume than other modifications and can be prefabricated as a package unit for flows of 0.05 to 1.0 MGD.
   - A disadvantage is that common process control calculations do not provide usable information.

6. Extended aeration:
   - Extended aeration does not require primary treatment.

TABLE 19.6
Activated Sludge Modifications

Parameter	Conventional	Contact Stabilization	Extended Aeration	Oxidation Ditch
Aeration time (hr)	4–8	0.5–1.5 (contact)	24	24
		3–6 (reaeration)		
Settling time (hr)	2–4	2–4	2–4	2–4
Return rate (% of influent flow)	25–100	25–100	25–100	25–100
MLSS (mg/L)	1500–4000	1000–3000	2000–6000	2000–6000
		3000–8000		
DO (mg/L)	1–3	1–3	1–3	1–3
SSV_{30} (ml/L)	400–700	400–700 (contact)	400–700	400–700
Food-to-mass ratio (lb BOD_5/lb MLVSS)	02–0.5	0.2–0.6 (contact)	0.05–0.15	0.05 – 0.15
MCRT (whole system) (days)	5–15	N/A	20–30	20 – 30
% Removal BOD_5	85–95	85–95	85–95	85–95
% Removal TSS	85–95	85–95	85–95	85–95
Primary treatment	Yes	No	No	No

- It is used frequently for small flows such as for schools and subdivisions.
- It uses 24-hour aeration.
- It produces low BOD effluent.
- It produces the least amount of waste activated sludge.
- The process is capable of achieving 95% or greater removal of BOD.
- It can produce effluent low in organic and ammonia nitrogen.

7. Oxidation ditch:
   - The oxidation ditch does not require primary treatment.
   - It is similar to the extended aeration process.

Table 19.6 lists the process parameters for each of the four most commonly used activated sludge modifications.

19.9.9 ACTIVATED SLUDGE PROCESS CONTROL PARAMETERS

When operating an activated sludge process, the operator must be familiar with the many important process control parameters that must be monitored frequently and adjusted occasionally to maintain optimal performance.

19.9.9.1 Alkalinity

Monitoring alkalinity in the aeration tank is essential to control of the process. Insufficient alkalinity will reduce organism activity and may result in low effluent pH and, in some cases, extremely high chlorine demand in the disinfection process.

19.9.9.2 Dissolved Oxygen (DO)

The activated sludge process is an aerobic process that requires some dissolved oxygen to be present at all times. The amount of oxygen required is dependent on the influ-ent food (BOD), the activity of the activated sludge, and the degree of treatment desired.

19.9.9.3 pH

Activated sludge microorganisms can be injured or destroyed by wide variations in pH. The pH of the aeration basin will normally be in the range of 6.5 to 9.0. Gradual variations within this range will not cause any major problems; however, rapid changes of one or more pH units can have a significant impact on performance. Industrial waste discharges, septic wastes, or significant amounts of stormwater flows may produce wide variations in pH. pH should be monitored as part of the routine process control testing schedule. Sudden changes in or abnormal pH values may indicate an industrial discharge of strongly acidic or alkaline wastes. Because these wastes can upset the environmental balance of the activated sludge, the presence of wide pH variations can result in poor performance. Processes undergoing nitrification may show a significant decrease in effluent pH.

19.9.9.4 Mixed Liquor Suspended Solids, Mixed Liquor Volatile Suspended Solids, and Mixed Liquor Total Suspended Solids

The mixed liquor suspended solids (MLSS) or mixed liquor volatile suspended solids (MLVSS) can be used to represent the activated sludge or microorganisms present in the process. Process control calculations, such as sludge age and sludge volume index, cannot be calculated unless the MLSS is determined. Adjust the MLSS and MLVSS by increasing or decreasing the waste sludge rates. The mixed liquor total suspended solids (MLTSS) are an important activated sludge control parameter. To increase the MLTSS, for example, the operator must decrease the waste rate or increase the MCRT. The MCRT must be decreased to prevent the MLTSS from changing when the number of aeration tanks in service is reduced.

Note: When performing the Gould sludge age test, assume that the source of the MLTSS in the aeration tank is influent solids.

19.9.9.5 Return Activated Sludge Rate and Concentration

The sludge rate is a critical control variable. The operator must maintain a continuous return of activated sludge to the aeration tank or the process will show a drastic decrease in performance. If the rate is too low, solids remain in the settling tank, resulting in solids loss and a septic return. If the rate is too high, the aeration tank can become hydraulically overloaded, causing reduced aeration time and poor performance. The return concentration is also important because it may be used to determine the return rate required to maintain the desired MLSS.

19.9.9.6 Waste Activated Sludge Flow Rate

Because the activated sludge contains living organisms that grow, reproduce, and produce waste matter, the amount of activated sludge is continuously increasing. If the activated sludge is allowed to remain in the system too long, the performance of the process will decrease. If too much activated sludge is removed from the system, the solids become very light and will not settle quickly enough to be removed in the secondary clarifier.

19.9.9.7 Temperature

Because temperature directly affects the activity of the microorganisms, accurate monitoring of temperature can be helpful in identifying the causes of significant changes in organization populations or process performance.

19.9.9.8 Sludge Blanket Depth

The separation of solids and liquid in the secondary clarifier results in a blanket of solids. If solids are not removed from the clarifier at the same rate they enter, the blanket will increase in depth. If this occurs, the solids may carryover into the process effluent. The sludge blanket depth may be affected by other conditions, such as temperature variation, toxic wastes, or sludge bulking. The best sludge blanket depth is dependent on such factors as hydraulic load, clarifier design, and sludge characteristics. The best blanket depth must be determined on an individual basis by experimentation.

Note: When measuring sludge blanket depth, it is general practice to use a 15- to 20-ft-long clear plastic pipe marked at 6-in. intervals; the pipe is equipped with a ball valve at the bottom.

19.9.10 ACTIVATED SLUDGE OPERATIONAL CONTROL LEVELS

(Much of the information in this section is based on *Activated Sludge Process Control*; see VWCB, 1990.) The operator has two methods available to operate an activated sludge system. The operator can wait until the process performance deteriorates and make drastic changes, or the operator can establish *normal* operational levels and make minor adjustments to keep the process within the established operational levels.

Note: Control levels can be defined as the upper and lower values for a process control variable that can be expected to produce the desired effluent quality.

Although the first method will guarantee that plant performance is always maintained within effluent limitations, the second method has a much higher probability of achieving this objective. This section discusses methods used to establish *normal* control levels for the activated sludge process. Several major factors should be considered when establishing control levels for the activated sludge system, including:

- Influent characteristics
- Industrial contributions
- Process sidestreams
- Seasonal variations
- Required effluent quality

19.9.10.1 Influent Characteristics

Influent characteristics were discussed earlier; however, major factors to consider when evaluating influent characteristics are the nature and volume of industrial contributions to the system. Waste characteristics (BOD, solids, pH, metals, toxicity, and temperature), volume, and discharge pattern (e.g., continuous, slug, daily, weekly) should be evaluated when determining if a waste will require pretreatment by the industry or adjustments to operational control levels.

19.9.10.2 Industrial Contributions

One or more industrial contributors produce a significant portion of the plant loading (in many systems). Identifying and characterizing all industrial contributors is important. Remember that the volume of waste generated may not be as important as the characteristics of the waste. Extremely high-strength wastes can result in organic overloading or poor performance because of insufficient nutrient availability. A second consideration is the presence of materials that even in small quantities are toxic to the process microorganisms or that create a toxic condition in the plant effluent or plant sludge. Industrial contributions to a biological treatment system should be thoroughly characterized prior to acceptance, monitored frequently, and controlled by either local ordinance or implementation of a pretreatment program.

19.9.10.3 Process Sidestreams

Process sidestreams are flows produced in other treatment processes that must be returned to the wastewater system for treatment prior to disposal. Examples of process sidestreams include the following:

- Thickener supernatant
- Aerobic and anaerobic digester supernatant
- Liquids removed by sludge dewatering processes (filtrate, centrate, and subnate)
- Supernatant from heat treatment and chlorine oxidation sludge treatment processes

Testing these flows periodically to determine both their quantity and strength is important. In many treatment systems, a significant part of the organic or hydraulic loading for the plant is generated by sidestream flows. The contribution of the plant sidestream flows can significantly change the operational control levels of the activated sludge system.

19.9.10.4 Seasonal Variations

Seasonal variations in temperature, oxygen solubility, organism activity, and waste characteristics may require several normal control levels for the activated sludge process; for example, during cold months of the year, aeration tank solids levels may have to be maintained at significantly higher level than are required during warm weather. Likewise, the aeration rate may be controlled by the mixing requirements of the system during the colder months and by the oxygen demand of the system during the warm months.

19.9.10.5 Control Levels at Startup

Control levels for an activated sludge system during startup are usually based on design engineer recommendations or information available from recognized reference sources. Although these levels provide a starting point, both the process control parameter sensitivity and control levels should be established on a plant-by-plant basis. During the first 12 months of operation, it is important to evaluate all potential process control options to determine the following:

- Sensitivity to effluent quality changes
- Seasonal variability
- Potential problems

19.9.11 Visual Indicators for Influent or Aeration Tanks

Wastewater operators are required to monitor or to make certain observations of treatment unit processes to ensure optimum performance and to make adjustments when required. When monitoring the operation of an aeration tank, the operator should look for the three physical parameters—turbulence, surface foam and scum, and sludge color and odor—that aid in determining how the process is operating and indicate if any operational adjustments should be made. This information should be recorded each time operational tests are performed. Aeration tank and secondary settling tank observations are summarized in the following sections. Remember that many of these observations are very subjective and must be based on experience. Plant personnel must be properly trained on the importance of ensuring that recorded information is consistent throughout the operating period.

19.9.11.1 Turbulence

Normal operation of an aeration basin includes a certain amount of turbulence. This turbulent action is, of course, required to ensure a consistent mixing pattern; however, whenever excessive, deficient, or nonuniform mixing occurs, adjustments to air flow may be necessary, or the diffusers may require cleaning or replacement.

19.9.11.2 Surface Foam and Scum

The type, color, and amount of foam or scum present may indicate the required wasting strategy to be employed. Types of foam include the following:

- *Fresh, crisp, white foam*—Moderate amounts of crisp white foam are usually associated with activated sludge processes producing an excellent final effluent. *Adjustment:* None, normal operation.
- *Thick, greasy, dark tan foam*—A thick, greasy, dark tan or brown foam or scum normally indicates an old sludge that is overoxidized with a high mixed liquor concentration and waste rate that is too high. *Adjustment:* More wasting required for old sludge.
- *White billowing foam*—Large amounts of a white, soapsuds-like foam indicate a very young, underoxidized sludge. *Adjustment:* Less wasting required for young sludge.

19.9.11.3 Sludge Color and Odor

Though not as reliable an indicator of process operations as foam, sludge color and odor are also useful indicators. Colors and odors that are important include the following:

- *Chocolate brown/earthy odor* indicates normal operation.
- *Light tan or brown/no odor* indicates sand and clay from infiltration and inflow. *Adjustment:* Decrease wasting for extremely young sludge.
- *Dark brown/earthy odor* indicates old sludge, high solids. *Adjustment:* Increase wasting.
- *Black color/rotten-egg odor* indicates septic conditions, low dissolved oxygen concentration, and rate of air flow that is too low. *Adjustment:* Increase aeration.

19.9.11.4 Mixed Liquor Color

A light chocolate brown mixed liquor color indicates a well-operated activated sludge process.

19.9.12 FINAL SETTLING TANK (CLARIFIER) OBSERVATIONS

Settling tank observations include flow pattern (normally uniform distribution), settling, amount and type of solids leaving with the process effluent (normally very low), and the clarity or turbidity of the process effluent (normally very clear). Observations should include the following:

- *Sludge bulking* occurs when solids are evenly distributed throughout the tank and leaving over the weir in large quantities.
- *Sludge solids washout* occurs when the sludge blanket is down but solids are flowing over the effluent weir in large quantities. Control tests indicate good quality sludge.
- *Clumping* occurs when large clumps or masses of sludge (several inches or more) rise to the top of the settling tank.
- *Ashing* occurs when fine particles of gray to white material are flowing over the effluent weir in large quantities.
- *Straggler floc* is comprised of small, almost transparent, very fluffy, buoyant solids particles (1/8 to 1/4 in. in diameter) that rise to the surface; usually accompanied by a very clean effluent. New growth is most notable in the early morning hours. Sludge age is slightly below optimum.
- *Pin floc* is comprised of very fine solids particles, usually less than 1/32 in. in diameter, that are suspended throughout lightly turbid liquid; usually the result of an overoxidized sludge.

19.9.13 PROCESS CONTROL SAMPLING AND TESTING

The activated sludge process generally requires more sampling and testing to maintain adequate process control than any of the other unit processes in the wastewater treatment system. During periods of operational problems, both the parameters tested and the frequency of testing may increase substantially. Process control testing may include settleability testing to determine the settled sludge volume, suspended solids testing to determine influent and mixed liquor suspended solids, return activated sludge solids and waste activated sludge concentrations, determination of the volatile content of the mixed liquor suspended solids, dissolved oxygen and pH of the aeration tank, BOD_5 or chemical oxygen demand (COD) of the aeration tank influent and process effluent, and microscopic evaluation of the activated sludge to determine the predominant organism. The following sections describe most of the common process control tests.

19.9.13.1 Aeration Influent and Effluent Sampling

19.9.13.1.1 pH

pH is tested daily with a sample taken from the aeration tank influent and process effluent. pH is normally close to 7.0 (normal), with the best pH range being 6.5 to 8.5 (a pH range of 6.5 to 9.0 is satisfactory). A pH of >9.0 may indicate toxicity from an industrial waste contributor. A pH of <6.5 may indicate loss of flocculating organisms, potential toxicity, industrial waste contributors, or acid storm flow. Keep in mind that the effluent pH may be lower because of nitrification.

19.9.13.1.2 Temperature

Temperature is important because as the:

Temperature increases:	Organism activity increases.
	Aeration efficiency decreases.
	Oxygen solubility decreases.
Temperature decreases:	Organism activity decreases.
	Aeration efficiency increases.
	Oxygen solubility increases.

19.9.13.1.3 Dissolved Oxygen

The content of dissolved oxygen (DO) in the aeration process is critical to performance. DO should be tested at least daily (peak demand). Optimum is determined for individual plants, but normal is from 1 to 3 mg/L. If the system contains too little DO, the process will become septic. If it contains too much DO, energy and money are wasted.

19.9.13.1.4 Settled Sludge Volume (Settleability)

Settled sludge volume (SSV) is determined at specified times during sample testing. Both 30- and 60-minute observations are used for control. Subscripts (e.g., SSV_{30} and SSV_{60}) indicate the settling time. The test is performed on aeration tank effluent samples.

$$SSV = \frac{Settled\ sludge\ (mL) \times 1000\ mL/L}{Sample\ (mL)} \quad (19.37)$$

$$\%SSV = \frac{\text{Settled sludge (mL)} \times 100}{\text{Sample (mL)}} \quad (19.38)$$

Under normal conditions, sludge settles as a mass, producing clear supernatant with SSV_{60} in the range of 400 to 700 ml/L. When higher values are indicated, this may indicate excessive solids (old sludge) or bulking conditions. Rising solids (if sludge is well oxidized) may rise after 2 or more hours; however, rising solids in less than 1 hour indicate a problem.

Note: Running the settleability test with a diluted sample can assist in determining if the activated sludge is old (too many solids) or bulking (not settling). Old sludge will settle to a more compact level when diluted.

19.9.13.1.5 Centrifuge Testing

The centrifuge test provides a quick, relatively easy control test for the solids level in the aerator but does not usually correlate with MLSS results. Results are directly affected by variations in sludge quality.

19.9.13.1.6 Alkalinity

Alkalinity is essential to biological activity. Nitrification requires 7.3 mg/L alkalinity per mg/L total Kjeldahl nitrogen.

19.9.13.1.7 BOD$_5$

Testing showing an increase in BOD$_5$ indicates increased organic loading; a decrease in BOD$_5$ indicates decreased organic loading.

19.9.13.1.8 Total Suspended Solids

An increase in total suspended solids indicates an increase in organic loading; a decrease in TSS indicates a decrease in organic loading.

19.9.13.1.9 Total Kjeldahl Nitrogen

Total Kjeldahl nitrogen determination is required to monitor process nitrification status and to determine alkalinity requirements.

19.9.13.1.10 Ammonia Nitrogen

Determination of ammonia nitrogen is required to monitor process nitrification status.

19.9.13.1.11 Metals

Metal contents are measured to determine toxicity levels.

19.9.13.2 Aeration Tank

19.9.13.2.1 pH

Normal pH range in the aeration tank is 6.5 to 9.0. Decreases in pH indicate process sidestreams or that insufficient alkalinity is available.

19.9.13.2.2 Dissolved Oxygen

The normal range of dissolved oxygen in an aeration tank is from 1 to 3 mg/L. Decreases in dissolved oxygen levels may indicate increased activity, increased temperature, increased organic loading, or decreased MLSS/MLVSS. An increase in dissolved oxygen could be indicative of decreased activity, decreased temperature, decreased organic loading, increased MLSS/MLVSS, or influent toxicity.

19.9.13.2.3 Dissolved Oxygen Profile

All dissolved oxygen profile readings should be >0.5 mg/L. Readings of <0.5 mg/L indicate inadequate aeration or poor mixing.

19.9.13.2.4 Mixed Liquor Suspended Solids

The range of mixed liquor volatile suspended solids is determined by the process modification used. When MLSS levels increase, more solids, organisms, and an older, more oxidized sludge are typical.

19.9.13.2.5 Microscopic Examination

The activated sludge process cannot operate as designed without the presence of microorganisms. Thus, microscopic examination of an aeration basin sample to determine the presence and the type of microorganisms is important. Different species prefer different conditions; therefore, the presence of various species can indicate process condition.

Note: It is important to point out that, during microscopic examination, identifying all organisms present is not required, but identification of the predominant species is.

Table 19.7 lists process conditions indicated by the presence and population of certain microorganisms.

19.9.13.3 Interpretation

Routine process control identification can be limited to the general category of organisms present. For troubleshooting more difficult problems, a more detailed study of organism distribution may be required (the knowledge required to perform this type of detailed study is beyond the scope of this text). The major categories of organisms found in the activated sludge are:

- Protozoa
- Rotifers
- Filamentous organisms

Note: Bacteria are the most important microorganisms in the activated sludge. They perform most of the stabilization or oxidation of the organic matter and are normally present in extremely large numbers. They are not, however, normally visible with a

TABLE 19.7
Process Condition vs. Organisms Present

Process Condition	Organism Population
Poor BOD$_5$ and TSS removal	Predominance of amoebae and flagellates
No floc formation	Mainly dispersed bacteria
Very cloudy effluent	A few ciliates present
Poor-quality effluent	Predominance of amoebae and flagellates
Dispersed bacteria	Some free-swimming ciliates
Some free-swimming ciliates	
Some floc formation	
Cloudy effluent	
Satisfactory effluent	Predominance of free-swimming ciliates
Good floc formation	Few amoebae and flagellates
Good settleability	
Good clarity	
High-quality effluent	Predominance of stalked ciliates
Excellent floc formation	Some free-swimming ciliates
Excellent settleability	A few rotifers
High effluent clarity	A few flagellates
High effluent TSS and low BOD$_5$	Predominance of rotifers
High settled sludge volume	Large numbers of stalked ciliates
Cloudy effluent	A few free-swimming ciliates; no flagellates

conventional microscope operating at the recommended magnification and are not included in the Table 19.7 list of indicator organisms.

Note: The presence of free-swimming and stalked ciliates, some flagellates, and rotifers in mixed liquor indicates a balanced, properly settling environment.

19.9.13.3.1 Protozoa

Protozoa are secondary feeders in the activated sludge process (secondary as feeders, but nonetheless definitely important to the activated sludge process). Their principal functions are to remove (eat or crop) dispersed bacteria and to help produce a clear process effluent. To help gain an appreciation for the role of protozoa in the activated sludge process, consider the following explanation.

The activated sludge process is typified by the successive development of protozoa and mature floc particles. This succession can be evidenced by the type of dominant protozoa present. At the start of the activated process (or recovery from an upset condition), the amoebae dominate.

Note: Amoebae have very flexible cell walls and move by shifting fluids within the cell wall. Amoebae predominate during process startup or during recovery from severe plant upsets.

As the process continues uninterrupted or without upset, small populations of bacteria begin to grow in logarithmic fashion, which, as the population increases, develop into mixed liquor. When this occurs, the flagellates dominate.

Note: Flagellated protozoa typically have single hair-like flagella, or tail, that they use for movement. The flagellate predominates when the MLSS and bacterial populations are low and organic load is high. As the activated sludge grows older and more dense, the flagellates decrease until they are seldom used.

When the sludge attains an age of about 3 days, lightly dispersed floc particles begin to form (flocculation "grows" fine solids into larger, more settleable solids), and bacteria increase. At this point, free-swimming ciliates dominate.

Note: The free-swimming ciliated protozoa have hair-like projections (*cilia*) that cover all or part of the cell. The cilia are used for motion and create currents that carry food to the organism. The free-swimming ciliates are sometimes divided into two subcategories: *free swimmers* and *crawlers*. The free swimmers are usually seen moving through the fluid portion of the activated sludge, while the crawlers appear to be walking or grazing on the activated sludge solids. The free-swimming ciliated protozoa usually predominate when large numbers of dispersed bacteria are present that can be used as food. Their predominance indicates a process nearing optimum conditions and effluent quality.

The process continues with floc particles beginning to stabilize, taking on irregular shapes, and beginning to show filamentous growth. At this stage, the crawling ciliates dominate. Eventually, mature floc particles develop and increase in size, and large numbers of crawling and stalked ciliates are present. When this occurs, the succession process has reached its terminal point. The succession of protozoan and mature floc particle development just described details the occurrence of phases of development in a step-by-step progression. Protozoan succession is also based on other factors, including dissolved oxygen and food availability.

Probably the best way to understand protozoan succession based on dissolved oxygen and food availability is to view the aeration basin of a wastewater treatment plant as a "stream within a container." Using the *saprobity system* to classify the various phases of the activated sludge process in relation to the self-purification process that takes place in a stream, we can see a clear relationship between the two processes based on available dissolved oxygen and food supply. Any change in the relative numbers of bacteria in the activated sludge process causes a corresponding change in the population of the microorganism. Decreases in bacteria increase competition between protozoa and result in the secession of dominant groups of protozoa.

The success or failure of protozoa to capture bacteria depends on several factors. Those with more advanced locomotion capability are able to capture more bacteria. Individual protozoan feeding mechanisms are also important in the competition for bacteria. At the beginning of the activated sludge process, amoebae and flagellates are the first protozoan groups to appear in large numbers. They can survive on smaller quantities of bacteria because their energy requirements are lower than other protozoan types. Because few bacteria are present, competition for dissolved substrates is low; however, as the bacteria population increases, these protozoa are not able to compete for available food. This is when the next group of protozoa (the free-swimming protozoa) enters the scene.

The free-swimming protozoa take advantage of the large populations of bacteria because they are better equipped with food-gathering mechanisms than the amoebae and flagellates. The free swimmers are important for their insatiable appetites for bacteria and also in floc formation. Secreting polysaccharides and mucoproteins that are absorbed by bacteria—which make the bacteria "sticky" through biological agglutination (biological gluing together)—allows them to stick together and, more importantly, to stick to floc. Thus, large quantities of floc are prepared for removal from secondary effluent and are either returned to aeration basins or wasted. The crawlers and stalked ciliates succeed the free swimmers.

Note: Stalked ciliated protozoa are attached directly to the activated sludge solids by a stalk. In some cases, the stalk is rigid and fixed in place; in others, the organism can move (contract or expand the stalk) to change its position. The stalked ciliated protozoa normally have several cilia that are used to create currents, which carry bacteria and organic matter to it. The stalked ciliated protozoa predominate when the dispersed bacteria population decreases and does not provide sufficient food for the free swimmers. Their predominance indicates a stable process, operating at optimum conditions.

The free swimmers are replaced in part because the increasing level of mature floc retards their movement. Additionally, the type of environment that is provided by the presence of mature floc is more suited to the needs of the crawlers and stalked ciliates. The crawlers and stalked ciliates also aid in floc formation by adding weight to floc particles, thus enabling removal.

19.9.13.3.2 Rotifers
Rotifers are a higher life form normally associated with clean, unpolluted waters. Significantly larger than most of the other organisms observed in activated sludge, rotifers can utilize other organisms, as well as organic matter, as their food source. Rotifers are usually the predominant organism; the effluent will usually be cloudy (pin or ash floc) and will have very low BOD_5.

19.9.13.3.3 Filamentous Organisms
Filamentous organisms (e.g., bacteria, fungi) occur whenever the environment of the activated sludge favors their predominance. They are normally present in small amounts and provide the basic framework for floc formation. When the environmental conditions (e.g., pH, nutrient levels, DO) favor their development, they become the predominant organisms. When this occurs, they restrict settling, and the condition known as *bulking* occurs.

Note: A microorganism examination of activated sludge is a useful control tool. In attempting to identify the microscopic contents of a sample, the operator should try to identify the predominant groups of organisms.

Note: During microscopic examination of the activated sludge, a predominance of amoebae indicates that the activated sludge is very young.

19.9.13.4 Settling Tank Influent
19.9.13.4.1 Dissolved Oxygen
The dissolved oxygen level of the activated sludge settling tank should be 1 to 3 mg/L; lower levels may result in rising sludge.

19.9.13.4.2   pH

Normal pH range in an activated sludge settling tank should be maintained between 6.5 to 9.0. Decreases in pH may indicate alkalinity deficiency.

19.9.13.4.3   Alkalinity

A lack of alkalinity in an activated sludge settling tank will prevent nitrification.

19.9.13.4.4   Total Suspended Solids

Mixed liquor suspended solids sampling and testing are required for determining solids loading, mass balance, and return rates.

19.9.13.4.5   Settled Sludge Volume (Settleability)

Settled sludge volume (SSV) is determined at specified times during sample testing:

- *Normal operation*—When the process is operating properly, the solids will settle as a blanket (a mass), with a crisp or sharp edge between the solids and the liquor above. The liquid over the solids will be clear, with little or no visible solids remaining in suspension. Settled sludge volume at the end of 30 to 60 min will be in the range of 400 to 700 mL.
- *Old or overoxidized activated sludge*—When the activated sludge is overoxidized, the solids will settle as discrete particles. The edge between the solids and liquid will be fuzzy, with a large number of visible solids (e.g., pin floc, ash floc) in the liquid. The settled sludge volume at the end of 30 or 60 min will be greater than 700 mL.
- *Young or underoxidized activated sludge*—When the activated sludge is underoxidized, the solids settle as discrete particles, and the boundary between the solids and the liquid is poorly defined. Large amounts of small visible solids are suspended in the liquid. The settled sludge volume after 30 to 60 min will usually be less than 400 mL.
- *Bulking activated sludge*—When the activated sludge is experiencing a bulking condition, very little or no settling is observed:

$$SSV = \frac{\text{Settled sludge (mL)} \times 1000 \text{ mL/L}}{\text{Sample (mL)}} \quad (19.39)$$

$$\%SSV = \frac{\text{Settled sludge (mL)} \times 100}{\text{Sample (mL)}} \quad (19.40)$$

Note: Running the settleability test with a diluted sample can assist in determining if the activated sludge is old (too many solids) or bulking (not settling). Old sludge will settle to a more compact level when diluted.

19.9.13.4.6   Flow

Monitoring flow in settling tank influent is important for determination of mass balance.

19.9.13.4.7   Jar Tests

Jar tests are performed as required on settling tank influent and are beneficial in determining the best flocculant aid and appropriate doses to improve solids capture during periods of poor settling.

19.9.13.5   Settling Tank

19.9.13.5.1   Sludge Blanket Depth

As mentioned, sludge blanket depth refers to the distance from the surface of the liquid to the solids–liquid interface or the thickness of the sludge blanket as measured from the bottom of the tank to the solids–liquid interface. Part of the operator's sampling routine, this measurement is taken directly in the final clarifier. Sludge blanket depth is dependent on hydraulic load, return rate, clarifier design, waste rate, sludge characteristics, and temperature. If all other factors remain constant, the blanket depth will vary with amount of solids in the system and the return rate; thus, it will vary throughout the day.

Note: Depth of sludge blanket provides an indication of sludge quality; it is used as a trend indicator. Many factors affect test results.

19.9.13.5.2   Suspended Solids and Volatile Suspended Solids

Suspended solids and volatile suspended solids concentrations of the mixed liquor (i.e., mixed liquor suspended solids), the return activated sludge (RAS), and waste activated sludge (WAS) are routinely sampled and tested because they are critical to process control.

19.9.13.6   Settling Tank Effluent

19.9.13.6.1 BOD_5 and Total Suspended Solids

BOD_5 and total suspended solids testing is conducted variably (daily, weekly, and monthly). Increases indicate that treatment performance is decreasing; decreases indicate that treatment performance is increasing.

19.9.13.6.2   Total Kjeldahl Nitrogen

Total Kjeldahl nitrogen (TKN) sampling and testing are variable. An increase in TKN indicates that nitrification is decreasing; a decrease in TKN indicates that nitrification is increasing.

19.9.13.6.3   Nitrate Nitrogen

Nitrate nitrogen sampling and testing are variable. Increases in nitrate nitrogen indicate increasing nitrification or industrial contribution of nitrates; a decrease indicates reduced nitrification.

19.9.13.6.4   Flow

Settling tank effluent flow is sampled and tested daily. Results are required for several process control calculations.

19.9.13.7   Return Activated Sludge and Waste Activated Sludge

19.9.13.7.1   Total Suspended Solids and Volatile Suspended Solids

Total suspended solids and total volatile suspended solids concentrations of the mixed liquor (mixed liquor suspended solids), return activated sludge (RAS), and waste activated sludge (WAS) are routinely sampled (using grab or composite samples) and tested, because they are critical to process control. The results of the suspended and volatile suspended tests can be used directly or to calculate such process control figures as mean cell residence time (MCRT) or food-to-microorganism ratio (F/M). In most situations, increasing the MLSS produces an older, denser sludge, while decreasing MLSS produces a younger, less dense sludge.

Note: Control of the sludge wasting rate by constant MLVSS concentration involves maintaining a certain concentration of volatile suspended solids in the aeration tank.

Note: The activated sludge aeration tank should be observed daily. Included in this daily observation should be a determination of the type and amount of foam, mixing uniformity, and color.

19.9.13.7.2   Flow

The flow of return activated sludge is tested daily. Test results are required to determine mass balance and for control of the sludge blanket, MLSS, and MLVSS. For waste activated sludge, flow is sampled and tested whenever sludge is wasted. Results are required to determine mass balance and to control solids levels in the process.

19.9.13.8   Process Control Adjustments

In the routine performance of their duties, wastewater operators make process control adjustments to various unit processes, including the activated sludge process. In the following a summary is provided of the process controls available for the activated sludge process and the result that will occur from adjustment of each.

19.9.13.8.1   Return Rate

Condition: Return rate is too high.

Result:

- Hydraulic overloading of aeration and settling tanks
- Reduced aeration time
- Reduced settling time
- Loss of solids over time

Condition: Return rate is too low.

Result:

- Septic return
- Solids buildup in settling tank
- Reduced MLSS in aeration tank
- Loss of solids over weir

19.9.13.8.2   Waste Rate

Condition: Waste rate is too high.

Result:

- Reduced MLSS
- Decreased sludge density
- Increased SVI
- Decreased MCRT
- Increased F/M ratio

Condition: Waste rate is too low.

Result:

- Increased MLSS
- Increased sludge density
- Decreased SVI
- Increased MCRT
- Decreased F/M ratio

19.9.13.8.3   Aeration Rate

Condition: Aeration rate is too high.

Result:

- Wasted energy
- Increased operating cost
- Rising solids
- Breakup of activated sludge

Condition: Aeration rate is too low.

Result:

- Septic aeration tank
- Poor performance
- Loss of nitrification

19.9.13.9   Troubleshooting Operational Problems

Without a doubt, the most important dual function performed by the wastewater operator is the identification of process control problems and implementing the appropriate actions to correct the problems. Table 19.8 shows typical aeration system operational problems with their symptoms, causes, and the appropriate corrective actions necessary to restore the unit process to a normal or optimal performance level.

19.9.13.10   Process Control Calculations

As with other wastewater treatment unit processes, process control calculations are important tools used by the operator to optimize and control process operations. In this section, we review the most frequently used activated sludge calculations.

TABLE 19.8
Symptoms, Causes, and Corrective Actions for Aeration System Problems

Symptom	Cause	Corrective Action
The solids blanket is flowing over the effluent weir (classic bulking); settleability test shows no settling.	Organic overloading	Reduce organic loading.
	Low pH	Add alkalinity.
	Filamentous growth	Add nutrients; add chlorine or peroxide to return.
	Nutrient deficiency	Add nutrients.
	Toxicity	Identify source; implement pretreatment.
	Overaeration	Reduce aeration during low-flow periods.
Solids settled properly in the settleability test but large amounts of solids are being lost over the effluent weir.	Billowing solids due to short-circuiting	Identify short-circuiting cause and eliminate if possible.
Large amounts of small pinhead-sized solids are leaving the settling tank.	Old sludge	Reduce sludge age (gradual change is best); increase waste rate.
Large amounts of light floc (low BOD_5 and high solids) are leaving the settling tank.	Excessive turbulence	Decrease turbulence (adjust aeration during low flows).
	Extremely old sludge	Reduce age; increase waste.
Large amounts of small translucent particles (1/16 to 1/8 in.) are leaving the settling tank.	Rapid solids growth	Decrease waste.
	Slightly young activated sludge	Increase sludge age.
Solids are settling properly but rise to the surface within a short time. Many clumps of solids, both small (1/4-in. diameter) and large (several feet in diameter), are on the surface of the settling tank.	Denitrification	Increase rate of return; adjust sludge age to eliminate nitrification.
	Overaeration	Reduce aeration.
Return activated sludge has a rotten-egg odor.	Septic return	Increase aeration rate.
	Return rate too low	Increase rate of return.
Activated sludge organisms die within a short time.	Toxic material in the influent	Isolate activated sludge (if possible); return all available solids; stop wasting; increase return rate; implement pretreatment program.
The surface of the aeration tank is covered with thick, greasy foam.	Extremely old activated sludge	Reduce activated sludge age; increase wasting; use foam-control sprays.
	Excessive grease and oil in system	Improve grease removal; use foam-control sprays; implement pretreatment program.
	Froth-forming bacteria	Remove froth-forming bacteria.
Large clouds of billowing white foam are on the surface of the aeration tank.	Young activated sludge	Increase sludge age; decrease wasting; use foam-control sprays.
	Low solids in aeration tank	Increase sludge age; decrease wasting; use foam-control sprays.
	Surfactants (detergents)	Eliminate surfactants; use foam-control sprays; add antifoam.

19.9.13.10.1 Settled Sludge Volume

Settled sludge volume (SSV) is the volume that a settled activated sludge occupies after a specified time. The settling time may be shown as a subscript (e.g., SSV_{60} indicates the reported value was determined at 60 min). The settled sludge volume can be determined for any time interval; however, the most common values are the 30-min reading (SSV_{30}) and 60-min reading (SSV_{60}). The settled sludge volume can be reported as milliliters of sludge per liter of sample (mL/L) or as a percent settled sludge volume:

$$\text{Settled sludge (mL/L)} = \frac{\text{Settled sludge vol. (mL)}}{\text{Sample vol. (L)}} \quad (19.41)$$

Note: 1000 milliliters = 1 liter

$$\text{Sample vol. (L)} = \text{Sample vol. (mL)}/1000 \text{ mL/L} \quad (19.42)$$

$$\%\text{SSV} = \frac{\text{Settled sludge volume (mL)} \times 100}{\text{Sample volume (mL)}} \quad (19.43)$$

■ EXAMPLE 19.38

Problem: Using the information provided below, calculate the SSV_{30} and the $\%SSV_{60}$:

Start	2500 mL
15 min	2250 mL
30 min	1800 mL
45 min	1700 mL
60 min	1600 mL

Solution:

$$\text{Settled sludge vol. }(SSV_{30}) = \frac{1800\text{ mL}}{2.5\text{ L}} = 720\text{ mL/L}$$

$$\%\text{ Settled sludge vol. }(SSV_{60}) = \frac{1600\text{ mL}\times100}{2500\text{ mL}} = 64\%$$

19.9.13.10.2 Estimated Return Rate

Many different methods are available for estimating the proper return sludge rate. A simple method described in the *Operation of Wastewater Treatment Plants, Field Study Program* (see Table 19.1) was developed by the California State University, Sacramento. It uses the 60-min percent settled sludge volume ($\%SSV_{60}$), which can provide an approximation of the appropriate return activated sludge rate. The results of this calculation can then be adjusted based on sampling and visual observations to develop the optimum return sludge rate.

Note: The $\%SSV_{60}$ must be converted to a decimal percent and total flow rate (i.e., wastewater flow and current return rate in million gallons per day must be used).

$$\text{Est. return rate(MGD)} = \left[\begin{pmatrix}\text{Influent flow (MGD)}\\+\text{ current return flow (MGD)}\end{pmatrix}\\\times\%SSV_{60}\right]$$

- Assumes $\%SSV_{60}$ is representative.
- Assumes return rate, in percent equals $\%SSV_{60}$.

Actual return rate is normally set slightly higher to ensure that organisms are returned to the aeration tank as quickly as possible. The rate of return must be adequately controlled to prevent the following:

- Aeration and settling hydraulic overloads
- Low MLSS levels in the aerator
- Organic overloading of aeration
- Solids loss due to excessive sludge blanket depth

■ EXAMPLE 19.39

Problem: The influent flow rate is 4.2 MGD and the current return activated sludge flow rate is 1.5 MGD. The SSV_{60} is 38%. Based on this information, what should be the return sludge rate in million gallons per day (MGD)?

Solution:

$$\text{Return} = (4.2\text{ MGD}+1.5\text{ MGD})\times0.38 = 2.2\text{ MGD}$$

19.9.13.10.3 Sludge Volume Index

Sludge volume index (SVI) is a measure of the settling quality (a quality indicator) of the activated sludge. As the SVI increases, the sludge settles more slowly, does not compact as well, and is likely to result in an increase in effluent suspended solids. As the SVI decreases, the sludge becomes more dense, settling is more rapid, and the sludge is becoming older. SVI is the volume in milliliters occupied by 1 gram of activated sludge. The settled sludge volume (SSV) (mL/L) and the mixed liquor suspended solids (MLSS) (mg/L) are required for this calculation.

$$SVI = \frac{SSV\text{ (mL/L)}\times1000}{MLSS\text{ (mg/L)}} \qquad (19.44)$$

■ EXAMPLE 19.40

Problem: The SSV_{30} is 365 mL/L and the MLSS is 2365 mg/L. What is the SVI?

Solution:

$$SVI = \frac{365\text{ mL/L}\times1000}{2365\text{ mg/L}} = 154.3$$

The SVI equals 154.3. What does this mean? It means that the system is operating normally with good settling and low effluent turbidity. How do we know this? Another good question. We know this because we can compare the 154.3 result with the expected conditions listed below:

SVI Value	Expected Condition
Less than 100	Old sludge; possible pin floc; effluent turbidity increasing
100 to 200	Normal operation; good settling; low effluent turbidity
Greater than 250	Bulking sludge; poor settling; high effluent turbidity

The SVI is best used as a trend indicator to evaluate what is occurring compared to previous SVI values. Based on this evaluation, the operator may determine if the SVI trend is increasing or decreasing (refer to the following chart):

SVI Value	Result	Adjustment
Increasing	Sludge is becoming less dense.	Decrease waste rate.
	Sludge is either younger or bulking.	Increase return rate.
	Sludge will settle more slowly.	
	Sludge will compact less.	
Decreasing	Sludge is becoming more dense.	Increase waste rate.
	Sludge is becoming older.	
	Sludge will settle more rapidly.	Decrease return rate.
	Sludge will compact more with no other process changes.	
Holding constant	Sludge should maintain its current characteristics.	No changes are indicated.

19.9.13.10.4 Waste Activated Sludge

The quantity of solids removed from the process as waste activated sludge (WAS) is an important process control parameter that operators need to be familiar with and, more importantly, should know how to calculate:

$$\text{Waste (lb/day)} = \begin{bmatrix} \text{WAS conc. (mg/L)} \\ \times \text{WAS flow (MGD)} \\ \times 8.34 \text{ lb/MG/mg/L} \end{bmatrix} \quad (19.45)$$

■ EXAMPLE 19.41

Problem: The operator wastes 0.44 MGD of activated sludge. The waste activated sludge has a solids concentration of 5540 mg/L. How many pounds of waste activated sludge are removed from the process?

Solution:

$$\text{Waste (lb/day)} = \begin{bmatrix} 5540 \text{ mg/L} \times 0.44 \text{ MGD} \\ \times 8.34 \text{ lb/MG/mg/L} \end{bmatrix}$$

$$= 20,329.6 \text{ lb/day}$$

19.9.13.10.5 Food-to-Microorganism Ratio

The food-to-microorganism ratio (F/M ratio) is a process control calculation used in many activated sludge facilities to control the balance between available food materials (BOD or COD) and available organisms (mixed liquor volatile suspended solids, or MLVSS). The chemical oxygen demand (COD) test is sometimes used, because the results are available in a relatively short period of time. To calculate the F/M ratio, the following information is required:

• Aeration tank influent flow rate (MGD)
• Aeration tank influent BOD or COD (mg/L)
• Aeration tank MLVSS (mg/L)
• Aeration tank volume (MG)

$$\text{F/M ratio} = \frac{\begin{array}{c} \text{Primary effluent COD/BOD (mg/L)} \\ \times \text{flow (MGD)} \times 8.34 \text{ lb/MG/mg/L} \end{array}}{\begin{array}{c} \text{MLVSS (mg/L)} \times \text{aerator volume (MG)} \\ \times 8.34 \text{ lb/MG/mg/L} \end{array}} \quad (19.46)$$

Typical F/M ratios for activated sludge processes are shown below:

Process	BOD₅ (lb)/ MLVSS (lb)	COD (lb)/ MLVSS (lb)
Conventional	0.2–0.4	0.5–1.0
Contact stabilization	0.2–0.6	0.5–1.0
Extended aeration	0.05–0.15	0.2–0.5
Oxidation ditch	0.05–0.15	0.2–0.5
Pure oxygen	0.25–1.0	0.5–2.0

■ EXAMPLE 19.42

Problem: Given the following data, what is the F/M ratio?

Primary effluent flow = 2.5 MGD
Primary effluent BOD = 145 mg/L
Primary effluent TSS = 165 mg/L
Effluent flow = 2.2 MGD
Effluent BOD = 22 mg/L
Effluent TSS = 16 mg/L
Aeration volume = 0.65 MG
Settling volume = 0.30 MG
MLSS = 3650 mg/L
MLVSS = 2550 mg/L
% Waste volatile = 71%
Desired F/M = 0.3

Solution:

$$\text{F/M ratio} = \frac{145 \text{ mg/L} \times 2.2 \text{ MGD} \times 8.34 \text{ lb/MG/mg/L}}{2550 \text{ mg/L} \times 0.65 \text{ MG} \times 8.34 \text{ lb/MG/mg/L}}$$

$$= 0.19 \text{ lb BOD/lb MLVSS}$$

Note: If the MLVSS concentration is not available, it can be calculated if the percent volatile matter (%VM, expressed as a decimal) of the mixed liquor suspended solids (MLSS) is known:

$$\text{MLVSS} = \text{MLSS} \times \%\text{VM} \quad (19.47)$$

Note: The "F" value in the F/M ratio for computing loading to an activated sludge process can be either BOD or COD. Remember that the reason for sludge production in the activated sludge process is to convert BOD to bacteria. One advantage of using COD over BOD for analysis of organic load is that COD is more accurate.

■ EXAMPLE 19.43

Problem: The aeration tank contains 2985 mg/L of MLSS. Laboratory tests indicate the MLSS is 66% volatile matter. What is the MLVSS concentration in the aeration tank?

Solution:

MLVSS (mg/L) = 2985 mg/L × 0.66 = 1970 mg/L

F/M Ratio Control

Maintaining the F/M ratio within a specified range can be an excellent control method. Although the F/M ratio is affected by adjustment of the return rates, the most practical method for adjusting the ratio is through waste rate adjustments.

Increasing the rate will:

• Decrease the MLVSS
• Increase the F/M ratio

Decreasing the waste rate will:

- Increase the MLVSS
- Decrease the F/M ratio

The desired F/M ratio must be established on a plant-by-plant basis. Comparison of F/M ratios with plant effluent quality is the primary means for identifying the most effective range for individual plants, when the range of F/M values that produce the desired effluent quality has been established.

Required MLVSS Quantity (Pounds)

The pounds of MLVSS required in the aeration tank to achieve the optimum F/M ratio can be determined from the average influent food (BOD or COD) and desired F/M ratio:

$$\text{MLVSS (lb)} = \frac{\text{Primary effluent BOD or COD} \times \text{flow (MGD)} \times 8.34}{\text{Desired F/M ratio}} \quad (19.48)$$

The required pounds of MLVSS determined by this calculation can then be converted to a concentration value by:

$$\text{MLVSS (mg/L)} = \frac{\text{Desired MLVSS (lb)}}{\text{Aeration volume (MG)} \times 8.34} \quad (19.49)$$

■ EXAMPLE 19.44

Problem: The aeration tank influent flow rate is 4.0 MGD, and the influent COD is 145 mg/L. The aeration tank volume is 0.65 MG. The desired F/M ratio is 0.3 lb COD/lb MLVSS. (1) How many pounds of MLVSS must be maintained in the aeration tank to achieve the desired F/M ratio? (2) What is the required concentration of MLVSS in the aeration tank?

Solution:

$$\text{MLVSS (lb)} = \frac{145 \text{ mg/L} \times 4.0 \text{ MGD} \times 8.34 \text{ lb/gal}}{0.3 \text{ lb COD/lb MLVSS}}$$
$$= 16,124 \text{ lb MLVSS}$$

$$\text{MLVSS (mg/L)} = \frac{16,124 \text{ MLVSS}}{0.65 \text{ MG} \times 8.34} = 2974 \text{ mg/L MLVSS}$$

Calculating Waste Rates Using F/M Ratio

Maintaining the desired F/M ratio is accomplished by controlling the MLVSS level in the aeration tank. This may be accomplished by adjustment of return rates; however, the most practical method is by proper control of the waste rate:

$$\text{Waste volatile solids (lb/day)} = \begin{bmatrix} \text{Actual MLVSS (lb)} \\ -\text{desired MLVSS (lb)} \end{bmatrix} \quad (19.50)$$

If the desired MLVSS is greater than the actual MLVSS, wasting is stopped until the desired level is achieved.

Practical considerations require that the required waste quantity be converted to a required volume to waste per day. This is accomplished by converting the waste pounds to flow rate in million gallons per day or gallons per minute.

$$\text{Waste (MGD)} = \frac{\text{Waste volatile (lb/day)}}{\text{Waste volatile concentration (mg/L)} \times 8.34} \quad (19.51)$$

Note: When the F/M ratio is used for process control, the volatile content of the waste activated sludge should be determined.

■ EXAMPLE 19.45

Problem: Given the following information, determine the required waste rate in gallons per minute to maintain an F/M ratio of 0.17 lb COD/lb MLVSS:

Primary effluent COD = 140 mg/L
Primary effluent flow = 2.2 MGD
MLVSS (mg/L) = 3549 mg/L
Aeration tank volume = 0.75 MG
Waste volatile concentration = 4440 mg/L

Solution:

$$\text{Actual MLVSS (lb)} = 3.549 \text{ mg/L} \times 0.75 \text{ MG} \times 8.34$$
$$= 22,199 \text{ lb}$$

$$\text{Required MLVSS (lb)} = \frac{140 \text{ mg/L} \times 2.2 \text{ MGD} \times 8.34}{0.17 \text{ lb COD/lb MLVSS}}$$
$$= 15,110 \text{ lb}$$

$$\text{Waste (lb/day)} = 22,199 \text{ lb} - 15,110 \text{ lb} = 7089 \text{ lb}$$

$$\text{Waste (MGD)} = \frac{7089 \text{ lb/day}}{(4440 \text{ mg/L} \times 8.34)} = 0.19 \text{ MGD}$$

$$\text{Waste (gpm)} = \frac{0.19 \text{ MGD} \times 1,000,000 \text{ gpd/MGD}}{1440 \text{ min/day}}$$
$$= 132 \text{ gpm}$$

19.9.13.10.6 Mean Cell Residence Time

Mean cell residence time (MCRT), sometimes referred to as *sludge retention time*, is a process control calculation used for activated sludge systems. The MCRT calculation illustrated in Example 19.46 uses the entire volume of the activated sludge system (aeration and settling):

$$\text{MRCT (days)} = $$
$$\frac{\begin{bmatrix} \text{MLSS (mg/L)} \times \text{aeration vol. (MG)} \\ + \text{clarifier vol. (MG)} \times 8.34 \end{bmatrix}}{\begin{bmatrix} (\text{WAS (mg/L)} \times \text{WAS flow (MGD)} \times 8.34) \\ + (\text{TSS out (mg/L)} \times \text{flow} \times 8.34) \end{bmatrix}} \quad (19.52)$$

Note: Due to the length of the MCRT equation, the units for the conversion factor 8.34 have not been included. The dimensions for the 8.34 conversion factor are lb/MG/mg/L.

Note: MCRT can be calculated using only the aeration tank solids inventory. When comparing plant operational levels to reference materials, it is important to determine which calculation the reference manual uses to obtain its example values. Other methods are available to determine the clarifier solids concentration; however, the simplest method assumes the average suspended solids concentration is equal to the solids concentration in the aeration tank.

■ **EXAMPLE 19.46**

Problem: Given the following data, what is the MCRT?

    Influent flow = 4.2 MGD
    Influent BOD = 135 mg/L
    Influent TSS = 150 mg/L
    Effluent flow = 4.2 MGD
    Effluent BOD = 22 mg/L
    Effluent TSS = 10 mg/L
    Aeration volume = 1.20 MG
    Settling volume = 0.60 MG
    MLSS = 3350 mg/L
    Waste rate = 0.080 MGD
    Waste concentration = 6100 mg/L
    Desired MCRT = 8.5 days

Solution:

$$\text{MRCT} = \frac{[3350 \text{ mg/L} \times (1.2 \text{ MG} + 0.6 \text{ MG}) \times 8.34]}{\begin{bmatrix} (6100 \text{ mg/L} \times 0.08 \text{ MGD} \times 8.34) \\ + (10 \text{ mg/L} \times 4.2 \text{ MGD} \times 8.34) \end{bmatrix}} = 11.4$$

Mean Cell Residence Time Control
Because it provides an accurate evaluation of the process condition and considers all aspects of the solids inventory, the MCRT is an excellent process control tool. Increases in the waste rate will decrease the MCRT, as will large losses of solids over the effluent weir. Reductions in waste rate will result in increased MCRT values.

Note: You should remember these important process control parameters.

Process Parameters and Impact on MCRT
- To increase F/M, decrease MCRT.
- To increase MCRT, decrease waste rate.
- If the MCRT is increased, the MLTSS and 30-min settling volume increase.
- Return sludge rate has no impact on MCRT.
- MCRT has no impact on F/M when the number of aeration tanks in service is reduced.

Typical MCRT Values
The following chart lists the various aeration process modifications and associated MCRT values:

Process	MCRT (Days)
Conventional	5–15
Step aeration	5–15
Contact stabilization (contact)	5–15
Extended aeration	20–30
Oxidation ditch	20–30
Pure oxygen	8–20

Control Values for MCRT
Control values for the MCRT are normally established based on effluent quality. When the MCRT range required to produce the desired effluent quality has been established, it can be used to determine the waste rate required to maintain it.

Waste Quantities/Requirements
Using the MCRT for process control requires determination of the optimum range for MCRT values. This is accomplished by comparison of the effluent quality with MCRT values. When the optimum MCRT is established, the quantity of solids to be removed (wasted) is determined by:

$$\underset{\text{(lb/day)}}{\text{Waste}} = \begin{bmatrix} \left(\dfrac{\text{MLSS} \times (\text{aerator (MG)} + \text{clarifier (MG)})}{\text{Desired MCRT}} \times 8.34 \right) \\ - (\text{TSS}_{\text{out}} \times \text{flow} \times 8.34) \end{bmatrix} \quad (19.53)$$

■ **EXAMPLE 19.47**

$$\text{Waste} = \begin{bmatrix} \left(\dfrac{3400 \text{ mg/L} \times (1.4 \text{ MG} + 0.50 \text{ MG}) \times 8.34}{8.6 \text{ days}} \right) \\ - (10 \text{ mg/L} \times 5.0 \text{ MGD} \times 8.34) \end{bmatrix}$$

$$= 5848 \text{ lb/day}$$

Waste Rate in Million Gallons/Day
When the quantity of solids to be removed from the system is known, the desired waste rate in million gallons per day can be determined. The unit used to express the rate (MGD, gpd, and gpm) is a function of the volume of waste to be removed and the design of the equipment.

$$\text{Waste (MGD)} = \frac{\text{Waste (lb/day)}}{\text{WAS conc. (mg/L)} \times 8.34} \quad (19.54)$$

$$\text{Waste (gpm)} = \frac{\text{Waste (MGD)} \times 1,000,000 \text{ gpd/MGD}}{1440 \text{ min/day}} \quad (19.55)$$

■ **EXAMPLE 19.48**

Problem: Given the following data, determine the required waste rate to maintain an MCRT of 8.8 days:

> MLSS = 2500 mg/L
> Aeration volume = 1.20 MG
> Clarifier volume = 0.20 MG
> Effluent TSS = 11 mg/L
> Effluent flow = 5.0 MGD
> Waste concentrations = 6000 mg/L

Solution:

$$\text{Waste (lb/day)} = \frac{2500 \text{ mg/L} \times (1.20 + 0.20) \times 8.34}{8.8 \text{ days}}$$
$$- (11 \text{ mg/L} \times 5.0 \text{ MGD} \times 8.34)$$
$$= 3317 \text{ lb/day} - 459 \text{ lb/day} = 2858 \text{ lb/day}$$

$$\text{Waste (MGD)} = \frac{2858 \text{ lb/day}}{(6000 \text{ mg/L} \times 8.34)} = 0.057 \text{ MGD}$$

$$\text{Waste (gpm)} = \frac{0.057 \text{ MGD} \times 1,000,000 \text{ gpd/MGD}}{1440 \text{ min/day}}$$
$$= 40 \text{ gpm}$$

19.9.13.10.7 Mass Balance

Mass balance is based on the fact that solids and BOD are not lost in the treatment system. In simple terms, the mass balance concept states that "what comes in must equal waste that goes out." The concept can be used to verify operational control levels and to determine if potential problems exist within the plant's process control monitoring program.

Note: If influent values and effluent values do not correlate within 10 to 15%, it usually indicates either a sampling or testing error or a process control discrepancy.

Mass balance procedures for evaluating the operation of a settling tank and a biological process are described in this section. Operators should recognize that, although the procedures are discussed in reference to the activated sludge process, the concepts can be applied to any settling or biological process.

Mass Balance: Settling Tank Suspended Solids

The settling tank mass balance calculation assumes that no suspended solids are produced in the settling tank. Any settling tank operation can be evaluated by comparing the solids entering the unit with the solids leaving the tank as effluent suspended solids or as sludge solids. If sampling and testing are accurate and representative, and process control and operation are appropriate, the quantity of sus-

pended solids entering the settling tank should equal (±10%) the quantity of suspended solids leaving the settling tanks as sludge, scum, and effluent total suspended solids.

Note: In most instances, the amount of suspended solids leaving the process as scum is so small that it is ignored in the calculation.

Mass Balance Calculations

$$\text{TSS}_{in} \text{ (lb)} = \text{TSS}_{in} \times \text{flow (MGD)} \times 8.34$$

$$\text{TSS}_{out} \text{ (lb)} = \text{TSS}_{out} \times \text{flow (MGD)} \times 8.34$$

$$\text{Sludge solids} = \left[\begin{array}{c} \text{Sludge pumped (gal)} \\ \times \% \text{ solids} \times 8.34 \end{array} \right] \tag{19.55}$$

$$\% \text{ Mass balance} = \frac{\left[\text{TSS}_{in} \text{ (lb)} - \left(\begin{array}{c} \text{TSS}_{out} \text{ (lb)} \\ + \text{sludge solids (lb)} \end{array} \right) \right] \times 100}{\text{TSS}_{in} \text{ (lb)}}$$

Explanation of Results

- If the mass balance is ±15% or less, the process is considered to be in balance. Sludge removal should be adequate and the sludge blanket depth should be remaining stable. Sampling is considered to be producing representative samples that are being tested accurately.

- If the mass balance is greater than ±15%, more solids are entering the settling tank than are being removed. Sludge blanket depth should be increasing, effluent solids may also be increasing, and effluent quality is decreasing. If the changes described are not occurring, the mass balance may indicate that sample type, location, times, or procedures and testing procedures are not producing representative results.

- If the mass balance is greater than 15%, fewer solids are entering the settling tank than are being removed. The sludge blanket depth should be decreasing; sludge solids concentration may also be decreasing. This could adversely impact sludge treatment processes. If the changes described are not occurring, the mass balance may indicate that sample type, location, times, or procedures and testing procedures are not producing representative results.

■ **EXAMPLE 19.49**

Problem: Given the following data, determine the solids mass balance for the settling tank:

> Influent flow = 2.6 MGD
> Influent TSS = 2445 mg/L

Effluent flow = 2.6 MGD
Effluent TSS = 17 mg/L
Return flow = 0.5 MGD
Return TSS = 8470 mg/L

Solution:

$$TSS_{in} \text{ (lb/day)} = 2445 \text{ mg/L} \times 2.6 \text{ MGD} \times 8.34$$
$$= 53,017 \text{ lb/day}$$

$$TSS_{out} \text{ (lb/day)} = 17 \text{ mg/L} \times 2.6 \text{ MGD} \times 8.34$$
$$= 369 \text{ lb/day}$$

$$\text{Sludge solids out (lb/day)} = 8470 \text{ mg/L} \times 0.5 \text{ MGD} \times 8.34$$
$$= 35,320 \text{ lb/day}$$

$$\% \text{ Mass balance} = \frac{\left[53,017 \text{ lb/day} - \left(\begin{array}{c} 369 \text{ lb/day} \\ + 35,320 \text{ lb/day} \end{array} \right) \right] \times 100}{53,017 \text{ lb/day}}$$
$$= 32.7\%$$

This value indicates that:

- The sampling point, collection procedure, or laboratory procedure is producing inaccurate data upon which to make process control decisions.
- Or, more solids are entering the settling tank each day than are being removed. This should result in either (1) a solids buildup in the settling tank, or (2) a loss of solids over the effluent weir.

Investigate further to determine the specific cause of the imbalance.

Mass Balance: Biological Process
Solids are produced whenever biological processes are used to remove organic matter from wastewater. Mass balance for an aerobic biological process must take into account both the solids removed by physical settling processes and the solids produced by biological conversion of soluble organic matter to insoluble suspended matter or organisms. Research has shown that the amount of solids produced per pound of BOD_5 removed can be predicted based on the type of process being used. Although the exact amount of solids produced can vary from plant to plant, research has developed a series of K factors that can be used to estimate the solids production for plants using a particular treatment process. These average factors provide a simple method to evaluate the effectiveness of a facility's process control program. The mass balance also provides an excellent mechanism to evaluate the validity of process control and effluent monitoring data generated. Table 19.9 lists average K factors

TABLE 19.9
K Conversion Factors

Process	lb Solids/lb BOD_5 Removed
Primary	1.7
Activated sludge with primary	0.7
Activated sludge without primary	
Conventional	0.85
Step feed	0.85
Extended aeration	0.65
Oxidation ditch	0.65
Contact stabilization	1.00
Trickling filter	1.00
Rotating biological contactor	1.00

in pounds of solids produced per pound of BOD removed for selected processes.

Conversion Factor
Conversion factors depend on the activated sludge modification involved in the process. Factors generally range from 0.5 to 1.0 lb of solids per pound of BOD removed (see Table 19.9).

Mass Balance Calculation

$$BOD_5 \text{ in (lb)} = \left[\begin{array}{c} BOD_{in} \text{ (mg/L)} \times \text{flow (MGD)} \\ \times 8.34 \end{array} \right]$$

$$BOD_5 \text{ out (lb)} = \left[\begin{array}{c} BOD_{out} \text{ (mg/L)} \times \text{flow (MGD)} \\ \times 8.34 \end{array} \right]$$

$$\text{Solids produced (lb/day)} = (BOD_{in} - BOD_{out}) \times K$$

$$TSS_{out} \text{ (lb/day)} = \left[\begin{array}{c} TSS_{out} \text{ (mg/L)} \times \text{flow (MGD)} \\ \times 8.34 \end{array} \right] \quad (19.56)$$

$$\text{Waste (lb/day)} = \left[\begin{array}{c} \text{Waste (mg/L)} \times \text{flow (MGD)} \\ \times 8.34 \end{array} \right]$$

$$\text{Solids removed (lb/day)} = \left[\begin{array}{c} TSS_{out} \text{ (lb/day)} \\ + \text{waste (lb/day)} \end{array} \right]$$

$$\% \text{ Mass balance} = \frac{(\text{Solids produced} - \text{solids removed}) \times 100}{\text{Solids produced}}$$

Explanation of Results
If the mass balance is ±15% or less, the process sampling, testing, and process control are within acceptable levels. If the balance is greater than ±15%, investigate further to determine if the discrepancy represents a process control problem or is the result of nonrepresentative sampling and inaccurate testing.

Sludge Waste Based on Mass Balance

The mass balance calculation predicts the amount of sludge that will be produced by a treatment process. This information can then be used to determine what that waste rate must be, under current operating conditions, to maintain the current solids level.

$$\text{Waste rate (MGD)} = \frac{\text{Solids produced (lb/day)}}{\text{Waste concentration} \times 8.34} \quad (19.57)$$

■ EXAMPLE 19.50

Problem: Given the following data for extended aeration (no primary), determine the mass balance of the biological process and the appropriate waste rate to maintain current operating conditions:

> Influent flow = 1.1 MGD
> Influent BOD_5 = 220 mg/L
> Influent TSS = 240 mg/L
> Effluent flow = 1.5 MGD
> Effluent flow BOD_5 = 18 mg/L
> Effluent flow TSS = 22 mg/L
> Waste flow = 24,000 gpd
> Waste TSS = 8710 mg/L

Solution:

$$BOD_5 \text{ in} = 220 \text{ mg/L} \times 1.1 \text{ MGD} \times 8.34 = 2018 \text{ lb/day}$$

$$BOD_5 \text{ out} = 18 \text{ mg/L} \times 1.1 \text{ MGD} \times 8.34 = 165 \text{ lb/day}$$

$$BOD_5 \text{ removed} = 2018 \text{ lb/day} - 165 \text{ lb/day} = 1853 \text{ lb/day}$$

$$\text{Solids produced} = 1853 \text{ lb/day} \times 0.65 \text{ lb/lb } BOD_5$$
$$= 1204 \text{ lb/day}$$

$$\text{Solids out (lb/day)} = 22 \text{ mg/L} \times 1.1 \text{ MGD} \times 8.34$$
$$= 202 \text{ lb/day}$$

$$\text{Sludge out (lb/day)} = 8710 \text{ mg/L} \times 0.024 \text{ MGD} \times 8.34$$
$$= 1743 \text{ lb/day}$$

$$\text{Solids removed (lb/day)} = (292 \text{ lb/day} + 1743 \text{ lb/day})$$
$$= 1945 \text{ lb/day}$$

$$\% \text{ Mass balance} = \frac{(1204 \text{ lb/day} - 1945 \text{ lb/day})}{1204 \text{ lb/day}} \times 100 = 62\%$$

The mass balance indicates that:

- The sampling points, collection methods, or laboratory testing procedures are producing nonrepresentative results.
- The process is removing significantly more solids than is required. Additional testing should be performed to isolate the specific cause of the imbalance.

To assist in the evaluation, the waste rate based on the mass balance information can be calculated:

$$\text{Waste (GPD)} = \frac{\text{Solids produced (lb/day)}}{\text{Waste TSS (mg/L)} \times 8.34} \quad (19.58)$$

$$\text{Waste (GPD)} = \frac{1204 \text{ lb/day} \times 1,000,000}{8710 \text{ mg/L} \times 8.34} = 16,575 \text{ gpd}$$

19.9.14 SOLIDS CONCENTRATION: SECONDARY CLARIFIER

The solids concentration in the secondary clarifier can be assumed to be equal to the solids concentration in the aeration tank effluent. It may also be determined in the laboratory using a core sample taken from the secondary clarifier. The secondary clarifier solids concentration can be calculated as an average of the secondary effluent suspended solids and the return activated sludge suspended solids concentration.

19.9.15 ACTIVATED SLUDGE PROCESS RECORDKEEPING REQUIREMENTS

Wastewater operators soon learn that recordkeeping is a major requirement and responsibility of their jobs. Records are important (essential) for process control, providing information on the cause of problems, providing information for making seasonal changes, and compliance with regulatory agencies. Records should include sampling and testing data, process control calculations, meter readings, process adjustments, operational problems and corrective action taken, and process observations.

19.10 DISINFECTION OF WASTEWATER

Like drinking water, liquid wastewater effluent is disinfected. Unlike drinking water, wastewater effluent is disinfected not to directly (direct end-of-pipe connection) protect a drinking water supply but instead to protect public health in general. This is particularly important when the secondary effluent is discharged into a body of water used for swimming or as a downstream water supply. In the treatment of water for human consumption, the treated water is typically chlorinated (although ozonation is also currently being applied in many cases). Chlorination is the preferred disinfection in potable water supplies because of the unique ability of chlorine to provide a residual. This chlorine residual is important because when treated water leaves the waterworks facility and enters the distribution system the possibility of contamination is increased. The residual works to continuously disinfect water right up to the consumer's tap.

In this section, we discuss basic chlorination and dechlorination. In addition, we describe ultraviolet (UV) irradiation, ozonation, bromine chlorine, and no disinfection. Keep in mind that much of the chlorination material presented in the following is similar to the chlorination information presented in Chapter 18.

19.10.1 Chlorine Disinfection

Chlorination for disinfection, as shown in Figure 19.1, follows all other steps in conventional wastewater treatment. The purpose of chlorination is to reduce the population of organisms in the wastewater to levels low enough to ensure that pathogenic organisms will not be present in sufficient quantities to cause disease when discharged.

Note: Chlorine gas is heavier than air (vapor density of 2.5); therefore, exhaust from a chlorinator room should be taken from floor level.

Note: The safest action to take in the event of a major chlorine container leak is to call the fire department.

Note: You might wonder why it is that chlorination of critical waters such as natural trout streams is not normal practice. This practice is strictly prohibited because chlorine and its byproducts (e.g., chloramines) are extremely toxic to aquatic organisms.

19.10.1.1 Chlorination Terminology

Several terms are pertinent to a discussion of disinfection by chlorination. Because it is important for the operator to be familiar with these terms, we are repeating those key terms here:

Chlorine—A strong oxidizing agent that has strong disinfecting capability. It is a yellow–green gas that is extremely corrosive and is toxic to humans in extremely low concentrations in air.

Contact time—The length of time the disinfecting agent and the wastewater remain in contact.

Demand—The chemical reactions that must be satisfied before a residual or excess chemical will appear.

Disinfection—The selective destruction of disease-causing organisms. All the organisms are not destroyed during the process. This differentiates disinfection from *sterilization*, which is the destruction of all organisms.

Dose—The amount of chemical being added in milligrams/liter.

Feed rate—The amount of chemical being added in pounds per day.

Residual—The amount of disinfecting chemical remaining after the demand has been satisfied.

Sterilization—The removal of all living organisms.

19.10.1.2 Wastewater Chlorination Facts

19.10.1.2.1 Chlorine Facts

- Elemental chlorine (Cl_2, gaseous) is a yellow–green gas, 2.5 times heavier than air.
- The most common use of chlorine in wastewater treatment is for disinfection. Other uses include odor control and activated sludge bulking control. Chlorination takes place prior to the discharge of the final effluent to the receiving waters (see Figure 19.1).
- Chlorine may also be used for nitrogen removal through a process known as *breakpoint chlorination*. For nitrogen removal, enough chlorine is added to the wastewater to convert all the ammonium nitrogen gas. To do this, approximately 10 mg/L of chlorine must be added for every 1 mg/L of ammonium nitrogen in the wastewater.
- For disinfection, chlorine is fed manually or automatically into a chlorine contact tank or basin, where it contacts flowing wastewater for at least 30 min to destroy any disease-causing microorganisms (pathogens) present.
- Chorine may be applied as a gas or a solid or in liquid hypochlorite form.
- Chorine is a very reactive substance. It has the potential to react with many different chemicals (including ammonia), as well as with organic matter. When chlorine is added to wastewater, several reactions occur:
  1. Chlorine will react with any reducing agent (e.g., sulfide, nitrite, iron, thiosulfate) present in wastewater. These reactions are known as *chlorine demand*. The chlorine used for these reactions is not available for disinfection.
  2. Chlorine also reacts with organic compounds and ammonia compounds to form chlororganics and chloramines. Chloramines are part of the group of chlorine compounds that have disinfecting properties and show up as part of the chlorine residual test.
  3. After all of the chlorine demands are met, addition of more chlorine will produce free residual chlorine. Producing free residual chlorine in wastewater requires very large additions of chlorine.

19.10.1.2.2 Hypochlorite Facts

Hypochlorite is relatively safe to work with, although some minor hazards are associated with its use (skin irritation, nose irritation, and burning eyes). It is normally available in dry form as a white powder, pellet, or tablet or in liquid form. It can be added directly using a dry chemical feeder or dissolved and fed as a solution.

Note: In most wastewater treatment systems, disinfection is accomplished by means of combined residual.

19.10.1.3 Water Chlorination Process Description

Chlorine is a very reactive substance. Chlorine is added to wastewater to satisfy all chemical demands—that is, to react with certain chemicals (such as sulfide, sulfite, or ferrous iron). When these initial chemical demands have been satisfied, chlorine will react with substances such as ammonia to produce chloramines and other substances that, although not as effective as chlorine, have disinfecting capability. This produces a combined residual, which can be measured using residual chlorine test methods. If additional chlorine is added, free residual chlorine can be produced. Due to the chemicals normally found in wastewater, chlorine residuals are normally combined rather than free residuals. Control of the disinfection process is normally based on maintaining total residual chlorine of at least 1.0 mg/L for a contact time of at least 30 min at design flow.

Note: Residual level, contact time, and effluent quality affect disinfection. Failure to maintain the desired residual levels for the required contact time will result in lower efficiency and increased probability that disease organisms will be discharged.

Based on water quality standards, total residual limitations on chlorine are:

- Freshwater—Less than 11 ppb total residual chlorine
- Estuaries—Less that 7.5 ppb for halogen-produced oxidants
- Endangered species—Use of chlorine is prohibited

19.10.1.4 Chlorination Equipment

19.10.1.4.1 Hypochlorite Systems

Depending on the form of hypochlorite selected for use, special equipment to control the addition of hypochlorite to the wastewater is required. Liquid forms require the use of metering pumps, which can deliver varying flows of hypochlorite solution. Dry chemicals require the use of a feed system designed to provide variable doses of the form used. The tablet form of hypochlorite requires the use of a tablet chlorinator designed specifically to provide the desired dose of chlorine. The hypochlorite solution or dry feed systems dispense the hypochlorite, which is then mixed with the flow. The treated wastewater then enters the contact tank to provide the required contact time.

19.10.1.4.2 Chlorine Systems

Because of the potential hazards associated with the use of chlorine, the equipment requirements are significantly greater than those associated with hypochlorite use. The system most widely used is a solution feed system. In this system, chlorine is removed from the container at a flow rate controlled by a variable orifice. Water moving through the chlorine injector creates a vacuum, which draws the chlorine gas to the injector and mixes it with the water. The chlorine gas reacts with the water to form hypochlorous and hydrochloric acid. The solution is then piped to the chlorine contact tank and dispersed into the wastewater through a diffuser. Larger facilities may withdraw the liquid form of chlorine and use evaporators (heaters) to convert to the gas form. Small facilities will normally draw the gas form of chlorine from the cylinder. As gas is withdrawn, liquid will be converted to the gas form. This requires heat energy and may result in chlorine line freeze-up if the withdrawal rate exceeds the available energy levels.

19.10.1.5 Chlorination Operation

In either type of system, normal operation requires adjustment of feed rates to ensure that required residual levels are maintained. This normally requires chlorine residual testing and adjustment based on the results of the test. Other activities include removal of accumulated solids from the contact tank, collection of bacteriological samples to evaluate process performance, and maintenance of safety equipment (respirator/air pack, safety lines, etc.). Hypochlorite operation may also include making solutions (solution feed systems) or adding powder or pellets to the dry chemical feeder or tablets to the tablet chlorinator.

Chlorine operations include adjustment of chlorinator feed rates, inspection of mechanical equipment, testing for leaks using ammonia swabs (white smoke indicates the presence of leaks), changing containers (which requires more than one person for safety), and adjusting the injector water feed rate when required. Chlorination requires routine testing of plant effluent for total residual chlorine and may also require collection and analysis of samples to determine the fecal coliform concentration in the effluent.

19.10.1.6 Troubleshooting Operation Problems

On occasion, operational problems with the disinfection process develop. The wastewater operator must be able to recognize these problems and correct them. For proper operation, the chlorination process requires routine observation, meter readings, process control and testing, and various process control calculations. Comparison of daily results with expected normal ranges is the key to identifying problems during the troubleshooting process and to taking the appropriate corrective action (if required). In this section, we review normal operational and performance factors, point out various problems that can occur with the disinfection process, identify causes, and suggest corrective actions that should be taken.

19.10.1.6.1 Operator Observations

The operator should consider the following items:

- *Flow distribution*—The operator monitors the flow to ensure that it is evenly distributed between all units in service and that the flow through each individual unit is uniform, with no indication of short-circuiting.
- *Contact tank*—The contact tanks or basins must be checked to ensure that no excessive accumulation of scum is present on the surface, that no solids are accumulating on the bottom, and that mixing appears to be adequate.
- *Chlorinator*—The operator should check to ensure that no evidence of leakage is apparent, operating pressure and vacuum are within specified levels, current chlorine feed settling is within expected levels, inline cylinders have sufficient chlorine to ensure continuous feed, and the exhaust system is operating as designed.

19.10.1.6.2 Factors Affecting Performance

Operators must be familiar with those factors that affect chlorination performance. Any item that interferes with the chlorine reactions or increases the demand for chlorine can affect performance and, in turn, produce nondisinfectant products. Factors affecting chlorination performance include:

- *Effluent quality*—Poor-quality effluents have higher chlorine demands, high concentrations of solids prevent chlorine–organism contact, and incomplete nitrification can cause extremely high chlorine demand.
- *Mixing*—To be effective, chlorine must be in contact with the organisms. Poor mixing results in poor chlorine distribution. Installation of baffles and utilizing a high length-to-width ratio will improve mixing and contact.
- *Contact time*—The chlorine disinfection process is time dependent. As the contact time decreases, process effectiveness decreases. A minimum of 30 minutes of contact must be available at design flow.
- *Residual levels*—The chlorine disinfection process is dependent on the total residual chlorine (TRC). The concentration of residual must be sufficient to ensure that the desired reactions

occur. At the design contact time, the required minimum total residual chlorine concentration is 1.0 mg/L.

19.10.1.6.3 Process Control Sampling and Testing

The process performance evaluation is based on the bacterial content (fecal coliform) of the final effluent. To ensure proper operation of the chlorination process, the operator must perform process control testing for the chlorination process. Process control testing consists of performing a *total chlorine residual test* on chlorine contact effluent. The frequency of the testing is specified in the plant permit. The normal expected range of results is also specified in the plant permit.

19.10.1.6.4 Troubleshooting

Table 19.10 presents common operational problems, symptoms, casual factors, and corrective actions associated with chlorination system use in wastewater treatment.

19.10.1.6.5 Dechlorination

The purpose of *dechlorination* is to remove chlorine and reaction products (chloramines) before the treated wastestream is discharged into its receiving waters. Dechlorination follows chlorination—usually at the end of the contact tank to the final effluent. Sulfur dioxide gas, sodium sulfate, sodium metabisulfate, or sodium bisulfates are the chemicals used to dechlorinate. No matter which chemical is used to dechlorinate, its reaction with chlorine is instantaneous.

19.10.1.7 Chlorination Environmental Hazards and Safety

Chlorine is an extremely toxic substance that can cause severe damage when released to the environment. For this reason, most state regulatory agencies have established chlorine water quality standards; for example, in Virginia, the standard is 0.011 mg/L in freshwaters for total residual chlorine and 0.0075 mg/L for chlorine produced oxidants in saline waters. Studies have indicated that above these levels chlorine can reduce shellfish growth and destroy sensitive aquatic organisms. Such standards have forced many treatment facilities to add an additional process to remove the chlorine prior to discharge. The process, *dechlorination*, uses chemicals that react quickly with chlorine to convert it to a less harmful form. Elemental chlorine is a chemical with potentially fatal hazards associated with it. For this reason, many different state and federal agencies regulate the transport, storage, and use of chlorine. All operators required to work with chlorine should be trained in proper handling techniques. They should also be trained to ensure that all procedures for storage transport, handling, and use of chlorine are in compliance with appropriate state and federal regulations.

TABLE 19.10
Symptoms, Causes, and Corrective Actions for Chlorination System Problems

Symptom	Cause	Corrective Action
Coliform count fails to meet required standards for disinfection.	Inadequate chlorination equipment capacity	Replace equipment as necessary to provide treatment based on maximum flow through the pipe.
	Inadequate chlorine residual control	Use chlorine residual analyzer to monitor and control chlorine dosage automatically.
	Short-circuiting in chlorine contact chamber	Install baffling in the chlorine contact chamber; install mixing device in chlorine contact chamber.
	Solids buildup in contact chamber	Clean contact chamber.
	Chlorine residual too low	Increase contact time or increase chlorine feed rate, or both.
Chlorine gas pressure at the chlorinator is low.	Insufficient number of cylinders connected to the system	Connect enough cylinders to system so feed rate does not exceed recommended withdrawal rate for cylinders.
	Stoppage or restriction of flow between cylinders and chlorinator	Disassemble chlorine header system at point where cooling begins; locate stoppage and clean with solvent.
Chlorinator has no chlorine gas pressure.	Chlorine cylinders empty or not connected to the system	Replace empty cylinders or connect cylinders.
	Plugged or damaged pressure-reducing valve	Repair reducing valve after shutting cylinder valves and decreasing gas in the header system.
Chlorinator will not feed any chlorine.	Dirty pressure-reducing valve in chlorinator	Disassemble chlorinator and clean valve stem and seat; precede valve with filter/sediment trap.
	Chlorine cylinder hotter than chlorine control apparatus (chlorinator)	Reduce temperature in cylinder area; do not connect a new cylinder that has been sitting in the sun.
Chlorine gas is escaping from the chlorine pressure-reducing valve (CPRV).	Main diaphragm of CPRV ruptured	Disassemble valve and diaphragm; inspect chlorine supply system for moisture intrusion.
Chlorine feed rate cannot be maintained without icing of chlorine system.	Insufficient evaporator capacity	Reduce feed rate to 75% of evaporator capacity. If this eliminates the problem, then the main diaphragm of the CPRV is ruptured.
Maximum feed rate from chlorinator cannot be obtained.	Inadequate chlorine gas pressure	Increase pressure; replace empty or low cylinders.
	Water pump injector clogged with deposits	Clean injector parts using muriatic acid; rinse parts with freshwater and place back in service.
	Leak in vacuum relief valve	Disassemble vacuum relief valve and replace all springs.
	Vacuum leak in joints, gaskets, tubing, etc. in chlorinator system	Repair all vacuum leaks by tightening joints, replacing gaskets, and replacing tubing or compression nuts.

19.10.1.7.1 Safe Work Practices for Chlorine

Because of the inherent dangers involved with handling chlorine, each facility using chlorine (for any reason) should be sure to have a written safe work practice in place and followed by plant operators. A sample safe work practice for handling chlorine is provided below:

1. Plant personnel *must* be trained and instructed on the use and handling of chlorine, chlorine equipment, chlorine emergency repair kits, and other chlorine emergency procedures.
2. Use extreme care and caution when handling chlorine.
3. Lift chlorine cylinders only with an approved and load-tested device.
4. Secure chlorine cylinders into position immediately. *Never* leave a cylinder suspended.
5. Avoid dropping chlorine cylinders.
6. Avoid banging chlorine cylinders into other objects.
7. Store 1-ton chlorine cylinders in a cool, dry place away from direct sunlight or heating units. Railroad tank cars compensate for direct sunlight.
8. Store 1-ton chlorine cylinders on their sides only (horizontally).
9. Do not stack unused or used chlorine cylinders.
10. Provide positive ventilation to the chlorine storage area and chlorinator room.
11. *Always* keep chlorine cylinders at ambient temperature. *Never* apply direct flame to a chlorine cylinder.
12. Use the oldest chlorine cylinder in stock first.
13. Always keep valve protection hoods in place until the chlorine cylinders are ready for connection.

TABLE 19.10 (cont.)
Symptoms, Causes, and Corrective Actions for Chlorination System Problems

Symptom	Cause	Corrective Action
Chlorinator system is unable to maintain sufficient water bath temperature to keep external CPRV open.	Clogged external CPRV cartridge	Flush and clean cartridge.
	Heating element malfunction	Remove and replace heating element.
Adequate chlorine feed rate cannot be maintained.	Malfunction or deterioration of chlorine water supply pump	Overhaul pump; if turbine pump is used, try closing valve to maintain proper discharge pressure.
Chlorine residual is too high in plant effluent to meet requirements.	Chlorine residual too high	Install dechlorination facilities.
Chlorine residual produced in the effluent varies widely.	Inadequate chlorine flow proportion meter capacity to meet plant flow rates	Replace with higher capacity chlorinator meter.
	Malfunctioning controls	Call manufacturer technical representative.
	Solids settled in chlorine contact chamber	Clean chlorine contact tank.
	Flow proportioning control device not zeroed or spanned correctly	Re-zero and span the device in accordance with the manufacturer's instructions.
Chlorine residual cannot be obtained.	High chemical demand	Locate and correct the source of the high demand.
	Test interference	Add sulfuric acid to samples to reduce interference.
Chlorine residual analyzer, recorder, and controller do not control chlorine residual properly.	Electrodes fouled	Clean electrodes.
	Loop time too long	Reduce control loop time by: (1) moving the injector closer to the point of application; (2) increasing the velocity in the sample line to the analyzer; (3) moving the cell closer to the sample point; or (4) moving the sample point closer to the point of application.
	Insufficient potassium iodide being added for the amount of residual being measured	Adjust potassium iodide feed to correspond with the chlorine residual being measured.
	Malfunctioning buffer additive system	Repair buffer additive system.
	Malfunctioning analyzer cell	Call authorized service personnel to repair electrical components.
	Poor mixing of chlorine at point of application	Install mixing device to cause turbulence at point of application.
	Improperly set rotameter tube range	Replace rotameter with a proper range of feed rate.

14. Except to repair a leak, do not tamper with the fusible plugs on chlorine cylinders.
15. Wear self-contained breathing apparatus (SCBA) whenever changing a chlorine cylinder, and have at least one other person with a standby SCBA unit outside the immediate area.
16. Inspect all threads and surfaces of chlorine cylinder and have at least one other person with a standby SCBA unit outside of the immediate area.
17. Use new lead gaskets each time a chlorine cylinder connection is made.
18. Use only the specified wrench to operate chlorine cylinder valves.
19. Open chlorine cylinder valves slowly, no more than one full turn.
20. Do not hammer, bang, or force chlorine cylinder valves under any circumstances.
21. Check for chlorine leaks as soon as the chlorine cylinder connection is made by gently expelling ammonia mist from a plastic squeeze bottle filled with approximately 2 ounces of liquid ammonia solution. Do not put liquid ammonia on valves or equipment.
22. Correct all minor chlorine leaks at the chlorine cylinder connection immediately.
23. Except for automatic systems, draw chlorine from only one manifolded chlorine cylinder at a time. *Never* simultaneously open two or more chlorine cylinders connected to a common manifold pulling liquid chlorine. Two or more cylinders connected to a common manifold pulling gaseous chlorine are acceptable.
24. Wear SCBA and chemical protective clothing that covers the face, arms, and hands before entering an enclosed chlorine area to investigate

a chlorine odor or chlorine leak; follow the two-person rule.

25. Provide positive ventilation to a contaminated chlorine atmosphere before entering whenever possible.

26. Have at least two people present before entering a chlorine atmosphere: one person to enter the chlorine atmosphere, the other to observe in the event of an emergency. *Never* enter a chlorine atmosphere unattended. Remember that the Occupational Safety and Health Administration (OSHA) mandates that only fully qualified Level III HAZMAT responders are authorized to aggressively attack a hazardous materials leak such as chlorine.

27. Use supplied-air breathing equipment when entering a chlorine atmosphere. *Never* use canister-type gas masks when entering a chlorine atmosphere.

28. Be sure that the supplied-air breathing apparatus has been properly maintained in accordance with the plant's self-contained breathing apparatus inspection guidelines as specified in the plant's respiratory protection program.

29. Stay upwind from all chlorine leak danger areas unless involved with making repairs. Look to plant windsocks for wind direction.

30. Contact trained plant personnel to repair chlorine leaks.

31. Roll uncontrollable leaking chlorine cylinders so the chlorine escapes as a gas, not as a liquid.

32. Stop leaking chlorine cylinders or leaking chlorine equipment (by closing off valves, if possible) prior to attempting repair.

33. Connect uncontrollable leaking chlorine cylinders to the chlorination equipment and feed the maximum chlorine feed rate possible.

34. Keep leaking chlorine cylinders at the plant site. Chlorine cylinders received at the plant site must be inspected for leaks prior to taking delivery from the shipper. *Never* ship a leaking chlorine cylinder back to the supplier after it has been accepted from the shipper (bill of lading has been signed by plant personnel); instead, repair or stop the leak first.

35. Keep moisture away from a chlorine leak. *Never* put water onto a chlorine leak.

36. Call the fire department or rescue squad if a person is incapacitated by chlorine.

37. Administer CPR (use barrier mask if possible) immediately to a person who has been incapacitated by chlorine.

38. Take only shallow breaths when exposed to chlorine without the appropriate respiratory protection.

39. Place a person who does not have difficulty breathing and is heavily contaminated with chlorine into a deluge shower. Remove the person's clothing under the water and flush all body parts that were exposed to chlorine.

40. Flush eyes contaminated with chlorine with copious quantities of lukewarm running water for at least 15 minutes.

41. Drink milk if throat is irritated by chlorine.

42. *Never* store other materials in chlorine cylinder storage areas; substances such as acetylene and propane are not compatible with chlorine.

19.10.1.8 Chlorination Process Calculations

Several calculations are useful when operating a chlorination system. Many of these calculations are discussed and illustrated in this section.

19.10.1.8.1 Chlorine Demand

Chlorine demand is the amount of chlorine in milligrams per liter that must be added to the wastewater to complete all of the chemical reactions that must occur prior to producing a residual:

$$\text{Chlorine demand} = \begin{bmatrix} \text{Chlorine dose (mg/L)} \\ -\text{ chlorine residual (mg/L)} \end{bmatrix} \quad (19.59)$$

■ EXAMPLE 19.51

Problem: The plant effluent currently requires a chlorine dose of 7.1 mg/L to produce the required 1.0-mg/L chlorine residual in the chlorine contact tank. What is the chlorine demand in milligrams per liter?

Solution:

$$\text{Chlorine demand} = 7.1 \text{ mg/L} - 1.0 \text{ mg/L} = 6.1 \text{ mg/L}$$

19.10.1.8.2 Chlorine Feed Rate

The chlorine feed rate is the amount of chlorine added to the wastewater in pounds per day.

$$\text{Chlorine feed rate} = \begin{bmatrix} \text{Dose (mg/L)} \times \text{flow (MGD)} \\ \times 8.34 \text{ lb/MG/mg/L} \end{bmatrix} \quad (19.60)$$

■ EXAMPLE 19.52

Problem: The current chlorine dose is 5.55 mg/L. What is the feed rate in pounds per day if the flow is 22.89 MGD?

Solution:

$$\text{Chlorine feed rate} = \begin{bmatrix} 5.55 \text{ mg/L} \times 22.89 \text{ MGD} \\ \times 8.34 \text{ lb/MG/mg/L} \end{bmatrix}$$

$$= 1060 \text{ lb/day}$$

19.10.1.8.3 Chlorine Dose

Chlorine dose is the concentration of chlorine being added to the wastewater. It is expressed in milligrams per liter:

$$\text{Dose (mg/L)} = \frac{\text{Chlorine feed rate (lb/day)}}{\text{Flow (MGD)} \times 8.34 \text{ lb/MG/mg/L}} \quad (19.61)$$

■ **EXAMPLE 19.53**

Problem: Each day, 320 lb of chlorine are added to a wastewater flow of 5.60 MGD. What is the chlorine dose in milligrams per liter?

Solution:

$$\text{Dose (mg/L)} = \frac{320 \text{ lb/day}}{5.60 \text{ MGD} \times 8.34 \text{ lb/MG/mg/L}}$$

$$= 6.9 \text{ mg/L}$$

19.10.1.8.4 Available Chlorine

When hypochlorite forms of chlorine are used, the available chlorine is listed on the label. In these cases, the amount of chemical added must be converted to the actual amount of chlorine using the following calculation:

$$\text{Available chlorine} = \begin{bmatrix} \text{Amount of hypochlorite} \\ \times \% \text{ available chlorine} \end{bmatrix} \quad (19.62)$$

■ **EXAMPLE 19.54**

Problem: The calcium hypochlorite used for chlorination contains 62.5% available chlorine. How many pounds of chlorine are added to the plant effluent if the current feed rate is 30 pounds of calcium hypochlorite per day?

Solution:

$$\text{Quantity of chlorine} = 30 \text{ lb} \times 0.625 = 18.75 \text{ lb}$$

19.10.1.8.5 Required Quantity of Dry Hypochlorite

This calculation is used to determine the amount of hypochlorite required to achieve the desired dose of chlorine:

$$\text{Hypochlorite quantity (lb/day)} = \frac{\text{Required chlorine dose (mg/L)} \times \text{flow (MGD)} \times 8.34 \text{ lb/MG/mg/L}}{\% \text{ Available chlorine}} \quad (19.63)$$

■ **EXAMPLE 19.55**

Problem: The laboratory reports that the chlorine dose required to maintain the desired residual level is 8.5 mg/L. Today's flow rate is 3.25 MGD. The hypochlorite powder used for disinfection is 70% available chlorine. How many pounds of hypochlorite must be used?

Solution:

$$\text{Hypochlorite quantity} = \frac{8.5 \text{ mg/L} \times 3.25 \text{ MGD} \times 8.34 \text{ lb/MG/mg/L}}{0.70}$$

$$= 329 \text{ lb/day}$$

19.10.1.8.6 Required Quantity of Liquid Hypochlorite

$$\text{Hypochlorite required (gal/day)} = \frac{\text{Chlorine dose (mg/L)} \times \text{flow (MGD)} \times 8.34 \text{ lb/MG/mg/L}}{\% \text{ Available chlorine} \times 8.34 \text{ lb/MG/mg/L} \times \text{hypochlorite solution SG}} \quad (19.64)$$

■ **EXAMPLE 19.56**

Problem: The chlorine dose is 8.8 mg/L and the flow rate is 3.28 MGD. The hypochlorite solution is 71% available chlorine and has a specific gravity of 1.25. How many pounds of hypochlorite must be used?

Solution:

$$\text{Hypochlorite quantity} = \frac{8.8 \text{ mg/L} \times 3.28 \text{ MGD} \times 8.34 \text{ lb/MG/mg/L}}{0.71 \times 8.34 \text{ lb/gal} \times 1.25}$$

$$= 32.5 \text{ gal/day}$$

19.10.1.8.7 Ordering Chlorine

Because disinfection must be continuous, the supply of chlorine must never be allowed to run out. The following calculation provides a simple method for determining when additional supplies must be ordered. The process consists of three steps:

1. Adjust the flow and use variations if projected changes are provided.
2. If an increase in flow or required dosage is projected, the current flow rate or dose must be adjusted to reflect the projected change.
3. Calculate the projected flow and dose:

$$\text{Projected flow} = \begin{bmatrix} \text{Current flow (MGD)} \\ \times (1.0 + \% \text{ change}) \end{bmatrix}$$

$$\text{Projected dose} = \begin{bmatrix} \text{Current dose (mg/L)} \\ \times (1.0 + \% \text{ change}) \end{bmatrix} \quad (19.65)$$

■ **EXAMPLE 19.57**

Problem: Based on the available information for the past 12 months, the operator projects that the effluent flow rate will increase by 7.5% during the next year. If the average daily flow has been 4.5 MGD, what will be the projected flow for the next 12 months?

Solution:

Projected flow = 4.5 MGD × (1.0 + 0.075) = 4.84 MGD

To determine the amount of chlorine required for a given period:

Chlorine required = Feed rate (lb/day) × no. days required

■ **EXAMPLE 19.58**

Problem: The plant currently uses 90 lb of chlorine per day. The town wishes to order enough chlorine to supply the plant for 4 months (assume 31 days/month). How many pounds of chlorine should be ordered to provide the needed supply?

Solution:

Chlorine required = 90 lb/day × 124 days = 11,160 lb

Note: In some instances, projections for flow or dose changes are not available but the plant operator wishes to include an extra amount of chlorine as a safety factor. This safety factor can be stated as a specific quantity or as a percentage of the projected usage. A safety factor as a specific quantity can be expressed as:

Total required Cl_2 = Chlorine required (lb) + safety factor

Note: Because chlorine is only shipped in full containers, unless asked specifically for the amount of chlorine actually required or used during a specified period, all decimal parts of a cylinder are rounded up to the next highest number of full cylinders.

19.10.2 ULTRAVIOLET IRRADIATION

Although ultraviolet (UV) disinfection was recognized as a method for achieving disinfection in the late 19th century, its application virtually disappeared with the evolution of chlorination technologies. In recent years, however, there has been a resurgence in its use in the wastewater field, largely as a consequence of concern for discharge of toxic chlorine residual. Even more recently, UV has gained more attention because of the tough new regulations on chlorine use imposed by both OSHA and the U.S. Environmental Protection Agency. Because of this rela-

tively recent increased regulatory pressure, many facilities are actively engaged in substituting chlorine for other disinfection alternatives. Moreover, UV technology itself has undergone many improvements that make UV attractive as a disinfection alternative. Ultraviolet light has very good germicidal qualities and is very effective in destroying microorganisms. It is used in hospitals, biological testing facilities, and many other similar locations. In wastewater treatment, the plant effluent is exposed to ultraviolet light of a specified wavelength and intensity for a specified contact period. The effectiveness of the process is dependent on:

- UV light intensity
- Contact time
- Wastewater quality (turbidity)

The Achilles' heel of UV for disinfecting wastewater is turbidity. If the wastewater quality is poor, the ultraviolet light will be unable to penetrate the solids, and the effectiveness of the process decreases dramatically. For this reason, many states limit the use of UV disinfection to facilities that can reasonably be expected to produce an effluent containing ≤30 mg/L BOD_5 and total suspended solids.

In the operation of UV systems, UV lamps must be readily available when replacements are required. The best lamps are those with a stated operating life of at least 7500 hr and those that do not produce significant amounts of ozone or hydrogen peroxide. The lamps must also meet technical specifications for intensity, output, and arc length. If the UV light tubes are submerged in the wastestream, they must be protected inside quartz tubes, which not only protect the lights but also make cleaning and replacement easier.

Contact tanks must be used with UV disinfection. They are designed with the banks of UV lights in a horizontal position, either parallel or perpendicular to the flow, or with the banks of lights placed in a vertical position perpendicular to the flow.

Note: The contact tank must provide, at a minimum, a 10-second exposure time.

We stated earlier that turbidity problems have been a problem with UV wastewater treatment—and this is the case. However, if turbidity is its Achilles' heel, then the need for increased maintenance (as compared to other disinfection alternatives) is the toe of the same foot. UV maintenance requires that the tubes be cleaned on a regular basis or as needed. In addition, periodic acid washing is also required to remove chemical buildup.

Routine monitoring of UV disinfection systems is required. Checking on bulb burnout, buildup of solids on quartz tubes, and UV light intensity is necessary.

Note: UV light is extremely hazardous to the eyes. Never enter an area where UV lights are in operation without proper eye protection. Never look directly into the ultraviolet light.

19.10.3 OZONATION

Ozone is a strong oxidizing gas that reacts with most organic and many inorganic molecules. It is produced when oxygen molecules separate, collide with other oxygen atoms, and form a molecule consisting of three oxygen atoms. For high-quality effluents, ozone is a very effective disinfectant. Current regulations for domestic treatment systems limit the use of ozonation to filtered effluents unless the effectiveness of the system can be demonstrated prior to installation.

Note: Effluent quality is the key performance factor for ozonation.

For ozonation of wastewater, the facility must have the capability to generate pure oxygen along with an ozone generator. A contact tank with ≥10-minute contact time at design average daily flow is required. Off-gas monitoring for process control is also required. In addition, safety equipment capable of monitoring ozone in the atmosphere and a ventilation system to prevent ozone levels exceeding 0.1 ppm are necessary.

The actual operation of the ozonation process consists of monitoring and adjusting the ozone generator and monitoring the control system to maintain the required ozone concentration in the off-gas. The process must also be evaluated periodically using biological testing to assess its effectiveness.

Note: Ozone is an extremely toxic substance. Concentrations in air should not exceed 0.1 ppm. It also has the potential to create an explosive atmosphere. Sufficient ventilation and purging capabilities should be provided.

Note: Ozone has certain advantages over chlorine for disinfection of wastewater, in that: (1) ozone increases DO in the effluent, (2) ozone has a briefer contact time, (3) ozone has no undesirable effects on marine organisms, and (4) ozone decreases turbidity and odor.

19.10.4 BROMINE CHLORIDE

Bromine chloride is a mixture of bromine and chlorine. It forms hydrocarbons and hydrochloric acid when mixed with water. Bromine chloride is an excellent disinfectant that reacts quickly and normally does not produce any long-term residuals.

Note: Bromine chloride is an extremely corrosive compound in the presence of low concentrations of moisture.

The reactions occurring when bromine chloride is added to the wastewater are similar to those occurring when chlorine is added. The major difference is the production of bromamine compounds rather than chloramines. The bromamine compounds are excellent disinfectants but are less stable and dissipate quickly. In most cases, the bromamines decay into other, less toxic compounds rapidly and are undetectable in the plant effluent. The factors that affect performance are similar to those affecting the performance of the chlorine disinfection process. Effluent quality, contact time, etc. have a direct impact on the performance of the process.

19.10.5 NO DISINFECTION

In a very limited number of cases, treated wastewater discharges without disinfection are permitted. These are approved on a case-by-case basis. Each request must be evaluated based on the point of discharge, the quality of the discharge, the potential for human contact, and many other factors.

19.11 ADVANCED WASTEWATER TREATMENT

Advanced wastewater treatment is defined as the methods and processes that remove more contaminants (suspended and dissolved substances) from wastewater than are taken out by conventional biological treatment. Put another way, advanced wastewater treatment is the application of a process or system that follows secondary treatment or that includes phosphorus removal or nitrification in conventional secondary treatment.

Advanced wastewater treatment is used to augment conventional secondary treatment because secondary treatment typically removes only between 85 and 95% of the biochemical oxygen demand (BOD) and total suspended solids (TSS) in raw sanitary sewage. Generally, this leaves 30 mg/L or less of BOD and TSS in the secondary effluent. To meet stringent water quality standards, this level of BOD and TSS in secondary effluent may not prevent violation of water quality standards—the plant may not make permit. Thus, advanced wastewater treatment is often used to remove additional pollutants from treated wastewater.

In addition to meeting or exceeding the requirements of water quality standards, treatment facilities use advanced wastewater treatment for other reasons, as well; for example, conventional secondary wastewater treatment is sometimes not sufficient to protect the aquatic environment. In a stream, for example, when periodic flow events occur, the stream may not provide the amount of dilution of effluent required to maintain the necessary dissolved oxygen levels for aquatic organism survival.

Secondary treatment has other limitations. It does not significantly reduce the effluent concentration of nitrogen and phosphorus (important plant nutrients) in sewage. An overabundance of these nutrients can overstimulate plant and algae growth such that they create water quality problems. If they are discharged into lakes, for example, these nutrients contribute to algal blooms and accelerated eutrophication (lake aging). Also, the nitrogen in the sewage effluent may be present mostly in the form of ammonia compounds. At high enough concentrations, ammonia compounds can be toxic to aquatic organisms. Yet another problem with these compounds is that they exert a *nitrogenous oxygen* demand in the receiving water, as they convert to nitrates. This process is called *nitrification*.

Note: The term *tertiary treatment* is commonly used as a synonym for advanced wastewater treatment; however, these two terms do not have precisely the same meaning. Tertiary suggests a third step that is applied after primary and secondary treatment.

Advanced wastewater treatment can remove more than 99% of the pollutants from raw sewage and can produce an effluent of almost potable (drinking) water quality. Obviously, however, advanced treatment is not cost free. The cost of advanced treatment—costs of operation and maintenance, as well as retrofit of existing conventional processes—is very high (sometimes doubling the cost of secondary treatment). A plan to install advanced treatment technology calls for careful study—the benefit-to-cost ratio is not always significant enough to justify the additional expense.

Even considering the expense, application of some form of advanced treatment is not uncommon. These treatment processes can be physical, chemical, or biological. The specific process used is based on the purpose of the treatment and the quality of the effluent desired.

19.11.1 CHEMICAL TREATMENT

The purpose of chemical treatment is to remove:

- Biochemical oxygen demand (BOD)
- Total suspended solids (TSS)
- Phosphorus
- Heavy metals
- Other substances that can be chemically converted to a settleable solid

Chemical treatment is often accomplished as an "add-on" to existing treatment systems or by means of separate facilities specifically designed for chemical addition. In each case, the basic process necessary to achieve the desired results remains the same:

- Chemicals are thoroughly mixed with the wastewater.
- The chemical reactions that occur form solids (coagulation).
- The solids are mixed to increase particle size (flocculation).
- Settling or filtration (separation) then removes the solids.

The specific chemical used depends on the pollutant to be removed and the characteristics of the wastewater. Chemicals may include the following:

- Lime
- Alum (aluminum sulfate)
- Aluminum salts
- Ferric or ferrous salts
- Polymers
- Bioadditives

19.11.1.1 Operational Observations, Problems, and Troubleshooting

Operation and observation of performance of chemical treatment processes are dependent on the pollutant being removed and on process design. Operational problems associated with chemical treatment processes used in advanced treatment usually revolve around problems with floc formation, settling characteristics, removal in the settling tank, and sludge (in settling tank) turning anaerobic. To correct these problems, the operator must be able to recognize the applicable problem indicators through proper observation. Table 19.11 lists common indicators and observations of operational problems, along with the applicable causal factors and corrective actions.

19.11.2 MICROSCREENING

Microscreening (also referred to as *microstraining*) is an advanced treatment process used to reduce suspended solids. The microscreens are composed of specially woven steel wire fabric mounted around the perimeter of a large revolving drum. The steel wire cloth acts as a fine screen, with openings as small as 20 μm (or millionths of a meter)—small enough to remove microscopic organisms and debris. The rotating drum is partially submerged in the secondary effluent, which must flow into the drum then outward through the microscreen. As the drum rotates, captured solids are carried to the top where a high-velocity water spray flushes them into a hopper or backwash tray mounted on the hollow axle of the drum. Backwash solids are recycled to plant influent for treatment. These units have found greatest application in treatment of industrial waters and final

TABLE 19.11
Symptoms, Causes, and Corrective Actions for Chemical Treatment Problems

Symptom	Cause	Corrective Action
Poor floc formation and settling characteristics	Insufficient chemical dispersal during rapid mix	Increase speed of rapid mixer.
	Excessive detention time in rapid mix	Reduce detention time to 15 to 60 sec.
	Improper coagulant dosage	Correct dosage (determine by jar testing).
	Excessive flocculator speed	Reduce flocculator speed.
Good floc formation, poor removal in settling tank	Excessive velocity between flocculation and settling	Reduce velocity to acceptable range.
	Settling tank operational problem	Correct settling tank operational problems.
Settling tank sludge turning anaerobic	Development of a sludge blanket in the settling tank	Increase sludge withdrawal to eliminate blanket.
	Excessive organic carryover from secondary treatment	Correct secondary treatment operational problems.

TABLE 19.12
Symptoms, Causes, and Corrective Actions for Microscreening Treatment Problems

Symptom	Cause	Corrective Action
Decrease in throughput rate (due to slime growth)	Inadequate cleaning	Increase backwash pressure (60–120 psi); add hypochlorite upstream of the unit.
	Spray nozzles plugged	Unclog nozzles.
Decreased performance due to leakage at the ends of the drum	Defective/leaking units	Tighten tension on sealing bands; replace sealing bands if excessive tension is required.
Screen capacity reduced after shutdown period	Fouled screen	Clean screen prior to shutdown; clean screen with hypochlorite.
Drive system running hot or noisy	Inadequate lubrication	Fill to specified level with recommended oil.
Erratic drum rotation	Improper drive belt adjustment	Adjust tension to specified level.
	Drive belts worn out	Replace drive belts.
Sudden increase in effluent solids	Hole in screen fabric	Repair fabric.
	Loose screws that secure fabric	Tighten screws.
	Overflowing solids collection trough	Reduce microscreen influent flow rate.
Decreased screen capacity after high-pressure washing	Iron or manganese oxide film on fabric	Clean screen with inhibited acid cleaner; follow manufacturer's instruction

polishing filtration of wastewater effluents. Expected performance for suspended solids removal is 95 to 99%, but the typical suspended solids removal achieved with these units is about 55%. The normal range is from 10 to 80%.

According to Metcalf & Eddy (2003), the functional design of the microscreen unit involves the following considerations: (1) characterization of the suspended solids with respect to the concentration and degree of flocculation, (2) selection of unit design parameter values that will not only ensure capacity to meet maximum hydraulic loadings with critical solids characteristics but also provide desired design performance over the expected range of hydraulic and solids loadings, and (3) provision of backwash and cleaning facilities to maintain the capacity of the screen.

19.11.2.1 Operational Observations, Problems, and Troubleshooting

Microscreen operators typically perform sampling and testing on influent and effluent TSS and monitor screen operation to ensure proper operation. Operational problems generally consist of gradual decreases in throughput rate, leakage at the ends of the drum, reduced screen capacity, hot or noisy drive systems, erratic drum rotation, and sudden increases in effluent solids (see Table 19.12).

19.11.3 FILTRATION

The purpose of *filtration* processes used in advanced treatment is to remove suspended solids. The specific operations associated with a particular filtration system

TABLE 19.13

Symptoms, Causes, and Corrective Actions for Filtration Treatment Problems

Symptom	Cause	Corrective Action
High effluent turbidity	Filter requires backwashing.	Backwash unit as soon as possible.
	Prior chemical treatment was inadequate.	Adjust and control chemical dosage properly.
High head loss through the filter	Filter requires backwashing	Backwash unit as soon as possible.
High head loss through unit right after backwashing	Backwash cycle was insufficient.	Increase backwash time.
	Surface scour/wash arm was inoperative.	Repair air scour or surface scrubbing arm.
Backwash water requirement that exceeds 5%	Excessive solids are in filter influent.	Improve treatment prior to filtration.
	Filter aid dosage is excessive.	Reduce control/filter aid dose rates.
	Surface washing/air scour is not operating.	Repair mechanical problem.
	Surface washing/air scour is not operating long enough during backwash cycle.	Increase surface wash cycle time.
	Excessive backwash cycle is being used.	Adjust backward cycle length.
Filter surface clogging	Prior treatment was inadequate (single-media filters).	Improve prior treatment; replace single media with dual/mixed media.
	Filter aid dosage is excessive (dual- or mixed-media filters).	Reduce or eliminate filter aid.
	Surface wash cycle is inadequate.	Provide adequate surface wash cycle.
	Backwash cycle is inadequate.	Provide adequate backwash cycle.
Short filter runs	Head loss is high.	See "Filter surface clogging" entry (above).
Filter effluent turbidity that increases rapidly	Filter aid dosage is inadequate.	Increase chemical dosage.
	Filter aid system has experienced mechanical failure.	Repair feed system.
	Filter aid requirement has changed.	Adjust filter aid dose rate (do jar test).
Mud ball formation	Backwash flow rate is inadequate.	Increase backwash flow to specified levels.
	Surface wash is inadequate.	Increase surface wash cycle.
Gravel displacement	Air is entering the underdrains during backwash cycle.	Control backwash volume; control backwash water head; replace media (severe displacement).
Medium lost during backwash cycle	Backwash flows are excessive.	Reduce backwash flow rate.
	Auxiliary scour is excessive.	Stop auxiliary scour several minutes before end of backwash cycle.
	Air has attached to filter media, causing it to float.	Increase backwash frequency to prevent bubble displacement and/or maintain maximum operating water depth above filter surface.
Filter backwash cycle not effective during warm weather	Water viscosity is decreased due to higher temperatures.	Increase backwash rate until required bed expansion is achieved
Increasing premature head loss due to air binding	Air bubble is produced when an influent containing high dissolved oxygen levels is exposed to less than atmospheric pressure.	Increase backwash frequency.
	Pressure drops are occurring during changeover to backwash cycle.	Maintain maximum operating water depth.

are dependent on the equipment used. A general description of the process follows.

19.11.3.1 Filtration Process Description

Wastewater flows to a filter (either gravity or pressurized). The filter contains single media, dual media, or multimedia. Wastewater flows through the media, which remove solids. The solids remain in the filter. Backwashing the filter as needed removes these trapped solids. The backwash solids are then returned to the plant for treatment. Processes typically remove 95 to 99% of the suspended matter.

19.11.3.2 Operational Observations, Problems, and Troubleshooting

Operators routinely monitor filter operation to achieve optimum performance and to detect operational problems based on indications or observation of equipment malfunction or suboptimal process performance. Operational problems typically encountered in filter operations are listed in Table 19.13.

TABLE 19.14
Symptoms, Causes, and Corrective Actions for Biological Nitrification Treatment Problems

Symptom	Cause	Corrective Action
Decreased pH with loss of nitrification	Alkalinity available for the process is insufficient.	If process alkalinity is less than 30 mg/L, add lime or sodium hydroxide to process influent.
	Acid wastes are in the process influent.	Identify source and control of acid wastes.
Incomplete nitrification	Process is dissolved oxygen and temperature limited.	Increase process aeration rate.
	Influent nitrogen loading has increased.	Decrease process nitrogen loading; increase nitrifying bacteria population; put additional units in service; modify operations to increase nitrogen removal.
	Process has a low nitrifying bacteria population.	Decrease wasting or solids loss; add settled raw sewage to nitrification unit to increase biological solids.
	Peak hourly ammonium concentrations exceed available oxygen supplies.	Increase oxygen supply; install flow equalization to minimize peaks.
Very high nitrification sludge SVI (>250).	Nitrification is occurring in the first stage (BOD removal sludge).	Transfer sludge from first to second stages; operate first stage at lower ratio of mean cell residence time (MCRT) to sludge retention time (SRT).

19.11.4 BIOLOGICAL NITRIFICATION

Biological *nitrification* is the first basic step of *biological nitrification–denitrification*. In nitrification, the secondary effluent is introduced into another aeration tank, trickling filter, or biodisc. Because most of the carbonaceous BOD has already been removed, the microorganisms that drive this advanced step are the nitrifying bacteria *Nitrosomonas* and *Nitrobacter*. In nitrification, the ammonia nitrogen is converted to nitrate nitrogen, producing a *nitrified effluent*. At this point, the nitrogen has not actually been removed, only converted to a form that is not toxic to aquatic life and that does not cause an additional oxygen demand. The nitrification process can be limited (performance affected) by alkalinity (requires 7.3 parts alkalinity to 1.0 part ammonia nitrogen), pH, dissolved oxygen availability, toxicity (ammonia or other toxic materials), and process mean cell residence time (sludge retention time). As a general rule, biological nitrification is more effective and achieves higher levels of removal during the warmer times of the year.

19.11.4.1 Operational Observations, Problems, and Troubleshooting

Making sure that the nitrification process performs as per design requires the operator to monitor the process and to make routine adjustments. The loss of solids from the settling tank, rotating biological contactor (RBC), or a trickling filter is a common problem that the operator must be able to identify and take actions to correct. In these instances, obviously, the operator must be familiar with activated sludge systems, RBC, and trickling filter operations. The operator must also be familiar with other nitri-

fication operational problems and must be able to take the proper corrective actions. Table 19.14 lists typical nitrification operational problems and recommended corrective actions.

19.11.5 BIOLOGICAL DENITRIFICATION

Biological denitrification removes nitrogen from the wastewater. When bacteria come in contact with a nitrified element in the absence of oxygen, they reduce the nitrates to nitrogen gas, which escapes the wastewater. The denitrification process can be carried out in either an anoxic activated sludge system (suspended growth) or a column system (fixed growth). The denitrification process can remove up to 85% or more of nitrogen. After effective biological treatment, little oxygen demanding material is left in the wastewater when it reaches the denitrification process. The denitrification reaction will only occur if an oxygen demand source exists when no dissolved oxygen is present in the wastewater. An oxygen demand source is usually added to reduce the nitrates quickly. The most common demand source added is soluble BOD or methanol. Approximately 3 mg/L of methanol are added for every 1 mg/L of nitrate–nitrogen. Suspended growth denitrification reactors are mixed mechanically but only enough to keep the biomass from settling without adding unwanted oxygen. Submerged filters of different types of media may also be used to provide denitrification. A fine media downflow filter is sometimes used to provide both denitrification and effluent filtration. A fluidized sand bed where wastewater flows upward through media composed of sand or activated carbon at a rate to fluidize the bed may also be used. Denitrification bacteria grow on the media.

TABLE 19.15
Symptoms, Causes, and Corrective Actions for Biological Denitrification Treatment Problems

Symptom	Cause	Corrective Action
BOD_5 suddenly increases.	Excessive methanol or other organic matter is present.	Reduce methanol addition; install automated methanol control system; install aerated stabilization unit for removal of excess methanol.
Effluent nitrate concentration suddenly increases.	Methanol control is inadequate.	Identify/correct control problem.
	Denitrification pH is outside 7.0–7.5 range required for process.	Correct pH problem in nitrification process; adjust pH at process influent; correct denitrification sludge return; increase denitrification sludge waste rate; decrease denitrification sludge waste rate.
	Pump failure is causing a loss of solids from the denitrification process.	Transfer sludge from carbonaceous units to denitrification unit.
	Excessive mixing is introducing dissolved oxygen.	Reduce mixer speed; remove some mixers from service.
Head loss is high (packed-bed nitrification).	Unit has excessive solids.	Backwash unit 1 to 2 min, then return to service.
	Nitrogen gas is accumulating in the unit.	
Out-of-service packed-bed unit binds on startup.	Solids have floated to the top during the shutdown.	Backwash units before removing from service and immediately before placing in servicing.

19.11.5.1 Operational Observations, Problems, and Troubleshooting

Operators monitor the performance of a denitrification process by observing various parameters that indicate process malfunctions or suboptimal performance requiring corrective action. Table 19.15 lists several of these indicators of poor process performance, their causal factors, and corrective actions.

19.11.6 Carbon Adsorption

The main purpose of *carbon adsorption* used in advanced treatment processes is the removal of refractory organic compounds (non-BOD_5) and soluble organic materials that are difficult to eliminate by biological or physicochemical treatment. In the carbon adsorption process, wastewater passes through a container filled with either carbon powder or carbon slurry. Organics adsorb onto the carbon (i.e., organic molecules are attracted to the activated carbon surface and are held there) with sufficient contact time. A carbon system usually has several columns or basins used as contactors. Most contact chambers are open concrete gravity-type systems or steel pressure containers applicable to upflow or downflow operation. With use, carbon loses its adsorptive capacity. The carbon must then be regenerated or replaced with fresh carbon. As head loss develops in carbon contactors, they are backwashed with clean effluent in much the same way the effluent filters are backwashed. Carbon used for adsorption may be in a granular form or in a powdered form.

Note: Powdered carbon is too fine for use in columns and is usually added to the wastewater, then later removed by coagulation and settling.

19.11.6.1 Operational Observations, Problems, and Troubleshooting

With regard to carbon adsorption systems for advanced wastewater treatment, operators are primarily interested in monitoring the system to prevent excessive head loss, to reduce levels of hydrogen sulfide in the carbon contactor, to ensure that the carbon is not fouled, and to minimize corrosion of metal parts and damage to concrete in contactors. Table 19.16 lists several of these indicators of poor process performance, their causal factors, and corrective actions.

19.11.7 Land Application

The application of secondary effluent onto a land surface can provide an effective alternative to the expensive and complicated advanced treatment methods discussed previously and the biological nutrient removal (BNR) system discussed later. A high-quality polished effluent (i.e., effluent with high levels of TSS, BOD, phosphorus, and nitrogen compounds as well as refractory organics are reduced) can be obtained by the natural processes that occur as the effluent flows over the vegetated ground surface and percolates through the soil. Limitations are involved with land application of wastewater effluent. For example, the process requires large land areas. Soil type and climate are also critical factors in controlling the design and feasibility of a land treatment process.

TABLE 19.16

Symptoms, Causes, and Corrective Actions for Carbon Adsorption Treatment Problems

Symptom	Cause	Corrective Action
Excessive head loss	Influent is highly turbid.	Backwash unit vigorously. Correct problem in prior treatment steps.
	Biological solids are growing and accumulating in the unit.	Operate as an expanded upflow bed to remove solids continuously; increase frequently of backwashing for downflow beds; improve soluble BOD_5 removal in prior treatment steps; remove carbon from unit and wash out fines.
	Carbon fines are excessive due to deterioration during handling.	Replace carbon with harder carbon.
	Inlet or outlet screens are plugged.	Backflush screens.
Hydrogen sulfide in carbon contactor	Little or no dissolved oxygen or nitrate is present in the contactor influent.	Add air, oxygen, or sodium nitrate to unit influent.
	Influent BOD_5 concentrations are high.	Improve soluble BOD_5 removal in prior treatment steps; precipitate sulfides already formed with iron on chlorine.
	Detention time in carbon contactor is excessive.	Reduce detention time by removing one or more contactors from service; backwash units more frequently and more vigorously, using air scour or surface wash.
Large decrease in COD removed per lb of carbon regenerated	Carbon is fouled and losing efficiency.	Improve regeneration process performance.
Corrosion of metal parts or damage to concrete in the contactors	Hydrogen sulfide is in the carbon contactors.	See corrective actions for "Hydrogen sulfide in carbon contactor" entry (above).
	Holes in protective coatings allow exposure to dewatered carbon.	Repair protective coatings.

19.11.7.1 Types and Modes of Land Application

Three basic types or modes of land application or treatment are commonly used: *irrigation* (slow rate), *overland flow*, and *infiltration–percolation* (rapid rate). The basic objectives of these types of land applications and the conditions under which they can function vary. In irrigation (also called *slow rate*), wastewater is sprayed or applied (usually by ridge-and-furrow surface spreading or by sprinkler systems) to the surface of the land. Wastewater enters the soil. Crops growing on the irrigation area utilize available nutrients. Soil organisms stabilize the organic content of the flow. Water returns to the hydrologic cycle through evaporation or by entering the surface water or groundwater.

The irrigation land application method provides the best results (compared with the other two types of land application systems) with respect to advanced treatment levels of pollutant removal. Not only are suspended solids and BOD significantly reduced by filtration of the wastewater, but also biological oxidation of the organics occurs in the top few inches of soil. Nitrogen is removed primarily by crop uptake, and phosphorus is removed by adsorption within the soil. Expected performance levels for irrigation include:

- BOD_5, 98%
- Suspended solids, 98%
- Nitrogen, 85%
- Phosphorus, 95%
- Metals, 95%

The overland flow application method utilizes physical, chemical, and biological processes as the wastewater flows in a thin film down the relatively impermeable surface. In the process, wastewater sprayed over sloped terraces flows slowly over the surface. Soil and vegetation remove suspended solids, nutrients, and organics. A small portion of the wastewater evaporates. The remainder flows to collection channels. Collected effluent is discharged to surface waters. Expected performance levels for overland flow include:

- BOD_5, 92%
- Suspended solids, 92%
- Nitrogen, 70–90%
- Phosphorus, 40–80%
- Metals, 50%

In the infiltration–percolation application method, wastewater is sprayed or pumped to spreading basins (also known as recharge basins or large ponds). Some wastewater evaporates. The remainder percolates or infiltrates into the soil. Solids are removed by filtration. Water recharges the groundwater system. Most of the effluent percolates to the groundwater; very little of it is absorbed by vegetation. The filtering and adsorption action of the soil removes most of the BOD, TSS, and phosphorus from the effluent; however, nitrogen removal is relatively poor. Expected performance levels for infiltration–percolation include:

- BOD_5, 85–99%
- Suspended solids, 98%
- Nitrogen, 0–50%
- Phosphorus, 60–95%
- Metals, 50–95%

19.11.7.2 Operational Observations, Problems, and Troubleshooting

Performance levels are dependent on the land application process used. To be effective, operators must monitor the operation of the land application process employed. Experience has shown that these processes can be very effective, but problems arise when the flow contains potentially toxic materials that may become concentrated in the crops being grown on the land. Other common problems include ponding, deterioration of distribution piping systems, malfunctioning sprinkler heads, waste runoff, irrigated crop die-off, poor crop growth, and too great of a flow rate. Table 19.17 lists several of these indicators of poor process performance, their causal factors, and corrective actions.

19.11.8 Biological Nutrient Removal

Recent experience has shown that *biological nutrient removal* (BNR) systems are reliable and effective in removing nitrogen and phosphorus. The process is based on the principle that, under specific conditions, microorganisms will remove more phosphorus and nitrogen than is required for biological activity; thus, treatment can be accomplished without the use of chemicals. Not having to purchase and use chemicals to remove nitrogen and phosphorus potentially has numerous cost–benefit implications. In addition, because chemicals are not required, chemical waste products are not produced, thus reducing the need to handle and dispose of waste. Several patented processes are available for this purpose. Performance depends on the biological activity and the process employed.

19.11.9 Enhanced Biological Nutrient Removal

Removing phosphorus from wastewater in secondary treatment processes has evolved into innovative *enhanced biological nutrient removal* (EBNR) technologies. An EBNR treatment process promotes the production of phosphorus-accumulating organisms, which utilize more phosphorus in their metabolic processes than a conventional secondary biological treatment process (USEPA, 2007).

The average total phosphorus concentrations in raw domestic wastewater are usually between 6 to 8 mg/L, and the total phosphorus concentration in municipal wastewater after conventional secondary treatment is routinely reduced to 3 or 4 mg/L. EBNR incorporated into the secondary treatment system can often reduce total phosphorus concentrations to 0.3 mg/L and less. Facilities using EBNR have significantly reduced the amount of phosphorus to be removed through the subsequent chemical addition and tertiary filtration process. This improved the efficiency of the tertiary process and significantly reduced the costs of chemicals used to remove phosphorus. Facilities using EBNR reported that their chemical dosing was cut in half after EBNR was installed to remove phosphorus (USEPA, 2007).

Treatment provided by these EBNR processes also removes other pollutants that commonly affect water quality to very low levels (USEPA, 2007). Biochemical oxygen demand and total suspended solids are routinely less than 2 mg/L and fecal coliform bacteria less than 10 cfu/100 mL. Turbidity of the final effluent is very low, which allows for effective disinfection using ultraviolet light, rather than chlorination. Recent studies report finding that wastewater treatment facilities using EBNR also significantly reduced the amount of pharmaceuticals and healthcare products from municipal wastewater, as compared to removal accomplished by conventional secondary treatment. The following section describes some of the EBNR treatment technologies currently being used in the United States.

19.11.9.1 0.5-MGD Capacity Plant

- *Advanced phosphorus treatment technology*—Chemical addition, two-stage filtration
- *Treatment process (liquid side only)*—Screening and grit removal; extended aeration and secondary clarification (in a combined aeration basin/clarifier); chemical addition for flocculation using polyaluminum silicate sulfate (PASS) and filtration through two-stage Dyna-Sand® filters.

19.11.9.2 1.5-MGD Capacity Plant

- *Advanced phosphorus treatment technology*—BNR, chemical addition, tertiary settlers, filtration
- *Treatment process (liquid side only)*—Screening and grit removal in the headworks; activated sludge biological treatment; biological aerated filter (IDI Biofor™ for nitrification); chemical coagulation using alum; flocculation and clarification using a tube settler (IDI Densadeg™); filtration (single-stage Parkson Dyna-

TABLE 19.17

Symptoms, Causes, and Corrective Actions for Land Application Treatment Problems

Symptom	Cause	Corrective Action
Water ponding in irrigated areas	Application rate is excessive.	Reduce application rate to an acceptable level.
	Drainage is inadequate because of groundwater levels.	Irrigate in portions of site where groundwater is not a problem; store wastewater until condition is corrected.
	Drainage wells are damaged.	Repair drainage wells.
	Well withdrawal rates are inadequate.	Increase drainage-well pumping rates.
	Drain tiles are damaged.	Repair damaged drain tiles.
	Distribution system has a broken pipe.	Repair pipe.
Deterioration of distribution piping	Effluent remains in pipe for long periods.	Drain pipe after each use; coat steel valves.
	Different metals are used in same line.	Install cathodic/anodic protection.
No flow from source sprinkler nozzles	Nozzles are clogged	Repair/replace screen on irrigation pump inlet.
Wastes running off irrigation area	High sodium adsorption ratio has caused clay soil to become impermeable.	Feed calcium and magnesium to maintain sodium adsorption ratio (SAR) to less than 9.
	Solids have sealed the soil surface.	Strip crop area.
	Application rate is greater than the soil infiltration rate.	Reduce application rate to acceptable level.
	Distribution piping has a break in it.	Repair system.
	Soil permeability has decreased because of continuous application of wastewater.	Allow 2- to 3-day rest period between each application.
	Rain has saturated the soil.	Store wastewater until soil has drained.
Failure of irrigated crop	Too much or not enough water has been applied.	Adjust application rate to appropriate level.
	Wastewater contains toxic materials in toxic concentrations.	Eliminate source of toxicity.
	Excessive insecticide or herbicide has been applied.	Apply only as permitted or directed.
	Inadequate drainage has flooded the root zone of the crop.	
Poor crop growth	Crops lack sufficient nitrogen (N) or phosphorus (P).	Increase application rate to supply N and P; augment N and P of wastewater with commercial fertilizer applications.
	Timing of nutrient applications does not coincide with plant nutrient need.	Adjust application schedule to match crop need.
Normal irrigation pump psi but above-average flow rate	Main, riser, or lateral is broken.	Locate and repair problem.
	Gasket is leaking.	Locate and replace defective gasket.
	Sprinkler head or nozzle is missing.	Replace missing component.
	Too many distribution laterals are in service at one time.	Correct valving to adjust number of laterals in service.
Above-average irrigation pump psi but below-average flow	System has a blockage.	Locate and correct blockage
Below-average irrigation pump psi and flow rate	Impeller is worn.	Replace impeller.
	Pump inlet screen is partially clogged.	Clean screen.
Excessive erosion	Application rates are excessive.	Reduce application rate.
	Crop coverage is inadequate.	Improve crop coverage.
Odor complaints	Wastes are turning septic during transport to treatment/irrigation site.	Aerate or chemically treat wastes during transport; install cover over discharge point; collect and treat gases before release; improve pretreatment.
	Storage reservoirs are septic.	Aerate storage reservoirs.
Center pivot irrigation rigs stuck in mud	Application rates are excessive.	Reduce application rate.
	Improper rig or tires are being used.	Install tires with higher flotation capabilities.
	Drainage is poor.	Improve drainage.
Increasing nitrate in groundwater near irrigation site	Nitrogen application rate is not balanced with crop need.	Change to crop with higher nitrogen requirement.
	Applications are occurring during dormant periods.	Adjust schedule to apply only during active growth periods.
	Crop is not being properly harvested and removed.	Harvest and remove crop as required.

Sand® filters); disinfection; dechlorination. The DynaSand filter reject rate is reported to be about 15 to 20%. The DynaSand filters are configured in four two-cell units for a total of eight filters beds, each of which is 8 feet deep. Influent concentrations of total phosphorus are usually measured at about 6 mg/L (which is a typical value for untreated domestic wastewater). The aeration basins are operated with an anoxic zone to provide for biological removal of phosphorus. About 60% of the influent phosphorus is removed through the biological treatment process. Sodium sulfate is added to maintain alkalinity through the treatment process for phosphorus removal. Sodium sulfate is applied to the wastewater at a rate of approximately 100 to 120 mg/L just upstream of where alum is added. Alum is used to precipitate phosphorus. The alum dose is typically 135 mg/L and is used with 0.5 to 1.0 mg/L cationic polymer.

Note: The DynaSand® filter is a continuous backwash, upflow, deep-bed, granular media filter.

19.11.9.3   1.55-MGD Capacity Plant

- *Advanced phosphorus treatment technology—* Chemical addition, two-stage filtration
- *Treatment process (liquid side only)—* Screening and grit removal; extended aeration and secondary clarification; chemical addition for flocculation using aluminum chloride (added to the wastewater at both the secondary clarifiers and the distribution header for the DynaSand filters); filtration through two-stage DynaSand filters; disinfection with chlorine; and dechlorination with sulfur dioxide. Chlorine is added to the filter influent to control biological growth in the filters.

19.11.9.4   2-MGD Capacity Plant

- *Advanced phosphorus treatment technology—* BNR, chemical addition, two-stage filtration
- *Treatment process (liquid side only)—* Screening and grit removal; BNR activated sludge (five-stage BardenPho™ process: anaerobic basin, anoxic basin, oxidation ditch aeration basin, anoxic basin, reaeration basin); clarifiers (two parallel rectangular); chemical addition using alum and polymer; effluent polishing and filtration (using four USFilter's Memcor™ filter modules); and UV disinfection. The USFilter units utilize two-stage filtration in which the first stage is upflow through plastic media with

air scour. The second stage filtration is through a downflow, mixed media with backwash cleaning. The concentration of alum used for coagulation is 95 mg/L.

19.11.9.5   2.6-MGD Capacity Plant

- *Advanced phosphorus treatment technology—* BNR, chemical addition, tertiary settlers, filtration
- *Treatment process (liquid side only)—* Screening and grit removal; aeration basins; secondary clarification; chemical coagulation and flocculation using alum and polymer; tertiary clarification (rectangular sedimentation basins with plate settlers); mixed-media bed filters (5 ft deep); and disinfection (the filtration process removes enough fecal coliform so conventional disinfection is not normally required). The average alum dose is 70 mg/L in the wastewater and varies from 50 to 180 m/L. A greater dose of alum is applied during the winter period. The polymer dose concentration is about 0.1 mg/L.

19.11.9.6   3-MGD Capacity Plant

- *Advanced phosphorus treatment technology—* BNR, chemical addition, tertiary settlers, filtration
- *Treatment process (liquid side only)—* Screening and grit removal; BNR; chemical coagulation and flocculation using polymer and alum; clarification via tube settlers; filtration through mixed-media bed filters; disinfection with chlorine and dechlorination (using sodium bisulfate)

19.11.9.7   4.8-MGD Capacity Plant

- *Advanced phosphorus treatment technology—* Multipoint chemical addition, tertiary settling, filtration
- *Treatment process (liquid side only)—* Screening and grit removal; primary clarification; trickling filters; intermediate clarification (with polymer addition to aid settling); rotating biological contactors; secondary clarification; chemical addition using polyaluminum chloride; filtration through mixed-media traveling bed filters; and ultraviolet disinfection. The final effluent is discharged down a cascading outfall to achieve reaeration prior to mixing in the receiving water. Approximately 1 MGD per day of final effluent is utilized by the local power company for cooling water.

19.11.9.8 5-MGD Capacity Plant

- *Advanced phosphorus treatment technology—* BNR, filtration
- *Treatment process description (liquid side only)—* Screening and grinding; primary clarification; BNR in the contact basins; secondary clarification; filtration through single-pass Parkson DynaSand® filters (four cells with four filters per each cell); UV disinfection

19.11.9.9 24-MGD Capacity Plant

- *Advanced phosphorus treatment technology—* BNR, chemical addition, filtration
- *Treatment process (liquid side only)—* Screening and grit removal; primary clarification; biological treatment with enhanced biological nutrient removal; secondary clarification; chemical addition of alum and polymer for phosphorus removal; tertiary clarification; filtration through dual-media, gravity bed filters; and disinfection. Lime is added to the biological process to maintain pH and alkalinity. A two-stage fermenter is operated to produce volatile fatty acids, which are added to the biological contact basin. The enhanced biological nutrient removal process at times reduces total phosphorus to levels that are less than the 0.11-mg/L permit limitation; however, this performance is not achieved during the entire period when the seasonal phosphorus limitations are in effect. The tertiary treatment with chemical addition and filtration provides assurance that the final effluent is of consistently good quality. Some of the treated effluent is reclaimed for irrigation.

19.11.9.10 39-MGD Capacity Plant

- *Advanced phosphorus treatment technology—* Chemical addition, filtration
- *Treatment process (liquid side only)—* Screening and grit removal; alum addition; primary clarification; extended aeration; secondary clarification; flocculation using alum and polymer; tertiary clarification; filtration; disinfection with chlorine, and dechlorination. Wastewater is treated in two separate trains. Four 60-ft-diameter ClariCone® tertiary clarifiers are used on one treatment train to provide contact with six monomedia anthracite gravity bed filters. The other treatment train uses conventional clarifiers for tertiary settling followed by filtration through four dual-media, gravity bed filters. Phosphorus is removed in four locations within

this system: alum-enhanced removal in the primary clarifiers, biological removal in the aeration basins, chemical flocculation and removal in the tertiary clarifiers, and removal through filtration.

19.11.9.11 42-MGD Capacity Plant

- *Advanced phosphorus treatment technology—* Chemical (high lime) and tertiary filtration
- *Treatment process (liquid side only)—* Conventional treatment that removes 90% of most incoming pollutants; screening; grit removal; primary clarification; aerobic biological selectors; activated sludge aeration basins with nitrification and denitrification processes; and secondary clarification. Also, a chemical advanced treatment—the high lime process—reduces phosphorus to below 0.10 mg/L, captures organics from secondary treatment, precipitates heavy metals, and serves as a barrier to viruses. Lime slurry is added to rapid mix basins (to achieve pH of 11), and anionic polymer is added in flocculation basins, followed by chemical clarification, first-stage recarbonation to lower pH to 10, recarbonation clarifiers to collect precipitated calcium carbonate, second-stage recarbonation to lower pH to 7, and storage in ballast ponds. Physical advanced treatment to meet stringent limits for TSS (1 mg/L) and COD (10 mg/L) include alum or polymer addition, multimedia filters, and activated carbon contactors. Disinfection is by a chlorination and dechlorination process.

19.11.9.12 54-MGD Capacity Plant

- *Advanced phosphorus treatment technology—* BNR, multi-point chemical addition, tertiary settling, filtration
- *Treatment process description (liquid only)—* Treatment consists of screening and grit removal; primary settling with possible addition of ferric chloride and polymer; addition of methanol or volatile fatty acid to biological reactor basins to aid BNR; ferric chloride and polymer addition prior to secondary settling; alum addition and mixing; tertiary clarification with inclined plate settlers; dual-media gravity bed filtration; UV disinfection; and post aeration.

19.11.9.13 67-MGD Capacity Plant

- *Advanced phosphorus treatment technology—* BNR, chemical addition, tertiary clarification, filtration

- *Treatment process (liquid side only)*—Screening; primary clarification; biological treatment with enhanced biological nutrient removal; polymer addition as needed; secondary clarification; equalization and storage in retention ponds; tertiary clarification with ferric chloride addition to remove phosphorus; disinfection with sodium hypochlorite; filtration through dual-/monomedia gravity bed filters.

19.12 SOLIDS (SLUDGE/BIOSOLIDS) HANDLING

The wastewater treatment unit processes described to this point remove solids and BOD from the wastestream before the liquid effluent is discharged to its receiving waters. What remains to be disposed of is a mixture of solids and wastes, called *process residuals*, more commonly referred to as *sludge* or *biosolids*.

Note: Sludge is the commonly accepted name for wastewater solids; however, if wastewater sludge is used for beneficial reuse (e.g., as a soil amendment or fertilizer), it is commonly referred to as *biosolids*.

The most costly and complex aspect of wastewater treatment can be the collection, processing, and disposal of sludge, because the quantity of sludge produced may be as high as 2% of the original volume of wastewater, depending somewhat on the treatment process being used.

Because sludge can be as much as 97% water content and because the cost of disposal will be related to the volume of sludge being processed, one of the primary purposes or goals of sludge treatment (along with stabilizing it so it is no longer objectionable or environmentally damaging) is to separate as much of the water from the solids as possible. Sludge treatment methods may be designed to accomplish both of these purposes.

Note: Sludge treatment methods are generally divided into three major categories: *thickening*, *stabilization*, and *dewatering*. Many of these processes include complex sludge treatment methods such as heat treatment, vacuum filtration, and incineration.

19.12.1 SLUDGE: BACKGROUND INFORMATION

When we speak of *sludge* or *biosolids*, we are speaking of the same substance or material; each is defined as the suspended solids removed from wastewater during sedimentation and then concentrated for further treatment and disposal or reuse. The difference between the terms *sludge* and *biosolids* is determined by the way they are managed.

Note: The task of disposing of, treating, or reusing wastewater solids is *sludge* or *biosolids management*.

Sludge is typically seen as wastewater solids that are disposed of. Biosolids are the same substance but are managed for reuse, commonly called *beneficial reuse* (e.g., for land application as a soil amendment, such as biosolids compost). Note that even as wastewater treatment standards have become more stringent because of increasing environmental regulations, so has the volume of wastewater sludge increased. Note also that, before sludge can be disposed of or reused, it requires some form of treatment to reduce its volume, to stabilize it, and to inactivate pathogenic organisms.

Sludge forms initially as a 3 to 7% suspension of solids; with each person typically generating about 4 gal of sludge per week, the total quantity generated each day, week, month, and year is significant. Because of the volume and nature of the material, sludge management is a major factor in the design and operation of all water pollution control plants.

Note: Wastewater solids account for more than half of the total costs in a typical secondary treatment plant.

19.12.2 SOURCES OF SLUDGE

Wastewater sludge is generated in primary, secondary, and chemical treatment processes. In primary treatment, the solids that float or settle are removed. The floatable material makes up a portion of the solid waste known as *scum*. Scum is not normally considered sludge; however, it should be disposed of in an environmentally sound way. The settleable material that collects on the bottom of the clarifier is known as *primary sludge*. Primary sludge can also be referred to as *raw sludge* because it has not undergone decomposition. Raw primary sludge from a typical domestic facility is quite objectionable and has a high percentage of water, two characteristics that make handling difficult.

Solids not removed in the primary clarifier are carried out of the primary unit. These solids are known as *colloidal suspended solids*. The secondary treatment system (e.g., trickling filter, activated sludge) is designed to change those colloidal solids into settleable solids that can be removed. Once in the settleable form, these solids are removed in the secondary clarifier. The sludge at the bottom of the secondary clarifier is called *secondary sludge*. Secondary sludges are light and fluffy and more difficult to process than primary sludges—in short, secondary sludges do not dewater well.

The addition of chemicals and various organic and inorganic substances prior to sedimentation and clarification may increase the solids capture and reduce the amount of solids lost in the effluent. This *chemical addition* results in the formation of heavier solids, which trap the colloidal solids or convert dissolved solids to settleable solids. The resultant solids are known as *chemical sludges*. As chemical

TABLE 19.18
Typical Water Content of Sludges

Water Treatment Process	Percent Moisture of Sludge	lb Water/lb Sludge Solids Generated
Primary sedimentation	95	19
Trickling filter		
Humus, low rate	93	13.3
Humus, high rate	97	32.3
Activated sludge	99	99

Source: USEPA, *Operational Manual: Sludge Handling and Conditioning*, EPA-430/9-78-002, U.S. Environmental Protection Agency, Washington, D.C., 1978.

usage increases, so does the quantity of sludge that must be handled and disposed of. Chemical sludges can be very difficult to process; they do not dewater well and contain lower percentages of solids.

19.12.3 SLUDGE CHARACTERISTICS

The composition and characteristics of sewage sludge vary widely and can change considerably with time. Notwithstanding these facts, the basic components of wastewater sludge remain the same. The only variations occur in quantity of the various components as the type of sludge and the process from which it originated changes. The main component of all sludges is *water*. Prior to treatment, most sludge contains 95 to 99% water (see Table 19.18). This high water content makes sludge handling and processing extremely costly in terms of both money and time. Sludge handling may represent up to 40% of the capital costs and 50% of the operating costs of a treatment plant. As a result, the importance of optimum design for handling and disposal of sludge cannot be overemphasized. The water content of the sludge is present in a number of different forms. Some forms can be removed by several sludge treatment processes, thus allowing the same flexibility in choosing the optimum sludge treatment and disposal method. The various forms of water and their approximate percentages for a typical activated sludge are shown in Table 19.19. The forms of water associated with sludges include:

- *Free water*—Water that is not attached to sludge solids in any way and can be removed by simple gravitational settling.
- *Floc water*—Water that is trapped within the floc and travels with them; it can be removed by mechanical dewatering.
- *Capillary water*—Water that adheres to the individual particles and can be squeezed out of shape and compacted.
- *Particle water*—Water that is chemically bound to the individual particles and cannot be removed without inclination.

TABLE 19.19
Distribution of Water in Activated Sludge

Water Type	% Volume
Free water	75
Floc water	20
Capillary water	2
Particle water	2.5
Solids	0.5
Total	100

Source: USEPA, *Operational Manual: Sludge Handling and Conditioning*, EPA-430/9-78-002, U.S. Environmental Protection Agency, Washington, D.C., 1978.

From a public health view, the second and probably more important component of sludge is the *solids matter*. Representing from 1 to 8% of the total mixture, these solids are extremely unstable. Wastewater solids can be classified into two categories based on their origin—organic and inorganic. *Organic solids* in wastewater, simply put, are materials that were at one time alive and will burn or volatilize at 550°C after 15 min in a muffle furnace. The percent organic material within a sludge will determine how unstable it is.

The inorganic material within sludge will determine how stable it is. The *inorganic solids* are those solids that were never alive and will not burn or volatilize at 550°C after 15 min in a muffle furnace. Inorganic solids are generally not subject to breakdown by biological action and are considered stable. Certain inorganic solids, however, can create problems when related to the environment—for example, heavy metals such as copper, lead, zinc, mercury, and others. These can be extremely harmful if discharged.

Organic solids may be subject to biological decomposition in either an aerobic or anaerobic environment. Decomposition of organic matter (with its production of objectionable byproducts) and the possibility of toxic organic solids within the sludge compound the problems of sludge disposal.

Note: Before moving on to a discussion of the fundamentals of sludge treatment methods, it is important to begin by covering sludge pumping calculations. It is important to point out that it is difficult (if not impossible) to treat the sludge unless it is pumped to the specific sludge treatment process.

19.12.4 Sludge Pumping Calculations

Wastewater operators are often called upon to make various process control calculations. An important calculation involves sludge pumping. The sludge pumping calculations that the operator may be required to make during plant operations (and should know for licensure examinations) are covered in this section.

19.12.4.1 Estimating Daily Sludge Production

The calculation for *estimation of the required sludge-pumping rate* provides a method to establish an initial pumping rate or to evaluate the adequacy of the current withdrawal rate:

$$\text{Est. pump rate} = \frac{\begin{array}{c}\text{(Influent TSS conc.}\\ -\text{ effluent TSS conc.)}\\ \times \text{flow} \times 8.34\end{array}}{\begin{array}{c}\% \text{ Solids in sludge} \times 8.34\\ \times 1440 \text{ min/day}\end{array}} \quad (19.66)$$

■ EXAMPLE 19.59

Problem: The sludge withdrawn from the primary settling tank contains 1.4% solids. The unit influent contains 285 mg/L TSS and the effluent contains 140 mg/L TSS. If the influent flow rate is 5.55 MGD, what is the estimated sludge withdrawal rate in gallons per minute (assuming the pump operates continuously)?

Solution:

$$\text{Sludge rate} = \frac{(285 \text{ mg/L} - 140 \text{ mg/L}) \times 5.55 \times 8.34}{0.014 \times 8.34 \times 1440 \text{ min/day}}$$

$$= 40 \text{ gpm}$$

Note: The following information is used for Examples 18.60 to 18.65:

Operating time = 15 min/cycle
Frequency = 24 times per day
Pump rate = 120 gpm
Solids = 3.70%
Volatile matter = 66%

19.12.4.2 Sludge Pumping Time

The sludge pumping time is the total time (in minutes) that a pump operates during a 24-hr period:

$$\text{Pump operation time} = \begin{bmatrix}\text{Time (min/cycle)}\\ \times \text{frequency (cycles/day)}\end{bmatrix} \quad (19.67)$$

■ EXAMPLE 19.60

Problem: What is the pump operating time?

Solution:

$$\text{Pump operating time} = 15 \text{ min/cycle} \times 24 \text{ cycles/day}$$

$$= 360 \text{ min/day}$$

19.12.4.3 Gallons Sludge Pumped per Day

$$\text{Sludge (gpd)} = \begin{bmatrix}\text{Operating time (min/day)}\\ \times \text{pump rate (gpm)}\end{bmatrix} \quad (19.68)$$

■ EXAMPLE 19.61

Problem: What is the sludge pumped per day in gallons?

Solution:

$$\text{Sludge} = 360 \text{ min/day} \times 120 \text{ gpm} = 43,200 \text{ gpd}$$

19.12.4.4 Pounds Sludge Pumped per Day

$$\text{Sludge pumped (lb/day)} = \begin{bmatrix}\text{Sludge pumped (gal)}\\ \times 8.34 \text{ lb/gal}\end{bmatrix} \quad (19.69)$$

■ EXAMPLE 19.62

Problem: What is the sludge pumped per day in gallons?

Solution:

$$\text{Sludge pumped} = 43,200 \text{ gal/day} \times 8.34 \text{ lb/gal}$$

$$= 360,300 \text{ lb/day}$$

19.12.4.5 Pounds Solids Pumped per Day

$$\text{Solids pumped (lb/day)} = \begin{bmatrix}\text{Sludge pumped (gpd)}\\ \times \% \text{ solids}\end{bmatrix} \quad (19.70)$$

■ EXAMPLE 19.63

Problem: What are the solids pumped per day?

Solution:

$$\text{Solids pumped} = 360,300 \text{ lb/day} \times 0.0370$$

$$= 13,331 \text{ lb/day}$$

19.12.4.6 Pounds Volatile Matter (VM) Pumped per Day

$$VM \ (lb/day) = Solids \ pumped \ (lb/day) \times \%VM \quad (19.71)$$

■ EXAMPLE 19.64

Problem: What is the volatile matter in pounds per day?

Solution:

$$VM = 13{,}331 \ lb/day \times 0.66 = 8798 \ lb/day$$

Note: If we wish to calculate the pounds of solids or the pounds of volatile solids removed per day, the individual equations demonstrated above can be combined into a single calculation:

$$Solids \ (lb/day) = \begin{bmatrix} Pump \ time \ (min/cycle) \\ \times frequency \ (cycles/day) \\ \times rate \ (gpm) \\ \times 8.34 \ lb/gal \times solids \ volume \end{bmatrix}$$

$$(19.72)$$

$$VM \ (lb/day) = \begin{bmatrix} Time \ (min/cycle) \\ \times frequency \ (cycles/day) \\ \times rate \ (gpm) \times 8.34 \\ \times \% \ solids \times \%VM \end{bmatrix}$$

■ EXAMPLE 19.65

$$Solids = \begin{bmatrix} 15 \ min/cycle \times 24 \ cycles/day \\ \times 120 \ gpm \times 8.34 \times 0.0370 \end{bmatrix} = 13{,}331 \ lb/day$$

$$VM = \begin{bmatrix} 15 \ min/cycle \times 24 \ cycles/day \\ \times 120 \ gpm \times 8.34 \times 0.0370 \times .66 \end{bmatrix} = 8798 \ lb/day$$

19.12.4.7 Sludge Production in Pounds per Million Gallons

A common method of expressing sludge production is in pounds of sludge per million gallons of wastewater treated:

$$Sludge \ (lb/MG) = \frac{Total \ sludge \ production \ (lb)}{Total \ wastewater \ flow \ (MG)} \quad (19.73)$$

Problem: Records show that a treatment plant has produced 85,000 gal of sludge during the past 30 days. The average daily flow for this period was 1.2 MGD. What was the plant's sludge production in pounds per million gallons?

Solution:

$$Sludge \ (lb/MG) = \frac{85{,}000 \ gal \times 8.34 \ lb/gal}{1.2 \ MGD \times 30 \ days}$$

$$= 19{,}692 \ lb/MG$$

19.12.4.8 Sludge Production in Wet Tons per Year

Sludge production can also be expressed in terms of the amount of sludge (water and solids) produced per year. This is normally expressed in wet tons per year:

$$Sludge \ (wet \ tons/yr) = \frac{\begin{array}{c} Sludge \ produced \ (lb/MG) \\ \times average \ daily \ flow \ (MGD) \\ \times 365 \ days/yr \end{array}}{2000 \ lb/ton} \quad (19.74)$$

■ EXAMPLE 19.66

Problem: The plant is currently producing sludge at the rate of 16,500 lb/MG. The current average daily wastewater flow rate is 1.5 MGD. What will be the total amount of sludge produced per year in wet tons per year?

Solution:

$$Sludge = \frac{16{,}500 \ lb/MG \times 1.5 \ MGD \times 365 \ days/yr}{2000 \ lb/ton}$$

$$= 4517 \ wet \ tons/yr$$

Important Point: Release of wastewater solids without proper treatment could result in severe damage to the environment. Obviously, we must have a system to treat the volume of material removed as sludge throughout the system. Release without treatment would defeat the purpose of environmental protection. A design engineer can choose from many processes when developing sludge treatment systems. No matter what the system or combination of systems chosen, the ultimate purpose will be the same: the conversion of wastewater sludges into a form that can be handled economically and disposed of without damage to the environment or creating nuisance conditions. Leaving either condition unmet will require further treatment. The degree of treatment will generally depend on the proposed method of disposal. Sludge treatment processes can be classified into a number of major categories. In this handbook, we discuss the processes of *thickening*, *digestion* (or *stabilization*), *dewatering*, *incineration*, and *land application*. Each of these categories has then been further subdivided according to the specific processes that are used to accomplish sludge treatment. As mentioned, the importance of adequate, efficient sludge treatment cannot be overlooked when designing wastewater treatment facilities. The inadequacies of a sludge treatment system can severely affect the overall performance capabilities of a plant. The inability to remove and process solids as quickly as they accumulate in the process can lead to the discharge of large quantities of solids to receiving waters. Even with proper design and capabilities in place, no system can be effective unless it is properly operated. Proper operation requires proper operator performance. Proper operator performance begins and ends with proper training.

19.12.5 SLUDGE THICKENING

The solids content of primary, activated, trickling-filter, or even mixed sludge (i.e., primary plus activated sludge) varies considerably, depending on the characteristics of the sludge. Note that the sludge removal and pumping facilities and the method of operation also affect the solids content. *Sludge thickening* (or *concentration*) is a unit process used to increase the solids content of the sludge by removing a portion of the liquid fraction. By increasing the solids content, more economical treatment of the sludge can be accomplished. Sludge thickening processes include:

* Gravity thickeners
* Flotation thickeners
* Solids concentrators

19.12.5.1 Gravity Thickening

Gravity thickening is most effective on primary sludge. Solids are withdrawn from primary treatment (and sometimes secondary treatment) and pumped to the thickener. The solids buildup in the thickener forms a solids blanket on the bottom. The weight of the blanket compresses the solids on the bottom and squeezes the water out. By adjusting the blanket thickness, the percent solids in the underflow (solids withdrawn from the bottom of the thickener) can be increased or decreased. The supernatant (clear water) that rises to the surface is returned to the wastewater flow for treatment. Daily operations of the thickening process include pumping, observation, sampling and testing, process control calculations, maintenance, and housekeeping.

Note: The equipment employed in thickening depends on the specific thickening processes used.

Equipment used for gravity thickening consists of a *thickening tank* that is similar in design to the settling tank used in primary treatment. Generally, the tank is circular and provides equipment for continuous solids collection. The collector mechanism uses heavier construction than

that in a settling tank because the solids being moved are more concentrated. The gravity thickener pumping facilities (i.e., pump and flow measurement) are used for withdrawal of thickened solids.

Solids concentrations achieved by gravity thickeners are typically 8 to 10% solids from primary underflow, 2 to 4% solids from waste activated sludge, 7 to 9% solids from trickling filter residuals, and 4 to 9% from combined primary and secondary residuals. The performance of gravity thickening processes depends on various factors, including:

* Type of sludge
* Condition of influent sludge
* Temperature
* Blanket depth
* Solids loading
* Hydraulic loading
* Solids retention time
* Hydraulic detention time

19.12.5.2 Flotation Thickening

Flotation thickening is used most efficiently for waste sludges from suspended-growth biological treatment process, such as the activated sludge process. Recycled water from the flotation thickener is aerated under pressure. During this time, the water absorbs more air than it would under normal pressure. The recycled flow together with chemical additives (if used) are mixed with the flow. When the mixture enters the flotation thickener, the excess air is released in the form of fine bubbles. These bubbles become attached to the solids and lift them toward the surface. The accumulation of solids on the surface is called the *float cake*. As more solids are added to the bottom of the float cake, it becomes thicker and water drains from the upper levels of the cake. The solids are then moved up an inclined plane by a scraper and discharged. The supernatant leaves the tank below the surface of the float solids and is recycled or returned to the wastestream for treatment. Typically, flotation thickener performance is 3 to 5% solids for waste activated sludge with polymer addition and 2 to 4% solids without polymer addition.

The flotation thickening process requires pressurized air, a vessel for mixing the air with all or part of the process residual flow, a tank in which the flotation process can occur, and solids collector mechanisms to remove the float cake (solids) from the top of the tank and accumulated heavy solids from the bottom of the tank. Because the process normally requires chemicals to be added to improve separation, chemical mixing equipment, storage tanks, and metering equipment to dispense the chemicals at the desired dose are required. The performance of the dissolved air-thickening process depends on various factors:

TABLE 19.20
Symptoms, Causes, and Corrective Actions for Solids Concentrator Problems

Symptom	Cause	Corrective Action
Gravity thickener		
Odors and rising sludge	Sludge withdrawal rate is too low.	Increase sludge withdrawal rate.
	Overflow rate is too low.	Increase influent flow rate.
	Thickener is septic.	Add chlorine, permanganate, or peroxide to influent.
Thickened sludge below desired solids concentration	Overflow rate is too high.	Decrease influent sludge flow rate.
	Sludge withdrawal rate is too high.	Decrease pump rate for sludge withdrawal.
	Unit is short-circuiting.	Identify cause of short-circuiting and correct.
Torque alarm activated	Heavy sludge accumulation is occurring.	Agitate sludge blanket to decrease density; increase sludge withdrawal rate.
	Collector mechanism is jammed.	Attempt to locate and remove obstacle; dewater tank, if necessary, to remove obstacle.
Dissolved air flotation thickener		
Float solids concentration too low	Skimmer speed is too high.	Adjust skimmer speed to permit concentration to occur.
	Unit is overloaded.	Stop sludge flow through unit; purge unit with recycle.
	Polymer dose is insufficient.	Determine proper chemical dose and adjust.
	Air-to-solids ratio is excessive.	Reduce airflow to pressurization tank.
	Dissolved air levels are low.	Identify malfunction and correct.
Dissolved air concentration too low	Mechanical malfunction has occurred.	Identify cause and correct.
Excessive solids in effluent (subnatant) flow	Unit is overloaded.	Stop sludge flow through unit; purge unit with recycle.
	Chemical dose is too low.	Determine proper chemical dose and adjust.
	Skimmer is not operating.	Turn skimmer on; adjust skimmer speed.
	Solids-to-air ratio is low.	Increase airflow to pressurization system.
	Solids are building up in the thickener.	Remove sludge from tank.

- Bubble size
- Solids loading
- Sludge characteristics
- Chemical selection
- Chemical dose

19.12.5.3 Solids Concentrators

Solids concentrators (belt thickeners) usually consist of a mixing tank, chemical storage and metering equipment, and a moving porous belt. The process residual flow is chemically treated and then spread evenly over the surface of the moving porous belt. As the flow is carried down the belt (similar to a conveyor belt), the solids are mechanically turned or agitated and water drains through the belt. This process is primarily used in facilities where space is limited.

19.12.5.4 Operational Observations, Problems, and Troubleshooting

As with other unit treatment processes, proper operation of sludge thickeners depends on operator observation. The operator must make routine adjustment of sludge addition and withdrawal rates to achieve desired blanket thickness. Sampling and analysis of influent sludge, supernatant, and thickened sludge are also required. Sludge addition and withdrawal should be continuous if possible to achieve optimum performance. Mechanical maintenance is also required. Expected performance ranges for gravity and dissolved air flotation thickeners are:

- Primary sludge, 8–19% solids
- Waste activated sludge, 2–4% solids
- Trickling filter sludge, 7–9% solids
- Combined sludges, 4–9% solids

Typical operational problems with sludge thickeners include odors, rising sludge, thickened sludge below desired solids concentration, dissolved air concentration being too low, effluent flow containing excessive solids, and torque alarm conditions. Table 19.20 lists several indicators of poor process performance, their causal factors, and corrective actions.

19.12.5.5 Process Calculations (Gravity and Dissolved Air Flotation)

Sludge thickening calculations are based on the concept that the solids in the primary of secondary sludge are equal to the solids in the thickened sludge. Assuming a negligible amount of solids are lost in the thickener overflow, the solids are the same. Note that the water is removed to thicken the sludge and results in higher percent solids.

19.12.5.4.1 Estimating Daily Sludge Production

Equation 19.75 provides a method for establishing an initial pumping rate or evaluating the adequacy of the current pump rate:

$$\text{Est. pump rate} = \frac{\left(\begin{array}{c}\text{Influent TSS conc.} \\ -\text{effluent TSS conc.}\end{array}\right) \times \text{flow} \times 8.34}{\% \text{ Solids in sludge} \times 8.34 \times 1440 \text{ min/day}} \quad (19.75)$$

■ **EXAMPLE 19.67**

Problem: The sludge withdrawn from the primary settling tank contains 1.5% solids. The unit influent contains 280 mg/L TSS, and the effluent contains 141 mg/L. If the influent flow rate is 5.55 MGD, what is the estimated sludge withdrawal rate in gallons per minute (assuming the pump operates continuously)?

Solution:

$$\text{Sludge rate} = \frac{(280 \text{ mg/L} - 141 \text{ mg/L}) \times 5.55 \text{ MGD} \times 8.34}{0.015 \times 8.34 \times 1440 \text{ min/day}} = 36 \text{ gpm}$$

19.12.5.4.2 Surface Loading Rate (gpd/ft²)

The surface loading rate (surface settling rate) is hydraulic loading—the amount of sludge applied per square foot of gravity thickener:

$$\text{Surface loading (gpd/ft}^2) = \frac{\text{Sludge applied to thickener (gpd)}}{\text{Thickener area (ft}^2)} \quad (19.76)$$

■ **EXAMPLE 19.68**

Problem: The 70-ft-diameter gravity thickener receives 32,000 gpd of sludge. What is the surface loading in gallons per square foot per day?

Solution:

$$\text{Surface loading} = \frac{32,000 \text{ gpd}}{0.785 \times 70 \text{ ft} \times 70 \text{ ft}}$$

$$= 8.32 \text{ gpd/ft}^2$$

19.12.5.4.3 Solids Loading Rate (lb/day/ft²)

The solids loading rate is the pounds of solids per day being applied to 1 ft² of tank surface area. The calculation uses the surface area of the bottom of the tank. It assumes the floor of the tank is flat and has the same dimensions as the surface.

$$\text{Solids loading rate (lb/day/ft}^2) = \frac{\% \text{ Solids} \times \text{sludge flow (gpd)} \times 8.34 \text{ lb/gal}}{\text{Thickener area (ft}^2)} \quad (19.77)$$

■ **EXAMPLE 19.69**

Problem: The thickener influent contains 1.6% solids. The influent flow rate is 39,000 gpd. The thickener is 50 ft in diameter and 10 ft deep. What is the solids loading in pounds per day?

Solution:

$$\text{Solids loading rate} = \frac{0.016 \times 39,000 \text{ gpd} \times 8.34 \text{ lb/gal}}{0.785 \times 50 \text{ ft} \times 50 \text{ ft}}$$

$$= 2.7 \text{ lb/day/ft}^2$$

19.12.5.4.4 Concentration Factor

The concentration factor (CF) represents the increase in concentration due to the thickener:

$$\text{CF} = \frac{\text{Thickened sludge concentration (\%)}}{\text{Influent sludge concentration (\%)}} \quad (19.78)$$

■ **EXAMPLE 19.70**

Problem: The influent sludge contains 3.5% solids. The thickened sludge solids concentration is 7.7%. What is the concentration factor?

Solution:

$$\text{CF} = \frac{7.7\%}{3.5\%} = 2.2$$

19.12.5.4.5 Air-to-Solids Ratio

The air-to-solids ratio is the ratio between the pounds of air being applied to the pounds of solids entering the thickener:

$$\text{Air/solids ratio} = \frac{\text{Air flow (ft}^3/\text{min}) \times 0.075 \text{ lb/ft}^3}{\text{Sludge flow (gpm)} \times \% \text{ solids} \times 8.34 \text{ lb/gal}} \quad (19.79)$$

■ **EXAMPLE 19.71**

Problem: The sludge pumped to the thickener is 0.85% solids. The air flow is 13 cfm. What is the air-to-solids ratio if the current sludge flow rate entering the unit is 50 gpm?

Solution:

$$\text{Air/solids ratio} = \frac{13 \text{ cfm} \times 0.075 \text{ lb/ft}}{50 \text{ gpm} \times 0.0085 \times 8.34 \text{ lb/gal}}$$

$$= 0.28$$

19.12.5.4.6 Recycle Flow in Percent

The amount of recycle flow is expressed as a percent:

$$\% \text{ Recycle} = \frac{\text{Recycle flow rate (gpm)} \times 100}{\text{Sludge flow (gpm)}} \quad (19.80)$$

■ EXAMPLE 19.72

Problem: The sludge flow to the thickener is 80 gpm. The recycle flow rate is 140 gpm. What is the percent recycle?

Solution:

$$\% \text{ Recycle} = \frac{140 \text{ gpm} \times 100}{80 \text{ gpm}} = 175\%$$

19.12.6 SLUDGE STABILIZATION

The purpose of sludge stabilization is to reduce volume, stabilize the organic matter, and eliminate pathogenic organisms to permit reuse or disposal. The equipment required for stabilization depends on the specific process used. Sludge stabilization processes include:

- Aerobic digestion
- Anaerobic digestion
- Composting
- Lime stabilization
- Wet air oxidation (heat treatment)
- Chemical oxidation (chlorine oxidation)
- Incineration

19.12.6.1 Aerobic Digestion

Equipment used for aerobic digestion includes an aeration tank (digester), which is similar in design to the aeration tank used for the activated sludge process. Either diffused or mechanical aeration equipment is necessary to maintain the aerobic conditions in the tank. Solids and supernatant removal equipment is also required. In operation, process residuals (sludge) are added to the digester and aerated to maintain a dissolved oxygen (DO) concentration of 1 mg/L. Aeration also ensures that the tank contents are well mixed. Generally, aeration continues for approximately 20 days of retention time. Periodically, aeration is stopped and the solids are allowed to settle. Sludge and the clear liquid supernatant are withdrawn as needed to provide more room in the digester. When no additional volume is available, mixing is stopped for 12 to 24 hours before

TABLE 19.21
Aerobic Digester Normal Operating Levels

Parameter	Normal Levels
Detention time (days)	10–20
Volatile solids loading (lb/ft^3/day)	0.1–0.3
Dissolved oxygen (mg/L)	1.0
pH	5.9–7.7
Volatile solids reduction (%)	40–50

solids are withdrawn for disposal. Process control testing should include alkalinity, pH, percent solids, percent volatile solids for influent sludge, supernatant, digested sludge, and digester contents. Normal operating levels for an aerobic digester are listed in Table 19.21.

A typical operational problem associated with an aerobic digester is pH control. When pH drops, for example, this may indicate normal biological activity or low influent alkalinity. This problem is corrected by adding alkalinity (e.g., lime, bicarbonate).

19.12.6.2 Process Control Calculations for the Aerobic Digester

Wastewater operators who operate aerobic digesters are required to make certain process control calculations for the digesters. Moreover, licensing examinations typically include aerobic digester problems for determining volatile solids loading, digestion time, digester efficiency, and pH adjustment. These process control calculations are explained in the following sections.

19.12.6.2.1 Volatile Solids Loading

Volatile solids loading for the aerobic digester is expressed in pounds of volatile solids entering the digester per day per cubic foot of digester capacity:

$$\frac{\text{Volatile solids}}{\text{loading}} = \frac{\text{Volatile solids added (lb/day)}}{\text{Digester volume (ft}^3)} \quad (19.81)$$

■ EXAMPLE 19.73

Problem: The aerobic digester is 25 ft in diameter and has an operating depth of 24 ft. The sludge added to the digester daily contains 1350 lb of volatile solids. What is the volatile solids loading in pounds per day per cubic foot?

Solution:

$$\text{Volatile solids loading} = \frac{1350 \text{ lb/day}}{.785 \times 25 \text{ ft} \times 25 \text{ ft} \times 24 \text{ ft}}$$

$$= 0.11 \text{ lb/day/ft}^3$$

19.12.6.2.2   Digestion Time (Days)

Digestion time is the theoretical time the sludge remains in the aerobic digester:

$$\text{Digestion time (days)} = \frac{\text{Digester volume (gal)}}{\text{Sludge added (gpd)}} \quad (19.82)$$

■ EXAMPLE 19.74

Problem: Digester volume is 240,000 gal. Sludge is being added to the digester at the rate of 13,500 gpd. What is the digestion time in days?

Solution:

$$\text{Digestion time} = \frac{240,000 \text{ gal}}{13,500 \text{ gpd}} = 17.8 \text{ days}$$

19.12.6.2.3   Digester Efficiency (% Reduction)

To determine digester efficiency or the percent reduction, a two-step procedure is required. First the percent volatile matter reduction must be calculated and then the percent moisture reduction.

- Step 1: *Calculate volatile matter.* Because of the changes occurring during sludge digestion, the calculation used to determine percent volatile matter reduction is more complicated:

$$\%\text{VM reduction} = \frac{(\%\text{VM}_{in} - \%\text{VM}_{out}) \times 100}{\left[\%\text{VM}_{in} - (\%\text{VM}_{in} \times \%\text{VM}_{out})\right]} \quad (19.83)$$

■ EXAMPLE 19.75

Problem: Using the digester data provided below, determine the percent volatile matter reduction for the digester:

Raw sludge volatile matter = 71%
Digested sludge volatile matter = 53%

Solution:

$$\%\text{VM reduction} = \frac{(0.71 - 0.53) \times 100}{\left[0.71 - (0.71 \times 0.53)\right]} = 53.9 \text{ or } 54\%$$

- Step 2: *Calculate moisture reduction:*

$$\%\text{ Moisture reduction} = \frac{(\%\text{ Moisture}_{in} - \%\text{ moisture}_{out}) \times 100}{\left[\%\text{ Moisture}_{in} - \left(\%\text{ moisture}_{in} \times \%\text{ moisture}_{out}\right)\right]} \quad (19.84)$$

■ EXAMPLE 19.76

Problem: Using the digester data provided below, determine the percent moisture reduction for the digester.

Note: Percent moisture = 100% – percent solids.

Solution:

Raw sludge:
   Percent solids = 6%
   Percent moisture = 100% – 6% = 94%
Digested sludge:
   Percent solids = 15%
   Percent moisture = 100% – 15% = 85%

$$\%\text{ Moisture reduction} = \frac{(0.94 - 0.85) \times 100}{\left[0.94 - (0.94 \times 0.85)\right]} = 64\%$$

19.12.6.2.4   pH Adjustment

Occasionally, the pH of the aerobic digester will fall below the levels required for good biological activity. When this occurs, the operator must perform a laboratory test to determine the amount of alkalinity required to raise the pH to the desired level. The results of the lab test must then be converted to the actual quantity of chemical (usually lime) required by the digester.

$$\text{Chemical required (lb)} = \left[\left(\frac{\text{Chemical used in lab test (mg)}}{\text{Sample volume (L)}}\right) \times \text{digester volume (MG)} \times 8.34\right] \quad (19.85)$$

■ EXAMPLE 19.77

Problem: The lab reports that 225 mg of lime were required to increase the pH of a 1-L sample of the aerobic digester contents to pH 7.2. The digester volume is 240,000 gal. How many pounds of lime will be required to increase the digester pH to 7.2?

Solution:

$$\text{Chemical required} = \frac{225 \text{ mg} \times 240,000 \text{ gal} \times 3.785 \text{ L/gal}}{1 \text{ L} \times 454 \text{ g/lb} \times 1000 \text{ mg/g}} = 450 \text{ lb}$$

19.12.6.3   Anaerobic Digestion

Anaerobic digestion is the traditional method of sludge stabilization that involves using bacteria that thrive in the absence of oxygen. It is slower than aerobic digestion but has the advantage that only a small percentage of the wastes are converted into new bacterial cells. Instead, most of the organics are converted into carbon dioxide and methane gas.

Note: In an anaerobic digester, the entrance of air should be prevented because of the potential for an explosive mixture resulting from air mixing with gas produced in the digester.

Equipment used in anaerobic digestion includes a sealed digestion tank with either a fixed or a floating cover, heating and mixing equipment, gas storage tanks, solids and supernatant withdrawal equipment, and safety equipment (e.g., vacuum relief, pressure relief, flame traps, explosion proof electrical equipment).

In operation, process residual (thickened or unthickened sludge) is pumped into the sealed digester. The organic matter digests anaerobically by a two-stage process. Sugars, starches, and carbohydrates are converted to volatile acids, carbon dioxide, and hydrogen sulfide. The volatile acids are then converted to methane gas. This operation can occur in a single tank (single stage) or in two tanks (two stages). In a single-stage system, supernatant and digested solids must be removed whenever flow is added. In a two-stage operation, solids and liquids from the first stage flow into the second stage each time fresh solids are added. Supernatant is withdrawn from the second stage to provide additional treatment space. Periodically, solids are withdrawn for dewatering or disposal. The methane gas produced in the process may be used for many plant activities.

Note: The primary purpose of a secondary digester is to allow for solids separation.

Various performance factors affect the operation of the anaerobic digester; for example, the percent volatile matter in raw sludge, digester temperature, mixing, volatile acids/alkalinity ratio, feed rate, percent solids in raw sludge, and pH are all important operational parameters that the operator must monitor.

Along with being able to recognize normal and abnormal anaerobic digester performance parameters, wastewater operators must also know and understand normal operating procedures. Normal operating procedures include sludge additions, supernatant withdrawal, sludge withdrawal, pH control, temperature control, mixing, and safety requirements. Important performance parameters are listed in Table 19.22.

19.12.6.3.1 Sludge Additions

Sludge must be pumped (in small amounts) several times each day to achieve the desired organic loading and optimum performance.

TABLE 19.22
Anaerobic Digester Sludge Parameters

Raw Sludge Solids	Impact
<4% solids	Loss of alkalinity
	Decreased sludge retention time
	Increased heating requirements
	Decreased volatile acid-to-alkalinity ratio
4–8% solids	Normal operation
>8% solids	Poor mixing
	Organic overloading
	Decreased volatile acid-to-alkalinity ratio

Note: Keep in mind that in fixed cover operations additions must be balanced by withdrawals; if not, structural damage occurs.

19.12.6.3.2 Supernatant Withdrawal

Supernatant withdrawal must be controlled for maximum sludge retention time. When sampling, sample all drawoff points and select the level with the best quality.

19.12.6.3.3 Sludge Withdrawal

Digested sludge is withdrawn only when necessary—always leave at least 25% seed.

19.12.6.3.4 pH Control

The pH should be adjusted to maintain a range of 6.8 to 7.2 by adjusting the feed rate, sludge withdrawal, or alkalinity additions.

Note: The buffer capacity of an anaerobic digester is indicated by the volatile acid/alkalinity relationship. Decreases in alkalinity cause a corresponding increase in the ratio.

19.12.6.3.5 Temperature Control

If the digester is heated, the temperature must be controlled to a normal temperature range of 90 to 95°F. Never adjust the temperature by more than 1°F per day.

19.12.6.3.6 Mixing

If the digester is equipped with mixers, mixing should be accomplished to ensure that organisms are exposed to food materials.

19.12.6.3.7 Safety

Anaerobic digesters are inherently dangerous; several catastrophic failures have been recorded. To prevent such failures, safety equipment such as pressure relief and vacuum relief valves, flame traps, condensate traps, and gas collection safety devices is necessary. It is important that these critical safety devices be checked and maintained for proper operation.

Note: Because of the inherent danger involved with working inside anaerobic digesters, they are automatically classified as *permit-required confined spaces*; therefore, all operations involving internal entry must be made in accordance with OSHA's confined space entry standard.

19.12.6.4 Process Control Monitoring, Testing, and Troubleshooting

During operation, anaerobic digesters must be monitored and tested to maintain proper operation. Testing should be performed to determine supernatant pH, volatile acids, alkalinity, BOD or COD, total solids, and temperature. Sludge (in and out) should be routinely tested for percent solids and percent volatile matter. Normal operating parameters are listed in Table 19.23. As with all other unit

processes, the wastewater operator is expected to recognize problematic symptoms with anaerobic digesters and take the appropriate corrective actions. Table 19.24 lists several indicators of poor process performance, their causal factors, and corrective actions.

19.12.6.5 Anaerobic Digester Process Control Calculations

Process control calculations involved with anaerobic digester operation include determining the required seed volume, volatile acid-to-alkalinity ratio, sludge retention time, estimated gas production, volatile matter reduction, and percent moisture reduction in digester sludge. Examples on how to make these calculations are provided in the following sections.

19.12.6.5.1 Required Seed Volume in Gallons

$$\text{Seed volume (gal)} = \text{Digester volume} \times \% \text{ seed} \quad (19.86)$$

■ EXAMPLE 19.78

Problem: The new digester requires a 25% seed to achieve normal operation within the allotted time. If the digester volume is 266,000 gal, how many gallons of seed material will be required?

Solution:

$$\text{Seed volume} = 266,000 \times 0.25 = 66,500 \text{ gal}$$

19.12.6.5.2 Volatile Acids-to-Alkalinity Ratio

The volatile acids-to-alkalinity ratio can be used to control operation of an anaerobic digester:

$$\text{Ratio} = \frac{\text{Volatile acids concentration}}{\text{Alkalinity concentration}} \quad (19.87)$$

TABLE 19.23
Normal Operating Ranges for Anaerobic Digesters

Parameter	Normal Range
Sludge retention time	
Heated	30–60 days
Unheated	180+ days
Volatile solids loading	0.04–0.1 lb VM/day/ft³
Operating temperature	
Heated	90–95°F
Unheated	Varies with season
Mixing	
Heated, primary	Yes
Unheated, secondary	No
% Methane in gas	60–72%
% Carbon dioxide in gas	28–40%
pH	6.8–7.2
Volatile acids-to-alkalinity ratio	≤0.1
Volatile solids reduction	40–60%
Moisture reduction	40–60%

TABLE 19.24
Symptoms, Causes, and Corrective Actions for Anaerobic Digester Problems

Symptom	Symptom	Symptom
Digester gas production is reduced; pH drops below 6.8; volatile acids-to-alkalinity ratio increases.	Digester souring Organic overloading Inadequate mixing Low alkalinity Hydraulic overloading Toxicity Loss of digestion capacity	Add alkalinity (e.g., digested sludge, lime); improve temperature control; improve mixing; eliminate toxicity; clean digester.
Gray foam is oozing from the digester.	Rapid gasification Foam-producing organisms in digester	Reduce mixing; reduce feed rate; mix slowly by hand; clean all contaminated equipment.

TABLE 19.25
Volatile Acid-to-Alkalinity Ratios

Operating Condition	Volatile Acid-to-Alkalinity Ratio
Optimum	≤0.1
Acceptable range	0.1–0.3
Increase in % carbon dioxide in gas	≥0.5
Decrease in pH	≥0.8

■ EXAMPLE 19.79

Problem: The digester contains 240 mg/L volatile acids and 1860 mg/L alkalinity. What is the volatile acids-to-alkalinity ratio?

Solution:

$$\text{Ratio} = \frac{240 \text{ mg/L}}{1860 \text{ mg/L}} = 0.13$$

Note: Increases in the ratio normally indicate a potential change in the operation condition of the digester, as shown in Table 19.25.

19.12.6.5.3 Sludge Retention Time

Sludge retention time (SRT) is the length of time the sludge remains in the digester:

$$\text{SRT (days)} = \frac{\text{Digester vol. (gal)}}{\text{Sludge vol. added per day (gpd)}} \quad (19.88)$$

■ EXAMPLE 19.80

Problem: Sludge is added to a 525,000-gal digester at the rate of 12,250 gal per day. What is the SRT?

Solution:

$$\text{SRT} = \frac{525,000 \text{ gal}}{12,250 \text{ gpd}} = 42.9 \text{ days}$$

19.12.6.5.4 Estimated Gas Production in Cubic Feet/Day

The rate of gas production is normally expressed as the volume of gas (ft³) produced per pound of volatile matter destroyed. The total cubic feet of gas a digester will produce per day can be calculated by:

$$\text{Gas production (ft}^3) = \begin{bmatrix} \text{VM}_{in} \text{ (lb/day)} \\ \times \text{\%VM reduction} \\ \times \text{production rate (ft}^3/\text{lb)} \end{bmatrix} \quad (19.89)$$

■ EXAMPLE 19.81

Problem: The digester receives 11,450 lb of volatile matter per day. Currently, the volatile matter reduction achieved by the digester is 52%. The rate of gas production is 11.2 ft³ of gas per pound of volatile matter destroyed.

Solution:

$$\text{Gas production} = 11,450 \text{ lb/day} \times 0.52 \times 11.2 \text{ ft}^3/\text{lb}$$
$$= 66,685 \text{ ft}^3/\text{day}$$

19.12.6.5.5 Percent Volatile Matter Reduction

Because of the changes occurring during sludge digestion, the calculation used to determine percent volatile matter reduction is more complicated:

$$\text{\%VM reduction} = \frac{(\text{\%VM}_{in} - \text{\%VM}_{out}) \times 100}{\left[\text{\%VM}_{in} - (\text{\%VM}_{in} \times \text{\%VM}_{out})\right]} \quad (19.90)$$

■ EXAMPLE 19.82

Problem: Using the data provided below, determine the percent volatile matter reduction for the digester:

Raw sludge volatile matter = 74%
Digested sludge volatile matter = 55%

Solution:

$$\text{\%VM reduction} = \frac{(0.74 - 0.55) \times 100}{[0.74 - (0.74 \times 0.55)]} = 57\%$$

19.12.6.5.6 Percent Moisture Reduction in Digested Sludge

$$\text{\% Moisture reduction} = \frac{(\text{\% Moisture}_{in} - \text{\% moisture}_{out}) \times 100}{\left[\text{\% Moisture}_{in} - \left(\begin{array}{c}\text{\% moisture}_{in} \\ \times \text{\% moisture}_{out}\end{array}\right)\right]} \quad (19.91)$$

■ EXAMPLE 19.83

Problem: Using the digester data provide below, determine the percent moisture reduction and percent volatile matter reduction for the digester.

Solution:

Raw sludge percent solids = 6%
Digested sludge percent solids = 14%

Note: Percent moisture = 100% – percent solids.

$$\text{\% Moisture reduction} = \frac{(0.94 - 0.86) \times 100}{[0.94 - (0.94 \times 0.86)]} = 61\%$$

19.12.6.6 Other Sludge Stabilization Processes

In addition to aerobic and anaerobic digestion, other sludge stabilization processes include composting, lime stabilization, wet air oxidation, and chemical (chlorine) oxidation. These other stabilization processes are briefly described in this section.

19.12.6.6.1 Composting

The purpose of composting sludge is to stabilize the organic matter, reduce volume, and eliminate pathogenic organisms. In a composting operation, dewatered solids are usually mixed with a bulking agent (e.g., hardwood chips) and stored until biological stabilization occurs. The composting mixture is ventilated during storage to provide sufficient oxygen for oxidation and to prevent odors. After the solids are stabilized, they are separated from the bulking agent. The composted solids are then stored for curing and applied to farmlands or other beneficial uses. Expected performance of the composting operation for both percent volatile matter reduction and percent moisture reduction ranges from 40 to 60%.

19.12.6.6.2 Lime Stabilization

In the lime stabilization process, residuals are mixed with lime to achieve a pH of 12. This pH is maintained for at least 2 hr. The treated solids can then be dewatered for disposal or directly land applied.

19.12.6.6.3 Thermal Treatment

Thermal treatment (or wet air oxidation) subjects sludge to high temperature and pressure in a closed reactor vessel. The high temperature and pressure rupture the cell walls of any microorganisms present in the solids and causes chemical oxidation of the organic matter. This process substantially improves dewatering and reduces the volume of material for disposal. It also produces a very high-strength waste, which must be returned to the wastewater treatment system for further treatment.

19.12.6.6.4 Chlorine Oxidation

Chlorine oxidation also occurs in a closed vessel. In this process chlorine (100 to 1000 mg/L) is mixed with a recycled solids flow. The recycled flow and process residual flow are mixed in the reactor. The solids and water are separated after leaving the reactor vessel. The water is returned to the wastewater treatment system, and the treated solids are dewatered for disposal. The main advantage of chlorine oxidation is that it can be operated intermittently. The main disadvantage is production of extremely low pH and high chlorine content in the supernatant.

19.12.6.7 Stabilization Operation

Depending on the stabilization process employed, the operational components vary. In general, operations include pumping, observations, sampling and testing, process control calculations, maintenance, and housekeeping. Performance of the stabilization process will also vary with the type of process used. Generally, stabilization processes can produce 40 to 60% reduction of both volatile matter (organic content) and moisture.

19.12.6.7.1 Sludge Dewatering

Digested sludge removed from the digester is still mostly liquid. Sludge dewatering is used to reduce volume by removing the water to permit easy handling and economical reuse or disposal. Dewatering processes include sand drying beds, vacuum filters, centrifuges, filter presses (belt and plate), and incineration.

19.12.6.7.2 Sand Drying Beds

Drying beds have been used successfully for years to dewater sludge. Composed of a sand bed (consisting of a gravel base, underdrains, and 8 to 12 in. of filter-grade sand), drying beds also include an inlet pipe, splash pad containment walls, and a system to return filtrate (water) for treatment. In some cases, the sand beds are covered to provide drying solids protection from the elements. In operation, solids are pumped to the sand bed and allowed to dry by first draining off excess water through the sand and then by evaporation. This is the simplest and least expensive method for dewatering sludge. Moreover, no special training or expertise is required. The downside is that drying beds require a great deal of manpower to clean them, they can create odor and insect problems, and they can cause sludge buildup during inclement weather.

19.12.6.8 Stabilization Performance Factors

In sludge drying beds, various factors affect the length of time required to achieve the desired solids concentrations. The major factors and their impact on drying bed performance include the following:

- *Climate*—Drying beds in cold or moist climates will require significantly longer drying time to achieve an adequate level of percent solids concentrations in the dewatered sludge.
- *Depth of applied sludge*—The depth of the sludge drawn onto the bed has a major impact on the required drying time. Deeper sludge layers require longer drying times. Under ideal conditions, a well-digested sludge drawn to a depth of approximately 8 in. will require approximately 3 weeks to reach the desired 40 to 60% solids.
- *Type of sludge applied*—The quality and solids concentration of the drying media will affect the time requirements.
- *Bed cover*—Covered drying beds prevent rewetting of the sludge during storm events. In most cases, this reduces the average drying time required to reach the desired solids levels.

TABLE 19.26
Symptoms, Causes, and Corrective Actions for Drying Bed Problems

Symptom	Cause	Corrective Action
Sludge takes a long time to dewater.	Applied sludge is too deep.	Allow the bed to dry to the minimum acceptable percent solids and remove. Use the following procedure to determine the appropriate sludge depth: (1) Clean the bed and apply a smaller depth of sludge (e.g., 6 to 8 in.); (2) measure the decrease in depth (drawdown) at the end of 3 days of drying; and (3) use a sludge depth equal to twice the 3-day drawdown depth for future applications. After the sludge has dried, remove the sludge and add 0.5 to 1.0 in. of clean sand.
	Sludge was applied to a dirty bed.	Use external water source (with backflow prevention) to slowly flush underdrains.
	The drain system is plugged or broken.	Repair/replace underdrains as required; prevent damage to underdrains by draining during freezing weather.
	Design capacity is insufficient.	Use polymer to increase bed performance.
	Weather has been inclement or drying conditions have been poor.	Cover or enclose the beds.
Influent sludge is very thin.	Coning is occurring in the digester.	Reduce rate of sludge withdrawal.
Sludge feed lines plug frequently.	Solids and grit are accumulating in the feed lines.	Open lines fully at the start of each withdrawal cycle; flush lines at the end of each withdrawal cycle.
Flies are breeding in the drying sludge.	Sludge is being digested inadequately, and natural insect reproduction is occurring.	Break sludge crust, and apply a larvicide (e.g., borax); use insecticide (if approved) to remove adult insects; remove sludge as soon as possible.
Objectionable odors are apparent when sludge is applied to the bed.	Raw or partially digested sludge is being applied to the bed.	Add lime to the sludge to control odors and potential insect and rodent problems; remove the sludge as quickly as possible; identify and correct the digester problem.

19.12.6.9 Operational Observations, Problems, and Troubleshooting

Although drying beds involve two natural processes—drainage and evaporation—that normally work well enough on their own, a certain amount of preparation and operator attention are still required to maintain optimum drying performance. In the preparation stage, for example, all debris must be removed from the raked and leveled media surface, then all of the openings to the bed are sealed. After the bed is properly prepared, the sludge lines are opened, and sludge is allowed to flow slowly onto the media. The bed is filled to the desired operating level (8 to 12 in.). The sludge line is closed and flushed, and the bed drain is opened. Water begins to drain. The sludge remains on the media until the desired percent solids (40 to 60%) is achieved. Later, the sludge is removed. In most operations, manual removal is required to prevent damage to the underdrain system. The sludge is disposed of in an approved landfill or by land application as a soil conditioner. In the operation of a sludge drying bed, the operator observes the operations and looks for various indicators of operational problems and makes process adjustments as required. Table 19.26 lists several indicators of poor process performance, their causal factors, and corrective actions.

19.12.7 Rotary Vacuum Filtration

Rotary vacuum filters have also been used for many years to dewater sludge. The vacuum filter includes filter media (belt, cloth, or metal coils), media support (drum), vacuum system, chemical feed equipment, and conveyor belts to transport the dewatered solids. In operation, chemically treated solids are pumped to a vat or tank in which a rotating drum is submerged. As the drum rotates, a vacuum is applied to the drum. Solids collect on the media and are held there by the vacuum as the drum rotates out of the tank. The vacuum removes additional water from the captured solids. When solids reach the discharge zone, the vacuum is released and the dewatered solids are discharged onto a conveyor belt for disposal. The media are then washed prior to returning to the start of the cycle.

19.12.7.1 Types of Rotary Vacuum Filters

The three principal types of rotary vacuum filters are rotary drum, coil, and belt. The *rotary drum* filter consists of a cylindrical drum rotating partially submerged in a vat or pan of conditioned sludge. The drum is divided lengthwise into a number of sections that are connected through internal piping to ports in the valve body (plant) at the hub. This plate rotates in contact with a fixed valve plate with similar parts, which are connected to a vacuum supply, a compressed air supply, and an atmosphere vent. As the drum rotates, each section is thus connected to the appropriate service.

The *coil type* vacuum filter uses two layers of stainless steel coils arranged in corduroy fashion around the drum. After a dewatering cycle, the two layers of springs leave the drum bed and are separated from each other so the cake is lifted off the lower layer and is discharged from the upper layer. The coils are then washed and reapplied to the drum. The coil filter is used successfully for all types of sludges; however, sludges with extremely fine particles or ones that are resistant to flocculation dewater poorly with this system.

The media on a *belt filter* leave the drum surface at the end of the drying zone and pass over a small-diameter discharge roll to aid in cake discharge. Washing of the media occurs next. The media are then returned to the drum and to the vat for another cycle. This type of filter normally has a small-diameter curved bar between the point where the belt leaves the drum and the discharge roll. This bar primarily aids in maintaining belt dimensional stability.

19.12.7.1.1 Filter Media

Drum and belt vacuum filters use natural or synthetic fiber materials. On the drum filter, the cloth is stretched and secured to the surface of the drum. In the belt filter, the cloth is stretched over the drum and through the pulley system. The installation of a blanket requires several days. The cloth (with proper care) will last several hundred to several thousand hours. The life of the blanket depends on the cloth selected, the conditioning chemical, backwash frequency, and cleaning (e.g., acid bath) frequency.

19.12.7.1.2 Filter Drum

The filter drum is a maze of pipe work running from a metal screen and wooden skeleton and connecting to a rotating valve port at each end of the drum. The drum is equipped with a variable speed drive to turn the drum from 1/8 to 1 rpm. Normally, solids pickup is indirectly related to the drum speed. The drum is partially submerged in a vat containing the conditioned sludge. Submergence is usually limited to 1/5 or less of filter surface at a time.

19.12.7.1.3 Chemical Conditioning

Sludge dewatered using vacuum filtration is normally chemically conditioned just prior to filtration. Sludge conditioning increases the percentage of solids captured by the filter and improves the dewatering characteristics of the sludge; however, conditional sludge must be filtered as quickly as possible after chemical addition to obtain these desirable results.

19.12.7.2 Operational Observations, Problems, and Troubleshooting

In operation, the rotating drum picks up chemically treated sludge. A vacuum is applied to the inside of the drum to draw the sludge onto the outside of the drum cover. This porous outside cover or filter medium allows the filtrate or liquid to pass through into the drum and the filter cake (dewatered sludge) to stay on the medium. In the cake release/discharge mode, slight air pressure is applied to the drum interior. Dewatered solids are lifted from the medium and scraped off by a scraper blade. Solids drop onto a conveyor for transport for further treatment or disposal. The filtrate water is returned to the plant for treatment. The operator observes drum speed, sludge pickup, filter cake thickness and appearance, chemical feed rates, sludge depth in vat, and overall equipment operation. Sampling and testing are routinely performed on influent sludge solids concentration, filtrate BOD and solids, and sludge cake solids concentration. Table 19.27 lists several indicators of poor process performance, their causal factors, and corrective actions.

19.12.7.3 Process Control Calculations

19.12.7.3.1 Vacuum Filter Yield (lb/hr/ft²)

Probably the most frequent calculation that vacuum filter operators have to make is determining filter yield. Example 17.84 illustrates how this calculation is made.

■ EXAMPLE 17.84

Problem: Thickened, thermally conditioned sludge is pumped to a vacuum filter at a rate of 50 gpm. The vacuum area of the filter is 12 ft wide with a drum diameter of 9.8 ft. If the sludge concentration is 12%, what is the filter yield in lb/hr/ft²? Assume the sludge weighs 8.34 lb/gal.

Solution: First calculate the filter surface area:

$$\text{Area of a cylinder side} = 3.14 \times \text{diameter} \times \text{length}$$
$$= 3.14 \times 9.8 \text{ ft} \times 12 \text{ ft}$$
$$= 369.3 \text{ ft}^2$$

Next, calculate the pounds of solids per hour:

$$\frac{50 \text{ gpm}}{1 \text{ min}} \times \frac{60 \text{ min}}{1 \text{ hr}} \times \frac{8.34 \text{ lb}}{1 \text{ gal}} \times \frac{12\%}{100\%} = 3002.4 \text{ lb/hr}$$

Divide the two:

$$\frac{3002.4 \text{ lb/hr}}{369.3 \text{ ft}^2} = 8.13 \text{ lb/hr/ft}^2$$

TABLE 19.27

Symptoms, Causes, and Corrective Actions for Rotary Vacuum Filter Problems

Symptom	Cause	Corrective Action
High solids in filtrate	Coagulant dosage is incorrect.	Adjust coagulant dosage; recalibrate coagulant feeder.
	Filter media are binding.	Clean synthetic cloth with steam and detergent; clean steel coil with acid bath; clean cloth with water or replace cloth.
Thin filter cake and poor dewatering	Chemical dosage is incorrect.	Adjust coagulant dosage; recalibrate coagulant feeder.
	Filter media are binding.	Clean synthetic cloth with steam and detergent; clean steel coil with acid bath; clean cloth with water or replace cloth.
	Vacuum is inadequate.	Repair vacuum system.
	Drum speed is too high.	Reduce drum speed.
	Drum is submerged too low.	Increase drum submergence.
Vacuum pump not running	Power to drive motor is off.	Reset heater, breaker, etc.; restart.
	Seal water is not flowing.	Initiate seal water flow.
	Drive belt is broken.	Replace drive belt.
Drum not rotating	Power to drive motor is off.	Reset heater, breaker, etc.; restart.
Receiver vibrating	Filtrate pump is clogged.	Clear pump.
	Bolts and gasket around inspection plate are loose.	Tighten bolts and gasket.
	Ball check valve in filtrate pump is worn.	Replace ball check valve.
	Suction line has air leaks.	Seal leaks.
	Drum face is dirty.	Clean face with pressure hose.
	Seal strips are missing.	Replace missing seal strips.
High vat level	Chemical conditioning is incorrect.	Change coagulant dosage.
	Feed rate is too high.	Reduce feed rate.
	Drum speed is too slow.	Increase drum speed.
	Filtrate pump is off or clogged.	Turn on or clean pump.
	Drain line is plugged.	Clean drain line.
	Vacuum pump has stopped.	Turn on or clean pump.
	Seal strips are missing.	Replace seal strips.
Low vat level	Feed rate is too low.	Increase feed rate.
	Vat drain valve is open.	Close vat drain valve.
Vacuum pump drawing high amperage	Filtrate pump is clogged.	Clear pump clog.
	Improper chemical conditioning.	Adjust coagulant dosage.
	High vat level.	See "High vat level" entry (above).
	Cooling water flow to vacuum pump is too high.	Decrease cooling water flow rate.
Scale buildup on vacuum pump seals	Hard, unstable water	Add sequestering agent.

19.12.8 Pressure Filtration

Pressure filtration differs from vacuum filtration in that the liquid is forced through the filter media by a positive pressure instead of a vacuum. Several types of presses are available, but the most commonly used types are plate-and-frame presses and belt presses. *Filter presses* include the belt or plate-and-frame types. The belt filter includes two or more porous belts, rollers, and related handling systems for chemical makeup and feed, as well as supernatant and solids collection and transport.

The plate-and-frame filter consists of a support frame, filter plates covered with porous material, a hydraulic or mechanical mechanism for pressing plates together, and related handling systems for chemical makeup and feed, as well as supernatant and solids collection and transport. In the plate-and-frame filter, solids are pumped (sandwiched) between plates. Pressure (200 to 250 psi) is applied to the plates and water is squeezed from the solids. At the end of the cycle, the pressure is released; as the plates separate, the solids drop out onto a conveyor belt for transport to storage or disposal. Performance factors for plate-and-frame presses include feed sludge characteristics, type and amount of chemical conditioning, operating pressures, and the type and amount of precoat.

The belt filter uses a coagulant (polymer) mixed with the influent solids. The chemically treated solids are discharged between two moving belts. First, water drains from the solids by gravity. Then, as the two belts move

TABLE 19.28

Symptoms, Causes, and Corrective Actions for Pressure Filtration Problems

Symptom	Cause	Corrective Action
Plate press		
Plates fail to seal.	Poor alignment	Realign parts.
	Inadequate shimming	Adjust shimming of stay bosses.
Cake discharge is difficult.	Inadequate precoat	Increase precoat, feed at 25–40 psig.
	Improper conditioning	Change conditioner type or dosage (use filter leaf test to determine).
Filter cycle times are excessive.	Improper conditioning	Change chemical dosage.
	Low feed solids	Improve thickening operation.
Filter cake sticks to conveyors.	Improper conditioning chemical/dosage	Increase inorganic conditioner dose.
Precoat pressures gradually increase.	Improper sludge conditioning	Change chemical dosage.
	Improper precoat feed	Decrease feed for a few cycles, then optimize.
	Plugged filter media	Wash filter media.
	Calcium buildup in media	Wash media with inhibited hydrochloric acid.
Media are frequently binding.	Inadequate precoat	Increase precoat.
	Initial feed rate too high (no precoat)	Reduce feed rate; develop initial cake slowly.
Excessive moisture is found in cake.	Improper conditioning	Change chemical dosage.
	Filter cycle too short	Lengthen filter cycle.
Sludge is blowing out of the press.	Obstruction between plates	Shut down feed pump; hit press closure drive; restart feed pump; clean after cycle.
Leaks are apparent around the lower faces of the plates.	Wet cake soiling media on lower faces	See "Excessive moisture is found in cake" entry (above).
Belt press		
Filter cake discharge is difficult.	Improper chemical dosage	Change conditioning chemical.
	Changing sludge characteristics	Adjust chemical dosage.
	Wrong conditioning chemical selected	Change chemical or sludge.
	Wrong application point	Adjust application point.
Sludge is leaking from belt edges.	Excessive belt tension	Reduce belt tension.
	Belt speed too low	Increase belt speed.
	Excessive sludge feed rate	Reduce sludge feed rate.
Excessive moisture is found in filter cake.	Improper belt speed or drainage time	Adjust belt speed.
	Wrong conditioning chemical	Change conditioning chemical.
	Improper chemical dosage	Adjust chemical dosage.
	Inadequate belt washing	Clear spray nozzles and adjust sprays.
	Wrong belt weave or material	Replace belt.
Belt wear along edges is excessive.	Roller misalignment	Correct roller alignment.
	Improper belt tension	Correct tension.
	Tension/alignment control system	Repair tracking and alignment system controls.
Belt shifts or seizes.	Uneven sludge distribution	Adjust feed for uniform sludge distribution.
	Inadequate or uneven belt washing	Clean and adjust belt-washing sprays.

between a series of rollers, pressure squeezes additional water out of the solids. The solids are then discharged onto a conveyor belt for transport to storage or disposal. Performance factors for the belt press include sludge feed rate, belt speed, belt tension, belt permeability, chemical dosage, and chemical selection.

Filter presses have lower operation and maintenance costs than vacuum filters or centrifuges. They typically produce a good-quality cake and can be batch operated; however, construction and installation costs are high. Moreover, chemical addition is required and the presses must be operated by skilled personnel.

19.12.8.1　Operational Observations, Problems, and Troubleshooting

Most plate and filter press operations are partially or fully automated. Operation consists of observation, maintenance, and sampling and testing. Operation of belt filter presses consists of preparation of conditioning chemicals, chemical feed rate adjustments, sludge feed rate adjustments, belt alignment, belt speed and belt tension adjustments, sampling and testing, and maintenance. Table 19.28 lists several indicators of poor process performance, their causal factors, and corrective actions.

19.12.8.2 Filter Press Process Control Calculations

As part of the operating routine for filter presses, operators are called upon to make certain process control calculations. The process control calculation most commonly used in operating the belt filter press determines the hydraulic loading rate on the unit. The process control calculation most commonly used in the operation of plate and filter presses determines the pounds of solids pressed per hour. Both of these calculations are demonstrated below.

19.12.8.2.1 Hydraulic Loading Rate for Belt Filter Presses

■ **EXAMPLE 19.85**

Problem: A belt filter press receives a daily sludge flow of 0.30 gal. If the belt is 60 in. wide, what is the hydraulic loading rate on the unit in gallons per minute for each foot of belt width (gpm/ft)?

Solution:

$$\frac{0.30\,\text{MG}}{1\,\text{day}} \times \frac{1,000,000\,\text{gal}}{1\,\text{MG}} \times \frac{1\,\text{day}}{1440\,\text{min}} = \frac{208.3\,\text{gal}}{1\,\text{min}}$$

$$60\,\text{in.} \times \frac{1\,\text{ft}}{12\,\text{in.}} = 5\,\text{ft}$$

$$\frac{208.3\,\text{gal}}{5\,\text{ft}} = 41.7\,\text{gpm/ft}$$

19.12.8.2.2 Pounds of Solids Pressed Per Hour for Plate and Frame Presses

■ **EXAMPLE 19.86**

Problem: A plate and frame filter press can process 850 gal of sludge during its 120-min operating cycle. If the sludge concentration is 3.7%, and if the plate surface area is 140 ft², how many pounds of solids are pressed per hour for each square foot of plate surface area?

Solution:

$$850\,\text{gal} \times \frac{3.7\%}{100\%} \times \frac{8.34\,\text{lb}}{1\,\text{gal}} = 262.3\,\text{lb}$$

$$\frac{262.3\,\text{lb}}{120\,\text{min}} \times \frac{60\,\text{min}}{1\,\text{hr}} = 131.2\,\text{lb/hr}$$

$$\frac{131.2\,\text{lb/hr}}{140\,\text{ft}^2} = 0.94\,\text{lb/hr/ft}^2$$

TABLE 19.29
Expected Percent Solids for Centrifuge Dewatered Sludges

Type of Sludge	Percent Solids
Raw sludge	25–35%
Anaerobic digestion	15–30%
Activated sludge	8–10%
Heat treated	30–50%

19.12.9 CENTRIFUGATION

Centrifuges of various types have been used in dewatering operations for at lease 30 years and appear to be gaining in popularity. Depending on the type of centrifuge used, chemical makeup and feed equipment and support systems for removal of dewatered solids are required, in addition to centrifuge pumping equipment for solids feed and centrate removal.

19.12.9.1 Operational Observations, Problems, and Troubleshooting

Generally, in operation, the centrifuge spins at a very high speed. The centrifugal force it creates throws the solids out of the water. Chemically conditioned solids are pumped into the centrifuge. The spinning action throws the solids to the outer wall of the centrifuge. The centrate (water) flows inside the unit to a discharge point. The solids held against the outer wall are scraped to a discharge point by an internal scroll moving slightly faster or slower than the centrifuge speed of rotation. In the operation of a continuous-feed, solid-bowl, conveyor-type centrifuge (this is the most common type currently used), as well as other commonly used centrifuges, solid/liquid separation by gravity occurs as a result of rotating the liquid at high speeds.

In the solid-bowl type, the solid bowl has a rotating unit with a bowl and a conveyor. The unit has a conical section at one end that acts as a drainage device. The conveyor screw pushes the sludge solids to outlet ports and the cake to a discharge hopper. The sludge slurry enters the rotating bowl through a feed pipe leading into the hollow shaft of the rotating screw conveyor. The sludge is distributed through ports into a pool inside the rotating bowl. As the liquid sludge flows through the hollow shaft toward the overflow device, the fine solids settle to the wall of the rotating bowl. The screw conveyor pushes the solids to the conical section, where the solids are forced out of the water and the water drains back in the pool.

Expected percent solids for centrifuge dewatered sludge are in the range of 10 to 15%. The expected performance is dependent on the type of sludge being dewatered, as shown in Table 19.29. Centrifuge operation is dependent on various performance factors:

- Bowl design—length/diameter ratio; flow pattern
- Bowl speed
- Pool volume
- Conveyor design
- Relative conveyor speed
- Type and condition of sludge
- Type and amount of chemical conditioning
- Operating pool depth
- Relative conveyor speed (if adjustable)

Centrifuge operators often find that the operation of centrifuges can be simple, clean, and efficient. In most cases, chemical conditioning is required to achieve optimum concentrations. Operators soon discover that centrifuges are noise makers; units run at very high speed and produce high-level noise that can cause loss of hearing with prolonged exposure. When working in an area where a centrifuge is in operation, special care must be taken to provide hearing protection.

Actual operation of a centrifugation unit requires the operator to control and adjust chemical feed rates, observe unit operation and performance, control and monitor centrate returned to the treatment system, and perform required maintenance as outlined in the manufacturer's technical manual. The centrifuge operator must be trained to observe and recognize (as with other unit processes) operational problems that may occur with centrifuge operation. Table 19.30 lists several indicators of poor process performance, their causal factors, and corrective actions.

19.12.10 SLUDGE INCINERATION

Not surprisingly, incinerators produce the maximum solids and moisture reductions. The equipment required depends on whether the unit is a multiple-hearth or fluid-bed incinerator. Generally, the system will require a source of heat to reach ignition temperature, a solids feed system, and ash-handling equipment. It is important to note that the system must also include all required equipment (e.g., scrubbers) to achieve compliance with air pollution control requirements. Solids are pumped to the incinerator. The solids are dried and then ignited (burned). As they burn, the organic matter is converted to carbon dioxide and water vapor, and the inorganic matter is left behind as ash or fixed solids. The ash is then collected for reuse of disposal.

19.12.10.1 Process Description

The incineration process first dries then burns the sludge; the process involves the following steps:

- The temperature of the sludge feed is raised to 212°F.
- Water evaporates from the sludge.

- The temperature of the water vapor and air mixture increases.
- The temperature of the dried sludge volatile solids rises to the ignition point.

Note: Incineration will achieve maximum reductions if sufficient fuel, air, time, temperature, and turbulence are provided.

19.12.10.2 Incineration Processes

19.12.10.2.1 Multiple-Hearth Furnace

The *multiple-hearth furnace* consists of a circular steel shell surrounding a number of hearths. Scrappers (rabble arms) are connected to a central rotating shaft. Units range from 4.5 to 21.5 feet in diameter and have from four to 11 hearths. Dewatered sludge solids are placed on the outer edge of the top hearth. The rotating rabble arms move them slowly to the center of the hearth. At the center of the hearth, the solids fall through ports to the second level. The process is repeated in the opposite direction. Hot gases generated by burning on lower hearths are used to dry the solids. The dry solids pass to the lower hearths. The high temperature on the lower hearths ignites the solids. Burning continues to completion. Ash materials discharge to lower cooling hearths, where they are discharged for disposal. Air flowing inside the center column and rabble arms continuously cools internal equipment.

19.12.10.2.2 Fluidized Bed Furnace

The *fluidized bed* incinerator consists of a vertical circular steel shell (reactor) with a grid to support a sand bed and an air system to provide warm air to the bottom of the sand bed. The evaporation and incineration process takes place within the super-heated sand bed layer. Air is pumped to the bottom of the unit. The airflow expands (fluidizes) the sand bed inside. The fluidized bed is heated to its operating temperature (1200 to 1500°F). Auxiliary fuel is added when necessary to maintain operating temperature. The sludge solids are injected into the heated sand bed. Moisture immediately evaporates. Organic matter ignites and reduces to ash. Residues are ground to fine ash by the sand movement. Fine ash particles flow up and out of unit with exhaust gases. Ash particles are removed using common air pollution control processes. Oxygen analyzers in the exhaust gas stack control the airflow rate.

Note: Because these systems retain a high amount of heat in the sand, the system can be operated as little as 4 hr per day with little or no reheating.

19.12.10.3 Operational Observations, Problems, and Troubleshooting

The operator of an incinerator monitors various performance factors to ensure optimal operation. These performance

TABLE 19.30

Symptoms, Causes, and Corrective Actions for Pressure Filtration Problems

Symptom	Cause	Corrective Action
Poor centrate clarity	Feed rate too high	Adjust sludge feed rate.
	Wrong plate dam position	Increase pool depth.
	Worn conveyor flights	Repair/replace conveyor.
	Speed too high	Change pulley setting to obtain lower speed.
	High feed sludge solids concentration	Dilute sludge feed.
	Improper chemical conditioning	Adjust chemical dosage.
Solids cake not dry enough	Feed rate too high	Reduce sludge feed rate.
	Wrong plate dam position	Decrease pool depth to increase dryness.
	Speed too low	Change pulley setting to obtain higher speed.
	Excessive chemical conditioning	Adjust chemical dosage.
	Influent too warm	Reduce influent temperature.
Frequent tripping of torque control	Feed rate too high	Reduce flows.
	Feed solids concentration too high	Dilute flows.
	Foreign material (e.g., tramp iron) in machine	Remove conveyor; clear foreign materials.
	Gear unit misaligned	Correct gear unit alignment.
	Mechanical problem in gear unit	Repair gear unit.
Excess vibration	Improper lubrication	Lubricate according to the manufacturer's instructions.
	Improper adjustment of vibration isolators	Adjust isolators.
	Discharge funnels contacting centrifuge	Reposition slip joints at funnels.
	Portion of conveyor flights plugged (causing an imbalance)	Flush centrifuge.
	Gear box improperly aligned	Align gearbox.
	Pillow box bearings damaged	Replace bearings.
	Bowl out of balance	Return rotating parts to factory for rebalancing
	Parts not tightly assembled	Tighten parts.
	Uneven wear on conveyor	Resurface and rebalance.
Sudden increase in power consumption	Contact between bowl exterior and accumulated solids in case	Apply hard surfacing to areas with wear.
	Effluent pipe plugged	Clear solids discharge.
Gradual increase in power consumption	Conveyor blade wear	Replace blades.
Spasmodic surging of solids discharge	Pool depth too low	Increase pool depth.
	Conveyor helix rough	Refinish conveyor blade area.
	Feed pipe too near drainage deck	Move feed pipe to effluent end (if applicable).
	Excessive vibration	See "Excess vibration" entry (above).
Centrifuge shutting down or not starting	Blown fuses	Replace fuses.
	Tripped overload relay	Flush centrifuge; reset relay.
	Motor overheated or thermal protectors tripped	Flush centrifuge; reset thermal protectors.
	Tripped torque control	See "Frequent tripping of torque control" entry (above).
	Tripped vibration switch	See "Excess vibration" entry (above).

factors include feed sludge volatile matter content, feed sludge moisture content, operating temperature, sludge feed rate, fuel feed rate, and air feed rate.

Note: To ensure that the volatile material is ignited, the sludge must be heated to between 1400 and 1700°F.

To be sure that operating parameters are in the correct range, the operator monitors and adjusts sludge feed rate, airflow, and auxiliary fuel feed rate. All maintenance conducted on an incinerator should be in accordance with manufacturer's recommendations. The operator of a multiple-hearth or fluidized bed incinerator must be able to recognize operational problems using various indicators or through observations. Table 19.31 lists several indicators of poor process performance, their causal factors, and corrective actions.

19.12.11 Land Application of Biosolids

The purpose of land application of biosolids is to dispose of the treated biosolids in an environmentally sound manner by recycling nutrients and soil conditioners. To be land applied, wastewater biosolids must comply with state and federal biosolids management and disposal regulations. Biosolids must not contain materials that are dangerous to

TABLE 19.31

Symptoms, Causes, and Corrective Actions for Sludge Incineration Problems

Symptom	Cause	Corrective Action
Multiple hearth		
Incinerator temperature too high	Excessive fuel feed rate	Decrease fuel feed rate.
	Greasy solids	Reduce sludge feed rate; increase air feed rate.
	Thermocouple burned out	Replace thermocouple.
Furnace temperature too low	Increased moisture content of sludge	Increase fuel feed rate until dewatering operation improves.
	Fuel system malfunction	Establish proper fuel feed rate.
	Excessive air feed rate	Decrease air feed rate; increase sludge feed rate.
	Flame out	Relight furnace.
Oxygen content of stack gas too high	Sludge feed rate too low	Increase sludge feed rate.
	Sludge feed system blockage	Clear any feed system blockages.
	Air feed rate too high	Decrease air feed rate.
Oxygen content of stack gas too low	Increased volatile or grease content in sludge	Increase air feed rate; decrease sludge feed rate.
	Air feed rate too low	Increase air feed rate.
Furnace refractories deteriorated	Rapid startup/shutdown of furnace	Repair furnace refractories; follow specified startup and shutdown procedures.
Unusually high cooling effect	Air leak	Locate and repair leak.
Short hearth life	Uneven firing	Fire hearths equally on both sides.
Center shaft shear pin failure	Rabble arm dragging on hearth	Adjust rabble arm to eliminate rubbing.
	Debris caught under the arm	Remove debris.
Scrubber temperature too high	Low water flow to scrubber	Adjust water flow to proper level.
Stack gas temperatures too low	Inadequate fuel feed supply	Increase fuel feed rate.
	Excessive sludge feed rate	Decrease sludge feed rate.
Stack gas temperatures too high	Higher volatile content (heat value) in sludge	Increase air feed rate; decrease sludge feed rate.
	Excessive fuel feed rate	Decrease fuel feed rate.
Furnace burners slagging up	Burner design	Replace burners with newer designs that reduce slagging.
Rabble arms dropping	Excessive hearth temperatures	Maintain temperatures within proper range; discontinue injection of scum into the hearth.
	Loss of cooling air	Repair cooling air system immediately.
Excessive air pollutants in stack gas	Incomplete combustion, insufficient air	Raise air-to-fuel ratio.
	Air pollution control malfunction	Repair or replace broken equipment.
Flashing or explosions	Scum or grease additions	Remove scum or grease before incineration.
Fluidized bed		
Bed temperature falling	Inadequate fuel supply	Increase fuel supply; repair any fuel system malfunction.
	Excessive sludge feed rate	Decrease sludge feed rate.
	Excessive sludge moisture levels	Correct sludge dewatering process problem.
	Excessive air flow	Decrease airflow rate.
Low (<3%) oxygen in exhaust gas	Low air flow rate	Increase blower air feed rate.
	Fuel feed rate too high	Reduce fuel feed rate.
Fluidized bed excessive (>6%) oxygen in exhaust gas	Sludge feed rate too low	Increase sludge feed rate; adjust fuel feed rate to maintain steady bed temperature.
Erratic bed depth on control panel	Bed pressure taps plugged with solids	Tap a metal rod into pressure tap pipe when the unit is not in operation; apply compressed air to pressure tap while the unit is in operation (follow manufacturer's safety guidelines).
Preheat burner fails and alarm sounds	Pilot flame not receiving fuel	Correct fuel system problem.
	Pilot flame not receiving spark	Replace defective part.
	Defective pressure regulator	Replace defective regulator.
	Pilot flame ignition but malfunctioning flame scanner	Clear scanner sight glass; replace defective scanner.
Bed temperature too high	Bed gun fuel feed rate too high	Reduce bed gun fuel feed rate.
	Grease or high organic content in sludge (high heat value)	Increase airflow rate; decrease sludge fuel rate.
Bed temperature reading off scale	Thermocouple burned out	Replace thermocouple
High temperature in scrubber inlet	Water not flowing in scrubber	Open valves to provide water; correct system malfunction to provide required pressure.
	Plugged spray nozzles	Clear nozzles and strainers.
	Ash water not recirculating	Repair or replace recirculation pump; unclog scrubber discharge line.
Poor bed fluidization	Sand leakage through support plate during shutdown	Clear wind box; clean wind box at least once per month.

human health (e.g., toxicity, pathogenic organisms) or dangerous to the environment (e.g., toxicity, pesticides, heavy metals). Treated biosolids are land applied by either direct injection or application and plowing in (incorporation).

19.12.11.1 Process Control: Sampling and Testing

Land application of biosolids requires precise control to avoid problems. The quantity and the quality of biosolids applied must be accurately determined. For this reason, the operator's process control activities include biosolids sampling/testing functions. Biosolids sampling and testing includes determination of percent solids, heavy metals, organic pesticides and herbicides, alkalinity, total organic carbon (TOC), organic nitrogen, and ammonia nitrogen.

19.12.11.2 Process Control Calculations

Process control calculations include determining disposal cost, plant available nitrogen (PAN), application rate (dry tons and wet tons per acre), metals loading rates, maximum allowable applications based on metals loading, and site life based on metals loading.

19.12.11.2.1 Disposal Cost

The cost of disposal of biosolids can be determined by:

$$\text{Cost} = \text{Wet tons/yr} \times \% \text{ solids} \times \text{cost/dry ton} \quad (19.92)$$

■ EXAMPLE 19.87

Problem: The treatment system produces 1925 wet tons of biosolids for disposal each year. The biosolids are 18% solids. A contractor disposes of the biosolids for $28 per dry ton. What is the annual cost for sludge disposal?

Solution:

$$\text{Cost} = 1925 \text{ wet tons/yr} \times 0.18 \times \$28/\text{dry ton} = \$9702$$

19.12.11.2.2 Plant Available Nitrogen (PAN)

One factor considered when land applying biosolids is the amount of nitrogen in the biosolids available to the plants grown on the site. This includes ammonia nitrogen and organic nitrogen. The organic nitrogen must be mineralized for plant consumption. Only a portion of the organic nitrogen is mineralized per year. The mineralization factor (f_1) is assumed to be 0.20. The amount of ammonia nitrogen available is directly related to the time elapsed between applying the biosolids and incorporating (plowing) the sludge into the soil. We provide volatilization rates based upon this example below:

$$\underset{(\text{lb/dry ton})}{\text{PAN}} = \left\{ \begin{bmatrix} (\text{Organic nitrogen (mg/kg)} \times f_1) \\ + (\text{ammonia nitrogen (mg/kg)} \times V_1) \end{bmatrix} \right\} \quad (19.93)$$
$$\times 0.002 \text{ lb/dry ton}$$

where:

f_1 = Mineral rate for organic nitrogen (assume 0.20).
V_1 = Volatilization rate ammonia nitrogen.
 = 1.00 if biosolids are injected.
 = 0.85 if biosolids are plowed in within 24 hr.
 = 0.70 if biosolids are plowed in within 7 days.

■ EXAMPLE 19.88

Problem: The biosolids contain 21,000 mg/kg of organic nitrogen and 10,500 mg/kg of ammonia nitrogen. The biosolids are incorporated into the soil within 24 hr after application. What is the plant available nitrogen (PAN) per dry ton of solids?

Solution:

$$\text{PAN} = \begin{bmatrix} (21,000 \text{ mg/kg} \times 0.20) \\ + (10,500 \times 0.85) \end{bmatrix} \times 0.002$$

$$= 26.3 \text{ lb/dry ton}$$

19.12.11.2.3 Application Rate Based on Crop Nitrogen Requirement

In most cases, the application rate of domestic biosolids to crop lands will be controlled by the amount of nitrogen the crop requires. The biosolids application rate based on the nitrogen requirement is determined by the following:

* Step 1. Use an agriculture handbook to determine the nitrogen requirement of the crop to be grown.
* Step 2. Determine the amount of sludge in dry tons required to provide this much nitrogen:

$$\text{Dry tons/ac} = \frac{\text{Plant nitrogen requirement (lb/ac)}}{\text{Plant available nitrogen (lb/dry ton)}} \quad (19.94)$$

■ EXAMPLE 19.89

Problem: The crop to be planted on the land application site requires 150 lb of nitrogen per acre. What is the required biosolids application rate if the PAN of the biosolids is 30 lb/dry ton?

Solution:

$$\text{Application rate} = \frac{150 \text{ lb/ac}}{30 \text{ lb/dry ton}} = 5 \text{ dry tons/ac}$$

19.12.11.2.4 Metals Loading

When biosolids are land applied, metals concentrations are closely monitored and their loading on land application sites is calculated:

$$\text{Loading (lb/ac)} = \text{Metal conc. (mg/kg)}$$
$$\times 0.002 \text{ lb/dry ton} \qquad (19.95)$$
$$\times \text{application rate (dry tons/ac)}$$

■ EXAMPLE 19.90

Problem: The biosolids contain 14 mg/kg of lead. Biosolids are currently being applied to the site at a rate of 11 dry tons per acre. What is the metals loading rate for lead in pounds per acre?

Solution:

$$\text{Loading rate} = 14 \text{ mg/kg} \times 0.002 \text{ lb/dry ton} \times 11 \text{ dry tons}$$

$$= 0.31 \text{ lb/ac}$$

19.12.11.2.5 Maximum Allowable Applications Based on Metals Loading

If metals are present, they may limit the total number of applications a site can receive. Metals loading is normally expressed in terms of the maximum total amount of metal that can be applied to a site during its use:

$$\text{Applications} = \frac{\begin{array}{c}\text{Max. allowable cumulative}\\ \text{load for the metal (lb/ac)}\end{array}}{\text{Metal loading (lb/ac/application)}} \quad (19.96)$$

■ EXAMPLE 19.91

Problem: The maximum allowable cumulative lead loading is 48.0 lb/ac. Based on the current loading of 0.35 lb/ac, how many applications of biosolids can be made to this site?

Solution:

$$\text{Applications} = \frac{48.0 \text{ lb/ac}}{0.35 \text{ lb/ac}} = 137$$

19.12.11.2.6 Site Life Based on Metals Loading

The maximum number of applications based on metals loading and the number of applications per year can be used to determine the maximum site life:

$$\text{Site life (yr)} = \frac{\text{Max. allowable applications}}{\begin{array}{c}\text{Number of applications}\\ \text{planned per year}\end{array}} \quad (19.97)$$

■ EXAMPLE 19.92

Problem: Biosolids are currently applied to a site twice annually. Based on the lead content of the biosolids, the maximum number of applications is determined to be 135

applications. Based on the lead loading and the application rate, how many years can this site be used?

Solution:

$$\text{Site life} = \frac{135 \text{ applications}}{2 \text{ applications per yr}} = 68 \text{ yr}$$

Note: When more than one metal is present, the calculations must be performed for each metal. The site life would then be the lowest value generated by these calculations.

19.13 PERMITS, RECORDS, AND REPORTS

Permits, records, and reports play a significant role in wastewater treatment operations. In fact, with regard to permits, one of the first things any new operator quickly learns is the importance of "making permit" each month. In this chapter, we briefly cover National Pollutant Discharge Elimination System (NPDES) permits and other pertinent records and reports with which the wastewater operator must be familiar.

Note: The discussion that follows is general in nature; it does not necessarily apply to any state in particular but instead is an overview of permits, records, and reports that are an important part of wastewater treatment plant operations. For guidance on requirements for a specific locality, contact the state's water control board or other authorized state agency for information. In this handbook, the term *board* signifies the state-reporting agency.

19.13.1 DEFINITIONS

Several definitions should be understood before we discuss the permit requirements for records and reporting:

Average daily limitation—The highest allowable average over a 24-hour period, calculated by adding all of the values measured during the period and dividing the sum by the number of values determined during the period.

Average hourly limitation—The highest allowable average for a 60-minute period, calculated by adding all of the values measured during the period and dividing the sum by the number of values determined during the period.

Average monthly limitation—The highest allowable average over a calendar month, calculated by adding all of the daily values measured during the month and dividing the sum by number of daily values measured during the month.

Average weekly limitation—The highest allowable average over a calendar week, calculated by adding all of the daily values measured during the calendar week and dividing the sum by the number of daily values determined during the week.

Daily discharge—The discharge of a pollutant measured during a calendar day or any 24-hour period that reasonably represents the calendar for the purpose of sampling. For pollutants with limitations expressed in units of weight, the daily discharge is calculated as the total mass of the pollutant discharged over the day. For pollutants with limitations expressed in other units, the daily discharge is calculated as the average measurement of the pollutant over the day.

Discharge monitoring report—Forms used to report self-monitoring results of the permittee.

Discharge permit—State Pollutant Discharge Elimination System permit that specifies the terms and conditions under which a point source discharge to state waters is permitted.

Effluent limitation—Any restriction by the state board on quantities, discharge rates, or concentrations of pollutants which are discharged from point sources into state waters.

Maximum daily discharge—The highest allowable value for a daily discharge.

Maximum discharge—The highest allowable value for any single measurement.

Minimum discharge—The lowest allowable value for any single measurement.

Point source—Any discernible, defined, and discrete conveyance, including but not limited to, any pipe, ditch, channel, tunnel, conduit, well, discrete fissure, container, rolling stock, vessel, or other floating craft from which pollutants are or may be discharged. This definition does not include return flows from irrigated agricultural land.

19.13.2 NPDES PERMITS

In the United States, all treatment facilities that discharge to state waters must have a discharge permit issued by the state water control board or other appropriate state agency. This permit is known on the national level as the National Pollutant Discharge Elimination System (NPDES) permit and on the state level as the (State) Pollutant Discharge Elimination System (state-PDES) permit. The permit states the specific conditions that must be met to legally discharge treated wastewater to state waters. The permit contains general requirements (applying to every discharger) and specific requirements (applying only to the point source specified in the permit). A general permit is a discharge permit that covers a specified class of dischargers. It is developed to allow dischargers in a specified category to discharge under specified conditions. All discharge permits contain general conditions. These conditions are standard for all dischargers and cover a broad series of requirements. Read the general conditions of the treatment facility's permit carefully. Permittees must retain certain records.

19.13.2.1 Monitoring

- Date, time, and exact place of sampling or measurements
- Names of the individuals performing sampling or measurement
- Dates and times analyses were performed
- Names of the individuals who performed the analyses
- Analytical techniques or methods used
- Observations, readings, calculations, bench data, and results
- Instrument calibration and maintenance
- Original strip chart recordings for continuous monitoring
- Information used to develop reports required by the permit
- Data used to complete the permit application

Note: All records must be kept at least 3 years (longer at the request of the state board).

19.13.2.2 Reporting

Generally, reporting must be made under the following conditions or situations (requirements may vary depending on the state regulatory body with reporting authority):

- *Unusual or extraordinary discharge reports*— Must be submitted to the board by telephone within 24 hours of occurrence, with a written report being submitted within 5 days. The report must include:
  - Description of the noncompliance and its cause
  - Noncompliance dates, times, and duration
  - Steps planned or taken to reduce or eliminate the problem
  - Steps planned or taken to prevent reoccurrence
- *Anticipated noncompliance*—Must notify the board at least 10 days in advance of any changes to the facility or activity that may result in noncompliance.
- *Compliance schedules*—Must report compliance or noncompliance with any requirements contained in compliance schedules no later than 14 days following the scheduled date for completion of the requirement.
- *24-hour reporting*—Any noncompliance that may adversely affect state waters or may endanger

public health must be reported orally with 24 hours of the time the permittee becomes aware of the condition. A written report must be submitted within 5 days.

- *Discharge monitoring reports (DMRs)*—Self-monitoring data generated during a specified period (normally 1 month). When completing the DMR, remember:
  - More frequent monitoring must be reported.
  - All results must be used to complete reported values.
  - Pollutants monitored by an approved method but not required by the permit must be reported.
  - No blocks on the form should be left blank.
  - Averages are arithmetic unless noted otherwise.
  - Appropriate significant figures should be used.
  - All bypasses and overflows must be reported.
  - The licensed operator must sign the report.
  - Responsible official must sign the report.
  - Department must receive the report by the 10th day of the following month.

19.13.3 Sampling and Testing

The general requirements of the permit specify minimum sampling and testing that must be performed on the plant discharge. Moreover, the permit will specify the frequency of sampling, sample type, and length of time for composite samples. Unless a specific method is required by the permit, all sample preservation and analysis must be in compliance with the requirements set forth in the federal regulations *Guidelines Establishing Test Procedures for the Analysis of Pollutants Under the Clean Water Act* (40 CFR 136).

Note: All samples and measurements must be representative of the nature and quantity of the discharge.

19.13.3.1 Effluent Limitations

The permit sets numerical limitations on specific parameters contained in the plant discharge. Limits may be expressed as:

- Average monthly quantity (kg/day)
- Average monthly concentration (mg/L)
- Average weekly quantity (kg/day)
- Average weekly concentration (mg/L)
- Daily quantity (kg/day)
- Daily concentration (mg/L)
- Hourly average concentration (mg/L)
- Instantaneous minimum concentration (mg/L)
- Instantaneous maximum concentration (mg/L)

19.13.3.2 Compliance Schedules

If the facility requires additional construction or other modifications to fully comply with the final effluent limitations, the permit will contain a schedule of events to be completed to achieve full compliance.

19.13.3.3 Special Conditions

Any special requirements or conditions set for approval of the discharge will be contained in this section. Special conditions may include:

- Monitoring required to determine effluent toxicity
- Pretreatment program requirements

19.13.3.4 Licensed Operator Requirements

The permit will specify, based on the treatment system complexity and the volume of flow treated, the minimum license classification required to be the designated responsible charge operator.

19.13.3.5 Chlorination/Dechlorination Reporting

Several reporting systems apply to chlorination or chlorination followed by dechlorination. It is best to review this section of the specific permit for guidance. If confused, contact the appropriate state regulatory agency.

19.13.4 Reporting Calculations

Failure to accurately calculate report data will result in violations of the permit. The basic calculations associated with completing the DMR are covered below.

19.13.4.1 Average Monthly Concentration

The average monthly concentration (AMC) is the average of the results of all tests performed during the month:

$$\text{AMC (mg/L)} = \frac{\sum \begin{array}{c} \text{Test}_1 + \text{Test}_2 + \text{Test}_3 \\ + \ldots + \text{Test}_n \end{array}}{n \text{ (tests during month)}} \qquad (19.98)$$

19.13.4.2 Average Weekly Concentration

The average weekly concentration (AWC) is the results of all the tests performed during a calendar week. A calendar week must start on Sunday and end on Saturday and be completely within the reporting month. A weekly average is not computed for any week that does not meet these criteria:

$$AWC \ (mg/L) = \frac{\sum \genfrac{}{}{0pt}{}{Test_1 + Test_2 + Test_3}{+ \ldots + Test_n}}{n \ (tests \ during \ calendar \ week)} \quad (19.99)$$

19.13.4.3 Average Hourly Concentration

The average hourly concentration (AHC) is the average of all of the test results collected during a 60-minute period.

$$AHC \ (mg/L) = \frac{\sum \genfrac{}{}{0pt}{}{Test_1 + Test_2 + Test_3}{+ \ldots + Test_n}}{n \ (tests \ during \ 60\text{-}min \ period)} \quad (19.100)$$

19.13.4.4 Daily Quantity (kg/day)

Daily quantity is the quantity of a pollutant in kilograms per day discharged during a 24-hour period.

$$kg/day = \genfrac{}{}{0pt}{}{Concentration \ (mg/L) \times flow \ (MGD)}{\times 3.785 \ kg/MG/mg/L} \quad (19.101)$$

19.13.4.5 Average Monthly Quantity

Average monthly quantity (AMQ) is the average of all the individual daily quantities determined during the month.

$$AMQ \ (kg/day) = \frac{\sum \genfrac{}{}{0pt}{}{DQ_1 + DQ_2 + DQ_3}{+ \ldots + DQ_n}}{n \ (tests \ during \ month)} \quad (19.102)$$

19.13.4.6 Average Weekly Quantity

The average weekly quantity (AWQ) is the average of all the daily quantities determined during a calendar week. A calendar week must start on Sunday and end on Saturday and be completely within the reporting month. A weekly average is not computed for any week that does not meet these criteria.

$$AWQ \ (kg/day) = \frac{\sum \genfrac{}{}{0pt}{}{DQ_1 + DQ_2 + DQ_3}{+ \ldots + DQ_n}}{n \ (tests \ during \ calendar \ week)} \quad (19.103)$$

19.13.4.7 Minimum Concentration

The minimum concentration is the lowest instantaneous value recorded during the reporting period.

19.13.4.8 Maximum Concentration

The maximum concentration is the highest instantaneous value recorded during the reporting period.

19.13.4.9 Bacteriological Reporting

Bacteriological reporting is used for reporting fecal coliform test results. To make this calculation, the geometric mean calculation is used and all monthly geometric means are computed using all the test values. Note that weekly geometric means are computed using the same selection criteria discussed for average weekly concentration and quantity calculations. The easiest method used in making this calculation requires a calculator that can perform logarithmic (log) or nth root functions:

$$Geometric \ mean = Antilog \left[\frac{Log \ X_1 + \log X_2 + \log X_3 + \ldots \log X_n}{n \ (number \ of \ tests)} \right] \quad (19.104)$$

or

$$Geometric \ mean = \sqrt[n]{X_1 \times X_2 \times \ldots \times X_n}$$

CHAPTER REVIEW QUESTIONS

19.1   Who must sign the DMR?
19.2   What does the COD test measure?
19.3   Give three reasons for treating wastewater.
19.4   Name two types of solids based on physical characteristics.
19.5   Define organic and inorganic.
19.6   Name four types of microorganisms that may be present in wastewater.
19.7   When organic matter is decomposed aerobically, what materials are produced?
19.8   Name three materials or pollutants that are not removed by the natural purification process.
19.9   What do we call the used water and solids from a community that flow to a treatment plant?
19.10 Where do disease-causing bacteria in wastewater come from?
19.11 What does the term *pathogenic* mean?
19.12 What do we call wastewater that comes from the household?
19.13 What do we call wastewater that comes from industrial complexes?
19.14 A lab test indicates that a 500-g sample of sludge contains 22 g of solids. What are the percent solids in the sludge sample?
19.15 The depth of water in the grit channel is 28 in. What is the depth in feet?
19.16 The operator withdraws 5250 gal of solids from the digester. How many pounds of solids have been removed?
19.17 Sludge added to the digester causes a 1920-ft^3 change in the volume of sludge in the digester. How many gallons of sludge have been added?

19.18   The plant effluent contains 30 mg/L solids. The effluent flow rate is 3.40 MGD. How many pounds per day of solids are discharged?

19.19   The plant effluent contains 25 mg/L BOD_5. The effluent flow rate is 7.25 MGD. How many kilograms per day of BOD_5 are being discharged?

19.20   The operator wishes to remove 3280 pounds per day of solids from the activated sludge process. The waste activated sludge concentration is 3250 mg/L. What is the required flow rate in million gallons per day?

19.21   The plant influent includes an industrial flow that contains 240 mg/L BOD. The industrial flow is 0.72 MGD. What is the population equivalent for the industrial contribution in people per day?

19.22   The label of a hypochlorite solution states that the specific gravity of the solution is 1.1288. What is the weight of 1 gal of the hypochlorite solution?

19.23   What must be done to the cutters in a comminutor to ensure proper operation?

19.24   What is grit? Give three examples of materials considered to be grit.

19.25   The plant has three channels in service. Each channel is 2 ft wide and has a water depth of 3 ft. What is the velocity in the channel when the flow rate is 8.0 MGD?

19.26   The grit from the aerated grit channel has a strong hydrogen sulfide odor upon standing in a storage container. What does this indicate, and what action should be taken to correct the problem?

19.27   What is the purpose of primary treatment?

19.28   What is the purpose of the settling tank in the secondary or biological treatment process?

19.29   A circular settling tank is 90 ft in diameter and has a depth of 12 ft. The effluent weir extends around the circumference of the tank. The flow rate 2.25 MGD. What is the detention time in hours, surface loading rate in gpd/ft^2, and weir overflow rate in gpd/ft?

19.30   Give three classifications of ponds based on their location in the treatment system.

19.31   Describe the processes occurring in a raw sewage stabilization pond (facultative).

19.32   How do changes in the season affect the quality of the discharge from a stabilization pond?

19.33   What is the advantage of using mechanical or diffused aeration equipment to provide oxygen?

19.34   Name three classifications of trickling filters and identify the classification that produces the highest quality effluent.

19.35   Microscopic examination reveals a predominance of rotifers. What process adjustment does this indicate is required?

19.36   Increasing the wasting rate will _____ the MLSS, _____ the return concentration, _____ the MCRT, _____ the F/M ratio, and _____ the SVI.

19.37   The plant currently uses 45.8 lb of chlorine per day. Assuming the chlorine usage will increase by 10% during the next year, how many 2000-lb cylinders of chlorine will be needed for the year (365 days)?

19.38   A plant has six 2000-lb cylinders on hand. The current dose of chlorine being used to disinfect the effluent is 6.2 mg/L. The average effluent flow rate is 2.25 MGD. Allowing 15 days for ordering and shipment, when should the next order for chlorine be made?

19.39   The plant feeds 38 lb of chlorine per day and uses 150-lb cylinders. Chlorine use is expected to increase by 11% next year. The chlorine supplier has stated that the current price of chlorine ($0.170 per pound) will increase by 7.5% next year. How much money should the town budget for chlorine purchases for the next year (365 days)?

19.40   The sludge pump operates 30 min every 3 hr, and the pump delivers 70 gpm. If the sludge is 5.1% solids and has a volatile matter content of 66%, how many pounds of volatile solids are removed from the settling tank each day?

19.41   The aerobic digester has a volume of 63,000 gal. The laboratory test indicates that 41 mg of lime were required to increase the pH of a 1-L sample of digesting sludge from 6.0 to the desired 7.1. How many pounds of lime must be added to the digester to increase the pH of the unit to 7.4?

19.42   The digester has a volume of 73,500 gal. Sludge is added to the digester at the rate of 2750 gpd. What is the sludge retention time in days?

19.43   The raw sludge pumped to the digester contains 72% volatile matter. The digested sludge removed from the digester contains 48% volatile matter. What is the percent volatile matter reduction?

19.44   What does NPDES stand for?

19.45   How can primary sludge be freshened going into a gravity thickener?

19.46   A neutral solution has what pH value?

19.47   Why is the seeded BOD test required for some samples?

19.48   What is the foremost advantage of the COD over the BOD?

19.49 High mixed liquor concentration is indicated by a _____ aeration tank foam.

19.50 What typically happens to the activity level of bacteria when the temperature is increased?

19.51 List three factors other than food that affect the growth characteristics of activated sludge.

19.52 What are the characteristics of facultative organisms?

19.53 BOD measures the amount of _____ material in wastewater.

19.54 The activated sludge process requires _____ in the aeration tank to be successful.

19.55 The activated sludge process cannot be successfully operated with a _____ clarifier.

19.56 The activated biosolids process can successfully remove _____ BOD.

19.57 Successful operation of a complete mix reactor in the endogenous growth phase is possible or not possible?

19.58 The bacteria in the activated biosolids process are either _____ or _____.

19.59 Step feed activated biosolids processes have _____ mixed liquor concentrations in different parts of the tank.

19.60 An advantage of contact stabilization compared to complete mix is _____ aeration tank volume.

19.61 Increasing the _____ of wastewater increases the BOD in the activated biosolids process.

19.62 Bacteria need phosphorus to successfully remove _____ in the activated biosolids process.

19.63 The growth rate of microorganisms is controlled by the _____.

19.64 Adding chlorine just before the _____ can control alga growth.

19.65 What is the purpose of the secondary clarifier in an activated biosolids process?

19.66 The _____ growth phase should occur in a complete mix activated biosolids process.

19.67 The typical DO value for activated biosolids plants is between _____ and _____ mg/L.

19.68 In the activated biosolids process, what change would an operator normally expect to make when the temperature decreases from 25°C to 15°C?

19.69 In the activated biosolids process, what change must be made to increase the MLVSS?

19.70 In the activated biosolids process, what change must be made to increase the F/M?

19.71 What does the Gould sludge age assume to be the source of the MLVSS in the aeration tank?

19.72 What is one advantage of complete mix over plug flow?

19.73 The grit in the primary sludge is causing excessive wear on primary treatment sludge pumps. The plant uses an aerated grit channel. What action should be taken to correct this problem?

19.74 When the mean cell residence time (MCRT) increases, what does the mixed liquor suspended solids (MLSS) concentration in the aeration tank do?

19.75 Exhaust air from a chlorine room should be taken from where?

19.76 If chlorine costs $0.21/lb, what is the daily cost to chlorinate a 5-MGD flow rate at a chlorine feed rate of 2.6 mg/L?

19.77 What term describes a normally aerobic system from which the oxygen has temporarily been depleted?

19.78 The ratio that describes the minimum amount of nutrients theoretically required for an activated sludge system is 100:5:1. What are the elements that fit this ratio?

19.79 A flotation thickener is best used for what type of sludge?

19.80 Drying beds *are* or *are not* an example of a sludge stabilization process?

19.81 The minimum flow velocity in collection systems should be _____.

19.82 What effect will the addition of chlorine, acid, alum, carbon dioxide, or sulfuric acid have on the pH of wastewater?

19.83 An amperometric titrator is used to measure _____.

19.84 The normal design detention time for a primary clarifier is _____.

19.85 The volatile acids-to-alkalinity ratio in an anaerobic digester should be approximately _____.

19.86 The surface loading rate in a final clarifier should be approximately _____.

19.87 In a conventional effluent chlorination system, the chlorine residual measured is mostly in the form of _____.

19.88 For a conventional activated biosolids process, the food-to-microorganism (F/M) ratio should be in the range of _____.

19.89 Denitrification in a final clarifier can cause clumps of sludge to rise to the surface. The sludge flocs attach to small sticky bubbles of _____ gas.

19.90 An anaerobic digester is covered and kept under positive pressure to _____.

19.91 During the summer months, the major source of oxygen added to a stabilization pond is _____.

19.92 Which solids cannot be removed by vacuum filtration?

19.93 The odor recognition threshold for H$_2$S is reported to be as low as _____.

REFERENCES AND SUGGESTED READING

Culp, G.L. and Heim, N.F. 1978. *Field Manual for Performance Evaluation and Troubleshooting at Municipal Wastewater Treatment Facilities*. Washington, D.C.: U.S. Environmental Protection Agency.

Metcalf & Eddy. 1991. *Wastewater Engineering: Treatment, Disposal, Reuse*, 3rd ed. New York: McGraw-Hill.

Metcalf & Eddy. 2003. *Wastewater Engineering: Treatment, Disposal, Reuse*, 4th ed. New York: McGraw-Hill.

Spellman, F.R. 2007. *The Science of Water*, 2nd ed. Boca Raton, FL: CRC Press.

USEPA. 2007. *Advanced Wastewater Treatment to Achieve Low Concentration of Phosphorus*. Washington, D.C.: U.S. Environmental Protection Agency.

VWCB. 1990. *Activated Sludge Process Control*, Part II, 2nd ed. Richmond: Virginia Water Control Board.

Appendixes

Appendix A

Answers to Chapter Review Questions

CHAPTER 1 ANSWERS

1.1 A pattern or point of view that determines what is seen as reality.

1.2 A change in the way things are understood and done.

1.3 (1) Assessing and protecting drinking water sources
(2) Optimizing treatment processes
(3) Ensuring the integrity of distribution systems
(4) Effecting correct cross-connection control procedures
(5) Continuous monitoring and testing of the water before it reaches the tap

1.4 Water/wastewater operations are usually low-profile activities and much of water/wastewater infrastructure is buried underground.

1.5 Secondary

1.6 Privatization means allowing private enterprise to compete with government in providing public services, such as water and wastewater operations. Reengineering is the systematic transformation of an existing system into a new form to realize quality improvements in operations, systems capability, functionality, and performance at lower cost, improved schedule, and less risk to the customer.

1.7 A process for rigorously measuring performance vs. "best-in-class" operations and using the analysis to meet and exceed the best in class.

1.8 Planning, research, observation, analysis, adaptation

CHAPTER 2 ANSWERS

2.1 Operators are exposed to the full range of hazards and work under all weather conditions.

2.2 Plants are upgrading to computerized operations.

2.3 Computerized maintenance management system

2.4 HAZMAT emergency response technician 24-hour certification

2.5 Safe Drinking Water Act

CHAPTER 3 ANSWERS

3.1 Answers will vary

CHAPTER 4 ANSWERS

4.1 Matching answers

1.	o	14.	v
2.	c	15.	r
3.	t	16.	w
4.	j	17.	l
5.	s	18.	x
6.	p	19.	m
7.	d	20.	y
8.	i	21.	f
9.	e	22.	b
10.	q	23.	z
11.	u	24.	h
12.	k	25.	n
13.	a	26.	g

CHAPTER 5 ANSWERS

5.1 $(2.5 \text{ mg/L})(5.5 \text{ MGD})(8.34 \text{ lb/gal}) = 115 \text{ lb/day}$

5.2 $(7.1 \text{ mg/L})(4.2 \text{ MGD})(8.34 \text{ lb/gal}) = 249 \text{ lb/day}$

5.3 $(11.8 \text{ mg/L})(4.8 \text{ MGD})(8.34 \text{ lb/gal}) = 472 \text{ lb/day}$

5.4 $\dfrac{(10 \text{ mg/L})(1.8 \text{ MGD})(8.34 \text{ lb/gal})}{\text{lb/day}} = 0.65$

5.5 $(60 \text{ mg/L})(0.086 \text{ MGD})(8.34 \text{ lb/gal}) = 43 \text{ lb}$

5.6 $(2220 \text{ mg/L})(0.225)(8.34 \text{ lb/gal}) = 4166 \text{ lb}$

5.7 $\dfrac{(8 \text{ mg/L})(0.83 \text{ MGD})(8.34 \text{ lb/gal})}{0.65} = 85 \text{ lb/day}$

5.8 $(450 \text{ mg/L})(1.84 \text{ MGD})(8.34 \text{ lb/gal}) = 6906 \text{ lb/day}$

5.9 $(25 \text{ mg/L})(2.90 \text{ MGD})(8.34 \text{ lb/gal}) = 605 \text{ lb/day}$

5.10 $(260 \text{ mg/L})(5.45 \text{ MGD})(8.34 \text{ lb/gal}) = 11,818 \text{ lb/day}$

5.11 $(144 \text{ mg/L})(3.66 \text{ MGD})(8.34 \text{ lb/gal}) = 4396 \text{ lb/day}$

5.12 $(290 \text{ mg/L})(3.31 \text{ MGD})(8.34 \text{ lb/gal}) = 8006 \text{ lb/day}$

5.13 $(152 \text{ mg/L})(5.7 \text{ MGD})(8.34 \text{ lb/gal}) = 7226 \text{ lb/day}$

5.14   (188 mg/L)(1.92 MGD)(8.34 lb/gal) = 3010 lb/day SS

5.15   (184 mg/L)(1.88 MGD)(8.34 lb/day) = 2885 lb/day SS

5.16   (150 mg/L)(4.88 MGD)(8.34 lb/gal) = 6105 lb/day BOD

5.17   (205 mg/L)(2.13 MGD)(8.34 lb/gal) = 3642 lb/day solids

5.18   (115 mg/L)(4.20 MGD)(8.34 lb/gal) = 4028 lb/day BOD

5.19   (2230 mg/L)(0.40 MG)(8.34 lb/gal) = 7439 lb SS

5.20   (1890 mg/L)(0.41 MG)(8.34 lb/gal) = 6463 lb MLVSS

5.21   (3125 mg/L)(0.18 MG)(8.34 lb/gal) = 4691 lb MLVSS

5.22   (2250 mg/L)(0.53 MG)(8.34 lb/gal) = 9945 lb MLSS

5.23   (2910 mg/L)(0.63 MG)(8.34 lb/gal) = 15,290 lb MLSS

5.24   (6150 mg/L)(x MGD)(8.34 lb/gal) = 5200 lb/day
$x = 0.10$ MGD

5.25   (6200 mg/L)(x MGD)(8.34 lb/gal) = 4500 lb/day
a. $x = 0.09$
b.   90,000 gpd ÷ 1440 min/day = 62.5 gpm

5.26   (6600 lb/day)(x MGD)(8.34 lb/gal) = 6070 lb/day
$x = 0.11$ MGD
$x = 110,000$ gpd ÷ 1440 min/day = 76 gpm

5.27   (6350 mg/L)(x MGD)(8.34 lb/gal) = 7350 lb/day
$x = 0.14$ MGD
$x = 140,000$ gpd ÷ 1440 min/day = 97 gpm

5.28   (7240 mg/L)(x MGD)(8.34 lb/gal) = 5750 lb/day
$x = 0.10$ MGD
$x = 100,000$ gpd ÷ 1440 min/day = 69 gpm

5.29   (2.5 mg)(3.65 MGD)(8.34 lb/gal) = 76.1 lb/day

5.30   (17 mg/L)(2.10 MGD)(8.34 lb/gal) = 298 lb/day

5.31   (190 mg/L)(4.8 MGD)(8.34 lb/gal) = 7606 lb/day SS removed

5.32   (9.7 mg/L)(5.5 MGD)(8.34 lb/gal) = 445 lb/day

5.33   (305 mg/L)(3.5 MGD) (8.34 lb/gal) = 8903 lb/day

5.34 $\dfrac{(10 \text{ mg/L})(3.1 \text{ MGD})(8.34 \text{ lb/gal})}{0.65} = 398$ lb/day

5.35   (210 mg/L)(3.44 MGD)(8.34 lb/gal) = 6025 lb/day solids

5.36   (60 mg/L)(0.09 MG)(8.34 lb/gal) = 45 lb chlorine

5.37   (2720 mg/L)(0.52 MG)(8.34 lb/gal) = 11,796 lb MLSS

5.38   (5870 mg/L)(x MGD)(8.34 lb/gal) = 5480 lb/day
$x = 0.11$ MGD

5.39   (120 mg/L)(3.312 MGD)(8.34 lb/gal) = 3315 lb/day BOD

5.40   (240 mg/L)(3.18 MGD)(8.34 lb/gal) = 6365 lb/day BOD

5.41   (196 mg/L)(1.7 MGD)(8.34 lb/gal) = 2779 lb/day BOD removed

5.42   (x mg/L)(5.3 MGD)(8.34 lb/day) = 330 lb/day
$x = 7.5$ mg/L

5.43   (5810 mg/L)(x MGD)(8.34 lb/gal) = 5810 mg/L
$x = 0.12$ MGD
$x = 120,000$ gpd ÷ 1440 min/day = 83 gpm

5.44 $\dfrac{3,400,000 \text{ gpd}}{(0.785)(100 \text{ ft})(100 \text{ ft})} = 433 \text{ gpd/ft}^2$

5.45 $\dfrac{4,525,000 \text{ gpd}}{(0.785)(90 \text{ ft})(90 \text{ ft})} = 712 \text{ gpd/ft}^2$

5.46 $\dfrac{3,800,000 \text{ gpd}}{870,000 \text{ ft}^2} = 4.4 \text{ gpd/ft}^2$

5.47 $\dfrac{280,749 \text{ ft}^3 \text{ day}}{696,960 \text{ ft}} = 0.4 \text{ ft/day}$

     (0.4 ft/day)(12 in./ft) = 4.8 in./day

5.48 $\dfrac{5,280,000 \text{ gpd}}{(0.785)(90 \text{ ft})(90 \text{ ft})} = 830 \text{ gpd/ft}^2$

5.49 $\dfrac{4.4 \text{ ac-ft/day}}{20 \text{ ac}} = 0.22 \text{ ft/day} = 3 \text{ in./day}$

5.50 $\dfrac{2,050,000 \text{ gpd}}{(70 \text{ ft})(25 \text{ ft})} = 1171 \text{ gpd/ft}^2$

5.51 $\dfrac{2,440,000 \text{ gpd}}{(0.785)(60 \text{ ft})(60 \text{ ft})} = 863 \text{ gpd/ft}^2$

5.52 $\dfrac{3,450,000 \text{ gpd}}{(110 \text{ ft})(50 \text{ ft})} = 627 \text{ gpd/ft}^2$

5.53 $\dfrac{1,660,000 \text{ gpd}}{(25 \text{ ft})(70 \text{ ft})} = 949 \text{ gpd/ft}^2$

5.54 $\dfrac{2,660,000 \text{ gpd}}{(0.785)(70 \text{ ft})(70 \text{ ft})} = 691 \text{ gpd/ft}^2$

5.55 $\dfrac{2230 \text{ gpm}}{(40 \text{ ft})(20 \text{ ft})} = 2.8 \text{ gpm/ft}^2$

5.56 $\dfrac{3100 \text{ gpm}}{(40 \text{ ft})(25 \text{ ft})} = 3.1 \text{ gpm/ft}^2$

5.57 $\dfrac{2500 \text{ gpm}}{(26 \text{ ft})(60 \text{ ft})} = 1.6 \text{ gpm/ft}^2$

5.58 $\dfrac{1528 \text{ gpm}}{(40 \text{ ft})(20 \text{ ft})} = 1.9 \text{ gpm/ft}^2$

5.59 $\dfrac{2850 \text{ gpm}}{880 \text{ ft}^2} = 3.2 \text{ gpm/ft}^2$

5.60 $\dfrac{4750 \text{ gpm}}{(14 \text{ ft})(14 \text{ ft})} = 24 \text{ gpm/ft}^2$

5.61 $\dfrac{4900 \text{ gpm}}{(20 \text{ ft})(20 \text{ ft})} = 12 \text{ gpm/ft}^2$

5.62 $\dfrac{3400 \text{ gpm}}{(25 \text{ ft})(15 \text{ ft})} = 9 \text{ gpm/ft}^2$

5.63 $\dfrac{3300 \text{ gpm}}{(75 \text{ ft})(30 \text{ ft})} = 4.4 \text{ gpm/ft}^2$

5.64 $\dfrac{3800 \text{ gpm}}{(15 \text{ ft})(20 \text{ ft})} = 12.7 \text{ gpm/ft}^2$

5.65 $\dfrac{3,770,000 \text{ gal}}{(15 \text{ ft})(30 \text{ ft})} = 8378 \text{ gal/ft}^2$

5.66 $\dfrac{1,860,000 \text{ gal}}{(20 \text{ ft})(15 \text{ ft})} = 6200 \text{ gal/ft}^2$

5.67 $\dfrac{3,880,000 \text{ gal}}{(25 \text{ ft})(20 \text{ ft})} = 7760 \text{ gal/ft}^2$

5.68 $\dfrac{1,410,200 \text{ gal}}{(20 \text{ ft})(14 \text{ ft})} = 5036 \text{ gal/ft}^2$

5.69 $\dfrac{5,425,000 \text{ gal}}{(30 \text{ ft})(20 \text{ ft})} = 9042 \text{ gal/ft}^2$

5.70 $\dfrac{1,410,000 \text{ gpd}}{163 \text{ ft}} = 8650 \text{ gpd/ft}$

5.71 $\dfrac{2,120,000 \text{ gpd}}{(3.14)(60 \text{ ft})} = 11,253 \text{ gpd/ft}$

5.72 $\dfrac{2,700,00 \text{ gpd}}{240 \text{ ft}} = 11,250 \text{ gpd/ft}$

5.73 $\dfrac{(1400 \text{ gpm})(1440 \text{ min/day})}{(3.14)(80 \text{ ft})} = 8025 \text{ gpd/ft}$

5.74 $\dfrac{2785 \text{ gpm}}{189 \text{ ft}} = 14.7 \text{ gpm/ft}$

5.75 $\dfrac{(210 \text{ mg/L})(2.45 \text{ MGD})(8.34 \text{ lb/gal})}{25.1 \, 1000 \text{ ft}^3}$
$= 171 \text{ lb BOD/day/1000 ft}^3$

5.76 $\dfrac{(170 \text{ mg/L})(0.120 \text{ MGD})(8.34 \text{ lb/gal})}{3.5 \text{ ac}}$
$= 49 \text{ lb BOD/day/ac}$

5.77 $\dfrac{(120 \text{ mg/L})(2.85 \text{ MGD})(8.34 \text{ lb/gal})}{34 \, 1000 \text{ ft}^3}$
$= 84 \text{ lb/BOD/day/1000 ft}^3$

5.78 $\dfrac{(140 \text{ mg/L})(2.20 \text{ MGD})(8.34 \text{ lb/gal})}{900 \, 1000 \text{ ft}^2}$
$= 2.9 \text{ lb/BOD/day/1000 ft}^2$

5.79 $(0.785)(90 \text{ ft})(90 \text{ ft})(4 \text{ ft}) = 25,434$

$\dfrac{(150 \text{ mg/L}) (3.5 \text{ MGD}) (8.34 \text{ lb/gal})}{25.4 \, 1000 \text{ ft}^3}$
$= 172 \text{ lb}^3 \text{ BOD/day/1000 ft}$

5.80 $\dfrac{(200 \text{ mg/L})(3.42 \text{ MGD})(8.34 \text{ lb/gal})}{(1875 \text{ mg/L})(0.42 \text{ MG})(8.34 \text{ lb/gal})} = 0.9$

5.81 $\dfrac{(190 \text{ mg/L})(3.24 \text{ MGD})(8.34 \text{ lb/gal})}{(1710 \text{ mg/L})(0.28 \text{ MG})(8.34 \text{ lb/gal})} = 1.3$

5.82 $\dfrac{(151 \text{ mg/L})(2.25 \text{ MGD})(8.34 \text{ lb/gal})}{x \text{ lb MLVSS}} = 0.9$

$x = 3148 \text{ lb MLVSS}$

5.83 $\dfrac{(160 \text{ mg/L})(2.10 \text{ MGD})(8.34 \text{ lb/gal})}{(1900 \text{ mg/L})(0.255 \text{ MG})(8.34 \text{ lb/gal})} = 0.7$

5.84 $\dfrac{(180 \text{ mg/L})(3.11 \text{ MGD})(8.34 \text{ lb/gal})}{(x \text{ mg/L})(0.88 \text{ MG})(8.34 \text{ lb/gal})}$

$x = 1262 \text{ mg/L MLVSS}$

5.85 $\dfrac{\left[\begin{array}{c}(2650 \text{ mg/L})(3.60 \text{ MGD}) \\ \times (8.34 \text{ lb/gal})\end{array}\right]}{(0.785)(70 \text{ ft})(70 \text{ ft})} = 20.7 \text{ lb MLSS/day/ft}^2$

5.86 $\dfrac{\left[\begin{array}{c}(2825 \text{ mg/L})(4.25 \text{ MGD}) \\ \times (8.34 \text{ lb/gal})\end{array}\right]}{(0.785)(80 \text{ ft})(80 \text{ ft})} = 19.9 \text{ lb MLSS/day/ft}^2$

5.87 $\dfrac{\left[\begin{array}{c}(x \text{ mg/L})(3.61 \text{ MGD}) \\ \times (8.34 \text{ lb/gal})\end{array}\right]}{(0.785)(60 \text{ ft})(60 \text{ ft})} = 26 \text{ lb MLSS/day/ft}^2$

$x = 2441 \text{ mg/L MLSS}$

5.88 $\dfrac{\left[\begin{array}{c}(2210 \text{ mg/L})(3.3 \text{ MGD}) \\ \times (8.34 \text{ lb/gal})\end{array}\right]}{(0.785)(60 \text{ ft})(60 \text{ ft})} = 21.5 \text{ lb MLSS/day/ft}^2$

5.89 $\dfrac{\left[\begin{array}{c}(x \text{ mg/L})(3.11 \text{ MGD}) \\ \times (8.34 \text{ lb/gal})\end{array}\right]}{(0.785)(60 \text{ ft})(60 \text{ ft})} = 20 \text{ lb MLSS/day/ft}^2$

$x = 2174 \text{ mg/L MLSS}$

5.90 $\dfrac{12,110 \text{ lb VS/day}}{33,100 \text{ ft}^3} = 0.37 \text{ lb VS/day/ft}^3$

5.91 $\dfrac{\left[\begin{array}{c}(124,000 \text{ lb/day}) \\ \times (0.065)(0.70)\end{array}\right]}{(0.785)(60 \text{ ft})(60 \text{ ft})(25 \text{ ft})} = 0.08 \text{ lb VS/day/ft}^3$

5.92 $\dfrac{\left[\begin{array}{c}(141,000 \text{ lb/day}) \\ \times (0.06)(0.71)\end{array}\right]}{(0.785)(50 \text{ ft})(50 \text{ ft})(20 \text{ ft})} = 0.15 \text{ lb VS/day/ft}^3$

5.93 $\dfrac{\left[\begin{array}{c}(21,200 \text{ gpd})(8.34 \text{ lb/gal}) \\ \times (0.055)(0.69)\end{array}\right]}{(0.785)(40 \text{ ft})(40 \text{ ft})(16 \text{ ft})} = 0.33 \text{ VS/day/ft}^3$

5.94 $\dfrac{\left[\dfrac{(22{,}000 \text{ gpd})(8.6 \text{ lb/gal})}{(0.052)(0.70)}\right]}{(0.785)(50 \text{ ft})(50 \text{ ft})(20 \text{ ft})} = 0.18 \text{ lb VS/day/ft}^3$

5.95 $\dfrac{2050 \text{ lb VS added per day}}{32{,}400 \text{ lb VS}} = 0.06$

5.96 $\dfrac{620 \text{ lb VS added per day}}{(174{,}600 \text{ lb})(0.061)(0.65)} = 0.09$

5.97 $\dfrac{(63{,}200 \text{ lb/day})(0.055)(0.73)}{(115{,}000 \text{ gal})(8.34 \text{ lb/gal})(0.066)(0.59)} = 0.07$

5.98 $\dfrac{x \text{ lb VS added per day}}{(110{,}000 \text{ gal})(8.34 \text{ lb/gal})(0.059)(0.58)} = 0.08$

$x = 2511 \text{ lb/day VS}$

5.99 $\dfrac{(7900 \text{ gpd})(8.34 \text{ lb/gal})(0.048)(0.73)}{x \text{ lb VS}} = 0.06$

$x = 38{,}477 \text{ lb VS}$

5.100 1733 people/5.3 ac = 327 people/ac

5.101 4112 people/10 ac = 411 people/ac

5.102 $\dfrac{\left[\begin{array}{c}(1765 \text{ mg/L})(0.381 \text{ MGD}) \\ \times (8.34 \text{ lb/gal})\end{array}\right]}{0.2 \text{ lb/day}} = 28{,}040 \text{ people}$

5.103 $\dfrac{6000 \text{ people}}{x \text{ ac}} = 420 \text{ people/ac}, x = 14.3 \text{ ac}$

5.104 $\dfrac{\left[\begin{array}{c}(2210 \text{ mg/L})(0.100 \text{ MGD}) \\ \times (8.34 \text{ lb/gal})\end{array}\right]}{0.2 \text{ lb/day}} = 9216 \text{ people}$

5.105 $\dfrac{2{,}250{,}000 \text{ gpd}}{(0.785)(80 \text{ ft})(80 \text{ ft})} = 448 \text{ gpd/ft}^2$

5.106 $\dfrac{2960 \text{ gpm}}{190 \text{ ft}^2} = 15.6 \text{ gpm/ft}^2$

5.107 $\dfrac{2{,}100{,}000 \text{ gpd}}{(3.14)(80 \text{ ft})} = 8360 \text{ gpd/ft}$

5.108 $\dfrac{3{,}300{,}000 \text{ gpd}}{(0.785)(90 \text{ ft})(90 \text{ ft})} = 519 \text{ gpd/ft}^2$

5.109 $\dfrac{(161 \text{ mg/L})(2.1 \text{ MGD})(8.34 \text{ lb/gal})}{x \text{ lb MLVSS}} = 0.7$

$x = 4028 \text{ lb MLVSS}$

5.110 $\dfrac{500 \text{ lb/day VS added per day}}{(182{,}000 \text{ lb})(0.064)(0.67)} = 0.06$

5.111 $\dfrac{\left[\begin{array}{c}(2760 \text{ mg/L})(3.58 \text{ MGD}) \\ \times (8.34 \text{ lb/gal})\end{array}\right]}{(0.785)(80 \text{ ft})(80 \text{ ft})} = 16 \text{ lb/day/ft}^2$

5.112 $\dfrac{(115{,}000 \text{ lb/day})(0.071)(0.70)}{(0.785)(70 \text{ ft})(70 \text{ ft})(21 \text{ ft})} = 0.09$

5.113 $\dfrac{4.15 \text{ ac-ft/day}}{25 \text{ ac}} = 0.17 \text{ ft/day}$

$= (0.17 \text{ ft/day})(12 \text{ in./ft}) = 2.0 \text{ in./day}$

5.114 $\dfrac{(174 \text{ mg/L})(3.335 \text{ MGD})(8.3 \text{ lb/gal})}{(x \text{ mg/L})(0.287 \text{ MG})(8.34 \text{ lb/gal})} = 0.5$

$x = 4033 \text{ mg/L MLVSS}$

5.115 $\dfrac{2{,}000{,}000 \text{ gpd}}{(80 \text{ ft})(25 \text{ ft})} = 1000 \text{ gpd/ft}^2$

5.116 $\dfrac{1{,}785{,}000 \text{ gal}}{(25 \text{ ft})(20 \text{ ft})} = 3570 \text{ gal/ft}^2$

5.117 $\dfrac{(150 \text{ mg/L})(2.69 \text{ MGD})(8.34 \text{ lb/gal})}{(1920 \text{ mg/L})(0.31 \text{ MG})(8.34 \text{ lb/gal})} = 0.68$

5.118 $\dfrac{x \text{ lb VS added/day}}{(24{,}500 \text{ gal})(8.34 \text{ lb/gal})(0.055)(0.56)} = 0.09$

$x = 566 \text{ lb/day}$

5.119 $\dfrac{3083 \text{ gpm}}{(40 \text{ ft})(30 \text{ ft})} = 2.6 \text{ gpm/ft}^2$

5.120 $\dfrac{(115 \text{ mg/L})(3.3 \text{ MGD})(8.34 \text{ lb/gal})}{20.1 \ 1000 \text{ ft}^3}$

$= 157 \text{ lb BOD/day/1000 ft}^3$

5.121 $\dfrac{2{,}560{,}000 \text{ gpd}}{(3.14)(80 \text{ ft})} = 10{,}191 \text{ gpd/ft}$

5.122 1900 people/5.5 ac = 345 people/ac

5.123 $\dfrac{(140 \text{ mg/L})(2.44 \text{ MGD})(8.34 \text{ lb/gal})}{750 \ 1000 \text{ ft}^2}$

$= 3.8 \text{ lb BOD/day/1000 ft}^2$

5.124 $\dfrac{2882 \text{ gpm}}{(40 \text{ ft})(30 \text{ ft})} = 2.4 \text{ gpm/ft}^2$

5.125 $\dfrac{(30 \text{ ft})(16 \text{ ft})(8 \text{ ft})(7.48 \text{ gal/ft}^3)}{1007 \text{ gpm}} = 29 \text{ min}$

5.126 $\dfrac{(80 \text{ ft})(20 \text{ ft})(12 \text{ ft})(7.48 \text{ gal/ft}^3)}{75{,}000 \text{ gph}} = 1.9 \text{ hr}$

5.127 $\dfrac{(3 \text{ ft})(4 \text{ ft})(3 \text{ ft})(7.48 \text{ gal/ft}^3)}{(6 \text{ gpm})(60 \text{ min/hr})} = 0.75 \text{ hr}$

5.128 $\dfrac{(0.785)(80 \text{ ft})(80 \text{ ft})(10 \text{ ft})(7.48 \text{ gal/ft}^3)}{216{,}667 \text{ gpd}} = 1.7 \text{ hr}$

5.129 $\dfrac{(500 \text{ ft})(600 \text{ ft})(6 \text{ ft})(7.48 \text{ gal/ft}^3)}{222{,}500 \text{ gpd}} = 60.5 \text{ days}$

5.130 $\dfrac{12{,}300 \text{ lb MLSS}}{2750 \text{ lb/day}} = 4.5 \text{ days}$

5.131
$$\dfrac{\begin{bmatrix}(2820 \text{ mg/L MLSS})(0.49 \text{ MG}) \\ \times (8.34 \text{ lb/gal})\end{bmatrix}}{(132 \text{ mg/L})(0.988 \text{ MGD})(8.34 \text{ lb/gal})} = 10.6 \text{ days}$$

5.132
$$\dfrac{\begin{bmatrix}(2850 \text{ mg/L MLSS})(0.20 \text{ MG}) \\ \times (8.34 \text{ lb/gal})\end{bmatrix}}{(84 \text{ mg/L})(1.52 \text{ MGD})(8.34 \text{ lb/gal})} = 4.5 \text{ days}$$

5.133
$$\dfrac{\begin{bmatrix}(x \text{ mg/L MLSS})(0.205 \text{ MG}) \\ \times (8.34 \text{ lb/gal})\end{bmatrix}}{(80 \text{ mg/L})(2.10 \text{ MGD})(8.34 \text{ lb/gal})} = 6 \text{ days}$$

$x = 4917 \text{ mg/L MLSS}$

5.134 $\dfrac{x \text{ lb MLSS}}{1610 \text{ lb/day SS}} = 5.5 \text{ days}, \ x = 8855 \text{ lb MLSS}$

5.135
$$\dfrac{\begin{bmatrix}(3300 \text{ mg/L})(0.50 \text{ MG}) \\ \times (8.34 \text{ lb/gal})\end{bmatrix}}{\begin{bmatrix}(1610 \text{ lb/day wasted}) \\ + (340 \text{ lb/day in SE})\end{bmatrix}} = 7.1 \text{ days}$$

5.136
$$\dfrac{(2750 \text{ mg/L MLSS})(0.360 \text{ MG})(8.35 \text{ lb/gal})}{\begin{bmatrix}(5410 \text{ mg/L})(0.0192 \text{ MG})(8.34 \text{ lb/gal}) \\ + (16 \text{ mg/L SS})(2.35 \text{ MGD})(8.34 \text{ lb/gal})\end{bmatrix}}$$

$$\dfrac{8257 \text{ lb}}{866 \text{ lb/day} + 314 \text{ lb/day}} = 7.0 \text{ days}$$

5.137
$$\dfrac{(2550 \text{ mg/L MLSS})(1.8 \text{ MG})(8.34 \text{ lb/gal})}{\begin{bmatrix}(6240 \text{ mg/L SS})(0.085 \text{ MGD})(8.34 \text{ lb/gal}) \\ + (20 \text{ mg/L})(2.8 \text{ MGD})(8.34 \text{ lb/gal})\end{bmatrix}}$$

$$\dfrac{38,281 \text{ lb MLSS}}{4424 \text{ lb/day} + 467 \text{ lb/day}} = 8 \text{ days}$$

5.138
$$\dfrac{(x \text{ mg/L})(0.970 \text{ MG})(8.34 \text{ lb/gal})}{\begin{bmatrix}(6340 \text{ mg/L})(0.032 \text{ MGD})(8.34 \text{ lb/gal}) \\ + (20 \text{ mg/L})(2.6 \text{ MGD})(8.34 \text{ lb/gal})\end{bmatrix}}$$

$$\dfrac{(x \text{ mg/L})(0.970 \text{ MG})(8.34 \text{ lb/gal})}{1692 \text{ lb/day} + 434 \text{ lb/day}} = 8 \text{ days}$$

$$\dfrac{(x \text{ mg/L})(0.970 \text{ MG})(8.34 \text{ lb/gal})}{2126} = 8 \text{ days}$$

$x = 2100 \text{ mg/L MLSS}$

5.139 $\dfrac{(75 \text{ ft})(30 \text{ ft})(14 \text{ ft})(7.48 \text{ gal/ft}^3)}{68,333 \text{ gph}} = 3.5 \text{ hr}$

5.140 $\dfrac{12,600 \text{ lb MLSS}}{2820 \text{ lb/day}} = 4.5 \text{ days}$

5.141
$$\dfrac{\begin{bmatrix}(3120 \text{ mg/L MLSS})(0.48 \text{ MG}) \\ \times (8.34 \text{ lb/gal})\end{bmatrix}}{1640 \text{ lb/day wasted} + 320 \text{ lb/day}} = 6.4 \text{ days}$$

5.142 $\dfrac{(40 \text{ ft})(20 \text{ ft})(10 \text{ ft})(7.48 \text{ gal/ft}^3)}{1264 \text{ gpm}} = 47 \text{ min}$

5.143
$$\dfrac{(2810 \text{ Mg/L MLSS})(0.325 \text{ MG})(8.34 \text{ lb/gal})}{\begin{bmatrix}(6100 \text{ mg/L})(0.0189 \text{ MGD})(8.34 \text{ lb/gal}) \\ + (18 \text{ mg/L})(2.4 \text{ MGD})(8.34 \text{ lb/gal})\end{bmatrix}}$$

$$\dfrac{7617 \text{ lb MLSS}}{962 \text{ lb/day} + 360 \text{ lb/day}} = 5.8 \text{ days}$$

5.144 $\dfrac{(3250 \text{ mg/L})(0.33 \text{ MG})(8.34 \text{ lb/gal})}{(100 \text{ mg/L})(2.35 \text{ MGD})(8.34 \text{ lb/gal})} = 4.6 \text{ days}$

5.145
$$\dfrac{(2408 \text{ mg/L})(1.9 \text{ MG})(8.34 \text{ lb/gal})}{\begin{bmatrix}(6320 \text{ mg/L})(0.0712 \text{ MGD})(8.34 \text{ lb/gal}) \\ + (25 \text{ mg/L})(2.85 \text{ MGD})(8.34 \text{ lb/gal})\end{bmatrix}}$$

$$\dfrac{38,157 \text{ lb}}{(3753 \text{ lb/day}) + (594 \text{ lb/day})} = 9.8 \text{ days}$$

5.146 $\dfrac{(2610 \text{ mg/L})(0.15 \text{ MG})(8.34 \text{ lb/gal})}{(140 \text{ mg/L})(0.92 \text{ MGD})(8.34 \text{ lb/gal})} = 3 \text{ days}$

5.147 $\dfrac{(0.785)(6 \text{ ft})(6 \text{ ft})(4 \text{ ft})(7.48 \text{ gal/ft}^3)}{12 \text{ gpm}} = 70 \text{ min}$

5.148 $\dfrac{x \text{ lb MLSS}}{(140 \text{ mg/L})(2.14 \text{ MGD})(8.34 \text{ lb/gal})} = 6 \text{ days}$

$x = 14,992 \text{ lb MLSS}$

5.149 $\dfrac{(400 \text{ ft})(440 \text{ ft})(6 \text{ ft})(7.48 \text{ gal/ft}^3)}{200,000 \text{ gpd}} = 39.5$

5.150
$$\dfrac{(x \text{ mg/L MLSS})(0.64 \text{ MG})(8.34 \text{ lb/gal})}{\begin{bmatrix}(6310 \text{ mg/L})(0.034 \text{ MGD})(8.34 \text{ lb/gal}) \\ + (12 \text{ mg/L})(2.92 \text{ MGD})(8.34 \text{ lb/gal})\end{bmatrix}} = 8 \text{ days}$$

$$\dfrac{(x \text{ mg/L})(0.64 \text{ MG})(8.34 \text{ lb/gal})}{(1789 \text{ lb/day}) + (292 \text{ lb/day})} = 8 \text{ days}$$

$$\dfrac{(x \text{ mg/L})(0.64 \text{ MG})(8.34 \text{ lb/gal})}{2081 \text{ lb/day}} = 8 \text{ days}$$

$x = 3141 \text{ mg/L MLSS}$

5.151 $\dfrac{89 \text{ mg/L removed}}{110 \text{ mg/L}} \times 100\% = 81\%$

5.152 $\dfrac{216 \text{ mg/L removed}}{230 \text{ mg/L}} \times 100 = 94\%$

5.153 $\dfrac{200 \text{ mg/L removed}}{260 \text{ mg/L}} \times 100 = 77\%$

5.154 $\dfrac{175 \text{ mg/L removed}}{310 \text{ mg/L}} \times 100 = 56\%$

5.155 $4.9 = \dfrac{x \text{ lb/day solids}}{(3700 \text{ gal})(8.34 \text{ lb/gal})} \times 100$

$x = 1512 \text{ lb/day solids}$

5.156 $\dfrac{0.87 \text{ g sludge}}{12.87 \text{ g sludge}} \times 100 = 6.8\%$

5.157 $\dfrac{1450 \text{ lb/day solids}}{x \text{ lb/day sludge}} \times 100 = 3.3\%,\ x = 43{,}939 \text{ lb/day}$

5.158 $4.4 = \dfrac{258 \text{ lb/day}}{(x \text{ gpd})(8.34 \text{ lb/gal})} \times 100,\ x = 703 \text{ gpd}$

5.159 $3.6 = \dfrac{x \text{ lb/day solids}}{291{,}000 \text{ lb/day sludge}} \times 100$

$x = 10{,}476 \text{ lb/day solids}$

5.160 $\dfrac{\left[\dfrac{(3100 \text{ gpd})(8.34 \text{ lb/gal})(4.4)}{100}\right] + \left[\dfrac{(4100 \text{ gpd})(8.34 \text{ lb/gal})(3.6)}{100}\right]}{\left[\begin{array}{l}(3100 \text{ gpd})(8.34 \text{ lb/gal}) \\ + (4100 \text{ gpd})(8.34 \text{ lb/gal})\end{array}\right]} \times 100$

$= \dfrac{1138 \text{ lb/day} + 1231 \text{ lb/day}}{25{,}854 \text{ lb/day} + 34{,}194 \text{ lb/day}} \times 100$

$= \dfrac{2369 \text{ lb/day}}{60{,}048 \text{ lb/day}} \times 100 = 3.9\%$

5.161 $\dfrac{\left[\dfrac{(8100 \text{ gpd})(8.34 \text{ lb/gal})(5.1)}{100}\right] + \left[\dfrac{(7000 \text{ gpd})(8.34 \text{ lb/gal})(4.1)}{100}\right]}{\left[\begin{array}{l}(8100 \text{ gpd})(8.34 \text{ lb/gal}) \\ + (7000 \text{ gpd})(8.34 \text{ lb/gal})\end{array}\right]} \times 100$

$= \dfrac{3445 \text{ lb/day solids} + 2394 \text{ lb/day solids}}{\left(\begin{array}{l}67{,}554 \text{ lb/day sludge} \\ + 58{,}380 \text{ lb/day sludge}\end{array}\right)} \times 100$

$= \dfrac{5839 \text{ lb/day solids}}{125{,}934 \text{ lb/day sludge}} \times 100 = 4.6\%$

5.162 $\dfrac{\left[\dfrac{(4750 \text{ gpd})(8.34 \text{ lb/gal})(4.7)}{100}\right] + \left[\dfrac{(5250 \text{ gpd})(8.34 \text{ lb/gal})(3.5)}{100}\right]}{\left[\begin{array}{l}(4750 \text{ gpd})(8.34 \text{ lb/gal}) \\ + (5250 \text{ gpd})(8.34 \text{ lb/gal})\end{array}\right]} \times 100$

$= \dfrac{1862 \text{ lb/day} + 1532 \text{ lb/day}}{39{,}615 + 43{,}785} \times 100$

$= \dfrac{3394 \text{ lb/day solids}}{83{,}400 \text{ lb/day sludge}} \times 100 = 4.1\%$

5.163 $\dfrac{\left[\dfrac{(8925 \text{ gpd})(8.34 \text{ lb/gal})(4.0)}{100}\right] + \left[\dfrac{(11{,}340 \text{ gpd})(8.34 \text{ lb/gal})(6.6)}{100}\right]}{\left[\begin{array}{l}(8925 \text{ gpd})(8.34 \text{ lb/gal}) \\ + (11{,}340 \text{ gpd})(8.34 \text{ lb/gal})\end{array}\right]} \times 100$

$= \dfrac{2977 \text{ lb/day} + 6242 \text{ lb/day}}{74{,}435 \text{ lb/day} + 94{,}576 \text{ lb/day}} \times 100$

$= \dfrac{9219 \text{ lb/day}}{169{,}011 \text{ lb/day}} \times 100 = 5.5\%$

5.164 $(3250 \text{ lb/day solids})(0.65) = 2113 \text{ lb/day VS}$

5.165 $(4120 \text{ gpd})(8.34 \text{ lb/gal})(0.07)(0.70) = 1684 \text{ lb/day VS}$

5.166 $98 \text{ ft} - 91 \text{ ft} = 7 \text{ ft drawdown}$

5.167 $125 \text{ ft} - 110 \text{ ft} = 15 \text{ ft drawdown}$

5.168 $161 \text{ ft} - 144 \text{ ft} = 17 \text{ ft drawdown}$

5.169 $(3.7 \text{ psi})(2.31 \text{ ft/psi}) = 8.5 \text{ ft sounding line water depth}$

$112 \text{ ft} - 8.5 \text{ ft} = 103.5 \text{ ft}$

$103.5 \text{ ft} - 86 \text{ ft} = 17.5 \text{ ft}$

5.170 $(4.6 \text{ psi})(2.31 \text{ ft/psi}) = 10.6 \text{ ft sounding line water depth}$

$150 \text{ ft} - 10.6 \text{ ft} = 139.4 \text{ ft}$

$171 \text{ ft} - 139.4 \text{ ft} = 31.4 \text{ ft drawdown}$

5.171 $300 \div 20 = 15 \text{ gpm per ft of drawdown}$

5.172 $420 \text{ gal} \div 5 \text{ min.} = 84 \text{ gpm}$

5.173 $810 \text{ gal} \div 5 \text{ min.} = 162 \text{ gpm}$

5.174 $\dfrac{856 \text{ gal}}{5 \text{ min}} = 171 \text{ gpm}$

$(171 \text{ gpm})(60 \text{ min/hr}) = 10{,}260 \text{ gph}$

5.175 $\dfrac{\left[\begin{array}{l}(0.785)(1 \text{ ft})(1 \text{ ft})(12 \text{ ft}) \\ \times (7.48 \text{ gal/ft}^3)(12 \text{ round trips})\end{array}\right]}{5 \text{ min}} = 169 \text{ gpm}$

5.176 $750 \text{ gal} \div 5 \text{ min.} = 150 \text{ gpm}$

$(150 \text{ gpm})(60 \text{ min/hr}) = 9000 \text{ gph}$

$(9000 \text{ gph})(10 \text{ hr/day}) = 90{,}000 \text{ gal/day}$

5.177 $200 \text{ gpm} \div 28 \text{ ft} = 7.1 \text{ gpm/ft}$

5.178 $620 \text{ gpm} \div 21 \text{ ft} = 29.5 \text{ gpm/ft}$

5.179 $1100 \text{ gpm} \div 41.3 \text{ ft} = 26.6 \text{ gpm/ft}$

5.180 $\dfrac{x \text{ gpm}}{42.8 \text{ ft}} = 33.4 \text{ fpm/ft}$

$x = (33.4)(42.8) = 1430 \text{ gpm}$

5.181 $\left[\begin{array}{l}(0.785)(0.5 \text{ ft})(0.5 \text{ ft}) \\ \times (140 \text{ ft})(7.48 \text{ gal/ft}^3)\end{array}\right] = 206 \text{ gal}$

$\left[\begin{array}{l}(40 \text{ mg/L})(0.000206 \text{ MG}) \\ \times (8.34 \text{ lb/gal})\end{array}\right] = 0.07 \text{ lb chlorine}$

5.182 $\begin{bmatrix} (0.785)(1\ \text{ft})(1\ \text{ft}) \\ \times (109\ \text{ft})(7.48\ \text{gal/ft}^3) \end{bmatrix} = 640\ \text{gal}$

$\begin{bmatrix} (40\ \text{mg/L})(0.000640\ \text{MG}) \\ \times (8.34\ \text{lb/gal}) \end{bmatrix} = 0.21\ \text{lb chlorine}$

5.183 $\begin{bmatrix} (0.785)(1\ \text{ft})(1\ \text{ft}) \\ \times (109\ \text{ft})(7.48\ \text{gal/ft}^3) \end{bmatrix} = 633\ \text{gal}$

$\begin{bmatrix} (0.785)(0.67\ \text{ft})(0.67\ \text{ft}) \\ \times (40\ \text{ft})(7.48\ \text{gal/ft}^3) \end{bmatrix} = 105\ \text{gal}$

$633 + 105\ \text{gal} = 738\ \text{gal}$

$\begin{bmatrix} (110\ \text{mg/L})(0.000738\ \text{gal}) \\ \times (8.34\ \text{lb/gal}) \end{bmatrix} = 0.68\ \text{lb chlorine}$

5.184 $(x\ \text{mg/L})(0.000540\ \text{gal})(8.34\ \text{lb/gal}) = 0.48\ \text{lb}$

$x = \dfrac{0.48}{(0.000540)(8.34)}$

$x = 107\ \text{mg/L}$

5.185 $\dfrac{0.09\ \text{lb chlorine}}{5.25/100} = 1.5\ \text{lb}$

$\dfrac{1.5\ \text{lb}}{8.34\ \text{lb/gal}} = 0.18\ \text{gal}$

$(0.18\ \text{gal})(128\ \text{fluid oz./gal}) = 23\ \text{fl oz.}$

5.186 $\begin{bmatrix} (0.785)(0.5\ \text{ft})(0.5\ \text{ft}) \\ \times (120\ \text{ft})(7.48\ \text{gal/ft}^3) \end{bmatrix} = 176\ \text{gal}$

$\dfrac{\begin{bmatrix} (50\ \text{mg/L})(0.000176\ \text{MG}) \\ \times (8.34\ \text{lb/gal}) \end{bmatrix}}{65/100} = 0.1\ \text{lb calcium hypochlorite}$

$(0.1\ \text{lb})(16\ \text{oz./1 lb}) = 1.6\ \text{oz. calcium hypochlorite}$

5.187 $\begin{bmatrix} (0.785)(1.5\ \text{ft})(1.5\ \text{ft}) \\ \times (105\ \text{ft})(748\ \text{gal/ft}^3) \end{bmatrix} = 1387\ \text{gal}$

$\dfrac{\begin{bmatrix} (100\ \text{mg/L})(0.001387\ \text{MG}) \\ \times (8.34\ \text{lb/gal}) \end{bmatrix}}{25/100} = 4.6\ \text{lb chloride of lime}$

5.188 $\dfrac{\begin{bmatrix} (60\ \text{mg/L})(0.000240\ \text{MG}) \\ \times (8.34\ \text{lb/gal}) \end{bmatrix}}{5.25/100} = 2.3\ \text{lb}$

$\dfrac{2.3\ \text{lb}}{8.34\ \text{lb/gal}} = 0.3\ \text{gal}$

$(0.3\ \text{gal})(128\ \text{fl oz./gal}) = 38.4\ \text{fl oz. sodium hypochlorite}$

5.189 $(4.0\ \text{psi})\ (2.31\ \text{ft/psi}) = 9.2\ \text{ft}$

5.190 $(94\ \text{ft} + 24\ \text{ft}) + (3.6\ \text{psi})(2.31\ \text{ft/psi})$
$= 118\ \text{ft} + 8.3\ \text{ft} = 126.3\ \text{ft}$

5.191 $(400\ \text{ft})(110\ \text{ft})(14\ \text{ft})(7.48\ \text{gal/ft}^3) = 4{,}607{,}680\ \text{gal}$

5.192 $(400\ \text{ft})(110\ \text{ft})(30\ \text{ft} \times 0.4\ \text{average depth})$
$\times (7.48\ \text{gal/ft}^3)$

$= 3{,}949{,}440\ \text{gal}$

5.193 $\dfrac{(200\ \text{ft})(80\ \text{ft})(12\ \text{ft})}{43{,}560\ \text{ft}^3/\text{ac-ft}} = 4.4\ \text{ac-ft}$

5.194 $\dfrac{(320\ \text{ft})(170\ \text{ft})(16\ \text{ft})(0.4)}{43{,}560\ \text{ft}^3/\text{ac-ft}} = 8.0\ \text{ac-ft}$

5.195 $\dfrac{(0.5\ \text{mg/L chlorine})(20\ \text{MG})(8.34\ \text{lb/gal})}{25/100}$

$= 334\ \text{lb copper sulfate}$

5.196 $131.9\ \text{ft} - 93.5\ \text{ft} = 38.4\ \text{ft}$

5.197 $\dfrac{707\ \text{gal}}{5\ \text{min}} = 141\ \text{gpm}$

$(141\ \text{gpm})(60\ \text{min/hr}) = 8460\ \text{gph}$

5.198 $\dfrac{\begin{bmatrix} (0.785)(1\ \text{ft})(1\ \text{ft})(12\ \text{ft}) \\ \times (7.48\ \text{gal/ft}^3)(8\ \text{round trips}) \end{bmatrix}}{5\ \text{gpm}} = 113\ \text{gpm}$

5.199 $(3.5\ \text{psi})(2.31\ \text{ft/psi}) = 8.1\ \text{sounding line water depth}$

$= 167\ \text{ft} - 8.1\ \text{ft}$

$= 158.9\ \text{ft pumping water level}$

$\text{Drawdown (ft)} = 158.9\ \text{ft} - 141\ \text{ft} = 17.9\ \text{ft}$

5.200 $\dfrac{610\ \text{gpm}}{28\ \text{ft drawdown}} = 21.8\ \text{gpm/ft}$

5.201 $\begin{bmatrix} (0.785)(0.5\ \text{ft})(0.5\ \text{ft}) \\ \times (150\ \text{ft})(7.48\ \text{gal/ft}^3) \end{bmatrix} = 220\ \text{gal}$

$\begin{bmatrix} (55\ \text{mg/L})(0.000220\ \text{MG}) \\ \times (8.34\ \text{lb/gal}) \end{bmatrix} = 0.10\ \text{lb chlorine req.}$

5.202 $780\ \text{gal} \div 5\ \text{min} = 156\ \text{gpm}$

$\begin{bmatrix} (156\ \text{gal/min})(60\ \text{min/hr}) \\ \times (8\ \text{hr/day}) \end{bmatrix} = 74{,}880\ \text{gal/day}$

5.203 $\begin{bmatrix} (x\ \text{mg/L})(0.000610\ \text{MG}) \\ \times (8.34\ \text{lb/gal}) \end{bmatrix} = 0.47\ \text{lb}$

$x = \dfrac{0.47}{(0.000610)(8.34)} = 92.3\ \text{mg/L}$

5.204 $\left[\begin{array}{c} (0.785)(1\ \text{ft})(1\ \text{ft}) \\ \times (89)(7.48\ \text{gal/ft}^3) \end{array}\right] = 523\ \text{gal}$

$\left[\begin{array}{c} (0.785)(0.67)(0.67) \\ \times (45\ \text{ft})(7.48\ \text{gal/ft}^3) \end{array}\right] = 119\ \text{gal}$

$523\ \text{gal} + 119\ \text{gal} = 642\ \text{gal}$

$\left[\begin{array}{c} (100\ \text{mg/L})(0.000642\ \text{MG}) \\ \times (8.34\ \text{lb/gal}) \end{array}\right] = 0.54\ \text{lb chlorine}$

5.205 $\dfrac{0.3\ \text{lb chlorine}}{5.25/100} = 5.7\ \text{lb}$

$\dfrac{5.7\ \text{lb}}{8.34\ \text{lb/gal}} = 0.68\ \text{gal}$

$(0.68\ \text{gal})(128\ \text{fl oz./gal}) = 87\ \text{fl oz.}$

5.206 Volume = (4 ft)(5 ft)(3 ft)(7.48 gal/ft³) = 449 gal

5.207 Volume = (50 ft)(20 ft)(8 ft)(7.48 gal/ft³) = 59,840 gal

5.208 Volume = (40 ft)(16 ft)(8 ft)(7.48 gal/ft³) = 38,298 gal

5.209 42 in. ÷ 12 in./ft = 3.5 ft
Volume = (5 ft)(5 ft)(3.5 ft)(7.48 gal/ft³) = 655 gal

5.210 2 in. ÷ 12 in./ft = 0.17 ft
Volume = (40 ft)(25 ft)(9.17 ft)(7.48 gal/ft³) = 68,592 gal

CHAPTER 6 ANSWERS

6.1 Screwed joint
6.2 Welded joint
6.3 Soldered joint
6.4 Screwed glove valve
6.5 Globe valve
6.6 Screwed check valve
6.7 Pump
6.8 Flexible line
6.9 Check valve
6.10 Heat exchanger
6.11 Expansion joint
6.12 Vibration absorber
6.13 Battery cell
6.14 Motor
6.15 Relay
6.16 Voltmeter
6.17 Ammeter
6.18 Knife switch
6.19 Fuse
6.20 Transformer
6.21 Ground
6.22 Normally open contacts
6.23 Local note reference
6.24 Fillet weld
6.25 Square butt weld
6.26 Single hem
6.27 Single flange
6.28 Location of weld
6.29 Steel section
6.30 Bevel weld
6.31 V weld
6.32 J-groove weld
6.33 4 times true size
6.34 Drawing number
6.35 180
6.36 Break line
6.37 Object line
6.38 3-D pictorial
6.39 Shape; complexity
6.40 Limits
6.41 Center line; finished surface
6.42 Volute
6.43 Hem
6.44 Seam
6.45 Relief valve
6.46 Gas; liquid
6.47 Butt
6.48 Architectural
6.49 Plot plan
6.50 Line

CHAPTER 7 ANSWERS

7.1 26 ft
7.2 77 ft
7.3 Eccentric, segmental
7.4 Flow nozzle
7.5 Ultrasonic flowmeter
7.6 4937 gal
7.7 4.57
7.8 213,904 ft³
7.9 103 ft
7.10 8064 lb
7.11 Always constant
7.12 Pressure due to the depth of water
7.13 The line that connects the piezometric surface along a pipeline
7.14 0.28 ft
7.15 254.1 ft
7.16 6.2×10^{-8}
7.17 0.86 ft
7.18 Pressure energy due to the velocity of the water
7.19 A pumping condition where the size of the impeller of the pump and above the surface of the water from which the pump is running
7.20 The slope of the specific energy line

CHAPTER 8 ANSWERS

8.1   Alternator
8.2   The effect that causes current flow in a conductor moving across magnetic lines of force
8.3   Mechanical, electrical
8.4   Increases, decreases, decreases, increases
8.5   To protect an electrical circuit and load
8.6   0.2 ohms
8.7   Orbits or shells
8.8   Protons and neutrons
8.9   The value of the resistor, the length of the conductors, and the diameter of the conductors
8.10 Direct current flow does not change direction, whereas alternating current periodically changes direction.
8.11 The magnetic poles
8.12 The flux lines, or magnetic flux along which a magnetic force acts
8.13 Natural magnet, permanent magnets, and electromagnets
8.14 Chemistry
8.15 Battery, two
8.16 A series circuit has only one path for current flow, whereas a parallel circuit has more than one path.
8.17 Source voltage
8.18 Voltage drop
8.19 Counterclockwise
8.20 2 amps
8.21 12 volts
8.22 16 watts
8.23 80 watts
8.24 Less, more
8.25 Resistivity
8.26 Circular mil
8.27 Circular mil
8.28 Conductivity
8.29 Smaller
8.30 Doubles
8.31 It will withstand high voltages.
8.32 The two are directly proportional. As flux density increases, field strength also increases.
8.33 The type of material and the flux density
8.34 North pole
8.35 Increases
8.36 141.4 volts
8.37 A voltage is induced in the conductor
8.38 AC, cut, counter
8.39 Counter
8.40 Current has an associated magnetic field
8.41 Increase
8.42 Increase

CHAPTER 9 ANSWERS

9.1   Positive-displacement
9.2   High-viscosity
9.3   Positive-displacement
9.4   High
9.5   High
9.6   Eye
9.7   Static, dynamic
9.8   Shut off
9.9 $V^2/2g$
9.10 Total head
9.11 Head capacity, efficiency, horsepower demand
9.12 Water
9.13 Suction lift
9.14 Elevation head
9.15 Water hp and pump efficiency
9.16 Centrifugal force
9.17 Stuffing box
9.18 Impeller
9.19 Rings, impeller
9.20 Casing

CHAPTER 10 ANSWERS

10.1   A flexible piping component that absorbs thermal and/or terminal movement
10.2   Fluid
10.3   Fluid
10.4   Connected
10.5   Flow
10.6   Pressure loss
10.7   Increases
10.8   Automatically
10.9   Insulation
10.10 Leakage
10.11 Four times
10.12 Routine preventive maintenance
10.13 12
10.14 Schedule, thickness
10.15 Increases
10.16 Ferrous
10.17 Increases
10.18 Iron oxide
10.19 Cast iron
10.20 Iron
10.21 Corrosion
10.22 Decreases
10.23 Clay, concrete, plastic, glass, or wood
10.24 Corrosion-proof
10.25 Cement
10.26 Pressed
10.27 Turbulent, lower
10.28 Steel

10.29 Fusion
10.30 Flexible
10.31 Aluminum
10.32 Annealed
10.33 Fusion
10.34 Metals, plastics
10.35 Laminar flow
10.36 Reinforced nonmetallic
10.37 Wire-reinforced
10.38 Dacron®
10.39 Diameter
10.40 Flexibility
10.41 E.E.
10.42 Reinforced, pressure
10.43 Flexible
10.44 Expansion joint
10.45 Vibration dampener
10.46 Plain
10.47 Bends
10.48 Pressure
10.49 Plug
10.50 A long-radius elbow

CHAPTER 11 ANSWERS

11.1   Na
11.2 H_2SO_4
11.3   7
11.4   Base
11.5   Changes
11.6   Solid, liquid, gas
11.7   Element
11.8   Compound
11.9   Periodic
11.10 Solvent, solute
11.11 An atom or group of atoms that carries a positive or negative electrical charge as a result of having lost or gained one or more electrons
11.12 Colloid
11.13 Turbidity
11.14 Result of dissolved chemicals
11.15 Toxicity
11.16 Organic
11.17 0; 14
11.18 Ability of water to neutralize an acid
11.19 Calcium and magnesium
11.20 Base

CHAPTER 12 ANSWERS

12.1   Bacteria, viruses, protozoa
12.2   During rain storms
12.3   No
12.4   Binary fission
12.5   Spheres, rods, spirals
12.6   Typhoid, cholera, gastroenteritis
12.7   Amoebic dysentery, giardiasis
12.8   Cyst
12.9   Host
12.10 Plug screens, machinery; cause taste and odor problems
12.11 No; bacteria is Machiavellian—it is a survivor.

CHAPTER 13 ANSWERS

13.1   Ecosystem
13.2   Benthos
13.3   Periphyton
13.4   Plankton
13.5   Pelagic
13.6   Neuston
13.7   Immigration
13.8   Autotrophs
13.9   Lentic
13.10 Dissolved oxygen solubility

CHAPTER 14 ANSWERS

14.1   Secondary maximum contaminant levels
14.2   Transpiration
14.3   Surface water
14.4   Agriculture, municipal wastewater plants, habitat and hydrologic modifications, resource extraction, and urban runoff and storm sewers
14.5   Solids content
14.6   Turbidity
14.7   Universal solvent
14.8   Alkalinity
14.9   Neutral state
14.10 Lead

CHAPTER 15 ANSWERS

15.1   Muffle furnace, ceramic dishes, furnace tongs, and insulated gloves
15.2   15 minutes
15.3   A sample collected all at one time; representative of the conditions only at the time taken
15.4   For pH, dissolved oxygen, total residual chlorine, fecal coliform, and any test by NPDES permit for grab sample
15.5   A series of samples collected over a specified period of time in proportion to flow
15.6   Collect from well-mixed location; clearly mark sampling points; easy location to read; no large or unusual particles; no deposits, growths, or floating materials; corrosion-resistant containers; follow safety procedures; test samples as soon as possible.

15.7 Refrigerate at 4°C.

15.8 Absorption of water during cooling, contaminants, fingerprints, etc.

CHAPTER 16 ANSWERS

16.1 Cone of depression

16.2 12 in.

16.3 Concrete

16.4 Surface water, groundwater, GUDISW

16.5 Groundwater under the direct influence of surface water

16.6 Easily located; softer than groundwater

16.7 The study of the properties of water and its distribution and behavior

16.8 Zone of influence

16.9 GUDISW

16.10 Prevent large material from entering the intake

CHAPTER 17 ANSWERS

17.1 A potential reserve area, usually distinct from the treatment plant, where natural or artificial lakes are used for water storage, natural sedimentation, and seasonal pretreatment with or without disinfection

17.2 Collection area into which water drains

17.3 Either of two choices in water utility management—keep it out of the watershed or take it out during treatment

17.4 Control algae and in turn decrease taste and odor problems

17.5 Best management practices

17.6 True

17.7 True

17.8 True

17.9 False

17.10 False

CHAPTER 18 ANSWERS

18.1 Calcium (Ca) and magnesium

18.2 Buffer

18.3 Sodium hydroxide

18.4 Chlorine feed rate (lb/day) = Dose (mg/L) × flow (MGD) × 8.34
= 1.2 mg/L × 1.6 MGD × 8.34 = 16.0 lb/day

18.5 2.4 (60 ÷ 25 = 2.4)

18.6 Chlorine dose (mg/L) − chlorine residual (mg/L)
= 1.0 mg/L − 0.5 mg/L = 0.5 mg/L

18.7 Residual = Dose − demand
= 6.0 mg/L − 3.3 mg/L = 2.7 mg/L

18.8 $$Dose\ (mg/L) = \frac{220\ lb/day\ Cl_2}{3.1 \times 8.34} = 8.5\ mg/L$$

Chlorine residual = 8.5 mg/L − 6.9 mg/L = 1.6 mg/L

18.9 A description of the soil encountered during well construction, water quantity, well casing information, and well development and testing

18.10 Dug well

18.11 Disinfection residual, turbidity, coliform analysis

18.12 National Sanitation Foundation (NSF)

18.13 Fit for human consumption

18.14 First, determine the required chlorine feed rate:

Feed rate (lb/day) = Dose (mg/L) × flow (MGD) × 8.34 = 0.6 mg/L × 1 MGD × 8.34 = 5.0 lb/day

If we require 5 lb/day of chlorine, we will require more pounds of hypochlorite because it is not 100% chlorine; 68% of the hypochlorite is available chlorine, and 68% = 68/100 = 0.68. Next:

(Cl_2 fraction)(hypochlorite) = Available chlorine

(0.68)(x lb/day hypochlorite) = 5.00 lb/day Cl_2

x lb/day hypochlorite = 5.00/0.68

x = 7.36 lb/day hypochlorite

18.15 Public

18.16 The transport of water from one location to another

18.17 Acute

18.18 Reduction of pathogens to safe levels

18.19 Hypochlorites

18.20 Reduce the number of pathogens to safe levels in water before the contact time is completed.

18.21 Feed rate (lb/day) = Dose (mg/L) × flow (MGD) × 8.34
= 0.4 mg/L × 5/3 MGD × 8.34 = 17.68 lb/day Cl_2

18.22 Residual = Dose − demand
= 10 (mg/L) − 2.6 (mg/L) = 7.5 mg/L

18.23 Turbidity can entrap or shield microorganisms from the chlorine.

18.24 Feed rate (lb/day) = Dose (mg/L) × flow (MGD) × 8.34
= 0.8 × 2.6 × 8.34 = 17.35 lb/day soda ash

18.25 *Given:*

Flow 0.75 MGD
Shape Circular
Size Radius = 20 ft
Depth 10 ft

Find the detention time.

a. Find tank volume:
Volume = $\pi r^2 h$
Volume = $\pi \times (20\ ft)^2 \times 10\ ft = 12,560\ ft^3$

b. Flow = 0.75 MGD × 1,000,000 = 750,000 gpd

$$Detention\ time\ (hr) = \frac{12,560\ ft^3 \times 7.4824\ hr}{750,000\ gal/day} = 3\ hr$$

18.26 Chemical feed rate = Dose (mg/L) × flow (MGD) × 8.34
= 35 mg/L × 2 MGD × 8.34
= 70 × 8.34 = 584 lb/day of alum

18.27 Yes

18.28 Chlorine residual

18.29 A link that connects two systems and a force that causes liquids in a system to move

18.30 Moderate

18.31 Negative; low

18.32 Peristaltic metering pump

18.33 Purchase of buffer zone around a reservoir; inspection of construction sites; public education

18.34 *Given:*

No. of filters   3
Size (each)   10 ft × 7 ft
Operating   1 out of service
Filtration rate   280 gal/min (total capacity for both filters)

Find the filtration rate square foot of filter:

Area of each filter = 10 ft × 7 ft = 70 ft^2

Total area of filters = 70 ft^2 = 140 ft^2 total

$$\text{Filtration rate} = \frac{280 \text{ gal/min}}{140 \text{ ft}^2} = 2 \text{ gal/min/ft}^2 \text{ of filter}$$

18.35 *Given:*

Filter area   300 ft^2
Backwash rate   15 gal/ft^2/min
Backwash time   8 min

Find the amount of water for backwash.

We have been given information on per foot of filter but we want to find the total water required to backwash the entire filter.

a. Find total filtration rate:

$$300 \text{ ft}^2 \times \frac{15 \text{ gal}}{\text{ft}^2/\text{min}} = 4500 \text{ gal/min}$$

b. Gallons per 8-minute backwash time

$$\frac{4500 \text{ gal}}{\text{min}} \times 8 \text{ min} = 36,000 \text{ gal used}$$

18.36 Velocity = Distance traveled ÷ time
Velocity = 600 ft ÷ 5 min = 120 ft/min

18.37 Material Safety Data Sheets (MSDS)

18.38 Chlorination and filtration

18.39 Pump more than rated capacity

18.40 Hypochlorous acid

18.41 Protozoa

18.42 Removal/inactivation of most resistant pathogens

18.43 Corrosivity

18.44 Turbidity, paddles speed, pH

18.45 We want to find the velocity; therefore, we must rearrange the general formula to solve for velocity:

$$V = Q/A$$

Given:

Q = Rate of flow = 11.2 cfs
A = Area in square feet
Width = 2.5 ft
Depth = 1.4 ft

Find the average velocity.

Step 1.

Area = Width × depth = 2.5 ft × 1.4 ft = 3.5 ft^2

Step 2.

Velocity (ft/sec) = Flow rate (ft^3/sec) ÷ area (ft^2)
Velocity = 11.2 ft^3/sec ÷ 3.5 ft^2 = 32 ft/sec

18.46 *Given:*

Height   100 ft
Diameter   20 ft
Shape   Cylindrical

Find total gallons of water contained in the tank.

a. Find the volume in cubic feet:

Volume = 0.785 × (diameter)^2 × height
= 0.785 × (20 ft)^2 × 100 ft
= 0.785 × 400 ft^2 × 100 ft = 31,400 ft^3

b. Our problem asks how many *gallons* of water will it contain:

31,400 ft^3 × 7.48 gal/ft^3 = 234,872 gal

18.47 Rapid mix, flocculation, sedimentation

18.48 Removal of color, suspended matter, and organics

18.49 Transform soluble ions to insoluble compounds

18.50 3 to 4 times the theoretical amount

18.51 5%

18.52 *Given:*

Distance   1500 ft
Time   4 min

a. Find the velocity in ft/min:

Velocity = 1500 ft ÷ 4 min = 375 ft/min

b. Convert minutes to seconds:

375 ft/min × 1 min/60 sec = 6.25 ft/sec

18.53 Gate

18.54 Achieve optimum corrosion control

18.55 50%

18.56 Sodium fluoride (NaF)

18.57 Mottled teeth enamel

18.58 0.75 mg/L

18.59 Amount of chlorine to add for breakpoint chlorination; correct amount of coagulant to use for proper coagulation; length of flash mix; proper amount of mixing and settling time

18.60 Corrosion control technology

18.61 *Given:*

Flow   350 GPM
Pipe size   6 in.

Find the velocity (ft/sec) = Distance ÷ time

a. Convert gallons to ft^3:

350 gal/min ÷ 7.48 gal/ft^3 = 46.8 ft^3/min

b. Find cross-sectional area of pipe:

Area of circle = πr^2
= 3.14 × (3 in. × 3 in.) = 28.26 in.^{2}

c. Convert square inches to square feet:

28.26 in.^{2} ÷ 144 in.^{2}/ft^2 = 0.20 ft^2

d. Find ft/min:

46.8 ft³/min ÷ 0.20 ft² = 234 ft/min

e. Convert minutes to seconds:

234 ft/min × 1 min/60 sec = 3.9 ft/sec

18.62 Air, chlorine, or potassium permanganate

18.63 pH, alkalinity, hardness

18.64 Adsorption

18.65 Prior to the rapid mix basin

18.66 Before the backwash, water reaches the lip of the wash water trough.

18.67 Chlorine

18.68 True

18.69 True

18.70 79,269 gal

18.71 Powdered activated carbon

18.72 Iron and manganese

18.73 Copper

18.74 Soluble polyvalent cations

18.75 Gains an electron in going from the +2 oxidation state to the +3 form

18.76 Bicarbonate

18.77 Negative head

18.78 Gravity

18.79 Influent

18.80 Uniform

18.81 Maximize the conversion of organic carbon from the dissolved phase to the particulate phase; the removal of natural organic material; optimize the removal of DHP precursor material.

18.82 30 hours

18.83 Phenyl arsine oxide

18.84 *Given:*

Surface area of pond = 20 ac
Height of water collected = 2 in.

Find the number of gallons collected in the reservoir after the storm.

a. Convert acres to ft²:

20 ac × 43,560 ft²/ac = 871,200 ft²

b. Convert inches to feet:

2 in. × 1 ft/12 in. = 0.167 ft

c. Calculate volume of water collected:

Area × height = 871,200 ft² × 0.167 ft = 145,490 ft³

d. Convert ft³ to gallons:

145,590 ft³ × 7.48 gal/ft³ = 1,089,013 gal

18.85 20.0 lb/day Cl₂

18.86 70 ft × 0.4 = 28 ft
28 ft × 0.433 = 12.1 psi

18.87 Groundwater

18.88 Aeration, boiling, adsorption

18.89 Addition of powdered activated carbon

18.90 Permeability

18.91 Water table

18.92 Waterborne

18.93 Coagulant

18.94 Greater

18.95 Copper sulfate

18.96 Lime

18.97 Disinfected

18.98 Oxygen

18.99 Binary fission

CHAPTER 19 ANSWERS

19.1 The licensed operator and the responsible official

19.2 The amount of organic material in a sample that can be oxidized by a strong oxidizing agent

19.3 Prevent disease, protect aquatic organisms, protect water quality

19.4 Dissolved and suspended

19.5 Organic indicates matter that is made up mainly of carbon, hydrogen, and oxygen and will decompose into mainly carbon dioxide and water at 550°C; inorganic materials, such as salt, ferric chloride, iron, sand, gravel, etc.

19.6 Algae, bacteria, protozoa, rotifers, virus

19.7 Carbon dioxide, water, more organics, stable solids

19.8 Toxic matter, inorganic dissolved solids, pathogenic organisms

19.9 Raw effluent

19.10 From body wastes of humans who have disease

19.11 Disease-causing

19.12 Domestic waste

19.13 Industrial waste

19.14 4.4%

19.15 2.3 ft

19.16 5250 gal × 8.34 lb/gal = 43,785 lb

19.17 14,362 gal

19.18 850.7 lb/day

19.19 686 kg/day

19.20 0.121 MGD

19.21 8477 people

19.22 9.41 lb/gal

19.23 Cutter may be sharpened or replaced when needed. Cutter alignment must be adjusted as needed

19.24 Grit is heavy inorganic matter; sand, gravel, metal filings, egg shells, coffee grounds, etc.

19.25 0.7 fps

19.26 A large amount of organic matter is present in the gut. The aeration rate must be increased to prevent settling of the organic solids.

19.27 To remove settleable and flotable solids

19.28 To remove the settleable solids formed by the biological activity

19.29 7962 gpd/ft

19.30 Stabilization pond, oxidation pond, polishing pond

19.31 Settling, anaerobic digestion of settled solids, aerobic/anaerobic decomposition of dissolved and colloidal organic solids by bacteria producing stable solids and carbon dioxide, photosynthesis

19.32 Products of oxygen by algae; summer effluent is high in solids (algae) and low in BOD; winter effluent is low in solids and high in BOD.

19.33 Eliminates wide diurnal and seasonal variation in pond dissolved oxygen

19.34 Standard, high rate, roughing

19.35 Increase waste rate

19.36 Decrease, decrease, decrease, increase, increase

19.37 10 containers

19.38 88 days

19.39 103 cylinders; $2823.49

19.40 4716 lb/day

19.41 21.5 lb

19.42 27 days

19.43 64.1%

19.44 National Pollutant Discharge Elimination System

19.45 By increasing the primary sludge pumping rate or by adding dilution water

19.46 7.0 pH

19.47 Because the microorganisms have been killed or they are absent

19.48 The time to do the test, 3 hours vs. 5 days

19.49 Dark, greasy

19.50 Increases

19.51 Temperature, pH, toxicity, waste rate, aeration tank configuration

19.52 Can function with or without dissolved oxygen; prefer dissolved oxygen but can use chemically combined oxygen such as sulfate or nitrate

19.53 Organic

19.54 Living organisms

19.55 Final

19.56 Colloidal

19.57 Not possible

19.58 Aerobic, facultative

19.59 Different

19.60 Reduced

19.61 Temperature

19.62 BOD

19.63 F/M

19.64 Secondary clarifier weirs

19.65 To separate and return biosolids to the aeration tank

19.66 Declining

19.67 1.5 and 2.5 mg/L

19.68 Increased MLVSS concentration

19.69 Decreased waste rate

19.70 Decreased MCRT

19.71 Concentration of aeration influent solids

19.72 Complete mix is more resistant to shock loads

19.73 Decrease the grit channel aeration rate

19.74 Increase

19.75 Floor level

19.76 $22.77

19.77 Anoxic

19.78 C:N:P

19.79 Secondary

19.80 Are not

19.81 2 ft/sec

19.82 Lower

19.83 Chlorine residual

19.84 2 hr

19.85 0.1

19.86 800 gpd/ft^2

19.87 Monochloramine

19.88 0.2 to 0.5

19.89 Nitrogen

19.90 Decrease explosive hazard, decrease odor release, maintain temperature, collect gas

19.91 Algae

19.92 Dissolved solids

19.93 0.0005 ppm

APPENDIX B

Formulae

AREA

a. Rectangular tank:

$A = L \times W$

b. Circular tank:

$A = \pi r^2$

or

$A = 0.785 \times (\text{diameter})^2$

VOLUME

a. Rectangular tank:

$V = L \times W \times H$

b. Circular tank:

$V = \pi r^2 \times H$

or

$0.785 \times (\text{diameter})^2 \times H$

FLOW

Gal/day (gpd) = gal/min (gpm) $\times$ 1440 min/day

Gal/day (gpd) = gal/hr (gph) $\times$ 24 hr/day

Million gallons/day (MGD) = (gpd)/1,000,000

DOSE

lb = ppm $\times$ MG $\times$ 8.34 lb/gal

ppm = lb/(MG $\times$ 8.34 lb/gal)

EFFICIENCY

$$\text{Efficiency (\% removal)} = \frac{\text{Influent} - \text{effluent}}{\text{Influent}} \times 100$$

WEIR LOADING

$$\text{Weir loading (overflow rate)} = \frac{\text{Total gpd}}{\text{Length of weir}}$$

SURFACE SETTLING RATE

$$\text{Surface settling rate} = \frac{\text{Total gpd}}{\text{Surface area of tank}}$$

DETENTION TIME

$$\text{Detention time (hr)} = \frac{\text{Capacity of tank (gal)} \times 24 \text{ hr/day}}{\text{Flow rate (gpd)}}$$

HORSEPOWER

$$\text{Horsepower (hp)} = \frac{\text{gpm} \times \text{head (ft)}}{3960 \times \text{total efficiency}}$$

Index

A

abiotic materials, 474, 475, 476
aboveground outdoor equipment enclosures, 34
absolute pressure, 379, 416
absorption, 30, 508, 542, 650, 705
acceleration due to gravity, 276, 281
access systems, security, 40
accumulators, 263, 411, 415, 419
acid rain, 69, 440, 514
acids, 433, 511, 515
   ascorbic, 553, 554
   carbonic, 433, 481, 611
   dimethyl arsenic, 653
   ethylenediaminetetraacetic, 550
   fatty, 514, 745
   fluorosilicic, 131–134, 136
   haloacetic, 639
   humic, 476
   hydrochloric, 433, 569, 622, 642, 646, 735
   hydrofluoric, 132
   hydrofluorosilicic, 132, 133
   hypochlorous, 569, 642, 647, 649
   monomethyl arsenic, 653
   nitric, 433, 646
   organic, 434, 515, 635
   silicofluoric, 132
   sulfuric, 131, 132, 139, 306, 309, 430, 433, 622
   volatile, *see* volatile acids
acre-feet (ac-ft), defined, 69
activated alumina, 655, 656
activated biosolids, 103, 152–159, 172, 514, 517
   F/M ratios, typical, 154
   settleability test, 186, 188
activated carbon, 69, 70, 435, 510, 569, 605, 606, 633, 637, 638, 640, 650, 739, 740, 745
activated sludge, 69, 70, 452, 453, 515, 667, 668, 689, 704–726, 745
   flagellates in, 715
   food-to-microorganism ratio, 721–722
   formation, 707–708
   mass balance, 724–726

   mean cell residence time, 722–724
   microorganism examination of, 716
   samples, 546
   solids, 186
activated sludge process, 69, 73, 74, 667, 704, 705–726, 739, 746
   adjustments, 718
   bacteria, 715–716
   calculations, 718–726
   control levels, 711–712
   control parameters, 710–711
   equipment, 706
   factors affecting, 707
   modifications, 709–710
   performance of, 708–709
   protozoa in, 715
   recordkeeping, 726
   sampling and testing, 713–726
   startup, 712
   troubleshooting, 718
active security barriers, 36–38
actuators, 227, 261, 412–413
adsorbents, hydrophobic, 70
adsorption, 70, 119, 149, 435, 440, 605, 618, 654, 657, 705
advanced wastewater treatment, 69, 70, 667, 672, 735–746
aerated ponds, 691
aeration, 11, 70, 181, 436, 548, 569, 597, 605, 610, 614, 680, 682, 705, 706, 707, 708, 711, 712, 742, 744, 745, 753
   diffused air, 705
   extended, 709–710
   step, 709
aeration tank, 75, 101, 152, 153, 154, 155, 156, 704, 706, 707, 712–716, 718, 720, 721, 722, 723
   microscopic examination of, 714
aerobic bacteria, 444, 448, 705
aerobic, defined, 70, 667, 705
aerobic digester/digestion, 172, 753
   calculations, 753–754
   normal operating levels for, 753
   pH adjustments, 754
aerobic ponds, 691

aerobic protozoa, 450
aerobic treatment processes, 443, 452, 462
aestivation, 490
age, biosolids, 155, 157, 451
age, sludge, 82, 155, 705, 707, 710, 711, 713
agglomeration, 70, 436, 615
Aggressive Index (AI), 613
agricultural best management practices, 597
agricultural pollutants, 73
agricultural runoff, 10, 11, 583
agroecosystem, 474–475
air binding, 388, 619, 620
air conditioning and refrigeration (AC&R) drawings, 240–243, 262, 264
air core, 344
   electromagnet, 336
air distribution, schematic symbols for, 264
air flotation thickeners, 170–171
air gap, 70
air pressure, 267
air receivers, 419–420
air-to-solids ratio, 171, 752
alarms, 34–36, 53, 54, 59, 372
alderflies, 493, 497–498
aldicarb, 439
algae, 4, 10, 119, 126, 148, 432, 433, 435, 436, 440, 443, 444, 451, 452, 453, 461, 465, 471, 488, 490, 491, 492, 493, 495, 510, 514, 519, 530, 551, 556, 568, 596, 597, 614, 618, 631, 634, 637, 674, 675, 690, 691, 692
   as fuel, 18
   blooms, *see* algal blooms
   control, 107
   monitoring protocols for, 520–521
algaecides, 597
algal blooms, 70, 444, 452, 465, 467, 508, 514, 516, 553, 568, 583, 614, 736
alignment chart, 281
alkaline cell battery, 309